Topics in Mining, Metallurgy and Materials Engineering

Series Editor

Carlos P. Bergmann, Federal University of Rio Grande do Sul, Porto Alegre, Rio Grande do Sul, Brazil

"Topics in Mining, Metallurgy and Materials Engineering" welcomes manuscripts in these three main focus areas: Extractive Metallurgy/Mineral Technology; Manufacturing Processes, and Materials Science and Technology. Manuscripts should present scientific solutions for technological problems. The three focus areas have a vertically lined multidisciplinarity, starting from mineral assets, their extraction and processing, their transformation into materials useful for the society, and their interaction with the environment.

**** Indexed by Scopus (2020) ****

More information about this series at http://www.springer.com/series/11054

Abdel Salam Hamdy Makhlouf ·
Gomaa A. M. Ali
Editors

Waste Recycling Technologies for Nanomaterials Manufacturing

Springer

Editors
Abdel Salam Hamdy Makhlouf
Central Metallurgical Research
and Development Institute (CMRDI)
Cairo, Egypt

Engineering, Metallurgy, Coatings &
Corrosion Consultancy (EMC3)
Edinburg, TX, USA

Gomaa A. M. Ali
Chemistry Department, Faculty of Science
Al-Azhar University
Assiut, Egypt

ISSN 2364-3293 ISSN 2364-3307 (electronic)
Topics in Mining, Metallurgy and Materials Engineering
ISBN 978-3-030-68030-5 ISBN 978-3-030-68031-2 (eBook)
https://doi.org/10.1007/978-3-030-68031-2

This Springer imprint is published by the registered company Springer Nature Switzerland AG
The registered company address is: Gewerbestrasse 11, 6330 Cham, Switzerland

Preface

Nowadays, nanomaterials (NMs) are used in many areas and applications, including medicine, energy, and environment. The initial cost of the NMs is high; thus, finding cheap sources is required. In addition, waste accumulation is a serious environmental problem. Therefore, recycling waste into valuable NMs is highly required, where it has environmental and economic benefits.

Waste management is pressing hard to warn the industry. Humans always produce waste and discard it in some way, influencing the environment. At present, no spot on the earth is not exposed to some waste. These materials may cause immediate health risks to humans and animals. Other wastes persist for a long time in the environment until they reach damaging levels to ecosystems. Hence, the upsurge in waste generated by the industries and human activities needs to be managed. Various recycling methods have been developed and applied for the conversion of wastes into useful forms of materials and NMs. The standard methods applied to recover the generated wastes, including recycling, reducing, and reuse, still need more developments.

Information and techniques for investigations are minimal. Nonetheless, it is incredibly likely that NMs used in several items would be in the waste stream. Environmental risks related to the treatment of nanowastes remain unexplored. Another factor is whether items containing NMs, consisting of recycling processes, will affect the waste management capabilities/performance or not. In comparison, NMs may substitute certain substances that make products, e.g., smarter or more efficient, to get into waste management sooner and potentially play a role in waste reduction. Draw up an overview of nanomaterial and waste-related scientific, health, and environmental problems, and assess the available recycling issues are needed. The ultimate goal is to consider looking for identical statistics to compare the potential hazards associated with the existence of NMs in the waste.

This book provides in-depth studies about these challenges and covers these issues in four parts.

Part One: Fundamentals, Current Prospects, and Future Trends

In this part, we covered the basics of nanomaterials in terms of manufacturing, characteristics, and applications. Various techniques used to recycle waste have been discussed. In addition, this part highlights the fundamentals, current prospects, and future trends of the recovered nanomaterials.

Part Two: Electronics Waste Recycling Technologies

In this part, we highlighted the importance of recycling in terms of environmental and economic perspectives. We discussed the recycling techniques of electronic waste, including lithium-ion batteries, zinc–carbon batteries, etc. For example, hierarchical cobalt oxide nanostructure has been recovered from spent lithium-ion batteries using magnetic electrodeposition. In addition, MnO_2 nanoflower has obtained from zinc–carbon batteries using electrodeposition and other methods. The materials used for manufacturing lithium-ion batteries also recovered from various waste sources. The applications of the recovered materials for supercapacitors, batteries, electrocatalytic, and sensing have been discussed.

Part Three: Agriculture Waste Recycling Technologies

In this part, we covered the conversion of agricultural waste into nanomaterials, mainly carbon-based nanomaterials and their composites. The studied agriculture waste includes rice husk, rice husk ash, bamboo leaves, bio-waste sago bark, banana fibers, argania spinosa seeds, corn grains, sugarcane fibers, and oil palm shells, palm kernel shells, orange peel, wheat flour, etc. Various nanomaterials compositions and morphologies were obtained, such as pure activated carbon, hetero-atom-doped carbon materials, and metal oxides/carbon nanocomposites. The recovered materials have been studied for various applications, including water treatments, energy storage, and forensic medicine applications.

Part Four: Plastic and Polymeric Waste Recycling Technologies

Plastic is one of the most significant hazards to the environment. Plastic is a non-biodegradable material, and several toxic chemicals leach out of it and seep through the soil, water, plants, and animals. In this part, we introduced the topic of utilizing plastic wastes as a precursor for the fabrication of carbon-based materials. While also highlighting the factors affecting the efficiency of each process and the recent progress in this regard, this part also highlights recycling polyethylene terephthalate waste into a novel magnetic nanoadsorbent. Recent breakthroughs in carbon-based nanomaterials' science and technology use paraffinic waxes as a carbon source where it consists of not less than 18 carbon number per single paraffin crystal. This part also describes the separation of paraffinic petroleum wax, its purification, and characterization beside nanocarbon synthesis. Different nanomaterials can be synthesized from the waste plastics, such as polyvinyl chloride plastic is used as the carbon source for the fabrication of MoC_2 nanoparticles.

Cairo, Egypt
Edinburg, USA
Assiut, Egypt
November 2020

Abdel Salam Hamdy Makhlouf

Gomaa A. M. Ali

Contents

Editors and Contributors

About the Editors

Prof. Dr. Abdel Salam Hamdy Makhlouf, Ph.D. President of Engineering, Metallurgy, Coatings and Corrosion Consultancy (EMC3), Texas, USA

Full Professor: Central Metallurgical R&D Institute

Website: https://www.emc3.website/

E-mail: asalam85@yahoo.com

Professor Makhlouf is an internationally recognized leader in the field of materials science and engineering with more than 27 years of independent research project management, teaching, and consulting. He has been included in Stanford University's List of World's Top 2% of Scientists, USA, 2020. He has a blend of both industrial and academic leadership as a President of EMC3, Full Professor at Central Metallurgical Research and Development Institute, Egypt, and a Former Full Professor of Manufacturing Engineering at the University of Texas, USA. He is the recipient of numerous national and international prizes and awards including the Humboldt Research Award for Experienced Scientists, at Max Planck Institute, Germany; Fulbright Scholar, NSF, and Department of Energy Fellowships, USA; Shoman Award in Engineering Science; and the State Prize of Egypt in Advanced Science and Technology, and more. He is a member of TMS-USA, EPSRC-UK, European Science Foundation—College of Expert Reviewers, Fulbright Alumni, Alexander von Humboldt Alumni, Max Planck Institute Alumni, etc. He has served as both a

Senior Editor and board member of many international journals, as well as a reviewer for several international funding agencies. He has excellent knowledge of USA, EU, and UK research landscape. He is a Consultant and Reviewer for several universities, and Advisory Editor for Elsevier USA.

Dr. Makhlouf is the author of over 200 peer-reviewed journal and conference papers, 19 books and handbooks, 30 book-chapters, as well as +100 technical reports. The h-index is 37, with > 4570 citations. Many of his publications have been ranked among the World's Best in the fields of Nanostructures, Nanomaterials, Biomedical Engineering, Materials Science, Coatings, Environmental Science, Nuclear Materials.

Assis. Prof. Dr. Gomaa A. M. Ali, Ph.D. Assistant Professor at Chemistry Department, Faculty of Science Al-Azhar University, Assiut, Egypt

E-mail: gomaasanad@gmail.com; gomaasanad@azhar.edu.eg

Dr. Gomaa A. M. Ali is an Assistant Professor at the Chemistry Department, Faculty of Science, Al-Azhar University, Assiut, Egypt. He has 14 years of experience working in the research areas of materials science, nanocomposites, humidity sensing, graphene, supercapacitors, water treatment, and drug delivery. He was awarded his Ph.D. in Advanced Nanomaterials for Energy Storage from UMP, Malaysia. He is the recipient of some national and international prizes and awards such as TWAS-AREP (2018), Gold Medal (Archimedes, Russia, 2014), Green Technology Award (CITREX, Malaysia, 2015), Gold Medal (British Invention Show, UK, 2015). Dr. Gomaa has published over 100 journal articles and 6 book chapters on a broad range of cross-disciplinary research fields, including advanced multifunctional materials, nanotechnology, supercapacitor, water treatment, and humidity sensing, biosensing, corrosion, drug delivery, and materials for energy applications. So far, he has more than 1800 citations and h-index of 24. Dr. Gomaa has served as both Senior

Editor and board member of many international journals and a reviewer for more than 50 WoS journals. Dr. Gomaa is a member of some national and international scientific societies such as the American Chemical Society (ACS) and the Egyptian Young Academy of Sciences (EYAS).

Contributors

M. Abd Elkodous Department of Electrical and Electronic Information Engineering, Toyohashi University of Technology, Toyohashi, Aichi, Japan;
Center for Nanotechnology (CNT), School of Engineering and Applied Sciences, Nile University, Giza, Egypt

Sabah M. Abdelbasir Central Metallurgical Research and Development Institute, Helwan, Cairo, Egypt

Mohamed Abdelmottaleb Chemistry Department, Faculty of Science, Al-Azhar University, Assuit, Egypt

Eslam A. A. Aboelazm Institute of Basic and Applied Science, Egypt-Japan University of Science and Technology, New Borg El-Arab, Alexandria, Egypt

Abdelaal S. A. Ahmed Chemistry Department, Faculty of Science, Al-Azhar University, Assuit, Egypt;
State Key Laboratory of Advanced Technology for Materials Synthesis and Processing, Wuhan University of Technology, Luoshi Road, Wuhan, People's Republic of China

Omid Akbarzadeh Nanotechnology & Catalysis Research Centre (NANOCAT), Institute for Advanced Studies (IAS), University for Malaya (UM), Kuala Lumpur, Malaysia

Atika Alhanish Chemical Engineering Department, Faculty of Petroleum and Natural Gas Engineering, University of Zawia, Zawia, Libya

Gomaa A. M. Ali Chemistry Department, Faculty of Science, Al-Azhar University, Assiut, Egypt;
The Smart Materials Research Institute, Southern Federal University, Rostov-on-Don, Russian Federation

Shimaa Hosny Ali Department of Chemistry, Faculty of Science, New Valley University, New Valley, Egypt

Enas Amdeha Process Design and Development Department, Egyptian Petroleum Research Institute, Cairo, Egypt

Ibrahim Saana Amiinu State Key Laboratory of Silicate Materials for Architecture, Wuhan University of Technology, Wuhan, People's Republic of China

Hasna Aziam High Throughput Multidisciplinary Research Laboratory (HTMRL), Mohammed VI Polytechnic University (UM6P), Ben Guerir, Morocco; IMED-Lab, Cadi Ayyad University (UCA), Marrakesh, Morocco

Radwa Mahmoud Azmy Entomology Department, Faculty of Science, Ain Shams University, Cairo, Egypt

Rania Azouz Clinical Microbiology Unit, Clinical and Chemical Pathology Department, Faculty of Medicine, Beni Suef University, Beni Suef, Egypt;
Medical Administration, Beni Suef University, Beni Suef, Egypt

Zinab H. Bakr Physics Department, Faculty of Science, Assiut University, Assiut, Egypt

Maged F. Bekheet Fachgebiet Keramische Werkstoffe/Chair of Advanced Ceramic Materials, Technische Universität Berlin, Institut für Werkstoffwissenschaften und -technologien, Berlin, Germany

Aiman Awadh Bin Mokaizh Faculty of Chemical and Natural Resources Engineering, Universiti Malaysia Pahang, Gambang, Pahang, Malaysia

Arghya Chakravorty School of Bio Sciences and Technology, Vellore Institute of Technology, Vellore, India

Cik Norhazrin Che Hamzah Forensic Science Programme, School of Health Sciences, Universiti Sains Malaysia, Health Campus, Kubang Kerian, Kelantan, Malaysia

Kwok Feng Chong Faculty of Industrial Sciences and Technology, Universiti Malaysia Pahang, Gambang, Kuantan, Malaysia

Moharana Choudhury Voice of Environment (VoE), Guwahati, Assam, India

Himadri Tanaya Das Department of Chemical Engineering, National Taipei University of Technology, Taipei, Taiwan;
Center of Excellence for Advanced Materials and Applications, RUSA, Utkal University, Vanivihar, Bhubaneswar, Odisha, India

Asmaa M. El Shafey Faculty of Science and Arts, King Khalid University, Abha, Saudi Arabia

T. Elango Balaji Department of Chemistry, Bishop Heber College, Tiruchirappalli, Tamil Nadu, India

Ahmed I. El-Batal Drug Microbiology Lab, Drug Radiation Research Department, National Center for Radiation Research and Technology (NCRRT), Egyptian Atomic Energy Authority (EAEA), Cairo, Egypt

Heba H. El-Maghrabi Department of Refining, Egyptian Petroleum Research Institute, Cairo, Egypt

Mohamed E. Elmowafy Chemical Engineering Department, Military Technical College, Cairo, Egypt

Radwa A. El-Salamony Egyptian Petroleum Research Institute, Cairo, Egypt

Mohamed A. Elsayed Chemical Engineering Department, Military Technical College, Cairo, Egypt

Gharieb S. El-Sayyad Drug Microbiology Lab, Drug Radiation Research Department, National Center for Radiation Research and Technology (NCRRT), Egyptian Atomic Energy Authority (EAEA), Cairo, Egypt;
Chemical Engineering Department, Military Technical College (MTC), Egyptian Armed Forces, Cairo, Egypt

Mohammed Y. Emran Department of Chemistry, Faculty of Science, Al-Azhar University-Assiut Branch, Assiut, Egypt

Manal Fawzy Environmental Sciences Department, Faculty of Science, Alexandria University, Alexandria, Egypt;
Green Technology Group, Faculty of Science, Alexandria University, Alexandria, Egypt;
National Biotechnology Network of Expertise (NBNE), Academy of Scientific Research and Technology (ASRT), Cairo, Egypt

Mohamed Gobara Chemical Engineering Department, Military Technical College (MTC), Egyptian Armed Forces, Cairo, Egypt

Hassanien Gomaa Faculty of Science, Department of Chemistry, Al-Azhar University, Assiut, Egypt

A. Hamdy Department of Analysis and Evaluation, Egyptian Petroleum Research Institute, Cairo, Egypt

Zhanglian Hong State Key Laboratory of Silicon Material, School of Materials Science and Engineering, Zhejiang University, Hangzhou, China

Abdel-Rahman A. M. Ismail Faculty of Science, Department of Chemistry, Al-Azhar University, Assiut, Egypt

Mohd Rafie Johan Nanotechnology & Catalysis Research Centre (NANOCAT), Institute for Advanced Studies (IAS), University for Malaya (UM), Kuala Lumpur, Malaysia

Mohamed Ayman Kammoun Laboratoire Sciences des Matériaux et Environnement, Faculté des Sciences de Sfax, Université de Sfax, Sfax, Tunisia

Javaria Kanwal Department of Chemistry, The Women University Multan, Multan, Pakistan

Diab Khalafallah State Key Laboratory of Silicon Material, School of Materials Science and Engineering, Zhejiang University, Hangzhou, China;
Mechanical Design and Materials Department, Faculty of Energy Engineering, Aswan University, Aswan, Egypt

Chin Wei Lai Nanotechnology & Catalysis Research Centre (NANOCAT), Institute for Advanced Studies (IAS), University for Malaya (UM), Kuala Lumpur, Malaysia

Neeta Laxman Lala Voice of Environment (VoE), Guwahati, India

K. Mahendraprabhu Department of Chemistry, MEPCO Schlenk Engineering College (Autonomous), Sivakasi, Tamil Nadu, India

Alaa El Din Mahmoud Environmental Sciences Department, Faculty of Science, Alexandria University, Alexandria, Egypt;
Green Technology Group, Faculty of Science, Alexandria University, Alexandria, Egypt;
National Biotechnology Network of Expertise (NBNE), Academy of Scientific Research and Technology (ASRT), Cairo, Egypt

Tahany Mahmoud Petroleum Application Department, Egyptian Petroleum Research Institute, Cairo, Egypt

Abdel Salam Hamdy Makhlouf Engineering, Metallurgy, Coatings & Corrosion Consultancy (EMC3), Edinburg, TX, USA;
Central Metallurgical Research and Development Institute, Helwan, Cairo, Egypt

Jyoti Mehta Department of Environmental Sciences, Central University of Jharkhand, Ranchi, Jharkhand, India

Eslam A. Mohamed Petroleum Application Department, Egyptian Petroleum Research Institute, Cairo, Egypt

Nourhan Mohamed Department of Metallurgical and Materials Engineering, Istanbul Technical University, Maslak, Istanbul, Turkey

Seyedehmaryam Moosavi Nanotechnology & Catalysis Research Centre (NANOCAT), Institute for Advanced Studies (IAS), University for Malaya (UM), Kuala Lumpur, Malaysia

Mohamed M. Motawea Faculty of Science, Department of Chemistry, Al-Azhar University, Assiut, Egypt

Yasser M. Moustafa Department of Analysis and Evaluation, Egyptian Petroleum Research Institute, Cairo, Egypt

Sara Musaddiq Department of Chemistry, The Women University Multan, Multan, Pakistan

Kiran Mustafa Department of Chemistry, The Women University Multan, Multan, Pakistan

Amr A. Nada Department of Analysis and Evaluation, Egyptian Petroleum Research Institute, Cairo, Egypt;
Laboratoire Plasma et Conversion de l'Energie (LAPLACE), Université de Toulouse, CNRS, INPT, UPS, Toulouse, France

Nik Fakhuruddin Nik Hassan Forensic Science Programme, School of Health Sciences, Universiti Sains Malaysia, Health Campus, Kubang Kerian, Kelantan, Malaysia

Andrews Grace Nirmala Centre for Nanotechnology Research, Vellore Institute of Technology, Vellore, India

A. A. Ragab Petroleum Application Department, Egyptian Petroleum Research Institute, Cairo, Egypt

Premanjali Rai Environmental Chemistry Division, CSIR-Indian Institute of Toxicology Research, Lucknow, India

Revathi Rajan Forensic Science Programme, School of Health Sciences, Universiti Sains Malaysia, Health Campus, Kubang Kerian, Kelantan, Malaysia;
Forensic Science Programme, Department of Biotechnology, Faculty of Applied Sciences, UCSI University, Cheras, Kuala Lumpur, Malaysia

Rehab A. Rayan Department of Epidemiology, High Institute of Public Health, Alexandria University, Alexandria 21526, Egypt

Patrice Raynaud Laboratoire Plasma et Conversion de l'Energie (LAPLACE), Université de Toulouse, CNRS, INPT, UPS, Toulouse, France

Priyabrata Roy Centre for Interdisciplinary Studies, Barrackpore, Kolkatta, India;
Department of Molecular Biology and Biotechnology, University of Kalyani, Kalyani, India

Ibtisam Saeed College of Science, University of Duhok, Duhok, Iraq

Vahid Safarifard Department of Chemistry, Iran University of Science and Technology, Tehran, Iran

Mohamed F. Sanad Basic Science Departments, Modern Academy for Engineering and Technology, Maadi, Egypt;
Basic Science Department, British University in Egypt, Cairo, Egypt;
Chemistry Department, Faculty of Science, Ain-Shams University, Abbasia, Cairo, Egypt;
University of Texas at El Paso, El Paso, TX, USA

Mohamed A. Sayed Refining Department, Egyptian Petroleum Research Institute, Cairo, Egypt

Tarek A. Seaf El-Nasr Faculty of Science, Department of Chemistry, Jouf University, Sakaka, Aljouf, Saudi Arabia;
Faculty of Science, Department of Chemistry, Al-Azhar University, Assiut, Egypt

Hanaa Selim Department of Analysis and Evaluation, Egyptian Petroleum Research Institute, Cairo, Egypt

Jun Haslinda Binti Haji Shariffuddin Faculty of Chemical and Natural Resources Engineering, Universiti Malaysia Pahang, Gambang, Pahang, Malaysia

Kunwar P. Singh Environmental Chemistry Division, CSIR-Indian Institute of Toxicology Research, Lucknow, India

Fathi S. Soliman Department of Refining, Egyptian Petroleum Research Institute, Cairo, Egypt

S. Vinoth Department of Electronics and Communication Engineering, Manakulavinayagar Institute of Technology, Kalitheerthalkuppam, Puducherry, India

Suhad Yasin Chemistry Department, College of Science, University of Duhok, Kurdistan Region, Iraq

Yusmazura Zakaria Biomedicine Programme, School of Health Sciences, Universiti Sains Malaysia, Health Campus, Kubang Kerian, Kelantan, Malaysia

Xiujian Zhao State Key Laboratory of Advanced Technology for Materials Synthesis and Processing, Wuhan University of Technology, Wuhan, People's Republic of China

Mingjia Zhi State Key Laboratory of Silicon Material, School of Materials Science and Engineering, Zhejiang University, Hangzhou, China

Fundamentals, Current Prospects, and Future Trends

Fundamentals of Waste Recycling for Nanomaterial Manufacturing

Gomaa A. M. Ali and Abdel Salam Hamdy Makhlouf

Abstract Nowadays, nanomaterials are used in many areas and applications, including medicine, energy, and environment. The initial cost of the nanomaterials is high; thus, finding another cheap source is required. In addition, waste accumulation is a serious environmental problem. Therefore, recycling waste into valuable nanomaterials is highly required, where it has environmental and economic benefits.

Keywords Nanomaterials · Waste · Recycling · Nanotechnology · Manufacturing

List of Abbreviations

CVD	Chemical vapour deposition
mZVI	Meso-scale zero valent iron
NSMs	Nanostructured materials
PVD	Physical vapour deposition
ppb	Parts per billion
SG	Sol–Gel
TEOS	Tetraethylorthosilicate

G. A. M. Ali (✉)
Chemistry Department, Faculty of Science, Al–Azhar University, Assiut 71524, Egypt
e-mail: gomaasanad@azhar.edu.eg; gomaasanad@gmail.com

A. S. H. Makhlouf
Engineering, Metallurgy, Coatings & Corrosion Consultancy (EMC3), Edinburg, TX, USA

Central Metallurgical Research and Development Institute, Helwan, Cairo 11421, Egypt

A. S. H. Makhlouf and G. A. M. Ali (eds.), *Waste Recycling Technologies for Nanomaterials Manufacturing*, Topics in Mining, Metallurgy and Materials Engineering, https://doi.org/10.1007/978-3-030-68031-2_1

1 Fundamentals of Nanomaterials Manufacturing

1.1 Nanoscience and Nanotechnology

Nanoscience is one of the hottest areas and rapidly expanding research fields. It aims to synthesize, study, characterize, and evaluate new materials, for which one or more of their dimensions lies in the nanometer scale (1–100 nm). Nano refers to the 10^{-9} power or one billionth. Nanotechnology aims to fabricate these materials in the form of devices systems and other industrial products. Advances in nanoscience lead to new developments in nanotechnology. Materials of this size are particularly attractive because of their inherent physical and chemical characteristics because of their high surface-to-volume ratio. Nanoparticles are very sensitive to surface environments. Nanotechnology aims to better understanding these properties and to find new ways of utilizing them in our daily life.

Nanomaterials have received a great deal of research interest because of their unique catalytic [1, 2], optical [3, 4], electrical [5–7], magnetic [8, 9], sensing [10–12], storage [13–16], and mechanical and chemical properties. Nanomaterials have unique characteristics than when in bulk (micro or macro) form. Nanomaterials can be classified into several categories based on their structures such as nanoparticles, nanotubes, nanoflowers, nanocapsules, nanocoatings, nano-thin-films, nanorods, nanoflakes, nanofibers, nanocarriers, nanoceramics [13, 17–25], and nanocomposites. The nanomaterials' applications can also be classified as nanomagnetism, nanomedicine, nanotoxicology, nanoelectrochemistry, nanoengineering, nanoscience, etc. [26–29].

1.2 Types of Nanomaterials

Solid substances can be divided into metals, ceramics, semiconductors, composites, and polymers [6, 7, 10, 30–36]. When the size of these materials is decreased to the nano-range, they can be further subdivided into nanoparticles, nanocrystals, nanotubes, nanorods, and nanocomposites [14, 15, 37–40]. The first classification scheme of nanostructured materials (NSMs) was provided by Gleiter in 2000 [41], and other classification was extended by Pokropivny and Skorokhod [42] as shown in Fig. 1. 0D, 1D, 2D, and 3D are the main classifications of NSs based on their dimensionality.

1.3 Nanosized Structures

Nanomaterials are the bridge between bulk materials and atomic structures. The physical properties of the bulk materials are constant regardless of its size, but for

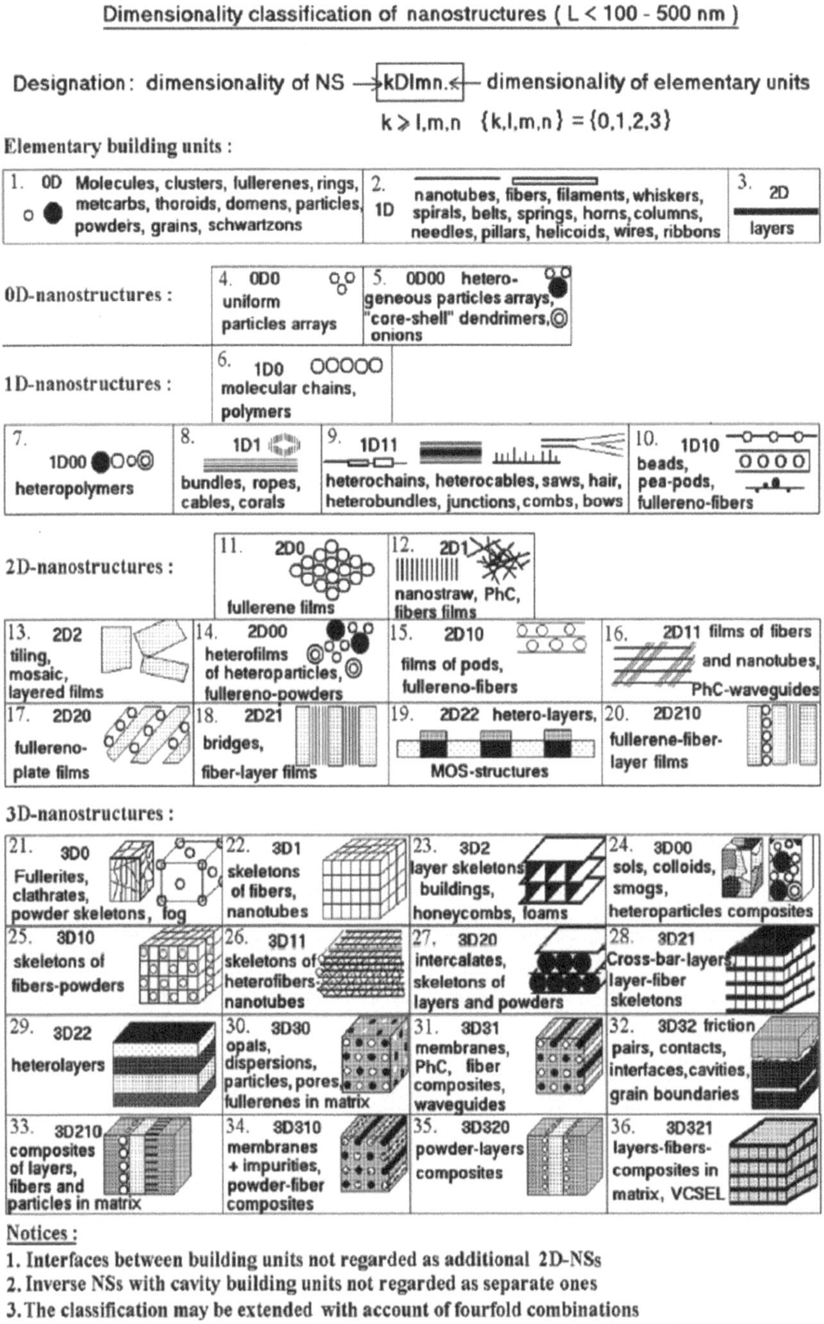

Fig. 1 Dimensionality classifications of nanostructured materials Adapted with permission from Ref. [42], Copyright 2007, Elsevier

the nanomaterials, the properties are varied with the variation in the particles size. Size-dependent properties are detected, such as quantum confinement [43] in semiconductor nanoparticles, surface plasmon resonance [44] in some metal nanoparticles, which are the fundamentals behind many colour-based biosensor applications and superparamagnetism [45] in magnetic nanomaterials. Nanocomposite materials can be considered as multiphase materials in which one or more from its components have nanometer dimensions [46]. One of its components is considering a guest phase, and the other is considered as the host one. The host phase acts as a matrix in which the guest phase is distributed [35], or as a shell coated the guest core [47, 48]. The nanocomposite materials exhibit properties of both the gust and host compounds [31, 49, 50].

2 Synthesis of Nanomaterials

The top-down and bottom-up are main approaches to nanomaterials synthesis. A typical top-down method is attrition or milling, whereas the bottom-up approach examples include colloidal dispersion and deposition. Synthesis routes play a crucial role in the properties of the target product. Many synthesis techniques have been evaluated and developed to obtain nanoscale materials with proper morphologies.

2.1 Vapor State Processing Routes

2.1.1 Physical Vapor Deposition

Thin-film nanomaterials could be easily prepared by physical vapor deposition (PVD). The process includes the generation of vapor phase species via sputtering [51], laser ablation [52], or plasma spray [53] (Fig. 2), and then the vapor is condensed onto a substrate followed by the nucleation and growth. The process has some limitations in the case of preparation of the multicomponent materials (nanocomposites) due to the differences in the evaporation temperature of the components because of the differences in vapor pressures of the evaporating species.

2.1.2 Chemical Vapor Deposition

The process in which the gaseous species decompose or react on a heated surface to make stable solid products is called chemical vapor deposition (CVD) [55]. Metallic, ceramic, and semiconducting thin films could be deposited using CVD. The process can be classified into thermally activated, laser-assisted, and

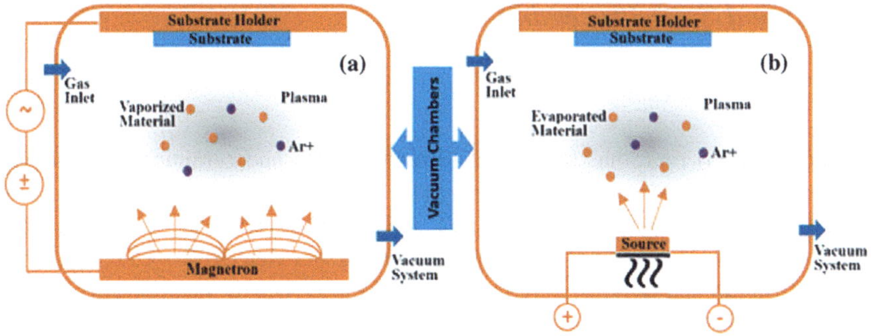

Fig. 2 Schematic illustration of conventional PVD processes (sputtering (**a**) and evaporating (**b**)). Adapted with permission from Ref. [54], Copyright 2018, MDPI

plasma-assisted CVD based on the activation sources for the chemical reactions [56]. Figure 3 shows the setup of a horizontal CVD and parallel-plate plasma-enhanced CVD reactors. In the case of metal chloride precursors, corrosive chlorine-containing by-products are formed [56]. CVD is a more complex method of forming thin films and coatings than PVD due to the surface and gas phase interactions.

2.1.3 Spray Conversion Processing

The chemical precursors could be atomized into aerosol droplets that are dispersed throughout a gas atmosphere, then moved into a heated reactor to form thin films or ultrafine particles. The atomization purposes use various aerosol generators, including pressure, electrostatic, or ultrasonic atomizer. Spray pyrolysis is the most commonly used aerosol processing method [58].

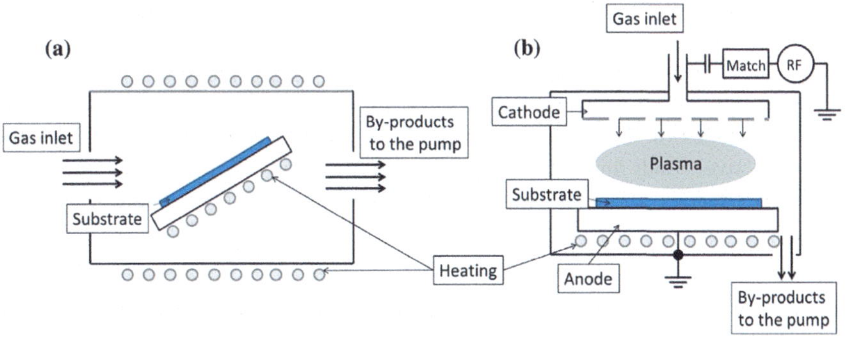

Fig. 3 Setup of a horizontal CVD (**a**) and a parallel-plate plasma-enhanced CVD (**b**) reactors. Adapted with permission from Ref. [57], Copyright 2019, Elsevier

2.2 *Liquid State Processing Routes*

2.2.1 Sol–Gel Method

The sol–gel (SG) route has been used for metal oxide and ceramic powders production with high homogeneity and purity [10, 35, 59, 60]. SG process involves the formation of a colloidal suspension (sol), which is converted to viscous gel and solid material. In the process, reactive metal precursors were hydrolyzed, followed by condensation and polymerization reactions. Many types of metal precursors can be used (alkoxides, carboxylic salts, chlorides, nitrates, etc. [61]). To achieve gel densification, the solvent removal and appropriate drying are required. Tetraethylorthosilicate (TEOS; $Si(OC_2H_5)_4$) is the most common studied metal alkoxides [62]. However, the conversion of metal precursor molecules needs acid or base catalysts such as sulphuric acid or ammonium hydroxide since the hydrolysis of silicon alkoxides is very slow. The conversion from a solution containing individual solvated particles to a colloidal three-dimensional network is represented in Fig. 4a. In addition, Fig. 4b, c shows the detailed synthesis process of CoO_x–SiO_2 system. The most advantages of the SG method are high purity, good homogeneity of the prepared materials, lower preparation temperature, precise composition control, versatile shaping, and preparation by cheap and straightforward apparatus compared with other methods.

2.2.2 Citrate-Gel-Pechini Process

In the citrate-gel-Pechini method, an organic network formed in precursor solutions; thus, fine oxide powders are obtained after a heating process stabilizes metal ions. This method can be used for the preparation of multicomponent compositions with good homogeneity and control of stoichiometry [64]. The citric acid is used as a capping agent because of its relatively strong multifunctional organic acid. Pechini route utilizes poly-chelates between the metal ions and C=O ligands of citric acid [65]. The chelating process takes place during the aggressive stirring of precursor solution containing metallic salts and CA. Figure 5 shows Pechini process for CuO/CeO_2 nanocomposites preparation. The simplicity of the method, homogeneous microstructure, and low crystallization temperature of the obtained material are the main advantages of the Pechini route [64]. Different types of precursors can be used. Moreover, both thin films and nanocomposite powders can be obtained [64, 66]. Higher surface area and lower particle size products can be obtained.

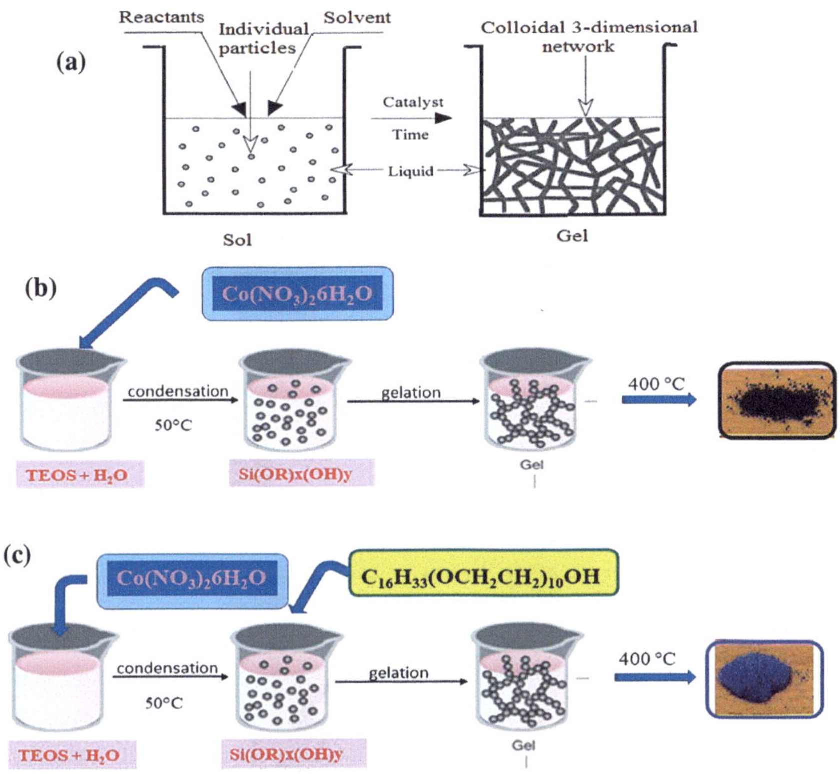

Fig. 4 Simple representation for sol-to-gel transformation (**a**) sol-gel process of CoO_x-SiO_2 preparation in presence of (**b**) hydrolytic alkoxide and non-ionic surfactant (**c**). Adapted with permission from Ref. [63], Copyright 2019, MPDI

2.2.3 Wet Chemical Synthesis

This simple and single-step method includes chemical reaction producing different shapes nanorods [68], and nanoparticles with approximate spherical shape [69]. Various nanostructured forms (nanoflowers, nanoneedles, and staking of flake-like structures) are achieved by controlling the experimental conditions [70]. Both organic and inorganic additives can yield different shapes of nanomaterials products. The shape of particles can be controlled by controlling the adsorption of some inorganic anions to particular crystal faces [71]. ZnO nanopencils have been synthesized using wet chemical method as shown in Fig. 6 together with its morphological investigations.

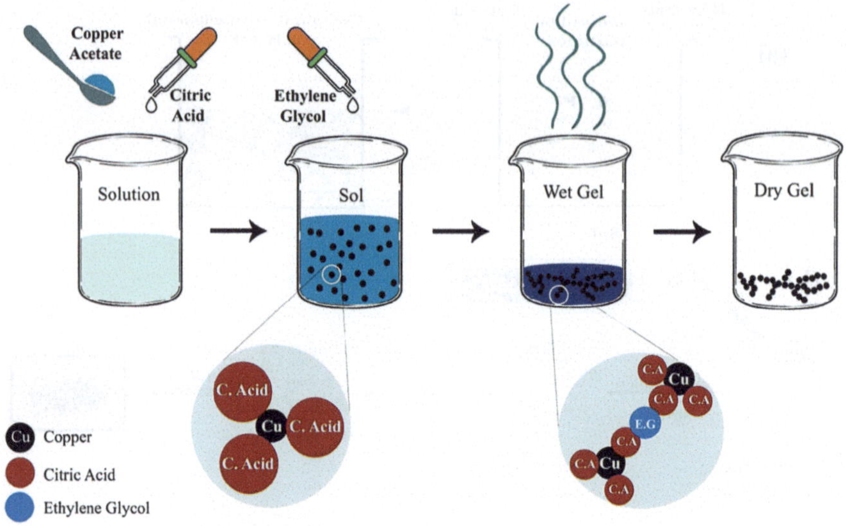

Fig. 5 Pechini process for CuO/CeO_2 nanocomposites preparation. Adapted with permission from Ref. [67], Copyright 2019, Elsevier

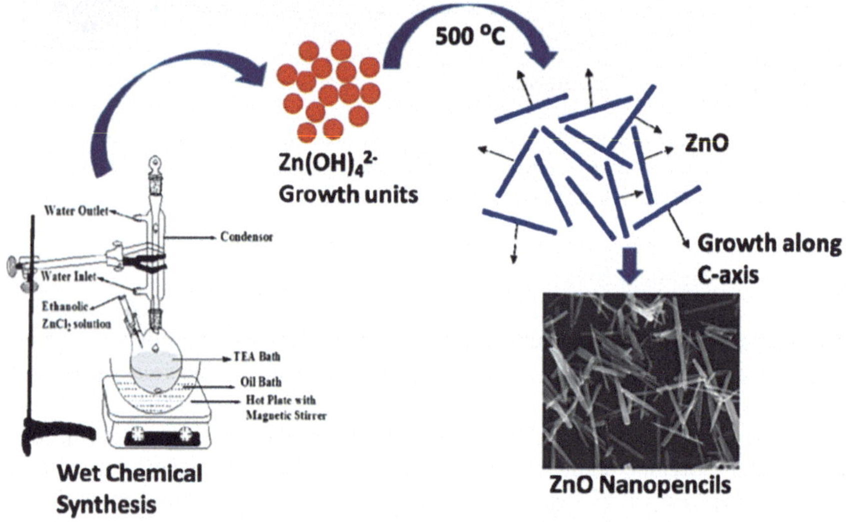

Fig. 6 Synthesis of ZnO nanopencils using wet chemical method. Adapted with permission from Ref. [72], Copyright 2016, Elsevier

2.3 *Solid-State Processing Routes*

2.3.1 Mechanical Milling

The particles' size of metal and ceramic materials can be reduced to the nanoscale by high-energy ball milling. Some factors must be taken in the account such as: (i) the powder-to-ball mass ratio, and (ii) addition of process control agent [73]. The ball milling techniques can successfully reduce the particle size to few nanometers, but phase transformation may occur [73, 74]. Present size reduction methods possess some disadvantages, such as contaminations from grinding media. Figure 7 shows the formation mechanism of meso-scale zero valent iron (mZVI) nanoparticles as an example by high-energy ball milling process.

2.3.2 Mechanochemical Preparation

The mechanochemical preparation process of the nanomaterials includes mechanical activation of solid-state displacement reactions in a ball mill. Thus, chemical

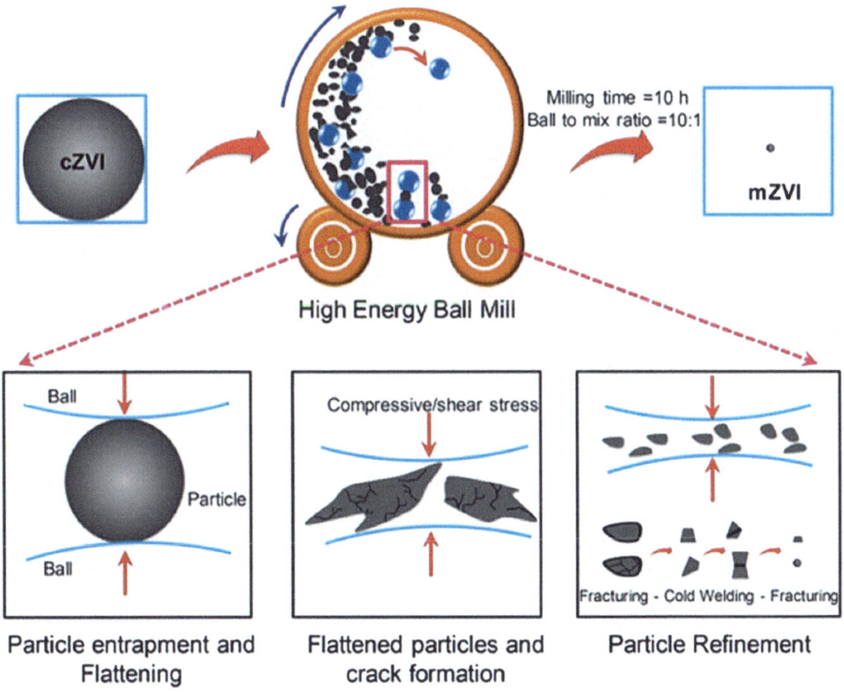

Fig. 7 Formation mechanism of meso-scale zero valent iron nanoparticles by high-energy ball milling process. Adapted with permission from Ref. [75], Copyright 2016, Elsevier

reactions are induced by mechanical energy. During milling or subsequent heat treatment, the chemical precursors (oxides, chlorides and/or metals) interact to form ultrafine particles. For example, Co_3O_4 has been prepared by the mechanochemical route [76] starting from cobalt nitrate and ammonium carbonate.

3 Properties of Nanomaterials

The properties of nanoparticles are highly size-dependent [77], where, as the particle size decreases, surface effects become increasingly important. The smaller particles have a large surface to volume ratio. This opens the door for their possible applications in diverse fields such as catalysis, electronics, optics, biology, and magnetism. One of the most critical properties of nanoparticles is the high surface-to-volume ratio making very high number of active centers. The central atom in a nanostructured system is surrounded by a first shell of 12, a second of 42, a third of 92 atoms, etc. The number of atoms in the nth shell is $10n^2 + 2$ (Table 1). Moreover, an increase in the surface-to-volume ratio leads to an increased fraction of the total crystal volume that consists of surface atoms. Some particularly exciting phenomenon in the study of nanomaterials and nanodevices is their ability to add value to materials and products through enhancement of specific properties.

Table 1 Surface to total number of atoms in shell cluster relationship [79]

Shells		Total atoms	Surface atoms (%)
One		13	92
Two		55	76
Three	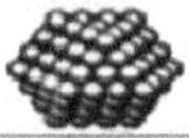	147	63
Four		307	52
Five	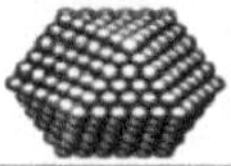	361	35
Seven		1415	45

3.1 Surface Area and Catalytic Activity

Compared with micro- or macro-scale size particles, nanosized particles have much greater surface area for the same volume of material. Preparation routes and chemical additives play an essential role in producing nanomaterials having high surface area [10, 46, 78]. High porosity and high surface area are interesting properties for using the materials as catalysts since they favor the accessibility to the active centers [52].

3.2 Electrical Properties

The nanosizing was found to enhance the DC conductivity of the nanomaterial. For example, the conductivity of NiO nanoparticle is about 8 orders of magnitude higher than that of bulk or single crystal [77]. Moreover, the activation energies for the nanomaterials are lower than for the bulk materials [77].

3.3 Energy Gap and Optical Properties

The optical properties of nanoparticles depend on their size. For example, the optical measurements of Co_3O_4 nanoparticles showed that by increasing the particles size from 7.3 to 14.2 nm, the optical bandgap was increased from 1.79 to 1.87 eV [4]. This shift phenomenon might be ascribed to the quantum effect [80].

3.4 Mechanical Strength

Nanomaterials' mechanical properties are size dependent that are significantly different from those of their coarse-grained counterparts. The pure metals are much stronger and less ductile than conventional ones. Recently, nanoparticles are being used as an advanced filler to improve the mechanical properties of the other materials [81]. It was proven that the mechanical properties of chlorinated poly (vinyl chloride) are improved by incorporating modified $CaCO_3$ nanoparticles as a filler [81]. Fine particles in the epoxy matrix influenced the deformation mechanism and improved all the investigated mechanical properties (tensile stress, elongation at break, toughness). Better mechanical properties are obtained for smaller particles than bigger particles [82].

3.5 *Melting Temperature and Thermodynamic Properties*

The nanomaterial size plays a critical role in the thermodynamic characteristics such as the melting temperature. For example, gold nanoparticles with diameters of smaller than 5 nm have a melting temperature of 300 °C compared to 1063 °C for the bulk particles [79]. The latent heat of fusion ΔH_m, as well as T_m, is particle size dependent. For bulk tin, the melting point is 232 °C [83], while with reducing the particle size to 5 nm, the T_m systematically decreases to about 70 °C.

3.6 *Color*

The visual characteristics of the nanomaterials depend on their size. For example, the colloidal gold solution containing uncoagulated particles (40 nm) has a beautiful ruby red color. For larger gold particles (coagulated), the color becomes blue, and absorption is accompanied by light scattering [66].

4 Applications of Nanomaterials

4.1 *Energy Applications*

Nowadays, humanity is in urgent need of energy generation and storage systems. The supercapacitor is one of the essential types of storage systems. The high cost of obtaining capacitor electrodes is the reason behind the researchers' attempts to find low-cost sources. A variety of single component and hybrid electrodes have been prepared by recycling various environmental wastes [84–90]. The obtained materials exhibited excellent behavior in the storage of electrical energy [84, 86–91]. For example, nanoflower manganese dioxide was prepared by electrodeposition of a solution prepared from the dry batteries and showed a specific capacitance of 208.5 F/g [90]. Carbon nanosphere was also prepared by recycling agricultural palm leaves residues and gave a specific capacitance of 309 F/g [92].

4.2 *Catalytic Applications*

Catalysis is a relevant field of nanomaterials application since it highly depends on the surface phenomena [93–97]. The intensely small particles size maximizes the exposed surface area (active centers), allowing more reactions to occur. The catalytic properties are improved when the size of catalyst reduced to the nanoscale. Nanocatalysts are usually heterogeneous catalysts. However, the thermal stability of these nanomaterials is limited by their critical sizes. The smaller the crystallite size is, the lower the thermal stability. Examples of metal oxides nanocatalysts are CeO_2, TiO_2, and ZnO.

4.3 Environmental Applications

Nanomaterials have been broadly applied for dye and heavy metal ions removals from wastewater [32, 33, 98–106]. Numerous novel nanoadsorbents have been evaluated for enhancing the efficiency and adsorption capacities [38, 107, 108]. The unique morphological and structural characteristics of the nanomaterials support their application as effective nanoadsorbents to solve several environmental problems. Adsorption processes using nanomaterials are highly effective and can be easily performed and employed for the removal of organic and inorganic pollutants [99–101, 105, 109–112].

4.4 Sensing Applications

Nanocomposite sensors are used extensively to detect and monitor a wide variety of gases, including toxic and explosive gases, organic and inorganic vapors, humidity, and odors. In many applications, sensors capable of monitoring parts per billion (ppb) concentrations are required. The most important characteristics of the sensor are sensitivity, reversibility, and selectivity. Sensitivity is the ability of the sensor to detect a given concentration of a gas. The sensitivity can be monitored by measuring the change in a physical or chemical property of the sensors such as electrical resistance [61], optical transmittance [113], etc. The changes in resistance with gas exposure should be reversible, where a reliable sensor should not show hysteresis on cycling between different ambient. An excellent sensor should respond quickly when the analyte is introduced, and it should recover immediately after its removal. Response and recovery times are, in part, determined by material topology relative to the molecular size of the sensed parameter, temperature, and binding energy between the material and analyte. In addition, the sensor element should be stable with time and possess repeated operational cycles. Metal oxide nanostructures exhibit gas-sensing capabilities of extraordinary sensitivity. Excellent sensing characteristics have been demonstrated with the scaling of physical feature size to the nanoscale [61].

Humidity is the amount of water vapor in the air. The vapor pressure and temperature are the main factors affecting the amount of water vapor. The vapor pressure is the pressure exerted by the water vapor molecules in the air. As the number of water vapor molecules increases, the vapor pressure increases. The saturated vapor pressure is the pressure at equilibrium condition when equal numbers of water molecules are entering and leaving a flat (plane) surface of pure liquid water.

5 Waste Recycling Technologies

5.1 Waste Classification

The waste could be classified into many types, according to their sources and nature. It can be classified into liquid and solid based on its physical state and can be categorized as agricultural, industrial, electronic, and medical wastes based on its source [114, 115].

5.1.1 Agricultural Waste

This waste is produced because of various agricultural activates and includes crop residue, fertilizers, pesticides and animal products or manure. Agriculture waste is considered as valuable source of carbon materials, and it can easily be converted into carbon-based nanomaterials applicable for many applications [86, 92, 116–118]. These wastes include oil palm biomass residues (leaves, fronds, trunks, empty fruit bunches, shells, and fibers), rice husk, sugarcane bagasse, etc. [119, 120]. Burning is the common practice in managing some of these residues.

5.1.2 Industrial Waste

Industrial waste is known as the waste generated by manufacturing or industrial processes. This waste contains cafeteria garbage, dirt and gravel, scrap metals, trash, masonry and concrete, oil, solvents, chemicals, weed grass and trees, wood and scrap lumber, etc. Industrial waste is classified into hazardous and nonhazardous waste. Manufacturing or industrial processes produce hazardous waste, while the non-hazardous industrial wastes are those that do not meet the EPA's definition of hazardous waste - and are not municipal waste [121]. Industrial waste may be toxic, burnable, corrosive, or reactive. If improperly managed, this waste can pose dangerous health and environmental consequences.

5.1.3 Electronic Waste

Home appliances, mobile phones, and computers are the main source for the e-waste. The increase in the electronics consumption can result in two major adverse ecological effects. First, it significantly increases mining and procurement for the materials needed for the gadgets production. Second, discarded devices produce large quantities of electronic waste. This e-waste could be reduced through reuse, repair, or resale. Spent batteries are the main form of the e-waste, and it could be recycled into valuable metals, metal hydroxides, and metal oxides [31, 84, 85, 87, 89, 90, 122].

5.2 Recycling Techniques

Recycling is the process of converting waste into other useful materials. It reduces the initial cost of materials, conserves energy, reduces greenhouse gases, reduces water and air pollution, and conserves natural resources. Protecting the environment is one of the most advantages of the recycling process. In addition, it also reduces the need for extraction such as mining, logging and quarrying and fair use of wealth. Agriculture, electronics, wood, paper, metals, cardboard, plastics, concrete, and green waste can all be recycled [123].

5.2.1 Pyrolysis Recycling

Pyrolysis is a recycling technique based on thermal and catalytic cracking to convert plastic waste into fuels, monomers, or other valuable materials. In other words, it thermally converts waste plastics into useful hydrocarbons liquids such as crude oil and diesel fuel. In addition, non-condensable gas products could be recycled by pyrolysis into bio-oils [124]. Moreover, pyrolysis method is a successful and effective way to convert agriculture waste into porous carbon nanomaterials [86, 88, 92, 104, 105, 112, 116–118, 125–128]. Plastic and rubber waste recycling is of increasing importance with complete exclusion of oxygen using melting vessels, rotary kilns, or fluidized bed processes [129].

5.2.2 Electrochemical Recycling

The metals and their oxides or hydroxide forms can be obtained in pure form by electrodeposition technique [85, 89]. In addition, electrocatalytic hydrogen evolution reaction could be used for recycling Ni and Co from Ni–MH batteries [130]. In addition, electrodeposition is an effective method for recycling metal oxides from spent batteries [31, 84, 85, 87, 122].

5.2.3 Chemical Recycling

Metal hydroxides and oxides could be recovered using the chemical precipitation of the leached spent batteries powder [130]. For example, the cathode material of the batteries was dissolved in HNO_3 under constant stirring, and then the solutions were filtered to remove insoluble graphite and plastics, and then to precipitate the metals [130].

6 Conclusion

Nanomaterials are used in many applications due to their matchless properties which are highly dependent on their particles size. Valuable nanomaterials could be obtained via waste recycling. Many recycling could be used to recover nanomaterials from waste precursors depending on the origin and type of waste. The recovered nanomaterials are applicable for many applications including medicine, energy, environment, etc. From environmental and economic points of view, the waste recycling into valuable products including nanomaterials is highly required.

References

1. Králik M, Biffis A (2001) Catalysis by metal nanoparticles supported on functional organic polymers. J Mol Catal A: Chem 177(1):113–138
2. Rogovin M, Neumann R (1999) Silicate xerogels containing cobalt as heterogeneous catalysts for the side-chain oxidation of alkyl aromatic compounds with tert-butyl hydroperoxide. J Mol Catal A: Chem 138(2–3):315–318
3. Kim KJ, Park YR (2003) Optical investigation of charge-transfer transitions in spinel Co_3O_4. Solid State Commun 127(1):25–28
4. Ali GAM, Fouad OA, Makhlouf SA (2013) Structural, optical and electrical properties of sol–gel prepared mesoporous Co_3O_4/SiO_2 nanocomposites. J Alloy Compd 579:606–611
5. Isshiki M, Fukuda Y, Igaki K (1984) Temperature dependence of the electrical resistivity of pure cobalt at low temperatures. J Phys F: Met Phys 14(12):3007–3013
6. Ali GAM, Fouad OA, Makhlouf SA, Yusoff MM, Chong KF (2016) Optical and electrochemical properties of Co_3O_4/SiO_2 nanocomposite. Adv Mater Res 1133:447–451
7. Ali GAM, Fouad Osama A, Makhlouf Salah A (2016) Electrical properties of cobalt oxide/ silica nanocomposites obtained by sol–gel technique. Am J Eng Appl Sci 9(1):12–16
8. Makhlouf SA (2004) Magnetic properties of Cr_2O_3 nanoparticles. J Magn Magn Mater 272–276:1530–1532
9. Sadegh H, Helmi H, Hamdy A, Masjedi A, Dastjerdi M, Shahryari-ghoshekandi R (2016) A developed simple, facile, economic and fast technique for preparation of high quality Fe_3O_4 magnetic nanoparticles. Chem Adv Mater 1(1):27–32
10. Fouad OA, Ali GAM, El-Erian MAI, Makhlouf SA (2012) Humidity sensing properties of cobalt oxide/silica nanocomposites prepared via sol–gel and related routes. Nano 7 (5):1250038
11. Makhlouf S (2003) Humidity sensing properties of NiO/Al_2O_3 nanocomposite materials. Solid State Ionics 164(1–2):97–106
12. Teo EYL, Ali GAM, Algarni H, Cheewasedtham W, Rujiralai T, Chong KF (2019) One-step production of pyrene-1-boronic acid functionalized graphene for dopamine detection. Mater Chem Phys 231:286–291
13. Aboelazm EAA, Ali GAM, Algarni H, Chong KF (2018) Flakes size-dependent optical and electrochemical properties of MoS_2. Curr Nanosci 14(2):1–5

14. Ali GAM, Fouad OA, Makhlouf SA, Yusoff MM, Chong KF (2014) Co_3O_4/SiO_2 nanocomposites for supercapacitor application. J Solid State Electrochem 18(9):2505–2512
15. Ali GAM, Lih Teo EY, Aboelazm EAA, Sadegh H, Memar AOH, Shahryari-Ghoshekandi R, Chong KF (2017) Capacitive performance of cysteamine functionalized carbon nanotubes. Mater Chem Phys 197:100–104
16. Ali GAM, Makhlouf SA, Yusoff MM, Chong KF (2015) Structural and electrochemical characteristics of graphene nanosheets as supercapacitor electrodes. Rev Adv Mater Sci 40 (1):35–43
17. Makhlouf ASH, Tiginyanu I (2011) Nanocoatings and ultra-thin films: technologies and applications. Elsevier
18. Makhlouf ASH, Scharnweber D (2015) Handbook of nanoceramic and nanocomposite coatings and materials. Butterworth-Heinemann
19. Makhlouf ASH, Abu-Thabit NY (2018) Stimuli responsive polymeric nanocarriers for drug delivery applications. Elsevier
20. Ghawanmeh AA, Ali GAM, Algarni H, Sarkar SM, Chong KF (2019) Graphene oxide-based hydrogels as a nanocarrier for anticancer drug delivery. Nano Res 12:973–990
21. Makhlouf ASH, Abu-Thabit NY (2018) Stimuli responsive polymeric nanocarriers for drug delivery applications. In: Volume 1: Types and triggers. Woodhead Publishing
22. Barhoum A, Bechelany M, Makhlouf ASH (2019) Handbook of Nanofibers. Springer International Publishing
23. Nekouei F, Kargarzadeh H, Nekouei S, Keshtpour F, Makhlouf ASH (2016) Novel, facile, and fast technique for synthesis of AgCl nanorods loaded on activated carbon for removal of methylene blue dye. Process Saf Environ Prot 103:212–226
24. Salam Hamdy A (2010) Corrosion protection performance via nano-coatings technologies. Recent Patents Mater Sci 3(3):258–267
25. Hamdy A (2010) The role of nanotechnology in designing high performance nano-ceramic coatings. Int Rev Chem Eng 2(2):256–262
26. Aliofkhazraei M, Makhlouf ASH (2016) Handbook of nanoelectrochemistry: electrochemical synthesis methods, properties, and characterization techniques. Springer
27. Makhlouf ASH, Barhoum A (2018) Fundamentals of nanoparticles: classifications, synthesis methods, properties and characterization. William Andrew
28. Abdal-Hay A, Makhlouf ASH, Vanegas P (2015) A novel approach for facile synthesis of biocompatible PVA-coated PLA nanofibers as composite membrane scaffolds for enhanced osteoblast proliferation. In: Handbook of nanoceramic and nanocomposite coatings and materials. Elsevier, pp 87–113
29. Abdal-hay A, Khalil KA, Hamdy AS, Al-Jassir FF (2017) Fabrication of highly porous biodegradable biomimetic nanocomposite as advanced bone tissue scaffold. Arab J Chem 10 (2):240–252
30. Hosseini M, Makhlouf ASH (2016) Industrial applications for intelligent polymers and coatings. Springer
31. Ali GAM, Yusoff MM, Algarni H, Chong KF (2018) One-step electrosynthesis of MnO_2/rGO nanocomposite and its enhanced electrochemical performance. Ceram Int 44(7):7799–7807
32. Abdel Ghafar HH, Ali GAM, Fouad OA, Makhlouf SA (2015) Enhancement of adsorption efficiency of methylene blue on Co_3O_4/SiO_2 nanocomposite. Desalin Water Treat 53 (11):2980–2989
33. Gupta VK, Agarwal S, Sadegh H, Ali GAM, Bharti AK, Hamdy AS (2017) Facile route synthesis of novel graphene oxide-β-cyclodextrin nanocomposite and its application as adsorbent for removal of toxic bisphenol A from the aqueous phase. J Mol Liq 237:466–472
34. Ali GAM, Thalji MR, Soh WC, Algarni H, Chong KF (2020) One-step electrochemical synthesis of MoS_2/graphene composite for supercapacitor application. J Solid State Electrochem 24(1):25–34
35. Fouad OA, Makhlouf SA, Ali GAM, El-Sayed AY (2011) Cobalt/silica nanocomposite via thermal calcination-reduction of gel precursors. Mater Chem Phys 128(1–2):70–76

36. Soliman H, Qian J, Tang S, Chen Y, Makhlouf ASH, Wan G (2020) Hydroxyquinoline/nano-graphene oxide composite coating of self-healing functionality on treated Mg alloys AZ31. Surf Coat Technol 385:125395
37. Barhoum A, Van Assche G, Makhlouf ASH, Terryn H, Baert K, Delplancke M-P, El-Sheikh SM, Rahier H (2015) A green, simple chemical route for the synthesis of pure nanocalcite crystals. Cryst Growth Des 15(2):573–580
38. Heibati B, Ghoochani M, Albadarin AB, Mesdaghinia A, Makhlouf ASH, Asif M, Maity A, Tyagi I, Agarwal S, Gupta VK (2016) Removal of linear alkyl benzene sulfonate from aqueous solutions by functionalized multi-walled carbon nanotubes. J Mol Liq 213:339–344
39. Ali GAM, Sadegh H, Yusoff MM, Chong KF (2019) Highly stable symmetric supercapacitor from cysteamine functionalized multi-walled carbon nanotubes operating in a wide potential window. Mater Today: Proc 16:2273–2279
40. Zhang C, Yan Y, Sheng Zhao Y, Yao J (2013) Synthesis and applications of organic nanorods, nanowires and nanotubes. Ann Rep Section "C" (Phys Chem) 109(0):211–239
41. Gleiter H (2000) Nanostructured materials: basic concepts and microstructure. Acta Mater 48(1):1–29
42. Pokropivny VV, Skorokhod VV (2007) Classification of nanostructures by dimensionality and concept of surface forms engineering in nanomaterial science. Mater Sci Eng, C 27(5–8):990–993
43. Sharma SN, Kohli S, Rastogi AC (2005) Quantum confinement effects of CdTe nanocrystals sequestered in TiO_2 matrix: effect of oxygen incorporation. Physica E 25(4):554–561
44. Zhang S, Berguiga L, Elezgaray J, Roland T, Faivre-Moskalenko C, Argoul F (2007) Surface plasmon resonance characterization of thermally evaporated thin gold films. Surf Sci 601(23):5445–5458
45. Haeiwa T, Segawa K, Konishi K (2007) Magnetic properties of isolated Co nanoparticles in SiO_2 capsule prepared with reversed micelle. J Magn Magn Mater 310(2):e809–e811
46. Khalil KMS, Makhlouf SA (2008) High surface area thermally stabilized porous iron oxide/silica nanocomposites via a formamide modified sol–gel process. Appl Surf Sci 254 (13):3767–3773
47. Fu W, Yang H, Hari B, Liu S, Li M, Zou G (2006) Preparation and characteristics of core–shell structure cobalt/silica nanoparticles. Mater Chem Phys 100(2–3):246–250
48. Yin XJ, Peng K, Hu AP, Zhou LP, Chen JH, Du YW (2009) Preparation and characterization of core–shell structured Co/SiO_2 nanosphere. J Alloy Compd 479(1–2):372–375
49. Marina PE, Ali GAM, See LM, Teo EYL, Ng E-P, Chong KF (2016) In situ growth of redox-active iron-centered nanoparticles on graphene sheets for specific capacitance enhancement. Arab J Chem 12(8):3883–3889
50. Ali GAM, Wahba OAG, Hassan AM, Fouad OA, Chong KF (2015) Calcium-based nanosized mixed metal oxides for supercapacitor application. Ceram Int 41(6):8230–8234
51. Deng Z, Tian Y, Yin X, Rui Q, Liu H, Luo Y (2008) Physical vapor deposited zinc oxide nanoparticles for direct electron transfer of superoxide dismutase. Electrochem Commun 10 (5):818–820
52. Li K-T, Lin P-H, Lin S-W (2006) Preparation of Ti/SiO_2 catalysts by chemical vapor deposition method for olefin epoxidation with cumene hydroperoxide. Appl Catal A 301 (1):59–65
53. Majima T, Yamamoto H, Kulinich S, Terashima K (2000) High-rate deposition of $LiNb_{1-x}Ta_xO_3$ films by thermal plasma spray CVD. J Cryst Growth 220(3):336–340
54. Baptista A, Silva F, Porteiro J, Míguez J, Pinto G (2018) Sputtering physical vapour deposition (PVD) coatings: a critical review on process improvement and market trend demands. Coatings 8(11):402
55. Makhlouf ASH, Barhoum A (2018) Emerging applications of nanoparticles and architectural nanostructures: current prospects and future trends. William Andrew
56. Goto T (2004) High-speed deposition of zirconia films by laser-induced plasma CVD. Solid State Ionics 172(1–4):225–229

57. Benelmekki M, Erbe A (2019) Nanostructured thin films–background, preparation and relation to the technological revolution of the 21st century. In: Benelmekki M, Erbe A (eds) Frontiers of nanoscience. Elsevier, pp 1–34
58. Wang Z-M, Yang G, Biswas P, Bresser W, Boolchand P (2001) Processing of iron-doped titania powders in flame aerosol reactors. Powder Technol 114(1–3):197–204
59. Lambert S, Tran KY, Arrachart G, Noville F, Henrist C, Bied C, Moreau JJ, Man MWC, Heinrichs B (2008) Tailor-made morphologies for Pd/SiO_2 catalysts through sol–gel process with various silylated ligands. Microporous Mesoporous Mater 115(3):609–617
60. Farag M, Liu H, Makhlouf ASH (2020) New nano-bioactive glass/magnesium phosphate composites by sol-gel route for bone defect treatment. Silicon. https://doi.org/10.1007/s12633-020-00485-3
61. Musat V, Fortunato E, do Rego AB, Monteiro R (2008) Sol–gel cobalt oxide–silica nanocomposite thin films for gas sensing applications. Thin Solid Films 516(7):1499–1502
62. Wagh P, Ingale S (2002) Comparison of some physico-chemical properties of hydrophilic and hydrophobic silica aerogels. Ceram Int 28(1):43–50
63. Esposito S (2019) "Traditional" sol–gel chemistry as a powerful tool for the preparation of supported metal and metal oxide catalysts. Materials 12(4):668
64. Popa M, Preda S, Fruth V, Sedlácková K, Balázsi C, Crespo D, Calderón-Moreno JM (2009) $BiFeO_3$ films on steel substrate by the citrate method. Thin Solid Films 517(8):2581–2585
65. Niranjan R, Hwang Y, Kim D-K, Jhung S, Chang J-S, Mulla I (2005) Nanostructured tin oxide: synthesis and gas-sensing properties. Mater Chem Phys 92(2–3):384–388
66. Patil M, Deshpande V, Ravi V (2006) Synthesis of nanocrystalline $Sn_{0.2}Zr_{0.8}TiO_4$ by the citrate gel method. Ceram Int 32(3):345–347
67. Zarei M, Borhani Zarandi M, Alizadeh M (2019) Preparation of CuO/CeO_2 composites by the Pechini method and investigation of their structural and electrical properties. Ceram Int 45(2, Part A):1991–1997
68. Roy P, Srivastava SK (2007) Low-temperature synthesis of CuS nanorods by simple wet chemical method. Mater Lett 61(8–9):1693–1697
69. Darezereshki E (2010) Synthesis of maghemite (γ-Fe_2O_3) nanoparticles by wet chemical method at room temperature. Mater Lett 64(13):1471–1472
70. Jana S, Das S, Das N, Chattopadhyay K (2010) CuO nanostructures on copper foil by a simple wet chemical route at room temperature. Mater Res Bull 45(6):693–698
71. Yu Z-G, Yang B-C (2008) Morphological investigation on cobalt oxide powder prepared by wet chemical method. Mater Lett 62(2):211–214
72. Shimpi NG, Jain S, Karmakar N, Shah A, Kothari DC, Mishra S (2016) Synthesis of ZnO nanopencils using wet chemical method and its investigation as LPG sensor. Appl Surf Sci 390:17–24
73. Lee J-S, Lee C-S, Oh S-T, Kim J-G (2001) Phase evolution of Fe_2O_3 nanoparticle during high energy ball milling. Scripta Mater 44(8–9):2023–2026
74. Li J, Ni X, Wang G (2007) Microstructure and magnetic properties of Co/Al_2O_3 nanocomposite powders. J Alloy Compd 440(1–2):349–356
75. Ambika S, Devasena M, Nambi IM (2016) Synthesis, characterization and performance of high energy ball milled meso-scale zero valent iron in Fenton reaction. J Environ Manage 181:847–855
76. Morsy SM, Shaban SA, Ibrahim AM, Selim MM (2009) Characterization of cobalt oxide nanocatalysts prepared by microemulsion with different surfactants, reduction by hydrazine and mechanochemical method. J Alloy Compd 486(1–2):83–87
77. Makhlouf SA, Kassem MA, Abdel-Rahim MA (2009) Particle size-dependent electrical properties of nanocrystalline NiO. J Mater Sci 44(13):3438–3444
78. Ethiraj AS, Uttam P, Varunkumar K, Chong KF, Ali GAM (2019) Photocatalytic performance of a novel semiconductor nanocatalyst: Copper doped nickel oxide for phenol degradation. Mater Chem Phys 242:122520
79. Tjong S, Chen H (2004) Nanocrystalline materials and coatings. Mater Sci Eng: R: Rep 45 (1–2):1–88

80. Salavati-Niasari M, Khansari A, Davar F (2009) Synthesis and characterization of cobalt oxide nanoparticles by thermal treatment process. Inorg Chim Acta 362(14):4937–4942
81. Haroun AAA, Rabie AGM, Ali GAM, Abdelrahim MYM (2019) Improving the mechanical and thermal properties of chlorinated poly (vinyl chloride) by incorporating modified $CaCO_3$ nanoparticles as a filler. Turk J Chem 43(3):750–759
82. Al-Turaif HA (2010) Effect of nano TiO_2 particle size on mechanical properties of cured epoxy resin. Prog Org Coat 69(3):241–246
83. Lai S, Guo J, Petrova V, Ramanath G, Allen L (1996) Size-dependent melting properties of small tin particles: nanocalorimetric measurements. Phys Rev Lett 77(1):99
84. Ali GAM, Yusoff MM, Shaaban ER, Chong KF (2017) High performance MnO_2 nanoflower supercapacitor electrode by electrochemical recycling of spent batteries. Ceram Int 43:8440–8448
85. Aboelazm EAA, Ali GAM, Chong KF (2018) Cobalt oxide supercapacitor electrode recovered from spent lithium-ion battery. Chem Adv Mater 3:67–74
86. Ali GAM, Divyashree A, Supriya S, Chong KF, Ethiraj AS, Reddy MV, Algarni H, Hegde G (2017) Carbon nanospheres derived from Lablab purpureus for high performance supercapacitor electrodes: a green approach. Dalton Trans 46(40):14034–14044
87. Ali GAM (2020) Recycled MnO_2 nanoflowers and graphene nanosheets for low-cost and high performance asymmetric supercapacitor. J Electron Mater 49(9):5411–5421
88. Ali GAM, Habeeb OA, Algarni H, Chong KF (2018) CaO impregnated highly porous honeycomb activated carbon from agriculture waste: symmetrical supercapacitor study. J Mater Sci 54:683–692
89. Aboelazm EAA, Ali GAM, Algarni H, Yin H, Zhong YL, Chong KF (2018) Magnetic Electrodeposition of the hierarchical cobalt oxide nanostructure from spent lithium-ion batteries: its application as a supercapacitor electrode. J Phys Chem C 122(23):12200–12206
90. Ali GAM, Tan LL, Jose R, Yusoff MM, Chong KF (2014) Electrochemical performance studies of MnO_2 nanoflowers recovered from spent battery. Mater Res Bull 60:5–9
91. Ali GAM, Yusoff MM, Chong KF (2016) Graphene electrochemical production and its energy storage properties. ARPN J Eng Appl Sci 11(16):9712–9717
92. Ali GAM, Abdul Manaf SA, Kumar A, Chong KF, Hegde G (2014) High performance supercapacitor using catalysis free porous carbon nanoparticles. J Phys D Appl Phys 47 (49):495307–495313
93. Solehudin M, Sirimahachai U, Ali GAM, Chong KF, Wongnawa S (2020) One-pot synthesis of isotype heterojunction g-C_3N_4-MU photocatalyst for effective tetracycline hydrochloride antibiotic and reactive orange 16 dye removal. Adv Powder Technol 31(5):1891–1902
94. Kamrani M, Seifpanahi-Shabani K, Seyed-Hakimi A, Al G, Agarwa S, Gupta V (2019) Degradation of cyanide from gold processing effluent by H_2O_2, NaClO and $Ca(ClO)_2$ combined with sequential catalytic process. Bul Chem Commun 51(3):384–393
95. Sharifi A, Montazerghaem L, Naeimi A, Abhari AR, Vafaee M, Ali GAM, Sadegh H (2019) Investigation of photocatalytic behavior of modified ZnS: Mn/MWCNTs nanocomposite for organic pollutants effective photodegradation. J Environ Manage 247:624–632
96. Giahi M, Pathania D, Agarwal S, Ali GAM, Chong KF, Gupta VK (2019) Preparation of Mg-doped TiO_2 nanoparticles for photocatalytic degradation of some organic pollutants. Stud Univ Babes-Bolyai, Chem 64(1):7–18
97. Abdal-hay A, Hamdy Makhlouf ASH, Khalil KA (2015) Novel, facile, single-step technique of polymer/TiO_2 nanofiber composites membrane for photodegradation of methylene blue. ACS Appl Mater Interfaces 7(24):13329–13341
98. Hamidreza S, Gomaa AMA (2019) Potential applications of nanomaterials in wastewater treatment: nanoadsorbents performance. In: Athar H, Sirajuddin A (eds) Advanced treatment techniques for industrial wastewater. IGI Global, Hershey, PA, USA, pp 51–61
99. Maazinejad B, Mohammadnia O, Ali GAM, Makhlouf ASH, Nadagouda MN, Sillanpää M, Asiri AM, Agarwal S, Gupta VK, Sadegh H (2020) Taguchi L9 (34) orthogonal array study based on methylene blue removal by single-walled carbon nanotubes-amine: adsorption

optimization using the experimental design method, kinetics, equilibrium and thermodynamics. J Mol Liq 298:112001
100. Sadegh H, Ali GAM, Makhlouf ASH, Chong KF, Alharbi NS, Agarwal S, Gupta VK (2018) MWCNTs-Fe_3O_4 nanocomposite for Hg(II) high adsorption efficiency. J Mol Liq 258:345–353
101. Sadegh H, Ali GAM, Gupta VK, Makhlouf ASH, Shahryari-ghoshekandi R, Nadagouda MN, Sillanpää M, Megiel E (2017) The role of nanomaterials as effective adsorbents and their applications in wastewater treatment. J Nanostruct Chem 7(1):1–14
102. Sadegh H, Ali GAM, Nia HJ, Mahmoodi Z (2019) Nanomaterial surface modifications for enhancement of the pollutant adsorption from wastewater. In: Nanotechnology Applications in Environmental Engineering, IGI Global, Hershey, PA, USA, pp 143–170
103. Seyed Arabi SM, Lalehloo RS, Olyai MRTB, Ali GAM, Sadegh H (2019) Removal of congo red azo dye from aqueous solution by ZnO nanoparticles loaded on multiwall carbon nanotubes. Physica E 106:150–155
104. Habeeb OA, Ramesh K, Ali GAM, Yunus RM (2017) Low-cost and eco-friendly activated carbon from modified palm kernel shell for hydrogen sulfide removal from wastewater: adsorption and kinetic studies. Desalin Water Treat 84:205–214
105. Habeeb OA, Ramesh K, Ali GAM, Yunus RM, Olalere OA (2017) Kinetic, isotherm and equilibrium study of adsorption of hydrogen sulfide from wastewater using modified eggshells. IIUM Eng J 18(1):13–25
106. Abed HO, Kanthasamy R, Ali GAM, Sethupathi S, Yunus RBM (2017) Hydrogen sulfide emission sources, regulations, and removal techniques: a review. Rev Chem Eng 34(6):837–854
107. Nekouei F, Kargarzadeh H, Nekouei S, Keshtpour F, Makhlouf ASH (2017) Efficient method for determination of methylene blue dye in water samples based on a combined dispersive solid phase and cloud point extraction using $Cu(OH)_2$ nanoflakes: central composite design optimization. Anal Bioanal Chem 409(4):1079–1092
108. Gupta VK, Moradi O, Tyagi I, Agarwal S, Sadegh H, Shahryari-Ghoshekandi R, Makhlouf ASH, Goodarzi M, Garshasbi A (2016) Study on the removal of heavy metal ions from industry waste by carbon nanotubes: effect of the surface modification: a review. Crit Rev Environ Sci Technol 46(2):93–118
109. Sadegh H, Ali GAM, Agarwal S, Gupta VK (2019) Surface Modification of MWCNTs with carboxylic-to-amine and their superb adsorption performance. Int J Environ Res 13(3):523–531
110. Agarwal S, Sadegh H, Majid Monajjemi, Makhlouf ASH, Ali GAM, Memar AOH, Shahryari-ghoshekandi R, Tyagi I, Gupta VK (2016) Efficient removal of toxic bromothymol blue and methylene blue from wastewater by polyvinyl alcohol. J Mol Liq 218:191–197
111. Sadegh H, Ali GAM, Abbasi Z, Nadagoud MN (2017) Adsorption of ammonium ions onto multi-walled carbon nanotubes. Stud Univ Babes-Bolyai, Chem 62(2):233–245
112. Habeeb OA, Ramesh K, Ali GAM, Yunus RM (2017) Application of response surface methodology for optimization of palm kernel shell activated carbon preparation factors for removal of H_2S from industrial wastewater. Jurnal Teknologi 79(7):1–10
113. Martucci A, Buso D, Guglielmi M, Zbroniec L, Koshizaki N, Post M (2004) Optical gas sensing properties of silica film doped with cobalt oxide nanocrystals. J Sol–Gel Sci Technol 32(1–3):243–246
114. Kaushika N, Reddy K, Kaushik K (2016) Sustainable energy and the environment: a clean technology approach. Springer
115. Ali GAM (2020) Supercapacitors by recycling technologies: economic and environmental aspects. ARID Int J Sci Technol 3(5):74–83
116. Hegde G, Abdul Manaf SA, Kumar A, Ali GAM, Chong KF, Ngaini Z, Sharma KV (2015) Biowaste sago bark based catalyst free carbon nanospheres: waste to wealth approach. ACS Sustain Chem Eng 5(9):2247–2253

117. Ali GAM, Supriya S, Chong KF, Shaaban ER, Algarni H, Maiyalagan T, Hegde G (2019) Superior supercapacitance behavior of oxygen self-doped carbon nanospheres: a conversion of Allium cepa peel to energy storage system. Biomass Convers Biorefinery
118. Ali GAM, Manaf SAA, Divyashree A, Chong KF, Hedge G (2016) Superior supercapacitive performance in porous nanocarbons. J Energy Chem 25(4):734–739
119. Rafatullah M, Ahmad T, Ghazali A, Sulaiman O, Danish M, Hashim R (2013) Oil palm biomass as a precursor of activated carbons: a review. Crit Rev Environ Sci Technol 43 (11):1117–1161
120. Chavalparit O, Rulkens W, Mol A, Khaodhair S (2006) Options for environmental sustainability of the crude palm oil industry in Thailand through enhancement of industrial ecosystems. Environ Dev Sustain 8(2):271–287
121. Polprasert C, Liyanage L (1996) Hazardous waste generation and processing. Resour Conserv Recycl 16(1–4):213–226
122. Ali GAM, Yusoff MM, Feng CK (2015) Electrochemical properties of electrodeposited MnO_2 nanoparticles. Adv Mater Res 1113:550–553
123. Sormunen P, Kärki T (2019) Recycled construction and demolition waste as a possible source of materials for composite manufacturing. J Build Eng 24:100742
124. Pala M, Marathe PS, Hu X, Ronsse F, Prins W, Kersten SRA, Lange J-P, Westerhof RJM (2020) Recycling of product gas does not affect fast pyrolysis oil yield and composition. J Anal Appl Pyrol 148:104794
125. Habeeb OA, Ramesh K, Ali GAM, and Yunus RM (2017) Experimental design technique on removal of hydrogen sulfide using CaO-eggshells dispersed onto palm kernel shell activated carbon: Experiment, optimization, equilibrium and kinetic studies. J Wuhan Univ Technol Mater Sci Ed 32(2):305–320
126. Habeeb OA, Ramesh K, Ali GAM, Yunus RM (2017) Isothermal modelling based experimental study of dissolved hydrogen sulfide adsorption from waste water using eggshell based activated carbon. Malays J Anal Sci 21(2):334–345
127. Habeeb OA, Ramesh K, Ali GAM, Yunus RM (2017) Optimization of activated carbon synthesis using response surface methodology to enhance H_2S removal from refinery wastewater. J Chem Eng Ind Biotechnol 1(1):1–17
128. Habeeb OA, Ramesh K, Ali GAM, Yunus RM, Thanusha TK, Olalere OA (2016) Modeling and optimization for H_2S adsorption from wastewater using coconut shell based activated carbon. Aust J Basic Appl Sci 10(17):136–147
129. Kaminsky W (1985) Thermal recycling of polymers. J Anal Appl Pyrol 8:439–448
130. Santos VEO, Celante VG, Lelis MFF, Freitas MBJG (2012) Chemical and electrochemical recycling of the nickel, cobalt, zinc and manganese from the positives electrodes of spent Ni–MH batteries from mobile phones. J Power Sources 218:435–444

Recycling, Management, and Valorization of Industrial Solid Wastes

Sabah M. Abdelbasir

Abstract Environmental pollution has been viewed as a serious issue all over the world. Waste management is pressing hard to warn the industry. Humans always produce waste and discard it in some way, influencing the environment. At present, there is no spot on the earth that is not exposed to some sort of waste. These materials may cause immediate health risks to humans and animals. Other wastes persist for a long time in the environment until they reach damaging levels to ecosystems. Hence, the upsurge in waste generated by the industries and human activities needs to be managed. To this end, various recycling methods have been developed and applied for the conversion of wastes into useful forms of materials and also nanomaterials. The common methods applied to recover the generated wastes, including recycling, reducing, and reuse, still need more developments. The main goal of this chapter is to discuss different waste recycling techniques and to provide a comprehensive review about the industrial waste recycling processes. Examples of recycling of particular types of waste are also discussed. Finally, the present status and economic considerations of waste recycling are also investigated and reviewed.

Keywords Waste recycling · Industrial solid waste · Environmental pollution · Disposal · Hazards

List of Abbreviations

3R	Reduce, recycle, and reuse
CNTs	Carbon nanotubes
REEs	Rare earth elements
EPA	Environmental protection agency
EU	European Union
ISW	Industrial solid waste
LiBs	Lithium-ion batteries

S. M. Abdelbasir (✉)
Central Metallurgical Research and Development Institute, P. O. Box: 87, Helwan, Cairo 11421, Egypt
e-mail: sfoda20@hotmail.com

A. S. H. Makhlouf and G. A. M. Ali (eds.), *Waste Recycling Technologies for Nanomaterials Manufacturing*, Topics in Mining, Metallurgy and Materials Engineering, https://doi.org/10.1007/978-3-030-68031-2_2

NMs	Nano materials
NOx	Nitrogen oxides
PCBs	Waste printed circuit boards
PMs	Precious metals
SW	Solid waste
SWM	Solid waste management
WEEE	Waste electric and electronic equipment

1 Introduction

Nowadays, increased population and rapid development of industries and lifestyle lead to more curtailment of virgin resources. Therefore, the increase in generated waste by human activities and the industries needs proper management. Scientists have focused on the ecological procedures, which progressively require the reduction, reuse, and recycling of waste to close the circle of material used all through the economy by providing waste-derived materials as a contribution to manufacture.

Solid waste management (SWM) has received academics attention in recent years due to its extreme importance to ecological, social, and commercial levels. SWM covers the concept of waste reduction and reuses in addition to recycling, valuation or recovery processes and techniques, composting treatment, heat treatment, energy recovery, fuel production, and landfill. Typically, solid waste (SW) is sorted as municipal, agricultural, and industrial solid waste. Generally, municipal solid waste refers to the integration of residential and commercial waste such as food waste, paper, textiles, and plastic products. On the other hand, industrial solid waste (ISW) refers to all wastes arising from different industries [1, 2]. Industrial solid waste contains valuable metals that could be valued for economic returns, along with diminishing its disastrous environmental effects that arise from mishandled waste components.

In general, ISW comes out from chemical and metallurgical plants, cement factories, food processing, paint, pharmaceutical, textile, and petroleum industries. ISW can be classified into two main classes: hazardous and non-hazardous waste. The non-hazardous waste is produced from food processing plants, paper mills, and textile industries. However, hazardous waste is produced from industries other than those above, and the most common examples are metals, chemicals, drugs, leather, electroplating, and rubbers.

Sustainable engineering progression and SWM keep up esteemed natural resources and forestall the harmful emissions, consequently, ensure safe wellbeing. The main objectives are to limit environmental impacts and generate financial resources. The solid-state reutilizing process turns an actual and fantastic practice to attain the green state establishing from recyclable waste to useful parts [3–6].

The created procedure can be considered as a typical ecologically friendly process. Its advantages are facile, cost-effective, energy-efficient, and eco-friendly.

2 Categorization of Industrial Solid Waste

Industrial solid waste can be identified, according to a variety of authors, as solid waste generated in industrial production activities. Many kinds of industrial production activities are potential sources for different sorts of solid wastes, like metallurgy, mining, oil refining, pharmaceuticals, transportation, light manufacture, automobiles, communication, electronics, building materials, and glass and metal processing. These sorts include wasted by-products, scraped materials, and equipment. Residues, sludge, and recyclable materials are also included. Two categories of industrial solid waste are known based on their origin. One is by-products resulting from different production plants as sewage, slag, sludge, and effluents [7]. The second type is expired crude materials or commodities suchlike waste leachates and solutions, discarded products, and ended-life equipment.

Industrial solid waste can be classified according to jeopardy into general, hazardous, and radioactive industrial waste. The solid wastes can be classified according to industrial activities they arise from, counting metallurgical industries, oil and chemicals, construction, electronic appliances, machine-driven fabrication, printing, papermaking, rubber and plastics, pharmaceutics, automobile, and many others. Another classification based on the structure is known as ferrous and non-ferrous metals, rare metals, and heavy metals. Also, they can be classified into inorganic and organic solid wastes based on the chemical category [8].

3 The Concept for Treatment of Solid Waste

Nowadays, the foremost attitude for the treatment and disposal of solid wastes is known as the reduce, recycle, and reuse (3R) [9]. The reduce applies to control the formation and release of waste at the origin to lessen hazards to the environment and public health and, above all, make full use of resources.

The reduce necessitates diminishing the volume and quantity of solid waste and the sorts of harmful wastes. To reduce the solid waste, various legislations and regulations have been established over the past years relying on the policy of whoever causes pollution must be responsible for the treatment. This policy necessitates generating less or no solid waste when industries are planning their technologies and products. It also means circular use in the case of generation to avoid the environmental burden of waste. Recycling means the use of technical methods to recycle materials and energy from solid waste using administrative and technological actions to accelerate the circulation of materials and energy, achieving economic value [5, 10–13]. Three approaches for recycling solid wastes are known [12]: (1) Recycling is a procedure of dealing with solid waste and recycling of secondary materials. (2) Conversion is making new forms of substances from waste, such as asphalt materials using rubber and waste glass, construction materials using blast furnace slag and fly ash as starting materials, and compost using sludge and organic trash. (3) Energy conversion is a process of recycling energy in waste treatment (as heat and electrical energy). For example, burning organic waste to restore heat and create electricity. The biogas produced from rubbish sludge is used to produce heat and electricity for industries and residents. Fuel oil and gas can be produced by the pyrolysis of waste plastics. The reuse refers to eco-friendly safe discarding or treating generated solid waste. Physical, chemical, or biological methods are used to detoxify the solid wastes, thus preventing and minimizing pollution generated from solid waste [12].

4 Solid Waste Management

Waste is simply defined as any substance or item discarded by the holder. It can be classified as hazardous (such as chemical waste) and non-hazardous waste (such as packaging waste) [14]. Consequently, waste dumping should be looked at as a last option. Waste disposal means throwing away valuable resources and energy and, consequently, the emission of methane from landfilled biodegradable waste. On another note, landfill zones are getting restricted. SWM, by definition, is the framework related to control of generation, stockpiling, assortment, transport, dumping, and processing of solid waste materials in a way that best suits wellbeing, preservation, monetary, engineering, and other environmental considerations. SWM includes arrangement, administrative, monetary, engineering, and legitimate issues. Its practices can contrast for ordinary and industrial makers, for different territories, and developed and non-developed nations. The managing of non-hazardous waste in civic zones is the job of local government specialists. Then again, the administration of hazardous waste materials is typically the commitment of the individuals producing it, as subject to local and even global specialists [15–18].

Lately, waste management has become a noteworthy topic for most ventures. All merchandise and items contain crude materials and energy. In the case that they are rejected, or life ended, this means getting rid of valuable common assets. Discarding waste can severely affect the surrounding air by contamination and greenhouse gasses. On that account, waste management can be seen as classification, transport, processing, control, and recycling of consumables that have been delivered by humans. It is mostly done to diminish the harmful impact on health and the environment and to restore resources [14].

The processing tactic for eco-friendly items implies a configuration process that represents ecological effects over the lifespan of those items. Consequently, ecological upgrades are linked to manufacturing processes that are linked to reduction, reuse, recycling, and remanufacturing. Nevertheless, eco-friendly comprises various items: purpose, extraction, manufacturing, building, maintenance, destruction, and financial strengthening [19]. Recently, sustainable, eco-friendly road development is progressively getting more consideration around the world. It makes our frameworks of street building clean and reduces ecological effects. It is additionally hastened by the progress in demanding sustainable cities and developments that are more naturally advantageous. Ecologically green roads and street development may have figured as a reaction of partners to the calls for sustainable advancement, which went up from the developing consciousness of the harmful effects of road development on our surrounding environments [20].

The quality of recycling wastes changes as a result of inadequate data on the properties of the manufacturing products, and the absence of affirmation of using recycled materials as input material in new development products, just as the absence of acknowledgment about the significant constituents and vital activities for recycling the wastes.

The challenges facing the recycling process are labor costs, lack of governmental consciousness, and support for recycling, constrained use of recycled materials, and assessing their quality. The main advantages of recycling are reducing the costs of transporting and disposing materials and protecting landfills to extend their lifetime and sometimes materials that have a lower cost than raw materials. Recycling saves natural resources and reduces their consumption, reduces energy use, and above all, reduces air pollution [8, 11, 21].

A growing in world plastic production and use due to increased advancement, overpopulation, and industrial development, increase the dispute arose by plastic wastes that constitute about 25% of municipal solid waste while polyethylene wastes constitute over 60% of plastic containers. Consequently, packing products are the most critical subscribers to environmental waste.

5 Valorization of Solid Waste

Increasingly strict guidelines concerning solid waste, such as expanding the interest for sustainable chemicals and fuels, are pushing the businessmen and the environment specialists forward toward more continuity to advance financial adequacy and meet clients' solicitations. Valorization of waste is one of the utmost critical research fields. It has drawn extraordinary attention as a possible option in contrast to the regular waste disposal of an extensive array of residues in landfill spots. The expanding development of environmental procedures to process such solid waste is an intriguing field of growing significance to our public community. The traditional landfill, incineration, and other different methods for dealing with solid wastes are well-known techniques for waste disposal. Still, these are not satisfying in managing solid waste [22].

Waste valorization is changing wasted materials into persistently important products counting substances, materials, and fuels. This idea arose a long time ago, and most of it was related to waste (Table 1). However, it was returned to the public with renewed concentration due to the rapid consumption of primary

Table 1 Diverse sorts of hazardous wastes engendered by industries and the metals present. Adapted with permission from Ref. [23], Copyright 2003, Elsevier

Waste type	Metals present
Waste batteries	Ni, Cd, Ag
Electronic waste	Cu, Sn, Au, Ag, Ni, Al, Zn
Waste X-ray films	Ag
MSW fly ash	Cu, Zn, Ni, Al, Cr, Pb
Petroleum spent catalyst	Ni, Co, Mo
Metal finishing industrial wastes	Cr, Ni, Cu, Zn, Au, Ag, Cd

resources, the creation of expanded waste and its dumping worldwide, and the urgent demand for more sustainable and cost-effective waste managing procedures [22, 23]. Waste valorization concerns the way of changing over waste materials into more advantageous products counting materials, fuels, and chemicals [24]. Such a tactic is generally interrelated to waste management for an extended period. Various valorization techniques are, currently, showing potential fulfilling of industrial demands. The application of flow chemical technology is one of such encouraging methodologies of waste processing to valued products. Serrano-Ruiz et al. [25] underlined the benefits of continuous stream valorization processes for biomass and food waste that included simplicity of upscaling, proficient reaction sequences giving a higher yield, better reaction control, and no need for catalyst separation.

6 Environmental Inducements for Industrial Waste Recycling

Environmental protection has been a real important worry, continually increasing in recent decades, with the growth of industrial activity in many nation-states. However, this situation is changing rapidly, as countries like China and India, with their immense population, have become major partners in industrial growth worldwide and have thus become more significant resource consumers and the right subscribers to environmental pollution. The metal and mineral industries have significant environmental impacts in two areas. The first one is the volume of industrial waste, effluents, and sludge. The general practice was to dispose of in landfills, causing serious health risks related to noxious metals in waste residues and landfills. Treating these mineral wastes and restoring valuable components and, in some cases, converting them into beneficial compounds will not only help reduce pressure on ponds and landfills but rather offset the cost of environmental protection. The second serious issue is the toxic emission of carbon dioxide, a large layer of gas that damages the ozone layer, which has led to constant climate change around the world [26, 27].

Recycling resources will reduce the energy required for raw materials and, in this way, lead to a lower rate of the environmental carbon dioxide increase in the atmosphere. The third ecological issue associated with the metallurgical industry is the release of harmful gases, such as sulfur dioxide, which is an unavoidable item in the extraction of base metals from sulfide minerals. Another waste gas is dioxin released from the thermal disintegration of organics related to raw material being handled. Technical uprisings such as catalytic adjustment of sulfur dioxide assist with diminishing the deleterious impacts of poisonous gas release. Recycling of metals from scrap assists with cutting the utilization of pure sulfide minerals and diminishes the immensity of sulfur dioxide outflow. Technical revolutions suchlike catalytic alteration of sulfur dioxide helps to lessen the harmful effects of noxious gas release. Recycling of resources from scrap helps to cut the use of virgin sulfide ores and reduce the magnitude of sulfur dioxide emission.

Recycling is the key to reducing the large volume of waste generated and is also benefit the environment. Furthermore, the increasing demand for metals can also be compensated by recycling and reuse of metals from secondary sources. In general, the treatment of solid wastes can confer at least two beneficial aspects: The first is the recovery of valuable metals, and the second is detoxification of the contaminated wastes [4, 28, 29]. Recycling will effectively reduce the loss of heavy metals to the environment and avoid new metals enter into circulation, thereby conserving energy resources and lowering the mining activities which disturb ecosystems. Aside from lowering gas emissions, recycling also diminishes air and water pollution accompanying new products manufactured from raw materials [30, 31].

Furthermore, the recovery of precious metals from waste will considerably affect the environment and the economy [31]. Other advantages of metals recovery from waste include a decrease in resource use, minimization of gaseous emissions, and toxic compounds, as well as consumption of much less energy. For instance, recycling of gold and Pd can lessen energy consumption to 65% and about 84% as compared with their production from primary resources [31, 32].

7 Traditional Methods of Industrial Waste Recycling

Due to extensive and uncontrolled use, nonrenewable resources are about exhaustion; hence, it is a pressing need to reduce the dependence on nonrenewable resources and to develop a cleaner technique based on renewable natural resources. Also, with the rising cost of energy and strict environmental regulations, there is an extensive loss of valuable metals due to wastes dumping [33, 34]. New resources of metals for tomorrow's need are themselves present in today's waste, which needs to be exploited and recovered with the aid of innovative technologies. Therefore, it has become mandatory to upsurge the retrieval of metals from secondary raw materials, which may contain metals of a much higher grade than ores themselves. Even with viewing as a promising metal resource, there are many sensible complications in exploiting secondary raw materials, concerning the monetary side, empowering limits, innovation, and the complicated nature of raw materials [35]. Various traditional methods are being applied for many years for the recovery of metals from waste. Some of these methods are advantageous over others, but sometimes, combinational methods appear to be the best choice from the perspective of economics as well as the environment and human wellness.

7.1 Pyrometallurgical Methods

Pyrometallurgical processing is a traditional method applied for the treatment of ores and wastes for metals recovery. It utilizes heat for metals separation out of other materials, employing the differences between their properties as densities, melting points, vapor pressures, oxidation potentials, and miscibility of their molten components. This method includes incineration or smelting in a blast or plasma arc furnace, melting, sintering, or drossing, for the treatment of wastes [36–38]. The heating is carried out in a blast furnace at temperatures above 1500 °C to change over waste to a refined configuration. Carbon, like coke or coal, is usually added as a reducing agent during heating of the oxide waste, where the metal's oxygen reacts with carbon forming CO_2 gas and metals' sulfides are converted into oxides that reduced then to metals. In the case of nonmetallic parts of waste, that is, gangue, it is removed by flux, which when heated with gangue forms a slag that floats on the top and can be skimmed or suctioned [39]. Despite many advantages of pyrometallurgical treatment including the efficient and economic recovery of maximum amounts of metals, it has some limitations, as the generation of hazardous emissions suchlike carcinogenic dioxins and furans due to the incineration of plastic or organic wastes. Metals such as Fe and Al end up in the waste slag as oxides. Along with this, incinerators also produce a high amount of greenhouse effect gases such as SO_2, Cl_2, hydrochloric acid (HCl), and NO_x and ash with a high concentration of heavy metals such as lead, arsenic, and cadmium, which are known to cause many serious health problems to humans. Metals are recovered partially and contain many impurities; thus, they need to be processed further to get the required purity by other treatments. Further, incomplete combustion releases carbon monoxide and other volatile compounds such as formaldehyde and acetaldehyde, whose treatment requires large capital expenses in equipment and advanced technologies [36, 40–46]. Overall, the waste is burnt, residue ash is left behind. It is assessed that almost one ton of ash is formed from three tons of incinerated waste. This ash is toxic and contains high concentrations of heavy metals and dioxins, which, when landfilled in the long run, leaked out into the soil, causing serious contamination to the environment. Many environmental concerns are related to smelting activities, as they eject very noxious pollutants into the air. Still, smelting with pollution control facilities is highly expensive, which contributes legitimately to significant expenses.

The pyrometallurgical methodologies (Fig. 1) are the basic successfully operating components of solid waste management frameworks [47–49]. They are depicted by higher temperatures and higher alteration rates than physicochemical or biochemical procedures. Their important advantageous effects are: (1) Significant decrease in waste volume for about 80 ~ 90% and mass (70 ~ 80%) [50]; (2) Short working and treatment period, from a few minutes to over 1 h or even a part of a day [51]; (3) Elimination of organic contaminants, for instance, dioxins and furans, phenols, and cyanides [52]; (4) Concentration and immobilization of inorganic contaminants, as heavy metals; which might be conveniently and securely utilized or probably discarded [53]; (5) Thorough decontamination, which is a decent strategy to get rid of pathogenic waste and organics dirtied waste; (6) Utilization of ferrous and non-ferrous metals from fly ash [54]; (7) Evading environmental burdens as verified through life cycle assessment findings [55, 56]. Waste-to-energy was appeared to have less ecological effects than practically some other sources of electric power [57]; (8) Possibility to utilize the sustainable energy of solid waste, such as electricity and heat [58]; yet, the expenses of the operation can be balanced somewhat by energy selling. Thermal treatment technology includes three main thermal conversion processes: combustion, gasification, and pyrolysis. Other thermal treatments include roasting (disposal of solid waste under the melting point), dewatering and drying, thermal decomposition, and sintering.

Incineration is the procedure of regulated and comprehensive burning of solid waste in an oxidizing atmosphere. At high temperatures, the solid waste decomposes and entirely oxidizes during the operation. By incinerations, we can recover eventual energy and destruct unsafe wastes, as clinics and hospital waste. The temperature in the incinerators fluctuates between 850 and 1200 °C. One of the most alluring points of interest is to reduce the starting volume of burnable waste by

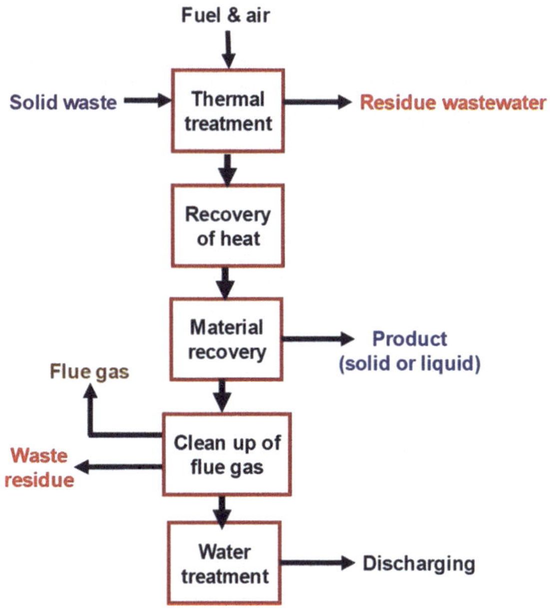

Fig. 1 Basic steps of thermal treatment processes

80–90%. When burning solid waste, its composition, characteristics, and heat value, should be taken into consideration. Solid waste incinerators can be sorted into three types: mechanical mesh, fluidized bed, and rotary kiln furnaces [59].

Pyrolysis is the chemical breakdown of organic waste into soot, liquid tar, and gases under reducing environments. Throughout pyrolysis, emanations are reduced by holding volatile constituents as alkalis and heavy metals (except Hg and Cd) and chlorine and sulfur contained inside the remains. Especially, pyrolysis largely prevents dioxins/furan formation and reduces thermal NO_x formation due to lower temperatures and reducing conditions [60]. Pyrolysis apparatuses can be sorted into a batch, semi-batch, and continual units based on the way of feeding and discharge. Another sorting relying on the flow, heat transfer, and reaction designs are fixed bed, fluidized bed, and screw oven ones. The goal of gasification is producing fuel, under incompletely oxidizing conditions to be stored and used in case of need; gasification is a waste-to-energy conversion system offering solutions to both waste discarding and energy issues. It is now one of the most significant thermal methods for disposing of solid waste.

7.2 Hydrometallurgical Methods

The hydrometallurgical process is well-defined as the chemical lixiviation in solutions aiming to recover valued metals from industrial solid waste [61–63]. In the hydrometallurgical method, chosen metals are separated from others depending on their solubility's differences and electrochemical properties while in aqueous solutions. Processed secondary materials are hydrometallurgically treated using strong acids or alkalis solutions targeting dissolution and precipitation of definite metals. The principle employed techniques during such treatment of secondary resources comprise lixiviation, ion exchange, solvent extraction, and precipitation varies contingent on the metals targeted for recovery. The main processes used for waste recycling are comparable to those used in the case of ores [64]. Hydrometallurgical pre-treatment is characteristically applied to recover important metals suchlike iron, steel, copper, manganese, and aluminum from electronic waste [65]. The followed procedure comprises three uninterrupted stages beginning with leaching, then the purification of solution and metals concentrate and ending with recovery of the desired metals [66–68]. This methodology has a low ecological effect, capital, and significant metal retrievals, just as being able to be applied for minor projects making it adaptable and productive. Table 2 illustrates various types of hydrometallurgical techniques for waste treatment.

The pregnant solution is separated, and metal is purified by removing impurities as gangue substances. The interested metals can then be precipitated as salts or compounds, extracted using solvents, adsorbed, or separated by ion exchange processes. Eventually, metals are probably retrieved from solution using electrorefining. Nevertheless, hydrometallurgical treatment of wastes engenders some hazardous gases such as chlorine, hydrogen cyanide, and other noxious fumes,

Table 2 Separation alternatives for waste treatment

Type of waste	Separation process
Soluble metals	Adsorption
	Cementation
	Electrowinning
	Ion exchange
	Membrane separations
	Precipitation
	Solvent extraction
Solid wastes	Biological separations
	Flotation
	Magnetic separations
	Pyrometallurgy
	Solvent partition

along with huge amounts of wastewater, which otherwise is not generated in furnace processes. However, as compared with pyrometallurgy, this treatment is more environmentally friendly. Gases formed during the process are not allowed to release, and solvents are locked up at normal temperatures, where they are prevented from producing dioxins or other greenhouse gases. Also, sulfur is presented as either a stable sulfate or natural sulfur as opposed to sulfur dioxide outflows.

Almost all waste components formed during or after treatment are segregated and retrieved for more recycling or reuse. Each component refining stage could be practiced in one procedure, lacking the need for redirection to another procedure.

Leaching processes yield residues and sludge, which can be sent for metal recovery. One of the limitations of hydrometallurgy is the slow and time-consuming process, which may affect the economy of the process when compared with pyrometallurgy. Some leaching agents employed for the process, for example, cyanide, are dangerous, and thus the process needs high safety standards. Also, valuable metals might be lost during various recovery steps [45, 69, 70].

Any hydrometallurgical process must include two essential steps: (1) the dissolution of metals from the solid material into an aqueous solution. This step is called leaching and frequently brings the concerned metals and other undesired constituents of the material into the leaching solution. (2) Separation of the wanted metals from others existing in solution.

Additional processing steps, along with these two primary steps, are usually needed. These were usually preconcentrate metals in the leachate in case of poor-quality materials.

The efficacy of the leaching process is influenced by various parameters such as type and concentration of the leaching agent, pH, temperature, lixiviation time, stirring speed, solid-to-liquid ratio, grain size, elements composition of the material, and many others [71]. Fast leaching kinetics at normal temperature is usually favored from an economic view, and slow kinetics for unwanted components is preferable as well. This makes precise leaching likely, leading to ease of subsequent processing and forming products with high pureness. Choosing suitable lixiviating agents affects the whole process and the consequent metal ions retrieval from the solution.

Albeit specific sorts of leachants might be less expensive and highly operative than others, they can cause problems in the treatment process if they are non-compatible with the chemical separation framework, for instance, coextracting or precipitating as unwanted compounds. The absolute most advantageous hydrometallurgical strategies for metal ions retrieval from leachates are solvent

Table 3 Merits and demerits for the separation of metal ions by some hydrometallurgical techniques. Adapted with permission from Ref. [72], Copyright 2016, Elsevier

Technique	Advantages	Disadvantages
Solvent extraction	• Gives highly pure products • Extractants are highly selective for various metals • Can adapt extractants or combine several extractants	• It consumes large chemicals • Some solvents are low-priced • The organic phase needs regeneration • Produces large amounts of wastes due to diverse steps • Needs complex equipment • Optimization and control of the process are not easy • Sensitive to solid impurities in the solution, forming scum or third phase • The product needs additional more processes to get solid metallic compounds
Ion exchange	• Gives very pure products can reach >99.999% • Highly selective • Can alter or advance ion exchangers having the favored necessities for separation • Can deal with complicated systems	• Not easy to scale up and high cost compared with precipitation and solvent extraction methods • The ion exchanger requires regeneration • Produces large amounts of secondary waste • Cannot process huge volumes in a short period, and its control is problematic than precipitation • Sensitive to large amounts of impurities in the solution • The product needs further processing to get final solid metallic compounds
Precipitation	• Simple and non-expensive • Easy to deal with and needs a simple apparatus • Yields products with extremely high purities • Generates less secondary waste • Can process huge volumes of in shorter periods • A solid product can be obtained	• Gives product of lower purity compared with other techniques • Coprecipitation issues • Hard to apply for complex aqueous solutions

extraction, ion exchange, and precipitation. A comparison between these techniques is revealed in Table 3.

Precipitation is suitable for systems in which the intrigued metals can be specifically precipitated from impurities, and coprecipitation is at least. For liquids having metal ions as contaminations, solvent extraction and ion exchange are relevant, knowing that solvent extraction is generally best suited for large-scale use. Full-scale solvent extraction separation utilizes counter-current equipment such as mixers and pulse columns [73]. The two types of equipment-tolerate multiple separation steps to get high purity standards. This can be made by using either column or by combining mixers.

Ion exchange methods, albeit costly and not easy to scale up, take into consideration extremely purities expected for certain products to be reused in highly advanced applications, for example, >99.99% for rare earth elements (REEs) [74]. Many of them have revealed great selectivity in metal ions separation from simple systems.

Nevertheless, when solutions contain many diverse chemical species, a set of different extractants, multiple extraction phases, or different extraction practices is required. One of the most important examples of this is the separation of individual REEs because of the great similarity in their chemical characteristics; individual separation requires highly selective extractants [75]. Usually, ultra-pureness is achieved using ion exchange resin.

8 Examples of Recycling Particular Types of Waste

8.1 Spent Hydroprocessing Catalyst Waste

Spent hydroprocessing catalysts are solid wastes resulting from petroleum refineries and incorporate several valuable and detrimental elements, suchlike V, Mo, Co, Fe, Ni, and As, along with elemental sulfur, carbon, and oils [76]. Hence, these spent catalysts recycling gained prominence due to environmental rules and protocols that consider them as unsafe materials owing to their noxious content and autonomous heating behavior. Spent hydroprocessing catalysts have been esteemed as hazardous materials by the environmental protection agency (EPA) of the USA [77]. In this manner, other than the ecological issues with discarding, spent catalysts can be looked at as secondary resources of raw materials. A number of recycling practices for metals retrieval from these catalysts were reported [78]. Direct smelting, calcination, chlorination, and salt roasting are examples of these applied practices [79, 80]. As well, numerous reagents, such as H_2SO_4, NH_3, NaOH, $(NH_4)_2SO_4$, oxalic acid with H_2O_2, and $Fe(NO_3)_2$, were experienced [78, 81–84].

Three principal metal retrieval varieties were considered by using pyro- and hydrometallurgical methods and their combination to take full advantage of the metal yield [85]. Ni, Mo, and V metals are retrieved as first-rate salts with more than 95% purity, while ultra-purity alumina is gained as boehmite. Metals can also be reclaimed through pyrometallurgical (roasting, heating) and hydrometallurgical (treating with organic and nonorganic solutions) routes (Fig. 2).

The retrieval of γ-alumina (boehmite) in addition to V, Mo, and Ni metals from the spent catalyst was found a cost affordable recycling waste and a better solution to the ecological case of the hazardous waste managing. The ammoniacal solution leaching exhibited poor metal separation and produced contaminated alumina. The Na_2CO_3 leaching method was found more effective than NaOH leaching, as molybdenum and vanadium are specifically separated over nickel and aluminum. The residue could be further isolated as nickel and solid alumina with high purity. For Mo and V extraction, inorganic acids were moderately not as effective as organic acids, yet they were similarly useful for Ni extraction. Oxalic acid was the most noteworthy efficient reagent for Mo and V metals but ineffectual for Ni retrieval. EDTA was found powerful reagent for recovering the three elements.

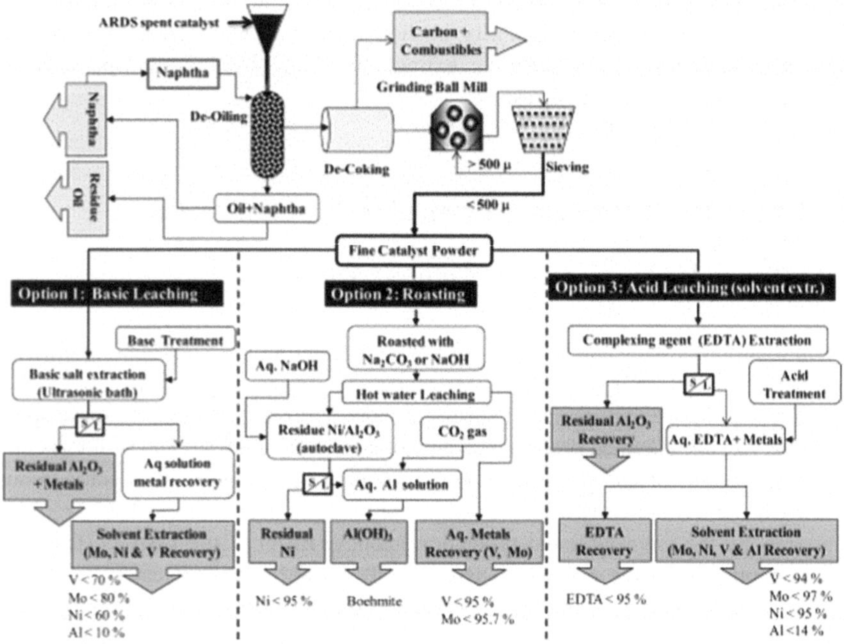

Fig. 2 Flow diagram of spent hydroprocessing catalyst recovery processes by using various pyrometallurgical options. Adapted with permission from Ref. [86], Copyright 2015, Royal Society of Chemistry

8.2 *Electronic Waste*

Waste electric and electronic equipment (WEEE) comprises cost-effective noteworthy amounts of precious, critical, and valuable elements, besides base metals and noxious compounds [11, 29, 87]. Based on comprehensive estimates, by the year 2021, the total global extent of WEEE possibly will surpass 50 million tons [88–90]. Risk-free discarding and weak ordinance are two major worries concerning the management of electronic waste, as the mainstream is exported to poor developing countries as China, India, and Africa [90, 91]. About 80% of American e-waste is exported to Asia [92]. Today, recycling and recovery of critical elements from electronic waste (e-waste) by a profitable technology are a matter of high priority in metallurgy owing to the speedy consumption of their innate resources. Wasted printed circuit boards (PCBs) constitute roughly 3% of approximately 50 Mt/ year of global electronic waste output [93, 94]. They are multifarious by their nature comprising of 40% metals, 30% ceramic, and 30% plastics. Of all these fractions, recovery of precious metals, especially Au, is the main motive behind recycling waste PCBs [65], which contain important metals suchlike Cu, Ni, and Fe in addition to precious metals (PMs) predominantly Ag, Au, and Pd [95]. Concentrations of PMs contained in PCBs are exceeding those present in their ores [96]. Hence, PCBs viewed as an excellent polymetallic secondary source of valued metals.

8.2.1 Waste Pre-treatment

Generally, the pre-treatment of waste PCBs comprise dismantling, removing solder, followed by physical separation techniques. Figure 3 overviews the order of these physical separation steps ending with obtaining the nonmetallic and metallic components out of waste PCBs. Physical pre-treatment is the foremost mandatory stage to recover the valued elements from e-waste. E-waste has significant varieties in materials content. This makes it tremendously difficult to be recycled for discerning retrieval of the enclosed elements due to its heterogeneity and the significant change in the composition as a consequence of the technology development over recent years [97–102]. Pre-treatment processes are usually required previous to the extraction processes for discerning retrieval of the demanded valuable or pollutant constituents, increasing the methodological efficacy and lowering the expenses of these processes [103].

Pre-treatment involves mainly dismantling, size decrease, and physical separation processes. These practices have been recognized as requisite for the common retrieval of metal fractions (magnetic and non-magnetic), glass, plastics, and others from e-waste. Nevertheless, part of precious constituents is lost throughout the pre-treatment procedures. This loss of elements in the overall recycling process is mainly because these substances end up in the released spouts (e.g., dust from shredding) [105, 96], suggesting that additional advancement of e-waste

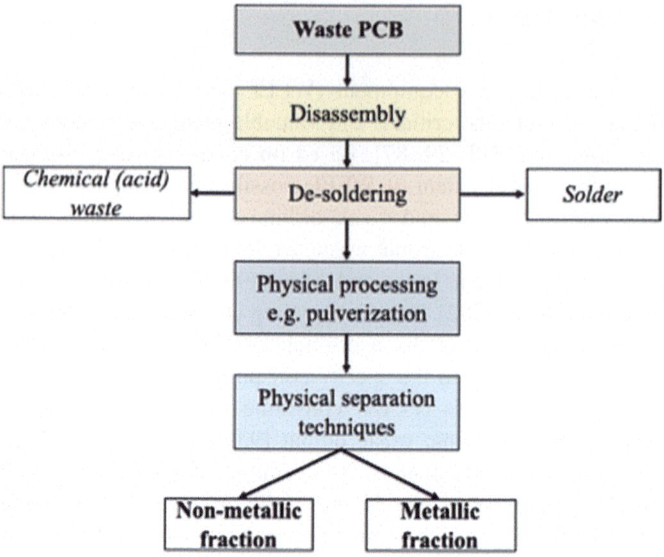

Fig. 3 A scheme of characteristic pre-treatment procedure for waste PCBs. Adapted with permission from Ref. [104], Copyright 2019, Royal Society of Chemistry

pre-handling should be done [106]. Chemical pre-treatment for the valorization of the enclosed values, elimination of harmful constituents, recovering energy, and so on can also be developed before extracting the metals. Overall, the pre-treatment phase involves manual dismantling, mechanical processing, and combined manual and mechanical preprocessing (Figs. 4 and 5).

8.2.2 Recovery of Metals

The essential incentive to recycle waste PCBs is the retrieval of metals. The primary paths of separating and retrieving metals are pyrometallurgy and hydrometallurgy. However, pyrometallurgy is still the most widely used classical technique on the industrial scale. Hydrometallurgy has gained attention with the growing advance of many practices, such as bioleaching [108]. Over 70% of waste PCBs are pyrometallurgical processed in smelters, in preference to mechanically treatment [45]. Crushed PCBs and the raw materials are added to smelters for separation and retrieval of copper, gold, and silver. Copper smelting processes can be classed as bath or flash smelting. In bath smelting, processing and transforming take place mainly in a molten bath where the concentrate contacts with the melted slag and matte by air injection into the molten bath. However, in flash smelting, processing takes place while the concentrate is widespread and suspended in the airflow. In Canada and Belgium, the Noranda and Umicore smelting processes are prime examples of efficient and most recognized industrially scale pyrometallurgical

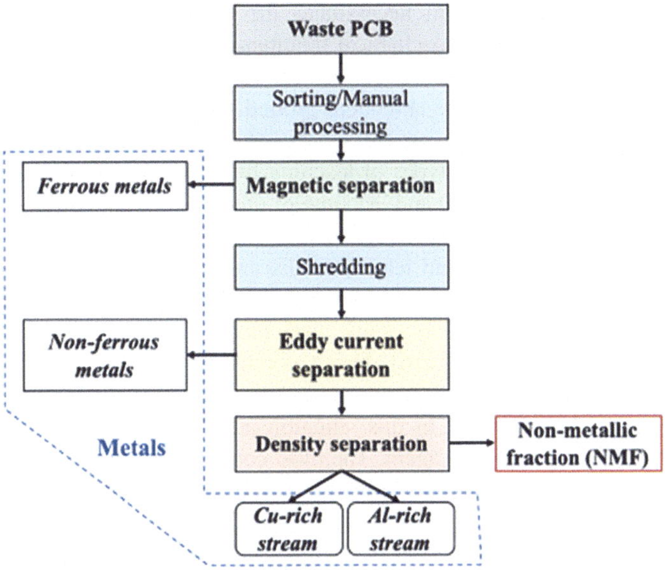

Fig. 4 Physical separation scheme for separation of metals from nonmetals fractions. Adapted with permission from Ref. [104], Copyright 2019, Royal Society of Chemistry

practices. The electronic scraps are initially introduced into an anode furnace then refined for metals recovery using electrowinning [42, 109–111]. The copper of the Umicore process is highly electro-refined after leaching; afterward, successive recovery steps were practiced for precious metal values recovery. Moreover, the process comprises an innovative practice that deters the release of noxious materials, thus diminishing human and environmental threats (Fig. 6).

On the technical side, integrated smelters recover only Cu metal leaving the Al and Fe concentrated in the slag. Also, the upgrading of metal value is limited owing to the incomplete separation of the metals. Thus, subsequent processing, as electrorefining, is essential for the complete separation of metals. The ceramics and glass components in PCBs lead to more slag formation and, consequently, more losses of precious and base metals. Finally, precious metals separation requires longer times and usually recovered by the end of the whole process [109]. Vacuum pyrometallurgy is another procedure of pyrometallurgy where metals having different vapor pressures can be separated and recovered through distillation and sequential condensation. Vacuum metallurgy separation for the recovery of heavy metals as Bi, Sb, Pb, and others has been proposed by many researchers [113]. Molten salt mixtures, as eutectic KOH–NaOH, at temperatures about 250 °C, have been shown to dissolve the organics along with the persistent and eventually recover Cu [113]. Techno-economic reports of pyrometallurgy have exposed that the e-waste recycling process based on copper smelting is economically feasible with a minimum plant capacity of 30 000 tons of e-waste per annum [114].

Hydrometallurgical treatment necessitates the detachment of the PCBs from most of the e-waste. At first, mechanical treatment is desirable owing to the multifaceted and encapsulated construction of metallic fractions of PCBs by plastics or resins. Generally, the PCBs are shredded, ground, and milled before leaching. At that stage, magnetic and eddy-current separation ease the retrieval of desired or undesired fractions [94]. Two types of leaching processes are notable: leaching of metals (as copper, zinc, and tin) and leaching of precious metals. Attributable to the precious metals' stability toward inorganic acids, selective leaching is most likely. Effectual leaching of copper and lead from discarded PCBs can be accomplished using nitric acid and subsequent electrowinning for metal recovery [94]. Adding hydrogen peroxide to inorganic acids, and raising the leaching temperature, enhances the efficiency [115, 116]. Using copper sulfate and sodium chloride, together with sulfuric acid, leads to increased efficiency in the leaching of copper, nickel, iron, and silver [117]. In this situation and the presence of oxygen, the copper ions act as an oxidizing agent. The most common process to recover copper from the leaching solutions is electrowinning, and better efficiencies are obtained using sulfuric acid leaching than aqua regia [118]. This is because the nitrate ions present in aqua regia leachates act as oxidizing agents, reducing the efficiency of the process.

For the acid leaching, three different reactions are represented. These reactions involve nitric acid, aqua regia, and sulfuric acid/H_2O_2 mechanisms [65, 119–123]. Acid leaching is currently the most common leaching method. Although it is quite corrosive, it has many vantages such as high leaching rate and fast kinetics. Cyanide leaching, the standard gold mining, is being gradually eliminated owing to its high noxiousness. Thiourea and thiosulfate are the least perilous leaching agents but are not as cost-effective. Thiourea is poorly stable while thiosulfate is kinetically slow, and leaching requires a large volume of both reagents. Intense leaching studying exposes several weaknesses associated with diverse types of acids and extraction reactions [65, 121–123]. Nitric acid (HNO_3) is the most common leaching solution used for metals dissolution, particularly Cu, Pb, and Sn (solder removal) from waste PCBs [124–127]. Researchers studied the use of aqua regia instead of HNO_3 for the precious metals extraction, stressing on Au [128, 129]. In general, highly concentrated acid solution and long leaching time result in higher metals extraction. Many researchers have also extracted and deposited copper from waste PCBs from nitric acid and aqua regia leaching media [130]. The results referred that nitric acid treatment is much preferred than the aqua regia as it gave higher Cu recovery. Nevertheless, the electrodeposition of Cu from this leaching solution could not be directly achieved due to the inhibition effect of the high acid content in the leaching solution [130]. Sulfuric acid also has shown to be efficient leaching, especially when mixed with an oxidizing agent, H_2O_2 [117, 121, 131–134]. Leaching of crushed waste PCBs (subjected to magnetic separation) in sulfuric acid solution (2 M H_2SO_4: 0.2 M H_2O_2) at 85 °C for 12 h yielded more than 95% Cu extraction [121]. The particle size of crushed PCBs substantially affects the Cu recovery, with reports exposed that crushing PCBs to less than 1 mm in size was optimum for Cu extraction [132].

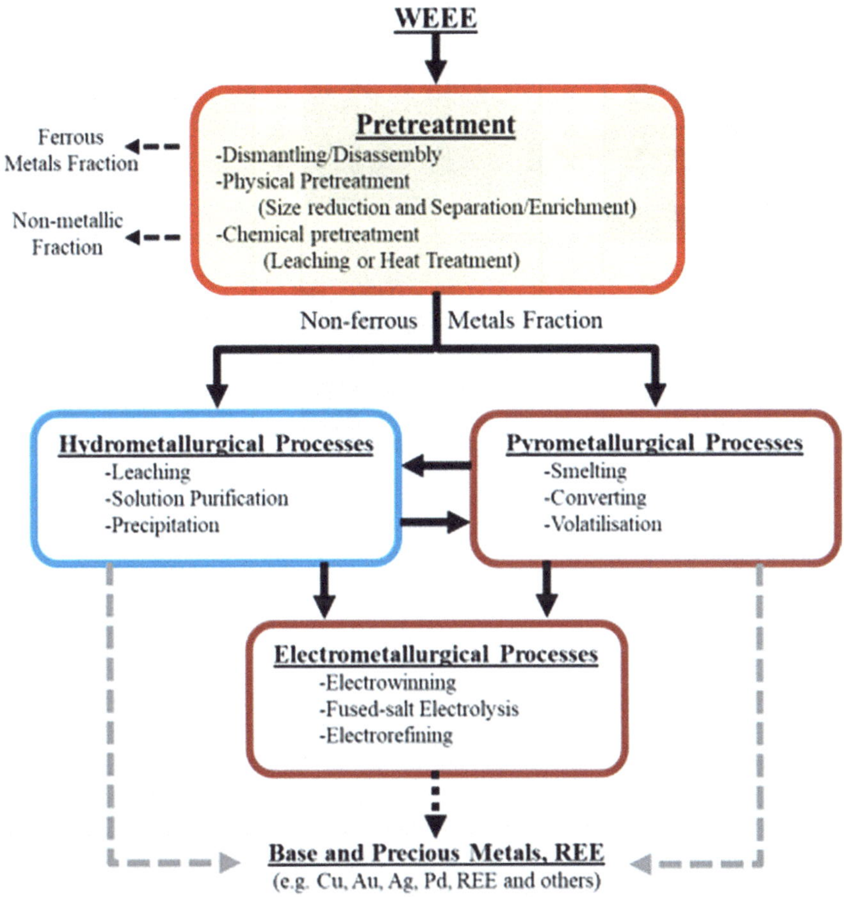

Fig. 5 Processes included the treatment of e-waste to recover valuable metals. Adapted with permission from Ref. [107], Copyright 2019, Taylor & Francis

Upon comparing the behavior of sulfuric acid, sulfuric acid, and hydrochloric acid, and aqua regia as leachants, unsurprisingly, aqua regia showed the highest percent of Cu extraction [135]. Aqua regia is known to be an effective solvent for metals leaching as Cu, Au, and Ag, but due to its highly corrosive and noxious nature, its industrial use is prohibited. Sulfuric acid, on the other hand, is somewhat lower in corrosivity, low-priced, and a lot easier to be industrially regenerated. Corrosion remains a problem, as sulfuric acid and hydrogen peroxide blend is corrosive; still, aqua regia is much more corrosive in comparison. Keeping these issues in mind, researchers have proposed using other leachants, namely ammonia leaching, which highly reacts with copper forming cupramine complexes, and their stability is controlled through ammonia concentration, pH, and oxidation potential.

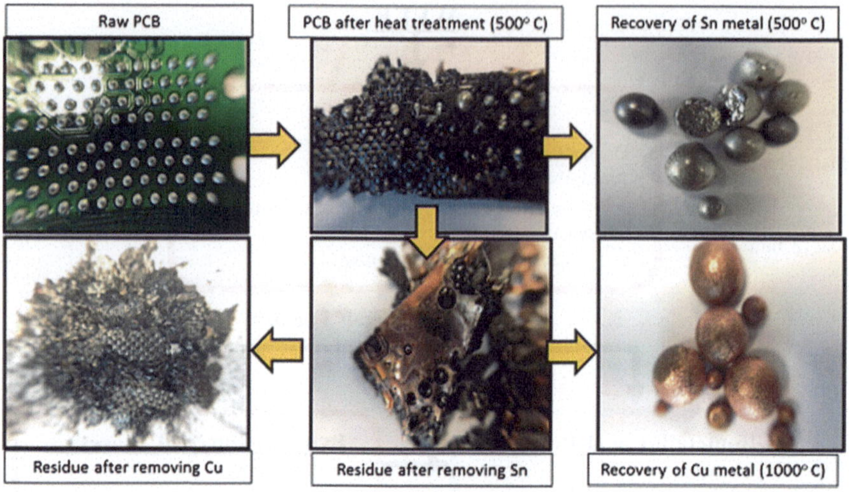

Fig. 6 Tin- and copper-based alloys recovered after thermal transformation at 500 and 1000 °C. Adapted with permission from Ref. [112], Copyright 2019, American Chemical Society

Ammonia-based leaching processes include ammonium carbonate, NH_3/NH_4Cl with $CuCl_2$ as an oxidant [136, 137].

Precious metal recovery is frequently accompanied by Cu recovery in a leaching process comprised of two various routes. The first one is the general acid leaching, and the second is frequently cyanide, thiosulfate, or thiourea leaching solutions for the extraction of precious metals. Cyanide leaching was traditionally used for Au and Ag mining, but it has been slowly ignored by time due to its toxicity [138]. Recent research has stressed using thiosulfate, thiourea solutions, and halide solutions for leaching of gold and silver [131, 139–143]. Some researchers have confirmed the validity of ammoniacal thiosulfate solutions to leach more than 90% Au from waste PCBs of mobile phones in the presence of ammonia and copper salt as a catalyst [144, 145]. Even though 95% Au and 100% Ag was extracted using 0.2 M $(NH_4)_2S_2O_3$, 0.02 M $CuSO_4$, and 0.4 M NH_4OH, the reaction took place in a considerably long time (24 to 48 h) [121]. In the same manner, leaching in thiourea and ferric ion as an oxidant has extracted about 86 and 71% for Au and Ag, respectively [131]. Nevertheless, the high cost and rapid consumption of thiourea are restrictive factors hinders process development and scaling-up [139, 146, 147].

8.2.3 Industrial-Scale Recycling

Regardless of the noteworthy number of studies overviewed above, the industrial pertinency of recycling methods to retrieve important metals out of e-waste is not as extensive. Most of the above-mentioned processes have not been widely used [148]. The retrieve problems are because of a few factors, for instance, collection, sorting, disassembly, difficult separation of the fractions of interest by classical milling activities, complicated and varied structure, low-grade of significant constituents, complex reagent processing, and so on. In many cases, specific data about processing steps are not usable due to competition and craft confidentiality. Umicore is one of the main recyclers of e-waste in Europe. Hydrometallurgical alongside pyro- and electrometallurgical processing is combined in its separation units. A broad assortment of metals is recovered, including common ones (cobalt, copper, nickel, lead, zinc, etc.), precious metals, indium, and so on [45]. These materials are inserted into a smelter where PCBs act as reducing agents and fuel alternatives. The separated copper which collects other melted metals in the smelter is additionally managed and refined via leaching and electrowinning. Then, metals are hydrometallurgically retrieved from each other and purified in precious metals' processing plant. Some industrial applications comprise both pyro- and hydrometallurgical processing [149]. During Umicore's smelting process, a Co–Cu–Ni–Fe alloy is obtained and processed using leaching with sulfuric acid followed by solvent extraction.

8.3 Lithium-Ion Batteries

A great number of rechargeable batteries were produced and used. Among these, lithium-ion batteries (LiBs) have increased quickly to outfit the electric energy source for transportable electronic appliances. Compared with other batteries (suchlike lead–acid and Ni–Cd batteries), LiBs are increasingly ecological because harmful heavy metals like lead and cadmium are not included in their structure [150]. The worldwide LiBs demand is expected by 6.9% each year. LiBs are extensively used in the field of different mobile phones, laptops, backup devices, power tools, digital products, and so on. It is assessed that by the year 2020, the amount of spent LiBs possibly will reach 200–500 tons per annum [150]. With this rate, wasted LIB is predicted to be 200–500 MT yearly, with 5–15 wt% cobalt and 2–7 wt% lithium metals content [151]. Recycling of LiBs has increased their significance due to the presence of combustible and harmful elements that may cause complications if not disposed of safely [63, 152, 153]. The retrieval of metals is one of the most important targets to recycle batteries to overcome the increasing use of batteries. The battery components are separated into several parts with a focus on the pure components. The EU battery directive considers battery recycling as a treatment for wasted batteries to recover products that can be used either in battery production or in any other applications.

The cathode materials in LiBs are lithium cobalt oxide, lithium manganese oxide, lithium nickel manganese oxide, lithium nickel cobalt aluminum oxide, and lithium titanate, and more of such materials are under development. Among these, the recycling of the electrode material ($LiCoO_2$) has good impacts, since its cobalt can be used in many applications. As of 2004, cobalt was reported to be twice costlier than nickel, and more costly than copper by 15 times [154, 155]. Besides, cobalt in LiBs has a low processing cost than its extraction from the ore. Although recycling batteries that have a small amount of lithium is high cost, it is expected that the battery usage will increase rapidly, which leads to an increase in demand for lithium in the foreseeable future. As a result, recycling has an economic return by recovering materials that will be future needed and avoiding disposal costs that are higher than collection and treatment costs.

The spent batteries can be recycled by various techniques, the basis of which is the physical and chemical methods. In the physical route, skinning, crust removing, crushing, sieving, and separation of materials are done to separate the cathode materials. Then these separated cathode materials are used to recover cobalt and other metals by a sequence of chemical procedures [156].

8.3.1 Physical Methods

After the spent LiBs have been categorized by type, usually the second stage is the physical processing [157, 158]. This pre-treatment is a must before performing the leaching and refining steps.

It is extremely difficult to achieve efficient, fast, and safe battery separation, especially for large batteries of electric cars. Plastic and iron scrapings are detached from the active electrode materials by physical processes. Plastics can be eliminated by flotation, and steel container chunks can be removed using magnetic separation [157, 159, 160].

After dismantling, a vibrating screen or dissolution with the help of ultrasound waves can be used to liberate the electrode materials from current collectors [161]. $LiCoO_2$ was then treated at a temperature range from 500 to 900 °C for burning the carbon and binder. Preceding leaching, the cathode material containing cobalt and lithium was ball milled and dried in a dryer at 60 °C for 48 h. Then, the obtained 15 μm sized powders were efficiently leached at 75 °C with nitric acid and hydrogen peroxide. Then, $LiCoO_2$ was prepared from the amorphous citrate precursor [162] (Fig. 7).

Chen et al. [164] presented a hydrometallurgical process to recycle spent LiBs and recover cobalt after dismantling to get rid of plastic and steel containers. Both electrodes materials were crushed to reach a size of 1–5 mm then heated at 150–200 °C for 2–3 h. The obtained powder was separated from copper and aluminum present in support materials. The dry mixture was ground for 30 min to achieve a fine powder (10–500 μm). That powder, consisting of 26.77% Co, 3.34% Li, 5.95% Al, 1.34% Cu, 3.76% Fe, 1.1% Mn, and 0.34% Ni (%w/w) was at that point leached in 5% sodium hydroxide for Al dissolution. The residue was calcined before reductive acid leaching. Performing such pyrolysis pre-treatment before hydrometallurgical processes break down the organics of the electrodes that can be problematic for the efficiency of the recovery process [165, 166].

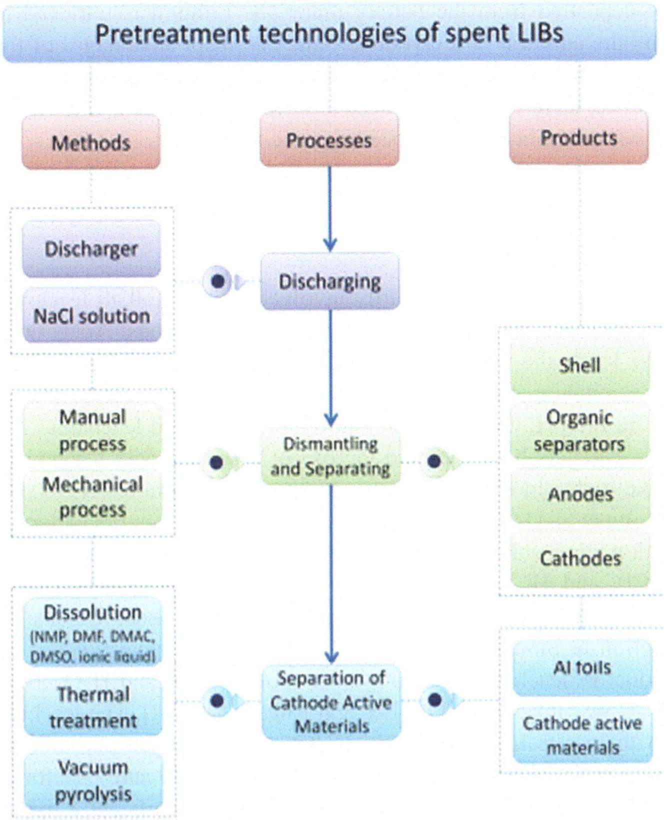

Fig. 7 Pre-treatment technologies of spent LiBs. Adapted with permission from Ref. [163], Copyright 2019, American Chemical Society

8.3.2 Thermal Methods

Thermal processes are interrelated to iron and its alloys as well as other metal alloys for industrial production. A method integrating many practices, mechanical, thermal, sol–gel, along with hydrometallurgical ones, has been reported to recuperate valuable materials out of used up LiBs [162]. In that method, heating in a muffle furnace (at 100–150 °C), dismantling in a high-speed shredder, and sorting according to size was performed. Then active electrode material was separated by vibrating screening, and the binder material was burned off at temperature ranging from 500 to 90 °C. The gained $LiCoO_2$ was lixiviated in acid solution (HNO_3); then, calcination was conducted to yield a fine residue. In a patented work by Lin and his team [167], pyro- and hydrometallurgical practices were used. The spent batteries were calcined and sifted to produce ash comprising metals and their oxides; the ash was dissolved and etched with HCl acid. Through electrolysis, Cu and Co are deposited from the battery materials. Lithium is retrieved as lithium carbonate by the addition of a carbonate salt. The main advantages of the thermal route are its simplicity and appropriateness. However, it fails to recover organic materials. Hence, there is a pressing need to have the equipment to purify the smoke and gases evolved from the combusted organic mixtures.

8.3.3 Chemical Methods

Chemical recycling processes primarily involves acid and alkali leaching, precipitation, extraction with solvents, and electrochemical methods. Acids or alkalis are employed to leach the cathode material so that metals can be retrieved using chemical precipitation, or through electrolysis [161]. The controlling factors in these steps are the pH, concentration, and type of the used leachants.

Several researchers have reported the use of inorganic acid leachants for spent LiBs suchlike hydrochloric (HCl), nitric (HNO_3), and sulfuric (H_2SO_4) acids. For instance, Zhang et al. [168] recovered lithium and cobalt from the battery electrodes by lixiviation in sulfurous acid (H_2SO_3), hydroxylamine hydrochloride ($HONH_2{\cdot}HCl$), and HCl. Hydroxylamine hydrochloride ($HONH_2{\cdot}HCl$) and HCl were more efficient than H_2SO_3, but HCl is preferable for cost reasons. A 4 mol/L of HCl solution at moderate temperature can lixiviate above 99% of cobalt and lithium. Reductive leaching of $LiCoO_2$ by 1 mol/L HNO_3 and H_2O_2 as a reducing agent at 75 °C was investigated by Lee and Rhee [151, 162]. The leaching efficacy increased with increasing temperature, HNO_3, and hydrogen peroxide concentrations. With no H_2O_2 addition, the leaching efficacy was 40 and 75% for cobalt and lithium, respectively, while 99% of both metals were significantly lixiviated with 1.7% by volume of H_2O_2. It should be mentioned that H_2O_2 lixiviates cobalt from the electrode's material containing divalent and trivalent Co ions, and the divalent ions are more readily dissolved in the HNO_3 solution (1 mol/L). This later solution can be utilized as the electrolyte for $LiCoO_2$ deposition on nickel sheets using fixed current [169]. Kang and his collaborators [170] reductively lixiviated powders containing 11.9% Co, 1.3% Li, 4.6% Cu, and 5.1% Al with 2 mol/L H_2SO_4, and 6% (by volume) H_2O_2 according to the following reaction:

$$LiCoO_2 + 1.5\ H_2O_2 + 1.5\ H_2SO_4 \rightarrow 0.5\ Li_2SO_4 + CoSO_4 + 3\ H_2O + O_2 \uparrow \quad (1)$$

Bearing in mind that cobalt is strongly united to oxygen, leaching of the cathode material ($LiCoO_2$) is not easy, and oxygen generated from H_2O_2 promotes cobalt oxide solvation by converting Co(III) to Co(II) [171]. Shin and coworkers [172] recycled $LiCoO_2$ battery cathodes. After physically separated, the powder was leached with 3 mol/L H_2SO_4 and 15% (by vol) H_2O_2 at a reasonable temperature of 75 °C. However, the organic matter of the binder causes filtration problems. Thus, burning the electrode powder at 900 °C before leaching eliminates the organics.

Unluckily, such treatment involves a substantial lessening of the cobalt leaching efficacy since its oxide surface was concealed by molten aluminum obstructing cobalt leaching for that, incineration is not endorsed here.

Chen et al. established a process relying on leaching and solvent extraction followed by precipitation aiming to recover cobalt metal from consumed LiBs [164]. After pretreating, the resultant powder with 26.77% Co, 3.34% Li, 5.95% Al, 3.76% Fe, Cu, and Mn around 1.3% was lixiviated by 5% NaOH solution to leach Al. The remainder was lixiviated with 4 mol/L H_2SO_4 and 10% (vol) H_2O_2 at a temperature of about 85 °C. Afterward, Fe (II) was precipitated as sodium jarosite ($NaFe_3(SO_4)_2(OH)_6$) at pH 3–3.5, and Mn was eliminated by reaction with ammonium persulfate at acidic pH of 4. In such circumstances, cobalt loss did not surpass 1.7%. Copper precipitated as hydroxide by NaOH at acidic pH ($\sim$5.5), accordingly, almost all Fe, Mn, and Cu were eliminated (98–99%), and concurrently, the loss of cobalt was below 2%. Li et al. [173] found that DL-malic acid ($C_4H_5O_6$), together with H_2O_2, could be used as effective leachants. At 40 °C, lithium, and cobalt were excellently lixiviated in DL-malic acid (1.5 mol/L) and H_2O_2 (2% by V) (Fig. 8).

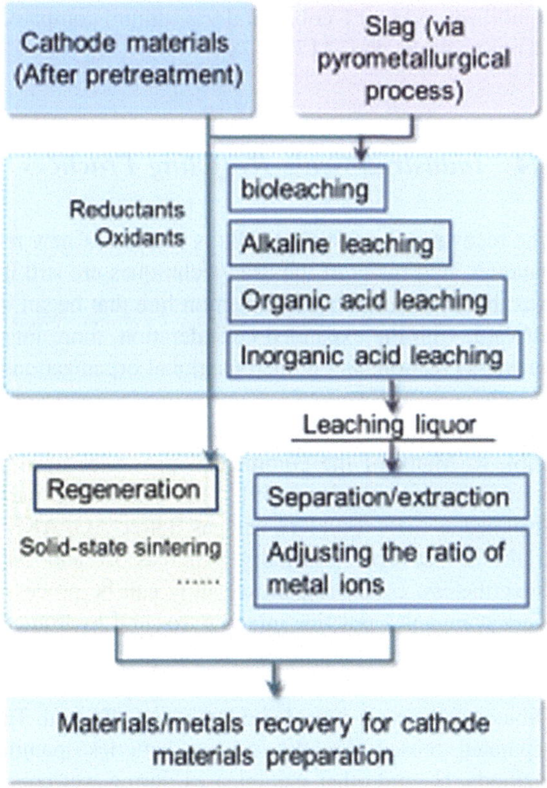

Fig. 8 Hydrometallurgical processes for LiBs. Adapted with permission from Ref. [174], Copyright 2018, American Chemical Society

In another study [175], H_2SO_4 (2 mol/L) and H_2O_2 (2.0%) were used to retrieve lithium and cobalt (87.5–96.3%) from $LiCoO_2$. CoC_2O_4. $2H_2O$ was precipitated by adding ammonium oxalate [$(NH_4)_2C_2O_4$], then separated and precipitated as Li_2CO_3 by adding Na_2CO_3. Eventually, Co and Li were retrieved with high purity (more than 94.0% and 71.0%, respectively).

Bioleaching is an intriguing approach retrieving valued metals from consumed lithium batteries. The establishment of inorganic acids as metabolous products by bacterial actions assists the metals lixiviation out of wastes and natural resources. Bioleaching of cobalt and lithium from consumed LiBs using chemolithotrophic and *acidithiobacillus ferrooxidans* bacterial strains was investigated [176]. Cobalt lixiviation by these bacteria is faster, and the addition of iron and sulfur to the medium advance the proficiency. The bioleaching mechanisms of $LiCoO_2$ electrodes were described [177] and were found reliant on the metal and the media. In the case of lithium, bioleaching is because of sulfuric acid generation, but in the case of cobalt, the dissolution is inferred to acid leachability and redox reactions of S and Fe.

Undeniably, because of Fe and S, the generated H_2SO_4 and Fe^{3+} lead to the direct dissolution of Co^{2+} and reduction–oxidation reactions forming Fe^{2+} encourage the solvation of insoluble Co^{3+}, respectively. Bioleaching of $LiCoO_2$ at low pH value ($\sim$2.5) with acidithiobacillus ferrooxidans, Fe and S (3 g/L and S) solubilizes over 60% cobalt and 9% lithium compared to only 20 and 5% of in case of not using bacteria [177, 178].

8.4 Industrial-Scale Recycling Practices

The recovery of consumed LiBs is somewhat new relative to the field of recycling batteries, and many of the used techniques are still in the beginning stages. Toxco was the first commercial production line that began in 1994 [179]. From that point forward, with the expanded consideration, more innovations have been applied in business creation, and numerous global organizations have perceived a total line to recycle spent LiBs.

Existing industrial-scale practices for the recycling of LiBs are outlined in Table 4. Many of the commercial processes used pyrometallurgical methods as secondary technologies [149]. This is because most of them are not originally designed as LiBs recyclers, such as Batrec AG, AKKUSER. Co, Ni, and Cu metals could be retrieved thermally, whereas Li and Al are eliminated in the slag. Nevertheless, cobalt and nickel alloy can be processed electrically. Thus, most of these companies practice mixed pyro- and hydrometallurgical methods.

Umicore process omits the discharging pre-treatment phase of spent LiBs and uses mechanical methods for materials separation. Compared to other processes, the Umicore process is considered facile and effectual. The Cu and Al foil, and steel are separated reused directly in new batteries manufacturing through mechanical methods. H_2 and other gas releases during wet crushing and thermal methods are

Table 4 Overview of top LiBs industrial technologies recyclers all over the world. Adapted with permission from Ref. [174], Copyright 2018, American Chemical Society

Company, country	Capacity (tons/year)	Used technology	Final output
SNAM, France	300	Sortation techniques; crushing then pyrolysis pre-treatment, crushing, and sieving	Cobaltiferous mix. "MELCO" Cobalt contents: 10–30%
INMETCO, USA	6000	Processing in a rotary furnace then refining in an electric arc furnace	Co/Ni/Fe alloy
Batrec Industrie AG, Switzerland	200	Shredding under CO_2 atmosphere, then treatment via the hydrometallurgical method	NA
Toxco Inc, Canada	4500	Shredding, milling, and screening, then leaching and precipitation	CoO and Li_2CO_3
AKKUSER, Finland	4000	Two-stage crushing; magnetic and separation methods; then smelting and lixiviation	Powdered metal
Umicore, USA	7000	Combined pyro-hydrometallurgical approaches	$CoCl_2$
Accurec GmbH, Germany	6000	Pyrolysis; mechanical separation of different materials; thermal treatment; slag treatment by hydrometallurgical methods	Co alloy
AEA Technology, UK	NA	Use of organic solvents to eliminate organics and electrolyte then, lixiviation-electrolysis of cathode material	LiOH and CoO
Shenzhen Green Eco Manufacture HiTech, China	2000	Leaching, purifying, and leaching-repreparation	Co powder
Glencore plc., Canada and Norway	7000	Combined pyro- and hydrometallurgical methods.	Co/Ni/Cu alloy
Bang Pu High-Tech Ni/Co, China	3600	Lixiviation, purification, and lixiviation—repreparation of the consumed cathode	Cathode material and Co_3O_4

*NA not applied

the outmost issues facing Toxco and Sumitomo-Sony. Other processes, including similar methods, also suffer from these issues. Conducting the recycling process under low temperature eases the disposal of flue gas and H_2, and treatment under CO_2 atmosphere also diminishes the jeopardy of mechanically treating the discharged batteries.

The Toxco process, Accurec, and other processes apply various hydrometallurgical methods. In these industries, mechanical treatment is performed first, and the slag left after is lixiviated in H_2SO_4 and NaClO to finally get $Co(OH)_3$. Li in the solution is usually precipitated as carbonate by CO_2 gas. In other pilots and corporations, the hydrometallurgical procedures comprise comparative approaches, suchlike leaching, solvent extraction, precipitation, and so on. Their main drawback is that they need a long time resulting in pollution and valuable metal losses. Some other organizations establish a closed circulatory system of recycling spent LiBs to regenerate cathode constituents or other invaluable materials in a short route [180].

8.5 Present Status and Economic Considerations

The valuation of wastes as a supply of value-added products is linked with effective utilization and recycling economics. One of the most crucial aspects faced in solid waste management (SWM) is the cost of treatment and handling the waste. Innovative approaches have been developed, which have reduced handling costs substantially. These approaches not only advance waste recovery options but also save costs [181]. Improved landfill and improved waste management are novel sustainable concepts for treating wastes [182]. A landfill is now seen as a non-permanent storage location for waste that has to be valued. Improved landfill grants a nice option to select the materials to be valued and could be used as a source of energy or product reliant on both waste and technology.

Solid waste management solutions ought to be monetarily maintainable, technically achievable, acceptable on a social and legal basis, and above all, ecologically safe. Today, waste valorization is a promising technological perspective through sorting them at the source and combining material recycling and energy generation methods. Nevertheless, sorting or discarding solid waste in landfill sites is not the best management. Thus, the selection of the dumping spot is extremely important for avoiding several social problems [183]. Efficient planning and development tactics concerning the extent and sorts of solid wastes are of outstanding importance to manage them sustainably. According to Senzige and his collaborators, quantization and characterization are the most important processes of proper waste managing [184]. At a specific point, information about the components and types of solid waste is necessary to integrate various waste management techniques, including recycling and recovery. Statistics can also help completely about infrastructure, develop the strategies used, and forecast any options related to the integrated waste management package [185].

Excessive disposal of waste is fading the landfill conception all around the World. Waste valorization is an interesting idea that is gaining prevalence in various regions due to the continuous growth in creating these wastes. For this reason, researchers are evolving valorization policies and concentrating on the use of safe materials and the design of clean technologies.

9 Conclusions

Industrial solid waste is an important environmental challenge. Improper waste management causes ecosystem alteration. Thus, it poses a tangible threat to public health. Reported studies witnessed that locals nearby industrial solid wastes services suffer low fetal weight at birth, genetic abnormalities, and more than one type of cancers. The progressive release of industrial solid wastes is a heavy burden on the civic budget. Population growth, rapid urban development, flourishing economy, and the rising quality of life have significantly enhanced the rate and amount of the industrial waste generated. Bio-disintegration of wastes with time is an important aspect governing the amount of materials to be recycled in a special way the organic contents. Recalling the current situation, our community would be better if we can efficiently reuse and valorize waste materials for the preparation and manufacture of valued materials and also nanomaterials. Recently, various waste materials have been utilized as a primary feedstock for producing important nanomaterials that can be used in many useful fields as for example in waste and wastewater treatment.

The complexity of electronic waste composition and the noxiousness of the materials contained in PCBs are significant challenges to the recycling practice. Upon the dismantling of PCBs, separation of metallic and nonmetallic fractions is performed by physical methods. Since the low-cost and high-tech ideas are controlling the development of new materials and designs for electronic equipment, these could weaken the concept of component recycling by physical procedures. Notwithstanding, virtually all current treatments for PCBs involve mechanical processing, even the highly advanced hydrometallurgical treatments. Pyrometallurgy is beneficial when related to the feed form of electronic scrap, but its problem is the problematic control of the product stream. The refining processes are not selective, leading to metals loss in the slag or the sludge. When compared to pyrometallurgy, hydrometallurgy is considered a greener selection and provides better management. Most hydrometallurgical methods aim to recover copper and precious metals. Leaching in acids is mostly used, where sulfuric and nitric acids are the superior reagents for Cu leaching, but they have poor leaching selectivity in subsequent processing steps.

Spent LiBs are a valued waste material containing important metals (Co, Li, Mn, Ni), and their recycling is economically beneficial. Recovery of cobalt and lithium is one of the most important goals of recycling spent LiBs since they are not widely distributed, valuable, and vital for many industries. Sustainable recycling of spent LiBs is extremely favored presently and for the coming future.

10 Future Perspectives

Waste valorization to renovate waste materials into more useful products has already existed long ago. However, it has been taken back to our society with restored attention due to the fast diminution of primary resources, the progressive waste generation, and the need for more maintainable and cost-adequate waste management procedures. The conversion of different wastes into products highlights the great potential of waste recycling and valorization strategies. The integration of recycling operations into future industries to produce value-added products and fuels will be a significant contribution toward achieving the world's highest priority aim, which is sustainable advancement. Nevertheless, the main and most important issue that has to be addressed for the next generations is society itself. The most common thought about waste as a problem, or as something of no value, should be replaced by a general acceptance by society of waste as a valuable resource. A resource, which is very complex, can simultaneously provide an infinite number of innovative solutions and alternatives to final products through revolutionary appraisement policies. These will require joint efforts from a range of disciplines, from engineering to chemistry, environmental science, legislation, and economics, to find innovative alternatives that we optimistically wish to see leading the way toward a more sustainable society.

It is imperative to recycle and valorize waste materials into products with high economic value and significant end use. In light of these facts, several new treatment methods have been explored to reduce the associated cost and the environmental impact of the recycling process. Waste can be used as resources to prepare value-added products including nanomaterials. Over the past few years, researchers have been synthesizing a variety of valuable compounds and nanomaterials from waste residues of industrial origin. Also, a number of researchers have employed and modified waste-derived NMs for removal of heavy metals, organic contaminants, and degradation of hazardous organic compounds from environmental effluents. Waste-derived carbon nanomaterials such as carbon nanodots and CNTs are extensively employed in different biomedical devices, including medical diagnosis and optical sensors. In the case of spent hydroprocessing catalysts, because of their hazardous nature, growing attention has been given to curtail their production at source along with developing safe and economically efficient procedures for recycling and metals recovery that are demonstrated in laboratory-scale studies. More research is still needed in this subject to develop these procedures from laboratory scale to marketable industrial scale.

At present, the processing approaches of PCBs is not sustainable; accordingly, new, eco-friendly approaches to recover invaluable constituents from waste PCBs are mandatory without delay. More studies on effective, efficient, and green metal separation and recovery techniques are necessary to defeat the challenges involved with PCB disposal and recycling.

For future research related to the recycling of spent LiBs, a complete recycling system ought to be distinguished and improved. Moreover, endeavors should also be focused on upgrading collection efficiency and diminishing the landfills of those consumed LiBs, which are considered dangerous to the ecosystem and living life forms on Earth. The inability to confine the recycling of used LiBs should be addressed. To wrap things up, a homogeneous worldwide system in manufacturing, evaluating, collecting, and recycling will significantly diminish the complexity of crude materials, which alternately lessens the energy utilization of the recycling procedure. These viewpoints have a lasting effect on accomplishing the monetary and eco-friendly recycling of spent LiBs.

References

1. Dixon N, Jones DRV (2005) Engineering properties of municipal solid waste. Geotext Geomembr 23(3):205–233
2. Prabakar D, Manimudi VT, Suvetha KS, Sampath S, Mahapatra DM, Rajendran K, Pugazhendhi A (2018) Advanced biohydrogen production using pretreated industrial waste: outlook and prospects. Renew Sustain Energy Rev 96:306–324
3. Ali GAM, Abdul Manaf SA, Kumar A, Chong KF, Hegde G (2014) High performance supercapacitor using catalysis free porous carbon nanoparticles. J Phys D Appl Phys 47 (49):495307–495313
4. Ali GAM, Divyashree A, Supriya S, Chong KF, Ethiraj AS, Reddy MV, Algarni H, Hegde G (2017) Carbon nanospheres derived from Lablab purpureus for high performance supercapacitor electrodes: a green approach. Dalton Trans 46(40):14034–14044
5. Ali GAM, Yusoff MM, Shaaban ER, Chong KF (2017) High performance MnO_2 nanoflower supercapacitor electrode by electrochemical recycling of spent batteries. Ceram Int 43:8440–8448
6. Ali GAM, Habeeb OA, Algarni H, Chong KF (2018) CaO impregnated highly porous honeycomb activated carbon from agriculture waste: symmetrical supercapacitor study. J Mater Sci 54:683–692
7. Han F, Le W (2019) Industrial solid waste recycling in western China. Springer Singapore
8. Sojobi AO, Nwobodo SE, Aladegboye OJ (2016) Recycling of polyethylene terephthalate (PET) plastic bottle wastes in bituminous asphaltic concrete. Cogent Eng 3(1)
9. Kumar V, Mishra AK, Mohan M (2016) Utilization of waste material in concrete paver blocks: a review. Int J Eng Sci Res Technol 4:1208–1210
10. Lu C, Zhang L, Zhong Y, Ren W, Tobias M, Mu Z, Ma Z, Geng Y, Xue B (2014) An overview of e-waste management in China. J Mater Cycles Waste Manage 17(1):1–12
11. Ali GAM, Tan LL, Jose R, Yusoff MM, Chong KF (2014) Electrochemical performance studies of MnO_2 nanoflowers recovered from spent battery. Mater Res Bull 60:5–9
12. Malinauskaite J, Jouhara H, Czajczyńska D, Stanchev P, Katsou E, Rostkowski P, Thorne RJ, Colón J, Ponsá S, Al-Mansour F, Anguilano L, Krzyżyńska R, López IC, Vlasopoulos A, Spencer N (2017) Municipal solid waste management and waste-to-energy in the context of a circular economy and energy recycling in Europe. Energy 141:2013–2044
13. Habeeb OA, Ramesh K, Ali GAM, Yunus RM (2017) Isothermal modelling based experimental study of dissolved hydrogen sulfide adsorption from waste water using eggshell based activated carbon. Malay J Anal Sci 21(2):334–345
14. Jassim AK (2017) Sustainable solid waste recycling. In: Skills development for sustainable manufacturing. InTech
15. Vesilind PA, Pereira NC (1980) Introduction to solid waste management. In: Wang LK, Pereira NC (eds) Solid waste processing and resource recovery. Humana Press, Totowa, NJ, pp 1–120

16. Ali GAM, Manaf SAA, Divyashree A, Chong KF, Hegde G (2016) Superior supercapacitive performance in porous nanocarbons. J Energ Chem 25(4):734–739
17. Habeeb OA, Ramesh K, Ali GAM, Yunus RM (2017) Experimental design technique on removal of hydrogen sulfide using CaO-eggshells dispersed onto palm kernel shell activated carbon: experiment, optimization, equilibrium and kinetic studies. J Wuhan Univ Technol Mater Sci Ed 32(2):305–320
18. Habeeb OA, Ramesh K, Ali GAM, Yunus RM (2017) Low-cost and eco-friendly activated carbon from modified palm kernel shell for hydrogen sulfide removal from wastewater: adsorption and kinetic studies. Desalin Water Treat 84:205–214
19. Rose CM, Stevels AB, Ishii K (2002) Method for formulating product end-of-life strategies for electronics industry. J Electron Manufact 11(02):185–196
20. Flynn A, Yu L, Feindt P, Chen C (2016) Eco-cities, governance and sustainable lifestyles: the case of the Sino-Singapore Tianjin Eco-City. Habitat Int 53:78–86
21. Hegde G, Abdul Manaf SA, Kumar A, Ali GAM, Chong KF, Ngaini Z, Sharma KV (2015) Biowaste sago bark based catalyst free carbon nanospheres: waste to wealth approach. ACS Sustain Chem Eng 5(9):2247–2253
22. Arancon RAD, Lin CSK, Chan KM, Kwan TH, Luque R (2013) Advances on waste valorization: new horizons for a more sustainable society. Energ Sci Eng 1(2):53–71
23. Cheremisinoff NP (2003) Municipal solid waste. In: Handbook of solid waste management and waste minimization technologies. Elsevier, pp 34–95
24. Kleerebezem R, Joosse B, Rozendal R, Van Loosdrecht MCM (2015) Anaerobic digestion without biogas? Rev Environ Sci Biotechnol 14(4):787–801
25. Serrano-Ruiz JC, Luque R, Campelo JM, Romero AA (2012) Continuous-flow processes in heterogeneously catalyzed transformations of biomass derivatives into fuels and chemicals. Challenges 3(2):114–132
26. Taylor DC (2000) Policy incentives to minimize generation of municipal solid waste. Waste Manag Res 18(5):406–419
27. Yang H-L, Innes R (2006) Economic incentives and residential waste management in taiwan: an empirical investigation. Environ Resour Econ 37(3):489–519
28. Elsayed DM, Abdelbasir SM, Abdel-Ghafar HM, Salah BA, Sayed SA (2020) Silver and copper nanostructured particles recovered from metalized plastic waste for antibacterial applications. J Environ Chem Eng 8(4):103826
29. Ali GAM, Yusoff MM, Algarni H, Chong KF (2018) One-step electrosynthesis of MnO_2/rGO nanocomposite and its enhanced electrochemical performance. Ceram Int 44(7):7799–7807
30. Needhidasan S, Samuel M, Chidambaram R (2014) Electronic waste—an emerging threat to the environment of urban India. J Environ Health Sci Eng 12(1)
31. Cayumil R, Khanna R, Rajarao R, Mukherjee PS, Sahajwalla V (2016) Concentration of precious metals during their recovery from electronic waste. Waste Manag 57:121–130
32. Wang X, Gaustad G (2012) Prioritizing material recovery for end-of-life printed circuit boards. Waste Manag 32(10):1903–1913
33. Shen R, Zhang G, Dell'Amico M, Brown P, Ostrovski O (2007) A feasibility study of recycling of manganese furnace dust
34. Xin B, Chen B, Duan N, Zhou C (2011) Extraction of manganese from electrolytic manganese residue by bioleaching. Biores Technol 102(2):1683–1687
35. Reuter M, Hudson C, Van Schaik A, Heiskanen K, Meskers C, Hagelüken C (2013) Metal recycling: opportunities, limits, infrastructure. In: A report of the working group on the global metal flows to the international resource panel
36. Sum EYL (1991) The recovery of metals from electronic scrap. JOM 43(4):53–61
37. Hoffmann JE (1992) Recovering precious metals from electronic scrap. JOM 44(7):43–48
38. Lee J-C, Song HT, Yoo J-M (2007) Present status of the recycling of waste electrical and electronic equipment in Korea. Resour Conserv Recycl 50(4):380–397
39. Singh S, Cameotra SS (2015) Anaerobic bioleaching by Acidophilic bacterial strains. In: Soil biology. Springer International Publishing, pp 179–201

40. Abdelbasir SM, Hassan SSM, Kamel AH, El-Nasr RS (2018) Status of electronic waste recycling techniques: a review. Environ Sci Pollut Res 25(17):16533–16547
41. Debnath B, Chowdhury R, Ghosh SK (2018) Sustainability of metal recovery from E-waste. Front Environ Sci Eng 12(6)
42. Hageluken C (2006) Improving metal returns and eco-efficiency in electronics recycling—a holistic approach for interface optimisation between pre-processing and integrated metals smelting and refining. In: Proceedings of the 2006 IEEE international symposium on electronics and the environment, IEEE
43. Chiang H-L, Lin K-H, Lai M-H, Chen T-C, Ma S-Y (2007) Pyrolysis characteristics of integrated circuit boards at various particle sizes and temperatures. J Hazard Mater 149 (1):151–159
44. Dalrymple I, Wright N, Kellner R, Bains N, Geraghty K, Goosey M, Lightfoot L (2007) An integrated approach to electronic waste (WEEE) recycling. Circuit World 33(2):52–58
45. Cui J, Zhang L (2008) Metallurgical recovery of metals from electronic waste: a review. J Hazard Mater 158(2–3):228–256
46. Khaliq A, Rhamdhani M, Brooks G, Masood S (2014) Metal extraction processes for electronic waste and existing industrial routes: a review and Australian perspective. Resources 3(1):152–179
47. Brunner PH, Rechberger H (2015) Waste to energy—key element for sustainable waste management. Waste Manag 37:3–12
48. Porteous A (2005) Why energy from waste incineration is an essential component of environmentally responsible waste management. Waste Manag 25(4):451–459
49. Psomopoulos CS, Bourka A, Themelis NJ (2009) Waste-to-energy: a review of the status and benefits in USA. Waste Manag 29(5):1718–1724
50. Consonni S, Giugliano M, Grosso M (2005) Alternative strategies for energy recovery from municipal solid waste. Waste Manag 25(2):123–135
51. McKay G (2002) Dioxin characterisation, formation and minimisation during municipal solid waste (MSW) incineration: review. Chem Eng J 86(3):343–368
52. Buekens A, Cen K (2011) Waste incineration, PVC, and dioxins. J Mater Cycles Waste Manage 13(3):190–197
53. Astrup T (2008) Management of APC residues from WtE plants. In: An overview of management options and treatment methods
54. Wiles CC (1996) Municipal solid waste combustion ash: state-of-the-knowledge. J Hazard Mater 47(1–3):325–344
55. Arena U, Mastellone ML, Perugini F (2003) The environmental performance of alternative solid waste management options: a life cycle assessment study. Chem Eng J 96(1–3):207–222
56. Azapagic A, Perdan S, Clift R (2004) Sustainable development in practice: case studies for engineers and scientists. John Wiley & Sons
57. Arena U (2012) Process and technological aspects of municipal solid waste gasification: a review. Waste Manage 32(4):625–639
58. Cheng H, Zhang Y, Meng A, Li Q (2007) Municipal solid waste fueled power generation in China: a case study of waste-to-energy in Changchun City. Environ Sci Technol 41 (21):7509–7515
59. Joseph A, Snellings R, Van den Heede P, Matthys S, De Belie N (2018) The use of municipal solid waste incineration ash in various building materials: a belgian point of view. Materials 11(1):141
60. Malkow T (2004) Novel and innovative pyrolysis and gasification technologies for energy efficient and environmentally sound MSW disposal. Waste Manag 24(1):53–79
61. Majed Al-Salem S, Constantinou A, Leeke GA, Hafeez S, Safdar T, Karam HJ, Al-Qassimi M, Al-Dhafeeri AT, Manos G, Arena U (2019) A review of the valorization and management of industrial spent catalyst waste in the context of sustainable practice: the case of the State of Kuwait in parallel to European industry. Waste Manage Res 37(11):1127–1141

62. Kentish S (2001) Innovations in separations technology for the recycling and re-use of liquid waste streams. Chem Eng J 84(2):149–159
63. Aboelazm EAA, Ali GAM, Algarni H, Yin H, Zhong YL, Chong KF (2018) Magnetic electrodeposition of the hierarchical cobalt oxide nanostructure from spent lithium-ion batteries: its application as a supercapacitor electrode. J Phys Chem C 122(23):12200–12206
64. Jha MK, Kumari A, Panda R, Kumar JR, Yoo K, Lee JY (2016) Review on hydrometallurgical recovery of rare earth metals. Hydrometallurgy 165:2–26
65. Tuncuk A, Stazi V, Akcil A, Yazici EY, Deveci H (2012) Aqueous metal recovery techniques from e-scrap: hydrometallurgy in recycling. Miner Eng 25(1):28–37
66. Brandl H, Bosshard R, Wegmann M (2001) Computer-munching microbes: metal leaching from electronic scrap by bacteria and fungi. Hydrometallurgy 59(2–3):319–326
67. Cerruti C, Curutchet G, Donati E (1998) Bio-dissolution of spent nickel–cadmium batteries using Thiobacillus ferrooxidans. J Biotechnol 62(3):209–219
68. Xu K, Deng T, Liu J, Peng W (2007) Study on the recovery of gallium from phosphorus flue dust by leaching with spent sulfuric acid solution and precipitation. Hydrometallurgy 86(3–4):172–177
69. La Brooy SR, Linge HG, Walker GS (1994) Review of gold extraction from ores. Miner Eng 7(10):1213–1241
70. Hilson G, Monhemius AJ (2006) Alternatives to cyanide in the gold mining industry: what prospects for the future? J Clean Prod 14(12–13):1158–1167
71. Ramachandra Rao S (2006) Resource recovery from process wastes. In: Resource recovery and recycling from metallurgical wastes. Elsevier, pp 375–457
72. Tunsu C, Retegan T (2016) Hydrometallurgical processes for the recovery of metals from WEEE. In: WEEE recycling. Elsevier, pp 139–175
73. Rydberg J (2004) Solvent extraction principles and practice, revised and expanded. CRC press
74. Tunsu C, Petranikova M, Ekberg C, Retegan T (2016) A hydrometallurgical process for the recovery of rare earth elements from fluorescent lamp waste fractions. Sep Purif Technol 161:172–186
75. Nash KL, Jensen MP (2001) Analytical-scale separations of the lanthanides: a review of techniques and fundamentals. Sep Sci Technol 36(5–6):1257–1282
76. Mishra D, Kim DJ, Ralph DE, Ahn JG, Rhee YH (2008) Bioleaching of spent hydro-processing catalyst using acidophilic bacteria and its kinetics aspect. J Hazard Mater 152(3):1082–1091
77. Marafi M, Stanislaus A (2008) Spent catalyst waste management: a review. Resour Conserv Recycl 52(6):859–873
78. Marafi M, Stanislaus A, Furimsky E (2017) Handbook of spent hydroprocessing catalysts. Elsevier
79. Busnardo RG, Busnardo NG, Salvato GN, Afonso JC (2007) Processing of spent NiMo and $CoMo/Al_2O_3$ catalysts via fusion with $KHSO_4$. J Hazard Mater 139(2):391–398
80. Gaballah I, Djona M, Mugica JC, Solozobal R (1994) Valuable metals recovery from spent catalysts by selective chlorination. Resour Conserv Recycl 10(1–2):87–96
81. Mulak W, Szymczycha A, Lesniewicz A, Zyrnicki W (2006) Preliminary results of metals leaching from a spent hydrodesulphurization (HDS) catalyst. Physicochem Probl Min Process 40:69–76
82. Rabah MA, Hewaidy IF, Farghaly FE (1997) Recovery of molybdenum and cobalt powders from spent hydrogenation catalyst. Powder Metall 40(4):283–288
83. Marafi M, Stanislaus A (2003) Options and processes for spent catalyst handling and utilization. J Hazard Mater 101(2):123–132
84. Mulak W, Miazga B, Szymczycha A (2005) Kinetics of nickel leaching from spent catalyst in sulphuric acid solution. Int J Miner Process 77(4):231–235
85. Al-Sheeha H, Marafi M, Raghavan V, Rana MS (2013) Recycling and recovery routes for spent hydroprocessing catalyst waste. Ind Eng Chem Res 52(36):12794–12801

86. Chauhan G, Pant KK, Nigam KDP (2015) Chelation technology: a promising green approach for resource management and waste minimization. Environ Sci Process Impacts 17 (1):12–40
87. Ali GAM, Yusoff MM, Feng CK (2015) Electrochemical properties of electrodeposited MnO_2 nanoparticles. Adv Mater Res 1113:550–553
88. Balde CP, Wang F, Kuehr R, Huisman J (2015) The global e-waste monitor 2014: quantities, flows and resources
89. Baldé A (2017) Quantities, flows, and resources. In: The global e-waste monitor. United Nations Univ
90. Robinson BH (2009) E-waste: an assessment of global production and environmental impacts. Sci Total Environ 408(2):183–191
91. Breivik K, Armitage JM, Wania F, Sweetman AJ, Jones KC (2015) Tracking the global distribution of persistent organic pollutants accounting for e-waste exports to developing regions. Environ Sci Technol 50(2):798–805
92. Perkins DN, Brune Drisse M-N, Nxele T, Sly PD (2014) E-waste: a global hazard. Ann Glob Health 80(4):286
93. Kaya M (2016) Recovery of metals and nonmetals from electronic waste by physical and chemical recycling processes. Waste Manag 57:64–90
94. Işıldar A, van de Vossenberg J, Rene ER, van Hullebusch ED, Lens PNL (2017) Biorecovery of metals from electronic waste, in sustainable heavy metal remediation. Springer International Publishing, pp 241–278
95. Anshu P, Hait S (2018) Comprehensive characterization of printed circuit boards of various end-of-life electrical and electronic equipment for beneficiation investigation. Waste Manag 75:103–123
96. Ebin B, Isik MI (2016) Pyrometallurgical processes for the recovery of metals from WEEE. In: WEEE recycling. Elsevier, pp 107–137
97. Cui J, Forssberg E (2003) Mechanical recycling of waste electric and electronic equipment: a review. J Hazard Mater 99(3):243–263
98. Stevens GC, Goosey M (2008) Materials used in manufacturing electrical and electronic products. In: Issues in environmental science and technology, p 40–74. Royal Society of Chemistry
99. Yazici EY, Deveci H (2015) Cupric chloride leaching ($HCl–CuCl_2–NaCl$) of metals from waste printed circuit boards (WPCBs). Int J Miner Process 134:89–96
100. Wu Z, Yuan W, Li J, Wang X, Liu L, Wang J (2017) A critical review on the recycling of copper and precious metals from waste printed circuit boards using hydrometallurgy. Front Environ Sci Eng 11(5)
101. Yamane LH, de Moraes VT, Espinosa DCR, Tenório JAS (2011) Recycling of WEEE: characterization of spent printed circuit boards from mobile phones and computers. Waste Manag 31(12):2553–2558
102. Kucuker M, Kerstin K (2018) Biomining–biotechnological systems for the extraction and recovery of metals from secondary sources. Glob NEST J 20(4):737–742
103. Chauhan G, Jadhao PR, Pant KK, Nigam KDP (2018) Novel technologies and conventional processes for recovery of metals from waste electrical and electronic equipment: challenges & opportunities—a review. J Environ Chem Eng 6(1):1288–1304
104. Hsu E, Barmak K, West AC, Park A-HA (2019) Advancements in the treatment and processing of electronic waste with sustainability: a review of metal extraction and recovery technologies. Green Chem 21(5):919–936
105. Marra A, Cesaro A, Belgiorno V (2018) Separation efficiency of valuable and critical metals in WEEE mechanical treatments. J Clean Prod 186:490–498
106. Chancerel P, Meskers CEM, Hagelüken C, Rotter VS (2009) Assessment of precious metal flows during preprocessing of waste electrical and electronic equipment. J Ind Ecol 13 (5):791–810
107. Sethurajan M, van Hullebusch ED, Fontana D, Akcil A, Deveci H, Batinic B, Leal JP, Gasche TA, Ali Kucuker M, Kuchta K, Neto IFF, Soares HMVM, Chmielarz A (2019)

Recent advances on hydrometallurgical recovery of critical and precious elements from end of life electronic wastes—a review. Crit Rev Environ Sci Technol 49(3):212–275
108. Teller M (2006) Recycling of electronic waste material. Sustainable metals management. Springer, Netherlands, pp 563–576
109. Cui J, Jørgen Roven H (2011) Electronic waste. In: Waste, pp 281–296. Elsevier
110. Gramatyka P, Nowosielski R, Sakiewicz P (2007) Recycling of waste electrical and electronic equipment. J Achievements Mater Manuf Eng 20(1–2):535–538
111. Moskalyk RR, Alfantazi AM (2003) Review of copper pyrometallurgical practice: today and tomorrow. Miner Eng 16(10):893–919
112. Hossain R, Nekouei RK, Mansuri I, Sahajwalla V (2019) Sustainable recovery of Cu and Sn from problematic global waste: exploring value from waste printed circuit boards. ACS Sustain Chem Eng 7(1):1006–1017
113. Zhan L, Xu Z (2009) Separating and recycling metals from mixed metallic particles of crushed electronic wastes by vacuum metallurgy. Environ Sci Technol 43(18):7074–7078
114. Ghodrat M, Rhamdhani MA, Brooks G, Masood S, Corder G (2016) Techno economic analysis of electronic waste processing through black copper smelting route. J Clean Prod 126:178–190
115. Deveci H, Yazici E, Aydin U, Yazici R, Akcil A (2010) Extraction of copper from scrap TV boards by sulphuric acid leaching under oxidising conditions. In: Proceedings of the going green-care innovation 2010 conference
116. Quinet P, Proost J, Van Lierde A (2005) Recovery of precious metals from electronic scrap by hydrometallurgical processing routes. Min Metall Explor 22(1):17–22
117. Yazici EY, Deveci H (2013) Extraction of metals from waste printed circuit boards (WPCBs) in H_2SO_4–$CuSO_4$–NaCl solutions. Hydrometallurgy 139:30–38
118. Veit HM, Bernardes AM, Ferreira JZ, Tenório JAS, Malfatti CdF (2006) Recovery of copper from printed circuit boards scraps by mechanical processing and electrometallurgy. J Hazard Mater 137(3):1704–1709
119. Pant D, Joshi D, Upreti MK, Kotnala RK (2012) Chemical and biological extraction of metals present in e-waste: a hybrid technology. Waste Manag 32(5):979–990
120. Xu Y, Li J, Liu L (2016) Current status and future perspective of recycling copper by hydrometallurgy from waste printed circuit boards. Procedia Environ Sci 31:162–170
121. Oh CJ, Lee SO, Yang HS, Ha TJ, Kim MJ (2003) Selective leaching of valuable metals from waste printed circuit boards. J Air Waste Manag Assoc 53(7):897–902
122. Zhang Y, Liu S, Xie H, Zeng X, Li J (2012) Current status on leaching precious metals from waste printed circuit boards. Procedia Environ Sci 16:560–568
123. Zhang L, Xu Z (2016) A review of current progress of recycling technologies for metals from waste electrical and electronic equipment. J Clean Prod 127:19–36
124. Mecucci A, Scott K (2002) Leaching and electrochemical recovery of copper, lead and tin from scrap printed circuit boards. J Chem Technol Biotechnol 77(4):449–457
125. Chaurasia A, Singh K, Mankhand T (2013) Extraction of tin and copper by acid leaching of PCBs. Int J Metall Eng 2(2):243–248
126. Kumari A, Jha MK, Singh RP (2016) Recovery of metals from pyrolysed PCBs by hydrometallurgical techniques. Hydrometallurgy 165:97–105
127. Neto IFF, Sousa CA, Brito MSCA, Futuro AM, Soares HMVM (2016) A simple and nearly-closed cycle process for recycling copper with high purity from end life printed circuit boards. Sep Purif Technol 164:19–27
128. Park YJ, Fray DJ (2009) Recovery of high purity precious metals from printed circuit boards. J Hazard Mater 164(2–3):1152–1158
129. Vijayaram R, Chandramohan K (2013) Studies on metal (Cu and Sn) extraction from the discarded printed circuit board by using inorganic acids as solvents. J Chem Eng Process Technol 4(2)
130. Maguyon MCC, Alfafara CG, Migo VP, Movillon JL, Rebancos CM (2012) Recovery of copper from spent solid printed-circuit-board (PCB) wastes of a PCB manufacturing facility

by two-step sequential acid extraction and electrochemical deposition. J Environ Sci Manag 15(1)
131. Behnamfard A, Salarirad MM, Veglio F (2013) Process development for recovery of copper and precious metals from waste printed circuit boards with emphasize on palladium and gold leaching and precipitation. Waste Manag 33(11):2354–2363
132. Yang H, Liu J, Yang J (2011) Leaching copper from shredded particles of waste printed circuit boards. J Hazard Mater 187(1–3):393–400
133. Xiao Y, Yang Y, van den Berg J, Sietsma J, Agterhuis H, Visser G, Bol D (2013) Hydrometallurgical recovery of copper from complex mixtures of end-of-life shredded ICT products. Hydrometallurgy 140:128–134
134. Silvas FPC, Jiménez Correa MM, Caldas MPK, de Moraes VT, Espinosa DCR, Tenório JAS (2015) Printed circuit board recycling: physical processing and copper extraction by selective leaching. Waste Manag 46:503–510
135. Vijayaram R, Nesakumar D, Chandramohan K (2013) Copper extraction from the discarded printed circuit board by leaching. Res J Eng Sci 2(1). ISSN 2278:9472
136. Koyama K, Tanaka M, Lee J-C (2006) Copper leaching behavior from waste printed circuit board in ammoniacal alkaline solution. Mater Trans 47(7):1788–1792
137. Sun ZHI, Xiao Y, Sietsma J, Agterhuis H, Yang Y (2016) Complex electronic waste treatment—an effective process to selectively recover copper with solutions containing different ammonium salts. Waste Manag 57:140–148
138. Fleming CA (1992) Hydrometallurgy of precious metals recovery. Hydrometallurgy 30(1–3):127–162
139. Birloaga I, De Michelis I, Ferella F, Buzatu M, Vegliò F (2013) Study on the influence of various factors in the hydrometallurgical processing of waste printed circuit boards for copper and gold recovery. Waste Manag 33(4):935–941
140. Alonso-Gómez AR, Lapidus GT (2009) Inhibition of lead solubilization during the leaching of gold and silver in ammoniacal thiosulfate solutions (effect of phosphate addition). Hydrometallurgy 99(1–2):89–96
141. Ficeriová J, Baláz P, Dutková E, Gock E (2008) Leaching of gold and silver from crushed Au–Ag wastes. Open Chem Eng J 2(1):6–9
142. Kejun L, Yen WT, Shibayama A, Miyazaki T, Fujita T (2004) Gold extraction from thiosulfate solution using trioctylmethylammonium chloride. Hydrometallurgy 73(1–2):41–53
143. Petter PMH, Veit HM, Bernardes AM (2014) Evaluation of gold and silver leaching from printed circuit board of cellphones. Waste Manag 34(2):475–482
144. Fotoohi B, Mercier L (2015) Recovery of precious metals from ammoniacal thiosulfate solutions by hybrid mesoporous silica: 2—a prospect of PGM adsorption. Sep Purif Technol 149:82–91
145. Ha VH, Lee J-C, Jeong J, Hai HT, Jha MK (2010) Thiosulfate leaching of gold from waste mobile phones. J Hazard Mater 178(1–3):1115–1119
146. Jing-ying L, Xiu-li X, Wen-quan L (2012) Thiourea leaching gold and silver from the printed circuit boards of waste mobile phones. Waste Manag 32(6):1209–1212
147. Gurung M, Adhikari BB, Kawakita H, Ohto K, Inoue K, Alam S (2013) Recovery of gold and silver from spent mobile phones by means of acidothiourea leaching followed by adsorption using biosorbent prepared from persimmon tannin. Hydrometallurgy 133:84–93
148. Graedel TE, Allwood J, Birat J-P, Buchert M, Hagelüken C, Reck BK, Sibley SF, Sonnemann G (2011) What do we know about metal recycling rates? J Ind Ecol 15(3):355–366
149. Chagnes A, Swiatowska J (2015) Lithium process chemistry: resources, extraction, batteries, and recycling. Elsevier
150. Vanitha M, Balasubramanian N (2013) Waste minimization and recovery of valuable metals from spent lithium-ion batteries—a review. Environ Technol Rev 2(1):101–115
151. Lee CK, Rhee K-I (2003) Reductive leaching of cathodic active materials from lithium ion battery wastes. Hydrometallurgy 68(1–3):5–10

152. Castillo S (2002) Advances in the recovering of spent lithium battery compounds. J Power Sources 112(1):247–254
153. Bankole OE, Gong C, Lei L (2013) Battery recycling technologies: recycling waste lithium ion batteries with the impact on the environment in-view. J Environ Ecol 4(1):14
154. Aboelazm EAA, Ali GAM, Chong KF (2018) Cobalt oxide supercapacitor electrode recovered from spent lithium-ion battery. Chem Adv Mater 3:67–74
155. Freitas MBJG, Garcia EM (2007) Electrochemical recycling of cobalt from cathodes of spent lithium-ion batteries. J Power Sources 171(2):953–959
156. Tenório JAS, Oliveira D, Chaves AP (1999) Carbon-zinc batteries treatment by ore processing methods. In: Proceedings: REWAS'99
157. Mantuano DP, Dorella G, Elias RCA, Mansur MB (2006) Analysis of a hydrometallurgical route to recover base metals from spent rechargeable batteries by liquid–liquid extraction with Cyanex 272. J Power Sources 159(2):1510–1518
158. Georgi-Maschler T, Friedrich B, Weyhe R, Heegn H, Rutz M (2012) Development of a recycling process for Li-ion batteries. J Power Sources 207:173–182
159. Xu J, Thomas HR, Francis RW, Lum KR, Wang J, Liang B (2008) A review of processes and technologies for the recycling of lithium-ion secondary batteries. J Power Sources 177 (2):512–527
160. Espinosa DCR, Bernardes AM, Tenório JAS (2004) An overview on the current processes for the recycling of batteries. J Power Sources 135(1–2):311–319
161. Li L, Lu J, Ren Y, Zhang XX, Chen RJ, Wu F, Amine K (2012) Ascorbic-acid-assisted recovery of cobalt and lithium from spent Li-ion batteries. J Power Sources 218:21–27
162. Lee CK, Rhee K-I (2002) Preparation of $LiCoO_2$ from spent lithium-ion batteries. J Power Sources 109(1):17–21
163. Yao Y, Zhu M, Zhao Z, Tong B, Fan Y, Hua Z (2018) Hydrometallurgical processes for recycling spent lithium-ion batteries: a critical review. ACS Sustain Chem Eng 6(11):13611–13627
164. Chen L, Tang X, Zhang Y, Li L, Zeng Z, Zhang Y (2011) Process for the recovery of cobalt oxalate from spent lithium-ion batteries. Hydrometallurgy 108(1–2):80–86
165. Sun L, Qiu K (2011) Vacuum pyrolysis and hydrometallurgical process for the recovery of valuable metals from spent lithium-ion batteries. J Hazard Mater 194:378–384
166. Sun L, Qiu K (2012) Organic oxalate as leachant and precipitant for the recovery of valuable metals from spent lithium-ion batteries. Waste Manag 32(8):1575–1582
167. Lin J-R, Fan C, Chang I-L, Shiu J-Y (2003) Clean process of recovering metals from waste lithium ion batteries. Google Patents
168. Zhang P, Yokoyama T, Itabashi O, Suzuki TM, Inoue K (1998) Hydrometallurgical process for recovery of metal values from spent lithium-ion secondary batteries. Hydrometallurgy 47 (2–3):259–271
169. Li L, Chen R, Sun F, Wu F, Liu J (2011) Preparation of $LiCoO_2$ films from spent lithium-ion batteries by a combined recycling process. Hydrometallurgy 108(3–4):220–225
170. Kang J, Senanayake G, Sohn J, Shin SM (2010) Recovery of cobalt sulfate from spent lithium ion batteries by reductive leaching and solvent extraction with Cyanex 272. Hydrometallurgy 100(3–4):168–171
171. Lain MJ (2001) Recycling of lithium ion cells and batteries. J Power Sources 97:736–738
172. Shin SM, Kim NH, Sohn JS, Yang DH, Kim YH (2005) Development of a metal recovery process from Li-ion battery wastes. Hydrometallurgy 79(3–4):172–181
173. Li L, Ge J, Chen R, Wu F, Chen S, Zhang X (2010) Environmental friendly leaching reagent for cobalt and lithium recovery from spent lithium-ion batteries. Waste Manag 30(12):2615–2621
174. Lv W, Wang Z, Cao H, Sun Y, Zhang Y, Sun Z (2018) A critical review and analysis on the recycling of spent lithium-ion batteries. ACS Sustain Chem Eng 6(2):1504–1521
175. Zhu S-G, He W-Z, Li G-M, Xu Z, X-J Zhang, Huang J-W (2012) Recovery of Co and Li from spent lithium-ion batteries by combination method of acid leaching and chemical precipitation. Trans Nonferrous Met Soc China 22(9):2274–2281

176. Mishra D, Kim D-J, Ralph D, Ahn J-G, Rhee Y-H (2008) Bioleaching of metals from spent lithium ion secondary batteries using *Acidithiobacillus ferrooxidans*. Waste Manag 28 (2):333–338
177. Xin B, Zhang D, Zhang X, Xia Y, Wu F, Chen S, Li L (2009) Bioleaching mechanism of Co and Li from spent lithium-ion battery by the mixed culture of acidophilic sulfur-oxidizing and iron-oxidizing bacteria. Biores Technol 100(24):6163–6169
178. Ou L, He R, Feng Q, Chen Y (2007) Investigation on the leaching of pyrite in a germ-free system. Min Metall Explor 24(2):111–114
179. Al-Thyabat S, Nakamura T, Shibata E, Iizuka A (2013) Adaptation of minerals processing operations for lithium-ion (LiBs) and nickel metal hydride (NiMH) batteries recycling: critical review. Miner Eng 45:4–17
180. Velázquez-Martínez O, Valio J, Santasalo-Aarnio A, Reuter M, Serna-Guerrero R (2019) A critical review of lithium-ion battery recycling processes from a circular economy perspective. Batteries 5(4):68
181. Ghiani G, Laganà D, Manni E, Musmanno R, Vigo D (2014) Operations research in solid waste management: a survey of strategic and tactical issues. Comput Oper Res 44:22–32
182. Rich C, Gronow J, Voulvoulis N (2008) The potential for aeration of MSW landfills to accelerate completion. Waste Manag 28(6):1039–1048
183. Hadjibiros K, Dermatas D, Laspidou C (2011) Municipal solid waste management and landfill site selection in Greece: irrationality versus efficiency. Glob Nest J 13:150–161
184. Senzige J, Nkansah-Gyeke Y, Makinde O, Njau K (2014) The potential for solid waste recycling in urban area of Tanzania: the case of Dar Es Salaam. Int J Environ Prot Policy 2330–7536(211):147–1522330
185. Abdalqader A, Hamad J (2012) Municipal solid waste composition determination supporting the integrated solid waste management in Gaza strip. Int J Environ Sci Dev 3:172–176

Environmental Susceptibility and Nanowaste

Priyabrata Roy and Moharana Choudhury

Abstract The appearance and fate of nanomaterials (NMs) also are new area of waste management. Information and techniques for investigations are minimal. Nonetheless, it is incredibly likely that nanomaterials used in several items or papers of one type would be in the waste stream. Environmental and environmental risks related to the treatment of nanowastes remain unexplored. Another factor is whether containing nanomaterials, consisting of recycling processes, will affect the waste management capabilities/performance. In comparison, nanomaterials may substitute certain substances that make products, for example, smarter or more efficient, to get into waste management sooner and potentially play a role in waste reduction. Draw up an overview of nanomaterial and waste-related scientific, health, and environmental problems, and assess the available recycling issues of environmental health significance are needed. One ultimate goal is to consider looking for identical statistics to compare the potential hazards associated with the existence of NMs in the waste. The emphasis is on eliminating consumer goods as waste and not creating waste anymore. Alternatively, instead of other residuals (e.g., cosmetics, containers, etc.), attention can be given to appliances and athletic equipment. Consciousness is usually on product forms and waste sources, where knowledge is at all available. Thus, the papers and studies that are indirectly available statistics are implicit delimitations; this research area is relatively new because of the reality. It has also sought to cover goods, however. Concerning incineration, its miles found it more relevant to observe the load and fate of particular NMs in respect of goods categories. After the initial activities, the spectrum can be similarly oriented and fabricated/designed nanomaterials to offer a selected

P. Roy (✉)
Centre for Interdisciplinary Studies, Barrackpore, Kolkatta 700123, India
e-mail: priyabrataroy2009@gmail.com

P. Roy
Department of Molecular Biology and Biotechnology, University of Kalyani, Kalyani 741235, India

M. Choudhury
Voice of Environment (VoE), Guwahati 781034, India
e-mail: moharanasil@gmail.com

A. S. H. Makhlouf and G. A. M. Ali (eds.), *Waste Recycling Technologies for Nanomaterials Manufacturing*, Topics in Mining, Metallurgy and Materials Engineering, https://doi.org/10.1007/978-3-030-68031-2_3

character in a product. Besides, the commonly considered nanomaterials within the context of the EU concept supported. Nevertheless, most of the evidence sources examined no longer detailed nanomaterials in exercise, and as a result, all known sources of information about nanomaterials were included in the waste.

Keywords Nanowaste · Environmental pollution · Nanomaterial · Nanotechnology · Toxicity of nanowaste · Nanohazards · Sustainability

List of Abbreviations

AFM	Atomic force microscopy
AgNPs	Silver nanoparticles
APS	Advanced photon source
BET	Brunauer-emmett-teller
CLSM	Controlled low strength materials
CPC	Condensation particle counter
CuNPS	Copper nanoparticles
DLS	Dynamic light scattering
EDX	Energy dispersive X-ray analysis
ENMs	Engineered nanomaterials
EOL	End of life cycle
EPA	Environmental protection agency
FFF	Fused filament fabrication
FS	Fluorescence spectroscopy
GC-MS	Gas chromatography mass spectrometry
GMD	Gestational diabetes mellitus (disease)
HDC	Hydrodynamic chromatography
HR-TEM	High-resolution transmission electron microscopy
ICP-MS	Inductively coupled plasma mass spectrometry
LC50	Lethal concentration 50%
MALS	Multiangle light scattering
NMs	Nanomaterials
NOx	Nitrogen oxides
NPs	Nanoparticles
POPs	Persistent organic pollutants
PPCPs	Pharmaceutical and personal care products
SAED	Semi-automatic external defibrillator
SAXS	Small angle X-ray scattering device
SCR	Selective catalytic reduction
SEC	Size-exclusion chromatography
SEM	Scanning electron microscopy
SNCR	Selective non-catalytic reduction
SNCR	Selective non-catalytic reduction technology
SP-ICP-MS	Single particle inductively coupled plasma mass spectroscopy

STP	Sewage treatment plant
TEM	Transmission electron microscopy
UV-Vis	Ultraviolet–visible spectroscopy
WHO	World health organization
XPS	X-ray photoelectron spectroscopy
XRD	X-ray diffraction

1 Introduction

Nanotechnology and nanoscience are a rapidly developing field with tremendous potential for advancing and benefiting many research and application fields. The latest development in material science is nanoscience, which deals with the characterization methods for tracking nanometer-scale materials (10^{-9} m) and the possibilities for understanding the interactions on an atomic level [1]. The growing production rate of different nanomaterials (NMs) increases; a new kind of waste containing NMs can be called nanowaste. Sometimes, without proper precautions or treatment, the waste is generating in the regular waste stream from manufacturing to recycling. No internationally accepted statistical model is available on the generation of nanowaste from different sources and processes. Therefore, the assessment of production and life cycle analyses of NMs is problematic. In some instances, engineered nanomaterials (ENMs) in the context of biocompatibility and potentially toxic nature are inconsistent if their disposal reaches the threshold point [2]. This condition means that nanowastes ongoing production could be calculated using data from an area manufacturing unit and based on the quantity of NMs-related technologies recorded in development.

2 Types of Nanomaterials and Their Uses

Nanoscience and nanotechnology are generally regarded as having tremendous potential for many fields of research and application, which rapidly draw investment from governments and companies in many parts of the world. This situation is also recognized as its implementation can pose new challenges in protection, regulation, or ethics that require debate in society. Nanomaterials could be organic or inorganic substances such as carbon, metals, and metal oxides [3–6]. Combinations of ingredients may also be present. There are several types of nanomaterials (Fig. 1). Nanomaterials can either be inadvertently or prepared deliberately in the sense of nanotechnology. While the ENMs have many benefits, they have tremendous potential for environmental degradation.

Fig. 1 Broad classification of engineered nanomaterials

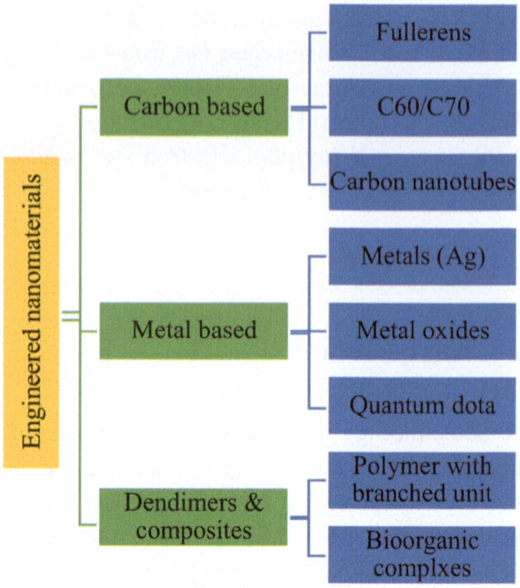

In the manufacturing of various materials/products the nanomaterials, like electrical, catalytic, structural, resonance surface, etc., have a wide range of applications, including sensors, thermo-electric components, photocatalysts, solar cells, and items such as paints, polycarbonate, sprays, cleaning materials, coatings, moisturizers, films, and packing. As these materials/products enter the end-of-life phase, they become extreme nanoresidues. Furthermore, it will be a significant concern to ensure its safe disposal. These waste materials come from several nano-based products, the manufacturing and production procedures such as metal waste/scrap, paper, and carton, plastics, fabrics and leather, electronic equipment, batteries, solar cells, tires, pharmaceuticals, municipal and industrial waste (Fig. 2). NMs in all three paths, such as solid, liquid, and gasses, can reach the environment at the end of its life. The waste product would be analyzed with a framework with a hierarchy, beginning from the acquisition of raw materials through to the production of supplies to the end of the processing by a customer. When NMs match this hierarchy of frameworks, they are dispersed into the waste stream, destined for the treatment plant.

On the other hand, ENMs are accumulated with solid waste and reach the site. The waste material in the nanoscale is known to have its end-stage as waste deposits. As a third step, the release of harmful vapors of NMs into the atmosphere and severe health disruptions during the processing of such goods occurs. Such nanowastes can be oxidized with other waste contaminants and released into the atmosphere as gas from the vent.

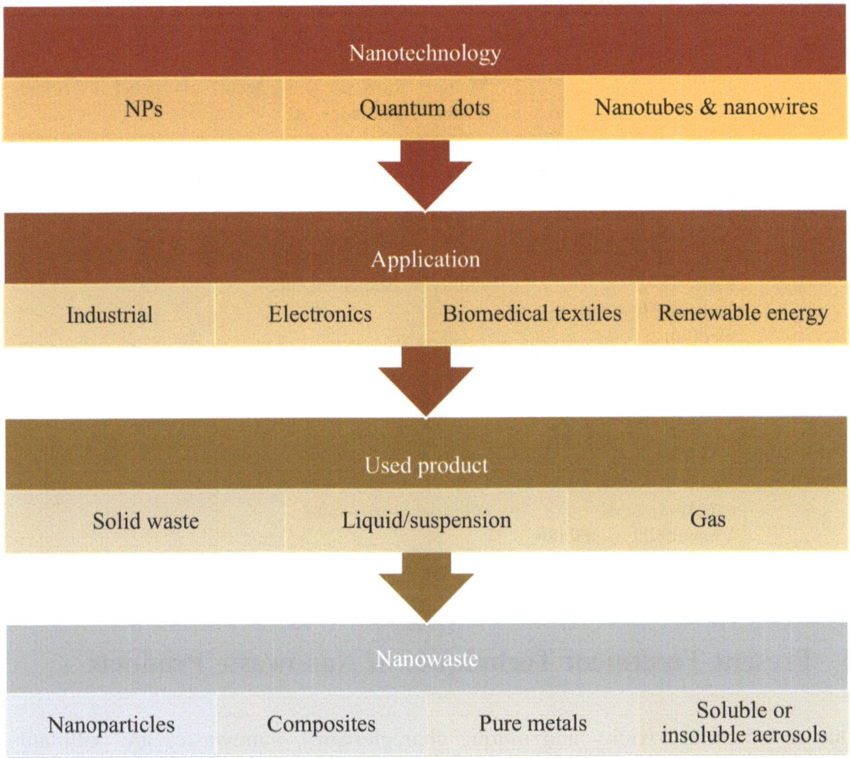

Fig. 2 Production pathway of nanowaste

3 Risk Description of Nanowaste

Specific references are not available on the current risk classification related to nanowaste of NMs, because of the multiple behaviors of various nanowaste of the same commodity and lack of risk assessment data. Nanowaste can be generally classified into five classes, as provided in Table 1 below. However, such classification is not adequate for nanoscale materials. Development, precautions, and approach of different nanowaste can perform in three ways: (i) isolation of waste during operation and management, (ii) storage, handling, and transportation, and (iii) appropriate treatment, recycling, and disposal.

Table 1 Nanowaste classifications as part of toxicity and sensitivity to the nanomaterials

Classes	Toxicity description	Examples
Class I	Risk: nontoxic Vulnerability: low to high	Memory chips, polishing agent, solar panel, TV screen display backplanes, etc.
Class II	Risk: harmful or toxic Vulnerability: low to medium	Paints and coating, polishing agents, etc.
Class III	Risk: toxic to very toxic Vulnerability: low to medium	Food packaging, personal care products, food additives, pesticides, etc.
Class IV	Risk: toxic to very toxic Vulnerability: medium to high	Paints and coating, pesticides, etc.
Class V	Risk: very toxic to extremely toxic Vulnerability: medium to high	Lotions, pesticides, food items containing fullerenes in colloidal suspensions

4 Present Treatment Techniques of Nanowaste Products

Due to potential toxicity and unique characteristics, nanowastes are pollutants. Engineered and synthesized nanowastes can be disposed of in three different ways such as (1) inhalation or atmospheric emission during incineration, (2) through landfilling, and (3) through liquid waste (production and purification).

5 Nanomaterial in Pollution Control and Recycling

Both hazardous waste sources are responded to by existing environmental remediation techniques: incineration, microbial treatment, biosorption, and advanced oxidation methods. Nevertheless, the environment includes some dangerous waste (like trace-toxic metal ions, and synthetic chemicals). Conventional forms of treatment did not eliminate those. Many emerging pollutants cannot be easily extracted, and, therefore, the increasingly strict water quality standards are difficult to meet [7].

Large quantities of wastewater and hydrocarbons are produced worldwide. Heavy metal ions and organic matter typically taint clean water. Nevertheless, some 780 million people worldwide lack access to modern facilities for water consumption. The contamination of rivers is dangerous for human beings and has implications for the environment. Clean water for all living species is an essential resource, and toxins such as this are urgently extracted from contaminated waters.

Furthermore, flue gas typically includes volatile organic compounds, NO_x, sulfur oxide, CO_2, etc. Air pollution control is also essential but very difficult. Nevertheless, waste should not be wasted as an undervalued tool. Organic matter is abundant, and in treatment, facilities are usually discarded or destroyed. To date, biological wastewater treatment contributes nearly 15 GW annually, or approximately 3% of the total electricity generated in the USA [8]. Over recent decades, several approaches have been developed to control the polluted environment. Various physical techniques can significantly improve the efficiency of treatment and the use of nanomaterials and nanotechnology [7]. Adsorption refers to a method of high performance, fast operation, and ease of use [9–12]. Biosorbents are created from natural substances, such as cellular products. Nano-adsorbents have gained considerable attention because of their excellent adsorption performance [5, 13–16]. Given some drawbacks, the ads alternative is soon to be regarded as a superior water treatment technology. There are ways to reduce emissions that are addressed in the following section: regulation of organic contaminants, law, and reuse of high metals, NO_x management, and CO_2 reduction. Chemical pollutants: Multiple organic contaminants severely impact the natural environment and associated human health. Biological pollutants, in general, can be classified into three main classes: primary pollutants, organic pollutants, and surface contaminants [17, 18]. This portion will briefly discuss the ways to remove persistent organic pollutants and toxic elements. Low biodegradability of persistent organic pollutants (POPs) makes them stackable and widely spread in both water and soil. Because of the widespread delivery coupled with their carcinogenicity, POPs raised various environmental problems. Polyhalogenated phenols have belonged to a group of XAr, and a few of them have also been classified as priority contaminants by the United States Environmental Protection Agency (EPA). These compounds appear to collect in fats rather than liquids and eventually remain in different organisms' fatty tissues. Hence, they are listed by the International Organization for Cancer Research as a category 2B human carcinogen. POPs typically resist the normal processes of photolytic, chemical, and biological degradation [19, 20]. To date, AOPs have increasingly preferred in remediating polluted water or soil to treat pentachlorophenol recalcitrant, and other XAr pollutants [21]. The hydroxyl radical (•OH) is the most reactive radical precursor produced at relatively close ambient temperature and pressure in these environmentally friendly AOP systems, which concentrate on hydrogen peroxide and ozone (O_3, O_3/UV). The existence of emerging pollutants in the atmosphere is primarily because of the wastewater treatment plants discharged. Traditional wastewater treatment techniques, including biologically mediated coagulation processes, flocculation, sedimentation, filtration, and sludge, are unable to remove these contaminants. These pollutants typically generated either naturally or synthetically and usually occur at relatively low concentrations [22]. One of the most significant potential pollutants is pharmaceutical and personal care products (PPCPs). A recently published report estimated that 20 million tonnes and the use of PPCP continue to grow, of PPCPs, produced commercially last year [23]. Endocrine disruptors (so-called environment hormones) are exogenous substances that can inhibit the body's regulatory function and its specific endocrine activity. There are two required fields of environmental behavior analysis and the treatment of endocrine disruptors. These micropollutants not removed from wastewater and drinking water by existing therapy methods typically bioaccumulated in invertebrates, other marine foodstuffs, and human beings [24]. A necessary and demanding role of environmental restoration continues to be a successful treatment of PPCP-polluted water. PPCP processes typically include adsorption, chemical reactions, and multiple treatments. The choice of a specific method of treatment depends on a variety of factors, in particular, the costs of the operation, quantity, and wastewater concentration to be used.

Metal ion management and utilization: Heavy metals in water and soil face significant health and environmental hazards. Most metals are not biodegradable, such as organic pollutants, and can move through levels tropically and reside in different body parts of humans, plants. While some metal ions are essential trace elements in life forms, they can cause high-dose acute and chronic toxicity. Stringent limits are placed on the disposal of different metal ions to wastewater. However, metal streams are often a vital commodity to rehabilitate valuable and scarce elements.

Chromium (Cr) has been used for various production processes, including metal electroplating production, chemical processing, metallurgy, and pigment Cr (VI), which is highly toxic to living organisms that have mutagenic and carcinogenic effects than the Cr (III) and appear water-soluble at maximum pH. According to the US EPA guidelines, the maximum allowable Cr in drinking water is 0.1 mg L^{-1} [25]. In crystals, galena, battery chargers, alloys, cable spikes, soldering machines, radiation protection, and X-ray equipment lead (Pb) have also been found. Pb(II) pollution, due to its persistence in polluted water and the nature of the biological toxicity process [26], is one of the environmental issues. Only small levels of Pb(II), including pulmonary and sinuses [27], lead to respiratory toxicity and many other fatal diseases. The application of Pb(II) for antiknock has also contributed to the effect of petroleum and paints and coatings on the atmosphere and health. Pb(II) is associated with water with mining and mineral processing, industrial wastewater, etc. Mercury (Hg) is one of the dangerous contaminants with eventual conversion into hypertoxic methyl mercury (MeHg) due to its volatility, persistence, and biotransformation [28]. More than 140 nations have signed a new treaty to fight mercury release into the atmosphere-the Mercury Minamata Convention. Soluble Hg(II)'s water ion must contain chloride, sulfide, or organic acid [29].

The widespread use of huge quantities of coal as a primary energy source in Asian countries, such as China and India, means that coal-fired power plants are one of the largest human activities for atmospheric mercury pollution (35%). Copper (Cu), which has been used in living cells, is a significant micronutrient. However, the chance of acute exposure to short- or long-term feed, GMD, diarrhea, copper immune function, and hepatic toxicity is considerably riskier to human or marine animals [30]. Cu is used in the mining, smelting, microprocessing, metallurgy, electrolysis, drawing, and scrubbing of copper. The Cu component levels are

1.5 mg L^{-1} for drinking water and 1.3 mg L^{-1} for hazardous waste, as defined by the World Health Organization (WHO) and the US EPA. Various physics, chemistry, and biology methods, including precipitation, membrane filtration, ion exchange, and electrochemistry, were developed to extract or trap heavy metal ions from waste [31]. Different methods have reported different metal ion tests. Reduction of carbon dioxide: one of the strongest greenhouse gases is carbon dioxide (CO_2). The rapid growth in CO_2 emission in 2016 with an atmospheric CO_2 level above 400 ppm was due to the rapidly increasing combustion of fossil fuels (coal, oil, natural gas) [32, 33]. Increasing atmospheric CO_2 will contribute to global warming and subsequent severe events of extreme weather that are a significant threat to human beings. Some attempts have, however, been made to reduce CO_2. Radiochemical, thermochemical, biochemical, photochemical, and electrochemical approaches are the latest techniques in CO_2 production [34, 35]. The reduction of electrocatalytic CO_2 emulates the effect of the photosynthesis process on the atmosphere and is therefore of great concern for the use of solar energy. The elimination of half a reaction only removes the electrons. Half-reaction oxidation in an ideal case will lead to the development of oxygen. Electrocatalytic CO_2 reduction represents an outstanding opportunity to reduce and convert CO_2 emissions into added-value chemicals and fuels.

NOx emissions from fossil fuel combustion had a significant impact on the ecosystem [32]. For open burning exhaust gases, nitric oxide (NO) produces 90% of NO_x. NO emissions also constitute the principal causes of acidic rain, i.e., when combined with water vapor in the clouds, they create nitric acid. Air-based NO_x contaminants are produced thermally in the N_2 and O_2 combustion cycles. Therefore, NO_x makes hydrocarbon reactions simpler in the production of complex chemical ozone (smog) and causes environmental harm after that. Common post-combustion NO_x controls are conventional selective catalytic reduction (SCR) and selective non-catalytic reduction (SNCR). If a molecule limits the nitrogen in a chemical reaction, NO_x is often produced easily at far low temperatures by an oxidative cycle. The NO_x produced from bound nitrogen can be regulated through rich combustion. Many future developments in combustion, like low-NOx heaters, would significantly reduce the thermal NOx emissions by flame air, fire, and water or steam. Apply NO_x (NO_x SNCR) of between 30 and 60% to the direct application of ammonia or urea to flute or combustion gas. Catalytic is also required after a procedure to achieve a high removal rate.

Recently, considerable advances in nanotechnology have led to a strong focus on air protection and pollutant handling. The most significant is to take advantage of nanomaterials' exciting features like high-specific surface areas, special electrical and optical properties, etc. Nanomaterials can also be used as excellent sensors, adsorbents, photographic/electrical catalysts, and disinfectants.

6 Toxicity of Nanowaste to the Environment

A nanotoxicity literature review revealed the first toxicity conference in Sweden in 2007, which was expressly devoted to the use of NM-based products. Since then, several research studies have been conducted on nanotoxicity and ecotoxicity. Although focusing on potential threats in mammalian model systems with minimal work in other ecological systems, it is suggested that most evolved NMs have effects on human and environmental health [1, 36–41]. The design, structure, and working features of NMs have been proved in vitro and in vivo risk analysis studies. NMs exposure levels and the vulnerability of species to NMs are both critical to bioavailability and the response to threats of NMs (Fig. 3) as well as to their fate and environmental cycle [42, 43]. In a comprehensive applied scenario, we can conclude that the paths that provide risk analysis for NMs in environmental streams are a crucial element, like: (1) Uncertain NMs or their EOL process fabrication exposure scenarios; (2) The presence or lack of NMs coating surfaces and the simplicity of removing nanotechnological NMs related to the amount of NMs releasing in the environment and its effect on the conversion and bioavailability mechanisms and the absorption mechanisms involved; and (3) A wide range of routes for nanowaste removal (e.g., water bodies, landfills, and recycling/burning); (4) Antagonistic or synergistic effects on the individual NMs population may occur or persist in a waste stream without specific contaminants.

The toxicity studies of NMs show that the potential threats posed to public health and the environment by these nanoscale materials are varied [44–50]. For many of the toxicity studies published, there has been no work into the possible hazard of

Fig. 3 Toxicity controlling properties of NPs

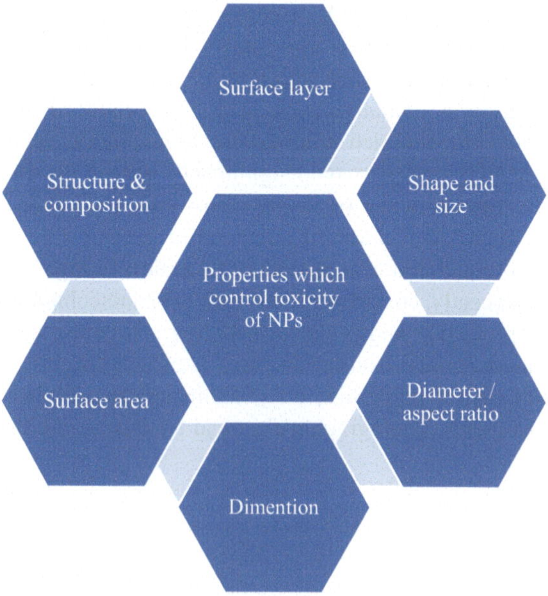

NMs in the mammalian model systems in another organ system, for example, in marine animals, microorganisms, plants, reptiles, and amphibians. However, the ecotoxicity of NMs in marine environments is currently inadequate. This situation means that current NMs ecotoxicity evidence in both estuarine and marine species are minimal or unaware of their association in risk assessments with aquatic freshwater organisms. The findings indicate that most toxicity studies in NMs are limited to their completeness, carbon nanotubes, and some metal oxides. Complexity and diversity of NMs make it difficult for various biological test organisms and endpoints to classify their exposure levels. This situation presents a challenge to determine which NMs properties cause the test organisms toxicity. This situation is unworkable because there are inadequate analytical techniques because of the limitations, including expense, difficulty, and feasibility of performing such a study. Also, for the part of the same bulk source, nature NMs determines the labeling of nanowaste classification.

Interactions of NMs with biological systems indicate that their sub-products can have a deleterious effect, whereas distilled NMs have no adverse effects on test organisms [51]. Exposure to pure carbon nanotubes, for example, does not present chronic threats of toxicity to sensitive Estuvial copepod crustaceans, even at environmentally unrealized levels of close to 10 mg L^{-1}. It was found that the presence of fullerene did not affect the toxicity of methyl parathion but had been detected 19 times higher than that of pentachlorophenol [52]. Phenanthrene, due to the presence of fullerene aggregates, has increased its toxicity by 85% in fullerene molecules by 60%. These findings indicate that sorbed phenanthrene has been bioavailable by research species. Several studies indicate that the sorption of contaminants in NMs is mostly due to their entire surface area [53–55]. These aspects, explicitly concerning their potential consequences for nanomaster management, are not adequately explained at the moment, and these research findings are far enough to offer a convincing argument for generalizing the fate and comportment of NMs in the ecosystem. The findings of Oberdörster et al. [47] also show different harmful effects on an organism when concentrations of up to 5 mg L^{-1} were observed to be too low for the assessment of LC50 levels with regards to the same material and physicochemical property when stirring the milli-Q suspension water process as a dispersion medium. This condition makes it more complex to interpret the ecotoxicity data because of restrictions on the dispersion methods used.

Nevertheless, this kind of toxicity is not discussed in many research articles, and its effect on the recorded data cannot be isolated. Due to the absence of consistent study findings due to the lack of accepted reference materials, these problems were considered. As a consequence, a great deal of ambiguity can occur concerning risk assessment by nanowastes of NMs in actual environmental factors. However, the relevant evidence was applied to the conceptualization of a classification system for nanowaste sources to ensure development at different stages (generation, handling, transport, and disposal).

Nonetheless, there is limited scientific evidence available about the nanowaste generated and the determination of all circumstances. Scenarios are likely to result in a full range of generated NMs being exposed every day and the absence of sufficient and reliable tools for in situ identification, sampling, classifying monitoring, and measurement [56]. Research approaches to avoid accidental release of NMs into the environment, and their unintended long-term toxic effects are urgently needed from this perspective. Nano-labeled products will cause nanopollution to the environment by nanowaste and therefore return to the human being.

The first debate in Sweden in 2007 was on the use of nano-based material and nanosubstances. The studies on nanotoxicity and ecotoxicity have been relatively limited and theoretical. With an emphasis on animal model systems dealing with biological systems (plants, insects, animals, and bacteria), many synthesized NMs are estimated to have environmental and human risks [40, 57, 58].

It can be stated that in different streams of environmental systems, risk assessment of NMs is an essential factor influenced by different ways: (1) Exposure during production and manufacturing of NMs, (2) Technological disadvantages of protection during operation of synthesis of NMs, (3) The protocol of NMs synthesis or production and the presence and absence of coating, (4) The different pathways of nanowaste disposal, and (5) Toxicological effects may be altered due to the presence of other contaminants in the waste system.

According to Boldrin et al. [56], limited scientific knowledge and lack of presence of precise tools for the characterization of nanowaste, it is difficult to predict all the situations which have more significant effects and lead to exposure to different types of NMs. There is an urgent need to develop strategies, research, and particular aims to prevent the long-term toxic effect of NMs into the environment (Table 2).

Therefore, the assessment after NMs generation and biocompatibility after interaction with living cells, which affects different environmental conditions, has also been another major problem considered in the impact of nanotoxicology. In this present case, a nano-based product is an innovative concept of the lifetimes of a

Table 2 A summarized view of nano-based technologies and materials and their predicted hazardous risk [59, 60]

Nanomaterials	Applications	Potential risk and remarks
AgNPs	R&D, Antimicrobial agent, food, textile, personal care product, paint, etc.	Landfill leaches, highly influential in the biochemical process, toxicity effect on the environment and biological systems
Carbon nanotubes	R&D, solar cell, electronics	Irritation, changes in brain pathology, changing in behavior
CuNPs	Agriculture, R&D	Toxic to planktons and other organisms living in the water
Quantum dot	Photocatalysis, solar cells	A few toxicity report
TiO_2, ZnO NPs	UV protections, paints, cosmetics, battery,cement, etc.	Direct release into the environment, disposal of containers
SiO_2, Fe_3O_4	Bioimaging, sensors, fire-proof glasses, pharmaceuticals products, contaminant purification, etc.	Contaminated nanowaste and unknown exposure to wildlife and plants

particular nano-application. NMs release process unchecked, environmental consequences, fate, and bioaccumulation. Bioavailability should be independently tested based on theoretical and experimental data. Additionally, the safety of the working environment must be maintained by successful control measures aimed at reducing NMs (air, fluid, and solids) emissions. Concentration, growth, time, and frequency, since all the materials with nanocomposites, can be a source of nanowaste for the human atmosphere.

7 Nanowaste Identification and Characterization of Analytical Tools

Various analytical tools (Table 3) can be applied to characterize and quantify the NMs in various studies. Analytical tool-based approaches are essential to developing accurate assessments for monitoring nanowaste in various environmental and biological and physical systems (Table 3). There are three simple steps for nanowaste monitoring, namely (i) sampling, (ii) separation, and (iii) characterization.

8 Fate and the Environmental Behavior of Nanomaterials

In the environment, NMs can go through several chemical processes and depend on various factors (like pH, salinity, temperature, organic and inorganic materials, etc.). The properties of NMs play an essential role. Depending on the bioavailability of the MNs, the toxicity of the NMs can be defined. However, the available research is not sufficient as different NMs possess different characters. Briefly, the fate and behavior of the NMs can be summarized below:

Table 3 Analytical tools for monitoring NMs in different systems

Parameters to be monitored of NMs	Analytical tools
Composition	EDX, ICP-MS, GC-MS, XPS
Particle size	AFM, DLS, FS, SEM, CPC, XRD, HDC, SP-ICP-MS, SEC
Size distribution	AFM, APS, CLSM, MALS, FFF, XRD, DMASEM
Shape	AFM, SEM, TEM, FFF
Mass or particle number composition	FS, GC-MS, ICP-MS, UV/vis
Surface area	AFM, SEM, XPS, BET
Structure	HR-TEM, SAXS, SAED

8.1 Air

In the atmosphere, the NMs can move from one zone to another zone of lower or higher concentration by diffusion. By rapid distribution of the particle, they can migrate great distances from the source. Due to their size distribution, it is challenging to detect nanoparticles in the air. Particular diameter, deposition character, and airspeed regulate the action and effectiveness of NMs on the system in particular. Despite their smaller diameters, nanoparticles from the air are released much slower than bigger particles.

8.2 Water

NMs usually behave like colloids in the mud. These colloids are the finely dispersed droplets or particles in a liquid. We are unstable and fall due to tidal force due to electrostatic forces. Natural bodies of water usually contain dissolved or dispersed natural NMs. Diverse physiochemical variables, such as salinity, pH, alkalinity, etc. Organic compounds control NMs' fate and conduct in water.

8.3 Soil

In this segment, there are fewer data available about air and water. NMs are believed to bind to solids in soil and sediments. The bioavailability of soil species and the possible toxicity of the NMs are under investigation.

9 Risk Assessment and Approaches

Academics, business, and government agencies, as well as politicians, have been increasingly worried about the ecological and human health threats associated with ENMs. Some basic aspects of environmental and safety risk and the approaches that ENMs use to assess and analyze these risks have yet to be discussed by the scientific community [61, 62] pointed out, nothing is being done in designing short-term decision-taking tools/strategies. Ecotoxicological and exposure information for NMs was a major concern. Therefore, NM's assessment of chemical risk is complex, time-consuming, and expensive [61, 63, 64]. Another key problem is the general (chemical and physical) characteristics of ENMs. In the case of a more straightforward quantitative risk assessment with available feed data on the risk assessment device, this approach can take several decades to take shape [65]. Yet these methods of risk management, which in turn, become more difficult and daunting to handle as the number of ENMs in consumer products is expected to expand [66]. The ENMs environmental risk analysis and evaluation system, which combines whole or relevant sections of the frames for a particular contextual risk assessment, can be facilitated by an integrated multidimensional approach. This methodology enables the use of different system resources for a comprehensive risk evaluation [61, 62]. The applicable risk assessment components for the NMs methodology are as follows [67, 68].

9.1 Identification of Risk and Hazard

The step of the risk assessment initiates the evaluation successfully, first identifying the protection or environmental risk characteristics of the ENMs. The lack of physical–chemical data on the receptor of these NMs leads to a major source of uncertainty in NMs safety or risk assessment. When these ENMs exist as small particles, powders, liquids and emulsions, macroscopic solids, and aerosols, nanoparticles may be isolated from the receiver by different procedures (i.e., product) [67, 69]. It requires a thorough assessment of the discharge capacity since other products do not exist if product matrices are published [66]. However, after processing these ENMs products, effects on nuclear exposure can release ENMs from their product arrays, for example, recycling or improper waste disposal after performing their specified role. Therefore, a detailed review of each NMs should be given additional importance when defining and assessing risk. In terms of the exposure assessment, the information on ENMs release and transport routes is relevant. The aerosol is among the most widely investigated release pathways. Their functionality and sources can be effectively disrupted by these nano-aerosols, which typically consist of solid aggregated particles near to their source [70]. The observations of carbon-black agglomerates in waste were found in Kuhlbusch and Fissan [71]. This led to the rise in the plant region's ultra-fine particle rates by analyzing the carbon black manufacturing process' particle characteristics. Several new species of particles were introduced through progress on nanotechnology's new dimension [72–74]. Fictionalized NMs that complicated the risk assessment mechanism further and toxicological consequences have not been thoroughly studied. The fluid nature of NMs as aerosols and as a single particle complicates the task of defining and characterizing NMs themselves.

9.2 Exposure and Hazard Assessment

The two methods for screening behaviors are direct calculation and model-based projections. Various aspects of NMs features are deemed necessary to adequately characterize and monitor released NM, including the distribution of quantities, surface area, form and size, composure, and chemical reactivity. Further, selecting suitable health assessment [65, 75, 76] is a critical factor for the implementation of the ENMs surveillance strategy, technique, and technology on an occupational basis. Advancements in technology have made reachable devices that can measure numerous different NMs metrics online and off-line. Specific descriptions of calculated and compositional features can be given by specific instruments [71, 77, 78]. Readily existing information on a large distribution scale enables this metric to be used for other essential NMs metrics. Thus, it makes it one of the most useful records available in the quantity size distribution [67]. Nonetheless, (i) instrument portability and cost-effectiveness, (ii) capacity to view, and (iii) the upcoming challenges for nanomaterial tracking and risk evaluation are the extension of detection to classify properties.

This dynamic model could also help approximate potential NMs pathways and sinks like soil, surface waters, STP effluents, sediments, etc. A qualitative assessment of data should be performed to assist in the classification of NMs risks and build on the quantitative estimates considering their complexities. The use of the consumer product accounting system and the most likely last release compartment may be another analytical process for determining the ENMs releases to environmental matrices (air, water, or soil). A production stage where certain occupational activities require managing bulk nanomaterial raw in quantities 73 [75] is the most direct way of developing the potential exposures in consideration of the entire life cycle. Additional potential for exposure occurs in downstream activities where the substance is used as consumers, or environmental release is exposed to NM. However, although general models are available for evaluating exposure estimates, they still have to be reviewed for NM. Many additional aspects of NMs are also challenging to determine their exposure because the dermal exposure of NMs is, in some situations, deliberate, as is the case with nanosilver for injury dressings. And in some situations, the nature of the exposure potential may change as its status changes, for example, from liquid to solid NMs, in paints that turn from liquid to solid. Thus, their emission potential dramatically reduces (limited to weathering and erosion).

In addition, the inventory database used during the above assessment is not comprehensive, and therefore more research is needed to identify and quantify NMs in consumer products. NMs toxicity has been suggested as a primary pathway for oxidative stress that contributes to inflammation-causing oxidative stress. Exceptional susceptibility to certain NMs has been observed, leading to chronic oxidative damage, inflammatory effects, cancer, genotoxicity, fibrosis, and mutations [79]. The intensity of these consequences depends heavily on the concentration, type of delivery, duration of the exposure, and the NM's physicochemical properties. Through systemic risk evaluation, the following types of toxic exposure are addressed: (a) oral exposure, (b) dermal exposure, (c) genotoxicity, (d) carcinogenicity, and (e) reproductive toxicity. Criteria for assessing situation adaptation, as defined in different guidelines, recognize the difference between exposure conditions, intensity, respiratory volumes between staff operations and light, differences between and interspecies, extrapolation duration, magnitude, and database trust. To order to evaluate INELs, each of these variables will, where appropriate, be used. Epidemiological studies of workers exposed to NMs may contribute significantly to the quantifiable risk analysis of personal information [80].

9.3 *Characterization of Risk*

In order to recognize a comprehensive risk assessment of NMs, extensive work is required to address challenges, such as the lack of exposure thresholds, detection systems, and hazardous endpoints, taking into account their distinctive characteristics. However, the risk analysis's crucial component must be determined if the exposures that occur are within the safety cap. As already stated, the danger is the product of the probability of threat and exposure.

The most popular approach is the risk matrix to classify the hazard as extreme, medium, or low by the incident's frequency and severity. The threshold is used to assess the magnitude of the danger. The PCE for carbonate NMs, as stated by Aschberger et al. [68], was three to six orders below their permissible soil and water threshold in their NMs risk characterization analysis. For metal-based ENMs, both PEC and INEC were identical together; thus, the ratio of PEC to INEC is used as comparative calculation methods for risk characterization. The enhanced risk to human health through long-term inhalation, excess ingestion, or skin exposures are prone to workers or the general public during high-release times. A distinction should be made between the exposures and the risk severity equivalent to environmental risk classification.

New opportunities for exploration are opening up through the advent of nanotechnology and nanoscale materials. Although nanomaterials' production has improved overall living standards and has found a continuing place in nanomedicine, the uncontrolled insertion into living cells of these materials is not desirous. Research on non-toxicity has improved our awareness of possible risks to nanoparticles in health and the environment. A regulatory structure must be developed and enforced to test engineered nanomaterial before application to humans and the environment.

10 Environmental Processes Which Can Affect the NMs Properties

(a) Dissolution: NMs can disintegrate in a solvent.
(b) Sedimentation: NMs can be isolated from a solution or a suspension.
(c) Transformation: NMs can withstand specific chemical or biological processes.
(d) Diffusion: The transfer of NMs by molecular movement from higher-to-lower concentrations.

(e) Binding to the component biotic or abiotic: NMs may interact with different living and non-living materials.
(f) Deposition: NMs may be moved from air to water or from air to soil.

11 The Positive Nanotechnology Impact

Water quality can be improved with the help of nanotechnology. Other nanomaterials, such as zinc oxide (ZnO), titanium dioxide (TiO_2), serve as a photocatalyst to oxidize organic compounds into harmless materials. Nanoparticles made from silver have antimicrobial effects. Toxic gasses should be purified in the soil. A nanocontact sensor can detect heavy metal ions and radioactive elements.

12 Restriction of the New Nanowaste Management Regulatory System

Based on existing information, academics, governments, and the industry need to collaborate to ensure the secure release of NMs that contain environmental sources, and concentrate on the following actions: The evaluation of the risk of exposure and release mechanism of NMs under appropriate conditions in order, during development, manufacturing, use, storage, and handling, establishes adequate, realistically similar exposure mechanisms in a specific framework. Creation of reproducible methods for analyzing the NMs in complex waste matrices throughout their life cycle process in a consistent way and tracking them. Nanotracers or NMs with distinguishable isotopic relationships appear to be an extremely potential approach for fingerprints to better understand the peripheral properties of endpoints and their actions in waste streams. Increased scientific work on the premise of the nano-labeled products produced by different methods will enable the scientific community, policymakers, and producers to resolve the lack of knowledge on nanotoxicity. This is studied to examine the dynamic interrelationships in both the long- and short-term relationships among hazardous NMs characteristics, potential hazards, and biological responsibilities, which could accurately describe the class of NMs under varying environmental and susceptibility circumstances. The use of dissolution or immobilization, microemulsion resolvents, etc., to provide effective and efficient disposal of wastes will enhance and support the identification of unique legislative protocols to deactivate chemical reactive NMs recycle and reuse nanodisposal options. A natural, alternative synthesis of nanomaterials and toxicity knowledge by applying green chemistry concepts outlined by US-EPA is the most promising area of bringing about a new revolution in the development of green nanotechnology by measuring prospective advantages from unintended adverse effects of greener products.

13 Conclusion

Several regulatory authorities investigate the possibilities in the growing problem and attempt to develop efficient regulations and policies. A little knowledge is available about the fate and behavior and impact of nanowaste and NMs on the environment and biotic system. Model calculation and comprehensive risk assessment are need to be done. Ecotoxicological investigations show that some NMs have potential health hazards. Long-term studies are necessary to understand the prolonged effects and behavior of NMs and nanowaste on the biotic, abiotic, and environmental systems.

14 Future Perspectives

Nanomaterials are a very innovative field of research that improves available technologies and opens new, exciting possibilities for changing our reality for a brighter future for many manufacturing industries. Although the current increase in ecotoxicological data on some NMs is extremely complex and highly restricted information. It provides adequate guidance to monitor, manage, and adhere effectively to potentially hazardous NMs or their nanowaste. Some existing NMs and newly manufactured products pose possible environmental or human health hazards, and their toxicity depends on the dynamic environmental effects of NMs particulate materials and the scenario of exposure and NMs behavior. The risks of nano-based technology (in particular in the fields of climate, agriculture, and energy) were still unanswered during its long- and short-term life. Therefore, many were not listed and thus, qualitative and quantitative assessments were avoided. Currently, NMs classification and management systems indicate new challenges and constraints in the handling, processing, recovery, or recycling of nano. NMs were created by a lack of adequate pollutant-area detection techniques (soil, water, and air). Restriction of the new nanowaste management regulatory system.

References

1. Bhatia S (2016) Nanoparticles types, classification, characterization, fabrication methods and drug delivery applications. In: Natural polymer drug delivery systems. Springer International Publishing, pp 33–93
2. Mrowiec B (2016) Risk of nanowastes. Eng Prot Environ 19(4):469-478
3. Thalji MR, Ali GAM, Algarni H, Chong KF (2019) Al^{3+} ion intercalation pseudocapacitance study of $W_{18}O_{49}$ nanostructure. J Power Sources 438:227028
4. Ali GAM, Fouad OA, Makhlouf SA, Yusoff MM, Chong KF (2014) Co_3O_4/SiO_2 nanocomposites for supercapacitor application. J Solid State Electrochem 18(9):2505–2512

5. Ali GAM, Barhoum A, Gupta VK, Nada AA, El-Maghrabi H, Kanthasamy R, Shaaban ER, Algarni H, Chong KF (2020) High surface area mesoporous silica for hydrogen sulfide effective removal. Curr Nanosci 16(2):226–234
6. Ali GAM (2020) Recycled MnO_2 nanoflowers and graphene nanosheets for low-cost and high performance asymmetric supercapacitor. J Electron Mater 49(9):5411–5421
7. Qu X, Brame J, Li Q, Alvarez PJJ (2012) Nanotechnology for a safe and sustainable water supply: enabling integrated water treatment and reuse. Acc Chem Res 46(3):834–843
8. McCarty PL, Bae J, Kim J (2011) Domestic wastewater treatment as a net energy producer-can this be achieved? Environ Sci Technol 45(17):7100–7106
9. Luo X, Shen T, Ding L, Zhong W, Luo J, Luo S (2016) Novel thymine-functionalized MIL-101 prepared by post-synthesis and enhanced removal of Hg^{2+} from water. J Hazard Mater 306:313–322
10. Shayegan H, Ali GAM, Safarifard V (2020) Amide-functionalized metal–organic framework for high efficiency and fast removal of Pb(II) from aqueous solution. J Inorg Organomet Polym Mater 30:3170–3178
11. Agarwal S, Sadegh H, Majid Monajjemi, Makhlouf ASH, Ali GAM, Memar AOH, Shahryari-ghoshekandi R, Tyagi I, Gupta VK (2016) Efficient removal of toxic bromothymol blue and methylene blue from wastewater by polyvinyl alcohol. J Mol Liq 218:191–197
12. Habeeb OA, Ramesh K, Ali GAM, Yunus RM (2017) Experimental design technique on removal of hydrogen sulfide using CaO-eggshells dispersed onto palm kernel shell activated carbon: Experiment, optimization, equilibrium and kinetic studies. J Wuhan Univ Technol Mater Sci Ed 32(2):305–320
13. Luo S, Li X, Chen L, Chen J, Wan Y, Liu C (2014) Layer-by-layer strategy for adsorption capacity fattening of endophytic bacterial biomass for highly effective removal of heavy metals. Chem Eng J 239:312–321
14. Gupta VK, Agarwal S, Sadegh H, Ali GAM, Bharti AK, Hamdy AS (2017) Facile route synthesis of novel graphene oxide-β-cyclodextrin nanocomposite and its application as adsorbent for removal of toxic bisphenol A from the aqueous phase. J Mol Liq 237:466–472
15. Abed HO, Kanthasamy R, Ali GAM, Sethupathi S, Yunus RBM (2017) Hydrogen sulfide emission sources, regulations, and removal techniques: a review. Rev Chem Eng 34(6): 837–854
16. Habeeb OA, Ramesh K, Ali GAM, Yunus RM (2017) Low-cost and eco-friendly activated carbon from modified palm kernel shell for hydrogen sulfide removal from wastewater: adsorption and kinetic studies. Desalin Water Treat 84:205–214
17. Wang J, Chen B, Xing B (2016) Wrinkles and folds of activated graphene nanosheets as fast and efficient adsorptive sites for hydrophobic organic contaminants. Environ Sci Technol 50 (7):3798–3808
18. Xu Y, Liu T, Zhang Y, Ge F, Steel RM, Sun L (2017) Advances in technologies for pharmaceuticals and personal care products removal. J Mater Chem A 5(24):12001–12014
19. Gu Y, Dong W, Luo C, Liu T (2016) Efficient reductive decomposition of perfluorooctane-sulfonate in a high photon flux UV/sulfite system. Environ Sci Technol 50(19):10554–10561
20. Zhang C, Zhou M, Ren G, Yu X, Ma L, Yang J, Yu F (2015) Heterogeneous electro-Fenton using modified iron–carbon as catalyst for 2,4-dichlorophenol degradation: Influence factors, mechanism and degradation pathway. Water Res 70:414–424
21. Yu F, Zhou M, Zhou L, Peng R (2014) A novel electro-fenton process with H_2O_2 generation in a rotating disk reactor for organic pollutant degradation. Environ Sci Technol Lett 1 (7):320–324
22. Luo Y, Guo W, Ngo HH, Nghiem LD, Hai FI, Zhang J, Liang S, Wang XC (2014) A review on the occurrence of micropollutants in the aquatic environment and their fate and removal during wastewater treatment. Sci Total Environ 473–474:619–641
23. Wang J, Wang S (2016) Removal of pharmaceuticals and personal care products (PPCPs) from wastewater: a review. J Environ Manag 182:620–640

24. Petrie B, Barden R, Kasprzyk-Hordern B (2015) A review on emerging contaminants in wastewaters and the environment: current knowledge, understudied areas and recommendations for future monitoring. Water Res 72:3–27
25. Zhuang L, Li Q, Chen J, Ma B, Chen S (2014) Carbothermal preparation of porous carbon-encapsulated iron composite for the removal of trace hexavalent chromium. Chem Eng J 253:24–33
26. Chooto P, Wararatananurak P, Innuphat C (2010) Determination of trace levels of Pb(II) in tap water by anodic stripping voltammetry with boron-doped diamond electrode. Sci Asia 36 (2):143–157
27. Huang Y, Li L, Zhang Y, Zhang L, Ge S, Li H, Yu J (2017) Cerium dioxide-mediated signal "On–Off" by resonance energy transfer on a lab-on-paper device for ultrasensitive detection of lead ions. ACS Appl Mater Interfaces 9(38):32591–32598
28. Cao T, Li Z, Xiong Y, Yang Y, Xu S, Bisson T, Gupta R, Xu Z (2017) Silica–Silver nanocomposites as regenerable sorbents for Hg^0 removal from flue gases. Environ Sci Technol 51(20):11909–11917
29. Zhou J, Hou W, Qi P, Gao X, Luo Z, Cen K (2013) CeO_2–TiO_2 sorbents for the removal of elemental mercury from syngas. Environ Sci Technol 47(17):10056–10062
30. Xing HT, Chen JH, Sun X, Huang YH, Su ZB, Hu SR, Weng W, Li SX, Guo HX, Wu WB, He YS, Li FM, Huang Y (2015) NH_2-rich polymer/graphene oxide use as a novel adsorbent for removal of Cu(II) from aqueous solution. Chem Eng J 263:280–289
31. Nancharaiah YV, Venkata Mohan S, Lens PNL (2015) Metals removal and recovery in bioelectrochemical systems: a review. Biores Technol 195:102–114
32. Xia Y, Zhao J, Li M, Zhang S, Li S, Li W (2016) Bioelectrochemical reduction of Fe(II) EDTA–NO in a biofilm electrode reactor: performance, mechanism, and kinetics. Environ Sci Technol 50(7):3846–3851
33. Cai W, Liu W, Sun H, Li J, Yang L, Liu M, Zhao S, Wang A (2018) Ni_5P_4–NiP_2 nanosheet matrix enhances electron-transfer kinetics for hydrogen recovery in microbial electrolysis cells. Appl Energy 209:56–64
34. Kondratenko EV, Mul G, Baltrusaitis J, Larrazábal GO, Pérez-Ramírez J (2013) Status and perspectives of CO_2 conversion into fuels and chemicals by catalytic, photocatalytic and electrocatalytic processes. Energy Environ Sci 6(11):3112
35. Qiao J, Liu Y, Hong F, Zhang J (2014) A review of catalysts for the electroreduction of carbon dioxide to produce low-carbon fuels. Chem Soc Rev 43(2):631–675
36. Wang J, Zhang X, Chen Y, Sommerfeld M, Hu Q (2008) Toxicity assessment of manufactured nanomaterials using the unicellular green alga Chlamydomonas reinhardtii. Chemosphere 73(7):1121 1128
37. Musee N (2010) Nanotechnology risk assessment from a waste management perspective: are the current tools adequate? Hum Exp Toxicol 30(8):820–835
38. Musee N (2011) Nanowastes and the environment: potential new waste management paradigm. Environ Int 37(1):112–128
39. Gajewicz A, Rasulev B, Dinadayalane TC, Urbaszek P, Puzyn T, Leszczynska D, Leszczynski J (2012) Advancing risk assessment of engineered nanomaterials: application of computational approaches. Adv Drug Deliv Rev 64(15):1663–1693
40. Kwak JI, An Y-J (2016) The current state of the art in research on engineered nanomaterials and terrestrial environments: different-scale approaches. Environ Res 151:368–382
41. Iavicoli I, Leso V, Beezhold DH, Shvedova AA (2017) Nanotechnology in agriculture: opportunities, toxicological implications, and occupational risks. Toxicol Appl Pharmacol 329:96–111
42. Engelmann W, Weber S, Leal D, Hohendorff RV (2018) The nanotechnological revolution and the complexity of waste: the possibility of using the OECD protocol as an alternative to nanowaste risk management. Adv Recycl Waste Manag 02(03)
43. Purohit R, Mittal A, Dalela S, Warudkar V, Purohit K, Purohit S (2017) Social, environmental and ethical impacts of nanotechnology. Mater Today: Proc 4(4):5461–5467

44. Dick CAJ, Brown DM, Donaldson K, Stone V (2003) The role of free radicals in the toxic and inflammatory effects of four different ultrafine particle types. Inhalation Toxicol 15(1):39–52
45. Lam C-W, James JT, McCluskey R, Arepalli S, Hunter RL (2006) A review of carbon nanotube toxicity and assessment of potential occupational and environmental health risks. Crit Rev Toxicol 36(3):189–217
46. Maynard AD, Kuempel ED (2005) Airborne nanostructured particles and occupational health. J Nanopart Res 7(6):587–614
47. Oberdörster E, Zhu S, Blickley TM, McClellan-Green P, Haasch ML (2006) Ecotoxicology of carbon-based engineered nanoparticles: effects of fullerene (C60) on aquatic organisms. Carbon 44(6):1112–1120
48. Lovern SB, Klaper R (2006) Daphnia magna mortality when exposed to titanium dioxide and fullerene (C60) nanoparticles. Environ Toxicol Chem 25(4):1132
49. Roberts AP, Mount AS, Seda B, Souther J, Qiao R, Lin S, Ke PC, Rao AM, Klaine SJ (2007) In vivo biomodification of lipid-coated carbon nanotubes by daphnia magna. Environ Sci Technol 41(8):3025–3029
50. Blaise C, Gagné F, Férard JF, Eullaffroy P (2008) Ecotoxicity of selected nano-materials to aquatic organisms. Environ Toxicol 23(5):591–598
51. Templeton RC, Ferguson PL, Washburn KM, Scrivens WA, Chandler GT (2006) Life-cycle effects of single-walled carbon nanotubes (SWNTs) on an estuarine meiobenthic copepod†. Environ Sci Technol 40(23):7387–7393
52. Baun A, Hartmann NB, Grieger K, Kusk KO (2008) Ecotoxicity of engineered nanoparticles to aquatic invertebrates: a brief review and recommendations for future toxicity testing. Ecotoxicology 17(5):387–395
53. Yang K, Zhu L, Xing B (2006) Adsorption of polycyclic aromatic hydrocarbons by carbon nanomaterials. Environ Sci Technol 40(6):1855–1861
54. Knauer K, Sobek A, Bucheli TD (2007) Reduced toxicity of diuron to the freshwater green alga Pseudokirchneriella subcapitata in the presence of black carbon. Aquat Toxicol 83 (2):143–148
55. Sun H, Zhang X, Niu Q, Chen Y, Crittenden JC (2006) Enhanced accumulation of arsenate in carp in the presence of titanium dioxide nanoparticles. Water Air Soil Pollut 178(1–4):245–254
56. Boldrin A, Hansen SF, Baun A, Hartmann NIB, Astrup TF (2014) Environmental exposure assessment framework for nanoparticles in solid waste. J Nanopart Res 16(6)
57. Gambardella C, Gallus L, Gatti AM, Faimali M, Carbone S, Antisari LV, Falugi C, Ferrando S (2014) Toxicity and transfer of metal oxide nanoparticles from microalgae to sea urchin larvae. Chem Ecol 30(4):308–316
58. Sharma VK, Siskova KM, Zboril R, Gardea-Torresdey JL (2014) Organic-coated silver nanoparticles in biological and environmental conditions: fate, stability and toxicity. Adv Coll Interface Sci 204:15–34
59. Goswami L, Kim K-H, Deep A, Das P, Bhattacharya SS, Kumar S, Adelodun AA (2017) Engineered nano particles: nature, behavior, and effect on the environment. J Environ Manag 196:297–315
60. Bystrzejewska-Piotrowska G, Golimowski J, Urban PL (2009) Nanoparticles: their potential toxicity, waste and environmental management. Waste Manag 29(9):2587–2595
61. Grieger KD, Baun A, Owen R (2009) Redefining risk research priorities for nanomaterials. J Nanopart Res 12(2):383–392
62. Grieger KD, Linkov I, Hansen SF, Baun A (2011) Environmental risk analysis for nanomaterials: review and evaluation of frameworks. Nanotoxicology 6(2):196–212
63. Handy ROAR (2007) Viewpoint: formulating the problems for environmental risk assessment of nanomaterials. Environ Sci Technol 41(16):5582–5588
64. Choi J-Y, Ramachandran G, Kandlikar M (2009) The impact of toxicity testing costs on nanomaterial regulation. Environ Sci Technol 43(9):3030–3034
65. Maynard AD (2006) Nanotechnology: assessing the risks. Nano Today 1(2):22–33

66. Linkov I, Steevens JA (2008) Nanomaterials: risks and benefits. Springer Science & Business Media
67. Savolainen K, Alenius H, Norppa H, Pylkkänen L, Tuomi T, Kasper G (2010) Risk assessment of engineered nanomaterials and nanotechnologies—a review. Toxicology 269(2–3):92–104
68. Aschberger K, Micheletti C, Sokull-Klüttgen B, Christensen FM (2011) Analysis of currently available data for characterising the risk of engineered nanomaterials to the environment and human health-lessons learned from four case studies. Environ Int 37(6):1143–1156
69. Nel A (2006) Toxic potential of materials at the nanolevel. Science 311(5761):622–627
70. Rothenbacher S, Messerer A, Kasper G (2008) Fragmentation and bond strength of airborne diesel soot agglomerates. Part Fibre Toxicol 5(1):9
71. Kuhlbusch TAJ, Fissan H (2006) Particle characteristics in the reactor and pelletizing areas of carbon black production. J Occup Environ Hyg 3(10):558–567
72. Takagi A, Hirose A, Nishimura T, Fukumori N, Ogata A, Ohashi N, Kitajima S, Kanno J (2008) Induction of mesothelioma in p53± mouse by intraperitoneal application of multi-wall carbon nanotube. J Toxicol Sci 33(1):105–116
73. Poland CA, Duffin R, Kinloch I, Maynard A, Wallace WAH, Seaton A, Stone V, Brown S, MacNee W, Donaldson K (2008) Carbon nanotubes introduced into the abdominal cavity of mice show asbestos-like pathogenicity in a pilot study. Nat Nanotechnol 3(7):423–428
74. Nasibulin AG, Queipo P, Shandakov SD, Brown DP, Jiang H, Pikhitsa PV, Tolochko OV, Kauppinen EI (2006) Studies on mechanism of single-walled carbon nanotube formation. J Nanosci Nanotechnol 6(5):1233–1246
75. Maynard AD, Aitken RJ (2007) Assessing exposure to airborne nanomaterials: current abilities and future requirements. Nanotoxicology 1(1):26–41
76. Seipenbusch M, Binder A, Kasper G (2008) Temporal evolution of nanoparticle aerosols in workplace exposure. Ann Occup Hyg 52(8):707–716
77. Wake D, Mark D, Northage C (2002) Ultrafine aerosols in the workplace. Ann Occup Hyg 46 (suppl_1):235–238
78. Vincent JH (2007) Aerosol sampling: science, standards, instrumentation and applications. John Wiley & Sons
79. Ng C-T, Li JJ, Bay B-H, Yung L-YL (2010) Current studies into the genotoxic effects of nanomaterials. J Nucleic Acids 2010:1–12
80. Schulte PA, Schubauer-Berigan MK, Mayweather C, Geraci CL, Zumwalde R, McKernan JL (2009) Issues in the development of epidemiologic studies of workers exposed to engineered nanoparticles. J Occup Environ Med 51(3):323–335

Electronics Waste Recycling Technologies

Recycling of Cobalt Oxides Electrodes from Spent Lithium-Ion Batteries by Electrochemical Method

Eslam A. A. Aboelazm, Nourhan Mohamed, Gomaa A. M. Ali, Abdel Salam Hamdy Makhlouf, and Kwok Feng Chong

Abstract Energy storage electrode materials suffer from high-cost production. For example, cobalt oxide price was increased from 20 to 60 USD per kg in 1998 and 2017, respectively. Consequently, seeking low-cost production methods is essential. Over the years, the ownership of electronic devices has transformed from a human luxury to basic requirements. Following the exponential growth in the electronic gadget demand, the invention of lithium-ion batteries (LiBs) has been the most used energy storage devices in the electronic devices. However, the disposal of LiBs in electronic devices is obvious due to the limited cycle life. In this context, the disposal of LiBs could be a source of environmental calamity, if it will not be treated correctly because of the contained toxic materials and heavy metals such as cobalt, manganese, nickel, and lithium. Furthermore, the recovery of such valuable heavy metals before LiBs disposal is important from the economic and environmental viewpoints. Many processes have been used to extract the cobalt from spent LiBs, such as solvent extraction, acid leaching, chemical precipitation, bioleaching, and electrochemical recovery. The recycled materials were successfully used in many applications, such as supercapacitors. This chapter discusses the fundamental

E. A. A. Aboelazm
Institute of Basic and Applied Science, Egypt-Japan University of Science and Technology, New Borg El-Arab, Alexandria 21934, Egypt

N. Mohamed
Department of Metallurgical and Materials Engineering, Istanbul Technical University, 34469 Maslak, Istanbul, Turkey

G. A. M. Ali
Chemistry Department, Faculty of Science, Al–Azhar University, Assiut 71524, Egypt

A. S. H. Makhlouf
Engineering, Metallurgy, Coatings & Corrosion Consultancy (EMC3), Edinburg, TX, USA

Central Metallurgical Research and Development Institute, Helwan, Cairo 11421, Egypt

K. F. Chong (✉)
Faculty of Industrial Sciences and Technology, Universiti Malaysia Pahang, 26300 Gambang, Kuantan, Malaysia
e-mail: ckfeng@ump.edu.my

A. S. H. Makhlouf and G. A. M. Ali (eds.), *Waste Recycling Technologies for Nanomaterials Manufacturing*, Topics in Mining, Metallurgy and Materials Engineering, https://doi.org/10.1007/978-3-030-68031-2_4

of energy storage devices and illustrates the supercapacitor storage types based on electrical double-layer capacitors and the redox-based capacitors and their materials. Moreover, different approaches for recovering metals from LiBs as a supercapacitor electrode are discussed in detail.

Keywords Lithium-ion batteries · Waste recycling · Cobalt oxide · Nanomaterials · Supercapacitor

List of Abbreviations

AFM	Atomic force microscope
CoS_x	Cobalt sulfides
DMF	N-dimethyl formamide
DMSO	Dimethyl sulfoxide
ECs	Electrochemical capacitors
EDLs	Electric double layers
FCC	Face cantered cubic
HCP	Hexagonal close backed
HER	Hydrogen evolution reaction
LDHs	Layered double hydroxides
LiBs	Lithium-ion batteries
MHD	Magnetohydrodynamics
NiS_x	Nickel sulfides
NMP	N-methyl pyrrolidone
PANI	Polyaniline
PCs	Pseudocapacitors
PEDOT	3,4-ethylenedioxythiophene)
PPy	Polypyrrole
PVDF	Polyvinylidene fluoride
rGO	Reduced graphene oxide
TiN	Titanium nitride
TiN	Titanium nitride
VN	Vanadium nitride

1 Introduction

Investigating nanomaterials with high surface area and superior performance is a research trend set by today's challenges in many areas, including energy, information, health, and environment. To continue the progress toward more advanced nanotechnology, a continuous supply of raw materials is required. However, nanomaterials are as risky as beneficial; as shown in Table 1 [1], careful handling of these materials is necessary. Their release from products containing nanomaterials to the environment, such as air, water, and soil, causes a serious health hazard. Hence, waste management of the products containing nanomaterials is a sustainable concept to maintain a continuous supply of raw materials and to protect the environment against this threat.

Nanomaterials are the cornerstone of energy storage applications, such as rechargeable batteries and supercapacitors [2]. This is because of their capability to

Table 1 Nanomaterials and their advantages and risks in different applications

Applications	Benefits	Risks
Nanocrystals harvest light in photovoltaic devices	Green, renewable energy, new self-lighting displays for electronic devices	Light pollution in rural areas, the opportunity cost to fossil fuel economies
Antimicrobial wound dressings contain nanocrystalline silver	Improved healing in wounds and reduced risk of infection	Release of antimicrobials into the environment, a hazard to the natural microbial system
Sunscreens containing titanium dioxide nanomaterials are incredibly effective at absorbing ultraviolet light	Consumer preference for transparent but effective sun creams. Potential decrease in skin cancer due to increased sunscreen use	Titanium hazard to intertidal organisms and sandy shore ecosystems
Metal nanomaterial supplements to increase the burn efficiency of fuels	Less soot from diesel vehicles and urban air pollution. Burn efficient aviation fuels Economic saving for transport infrastructure on fuel costs. Reduced greenhouse gases	Respiratory exposure to nanomaterials in fuel exhausts. Long-range transport of particles in the atmosphere
Medical applications of hydroxyapatite and nano-silica applications in bone reconstruction	Structural repairs to teeth and bone using a natural material already in the body (no adverse immune response)	Durability-particles eroded from the surface may cause pathology in other internal organs in the long term
Nanomaterials in food packaging	Stronger lighter packaging to protect soft foods, antibacterial packing to improve shelf-life increased food safety	Unintended transfer of nanomaterial from the packaging to the food. Uncertain lifetime oral exposure risk
Use of carbon nanotubes to improve the strength and flexibility of sports equipment	The better product that lasts longer for the consumer. Reduced sports-related injuries	Life cycle analysis, what happens to the materials in landfills at the end of their use?
Use of nanomaterials as a catalyst in industrial processes such as coal liquefaction and producing gas	Improved efficiency and economy of industrial processes. Less industrial waste/ton of production	Inadvertent incorporation of toxic catalysts in consumer products, waste disposal of catalytic converters to landfill

improve the ionic transport and electronic conductivity compared to the bulk materials. They are also more tolerant of high currents, making them suitable for high-energy and high-power devices. Li-ion batteries (LiB) are one of the most important revolutionary technological inventions of the last century. LiBs contributed to the development of high-performance electronics because of their remarkable electrochemical properties, including the high operating voltage of 3.6 V, high energy density, and low weight and volume. Production of electric vehicles and advanced portable electronic devices strongly stimulate the rise of LiBs production. The vital task is to provide enough raw materials like Li, Co, Mn, and Ni to meet future demand [3]. Because any limitations in the number of raw materials may hinder the industry growth, many creative strategies that consider both the recovery of metals from batteries and the environmental impact of the inappropriate disposal of LiBs have been extensively studied through waste management [4–6].

The waste management of batteries that reached the end of life follows a hierarchal order that includes different management options: disposal, recovering, recycling, reusing, and prevention, ordered from less to most preferred. According to this order, reusing a battery in an application after its life ends in another application is beneficial and sustainable because it does not require further treatment. Recycling is another strategy that potentially helps decrease pollution and satisfies the need for raw materials to be used for any purpose not necessary for batteries again [7].

The main available recycling process is based on pyrometallurgical or hydrometallurgical methods to recover the metals, mainly cobalt, copper, manganese, nickel, lithium, and aluminum. In the pyrometallurgical process, the active materials are reduced into metallic alloy and the battery through smelting in a high-temperature furnace. This method results in toxic gas emissions, and it requires filters to trap gases. On the other hand, in the hydrometallurgical process, the metal is leached from the cathode using an aqueous solution such as a mixture of acid and reducing agents. The metal can be obtained in different ways, including extraction, precipitation, or electrodeposition. In the recovery process, the main concern should not be to obtain the highest possible yield of materials. However, it should be considered side by side with the quality and the properties of the recovered materials to be suitable for the next life without further treatment. For example, controlled size, size distribution, structure, and morphology of Co from spent LiBs can be determined during the recycling concerning the application of the next use. This chapter discusses the recycling of LiBs, recovery of Co, and its use for supercapacitors. Also, the advantages and limitations of the existing recovery techniques are critically presented and highlighted.

2 Electrochemical Energy Storage

The industry's rapid growth is the main reason behind global warming and the increasing demand for fossil fuels. Many efforts have been invested in fabricating efficient, clean, and sustainable energy conversion and storage devices. Fuel cells, batteries, and electrochemical capacitors (ECs) are three classes of the essential electrochemical energy conversion and storage devices [8–10]. These three classes have the same electrodes (cathode and anode) system connecting through the electrolyte. In fuel cells and batteries, the generation of electrical energy occurs by converting chemical energy via reduction and oxidation reactions at the cathode and anode. The oxidation reaction occurs at the anode, and the reduction reaction occurs at the cathode. The difference between supercapacitors, fuel cells, and batteries is correlated with their range of power densities and energy densities, as shown in Fig. 1. Batteries are closed systems with charge transformation and redox reactions. Fuel cells are an open system where anode and cathode are only charges, transfer media, and the redox reaction from the environment or outside the cell [11]. In ECs, the energy does not transfer by redox reaction as mentioned in batteries and fuel cells but by the formation and releasing of electric double layers (EDLs) caused by the electrolyte ions' orientation [11–15].

Ragone plot represents the difference in power capabilities of a storage device, which shows that capacitors have high energy density up to 10^6 Wh kg^{-1} but with lower power density. Fuel cells and batteries have high power density with very low energy density. Supercapacitors balance the gap between power density and energy density in capacitors and batteries, which give supercapacitors the ability to conduct a new application [9, 11].

Several types of ECs can be classified based on the energy storage mechanism and the active materials used. Firstly, electrical double-layer capacitors (EDLCs) follow non-Faradic behavior and commonly contain materials with very high surface areas such as carbon-based materials. Secondly, redox-based capacitors or

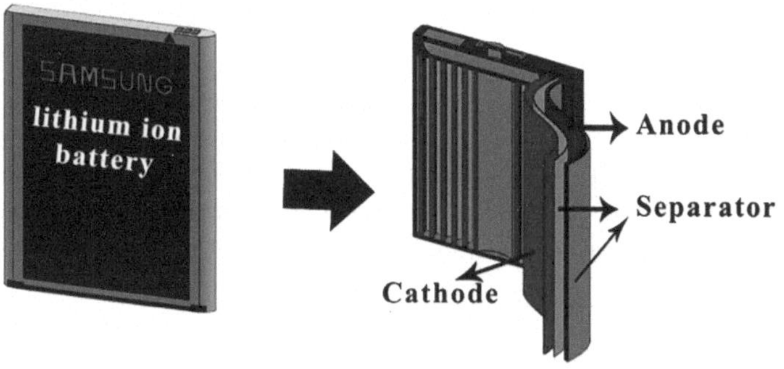

Fig. 1 Shape and components of lithium-ion batteries

pseudocapacitors (PCs) use high-speed and reversible surface or near-surface reactions for energy storage, such as transition metal oxides, including MnO_2, NiO, Co_3O_4, $NiCo_2O_4$, $W_{18}O_{49}$, etc. [16–25].

2.1 Electrical Double-Layer Capacitors

Electrical double-layer capacitors (EDLCs) store the charges electrostatically through the reversible adsorption of ions of the electrolyte onto electrode materials characterized by high specific surface area and electrochemical stability [18, 26, 27]. Charge separation occurs on polarization at the electrolyte–electrode interface leading to EDL formation, without charge transfer or chemical reactions, which is called the non-Faradic process. Due to their high chemical stability, EDLCs can perform high cyclic stability up to one million cycles [28–31]. This surface storage mechanism allows high-speed energy transmission with excellent power performance. Besides, this non-Faradic reaction excludes the swelling in the electrode material that happens during charge/discharge in batteries [18].

2.2 Redox-Based Capacitors (Pseudocapacitors)

Redox-based capacitors or pseudocapacitors (PCs) store charges based on the Faradaic process. Faradaic mechanisms include (1) redox pseudocapacitance by involving chemical reaction charge transfer through reduction and oxidation reactions, (2) underpotential deposition which happens when metal ions form an adsorbed layer at several metal's surfaces which is above their redox potential, and (3) intercalation pseudocapacitance which takes place when ions intercalate into the layers or tunnels of a redox-active material same like LiBs mechanism accompanied by a faradaic charge transfer without any changes in the crystallographic phase [32, 33].

There is a remarked difference between the cyclic voltammogram and the charge–discharge behavior of the EDLs capacitor and pseudocapacitance. Firstly, the EDL capacitor shows a rectangular shape of cyclic voltammogram, and the galvanic change discharge is always linear with time. This type of energy storage is a purely electrostatic process. It shows that the current reversed immediately with the reversing of the potential sweep. On the other hand, PCs with a charge accumulation that depends on the electrode potential show a deviation from the rectangular shape [34–36].

3 Pseudocapacitors Electrode Materials

The most common materials for redox pseudocapacitive electrodes are transition metal sulfides, metal nitrides, layered double hydroxide, conducting polymers, and transition metal oxides [9].

3.1 *Transition Metal Oxides*

Ruthenium (IV) oxide (RuO_2) is the first investigated material for PCs due to its high electrochemical reversible electron transfer. Moreover, it has a cyclic voltammogram with a quite broad shape called quasi-rectangular shape [36, 37]. Because of its excellent conductivity and the presence of structural water, it is widely used to produce very high capacitance that is close to its theoretical value. Hydrous RuO_2 showed a specific capacitance of 1340 F g^{-1}, and the theoretical value is 1400 F g^{-1} [38]. One of the most common characteristics of RuO_2 is the increased surface area of the electrode material to possess more active ions and, hence, enhance specific capacitance by pseudocapacitance effect. For example, a thick layer of MnO_2 provides a specific capacitance of 200 F g^{-1}, while an ultra-thin film of the MnO_2 supercapacitor electrode provides 1000 F g^{-1} [39, 40]. The transition metal oxides can be divided into two categories: firstly, the noble metal oxides such as RuO_2 and IrO_2 and secondly, the base transition metal oxides such as Mn_3O_4, Co_3O_4, Fe_3O_4, $W_{18}O_{49}$, and NiO [19, 41–43].

Recently, spinal metal oxides such as Mn_3O_4, Fe_3O_4, Co_3O_4, and NiO have been considered promising electrode materials for high capacitive performance with specific capacitance higher than 3000 F g^{-1} [25, 37, 44–46]. The following redox reactions represent the pseudocapacitance process of Co_3O_4, as shown in Eq. (1):

$$Co_3O_4 + OH^- + H_2O \leftrightarrows 3CoOOH + e^- \quad (1)$$

Many studies were performed to enhance the capacitive performance of Co_3O_4 supercapacitor and get specific capacitance near to its theoretical specific capacitance (3560 F g^{-1}) [44, 46–49]. Zhang et al. improved the electron transfer by growing nanostructured cobalt oxide on nickel foam by hydrothermal process. The resulting active material has a specific capacitance of 1160 F g^{-1} at 2 A g^{-1} [50]. Researchers have recently obtained a wide range of values of specific capacitance of cobalt oxide based on its morphology and size. Ultra-thin Co_3O_4 nanosheets with a thickness of 10 nm supercapacitor electrode prepared through hydrothermal reaction showed a specific capacitance of 1782 F g^{-1} at 1.8 A g^{-1} [47]. Moreover, the nanocubes cobalt oxide electrode synthesized by the solvothermal method showed ultra-high specific capacitance of 1913 F g^{-1} at 8 A g^{-1} [48].

Some other approaches have been used for cobalt oxide to reach the theoretical capacitance by combining the cobalt oxide with carbon-based materials like graphene [51–54]. For example, Fan et al. produced Co_3O_4 nanosheet arrays on three-dimensional (3D) porous graphene/nickel foam by a hydrothermal process. This system showed a very high specific capacitance of 3533 F g^{-1} at 1 A g^{-1}, which is very close to the theoretical specific capacitance of Co_3O_4 [55]. Another strategy to enhance Co_3O_4 is by doping with elements such as Mn, Cu, N, and C [56–58]. Recently, Li et al. prepared Mn-doped Co_3O_4 mesoporous nanoneedles that exhibited almost three times higher specific capacitance compared to the undoped Co_3O_4 [59].

3.2 Transition Metal Sulfides

Nickel sulfides (NiS_x) and cobalt sulfides (CoS_x) have received increasing attention as transition metal sulfides materials for PCs because of their unexpected performance in alkaline electrolytes [42, 60–64]. Tao et al. performed the first study on amorphous CoS_x as supercapacitor electrode material and showed a specific capacitance of 369 F g^{-1} at current density 50 mA cm^{-1} [65]. NiS_x is a notable material in the metal sulfides because of its cost-effectiveness, remarkable electrical conduction, slight toxicity, and the possibility of producing it using elementary synthesis methods such as chemical vapor deposition, hydrothermal technique, and electrochemical deposition. Gaikar et al. synthesized NiS film by a simple and inexpensive chemical bath deposition method. The results showed a significant electrochemical supercapacitor performance with a specific capacitance of 788 F g^{-1} at 1 mA cm^{-2} and 96% cyclic stability after 1000 charge/discharge cycles [66]. Amid metal sulfides, copper sulfide (CuS) is a crucial multi-property semiconductor material with various applications in energy storage, gas sensors, and solar energy devices. Wire-like CuS was synthesized via liquid–solid reactions and showed a tremendous reversible charge/discharge performance with a specific capacitance of 305 F g^{-1} at an energy density of 70 Wh kg^{-1} [67]. Many approaches have been proposed to enhance transition metal sulfides' capacitance, such as combining the nickel sulfide with reduced graphene oxide (rGO) to improve the capacitance up to 961 F g^{-1} at 15 A g^{-1} [68, 69]. Enhancing the morphology of $NiCo_2S_4$ hollow structure showed a very high porosity on carbon substrates as free-standing electrodes and produced high specific capacitance of 1418 F g^{-1} at 5 A g^{-1} [70].

3.3 Metal Nitrides

Metal nitrides are characterized by their chemical stability in bases and acids in addition to their high melting point [9]. Molybdenum nitride is the first metal nitride used as a supercapacitor electrode and showed a wide operation window and very high chemical stability in the acidic electrolyte [71]. Titanium nitride (TiN) and vanadium nitride (VN) received attention because of their attractive electrical conductivity, which allows them to reach the metal oxide electrode specific capacitance values. Choi et al. reported that VN had shown a specific capacitance of 1340 F g^{-1}, even with a relatively low mass loading [72]. Li et al. [73] successfully synthesized Mo_2N nanoparticles and studied their electrochemical performance. The as-prepared γ-Mo_2N showed good capacitive performance with an excellent specific capacitance of 172 F g^{-1} in a wide potential window from −0.6 V to 0.5 V. Balogun et al. found that Mo_2N showed a specific capacitance of 16 mF $cm^{-2,}$ which is higher than its corresponding oxide [73, 74]. Several research works have been performed on titanium nitride (TiN) and vanadium nitride (VN) by changing their oxidation states and involving them with carbon nanotubes, carbon nanofiber and rGO to improve their capacitive performance and cyclic stability [9]. The superlattice structure of a boron nitride/rGO composite showed a promising specific capacitance of 824 F g^{-1} at 4 A g^{-1} [75].

3.4 Layered Double Hydroxides

Layered double hydroxides (LDHs) are remarkable by their lamellar structure that can be described by $\left[M_{1-x}^{2+}M_x^{3+}(OH)_2\right][A^{n-}]_{x/n}{\cdot}zH_2O$, where M^{2+} and M^{3+} are related to the bivalent and trivalent cations, respectively, x is always between 0.2 and 0.4, and A^{n-} is the balancing anion [76–79]. Chen et al. synthesized nickel–cobalt LDHs nanosheets using a facile one-step method on nickel foam. Results showed a remarkable specific capacitance of 2682 F g^{-1} at 3 A g^{-1} and an energy density of 77.3 Wh kg^{-1} at 623 W kg^{-1} [80]. Guo et al. synthesized Ni–Mn LDHs nanosheets on nickel foam using facile one-step wet method. The optimized Ni_3Mn_1–LDH on Ni foam showed a high specific capacitance of 1511 F g^{-1} at 2.5 A g^{-1}, excellent cyclic stability with an efficiency loss of 8% after 3000 cycles with high Coulombic efficiency [81]. High specific capacitance was obtained by the hollow Ni–Al LDH microsphere and showed a value of 735 F g^{-1} at 2 A g^{-1} with good long cycling life [82]. Moreover, Zhang et al. reported a remarkable specific capacitance of 1255 F g^{-1} by synthesizing a composite of graphene nanosheet/Ni–Al LDH [83].

3.5 Conducting Polymers

Conducting polymers are a special kind of promising PCs electrode materials due to their outstanding electrochemical reversibility, rapid doping, and de-doping capability [9, 84, 85]. Among different conducting polymers, poly (styrene sulfonate), poly(3,4-ethylenedioxythiophene) (PEDOT), polyaniline (PANI), polypyrrole (PPy), polythiophene, and polymethyl methacrylate are the most studied conducting polymers used as supercapacitors electrodes. They showed high specific capacitance in the range of 500 to 3400 F g^{-1} which is considerably larger than carbon-based materials and comparable to metal oxides [86, 87]. The drawback of these conducting polymers materials has their very high electrochemical resistances. To overcome this problem, researchers combined them with other porous and highly conductive materials like graphene and carbon nanotube [87, 88]. Flexible polymer supercapacitor showed a specific capacitance of 202 F cm^{-3} at 0.54 A cm^{-3} using a thermal treatment of poly(3,4-ethylenedioxythiophene):-poly (4-styrenesulfonate) to form hydrogel flexible supercapacitor [89].

4 Lithium-Ion Batteries as a Source of Cobalt Oxide

4.1 Cobalt Production

Cobalt (Co) is a strategical metal in a broad spectrum of applications that dominate the global market [90]. Cobalt has excellent mechanical properties, high heat resistance, and ferromagnetic properties with a very high Curie temperature of 1121 °C, making it a potential candidate in various applications such as energy storage, sensing, and magnetism [91, 92]. The Co demand increased dramatically in the past two decades, and the main reason was developing the first commercial rechargeable battery based on $LiCoO_2$. The rising demand resulted in a progressive rise in the Co production from 34,000 tonnes in 2000 to 126,000 tonne in 2016. In the past 8 years, Co's average prices showed a continuous increase from 24,000 euro/tonne in 2012 to 65,000 euro/tonne in 2018.

On the other hand, some geopolitical conditions in the countries that hold Co mining can cause instability in its productions, such as the largest cobalt resources country globally, the Democratic Republic of Congo [93]. These conditions could cause a risk of instability in the supply of Co. Based on this information, the development of the LiBs can be significantly affected in the future if Co became deficient as a result of the price increase and geopolitical conditions. However, different strategies were considered to avoid this situation, including (i) recycling (ii) substitution of Co with other potential metals such as Ni. Recycling is essential as a waste management strategy that is critical to face the possible risk of shortage in Co's production.

4.1.1 Cobalt Production Processes

The production of cobalt depends on the ore type in which the place has occurred. In other words, Co is commonly found with other elements or minerals such as sulfide minerals associated with nickel, iron, and copper [91]. So, the production of Co depends on two main groups. The first one is recovered as a by-product of nickel extraction by going through three processes, from smelting to leaching processes, followed by electrowinning or hydrogen reduction [90, 94]. The second group extracts Co from copper–cobalt ores, mostly from the Republic of Congo and Zambia. This process requires leaching, solvent extraction, and finally, electrowinning [90].

4.1.2 Cobalt Production Drawbacks

In 2017, cobalt price raised to 70% to be 60,000 USD per ton because of the very high demand for LiBs in addition to the new electric car technology [95–97]. The price of cobalt is expecting to have an exponential raising by the coming years. There is a scary fact that the mineral sources are going to end one day. Another important problem is the price drop of by-product metals like copper and nickel because they are less in demand than cobalt, and this problem forced some mines to stop their work recently [95, 98].

4.2 Approaches to Recover Cobalt from Lithium-Ion Batteries

There are many advances in cobalt recycling sources such as hard metals, LiBs, aerospace alloys, and petrochemical catalysts. However, a big portion of this recycling will go to LiBs because it takes more than half of the world's cobalt demand [99, 100]. All batteries consist of three main essential components: cathode, anode, and the electrolyte mixture, as shown in Fig. 1. The cathode is an aluminum foil coated with a mixture of the active material such as $LiCoO_2$ or $LiNi_{1/3}Mn_{1/3}Co_{1/3}O_2$ depending on the battery type, but the commonly used is lithium cobalt oxide. The anode is a copper foil coated with graphite, which is commonly used in anode electrodes. The last component is the electrolyte mixture, which is responsible for traveling the lithium ions between the anode and the cathode, and the most common electrolytes are $LiPF_6$, $LiBF_4$, or $LiCO_4$ [4, 101–103].

The current recycling processes of valuable metals like cobalt and nickel from LiBs are going through two main processes: the physical processes and the chemical processes [104–106].

4.2.1 Physical Processes

Generally, the physical processes focus on the separation of the battery components, depending on its physical parameters such as density, magnetism, and conductivity [107]. The physical processes include thermal treatment, mechanical separation processes, dissolution processes, and mechanochemical processes.

Mechanical Separation Processes

The mechanical separation process is essential to avoid the explosion that could happen in the lithium batteries during the recovery process and reduce the recycled battery [108]. The mechanical separation process is usually used as a pre-treatment like removing the outer case and separation of the battery component, which will lead to continuing extracting the metals through the rest of the processes [109, 110]. Cobalt has been recovered by a hydrometallurgical process based on leaching, including the mechanical separation of the metal particles [73]. This process has a very critical disadvantage because it cannot fully recover the metals due to the organic and chemical salts such as $LiPF_6$, $LiBF_4$, and $LiClO_4$, which penetrate the metals [111]. For example, only 42.7 wt% of $LiCoO_2$ was recovered by Bertoul et al. using mechanical methods such as grinding, sieving, and elutriation [112].

Mechanochemical Process

The mechanochemical process, such as surface grinding treatment, is a kind of preparation and treatment of the electrode material to enhance its leaching processes to enable cobalt and lithium [104]. More than 90% of cobalt recovered using a planetary mill with and without aluminum oxide powder followed by leaching the scraped material in nitric acid (HNO_3) at room temperature [113]. Moreover, a new procedure was developed to recover cobalt and lithium from LiBs through cobalt chloride formation by co-grinding the cathode material with PVC in the air. PVC addition gives the grinding materials the chance to form chloride, soluble in water to extract the cobalt [114, 115]. It was reported that some of the grinding techniques could change the crystal structure of $LiCoO_2$ to ease its leaching at the room temperature [116, 117].

Thermal Treatment

This process is usually used to remove the organic salts and the insoluble adhesives by heating at a temperature range between 80 and 150°C [104]. Microcrystalline Co ferrite of 0.14 nm crystalline size was produced from a mixture of lithium cobalt oxide and iron oxide at 1000 °C for more than 4 h [118]. Kim et al. [109] used four steps recycling process to recover the cobalt, starting from mechanical to thermal, then going with a hydrometallurgical and sol–gel process.

Dissolution Process

This process is one of the most efficient recovering processes. It mainly breaks the polyvinylidene fluoride (PVDF) adhesive, which acts as a separator between the active materials (anode or cathode) and the current collectors. The selection of the solvents is very important to improve the dissolution process. Examples of the solvent are *N*, *N*-dimethyl formamide (DMF) and *N*-methyl pyrrolidone (NMP) [119, 120]. Dimethyl sulfoxide (DMSO) was used at 80 °C for more than 1.5 h, and it was counted as the most efficient solvent because of its environmental safety, low toxicity, and low prices [121]. Some approaches targeted the electrolyte extraction using organic solvents or isobutyl alcohol to decrease the toxicity of the recovering procedures in the following steps [122].

4.2.2 Chemical Processes

The chemical processes involve hydrometallurgical methods, including base and acid leaching, chemical precipitation, solvent extraction, electrochemical process bioprocess, or overlapping of these processes [5, 104, 117, 123, 124].

Acid Leaching

In this process, the leaching acid or solution can extract the metal by converting it from solid to the liquid phase and leaving the rest of the solids' original component. For instance, inorganic acids such as HCl, HNO_3, and H_2SO_4 were used to dissolve $LiCoO_2$ [125–127]. Also, environmentally-friendly organic acids, such as maleic acid, citric acid, ascorbic acid, and tartaric acid, are used [128–130]. Many studies have been performed to improve leaching processes' performance using several types of leaching solutions such as HCl, hydroxylamine acid, hydrochloride, and H_2SO_4. More than 99% of cobalt recovering was achieved using HCl at 80 °C for 1 h [131].

HNO_3 was used instead of HCl to produce a recycled lithium cobalt electrode with high performance [126, 132]. The particle size of the recovered metals was affected by the leaching acid and the effect of mechanical separation. Kim et al. reported a new procedure for cobalt recovery, which can be upgraded from laboratory scale to commercial recycling factories [109]. Li et al. recovered nearly 100% of lithium and more than 96% of cobalt from spent LiBs using citric acid as a leaching acid. Three acids have been tested, and the optimal performance of leaching was found to be 0.55 M H_2O_2 with 0.5 M of citric acid at 60 °C for 5 h [133]. Zeng et al. recovered 97% of cobalt from LiBs by using oxalic acid and proved that oxalic acid could be used alone without the assisting of H_2O_2. The optimal conditions are agitation of 400 rpm at 95 °C for 150 min; the solid–liquid ratio was 15 g L^{-1} [134].

Bioleaching

Bioleaching is a promising cost-effective technique due to its high efficiency in extracting cobalt at a low cost [135, 136]. In this process, the insoluble metal sulfates are changed to water-soluble through bio-oxidation conducted by microorganisms [137]. The mechanism of cobalt bioleaching from LiBs has been investigated by Xin et al. [138] by mixing iron-oxidizing, and sulfur-oxidizing bacteria, where the sulfur leads to dissolution and the iron leads to reduction of cobalt species [138]. A copper-catalyzed bioleaching process was developed to recycle cobalt from spent LiBs by Zeng et al. using *Acidithiobacillus ferrooxidans*. The influence of copper ions on the cobalt leaching was studied. The process achieved almost 99.9% of the cobalt dissolved in the solution after 6 days using copper ions. On the other hand, it takes more than 10 days to dissolve 43.1% without the assistance of copper ions [139].

Solvent Extraction

There are different routes and procedures for the solvent extraction method, widely used to extract and recover the metals from their leaching solutions such as Cyanex 272, PC-88A, and D2EHPA [73, 140, 141]. Swain et al. enhanced a hydrometallurgical method to extract cobalt sulfate solution from spent LiBs in purity. The optimum recovery condition was founded to be 2 M H_2SO_4 with 5 vol% of H_2O_2 at 75 °C for 30 min. The cobalt extraction was performed using Cyanex 272, with a concentration of 0.5 M [140]. Very high purity >99% of Co, Li, Ni, and Mn precipitates were recovered from LiBs using 20% Acorga M5640 in kerosene [142]. Chen et al. proposed a combined hydrometallurgical process (solvent extraction and precipitation) to recover the valuable metals from the LiBs in a leaching solution. The precipitation method was used to recover nickel and to extract cobalt and manganese using D2EHPA as a solvent extraction method. Consequentially, the cobalt was precipitated by ammonium oxalate solution as CoC_2O_4 and Li_2CO_3 [143].

Chemical Precipitation

This method is based on a precipitating agent's addition to precipitate the valuable metals such as cobalt present in the batteries. The key to achieving this method lies in the choosing of suitable chemical precipitation agents [104]. NaOH is the main solution to be added to perform the chemical precipitation and to form cobalt hydroxide. The precipitation of $Co(OH)_2$ begins at pH 6, and it should be completed when the pH value reaches 8. It was found that NaOH is the best agent to precipitate cobalt [108, 111, 114].

Electrochemical Process

The hydrometallurgical process starts with the dissolution of the cathode in concentrated acid, and then, the metals get recovered by chemical or electrochemical methods. This method is green compared with the pyrometallurgical process because it does not cause toxic gas emissions. The electrochemical process generally refers to selecting and separating the metal of interest from the leaching solution by electrodeposition. Electrodeposition is a desirable method because of ease of application, low cost, and ability to recover a metal with very high purity and controllable structure and morphology [104, 144, 145].

The process of battery recycling requires a sequence of physical and electrochemical steps. It starts by fully discharging the battery and then physically dismantling and separating its components. Afterward, the washed and dried cathode powder is dissolved in a leaching solution that contains acids, such as H_2SO_4 or HCL, as well as reducing agents. The next step is to selectively electrodeposit Co by an electrochemical technique such as cyclic voltammetry or chronoamperometry. The leaching solution's properties, including pH, temperature, and leaching duration, are critical for the electrochemical recovery step. Figure 2 shows that the amount of recovered Co increases with the leaching time until it reaches a plateau after almost 60 min which is the time required to dissolve the maximum amount of Co salt in the solution. It also clearly shows that the absence of the reducing agent such as H_2O_2 or ascorbic acid deaccelerates the leaching reaction that involves converting soluble Co^{3+} into the soluble $^{+2}$ (Eqs. 2 and 3), resulting in a lower amount of recovered Co.

$$2LiCoO_2 + 3H_2SO_4 + H_2O_2 = Li_2SO_4 + 2CoSO_4 + 4H_2O + O_2 \quad (2)$$

$$3H_2SO_4 + C_6H_8O_6 + 2LiCoO_2 = Li_2SO_4 + 2CoSO_4 + 4H_2O + C_6H_6O_6 \quad (3)$$

The properties of deposited Co depend on the parameters and the environment of the experiment, including (1) leaching time, (2) reduction potential of the metal, (3) pH, (4) charge density, and (5) electrolyte composition [127, 147–150]. There are several proposed mechanisms to explain the Co formation during electrodeposition as it is a complicated process. One of the mechanisms that explain the reduction of Co as a pH-dependent process shows that it involves the formation of $CoOH^+$ or $Co(OH)_2$ as intermediate species [149]. The formation of any of these species depends on the

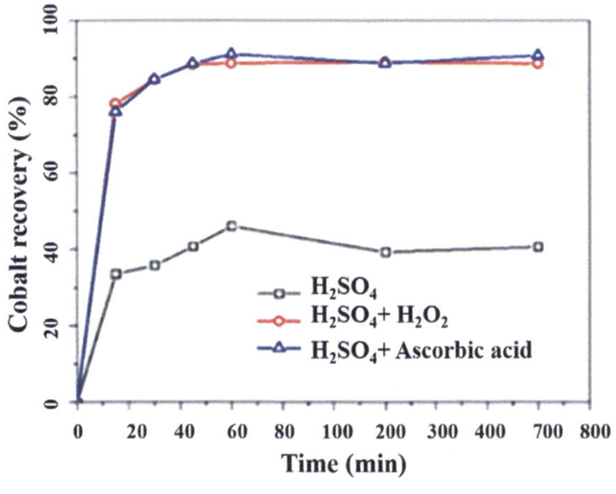

Fig. 2 Effect of leaching time at different leaching solutions on Co recovery at 65 °C in 1.5 M H_2SO_4, 0.15 M $C_6H_8O_6$, 0.636 M H_2O_2. Adapted from Ref. [146], Copyright 2019, MDPI

pH of the solution. At pH < 4, the Co^{+2} interacts with OH^- to form an unstable complex of $CoOH^+$, and then, it gets reduced and interacts with adsorbed hydrogen to form metallic Co as demonstrated in Eqs. 4-6. At pH > 4, the $Co(OH)_2$ is formed because of the Co^{+2} interaction with OH^-. Then, $Co(OH)_2$ was reduced into Co metal, as shown in Eqs. 7–8. This mechanism assumes that there is no hydrogen evolution reaction (HER) involved. Other studies showed that the electrodeposition of Co is accompanied with HER as a side reaction that changes the pH at the electrode surface in addition to decreasing the current efficiency during electrodeposition [151]. This can result in a modification of the kinetics of metal reduction and cause cobalt hydroxide precipitation instead of the metallic formation Co.

Using buffering agents such as boric acid in the electrodeposition solution could improve and control the electrodeposited Co properties. For example, boric acid acts as a buffering agent to prevent the H_2 evolution during electrodeposition. Moreover, it inhibits the reduction of Co^{2+} and shifts the cathodic reactions to a more negative potential, and it plays a leading role in controlling the morphology during electrodeposition. Ali et al. found that boric acid reduces the overpotential required for electrodeposition of Co complexes. The same author used a combination of sulfuric and ascorbic acid as a leaching solution since the ascorbic acid works as a leaching reducing agent and morphology modifier [146].

$$Co^{2+}_{(aq)} + 2e^- \rightarrow Co_{(s)} \tag{4}$$

$$H^+_{(aq)} + Co + e^- \rightarrow CoH_{(ads)} \tag{5}$$

$$H^+_{(aq)} + CoH_{(ads)} + e^- \rightarrow Co_{(s)} + H_{2(ads)} \tag{6}$$

$$Co^{2+}_{(aq)} + 2OH^-_{(aq)} \rightarrow Co(OH)_{2(s)} \tag{7}$$

$$Co(OH)_{2(s)} + 2e^- \rightarrow Co_{(s)} + 2OH_{(aq)} \tag{8}$$

Freitas et al. studied the charge efficiency as a function of the electrodeposition potential and pH values between 1.5 and 5.4, as shown in Fig. 3 [149]. The maximum charge efficiency is obtained at −1.00 V for all the tested pH values and was found to be 96.9, 88.2, and 17.2% for pH 1.5, 2.7, and 5.4, respectively. The highest efficiency was obtained at pH = 5.4, and this can be explained in the light of the hydrogen reduction reaction, where the contribution of HER is minimal. On the other hand, at low pH, the HER occurs and consumes part of the applied current resulting in lower efficiency.

The same study showed that initial stages of nucleation and growth of the electrodeposited Co are dependent on the pH under high efficiency conditions [149]. At pH 5.4, the electrodeposition follows the progressive nucleation model (Fig. 4) in which the nuclei are formed progressively at a constant rate in the initial instants, and the formed Co shows a large nucleus and grain size (three-dimensional growth). At lower pH 2.70, the electrodeposition follows the instantaneous

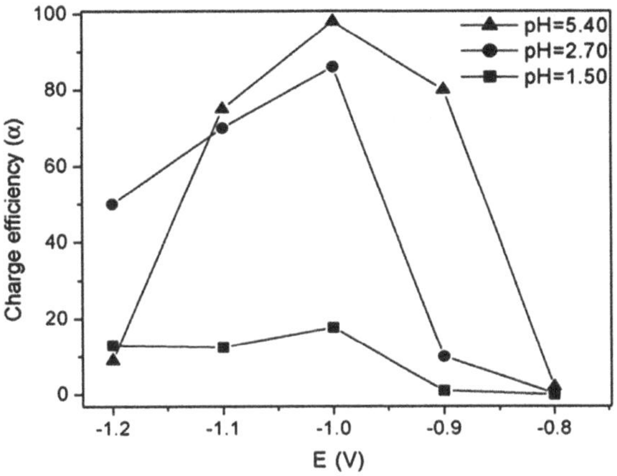

Fig. 3 Charge efficiency as a function of the electrodeposition potential under conditions of (0.10 mol L^{-1} Co, pH = 5.40, 2.70, and 1.50 in the presence of 0.10 mol L^{-1} H_3BO_3 buffer). Adapted from Ref. [149], Copyright 2007, Elsevier

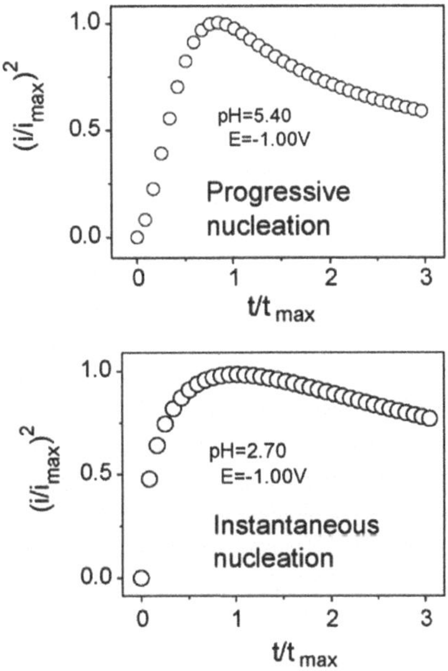

Fig. 4 Nucleation models of cobalt electrodeposited at −1.00 V and pH of 5.40 and 2.70, while the cobalt concentration is 0.10 mol L^{-1} in the presence of 0.10 mol L^{-1} H_3BO_3 buffer. Adapted from Ref. [149], Copyright 2007, Elsevier

nucleation model. In this approach, there is a high nucleus formation rate, and the formed Co exhibits a large number of small nuclei.

The structure of cobalt prepared by electrodeposition can be face-centered cubic (fcc) or hexagonal close-packed (hcp) or a combination [152, 153]. The structure is influenced by the nucleation and the growth process and is mainly controlled by the pH, the applied overpotential, and the electrolyte type. It was observed that at low pH ($\sim$1.6), the electrodeposited Co mostly consists of highly faulted fcc regions, while at high pH ($\sim$5.7), hcp phase with greatly reduced density of faults was observed [154–156]. The formation of the cobalt fcc is dominated by the adsorption and incorporation of atomic hydrogen that can diffuse during or after the electrodeposition to form a more stable hcp phase.

Freitas et al. studied the electrochemical recovery in HCl solution at different pH values (2.7 and 5.4) adjusted by NaOH and at different current densities of 10 and 50 C cm^{-2}. They showed that the hcp structure is obtained under both pH conditions [157]. However, at a low charge density of 10 C cm^{-2}, the [110] crystallographic direction is preferred at 2.7 pH, while the [002] direction is more preferred at a pH of 5.4. The crystal growth orientation can give considerable information about the growth of the deposited Co. The [110] preferred direction indicates that the Co grows layer by layer in a two-dimensional plane, while [002] shows that the growth is three dimensional. On the other hand, at a higher current density of 50 C/ cm^{-2}, the [002] and [110] show similar intensities at both pH values, but the reflections [100] and [101] are more pronounced in both cases, which can be explained by the hybrid 2D and 3D growth mechanism since [100] and [101] are assigned to the 2D and 3D growth, respectively.

Lupi et al. recovered nickel and cobalt, leaving less than 100 ppm of the metal in the leaching solution through two steps. The first step was the solvent extraction using 0.5 M CYANEX 272 in kerosene. The second step was carried out by potentiostat to electrodeposit cobalt and nickel on stainless steel and titanium net cathodes. Constant potentials of −0.85 to −0.9 V and of −1.20 to −1.50 V were applied in the case of Co and Ni, respectively [158].

A yield of 96.9% of cobalt was recovered from LiBs using HCl and H_2O_2 as a leaching mixture [149]. The electrochemical recovery was conducted in three-electrode systems using aluminum foil as a working electrode at a reduction potential of −0.5 to −1.2 V. The best performance of cobalt recovering was obtained at pH of 5.4 with a constant reduction potential of −1.0 V [149]. Garcia et al. followed the same procedures during their study on the recycled material as a supercapacitor and obtained a specific capacitance value of 601 F g^{-1} at 0.23 mA cm^{-1} [147]. Cobalt and copper multilayer films were collected on Al from spent LiBs by electrodeposition. The leaching solution was mainly composed of a mixture of H_2SO_4 (470 mL) and H_2O_2 (30 mL). The Cu–Co multilayers were conducted through potentiostatic steps with 50 s and a step potential of −1 and −0.3 V for the cobalt and the copper, respectively [159]. The Co deposit mechanism followed the instantons' mechanism at low pH (2.7) and the progressive mechanism at higher pH (5.4), which is similar to the individual deposition of Co.

Barbieri et al. conducted the electrodeposition of the cobalt on a conductive ITO glass substrate using HNO_3 as an acid leaching agent. The electrodeposition was performed through a three-electrode system with a constant voltage of −0.85 V and a pH of 6.5. The recovering performance was found to be 66.67%. The capacitive performance was studied for the recovered cobalt and showed a specific capacitance of 31.2 F g^{-1} at a scan rate of 1.0 mV s^{-1} [127].

Table 2 summarizes several studies of Co and Co compounds recovered from spent LiBs. The Table includes a column called capacitance, and it shows the specific capacitance of the electrodeposited Co compound as prepared or after treatment to convert it into electroactive materials for supercapacitor.

4.2.3 Magnetic Electrodeposition

Electrodeposition is a cost-effective and high-quality method for the production of nanostructured electrodeposited layers of controlled morphology, which can be used in a wide range of applications [164, 165]. The magnetic field can be applied during the electrodeposition process to enhance the electrodeposition rate. Bagard et al., at the end of the nineteenth century, investigated a new phenomenon after applying a low magnetic field on a metal deposition [166]. This idea receives more attention in terms of theoretical and practical studies by investigating the influence of homogeneous magnetic fields on mass transport in the electrodeposition processes. Furthermore, this study takes a part of Lorentz force, which has a significant contribution to mass transport and the morphology of the deposition layer [167]. Two magnetic forces occur when the electrochemical process takes place under the effect of the magnetic field. Firstly, the Lorentz force is shown by Eq. 9, where B is the magnetic induction and j represents the current density.

$$F_l = j \times B \tag{9}$$

To be more intense, through the electrochemical process, magnetic moments possess by the electrolyte ions. The magnetic field causes a torque, which makes the thermal motion in the fluid. This process called magnetohydrodynamics (MHD) [164]. MHD disturbs the surface during the electrodeposition causing agitation to the solution and increasing the ion transfer rate. Besides, the magnetic driving force makes a gradient in the ions concentration.

Aboelazm et al. enhanced the electrodeposition of Co by applying a magnetic field [160]. The electrodeposition was carried out in HCl and H_2O_2 leaching solution at −1 V for 20 s. Initially, the $Co(OH)_2$ was formed and reduced into metallic Co. After the electrodeposition, oxidation of the thin layer of Co into Co_2O_3 was observed due to the presence in the H_2O_2 medium. They explained the process according to the Eqs. 10–12. The Co_2O_3 prepared by this magnetic enhanced electrochemical recycling showed promising results as an active material for supercapacitors, where it showed a specific capacitance of 1273.0 F g^{-1} at 1 A g^{-1} with 96% capacitance retention at 5000 charge–discharge cycles.

$$Co^{2+}_{(aq)} + 2OH^-_{(aq)} \rightarrow Co(OH)_{2(s)} \tag{10}$$

$$Co(OH)_{2(s)} + 2e^- \rightarrow Co_{(s)} + OH^-_{(aq)} \tag{11}$$

$$Co_{(s)} + H_2O_{2(aq)} \rightarrow Co_3O_{4(s)} + H_2O_{(aq)} \tag{12}$$

Table 2 Electrodeposition conditions of Co and Co compounds from studies on the recovery of metals from spent LiB

	Electrodeposition solution	Electrochemical technique	Electrodeposition conditions	Temp.	pH	Capacitance	Ref.
Co and Li	$H_2SO_4 + H_2O_2$ + Ascorbic acid + Boric acid	Potentiostatic	Direct current "20 mA cm^{-2} for 600 s" Pulsed current "current density of 20 mA cm^{-2}" $T_{on} = T_{off} = 400$ ms	RT	–	NT	[146]
$Co(OH)_2$	$HNO_3 + KOH + H_3BO_3$	Potentiostatic	−0.85 V versus Ag/AgCl	RT	6.5	After heat treatment, a Co_3O_4 film formed and it shows 31.2 Fg^{-1} at 1 mV s^{-1}	[127]
Co	470 ml of HCl and 30 ml of H_2O_2	Potentiostatic under magnetic field	−1 V versus Ag/AgCl for 20 s	RT	5	H_2O_2 oxidize the Co immediately after electrodeposition to Co_3O_4; it shows 1273.0 F g^{-1} at 1 A g^{-1}	[160]
Co	3 M HCl and 30 mL of H_2O_2	Potentiostatic	−1 V versus Ag/AgCl for 50 s	RT	5	H_2O_2 oxidize the Co immediately after electrodeposition to Co_3O_4 with 87 F g^{-1} at 1 A g^{-1}	[161]
Co	HCL + $H_2O_2 + H_3BO_3$	Potentiostatic	−1 V versus Ag/AgCl/NaCl	25	5.4	NT	[157]
Li–Ni–Mn–Co Hydroxide	HCL + H_2O_2	Cyclic voltammetry	−1.2 to 0.2 V versus SCE for 5 cycles	–	–	951 F g^{-1} at 1 A g^{-1}	[162]
Co	$H_2SO_4 + H_2O_2$	Potentiostatic	−1.5 V versus Ag/AgCl	25	3	NT	[163]
Co and Cu	$H_2SO_4 + H_2O_2$	Potentiostatic steps	−1 V for Co −0.3 V for Cu versus Ag/AgCl/KCl	–	2.7 and 5.4	NT	[159]

NT refers to "Not tested for energy storage

4.2.4 Approaches in Magnetic Electrodeposition

Matsushima et al. studied the effect of the magnetic field on the iron electrodeposition by permanent magnets with a magnetic force range from 0, 1, 2, 3, 4, and 5 T. The effect of the magnetic field causes two significant changes; the first one is decreasing the current efficiency, and the second one is the changes in morphology round, angular shape to smooth roundish surface grains with small size grain distribution which became smaller with increasing the magnetic field [168]. Tanase et al. [169] studied the surface morphology and magnetic properties of Co–Ni–N films electrodeposited under magnetic flux lines generated by the electromagnet. Results showed a clear difference between the electrodeposition without the magnetic field, which revealed the inhomogeneous morphology of acicular and nodular grains and a well-defined acicular grain in the presence of a magnetic field with a quite homogeneous distribution. This change was explained in the light of the MHD convection of the ions at the electrode interface and the electrolyte induced by Lorentz force [169].

Yu et al. tested a magnetic field with a force of 1 T parallel to the electrode's surface in the cobalt electrodeposition. The deposition rate increased in the presence of the magnetic field because of the Lorentz forces as well as the interaction between the magnetic field and the electric current that causes MHD phenomenon. Then, the electrolyte agitation resulted in a decrease in the diffusion layer's thickness and, consequently, an increase in the deposition rate. A strong deviation was reported due to the instantaneous nucleation process regarding the micro-MHD flows [170].

Yu et al. applied a magnetic field parallel to the electric field for preparing a thin cobalt layer during the electrodeposition. The steady-state current and the deposition mass decreased gradually with the increase of the magnetic field. The gradient can explain this in concentrations of cobalt ions, which produce a magnetic driving force. This magnetic driving force pushes the cobalt ions away from the cathode and decreases the steady-state current. AFM images revealed that, a non-uniform nodular structure is detected in the absence of the magnetic field (Fig. 5a), while a hill-like morphology is observed in the presence of the magnetic field in Fig. 5b, c. The formation of hill-like morphology can be attributed to the cobalt's ferromagnetic property, which forces the ions to grow along the magnetic field lines and form the hill-like morphology [171].

Li et al. studied the effect of high magnetic field annealing up to 12 T in the pulsed electrodeposition of Co–Ni–p thin film. The mechanical properties were studied, and it was found that the optimal condition is the presence of a magnetic field with a magnetic force of 6 T. Under this condition, a clear modification happened to the deposited layer after applying the magnetic field's different values. From the FESEM images, it can be seen that Fig. 6a–d has almost the same

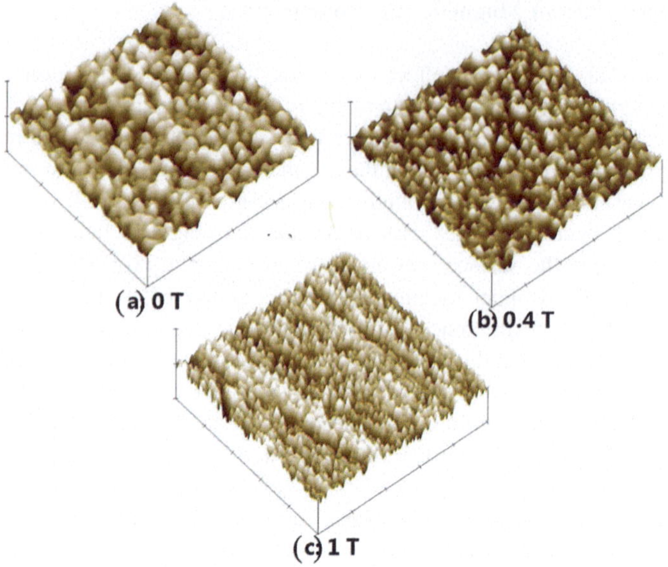

Fig. 5 AFM images of cobalt obtained by magnetic electrodeposition technique. Adapted from Ref. [171], Copyright 2015, ESG publisher

spherical cluster morphologies but with a bit different smoothness. On the other hand, Fig. 6c shows that the spherical clusters vanish, and an intestine-like morphology is formed. This finding can be attributed to the fall of the maximum saturation magnetization of the materials to a minimum value in the case of 6 T, which is recorded as the optimum magnetic field for the electrodeposition [55].

4.3 Advantages of Magnetic Electrodeposition of Cobalt from Lithium-Ion Batteries

Many approaches have been reported for the recycling of cobalt from the spent LiBs. However, there is a gap in the purity and quality of the manned commercial metals and the recovered metals from the waste such as LiBs. Moreover, the current recovering techniques request multi-steps to reach the end products.

The current electrodeposition technologies could recover the cobalt from spent LiBs. However, some limitations are related to uncontrolled properties such as morphology, which hinder its commercial and industrial growth. Considering the ferromagnetic properties of cobalt and cobalt ions' paramagnetic properties, the magnetic field was found to be an efficient process to enhance the controlled electrodeposition process to produce cobalt oxide with tailored morphology from

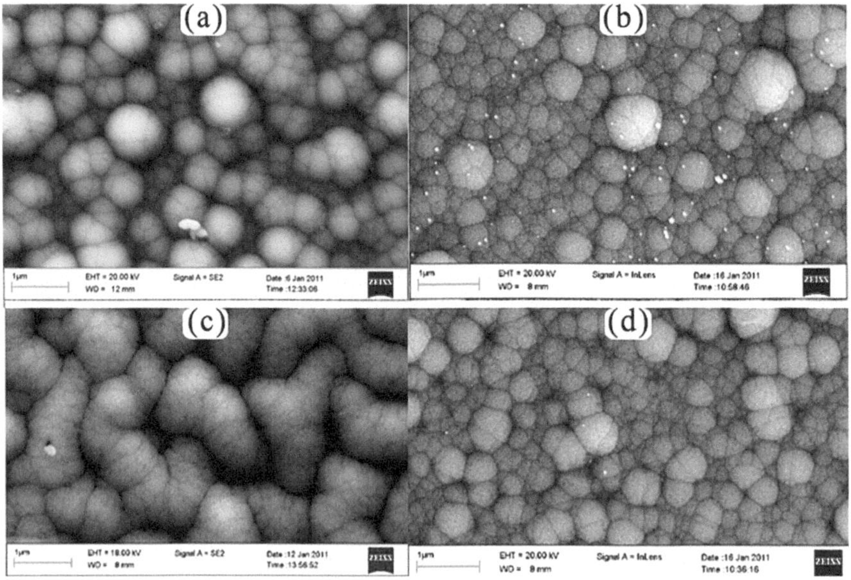

Fig. 6 FESEM images of Co–Ni–P film electrodeposition **a** before annealing and with magnetic annealing of **b** 0 T, **c** 6 T, and **d** 12 T. Adapted from Ref. [165], Copyright 2016, Hindawi Publishing Cooperation

spent LiBs. This could be one of the most promising electrodeposition processes to produce a high-quality cobalt layer at a low cost, which could be used in several industrial applications.

5 Conclusions

This chapter discusses the recycling of LiBs as a source of materials for electrochemical energy storage systems with more details in supercapacitors. Battery recycling is the most feasible future resource of metals such as Co and Li, considering the huge number of LiBs waste around the world. In addition, recycling is a very important strategy to avoid the negative environmental impact of batteries disposed of. The pseudocapacitor materials, such as cobalt oxide, are exciting because of their high theoretical specific capacitance. Moreover, cobalt production is challenging because of many problems related to its natural resources, purity, and price. Therefore, cobalt recovering from LiBs has received increasing attention as it consumes more than 50% of world cobalt demand. Battery recycling starts from the physical process, which includes (mechanical, mechanochemical, thermal, and dissolution treatments). Chemical processes include (acid leaching, bioleaching, solvent extraction, chemical precipitation, and electrochemical process).

The recovered processes focus only on the recovering yield without any attention on the recovered metals' performance.

The magnetic electrodeposition is one of the most promising techniques to enhance the morphology and the nucleation of the deposited materials. A combination between the electrochemical and the magnetic electrodeposition proved to be an effective process to extract cobalt and provide different characteristics of the recovered material to facilitate its use in a wide range of industrial applications.

6 Future Perspectives

The main challenge for the recovery of Co from LiBs is that its capacitive performance is still far from either the theoretical values or the commercial materials' capacitive performance. More research efforts are required to increase the efficiency of the recovered materials. Some interesting ideas include: (1) enhancing the material's electrodeposition from the leaching solution by controlling the conditions to enhance the deposition rate and morphology such as temperature, ultrasonication, or catalytic agents; (2) surface treatment of the electrode material before the electrodeposition seems to be very important. The deposited material properties could remarkably be improved if the energy of the surface increased by using plasma metal treatment or increasing the surface area by using an acid etching technique; (3) a composite of the electrode material or multilayer electrodeposition can be done with carbon-based materials such as activated carbon or rGO to increase the conductivity of the electrode, which could improve the capacitive performance.

References

1. Klaine SJ, Koelmans AA, Horne N, Carley S, Handy RD, Kapustka L, Nowack B, von der Kammer F (2012) Paradigms to assess the environmental impact of manufactured nanomaterials. Environ Toxicol Chem 31(1):3–14
2. Nagpure SC, Bhushan B (2012) Nanomaterials for electrical energy storage devices. In: Bhushan B (eds) Encyclopedia of nanotechnology. Springer Netherlands, Dordrecht, pp 1597–1608
3. Fan E, Li L, Wang Z, Lin J, Huang Y, Yao Y, Chen R, Wu F (2020) Sustainable recycling technology for li-ion batteries and beyond: challenges and future prospects. Chem Rev
4. Horn D, Zimmermann J, Gassmann A, Stauber R, Gutfleisch O (2019) Battery recycling: focus on li-ion batteries. In: Modern battery engineering: a comprehensive introduction, p 223
5. Santana IL, Moreira TFM, Lelis MFF, Freitas MBJG (2017) Photocatalytic properties of $Co_3O_4/LiCoO_2$ recycled from spent lithium-ion batteries using citric acid as leaching agent. Mater Chem Phys 190:38–44
6. Pinegar H, Smith YR (2019) Recycling of end-of-life lithium ion batteries, part i: commercial processes. J Sustain Metall 5(3):402–416

7. Harper G, Sommerville R, Kendrick E, Driscoll L, Slater P, Stolkin R, Walton A, Christensen P, Heidrich O, Lambert S, Abbott A, Ryder K, Gaines L, Anderson P (2019) Recycling lithium-ion batteries from electric vehicles. Nature 575(7781):75–86
8. Rekioua D (2020) Storage in hybrid renewable energy systems. In: Rekioua D (ed) Hybrid renewable energy systems: optimization and power management control. Springer International Publishing, Cham, pp 139–172
9. Wang J, Dong S, Ding B, Wang Y, Hao X, Dou H, Xia Y, Zhang X (2017) Pseudocapacitive materials for electrochemical capacitors: from rational synthesis to capacitance optimization. Nat Sci Rev 4(1):71–90
10. Thalji MR, Ali GAM, Algarni H, Chong KF (2019) Al^{3+} ion intercalation pseudocapacitance study of $W_{18}O_{49}$ nanostructure. J Power Sources 438:227028
11. Winter M, Brodd RJ (2004) What are batteries, fuel cells, and supercapacitors? Chem Rev 104(10):4245–4270
12. Horn M, Gupta B, MacLeod J, Liu J, Motta N (2019) Graphene-based supercapacitor electrodes: addressing challenges in mechanisms and materials. Curr Opin Green Sustain Chem 17:42–48
13. Ali GAM, Supriya S, Chong KF, Shaaban ER, Algarni H, Maiyalagan T, Hegde G (2019) Superior supercapacitance behavior of oxygen self-doped carbon nanospheres: a conversion of Allium cepa peel to energy storage system. Biomass Conv Bioref. https://doi.org/10.1007/s13399-019-00520-3
14. Marina PE, Ali GAM, See LM, Teo EYL, Ng E-P, Chong KF (2016) In situ growth of redox-active iron-centered nanoparticles on graphene sheets for specific capacitance enhancement. Arab J Chem. 12(8):3883–3889
15. Ali GAM, Yusoff MM, Chong KF (2016) Graphene Electrochemical production and its energy storage properties. ARPN J Eng Appl Sci 11(16):9712–9717
16. Yao B, Chandrasekaran S, Zhang J, Xiao W, Qian F, Zhu C, Duoss EB, Spadaccini CM, Worsley MA, Li Y (2019) Efficient 3D printed pseudocapacitive electrodes with ultrahigh MnO_2 loading. Joule 3(2):459–470
17. Xu R, Lin J, Wu J, Huang M, Fan L, Xu Z, Song Z (2019) A high-performance pseudocapacitive electrode material for supercapacitors based on the unique $NiMoO_4$/NiO nanoflowers. Appl Surf Sci 463:721–731
18. Simon P, Gogotsi Y (2008) Materials for electrochemical capacitors. Nat Mater 7(11):845–854
19. Thalji MR, Ali GAM, Liu P, Zhong YL, Chong KF (2021) $W_{18}O_{49}$ nanowires-graphene nanocomposite for asymmetric supercapacitors employing $AlCl_3$ aqueous electrolyte. Chem Eng J 409:128216
20. Ali GAM, Fouad OA, Makhlouf SA, Yusoff MM, Chong KF (2014) Co_3O_4/SiO_2 nanocomposites for supercapacitor application. J Solid State Electrochem 18(9):2505–2512
21. Ali GAM, Yusoff MM, Shaaban ER, Chong KF (2017) High performance MnO_2 nanoflower supercapacitor electrode by electrochemical recycling of spent batteries. Ceram Int 43:8440–8448
22. Ali GAM, Yusoff MM, Ng YH, Lim NH, Chong KF (2015) Potentiostatic and galvanostatic electrodeposition of MnO_2 for supercapacitors application: a comparison study. Curr Appl Phys 15(10):1143–1147
23. Ali GAM, Yusoff MM, Feng CK (2015) Electrochemical properties of electrodeposited MnO_2 nanoparticles. Adv Mater Res 1113:550–553
24. Ali GAM, Tan LL, Jose R, Yusoff MM, Chong KF (2014) Electrochemical performance studies of MnO_2 nanoflowers recovered from spent battery. Mater Res Bull 60:5–9
25. Ali GAM (2020) Recycled MnO_2 nanoflowers and graphene nanosheets for low-cost and high performance asymmetric supercapacitor. J Electron Mater 49(9):5411–5421
26. Ali GAM, Makhlouf SA, Yusoff MM, Chong KF (2015) Structural and electrochemical characteristics of graphene nanosheets as supercapacitor electrodes. Rev Adv Mater Sci 40(1):35–43

27. Ali GAM, Lih Teo EY, Aboelazm EAA, Sadegh H, Memar AOH, Shahryari-Ghoshekandi R, Chong KF (2017) Capacitive performance of cysteamine functionalized carbon nanotubes. Mater Chem Phys 197:100–104
28. Ren H, Tian H, Lee H-S, Park T, Leung FC, Ren T-L, Chae J (2015) Regulating the respiration of microbe: a bio-inspired high performance microbial supercapacitor with graphene based electrodes and its kinetic features. Nano Energy 15:697–708
29. Ali GAM, Manaf SAA, A D, Chong KF, Hegde G (2016) Superior supercapacitive performance in porous nanocarbons. J Energy Chem 25(4):734–739
30. Lee SP, Ali GAM, Hegazy HH, Lim HN, Chong KF (2021) Optimizing reduced graphene oxide aerogel for a supercapacitor. Energy & Fuels 35(5):4559–4569
31. Ali GAM, Divyashree A, Supriya S, Chong KF, Ethiraj AS, Reddy M, Algarni H, Hegde G (2017) Carbon nanospheres derived from Lablab purpureus for high performance supercapacitor electrodes: a green approach. Dalton Trans 46(40):14034–14044
32. Conway BE (1999) Electrochemical supercapacitors: scientific fundamentals and technological applications. Springer Science & Business Media
33. Augustyn V, Simon P, Dunn B (2014) Pseudocapacitive oxide materials for high-rate electrochemical energy storage. Energy Environ Sci 7(5):1597–1614
34. Frackowiak E, Béguin F (2001) Carbon materials for the electrochemical storage of energy in capacitors. Carbon 39(6):937–950
35. Boota M, Gogotsi Y (2019) MXene—conducting polymer asymmetric pseudocapacitors. Adv Energy Mater 9(7):1802917
36. Jiang Y, Liu J (2019) Definitions of pseudocapacitive materials: a brief review. Energy Environ Mater 2(1):30–37
37. Wang Y, Guo J, Wang T, Shao J, Wang D, Yang Y-W (2015) Mesoporous transition metal oxides for supercapacitors. Nanomaterials 5(4):1667–1689
38. Hu C-C, Chen W-C, Chang K-H (2004) How to achieve maximum utilization of hydrous ruthenium oxide for supercapacitors. J Electrochem Soc 151(2):A281–A290
39. Wang X, Wang X, Huang W, Sebastian P, Gamboa S (2005) Sol–gel template synthesis of highly ordered MnO_2 nanowire arrays. J Power Sources 140(1):211–215
40. Toupin M, Brousse T, Bélanger D (2004) Charge storage mechanism of MnO_2 electrode used in aqueous electrochemical capacitor. Chem Mater 16(16):3184–3190
41. Ali GAM, Yusoff MM, Algarni H, Chong KF (2018) One-step electrosynthesis of MnO_2/rGO nanocomposite and its enhanced electrochemical performance. Ceram Int 44(7):7799–7807
42. Ali GAM, Thalji MR, Soh WC, Algarni H, Chong KF (2020) One-step electrochemical synthesis of MoS_2/graphene composite for supercapacitor application. J Solid State Electrochem 24(1):25–34
43. Ali GAM, Habeeb OA, Algarni H, Chong KF (2018) CaO impregnated highly porous honeycomb activated carbon from agriculture waste: symmetrical supercapacitor study. J Mater Sci 54:683–692
44. Cheng M, Duan S, Fan H, Su X, Cui Y, Wang R (2017) Core@shell CoO@Co_3O_4 nanocrystals assembling mesoporous microspheres for high performance asymmetric supercapacitors. Chem Eng J 327:100–108
45. Yan X, Tian L, He M, Chen X (2015) Three-dimensional crystalline/amorphous Co/Co_3O_4 core/shell nanosheets as efficient electrocatalysts for the hydrogen evolution reaction. Nano Lett 15(9):6015–6021
46. Wang X, Hu A, Meng C, Wu C, Yang S, Hong X (2020) Recent advance in Co_3O_4 and Co_3O_4-containing electrode materials for high-performance supercapacitors. Molecules 25 (2):269
47. Yang Q, Lu Z, Sun X, Liu J (2013) Ultrathin Co_3O_4 nanosheet arrays with high supercapacitive performance. Sci Rep 3:3537

48. Pal M, Rakshit R, Singh AK, Mandal K (2016) Ultra high supercapacitance of ultra small Co_3O_4 nanocubes. Energy 103:481–486
49. Liu Z, Zhou W, Wang S, Du W, Zhang H, Ding C, Du Y, Zhu L (2019) Facile synthesis of homogeneous core-shell Co_3O_4 mesoporous nanospheres as high performance electrode materials for supercapacitor. J Alloy Compd 774:137–144
50. Zhang F, Yuan C, Lu X, Zhang L, Che Q, Zhang X (2012) Facile growth of mesoporous Co_3O_4 nanowire arrays on Ni foam for high performance electrochemical capacitors. J Power Sources 203:250–256
51. Li J, Zhang P, Zhao X, Chen L, Shen J, Li M, Ji B, Song L, Wu Y, Liu D (2019) Structure-controlled Co–Al layered double hydroxides/reduced graphene oxide nanomaterials based on solid-phase exfoliation technique for supercapacitors. J Colloid Interface Sci 549:236–245
52. Wang P, Zhou H, Meng C, Wang Z, Akhtar K, Yuan A (2019) Cyanometallic framework-derived hierarchical Co_3O_4–NiO/graphene foam as high-performance binder-free electrodes for supercapacitors. Chem Eng J 369:57–63
53. Babu IM, William JJ, Muralidharan G (2019) Ordered mesoporous Co_3O_4/CMC nanoflakes for superior cyclic life and ultra high energy density supercapacitor. Appl Surf Sci 480:371–383
54. Ali GAM, Sadegh H, Yusoff MM, Chong KF (2019) Highly stable symmetric supercapacitor from cysteamine functionalized multi-walled carbon nanotubes operating in a wide potential window. Mater Today: Proc 16:2273–2279
55. Fan H, Quan L, Yuan M, Zhu S, Wang K, Zhong Y, Chang L, Shao H, Wang J, Zhang J, Cao C-n (2016) Thin Co_3O_4 nanosheet array on 3D porous graphene/nickel foam as a binder-free electrode for high-performance supercapacitors. Electrochim Acta 188:222–229
56. Chen H, Wang J, Liao F, Han X, Zhang Y, Xu C, Gao L (2019) Uniform and porous Mn-doped Co_3O_4 microspheres: solvothermal synthesis and their superior supercapacitor performances. Ceram Int 45(9):11876–11882
57. Wang Y, Huo J (2020) C, N co-doped Co_3O_4 hollow spheres prepared via hydrothermal reaction for improved electrochemical activity. Int J Hydrogen Energy 45(1):1170–1176
58. Deng S, Xiao X, Chen G, Wang L, Wang Y (2016) Cd doped porous Co_3O_4 nanosheets as electrode material for high performance supercapacitor application. Electrochim Acta 196:316–327
59. Li G, Chen M, Ouyang Y, Yao D, Lu L, Wang L, Xia X, Lei W, Chen S-M, Mandler D (2019) Manganese doped Co_3O_4 mesoporous nanoneedle array for long cycle-stable supercapacitors. Appl Surf Sci 469:941–950
60. Tao F, Zhao Y-Q, Zhang G-Q, Li H-L (2007) Electrochemical characterization on cobalt sulfide for electrochemical supercapacitors. Electrochem Commun 9(6):1282–1287
61. Yu Z, Tetard L, Zhai L, Thomas J (2015) Supercapacitor electrode materials: nanostructures from 0 to 3 dimensions. Energy Environ Sci 8(3):702–730
62. Qin Q, Chen L, Wei T, Liu X (2019) MoS_2/NiS yolk–shell microsphere-based electrodes for overall water splitting and asymmetric supercapacitor. Small 15(29):1803639
63. Shinde SK, Ramesh S, Bathula C, Ghodake GS, Kim DY, Jagadale AD, Kadam AA, Waghmode DP, Sreekanth TVM, Kim HS, Nagajyothi PC, Yadav HM (2019) Novel approach to synthesize $NiCo_2S_4$ composite for high-performance supercapacitor application with different molar ratio of Ni and Co. Sci Rep 9(1):13717
64. Aboelazm EAA, Ali GAM, Algarni H, Chong KF (2018) Flakes size-dependent optical and electrochemical properties of MoS_2. Curr Nanosci 14(2):1–5
65. Zhu T, Wang Z, Ding S, Chen JS, Lou XWD (2011) Hierarchical nickel sulfide hollow spheres for high performance supercapacitors. RSC Adv 1(3):397–400
66. Gaikar P, Pawar SP, Mane RS, Nuashad M, Shinde D (2016) Synthesis of nickel sulfide as a promising electrode material for pseudocapacitor application. RSC Adv 6(113):112589–112593

67. Hsu Y-K, Chen Y-C, Lin Y-G (2014) Synthesis of copper sulfide nanowire arrays for high-performance supercapacitors. Electrochim Acta 139(Suppl C):401–407
68. Zhang H, Yu X, Guo D, Qu B, Zhang M, Li Q, Wang T (2013) Synthesis of bacteria promoted reduced graphene oxide-nickel sulfide networks for advanced supercapacitors. ACS Appl Mater Interfaces 5(15):7335–7340
69. Zhang R, Lu C, Shi Z, Liu T, Zhai T, Zhou W (2019) Hexagonal phase NiS octahedrons co-modified by 0D-, 1D-, and 2D carbon materials for high-performance supercapacitor. Electrochim Acta 311:83–91
70. Chen W, Xia C, Alshareef HN (2014) One-step electrodeposited nickel cobalt sulfide nanosheet arrays for high-performance asymmetric supercapacitors. ACS Nano 8(9):9531–9541
71. Liu TC, Pell W, Conway B, Roberson S (1998) Behavior of molybdenum nitrides as materials for electrochemical capacitors comparison with ruthenium oxide. J Electrochem Soc 145(6):1882–1888
72. Choi D, Blomgren GE, Kumta PN (2006) Fast and reversible surface redox reaction in nanocrystalline vanadium nitride supercapacitors. Adv Mater 18(9):1178–1182
73. Li L, Ge J, Wu F, Chen R, Chen S, Wu B (2010) Recovery of cobalt and lithium from spent lithium ion batteries using organic citric acid as leachant. J Hazard Mater 176(1):288–293
74. Balogun M-S, Qiu W, Wang W, Fang P, Lu X, Tong Y (2015) Recent advances in metal nitrides as high-performance electrode materials for energy storage devices. J Mater Chem A 3(4):1364–1387
75. Saha S, Jana M, Khanra P, Samanta P, Koo H, Murmu NC, Kuila T (2015) Band gap engineering of boron nitride by graphene and Its application as positive electrode material in asymmetric supercapacitor device. ACS Appl Mater Interfaces 7(26):14211–14222
76. Wang Y, Yang W, Yang J (2007) A Co–Al layered double hydroxides nanosheets thin-film electrode fabrication and electrochemical study. Electrochem Solid-State Lett 10(10):A233–A236
77. Prevot V, Forano C, Khenifi A, Ballarin B, Scavetta E, Mousty C (2011) A templated electrosynthesis of macroporous NiAl layered double hydroxides thin films. Chem Commun 47(6):1761–1763
78. Gao X, Liu X, Wu D, Qian B, Kou Z, Pan Z, Pang Y, Miao L, Wang J (2019) Significant role of Al in ternary layered double hydroxides for enhancing electrochemical performance of flexible asymmetric supercapacitor. Adv Func Mater 29(36):1903879
79. Gao X, Zhao Y, Dai K, Wang J, Zhang B, Shen X (2020) NiCoP nanowire@ NiCo-layered double hydroxides nanosheet heterostructure for flexible asymmetric supercapacitors. Chem Eng J 384:123373
80. Chen H, Hu L, Chen M, Yan Y, Wu L (2014) Nickel–Cobalt layered double hydroxide nanosheets for high-performance supercapacitor electrode materials. Adv Func Mater 24 (7):934–942
81. Guo XL, Liu XY, Hao XD, Zhu SJ, Dong F, Wen ZQ, Zhang YX (2016) Nickel-Manganese layered double hydroxide nanosheets supported on nickel foam for high-performance supercapacitor electrode materials. Electrochim Acta 194(Suppl C):179–186
82. Shao M, Ning F, Zhao Y, Zhao J, Wei M, Evans DG, Duan X (2012) Core–shell layered double hydroxide microspheres with tunable interior architecture for supercapacitors. Chem Mater 24(6):1192–1197
83. Zhang L, Wang J, Zhu J, Zhang X, San Hui K, Hui KN (2013) 3D porous layered double hydroxides grown on graphene as advanced electrochemical pseudocapacitor materials. J Mater Chem A 1(32):9046–9053
84. McNeill R, Siudak R, Wardlaw J, Weiss D (1963) Electronic conduction in polymers. I. The chemical structure of polypyrrole. Aust J Chem 16(6):1056–1075
85. Wang Y, Ding Y, Guo X, Yu G (2019) Conductive polymers for stretchable supercapacitors. Nano Res:1–10

86. Hall PJ, Mirzaeian M, Fletcher SI, Sillars FB, Rennie AJ, Shitta-Bey GO, Wilson G, Cruden A, Carter R (2010) Energy storage in electrochemical capacitors: designing functional materials to improve performance. Energy Environ Sci 3(9):1238–1251
87. Wang G, Zhang L, Zhang J (2012) A review of electrode materials for electrochemical supercapacitors. Chem Soc Rev 41(2):797–828
88. Ehsani A, Heidari AA, Shiri HM (2019) Electrochemical pseudocapacitors based on ternary nanocomposite of conductive polymer/graphene/metal oxide: an introduction and review to it in recent studies. Chem Rec 19(5):908–926
89. Yao B, Wang H, Zhou Q, Wu M, Zhang M, Li C, Shi G (2017) Ultrahigh-conductivity polymer hydrogels with arbitrary structures. Adv Mater 29(28):1700974-n/a
90. Pazik PM, Chmielewski T, Glass HJ, Kowalczuk PB (2016) World production and possible recovery of cobalt from the Kupferschiefer stratiform copper ore. E S Web of Conf 8:01063
91. Cheang CY, Mohamed N (2016) Removal of cobalt from ammonium chloride solutions using a batch cell through an electrogenerative process. Sep Purif Technol 162:154–161
92. Fouad OA, Ali GAM, El-Erian MAI, Makhlouf SA (2012) Humidity sensing properties of cobalt oxide/silica nanocomposites prepared via sol–gel and related routes. Nano 7 (5):1250038
93. Statista (2017) World cobalt reserves as of 2016 (cited 13 Aug 2017). Available from https://www.statista.com/statistics/264930/global-cobalt-reserves/
94. Crundwell FK (2011) Extractive metallurgy of nickel, cobalt and platinum group metals. Elsevier
95. Nelson E (2017) The hottest thing in the markets right now is an obscure metal mined in DR Congo. Batteries 2017 (cited 15 Aug 2017). Available from https://qz.com/969115/cobalt-is-the-unlikely-hero-of-financial-markets-right-now/
96. Randall T (2017) The electric-car boom is so real even oil companies say it's coming. Available from https://www.bloomberg.com/news/articles/2017-04-25/electric-car-boom-seen-triggering-peak-oil-demand-in-2030s
97. Nitta N, Wu F, Lee JT, Yushin G (2015) Li-ion battery materials: present and future. Mater Today 18(5):252–264
98. Times F (2017) Glencore deals mining blow to African copper producers (cited 12 Sept 2017). Available from https://www.ft.com/content/3f1d9664-562f-11e5-a28b-50226830d644
99. Raugei M, Winfield P (2019) Prospective LCA of the production and EoL recycling of a novel type of Li-ion battery for electric vehicles. J Clean Prod 213:926–932
100. van den Brink S, Kleijn R, Sprecher B, Tukker A (2020) Identifying supply risks by mapping the cobalt supply chain. Resour Conserv Recycl 156:104743
101. Tarascon JM, Armand M (2001) Issues and challenges facing rechargeable lithium batteries. Nature 414(6861):359–367
102. Li L, Dunn JB, Zhang XX, Gaines L, Chen RJ, Wu F, Amine K (2013) Recovery of metals from spent lithium-ion batteries with organic acids as leaching reagents and environmental assessment. J Power Sources 233:180–189
103. Bozich J, Hang M, Hamers R, Klaper R (2017) Core chemistry influences the toxicity of multicomponent metal oxide nanomaterials, lithium nickel manganese cobalt oxide, and lithium cobalt oxide to Daphnia magna. Environ Toxicol Chem 36(9):10
104. Ordoñez J, Gago EJ, Girard A (2016) Processes and technologies for the recycling and recovery of spent lithium-ion batteries. Renew Sustain Energy Rev 60:195–205
105. Porvali A, Aaltonen M, Ojanen S, Velazquez-Martinez O, Eronen E, Liu F, Wilson BP, Serna-Guerrero R, Lundström M (2019) Mechanical and hydrometallurgical processes in HCl media for the recycling of valuable metals from Li-ion battery waste. Resour Conserv Recycl 142:257–266
106. Dhiman S, Gupta B (2019) Partition studies on cobalt and recycling of valuable metals from waste Li-ion batteries via solvent extraction and chemical precipitation. J Clean Prod 225:820–832

107. Tenorio J (1999) Carbon-zinc batteries treatment by ore processing methods. In: REWAS'99: global symposium on recycling, waste treatment and clean technology
108. Contestabile M, Panero S, Scrosati B (1999) A laboratory-scale lithium battery recycling process. This work has been presented as an invited talk at the 4th international battery recycling congress, Hamburg, Germany, 1–3 July, 1998. J Power Sources 83(1):75–78
109. Kim D-S, Sohn J-S, Lee C-K, Lee J-H, Han K-S, Lee Y-I (2004) Simultaneous separation and renovation of lithium cobalt oxide from the cathode of spent lithium ion rechargeable batteries. J Power Sources 132(1):145–149
110. Bernardes AM, Espinosa DCR, Tenório JAS (2004) Recycling of batteries: a review of current processes and technologies. J Power Sources 130(1):291–298
111. Xu J, Thomas HR, Francis RW, Lum KR, Wang J, Liang B (2008) A review of processes and technologies for the recycling of lithium-ion secondary batteries. J Power Sources 177 (2):512–527
112. Bertuol DA, Toniasso C, Jiménez BM, Meili L, Dotto GL, Tanabe EH, Aguiar ML (2015) Application of spouted bed elutriation in the recycling of lithium ion batteries. J Power Sources 275:627–632
113. Zhang Q, Lu J, Saito F, Nagata C, Ito Y (2000) Room temperature acid extraction of Co from $LiCo_{0.2}Ni_{0.8}O_2$ scrap by a mechanochemical treatment. Adv Powder Technol 11 (3):353–359
114. Saeki S, Lee J, Zhang Q, Saito F (2004) Co-grinding $LiCoO_2$ with PVC and water leaching of metal chlorides formed in ground product. Int J Miner Process 74:S373–S378
115. Wang M, Zhang F (2017) Mechanochemical recycling of spent lithium-ion battery and reaction mechanisms clarification. Chinese J Environ Eng 11(2):1069–1074
116. Zhang Q, Saito F (1998) Non-thermal process for extracting rare earths from bastnaesite by means of mechanochemical treatment. Hydrometallurgy 47(2):231–241
117. Jha MK, Kumari A, Panda R, Rajesh Kumar J, Yoo K, Lee JY (2016) Review on hydrometallurgical recovery of rare earth metals. Hydrometallurgy 165:2–26
118. Bahgat M, Farghaly FE, Basir SMA, Fouad OA (2007) Synthesis, characterization and magnetic properties of microcrystalline lithium cobalt ferrite from spent lithium-ion batteries. J Mater Process Technol 183(1):117–121
119. Dai J, Teng X, Song Y, Ren J (2017) Effect of casting solvent and annealing temperature on recast Nafion membranes for vanadium redox flow battery. J Membr Sci 522:56–67
120. Zhou X, He W-Z, Li G-M, Zhang X-J, Zhu S-G, Huang J-W (2010) Recycling of electrode materials from spent lithium-ion batteries. In: 2010 4th International conference on bioinformatics and biomedical engineering (iCBBE). IEEE
121. Bankole OE, Gong C, Lei L (2013) Battery recycling technologies: Recycling waste lithium ion batteries with the impact on the environment in-view. J Environ Ecol 4(1):14–28
122. Bankole OE, Lei L (2014) Silicon exchange effects of glassware on the recovery of $LiPF_6$: Alternative route to preparation of Li_2SiF_6. J Solid Waste Technol Manag 39(4):254–259
123. Chen X, Ma H, Luo C, Zhou T (2017) Recovery of valuable metals from waste cathode materials of spent lithium-ion batteries using mild phosphoric acid. J Hazard Mater 326:77–86
124. Li J, Sollami Delekta S, Zhang P, Yang S, Lohe MR, Zhuang X, Feng X, Östling M (2017) Scalable fabrication and integration of graphene microsupercapacitors through full inkjet printing. ACS Nano 11(8):8249–8256
125. Mantuano DP, Dorella G, Elias RCA, Mansur MB (2006) Analysis of a hydrometallurgical route to recover base metals from spent rechargeable batteries by liquid–liquid extraction with Cyanex 272. J Power Sources 159(2):1510–1518
126. Castillo S, Ansart F, Laberty-Robert C, Portal J (2002) Advances in the recovering of spent lithium battery compounds. J Power Sources 112(1):247–254

127. Barbieri EMS, Lima EPC, Cantarino SJ, Lelis MFF, Freitas MBJG (2014) Recycling of spent ion-lithium batteries as cobalt hydroxide, and cobalt oxide films formed under a conductive glass substrate, and their electrochemical properties. J Power Sources 269:158–163
128. Nayaka G, Pai K, Santhosh G, Manjanna J (2016) Dissolution of cathode active material of spent Li-ion batteries using tartaric acid and ascorbic acid mixture to recover Co. Hydrometallurgy 161:54–57
129. Nayaka GP, Pai KV, Manjanna J, Keny SJ (2016) Use of mild organic acid reagents to recover the Co and Li from spent Li-ion batteries. Waste Manag 51:234–238
130. Nayaka G, Zhang Y, Dong P, Wang D, Zhou Z, Duan J, Li X, Lin Y, Meng Q, Pai K (2019) An environmental friendly attempt to recycle the spent Li-ion battery cathode through organic acid leaching. J Environ Chem Eng 7(1):102854
131. Zhang P, Yokoyama T, Itabashi O, Suzuki TM, Inoue K (1998) Hydrometallurgical process for recovery of metal values from spent lithium-ion secondary batteries. Hydrometallurgy 47 (2):259–271
132. Lee CK, Rhee K-I (2002) Preparation of $LiCoO_2$ from spent lithium-ion batteries. J Power Sources 109(1):17–21
133. Li L, Zhai L, Zhang X, Lu J, Chen R, Wu F, Amine K (2014) Recovery of valuable metals from spent lithium-ion batteries by ultrasonic-assisted leaching process. J Power Sources 262:380–385
134. Zeng X, Li J, Shen B (2015) Novel approach to recover cobalt and lithium from spent lithium-ion battery using oxalic acid. J Hazard Mater 295:112–118
135. Brandl H, Faramarzi MA (2006) Microbe-metal-interactions for the biotechnological treatment of metal-containing solid waste. China Particuology 4(2):93–97
136. Naseri T, Bahaloo-Horeh N, Mousavi SM (2019) Environmentally friendly recovery of valuable metals from spent coin cells through two-step bioleaching using *Acidithiobacillus thiooxidans*. J Environ Manage 235:357–367
137. Rohwerder T, Gehrke T, Kinzler K, Sand W (2003) Bioleaching review part A. Appl Microbiol Biotechnol 63(3):239–248
138. Xin B, Zhang D, Zhang X, Xia Y, Wu F, Chen S, Li L (2009) Bioleaching mechanism of Co and Li from spent lithium-ion battery by the mixed culture of acidophilic sulfur-oxidizing and iron-oxidizing bacteria. Biores Technol 100(24):6163–6169
139. Zeng G, Deng X, Luo S, Luo X, Zou J (2012) A copper-catalyzed bioleaching process for enhancement of cobalt dissolution from spent lithium-ion batteries. J Hazard Mater 199:164–169
140. Swain B, Jeong J, Lee J-c, Lee G-H, Sohn J-S (2007) Hydrometallurgical process for recovery of cobalt from waste cathodic active material generated during manufacturing of lithium ion batteries. J Power Sources 167(2):536–544
141. Nan J, Han D, Zuo X (2005) Recovery of metal values from spent lithium-ion batteries with chemical deposition and solvent extraction. J Power Sources 152:278–284
142. Nayl AA, Hamed MM, Rizk SE (2015) Selective extraction and separation of metal values from leach liquor of mixed spent Li-ion batteries. J Taiwan Inst Chem Eng 55:119–125
143. Chen X, Chen Y, Zhou T, Liu D, Hu H, Fan S (2015) Hydrometallurgical recovery of metal values from sulfuric acid leaching liquor of spent lithium-ion batteries. Waste Manag 38:349–356
144. Kalyani M, Emerson R (2019) Electrodeposition of nano crystalline cobalt oxide on porous copper electrode for supercapacitor. J Mater Sci: Mater Electron 30(2):1214–1226
145. Altimari P, Schiavi PG, Rubino A, Pagnanelli F (2019) Electrodeposition of cobalt nanoparticles: An analysis of the mechanisms behind the deviation from three-dimensional diffusion-control. J Electroanal Chem 851:113413

146. Rafsanjani-Abbasi A, Rahimi E, Shalchian H, Vahdati-Khaki J, Babakhani A, Hosseinpour S, Davoodi A (2019) Recycled cobalt from spent Li-ion batteries as a superhydrophobic coating for corrosion protection of plain carbon steel. Materials 12(1):90
147. Garcia EM, Tarôco HA, Matencio T, Domingues RZ, dos Santos JAF, Ferreira RV, Lorençon E, Lima DQ, de Freitas MBJG (2012) Electrochemical recycling of cobalt from spent cathodes of lithium-ion batteries: its application as supercapacitor. J Appl Electrochem 42(6):361–366
148. Garcia EM, Santos JS, Pereira EC, Freitas MBJG (2008) Electrodeposition of cobalt from spent Li-ion battery cathodes by the electrochemistry quartz crystal microbalance technique. J Power Sources 185(1):549–553
149. Freitas MBJG, Garcia EM (2007) Electrochemical recycling of cobalt from cathodes of spent lithium-ion batteries. J Power Sources 171(2):953–959
150. Myoung J, Jung Y, Lee J, Tak Y (2002) Cobalt oxide preparation from waste $LiCoO_2$ by electrochemical–hydrothermal method. J Power Sources 112(2):639–642
151. Gabe DR (1997) The role of hydrogen in metal electrodeposition processes. J Appl Electrochem 27(8):908–915
152. Pirota K, Béron F, Zanchet D, Rocha T, Navas D, Torrejón J, Vazquez M, Knobel M (2011) Magnetic and structural properties of fcc/hcp bi-crystalline multilayer Co nanowire arrays prepared by controlled electroplating. J Appl Phys 109(8):083919
153. Bubendorff J, Mény C, Beaurepaire E, Panissod P, Bucher J-P (2000) Electrodeposited cobalt films: Hcp versus fcc nanostructuring and magnetic properties. Eur Phys J B 17:635–643
154. Nakahara S, Mahajan S (1980) The influence of solution pH on microstructure of electrodeposited cobalt. J Electrochem Soc 127(2):283
155. Darques M, Piraux L, Encinas A, Bayle Guillem P, Popa A, Ebels U (2005) Appl Phys Lett 86:072508–072511
156. Vicenzo A, Cavallotti PL (2004) Growth modes of electrodeposited cobalt. Electrochim Acta 49(24):4079–4089
157. Freitas MBJG, Garcia EM, Celante VG (2009) Electrochemical and structural characterization of cobalt recycled from cathodes of spent Li-ion batteries. J Appl Electrochem 39 (5):601–607
158. Lupi C, Pasquali M, Dell'Era A (2005) Nickel and cobalt recycling from lithium-ion batteries by electrochemical processes. Waste Manag 25(2):215–220
159. Freitas MBJG, Celante VG, Pietre MK (2010) Electrochemical recovery of cobalt and copper from spent Li-ion batteries as multilayer deposits. J Power Sources 195(10):3309–3315
160. Aboelazm EAA, Ali GAM, Algarni H, Yin H, Zhong YL, Chong KF (2018) Magnetic electrodeposition of the hierarchical cobalt oxide nanostructure from spent lithium-ion batteries: its application as a supercapacitor electrode. J Phys Chem C 122(23):12200–12206
161. Aboelazm EAA, Ali GAM, Chong KF (2018) Cobalt oxide supercapacitor electrode recovered from spent lithium-ion battery. Chem Adv Mater 3(4):67–74
162. Allam NK, Mesbah YI, Ahmed N, and Ali BA (2020) Recycling of Li–Ni–Mn–Co hydroxide from spent batteries to high-performance supercapacitors with exceptional stability. ChemElectroChem
163. Garcia EM, Tarôco HA, Matencio T, Domingues RZ, dos Santos JAF, de Freitas MBJG (2011) Electrochemical recycling of cobalt from spent cathodes of lithium–ion batteries: its application as coating on SOFC interconnects. J Appl Electrochem 41(11):1373
164. Uhlemann M, Tschulik K, Gebert A, Mutschke G, Fröhlich J, Bund A, Yang X, Eckert K (2013) Structured electrodeposition in magnetic gradient fields. Eur Phys J Spec Top 220 (1):287–302
165. Li D, Wu C, Wang Q, Chopart J-P, He J, Franczak A, Levesque A (2016) Effects of high magnetic field postannealing on microstructure and properties of pulse electrodeposited Co–Ni–P films. Adv Mater Sci Eng 2016:6
166. Bagard H (1896) Phénomène de Hall dans les liquides. J Theor Appl Phys 5(1):499–508
167. Fahidy TZ (1983) Magnetoelectrolysis. J Appl Electrochem 13(5):553–563

168. Matsushima H, Nohira T, Mogi I, Ito Y (2004) Effects of magnetic fields on iron electrodeposition. Surf Coat Technol 179(2):245–251
169. Tanase SI, Tanase D, Sandu AV, Georgescu V (2012) Magnetic field effects on surface morphology and magnetic properties of Co–Ni–N thin films prepared by electrodeposition. J Supercond Novel Magn 25(6):2053–2057
170. Yu YD, Song ZL, Ge HL, Wei GY (2014) Influence of magnetic fields on cobalt electrodeposition. Surf Eng 30(2):83–86
171. Yu Y, Song Z, Ge H, Wei G, Jiang L (2015) Effects of magnetic fields on the electrodeposition process of cobalt. Int J Electrochem Sci 10:4812–4819

Recovery of Nanomaterials for Battery Applications

Hasna Aziam

Abstract In the last decades, many researchers were inspired to develop recycling technologies for nanomaterial manufacturing and manage the excessive generation of wastes (biomass, biological, plastic, and industrial wastes). Cost-efficient, sustainable process, and good material properties are the requirements to meet for a successful recycling route and, consequently, inducing huge economic and environmental benefits. Moreover, the use of the recycled nanomaterials for several applications was reported in the literature, such as catalysis, energy storage, and biomedical applications. This chapter will be devoted to reviewing the studies carried on the recycling of nanomaterials for battery applications, mainly alkali metal ion batteries (alkali metal: Li, Na, Mg, K), conventional secondary batteries, and alkaline batteries.

Keywords Recycling · Nanomaterials · Wastes · Cost-efficient · Sustainable · Energy storage · Batteries

List of Abbreviations

3D	Three-dimensional
ACN	Acetonitrile
BET	Brunauer–Emmett–Teller
CMC	Carboxymethyl Cellulose
EDS	Energy dispersive X-ray spectroscopy
EMD	Electrolytic Manganese Dioxide
LiBs	Lithium-ion batteries
Ni–MH	Nickel-metal hydride
NMP	*N*-methylpyrolidinone
PVP	Polyvinylpyrrolidone

H. Aziam (✉)
High Throughput Multidisciplinary Research Laboratory (HTMRL), Mohammed VI Polytechnic University (UM6P), Lot 660 Hay Moulay Rachid, Ben Guerir, Morocco
e-mail: hasna.aziam@um6p.ma

IMED-Lab, Cadi Ayyad University (UCA), Av. a. El Khattabi, P. B. 549, Marrakesh, Morocco

A. S. H. Makhlouf and G. A. M. Ali (eds.), *Waste Recycling Technologies for Nanomaterials Manufacturing*, Topics in Mining, Metallurgy and Materials Engineering, https://doi.org/10.1007/978-3-030-68031-2_5

REO	Rare earth oxide
SBR	Styrene-butadiene rubber
$scHHPCO_2$	Supercritical helium head pressurized carbon dioxide $scHHPCO_2$
SEM	Scanning electronic microscopy
SiO_{2slag}	Silica slag
Si_{slag}	Silicon slag
SWCNTs	Single-wall carbon nanotubes
TEM	Transmission electron microscope
XPS	X-ray photoelectron spectroscopy

1 Introduction

Recycling is converting waste material into a new material showing physicochemical properties suitable for a particular application. Billion tons of wastes, such as glass, biomass, electronics, metals, paper, plastic, textiles, batteries, electronics, and many more, are generated every day due to the population's growing needs and consumptions. Therefore, recycling these spent materials into the manufacture of new products would significantly impact the environment and the economy. Furthermore, our lifestyle nowadays is entirely dependent on energy, especially battery systems. A battery is needed to use mobile phones, work on a laptop, or even start a car engine. Now, the energy market's challenge is developing cost-efficient, high energy density, and safe electrode materials. Thus, nanomaterials (1–100 nm) have shown remarkable properties, which induced enhanced electrochemical performances for battery applications [1–14]. For example, nanomaterials demonstrate a high surface area, which provides a large contact area with the electrolyte inducing a high lithium-ion flux across the interface for lithium-ion batteries (LiBs).

Moreover, electron transport within the nanometer-sized particles is improved as well [15–18]. Reviewing some statistics, China alone is annually producing around 50–60 million units of waste lead batteries [19]. The annual consumption of batteries is around 8 billion units per year in the USA and Europe, 6 billion in Japan, and 1 billion in Brazil [20]. These batteries contain high value and a low abundance of chemical elements. LiBs are generally composed of 6.5% of Co, 5% of electrolyte, 1.6% of Al, 2.8% of Cu, and 0.8% of Li, where cobalt, lithium, and the electrolyte are the most valuable components. Government regulations oblige battery manufacturers to recycle the battery wastes for safety and health reasons [21].

This book chapter reviews waste recycling technologies and the recovery of nanomaterials suitable for battery applications. First, a brief overview is dedicated to explain the battery technology and emphasize the working principle of given examples of battery types. Second, the state of the art of nanomaterial recycling is detailed for each battery type: alkali ion batteries, conventional secondary batteries, and alkaline batteries.

2 A Brief Overview of Battery Technology

The battery concept is based on the electrochemical storage of energy in electrochemical cells connected in series and/or in parallel [22]. A single electrochemical cell is composed of a positive and a negative electrode, referred to as cathode and anode, respectively, separated with a separator to prevent short-circuiting and enable ion transportation. The electrochemical energy storage consists of converting the electrical energy into stored chemical energy and vice versa via a redox reaction happening between the positive and negative electrodes. Batteries are classified into two types: primary batteries (non-rechargeable) and secondary batteries (rechargeable).

Moreover, secondary batteries are differentiated as well into: (1) alkali metal ion secondary batteries (where alkali metal: Li^+, Na^+, K^+, Zn^{2+}, Mg^{2+}) [23–33], (2) conventional secondary batteries (e.g., lead–acid batteries, nickel-electrode batteries) [34–39], (3) molten salt batteries (e.g., sodium–nickel chloride batteries, sodium–sulfur batteries) [40–45], and (4) flow batteries (e.g., polysulfide–bromide batteries, vanadium redox batteries, zinc–bromine batteries) [46–48].

First, alkali metal ion batteries' operating principle consists of the intercalation/de-intercalation of the alkali metal ions between the negative/positive electrode systems. The electrolyte fills the space between the two electrodes to ensure proper ionic movement [49]. $LiCoO_2$ and $LiFePO_4$ are the most popular materials used for the positive electrode, while graphite is used for the negative side. Figure 1 shows a schematic illustration of the working principle in $LiCoO_2$//graphite cell. During charge, the lithium ions immigrate from $Li_{1-x}CoO_2$ through the electrolyte to settle in graphite sheets. During discharge, the reversible process occurs where lithium ions are de-intercalated from graphite.

Lead–acid secondary batteries are generally composed of several plates containing lead dioxide PbO_2 paste at the negative electrode and lead sponge at the positive part, immersed in an aqueous sulfuric acid electrolyte solution. In this case, the redox reaction converts both materials to lead sulfate $PbSO_4$, which remains essentially insoluble in the electrolyte [51, 52].

For the molten salt batteries, sodium–sulfur batteries, as shown in Fig. 2, contain molten elemental sodium in the negative electrode, molten sulfur in the positive electrode, and a solid beta alumina ceramic electrolyte. The temperature should be in the range of 300–350 °C to keep the electrodes in a molten state. Sodium–nickel chloride batteries are very similar to sodium–sulfur batteries. The negative electrode is composed of liquid sodium and the positive electrode of solid nickel chloride, separated by a solid beta alumina electrolyte. The difference is using a second liquid electrolyte of $NaAlCl_4$, which allows the fast transport of Na^+ ions [53].

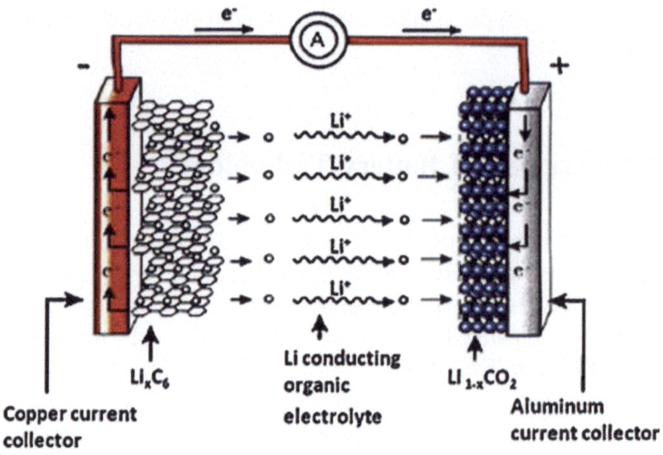

Fig. 1 Scheme of a common lithium-ion battery. Adapted with permission from Ref. [50], Copyright 2010, Elsevier

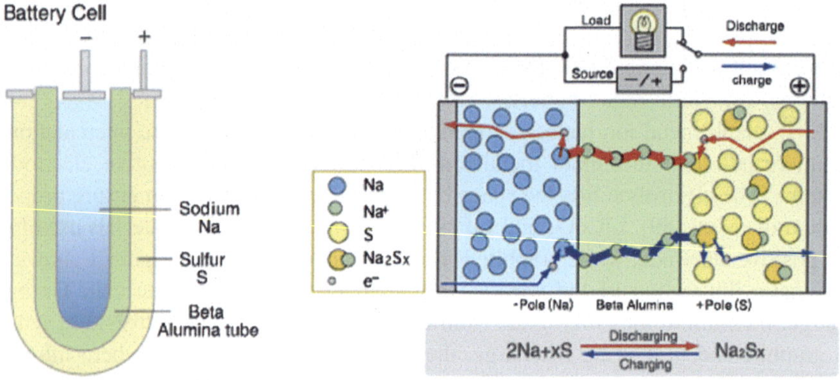

Fig. 2 Schematic diagram of a sodium–sulfur battery energy storage system. Adapted with permission from Ref. [54], Copyright 2017, MDPI

Last but not least, flow batteries or flowing electrolyte batteries are of a unique operating principle. Two electrolyte sets are pumped through separate loops, both negative and positive, and a third electrolyte separator. For example, in the zinc–bromine batteries, both electrolytes are aqueous solutions of zinc bromide $ZnBr_2$. The negative electrode originally contains a minimal quantity of zinc, and it is more plated during the charging process, while at the positive electrode, bromine is formed. During the inverse process (i.e., discharge), zinc dissolves into its aqueous state [22]. Figure 3 demonstrates a schematic illustration of the zinc–bromine battery system.

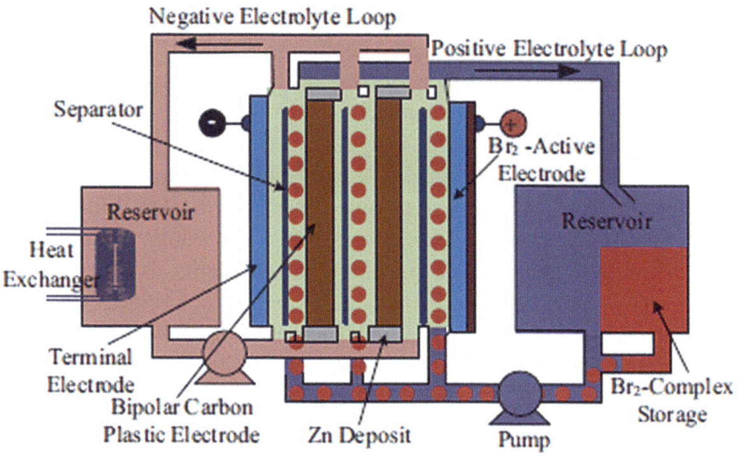

Fig. 3 Schematic illustration of zinc–bromine battery. Adapted with permission from Ref. [54], Copyright 2017, MDPI

3 Recovery of Nanomaterials for Alkali Metal Ion Batteries

3.1 Recovery of Graphite

Graphite is commonly used as a negative electrode material for commercial LiBs. It shows high thermal stability, good mechanical structure, high electrical conductivity; it is environmentally benign, abundant, and causes no lithium dendrites formation in the battery [55]. Therefore, recycling graphite from its various waste resources gained large attention as a candidate solution to meet the market's high demand [56–68]. Rothermel et al. [56] reported a study about the feasibility of recycling graphite recovered from spent LiBs via three different concepts: (1) thermal treatment of graphite without electrolyte recovery, (2) use of supercritical carbon dioxide as extractant, subsequently followed by the thermal treatment (3) use of subcritical carbon dioxide-assisted electrolyte extraction before thermal treatment. The first procedure consists of detaching the active material from copper foil, washing with distilled water, then NMP, and using ethanol. The graphite powder was dried overnight at 80 °C and then thermally treated at 1000 °C/ 5 h under argon atmosphere. The second method is based on the electrolyte removal using supercritical helium head pressurized carbon dioxide (scHHPCO_2) as extractant, and by the end, the same thermal treatment at 1000 °C was carried to recover graphite. The third approach is about removing electrolytes by applying a flow through subcritical carbon dioxide (liqCO_2) extraction with acetonitrile (ACN) admixed as an additional solvent. ACN is used to enhance conductive salt extraction. After electrolyte extraction, the electrodes were separated from each other, and graphite was detached from its current collector (i.e., copper foil) and thermally treated at 1000 °C in the same conditions followed in approach 1 and 2. Hence, the best electrochemical performances of the recycled graphite were demonstrated when approach 3 is applied.

Moreover, battery-grade graphite is synthesized from various biomass wastes such as cherry stone, olive stone, rice husk and banana peels, etc. For example, the carbon issued from chicken eggs is obtained in the form of N_2-rich mesoporous carbon. This material shows a reversible capacity of around 1780 mAh g^{-1} as anode for LiBs, one of the highest capacities obtained from carbonaceous materials. Another type of waste, i.e., waste tires, was studied to fabricate functionalized carbon black. The carbon prepared by this route revealed an ordered assembly of graphitic domains. Electrochemical tests demonstrated that this material exhibits a higher reversible capacity than graphite, around 390 mAh g^{-1} after 100 cycles [57]. Single-walled carbon nanotubes (abbreviated as SWCNTs) are another type of carbonaceous materials used in battery manufacturing. Schauerman et al. [69] studied the feasibility of recycling Single-walled carbon nanotubes (SWCNTs) recovered from spent LiBs electrodes and evaluated their potential reuse. The recycling procedure consists of (1) recovering the material by a mechanical separation or shredding process, (2) removing free carbon ad organic compounds from the active material via a thermal treatment, and (3) recovering valuable cathode metals and lithium-containing compounds via a chemical leaching using an acid (e.g., nitric acid, hydrochloric acid, sulfuric acid). The properties of pure and recycled SWCNTs were characterized using scanning electron microscopy, thermogravimetric analysis, Raman spectroscopy, optical absorption spectroscopy, and galvanostatic test. Moreover, the required energy to functionalize the recycled SWCNTs was measured and found to be less than half of the direct energy needed to synthesize new SWCNTs.

3.2 Recovery of Silicon

Silicon has been studied mainly as negative electrode material for alkali metal ion batteries (where alkali metal: Li, Na, K) [70–78]. Silicon shows interesting properties, mainly high abundance, non-toxicity, the low discharging potential of 0.2 V versus Li^+/Li, and a high theoretical capacity of around 4200 mAh g^{-1}, which is more than 10 times that of commercial graphite. Nevertheless, it demonstrates challenges like the huge volume expansion (>300%) during the lithiation/delithiation process, low electronic conductivity (around $\approx 10^{-3}$ S cm^{-1}, and it increases to $\approx 10^{2}$ S cm^{-1} after lithiation), and low lithium diffusion coefficient (around 10^{-14}–10^{-13} cm^2 S^{-1}). These drawbacks induce drastic capacity fade and poor cycling efficiency [79, 80]. In the literature, many studies reported the preparation of silicon nanoparticles by recycling several wastes such as rice husk [81–85], waste iron slag [86], silicon wafer waste [87], waste slicing slug [88–90], glass bottles [91], kerf powder [92], and silicon oxide [93]. In 2013, Jung et al. [81]

investigated 3D nanoporous silicon preparation from rice husk and its potential use as negative electrode material for LiBs. Firstly, the recycling procedure starts with acid leaching using HCl to remove the alkali metal impurities and then a thermal treatment at 650 °C for 3 h to remove organic components. Secondly, the magnesiothermic reduction step consists of adding magnesium and firing at 850 °C/3 h under argon atmosphere to obtain silicon following the chemical reaction: $2Mg + SiO_2 \rightarrow 2MgO + Si$. Last, high-purity 3D porous silicon is prepared by two-stage etching. The first is removing MgO by stirring the magnesia/silicon mixture in HCl solution (where the chemical reaction is: $MgO + 2HCl \rightarrow MgCl_2 + H_2O$) and then in HF solution to remove residual silica. This material shows a high capacity in the first cycle around 1554 mAh g^{-1} at 1 C current rate. Furthermore, it shows high capacity retention of about 100% after 200 cycles. Chun et al. [86] studied the preparation of highly mesoporous silicon using waste iron slag and evaluating its electrochemical performances as negative electrode material for LiBs. First, the silica slag (labeled as SiO_{2slag}) iron slag was ground to fine particles using a ball-miller at 120 rpm for 5 h. The slag powder was leached using HCl at 80 °C/5 h and then thermally treated at 700 °C/2 h. Second, the porous silicon slag (abbreviated as Si_{slag}) is prepared by modified magnesiothermic reduction. In water solution, Si_{slag} is mixed with NaCl, heated at 60 °C, and dried. Magnesium was added to the dried porous SiO_{2slag}/NaCl and heated at 650 °C/ 2.5 h under argon atmosphere. Two steps of etching using HCl and HF were carried to dissolve the magnesia and eliminate residual silica, respectively. Porous Si_{slag} shows a stable capacity of around 1500 mAh g^{-1} at 1000 mA g^{-1} current density for 100 cycles. Jang et al. [88] emphasized silicon nanoparticles' extraction from wafer slicing waste using an aerosol-assisted process. The Si sludge is washed using hydrochloric acid to eliminate metal impurities, dispersed in distilled water to form a colloidal suspension, and then nebulized using ultrasonic atomization to separate the silicon nanoparticles (smaller, lighter) from silicon carbide (large size, higher density). Dried argon flow was introduced to transport the aerosol droplets to a pre-heated furnace (300–500 °C), and by the end, dried Si particles are collected. The electrochemical properties of the aerosol recovered silicon nanoparticles showing superior capacity retention compared to commercial Si nanoparticles. Kim et al. [93] studied the feasibility of recycling silicon oxide SiO_x particles mixed with crystalline Si recovered from Si vapors in the ingot growing process. Afterward, the potential use of these particles as negative electrode material for LiBs was demonstrated. First, characterizations using SEM, TEM, EDS, and XPS techniques confirmed that the collected SiO_x particles are mixed with crystalline Si and amorphous SiO_2. Moreover, the average of these particles is distributed in two groups with 322.9 nm and 17.9 μm. The BET technique revealed that the specific surface area was 184.4 m^2 g^{-1}, and the average pore size was 7.4 nm. The electrochemical properties of this material were evaluated versus Li^+/Li, showing good coulombic efficiency up to 100 cycles around 99%. Recently, Eshraghi et al. [94] published a study on the recovery of solar-grade silicon (Si) from spent photovoltaic panels and their reuse as anodes for LiBs. The polycrystalline silicon wafer scraps were leached in two media. The first was in KOH solution to remove Al impurities, whereas the second was in HNO_3 to remove Pbb and Ag. Nano-sized Si was obtained by 10 h ball-milling delivering capacities as high as 1400 mAh g^{-1}.

3.3 *Recovery of Valuable Chemical Elements*

The increasing demand and use of LiBs are inducing large quantities of spent batteries containing valuable chemical elements such as Li, Co, Ni, and Mn. Governmental regulations oblige LiBs manufacturers to recycle it as considering the dangerous environmental impacts and beneficial economic/environmental revenue. In addition, these chemicals are of high cost and less abundant in the earth's crust. Therefore, this has inspired many researchers to dig into studying their recovery and reuse. These elements are recovered, in general, in the form of metallic alloys, compounds, or solutions containing metallic ions [57]. In 1999, Contestabile et al. [95] described a multi-step process for recovering spent Li// MnO_2 primary batteries. First, the batteries are opened by cutting the steel cases after cooling them to avoid any explosions caused by the components' high reactivity. The metallic lithium, separator, cathode material was isolated. Then, the cathode material MnO_2/C is separated from the steel casings via mechanical riddling and treated with concentrated sulfuric acid. Later, manganese is recovered as electrolytic manganese dioxide (EMD) by electrolysis of a neutralized and purified solution of $MnSO_4$ with H_2SO_4. The obtained EMD demonstrates high-purity, battery-grade, and can be reused as raw material to manufacture high energy density LiBs. Castillo [96] reported a recycling process for spent LiBs by recovering $LiCoO_2$ electrodes. First, the electrodes were crushed and sieved, and then soaked in *N*-methylpyrolidinone (NMP) at 60 °C for 1 h to dissolve the binder and filtered to detach the active material (i.e., $LiCoO_2$) from the aluminum foil used as a current collector. The recovered $LiCoO_2$ was leached using sulfuric acid for 1 h at 80 °C, and then, cobalt was precipitated in the form of cobalt hydroxide $Co(OH)_2$ by adding sodium hydroxide. Kim and Shin [97] investigated the re-synthesis feasibility and reuse of spent $LiFePO_4$ cathode material for LiBs. For this study, the electrodes were separated from aluminum foil and heated to decompose the CMC and SBR, used as binders. Three carbonization temperatures were carried (400 °C, 500 °C, and 600 °C), and their effect on the electrochemical properties of the recycled $LiFePO_4$ was evaluated in comparison to the newly synthesized $LiFePO_4$. Surprisingly, the recycled $LiFePO_4$ at 500 °C showed higher specific capacity at 1 C and 0.1 C current rates compared to that shown for the new $LiFePO_4$. This finding is probably assigned to the enhanced electronic conductivity of recycled $LiFePO_4$ due to the carbonization of CMC and SBR binders. Li et al. [98] studied the recovery of Li, Fe, and P from spent $LiFePO_4$ electrode materials using a simple, efficient, and cost-effective process. This procedure involves the use of sulfuric acid at low concentration as leachant and H_2O_2 as oxidant to leach Li into the solution and Fe/P form as $FePO_4$ in the leaching residue as stated in Eq. (1). Li is recovered by adding Na_3PO_4 to Li solution, and Li_3PO_4 precipitate was obtained following the chemical Eq. (2), whereas the leaching residue is thermally treated at 600 °C/4 h to obtain $FePO_4$. The chemical reactions are presented as follows:

$$2LiFePO_4 + H_2SO_4 + H_2O_2 \rightarrow Li_2SO_4 + 2FePO_4 \downarrow + 2H_2O \quad (1)$$

$$3Li_2SO_4 + 2Na_3PO_4 \rightarrow 3Na_2SO_4 + 2Li_3PO_4 \downarrow \quad (2)$$

Moreover, several companies all over the globe are interested in recycling of Li, Co, Mn, Ni, and their reuse in manufacturing LiBs. In Canada, Xstrata nickel international company is dealing with the recovery of Cu, Ni, and Co. In France, SNAM company is producing a coralliferous mixture from spent LiBs. Accuracy company (Germany) is recycling Li, Mn, and Co in lithium oxide, LiCl, and Co–Mn alloy. Umicore company (Belgium) is using a pyrometallurgical method to recycle Ni and cobalt as $Ni(OH)_2$ and $LiCoO_2$ [57].

4 Recovery of Nanomaterials for Conventional Secondary Batteries

4.1 Recovery of Nanomaterials for Ni–Cd and Ni–MH Batteries

The nickel–cadmium batteries' market is huge; they are used in many portable electronic devices, military and defense applications. However, these spent batteries are classified as hazardous wastes due to toxic cadmium. Yet, researchers craved to find good recycling solutions and minimize their negative impact on the environment [99–112]. For example, the electrodeposition technique is used to recover valuable metals. Freitas's group developed many research types on the recovery of Cd and Ni from spent Ni–Cd batteries [99–101]. In 2005, a study published by this group reported [59] the electrochemical recycling of cadmium from spent Ni–Cd batteries. Ionic cadmium is recovered from an acidic solution using a galvanostatic technique. First, the cadmium electrodes were leached using sulfuric acid. The suspension was left under constant stirring at 298 K. After the dissolution of cadmium hydroxide, the suspension was filtered. The leaching solution's pH and the conductivity were equal to 0.571 and 362.2 mS cm^2 mol^{-1}, respectively. A clear green solution was obtained due to the presence of ionic nickel from the cathode. The reference electrode was Ag/AgCl/NaCl. Cd electrodeposition was conducted at different current densities. Note that the electrodeposition charge efficiency and deposit morphology depend on current density. Freitas and Rosalém demonstrated that up to 85.0% in charge efficiency is reached for a current density in the range 5.0–10.0 mA cm^{-2}. Another study by Freitas et al. [100] investigated Cd's recovery from Ni–Cd batteries by chemical precipitation and electrodeposition. The electrode leaching process is carried here using H_2SO_4. The chemical precipitation occurs by adding NaOH to the leaching solution, and therefore, CdOH is obtained, whereas the electrodeposition route showed 95.0% in charge efficiency for a current density between 10.0 and 30.0 mA cm^{-2}. Reddy et al. [101] studied the feasibility of recycling Cd, Ni, and Co from spent Ni–Cd batteries using a hydrometallurgical route. This process relies on crushing the spent electrodes, followed by physical separation of the elements. The electrode materials are dissolved in an acidic solution (i.e., HCl leaching), and the metals are recovered by solvent extraction/ion exchange/cementation. In this study, Cyanex 923 is a mixture of four trialkyl-phosphine oxides ($R_3P = O$, $R_2R'P = O$, $R_3'P = O$, $R_2'R\ P = O$, where R = hexyl and R' = octyl) which was reported for the first time as selective Cd extractant. Moreover, Cyanex 272 (bis(2,4,4-trimethylpentyl)phosphinic acid) and TOPS 99, (an equivalent of di-2-ethylhexyl phosphoric acid) are used as extractants for Co and Ni, respectively. Chloride salts of cadmium, cobalt, and nickel were obtained in pure form with a recovery efficiency >99.95% with 0.6 M Cyanex 923, 0.03 M Cyanex 272, and 1 M TOPS 99. Moreover, a green route was explored to recycle Ni–Cd battery wastes. Lacerda et al. [103] investigated the separation of Ni and Cd from used Ni–Cd batteries by an aqueous two-phase route made of copolymer L35, Li_2SO_4, and water in the presence of potassium iodide (KI) as extracting agent. This liquid–liquid extraction is eco-friendly since it is water-based and does not require any toxic organic solvent. In this work, a maximum extraction of Cd (99.2 ± 3.1)% and Ni (10.6 ± 0.4)% is reached when the electrodes are leached with HCl using a concentration of KI equal to 50.00 mmol kg^{-1}, the mass ratio of the phases equal to 0.5 and a dilution factor of battery samples of 35.

The spent nickel–metal hydride battery, abbreviated as Ni–MH, is eco-friendly disposable compared to Ni–Cd battery. The cathode material is a Ni coated with NiOH, whereas the anode material is composed of a hydrogen storage alloy based on mischmetal (mainly cerium, lanthanum, praseodymium, and neodymium) and nickel, including substituents. As Ni–MH batteries contain around 36–42% of nickel and 3–4% of cobalt, their recycling and reuse were the concern of many researchers worldwide [113–117]. Müller and Friedrich [113] developed a study on Ni recycling from spent Ni–MH batteries using the slag systems $CaO–CaF_2$ and $CaO–SiO_2$. In a DC electric arc furnace, dismantled Ni–MH electrodes were melted, resulting in a Ni–Co alloy and a slag phase enriched with rare earth oxides. $CaO–CaF_2$ slag system showed better melting behavior, ensuring the recovery of Ni–Co alloy. Maaroufi et al. [114] presented a study on the recovery of rare earth oxides from end-of-life Ni–MH battery via thermal isolation. The anode materials are 54 wt% Ni, 23.7 wt% La, 6.7 wt% Ce, 5.4 wt% Co, 3.6 wt% Nd, and 3.4 wt% Mn. On the one hand, this material was heated firstly at 1000 °C for 1 h in the air to oxidize all the elements and secondly, at 1550 °C for 1 h 30 min. This study's reducing agent is a waste developer kit with Fe content >99 wt%. As a result, a ferronickel alloy was formed after reducing nickel and cobalt oxides and their diffusion into metallic iron. On the other hand, the anode material was combined with pure hematite, which plays an oxidizing agent in the oxidation–reduction process. The mixture was heated at 1550 °C for 90 min under argon atmosphere.

Both routes have shown that Fe-based metal (ferronickel) and rare earth oxide (REO) phases were successfully separated. Jiang et al. [115] demonstrated the extraction of valuable elements in Ni–MH battery electrode materials using the pyrometallurgical process. Ni–Co alloy and slag enriched with REO were obtained using selective gas redox and melting separation. The electrodes were heated with H_2/H_2O at 800 °C and 900 °C. Rare-earth elements, Al, and Mn (active elements) were oxidized, whereas Ni and Co (inert elements) were transformed into their elemental states. Later, SiO_2 and Al_2O_3 powders were added as fluxes. Then, they fired at 1550 °C to produce a nickel–cobalt alloy and an REO–SiO_2–Al_2O_3–MnO slag.

4.2 Recovery of Nanomaterials for Lead–Acid Batteries

Lead–acid batteries are commercialized for applications such as power/emergency supplies and lighting. Lead–acid batteries' market is in continuous growth due to their low cost and good performances with solid industrial recycling approaches [118–124]. The recovery of lead from spent lead–acid batteries was the focus of multiple published studies [118–121]. Ma and Qiu [118] presented the recovery of lead from the spent lead–acid battery using hydrometallurgical desulfurization and vacuum thermal reduction. The lead paste is composed of lead sulfate, lead oxides, and lead metal. First, lead sulfate $PbSO_4$ was desulfurized and transformed into lead carbonate $PbCO_3$ by Na_2CO_3, followed by thermal treatment at 315 °C to decompose $PbCO_3$ into PbO. Last, Pb was obtained by the reduction of PbO under vacuum using charcoal as a reducing agent. The reaction conditions were optimized to 850 °C temperature, 20 Pa pressure, 45 min reaction time, and 22.11×10^{-2} g cm^{-2} min^{-1} reduction rate. Zhu et al. [119] proposed a simple green process to recycle PbO from the lead paste in spent Pb/acid batteries. First, the desulfurization of the lead paste was carried by adding citric acid to produce lead citrate.

Second, the product was calcined at low temperatures to form ultra-fine lead oxide. At 370 °C, the resultant PbO shows a particle size of 100–500 nm. Pan et al. [120] reported a new route to recover high-purity lead from lead–acid batteries waste. Electrolyzing PbO directly produces metallic lead in NaOH alkaline solution, where the sodium ionic exchange membrane prevents the oxidation of $HPbO_2^-$ to PbO_2. This study showed that the electrolytic bath's cell voltage is 1.23 V, the current efficiency is 99.9%, and the lead recovery efficiency is 99.8% at a current density of 20 mA cm^{-2}.

Furthermore, other studies discussed the potential use of recycled PbO as negative electrode materials for energy storage. Pan et al. [122] worked on the fabrication of pure lead oxide PbO from spent lead–acid batteries using a hydrometallurgical route. This process involves the catalytic conversion of waste lead pastes into $PbSO_4$ and desulfurization–recrystallization in NaOH solution. Moreover, the electrochemical tests revealed that the prepared PbO delivers slightly superior electrochemical performances compared to other PbO prepared by the

Shimadzu ball-milling method. Hu et al. [123] studied the preparation of PbO@C composite by pyrolyzing the lead citrate precursor recovered from the spent lead paste of lead-acid batteries. The particles of the obtained composites PbO@C are spherical with 10–50 nm diameter. The electrochemical performances were enhanced due to the chemical bond between carbon and lead oxide. The hydrophilic character of carbon acts as a 3D electroosmotic pump, which eases the electrolyte's absorption and transportation. Zhou et al. [124] emphasized a recycling procedure using spent lead paste and the synthesis of battery-grade lead oxide. First, the lead oxide powder was dissolved in acetate acid solution and diluted to obtain a Pb $(CH_3COO)_2$ stock solution. Later, polyvinylpyrrolidone (PVP) was added to Pb $(CH_3COO)_2$ and then dried. Last, the PVP/Pb$(CH_3COO)_2$ mixture was calcined in a muffle furnace at 350 °C, 450 °C, and 550 °C, respectively, to decompose lead acetate and produce PbO, with an average size of about 200 nm. The prepared lead oxide PbO material demonstrated enhanced with a discharge capacity of 159 mAh g^{-1} at 50 mA g^{-1} current density. Moreover, Eco-Bat Technologies is a world's leading company and the first group dealing with lead–acid recycling batteries in a closed recycling loop, i.e., the recovered materials are directly used in the fabrication of new Pb/acid batteries [57].

5 Recovery of Nanomaterials for Alkaline Batteries

5.1 Recovery of Nanomaterials for Rechargeable Zn//MnO$_2$ Batteries

The annual market of portable batteries using Zn–Mn battery technology is estimated by over 90% of China, Japan, Korea, and the USA. The excessive use of these batteries produces tons of wastes. Consequently, numerous studies reported the recycling spent zinc–manganese dioxide batteries and the preparation of nanomaterials subjected to several applications. Moreover, special attention was dedicated to synthesizing zinc–manganese ferrites, zinc oxide, manganese oxide, and zinc from spent Zn–MnO_2 alkaline batteries [125–133]. In 2004, Xi et al. [125] studied manganese–zinc ferrites' fabrication from spent Zn–Mn batteries. Therefore, the spent electrodes were dissolved in sulfuric acid containing 2.4 wt% H_2O_2. The acid solution was filtered after complete dissolution, and the filtrate was heated to the boiling point for 1 h to remove H_2O_2. The pH of the filtrate was adjusted to 3–5 by adding NaOH solution. $MnSO_4$ and Fe powder were added to the filtrate to obtain the chemical composition Fe_2O_3/MnO/ZnO = 1:0.6:0.4 (in mole). Furthermore, the pH was adjusted to 6.0–9.5 by adding $NH_3.H_2O$, and NH_4HCO_3 was to precipitate the metals. Afterward, the precipitate was filtered, washed, dried, and calcined in air at the range of 1000–1150 °C. Thus, single spinel phase $Mn_{0.6}Zn_{0.4}Fe_2O_4$ was obtained in the pH co-precipitation range 7.0–7.5, 1100–1150 °C, with a mean crystallite size of 22.4 nm. Nan et al. [126]

investigated the preparation as well of $Mn_{0.26}Zn_{0.24}FeO_2$ from exhausted Zn–MnO_2 batteries using co-precipitation with ammonium oxalate and 850–1250 °C calcination temperature. In 2011, Deep et al. [128] investigated the recovery of pure ZnO nanoparticles from spent Zn–MnO_2 alkaline batteries. The electrode material was recovered, washed with water to remove the electrolyte, dried, and milled to produce fine particles. Zinc was leached using hydrochloric acid at 70 °C for 2 h, then filtered and washed with distilled water. The liquid–liquid extraction procedure followed using the extractant solution of Cyanex 923 in n-hexane. Zn(II) was distributed in the organic phase as complex $ZnCl_2$.2R (R = Cyanex 923 molecule) and was calcined at 600 °C where pure ZnO nanoparticles with a size in the range of 40–50 nm. In 2016, the same research group proposed the recovery of nanosized pure ZnO using a one-pot solvothermal process. TEM technique revealed that the size of the obtained ZnO nanoparticles is 5 nm. Xiang et al. [130] demonstrated zinc nanoparticles' fabrication from spent zinc–manganese batteries via vacuum separation and inert gas condensation. The separation efficiency of 99.68% of uniform zinc hexagonal prisms with 100–300 nm diameter was reached by optimizing the conditions to 1000 Pa inert pressure, 1073 K heating temperature, lower than 473 K condensing temperature, and 10–30 cm condensation distance.

5.2 *Recovery of Nanomaterials for Primary Zinc–Carbon Batteries*

The zinc–carbon battery is a nonchargeable battery type, widely used for alarm clocks, remote controls, radios, etc. Several research types were dedicated to the recycling of the spent Zn–C batteries [109–111, 132, 133]. Ferella et al. [134] proposed the recovery of metallic zinc and mixed manganese oxides from spent alkaline and Zn–C batteries. The electrodes are leached in 1.5 M sulfuric acid, where Zn is dissolved within 3 h at 80 °C. Later, metallic zinc (purity 99.6%) is obtained by electrolysis, and the manganese oxide mixture Mn_3O_4 and Mn_2O_3 (70% grade of Mn) are formed by thermally treating the solid residue at 900 °C for 30 min. Kim et al. [135] demonstrated the recovery of zinc–manganese ferrite from spent zinc–carbon batteries. Sulfuric acid leaching with H_2O_2 and oxidative alkaline co-precipitation with the O_2 approach were followed. In fact, 97.9% Mn, 98.0% Zn, and 55.2% Fe were extracted within 1 h at a solid/liquid ratio of 1:10. Furthermore, the stoichiometry was adjusted to form nano-sized $MnZnFe_4O_8$ (around ~20 nm diameter) by adding metal sulfates (Metal: Mn, Zn, Fe). Gabal's group [136] worked as well on the recycle of zinc–carbon battery wastes into the preparation of $Mn_{1-x}Zn_xFe_2O_4$ ($x = 0.2 - 0.8$) nanocrystals. First, the electrodes were leached using a mixture of 2 M HNO_3 and 2 wt% H_2O_2. After complete dissolution, the solution was filtered, and amounts of metal nitrates were added to the filtrate to obtain the targeted stoichiometry. Mahandra et al. [104] presented a study on the recovery of nano-sized Zn from spent Zn–C batteries in the form of zinc oxide. The dismantled electrodes were leached in 5.0 mol L^{-1} HCl for 1 h at 70 °C. Then, the solution was cooled, filtered, and diluted to 100 mL to maintain the acidity at 0.5 mol L^{-1} HCl. In this work, Cyphos IL 104 (trihexyl(tetradecyl)-phosphonium bis(2,4,4-trimethylpentyl) phosphinate) was used as extracting agent. The leach liquor and Cyphos IL 104 were mixed and left to separate. Zinc oxide was recovered from the loaded organic phase after precipitation followed by thermal treatment at 400 °C with a quantitative percentage of more than 99.9%.

6 Conclusions

This book chapter provides an overview of the state of waste recycling technologies for nanomaterial manufacturing and their potential use for battery applications. Most commercial LiBs contain graphite as the negative electrode material. Battery-grade graphite is prepared by recycling various biomass wastes such as olive stone, charcoal, rice husk, banana peels, and nonedible chicken egg-based. Graphite is recovered as well by recycling spent LiBs. The preparation of silicon nanoparticles, another type of negative electrode material, was reported by recycling several wastes such as rice husk, waste iron slag, silicon wafer waste, waste slicing slug, glass bottles, kerf powder, and silicon oxide. Lithium, cobalt, nickel, manganese, and copper are valuable chemical elements largely used in the battery industry. Several studies were published investigating their recovery using chemical routes in the form of hydroxides, oxides, or as Li_3PO_4. Other studies focused on the recovery of $LiFePO_4$ from end-of-life batteries delivering enhanced electrochemical properties. First, $LiFePO_4$ was separated from aluminum foil, mixed them with CMC and SBR binders, and then decomposed. Furthermore, worldwide companies such as Xstrata nickel international company (Canada), SNAM (France), Accurec (Germany), and Umicore (Belgium) run a huge recycling business on the industrial level of battery processing. For Ni–Cd, Ni–MH, lead–acid batteries, the recovery of Ni, Cd, Co, Pb, and PbO was reported using electrodeposition and hydrometallurgical routes in the form of metals, oxides, or alloys. The recovered elements were used as raw materials for the manufacturing of new batteries. Eco-Bat technology developed a closing recycling loop for lead–acid batteries. Finally, zinc–manganese ferrites, zinc oxides, manganese oxides, and metallic zinc are restored from spent alkaline batteries using acid leaching technique co-precipitation.

7 Future Perspectives

The state of the art of nanomaterial recovery and their use for battery fabrication is still in a growing process. Nanomaterials show interesting properties, which meet the requirements of cost-efficient and high energy density battery systems such as a high surface area, enabling a large contact area with the electrolyte. Consequently, more efforts should be deployed in the future to investigate the beneficial effect of the use of this technology waste on the economy, environment, and energetic performances.

Acknowledgements The author would like to give special thanks to Dr. Gomaa A. M. Ali for his unlimited guidance and help. Thank you for your remarks and suggestions during the writing process of this chapter.

References

1. Bruce PG, Scrosati B, Tarascon J-M (2008) Nanomaterials for rechargeable lithium batteries. Angew Chem Int Ed 47(16):2930–2946
2. Jiang C, Hosono E, Zhou H (2006) Nanomaterials for lithium ion batteries. Nano Today 1 (4):28–33
3. Lee KT, Cho J (2011) Roles of nanosize in lithium reactive nanomaterials for lithium ion batteries. Nano Today 6(1):28–41
4. Yaroslavtsev AB, Kulova TL, Skundin AM (2015) Electrode nanomaterials for lithium-ion batteries. Russ Chem Rev 84(8):826–852
5. Sun Y, Liu N, Cui Y (2016) Promises and challenges of nanomaterials for lithium-based rechargeable batteries. Nat Energy 1(7)
6. Xu Z-L, Kim J-K, Kang K (2018) Carbon nanomaterials for advanced lithium sulfur batteries. Nano Today 19:84–107
7. Devaraju MK, Honma I (2012) Hydrothermal and solvothermal process towards development of $LiMPO_4$ (M=Fe, Mn) nanomaterials for lithium-ion batteries. Adv Energy Mater 2 (3):284–297
8. Mei J, Liao T, Kou L, Sun Z (2017) 2D metal oxides: two-dimensional metal oxide nanomaterials for next-generation rechargeable batteries. Adv Mater 29(48):1770344
9. Xia Q, Li W, Miao Z, Chou S, Liu H (2017) Phosphorus and phosphide nanomaterials for sodium-ion batteries. Nano Res 10(12):4055–4081
10. Yuan YF, Tu JP, Wu HM, Li Y, Shi DQ, Zhao XB (2006) Effect of ZnO nanomaterials associated with $Ca(OH)_2$ as anode material for Ni–Zn batteries. J Power Sources 159 (1):357–360
11. Lu Y, Lu Y, Niu Z, Chen J (2018) Graphene-based nanomaterials for sodium-ion batteries. Adv Energy Mater 8(17):1702469
12. Jin T, Han Q, Wang Y, Jiao L (2017) 1D nanomaterials: design, synthesis, and applications in sodium-ion batteries. Small 14(2):1703086
13. Zhang J, Gu P, Xu J, Xue H, Pang H (2016) High performance of electrochemical lithium storage batteries: ZnO-based nanomaterials for lithium-ion and lithium–sulfur batteries. Nanoscale 8(44):18578–18595
14. Ali GAM, Fouad OA, Makhlouf SA, Yusoff MM, Chong KF (2016) Optical and electrochemical properties of Co_3O_4/SiO_2 nanocomposite. Adv Mater Res 1133:447–451
15. Ali GAM, Wahba OAG, Hassan AM, Fouad OA, Chong KF (2015) Calcium-based nanosized mixed metal oxides for supercapacitor application. Ceram Int 41(6):8230–8234
16. Ali GAM, Fouad OA, Makhlouf SA, Yusoff MM, Chong KF (2014) Co_3O_4/SiO_2 nanocomposites for supercapacitor application. J Solid State Electrochem 18(9):2505–2512
17. Ali GAM, Habeeb OA, Algarni H, Chong KF (2018) CaO impregnated highly porous honeycomb activated carbon from agriculture waste: symmetrical supercapacitor study. J Mater Sci 54:683–692
18. Ali GAM, Fouad OA, Makhlouf SA (2013) Structural, optical and electrical properties of sol-gel prepared mesoporous Co_3O_4/SiO_2 nanocomposites. J Alloy Compd 579:606–611

19. Samaddar P, Ok YS, Kim K-H, Kwon EE, Tsang DCW (2018) Synthesis of nanomaterials from various wastes and their new age applications. J Clean Prod 197:1190–1209
20. Xu J, Thomas HR, Francis RW, Lum KR, Wang J, Liang B (2008) A review of processes and technologies for the recycling of lithium-ion secondary batteries. J Power Sources 177 (2):512–527
21. Chagnes A, Pospiech B (2013) A brief review on hydrometallurgical technologies for recycling spent lithium-ion batteries. J Chem Technol Biotechnol 88(7):1191–1199
22. Bradbury K (2010) Energy storage technology review. Duke University, pp 1–34
23. Rajagopalan R, Tang Y, Ji X, Jia C, Wang H (2020) Advancements and challenges in potassium ion batteries: a comprehensive review. Adv Func Mater 30(12):1909486
24. Xu X, Xiong F, Meng J, Wang X, Niu C, An Q, Mai L (2020) Vanadium-based nanomaterials: a promising family for emerging metal-ion batteries. Adv Func Mater 30 (10):1904398
25. Rashad M, Asif M, Wang Y, He Z, Ahmed I (2020) Recent advances in electrolytes and cathode materials for magnesium and hybrid-ion batteries. Energy Storage Mater 25:342–375
26. Massé RC, Uchaker E, Cao G (2015) Beyond Li-ion: electrode materials for sodium- and magnesium-ion batteries. Sci China Mater 58(9):715–766
27. Huie MM, Bock DC, Takeuchi ES, Marschilok AC, Takeuchi KJ (2015) Cathode materials for magnesium and magnesium-ion based batteries. Coord Chem Rev 287:15–27
28. Kim H, Kim H, Ding Z, Lee MH, Lim K, Yoon G, Kang K (2016) Recent progress in electrode materials for sodium-ion batteries. Adv Energy Mater 6(19):1600943
29. Wang LP, Yu L, Wang X, Srinivasan M, Xu ZJ (2015) Recent developments in electrode materials for sodium-ion batteries. J Mater Chem A 3(18):9353–9378
30. Dahbi M, Yabuuchi N, Kubota K, Tokiwa K, Komaba S (2014) Negative electrodes for Na-ion batteries. Phys Chem Chem Phys 16(29):15007
31. Shea JJ, Luo C (2020) Organic electrode materials for metal ion batteries. ACS Appl Mater Interfaces 12(5):5361–5380
32. Ohzuku T, Brodd RJ (2007) An overview of positive-electrode materials for advanced lithium-ion batteries. J Power Sources 174(2):449–456
33. Kubota K, Yabuuchi N, Yoshida H, Dahbi M, Komaba S (2014) Layered oxides as positive electrode materials for Na-ion batteries. MRS Bull 39(5):416–422
34. Sakai T, Uehara I, Ishikawa H (1999) R&D on metal hydride materials and Ni–MH batteries in Japan. J Alloy Compd 293–295:762–769
35. Zhu WH, Zhu Y, Davis Z, Tatarchuk BJ (2013) Energy efficiency and capacity retention of Ni–MH batteries for storage applications. Appl Energy 106:307–313
36. Geng M (2003) Development of advanced rechargeable Ni/MH and Ni/Zn batteries. Int J Hydrogen Energy 28(6):633–636
37. Ruetschi P (1977) Review on the lead—acid battery science and technology. J Power Sources 2(1):3–120
38. Yang J, Hu C, Wang H, Yang K, Liu JB, Yan H (2016) Review on the research of failure modes and mechanism for lead-acid batteries. Int J Energy Res 41(3):336–352
39. Feng F (2001) Electrochemical behaviour of intermetallic-based metal hydrides used in Ni/metal hydride (MH) batteries: a review. Int J Hydrogen Energy 26(7):725–734
40. Zhang J, Zhang G, Chen Z, Dai H, Hu Q, Liao S, Sun S (2020) Emerging applications of atomic layer deposition for lithium-sulfur and sodium-sulfur batteries. Energy Storage Mater 26:513–533
41. Wen Z, Cao J, Gu Z, Xu X, Zhang F, Lin Z (2008) Research on sodium sulfur battery for energy storage. Solid State Ionics 179(27–32):1697–1701
42. Benato R, Cosciani N, Crugnola G, Dambone Sessa S, Lodi G, Parmeggiani C, Todeschini M (2015) Sodium nickel chloride battery technology for large-scale stationary storage in the high voltage network. J Power Sources 293:127–136
43. Sudworth J (2001) The sodium/nickel chloride (ZEBRA) battery. J Power Sources 100(1–2):149–163

44. Vanzyl A (1996) Review of the zebra battery system development. Solid State Ionics 86–88:883–889
45. Ottaviani M, Turconi A, Basso D (2020) The sodium/nickel chloride battery. In: Encyclopedia of Electrochemistry. Wiley, pp 1–22
46. Weber AZ, Mench MM, Meyers JP, Ross PN, Gostick JT, Liu Q (2011) Redox flow batteries: a review. J Appl Electrochem 41(10):1137–1164
47. Alotto P, Guarnieri M, Moro F (2014) Redox flow batteries for the storage of renewable energy: a review. Renew Sustain Energy Rev 29:325–335
48. Cunha Á, Martins J, Rodrigues N, Brito FP (2014) Vanadium redox flow batteries: a technology review. Int J Energy Res 39(7):889–918
49. Zhang X, Ji L, Toprakci O, Liang Y, Alcoutlabi M (2011) Electrospun nanofiber-based anodes, cathodes, and separators for advanced lithium-ion batteries. Polym Rev 51(3):239–264
50. Scrosati B, Garche J (2010) Lithium batteries: Status, prospects and future. J Power Sources 195(9):2419–2430
51. Zabaniotou A, Kouskoumvekaki E, Sanopoulos D (1999) Recycling of spent lead/acid batteries: the case of Greece. Resour Conserv Recycl 25(3–4):301–317
52. Huang P-H, Kuo J-K, Huang C-Y (2015) A new application of the ultra battery to hybrid fuel cell vehicles. Int J Energy Res 40(2):146–159
53. Molina MG (2017) Energy storage and power electronics technologies: a strong combination to empower the transformation to the smart grid. Proc IEEE 105(11):2191–2219
54. Akinyele D, Belikov J, Levron Y (2017) Battery storage technologies for electrical applications: impact in stand-alone photovoltaic systems. Energies 10(11):1760
55. Zhang W-J (2011) A review of the electrochemical performance of alloy anodes for lithium-ion batteries. J Power Sources 196(1):13–24
56. Rothermel S, Evertz M, Kasnatscheew J, Qi X, Grützke M, Winter M, Nowak S (2016) Graphite recycling from spent lithium-ion batteries. Chemsuschem 9(24):3473–3484
57. Moradi B, Botte GG (2015) Recycling of graphite anodes for the next generation of lithium ion batteries. J Appl Electrochem 46(2):123–148
58. Sabisch JEC, Anapolsky A, Liu G, Minor AM (2018) Evaluation of using pre-lithiated graphite from recycled Li-ion batteries for new LiB anodes. Resour Conserv Recycl 129:129–134
59. Yang Y, Song S, Lei S, Sun W, Hou H, Jiang F, Ji X, Zhao W, Hu Y (2019) A process for combination of recycling lithium and regenerating graphite from spent lithium-ion battery. Waste Manag 85:529–537
60. Zhao L, Liu X, Wan C, Ye X, Wu F (2018) Soluble graphene nanosheets from recycled graphite of spent lithium ion batteries. J Mater Eng Perform 27(2):875–880
61. Yu J, He Y, Ge Z, Li H, Xie W, Wang S (2018) A promising physical method for recovery of $LiCoO_2$ and graphite from spent lithium-ion batteries: grinding flotation. Sep Purif Technol 190:45–52
62. He Y, Zhang T, Wang F, Zhang G, Zhang W, Wang J (2017) Recovery of $LiCoO_2$ and graphite from spent lithium-ion batteries by Fenton reagent-assisted flotation. J Clean Prod 143:319–325
63. Chen Y, Liu C, Sun X, Ye H, Cheung C, Zhou L (2015) Recycled diesel carbon nanoparticles for nanostructured battery anodes. J Power Sources 275:26–31
64. Zhang G, He Y, Feng Y, Wang H, Zhu X (2018) Pyrolysis-ultrasonic-assisted flotation technology for recovering graphite and $LiCoO_2$ from spent lithium-ion batteries. ACS Sustain Chem Eng 6(8):10896–10904
65. Zhang W, Liu Z, Xia J, Li F, He W, Li G, Huang J (2017) Preparing graphene from anode graphite of spent lithium-ion batteries. Front Environ Sci Eng 11(5)
66. Xiao J, Li J, Xu Z (2017) Recycling metals from lithium ion battery by mechanical separation and vacuum metallurgy. J Hazard Mater 338:124–131

67. Guo Y, Li F, Zhu H, Li G, Huang J, He W (2016) Leaching lithium from the anode electrode materials of spent lithium-ion batteries by hydrochloric acid (HCl). Waste Manag 51:227–233
68. Huang B, Pan Z, Su X, An L (2018) Recycling of lithium-ion batteries: recent advances and perspectives. J Power Sources 399:274–286
69. Schauerman CM, Ganter MJ, Gaustad G, Babbitt CW, Raffaelle RP, Landi BJ (2012) Recycling single-wall carbon nanotube anodes from lithium ion batteries. J Mater Chem 22 (24):12008
70. Chou C-Y, Lee M, Hwang GS (2015) A comparative first-principles study on sodiation of Silicon, Germanium, and Tin for Sodium-Ion batteries. J Phys Chem C 119(27):14843–14850
71. Dimov N, Kugino S, Yoshio M (2003) Carbon-coated silicon as anode material for lithium ion batteries: advantages and limitations. Electrochim Acta 48(11):1579–1587
72. Yoshio M, Tsumura T, Dimov N (2005) Electrochemical behaviors of silicon based anode material. J Power Sources 146(1–2):10–14
73. Park M-H, Kim MG, Joo J, Kim K, Kim J, Ahn S, Cui Y, Cho J (2009) Silicon nanotube battery anodes. Nano Lett 9(11):3844–3847
74. Fang S, Tong Z, Nie P, Liu G, Zhang X (2017) Raspberry-like nanostructured silicon composite anode for high-performance lithium-ion batteries. ACS Appl Mater Interfaces 9 (22):18766–18773
75. Wang H, Tang W, Ni L, Ma W, Chen G, Zhang N, Liu X, Ma R (2020) Synthesis of silicon nanosheets from kaolinite as a high-performance anode material for lithium-ion batteries. J Phys Chem Solids 137:109227
76. Pan Q, Zhao J, Du Y, Liu R, Li N, Xing B, Jiang S, Pang M, Qu W, Liang W, Li Z, Cao F (2020) Coal-based kaolin derived porous silicon nanoparticles as anode materials for Li-ion batteries. Microporous Mesoporous Mater 294:109918
77. Loaiza LC, Monconduit L, Seznec V (2020) Si and Ge-based anode materials for Li-, Na-, and K-ion batteries: a perspective from structure to electrochemical mechanism. Small 16 (5):1905260
78. Wen Z, Lu G, Mao S, Kim H, Cui S, Yu K, Huang X, Hurley PT, Mao O, Chen J (2013) Silicon nanotube anode for lithium-ion batteries. Electrochem Commun 29:67–70
79. Su X, Wu Q, Li J, Xiao X, Lott A, Lu W, Sheldon BW, Wu J (2013) Silicon-based nanomaterials for lithium-ion batteries: a review. Adv Energ Mater 4(1):1300882
80. Yin Y, Wan L, Guo Y (2012) Silicon-based nanomaterials for lithium-ion batteries. Chin Sci Bull 57(32):4104–4110
81. Jung DS, Ryou MH, Sung YJ, Park SB, Choi JW (2013) Recycling rice husks for high-capacity lithium battery anodes. Proc Natl Acad Sci 110(30):12229–12234
82. Zhang Y-C, You Y, Xin S, Yin Y-X, Zhang J, Wang P, Zheng X-s, Cao F-F, Guo Y-G (2016) Rice husk-derived hierarchical silicon/nitrogen-doped carbon/carbon nanotube spheres as low-cost and high-capacity anodes for lithium-ion batteries. Nano Energy 25:120–127
83. Song J, Guo S, Ren D, Liu H, Kou L, Su J, Zheng P (2020) Rice husk-derived SiO_x@carbon nanocomposites as a high-performance bifunctional electrode for rechargeable batteries. Ceram Int 46(8):11570–11576
84. Chu H, Wu Q, Huang J (2018) Rice husk derived silicon/carbon and silica/carbon nanocomposites as anodic materials for lithium-ion batteries. Coll Surf A 558:495–503
85. Wang L, Xue J, Gao B, Gao P, Mou C, Li J (2014) Rice husk derived carbon–silica composites as anodes for lithium ion batteries. RSC Adv 4(110):64744–64746
86. Chun J, An S, Lee J (2015) Highly mesoporous silicon derived from waste iron slag for high performance lithium ion battery anodes. J Mater Chem A 3(43):21899–21906
87. Lee S, Kim T-H, Kim D-W, Park D-W (2017) Preparation of silicon nanopowder by recycling silicon wafer waste in radio-frequency thermal plasma process. Plasma Chem Plasma Process 37(4):967–978

88. Jang HD, Kim H, Chang H, Kim J, Roh KM, Choi J-H, Cho B-G, Park E, Kim H, Luo J, Huang J (2015) Aerosol-assisted extraction of silicon nanoparticles from wafer slicing waste for lithium ion batteries. Sci Rep 5(1)
89. Bao Q, Huang Y-H, Lan C-K, Chen B-H, Duh J-G (2015) Scalable upcycling silicon from waste slicing sludge for high-performance lithium-ion battery anodes. Electrochim Acta 173:82–90
90. Chou C-Y, Kuo J-R, Yen S-C (2018) Silicon-based composite negative electrode prepared from recycled silicon-slicing slurries and lignin/lignocellulose for Li-ion cells. ACS Sustain Chem Eng 6(4):4759–4766
91. Li C, Liu C, Wang W, Mutlu Z, Bell J, Ahmed K, Ye R, Ozkan M, Ozkan CS (2017) Silicon derived from glass bottles as anode materials for lithium ion full cell batteries. Sci Rep 7(1)
92. Wagner NP, Tron A, Tolchard JR, Noia G, Bellmann MP (2019) Silicon anodes for lithium-ion batteries produced from recovered kerf powders. J Power Sources 414:486–494
93. Kim J, Kim SY, Yang C-M, Lee GW (2019) Possibility of recycling SiO_x particles collected at silicon ingot production process as an anode material for lithium ion batteries. Sci Rep 9 (1)
94. Eshraghi N, Berardo L, Schrijnemakers A, Delaval V, Shaibani M, Majumder M, Cloots R, Vertruyen B, Boschini F, Mahmoud A (2020) Recovery of nano-structured silicon from end-of-life photovoltaic wafers with value-added applications in lithium-ion battery. ACS Sustain Chem Eng 8(15):5868–5879
95. Contestabile M, Panero S, Scrosati B (1999) A laboratory-scale lithium battery recycling process. This work has been presented as an invited talk at the 4th international battery recycling congress, Hamburg, Germany, 1–3 July, 1998. J Power Sources 83(1–2):75–78
96. Castillo S (2002) Advances in the recovering of spent lithium battery compounds. J Power Sources 112(1):247–254
97. Kim HS, Shin EJ (2013) Re-synthesis and electrochemical characteristics of $LiFePO_4$ cathode materials recycled from scrap electrodes. Bull Korean Chem Soc 34(3):851–855
98. Li H, Xing S, Liu Y, Li F, Guo H, Kuang G (2017) Recovery of lithium, iron, and phosphorus from spent $LiFePO_4$ batteries using stoichiometric sulfuric acid leaching system. ACS Sustain Chem Eng 5(9):8017–8024
99. Freitas MBJG, Rosalém SF (2005) Electrochemical recovery of cadmium from spent Ni–Cd batteries. J Power Sources 139(1–2):366–370
100. Freitas MBJG, Penha TR, Sirtoli S (2007) Chemical and electrochemical recycling of the negative electrodes from spent Ni–Cd batteries. J Power Sources 163(2):1114–1119
101. Rozário A, Silva e Silva RK, Freitas MBJG (2006) Recycling of nickel from NiOOH/Ni $(OH)_2$ electrodes of spent Ni–Cd batteries. J Power Sources 158(1):754–759
102. Reddy BR, Neela Priya D, Venkateswara Rao S, Radhika P (2005) Solvent extraction and separation of Cd(II), Ni(II) and Co(II) from chloride leach liquors of spent Ni–Cd batteries using commercial organo-phosphorus extractants. Hydrometallurgy 77(3–4):253–261
103. Lacerda VG, Mageste AB, Santos IJB, da Silva LHM, da Silva MdCH (2009) Separation of Cd and Ni from Ni–Cd batteries by an environmentally safe methodology employing aqueous two-phase systems. J Power Sources 193(2):908–913
104. Mahandra H, Singh R, Gupta B (2018) Recycling of Zn–C and Ni–Cd spent batteries using Cyphos IL 104 via hydrometallurgical route. J Clean Prod 172:133–142
105. Huang K, Li J, Xu Z (2010) Characterization and recycling of cadmium from waste nickel–cadmium batteries. Waste Manag 30(11):2292–2298
106. Assefi M, Maroufi S, Mayyas M, Sahajwalla V (2018) Recycling of Ni–Cd batteries by selective isolation and hydrothermal synthesis of porous NiO nanocuboid. J Environ Chem Eng 6(4):4671–4675
107. Assefi M, Maroufi S, Yamauchi Y, Sahajwalla V (2020) Pyrometallurgical recycling of Li-ion, Ni–Cd and Ni–MH batteries: a minireview. Curr Opin Green Sustain Chem 24:26–31
108. Aboelazm EAA, Ali GAM, Chong KF (2018) Cobalt oxide supercapacitor electrode recovered from spent lithium-ion battery. Chem Adv Mater 3:67–74

109. Ali GAM, Tan LL, Jose R, Yusoff MM, Chong KF (2014) Electrochemical performance studies of MnO_2 nanoflowers recovered from spent battery. Mater Res Bull 60:5–9
110. Ali GAM, Yusoff MM, Feng CK (2015) Electrochemical properties of electrodeposited MnO_2 nanoparticles. Adv Mater Res 1113:550–553
111. Ali GAM, Yusoff MM, Shaaban ER, Chong KF (2017) High performance MnO_2 nanoflower supercapacitor electrode by electrochemical recycling of spent batteries. Ceram Int 43:8440–8448
112. Aboelazm EAA, Ali GAM, Algarni H, Yin H, Zhong YL, Chong KF (2018) Magnetic electrodeposition of the hierarchical cobalt oxide nanostructure from spent lithium-ion batteries: its application as a supercapacitor electrode. J Phys Chem C 122(23):12200–12206
113. Müller T, Friedrich B (2006) Development of a recycling process for nickel–metal hydride batteries. J Power Sources 158(2):1498–1509
114. Maroufi S, Nekouei RK, Hossain R, Assefi M, Sahajwalla V (2018) Recovery of rare earth (i.e., La, Ce, Nd, and Pr) oxides from end-of-life Ni–MH battery via thermal isolation. ACS Sustain Chem Eng 6(9):11811–11818
115. Jiang Y-J, Deng Y-C, Bu W-G (2015) Pyrometallurgical extraction of valuable elements in Ni–Metal hydride battery electrode materials. Metall Mater Trans B 46(5):2153–2157
116. Agarwal V, Khalid MK, Porvali A, Wilson BP, Lundström M (2019) Recycling of spent NiMH batteries: Integration of battery leach solution into primary Ni production using solvent extraction. Sustain Mater Technol 22:e00121
117. Mantuano DP, Dorella G, Elias RCA, Mansur MB (2006) Analysis of a hydrometallurgical route to recover base metals from spent rechargeable batteries by liquid–liquid extraction with Cyanex 272. J Power Sources 159(2):1510–1518
118. Ma Y, Qiu K (2015) Recovery of lead from lead paste in spent lead acid battery by hydrometallurgical desulfurization and vacuum thermal reduction. Waste Manag 40:151–156
119. Zhu X, Li L, Sun X, Yang D, Gao L, Liu J, Kumar RV, Yang J (2012) Preparation of basic lead oxide from spent lead acid battery paste via chemical conversion. Hydrometallurgy 117–118:24–31
120. Pan J, Zhang C, Sun Y, Wang Z, Yang Y (2012) A new process of lead recovery from waste lead-acid batteries by electrolysis of alkaline lead oxide solution. Electrochem Commun 19:70–72
121. Yang J, Zhu X, Kumar RV (2011) Ethylene glycol-mediated synthesis of PbO nanocrystal from PbSO4: a major component of lead paste in spent lead acid battery. Mater Chem Phys 131(1–2):336–342
122. Pan J, Zhang X, Sun Y, Song S, Li W, Wan P (2016) Preparation of high purity lead oxide from spent lead acid batteries via desulfurization and recrystallization in sodium hydroxide. Ind Eng Chem Res 55(7):2059–2068
123. Hu Y, Yang J, Hu J, Wang J, Liang S, Hou H, Wu X, Liu B, Yu W, He X, Kumar RV (2018) Lead-carbon batteries: synthesis of nanostructured PbO@C composite derived from spent lead-acid battery for next-generation lead-carbon battery. Adv Func Mater 28 (9):1870056
124. Zhou H, Su M, Lee P-H, Shih K (2017) Synthesis of submicron lead oxide particles from the simulated spent lead paste for battery anodes. J Alloy Compd 690:101–107
125. Xi G, Li Y, Liu Y (2004) Study on preparation of manganese–zinc ferrites using spent Zn–Mn batteries. Mater Lett 58(7–8):1164–1167
126. Nan J, Han D, Cui M, Yang M, Pan L (2006) Recycling spent zinc manganese dioxide batteries through synthesizing Zn–Mn ferrite magnetic materials. J Hazard Mater 133(1–3):257–261
127. Kanemaru T, Iwasaki T, Suda S, Kitagawa T (1998) Preparation of ferrite from used dry cells. Google Patents
128. Deep A, Kumar K, Kumar P, Kumar P, Sharma AL, Gupta B, Bharadwaj LM (2011) Recovery of pure ZnO nanoparticles from spent Zn–MnO_2 alkaline batteries. Environ Sci Technol 45(24):10551–10556

129. Deep A, Sharma AL, Mohanta GC, Kumar P, Kim K-H (2016) A facile chemical route for recovery of high quality zinc oxide nanoparticles from spent alkaline batteries. Waste Manag 51:190–195
130. Xiang X, Xia F, Zhan L, Xie B (2015) Preparation of zinc nano structured particles from spent zinc manganese batteries by vacuum separation and inert gas condensation. Sep Purif Technol 142:227–233
131. Deng J, Wang X, Duan X, Liu P (2015) Facile preparation of MnO_2/graphene nanocomposites with spent battery powder for electrochemical energy storage. ACS Sustain Chem Eng 3(7):1330–1338
132. Ali GAM, Yusoff MM, Algarni H, Chong KF (2018) One-step electrosynthesis of MnO_2/rGO nanocomposite and its enhanced electrochemical performance. Ceram Int 44(7):7799–7807
133. Ali GAM (2020) Recycled MnO_2 nanoflowers and graphene nanosheets for low-cost and high performance asymmetric supercapacitor. J Electron Mater 49(9):5411–5421
134. Ferella F, De Michelis I, Vegliò F (2008) Process for the recycling of alkaline and zinc–carbon spent batteries. J Power Sources 183(2):805–811
135. Kim TH, Senanayake G, Kang JG, Sohn JS, Rhee KI, Lee SW, Shin SM (2009) Reductive acid leaching of spent zinc–carbon batteries and oxidative precipitation of Mn–Zn ferrite nanoparticles. Hydrometallurgy 96(1–2):154–158
136. Gabal MA, Al-luhaibi RS, Al Angari YM (2014) Recycling spent zinc–carbon batteries through synthesizing nano-crystalline Mn–Zn ferrites. Powder Technol 258:32–37

Cost-Effective Nanomaterials Fabricated by Recycling Spent Batteries

Himadri Tanaya Das, T. Elango Balaji, K. Mahendraprabhu, and S. Vinoth

Abstract The renewable energies have become affordable and accessible door to door due to well-equipped energy storage devices. Lithium-ion batteries (LiBs) and supercapacitors (SCs) are the prominent energy storage devices in the market. Recycling electronic wastes are one of the steps to build a greener and cleaner environment. In this chapter, reusing of disposed spent batteries components to generate electrode materials for LiBs applications has been discussed. Batteries developed from waste-materials are expected to fulfill the demand of people for energy storage devices with the high energy density in consumer electronics. Recycling LiBs will make the energy storage device, market cost-effective. Thus, apt energy storage devices are creating a new generation of electronics with excellent potential to improve the quality of human life. Owing to the advancement in or advanced, the researchers or industries are focused on developing electrodes for smart electronics by fabricating a potential device from the waste materials which is quite a challenging technique. In the context of making the environment clean and free from electronic wastages, recycling the spent batteries could be an estimable step. In this chapter, the basic recycling methods to extract the electrode materials such as different nanostructured oxides and carbon materials will be

H. T. Das (✉)
Department of Chemical Engineering, National Taipei University of Technology, Taipei-10607, Taiwan
e-mail: himadritanaya1991@gmail.com

H. T. Das
Center of Excellence for Advanced Materials and Applications, RUSA, Utkal University, Vanivihar, Bhubaneswar 751004, Odisha, India

T. Elango Balaji
Department of Chemistry, Bishop Heber College, Tiruchirappalli, Tamil Nadu, India

K. Mahendraprabhu
Department of Chemistry, MEPCO Schlenk Engineering College (Autonomous), Sivakasi, Tamil Nadu, India

S. Vinoth
Department of Electronics and Communication Engineering, Manakulavinayagar Institute of Technology, Kalitheerthalkuppam, Puducherry, India

A. S. H. Makhlouf and G. A. M. Ali (eds.), *Waste Recycling Technologies for Nanomaterials Manufacturing*, Topics in Mining, Metallurgy and Materials Engineering, https://doi.org/10.1007/978-3-030-68031-2_6

discussed along with the configuration of various electrodes and electrolytes in batteries. Such extensive discussion can bring out a summary on operational parameters such as stability, storage capacity, energy density, and cycle life of the LiBs developed from spent batteries. The discussions in-depth, on the general pros and cons of conventional energy storage devices, can lead to the next phase for generating advance battery technology from recycling the wastage.

Keywords E-wastes · Spent batteries · Recycling · Metal oxides · Energy storage devices · Lithium-ion batteries

List of Abbreviations

EPA	Environmental protection agency
EC	Ethylene carbonate
IL	Ionic liquids
ITO	Indium-doped tin oxide
LFO	Lithium Ferrous oxide
LFO	Lithium ferrous oxide
LiBs	Lithium-ion batteries
LMO	Lithium magnesium oxide
LMO	Lithium magnesium oxide
NMP	N-methyl pyrrolidone
PC	Propylene carbonates
SC	Supercapacitors
$scCO_2$	Supercritical carbon dioxide
WEEE	Waste Electric and Electronic Equipment

1 Introduction

Energy is a primary requirement of human civilization. At present, the energy crisis being a worldwide concern, we have to address it with a careful balance of acquired knowledge as well as innovations in research and technology. Tackling the energy crisis must be cost-effective for it to benefit civilization [1]. Fossil fuels such as petroleum products and coal have been utilized for power production in large quantities, leading to wide-spread pollution and the depletion of petroleum reserves. Thus, it is important to seek alternative, renewable, and green energy technologies to address the energy crisis [2]. Currently, renewable energy resources have become affordable and accessible door to door due to well-equipped energy storage devices. Lithium-ion batteries (LiBs) and supercapacitors (SCs) are the premier energy storage devices in the market [3–6]. The efficient recycling of electronic waste is crucial in building a greener and cleaner environment. In this chapter, reusing spent disposed batteries to generate electrode materials for LiBs applications has been discussed.

The process of developing new batteries generates a higher amount of electronic wastes. USA contributes about 180,000 tons of electronic wastes in which 86,000 tons of alkaline-type batteries and 14,000 tons of secondary-type batteries each year as per the reports from the Environmental Protection Agency (EPA). In 2013, China alone generated 570,000 tons of battery wastes [8]. The cost of 1 tonne of spent LiBs worth \$7708, among spent LiBs parts, the share of the cathode material is expected to worth \$6101 [9]. From Fig. 1, we can observe that the amount of spent LiBs increased rapidly since 2010. Apart from the production of efficient batteries technology, the generation of waste has raised its head as a major issue for our current situation. So many researchers, as well as industries, have come together to face the challenge of accumulation of e-waste and its recycling. This shows that, we are concerned with preserving the Earth and prevent pollution of the environment. However, it has been found that recycling is also quite an expensive and difficult process. Therefore, across the globe, it has garnered the researcher to discover new and cost-effective technologies.

For the batteries, developed from waste material are expected to fulfill the fast-increasing demand for energy storage devices in the consumer electronics industry [10]. Thus, this chapter will bring out insights on recycling spent batteries to LiBs to make the energy storage device market cost-effective. Thus, innovative and environment-friendly storage devices are creating a new generation of electronics with excellent potential to improve the quality of life around the globe.

In the modern world, electronic wastes are a concern for being the primary source of various pollutions. The components of batteries leaching out are a major cause of soil and water pollution as they consist of heavy metals such (Mn, Co, Ni, Cd, etc.) [11–14]. These hazardous electrode materials create an imbalance in the ecosystem. Thus, we must extract all the possible components from used batteries. Recycling batteries is not enough unless recovered materials are reused [12, 13, 15].

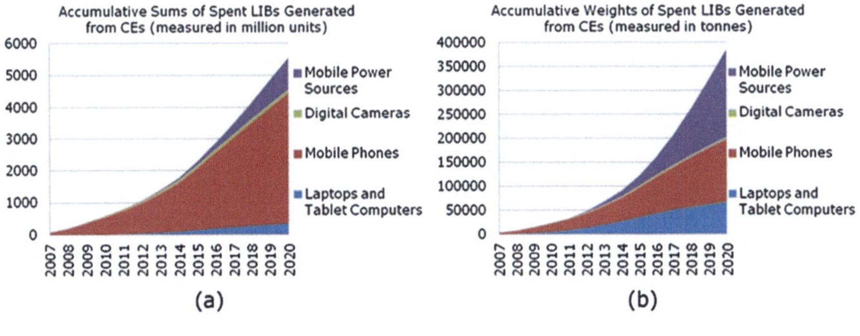

Fig. 1 Estimated amounts of spent LiBs generated from related consumer electronics: **a** accumulative sums, and **b** accumulative weights. Reproduced with permission from Ref. [7], Copyright 2017, Elsevier

Also, reusing those as energy storage devices will give rebirth to some active components; so, researchers need to identify the universal truth of making art from waste.

2 Overview of Batteries, Its Components, and Their Harmful Effects

Batteries are electrochemical cells whose redox reactions (reduction and oxidation) create a flow of electrons in a circuit. Batteries are made up of main components such as suitable electrodes (an anode and a cathode) and electrolyte. When electrodes of a battery are connected to a circuit, a redox reaction will take place between the anode and the electrolyte which causes the flow of electrons through the circuit (oxidation) and back into the cathode where reduction would take place [16]. When electrode materials are consumed, the battery is unable to generate electricity, and that is when the battery is said to be dead. Those batteries that should be thrown away after once used are called primary batteries. Those batteries that can be recharged after use, repeatedly, are said to be secondary batteries. Some examples of primary batteries are alkaline, Zn–C, and manganese dry cells [6, 12–14, 17–19]. The primary batteries are still in usage in various products such as watches, remotes, wall clocks, wireless mouse and keyboards, and torchlights which result in higher accumulation of battery waste. The spent primary batteries will have 5–7% of lithium, 5–10% of cobalt, 5–10% of nickel, 15% of suitable organic reagents, and 7% of plastics. These proportions may change with different manufacturers [20].

The lead–acid battery technology, consisting of lead peroxide as a cathode, lead as an anode, and sulfuric acid (6 M) as an electrolyte, has opened the way to develop rechargeable secondary batteries in the eighteenth century [21]. Since then, there is a high demand for lead–acid batteries in various applications such as vehicles, but this demand means, battery waste is continuously added to the environment. Lead has been declared hazardous waste by the Environment Protection Agency (EPA), and it is also reported that used lead–acid batteries formed in the USA are effectively recycled (more than 97%) [22]. The chemicals exposed to the environment take considerable time to decompose. Owing to the detrimental environmental impact of Ni–Cd batteries, their practical usage truncated rapidly. Although nickel hydride batteries replaced it, it was neither an effective nor an environmentally safe alternative. Numerous research reports have proved that the harmful effect of nickel on the environment and especially the health of human, including DNA damage and possibly premature aging [23]. Besides that, cadmium is one of the most deadly for human life, causing kidney, bone, and pulmonary damages [24].

Indeed, in the nineteenth century, LiBs conquered the market of battery technology and made electronics easily accessible. Figure 2 is an online survey data, and it shows the ownership of various electronics things among people these

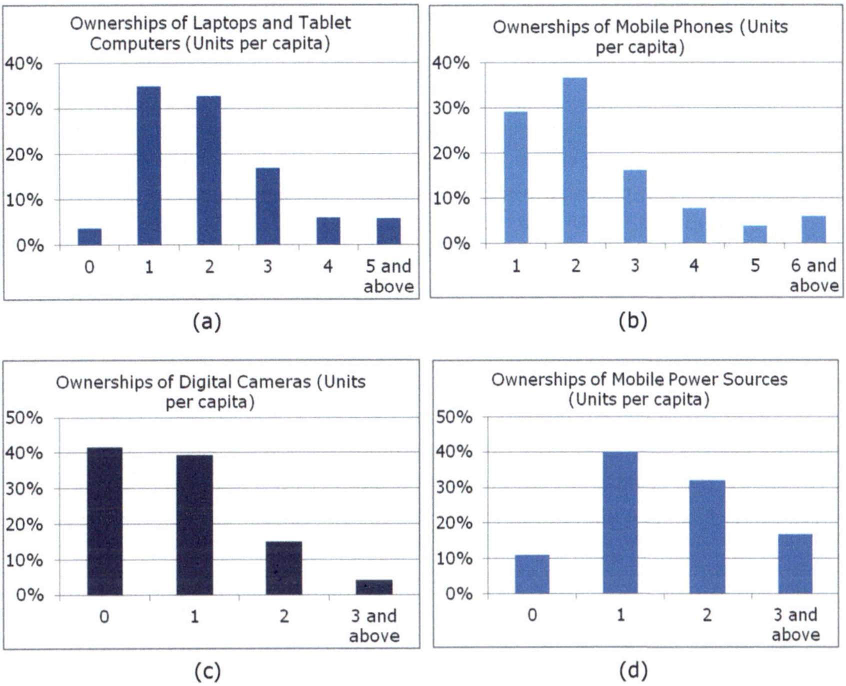

Fig. 2 **a** Ownership of laptops and tablets, **b** ownership of cellular phones, **c** ownership of digital cameras and **d** ownership of mobile power sources. Reproduced with permission from [7], Copyright 2017, Elsevier

electronics things will work only in the presence of a power source and the power source for all of these are batteries. There are several advantages to having LiBs as energy storage devices in electronics: they are small, compatible & portable and deliver higher energy density with a longer lifetime at a high potential (3.7 V) [25]. In search of a potential electrode or a high-performance secondary electrode battery, we have landed on lithium-based electrode materials, as lithium electrochemical equivalence is 3860 mA h g^{-1} at the lowest potential of −3.05 V [26]. Hence, lithium being an alkali metal is the lightest among all and has the highest theoretical specific capacity with a wide potential window.

Along with lithium-based cathode (positive electrode), LiBs consist of graphite as the anode along with the ionic electrolyte (lithium–ion rich) [27]. Usually, $LiCoO_2$ with theoretical capacity 274 mA h g^{-1} used against the graphite (372 mA h g^{-1} in organic solvent $LiPF_6$ as the electrolyte) for today's market of battery technology. In 1991, Sony Corporation announced the commercialization of LiBs. Despite the excellent performance of existing cathodes, LiBs encountered several inadequacies such as the parasitic reaction of Li-metal with electrolytes resulting in dendrite formation, which further created safety concerns such as volume expansion, thermal breakaways, and explosion.

2.1 Nanomaterials Used as Cathodes in Lithium-Ion Batteries

In general, $LiMO_2$ is the layered metal oxides that used as cathodes in LiBs. Lithium cobalt oxide ($LiCoO_2$), for which Goodenough was awarded Nobel for Chemistry 2019, still prevails in the market and is used in various devices. Owing to their exceptional cycle life and ease of bulk manufacturing, $LiCoO_2$ is undoubtedly the most extensively used anode material for LiBs [6, 14, 21]. Interestingly, the layered structures help in the intercalation/de-intercalation processes of Li-ions (during the charging and discharging). The coating of various metal oxides (MgO, Al_2O_3, ZnO) and metal phosphates (e.g., $AlPO_4$, $FePO_4$) has been reported as an effective alternative to prevent phase instability and improve cyclability of $LiCoO_2$ [28]. However, owing to factors such as high power and high energy density, scarce availability and material toxicity, their use in hybrid vehicles is restricted. Of late, the higher input costs associated with the use of $LiCoO_2$ were compensated by replacing Co with other metals, namely Ni, Mn, or Al within the lattice structure of $LiMO_2$, due to identical crystal lattice pattern with $LiCoO_2$. The advent of $LiNiO_2$ was touted to be an ideal replacement for $LiCoO_2$ anodes owing to their excellent electrical conductivity, low cost, and better discharge capacity (200 mA h g^{-1}) that is higher (almost 40%) than that of $LiCoO_2$ [29]. After the commercialization of LiBs, $LiCoO_2$ continued to dominate the LiBs market for a long time. The superior theoretical capacity of $LiMnO_2$ is remarkable (285 mA h g^{-1}), and environment-friendly character makes them an exciting candidate among the layered $LiMO_2$ cathode materials [30]. To overcome the shortfalls associated with the $LiMn_2O_4$ spinel, researchers have put many potential alternatives forward. Among all the composites, $LiNi_{0.5}Mn_{1.5}O_4$ showed the best overall electrochemical performances, which means it can serve as ideal high-power lithium secondary batteries in the automobile industry [31]. In recent years, the olivine type compound lithium iron phosphate ($LiFePO_4$), commonly referred to as, ($LiFePO_4$) gained significant interest as a cathode material in LiBs owing to its extremely low cost, environment-friendliness, and structural stability at harsh conditions. Despite offering several advantages such as stability at elevated temperature and long cycle life, $LiFePO_4$ suffers from inherent poor electronic and ionic conductivity, delivering a relatively lower capacity [31]. So few other materials such as V_2O_5, $LiTi_2O_4$, $LiMnCoO_4$, $Li_2Fe_2(SO_4)_3$, $Li_xTiNb(PO_4)_3$, and Li_xTiS_2, are being explored as electrodes for LiBs [26, 32, 33]. Such materials have small cyclic life and low power densities to incorporate in applications like electric vehicles. Therefore, the search for various anode materials also continues.

2.2 Nanomaterials Used as Anodes in Lithium-Ion Batteries

On the other hand, to achieve higher capacity and stability, there is always a search for a substitute anode material. Different metal oxides, carbides, or nitrides along with metal chalcogenides, are used as anode materials. The metal oxides/chalcogenides are nanocomposites with carbon-based materials or with other nanostructured materials for higher electrochemical performance. Carbonaceous materials are preferred as anodes in LiBs, because of their enhanced specific charges, longer stability, and high negative redox potentials. The morphology, porous structure, and different dimension (0D, 1D, 2D, 3D) can be easily tuned to complement the performance of LiBs. Carbon-based materials are categorized in different forms like amorphous, activated, graphite, soft carbon (non-graphitized carbon), and hard carbon which have different structures and electrochemical properties. The advantages of graphite are layered structures of carbon hexagon networks with high surface areas in LiBs. However, several other important factors such as SEI formation and solvated lithium insertion affect the reversibility of the intercalation, which increases cell resistance led to a decrease in the power density with the different cycles. In addition, metals such as Co, Ni, Mn, Fe, Si, and Sn can be used as nanocomposites electrode materials are hazardous for the environment.

2.3 Electrolytes in Lithium-Ion Batteries

Along with electrodes, there is also development in electrolyte research for the ideal performance of LiBs. Thus, the selection of suitable electrolyte material is another important thing. Usually, organic solvents (aprotic type) mixed with lithium salts are used as electrolytes, e.g., alkyl carbonates (ethylene carbonate (EC)), propylene carbonates (PC)) etc., mixed with hexafluorophosphate ($LiPF_6$) or lithium perchlorate ($LiClO_4$) [34]. Other important parameters such as dielectric constants and viscosity for dissolving solute to get a stable rechargeable cell structure. The operating cell voltage of LiBs mainly relies on the stable electrochemical potential window of the electrolytes, if electrodes are stable within the range of working voltage. The potential window of aqueous electrolyte is about 1.23 V (H_2/O_2 evolutions at ambient conditions), while the organic electrolyte-based and ionic liquid (IL, such as $[CMMe_3N^+][NTf^{2-}]$, ($[CMMeIm^+][NTf^{2-}]$, etc., found in LiBs have potential windows of 2.5–2.7 V and 3.5–4.0 V, respectively [35]. In addition to the determining role of the operating voltage window in ES energy density, electrolytes/solutions also play a vital role in establishing other essential properties such as power density, internal resistance, rate performance, operating temperature range, cycling lifetime, self-discharge, and toxicity. Moreover, the exploration of efficient solid or semi-solid electrolytes has led to the invention of flexible or solid-state LiBs, which are claimed to have no potential leakage issues especially in the liquid electrolyte-based LiBs [36].

3 Effect of Lithium-Ion Batteries Development on the Environment

LiBs are fast-growing in consumer electronics and find applications in many other fields. This popularity of LiBs is attributed to their extremely low weight, wide working potential window (up to 3.6 V), high theoretical energy density (150 W h kg^{-1}), and power density (400 W h kg^{-1}). Also, LiBs are characterized by a very low self-discharge rate of 2–8% per month, absence of memory effect, high coulombic efficiency over longer cycle life of more than 1000–3000 cycles, capacities. Many industries and researchers are investing in millions for advancement in LiBs, in the worldwide market. Besides all the advantages of LiBs, shortcomings such as the safety of LiBs, low power density, capacity retention, and hazardous chemical components are still a matter of concern. LiBs are available in the market in various sizes and shapes with a life cycle of 2–3 years [37]. The cost of LiBs is high, and they get ruined if they are completely discharged. Besides, there is a minor risk of LIBs, getting bursted into flames, when the battery is manufactured with defects in the fabrication process, and the risk is as low as 2–3 packs produced per million batteries. From this overview of LiBs, we understand the components being used in the LiBs and their role in our recycling goal.

From Fig. 3, we can see that the most recycled components in LiBs are the electrode materials since they have many adverse environmental effects when thrown-out or dumped in landfills. Ni–Cd and NiMH cannot be landfilled due to the reason that the leaching of heavy metal results in the contamination of groundwater [39]. As per the survey in 2012, the market was occupied by batteries, which made up of more than half of the cathode material by $LiCoO_2$ and $LiNi_{0.33}Co_{0.33}Mn_{0.33}O_2$, metals containing hazardous heavy metal (cobalt). Although cobalt is not considered as harmful metal yet, it raises flags as it is soluble in biological fluids and uptake of cobalt in our biological system leads to a rapid urinary elimination lasting for about 2 days. However, when the metal is entering our biological system, it may affect our normal well-being as well as an excess of this can lead to toxicity [40, 41]. Lead–acid batteries contain lead, and this lead is harmful to human health. A recent study in Bangladesh showed that there is an increased amount of lead in the blood level of employees who had worked with lead-acid batteries and these increased levels of lead resulted in hematopoietic, gastrointestinal, nervous, renal and reproductive problems which may accelerate the onset of various diseases [42].

4 Recycling Nanomaterials from Lithium-Ion Batteries

Recycling all possible active components of LiBs in spent electronic devices by adopting efficient methods became mandatory for reducing pollution and also to meet resource limitations. The components in spent electronic devices possess

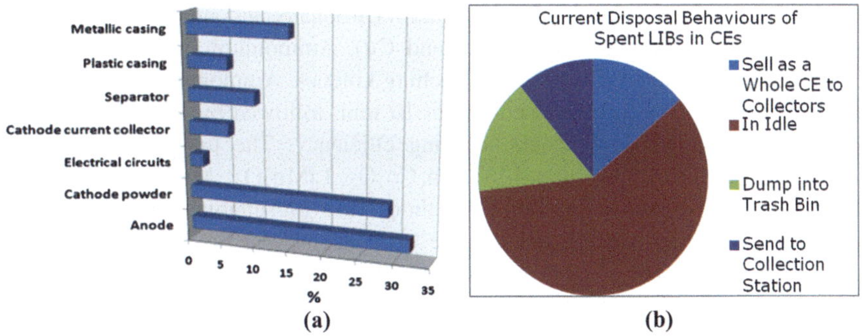

Fig. 3 **a** Breakdown of components in LiBs. Reproduced with permission from Ref. [38], Copyright 2017, Elsevier. **b** current disposal behaviors of spent LiBs in consumer electronics. Reproduced with permission from Ref. [38], Copyright 2015, Elsevier

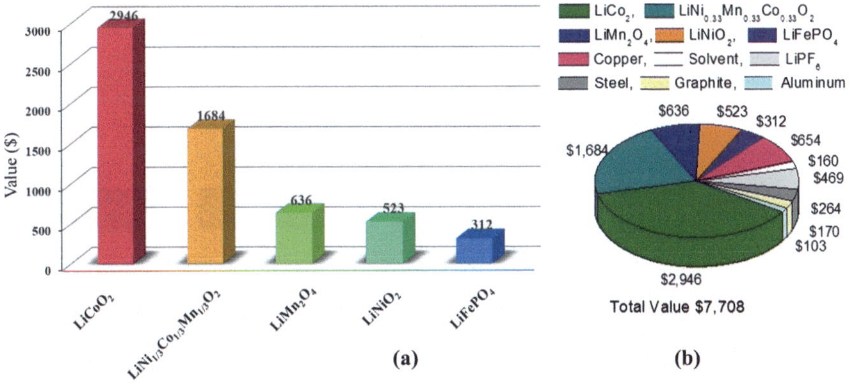

Fig. 4 Economic value of components in LiBs: **a** value of cathode materials used in LiBs and **b** economic value of components of LiBs. Reproduced with permission from Ref. [9], Copyright 2018, Elsevier, and Ref. [43], Copyright 2014, Elsevier

materials of significant economic value. Figure 4 summarizes the value of active components in spent LiBs.

It is reported that the total value of all components is about \$7708 per ton of LiBs. Gratz et al. recovered mixed cathode materials ($LiNi_xMn_yCo_zO_2$) efficiently (about 90%), and with high purity (Al, Cu, and Fe impurities are observed in less than 50 ppm) using the low-temperature hydrometallurgical process (the steel casing is removed by magnetic separation). This process resulted with a recovery efficiency of about 90% (with the purity of the compound containing Al, Cu, and Fe impurities are less than 5 ppm) and another interesting feature is that \$5013 profit margin can be achieved from one ton of waste batteries [43].

Castillo et al. elaborated various cost-effective chemical treatments using acids, etc., for the extraction of the active components of used LiBs [44]. Ku et al. used leaching agents such as ammonia, ammonium carbonate, and ammonium sulfite to extract active metals (Ni, Co, Mn, Al, and Cu). Ammonium sulfite used as the reducing agent and also enhancing the leaching kinetics. Ammonium carbonate acts as a buffer. Copper is fully leached within 10 min, followed by Ni and Co within 40 min. Mn and Al showed less leaching efficiency. The final product of the leached active materials such as $LiNi_xMn_yCo_zO_2$, $LiMn_2O_4$, Al_2O_3, $MnCO_3$, and Mn oxides [25]. Various dissolution methods (use of suitable organic solvents, which are non-toxic) used to dissolve the toxic nature of PVDF, which acts as a binder present in spent LiBs to minimize pollutions (when burned it releases HF gases) are reviewed as shown in Fig. 5 [34]. By incorporating pyrolysis before leaching has provided an environmentally favorable method of treating the LiBs and also washing the inner materials which are present inside the batteries with suitable organic solvents for extracting electrolytes to minimize pollution is suggested to avoid the generation of wastes [34]. Boyden et al. verified that more

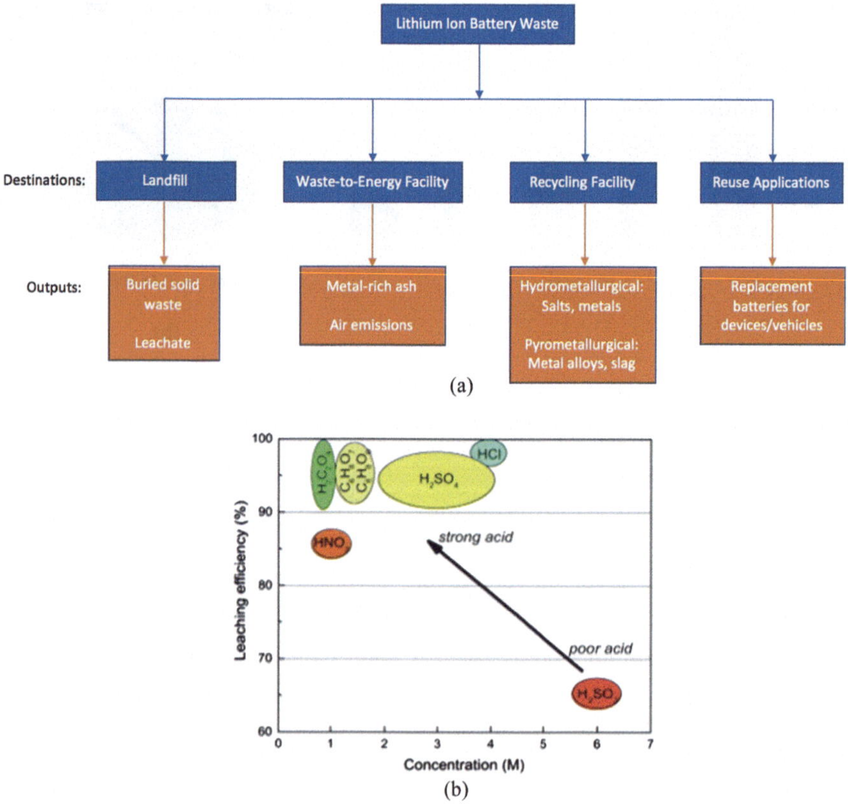

Fig. 5 **a** Different methods for depositing of spent LiBs. Reproduced with permission from Ref. [22], Copyright 2018 Elsevier, and **b** the variation of leaching efficiency on weak acid to strong acid. Reproduced with permission from Ref. [45], Copyright 2015, Elsevier

materials recovered by the hydrometallurgical process than the pyrometallurgical process on average and the recovered materials are copper, nickel, and cobalt. The pyrometallurgical process has some adverse effects like plastic incineration, which causes global warming and terrestrial ecotoxicity. The impacts of these processes, when compared to the hydrometallurgy process, showed more effect on the environment than pyrometallurgy and landfills method within the potential global warming type.

4.1 Recycled Nanomaterials from Lithium-Ion Batteries

Furthermore, another problem has great potential for increasing the toxicity, which is transporting the waste batteries for processing. For instance, transporting batteries led to global warming potential by 45% by the pyro-hydrometallurgical process and about 50% increase of human toxicity potential for the hydrometallurgical process [46]. Pinna et al. investigated that the mixed $LiCoO_2$ in cathodes of spent LiBs is recovered by reductive dissolution process by adding phosphoric acid. Specific parameters such as concentrations (leaching and reducing agent), reaction time, reaction temperature, and speed of the stirrer are studied. This process resulted in the dissolution value near 100% with an output of lithium and cobalt.

The composition of the LiBs was investigated, and about 29% of the mass of LiBs is made up of 83% $LiCoO_2$, 14.5% C, and less than 2.5% was made up of Al, Al_2O_3, and Co_3O_4. When this spent cathode material was used for the degradation of dyes, it showed an increase in degradation of methylene blue in the dark condition by 200 times which is due to decrease in activation energy (26 kJ mol^{-1} from 83 kJ mol^{-1}). With this, two potentially harmful effects can be reduced, the hazardous environmental effects of spent LiBs and toxicity of organic dyes [38]. Zeng et al. adopted the oxalic acid leaching and filtering method for the extraction of $LiCoO_2$. Combined processes in which leaching of chemicals then precipitating and finally filtering had been followed. Lithium and cobalt (about 98% and 97%, respectively) were recovered by using oxalic acid leaching method [45].

Li et al. recovered graphite, carbonates of cobalt and lithium using in situ oxygen-free roasting and wet magnetic separation. The change of waste into resources from components (waste + waste → resources), from this process, about 95.72% of Co, 98.93% of Li, and 91.05% of graphite were successfully recovered. This method of recovery does not use any chemicals which preserve the cost of treating secondary pollution [47]. Barbieri et al. reported that cobalt oxide and cobalt hydroxide are recovered using the solution leaching method. $Co(OH)_2$ acted as starting material for the synthesis of Co_3O_4 and then it was electrodeposited on the indium-doped tin oxide (ITO). The process showed a high conversion ($Co(OH)_2$ to Co_3O_4) efficiency of 64.29%. The recovered material shows capable applications in the field of electrocatalysis and as pseudocapacitors as the electrode material showed the charge efficiency (77%) for the first 10 cycles and a specific capacitance of 10.5 F g^{-1} at 10 mV s^{-1} [48]. Dewulf et al. reported that the recycling of spent

LiBs led to saving of natural resource by 51.3%, the scenario due to minimizing the dependency on mineral ore and also due to reduced fossil resource (reduction by 45.3%) and nuclear energy demand (57.2%). As of 2007 data reports, about 25% of the Co demand all over the world due to battery applications. A comprehensive analysis to recover metals such as Co and Ni from the mixed metal oxide battery, then it is reused for the battery application related to the synthesis using virgin materials [49]. Li et al. reported the leaching process is an eco-friendly manner to recover Co and Li in which organic acid DL-malic acid ($C_4H_5O_6$) used as a leaching reagent which is easily degradable. Various essential parameters were studied to recover more than 90 wt% of Co and 100 wt% of lithium by experimentally changing the concentration of leaching agent, time, and temperature. Also, the solid-to-liquid ratio was investigated that hydrogen peroxide in DL-malic acid solution, which enhanced the efficiency of leaching action, making it an effective reducing agent. Leaching was done with DL-malic acid (1.5 M), 2.0 vol.% of hydrogen peroxide which helped to recover metals efficiently within 40 min at 90 °C and an increase in temperature and time also had the significant key role in increasing the efficiency of leaching. It is also noted that hydrogen peroxide accelerated the dissolution [50].

Freitas et al. recovered Co from spent cellular telephone LiBs by electrochemical techniques. The composition of cathodes of spent LiBs is $LiCoO_2$, Co_3O_4, C, and Al as per the XRD data. It confirmed the presence of Co_3O_4 in cathode due to the active material ($LiCoO_2$) and the maximum charge efficiency was found to be 96.90% at pH 5.4 at a current density (10.0 C cm^{-2}). It is observed that there is the decrease in charge efficiency with a decrease in pH for electrochemical recycling of cobalt [51]. Garcia et al. recycled cobalt from spent cathodes of LiBs using the electrochemical technique for supercapacitor application. Cobalt is recovered from $LiCoO_2$ by the acidic dissolution of lithium cobalt oxide. The electrochemical studies of the directly applied recycled material showed the high specific capacitance (601 F g^{-1}) at the current density of 0.23 mA cm^{-2} in 6 M KOH. Also, it was found that recycled cobalt is compatible with SCs materials [52]. Rothermel et al. investigated the anode recycling along with three different electrolytes. In the first process, the graphite is heated without recovering the electrolyte. The second method adopted the electrolyte extraction using subcritical carbon dioxide ($scCO_2$)-assisted extraction. Finally, the use of $scCO_2$ as extraction was done followed by the thermal treatment. While using $scCO_2$-for electrolyte extraction, recycled graphite anodes achieved the best performance. When comparing with the freshly synthesized graphite anode with recycled graphite, the recycled graphite anode shows a better electrochemical performance than the synthesized graphite (TIMREX SLP50) [53]. Kevin et al. utilized the study of life cycle impact assessments to recycle certain components in LiBs, which results in fewer air emissions. The spent LiBs can ignite or release hazardous chemicals spontaneously under landfill conditions [22]. He et al. prepared porous carbon monoliths with hierarchical structure by recycling the spent lead–acid secondary battery using the hydrometallurgical process.

The recovered material showed a large micropore size of 2 nm with a volume of 0.0248 $cm^3\ g^{-1}$. The surface area was found to be 138.5 $m^2\ g^{-1}$ from BET analysis. It can be noted that the high charge–discharge capacity (217 $mA\ h\ g^{-1}$) was maintained at high current density (5000 $mA\ g^{-1}$) [54]. Chagnes et al. reviewed the recycling of various components in spent LiBs using hydrometallurgical technologies. Also, solvent extraction strategies from spent LiBs are reviewed. Cobalt and lithium are the main objectives of recovery due to the high price, and cobalt is precious and rare when compared to other components of the LiBs. There is very little research on the recovery of polymetallic oxides, which is the next generation of LiBs, using the solvent extraction method selective recovery of the metal. Hence, more research should be focused on the development of selective extractants to recycle polymetallic electrodes [55] (Fig. 6).

Hannan et al. highlighted the energy management strategy of LiBs in electric vehicles in which the need for alternatives for fossil fuels is emphasized. It has been stated that LiBs hurt the environment if it is not disposed of properly, and recycling is the best way to minimize CO_2 and GHGs emissions. The recycling process develops the preservation and reuse of Co, Ni, and Li by recovering it. By doing this, the life cycle of the LiBs is improved the processes now incorporated for the recycling are a hydrometallurgical method, pyrometallurgical method, and direct

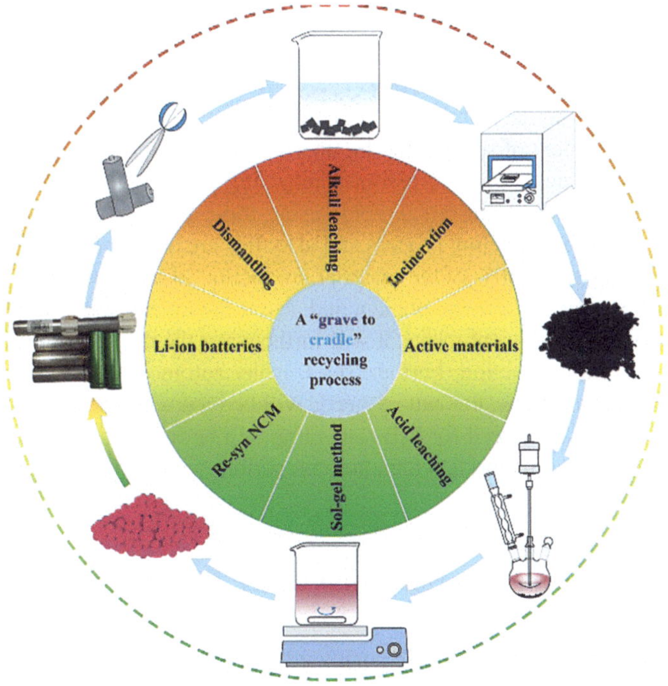

Fig. 6 Schematic view of the recycling process. Reproduced with permission from Ref. [9], Copyright 2018, Elsevier

recycling method. It has to be noted that direct recycling has a higher potential for recovering LiBs materials [56]. Li et al. obtained $LiCoO_2$ films by combined recycling processes such as dismantling of components, detachment with N-methyl pyrrolidone (NMP), acid leaching and resynthesis of $LiCoO_2$. $LiCoO_2$ crystals were regenerated on Ni plate with electrochemical deposition technology. Initially, the charge and discharge capacities were found to be 130.8 and 127.2 mA h g^{-1}, respectively. The capacity had decreased after 30 cycles by less than 4% compared with the first cycle. The recovery/regeneration method is more economical as well as eco-friendly one [57]. Nie et al. used the green method to recycle active components in spent LiBs. It is reported that $LiCoO_2$ regenerated at 900 °C, which could meet the commercial requirements for the reuse in LiBs. The structure of $LiCoO_2$ remains unchanged, and binders, conductive polymers are removed during the recycling process. The reported process in an eco-friendly, cost-efficient, and suitable for industrial application [58].

Lee et al. adopted the set of recycling processes (mechanical, thermal, hydrometallurgical, and sol–gel method to recycle cobalt and lithium and to recover $LiCoO_2$. The recovery of Co and Li by nitric acid is investigated. Among many processes, amorphous citrate precursor is adopted to obtain $LiCoO_2$. The precursor is calcined at 950 °C for 24 h to obtain pure crystalline $LiCoO_2$. The particle size and specific surface area were found to be 20 μm and 30 cm^2 g^{-1}, respectively. It is noted that the charge and discharge capacities of $LiCoO_2$ are 165 and 154 mA h g^{-1}, respectively [11]. Gaines et al. emphasized the lifecycle burdens of LiBs and stressed the significances of constituent-material production and the subsequent manufacturing of batteries.

Figure 7 shows some of the best recycling process used until now. Figure 7a shows the pretreatment of spent LiBs. Figure 7b shows the recovering of Co and Li through DL-malic acid. Figure 7c shows recycling of LiBs through maleic acid (strong acid) and acetic acid (weak acid) using these types of organic acids will reduce the cost of the recycling process and also they are environment-friendly. The recycling of LiBs saves 50% of the material production, if the production efficiency of the recycling process is more efficient and also if the money spent on the process is also economically efficient means it will further save the use of fresh materials in the manufacturing production. Further, soon, many automobile vehicles are going to be introduced; after this, the battery needs will be increased by 200–500%. So in the upcoming years, the need for LiBs recycling will be increasing, and we have to develop the recycling methods which are compatible with our needs [60].

4.2 *Recycled Nanomaterials from Other Battery Cathodes*

Apart from $LiCoO_2$ cathodes in LiBs, many other electrodes are also considered for recycling and reuse. Many researchers and industries have recycled Ni–Cd, Ni-metal hydride, and zinc batteries in an effective way. Huang et al. recovered Cd from the spent Ni–Cd battery by the separation method of vacuum metallurgy with

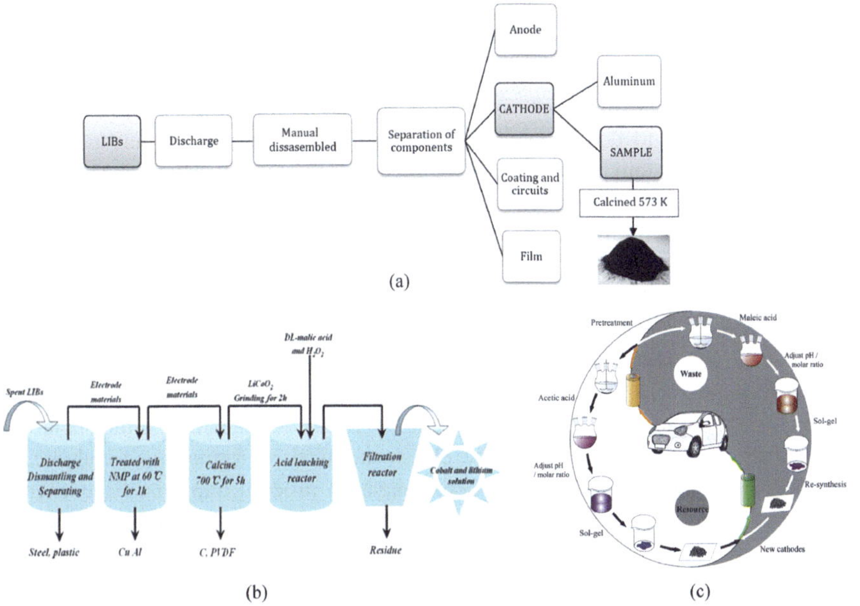

Fig. 7 Some of the best processes used in recycling of LiBs **a** Reproduced with permission from Ref. [3], Copyrights 2017, Elsevier, sample preparation and pretreatment steps. Reproduced with permission from Ref. [50], Copyrights 2010, Elsevier, **b** recover and reuse of cobalt and lithium using DL-malic acid, and **c** recycling of LiBs through maleic acid and acetic acid. Reproduced with permission from Ref. [59], Copyrights 2018, Elsevier

a recycling efficiency of 99.98%. Five different types of battery models and brands were collected, and orthogonal design is used to optimize the parameters, which affect the recycling of cadmium. The orthogonal design is a convenient tool in evaluating the parameters impacting the recycling of cadmium by the thermodynamics theory and numerical analysis [61]. Vassura et al. recovered materials in mixed batteries (Ni–Cd, Ni-metal hydride, and LiBs). It is reported that about 250 g of Co, 110 g of Ni, 120 g of Cu from the spent LiBs. The leaching test results show that only very few metals were released into the water. From this, we can understand that only a limited amount of contamination can happen when these materials are kept in contact with water [62]. Mariana et al. recycled spent cathode and used as a catalyst in the degradation of the toxins in the organic dye; this would aid the safe disposal of batteries. Reitas et al. reported that recovering of Zn from spent Zn–MnO_2 batteries. Zinc can be recovered using the galvanostatic technique in acidic or alkaline solutions. The optimum current density is in between 10 mA cm^{-2} and 25 mA cm^{-2}, which was obtained for the recovery of zinc ions in acidic solutions. The optimum current density is 15 mA cm^{-2} for electrodeposition of zinc in an alkaline solution. The current density and charge efficiency are inversely proportional. Charge efficiency was found out to be 80%, and it decreases with increasing current density [63]. Kim et al. proposed a new method (electrochemical λ-MnO_2/

BDD system) to recover the lithium metal present in wastewater from battery recycling plants from this approximately 1900 mg L^{-1} of lithium is discarded. The wastewater has some amount of organic pollutants (approximately 300 mg L^{-1} of dissolved carbon). In this work, an electrochemical system is proposed to recycle lithium as well as to decompose various organic pollutants simultaneously. The electrochemical system made up of two electrodes, such as a lithium-recovering electrode and the oxidant-generating electrode. The system successfully recovered 98.6% lithium and the reduction of organic pollutants by 65% [64]. Fouad et al. synthesized $LiAlO_2$ from spent LiBs by thermal treatment at 900 °C supported by Al sheets. The SEM analysis shows the γ-$LiAlO_2$ has a coconut like grain morphology [65].

Similarly, it is noteworthy that Li^+ (88%) is recovered as Li_3PO_4 with the purity of 98.3% and also Co^{2+} (99%) is recovered as CoC_2O_4 with the purity of 97.8% [3]. Ganter et al. adopted the process of cathode refunctionalization, which enables remanufacturing cathode to retain the electrochemical performance of cathode. It demonstrated that the $LiFePO_4$ refunctionalization process minimized the energy requirement (by 50%) when compared to the production of the cathode from raw materials. The cathode refunctionalized using electrochemical and chemical lithiation method by incorporating this process, the material retained the conventional capacity (150–155 mA h $g^{-1)}$ [10].

Chen et al. adopted the green recycling process for the first time to recycle cathode powders where no harmful reagents are used, no harmful gases are evolved during the reaction, and no wastewater is leaked into the environment. Considering the disposal of $LiFePO_4$, if thrown away without proper disposal or recycling, it causes severe environmental problems. Fragments of $LiFePO_4$ host particles decomposed into different compounds such as $FePO_4$, Fe_2O_3, P_2O_5, and Li_3PO_4. The material showed poor electrochemical performance after performing some charge–discharge cycles, which may be due to the residue of PVDF binders. The cathode powders are finely recycled and showed better tap densities and discharge capabilities as compared to the low temperature treated material with heat treatment at 650 °C. This research provides not only a green process for recycling but also, the small-scale model line is developed [66]. Renault et al. adopted green solvents as an eco-friendly process and subsequent thermal treatments to recycle lithium. It is noted that dimethyl carbonate/lithium bis(trifluoromethane sulfonyl) imide was not decomposed in water. This method gives up to 99% of the capacity restored. An only small amount of lithium is lost in the whole process. The excellent recovery rate is due to the electrolyte salt and solvent [67]. Ciez et al. compared different recycling processes (pyrometallurgical process and hydrometallurgical recycling process) which reduce the cells into elemental products. It is reported that pyrometallurgical and hydrometallurgical processes have not reduced greenhouse gas emissions significantly, but the direct recycling process reduced emissions with also economically competitive [68]. Poyraz et al. adopted the thermal regeneration method, which created new possibilities in electrochemical energy storage devices with long life cycles. There are many types of research done on recycling of the LiBs, but they are multiple-step processes where only the base material is recycled, and the capacity of the material fades out due to structural regeneration. Herein, a new process was incorporated, which is attracted by thermal regeneration strategies. In this process, the removal of cathodes had undergone heat treatment and then inserted into the cell with restoring capacity and cyclic stability. To test this α-MnO_2 (binder-free self-supporting) electrode was synthesized, and the electrochemical activities of these electrode materials can be restored many times using thermal regeneration method. It is noted that the process restored the average oxidation state, crystallinity, and electrochemical performance. This process can be adopted for all electrode materials with nanowire morphology [69].

Researchers of Worcester Polytechnic Institute have been developed new technologies to recover $LiNi_xMn_yCo_zO_2$ cathode material by the hydrometallurgical process, which is more economically viable. Recycling of LiBs must be done by introducing the flexible recycling method, which is named as closed-loop LiBs recycling system. There has been a good development in the LiBs, but still, the recycling industry is struggling, and about 95% of the LiBs are landfilled. Previously, there were some methods reported for direct synthesis of cathode material from spent LiBs. However, by incorporating this method the cathode material for a battery can be directly synthesized regardless of size, shape, the chemistry of cathode materials and most importantly, it can recover active materials which account for the 70% of the battery value. The other advantage of this method is that it does not include a pyrometallurgical method, so no carbon is emission occurs during the process. With this process, we can recover all the material except the electrolyte and the solvent used, also after the process is over, some of the wastewater remains and should be disposed of safely [70]. Li et al. developed an economically effective method to recover spent $LiNi_{1/3}Co_{1/3}Mn_{1/3}O_2$ with different organic acids in which 98% Li, Co, Ni, and Mn can be leached out and recover $LiNi_{1/3}Co_{1/3}Mn_{1/3}O_2$. Organic acids are chosen based on their chelating tendencies (acetic acid has weak chelation, but maleic acid has strong chelation). The leaching mechanism is studied in detail (in both micro- and macro-scales). The sol–gel method is adopted, and the performance of acetic acid (NCM–Ac) and maleic acid (NCM–Ma) is studied. The obtained results showed that acetic acid leaching with H_2O_2 showed a leaching efficiency of 98.39% for Li, 97.72% for Co, 97.27% for Ni, and 97.07% for Mn. When leached with maleic acid with H_2O_2, it showed a leaching efficiency of 98.24% for Li, 98.41% for Co, 98.05% for Ni, and 98.06% for Mn, respectively. The recovered materials from the two leachates were characterized, the electrochemical performance was compared, and it was concluded that maleic acid leaching showed the layered structure without impurities. It attained a higher capacity (151.6 mA h g^{-1}) which is better performance than the material leached from acetic acid.

Also, considering the economy of this organic acid leaching, it is proved that the cost of the chemical used was less, and it was a green process, and it has good applicability for using in electric vehicles [59]. Li et al. proposed the grave-to-cradle process to synthesize the mixture of cathode materials ($LiMn_2O_4$, $LiCoO_2$, and $LiCo_{1/3}Ni_{1/3}Mn_{1/3}O_2$) by acids. Figure 6 illustrates the process involved to recover active materials using oxalic acid and hydrogen peroxide. It is

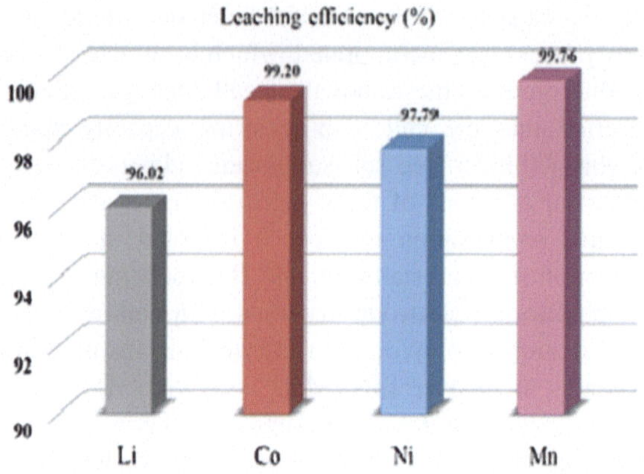

Fig. 8 Leaching efficiencies of metals (Li, Co, Ni, and Mn) in optimized conditions. Reproduced with permission from Ref. [9], Copyright 2018, Elsevier

noted that the efficiencies exceeded 95% in recovering $LiCo_{1/3}Ni_{1/3}Mn_{1/3}O_2$ material using the sol–gel method. From Fig. 7, we can observe that Mn shows a high leaching efficiency of 99.76%. Thus, obtained material showed better performance than the directly obtained from original ones, and it was found that it was due to Al doping [9] (Fig. 8).

Pagnanelli et al. adopted physical pretreatments such as milling, crushing, sieving, second granulation, and sieving. It is noted that 50% of initial LIBs wastes were recycled in the first treatment than that of second treatment in which only 37% can be successfully recovered (with size <1 mm). Solid/liquid ratios and the concentration of sulfuric acid were changed. There were two samples evaluated, sample No. 1 has 1/10 g mL^{-1} (solid/liquid ratio) and excess of acid (+50% stoichiometric) which resulted in >90% extraction of Co and Li from sample No. 1. In sample No. 2, the same solid/liquid ratio is maintained and +100% acid excess, and it resulted in 96% of Co and 86% of Li [71].

5 Quantitative Analysis of Recycling Various Lithium-Ion Batteries Electrodes

It was reported that China is the major hub for the transportation of Waste Electric and Electronic Equipment (WEEE) illegally, and those cannot be assessed thoroughly since they operate underground. It is roughly estimated that 6,00,000 tons of WEEE were imported in the year 2014, and it consisted of electronic items at large scales. To avoid this illegal transportation and also to collect the spent LiBs from respondents who do not know where to deposit these, proper regulations with labels should be incorporated and spent LiBs must be classified into hazardous wastes [7]. Gu et al. overviewed the present status of numerous methods to recover materials spent LiBs in China, including consumer behavior, recycling processes, and recycling rate. Policies and action plans were also suggested to enhance recycling rates of spent LiBs. China is the main manufacturer and consumer of LiBs, the collection, generation, and recycling of spent LiBs, examined using combined methodologies such as online surveys, field investigations, and estimations. It is noteworthy that respondents have more willingness toward recycling the spent LiBs, but they did not have an awareness of where to transport the spent LiBs. It should be noted that spent LiBs stored at home as responded by 59.6% of the people involved in the survey and 93.8% of them are unknown where they have to dispose of the spent LiBs; it impacts the current recycling industry since the supply of spent LiBs is inadequate. It has been roughly assessed that only 10% of the LiBs used by the consumers are sent to recycling plants, while others end up idle or deposited in scrapyards and landfills.

Wang et al. reported the case study, which helps to enhance end-of-life management for spent LiBs. An optimized model is utilized to study the chemistry of cathode and overall battery waste stream. The profitability of recycling process depends on the chemistry of cathodes because of the fluctuation of potential from $360 per ton for $LiMn_2O_4$ to $8900 per ton for $LiCo_2O_4$. As the current research is focused on finding low-cost cathode materials, the focus toward the recycling of those batteries will weaken. Further, if the fraction of $LiCoO_2$ cathode lies below 21% of the total scrap stream, the recycling process will not be profitable. This study also displays that the rate of the current collection of spent LiBs probably less than 10%. If the rate increases, local recycling of spent LiBs will also increase which reduces the transport cost, and also if the rate or the recycling rates of the valuable metals increase in the future, it will serve as further motivation to expand the recycling process [72]. Table 1 summarizes various methodologies for the recycling, reagents used in recycling method, and the end products from the recycling process.

6 Conclusion

In the context of making the environment clean and pollution-free from electronic wastages, recycling spent batteries could be an estimable step. Application of LiBs in various electronic devices leads to the large-scale accumulation of spent LiBs. So, the proper disposal methods of spent LiBs have become mandatory to sustain a greener environment. This study reviewed various methods to recover active materials from spent LiBs such as chemical treatments, mechanical separation, magnetic separation, hydrometallurgy, and vacuum metallurgy separation. Recycling methods to recover components of LiBs must be greener and cleaner. Chemical treatments involve the use of mineral acids, alkalis, and suitable solvents to recover metal and metal compounds. Mechanical and magnetic separation methods are better than chemical treatments for their cleaner perspectives. Numerous methods have already been reported in past literatures, but novel methods with cost-effective and greener methods to achieve maximum efficiency are yet to be developed. In future, only simple, low cost, greener methods would be

Table 1 Summary of recycling methods and conditions to recover metal from spent LiBs

Methods	Chemicals/reagents	Outcomes	Ref.
Chemical treatment	Trifluoroacetic acid	$LiNi_{1/3}Co_{1/3}Mn_{1/3}O_2$	[73]
Chemical treatment and mechanochemical method	Polyvinylchloride	Cathode materials (C/$LiCoO_2$)	[74]
Mechanical separation and vacuum metallurgy	–	Mn_3O_4, Li_2CO_3 binder, graphite and $LiMn_2O_4$	[75]
Mechanochemical process	Iron powders for reduction	Li, Co, Mn, and Ni	[76]
Metallurgical and mechanical methods	Acids, leaching agents and solvents	Electrodes and electrolyte	[77]
Electrochemical processes	Manganese and $(NH_4)_2SO_4$	Ni and Co	[78]
Mechanical pretreatment and solvent extraction operations	Leaching agents and solvents	Li, Ni, and Co	[79]
Chemical treatment	Citric acid	Co_3O_4/$LiCoO_2$	[80]
Chemical deposition and solvent extraction	Acid and alkali	Co and Cu	[81]
Chemical treatments	Solvents, acids, and alkalis	Co, Li, and Mn compounds	[82]
Acid and thermal treatments	Acids	Single-wall carbon nanotube	[83]
Chemical treatments	Solvents, acids, and alkalis	Electrodes and electrolyte	[84]
Reductive ammonia leaching method	Sodium sulfite	Cathode material	[85]
Acid and thermal treatments	Hydrogen peroxide and HNO_3	Cathode material	[86]
Thermal decomposition and mechanical separation	–	Electrodes and electrolyte	[87]
Calcination, solvent dissolution, and basic solution dissolution	Solvents and alkalis	$Li(Ni_{1/3}Co_{1/3}Mn_{1/3})O_2$	[88]
Crushing methods (wet and dry)	Water	Electrodes and electrolyte	[89]
Vacuum pyrolysis, oxalate leaching, and precipitation	Acid oxalate and acid	$LiCoO_2$, Li, C, and Co	[90]
Chemical and thermal treatments	H_2SO_4 and H_2O_2	Lithium and cobalt	[91]
Hydrometallurgical process	Solvents, carbon, DMG	Li and Li_2CO_3	[92]
Hydrometallurgical process and chemical treatments	Solvents and H_2SO_4	Co and $CoCO_3$	[93]
Chemical treatments	Carbon black powder and sodium carbonate	Co and Li	[94]
Chemical treatments	Iminodiacetic acid (IDA) and maleic acid (MA)	Co and Li	[95]

(continued)

Table 1 (continued)

Methods	Chemicals/reagents	Outcomes	Ref.
Chemical and thermal treatments	H_2SO_4 and H_2O_2	Lithium and cobalt	[91]
Hydrometallurgical process	Solvents, carbon, DMG	Li and Li_2CO_3	[92]
Hydrometallurgical process and chemical treatments	Solvents and H_2SO_4	Co and $CoCO_3$	[93]
Chemical treatments	Carbon black powder and sodium carbonate	Co and Li	[94]
Chemical treatments	Iminodiacetic acid (IDA) and maleic acid (MA)	Co and Li	[95]
Chemical treatments	H_2SO_4 and NaOH	Li, Mn, Co, Ni, and $LiNi_{1/3}Co_{1/3}Mn_{1/3}O_2$	[96]
Chemical treatments	Citric acid	Co and Li	[97]
Hydrometallurgical method	Glycine and ascorbic acid	Cobalt	[98]
Chemical treatments	NH_4OH, H_2SO_4, and H_2O_2	$MnCO_3$, $NiCO_3$, and Li_2CO_3	[99]
Chemical treatments	H_2SO_4	$LiCoO_2$, Co, and carbon	[100]
Vacuum pyrolysis and hydrometallurgical process	2 M H_2SO_4	Li and Co	[101]
Leaching and precipitation method	0.7 M H_3PO_4.	Li and Co	[102]

appreciated to recover components of LiBs. Thus, the focus of recycling the spent batteries is the reuse of extracted materials for developing cost-effective LiBs for future generation and a cleaner environment.

7 Future Perspective

Recycling of LiBs gives a potential profit of nearly \$5013 per ton [102], and this mainly depends on the cathode chemistry, for example, recycling of $LiMn_2O_4$ gives a profit of \$360 per ton, and $LiCo_2O_4$ gives \$8900 per ton [72]. In the year 2014, the global batteries market was worth approximately 62 billion USD [8]. In 2012, the total sales of LiBs were \$11.8 billion in which portable battery contributed to 60% of the total sales [43].

Figure 9 shows the most important reasons for replacements and the perception and attitude among people toward recycling. From this, we can understand that there are more number of batteries that are being replaced due to some of the reason like malfunction, battery degradation as mentioned in Fig. 9(a), (c), (d) shows that the maximum number of people do not have an awareness of recycling, so they are not selling it for recycling. So maximum awareness should be created among people about the recycling of LiBs.

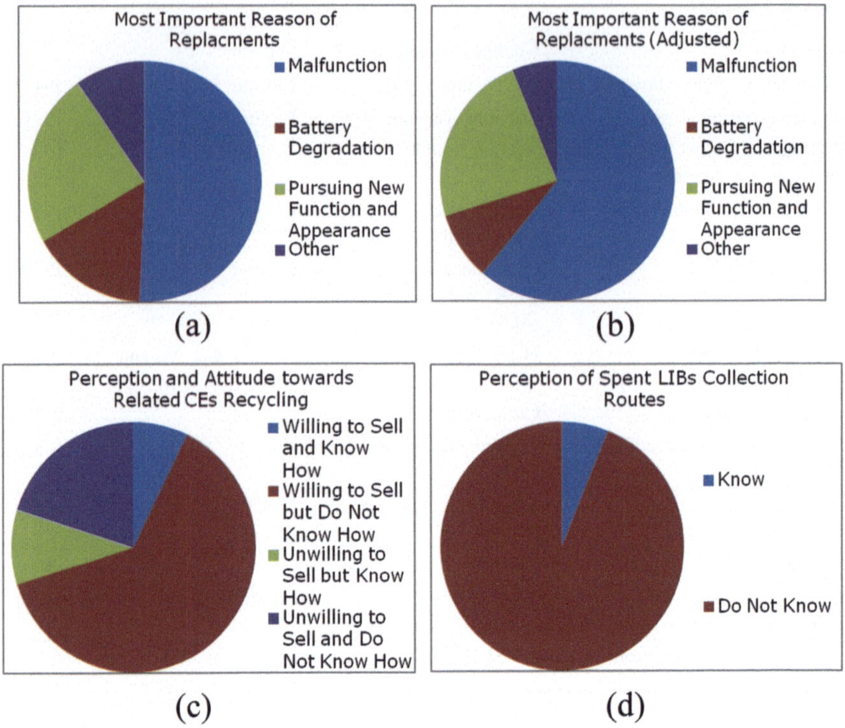

Fig. 9 Need for new batteries and awareness of recycling among people: **a** important reasons for replacement of batteries in consumer electronics and **b** important reasons for replacements after post-stratification, **c** peoples' attitude toward consumer electronics related recycling and **d** awareness of recycling spent LiBs among people. Reproduced with permission from Ref. [7], Copyright 2017, Elsevier

As of now, research is focused on developing cost-effective cathode material for LiBs. If the cost of the cathode materials gets decreased means, then the value of current recycling method will also get decreased because using expensive process and getting very low-cost materials will result in depletion in the recycling industry. For example, the cost of the recovered material is \$100, and the cost of the process is \$120 means the incentive toward recycling will decrease. To resolve this issue, new recycling methods have to be implemented, which should be cost-effective and also environmentally favorable. In situ transformation of waste LiBs should be increased (waste + waste $\rightarrow$ resources) [47], by doing this type of process, we can save many resources used in the process. The cost of some of the precious metals like cobalt is gradually increasing so; the recycling process is highly beneficial for the present and future generations. Also, the choice of cathode chemistry should be considered while choosing a battery to manufacture, for example, lithium magnesium oxide (LMO) battery and lithium ferrous oxide (LFO) battery show similar efficiency but the environmental impacts vary, CO_2 emissions of LFO battery is 26 and 24% of total equivalent emissions from the manufacturing and recycling processes.

In contrast, the CO_2 emission of LMO is 14 and 11% of CO_2 emissions from manufacturing and recycling. Hence, the manufacturer should consider cathode chemistry before manufacturing. The recycling process should not be a closed loop, and we should use the recovered material as a photocatalyst, electrocatalyst, among others; by doing so, this recycling process is never going to get saturated.

References

1. Das HT, Mahendraprabhu K, Maiyalagan T, Elumalai P (2017) Performance of solid-state hybrid energy-storage device using reduced graphene-oxide anchored sol-gel derived Ni/NiO nanocomposite. Sci Rep 7(1):1–14
2. Duraisamy E, Das HT, Sharma AS, Elumalai P (2018) Supercapacitor and photocatalytic performances of hydrothermally-derived Co_3O_4/CoO@ carbon nanocomposite. New J Chem 42(8):6114–6124
3. Pinna EG, Ruiz MC, Ojeda MW, Rodriguez MH (2017) Cathodes of spent Li-ion batteries: dissolution with phosphoric acid and recovery of lithium and cobalt from leach liquors. Hydrometallurgy 167:66–71
4. Rajkumar S, Elanthamilan E, Balaji TE, Sathiyan A, Jafneel NE, Merlin JP (2020) Recovery of copper oxide nanoparticles from waste SIM cards for supercapacitor electrode material. J Alloy Compd 156582
5. Ali GAM, Lih Teo EY, Aboelazm EAA, Sadegh H, Memar AOH, Shahryari-Ghoshekandi R, Chong KF (2017) Capacitive performance of cysteamine functionalized carbon nanotubes. Mater Chem Phys 197:100–104
6. Aboelazm EAA, Ali GAM, Algarni H, Yin H, Zhong YL, Chong KF (2018) Magnetic electrodeposition of the hierarchical cobalt oxide nanostructure from spent lithium-ion batteries: its application as a supercapacitor electrode. J Phy Chem C 122(23):12200–12206
7. Gu F, Guo J, Yao X, Summers PA, Widijatmoko SD, Hall P (2017) An investigation of the current status of recycling spent lithium-ion batteries from consumer electronics in China. J Clean Prod 161:765–780
8. Rarotra S, Sahu S, Kumar P, Kim KH, Tsang YF, Kumar V, Kumar P, Srinivasan M, Veksha A, Lisak G (2020) Progress and challenges on battery waste management: a critical review. ChemistrySelect 5(20):6182–6193
9. Li L, Bian Y, Zhang X, Guan Y, Fan E, Wu F, Chen R (2018) Process for recycling mixed-cathode materials from spent lithium-ion batteries and kinetics of leaching. Waste Manag 71:362–371
10. Ganter MJ, Landi BJ, Babbitt CW, Anctil A, Gaustad G (2014) Cathode refunctionalization as a lithium ion battery recycling alternative. J Power Sources 256:274–280
11. Lee CK, Rhee K-I (2002) Preparation of LiCoO2 from spent lithium-ion batteries. J Power Sources 109(1):17–21
12. Ali GAM, Tan LL, Jose R, Yusoff MM, Chong KF (2014) Electrochemical performance studies of MnO_2 nanoflowers recovered from spent battery. Mater Res Bull 60:5–9
13. Ali GAM, Yusoff MM, Shaaban ER, Chong KF (2017) High performance MnO_2 nanoflower supercapacitor electrode by electrochemical recycling of spent batteries. Ceram Int 43:8440–8448
14. Aboelazm EAA, Ali GAM, Chong KF (2018) Cobalt oxide supercapacitor electrode recovered from spent lithium-ion battery. Chem Adv Mater 3:67–74

15. Ali GAM (2020) Recycled MnO_2 nanoflowers and graphene nanosheets for low-cost and high performance asymmetric supercapacitor. J Electron Mater 49(9):5411–5421
16. Ordoñez J, Gago EJ, Girard A (2016) Processes and technologies for the recycling and recovery of spent lithium-ion batteries. Renew Sustain Energy Rev 60:195–205
17. Tian X, Gong Y, Wu Y, Agyeiwaa A, Zuo T (2014) Management of used lead acid battery in China: secondary lead industry progress, policies and problems. Resour Conserv Recycl 93:75–84
18. Das HT, Saravanya S, Elumalai P (2018) Disposed dry cells as sustainable source for generation of few layers of graphene and manganese oxide for solid-state symmetric and asymmetric supercapacitor applications. ChemistrySelect 3(46):13275–13283
19. Ali GAM, Yusoff MM, Algarni H, and Chong KF (2018) One-step electrosynthesis of MnO_2/rGO nanocomposite and its enhanced electrochemical performance. Ceramics Int
20. Zeng X, Li J, Singh N (2014) Recycling of spent lithium-ion battery: a critical review. Crit Rev Environ Sci Technol 44(10):1129–1165
21. Zubi G, Dufo-López R, Carvalho M, Pasaoglu G (2018) The lithium-ion battery: state of the art and future perspectives. Renew Sustain Energy Rev 89:292–308
22. Winslow KM, Laux SJ, Townsend TG (2018) A review on the growing concern and potential management strategies of waste lithium-ion batteries. Resour Conserv Recycl 129:263–277
23. Richa K, Babbitt CW, Gaustad G (2017) Eco-efficiency analysis of a lithium-ion battery waste hierarchy inspired by circular economy. J Ind Ecol 21(3):715–730
24. Godt J, Scheidig F, Grosse-Siestrup C, Esche V, Brandenburg P, Reich A, Groneberg DA (2006) The toxicity of cadmium and resulting hazards for human health. J Occup Med Toxicol 1(1):22
25. Ku H, Jung Y, Jo M, Park S, Kim S, Yang D, Rhee K, An E-M, Sohn J, Kwon K (2016) Recycling of spent lithium-ion battery cathode materials by ammoniacal leaching. J Hazard Mater 313:138–146
26. Mishra A, Mehta A, Basu S, Malode SJ, Shetti NP, Shukla SS, Nadagouda MN, Aminabhavi TM (2018) Electrode materials for lithium-ion batteries. Mater Sci Energy Technol 1(2):182–187
27. Li X, Liu J, Meng X, Tang Y, Banis MN, Yang J, Hu Y, Li R, Cai M, Sun X (2014) Significant impact on cathode performance of lithium-ion batteries by precisely controlled metal oxide nanocoatings via atomic layer deposition. J Power Sources 247:57–69
28. Wiedemann AH, Goldin GM, Barnett SA, Zhu H, Kee RJ (2013) Effects of three-dimensional cathode microstructure on the performance of lithium-ion battery cathodes. Electrochim Acta 88:580–588
29. Uddin J, Bryantsev VS, Giordani V, Walker W, Chase GV, Addison D (2013) Lithium nitrate as regenerable SEI stabilizing agent for rechargeable Li/O_2 batteries. J Phys Chem Lett 4(21):3760–3765
30. Nagasubramanian A, Yu DYW, Hoster H, Srinivasan M (2014) Enhanced cycling stability of o-$LiMnO_2$ cathode modified by lithium boron oxide coating for lithium-ion batteries. J Solid State Electrochem 18(7):1915–1922
31. Hua AC-C, Syue BZ-W (2010) Charge and discharge characteristics of lead-acid battery and LiFePO4 battery. In: The 2010 international power electronics conference-ECCE ASIA. IEEE
32. Prasad KH, Vinoth S, Ratnakar A, Venkateswarlu M, Satyanarayana N (2016) Structural and electrical conductivity studies of spinel $LiMn_2O_4$ cathode films grown by RF sputtering. Mater Today Proc 3(10):4046–4051
33. Prasad KH, Arumugama R, Srinadhu E, Satyanarayana N. AC conductivity and electric modulus studies of $LiNi_{1/3}Mn_{1/3}Co_{1/3}O_2$ cathode thin films grown by pulsed laser deposition technique
34. Bankole OE, Gong C, Lei L (2013) Battery recycling technologies: recycling wastelithium ion batteries with the impact on the environment in-view. J Environ Ecol 4:14–28

35. Lewandowski A, Świderska-Mocek A (2009) Ionic liquids as electrolytes for Li-ion batteries —an overview of electrochemical studies. J Power Sources 194(2):601–609
36. Kato Y, Hori S, Saito T, Suzuki K, Hirayama M, Mitsui A, Yonemura M, Iba H, Kanno R (2016) High-power all-solid-state batteries using sulfide superionic conductors. Nat Energy 1 (4):1–7
37. Fergus JW (2010) Recent developments in cathode materials for lithium ion batteries. J Power Sources 195(4):939–954
38. Gonçalves MCA, Garcia EM, Taroco HA, Gorgulho HF, Melo JO, Silva RR, Souza AG (2015) Chemical recycling of cell phone Li-ion batteries: application in environmental remediation. Waste Manag 40:144–150
39. Richa K, Babbitt CW, Gaustad G, Wang X (2014) A future perspective on lithium-ion battery waste flows from electric vehicles. Resour Conserv Recycl 83:63–76
40. Huang B, Pan Z, Su X, An L (2018) Recycling of lithium-ion batteries: recent advances and perspectives. J Power Sources 399:274–286
41. Scansetti G, Lamon S, Talarico S, Botta G, Spinelli P, Sulotto F, Fantoni F (1985) Urinary cobalt as a measure of exposure in the hard metal industry. Int Arch Occup Environ Health 57(1):19–26
42. Ahmad S, Khan MH, Khandker S, Sarwar A, Yasmin N, Faruquee M, Yasmin R (2014) Blood lead levels and health problems of lead acid battery workers in Bangladesh. Sci World J
43. Gratz E, Sa Q, Apelian D, Wang Y (2014) A closed loop process for recycling spent lithium ion batteries. J Power Sources 262:255–262
44. Castillo S, Ansart F, Laberty-Robert C, Portal J (2002) Advances in the recovering of spent lithium battery compounds. J Power Sources 112(1):247–254
45. Zeng X, Li J, Shen B (2015) Novel approach to recover cobalt and lithium from spent lithium-ion battery using oxalic acid. J Hazard Mater 295:112–118
46. Boyden A, Soo VK, Doolan M (2016) The environmental impacts of recycling portable lithium-ion batteries. Procedia Cirp 48:188–193
47. Li J, Wang G, Xu Z (2016) Environmentally-friendly oxygen-free roasting/wet magnetic separation technology for in situ recycling cobalt, lithium carbonate and graphite from spent $LiCoO_2$/graphite lithium batteries. J Hazard Mater 302:97–104
48. Barbieri E, Lima E, Cantarino S, Lelis M, Freitas M (2014) Recycling of spent ion-lithium batteries as cobalt hydroxide, and cobalt oxide films formed under a conductive glass substrate, and their electrochemical properties. J Power Sources 269:158–163
49. Dewulf J, Van der Vorst G, Denturck K, Van Langenhove H, Ghyoot W, Tytgat J, Vandeputte K (2010) Recycling rechargeable lithium ion batteries: critical analysis of natural resource savings. Resour Conserv Recycl 54(4):229–234
50. Li L, Ge J, Chen R, Wu F, Chen S, Zhang X (2010) Environmental friendly leaching reagent for cobalt and lithium recovery from spent lithium-ion batteries. Waste Manag 30(12):2615–2621
51. Freitas M, Garcia E (2007) Electrochemical recycling of cobalt from cathodes of spent lithium-ion batteries. J Power Sources 171(2):953–959
52. Garcia EM, Tar HA, Matencio T, Domingues RZ, dos Santos JA, Ferreira RV, Lorenšon E, Lima DQ, de Freitas MB (2012) Electrochemical recycling of cobalt from spent cathodes of lithium-ion batteries: its application as supercapacitor. J Appl Electrochem 42(6):361–366
53. Rothermel S, Evertz M, Kasnatscheew J, Qi X, Grützke M, Winter M, Nowak S (2016) Graphite recycling from spent lithium-ion batteries. Chemsuschem 9(24):3473–3484
54. He X, Peng X, Zhu Y, Lai C, Ducati C, Kumar RV (2015) Producing hierarchical porous carbon monoliths from hydrometallurgical recycling of spent lead acid battery for application in lithium ion batteries. Green Chem 17(9):4637–4646
55. Chagnes A, Pospiech B (2013) A brief review on hydrometallurgical technologies for recycling spent lithium-ion batteries. J Chem Technol Biotechnol 88(7):1191–1199
56. Hannan MA, Hoque MM, Hussain A, Yusof Y, Ker PJ (2018) State-of-the-art and energy management system of lithium-ion batteries in electric vehicle applications: issues and recommendations. IEEE Access 6:19362–19378

57. Li L, Chen R, Sun F, Wu F, Liu J (2011) Preparation of $LiCoO_2$ films from spent lithium-ion batteries by a combined recycling process. Hydrometallurgy 108(3–4):220–225
58. Nie H, Xu L, Song D, Song J, Shi X, Wang X, Zhang L, Yuan Z (2015) $LiCoO_2$: recycling from spent batteries and regeneration with solid state synthesis. Green Chem 17(2):1276–1280
59. Li L, Bian Y, Zhang X, Xue Q, Fan E, Wu F, Chen R (2018) Economical recycling process for spent lithium-ion batteries and macro-and micro-scale mechanistic study. J Power Sources 377:70–79
60. Gaines L, Sullivan J, Burnham A, Belharouak I (2011) Life-cycle analysis of production and recycling of lithium ion batteries. Transp Res Rec 2252(1):57–65
61. Huang K, Li J, Xu Z (2010) Characterization and recycling of cadmium from waste nickel–cadmium batteries. Waste Manag 30(11):2292–2298
62. Vassura I, Morselli L, Bernardi E, Passarini F (2009) Chemical characterisation of spent rechargeable batteries. Waste Manag 29(8):2332–2335
63. Freitas M, De Pietre M (2004) Electrochemical recycling of the zinc from spent $Zn–MnO_2$ batteries. J Power Sources 128(2):343–349
64. Kim S, Kim J, Kim S, Lee J, Yoon J (2018) Electrochemical lithium recovery and organic pollutant removal from industrial wastewater of a battery recycling plant. Environ Sci Water Res Technol 4(2):175–182
65. Fouad O, Farghaly F, Bahgat M (2007) A novel approach for synthesis of nanocrystalline γ-$LiAlO_2$ from spent lithium-ion batteries. J Anal Appl Pyrol 78(1):65–69
66. Chen J, Li Q, Song J, Song D, Zhang L, Shi X (2016) Environmentally friendly recycling and effective repairing of cathode powders from spent $LiFePO_4$ batteries. Green Chem 18 (8):2500–2506
67. Renault S, Brandell D, Edström K (2014) Environmentally-friendly lithium recycling from a spent organic Li-ion battery. Chemsuschem 7(10):2859–2867
68. Ciez RE, Whitacre J (2019) Examining different recycling processes for lithium-ion batteries. Nat Sustain 2(2):148–156
69. Poyraz AS, Huang J, Cheng S, Bock DC, Wu L, Zhu Y, Marschilok AC, Takeuchi KJ, Takeuchi ES (2016) Effective recycling of manganese oxide cathodes for lithium based batteries. Green Chem 18(11):3414–3421
70. Heelan J, Gratz E, Zheng Z, Wang Q, Chen M, Apelian D, Wang Y (2016) Current and prospective Li-ion battery recycling and recovery processes. JOM 68(10):2632–2638
71. Pagnanelli F, Moscardini E, Altimari P, Atia TA, Toro L (2017) Leaching of electrodic powders from lithium ion batteries: optimization of operating conditions and effect of physical pretreatment for waste fraction retrieval. Waste Manag 60:706–715
72. Wang X, Gaustad G, Babbitt CW, Richa K (2014) Economies of scale for future lithium-ion battery recycling infrastructure. Resour Conserv Recycl 83:53–62
73. Zhang X, Xie Y, Cao H, Nawaz F, Zhang Y (2014) A novel process for recycling and resynthesizing $LiNi_{1/3}Co_{1/3}Mn_{1/3}O_2$ from the cathode scraps intended for lithium-ion batteries. Waste Manag 34(9):1715–1724
74. Wang MM, Zhang CC, Zhang FS (2017) Recycling of spent lithium-ion battery with polyvinyl chloride by mechanochemical process. Waste Manag 67:232–239
75. Xiao J, Li J, Xu Z (2017) Recycling metals from lithium ion battery by mechanical separation and vacuum metallurgy. J Hazard Mater 338:124–131
76. Guan J, Li Y, Guo Y, Su R, Gao G, Song H, Yuan H, Liang B, Guo Z (2017) Mechanochemical process enhanced cobalt and lithium recycling from wasted lithium-ion batteries. ACS Sustain Chem Eng 5(1):1026–1032
77. Yun L, Linh D, Shui L, Peng X, Garg A, Le MLP, Asghari S, Sandoval J (2018) Metallurgical and mechanical methods for recycling of lithium-ion battery pack for electric vehicles. Resour Conserv Recycl 136:198–208
78. Lupi C, Pasquali M, Dell'Era A (2005) Nickel and cobalt recycling from lithium-ion batteries by electrochemical processes. Waste Manag 25(2):215–220

79. Granata G, Pagnanelli F, Moscardini E, Takacova Z, Havlik T, Toro L (2012) Simultaneous recycling of nickel metal hydride, lithium ion and primary lithium batteries: accomplishment of European guidelines by optimizing mechanical pre-treatment and solvent extraction operations. J Power Sources 212:205–211
80. Santana I, Moreira T, Lelis M, Freitas M (2017) Photocatalytic properties of Co_3O_4/$LiCoO_2$ recycled from spent lithium-ion batteries using citric acid as leaching agent. Mater Chem Phys 190:38–44
81. Nan J, Han D, Zuo X (2005) Recovery of metal values from spent lithium-ion batteries with chemical deposition and solvent extraction. J Power Sources 152:278–284
82. Paulino JF, Busnardo NG, Afonso JC (2008) Recovery of valuable elements from spent Li-batteries. J Hazard Mater 150(3):843–849
83. Schauerman CM, Ganter MJ, Gaustad G, Babbitt CW, Raffaelle RP, Landi BJ (2012) Recycling single-wall carbon nanotube anodes from lithium ion batteries. J Mater Chem 22 (24):12008–12015
84. Lain MJ (2001) Recycling of lithium ion cells and batteries. J Power Sources 97:736–738
85. Zheng X, Gao W, Zhang X, He M, Lin X, Cao H, Zhang Y, Sun Z (2017) Spent lithium-ion battery recycling–reductive ammonia leaching of metals from cathode scrap by sodium sulphite. Waste Manag 60:680–688
86. Lee CK, Rhee K-I (2003) Reductive leaching of cathodic active materials from lithium ion battery wastes. Hydrometallurgy 68(1–3):5–10
87. Hanisch C, Loellhoeffel T, Diekmann J, Markley KJ, Haselrieder W, Kwade A (2015) Recycling of lithium-ion batteries: a novel method to separate coating and foil of electrodes. J Clean Prod 108:301–311
88. Zhang X, Xue Q, Li L, Fan E, Wu F, Chen R (2016) Sustainable recycling and regeneration of cathode scraps from industrial production of lithium-ion batteries. ACS Sustain Chem Eng 4(12):7041–7049
89. Zhang T, He Y, Ge L, Fu R, Zhang X, Huang Y (2013) Characteristics of wet and dry crushing methods in the recycling process of spent lithium-ion batteries. J Power Sources 240:766–771
90. Sun L, Qiu K (2012) Organic oxalate as leachant and precipitant for the recovery of valuable metals from spent lithium-ion batteries. Waste Manag 32(8):1575–1582
91. Jha MK, Kumari A, Jha AK, Kumar V, Hait J, Pandey BD (2013) Recovery of lithium and cobalt from waste lithium ion batteries of mobile phone. Waste Manag 33(9):1890–1897
92. Chen X, Chen Y, Zhou T, Liu D, Hu H, Fan S (2015) Hydrometallurgical recovery of metal values from sulfuric acid leaching liquor of spent lithium-ion batteries. Waste Manag 38:349–356
93. Pagnanelli F, Moscardini E, Altimari P, Atia TA, Toro L (2016) Cobalt products from real waste fractions of end of life lithium ion batteries. Waste Manag 51:214–221
94. Barik S, Prabaharan G, Kumar B (2016) An innovative approach to recover the metal values from spent lithium-ion batteries. Waste Manag 51:222–226
95. Nayaka GP, Pai KV, Manjanna J, Keny SJ (2016) Use of mild organic acid reagents to recover the Co and Li from spent Li-ion batteries. Waste Manag 51:234–238
96. Yang Y, Xu S, He Y (2017) Lithium recycling and cathode material regeneration from acid leach liquor of spent lithium-ion battery via facile co-extraction and co-precipitation processes. Waste Manag 64:219–227
97. Li L, Ge J, Wu F, Chen R, Chen S, Wu B (2010) Recovery of cobalt and lithium from spent lithium ion batteries using organic citric acid as leachant. J Hazard Mater 176(1–3):288–293
98. Nayaka G, Pai K, Santhosh G, Manjanna J (2016) Recovery of cobalt as cobalt oxalate from spent lithium ion batteries by using glycine as leaching agent. J Environ Chem Eng 4 (2):2378–2383
99. Nayl A, Elkhashab R, Badawy SM, El-Khateeb M (2017) Acid leaching of mixed spent Li-ion batteries. Arabian J Chem 10:S3632–S3639
100. Wang J, Chen M, Chen H, Luo T, Xu Z (2012) Leaching study of spent Li-ion batteries. Procedia Environ Sci 16:443–450

101. Sun L, Qiu K (2011) Vacuum pyrolysis and hydrometallurgical process for the recovery of valuable metals from spent lithium-ion batteries. J Hazard Mater 194:378–384
102. Chen X, Ma H, Luo C, Zhou T (2017) Recovery of valuable metals from waste cathode materials of spent lithium-ion batteries using mild phosphoric acid. J Hazard Mater 326: 77–86

Recycled Nanomaterials for Energy Storage (Supercapacitor) Applications

Gomaa A. M. Ali, Zinab H. Bakr, Vahid Safarifard, and Kwok Feng Chong

Abstract Nowadays, humankind is in urgent need of energy generation and storage systems. The supercapacitor is one of the essential types of storage systems. The high cost of obtaining capacitor electrodes is the reason behind the researchers' attempts to find low-cost sources. The need for the development of efficient energy storage systems is paramount in meeting the world's future energy targets, especially when the energy costs are on the increase in addition to the escalating demand. Energy storage technologies can improve efficiencies in supply systems by storing the energy when it is in excess, and then release it timely. Batteries are slowly becoming obsolete due to their poor cyclability (limited to a few thousand) and long charge time (tens of minutes) in comparison to supercapacitors. On the other hand, supercapacitors have a long lifetime and fast charging times. Nowadays, the research focuses on advanced suitable electrode materials that directly reflect in supercapacitor technology enhancement. The researchers have prepared a variety of single components and hybrid electrodes by recycling various environmental wastes. The recycled materials include metal oxides (MnO_2, Co_3O_4, etc.), carbon materials (carbon nanosphere, porous carbon nanoparticles, activated carbon), and hybrid materials (MnO_2/graphene, CaO/AC). The obtained materials exhibited interesting structural and morphological properties as well as excellent energy storage behavior. The recycling technique provides a unique alternative cheap way

G. A. M. Ali (✉)
Chemistry Department, Faculty of Science, Al–Azhar University, Assiut 71524, Egypt
e-mail: gomaasanad@azhar.edu.eg; gomaasanad@gmail.com

Z. H. Bakr
Physics Department, Faculty of Science, Assiut University, Assiut 71516, Egypt

V. Safarifard
Department of Chemistry, Iran University of Science and Technology, 16846-13114 Tehran, Iran

K. F. Chong (✉)
Faculty of Industrial Sciences and Technology, Universiti Malaysia Pahang, 26300 Kuantan, Malaysia
e-mail: ckfeng@ump.edu.my

A. S. H. Makhlouf and G. A. M. Ali (eds.), *Waste Recycling Technologies for Nanomaterials Manufacturing*, Topics in Mining, Metallurgy and Materials Engineering, https://doi.org/10.1007/978-3-030-68031-2_7

for getting supercapacitor electrode materials, as well as it helps to maintain a clean environment.

Keywords Energy storage · Supercapacitors · Electrochemical capacitors · EDLC · Pseudocapacitors · Recycling · Environmental waste

List of Abbreviations

2ES	Two-electrode system
3ES	Three-electrode system
AC	Activated carbon
ACPKS	Activated carbon palm kernel shell
ASSCs	Asymmetric supercapacitors
C	Capacitance
CNTs	Carbon nanotubes
C_s	Specific capacitance
CVD	Chemical vapor deposition
ECs	Electrochemical capacitors
EDL	Electrochemical double-layer
EDLCs	Electrochemical double-layer capacitors
EMIMBF4	1-ethyl-3-methylimidazolium tetrafluoroborate
HCs	Hybrid capacitors
LiBs	Lithium-ion batteries
MWCNTs	Multi-walled carbon nanotubes
PCs	Pseudocapacitors
PKS	Palm kernel shell
SWCNTs	Single-walled carbon nanotubes
SWCNHs	Single-walled carbon nanohorns
SSCs	Symmetric supercapacitors
TEA–BF4	N,N,N,N-tetraethylammonium tetrafluoroborate
USEPA	United States Environmental Protection Agency
Zn–C	Zinc–carbon

1 Introduction

The capacitor is a two-terminal electrical component used to store electrical energy. The practical capacitors have many forms, but each one contains at least two electrical conductors separated by a dielectric. The capacitor can be used like a temporary battery, or like other types of the rechargeable energy storage system, where it can store electric energy when disconnected from its charging circuit. The capacitance (C) of an object is a measure of the number of charges that an object can hold without discharging. The capacitance depends on the separation width between conductors and the conductor's surface area [1]. The capacitance in Farad (F) can be simply calculated as C = Q/V where Q is the charge in coulomb and V is the voltage in volt.

Electrochemical capacitors (ECs) are also known as ultracapacitors or supercapacitors. It can be classified into two main categories based on energy storage mechanism, pseudocapacitors (PCs), and electrochemical double-layer capacitors (EDLCs). In addition, ECs are classified into three groups based on the type of electrode materials, EDLCs, PCs, and hybrid capacitors (HCs). In where there are some differences between supercapacitors and conventional physical capacitors such as supercapacitors; the polarization/depolarization takes place at the electrode surface, and the insulating barrier is replaced by electrolyte [2]. Consequently, the supercapacitors can store more charges and have very high capacitance value than a conventional physical capacitor. PCs store electrical energy Faradically by electron charge transfer between electrode and electrolyte. Metal oxides and conducting polymers are used as electrode materials for PCs. In EDLCs, a double layer of electrolyte ions is formed on the surface of an electrode material, which arises from the potential-dependence of the surface density of charges stored electrostatically. The electrode materials for EDLCs include all carbon-based materials. Supercapacitors could be used in many applications because of their higher energy output as compared to conventional capacitors and higher power than batteries, in addition to their miniature size. Various types of electrode materials can be used in supercapacitors, including carbon-based materials, conducting polymers, and metal oxides. In addition, the electrolyte could be an aqueous, organic, or an ionic liquid. In the case of an aqueous electrolyte, the operating voltage is limited to 1 V (due to the electrochemical decomposition of water at 1.23 V). In contrast, an organic electrolyte can achieve a voltage range of 2.5–3.5 V [3]. A higher voltage of up to 4.0 V can be obtained for the ionic liquid.

2 Supercapacitors, Batteries, and Fuel Cells

There are three types of electrochemical systems: supercapacitors, batteries, and fuel cells, and there are some electrochemical similarities and differences between these three systems. The three systems consist of two electrodes separated from each other and in contact with the electrolyte. The main difference is that ECs store the charge physically with no chemical reaction or phase change, so the charge–discharge could typically be cycled many times (millions of times) compared to batteries (few thousand at best) [1, 4]. In batteries and fuel cells, the chemical energy is converted into electrical energy via redox reaction of the electrode materials [5]; therefore, there is a volume changed, leading to a limitation of cyclability. Moreover, there are some other limitations in the case of batteries, such as reaction kinetics and mass transport, limiting their power performance. These limitations do not occur in supercapacitors. The state of charge for batteries cannot precisely be determined due to the thermodynamics of batteries reacted species giving constant operating voltage.

In contrast, it is easily identified in the case of ECs due to the linearity of the voltage change with time. Another critical difference between ECs and batteries is the time of response. In ECs, the formation of an electrochemical double layer (EDL) takes place very fast within 10^{-8} s, while the chemical reaction in batteries is much slower and needs more time (10^{-2}–10^{-4} s) [5].

The energy density is the Faradic charge withdrawable from the system at a potential (V) per unit of mass of active material [6]. The characterization of the system by its energy and power densities is important for the evaluation of the electrochemical performance. ECs have a high energy density compared to conventional capacitors and higher power ratings than batteries [7]. Therefore, supercapacitors have replaced batteries in such applications. Ragone plot, which shows the relation between energy and power densities for various systems [4–6], is shown in Fig. 1.

3 Supercapacitors Applications

Supercapacitors give access to new power electronic and industrial storage applications due to their higher energy density and a long lifetime, as well as their ability to turn on instantaneously, charge quickly, and require less complicated charging circuits. Supercapacitors could be applied in many fields such as transportation, including buses, hybrid electric vehicles, as well as metro trains and tramways [4]. Supercapacitors' reliability now sees them across batteries in many electric vehicles or completely replacing batteries in the Toyota racing car. Moreover, they could be used in emergency doors in Airbus A380 planes [4]. The automobile industry requires ECs with an energy density of 16 W h kg^{-1} [6]. In addition, they could be used in toys, electronics, PC–board, military, and medical applications. ECs could

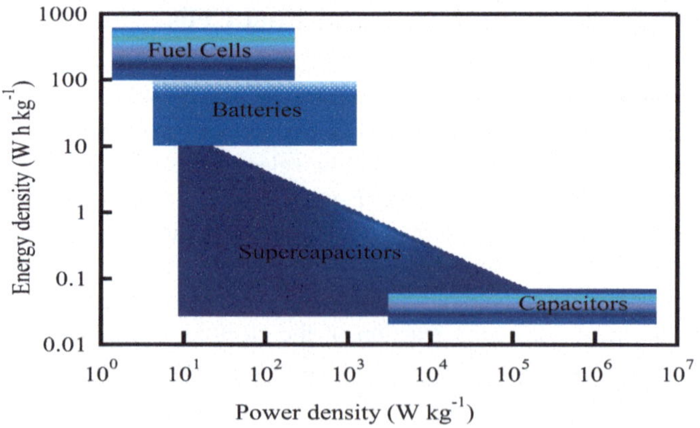

Fig. 1 Ragone plot of various energy storage systems. Adapted with permission from Ref. [5], Copyright 2004, American Chemical Society

be used as standby power for random access memory devices or telephone equipment [6]. The high power densities and very long lifetimes are strongly supported using supercapacitors in smartphone technologies.

Particularly in medicine and health, patients with cardiac pacemakers and other implantable electronic medical devices end up requiring reimplantations when the batteries run out. While we have seen the slow but steady progress of battery technology, there is a great deal of opportunity for harvesting electrical energy from the body and using it to power implants. To this end, a "biological supercapacitor" is capable of storing electrical energy inside the body using safe, non-toxic components, including the body's fluids [8]. Consisting of sheets of graphene, a material made of a flat sheet of carbon atoms, layered with a protein produced by the human body that acts as the electrode, a single capacitor is only a micrometer thick. It can be connected in bunches to provide more electricity storage.

4 Energy Storage Mechanisms

Two types of energy storage mechanisms have been reported. The first is the EDLCs in which the energy is stored and released by nanoscopic charge separation at the electrochemical interface between the electrode and the electrolyte [9, 10]. Electrodouble layer materials include all carbon-based materials such as: graphene, carbon nanotubes (CNTs), carbon nanoparticles, and other carbon-based materials [10–12], and the second is pseudocapacitive in which the materials store the energy based on redox reaction of the electrode material. These materials include metal oxides [4, 13–15] and conducting polymers [16].

In EDLCs, a double layer of electrolyte ions is formed on the surface of an electrode material, which arises from the potential-dependence of the surface density of charges stored electrostatically. Charge separation takes place on either side of the interface leading to the formation of an electrodouble layer. No charge transfer from chemical reactions in the electrode takes place. The charge distribution on the surface of the electrode depends on the porosity and the crystal structure of the electrode material. The electrodouble layer materials should have a high surface area for high charge accumulation and a suitable pore structure to allow a rapid motion of the electrolyte ions [7, 9, 17, 18]. Helmholtz, Gouy-Chapman, and Stern [17] models have illustrated the electrodouble layer formation. Helmholtz model considered the concept of the charge separation at the interface between a metallic electrode and an electrolyte solution. The charge on the electrode is balanced by redistribution of the ions in the solution by an equal but oppositely charged amount of it. Gouy and Chapman's model takes into account the thermal motion of ions near a charged surface. In contrast, Stern model combined the two previous models by adapting the compact layer of ions used by Helmholtz and next to the diffuse layer of the Gouy-Chapman model extending into the bulk solution. He took into account the fact that ions have a finite size. A simple representation of the electrodouble layer in charging and discharging states is shown in Fig. 2.

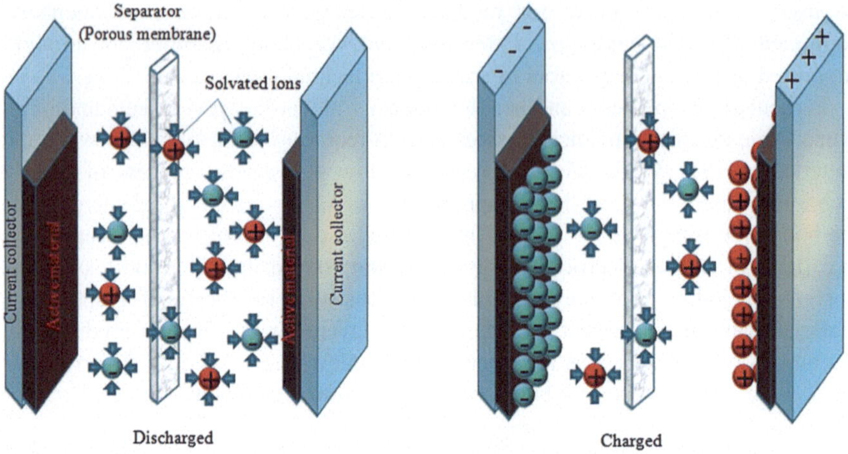

Fig. 2 Schematic diagrams of EDLCs in discharged (left) and charged (right) states. Adapted with permission from Ref. [19], Copyright 2013, Elsevier

The electrodouble layer characteristics depend on the electrode surface, the composition of the electrolyte, and the potential field between the charges at the interface [5, 19].

On the other hand, PCs store electrical energy Faradically by electron charge transfer between electrode and electrolyte. This is accomplished through electrosorption, oxidation-reduction reactions (redox), and intercalation processes [2, 15]. Pseudocapacitive might be caused by electrosorption of H^+ or metal ions and redox reactions of electroactive species. Therefore, it involves the passage of charge across the double layer, as in battery charging or discharging [6]. Pseudocapacitive arises from the charge transfer between oxidized and reduced species. PCs give higher C_s than EDLCs, but they suffer from the lack of stability due to the redox reaction occurring [5]. Three processes give rise to pseudocapacitive: redox reactions, electrochemical intercalation, and under-potential deposition [6]. In detail, the electrochemical storage of the electrical energy is achieved by redox reactions with the adsorbed electrolyte ions, intercalation of atoms in the layer lattice or electrosorption, and underpotential deposition of hydrogen or metal atoms in surface lattice sites with all results in a reversible Faradic charge transfer.

Usually, EDLCs exhibit around 1–5% of their capacitance as pseudocapacitance due to the Faradic reactivity of surface oxygen functionalities and PCs exhibit some EDLCs component (5–10%), which is proportional to their electrochemically accessible interfacial areas [6]. Cyclic voltammograms can clearly show the differences between EDLCs and PCs where the cyclic voltammogram for EDLCs is rectangular. Once the potential sweep is reversed, the sign of current is also immediately reversed. On the other hand, the PCs show a deviation from the rectangular shape combined with redox peaks because of the redox reaction of the electrode material. A parallelogram shape could be demonstrated for double-layer

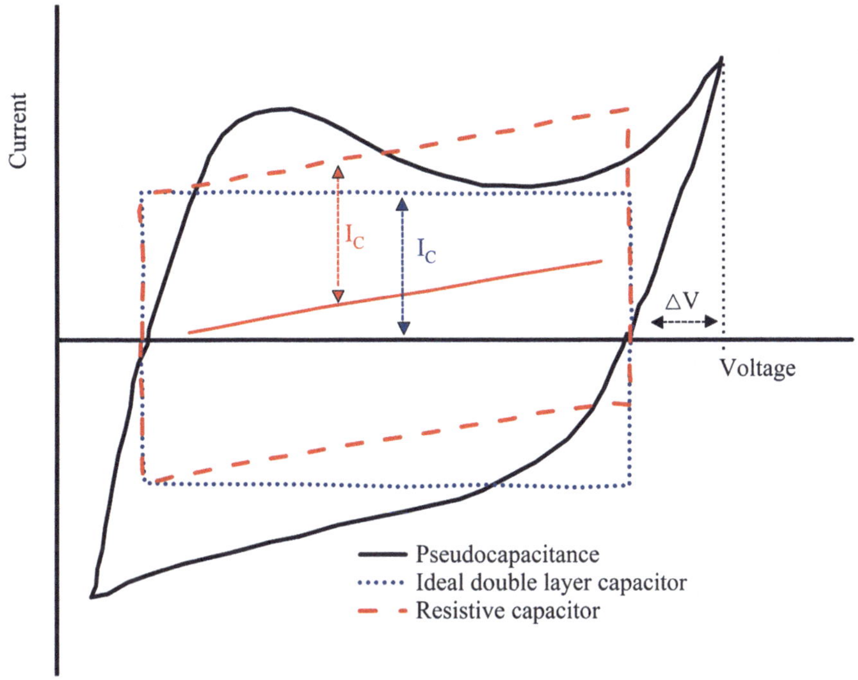

Fig. 3 Cyclic voltammograms of electrochemical capacitors. Adapted with permission from Ref. [20], Copyright 2001, Elsevier

capacitors with resistive losses [20]. A simple illustration of these cyclic voltammogram shapes is shown in Fig. 3.

5 Supercapacitors Components

5.1 Electrode Materials

The kinetics of ion and electron transport in the electrodes and at the electrode/electrolyte interface play an important role in the electrochemical performance of supercapacitors. Therefore, electrodes with proper pore structure and good electrical conductivity are highly desirable [5].

5.1.1 Metal Oxides

Metal oxide electrodes are excellent candidates for electrical storage devices due to their high capacitance value. Hydrous ruthenium oxide ($RuO_2.xH_2O$) is a typical example of metal oxide giving pseudocapacitive properties. The electrochemical reactions of RuO_2 involve the injection of protons and electrons into the surface of hydrous RuO_2 as well as conversion between Ru(IV) and Ru(III) species according to the following reactions (Eqs. (1–2)) [21]:

$$RuO_x(OH)_y + \delta H^+ + \delta e^- \leftrightarrows RuO_{x-u}(OH)_{y+\delta} \quad (1)$$

$$Ru(IV) - O^{-2} + H^+ + e^- \leftrightarrows Ru(III) - OH^- \quad (2)$$

The theoretical pseudocapacitance (C_{theo}) of the metal oxides can be calculated using Eq. (3) [22]:

$$C_{theo} = \frac{nF}{M\Delta E} \quad (3)$$

where n is the number of the electrons transferred in the redox reaction, F is the Faraday constant (F = 96,500 C), M and ΔE are the molar mass and the redox potential of the material, respectively. The theoretical C_s of $RuO_2.0.5H_2O$ was estimated to ca. 2000 F g^{-1} [22]. Hu obtained 1580 F g^{-1} for a thin film of 10% gold AC–RuO_2 derived from sol–gel and deposited on stainless steel substrate, which is around 80% of its theoretical capacitance [22]. It was found that the C_s depends on the crystallinity of RuO_2, where the amorphous RuO_2 showed 720 F g^{-1}. In comparison, the crystalline RuO_2 explained only 530 F g^{-1} [23], where the amorphous structure allows the proton to diffuse faster inside the oxide particle as compared to the crystalline one [24].

Also, many other transition metal oxides, hydroxides, sulfides, and their composites were studied for supercapacitive applications, including NiO, CuO, Fe_3O_4, V_2O_5, MnO_2, MoS_2, and Co_3O_4 [13, 25–32]. Among all the transition of metal oxides, MnO_2 and Co_3O_4 show high theoretical pseudocapacitance of 1360 and 3560 F g^{-1}, respectively, and have multivalance states [27, 33]. Ultralayered Co_3O_4 structure, which has suitable redox property, showed high C_s of 548 F g^{-1} [26].

In recent years, MnO_2 has drawn increasing attention for supercapacitors application, mainly due to the high abundance of manganese [34] that contributes to low material cost as compared to the expensive ruthenium metal. Pang et al. reported high specific capacitance (C_s) (700 F g^{-1}) for MnO_2 thin films in the year 2000, and their findings had sparked intense interest among the energy research community for its application in supercapacitor electrode [35]. Such high capacitance value arises from the ions insertion/desertion within MnO_2 structure, and it depends crucially on the particle size, surface area, and porosity. Since then, in achieving optimized conditions for the properties above, MnO_2 with different morphologies have been developed, such as nanoflakes [36], nanorods [37], nanowires [38], nanopetals [39], and nanosheets [40]. The synthesis route plays a vital role in determining its morphology.

5.1.2 Carbon-Based Materials

Carbon-based materials possess high surface area as the electrode material, and the capacitive originates from the charge accumulation at the interface between electrode and electrolyte [12]. PCs employ transition metal oxides or conductive polymers [14, 16] as the electrode material. Though the energy densities in PCs are higher than that of EDLCs, the faradic reactions within PCs could lead to phase changes and limit their lifetime [41]. Graphene, with its high surface area and nanosheets morphology and carbon nanoparticles with the porous structure, is promising materials from energy storage applications.

Carbon materials are used widely as electrode materials for energy storage applications for several reasons, such as abundance, high surface area, good electrical conductivity, and low production cost [4, 42, 43]. Many carbon materials have been tested as EDL electrodes such as activated carbon (AC), C60, single-walled carbon nanotubes (SWCNTs), multiwalled carbon nanotubes (MWCNTs), single-walled carbon nanohorns (SWCNHs), carbon nanospheres, carbon nanoparticles, carbon nano-onion, and graphene [9, 18, 44–54].

Treatment of AC with KOH leads to getting a high specific surface area in the range of 1300–1500 $m^2\ g^{-1}$ and high C_s of 200 and 150 F g^{-1} in aqueous and organic electrolytes, respectively [18]. C60–AC composite shows a C_s of 172 F g^{-1} at 1 wt% C60 loading [46]. SWCNHs are a new material showing EDL performance with C_s of 114 F g^{-1} [47]. Tashima and his group prepared carbon nanospheres with a very high specific surface area of 3301 $m^2\ g^{-1}$. The obtained carbon nanospheres showed EDLCs performance in 0.5 M H_2SO_4 with C_s up to 219 F g^{-1} [45]. CNTs have been widely studied and attracted extensive attention was found due to their intriguing and potentially useful structural, electrical, and mechanical properties. One of the crucial applications of SWCNTs is as supercapacitors, where it showed a C_s of 180 F g^{-1} and a power density of 20 kW kg^{-1} [44].

Graphene is a single carbon layer, describing its nature by analogy to a polycyclic aromatic hydrocarbon of quasi-infinite size. Graphene exhibits many interesting electronic, optical, and mechanical properties due to its 2D crystal structure. The charge carriers move ballistically in the 2D crystal lattice of graphene. Its high surface area and superb electronic properties also contribute to energy storage capability, especially in supercapacitor and other applications such as optoelectronics and sensing [11, 54, 55]. Graphite could be oxidized to graphite oxide, then by the exfoliation of the later, graphene oxide is obtained. Graphene can be prepared by chemical, electrochemical, or thermal reduction of graphene oxide [9, 10]. Moreover, graphene can be prepared by chemical vapor deposition (CVD) [56]. Graphene has a high theoretical specific surface area of 2630 $m^2\ g^{-1}$ and a very high electrical conductivity as well as high mechanical strength and chemical stability [9]. Various chemically modified graphene-based materials derived from graphene oxide have reported high-end capacitance values of 135 F g^{-1} in aqueous (KOH), 99 F g^{-1} in organic (TEA-BF4/AN), and 154.1 F g^{-1} in ionic (EMIMBF4) electrolytes [9, 55]. Graphene is an ideal material for EDLCs due to its interesting nanosheets structure, which makes all area is easily reachable by the electrolyte ions and is capable of storing a value of 550 F g^{-1} [55].

5.1.3 Polymeric Materials

Conducting polymers are organic materials that can offer high electrical conductivity. The electrical conductivity of these polymeric materials rises from π-orbital conjugation and delocalized electrons along the polymer chains. Conducting polymers such as polyaniline, polypyrrole, and polythiophene have been studied widely for supercapacitive applications due to their excellent capacity for energy storage, easy synthesis, higher conductivity, and lower cost [57]. The process that involves both oxidation and reduction of the polymer backbone and the concomitant changes in the electronic structure is known as p-doping and n-doping, respectively [6]. In terms of energy and power, the capacitors with n-doped as positive electrode and p-doped as the negative electrode are promising. Electric energy can be stored and delivered in conducting polymers as delocalized π-electrons are accepted and released during electrochemical doping/undoping, respectively. Poly(3-methylthiophene) shows a high capacitance of 220 F g^{-1} and low electrical resistance of 2 Ω cm^2 [58]. In addition, polyaniline nanofibers exhibit a large C_s of 428 F g^{-1} at a current density of 2 mA cm^{-2} in 1 M H_2SO_4 solution [59]. Poly(tris(4-(thiophen-2-yl)phenyl)amine) nanotubes in organic electrolytes showed high C_s of 990 F g^{-1} [60].

5.1.4 Hybrid Materials

Although metal oxides have high specific capacitance, they have low stability. On the other hand, carbon materials have high stability with good electrical conductivity. Therefore, the focus nowadays is on hybrid materials, which exhibit the properties of both components. Different hybrid systems from (i) carbon/carbon materials [61], (ii) carbon materials/metal oxides [62, 63], (iii) carbon materials/polymers [64], (iv) metal oxides/metal oxides [65], and (v) metal oxides/polymer [66] have been studied for supercapacitor applications. Liu and co-workers maximized the capacitance of graphene/MWCNTs by using a 1:1 weight ratio of graphene to MWCNTs, where it maximized the accessible surface and further enhanced the charge collection [67]. Graphene/MnO_2 nanocomposite, which has been prepared by anodic electrodeposition and electrophoresis, showed a multilayered structure and a high C_s of 252 F g^{-1} at 2 A g^{-1} [68]. Although polyaniline has poor stability during the CDC process, the composite of polyaniline and graphene exhibited superior capacitive behavior [64]. Liu and his co-authors studied the effect of Co:Ru ratio in $Co_3O_4/RuO_2{\cdot}xH_2O$ system, where they obtained the highest C_s of 642 F g^{-1} with a molar ratio of Co:Ru = 1:1 [69]. In addition, Co_3O_4@MnO_2 core–shell arrays, which had been prepared by the hydrothermal method, showed a high C_s of 560 F g^{-1} at 0.2 A g^{-1}, energy density of 17.7 W h kg^{-1}, and a power density of 158 kW kg^{-1} [65].

5.2 Electrolytes

Three types of electrolytes could be used in supercapacitors, namely aqueous, organic, and ionic liquid electrolytes. The aqueous electrolyte has much higher ionic conductivity (0.5–1 S cm^{-1}) but limits the potential operating window of about 1.0 V, determined by the thermodynamic electrochemical window of water (1.23 V) [70]. For example, H_2SO_4, Na_2SO_4, and KOH are used as acidic, neutral, and basic aqueous electrolytes, respectively. It has been found that both acidic and basic electrolytes originate pseudocapacitive in EDLC materials as a result of the interaction of H^+ and OH^- ions with the material [53, 71, 72]. Choosing the electrolyte concentration is another important factor in supercapacitive testing because it affects the life cycle and restricts the range of possible electrode materials because most of the metal oxides degrade dramatically in concentrated solutions. Organic electrolytes such as N,N,N,N-tetraethylammonium tetrafluoroborate (TEA-BF4) dissolved in high propylene carbonate has been used for the high voltage of 2.7 V [9]. Ionic liquid electrolytes such as 1-ethyl-3-methylimidazolium tetrafluoroborate (EMIMBF4) have a broader electrochemical window up to 4 V [55]. Many aspects must be taken into account in choosing the proper electrolyte, such as corrosion of electrode structures, which depends on the nature of the solution and the pH and anion type (for aqueous media).

5.3 Separators

Separators are inert, porous, and very thin materials. They are used to separate the two electrodes, prevent short circuits, and allow the transfer of charged ions. Separator type depends on the electrolyte, where fiberglass or ceramic separators are suitable for aqueous electrolytes, while polymer or paper separators are used in the case of organic electrolytes.

5.4 Supercapacitors Cell Assembly

Symmetric supercapacitors (SSCs) and asymmetric supercapacitors (ASSCs) are two categories of supercapacitors often used depending on types of the cell assembly. In SSCs, two identical electrodes from the same active materials (often EDLCs materials) are used as positive and negative electrodes, while in ASSCs, two different electrodes (usually one is carbon material and the other metal oxide) are used for the former purpose. Thus, the hybrid of an electric double layer system and a Faradic pseudocapacitive system could be a good candidate for a supercapacitor with high C_s and energy density. Symmetric (NiO/NiO) and asymmetric (NiO/activated carbon) supercapacitors showed that the asymmetric configuration has higher C_s and higher relaxation time constant [73, 74]. Asymmetric MWCNT–NiO/MWCNT in H_2SO_4 solution showed very high C_s of 950 F g^{-1} compared to symmetric MWCNT–NiO/NiO–MWCNT, which only showed 80 F g^{-1} [75].

5.5 Cells Setup

Two cell designs are used for supercapacitive testing, two-electrode system (2ES), and three-electrode system (3ES). The 3ES consists of a working electrode (the active material), reference electrode (Ag/AgCl or calomel), and counter electrode (Pt). In 2ES, two electrodes of the active material are electrically isolated from each other by porous membrane pre-soaked with the electrolyte solution. It was then sandwiched and pressed into a suitable design.

6 Supercapacitors Electrodes by Waste Recycling

Supercapacitor electrode material resources suffer from high-cost production. For example, cobalt oxide price enlarged from 20 to 60 USD per kg in 1998 and 2017, respectively [76]. Consequently, seeking for low-cost production method is essential. Over the years, the ownership of electronic devices has transformed from a personal luxury to essential requirements.

6.1 Recycled Metal Oxides

Battery recycling can be divided into a pyrometallurgical process and hydrometallurgical process [77]. The former involves selective volatilization of scrapped battery at elevated temperature followed by condensation for metal recovery. It is the most popular battery recycling process in the industry due to its simplicity, as battery dismantling is not required [78–80]. On the other hand, the hydrometallurgical process involves dismantling, pre-treatment followed by metal ions leaching, and precipitation. The hydrometallurgical process is more efficient in metal recovery and environmentally friendly as its energy consumption is lower [81, 82]. A comparison of these processes is reported elsewhere [83], with detail technical information about battery treatment. However, battery recycling activities are not favored all the time as economic interests usually supersede environmental obligations. Therefore, recycling spent battery into a valuable product is expected to be a solution in this context.

MnO_2 has been identified as a promising pseudocapacitive material to replace toxic and costly substances, especially ruthenium oxide. Though manganese sources can be found abundantly in nature, it is imperative to stop exploiting nature for the advancement of technology. Instead, the recovery of manganese from waste sources could be an alternative to obtain MnO_2. According to the United States Environmental Protection Agency (USEPA) analysis, an average of 8 disposable batteries is consumed by an individual annually. Annually, around 160,000 tonnes of batteries are placed on the market, and about 20,000 tonnes per year of manganese could be recovered [84]. Thus, figures raise the alarm on the disposal issue where the common practice in handling spent batteries is landfill, which could potentially harm the environment. Furthermore, a high percentage of manganese in spent batteries could be a motivation in manganese recovery from batteries to be used as a supercapacitor electrode [85].

6.1.1 MnO_2 by Recycling Spent Zinc–Carbon Batteries

Zinc–carbon (Zn–C) battery is frequently used in electronic and electrical appliances as it is the least expensive battery among primary batteries. Fresh Zn–C battery consists of Zn metal as an anode and MnO_2 powder as a cathode. In discharged form, Zn is present as ZnO, while Mn is present as Mn_2O_3 and Mn_3O_4 [80, 85, 86]. Research shows that Zn and Mn contents are 28.3% and 26.3%, respectively, of the total mass in a spent Zn–C battery [85]. Such high Zn and Mn contents highlight the importance of battery recycling, both from the economy and environment perspectives.

In a detailed study, a spent Zn–C battery (EVEREADY® D cell) was disassembled, and the cathode black paste was recycled into MnO_2 nanoflowers [29]. The process involves a combination of leaching (Eqs. (4–6)) and electrowinning (Eqs. (7, 8)) processes, as shown below [29, 87]:

$$Mn_2O_3 + H_2SO_4 \rightarrow MnO_2 + MnSO_4 + H_2O \tag{4}$$

$$Mn_3O_4 + 2H_2SO_4 \rightarrow MnO_2 + 2MnSO_4 + 2H_2O \tag{5}$$

$$MnO_2 + H_2C_2O_4 + H_2SO_4 \rightarrow MnSO_4 + 2CO_2 + 2H_2O \tag{6}$$

$$2Mn^{+2} \rightarrow 2Mn^{+3} + 2e^- \tag{7}$$

$$2Mn^{+3} + 2H_2O \rightarrow Mn^{+2} + MnO_2 + 4H^+ \tag{8}$$

The recovered MnO_2 nanoflowers showed an excellent electrochemical performance and exhibited a high specific capacitance of 294 F g^{-1} in 1 M Na_2SO_4 electrolyte, with stable electrochemical cycling of 88%. The nanoflowers' morphology and CV curves for the recovered MnO_2 are shown in Fig. 4.

Cyclic voltammetry conversion is another electrochemical method for Zn–C spent batteries recycling into porous MnO_2 nanoflowers [88]. The process involves the electrochemical conversion of heat treatment (900 °C) of spent Zn–C batteries cathode powder by CV within a potential window range of −0.1 to 0.9 V for 1000

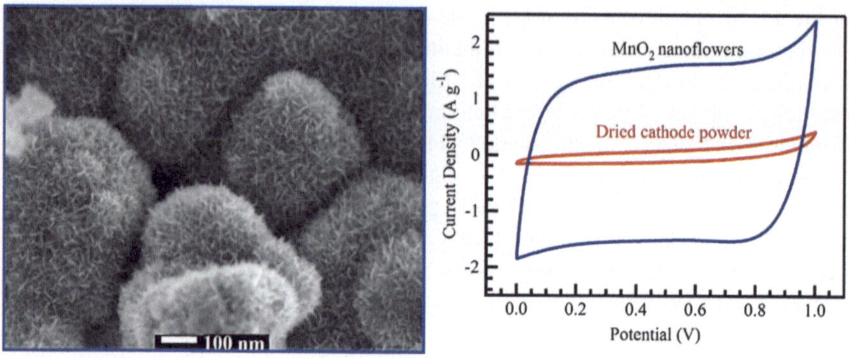

Fig. 4 Simple schematic representation of recycling Zn–C spent batteries into MnO_2 nanoflowers. Adapted with permission from Ref. [29], Copyright 2014, Elsevier

cycles at 50 mV s^{-1}. A specific capacitance of 309 F g^{-1} (six times higher than the heat-treated cathode powder) and high stability of 93% were obtained, as shown in Fig. 5. In addition, MnO_2 nanoflowers were recycled from spent Zn–C batteries in the form of nanocomposite with graphene using a one-step MnO_2/rGO nanocomposite that has been prepared by one-step electrochemical approach [89]. MnO_2/rGO exhibited 473 F $g^{-1,}$ which is remarkably three times higher than prior conversion. In addition, it showed excellent stability (95%) and the electroactive surface area.

6.1.2 Co_3O_4 by Recycling Spent Lithium–Ion Batteries

Following the exponential growth in the electronic gadget demand, the invention of lithium–ion batteries (LiBs) should be extended as it is still the most generally used energy storage device in the electronic devices [90, 91]. However, the disposal of LiBs in the electronic device is obvious due to the limited cycle life of LiBs. In this context, the disposal of LiBs could be a source of incredible environmental calamity, if it is not treated correctly because of the contained toxic materials and heavy metals such as cobalt, manganese, nickel, and lithium [92]. Furthermore, it would be a fault from the economic and environmental viewpoints if the heavy metals are not recovered before LiBs disposal. Many processes and procedures were done to extract the cobalt from spent LiBs, such as solvent extraction, acid leaching, chemical precipitation, bioleaching, and electrochemical recovery [93], which could recover more than 99% [93–95]. Alternatively, LiBs should be more excellent recycling by linking the recycled material with current applications such as supercapacitors [96–99].

The current electrodeposition approaches could recover the cobalt from spent LiBs, though at the uncontrolled morphology, restricting its technological applications [98–101]. Based on the ferromagnetic properties of cobalt, the magnetic field has been used in the electrodeposition to improve the roughness and enhance the morphology of the deposited thin film.

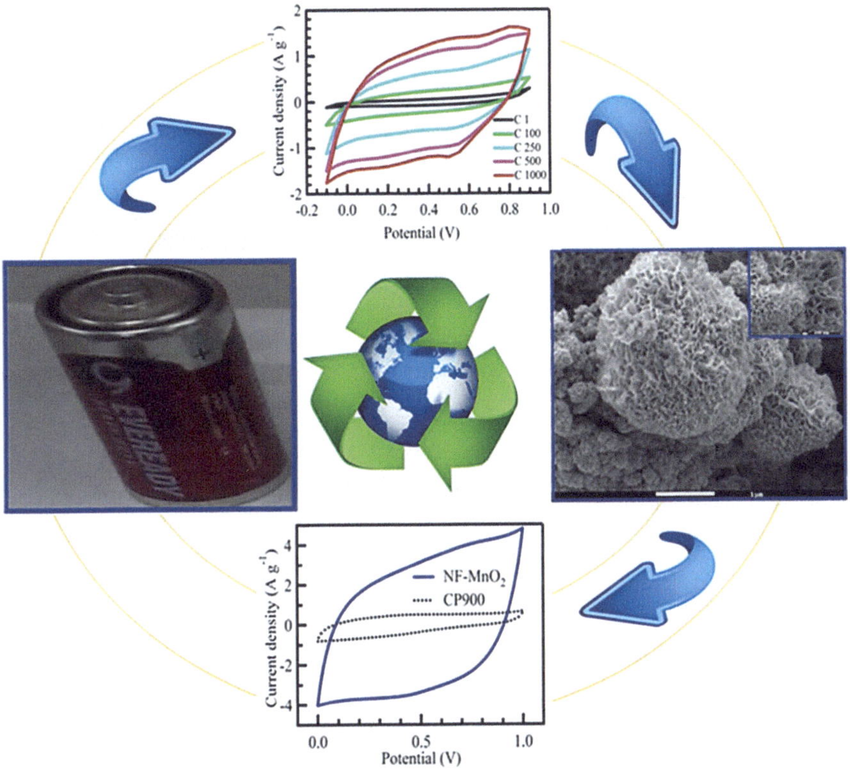

Fig. 5 Simple schematic representation of recycling Zn–C spent batteries into MnO_2 nanoflowers. Adapted with permission from Ref. [88], Copyright 2017, Elsevier

The new recycling processes of the valuable metals like cobalt and nickel from LiBs are going through two main methods. Firstly, the physical processes include thermal treatment, mechanical separation processes, dissolution processes, and mechanochemical processes. Secondly, the chemical processes are acid leaching, chemical precipitation, solvent extraction, electrochemical processes, and bioleaching [93]. The current electrodeposition technologies could recover the cobalt from spent LiBs, however, at the uncontrolled morphology, limiting its technological applications. Motivated by the ferromagnetic properties of cobalt, the magnetic field can be used to enhance the electrodeposition process to produce controlled morphology cobalt oxide from spent LiBs. The electrochemical recovering of cobalt will be used to the effect of the magnetic field, which was mentioned in the literature, promising to get a few steps enhanced cobalt layer to be used in serval of applications.

Well-defined hierarchical Co_3O_4 nanostructures with the higher electroactive surface area were formed during the electrodeposition process of spent LiBs in the presence of a magnetic field. The recovered Co_3O_4 nanostructures exhibited

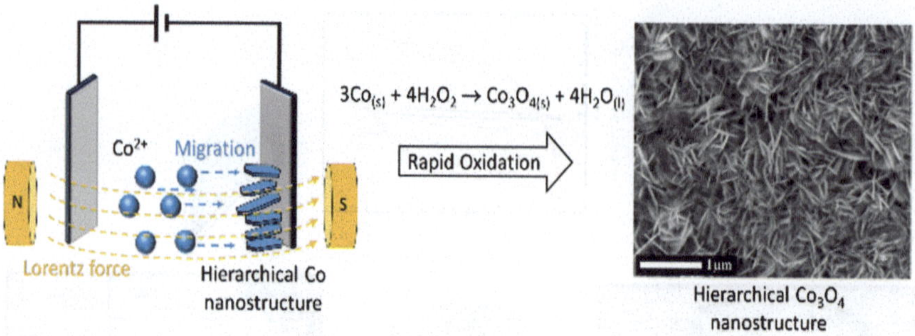

Fig. 6 Simple schematic representation of recycling lithium–ion batteries into hierarchical cobalt oxide nanostructure. Adapted with permission from Ref. [88, 102], Copyright 2018, American Chemical Society

excellent charge storage capabilities of 1273 F g^{-1} and high cycling stability of 96% [102]. The schematic representation of the magnetic electrodeposition process and the morphology of the recovered material are shown in Fig. 6.

6.2 *Recycled Carbon Materials*

6.2.1 Carbon Materials from Agriculture Waste

Carbon-based materials are the most widely used materials in a commercial supercapacitor. However, activated carbon possesses the problem of achieving high Cs and thus limiting its wide application in a supercapacitor. Graphene possesses high surface area, stable structure and exhibits many interesting electronic, optical, and mechanical properties due to its 2D crystal structure. Graphene could be the solution to this problem. On the other hand, as a move to preserve the environment as well as maintain low-cost material, waste precursors could be the potential source for the production of carbon-based materials. This includes oil palm biomass residues (leaves, fronds, trunks, empty fruit bunches, shells, and fibers) that constitute biomass waste produced from oil palm industries. A common practice in managing oil palm residues is burning, which gives rise to environmental issues. Furthermore, it is composed of high carbon content (about 18 wt%) and could be the potential source for the production of carbon-based material for supercapacitor electrode construction.

Carbon materials have synthesized from different waste sources such as banana fibers, argan (Argania spinosa) seed shells, corn grains, camellia oleifera shell, sugar cane bagasse, oil palm (empty fruit bunches and leaves), and palm kernel shells [99, 103–111]. AC prepared from camellia oleifera shell showed a C_s of 374 and 266 F g^{-1} in 1 M H_2SO_4 and 6 M KOH electrolytes, respectively [99]. In addition, the symmetric sugar cane bagasse carbons capacitor in 1 M H_2SO_4

exhibited a specific energy up to 10 Wh kg^{-1} and C_s close to 300 F g^{-1} [107]. Subramanian and his group prepared studied the electrochemical performance of the carbon material derived from banana fibers by chemical activation processes involving KOH and $ZnCl_2$ in a neutral electrolyte (1 M Na_2SO_4). Where in neutral electrolytes, the C_s is mainly governed by the non-faradic electrostatic sorption of ions at the double layer [103]. All preparation methods of carbon materials involve a template, a catalyst, or an oxidizing agent [43, 99, 112].

Catalyst-free porous carbon nanospheres (particle size ranging from 40 to 70 nm) were obtained from biowaste sago bark using one-step pyrolysis techniques. The electrochemical investigation showed a specific capacitance value of 180 F g^{-1} and the cycling stability up to 1700 cycles [48]. In addition, Lablab purpureus was used as another biowaste precursor for carbon nanospheres preparation using a green approach at different temperatures. The single electrode showed a specific capacitance of 300 F g^{-1} in 5 M KOH electrolyte [51] and 94% capacitance retention even after 5200 charge/discharge cycles entailing excellent recycling durability (as shown in Fig. 7). Moreover, a practical symmetrical supercapacitor was fabricated from the recovered carbon nanosphere. It showed unique storage characteristics under a wide potential window up to 1.7 V. It showed a high energy density of 17.9 W h kg^{-1}.

Palm kernel shells are widely used as an agriculture waste precursor for carbon recovery [111, 113–118]. Misnon et al. recovered activated carbon from palm kernel shell (PKS) and applied for supercapacitor in three different aqueous electrolytes (1 M H_2SO_4, 1 M Na_2SO_4, and 6 M KOH) [114]. The operating potential was found to be dependent on the electrolyte where it was 1.0, 1.2, and 2.0 for H_2SO_4, KOH, and Na_2SO_4, respectively. The device showed an energy density of 7.4 Wh kg^{-1} in Na_2SO_4 electrolyte [114]. Further improvement was performed on the carbon recovered from PKS using CaO impregnation obtained

Fig. 7 Conversion Lablab purpureus into carbon nanospheres. Adapted with permission from Ref. [51], Copyright 2017, Royal Society of Chemistry

Fig. 8 Conversion palm kernel shells and eggs shells into CaO impregnated highly porous honeycomb activated carbon. Adapted with permission from Ref. [113], Copyright 2019, Springer-Nature

from chicken eggshell waste. The obtained nanocomposite (CaO/ACPKS) showed a highly porous honeycomb structure with a homogeneous distribution of CaO nanoparticles (30–50 nm in size) [113]. The electrochemical investigations showed a specific capacitance value of 222 F g^{-1} for CaO/ACPKS, which is around three times higher than that for ACPKS. Schematic representation of the recycling process and the morphology of the recovered material are shown in Fig. 8. Table 1 summarizes different waste precursors used for nanomaterials recycling. It compares the capacitance and stability values together with their morphological structures.

6.2.2 Carbon Materials from Other Waste

Engine oil waste could be used as a precursor for recycling carbon hierarchical porous carbon nanosheet [119]. Engine oil waste is a toxic and hazardous waste, and it is accumulated with a high amount (about 45 million tons per year worldwide). Therefore, the recycling process, including treatment with H_2SO_4 and pyrolysis at 600–800 °C, was proposed to convert this waste into useful carbon material [119]. The obtained hierarchical porous carbon nanosheets showed a high specific surface area (up to 2276 $m^2 g^{-1}$), high specific capacitance reached 352 F g^{-1}, and 87.7% of initial capacitance was retained observed after 5000 charge and discharge cycles. In addition, a hybrid supercapacitor was fabricated from MnO_2/porous carbon flakes nanocomposite in which the porous carbon flakes were recycled from polystyrene waste through template method [120]. The symmetric supercapacitor device showed suburb electrochemical performance (high and long stability with high capacitance). Moreover, spent Zn–Mn batteries, which is considered as a serious dangerous waste on ecosystem and human health because of the toxic heavy metal ions, could be used for carbon recovery where its positive electrode is carbon package [121]. Much other waste precursors were used to recover carbon materials for supercapacitor applications including crab shell [122], vehicle exhaust [123], polystyrene [124], sludge [125], and waste filter in household water purifier [126]. The summary of some carbon waste precursors is given in Fig. 9.

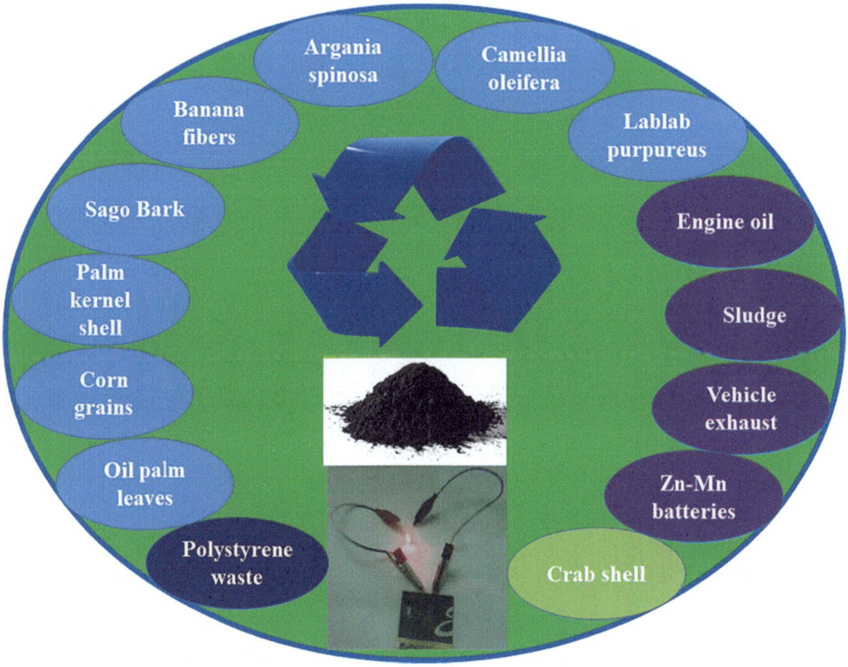

Fig. 9 Summary of some carbon waste precursors

Table 1 Comparison of reported capacitance properties of the recovered nanomaterials by recycling the environmental wastes

Waste precursor	Recovered material	Preparation method	Morphology	Specific capacitance ($F\ g^{-1}$)	Stability (%) @ cycles	Reference
Banana fibers	Activated carbon	Treatment with $KOH/ZnCl_2$ and pyrolysis	Tube-like structures	74 @ 0.5 $A\ g^{-1}$	88 @ 500	[103]
Sago bark	Carbon nanospheres	Catalyst-free, one-step and straightforward pyrolysis	Carbon nanospheres	180 @ 2 $mV\ s^{-1}$ 113 @ 0.02 $A\ g^{-1}$	94 @ 1700	[48]
Palm kernel shell and chicken eggshell	CaO/carbon	Pyrolysis and impregnation	Nanoparticles/porous honeycomb structure	222 @ 0.025 $A\ g^{-1}$	93 @ 3300	[113]
Corn grains	Activated carbon	Chemical activation	Monolithic forms	257 @ 0.125 $A\ g^{-1}$	–	[105]
Argania spinosa seed shells	Activated carbon	KOH activation	Microporous carbons	259 @ 0.1 $A\ cm^{-2}$	52	[104]
Camellia oleifera shell	Activated carbon	Pyrolysis and impregnation	Large thin sheets	266 @ 0.2 $A\ g^{-1}$	91.3 @ 5000	[99]
Lablab purpureus	Carbon nanospheres	Pyrolysis at different temperatures	Carbon nanospheres	300 @ 0.1 $A\ g^{-1}$	94 @ 5200	[51]
Oil palm fronds	Carbon nanoparticles	Single step pyrolysis in N_2	Porous carbon nanoparticles	343 @ 5 $mV\ s^{-1}$ 309 @ 0.06 $A\ g^{-1}$	90 @ 2500	[108]
Engine oil	Carbon nanosheet	Acid treatment and pyrolysis	Hierarchical porous carbon nanosheet	352 @ 0.5 $A\ g^{-1}$	87.7 @ 5000	[119]
Oil palm leaves	Nanocarbons	Simple pyrolysis	Porous nanocarbons	368 @ 0.06 $A\ g^{-1}$	96 @ 1700	[109]
LiB	Co_3O_4	Chemical precipitation and heat-treatement	Agglomerated crystals	13 @ 1 $mV\ s^{-1}$	–	[96]
LiB	Co_3O_4	Electrodeposition	Sheet-like	143 @ 0.5 $A\ g^{-1}$	–	[127]
LiB	Co_3O_4	Magnetic electrodeposition	Well-defined hierarchical nanostructures	1273 @ 1 $A\ g^{-1}$	81 @ 5000	[102]

(continued)

Table 1 (continued)

Waste precursor	Recovered material	Preparation method	Morphology	Specific capacitance ($F\ g^{-1}$)	Stability (%) @ cycles	Reference
Zn–C Batteries	MnO_2	Leaching and electrowinning	Nanoflower	294 @ 10 $mV\ s^{-1}$ 208.5 @ 0.1 $A\ g^{-1}$	88 @ 900	[29]
Polystyrene waste	MnO_2/porous carbon	Pyrolysis	Nanosheets/porous carbon flakes	308 @ 1 $mV\ s^{-1}$ 247 @ 1 $A\ g^{-1}$	93.4 @ 10,000	[120]
Zn–C Batteries	MnO_2	Cyclic voltammetry conversion	Nanoflower	309 @ 0.1 $A\ g^{-1}$	93 @ 1650	[88]
Zn–C Batteries	MnO_2/rGO	Cyclic voltammetry conversion	Nanoflower/nanosheets	473 @ 0.25 $A\ g^{-1}$	95 @ 2000	[89]

7 Conclusions

Environmental wastes recycling is one of the alternative ways to obtain promising candidates for supercapacitors electrode materials. Recycling techniques have another environmental aspect where they lead to a cleaner environment. A variety of single components and hybrid electrodes can be prepared from waste. The recycled materials include metal oxides (MnO_2, Co_3O_4, etc.), carbon materials (carbon nanosphere, porous carbon nanoparticles, activated carbon), and hybrid materials (MnO_2/graphene, CaO/AC). The obtained materials exhibited interesting structural and morphological properties as well as excellent energy storage behavior. Further improvement of recycling technologies and process parameters control leads to obtain superior materials with unique characteristics, making them promising candidates for supercapacitors application.

8 Future Prospectives

The present study focused on the successful synthesis of different nanomaterials by recycling techniques. The methods were performed on the laboratory scale giving small product weight; the further step is to upscale these methods onto the industrial scale. Therefore, the scalability of the process is one of the immediate attention. The electrochemical (supercapacitive) properties have been tested using laboratory-made electrodes, so this study recommends that fabrication of these materials in a more suitable form, and it can be employed in supercapacitors technology.

References

1. Rötz R, Carlen M (2000) Principles and applications of electrochemical capacitors. Electrochim Acta 45:2483–2498
2. Liu C, Li F, Ma LP, Cheng HM (2010) Advanced materials for energy storage. Adv Mater 22(8):28–62
3. Syzdek J, Marcinek M, Kostecki R (2014) Electrochemical activity of carbon blacks in LiPF6-based organic electrolytes. J Power Sources 245:739–744
4. Simon P, Gogotsi Y (2008) Materials for electrochemical capacitors. Nat Mater 7:845–854
5. Winter M, Brodd RJ (2004) What are batteries, fuel cells, and supercapacitors. Chem Rev 104:4245–4269
6. Conway BE (1999) Electrochemical supercapacitors scientific fundamentals and technological applications. Kluwer Academic/Plenum Press, New York
7. Lota G, Fic K, Frackowiak E (2011) Carbon nanotubes and their composites in electrochemical applications. Energy Environ Sci 4(5):1592–1605
8. Mosa IM, Pattammattel A, Kadimisetty K, Pande P, El-Kady MF, Bishop GW, Novak M, Kaner RB, Basu AK, Kumar CV (2017) Ultrathin graphene-protein supercapacitors for miniaturized bioelectronics. Adv Energy Mater 7(17):1700358

9. Stoller MD, Park S, Zhu Y, An J, Ruoff RS (2008) Graphene-based ultracapacitors. Nano Letter 8(10):3498–3502
10. Lee SP, Ali GAM, Hegazy HH, Lim HN, Chong KF (2021) Optimizing reduced graphene oxide aerogel for a supercapacitor. Energy & Fuel 35(5):4559–4569
11. Zhu Y, Murali S, Cai W, Li X, Suk JW, Potts JR, Ruoff RS (2010) Graphene and graphene oxide: synthesis, properties, and applications. Adv Mater 22(35):3906–39024
12. Portet C, Taberna PL, Simon P, Flahaut E, Laberty-Robert C (2005) High power density electrodes for carbon supercapacitor applications. Electrochim Acta 50(20):4174–4181
13. Nathan T, Aziz A, Noor AF, Prabaharan SRS (2008) Nanostructured NiO for electrochemical capacitors: synthesis and electrochemical properties. J Solid State Electrochem 12(7–8):1003–1009
14. Xie L, Li K, Sun G, Hu Z, Lv C, Wang J, Zhang C (2012) Preparation and electrochemical performance of the layered cobalt oxide (Co_3O_4) as supercapacitor electrode material. J Solid State Electrochem 17(1):55–61
15. Thalji MR, Ali GAM, Algarni H, Chong KF (2019) Al^{3+} ion intercalation pseudocapacitance study of $W_{18}O_{49}$ nanostructure. J Power Sources 438:227028
16. Patil DS, Pawar SA, Devan RS, Gang MG, Ma YR, Kim JH, Patil PS (2013) Electrochemical supercapacitor electrode material based on polyacrylic acid/polypyrrole/silver composite. Electrochim Acta 105:569–577
17. Endo M, Takeda T, Kim YJ, Koshiba K, Ishii K (2001) High power electric double layer capacitor (EDLC's); from operating principle to pore size control in advanced activated carbons. Carbon Sci 1(2–3):117–128
18. Lota G, Centeno TA, Frackowiak E, Stoeckli F (2008) Improvement of the structural and chemical properties of a commercial activated carbon for its application in electrochemical capacitors. Electrochim Acta 53(5):2210–2216
19. Chen T, Dai L (2013) Carbon nanomaterials for high–performance supercapacitors. Mater Today 16(7–8):272–280
20. Frackowiak E, Béguin F (2001) Carbon materials for the electrochemical storage of energy in capacitors. Carbon 39:937–950
21. McKeown DA, Hagans PL, Carette LPL, Russell AE, Swider KS, Rolison DR (1999) Structure of hydrous ruthenium oxides implications for charge storage. J Phys Chem B 103:4825–4832
22. Hu CC, Chen WC (2004) Effects of substrates on the capacitive performance of $RuO_x{\cdot}nH_2O$ and activated carbon–RuO_x electrodes for supercapacitors. Electrochim Acta 49(21):3469–3477
23. Zheng JP, Cygan PJ, Jow TR (1995) Hydrous ruthenium oxide as an electrode material for electrochemical capacitors. J Electrochem Soc 142(8):2699–2703
24. Kim H, Popov BN (2003) A mathematical model of oxide/carbon composite electrode for supercapacitors. J Electrochem Soc 150(9):A1153–A1160
25. Wee G, Soh HZ, Cheah YL, Mhaisalkar SG, Srinivasan M (2010) Synthesis and electrochemical properties of electrospun V_2O_5 nanofibers as supercapacitor electrodes. J Mater Chem 20(32):6720–6725
26. Meher SK, Rao GR (2011) Ultralayered Co_3O_4 for high–performance supercapacitor applications. J Phy Chem C 115(31):15646–15654
27. Yang P, Ding Y, Lin Z, Chen Z, Li Y, Qiang P, Ebrahimi M, Mai W, Wong CP, Wang ZL (2014) Low–cost high–performance solid–state asymmetric supercapacitors based on MnO_2 nanowires and Fe_2O_3 nanotubes. Nano Lett 14(2):731–736
28. Ali GAM, Fouad OA, Makhlouf SA, Yusoff MM, Chong KF (2014) Co_3O_4/SiO_2 nanocomposites for supercapacitor application. J Solid State Electrochem 18(9):2505–2512
29. Ali GAM, Tan LL, Jose R, Yusoff MM, Chong KF (2014) Electrochemical performance studies of MnO_2 nanoflowers recovered from spent battery. Mater Res Bull 60:5–9
30. Ali GAM (2020) Recycled MnO_2 nanoflowers and graphene nanosheets for low-cost and high performance asymmetric supercapacitor. J Electron Mater 49:5411–5421

31. Ali GAM, Wahba OAG, Hassan AM, Fouad OA, Chong KF (2015) Calcium-based nanosized mixed metal oxides for supercapacitor application. Ceram Int 41(6):8230–8234
32. Aboelazm EAA, Ali GAM, Algarni H, Chong KF (2018) Flakes size-dependent optical and electrochemical properties of MoS_2. Curr Nanosci 14(2):1–5
33. Ali GAM, Fouad OA, Makhlouf SA, Yusoff MM, Chong KF (2016) Optical and electrochemical properties of Co_3O_4/SiO_2 nanocomposite. Adv Mater Res 1133:447–451
34. Jang H, Suzuki S, Miyayama M (2012) Self-reassembled MnO_2 nanosheets for electrochemical capacitors in neutral aqueous solution. J Electrochem Soc 159(9):A1425–A1430
35. Pang SC, Anderson MA, Chapman TW (2000) Novel electrode materials for thin-film ultracapacitors comparison of electrochemical properties of sol-gel-derived and electrodeposited manganese dioxide. J Electrochem Soc 147(2):444–450
36. Chou SL, Cheng FY, Chen J (2006) Electrodeposition synthesis and electrochemical properties of nanostructured γ–MnO_2 films. J Power Sources 162(1):727–734
37. Yousefi T, Golikand AN, Hossein Mashhadizadeh M, Aghazadeh M (2012) Facile synthesis of α-MnO_2 one-dimensional (1D) nanostructure and energy storage ability studies. J Solid State Chem 190:202–207
38. Yousefi T, Davarkhah R, Golikand AN, Mashhadizadeh MH (2013) Synthesis, characterization, and supercapacitor studies of manganese (IV) oxide nanowires. Mater Sci Semicond Process 16(3):868–876
39. Yang G, Wang B, Guo W, Bu Z, Miao C, Xue T, Li H (2012) Liquid crystalline phase synthesis of nanoporous MnO_2 thin film arrays as an electrode material for electrochemical capacitors. Mater Res Bull 47(11):3120–3123
40. Yan D, Guo Z, Zhu G, Yu Z, Xu H, Yu A (2012) MnO_2 film with three-dimensional structure prepared by hydrothermal process for supercapacitor. J Power Sources 199:409–412
41. Compton OC, Nguyen ST (2010) Graphene oxide, highly reduced graphene oxide, and graphene: versatile building blocks for carbon–based materials. Small 6(6):711–723
42. Ali GAM, Makhlouf SA, Yusoff MM, Chong KF (2015) Structural and electrochemical characteristics of graphene nanosheets as supercapacitor electrodes. Rev Adv Mater Sci 40 (1):35–43
43. Barhoum A, Shalan AE, El-Hout SI, Ali GAM, Abdelbasir SM, Abu Serea ES, Ibrahim AH, Pal K (2019) A broad family of carbon nanomaterials: classification, properties, synthesis, and emerging applications. In: Barhoum A, Bechelany M, Makhlouf A (eds) Handbook of nanofibers. Springer International Publishing, Cham, pp 1–40
44. Pan H, Li J, Feng Y (2010) Carbon nanotubes for supercapacitor. Nanoscale Res Lett 5 (3):654–668
45. Tashima D, Yamamoto E, Kai N, Fujikawa D, Sakai G, Otsubo M, Kijima T (2011) Double layer capacitance of high surface area carbon nanospheres derived from resorcinol–formaldehyde polymers. Carbon 49(14):4848–4857
46. Okajima K, Ikeda A, Kamoshita K, Sudoh M (2005) High rate performance of highly dispersed C60 on activated carbon capacitor. Electrochim Acta 51(5):972–977
47. Yang CM, Kim YJ, Endo M, Kanoh H, Yudasaka M, Iijima S, Kaneko K (2007) Nanowindow-regulated specific capacitance of supercapacitor electrodes of single-wall carbon nanohorns. J Am Chem Soc 129:20–21
48. Hegde G, Abdul Manaf SA, Kumar A, Ali GAM, Chong KF, Ngaini Z, Sharma KV (2015) Biowaste sago bark based catalyst free carbon nanospheres: waste to wealth approach. ACS Sustain Chem Eng 5(9):2247–2253
49. Marina PE, Ali GAM, See LM, Teo EYL, Ng E-P, Chong KF (2016) In situ growth of redox-active iron-centered nanoparticles on graphene sheets for specific capacitance enhancement. Arab J Chem 12(8):3883–3889
50. Ali GAM, Yusoff MM, Chong KF (2016) Graphene electrochemical production and its energy storage properties. ARPN J Eng Appl Sci 11(16):9712–9717

51. Ali GAM, Divyashree A, Supriya S, Chong KF, Ethiraj AS, Reddy M, Algarni H, Hegde G (2017) Carbon nanospheres derived from lablab purpureus for high performance supercapacitor electrodes: a green approach. Dalton Trans 46(40):14034–14044
52. Ali GAM, Teo EYL, Aboelazm EAA, Sadegh H, Memar AOH, Shahryari-Ghoshekandi R, Chong KF (2017) Capacitive performance of cysteamine functionalized carbon nanotubes. Mater Chem Phys 197:100–104
53. Ali GAM, Megiel E, Romański J, Algarni H, Chong KF (2018) A wide potential window symmetric supercapacitor by TEMPO functionalized MWCNTs. J Mol Liq 271:31–39
54. Teo EYL, Ali GAM, Algarni H, Cheewasedtham W, Rujiralai T, Chong KF (2019) One-step production of pyrene-1-boronic acid functionalized graphene for dopamine detection. Mater Chem Phys 231:286–291
55. Liu C, Yu Z, Neff D, Zhamu A, Jang BZ (2010) Graphene–based supercapacitor with an ultrahigh energy density. Nano Letter 10(12):4863–4868
56. Chung TF, Shen T, Cao H, Jauregui LA, Wu WEI, Yu Q, Newell D, Chen YP (2013) Synthetic graphene grown by chemical vapor deposition on copper foils. Int J Mod Phys B 27(10):1341002–1341016
57. Shi Y, Pan L, Liu B, Wang Y, Cui Y, Bao Z, Yu G (2014) Nanostructured conductive polypyrrole hydrogels as high-performance, flexible supercapacitor electrodes. J Mater Chem A 2(17):6086–6091
58. Mastragostino M, Arbizzani C, Soavi F (2002) Conducting polymers as electrode materials in supercapacitors. Solid State Ionics 148:493–498
59. Mi H, Zhang X, Yang S, Ye X, Luo J (2008) Polyaniline nanofibers as the electrode material for supercapacitors. Mater Chem Phys 112(1):127–131
60. Roberts ME, Wheeler DR, McKenzie BB, Bunker BC (2009) High specific capacitance conducting polymer supercapacitor electrodes based on poly(tris(thiophenylphenyl)amine). J Mater Chem 19(38):6977–6979
61. Ma J, Guo Q, Gao HL, Qin X (2015) Synthesis of C_{60}/graphene composite as electrode in supercapacitors. Fullerenes, Nanotubes, Carbon Nanostruct 23(6):477–482
62. Wang JG, Yang Y, Huang ZH, Kang F (2012) Synthesis and electrochemical performance of MnO_2/CNTs-embedded carbon nanofibers nanocomposites for supercapacitors. Electrochim Acta 75:213–219
63. Ali GAM, Thalji MR, Soh WC, Algarni H, Chong KF (2020) One-step electrochemical synthesis of MoS_2/graphene composite for supercapacitor application. J Solid State Electrochem 24(1):25–34
64. Wang L, Ye Y, Lu X, Wen Z, Li Z, Hou H, Song Y (2013) Hierarchical nanocomposites of polyaniline nanowire arrays on reduced graphene oxide sheets for supercapacitors. Sci Rep 3:3568–3577
65. Huang M, Zhang Y, Li F, Zhang L, Wen Z, Liu Q (2014) Facile synthesis of hierarchical Co_3O_4@MnO_2 core–shell arrays on Ni foam for asymmetric supercapacitors. J Power Sources 252:98–106
66. Meng F, Yan X, Zhu Y, Si P (2013) Controllable synthesis of MnO_2/polyaniline nanocomposite and its electrochemical capacitive property. Nanoscale Res Lett 8(1):1–8
67. Liu JZ, Mirri F, Notarianni M, Pasquali M, Motta N (2015) High performance all-carbon thin film supercapacitors. J Power Sources 274:823–830
68. Wu MS, Lin CJ, Ho CL (2012) Multilayered architecture of graphene nanosheets and MnO_2 nanowires as an electrode material for high-performance supercapacitors. Electrochim Acta 81:44–48
69. Liu Y, Zhao W, Zhang X (2008) Soft template synthesis of mesoporous $Co_3O_4/RuO_2{\cdot}xH_2O$ composites for electrochemical capacitors. Electrochim Acta 53(8):3296–3304
70. Tanahashi I, Yoshida A, Nishino A (1990) Comparison of the electrochemical properties of electric double-layer capacitors with an aqueous electrolyte and with a nonaqueous electrolyte. Bulletin Chem Soc Japan 63(12):3611–3614
71. Hulicova D, Yamashita J, Soneda Y, Hatori H, Kodama M (2005) Supercapacitors prepared from melamine–based carbon. Chem Mater 17(5):1241–1247

72. Oh YJ, Yoo JJ, Kim YI, Yoon JK, Yoon HN, Kim JH, Park SB (2014) Oxygen functional groups and electrochemical capacitive behavior of incompletely reduced graphene oxides as a thin-film electrode of supercapacitor. Electrochim Acta 116:118–128
73. Ganesh V, Pitchumani S, Lakshminarayanan V (2006) New symmetric and asymmetric supercapacitors based on high surface area porous nickel and activated carbon. J Power Sources 158:1523–1532
74. Yin J, Park J (2015) Asymmetric supercapacitors based on the in situ–grown mesoporous nickel oxide and activated carbon. J Solid State Electrochem 1–8
75. Adekunle AS, Ozoemena KI, Mamba BB, Agboola BO, Oluwatobi OS (2011) Supercapacitive properties of symmetry and the asymmetry two electrode coin type supercapacitor cells made from MWCNTS/nickel oxide nanocomposite. Int J Electrochem Sci 6:4760–4774
76. Freitas M, Garcia E (2007) Electrochemical recycling of cobalt from cathodes of spent lithium-ion batteries. J Power Sources 171(2):953–959
77. Zhang X, Xie Y, Lin X, Li H, Cao H (2013) An overview on the processes and technologies for recycling cathodic active materials from spent lithium-ion batteries. J Mater Cycles Waste Manage 15(4):420–430
78. Li Y, Xi G (2005) The dissolution mechanism of cathodic active materials of spent Zn–Mn batteries in HCl. J Hazard Mater 127(1–3):244–248
79. Nan J, Han D, Cui M, Yang M, Pan L (2006) Recycling spent zinc manganese dioxide batteries through synthesizing Zn–Mn ferrite magnetic materials. J Hazard Mater 133(1–3):257–261
80. Rácz R, Ilea P (2013) Electrolytic recovery of Mn_3O_4 and Zn from sulphuric acid leach liquors of spent zinc-carbon-MnO_2 battery powder. Hydrometallurgy 139:116–123
81. Ferella F, De Michelis I, Vegliò F (2008) Process for the recycling of alkaline and zinc–carbon spent batteries. J Power Sources 183(2):805–811
82. Freitas MBJG, Pegoretti VC, Pietre MK (2007) Recycling manganese from spent Zn–MnO_2 primary batteries. J Power Sources 164(2):947–952
83. Bernardes AM, Espinosa DCR, Tenório JAS (2004) Recycling of batteries: a review of current processes and technologies. J Power Sources 130(1–2):291–298
84. Gallegos MV, Falco LR, Peluso MA, Sambeth JE, Thomas HJ (2013) Recovery of manganese oxides from spent alkaline and zinc–carbon batteries. An application as catalysts for VOCs elimination. Waste Manag 33(6):1483–1490
85. Sayilgan E, Kukrer T, Civelekoglu G, Ferella F, Akcil A, Veglio F, Kitis M (2009) A review of technologies for the recovery of metals from spent alkaline and zinc-carbon batteries. Hydrometallurgy 97(3–4):158–166
86. Michelis ID, Ferella F, Karakaya E, Beolchini F, Vegliò F (2007) Recovery of zinc and manganese from alkaline and zinc–carbon spent batteries. J Power Sources 172(2):975–983
87. Souza CCBM, Tenório JAS (2004) Simultaneous recovery of zinc and manganese dioxide from household alkaline batteries through hydrometallurgical processing. J Power Sources 136(1):191–196
88. Ali GAM, Yusoff MM, Shaaban ER, Chong KF (2017) High performance MnO_2 nanoflower supercapacitor electrode by electrochemical recycling of spent batteries. Ceram Int 43 (11):8440–8448
89. Ali GAM, Yusoff MM, Algarni H, Chong KF (2018) One-step electrosynthesis of MnO_2/rGO nanocomposite and its enhanced electrochemical performance. Ceram Int 44(7):7799–7807
90. Hu J, Zhang J, Li H, Chen Y, Wang C (2017) A promising approach for the recovery of high value-added metals from spent lithium-ion batteries. J Power Sources 351:192–199
91. Xiao J, Li J, Xu Z (2017) Recycling metals from lithium ion battery by mechanical separation and vacuum metallurgy. J Hazard Mater 338:124–131
92. Li L, Ge J, Chen R, Wu F, Chen S, Zhang X (2010) Environmental friendly leaching reagent for cobalt and lithium recovery from spent lithium-ion batteries. Waste Manag 30(12):2615–2621

93. Ordoñez J, Gago EJ, Girard A (2016) Processes and technologies for the recycling and recovery of spent lithium-ion batteries. Renew Sustain Energy Rev 60:195–205
94. Contestabile M, Panero S, Scrosati B (1999) A laboratory-scale lithium battery recycling process. J Power Sources 83(1):75–78
95. Castillo S, Ansart F, Laberty-Robert C, Portal J (2002) Advances in the recovering of spent lithium battery compounds. J Power Sources 112(1):247–254
96. Barbieri EMS, Lima EPC, Lelis MFF, Freitas MBJG (2014) Recycling of cobalt from spent Li-ion batteries as β-$Co(OH)_2$ and the application of Co_3O_4 as a pseudocapacitor. J Power Sources 270:158–165
97. Barbieri EMS, Lima EPC, Cantarino SJ, Lelis MFF, Freitas MBJG (2014) Recycling of spent ion-lithium batteries as cobalt hydroxide, and cobalt oxide films formed under a conductive glass substrate, and their electrochemical properties. J Power Sources 269:158–163
98. Garcia EM, Tarôco HA, Matencio T, Domingues RZ, dos Santos JAF, Ferreira RV, Lorençon E, Lima DQ, de Freitas MBJG (2012) Electrochemical recycling of cobalt from spent cathodes of lithium-ion batteries: its application as supercapacitor. J Appl Electrochem 42(6):361–366
99. Zhang J, Gong L, Sun K, Jiang J, Zhang X (2012) Preparation of activated carbon from waste Camellia oleifera shell for supercapacitor application. J Solid State Electrochem 16 (6):2179–2186
100. Freitas MBJG, Celante VG, Pietre MK (2010) Electrochemical recovery of cobalt and copper from spent Li–ion batteries as multilayer deposits. J Power Sources 195(10):3309–3315
101. Garcia EM, Santos JS, Pereira EC, Freitas MBJG (2008) Electrodeposition of cobalt from spent Li–ion battery cathodes by the electrochemistry quartz crystal microbalance technique. J Power Sources 185(1):549–553
102. Aboelazm EAA, Ali GAM, Algarni H, Yin H, Zhong YL, Chong KF (2018) Magnetic electrodeposition of the hierarchical cobalt oxide nanostructure from spent lithium–ion batteries: its application as a supercapacitor electrode. J Phys Chem C 122(23):12200–12206
103. Subramanian V, Luo C, Stephan AM, Nahm KS, Thomas S, Wei B (2007) Supercapacitors from activated carbon derived from banana fibers. J Phys Chem C 111(20):7527–7531
104. Elmouwahidi A, Zapata-Benabithe Z, Carrasco–Marin F, Moreno-Castilla C (2012) Activated carbons from KOH–activation of argan (Argania spinosa) seed shells as supercapacitor electrodes. Biores Technol 111:185–190
105. Balathanigaimani MS, Shim WG, Lee MJ, Kim C, Lee JW, Moon H (2008) Highly porous electrodes from novel corn grains–based activated carbons for electrical double layer capacitors. Electrochem Commun 10(6):868–871
106. Farma R, Deraman M, Awitdrus A, Talib IA, Taer E, Basri NH, Manjunatha JG, Ishak MM, Dollah BN, Hashmi SA (2013) Preparation of highly porous binderless activated carbon electrodes from fibres of oil palm empty fruit bunches for application in supercapacitors. Biores Technol 132:254–261
107. Rufford TE, Hulicova-Jurcakova D, Khosla K, Zhu Z, Lu GQ (2010) Microstructure and electrochemical double-layer capacitance of carbon electrodes prepared by zinc chloride activation of sugar cane bagasse. J Power Sources 195(3):912–918
108. Ali GAM, Abdul Manaf SA, Kumar A, Chong KF, Hegde G (2014) High performance supercapacitor using catalysis free porous carbon nanoparticles. J Phys D-Appl Phys 47 (49):495307–495313
109. Ali GAM, Manaf SAA, Divyashree A, Chong KF, Hegde G (2016) Superior supercapacitive performance in porous nanocarbons. J Energy Chem 25(4):734–739
110. Habeeb OA, Ramesh K, Ali GAM, Yunus RM (2017) Experimental design technique on removal of hydrogen sulfide using CaO-eggshells dispersed onto palm kernel shell activated carbon: experiment, optimization, equilibrium and kinetic studies. J Wuhan Univ Technol Mater Sci 32(2):305–320
111. Habeeb OA, Ramesh K, Ali GAM, Yunus RM (2017) Low-cost and eco-friendly activated carbon from modified palm kernel shell for hydrogen sulfide removal from wastewater: adsorption and kinetic studies. Desalin Water Treat 84:205–214

112. Abdul Manaf SA, Roy P, Sharma KV, Ngaini Z, Malgras V, Aldalbahi A, Alshehri SM, Yamauchi Y, Hegde G (2015) Catalyst-free synthesis of carbon nanospheres for potential biomedical applications: waste to wealth approach. RSC Adv 5(31):24528–24533
113. Ali GAM, Habeeb OA, Algarni H, Chong KF (2019) CaO impregnated highly porous honeycomb activated carbon from agriculture waste: symmetrical supercapacitor study. J Mater Sci 54:683–692
114. Misnon II, Zain NKM, Jose R (2018) Conversion of oil palm kernel shell biomass to activated carbon for supercapacitor electrode application. Waste Biomass Valorization. https://doi.org/10.1007/s12649-018-0196-y
115. Habeeb OA, Ramesh K, Ali GAM, Yunus RM, Thanusha TK, Olalere OA (2016) Modeling and optimization for H_2S adsorption from wastewater using coconut shell based activated carbon. Aust J Basic Appl Sci 10(17):136–147
116. Habeeb OA, Ramesh K, Ali GAM, Yunus RM (2017) Application of response surface methodology for optimization of palm kernel shell activated carbon preparation factors for removal of H_2S from industrial wastewater. Jurnal Teknologi 79(7):1–10
117. Habeeb OA, Ramesh K, Ali GAM, Yunus RM (2017) Isothermal modelling based experimental study of dissolved hydrogen sulfide adsorption from waste water using eggshell based activated carbon. Malays J Anal Sci 21(2):334–345
118. Habeeb OA, Ramesh K, Ali GAM, Yunus RM, Olalere OA (2017) Kinetic, isotherm and equilibrium study of adsorption of hydrogen sulfide from wastewater using modified eggshells. IIUM Eng J 18(1):13–25
119. Li Y, Zhang D, He J, Wang Y, Zhang X, Zhang Y, Liu X, Wang K, Wang Y (2019) Hierarchical porous carbon nanosheet derived from waste engine oil for high-performance supercapacitor application. Sustain Energy Fuels 3(2):499–507
120. Min J, Zhang S, Li J, Klingeler R, Wen X, Chen X, Zhao X, Tang T, Mijowska E (2019) From polystyrene waste to porous carbon flake and potential application in supercapacitor. Waste Manag 85:333–340
121. Guo X, Luan Z, Lu Y, Fu L, Gai L (2019) $NaTi_2(PO_4)_3/C\|$carbon package asymmetric flexible supercapacitors with the positive material recycled from spent Zn–Mn dry batteries. J Alloys Compd 576–585
122. Fu M, Chen W, Zhu X, Yang B, Liu Q (2019) Crab shell derived multi-hierarchical carbon materials as a typical recycling of waste for high performance supercapacitors. Carbon 141:748–757
123. Li X, Yin X, Wang W, Zhao H, Liu D, Zhou L, Zhang C, Wang J (2019) Carbon captured from vehicle exhaust by triboelectric particular filter as materials for energy storage. Nano Energy 56:792–798
124. de Paula FGF, de Castro MCM, Ortega PFR, Blanco C, Lavall RL, Santamaría R (2018) High value activated carbons from waste polystyrene foams. Microporous Mesoporous Mater 267:181–184
125. Zhang JJ, Fan HX, Dai XH, Yuan SJ (2018) Digested sludge-derived three-dimensional hierarchical porous carbon for high-performance supercapacitor electrode. Royal Soc Open Sci 5(4):1–12
126. Li G, Li Q, Ye J, Fu G, Han JJ, Zhu Y (2017) Activated carbon from the waste water purifier for supercapacitor application. J Solid State Electrochem 21(11):3169–3177
127. Aboelazm EAA, Ali GAM, Chong KF (2018) Cobalt oxide supercapacitor electrode recovered from spent Lithium–Ion battery. Chem Adv Mater 3(4):67–73

Recovery of Metal Oxide Nanomaterials from Electronic Waste Materials

Heba H. El-Maghrabi, Amr A. Nada, Fathi S. Soliman, Patrice Raynaud, Yasser M. Moustafa, Gomaa A. M. Ali, and Maged F. Bekheet

Abstract The exploitation of spent battery and electronic waste for the recovery and preparation of metal oxide nanomaterials (MONMs) is vital for technology, economical, sustainable, and environmental research. The recovery of MONMs through recycling waste materials reduces environmental pollutions and saves the primary resources due to industrial consumption. However, the economic benefits of recycling electronic waste for the recovery of these high-value MONMs have still been debated because of the low purity and stability of recovered materials restricting their commercial use. In this chapter, we discuss the motivation and importance of waste recycling for the recovery of nanomaterials, focusing on the possible techniques that can be applied for the efficient synthesis of commercial-grade MONMs (e.g., ferrites, zinc oxides, indium oxides, tin oxides, etc.) with high purity at a minimal cost. Besides, the profit of recovered MONMs in

H. H. El-Maghrabi (✉) · F. S. Soliman
Department of Refining, Egyptian Petroleum Research Institute, Cairo 11727, Egypt
e-mail: hebachem@yahoo.com

A. A. Nada (✉) · Y. M. Moustafa
Department of Analysis and Evaluation, Egyptian Petroleum Research Institute, Cairo 11727, Egypt
e-mail: chem_amr@yahoo.com; amr.nada@epri.sci.eg

A. A. Nada · P. Raynaud
Laboratoire Plasma et Conversion de l'Energie (LAPLACE), Université de Toulouse, CNRS, INPT, UPS, Toulouse, France

G. A. M. Ali
Chemistry Department, Faculty of Science, Al-Azhar University, Assiut 71524, Egypt

G. A. M. Ali
The Smart Materials Research Institute, Southern Federal University, Sladkova Str. 178/24, Rostov-on-Don, Russian Federation

M. F. Bekheet
Fachgebiet Keramische Werkstoffe/Chair of Advanced Ceramic Materials, Technische Universität Berlin, Institut für Werkstoffwissenschaften und -technologien, Hardenbergstraße 40, 10623 Berlin, Germany

A. S. H. Makhlouf and G. A. M. Ali (eds.), *Waste Recycling Technologies for Nanomaterials Manufacturing*, Topics in Mining, Metallurgy and Materials Engineering, https://doi.org/10.1007/978-3-030-68031-2_8

potential applications for wastewater remediation and renewable energy production are addressed.

Keywords Nanomaterials · Metal oxides · Waste recycling · Environmental pollution · Renewable energy

List of Abbreviations

AFM	Atomic force microscope
CF	Color filter
DTA	Differential thermal analysis
FTIR	Fourier-transform infra-red
Gg	Gigagram
GO	Graphene oxide
ICP	Inductively coupled plasma
ITO	Indium tin oxide
LCD	Liquid crystal display
MONMs	Metal oxide nano materials
NPs	Nanoparticles
PVC	Poly vinyl chloride
RGO	Reduced graphene oxide
SEM	Scanning electron microscope
TEM	Transmission electron microscopy
TFT	Thin film transistor
TGA	Thermal gravimetric analysis
VSM	Value stream mapping
XPS	X-ray photoelectron spectroscopy
XRD	X-ray diffraction

1 Introduction

Recycling waste materials to recover high-value end-products is probably the most effective environmental preservation [1]. The goals of recycling are to achieve less natural spent consumption and reduce the amount of trash, as illustrated in Fig. 1. It is important to recycle waste materials with the lowest possible energy consumption [2]. Thus, the raw materials can be recovered several times through the recycling processes. Recycling is designed to reduce the consumption of energy and save natural resources. Up to date, we are far uptake from a locked-loop material cycle because we are not able to recycle all the waste materials (Fig. 2). A closed-loop platform has three fundamentals as following: gathering, processing, and purchase. The gathering is rerouting and collecting spent wastes for recycling and treatment. In the processing, the collected spent wastes are re-manifested for fabricating a new

Fig. 1 Recycling and recovery of waste

Fig. 2 Closed-loop materials cycle

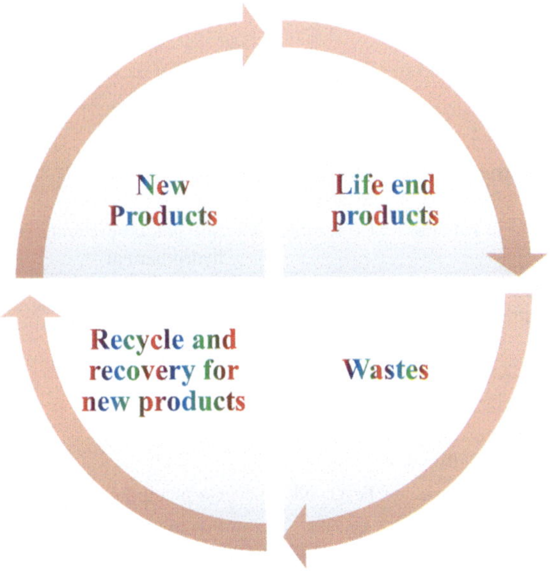

output. Finally, the freshly synthetic output is then sold to a tertiary party before reprocessing and repeat the operation once more. Regarding the entire closing of the spent recycle, the most crucial aspect is to beat several restrictions enjoined in reprocess like output layout, process thermodynamics, recovery methods, social behavior, and economics [3].

It could be announced that recovery and the up-scale recycling for several spent wastes (e.g., newspapers, glass bottles, cans, polymers, electronics, and batteries) are recently echoed in many countries of all the world [4]. Converting waste into fresh, valuable, and vendible products might be an essential strategy in boosting earnest recovery efforts to prospect desirable economic feedback with polity subsidization [5–10]. Several nanomaterials (metals, metal oxides, and hydroxides) such as ferrites, TiO_2, Ag, Fe, Fe–O, Au, ZnO, SiO_2, and hydroxyapatite have been recovered from waste materials by different recycling processes [11, 12]. These recovered nanomaterials could be extensively used in many applications, such as in renewable energy solar cells, catalysis, water remediation, and supercapacitors [13–15]. Many researchers have reported nanomaterials recovery from electronic waste by different recycling processes under available conditions in most research laboratories, as revealed in Table 1.

The spent batteries are the most predominant waste materials recycled to recover different nanomaterials [16–21]. Nowadays, batteries are widely used as mobile sources of energy for many purposes. Annual battery consumption is estimated at 8, 6, and 1 billion units annually in Europe/United States, Japan, and Brazil, respectively [22]. Despite these batteries made of many valuable materials that can be recovered, they are usually disposed as spent waste after their end life. Nevertheless, scientists recently made some attempts to recover these valuables metal and metal oxides from the spent batteries waste, such as wastes of Zn–C,

Table 1 Nano-metal-oxides recovered from recyclable waste

Waste	Oxides	Methods	Recovery (%)	Ref.
Zn–C, Zn–Mn batteries	Ni–Mn–Zn/ Mn–Zn ferrites	Co-precipitation, Sol–gel, and leaching	56 Fe and 97 of Mn and Zn	[16, 36, 37]
Li–ion batteries	Co ferrites	Sol–gel, co-precipitation, hydrothermal	–	[38–40]
MnO_2–Zn dry batteries	ZnO NPs	Liquid extraction, precipitation, and hydrometallurgy-	99	[41–43]
Zn–MnO_2 batteries	$Zn_xMn_{1-x}O$ NPs	Hydrometallurgy	57	[44]
Furnace slag, red mud, fly ash	TiO_2, SiO_2	–	–	[45]
Pickling waste	Fe_3O_4 NPs	Co-precipitation	–	[46]
Waste silver chloride	Ag NPs	Precipitation and reduction	–	[47]

Li–ion, Ni–MH, and zinc–manganese oxide alkaline batteries [8, 23–25]. Several nanoparticles (NPs) such as Fe, Co, Mn, Li, Ag, Au, Zn, Ti, Pd, Ni, Cu, and rare earth elements could be recovered from the spent batteries wastes [8, 24, 26, 27]. Analogous with these, various oxide nanomaterials such as Co_3O_4, MnO_2, ZnO, $ZnMn_2O_4$, Ni–Mn–Co–Zn ferrites, and MnO_2/RGO could also be synthesized from recovered materials from the spent batteries wastes [8, 23, 24, 27, 28].

Dry cell batteries (i.e., alkaline batteries) represent about 89% of commercially available batteries globally, whereas hundreds of million pieces of dry batteries are manufactured per year [29, 30]. Thus, the disposal of the spent dry batteries is a huge issue that cannot be neglected [23, 30, 31]. In various countries, gathering ways were established to manage the spent dry batteries wastes. However, a considerable quantity is still left to be castaway along with household spent. Burying, burning, and recycling are the three main administration routes for the spent dry batteries [32]. The highest practical strategies for the finite availability of sites were incineration and recycling. Recycling is also a viable solution for addressing these serious problems associated with spent waste [33–35].

Accordingly, this chapter aims to discuss the most novel and environmentally friendly technologies for recycling various batteries and electronic wastes to recover valuable metals oxide nanomaterials such as metal ferrites, zinc oxide, and indium tin oxide. The possible applications of recovered nanomaterials in different fields such as catalysis, water treatment, photocatalyst, electrocatalysis, and supercapacitors will also be demonstrated.

2 Recent Categories and Strategies of Metal Oxide Recovery

In the present subsections, we discuss a vast range of waste items regenerated and recovered to recuperate precious metals and metal oxides such as electrical machines, batteries, and sludges. As illustrated in Fig. 3, these waste materials can be categorized into three elementary classes, e.g., batteries (Category A), used instruments (Category B), and wastes and sludges (Category C). Several metal NPs, ferrites, and metal oxides could be recovered by recycling these three categories of waste materials using several step methods of solid/liquid spent, as discussed in several literature studies. However, the hydrometallurgical and pyrometallurgical approaches are the most popular processes that yield reusable oxides or ferrites at a considerable quantity. They work on co-precipitation methods and thermal handling. Figure 4 depicts and summarizes the main steps of the recycling and recovery processes of ultra-pure NPs from combustible wastes such as dry batteries and electronics using hydrometallurgical, pyrometallurgical or physical separation approaches. The main fundamentals of those pathways are discussed in the next sections.

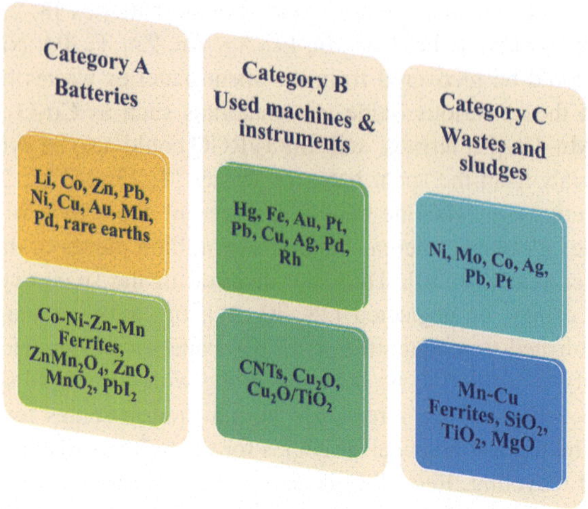

Fig. 3 Categories of recycling and fabrication of nanosized metals oxides and metals from spent/waste

2.1 *Hydrometallurgical Approach Pathway*

The hydrometallurgical process is one of the most powerful ways of recycling nanosized metals and metal oxides. As illustrated in Fig. 4, the hydrometallurgical process is divided into several steps, including the dismantlement of the life end products, washing and cleaning with water to uptake all undesirable digestion, alkali/alkaline salts, extraction, thermal treatment, and filtration [48]. The selective

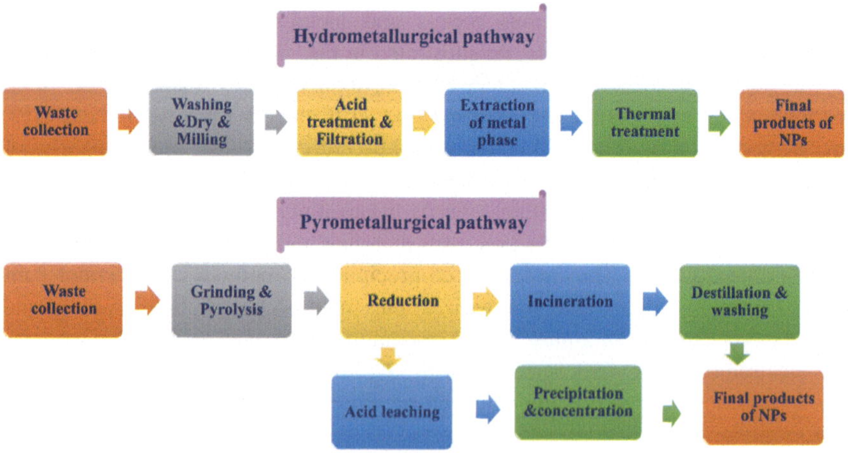

Fig. 4 Strategies of metal oxide recovery from spent/waste

leaching of metals could be done in acid (e.g., HCl and H_2SO_4) or alkaline (e.g., NH_3/NaOH) solutions. Similarly, efficient mineral extraction depends on several parameters such as the acid and base concentration, saturation solubility of the metal, solution pH, and reaction temperature. Furthermore, the effect of oxidants agents (e.g., H_2O_2, $KMnO_4$) and reductants agents (e.g., citric acid, acidic oxalate, and ascorbic acid) on the selective co-precipitation of metal ions was also inquired [49, 50]. Toro et al. [51] have reported a patented hydrometallurgical method for the large-scale production of the most valuable products using spent alkaline batteries. This method was based on alkaline batteries dissociation to split up fine (mostly manganese, graphite, and zinc) and coarse materials (paper, plastic, and steel), followed by washing and liquidation co-precipitation, cementing, and recrystallization. An agriculture fertilizer such as potassium sulfate was recovered [52], while the leftover reagents, including acid, reagents, and water, were directed back to leaching's reactor. The residue reagents were regenerated via simple strategies to reduce the wastes generated through nanomaterial fabrication. In this case study, the reagent solvents utilized to generate the nanomaterials were recycled and used later several times in the same preparation. Several other modern examples of laboratory-bench scale hydrometallurgical preparation of NPs exist as MnO_2/graphene nanocomposites [20], ZnO NPs [41, 42], $Zn_xMn_{1-x}O$ NPs [44], Zn substituted $MnFe_2O_4$ NPs [37], Co ferrites [37, 40], and Ni–Zn–Mn ferrites [53]. In the hydrometallurgy, we mostly use acid to leach the waste batteries and form soluble salts and then disinfect the product by adjusting the pH, extraction, co-precipitation, etc. [54]. Similarly, some valuable Zn and Mn could be extracted from spent Zn–Mn batteries using bio-hydrometallurgy, whereas bioleaching bacteria species were used to reduce the extraction time to 24 h regardless of energy source types [55].

2.2 *Pyrometallurgical Approach Pathway*

The pyrometallurgical approach pathway is an alternative conventional operation for the recycling and recovering precious metals and metal oxides from the spent electronic equipment wastes. Pyrometallurgical recovery and regenerating operations include mainly the following sequences: pyrolysis, reduction, distillation, and incineration [56]. Unlike hydrometallurgy, pyrometallurgical processes need high temperatures to generate metals and metal oxides. For instance, Umicore chemical company (Belgium) could regenerate seventeen different metals from spent electrical and electronic equipment by extracting the metal slags at high temperatures (i.e., 1200 °C).

Nevertheless, the formation of toxic or harmful fumes and gases meets a major barrier in the current period [57]. The pyrometallurgy recovery process can be performed under vacuum (i.e., vacuum metallurgy) or different atmospheric pressure (i.e., pressurized metallurgy). Although pressurized metallurgy has been used to recycle and recover zinc oxide from spent waste batteries, it is highly costly due to the latest tail gas treatment equipment. Thus, the vacuum metallurgy recovery process has recently received more attention due to its environmentally friendly advantages [58–63]. Because of the metals lower boiling points during vacuum, metal recycling by vacuum mining reduces energy compared to traditional thermal mining operations (pyrometallurgy processes) [64, 65].

2.3 *Physical Separation Approach*

Although wet recycling has attracted complete attention due to its mild reaction conditions, it shows high energy consumption and serious subaltern contamination, limiting its widespread application [66]. Subsequently, low-cost, high-efficiency, and environmentally friendly recovery processes should be considered. Physical separation is an alternative process applied to retrieve useful components from the spent fluorescent lamp, printed circuit board, and batteries. For instance, rare earth phosphorus could be recovered from the spent fluorescent lamps with a recovery efficiency of about 70–90% by the flotation method [67]. Similarly, polyvinyl chloride (PVC) could be recovered completely (i.e., 100%) from spent electronic plastics by froth flotation separation after treating the E-waste with Ca/CaO composite to hydrophilize the surface of PVC [68]. The electrode materials (e.g., $LiCoO_2$ and graphite) could also be recovered from the spent dry batteries wastes with a recovery rate of 98.99% via Fenton reagent-assisted flotation [69].

3 Recovery of Ferrites

Ferrites are a ceramic-like substance that has magnetic properties useful in many types of electronic devices. Ferrites are iron-based materials, generally gray or black, and exist in a polycrystalline form of many small crystals. Several other transition metals can be present in ferrites materials. For instance, Zn–Mn–Co ferrites consist of iron oxide and one or more other cations of these transition metals. Magnetic ferrite is widely valued in different fields due to its high electrical resistance, excellent magnetic properties, chemical stability, mechanical rigidity, and relatively low cost; these materials' properties vary significantly from the properties of bulk materials [70–74]. As illustrated in Table 1, the recycling and recovery of ferrites NPs from the spent wastes started in the last decade, approximately in 2004. Xi et al. are among the pioneering groups that established Zn–Mn magnetic ferrite NPs recovery from dry Zn–MnO_2 batteries after end life via co-precipitation process [75]. They found that the recovery process could be affected by several factors, such as co-precipitation pH and temperature of co-precipitation and calcination processes. Their studies showed that Zn–Mn magnetic ferrite NPs could be appropriately recovered using a co-precipitation pH of about 7.0–7.5 at 50 °C and calcination temperatures between 1100 and 1150 °C. As illustrated in Fig. 5, magnetic manganese ferrite ($MnFe_2O_4$) could be

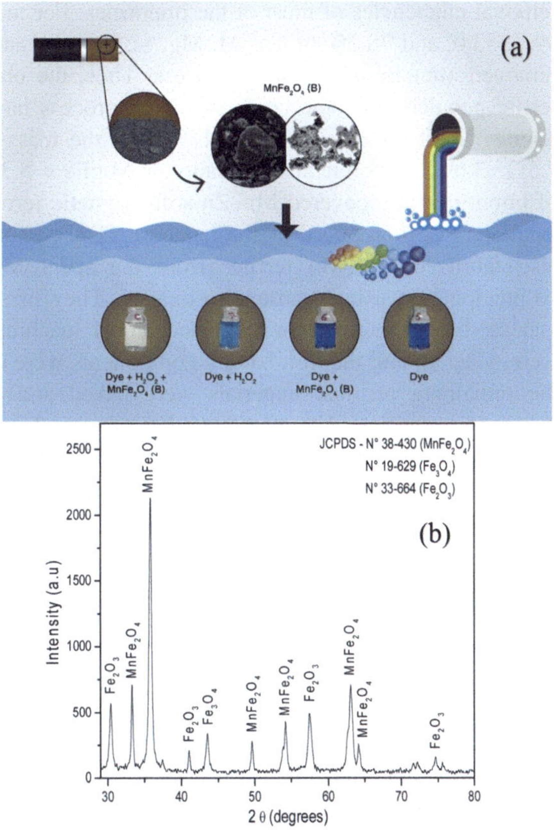

Fig. 5 (**a**) schematic representation of efficiency of $MnFe_2O_4$ for water remediation, (**b**) X-ray diffractogram of $MnFe_2O_4$ synthesized from spent batteries. Adapted with permission from Ref. [19], Copyright 2020, Elsevier

successfully synthesized from manganese solution recycled and recovered by leaching the cathode material of spent wastes of zinc–manganese oxide batteries with 0.5 mol L^{-1} nitric acid and 30% (v/v) H_2O_2 at 80 °C, followed by reaction with ferric chloride and calcination of obtained co-precipitate at 450 °C [19]. The obtained $MnFe_2O_4$ materials were a promising catalyst for heterogeneous photo-Fenton photodegradation of methylene blue dye, which accomplished 92% dye decolorization within 120 min irradiation [19].

Peng et al. [76] have reported the synthesis of Mn–Zn light magnetic ferrite NPs from the waste of Zn–Mn batteries (e.g., spent scrap iron and pyrolusite MnO_2) via leaching, purification, and co-precipitation processes. In the first step, the leaching of the spent materials in H_2SO_4 has yielded about 92.02, 96.14, and 98.34% of Fe, Mn, and Zn, respectively. They have then washed the leached solutions with ammonia, ammonium sulfhydrate, and ammonium sulfide to remove the heavy metals, $Ca^{2+,}$ and Mg^{2+} by vulcanized and fluorination precipitations (i.e., double salt precipitation deep purification). This deep purification process has led to high removal efficiencies of most of the impurities. For instance, 99.7, 96.5, 92.3, 63.6, 99.9, 53.0, and 78.7% of Ca, Al, Mg, Si, Cu, Cd, and Ni, respectively, have been removed after the purification process. Thus, the obtained Mn–Zn soft magnetic ferrite powders after the co-precipitation process have been found to contain Zn 4.49%, Mn 13.92%, and Fe 41.41% with the mass ratio of Mn:Zn:Fe equals to 7.5:23.3:69.2 (theoretical mass ratio Zn:Mn:Fe = 8.3:24.4:67.3). The low content of impurities in recovered Mn–Zn soft magnetic ferrite made them better than the synthetic PC30 produced by the Japanese TDK company [76]. Nan et al. [36] have also fabricated Zn–Mn ferrites from the spent Mn–Zn batteries by chemical co-precipitation using ammonium oxalate. They first dismantled spent zinc–manganese dioxide batteries with a mechanical machine. The dismantling materials were watered, and the iron battery components were collected using a magnet, and the remaining residual materials were baked and griddled to obtain zinc and manganese species. The separated Fe-, Zn- and Mn-species were dissolved in H_2SO_4 to obtain $FeSO_4$, $ZnSO_4$, and $MnSO_4$ solution, respectively, which were used in the synthesis of Zn–Mn ferrite magnetic materials with the composition of $Mn_{0.26}Zn_{0.24}FeO_2$ after the calcination of the co-precipitated precursor at 850–1250 °C [36]. Xi et al. [18] have fabricated nanocrystalline ferrites from the spent alkaline Zn–Mn batteries using sol-gel and combustion methods. The gel containing Fe, Mn, and Zn cations was first prepared by dissolving the spent battery materials in an HNO_3 solution containing 2.5 wt% H_2O_2, followed by the addition of citric acid and ammonia and drying at 135 °C for 2 h. The XRD, TGA, DTA, SEM, ICP, and FTIR characterizations showed that the dried gels are self-propagating combustion led to the nanocrystalline formation of manganese zinc ferrites with the stoichiometry $MnZn_{0.85}Fe_2O_4$ [18]. However, it is worth mentioning that they have used analytical grade manganese nitrate, zinc nitrate, and iron nitrates to adjust the concentration of Mn, Zn, and Fe to obtain this stoichiometric spinel.

Similarly, Mylarappa et al. [17] have also prepared nanocrystalline zinc–manganese ferrites from the spent Zn–C batteries as raw materials by dissolution in H_2SO_4 followed by co-precipitation in NaOH at 100 °C. They used the obtained zinc–manganese ferrites to decorate reduced graphene oxide (rGO) by using a solvothermal route. Their GO/Mn–Zn ferrite nanocomposite showed excellent photoactivity performance to decompose acid orange 88 dye in wastewater under ultraviolet irradiated, as illustrated in Fig. 6 [17].

Several harmful byproducts can be produced in the stainless-steel industries, such as fly ash and ultrafine particle, which can easily bio-accumulate leak and snuff, and pose a critical menace to the humans and environment of the world [77–80]. Magnetic ferrites have also been recovered and purified from industrial fly ash wastes. The magnetic ferrites (e.g., γ-Fe_2O_3 and Fe_3O_4) from fly ash waste depend on the interactions between magnetic ferrites with the applied magnetic field [79]. Lin et al. have utilized a magnetic separator to separate magnetic Ni-ferrites NPs from a steel industrial fly ash (RFSIF) with a good separation factor of 74%. These Ni-ferrites NPs were found suitable for various applications, especially as a catalyst

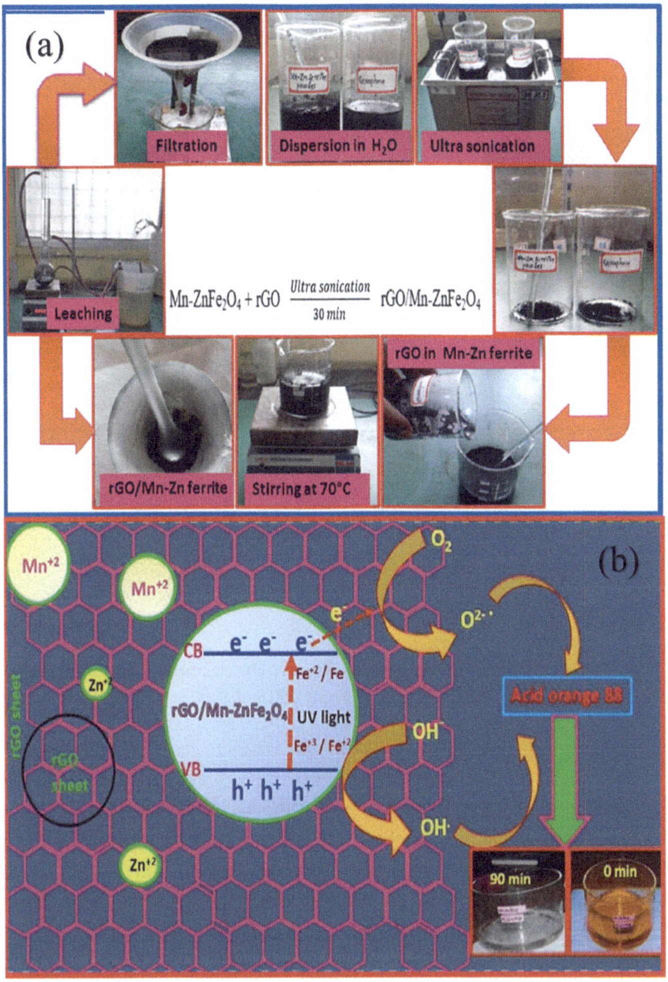

Fig. 6 (**a**) Schematic representation of recovery of nanosized Mn–Zn ferrites using Zn–C dry batteries, (**b**) mechanism of acid orange 88 decomposition for Mn–Zn ferrite-rGO nanocomposite. Adapted with permission from Ref. [17], Copyright 2019, Elsevier

for the hydrolysis and methanation of CO_2 using fluid hydrogen gas, as shown in Fig. 7 [81].

Magnetic cobalt ferrites are another class of ferrites materials with distinct chemical, physical, and mechanical properties [40, 82, 83]. For instance, they typically demonstrate a low-frequency magnetic permeability and a high magnetic polarization angle in visible light at ambient temperature [40, 82, 83]. Cobalt-ferrite also has excellent chemical consistency and high wear resistance to corrosion [84]. Magnetic cobalt ferrites have also been applied as a microwave absorber [85]. As shown in Fig. 8, spinel cobalt ferrites, NPs with the composition of $Co_{0.8}Fe_{2.2}O_4$,

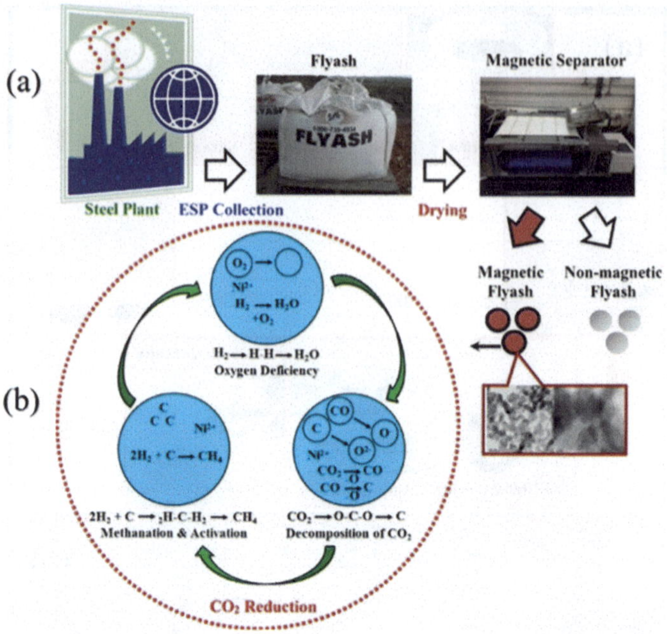

Fig. 7 (**a**) Recovery of magnetic ferrite from fly ash (**b**) decomposition/methanation of CO_2. Adapted with permission from Ref. [81], Copyright 2018, Elsevier

have been fabricated from the spent cathode of lithium–ion batteries as raw materials by sol-gel route using ethylene glycol as a gelling agent [40]. TEM, XRD, XPS, FTIR, SEM, and VSM characterizations showed that the synthesis parameters could influence the cobalt ferrites' chemical structure, morphology, and magnetic properties. Spinel $Co_{0.8}Fe_{2.2}O_4$ NPs with a saturated magnetization of around 61.96 emu g L^{-1} could be obtained at the optimum synthesis parameters of metal ion to ethylene glycol ratio $\sim$1:0.8, annealing temperature $\sim$800 °C, and the annealing time about 6 h [40].

4 Recovery of Zinc Oxide

ZnO has shown remarkable implementation prospects in treating environmental contamination, particularly in organic wastewater remediation [86]. Researchers worldwide have proposed different recovery methods of Zn-based materials, which can mainly be divided into pyrometallurgy and hydrometallurgy. In China, about 373 Gg of ZnO have been utilized to manufacture batteries in 2003 and have increased to 535 Gg in 2005 [87]. NPs could be effectively prepared from the spent Zn–Mn dry batteries by high-temperature evaporation-separation and oxygen

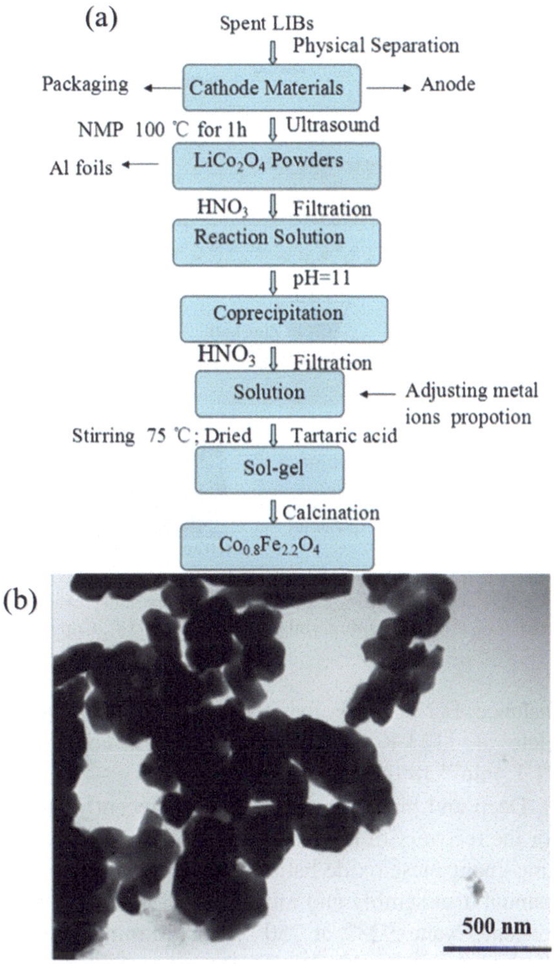

Fig. 8 (**a**) Schematic chart of magnetic cobalt-ferrite spinel prepared from spent Li–ion batteries, (**b**) TEM analysis of the magnetic cobalt-ferrite spinel elaborated from Li–ion batteries. Adapted with permission from Ref. [40], Copyright 2016, Elsevier

control oxidation using air as both carrier gas and oxidizer [88]. Although carbon powder or lead powder were used as additional materials to isolate the oxygen, zinc's oxidation rate was still fast. However, Zn could be first evaporated and recovered from the zinc hull covered with fiber mat with a recovery efficiency of ~98.99% and then oxidized into nanotetrapod ZnO by using a heating temperature of 1123 K, air pressure of 3 kPa, and blowing airflow at 723 K [88]. Nevertheless, due to the unstable content and flow rate of oxygen, the morphology of the substance obtained in this way is not standardized. ZnO NPs with variable morphologies have been obtained from the spent Zn–Mn battery by vacuum evaporation and oxygen control method [89]. In this method, the morphology of ZnO NPs could be controlled by adjusting the flow rate of the N_2 conveyor, the content of O_2 oxidizer, calcination temperature, used substrate, and condensing

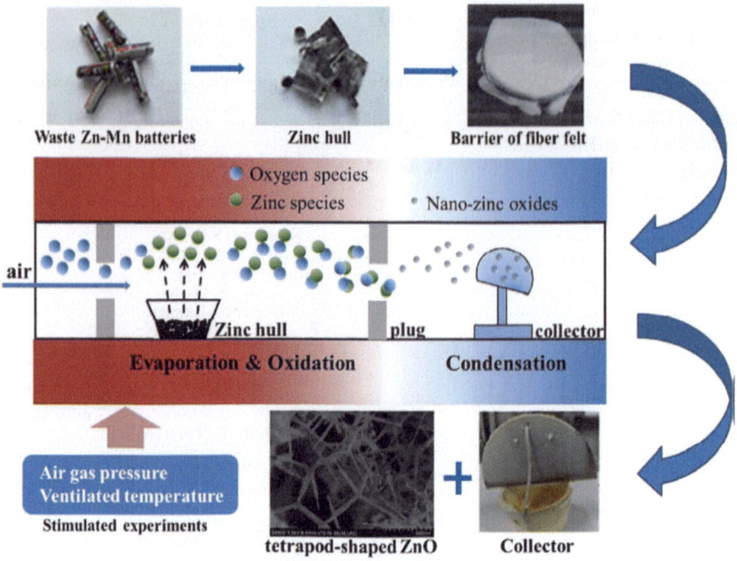

Fig. 9 Recovery of zinc oxide by high-temperature evaporation–separation and oxygen control oxidation. Adapted with permission from Ref. [88], Copyright 2018, American Chemical Society

distance. For instance, nanotetrapod ZnO photocatalyst could be formed on a glass plate at 1123 K under 12.5% oxygen content and a nitrogen flow rate of 21 L min^{-1} [89] (Fig. 9).

Deep and his co-worker [42] have recently proposed a novel and easy method for the recovery of highly pure ZnO NPs from the electrode waste of spent alkaline zinc–manganese oxide batteries. The spent electrode materials were gathered using manual disassembly and mixed with 5 M HCl to interact with the phosphine oxide reagent Cyanex923® at 250 °C for 30 min to form Zn–Cyanex 923 complex, as illustrated in Fig. 10. With approximately 5 nm diameter, ZnO NPs were then prepared from the Zn–Cyanex 923 complex by an ethanolic precipitation process. Therefore, the proposed process provides an easy and efficient way to regenerate ultra-pure ZnO NPs from the spent dry waste batteries compared with other high-temperature evaporation-separation techniques oxygen control oxidation.

5 Recovery of Indium Tin Oxide

Recently, the world's indium reserves are around 12,000–13,000 tons, representing only one-sixth of the gold reserves; access to 0.03% indium content is considered a mining value [90, 91]. The binary and ternary indium oxides have recently attracted much scientific and technological interest among indium compounds due to their fascinating structural chemistry and promising functionalities [92–95].

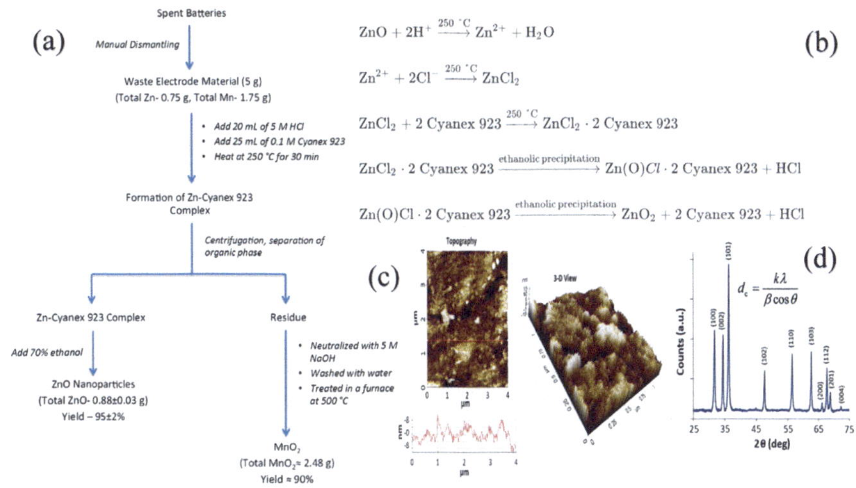

Fig. 10 (**a**) Schematic flowchart of ZnO NPs recovery from spent alkaline batteries, (**b**) equation mechanism of ZnO NPs recovery, (**c**) AFM topography, line analysis, and 3D view of the recovered ZnO NPs, (**d**) XRD analysis of the recovered ZnO NPs. Adapted with permission from Ref. [42], Copyright 2016, Elsevier

Indium-based materials are widely applied in radio-electronic, spintronic, and semiconductor manufacturing processes due to its special electrical, chemical, and physical properties [96–100]. Thus, it is considered an essential strategic resource in recent industrial growth [101]. More than 80% of extracted indium in the world is employed in the fabrication of indium tin oxide (ITO) materials [102] that contain more than 90% indium oxide and less than 10% tin oxide [103]. ITO is widely used to fabricate the touch display of smartphones and liquid-crystal display panels (LCD) [104]. Thus, the amount of indium in the world is reduced with time, and it is getting expansive due to its scarcity on the earth [105]. The recovery of indium from the wastes of LCD devices should be urgently regulated [106].

As illustrated in Fig. 11, the board of LCD consists of two polarizing films, two glasses substrate covered by the ITO layer, a color filter film, and a liquid-crystal band. Because of the large indium content in an ITO film, the waste LCD is considered a potential secondary source of indium [107]. However, the recovery of In from LCD screens faces some problems due to the harmful impurities and components in the display (e.g., liquid crystals, plastics, and mercury) that can be released during the dismantling process of the LCD [108]. To date, most of the studies on the recovery of indium from waste LCD panels focused on the aqueous hydrometallurgical pathway. For instance, Li and his co-workers [109] have reported the recovery of indium from the waste LCD screens using a multi-step process, including the removal of LCD polarizing film and liquid crystals by thermal shock at 230–240 °C and the ultrasonic cleaning at 40 kHz, respectively, and the recovery of indium from ITO glass by acid dissolution. This multi-step process led to the recovery of 92 wt% of indium metal when a solution mixture of

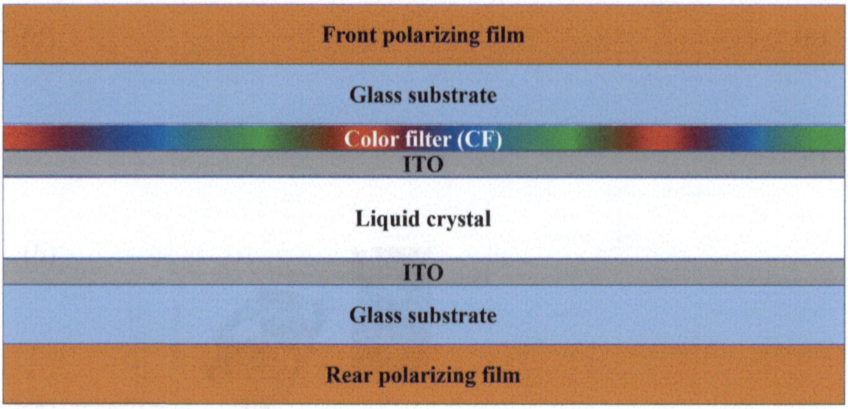

Fig. 11 Configuration of the LCD panel

HCl and HNO_3 acids with the ratio of $HCl:HNO_3:H_2O$ = 45:5:50, and aging temperature and time of 60 °C and 30 min, respectively, are used. Lee et al. [110] have reported that the particle size of ITO materials crushed by high energy ball milling and the amount of HCl in the acidic solution could strongly affect the recovery efficiency of indium from the waste LCD panels. They could rapidly recover 86% of the indium in the waste materials after 1 min of ball milling using the acid solution ($HCl:H_2O$ = 50:50). Solvent extraction, which is based on sulfuric acid leaching, di(2-ethylhexyl)phosphoric acid extraction, and HCl back extraction, is another alternative approach to recover indium from thin-film-transistor liquid-crystal display (TFT-LCDs) waste [111]. In this method, 97% of indium in the raw materials could be recovered by leaching the waste materials H_2SO_4 (1:1, v/v) acid solution, followed by two extraction process using 30% of di (2-ethylhexyl)phosphoric acid solution (O/A = 1:5) for 5 min, and 4 M HCl acid solution. Hsieh and his co-worker [112] have reported the recovery of highly pure sponge indium (>99%) from spent ITO waste using hydrometallurgical in 30 vol.% HCl solution at 80 °C for 30 min, followed by hot immersion in KOH solution at 300 °C for 10 min.

In addition to the hydrometallurgical method, chlorination, vacuum carbon reduction, sub-critical water have also been applied to recover indium from ITO waste. For instance, Ma and Xu [113] have recovered 99.97% of indium in the ITO waste as indium chloride by using vacuum pyrolysis at 300 °C and 50 Pa to separate oil and organic materials, followed by vacuum chlorinated separation using 50 wt% NH_4Cl at 450 °C. Similarly, 66.7 and 54.1% of indium could be recovered from LCD waste in the form of indium chloride by chloride volatilization process under air and N_2, respectively, using polyvinyl chloride (PVC) as chlorine agent at 350 °C and a Cl/In a molar ratio of 11 [114]. Zhang and his co-worker [115] could also prepare indium chloride with 36% of indium from waste LCD panels containing 0.02 wt% in a multi-step process including mechanical stripping, a series of pyrolysis separation, chlorinated vacuum separation, and substitution reaction.

He et al. [116] have reported that 90 wt% of indium could be recovered from waste LCD panel by vacuum carbon reduction using 30 wt% carbon as additives at 1223 K and 1 Pa for 30 min. Yoshida et al. [117] have demonstrated that 83 and 10% of indium could be recovered from stained glass (CF) and TFT glass of LCD waste, respectively, using semi-critical water 360 °C. They also found that NaOH addition during the recovery process could increase the recovery efficiency of indium from both CF and TFT glasses to 99 and 95%, respectively, at 220 °C after 5 min. Indium could also be recovered as In^{3+} ions from LCD waste via the microbial adsorption method, whereas a gram-negative bacterium could adsorb In^{3+} ions generated by leaching the spent LCD panel in HCl solution under hydrothermal conditions (120 °C, 0.198 MPa, 5 min) [118].

Even if indium tin oxide is the target, the CF glass is the primary medium because ITO adheres to the CF, easily removed by crushing from the glass (Fig. 11). The flotation process has also been utilized to recover ITO from color filter (CF) glass in waste LCD panels, as illustrated in Fig. 12 [119]. ITO could be easily separated from glass for indium recycling by the floatability of CF due to its hydrophobicity caused by hydrophobic methyl groups and aromatic ring. The separation of CF containing ITO from the glass substrate was found to depend on the feed size of crushed materials, and the recovery of indium and ITO increased with decreasing feed size. For instance, 85.21% of ITO containing 0.61% indium could be recovered for particles with a size smaller than 0.025 mm. In addition, the indium oxide concentration and the recovery of indium tin oxide increased with increasing grinding time. The 92.51% of ITO containing 0.60% indium was recovered after 80 min of milling.

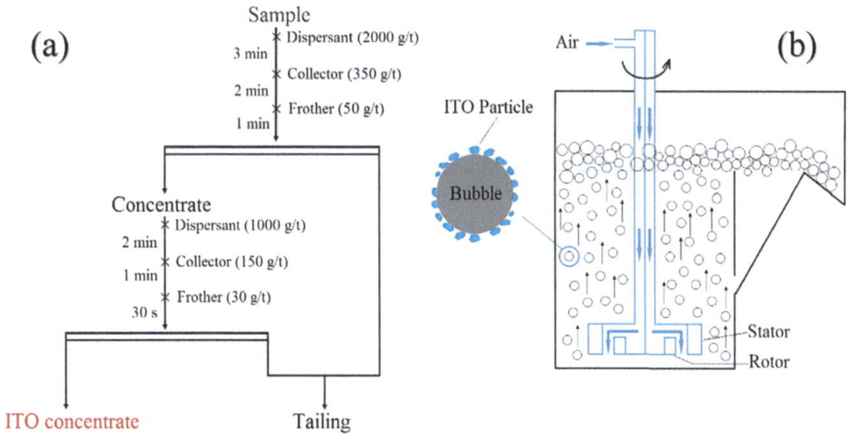

Fig. 12 (**a**) Flowsheet of ITO recovery by flotation, (**b**) schematic diagram for ITO and glass separation by flotation. Adapted with permission from Ref. [119], Copyright 2018, Elsevier

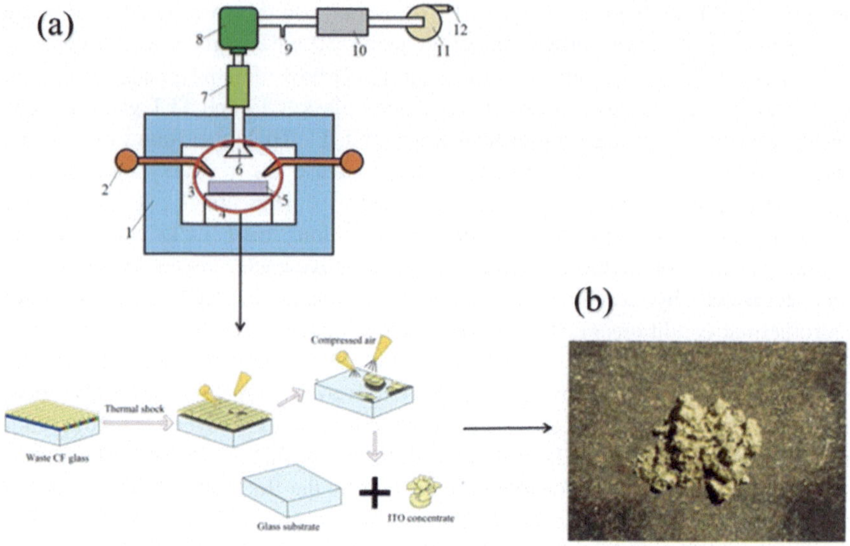

Fig. 13 (**a**) Recovery of ITO by thermal decomposition of waste color filter glass: 1) muffle furnace; 2) gas storage; 3) nozzle; 4) sample holder; 5) CF glass panel; 6) dust collecting cover; 7) condenser; 8) dust collector; 9) gas sampling point 1; 10) activated carbon box; 11) induced draft fan; and 12) gas sampling point 2. (**b**) optical image of recovery ITO powder Adapted with permission from Ref. [120], Copyright 2020, Elsevier

Thermal decomposition is another approach that has been applied to recover ITO from the waste color filter (CF) glass, as illustrated in Fig. 13 [120]. The ITO layer could be first separated from the glass substrate after the color layer's thermal oxidation. Although the yield of recovered ITO was found to decrease with increasing temperature and time of the thermolysis process, the recovery and enrichment ratio was increased. After thermolysis at 600 °C for 8 min, 98% of ITO could be efficiently recovered with yield and enrichment ratio of 0.06% and 1669, respectively. Moreover, the crystal structure of recovered ITO has not been changed during the thermolysis process. It is worth noting that the authors have used activated carbon during thermolysis to adsorb the released toxic mixed flue gas that might harm the atmospheric environment [120].

6 Conclusions

There is no dispute concerning the significance of recycling spent waste and the recovery and regeneration of high-value products, including minerals, NPs, and other materials. Nevertheless, the literature in this work remains to be explained inappropriate technology due to the limitation of cost-effectiveness. Moreover, although recycling is beneficial from an environmental point of view, economic sustainability cannot often be achieved for an advanced process without subsidizing the country due to low or non-existent profits. However, new methods appear more favorable, perhaps for modern trade models to become more profitable than debris. With the scheme increasing by 21% annually in the global depletion of metal oxides by 2019, there is a critical need to generate potential metal oxide supplies. Conventional processes for converting bulk materials into metal oxides NPs can increase supply slightly. Recycling such materials from recycled expenses thus opens up the possibility that some of the most important materials will create a secondary supply chain. Overall, we have found that most of the currently available recycling technologies need more refinement to produce finished commercial goods.

7 Future Prospective

Furthermore, the reproduction of specific recovery methods and stability should be dealt with more widely. Reuse of materials that have been produced or degraded after application for various treatment purposes is crucial, especially as production costs for NPs with high-purity metal oxides are generally high. The replication and degradability or reuse of recycled metal oxides should be given greater importance to close existing research gaps. For the nanostructures to survive, a systematic approach is also required to determine risks associated with the use and disposal of the major metal oxides associated with the product. The evolution of the process for assessing toxicity that predicts potential impacts on human health and the environment is still critical. Finally, applying a green process may significantly raise the benefits of pilot projects, and it should be taken into consideration.

References

1. Al-Salem SM, Lettieri P, Baeyens J (2009) Recycling and recovery routes of plastic solid waste (PSW): a review. Waste Manag 29(10):2625–2643
2. Ordoñez J, Gago EJ, Girard A (2016) Processes and technologies for the recycling and recovery of spent lithium-ion batteries. Renew Sustain Energy Rev 60:195–205
3. Reck BK, Graedel TE (2012) Challenges in metal recycling. Science 337(6095):690–695
4. Izatt RM, Izatt SR, Bruening RL, Izatt NE, Moyer BA (2014) Challenges to achievement of metal sustainability in our high-tech society. Chem Soc Rev 43(8):2451–2475
5. Söderholm P, Ekvall T (2019) Metal markets and recycling policies: impacts and challenges. Miner Econ 33(1–2):257–272
6. Ali GAM, Supriya S, Chong KF, Shaaban ER, Algarni H, Maiyalagan T, Hegde G (2019) Superior supercapacitance behavior of oxygen self-doped carbon nanospheres: a conversion of Allium cepa peel to energy storage system. Biomass Conv Bioref. https://doi.org/10.1007/s13399-019-00520-3
7. Ali GAM, Habeeb OA, Algarni H, Chong KF (2018) CaO impregnated highly porous honeycomb activated carbon from agriculture waste: symmetrical supercapacitor study. J Mater Sci 54:683–692

8. Aboelazm EAA, Ali GAM, Algarni H, Yin H, Zhong YL, Chong KF (2018) Magnetic electrodeposition of the hierarchical cobalt oxide nanostructure from spent lithium-ion batteries: its application as a supercapacitor electrode. J Phys Chem C 122(23):12200–12206
9. Habeeb OA, Ramesh K, Ali GAM, Yunus RM (2017) Low-cost and eco-friendly activated carbon from modified palm kernel shell for hydrogen sulfide removal from wastewater: adsorption and kinetic studies. Desalin Water Treat 84:205–214
10. Habeeb OA, Ramesh K, Ali GAM, Yunus RM (2017) Experimental design technique on removal of hydrogen sulfide using CaO-eggshells dispersed onto palm kernel shell activated carbon: experiment, optimization, equilibrium and kinetic studies. J Wuhan University Technol-Mater Sci Ed 32(2):305–320
11. Bensalah H, Bekheet MF, Younssi SA, Ouammou M, Gurlo A (2018) Hydrothermal synthesis of nanocrystalline hydroxyapatite from phosphogypsum waste. J Environ Chem Eng 6(1):1347–1352
12. Soltani N, Simon U, Bahrami A, Wang X, Selve S, Epping JD, Pech-Canul MI, Bekheet MF, Gurlo A (2017) Macroporous polymer-derived SiO_2/SiOC monoliths freeze-cast from polysiloxane and amorphous silica derived from rice husk. J Eur Ceram Soc 37(15):4809–4820
13. Witik RA, Teuscher R, Michaud V, Ludwig C, Månson J-AE (2013) Carbon fibre reinforced composite waste: an environmental assessment of recycling, energy recovery and landfilling. Compos A Appl Sci Manuf 49:89–99
14. Bensalah H, Younssi SA, Ouammou M, Gurlo A, Bekheet MF (2020) Azo dye adsorption on an industrial waste-transformed hydroxyapatite adsorbent: kinetics, isotherms, mechanism and regeneration studies. J Environ Chem Eng 8(3):103807
15. Abdullayev A, Bekheet MF, Hanaor DA, Gurlo A (2019) Materials and applications for low-cost ceramic membranes. Membranes 9(9):105
16. Kim TH, Senanayake G, Kang JG, Sohn JS, Rhee KI, Lee SW, Shin SM (2009) Reductive acid leaching of spent zinc–carbon batteries and oxidative precipitation of Mn–Zn ferrite nanoparticles. Hydrometallurgy 96(1–2):154–158
17. Mylarappa M, Venkata Lakshmi V, Vishnu Mahesh KR, Nagaswarupa HP, Raghavendra N (2019) Recovery of Mn–Zn ferrite from waste batteries and development of rGO/Mn–Zn ferrite nanocomposite for water purification. Mater Today Proc 9:256–265
18. Xi G, Yang L, Lu M (2006) Study on preparation of nanocrystalline ferrites using spent alkaline Zn–Mn batteries. Mater Lett 60(29–30):3582–3585
19. Morais VS, Barrada RV, Moura MN, Almeida JR, Moreira TFM, Gonçalves GR, Ferreira SAD, Lelis MFF, Freitas MBJG (2020) Synthesis of manganese ferrite from spent Zn–MnO_2 batteries and its application as a catalyst in heterogeneous photo-Fenton processes. J Environ Chem Eng 8(3):103716
20. Deng J, Wang X, Duan X, Liu P (2015) Facile preparation of MnO_2/graphene nanocomposites with spent battery powder for electrochemical energy storage. ACS Sustain Chem Eng 3(7):1330–1338
21. Duan X, Deng J, Wang X, Guo J, Liu P (2016) Manufacturing conductive polyaniline/graphite nanocomposites with spent battery powder (SBP) for energy storage: a potential approach for sustainable waste management. J Hazard Mater 312:319–328
22. Xu J, Thomas HR, Francis RW, Lum KR, Wang J, Liang B (2008) A review of processes and technologies for the recycling of lithium-ion secondary batteries. J Power Sources 177 (2):512–527
23. Ali GAM, Tan LL, Jose R, Yusoff MM, Chong KF (2014) Electrochemical performance studies of MnO_2 nanoflowers recovered from spent battery. Mater Res Bull 60:5–9
24. Ali GAM, Yusoff MM, Shaaban ER, Chong KF (2017) High performance MnO_2 nanoflower supercapacitor electrode by electrochemical recycling of spent batteries. Ceram Int 43:8440–8448
25. Aboelazm EAA, Ali GAM, Chong KF (2018) Cobalt oxide supercapacitor electrode recovered from spent lithium-ion battery. Chem Adv Mater 3:67–74

26. Ali GAM (2020) Recycled MnO_2 nanoflowers and graphene nanosheets for low-cost and high performance asymmetric supercapacitor. J Electron Mater 49:5411–5421
27. Ali GAM, Yusoff MM, Algarni H, Chong KF (2018) One-step electrosynthesis of MnO_2/rGO nanocomposite and its enhanced electrochemical performance. Ceram Int 44:7799–7807
28. Song M, Cho Y, Yoon D, Nahm S, Oh S, Woo K, Ko J, Cho W (2014) Solvothermal synthesis of $ZnMn_2O_4$ as an anode material in lithium ion battery. Electrochim Acta 137:266–272
29. Smith WN, Arutunian M, Swoffer S (2013) Process for recycling alkaline batteries. U.S. Patent 8,586,218.
30. Ma Y, Cui Y, Zuo X, Huang S, Hu K, Xiao X, Nan J (2014) Reclaiming the spent alkaline zinc manganese dioxide batteries collected from the manufacturers to prepare valuable electrolytic zinc and $LiNi_{0.5}Mn_{1.5}O_4$ materials. Waste Manag 34(10):1793–1799
31. Ali GAM, Yusoff MM, Feng CK (2015) Electrochemical properties of electrodeposited MnO_2 nanoparticles. Adv Mater Res 1113:550–553
32. Xará S, Almeida MF, Costa C (2015) Life cycle assessment of three different management options for spent alkaline batteries. Waste Manag 43:460–484
33. Ali GAM, Manaf SAA, Divyashree A, Chong KF, Hegde G (2016) Superior supercapacitive performance in porous nanocarbons. J Energy Chem 25(4):734–739
34. Hegde G, Abdul Manaf SA, Kumar A, Ali GAM, Chong KF, Ngaini Z, Sharma KV (2015) Biowaste sago bark based catalyst free carbon nanospheres: waste to wealth approach. ACS Sustain Chem Eng 5(9):2247–2253
35. Ali GAM, Abdul Manaf SA, Kumar A, Chong KF, Hegde G (2014) High performance supercapacitor using catalysis free porous carbon nanoparticles. J Phys D Appl Phys 47 (49):495307–495313
36. Nan J, Han D, Cui M, Yang M, Pan L (2006) Recycling spent zinc manganese dioxide batteries through synthesizing Zn–Mn ferrite magnetic materials. J Hazard Mater 133(1–3):257–261
37. Gabal MA, Al-Harthy EA, Al Angari YM, Abdel Salam M, Asiri AM (2016) Synthesis, characterization and magnetic properties of MWCNTs decorated with Zn-substituted $MnFe_2O_4$ nanoparticles using waste batteries extract. J Magn Magn Mater 407:175–181
38. Yao L, Xi Y, Xi G, Feng Y (2016) Synthesis of cobalt ferrite with enhanced magnetostriction properties by the sol−gel−hydrothermal route using spent Li–ion battery. J Alloy Compd 680:73–79
39. Xi G-x, Jiao Y-z, Lu M-x (2008) Preparation of $CoFe_2O_4$ nanocrystal from spent lithium-ion batteries with coprecipitation method. Electron Compon Mater 27(5):19
40. Yang L, Xi G, Lou T, Wang X, Wang J, He Y (2016) Preparation and magnetic performance of $Co_{0.8}Fe_{2.2}O_4$ by a sol–gel method using cathode materials of spent Li–ion batteries. Ceram Int 42(1):1897–1902
41. Deep A, Kumar K, Kumar P, Kumar P, Sharma AL, Gupta B, Bharadwaj LM (2011) Recovery of pure ZnO nanoparticles from spent Zn–MnO_2 alkaline batteries. Environ Sci Technol 45(24):10551–10556
42. Deep A, Sharma AL, Mohanta GC, Kumar P, Kim K-H (2016) A facile chemical route for recovery of high quality zinc oxide nanoparticles from spent alkaline batteries. Waste Manag 51:190–195
43. Mantuano DP, Dorella G, Elias RCA, Mansur MB (2006) Analysis of a hydrometallurgical route to recover base metals from spent rechargeable batteries by liquid–liquid extraction with Cyanex 272. J Power Sources 159(2):1510–1518
44. Qu J, Feng Y, Zhang Q, Cong Q, Luo C, Yuan X (2015) A new insight of recycling of spent Zn–Mn alkaline batteries: synthesis of $Zn_xMn_{1-x}O$ nanoparticles and solar light driven photocatalytic degradation of bisphenol A using them. J Alloy Compd 622:703–707
45. Bennett JA, Wilson K, Lee AF (2016) Catalytic applications of waste derived materials. J Mater Chem A 4(10):3617–3637

46. Tang B, Yuan L, Shi T, Yu L, Zhu Y (2009) Preparation of nano-sized magnetic particles from spent pickling liquors by ultrasonic-assisted chemical co-precipitation. J Hazard Mater 163(2–3):1173–1178
47. Khaloo SS, Torabbeigi M, Jazani RK, Douraghi M, Ghalavand Z (2013) Laboratory waste minimization by recovery of silver as nano-silver colloidal dispersion from waste silver chloride. J Mater Cycles Waste Manage 15(3):342–347
48. Buzatu T, Popescu G, Birloaga I, Săceanu S (2013) Study concerning the recovery of zinc and manganese from spent batteries by hydrometallurgical processes. Waste Manag 33 (3):699–705
49. Sayilgan E, Kukrer T, Yigit NO, Civelekoglu G, Kitis M (2010) Acidic leaching and precipitation of zinc and manganese from spent battery powders using various reductants. J Hazard Mater 173(1–3):137–143
50. Provazi K, Campos BA, Espinosa DCR, Tenório JAS (2011) Metal separation from mixed types of batteries using selective precipitation and liquid–liquid extraction techniques. Waste Manag 31(1):59–64
51. Toro L, Veglio F, Beolchini F, Pagnanelli F, Zanetti M, Furlani G (2006) Process and plant for the treatment of run-down batteries. European Patent Appl EP 1:684
52. Giacinto V (1967) Process for recovering potassium sulfate from final mother liquors of kaineite. U.S. Patent 3,348,913.
53. Wen X, Qiao X, Han X, Niu L, Huo L, Bai G (2015) Multifunctional magnetic branched polyethylenimine nanogels with in-situ generated Fe_3O_4 and their applications as dye adsorbent and catalyst support. J Mater Sci 51(6):3170–3181
54. Lannoo S, Vilas-Boas A, Sadeghi SM, Jesus J, Soares HMVM (2019) An environmentally friendly closed loop process to recycle raw materials from spent alkaline batteries. J Clean Prod 236:117612
55. Xin B, Jiang W, Aslam H, Zhang K, Liu C, Wang R, Wang Y (2012) Bioleaching of zinc and manganese from spent Zn–Mn batteries and mechanism exploration. Biores Technol 106:147–153
56. Cui J, Zhang L (2008) Metallurgical recovery of metals from electronic waste: a review. J Hazard Mater 158(2–3):228–256
57. Oliveira CRd, Bernardes AM, Gerbase AE (2012) Collection and recycling of electronic scrap: a worldwide overview and comparison with the Brazilian situation. Waste Manag 32 (8):1592–1610
58. Zhan L, Xu Z (2008) Application of vacuum metallurgy to separate pure metal from mixed metallic particles of crushed waste printed circuit board scraps. Environ Sci Technol 42 (20):7676–7681
59. Zhan L, Xu Z (2011) Separating and recovering Pb from copper-rich particles of crushed waste printed circuit boards by evaporation and condensation. Environ Sci Technol 45 (12):5359–5365
60. Zhan L, Xu Z (2009) Separating and recycling metals from mixed metallic particles of crushed electronic wastes by vacuum metallurgy. Environ Sci Technol 43(18):7074–7078
61. Xing M, Zhang F-S (2011) Nano-lead particle synthesis from waste cathode ray-tube funnel glass. J Hazard Mater 194:407–413
62. Saitoh S, Melber A, Ohbayashi H, Nakai S, Matsuoka M, Hosomi M (2008) Simplified heating time calculation using the schmidt graphical method for PCB-contaminated capacitors undergoing the vacuum thermal recycling process. J Mater Cycles Waste Manage 10(1):71–78
63. Ohbayashi H, Hosomi M, Kanbe H, Melber A, Bruckamp J (2002) Removal of polychlorinated biphenyls from capacitors and pressure-sensitive paper by vacuum thermal recycling. Waste Manag 22(1):91–98
64. Saotome Y, Nakazawa Y, Yamada Y (1999) Disassembling and materials recovering process of alkaline manganese dry batteries by vacuum-aided recycling systems technology (VARS Tech.). Vacuum 53(1–2):101-104

65. Huang K, Li J, Xu Z (2009) A novel process for recovering valuable metals from waste nickel−cadmium batteries. Environ Sci Technol 43(23):8974–8978
66. Kaya M (2016) Recovery of metals and nonmetals from electronic waste by physical and chemical recycling processes. Waste Manag 57:64–90
67. Hirajima T, Bissombolo A, Sasaki K, Nakayama K, Hirai H, Tsunekawa M (2005) Floatability of rare earth phosphors from waste fluorescent lamps. Int J Miner Process 77 (4):187–198
68. Mallampati SR, Heo JH, Park MH (2016) Hybrid selective surface hydrophilization and froth flotation separation of hazardous chlorinated plastics from E-waste with novel nanoscale metallic calcium composite. J Hazard Mater 306:13–23
69. He Y, Zhang T, Wang F, Zhang G, Zhang W, Wang J (2017) Recovery of $LiCoO_2$ and graphite from spent lithium–ion batteries by Fenton reagent-assisted flotation. J Clean Prod 143:319–325
70. Kachniarz M, Salach J, Szewczyk R, Bieńkowski A (2015) Temperature Influence on the magnetic characteristics of Mn–Zn ferrite materials. In: Advances in intelligent systems and computing, Springer International Publishing, pp 121–127.
71. Karpova TS, Vasil'ev VG, Vladimirova EV, Nosov AP (2012) Synthesis of ferrite spinel $NiFe_2O_4$ by thermal hydrolysis and its magnetic properties. Inorg Mater Appl Res 3(2):107–112
72. Cao C, Xia A, Liu S, Tong L (2013) Synthesis and magnetic properties of hydrothermal magnesium–zinc spinel ferrite powders. J Mater Sci: Mater Electron 24(12):4901–4905
73. Tudorache F, Petrila I (2014) Effects of partial replacement of iron with tungsten on microstructure, electrical, magnetic and humidity properties of Copper–Zinc ferrite material. J Electron Mater 43(9):3522–3526
74. Rashad M, Soltan S, Ramadan A, Bekheet M, Rayan D (2015) Investigation of the structural, optical and magnetic properties of $CuO/CuFe_2O_4$ nanocomposites synthesized via simple microemulsion method. Ceram Int 41(9):12237–12245
75. Xi G, Li Y, Liu Y (2004) Study on preparation of manganese–zinc ferrites using spent Zn–Mn batteries. Mater Lett 58(7–8):1164–1167
76. Peng C-h, Bai B-s, Chen Y-f (2008) Study on the preparation of Mn–Zn soft magnetic ferrite powders from waste Zn–Mn dry batteries. Waste Manag 28(2):326–332
77. Lu L, Fang Y, Huang Z, Huang Y, Ren ZJ (2016) Self-sustaining carbon capture and mineralization via electrolytic carbonation of coal fly ash. Chem Eng J 306:330–335
78. Huang TY, Chiueh PT, Lo SL (2017) Life-cycle environmental and cost impacts of reusing fly ash. Resour Conserv Recycl 123:255–260
79. Li S, Wang X-m (2016) Fly-ash-based magnetic coagulant for rapid sedimentation of electronegative slimes and ultrafine tailings. Powder Technol 303:20–26
80. Li S, Wang X-m, Zhang Q-l (2016) Dynamic experiments on flocculation and sedimentation of argillized ultrafine tailings using fly-ash-based magnetic coagulant. Trans Nonferrous Met Soc China 26(7):1975–1984
81. Lin K-S, Chiang C-L, Hsu P-J, Bat-Erdene K, Tang C-Y, Wu C-M (2018) Magnetic separation and recycling of ferrite nanocatalysts for CO_2 decomposition with CH_4 recovery from steel industrial flyash. Catal Today 307:260–271
82. Kim G-Y, Jeon J-H, Kim M-H, Suvorov D, Choi S-Y (2013) Microstructural development of cobalt ferrite ceramics and its influence on magnetic properties. Met Mater Int 19 (6):1209–1213
83. Emamian HR, Honarbakhsh-Raouf A, Ataie A, Mirzaee O (2015) Characterization of mesoporous nanostructured cobalt ferrite foams fabricated by microwave and conventional sintering routes. J Porous Mater 22(3):603–611
84. Amiri M, Salavati-Niasari M, Akbari A, Razavi R (2017) Sol–gel auto-combustion synthesize and characterization of a novel anticorrosive cobalt ferrite nanoparticles dispersed in silica matrix. J Mater Sci: Mater Electron 28(14):10495–10508
85. Khan K (2014) Microwave absorption properties of radar absorbing nanosized cobalt ferrites for high frequency applications. J Supercond Novel Magn 27(2):453–461

86. Zhou J-p, Fu W-l, Qiu K-q, Chen Q-y (2007) Photocatalytic properties of nano tetrapod-like zinc oxide whiskers. J Funct Mater Devices 13(3):195
87. Guo X, Zhong J, Song Y, Tian Q (2010) Substance flow analysis of zinc in China. Resour Conserv Recycl 54(3):171–177
88. Zhan L, Li O, Wang Z, Xie B (2018) Recycling zinc and preparing high-value-added nanozinc oxide from waste zinc–manganese batteries by high-temperature evaporation-separation and oxygen control oxidation. ACS Sustain Chem Eng 6 (9):12104–12109
89. Zhan L, Li O, Xu Z (2020) Preparing nano-zinc oxide with high-added-value from waste zinc manganese battery by vacuum evaporation and oxygen-control oxidation. J Clean Prod 251:119691
90. Xie Y, Wang S, Tian X, Che L, Wu X, Zhao F (2019) Leaching of indium from end-of-life LCD panels via catalysis by synergistic microbial communities. Sci Total Environ 655:781–786
91. Lahtela V, Virolainen S, Uwaoma A, Kallioinen M, Kärki T, Sainio T (2019) Novel mechanical pre-treatment methods for effective indium recovery from end-of-life liquid-crystal display panels. J Clean Prod 230:580–591
92. Bekheet MF, Schwarz MR, Lauterbach S, Kleebe HJ, Kroll P, Riedel R, Gurlo A (2013) Orthorhombic In_2O_3: a metastable polymorph of indium sesquioxide. Angew Chem Int Ed 52(25):6531–6535
93. Bekheet MF, Schwarz MR, Lauterbach S, Kleebe H-J, Kroll P, Stewart A, Kolb U, Riedel R, Gurlo A (2013) In situ high pressure high temperature experiments in multi-anvil assemblies with bixbyite-type In_2O_3 and synthesis of corundum-type and orthorhombic In_2O_3 polymorphs. High Press Res 33(3):697–711
94. Bekheet MF, Schwarz MR, Kroll P, Gurlo A (2015) Kinetic control in the synthesis of metastable polymorphs: bixbyite-to-Rh_2O_3 (II)-to-corundum transition in In_2O_3. J Solid State Chem 229:278–286
95. De Boer T, Bekheet M, Gurlo A, Riedel R, Moewes A (2016) Band gap and electronic structure of cubic, rhombohedral, and orthorhombic In_2O_3 polymorphs: experiment and theory. Phys Rev B 93(15):155205
96. Boundy T, Boyton M, Taylor P (2017) Attrition scrubbing for recovery of indium from waste liquid crystal display glass via selective comminution. J Clean Prod 154:436–444
97. Bekheet MF, Miehe G, Fasel C, Gurlo A, Riedel R (2012) Low temperature synthesis of nanocrystalline $MnIn_2O_4$ spinel. Dalton Trans 41(12):3374–3376
98. Schlicker L, Bekheet MF, Gili A, Doran A, Gurlo A, Ploner K, Schachinger T, Penner S (2018) Hydrogen reduction and metal-support interaction in a metastable metal-oxide system: Pd on rhombohedral In_2O_3. J Solid State Chem 266:93–99
99. Bekheet MF, Svoboda I, Liu N, Bayarjargal L, Irran E, Dietz C, Stark RW, Riedel R, Gurlo A (2016) Ferroelectric $InMnO_3$: growth of single crystals, structure and high-temperature phase transitions. J Solid State Chem 241:54–63
100. Bekheet MF, Dubrovinsky L, Gurlo A (2015) Compressibility and structural stability of spinel-type $MnIn_2O_4$. J Solid State Chem 230:301–308
101. Zhuang XN, Ye YY, He WZ, Li GM, Huang JW (2014) Hydrothermal treatment of liquid crystal using a batch reactor. Adv Mater Res 878:563–568
102. Virolainen S, Ibana D, Paatero E (2011) Recovery of indium from indium tin oxide by solvent extraction. Hydrometallurgy 107:56–61
103. Pa P-s (2009) Effective tool design of three-rank form as precision removal-process of ITO thin-films. Trans Nonferrous Met Soc China 19:s232–s237
104. Inoue K, Nishiura M, Kawakita H, Ohto K, Harada H (2008) Recovery of indium from spent panel of liquid crystal display panels. Kagaku Kogaku Ronbunshu 34(2):282–286
105. Rocchetti L, Amato A, Beolchini F (2016) Recovery of indium from liquid crystal displays. J Clean Prod 116:299–305
106. Zhang K, Wu Y, Wang W, Li B, Zhang Y, Zuo T (2015) Recycling indium from waste LCDs: a review. Resour Conserv Recycl 104:276–290

107. Wang S, He Y, Zhang T, Zhang G (2017) Recovery of valuable components from waste LCD panel through a dry physical method. Waste Manag 64:255–262
108. Savvilotidou V, Hahladakis JN, Gidarakos E (2014) Determination of toxic metals in discarded liquid crystal displays (LCDs). Resour Conserv Recycl 92:108–115
109. Li J, Gao S, Duan H, Liu L (2009) Recovery of valuable materials from waste liquid crystal display panel. Waste Manag 29(7):2033–2039
110. Lee C-H, Jeong M-K, Fatih Kilicaslan M, Lee J-H, Hong H-S, Hong S-J (2013) Recovery of indium from used LCD panel by a time efficient and environmentally sound method assisted HEBM. Waste Manag 33(3):730–734
111. Ruan J, Guo Y, Qiao Q (2012) Recovery of indium from scrap TFT-LCDs by solvent extraction. Procedia Environ Sci 16:545–551
112. Hsieh S-J, Chen C-C, Say WC (2009) Process for recovery of indium from ITO scraps and metallurgic microstructures. Mater Sci Eng, B 158(1–3):82–87
113. Ma E, Xu Z (2013) Technological process and optimum design of organic materials vacuum pyrolysis and indium chlorinated separation from waste liquid crystal display panels. J Hazard Mater 263:610–617
114. Park K-S, Sato W, Grause G, Kameda T, Yoshioka T (2009) Recovery of indium from In_2O_3 and liquid crystal display powder via a chloride volatilization process using polyvinyl chloride. Thermochim Acta 493(1–2):105–108
115. Zhang L, Wu B, Chen Y, Xu Z (2017) Energy and valuable resource recovery from waste liquid crystal display panels by an environment-friendly technological process: pyrolysis of liquid crystals and preparation of indium product. J Clean Prod 162:141–152
116. He Y, Ma E, Xu Z (2014) Recycling indium from waste liquid crystal display panel by vacuum carbon-reduction. J Hazard Mater 268:185–190
117. Yoshida H, Izhar S, Nishio E, Utsumi Y, Kakimori N, Asghari Feridoun S (2014) Recovery of indium from TFT and CF glasses in LCD panel wastes using sub-critical water. Sol Energy Mater Sol Cells 125:14–19
118. Higashi A, Saitoh N, Ogi T, Konishi Y (2011) Recovery of indium by biosorption and its application to recycling of waste liquid crystal display panel. J Jpn Inst Met 75(11):620–625
119. Wang S, He Y, Yang J, Feng Y (2018) Enrichment of indium tin oxide from colour filter glass in waste liquid crystal display panels through flotation. J Clean Prod 189:464–471
120. Wang S, He Y, Feng Y, Zhang F, Zhang T (2020) Indium tin oxide recycling from waste colour filter glass via thermal decomposition. J Hazard Mater 392:122503

Nanosensors and Nanobiosensors for Monitoring the Environmental Pollutants

Alaa El Din Mahmoud and Manal Fawzy

Abstract Nowadays, both developed and developing countries are more concerned about their sensitive natural resources, especially water resources. Therefore, conservation and sustainable utilization of available and limited water resources are a must. Globally, agriculture is the primary water-using sector. Usually, water and soil qualities are monitored and recorded for agricultural purposes through traditional analytical techniques that require planning and effort for sampling. The need for more sensitive and simple techniques is the main driving force for using nanobiosensors. Recycled nanomaterials can be integrated into nanobiosensors so that various nanobiosensors can achieve the same goals with high sensitivity and without sample preparations. The new nanobiosensors can easily operate at low cost and function at a wide range of detection scales. Nanobiosensors can monitor and detect either physicochemical parameters or microbes in remote areas and monitor environmental conditions. Real-time nanobiosensors can boost agricultural production by monitoring the temperature and humidity and regulating the usage of fertilizers/pesticides at a specific time in a targeted location. This chapter provides an insight into the concepts and parameters of nanobiosensors, emphasizing the recent progress of their applications in organic and inorganic pollutants sensing. The chapter also summarizes the statistics of the last two decades for using nanosensors and nanobiosensors for environmental monitoring.

Keywords Nanosensors · Nanobiosensors · Recycled nanomaterials · Conservation · Sustainable · Agriculture · Water pollution · Statistics

Alaa El Din Mahmoud (✉) · M. Fawzy
Environmental Sciences Department, Faculty of Science, Alexandria University, Alexandria 21511, Egypt
e-mail: alaa-mahmoud@alexu.edu.eg

Alaa El Din Mahmoud · M. Fawzy
Green Technology Group, Faculty of Science, Alexandria University, Alexandria 21511, Egypt

Alaa El Din Mahmoud · M. Fawzy
National Biotechnology Network of Expertise (NBNE), Academy of Scientific Research and Technology (ASRT), Cairo, Egypt

A. S. H. Makhlouf and G. A. M. Ali (eds.), *Waste Recycling Technologies for Nanomaterials Manufacturing*, Topics in Mining, Metallurgy and Materials Engineering, https://doi.org/10.1007/978-3-030-68031-2_9

List of Abbreviations

APTES	3-aminopropyltriethoxysilane
Au NRs	Gold nanorods
BPEI-CQDs	Branched polyethylenimine- capped Carbon Quantum Dots
CNTs	Carbon nanotubes
CQDs	Carbon quantum dots
FRET	Fluorescence resonance energy transfer
Hg	Mercury
LOD	Limit of detection
MNPs	Magnetic nanoparticles
MRI	Magnetic resonance imaging
MRS	Magnetic relaxation switch
MWCNTs	Multiwalled carbon nanotubes
PL	Photoluminescence
QDs	Quantum dots
SPR	Surface plasmon resonance
QD	Quantum dot

1 Introduction

The world's population is estimated to grow from 7 billion (2020) to 9.7 billion (2050). Such dynamic changes have pressures on our environmental resources, which lead to threatening resources sustainability and food security [1]. On the other hand, agricultural productivity must increase in parallel to fulfill the needs of developing and developed countries. It is projected that the global agricultural sector will consume 1405 billion cubic meters of water in 2040 [2].

Freshwater scarcity is a major issue for arid and semiarid countries, threatening socioeconomic development. In Africa, about 300 million people out of 800 million live in water-scarce environments [3]. It is estimated that the amount of freshwater resources is less than 500 m^3 $capita^{-1}$ $year^{-1}$ in most of the Middle Eastern and North African countries [4]. However, the practices of the agricultural sector are still unsustainable, using traditional methods of irrigation which consume huge amounts of freshwater and poor water resources management systems [5].

The agricultural sector is the largest sector for consuming the available water, as shown in Fig. 1. The available water consumption reaches 68% globally (Fig. 1a), 80% in Africa (Fig. 1b), and 85% in Egypt (Fig. 1c). This can be due to the irrigated Egyptian areas increased from 3.61 million hectare in 2011 to 3.78 million hectare in 2015 [6].

Conservation of available water resources and its sustainable usage can be controlled through regular environmental monitoring for organic and inorganic pollutants released in our environment (air, soil, and water). Nanotechnology can

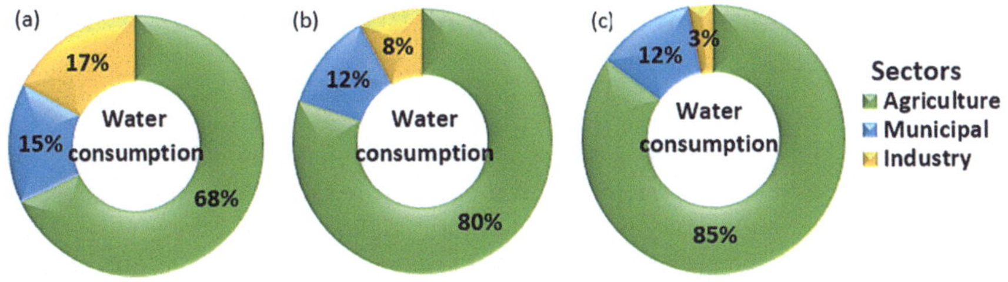

Fig. 1 Water consumption percentage by different sectors: **a** worldwide, **b** Africa, and **c** Egypt as an example

assist in this direction of application [7–9]. We highlight the preparation techniques of nanomaterials that can be involved in nanobiosensors in the next section.

First of all, the sensor is an analytical device utilized to convert the received signal as heat, motion, sound, light or electric effect into an electric signal with information or as a measurement of physical or chemical parameters in any type of sample [10–12]. When nanomaterials are integrated into the components of sensors, they are called "nanosensors," which are built with a dimension of about 10 nm. They are small enough to be lightweight, portable, and require low power consumption. They are on focus in this chapter because they could be integrated into portal devices and be reproducible on large-scale production when they are cost-effective.

Nanobiosensors are the same nanosensor concept but with a biologically active element to generate a measurable signal proportional to the concentration of chemical species in any type of sample. They are based on biological recognition and sensing. Therefore, they show considerably improved biocompatibility, sensitivity, and specificity over most conventional sensing systems [13, 14]. The integration of nanosensors/nanobiosensors in the water system will enable the users to: (a) remotely monitor and identify pollution problems, and (b) comply with regulations and policies on water quality and conservation [15]. Besides, nanosensors and nanobiosensors could be the key to apply the sustainable target number 6.4 (more efficient water use).

This chapter aims to provide an insight into the concepts and parameters of nanosensors and nanobiosensors with recycled nanomaterials with emphasis on the recent progress of their applications in aquatic ecosystems and organic/inorganic pollutants sensing.

2 Recent Preparation Techniques of Recycled Nanomaterials

Nanomaterials play an essential role in either nanosensors or nanobiosensors as they demonstrate unique physiochemical properties that vary from the bulk materials [16–20]. As an example, their strength and high surface–volume ratios [21–25]. The synthesis of nanomaterials can be conducted using eco-friendly routes without the need for harsh and corrosive chemicals [26]. The opportunity of recycling the agricultural wastes and incorporated in the nanosynthesis route are applicable at a low cost [27, 28]. Alternative reducing agents and capping agents were searched to apply the principles of green chemistry.

Mhamane et al. [29] synthesized highly water dispersible and stable functionalized graphene nanosheets by using the plant extracts of *Potamogeton pectinatus* L., *Lemna gibba, Ceratophyllum demersum*, and *Cyperus difformis.* The measured current-voltage (I–V) of the prepared recycled graphene nanosheets were higher $\sim 10^3$ times compared to that with graphene oxide films. Mahmoud [30] developed a new technique to prepare reduced graphene oxide from agricultural byproducts and aquatic macrophytes. The proposed technique could be used to form nanocomposites for energy or environmental applications to be integrated into nanobiosensors. Akhavan et al. [31] synthesized graphene from different natural carbonaceous waste materials and industrial soot powder. The synthesized graphene materials were comparable to those obtained through Hummers' method. Furthermore, metallic nanoparticles could be synthesized using plant extracts. Gold (Au), silver (Ag), and Au–Ag bimetallic nanoparticles were synthesized using *Potamogeton pectinatus* L. [32].

Recently, the green synthesis of carbon quantum dots (CQDs) was successfully prepared via the hydrothermal method by using the extract of *Echinops persicus* plant [33]. Singh et al. [34] synthesized biogenic zinc oxide (ZnO) quantum dots using *Eclipta alba* leaf extract as a reducing agent. Another study focused on the preparation of CdS quantum dots with *Camellia sinensis* [35].

3 Applications of Nanosensors/Nanobiosensors for Environmental Monitoring

Nanotechnology has a significant role in the development and enhancement of nanosensors/nanobiosensors-based applications. Nanotechnology-enabled sensors allow being integrated into miniaturized and automated devices [36]. Hence, the market of nanosensors is expected to grow by 3.2$ billion during 2018–2022 [37]. This enormous growth point to the current significance of nanosensors/nanobiosensors as they are very small, highly sensitive, more selective, and easier to use.

Table 1 Parameters to be considered in nanosensors fabrication [10, 39]

Parameters	Explanation
Resolution	-The smallest change in measured analyte values detected by the nanosensor. -High resolution is preferable.
Sensitivity	-The change in output value per unit change in the input variable. -High sensitivity means more accuracy and possible detection of a single analyte. -Example of optical nanosensors due to the unique interactions between used nanomaterials and light waves. -Nanogold-based nanosensor for quick detection of DNA and protein biomarkers for cancer and other diseases.
Selectivity	-Detecting a specific analyte despite the existence of interference in the output.
Response time	-The time required for the sensor to attain 90% of its signal for an analyte concentration. -Real-time monitoring reflecting how quickly the nanosensor responds to the change in the detected signal. -More detection speed requires a low response time.
Repeatability	-The reproducibility of the detected analyte values at certain measurements.
Drift	-The shift in the sensor calibration while using over time.
Stability	-The performance of the nanosensor over a certain time with no need for recalibration.

The performance of a nanobiosensor depends on its parameters. Table 1 summarizes these parameters to be more coherent. The input values and output signals of a nanosensor usually refer to the minimum and maximum range that can affect the sensitivity [38]. However, nanobiosensors usually have constant sensitivity over the determined range and can be estimated from the slope of the straight-line plot.

Environmental nanosensors/nanobiosensors help real-time onsite monitoring of moisture, temperature, nutrients, and water level. Henceforth, the development of real-time monitoring nanosensors/nanobiosensors is more efficient and reliable for any field applications. Various kinds of nanosensors, such as chemical, electrical, and biosensors, have been used in various sectors.

Generally, any nanosensor involves: (i) single or composite nanomaterials (metals, magnetic, quantum dots, carbonaceous), (ii) targeting analytes, and (iii) transducer (Fig. 2). The ability of nanosensors to interact with and recognize the analyte is considered as the detection system, the first part of the nanosensor [40]. After the detection system, there is the translation system where the transducer operates as a sensor fabricated with the nanomaterials mentioned above to convert the interaction between the analyte and the detector into a readable output signal. The categories of either nanosensors or nanobiosensors are based on transduction mechanisms for generating output like optical, electrochemical, and mass nanosensors [41], as shown in Table 2.

Nanosensors/nanobiosensors are applied in various environmental, military, aerospace, and industrial applications. Figure 3 focuses on environmental applications of nanobiosensors with an emphasis on organic/inorganic pollutants sensing.

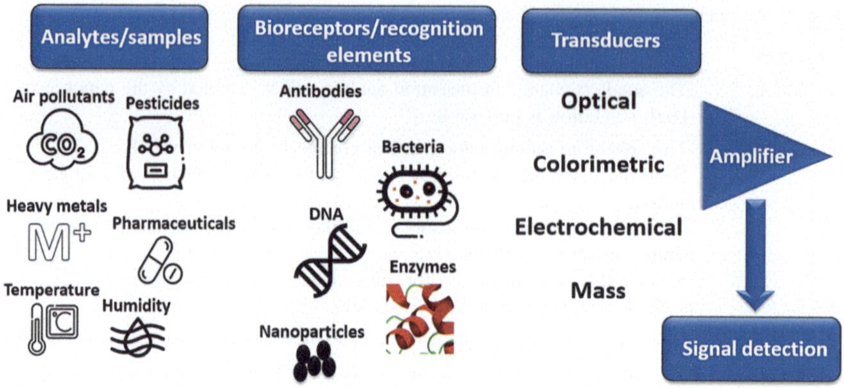

Fig. 2 Components of a typical nanobiosensors starting from sample analyte to bioreceptor, transducer with nanoparticles, and finally, detectors

The family of the nanosensors can include nanofibers, nanotubes, nanowires, nanograins, nanorods, nanocomposites and quantum dots that have the potential for environmental monitoring (Fig. 4). Each one has merits [46]; (i) nanotubes can be functionalized and be better electrical transducer; (ii) nanowires are very versatile and have better charge conduction, (iii) quantum dots (QDs) have fluorescence property and size-tunable band energy. The bandgap of the QDs depends on the size of the nanocrystal. This is due to the high difference in the energy levels of the smaller nanocrystals that lead to the wider the energy gap and the shorter the wavelength of the fluorescence [47]. The advantage of QDs is using multiplex assays with single excitation sources [47]. Cheng et al. [48] reported the preparation of carbon QDs through the carbonization of walnut shells as a facile technique. Carbon nanotubes (CNT)/graphene hybrid were prepared by the hydrothermal method [49].

Pollutants monitoring and detection are essential to regulate water quality in specific areas. It is usually conducted by conventional analytical techniques using atomic absorption, HPLC, and mass spectrometry after collecting samples from the environment [51]. However, these techniques have limitations such as expensive analysis, time-consuming for sample preparation, complexity, and trained technicians [52]. Furthermore, these techniques are not adequate for onsite applications because they are usually performed in the laboratory and the long-time delay between sampling and measuring.

Currently, research directions are oriented to label-free nanosensors due to the ability of the direct detection of the analyte of interest without the amplification step. So the development of label-free and low-cost nanosensors is necessary for the sustainable use of water, fertilizers, pesticides, and nutrients to enhance crop

Table 2 Classification of nanobiosensors based on transduction mechanism and detection levels

Classification	Explanations	Detection levels*	Ref.
Optical	–It is ascribed with the enhanced optical property of the metallic nanoparticle, and quantum dots. –Depends on: Fluorescence output signals, Colourimetric output signals, Surface-enhanced Raman spectroscopy (SERS) that is promising in the field of biology (miRNA analysis) since SERS is an ultra-sensitive and non-destructive technique. –Easily portable. –Enable the production of large-scale sensors.	pM (fluorescence) nM (colorimetric) zM (SERS)	[40, 42]
Electrochemical	–Rely on metal oxide nanomaterials, magnetic nanomaterials, graphene sheets or carbon nanotubes-based electrode. –Facilitate or analyze the biochemical reactions with the help of improved electrical means and metallic nanoparticles, which significantly help in achieving immobilization of one of the reactants. –Measure the change in the voltage (potentiometry), current (voltammetry), impedance or conductance resulting from a chemical reaction that transfers or separates electric charge with sufficient selectivity and sensitivity.	fM	[43, 44]
Mass/ Mechanical	–Rely on the mechanical (acoustic) and piezoelectric properties of microcantilever and crystals. –As an example, microcantilever could be used to sense changes in temperature and relative humidity.	fM	[45]

*pM: pico (10^{-12}), nM: nano (10^{-9}), zM: zepto (10^{-21}), fM: femto (10^{-15})

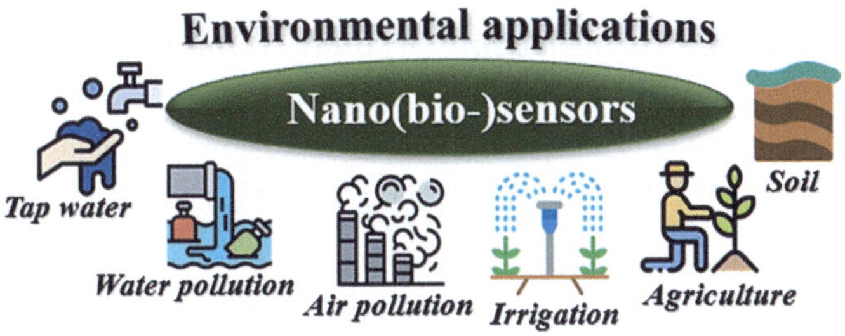

Fig. 3 Potential environmental applications of nanobiosensors

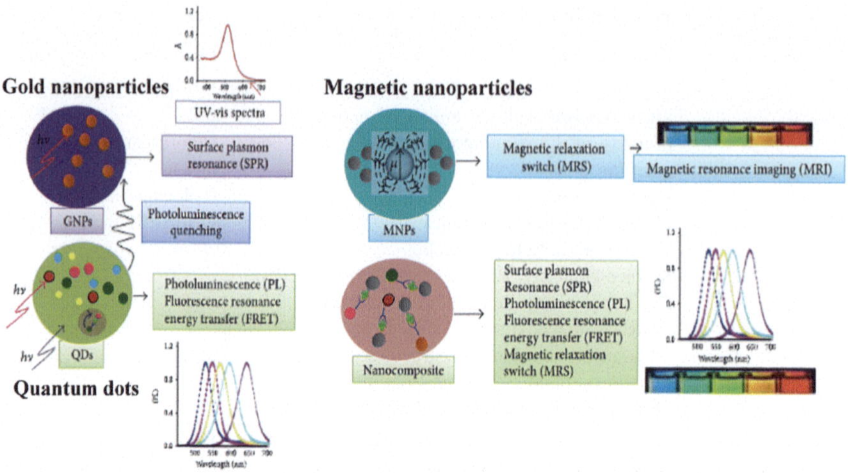

Fig. 4 Diagram of various nanoparticles and nanocomposites possessing optical properties that can confer multi-functionalities to provide distinct advantages for environmental monitoring. Adapted with permission from Ref. [50], Copyright 2015, Hindawi Publishing Corporation

productivity with environmental safety manner and economic stability [53, 54]. Label-free detection allows the sensors to detect the molecule of interest directly. In addition, environmental nanosensors help real-time onsite monitoring of moisture, temperature, nutrients, and water level.

Recently, DNA-based nanobiosensors are very promising in environmental applications because DNA serves as a building block for the construction of refined nanostructures [36]. Apart from the role of DNA in species identification and classification, it can be a recognition element (Fig. 2) with many pros: robustness, versatility, cost-effective mass synthesis. Furthermore, there is a nanobarcode technology that can interpret the encoded information or data in the form of a map [41]. For instance, DNA barcode/biobarcode is used to identify single species of plants, microbes, and animals by taking a small stretch of a single gene. Further details are provided in Srivastava et al. [41].

3.1 Nanosensors for Detecting Organic Pollutants

Precision farming is an important technique used for boosting crop productivity by monitoring environmental variables and applying targeted action [55]. It includes nanosensors which can assist the management of fertilizers/pesticides usages and the management of the inputs and time conditions [56].

Pesticides and fertilizers detections attract the major interest of the agricultural sector and farmers because they help farmers in protecting the crops from pests and

Table 3 Classification of major pesticides and growth hormones

Main groups	Examples with structures	
Insecticides	Tetradifon, 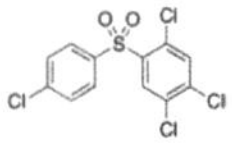	Heptachlor, 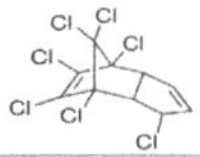
	Dichloro-diphenyl-trichloroethane (DDT), 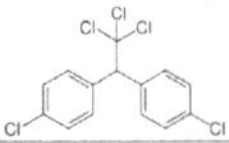	Chlordane, 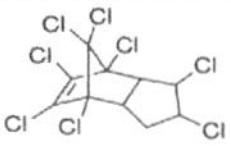
	Aldrin (used until 1999 then banned), 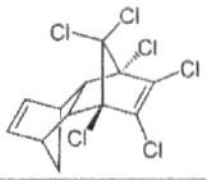	Parathion, 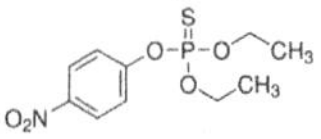
	Carbofuran, 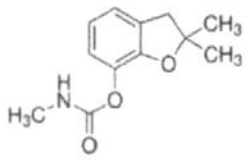	Endosulfan sulfate 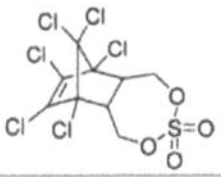
Herbicides	Atrazine, 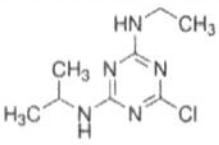	Imazapyr, 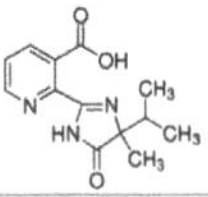
	Simazine, 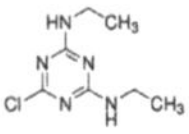	Pyrazon, 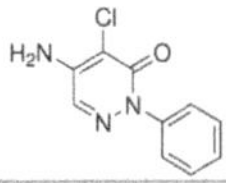
	Metolachlor, 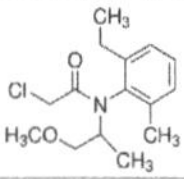	Alachlor 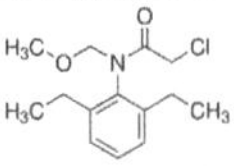
	Chlorthiamid, 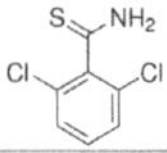	
Fungicides	Captafol, 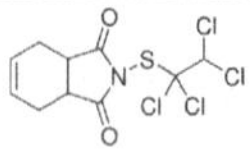	Dicloran, 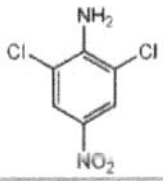
	Chlorothalonil, 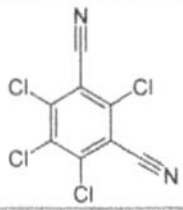	Benomyl 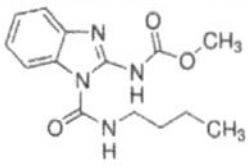
Growth hormones	Auxins 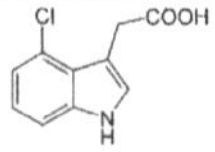	Gibberellins 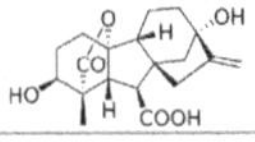
	Cytokinins (i.e., Zeatin) 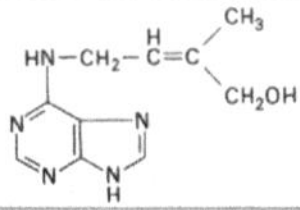	

provide nutrients to the crops [57]. For instance, fertilizers contribute to 37.5% of crop productivity. Their classifications are shown in Table 3, collecting major pesticides and growth hormones used in agriculture.

Table 4 Recent literature of nanobiosensors applied for organic pollutants detection

Analyte	Nano(bio-)sensor	LOD*	Detection range	Ref
Acetamiprid	Apta-nanosensor (peroxidase-hemin-functionalized reduced graphene oxide)	40 nM	0.10–100 μM	[63]
Dinosulfon	Zinc oxide (ZnO)/Multiwalled Carbon Nanotubes (MWCNTs)	12.39 nM	0.054–161.9 μM	[58]
1-hydroxypyrene	Graphene modified glassy carbon electrode	0.84 nM	0.005–0.3 μM	[64]
4-Aminophenol	Cobalt metal-organic framework	15 nM	–	[65]
4-Methylphenol		35 nM	–	
Phenol		60 nM	–	
4-Chlorophenol		60 nM	–	
Organophosphate pesticides	Copper Oxide Nanorods	2 nM	–	[66]
Imazapyr	APTES-coated ytterbium oxide (Yb_2O_3)	–	0.76 μM	[61]

*LOD: Limit of detection is the smallest concentration which the nanosensor can detect

Table 4 summarizes the literature that investigated nanosensors in either the agricultural sector or water pollutants. Electrochemical nanosensor composed of zinc oxide (ZnO)/multiwalled carbon nanotubes (MWCNTs) was used for dinosulfon in water samples. Reddy et al. [58] selected ZnO/MWCNTs nanocomposite because it is a very effective electrocatalysis for the nanosensors and may improve the electrochemical performance. The prepared nanosensor had a wide linear detection range at pH = 8.0.

Fluorescence nanosensor based on cadmium selenium/zinc sulfide (CdSe/ZnS) quantum dots was successfully applied for the detection of anthracene, phenanthrene, and pyrene in water [59]. The trace level of these polycyclic aromatic hydrocarbons could be enhanced, and the low detection limit of the nanosensor was at 5 nM.

Sahoo et al. [60] tested zinc oxide quantum dot (ZnO_2 QD) nanosensor for sensing and degrading four pesticides widely used in agriculture, namely aldrin, tetradifon, glyphosate, and atrazine. They found there are covalent bonds between QDs and aldrin and tetradifon because of the existence of the primary amine group of QDs that can substitute the leaving groups (-Cl) in both pesticides. In the case of glyphosate, ionic interaction has a role in binding with QDs.

Imazapyr, as herbicide, is frequently utilized for regulating the growth of grasses in the crop fields of corn, sunflower, and sugarcane. Hence, Kumar et al. [61] fabricated APTES coated Yb_2O_3 nanosensor for imazapyr detection. They verified the sensitivity and selectivity of this sensor with simple fluorescence spectroscopy. H-bonding can explain this, and electrostatic interactions between APTES-coated Yb_2O_3 and the imazapyr. This is due to the interaction between the carboxylic group of imazapyr (see Table 3) and the amine group of APTES. There is also another type called APTA-nanosensors that were used for the detection of different pesticides. Verdian [62] summarized these sensors for the detection of acetamiprid.

3.2 Nanosensors for Detecting Inorganic Pollutants

Environmental pollution by heavy metals considers a global issue because of their impact on natural resources (biodiversity, agricultural productivity) and our health [67–69]. Therefore, testing water quality factors are necessary for drinking and many industrial applications. Various parameters can be monitored using nanosensors, for instance, physical parameters (temperature, turbidity, conductivity), chemical parameters (pH, heavy metals, organics, dissolved oxygen, nitrates), and biological (biological oxygen demand, bacterial content).

Recent literature on the application of nanosensors in heavy metals detection is shown in Table 5. Carbon dot-based nanosensor was first used for fluorescence detection of chromium (VI) in water with a detection limit of 2.3 nM [70]. The advantages of this nanosensor are its low toxicity and excellent photostability. Maric et al. [71] demonstrated platinum–halloysite nanoclay nanojets (Pt–Halloysite) as a nanosensor for the detection of heavy metals such as Cd^{2+}, Zn^{2+}, Pb^{2+}, and Hg^{2+}. The selectivity of the nanosensor was found in the order of $Hg^{2+} > Pb^{2+} > Zn^{2+} > Cd^{2+}$ because of the electronegativity and atomic radius of these metals. Further details can be found in Mahmoud [21]. In summary, Hg^{2+} and Pb^{2+} possess high electronegativity, which causes strong attachment to the Pt surface of the nanojets.

Thatai et al. [72] applied gold nanorods (Au NRs) as a nanosensor for the detection of Fe^{3+} in aqueous solutions. They observed that Fe^{3+} ions adsorbed on Au NRs where oxidation of Au NRs occurred, and Au NRs are sensitive to Fe^{3+} ions at a level of 100 ppb. Another colorimetric nanosensor of dithioacetal-mechanized mesoporous silica was developed for Hg^{2+} detection [73].

Table 5 Recent nanobiosensors literature used for heavy metals detection

Analytes	Nanobiosensors	LOD*	Detection range	Ref.
Cu^{2+}	Branched polyethylenimine-capped carbon quantum dots (BPEI-CQDs)	115 nM	0.33–66.6 μM	[74]
Fe^{3+}	Highly fluorescent carbon dots	6.4 nM	–	[75]
Hg^{2+}	Platinum–Halloysite Nanoclay Nanojets	16.2 nM	1.24–4985.3 nM	[71]
Hg^{2+}	Dithioacetal-mechanized mesoporous nanosensor	60 nM	0.154–31 nM	[73]

*LOD: Limit of detection is the smallest concentration which the nanobiosensor can detect

The mechanism of detection can be that Hg^{2+} disrupts the linkages of dithioacetal. The developed nanosensor displayed high storage stability for 2 months of storage at room temperature.

4 Other Applications of Nanobiosensors

Smart packages are still under development to ensure and preserve food quality through nanobiosensors. The role of nanobiosensors is to indicate temperature, ripeness, freshness, and pathogen status on the food packages [56]. In 2016, the first intelligent sensor was introduced to the fruit market to indicate especially the ripeness of pears fruits in New Zealand [76]. It is a label that changes its color from dark red gradually till pale yellow, as shown in Fig. 5. Its mechanism depends on the released aromas of the fruit, which reacts with the nanomaterials.

The same concept applied to milk or meat to inform the consumers about the freshness of the purchased products by the color change when oxidation occurs in the package. Gold nanoparticles have been used to detect fast the presence of melamine in milk products [77]. When melamine exists, the mixture changes from pink to blue. Moreover, nanobiosensors can timely detect the leakage of oxygen that causes food spoilage besides carbon dioxide and moisture [78]. For instance, nanoclays can increase the durability of milk production from 3 to 9 days.

The main challenges of nanobiosensors applications in food packages are their sensitive operation in the presence of many microorganisms and other components that might interfere with the nanosensor accuracy [12]. Furthermore, gaining public acceptance for using either nanobiosensors or nanomaterials in the food packages is still resistive.

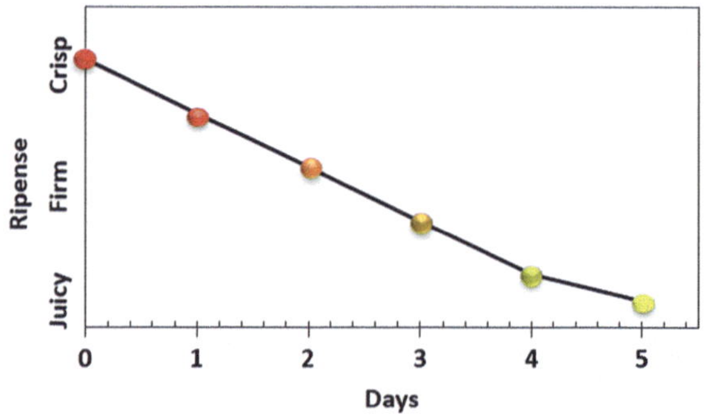

Fig. 5 Color change of label nanosensor to indicate the ripeness of pears fruits

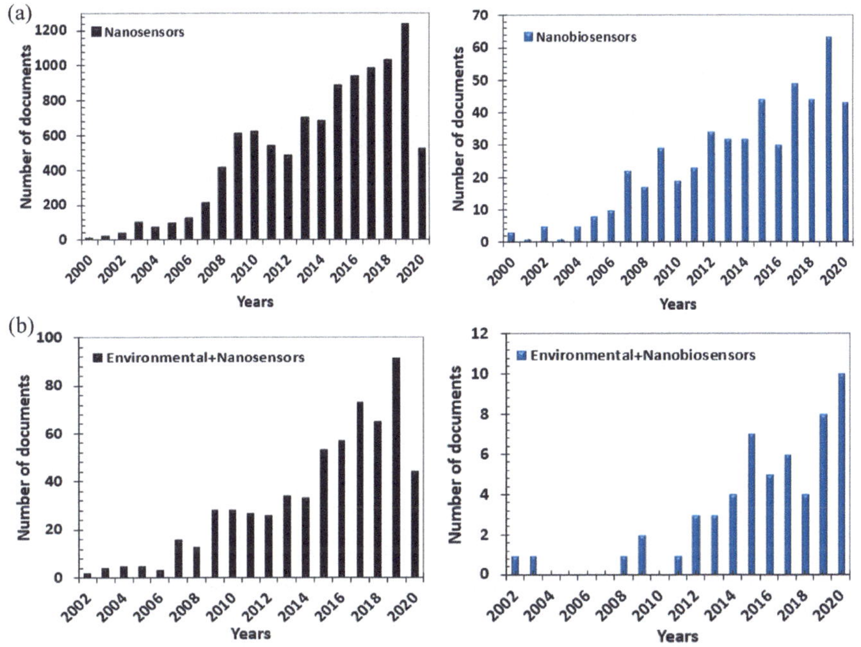

Fig. 6 Number of documents indexed in Scopus for keywords **a** "Nanosensors"/ "Nanobiosensors" and **b** "Environmental" and "Nanosensors"/"Nanobiosensors" during 2000–2020

5 Statistics for Environmental Nanobiosensors

It has been found that the number of published nanobiosensors documents is increasing rapidly over time since 2000 (Fig. 6). In total, 10393 and 514 published documents of nanosensors and nanobiosensors are recorded in the Scopus database, respectively (Fig. 6a). However, the documents of environmental nanosensors and environmental nanobiosensors represent 5.84 and 0.11% of the total nanosensors/ nanobiosensors documents, which require further investigations in the future (Fig. 6b). The two leading countries are the USA and China, which possess 28.7 and 22.6% of the published documents, respectively.

6 Conclusions

The development of cost-effective and label-free nanobiosensors is necessary to mitigate the impacts of water scarcity and pollution. Agriculture productivity is enhanced because the nanosensors can control the environmental conditions by minimizing the inputs and enabling the crops to use pesticides and water more efficiently. The nanosensors are also very useful for irrigation in precision and can digitize the agriculture concept. However, they should provide the following aspects: cost-effective, high sensitivity, portable, long-term, and onsite monitoring for environmental conditions (soil, water), fast and accurate detection of the desired analyte, selectivity, and remote operation (wireless) to be applied on a wide scale. The average limit of detection (LOD) of the mentioned nanosensors for heavy metals ($n = 3$) is 63.72 nM. On the other hand, the average LOD of pesticides ($n = 7$) is 23.59 nM. Further research is needed to investigate more analytes with higher sensitivity. Various nanomaterials or nanocomposites can enhance and update the next generation of the nanosensors. The advancement of nanobiosensors can be achieved through multidisciplinary backgrounds from science, biology, engineering, and physics.

7 Future Perspectives

It is recommended that nanobiosensors should detect multiple heavy metals and pesticides with higher sensitivity. When recycled nanomaterials from agricultural wastes are combined in the fabrication of sensors and biosensors, those nanosensors will be the key for sustainable development, especially in water monitoring, agriculture, food, and industry. Nanobiosensors can facilitate the establishment of networked analytical stations in different natural environments. This gives an accurate, robust monitoring, and quick actions for global environmental issues. For the implementation of nanobiosensors in different sectors, more concern is needed for raising people's awareness about the crucial role of nanotechnology and sensors in the agricultural and industrial sectors. The large-scale application of environmental nanosensors requires synergy among farmers, scientists, the private sector, NGOs, and government.

Acknowledgements The authors thank "Green Technology Group" in the Environmental Sciences Department. Furthermore, we acknowledge the support of ERANETMED 3 for the research project entitled: "Smart wireless sensor network to detect and purify water salinity and pollution for agriculture irrigation" (SMARTWATIR) with ID: 227. In addition, the support of Science, Technology and Innovation Funding Authority (STDF-STIFA), Egypt, for the project ID: 42961 and the support of Science and Technology Development Fund-Support of Scientific Events (STDF-SSE Youth), Egypt, for the workshop ID: 43512.

References

1. Garg D, Payasi DK (2020) Nanomaterials in agricultural research: an overview. In: Environmental nanotechnology, vol 3, pp 243–275. Springer
2. Statista. (2019) Projected water consumption from 2014 to 2040 by sector. Available from: https://www.statista.com/statistics/1012228/global-consumption-for-water-by-sector/

3. Radingoana MP, Dube T, Mazvimavi D (2020) Progress in greywater reuse for home gardening: opportunities, perceptions and challenges. Phys Chem Earth Parts A/B/C 102853
4. Mizyed NR (2013) Challenges to treated wastewater reuse in arid and semi-arid areas. Environ Sci Policy 25:186–195
5. Ziarati P, El-Esawi M, Sawicka B, Umachandran K, Mahmoud AED, Hochwimmer B, Vambol S, Vambol V (2019) Investigation of prospects for phytoremediation treatment of soils contaminated with heavy metals. J Med Discov 4
6. CAPMAS (2018) Available from: http://www.capmas.gov.eg/
7. Mahmoud AED (2020) Graphene-based nanomaterials for the removal of organic pollutants: insights into linear versus nonlinear mathematical models. J Environ Manage 270:110911
8. Mahmoud AED, Stolle A, Stelter M (2018) Sustainable synthesis of high-surface-area graphite oxide via dry ball milling. ACS Sustain Chem Eng 6(5):6358–6369
9. Mahmoud AED, Franke M, Stelter M, Braeutigam P (2020) Mechanochemical versus chemical routes for graphitic precursors and their performance in micropollutants removal in water. Powder Technol 366:629–640
10. Khanna VK (2011) Nanosensors: physical, chemical, and biological. CRC Press
11. Singh K (2020) Nanosensors for food safety and environmental monitoring, in nanotechnology for food, agriculture, and environment, pp 63–84. Springer
12. Caon T, Martelli SM, Fakhouri FM (2017) New trends in the food industry: application of nanosensors in food packaging. In: Nanobiosensors, pp 773–804. Elsevier
13. Wang J, Qu X (2013) Recent progress in nanosensors for sensitive detection of biomolecules. Nanoscale 5(9):3589–3600
14. Touhami AJN (2014) Biosensors and nanobiosensors: design and applications 15:374–403
15. Coles C (2020) Sensors in the water and wastewater treatment industries 2020–2030. IDTechEX
16. Fouad OA, Ali GAM, El-Erian MAI, Makhlouf SA (2012) Humidity sensing properties of cobalt oxide/silica nanocomposites prepared via sol-gel and related routes. Nano 7(5):1250038
17. Teo EYL, Ali GAM, Algarni H, Cheewasedtham W, Rujiralai T, Chong KF (2019) One-step production of pyrene-1-boronic acid functionalized graphene for dopamine detection. Mater Chem Phys 231:286–291
18. Pourtaheri E, Taher MA, Ali GAM, Agarwal S, Gupta VK (2019) Electrochemical detection of gliclazide and glibenclamide on $ZnIn_2S_4$ nanoparticles-modified carbon ionic liquid electrode. J Mol Liq 289:111141
19. Pourtaheri E, Taher MA, Ali GAM, Agarwal S, Gupta VK (2019) Low-cost and highly sensitive sensor for determining atorvastatin using PbTe nanoparticles-modified graphite screen-printed electrode. Int J Electrochem Sci 14:9622–9632
20. Salehi Rozveh Z, Kazemi S, Karimi M, Ali GAM, Safarifard V (2020) Effect of functionalization of metal-organic frameworks on anion sensing. Polyhedron 183:114514
21. Mahmoud AED (2020) Nanomaterials: green synthesis for water applications. In: Kharissova OV, Martínez LMT, Kharisov BI (eds) Handbook of nanomaterials and nanocomposites for energy and environmental applications. Springer International Publishing, Cham, pp 1–21
22. Mahmoud AED, Stolle A, Stelter M, Braeutigam P (2018) Adsorption technique for organic pollutants using different carbon materials. In: Abstracts of papers of the american chemical society. AMER Chemical Soc 1155 16TH ST, NW, Washington, DC 20036 USA
23. Abdal-hay A, Hamdy Makhlouf AS, Khalil KA (2015) Novel, facile, single-step technique of polymer/TiO_2 nanofiber composites membrane for photodegradation of methylene blue. ACS Appl Mater Interfaces 7(24):13329–13341
24. Farag M, Liu H, Makhlouf A (2020) New nano-bioactive glass/magnesium phosphate composites by sol-gel route for bone defect treatment. Silicon https://doi.org/10.1007/s12633-020-00485-3
25. Nekouei F, Kargarzadeh H, Nekouei S, Keshtpour F, Makhlouf ASH (2016) Novel, facile, and fast technique for synthesis of AgCl nanorods loaded on activated carbon for removal of methylene blue dye. Process Saf Environ Prot 103:212–226

26. Barhoum A, Van Assche G, Makhlouf ASH, Terryn H, Baert K, Delplancke M-P, El-Sheikh SM, Rahier H (2015) A green, simple chemical route for the synthesis of pure nanocalcite crystals. Cryst Growth Des 15(2):573–580
27. Ali GAM (2020) Recycled MnO_2 nanoflowers and graphene nanosheets for low-cost and high performance asymmetric supercapacitor. J Electron Mater 49(9):5211-5221
28. Ali GAM, Yusoff MM, Shaaban ER, Chong KF (2017) High performance MnO_2 nanoflower supercapacitor electrode by electrochemical recycling of spent batteries. Ceram Int 43:8440–8448
29. Mhamane D, Ramadan W, Fawzy M, Rana A, Dubey M, Rode C, Lefez B, Hannoyer B, Ogale S (2011) From graphite oxide to highly water dispersible functionalized graphene by single step plant extract-induced deoxygenation. Green Chem 13(8):1990–1996
30. Mahmoud AED (2020) Eco-friendly reduction of graphene oxide via agricultural byproducts or aquatic macrophytes. Mater Chem Phys 253:123336
31. Akhavan O, Bijanzad K, Mirsepah A (2014) Synthesis of graphene from natural and industrial carbonaceous wastes. RSC Adv 4(39):20441–20448
32. AbdelHamid AA, Al-Ghobashy MA, Fawzy M, Mohamed MB, Abdel-Mottaleb MMSA (2013) Phytosynthesis of Au, Ag, and Au–Ag bimetallic nanoparticles using aqueous extract of sago pondweed (*Potamogeton pectinatus* L.). ACS Sustain Chem Eng 1(12):1520–1529
33. Nasseri MA, Keshtkar H, Kazemnejadi M, Allahresani A (2020) Phytochemical properties and antioxidant activity of *Echinops persicus* plant extract: green synthesis of carbon quantum dots from the plant extract. SN Appl Sci 2(4):670
34. Singh AK, Pal P, Gupta V, Yadav TP, Gupta V, Singh SP (2018) Green synthesis, characterization and antimicrobial activity of zinc oxide quantum dots using Eclipta alba. Mater Chem Phys 203:40–48
35. Shivaji K, Mani S, Ponmurugan P, De Castro CS, Lloyd Davies M, Balasubramanian MG, Pitchaimuthu S (2018) Green-synthesis-derived CdS quantum dots using tea leaf extract: antimicrobial, bioimaging, and therapeutic applications in lung cancer cells. ACS Appl Nano Mater 1(4):1683–1693
36. Soukarié D, Ecochard V, Salomé L (2020) DNA-based nanobiosensors for monitoring of water quality. Int J Hyg Environ Health 226:113485
37. Technavio (2018) Global nanosensors market 2018–2022
38. McGrath MJ, Scanaill CN (2013) Sensing and sensor fundamentals, in sensor technologies, pp 15–50. Springer
39. Swierczewska M, Liu G, Lee S, Chen X (2012) High-sensitivity nanosensors for biomarker detection. Chem Soc Rev 41(7):2641–2655
40. Schoolaert E, Hoogenboom R, De Clerck K (2017) Colorimetric nanofibers as optical sensors. Adv Func Mater 27(38):1702646
41. Srivastava AK, Dev A, Karmakar S (2018) Nanosensors and nanobiosensors in food and agriculture. Environ Chem Lett 16(1):161–182
42. Sun Y, Shi L, Mi L, Guo R, Li T (2020) Recent progress of SERS optical nanosensors for miRNA analysis. J Mater Chem B
43. Malik P, Katyal V, Malik V, Asatkar A, Inwati G, Mukherjee TK (2013) Nanobiosensors: concepts and variations. ISRN Nanomater 2013:327435
44. Wei D, Bailey MJ, Andrew P, Ryhänen T (2009) Electrochemical biosensors at the nanoscale. Lab Chip 9(15):2123–2131
45. Etayash H, Thundat T (2014) Microcantilever chemical and biological sensors, in encyclopedia of nanotechnology. In: Bhushan B (ed), Dordrecht, Springer Netherlands, pp 1–9
46. Kuswandi B (2019) Nanobiosensor approaches for pollutant monitoring. Environ Chem Lett 17(2):975–990
47. Long F, Zhu A, Gu C, Shi H (2013) Recent progress in optical biosensors for environmental applications, in state of the art in biosensors: environmental medical applications. In: Rinken T (ed) Intechopen, pp 4–28

48. Cheng C, Shi Y, Li M, Xing M, Wu Q (2017) Carbon quantum dots from carbonized walnut shells: structural evolution, fluorescence characteristics, and intracellular bioimaging. Mater Sci Eng C 79:473–480
49. Kou L, Huang T, Zheng B, Han Y, Zhao X, Gopalsamy K, Sun H, Gao C (2014) Coaxial wet-spun yarn supercapacitors for high-energy density and safe wearable electronics. Nat Commun 5(1):1–10
50. Koedrith P, Thasiphu T, Weon J-I, Boonprasert R, Tuitemwong K, Tuitemwong P (2015) Recent trends in rapid environmental monitoring of pathogens and toxicants: potential of nanoparticle-based biosensor and applications. The Scientific World Journal 2015
51. Mahmoud AED, Fawzy M (2015) Statistical methodology for Cadmium (Cd (II)) removal from wastewater by different plant biomasses. J Bioremediat Biodeg 6:1–7
52. Gheorghe I, Czobor I, Lazar V, Chifiriuc MC (2017) Present and perspectives in pesticides biosensors development and contribution of nanotechnology. In: New pesticides and soil sensors, pp 337–372. Elsevier
53. Kaushal M, Wani SP (2017) Nanosensors: frontiers in precision agriculture. In: Nanotechnology, pp 279–291. Springer
54. Pandey G (2018) Challenges and future prospects of agri-nanotechnology for sustainable agriculture in India. Environ Technol Innovation 11:299–307
55. Chen H, Yada R (2011) Nanotechnologies in agriculture: new tools for sustainable development. Trends Food Sci Technol 22(11):585–594
56. Sekhon BS (2014) Nanotechnology in agri-food production: an overview. Nanotechnol, Sci Appl 7:31
57. Bhatt J, Jangid M, Kapoor N, Ameta R, Ameta SC (2020) Detection and degradation of pesticides using nanomaterials, in biogenic nano-particles and their use in agro-ecosystems, pp 431–455. Springer
58. Reddy CN, Prasad PR, Sreedhar N (2016) ZnO/MWCNTs Nanosensor for differential pulse voltammetric determinations of dinosulfon in water samples. J Chem Chem Sci 6(11):997–1005
59. Duong HD, Reddy CG, Rhee JI, Vo-Dinh T (2011) Amplification of fluorescence emission of CdSe/ZnS QDs entrapped in a sol–gel matrix, a new approach for detection of trace level of PAHs. Sens Actuators B Chem 157(1):139–145
60. Sahoo D, Mandal A, Mitra T, Chakraborty K, Bardhan M, Dasgupta AK (2018) Nanosensing of pesticides by zinc oxide quantum dot: an optical and electrochemical approach for the detection of pesticides in water. J Agri Food Chem 66(2):414–423
61. Kumar S, Sachdeva S, Chaudhary S, Chaudhary GR (2020) Assessing the potential application of bio-compatibly tuned nanosensor of Yb_2O_3 for selective detection of imazapyr in real samples. Colloids Surf A: Physicochemical Eng Aspects 593:124612
62. Verdian A (2018) Apta-nanosensors for detection and quantitative determination of acetamiprid—a pesticide residue in food and environment. Talanta 176:456–464
63. Yang Z, Qian J, Yang X, Jiang D, Du X, Wang K, Mao H, Wang K (2015) A facile label-free colorimetric aptasensor for acetamiprid based on the peroxidase-like activity of hemin-functionalized reduced graphene oxide. Biosens Bioelectron 65:39–46
64. Pang Y, Zhang Y, Sun X, Ding H, Ma T, Shen X (2019) Synergistical accumulation for electrochemical sensing of 1-hydroxypyrene on electroreduced graphene oxide electrode. Talanta 192:387–394
65. Bigdeli F, Rouhani F, Morsali A, Ramazani A (2020) Ultrasonic-assisted synthesis of the nanostructures of a Co (II) metal organic framework as a highly sensitive fluorescence probe of phenol derivatives. Ultrason Sonochem 62:104862
66. Chongsuebsirikul J, Sakdarat P, Phongphut A, Klinthingchai Y, Prichanont S, Thanachayanont C, Pungetmongkol P (2019) Copper oxide nanorods electrodes for organophosphate pesticide sensor. In: 2019 IEEE 14th international conference on nano/micro engineered and molecular systems (NEMS). IEEE

67. El Din Mahmoud A, Fawzy M (2016) Bio-based methods for wastewater treatment: green sorbents. In: Ansari AA, Gill SS, Gill R, Lanza GR, Newman L (eds) Phytoremediation: management of environmental contaminants, vol 3. Springer International Publishing, Cham, pp 209–238
68. Mahmoud AED, Fawzy M, Radwan A (2016) Optimization of cadmium (Cd^{2+}) removal from aqueous solutions by novel biosorbent. Int J Phytorem 18(6):619–625
69. Nasr M, Mahmoud AED, Fawzy M, Radwan A (2017) Artificial intelligence modeling of cadmium (II) biosorption using rice straw. Appl Water Sci 7(2):823–831
70. Zhang S, Jin L, Liu J, Wang Q, Jiao L (2020) A label-free yellow-emissive carbon dot-based nanosensor for sensitive and selective ratiometric detection of chromium (VI) in environmental water samples. Mater Chem Phys 122912
71. Maric T, Mayorga-Martinez CC, Nasir MZM, Pumera M (2019) Platinum–halloysite nanoclay nanojets as sensitive and selective mobile nanosensors for mercury detection. Adv Mater Technol 4(2):1800502
72. Thatai S, Khurana P, Prasad S, Kumar D (2014) A new way in nanosensors: gold nanorods for sensing of Fe(III) ions in aqueous media. Microchem J 113:77–82
73. Jimenez-Falcao S, Villalonga A, Parra-Nieto J, Llopis-Lorente A, Martinez-Ruiz P, Martinez-Mañez R, Villalonga R (2020) Dithioacetal-mechanized mesoporous nanosensor for Hg(II) determination. Microporous Mesoporous Mater 297:110054
74. Liu Y, Zhao Y, Zhang Y (2014) One-step green synthesized fluorescent carbon nanodots from bamboo leaves for copper (II) ion detection. Sens Actuators B Chem 196:647–652
75. Vandarkuzhali SAA, Jeyalakshmi V, Sivaraman G, Singaravadivel S, Krishnamurthy KR, Viswanathan B (2017) Highly fluorescent carbon dots from pseudo-stem of banana plant: applications as nanosensor and bio-imaging agents. Sens Actuators B Chem 252:894–900
76. RipeSense. Available from: www.ripesense.co.nz/
77. Wei F, Lam R, Cheng S, Lu S, Ho D, Li N (2010) Rapid detection of melamine in whole milk mediated by unmodified gold nanoparticles. Appl Phys Lett 96(13):133702
78. Maksimović M, Omanović-Mikličanin E, Badnjević A (2019) Is nanofood safe? In: Nanofood and internet of nano things, pp 87–137. Springer

Waste-Recovered Nanomaterials for Emerging Electrocatalytic Applications

Abdelaal S. A. Ahmed, Ibrahim Saana Amiinu, Xiujian Zhao, and Mohamed Abdelmottaleb

Abstract Energy is essential and affects all aspects of our society, including the economy and modern living. However, the unparalleled rise in the global population, technological advancements, and changes in the scope of energy resources are all affecting the present energy landscape. With the increasing demands for energy and over-consumption of fossil energy, CO_2 emission is anticipated to rise over the next decades with devastating consequences on the environment and humans' lives. To avoid future eventualities, clean energy technologies have evolved with the expectation to diversify the global energy resources. Alternative energies are likely to show a crucial role in meeting not just the future energy needs but to remedy the escalating negative impact of fossil energy. Various clean energy systems, including fuel cells, electrolytic cells, rechargeable batteries, solar cells, etc., have emerged as viable renewable energy systems with even a wider range of applications and less impact on the environment. The efficiency of these energy systems is critical but is dependent on several technical factors, including electrochemical hydrogen evolution reaction (HER), oxygen evolution reaction (OER), and oxygen reduction reaction (ORR). An efficient electrocatalyst is required to drive the kinetics of these electrochemical processes effectively. However, developing practically efficient electrocatalyst is a significant challenge in terms of striking a balance between cost, performance, and sustainability of the active materials. Irrespective of any challenges, developing cost-effective and efficient electrode materials is vital for large-scale implementations of these energy systems. This chapter discusses the alternatives, recent progress, and future trends of using various waste materials for the development of advanced electrodes for various electrochemical systems.

A. S. A. Ahmed (✉) · M. Abdelmottaleb
Chemistry Department, Faculty of Science, Al-Azhar University, Assuit 71524, Egypt
e-mail: abdelaalsaiyd@gmail.com

I. S. Amiinu
State Key Laboratory of Silicate Materials for Architecture, Wuhan University of Technology, Luoshi Road, Wuhan 430070, People's Republic of China

A. S. A. Ahmed · X. Zhao
State Key Laboratory of Advanced Technology for Materials Synthesis and Processing, Wuhan University of Technology, Luoshi Road, Wuhan 430070, People's Republic of China

A. S. H. Makhlouf and G. A. M. Ali (eds.), *Waste Recycling Technologies for Nanomaterials Manufacturing*, Topics in Mining, Metallurgy and Materials Engineering, https://doi.org/10.1007/978-3-030-68031-2_10

Keywords Waste-recovered nanomaterials · Electrochemical water splitting · Fuel cell · Metal–air batteries · Dye-sensitized solar cell

List of Abbreviations

AFM	Atomic force microscopy
BCMs/a-BCMs	Biochar microspheres/activated biochar microspheres
BET	Brunauer–Emmett–Teller
CE	Counter electrode
CNFs	Carbon nanofibers
CSEM	Carbonized sucrose-coated eggshell membrane
DSSCs	Dye-sensitized solar cells
1D	One-dimensional
3D	Three-dimensional
E^o	The standard potential
FESEM	Field emission scanning electron microscope
HER	Hydrogen evolution reaction
HPNS	Hierarchical porous nanosheets
HRTEM	High-resolution TEM
LiBs	Lithium-ion batteries
LSV	Linear sweep voltammetry
MFCs	Microbial fuel cells
OER	Oxygen evolution reaction
ORR	Oxygen reduction reaction
PCE	Power conversion efficiency
PV	Photovoltaics
REN21	Renewable Energy Policy Network for the twenty-first Century
TEM	Transmission electron microscopy
TFSCs	Thin-film solar cells
XRD	X-ray powder diffraction
ΔG^o	The free energy change for the reaction
η	Overpotential

1 Introduction

By advent, the industrial revolution in the last century combined with the growing global population, the global energy consumption has risen from 5.52×10^{20} J in 2010 to 6.07×10^{20} J in 2015 and is projected to reach 6.97×10^{20} J by 2030 [1]. Electricity is the most significant form of end-use energy, with an annual growth of about 1% [2]. Currently, fossil fuels supply about 80% of the total global energy, with about 34.3% share for oil, 20.9% for natural gas, and 25.1% for coal. The estimated reserves of these resources are projected to deplete in about 40 years for oil, 60 years for natural gas, and 200 years for coal [3]. The burning of fossil fuels emits greenhouse gases, including CO_2, as the main component driving for global climate change and environmental pollution [4]. Therefore, governments and various institutions have begun mobilization efforts to find sustainable energy resources to meet future energy demands. Renewable energy storage and conversion systems are considered to be sustainable without harmful emissions. As reported by Renewable Energy Policy Network for the twenty-first Century (REN21), 146 countries have national goals for renewable energy power in different fields. For instance, Denmark has a target of 100% use of renewable energy [5]. The European Commission has set a strategic target to use at least 20% energy from renewable sources by 2020 and 32% by 2030 [6]. Recently, electrocatalytic-based techniques have received attention as promising candidates for sustainable energy storage and conversion applications [7]. As a result of the modernization and increasing human activities, a vast amount (millions to hundreds of millions of tons per annum scale) of unwanted residues as waste are produced continuously [8, 9]. The releases of these wastes with limited or no recycling have a significant impact on the environment and human health. Moreover, the disposal of these waste materials through improper management systems makes their impact more detrimental [10]. Therefore, over the past two decades, the treatment of waste materials and recycling has received increasing attention from the economic and environmental protection point of views [8, 11–19]. Nanomaterials derived from waste have been particularly used in many modern technological applications [20–23]. This chapter highlights the recent progress on the recovery of waste materials for various emerging electrocatalytic applications.

2 Electrochemical Water Splitting

Due to the increasing rate of depletion and the environmental problems related to the usage of fossil fuels, research efforts are being geared at developing eco-friendly, abundant, and renewable energy resources. Several innovative techniques have been recommended as potential energy conversion systems. One of these techniques is the fuel cell technology for generating eco-friendly energy [24]. A fuel cell works using hydrogen (H_2) and oxygen (O_2) gases as fuels while emitting zero carbon and water as waste. Typically, steam reformation of hydrocarbons and water electrolysis are the common ways to produce H_2 for fuel cells [24, 25]. Steam reformation usually emits CO_2 along with H_2 gas, which reduces the purity of the obtained H_2, and thus the life cycle, as well as the efficiency of assembled fuel cells, will be affected. Electrochemical water splitting [26] has been considered the best alternative pathway for renewable hydrogen production since the work of Troostwijk and Deinman in 1789 [27]. The advantages of electrochemical water splitting include (i) high-purity H_2 production, (ii) the abundance of water as a fuel source, and (iii) no need for high-temperature and/or high-pressure reactors. Until recently, electrochemical water splitting is limited and contributes

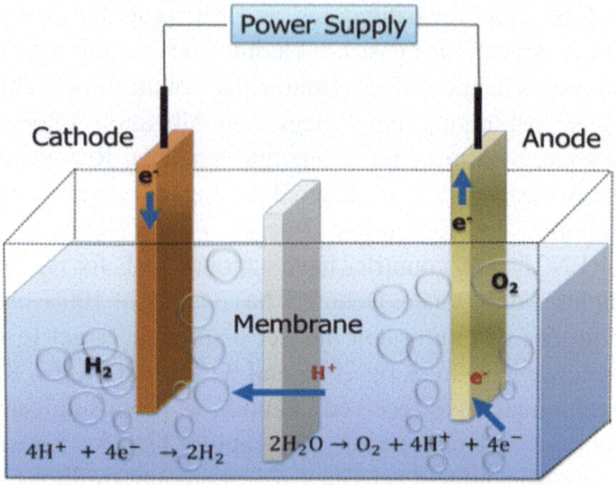

Fig. 1 Schematic illustration of electrochemical water splitting

only about 4% of the global production [28, 29]. The water-splitting process is a double electrochemical process that involves two half-reactions (Fig. 1). A reduction process at the cathode is called hydrogen evolution reaction (HER), while the oxidation reaction process at the anode is known as oxygen evolution reaction (OER). The cathodic and anodic reactions are strongly dependent on the pH of the solution according to the following equations [30].

(a) The overall electrocatalytic water splitting is represented by Eq. 1

$$2H_2O \rightarrow 2H_2 + O_2\ \Delta E^o = 1.23\ V,\ \Delta G^\circ = 237.2\ kJ\ mol^{-1} \tag{1}$$

(b) Under acidic conditions, the cathodic and anodic reactions are represented by Eqs. 2 and 3, respectively.

$$4H^+ + 4e^- \rightarrow 2H_2 \quad E_c^o = 0.00\ V \tag{2}$$

$$2H_2O_l \rightarrow O_{2g} = +4H^+ + 4e^- \quad E_a^o = 1.23\ V \tag{3}$$

(c) Under alkaline conditions, the cathodic and anodic reactions are represented by Eq. 4 and 5, respectively.

$$4H_2O + 4e^- \rightarrow 2H_2 + 4OH^- \quad E_c^o = -0.83\ V \tag{4}$$

$$4OH^- \rightarrow 2H_2 + 2H_2O_l + 4e^- \quad E_a^o = -0.40\ V \tag{5}$$

where E^o is the standard potential (V) and ΔG^o is the free energy change for the reaction ($kJ\ mol^{-1}$).

The theoretical minimum energy input required for the catalytic splitting of water is $\Delta G^{o} = 237.1\ kJ\ mol^{-1}$, which relates to a voltage of 1.23 V to deliver the thermodynamic driving force for the uphill reaction [31]. However, the experimental value is usually higher due to the practical overpotential (η) correlated with the sluggish reaction kinetics, particularly for the OER [32, 33]. Therefore, intensive research has been done to increase the reaction rate and reduce the overpotential of this process by using various electrocatalytic materials [34]. Nobel-based metals such as Pt and Ru are the benchmark catalysts for HER and OER, respectively [35–37]. However, their rarity, high cost of Pt (≈$50.000/kg), and poor durability are the main challenges that are preventing them from large-scale applications. Recently, tremendous efforts have been devoted to develop alternatives, catalytic materials that display high performance, and are more economically viable. These materials include sulfides [38–44], phosphides [45–48], and heteroatom-doped nanomaterials [49]. However, most of them are obtained by using toxic, hazardous, and expensive chemicals that are not suitable for large-scale production [9, 26]. To assure sustainable hydrogen production, it is highly desired to develop renewable and inexpensive alternatives. In the following section, the basic principles for overall water splitting, HER and OER reactions, and the recent progress of using recovered waste nanomaterials as electrocatalysts will be discussed.

2.1 *Electrocatalytic Reaction*

Generally, the electrocatalytic study is essential to investigate the kinetic parameters that define the electron transfer reactions at the interfaces. As water splitting is a typical electrocatalytic reaction, it is important to briefly discuss the basics of such reactions to give the reader a general overview. Generally, the electrochemical reaction is the process involving charge transfer ($O + ne^{-} \leftrightarrow R$) facilitated by an electrocatalyst. The key role of the electrocatalyst is to adsorb the reactants on its surface to form adsorbed intermediates and to facilitate the charge transfer between the electrode and reactants [50]. The performance of the electrocatalytic reactions is affected by many factors such as the overpotential (η) exchange current density (i_o) and Tafel slope (b), and these parameters will be briefly introduced in the following subsections.

2.1.1 The Overpotential

The overpotential (η) is a critical parameter to investigate the performance of the electrocatalyst. Typically, the varying of applied potential to drive an electrochemical reaction should be equal to the potential at equilibrium. However, the real applied potential is much higher. From the Nernst equation, the applied potential (E) can be expressed as Eq. 6.

$$E = E^{o\prime} + \frac{RT}{nF} \ln \frac{C_o}{C_R} \tag{6}$$

where $E^{o\prime}$ is the potential of the overall reaction (V), R is the ideal gas constant (8.314 J mol^{-1} K^{-1}), T is the Kelvin temperature, n is the number of transferred electrons in the reaction, and F is the Faraday constant (96485.33 C mol^{-1}). C_O and C_R are the bulk concentrations for the oxidized and reduced species (mol L^{-1}), respectively.

The overpotential is defined as the difference between the applied potential and the equilibrium potential (E_{eq}) as in Eq. 7 [51].

$$\eta = E - E_{eq} \tag{7}$$

Typically, the overpotential is the additional potential needed to drive an electrochemical reaction, and the electrocatalytic performance of catalytic materials is inversely proportional to the η.

2.1.2 Exchange Current Density

The total current (j) of an electrochemical reaction is the sum of anodic (j_a) and cathodic (j_c) currents (Eq. 8).

$$j = j_a + j_c \tag{8}$$

The anodic current is repented in Eq. 9, and the cathodic current in Eq. 10.

$$j_a = nFk_a[C_R] \exp\left(\frac{\alpha_a nFE}{RT}\right) \tag{9}$$

$$j_c = nFk_c[C_O] \exp\left(-\frac{\alpha_C nFE}{RT}\right) \tag{10}$$

Under equilibrium conditions (η = 0; E = E_{eq}), and j_a = j_c, thus the net current is zero.

Usually, the current is divided by the surface area of the electrode (A) to obtain the i_o, which is commonly used.

$$i_o = \frac{j_o}{A} \tag{11}$$

The rate of the electrochemical reaction is directly correlated with i_o. Thus, the smaller η indicates a higher $i_{o,}$ which is desired for efficient catalytic materials [35–37].

2.1.3 Tafel Equation and Tafel Slope

The Tafel plot is an essential technique in the electrochemical studies to estimate the current exchange density of the reaction kinetics. The relation between the i_o and the applied η of a simple electrochemical redox reaction can be labeled by the Butler–Volmer equation (Eq. 12). [50]

$$i = i_o \left[\exp\left(\frac{\alpha_a n\mathrm{FE}}{\mathrm{RT}}\right) + \exp\left(\frac{\alpha_c n\mathrm{FE}}{\mathrm{RT}}\right)\right] \tag{12}$$

where α is the transfer coefficient and α is the symmetry factor ($\approx$0.5).

At high anodic overpotential, the net current attributed to the anodic end while the impact of the cathodic part is insignificant, and the Eq. 12 can be simplified as Eq. 13.

$$i \approx i_o \exp\left(\frac{\alpha_a n\mathrm{F}\eta}{\mathrm{RT}}\right) \tag{13}$$

According to the logarithmic form, the Tafel relationship is revealed as Eq. 14.

$$\log(i) = \log(i_o) + \frac{\eta}{b} \tag{14}$$

Also, Eq. (2.8) can be expressed in the original Tafel format

$$\eta = a + b\log(i) \tag{15}$$

where a is a constant and b is the Tafel slope, which can be represented by Eq. 16.

$$b = \frac{2.303\ \mathrm{RT}}{\alpha\mathrm{F}} \tag{16}$$

The value of b can be extracted from plots η versus $\log(i)$.

A smaller b indicates that i is significantly enhanced with a small η change. This indicates a faster reaction rate constant and implies good electrocatalytic kinetics.

Another important function of the Tafel slope is to determine the pathway for an electrochemical reaction as follows:

(a) For a single-electron transfer reaction, α is usually represented by the symmetry factor (β) and can be recognized as Eq. 17.

$$\alpha = \beta = \frac{1}{2} + \frac{\eta}{\lambda} \tag{17}$$

β describes a symmetrical potential energy barrier and is generally close to 0.5 because the η is usually much smaller than the reorganization energy (λ). A value of 120 mV dec^{-1} implies that the rate-determining step is controlled by the single-electron transfer step.

(b) For a multi-step electron transfer reaction, which involves a series of consecutive reaction steps, α can be represented by Eq. 18.

$$\alpha_a = \frac{n_b}{v} + n_r\beta \tag{18}$$

where n_b is the number of electrons transferred before the rate-determining step, v is the number of times the rate-determining step occurs for one repetition of the overall reaction, and n_r is the number of electrons that participate in the rate-determining step.

According to Guidelli et al. [52], it is not possible to have more than one electron transferred simultaneously; thus, n_b is either 1 or 0 when the rate-determining step is the first electron transfer step or the chemical reaction, respectively.

From Eq. 18, the rate-determining step for electrochemical reactions can be predicted as follows: (i) If the first electron transfer reaction is the rate-determining step, the n_b and v are equal to 0, while n_r is equal to 1, $\beta = \alpha_a$ is equal to 0.5, and the corresponding Tafel slope is 120 mV dec^{-1} which is close to a single-electron transfer process, and (ii) if the second electron transfer step is a rate-determining step, the n_b is equal to 1 and α_a is equal to 1.5. Assuming v is equal to 1, the estimated Tafel slope is 40 mV dec^{-1}; (iii) if the chemical reaction is the rate-determining step, or the rate-determining step includes a chemical step successive to the first electron transfer step, the n_b and v are equal to 1, while n_r is 0. Consequently, α_a is unity and the estimated Tafel slope is 60 mV dec^{-1}, and (iv) in more complex systems such as OER, if the rate-determining step is the third electron transfer step, n_b and v are equal to 2 and 1 (n_r and β are 0), respectively. Thus, α_a is equal to 2 and the estimated Tafel slope is 30 mV dec^{-1}. Thus, Tafel slope is a strong indicator for the rate-determining step. In a consecutive reaction, a smaller Tafel slope reveals that the rate-determining step is at the end part of the multiple-electron transfer process, which is usually a sign of a suitable electrocatalyst.

2.2 *Recovered Nanomaterials for Hydrogen Evolution Reaction*

H_2 gas is one of the best options with zero end-use emissions and continually replenished resources to reduce the emissions of environmental hazards associated with the use of fossil fuels [53, 54]. H_2 is used as a fuel in fuel cells to generate electricity, which is a source of clean carbon-free chemical energy. Thus, H_2 gas has great potential as an energy carrier due to the following reasons: (i) compatibility with fuel cell technology and recyclability, (ii) H_2 gas can be produced via eco-friendly water splitting without CO_2 or other toxic emissions [46], (iii) the energy density of H_2 gas is higher than that of natural gas [53], and (iv) H_2 gas can be stored in gaseous or liquid forms, also together with metal hydrides [53]. Although H_2 gas can be produced from fossil fuels or nuclear energy, these routes display many environmental problems. Therefore, considerable efforts have been made to develop cost-effective methods from sustainable resources [31, 53]. The cathodic reaction in electrochemical water splitting (HER) is a possible way to store the obtained energy in the chemical bonds [55]. HER ($2H^+ + 2e^- \rightarrow H_2$) is a multi-step electrochemical process occurring on the electrode surface. Typically, pure water (pH $\approx$ 7) is a poor ionic conductor with high Ohmic overpotential and therefore is seldomly utilized for H_2 production. For efficient water splitting, the voltage and conductivity of the water should be improved by adding acids or alkalis. The aqueous acidic and alkaline solutions offer high ionic concentrations, thus possessing a low electrical resistance. At acidic conditions, HER is superficial as abundantly of protons are available, and it proceeds by a multi-step reaction with two workable mechanisms. The possible electrochemical mechanism of HER in both acidic and alkaline solutions is composed of two steps depending on the Tafel slope [54].

(a) The first step is the electrochemical hydrogen adsorption, where the dehydrated proton (H^+) associates with an electron (e^-) and chemisorbed on the electrode surface (M) to form adsorbed hydrogen (H^*) as Eqs. (19 and 20). This process is known as Volmer reaction, and its Tafel slope (b_V) is expressed in Eq. 21.

$$H^+ + M + e^- \leftrightarrows M - H^* \text{(acid solution)} \tag{19}$$

$$H_2O + M + e^- \leftrightarrows M - H^* + OH^- \text{(alkaline solution)} \tag{20}$$

$$b_V = \frac{2.303\,RT}{\alpha F} = 120 \text{ mV dec}^{-1} \tag{21}$$

where H* is the hydrogen atom chemisorbed on the active site of the electrode surface (M).

(b) The second step is desorption of H_2 from the active site of the electrode surface to generate H_2 gas. Based on the coverage of H^*, this process can take place by two possible pathways known as electrochemical or chemical desorption as follows:

(i) In the case of low coverage H^* with enough active sites on the electrode surface, the H^* will connect with a proton and an electron simultaneously to evolve a molecule of H_2 (Eqs. 22 and 23). This is known as the electrochemical desorption of H_2 and usually called a Heyrovsky reaction. Hence, the HRE follows a *Volmer–Heyrovsky* mechanism. The Tafel slope (b_H) for this process can be expressed by Eq. 6

$$M - H^* + H^+ + e^- \leftrightarrows M + H_2 \qquad \text{acid solution} \tag{22}$$

$$M - H^* + H_2O + e^- \leftrightarrows M + OH^- + H_2 \qquad \text{alkaline solution} \tag{23}$$

$$b_H = \frac{2.303\ RT}{(1+\alpha)F} = 40\ \text{mV dec}^{-1} \tag{24}$$

(ii) In case of high coverage H^*, two adjacent H^* will chemically join together and evolve H_2 molecule. This is known as chemical desorption of H_2 (Eq. 25), otherwise called a Tafel reaction; thus, the HER follows a *Volmer–Tafel* mechanism. The Tafel slope (b_T) for this process can be expressed by Eq. 26.

$$2M - H^* \leftrightarrows 2M + H_2 \qquad \text{acidic and alkaline solutions} \tag{25}$$

$$b_T = \frac{2.303\ RT}{2F} = 30\,\text{mV dec}^{-1} \tag{26}$$

As the progress of the HER depends on the adsorption and desorption of proton on the surface of the catalyst, the bonding between HER catalyst and H^* should be appropriately strong to facilitate the proton–electron transfer and weak enough to facile bond breaking and the evolvement of gaseous H_2 product [46, 54]. The overall HER also depends on the change in free energy of H^* adsorption on a catalyst surface (ΔGH^*), which is zero for an optimum HER catalyst. The sufficient HER catalyst should display a lower overpotential with a high ability to promote the reaction rate in acidic and alkaline solutions. Pt-based materials are the benchmark for HER electrocatalyst due to their matchless electrocatalytic performance. However, the high cost and insufficient Pt reserves limit their commercial applications [54]. Therefore, a significant research work has been dedicated to develop alternative catalysts that exhibit high performance and are more economically viable [26]. Various nanomaterials include sulfides [56], phosphides [57, 58], and chalcogenides [31, 59]. In this regard, great efforts have been put to explore cheaper electrocatalytic materials without sacrificing the catalytic performances. Waste-derived nanomaterials showed promising potential for H_2 production [9]. An overview of recent progress concerning the waste nanomaterials HER catalysts is given below.

Despite the potential catalytic activity of the traditional carbon materials (e.g. graphene, carbon nanotubes, etc.), their high cost and sometimes using toxic reagents during their synthetic process restrict their large-scale application [60]. Therefore, developing carbonaceous materials from renewable and highly abundant resources is highly desired to push sustainable renewable energy systems. Carbon is a highly abundant element in our universe and attached with hydrogen, oxygen, nitrogen, and sulfur can form the energy mass (e.g., carbohydrates, fats, and protein) and structural tissue (e.g., bones, muscles, stems, and leaves) for organisms [60]. However, a vast amount of wasted biomass from agriculture, stock farming, ocean fisheries, and human activities are obtained each year which leads to pollution of our environment and releases CO_2 [61, 62]. Thus, conversion biomass to carbonaceous materials has received an intense interest as a promising catalyst in emerging renewable energy technologies; this mainly can be assigned to their sustainability, broad availability with low cost, and their high ability to design various heteroatoms (N, S, P)–doped carbon metal-free catalysts [60]. Every day, a huge amount of animal biowastes such as stock farming, ocean fisheries, and other human activities are obtained. The conversion of this unwanted waste material to catalytic-based nanomaterials is an encouraging pathway. For instance, Liu et al. synthesized N-doped porous carbon from Bombyx mori silk cocoons with KCl activation at 400–900 °C [63]. The obtained carbonized silk fibers showed a honeycomb-like structure, and the sample at 900 °C (KCl–900) displayed 4.7% nitrogen contents with a BET surface area of 349.3 m^2 g^{-1}. The electrochemical analysis of KCl-900 showed promising electrocatalytic ability toward HER in acid solution (Fig. 2c) in terms of low onset potential (63.3 mV), low overpotential potential (137 mV at 10 mA cm^{-2}), and small Tafel slope (131.6 mV dec^{-1}) as well as good electrochemical durability. It has been reported that heteroatom-doped carbon displayed very poor catalytic ability toward water splitting [64]. Codoping (more than one type of heteroatoms is incorporated) in the carbon network can enhance the catalytic performance for water splitting [65, 66]. Most of the developed catalytic materials have been synthesized through a multi-step process and using some toxic reagents which are not suitable for economic and environmental aspects. Thus, intensive research has been done to find cheaper materials as a precursor for carbon catalysts. Recently, Prabu et al. utilized spathe-pollen waste of palm plant to prepare graphene-like hierarchical porous nanosheets (HPNs) [67]. The obtained HPNs displayed a dual micro-/mesoporous structure with a BET surface area of 1297 m^2 g^{-1}. As a HER catalyst in acidic conditions (Fig. 2e), HPNs showed a promising performance with an overpotential of 330 mV and a Tafel slope of 63 mV dec^{-1}.

Furthermore, the carbon derived from wastes can be utilized as a catalyst to support composite catalytic system. For instance, Mir et al. utilized waste polythene as a carbon source to support molybdenum carbide (Mo_2C) in C–Mo_2C nanocomposites [59]. By the further introduction of nitrogen, the obtained composite (C/N–Mo_2C) showed higher HER performances with high electrochemical stability over 2000 cyclic voltammetry cycles. Also, Yan et al. utilized waste walnut shells as the carbon source and carbon support to synthesize Mo_2C@C catalyst (Fig. 3a) [68]. Different mass ratios of the walnut shell-derived carbon and

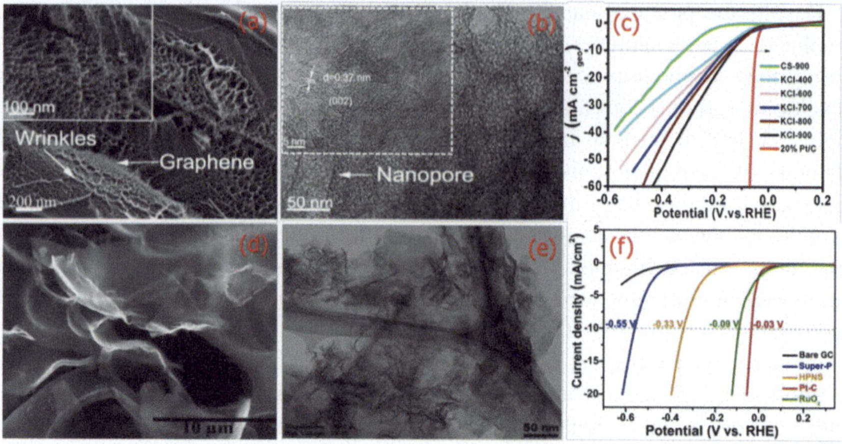

Fig. 2 **a** FESEM, **b** TEM images (inside is HRTEM) of the KCI-900 sample, and **c** linear sweep voltammetry (LSV) curves of the various catalysts in 0.5 M H_2SO_4 solution. Adapted from Ref. [63] with permission, Copyright 2017, Elsevier. **d** FESEM, **e** TEM images of HPNS and **f** LSV HER LSV curves of the various catalysts in 0.5 M H_2SO_4 solution. Adapted from Ref. [67] with permission, Copyright 2017, Elsevier

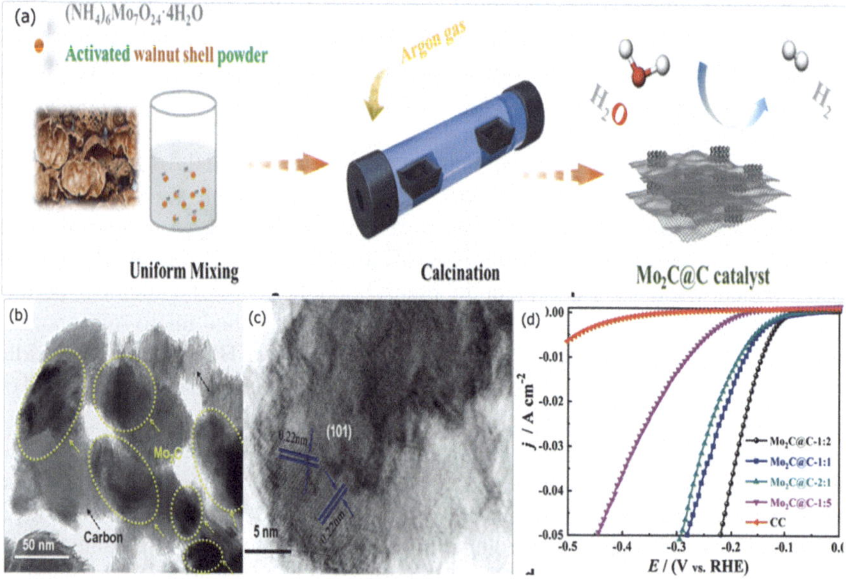

Fig. 3 **a** Schematic synthesis of Mo_2C@C catalysts, **b** TEM, **c** HRTEM image of Mo_2C@C-1:2, **d** HER LSV curves of the various catalysts in 0.5 M H_2SO_4 solution. Adapted from Ref. [68] with permission, Copyright 2017, Elsevier

ammonium heptamolybdate were physically mixed at an appropriate temperature under argon atmosphere. The obtained Mo_2C@C–1:2 displayed the high HER performance with the small overpotential (140 mV) and a Tafel slope of 63 mV dec^{-1} in acidic solution. Duckweed (DW) famous kind of biomass with high nitrogen, sulfur, and carbon contents was utilized to prepare N, S-doped carbon to support bimetallic NiFe alloy nanoparticle [69]. The un-annealed electrocatalyst displayed high performance toward HER performance with an overpotential of 106 mV at 10 mA cm^2 in 1 M KOH.

2.3 Recovered Nanomaterials for Oxygen Evolution Reaction

The anodic electrocatalytic oxygen evolution reaction (OER) of water electrolysis is an eco-friendly method for oxygen (O_2) production in large scale from renewable resources [70]. In the earth's crust, O_2 is among the most abundant elements and is essential for all living organisms to produce energy for vital metabolic processes. OER process involves four consecutive proton–electron transfers per oxygen molecule and needs a thermodynamic potential higher than 1.23 V versus of the reversible hydrogen electrode (RHE) [51]. This overpotential consumes more energy and decreases the energy conversion efficiency. Accordingly, the inefficiency of OER is a major bottleneck to the water-splitting systems for commercial applications. The proposed mechanism of OER depends on the adsorption/desorption of oxygen-containing intermediates (e.g. O, OH, and OOH).

The possible mechanism in the acidic condition is represented by the following steps: (i) The first step (Eq. 27) is a charge-transfer process involving the formation of adsorbed hydroxy species (OH) onto an active surface site (M). The Tafel slope of this step is 120 mV dec^{-1}. (ii) The second step can be occurred by either an electrochemical oxide path with a second electron transfer and Tafel slope of 40 mV dec^{-1} (Eq. 28) or an oxide path with a recombination step and Tafel slope of 30 mV dec^{-1} (Eq. 29). (iii) The third step involves the formation of O_2, and two active sites occur with a Tafel slope of 15 mV dec^{-1}.

Acidic condition

$$M + H_2O_l \rightarrow M - OH + H^+ e^- \tag{27}$$

$$M - OH + OH^- \rightarrow M - O + H_2O_l + e^- \tag{28}$$

$$2M - O \rightarrow 2M + O_{2g} \tag{29}$$

$$M - O + H_2O_l \rightarrow M - OOH + H^+ + e^- \tag{30}$$

$$M - OOH + H_2O_l \rightarrow M + O_2 + H^+ + e^- \tag{31}$$

The following equations represent the reaction mechanism in alkaline solutions [71].

Alkaline condition

$$M + OH^- \rightarrow M - OH \tag{32}$$

$$M - OH + OH^- \rightarrow M - O + H_2O_l \tag{33}$$

$$2M - O \rightarrow 2M + O_{2g} \tag{34}$$

$$M - O + OH^- \rightarrow MOOH + e^- \tag{35}$$

$$M - OOH + OH^- \rightarrow M + O_{2g} + H_2O_l \tag{36}$$

As shown in Fig. 4, the O_2 can be obtained from the Oxo (MO) intermediate in two different ways [51]. The first is represented by the green line, which involves the production of O_2 by direct combination of two MO groups, while the second involves the formation of peroxide (MOOH) intermediates followed by decomposition to form O_2. The OER is a heterogeneous reaction in which the bonding interactions (M–O) within the oxygen-containing intermediates play a crucial role in the overall electrocatalytic ability.

The unfavorable thermodynamic and kinetic conditions for removing $4e^-$ during the splitting of water to form oxygen–oxygen double bond limit the performance of the water-splitting system. Therefore, intensive efforts have been devoted to

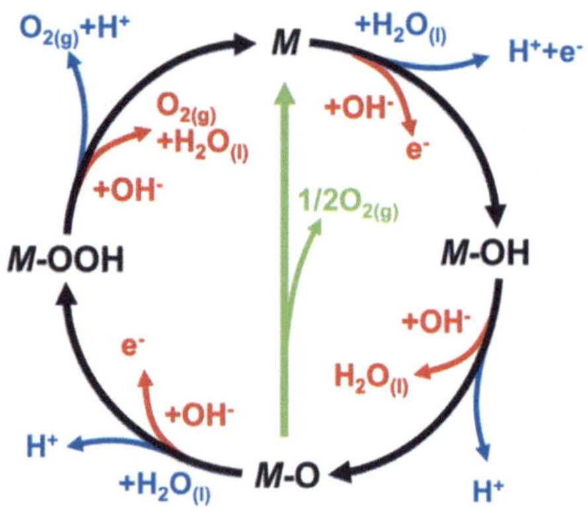

Fig. 4 OER reaction mechanism in acid (blue) and alkaline (red) conditions. The black line represents the oxygen evolution with the formation of M–OOH intermediate while the green route for direct reaction of two adjacent M–O intermediates to produce oxygen. Adapted from Ref. [51] with permission, Copyright 2017, Royal Society of Chemistry

develop efficient OER materials with valuable rates of oxygen evolution at the lowest overpotential. Metal oxides have been considered as suitable material toward OER reaction [72]. However, their poor electrical conductivity limits their electrocatalytic enhancement [30]. At present, oxides of RuO_2 and iridium-type IrO_2 with a low overpotential of ≈220 and 280 mV, respectively, are considered as the benchmark for OER [73]. However, under high anodic potential, RuO_2 displayed poor stability; thus, it will be oxidized to RuO_4 and dissolved into the solution. In comparison with RuO_2 catalyst, not only is IrO_2 more stable but can also be sustained at a higher anodic potential [51]. However, IrO_2 displays similar problems as RuO_2 catalyst and oxidizes to IrO_3 during OER catalytic conditions [74]. Also, both RuO_2 and IrO_2 are precious metals with a high cost, which limits their large-scale applications [75]. Accordingly, in the last decades, intensive research has been done to develop abundant, low-cost alternatives toward promoting the commercialization of water-splitting-related devices [75]. Recently, a lot of materials have been tested for catalyzing the OER, including carbon-based materials [76–78], transition metal oxides [79], selenides [80], and phosphides [81]. Among these materials, wastes recovered nanomaterials are widely explored as an alternative. Hence, recent progress on waste-recovered nanomaterials as OER catalysts is discussed below. Steel is one of the most wildly used materials in several aspects of our life. Each year, in the world millions of tons of scrap steel is created. The high content of Fe (~98%), excellent stability, good conductivity, low cost, and wide availability are encouraging factors to extract iron nanoparticles. Devi et al. extracted iron nanosheets from waste tin-plated steel as anode for water splitting [82] using a single-step hydrometallurgy process involving treatment with concentrated hydrochloric acid at different times up to 120 min at room temperature. The extracted iron nanosheets are in 20–25 nm thick range with a dimension of 1.4–1.8 μm. The electrochemical analysis showed that the extracted iron nanosheets by acid treatment at 10 min showed the best OER performance with a low overpotential (329 mV), small Tafel slope (60 mV dec^{-1}), and a high exchange current density (17.7 μA cm^{-1}) in alkaline medium. The high consumption of electronic devices has led to the rapid obsolescence of many electric and electronic appliances [83]. The short lifespan of these devices results in rapid increase in related waste materials with a wide-ranging negative impact on the environment [84]. Therefore, finding a new use of these materials is highly desired from both environmental and economic points of view. The electric Cu cable wires can be completely recyclable with the same physical and chemical properties. Therefore, recovering Cu from spent electronics not only is environmentally required but also keeps the resource for the coming generations [83]. Great efforts have been done to recycle Cu. For instance, Babar et al. employed waste copper Cu wire as an electron collector to enhance the electrocatalytic performance of OER [84]. In this study, a chemical oxidation method was utilized to deposit $Cu(OH)_2$ nanowires on the surface of Cu wire followed by electrodeposition of nickel–iron hydroxide (NiFe LDH) as described in Fig. 5a. The assembled core–shell NiFe LDH/Cu $(OH)_2$/Cu electrode exhibits significant OER performance in terms of low overpotential

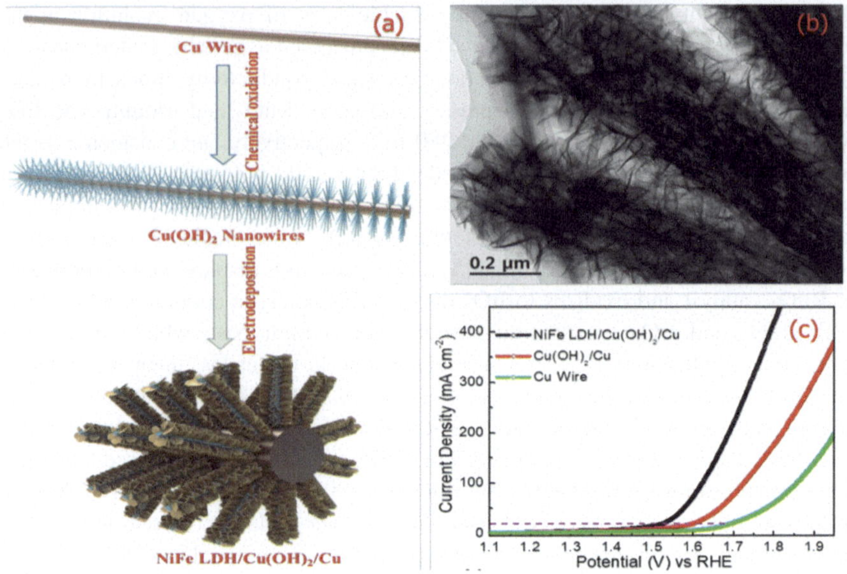

Fig. 5 **a** Schematic synthesis of core–shell NiFe LDH/$Cu(OH)_2$/Cu on waste Cu cable wire, **b** TEM image of NiFe LDH/$Cu(OH)_2$/Cu, and **c** OER LSV curves of the various catalysts in 1 M KOH solution at 5 mV s^{-1}. Adapted from Ref. [84] with permission, Copyright 2017, Elsevier

(275 mV) and 390 mV to achieve 20 and 100 mA cm^{-2} current densities, respectively, in addition to its excellent durability in alkaline media.

Since the last three decades, lithium-ion batteries (LiBs) have been considered the most usable storage devices for most of our daily electronic devices such as mobile phones, laptops, and more recently for hybrid electric vehicles and electric vehicles [85]. However, tremendous consumptions and lack of efficient recycling systems make spent LiBs a serious environmental problem as they contain flammable and toxic substances [86, 87]. Typically, LiBs are composed of Al foil coated with Li-based metal oxides such as $LiCoO_2$, $LiMn_2O_4$, $LiNiO_2$, and $LiNi_xCo_yMnzO_2$ as a cathode, and Cu foil coated with graphite as anode and a polymer as a separator. The content of the active materials in the cathode is usually high in some spent LiBs. Therefore, utilizing the waste LiBs as a source to recover these valuable metals keeps our environment not only safe but also economically beneficial [88]. Recently, many studies have been done on recycling the spent LiBs. For instance, Chen et al. recovered $LiCoO_2$ from the spent LiBs to be an OER electrocatalyst [89]. The electrochemical analysis exposed that the catalytic performance of the $LiCoO_2$ was considerably enhanced with increasing the number of LIB cycles from 100 to 500 cycles. The $LiCoO_2$ from 500 cycles displayed the best OER activity with a slightly smaller Tafel slope (67.41 mV dec^{-1}) than that for the original (72.58 mV dec^{-1}). Natarajan et al. [85] utilized Mn and Co from spent LiBs for recovering spinel $MnCo_2O_4$ oxides with a surfactant-free

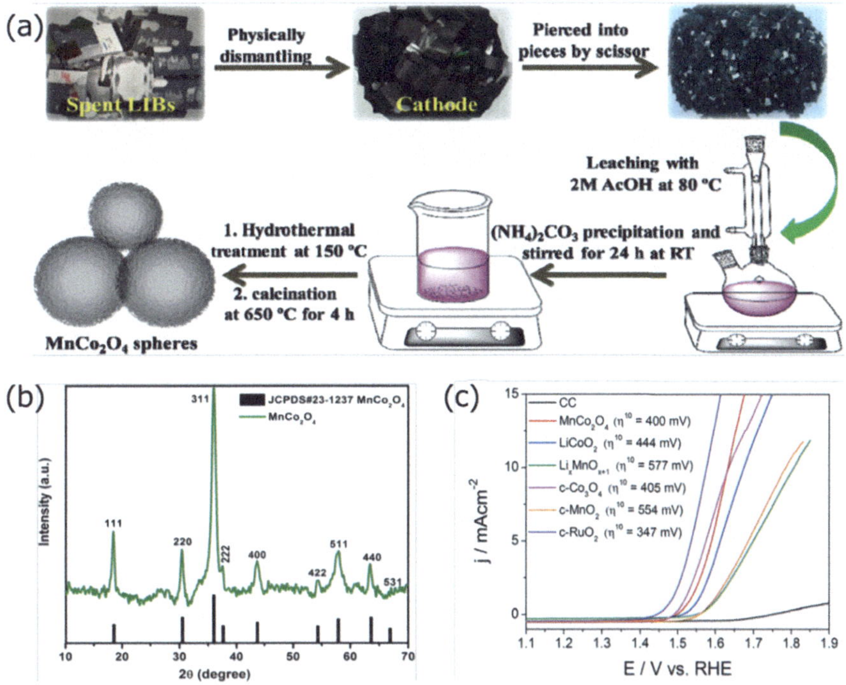

Fig. 6 **a** Scheme of recovery of spinel $MnCo_2O_4$ from the spent LiBs, **b** XRD pattern of spinel $MnCo_2O_4$ synthesized by thermal annealing of the carbonate precursor after hydrothermal treatment, and **c** OER LSV curves of the various catalysts in 1 M KOH solution at 5 mV s^{-1}. Adapted from Ref. [85] with permission, Copyright 2017, Royal Society of Chemistry

hydrometallurgical route, as shown in Fig. 6a. The recovered spinal $MnCo_2O_4$ as an ORE electrocatalyst in 1 M KOH showed a superior catalytic performance than that of recovered $LiCoO_2$, Li_XMnO_{X+1}, c–Co_3O_4, c–MnO_2, and c–RuO_2 (Fig. 6c). This resulted in a lower overpotential (400 mV) at 10 mA cm^{-2} and smaller Tafel slope (80 mV dec^{-1}).

The hydrometallurgical process showed some limitations, such as producing secondary waste, consuming time, and using some hazardous chemical reagents. To overcome these limitations, Yang et al. utilized direct mechanochemical activation for recovering of Li_2CO_3 and $Ni_{0.5}Mn_{0.3}Co_{0.2}(OH)_2$ from spent $LiNi_{0.5}Mn_{0.3}Co_{0.2}O_2$ batteries [90]. The electrocatalytic performance of the recovered $Ni_{0.5}$ $Mn_{0.3}$ $Co_{0.2}(OH)_2$ toward OER reaction was comparable to IrO_2 (Fig. 7d). The estimated Tafel slope of $Ni_{0.5}$ $Mn_{0.3}$ $Co_{0.2}(OH)_2$, $LiNi_{0.5}Mn_{0.3}Co_{0.2}O_2$, and IrO_2 was 6.79, 18.61, and 5.56 mV dec^{-1}, respectively.

Pegoretti et al. recycled high-temperature (HT) $LiCoO_2$ from spent LiBs as an electrocatalyst for OER in alkaline media [91]. Firstly, cobalt was recycled as Co$(OH)_2$ and served as a precursor material for synthesizing HT $LiCoO_2$ using Li_2CO_3 as a reagent. The OER reaction by HT $LiCoO_2$ starts at lower potential compared

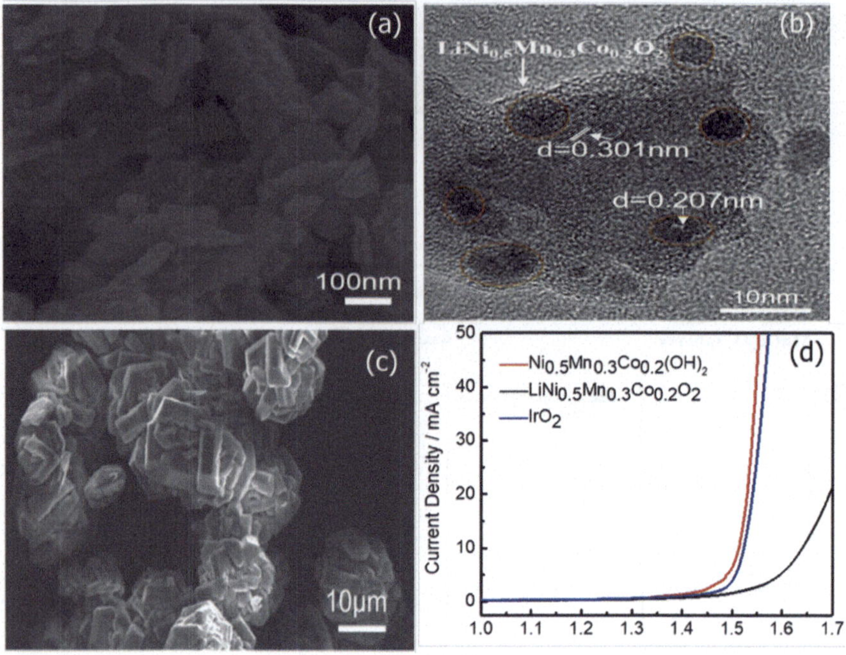

Fig. 7 **a** SEM and **b** TEM images of $LiNi_{0.5}Mn_{0.3}Co_{0.2}O_2$ powder after MA treatment. **c** SEM images of the recycled Li_2CO_3 sample and **d** OER LSV curves of the various catalysts in 1 M KOH solution at 5 mV s^{-1}. Adapted from Ref. [90] with permission, Copyright 2017, Elsevier

with the Ni and Pt, indicating promising catalytic ability. Mei et al. utilized spent $LiFePO_4$ and $LiCoO_2$ batteries to extract Fe and Co ions by oxalic acid as a leaching agent [92]. The Fe extracted from $LiFePO_4$ by chemical leaching process is electrochemically deposited on Fe-doped tin oxide (FTO) followed by annealing in air at 770 °C to form hematite film (Fig. 8a–b). Moreover, the cobalt phosphate (CoPi) extracted by hydrothermal leaching process and the obtained CoPi are electrochemically deposited on the hematite film (Fig. 8c–d). The hematite and hematite/CoPi/FTO photoanode power conversion efficiencies for water oxidation using hematite/FTO and hematite/CoPi photoanodes were 0.053 (at1.14 V) and 0.13% (at 1.10 V) in 1 M NaOH, respectively (Fig. 8f). Mu's group recycled a sheep horn as a rich source of N and S for the one-step preparation of N,S-codoped 3D porous graphene which showed a relatively high electrocatalytic activity toward OER and even oxygen reduction performance higher to their counterparts obtained via toxic chemicals and/or multi-step routes [93].

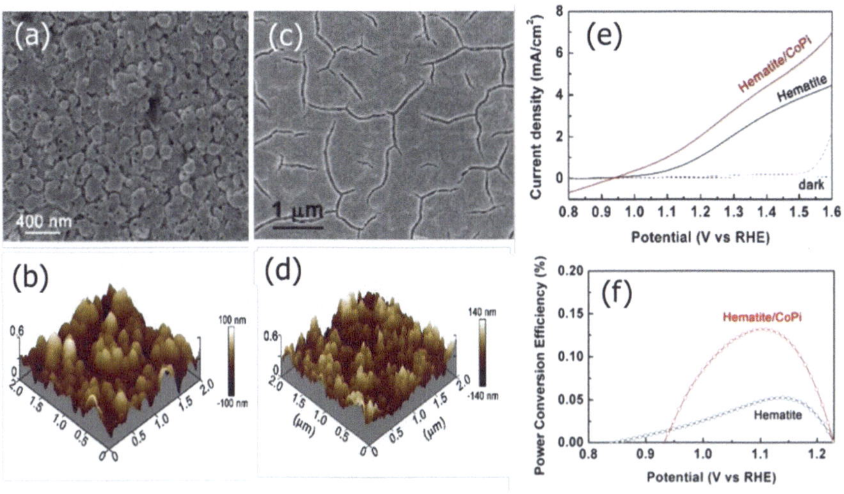

Fig. 8 **a** SEM and **b** 3D AFM images of the hematite films annealed at 770 °C for 3 min. **c** SEM and **d** 3D AFM of hematite/CoPi film. **e** J–V curves of hematite and hematite/CoPi and **d** power conversion efficiency of the hematite and hematite/CoPi photoanodes for water splitting. Adapted from Ref. [92] with permission, Copyright 2017, American Chemical Society

2.4 Recovered Nanomaterials for Electrocatalytic Overall Water Splitting

As mentioned above, overall water spiting requires high overpotential. Therefore, developing efficient HER and OER catalysts is highly desired. Most of the designed materials can be separately utilized as HER or OER catalysts. However, there are growing interest materials with bifunctional properties toward both OER and HER in the same electrolytic system and pH range [94–96]. However, almost all the best OER catalysts work well in neutral (pH $\approx$ 7) or alkaline media, whereas most of the HER catalysts are only good in acidic media. In this section, a brief overview of the current progress of utilizing waste-recovered nanomaterials for overall water splitting is discussed. Recently, Jothi et al. [83] used scrap Cu wire (SCW) recovered from waste electric cables as a substrate to grow nickel–cobalt phosphide (NiCoP) by an electrodeposition process. The assembled NiCoP/SCW displayed remarkable OER performance (Fig. 9b) with 220 mV overpotential at 10 mA cm^{-2}. The NiCoP/SCW also displayed high catalytic activity toward H_2 evolution in alkaline solution (Fig. 9c). Moreover, the electrodeposited NiCoP/SCW showed considerable electrocatalytic ability for full electrochemical water splitting in alkaline condition at a low cell voltage of 1.59 V and displayed a stable current density of 10 mA cm^{-2} during 24 h.

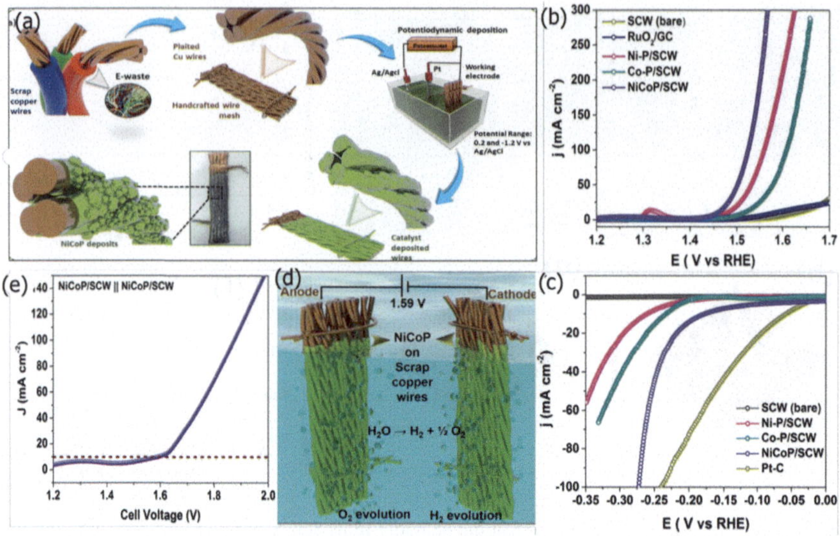

Fig. 9 **a** Schematic synthesis of NiCoP/SCW. **b** OER LSV curves and **c** HER LSV curves of the various electrodes recorded at 5 mV s^{-1} in 1 M KOH. **d** Schematic representation of the two-electrode cell configuration and **e** LSV curve of NiCoP/SCW coupled electrodes in a two-electrode setup. Adapted from Ref. [83] with permission, Copyright 2017, Wiley-VCH

Recently, Tiwari et al. [97] utilized multi-heteroatoms (nitrogen, sulfur, and phosphorus)-doped carbon (MHC) catalysts from waste yeast biomass as catalyst supports for ruthenium single atoms (RuSAs) along with Ru nanoparticles (RuNPs) as anode and Fe_3O_4 as a cathode for water electrolysis. The RuSAs + RuNPs@MHC displayed excellent HER catalytic ability comparable to the standard platinum on carbon catalyst. Moreover, Fe_3O_4@MHC catalyst displayed outstanding OER catalysts compared with IrO_2 in terms of overpotential, exchange current density, and Tafel slope. The as-prepared catalysts for overall water splitting need a solar voltage of 1.74 V to drive ~30 mA with remarkable long-term stability in the presence and absence of solar energy (Fig. 10).

3 Oxygen Reduction Reaction

Due to its availability in the atmosphere, O_2 is considered the most common oxidant. The reduction of O_2 by an ORR has captured substantial attention as the most important cathodic reaction in many fields such as corrosion and most of the sustainable energy conversion devices (e.g., fuel cells, metal–air batteries, solar cells, etc.). Typically, ORR is six or more times slower than anodic hydrogen oxidation reaction (HOR), which is an obstacle to their total performance [98]. In aqueous solutions, ORR can be done by two possible pathways: (i) transfer of two

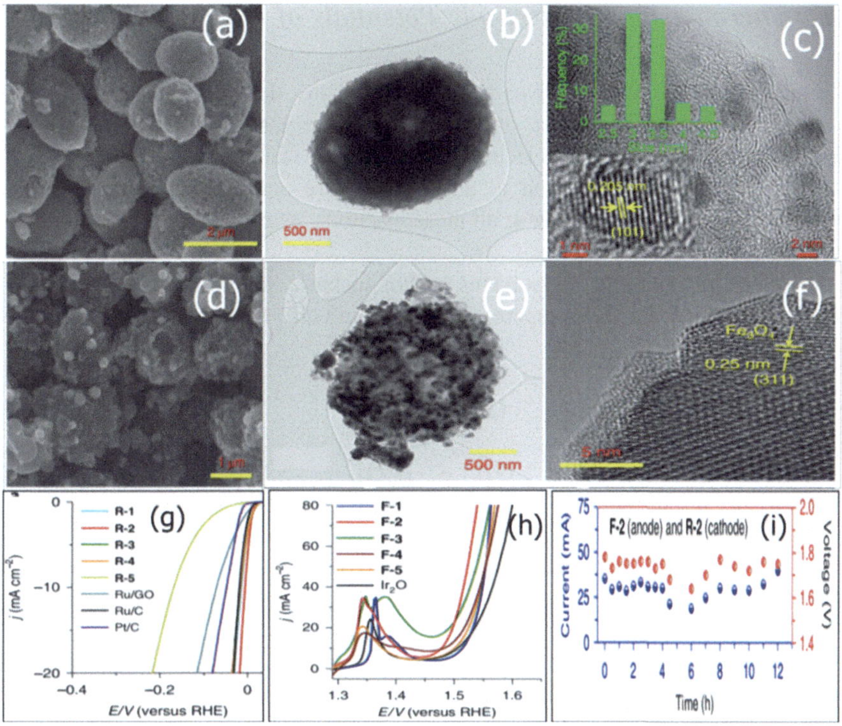

Fig. 10 **a** SEM, **b** TEM, and **c** HRTEM images of Ru–NP (R–2). **d** SEM, **e** TEM, and **f** HRTEM images of Fe_3O_4 NPs on the surface of carbonized yeast cells (F–2). **g** HER LSV curves of various catalysts in 1 M KOH at 2 mV s^{-1} [R–1 (35 mg $RuCl_3.xH_2O$), R–2 (40 mg $RuCl_3.xH_2O$), R–3 (45 mg $RuCl_3.xH_2O$), R–4 (40 mg $RuCl_3.xH_2O$, without glutaraldehyde), and R–5 (40 mg $RuCl_3.xH_2O$, without hydrothermal process)], **h** OER LSV curves of various catalysts in 1 M KOH at 2 mV s^{-1}; F–(1–5)/NF and IrO_2/NF [F–1 (1.2 g $Fe(NO_3)_3{\cdot}9H_2O$), F–2 (1.4 g $Fe(NO_3)_3{\cdot}9H_2O$), F–3 (1.6 g $Fe(NO_3)_3{\cdot}9H_2O$), F–4 (1.4 g $Fe(NO_3)_3{\cdot}9H_2O$, without glutaraldehyde), and F–5 (1.4 g $Fe(NO_3)_3{\cdot}9H_2O$ without hydrothermal treatment)], **i** solar-panel-driven activity and stability of catalysts in a two-electrode setup. Adapted from Ref. [97] with permission, Copyright 2017, Nature

electrons to produce hydrogen peroxide (H_2O_2) or (ii) direct transfer of four electrons to produce H_2O [99]. In a fuel cell, ORR has to proceed through the 4e-pathway, since a 2e-pathway results in a formation of stable H_2O_2 intermediate species, thus decreasing the catalytic activity [100]. Moreover, in non-aqueous solvents and/or in alkaline solution, the one-electron transfer pathway from O_2 to superoxide (O_2^-) could also occur, followed by reduction to O_2^{2-}. In the following sections, the thermodynamics of ORR process and recent progress on utilizing wastes-recovered nanomaterials for fuel cells and metal–air batteries are presented.

3.1 Thermodynamic Electrode Potentials of ORR

The mechanism of ORR involves many intermediates depending on the nature of the electrode material, catalyst, and electrolyte. The various ORR processes and their related thermodynamic electrode potentials at standard conditions (zero in 1.0 M proton aqueous solution at any temperature and 1.0 atm. hydrogen gas pressure) are presented in Table 1; all these reactions are called half electrochemical cathodic reaction.

Moreover in non-aqueous solvents and/or in basic solutions, the one-electron transfer pathway from O_2 to superoxide (O_2^-) could also happen (Eq. 43), followed by reduction to O_2^{2-} (Eq. 44).

$$O_2 + e^- \rightarrow O_2^- \tag{43}$$

$$O_2^- + e^- \rightarrow O_2^{2-} \tag{44}$$

The various electron transfer processes (1-, 2-, and 4-e) can be utilized in different applications [102]. The 4-electron direct pathway is preferred in fuel cell processes. The 2-electron pathway is used for H_2O_2 production, during the 1-electron pathway for exploration of the ORR mechanism.

In the case of nonstandard conditions, the electrode potentials can be expressed by the Nernst equations. Taking Eq. 37 in Table 1 as an example,

$$O_2 + 4H^+ + 4e^- \rightarrow H_2O \qquad E^o_{O_2/H_2O} = 1.229\,V \tag{37}$$

The thermodynamic reversible electrode potential for cathode at nonstandard conditions and assuming O_2 in gas phase and H_2O in liquid phase can be expressed by Eq. 45.

$$E^r_{O_2/H_2O} = E^o_{O_2/H_2O} + \frac{RT}{n_{O_2}F}\ln\left(P_{O_2}C^4_{H^+}\right) \tag{45}$$

Table 1 Thermodynamic electrode potentials of ORR in aqueous electrolyte solutions at standard conditions (25 °C, 1.0 atm.) [101, 102]

Electrolyte	ORR reactions	Thermodynamic electrode potential (V vs. SHE)
Acidic condition	$O_2 + 4H^+ + 4e^- \rightarrow H_2O$ (37)	1.23 V
	$O_2 + 2H^+ + 2e^- \rightarrow H_2O_2$ (38)	0.70 V
	$H_2O_2 + 2H^+ + 2e^- \rightarrow 2H_2O$ (39)	1.76 V
Alkaline condition	$O_2 + H_2O + 4e^- \rightarrow 4OH^-$ (40)	0.401
	$O_2 + H_2O + 2e^- \rightarrow HO_2^- + OH^-$ (41)	−0.065
	$HO_2^- + H_2O + 2e^- \rightarrow 3OH^-$ (42)	0.867

where $E^o_{O_2/H_2O}$ is the thermodynamic electrode potential as in Table 4.1, P_{O_2} is the partial pressure of O_2 gas (atmospheres), $C^4_{H^+}$ is the concentration of proton (moles per cubic decimeter), and n_{O2} is the number of electron transfer (here = 2).

The other half electrochemical reactions in Table 1 can also be expressed by the Nernst equation, as described above. In any electrochemical cell, there are two half-reactions, the anode and the cathode reactions. For example, in the fuel cell with H_2/O_2 supplies, the two half-reactions are

$$H_2 \leftrightarrow 2H^+ + 2e^- \qquad E^o_{H_2/H^+} = 0.000\ V \tag{46}$$

$$\frac{1}{2}O_2 + 2H^+ + 2e^- \leftrightarrow H_2O \qquad E^o_{O_2/H_2O} = 1.229\ V \tag{47}$$

Thus, the cathodic half-reaction has a lower overpotential and a higher kinetic rate, whereas the ORR is inactive, the overall cell reaction as in Eq. 48.

$$\frac{1}{2}O_2 + H_2 \leftrightarrow H_2O \qquad E^o_{cell} = 1.23\ V \tag{48}$$

This means that the overall electrochemical reaction has a thermodynamic cell voltage (E^o_{cell}) of 1.23 V at reversible conditions. Accordingly, the standard free energy change of the overall reaction is $\Delta G^o_{cell} = -nFE^o_{cell}$, where n is the number of electron transfer for the overall reaction (here = 2).

The thermodynamic reversible electrode potential for anode at nonstandard conditions and assuming O_2 in gas phase and H_2O in liquid that can be expressed by Eq. 49.

$$E^r_{H_2/H^+} = E^o_{H_2/H^+} + \frac{RT}{n_{H_2}F}\ln\left(\frac{C^2_{H^+}}{P_{H_2}}\right) \tag{49}$$

where $E^o_{H_2/H^+}$ is the thermodynamic electrode potential of the anode reaction and P_{H_2} is the partial pressure of H_2 gas (atmosphere).

By combining the thermodynamic reversible electrode potential for anode and cathode, the overall thermodynamic/fuel cell voltage/theoretical open-circuit voltage (E^o_{cell}) can be obtained as Eq. 50.

$$E^o_{cell} = E^r_{O_2/H_2O} - E^r_{\frac{H_2}{H^+}} = E^o_{\frac{O_2}{H_2O}} - E^o_{\frac{H_2}{H^+}} + \frac{RT}{nF}\ln\left(P^{\frac{1}{2}}_{O_2}P_{H_2}\right) \tag{50}$$

where n is the electron transfer number of the fuel cell reaction (=2) and other symbols have the same meanings as described above.

3.2 Waste-Recovered Nanomaterials for ORR in Fuel Cells

In the last decades, fuel cells have attracted attention as a potential device to directly convert the chemical energy from fuels to electrical energy [103]. The proton exchange membrane (PEM) fuel cells composed of anode and cathode separated with a solid electrolyte are shown in Fig. 11. The pumped fuel (H_2 gas) at the anode split into electrons and protons as in Eq. 46. The electrons flow out to deliver electrical power and end up at the cathode to reduce oxygen through (ORR) as in Eq. 47, while the protons diffuse through the solid electrolyte membrane toward the cathode to combine with the reduced oxygen forming water (Eq. 48) [102]. Although ORR plays a critical role in the work function of fuel cells, its sluggish kinetics leads to poor energy conversion efficiency and low output power density, which remain a major drawback restricting the widespread application of fuel cells [104].

To enhance the ORR kinetics for practical fuel cell, developing high cathode ORR catalyst is necessary. Currently, Pt/C catalysts have been widely fulfilled as cathodic catalysts due to its tremendous electrocatalytic activity for the ORR [105]. However, the high cost, inadequacy, poor stability, and susceptibility to methanol crossover in acidic medium hinder their large-scale application [99, 106]. Therefore, significant efforts have been devoted to designing cost-effective catalysts with excellent ORR activity and high accessibility. These developed materials include Pt-based alloys [107, 108] and carbon-based materials [109–112]. Waste-recovered nanomaterials also show high potentials as ORR catalysts for fuel

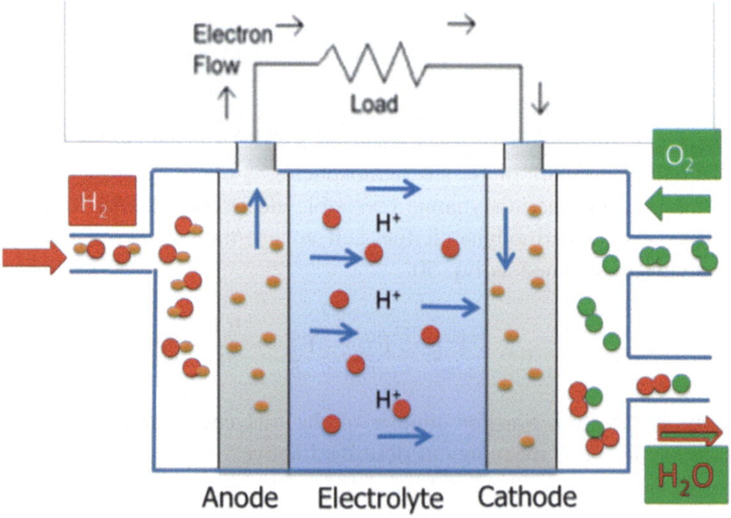

Fig. 11 Schematics of a fuel cell. Adapted from Ref. [98] with permission, Copyright 2017, Science

cells [104, 113–115]. Currently, many researchers have synthesized carbon-based material derived from biowastes. The animal biowastes displayed a promising candidate for preparing carbon-based nanomaterials. For instance, Guo et al. prepared hierarchically porous three-dimensional nanocarbons via pyrolysis process of fish-scale (FS) pyrolysis at 350 °C followed by activation with $ZnCl_2$ at 700–1000 °C [116]. Compared to the commercial Pt/C, the FS350-Z900 (the porous carbon obtained by $ZnCl_2$ activation followed by pyrolysis at 900 °C) with a BET surface area of 850 m^2 g^{-1} possesses the best ORR activity in both acidic and alkaline solutions. The estimated onset potentials (E_{ORR}) of the FS350-Z900 are 860 mV and 1000 mV for acidic and alkaline solutions, respectively. The relatively high content of N makes biowaste materials potential candidates to design N-doped carbon catalyst which is highly desired for ORR [117]. Zhoua et al. prepared N, S-codoped carbon nanofibers (CNFs) by pyrolysis natural spider silk (SS) as a precursor with 1D structure [118]. The obtained N, S-codoped CNFs exhibited high catalytic performance as ORR catalyst with onset potential of 850 mV. The assembled device showed a power density of 1800 mW m^{-2}, which is higher than that obtained by Pt/C cathode (1152 mW m^{-2}). Horn as animal biowaste was engaged as a precursor to prepare N, S-codoped 3D porous graphene (NSG) [93]. The resultant catalyst exhibits a promising ORR performance in both acid and alkaline media. Another study showed that the N-doped mesoporous carbon with ultra-high surface area was prepared by pyrolysis of sheep bones [119]. The ORR electrocatalytic activity of the obtained carbon material in alkaline solution is compared to commercial Pt/C. Moreover, cattle bones are utilized to synthesize N- and P-codoped hierarchically porous carbon (N,P-HPC) by pyrolyzing with phytic acid (PA) and dicyandiamide (DCDA) [120]. The obtained N, P-HPC exhibits remarkable ORR activity in alkaline electrolyte comparable to commercial Pt/C. Furthermore, human hair is converted to N, S-codoped carbons [121]. The obtained carbon exhibited as ORR catalyst displayed outstanding electrocatalytic performance with an onset potential of 950 mV and a high selectivity durability.

There are various plant waste materials utilized to prepare carbon-based nanomaterials as ORR catalysts. For example, pomelo is broadly cultivated and produces large amounts of pomelo peels as environmental waste. Yuan et al. prepared nitrogen-doped nanoporous carbon (NPC) catalyst by carbonizing pomelo peel waste at 800–1000 °C under NH_3 atmosphere [122]. The obtained catalyst (NPC-1000) with the highest BET surface area (Fig. 12b–c) displayed the highest ORR electrocatalytic performance in both acid and alkaline media with tolerance to methanol poisoning and high durability. To overcome utilizing NH_3 as a nitrogen source, Zhang et al. [104] used pomelo peels as a source for both carbon and nitrogen to prepare biochar microspheres (BCMs) and their activated counterpart (a-BCMs) via hydrothermal and thermal treatment, followed by annealing at 900 °C (Fig. 12d–e). The assembled microbial fuel cells (MFCs) with a-BCMs as cathode obtained power density of 907.2 mW m^{-2} operated continuously for 90 days.

To further enhance the ORR performance, Ma et al. used pomelo peel-derived carbon as a support to prepare Fe_3C/WC/GC nanocomposite [123]. First, pomelo

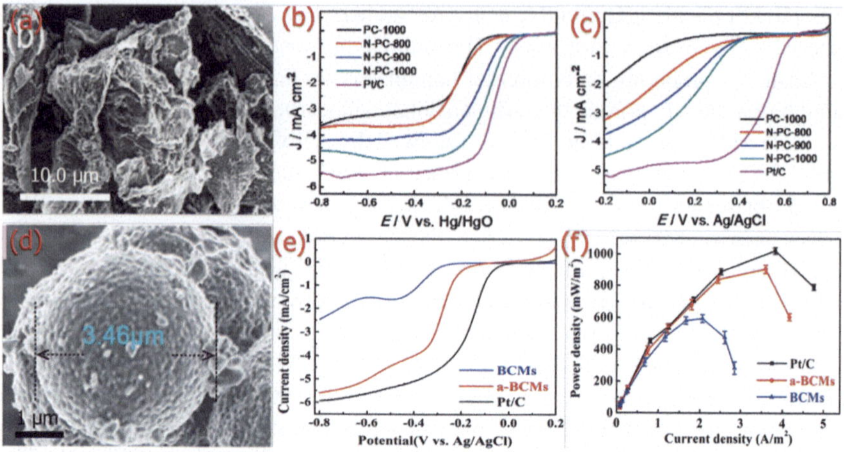

Fig. 12 **a** HRTEM image of NPC-1000. **b** LSV curves of the various catalysts in 1O_2-saturated 0.1 M KOH solution and **c** in O_2-saturated 0.5 M H_2SO_4 solution. Adapted from Ref. [122] with permission, Copyright 2017, Royal Society of Chemistry. **d** SEM image of a-BCMs, **e** LSV of OER LSV curves of the various catalysts in O_2-saturated 0.1 mol L^{-1} KOH solutions at 10 mV s^{-1}, and **f** power densities of the MFCs with different cathodes as a function of current density. Adapted from Ref. [104] with permission, Copyright 2017, Royal Society of Chemistry

peel was immersed in the K4Fe $(CN)_6$ and Na_2WO_4 precursors followed by carbonization at 600–1100 °C. The obtained Fe_3C/WC/GC nanocomposite exposed excellent ORR performance in the pH-neutral electrolyte (Fig. 13e). As an air cathode in MFCs, the assembled device displayed a power density of 1997 mW m^{-2}, which is about 67.82% higher than Pt/C (1190 mW m^{-2}), as shown in Fig. 13f. Moreover, the device displays negligible voltage decay during the operation for about 91 days.

Furthermore, Wang et al. inserted Fe_2N nanoparticles into pomelo peel-derived N-doped porous carbon (N-PPC) via a two-step carbonization process at 900 °C with $FeCl_3{\cdot}6H_2O$ and NH_3 atmosphere [124]. The obtained Fe_2N/N-PPC nanocomposite revealed a remarkable ORR activity in alkaline solution with an onset potential of 966 mV which is superior to those of PPC (890 mV), N-PPC (899 mV), Fe-PPC (940 mV), and 20% Pt/C (952 mV) catalysts. This was attributed to the high surface area and the synergistic effect of Fe_2N. Meng Li et al. prepared biochar by pyrolysis of corncob at 250–750 °C [125]. The obtained biochar at 650 °C (CC-650) showed higher BET surface area of 655.89 m^2 g^{-1} with an excellent catalytic ORR performance. As a cathode in MFCs, the device assembled with CC-650 cathode showed a maximum power density of 58.85 mW m^{-2}. Moreover, inserting metal ions is an effective strategy to enhance the catalytic ability for ORR [126]. For example, hierarchical carbon with honeycomb-like interconnected macro-mesoporous frameworks multi-doped with N, P, and Fe has been prepared by pyrolysis of livestock sewage sludge under nitrogen [113]. The assembled MFCs with the as-prepared N, P and Fe-doped

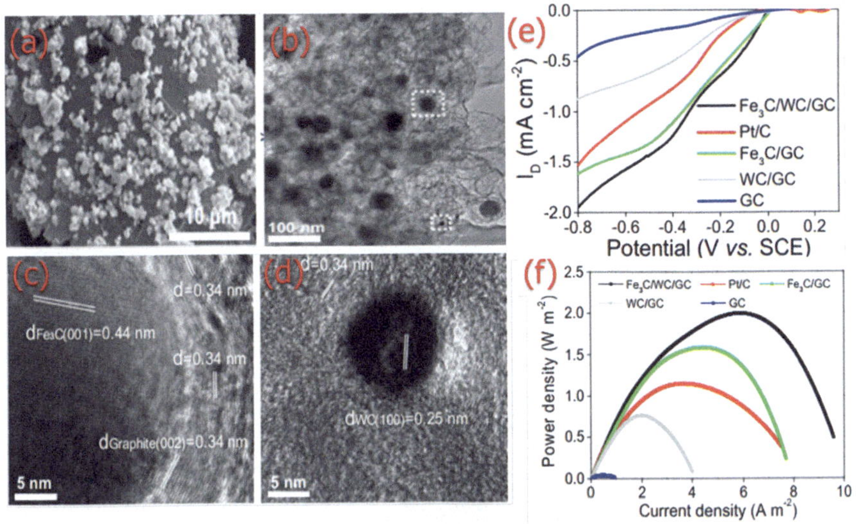

Fig. 13 **a** SEM, **b** TEM, and **c–d** HRTEM images of Fe_3C/WC/GC composite at 1000 °C. **e** SLV curves of different electrocatalysts in O_2-saturated and **f** power density of the MFCs with different cathodes as a function of current density. Adapted from Ref. [123] with permission, Copyright 2017, American Chemical Society

carbon cathode displayed a maximum power density of 1273 mW m^{-2} which is comparable to that obtained by the device constructed with Pt/C catalysts (1294 mW m^{-2}). After 90 days, the power density decreases by only 10.2% which is quite lower than that with Pt/C cathode (28.4%). Soybean straws are abundant waste with high contents of celluloses, hemicelluloses, and lignin which are desired for synthesizing ORR catalysts. Thus, Lu et al. [127] utilized carbon-derived soybean straw (SS) as supporting materials to prepare nitrogen and cobalt dual-doped activated carbon (Co-NASS) as in Fig. 14a. The as-prepared Co-NASS showed an excellent ORR activity in alkaline media with an onset potential of 870 mV which is comparable to that of the commercial Pt/C (940 mV).

Lu et al. utilized soybean straw to prepare Fe-NPC via a one-step pyrolysis (Fig. 15a) [128]. The obtained Fe-NPC catalyst displays outstanding ORR performance with an onset potential of 989 mV and 886 mV in alkaline and acid conditions, respectively. This was mainly attributed to the high BET surface area and the synergistic effect of Fe and N codoping. Chitosan is a highly abundant biopolymer existing in the exoskeletons of crabs, shrimp shells, etc. Although chitosan derives carbon display low ORR performance [129], its large amount of functional groups allowed to chelate with various metal ions strongly and thus can be converted to metal-NC (M-NC). Xie et al. synthesized Co- and N-codoped C (Co-NC) catalysts from chitosan at 900 °C [130]. The Co-NC-900 showed an excellent ORR performance in alkaline solution with onset negative potential of −44 mV that is comparable to that of Pt/C (−46 mV).

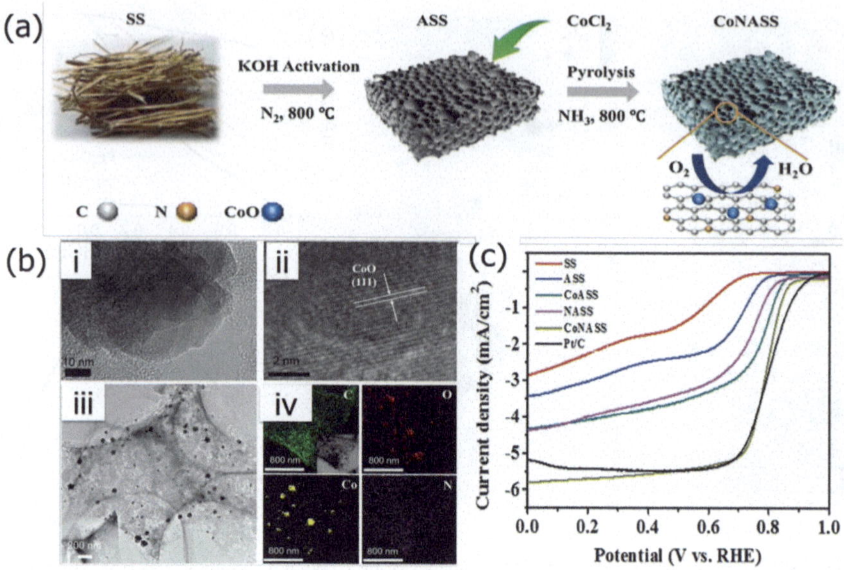

Fig. 14 **a** Schematic synthesis of N and Co codoping activated carbon (Co-NASS), **b** TEM (i) and HRTEM images (ii and iii) and HRTEM element mapping of Co-NASS (iv). **c** LSV curves of various catalysts in O_2-saturated 0.1 M KOH solution at 1600 rpm. Adapted from Ref. [127] with permission, Copyright 2017, Royal Society of Chemistry

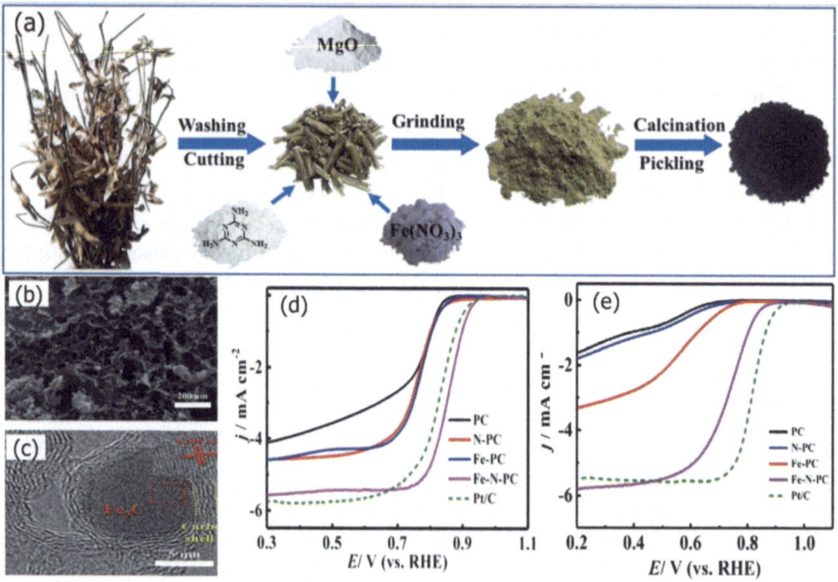

Fig. 15 **a** Schematic synthesis of Fe–N codoped porous carbon. **b** SEM image and **c** HRTEM image of Fe–NPC catalysts. **d** LSV curves of the series of soybean straw catalysts at a rotation speed of 1600 rpm in O_2-saturated 0.1 M KOH solution (**e**) and 0.1 M $HClO_4$ solution. Adapted from Ref. [128] with permission, Copyright 2017, Royal Society of Chemistry

3.3 Waste-Recovered Nanomaterials for Metal–Air Battery

Energy storage devices are indispensable for future energy systems to buffer the unpredictable energy generation and supply derived from renewable sources [131–135]. Currently, lithium-ion batteries (LiBs) remain the dominant player in portable electronic devices [136]. However, the intercalation chemistry of the electrode materials limits the maximum energy density of LiBs. Additionally, the supply of active LIB electrode materials, including cobalt and lithium metal, is limited and cannot meet future energy needs, particularly in an electric-vehicle-dominated market [131]. In this respect, many efforts have been devoted to find suitable alternatives such as metal–air batteries [136, 137]. Currently, metal–air batteries have attracted consideration due to their high energy density compared to that of other rechargeable batteries [36]. Generally, metal–air batteries can be divided into two kinds based on their electrolytes. One system is based on an aqueous electrolyte with high stability in moisture. The other is based on aprotic solvents and sensitive to moisture. Typical structure of metal–air battery consists of a metal anode (Li, Na, K, Mg, Zn, etc.), an air-breathing cathode with an open porous architecture and an electrolyte (aqueous or non-aqueous) according to the utilized anode. The theoretical energy density of these batteries is about 3-30 times greater than those of conventional LiBs [131]. Although metal–air batteries were developed in the early 1960s, their development faced many problems, typically consisting of metal anodes, air catalysts, and electrolytes [138]. Since then, aqueous metal–air batteries have been considered promising alternative energy storage devices due to their low cost, environmentally friendliness, and high power density. The air cathode architecture plays a key role in their performance [36]. As the ORR is a highly sluggish reaction, the air cathode is the performance-limiting electrode in metal–air batteries. As the cathode is interfaced with a liquid electrolyte and a gaseous O_2, the reaction at the air cathode takes place at the triple-phase boundary. Thus, it would be greatly beneficial to develop active air catalysts to expedite the ORR kinetics and to design proper electrode architecture to enlarge the triple-phase boundary to enhance the battery discharge performance. Although the noble-based metals are widely used for ORR, the relatively high cost with poor durability are the main challenges for large scale applications. In order to find low-cost candidates without sacrificing the catalytic performances, several materials have been investigated and reviewed as ORR catalysts [51]. Recently, nanomaterials recovered from waste exhibit a high promise as air cathodes in metal–air batteries. The recent progress on the application of these nanomaterials as an air-cathodes in metal-air batteries will be reviewed in the following section. Biowaste materials are used to prepare heteroatoms-doped carbons which displayed a promising potential as air cathode for metal–air batteries. For instance, Wang et al. [139] used the pine needle to prepare porous carbon via hydrothermal method and treatment with NH_3 at 900–1100 °C. The as-prepared carbon material at 1000 °C (N-SZ-1000) exhibits ORR performance superior to the commercial Pt/C. Moreover, the N-SZ-1000 cathode in aqueous zinc–air batteries shows 1.3 V discharge voltage at 10 mA cm^{-2}. Lei et al.

also prepared nitrogen-doped micropore carbon (N-PPC) derived from pine cone waste [140]. The pine cone was first pre-carbonized at low temperature and then post-treated in NH_3 at 800–1000 °C. The prepared N-PPC at 900 °C (N-PPC-900) showed the highest BET surface area (1555.25 m^2 g^{-1}) and exhibited overwhelming ORR performance than the commercial Pt/C in alkaline solution. Compared to the zinc–air battery assembled with Pt/C air cathode, the device with N-PPC-900 cathode showed higher discharge capacity at 50 mA cm^{-2} (Fig. 16c).

Another work by Ma et al. based on N, S-codoped porous carbon was synthesized by self-pyrolysis of garlic stems at 600–1000 °C (Fig. 17a) [141]. The as-prepared N, S-codoped carbon at 900 °C (GSC-900) displayed a higher BET surface area (991 m^2 g^{-1}) and the best ORR performance in terms of onset potential (860 mV) in alkaline solution. As an air cathode in zinc–air batteries, adding GSC-900 to FeCoOx enhances the conductivity and endows bifunctional ORR/OER properties. Choudhary et al. [142] successfully prepared a tungsten trioxide-modified carbon nanosheet decorated with palladium nanoparticles using cow dung as a carbon source. The as-prepared cow dung-derived nanodisc electrocatalyst (Pd@WO_3-NDs) showed excellent ORR performance in alkaline

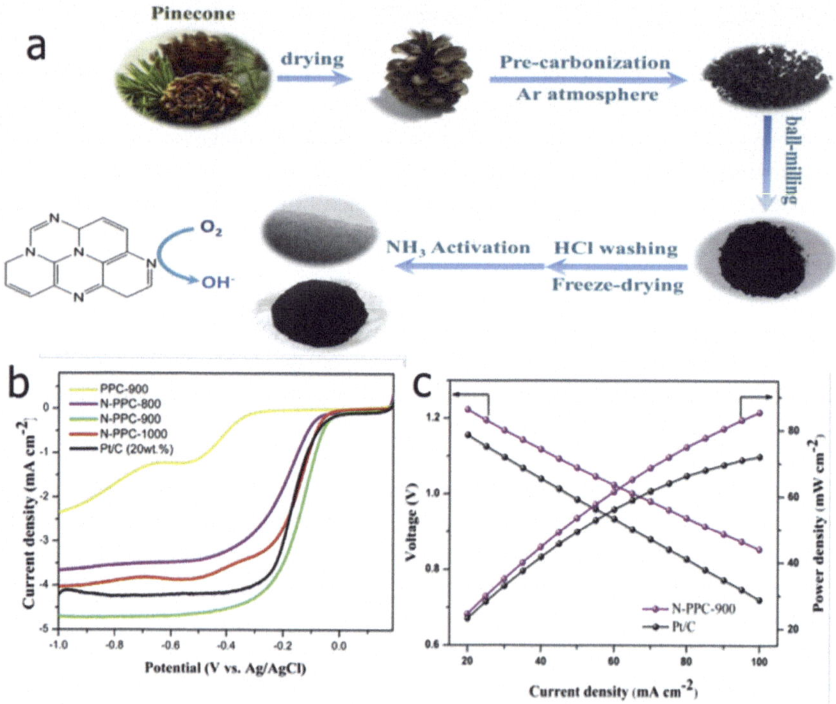

Fig. 16 **a** Preparation of the nitrogen-doped nanoporous carbon nanosheets. **b** LSV curves of different samples at a scan rate of 5 mV s^{-1} and **c** N–PPC–900 and Pt/C loaded air cathode discharge curves in zinc–air battery at 50 mA cm^{-2}. Adapted from Ref. [140] with permission, Copyright 2017, Elsevier

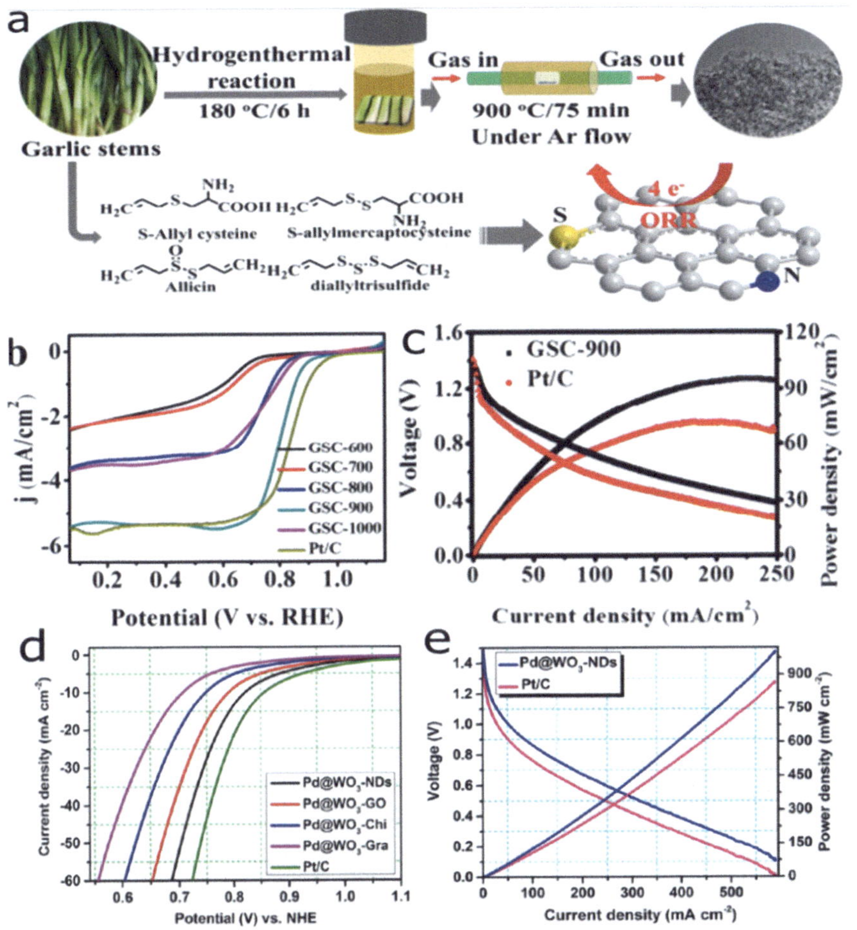

Fig. 17 **a** Preparation procedure of N, S-dual-doped porous carbon from garlic stems. **b** LSV curves of corresponding catalysts for ORR in O_2-saturated 0.1 M KOH at 1600 rpm and **c** polarization curve and corresponding power density plot of the battery prepared with GSC-900 as the cathode catalyst compared with the battery prepared using commercial Pt/C catalyst. Adapted from Ref. [141] with permission, Copyright 2018, Elsevier. **d** LSV curves of corresponding catalysts for ORR in O_2-saturated 0.1 M KOH at 1600 rpm and **e** polarization curve and corresponding power density plot of the battery prepared with Pd@WO_3-NDs as the cathode catalyst compared with the battery prepared using commercial Pt/C catalyst. Adapted from Ref. [142] with permission, Copyright 2017, American Chemical Society

solution (Fig. 17d) with onset potential of 195 mV at 10 mA cm^{-2}. In comparison with zinc–air batteries assembled with Pt/C cathode, the device with Pd@WO_3-NDs not only showed higher current density (250 mA cm^{-2}) at 0.6 V voltage, but also displayed higher power density at 1.0 V (Fig. 17e).

Yang et al. [143] utilized peanut shells as a carbon source to prepare FeNi alloy incorporated into nitrogen-doped carbon (NC). The as-prepared FeNi–NC catalyst displayed bifunctional ability toward ORR and OER comparable with the noble metals. The FeNi–NC catalyst for ORR in alkaline solution displayed the onset potential of 980 mV and a limiting current of 4.84 mA cm^{-2}, which is better than that of commercial Pt/C. This was mainly attributed to the synergistic effects of FeNi alloy and NC. Moreover, zinc–air battery assembled with FeNi–NC air catalyst not only produces discharging potential of 1.2 V and charging potential of 2.025 V at 8.0 mA cm^{-2}, but also produces a power density of 80.8 mW cm^{-2} at 115 mA cm^{-2} which is better than the 20% Pt/C-based device (72.9 mW cm^{-2} at 97.7 mA cm^{-2}). Nanomaterials derived from spent LiBs were also used as air cathodes in metal–air batteries. For instance, Wei et al. utilized Ni–Co–Mn oxides from spent LiBs as air electrode [144]. The Ni–Co–Mn oxide was recovered at 300–900 °C between 60 and 420 min. The ORR performance is significantly influenced by temperature and time. Accordingly, Ni–Co–Mn oxides heated at 600 °C for 300 min exhibited the best ORR performance with a limiting current density of 1.23 mA cm^{-2}. The zinc–air battery assembled with Ni–Co–Mn oxides, heated at 600 °C for 300 min, showed the best performance of discharge and charge capacity at 1.3 V.

4 Dye-Sensitized Solar Cells

In the last decades, intensive efforts have been made use of solar energy as a free ultimate renewable energy source. Each year, the earth receives about 5.29×10^{24} J of solar energy which is about 10 thousand times more than mankind consumes. Only about 3.78×10^{24} J solar energy strikes the surface of the earth, the earth receives about 4.30×10^{20} J solar energy per hour, which is more than the average global energy consumed per year (4.10×10^{20} J) [145]. If only 0.10% of the received sun is converted to electricity, it would be about four times the world's total energy capacity. Thus, solar energy received great attention as encouraging sustainable energy for future generations [146]. Solar cells or photovoltaics (PV) are devices that directly convert the incident sunlight into electrical energy through photovoltaic effect [147]. According to the basic materials, the developments of PV cells are divided into three generations. The first generation is the crystalline silicon (c-Si) PV cells demonstrated in 1954 by Chaplin, Fuller, and Pearson at the Bell Labs [148]. Although this kind of PV cells is the holder of the market (90%), the high cost of crystalline silicon restricts their widely spread. The second generation is thin-film solar cells (TFSCs) developed in the 1970s after the oil shock [149]. The TFSCs displayed ability to be designed in flexible shapes with attractive appearance which is suitable to be building-integrated PV cells. Despite the global interests for the first two generations, these devices are single junction obeyed the Shockley–Queisser limit with an estimated thermodynamic efficiency of 31% [150]. In contrast to such materials, a third generation of PV cells was

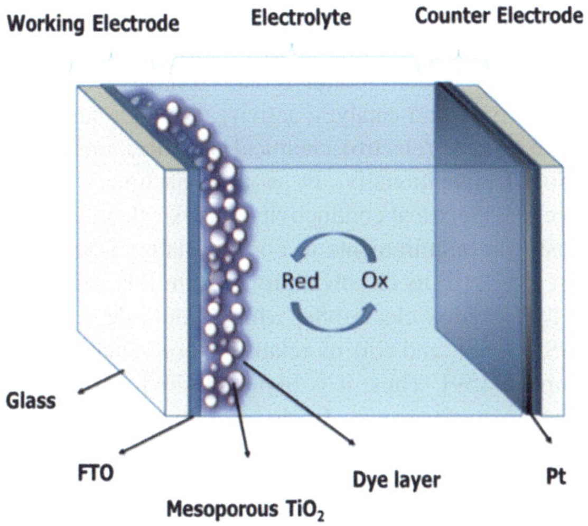

Fig. 18 Schematic structure of DSSC device. Adapted from Ref. [146] with permission, Copyright 2010, American Chemical Society

developed to overcome the limitations of the first two generations. In this regard, a third generation which called emergency solar cells was developed. The emergency solar cells include all systems which are not included into the first two generations such as dye-sensitized solar cells [151], quantum dot solar cells [152], and perovskite solar cells [153]. Here, we will discuss only the dye-sensitized solar cell (DSSC). In 1991, Grätzel et al. developed a member of the third-generation solar cells based on semiconductors known as DSSC or Grätzel cell [151, 154]. In the last decades, DSSCs have been considered the best alternatives to the traditional Si-based cells due to their simplistic assembly, cost-effectiveness, design in a flexibility, and high mechanical strength [154]. The typical architecture of a DSSC device is composed of a porous film of n-type TiO_2, a photosensitized dye, a redox couple consisting of I_3^-/I^- electrolyte, and a Pt counter electrode (CE) as illustrated in Fig. 18.

Under solar illumination, the excited dye molecule injects electrons into the conduction band of TiO_2. The oxidized dye subsequently stimulates the oxidation of iodide to triiodide in the electrolyte, and the electron is injected into the conduction band of TiO_2 and transported to the CE. Accordingly, the electrical properties and catalytic ability of the CE play a significant role in the DSSC process and the total cost of DSSCs [155].

4.1 WasteRecovered Nanomaterials as Catalyst for Dye-Sensitized Solar Cell

The main functions of the counter electrode (CE) in DSSC device are to transfer the photogenerated electrons from the external circuit to the redox couple and promote/

catalyze the reduction of the oxidized species of the redox couple, hence regenerating the oxidized form of the dye to its ground state [48]. An effective catalyst for DSSC application should satisfy some basic criteria such as (i) high electrical conductivity and catalytic activity, (ii) high surface area with high porosity [146], (iii) excellent (electro) chemical stability, and (iv) high abundance and low cost [156]. Conventionally, Pt is the benchmark catalyst for DSSCs owing to its excellent electrical conductivity and excellent catalytic activity. However, Pt cannot satisfy the requirements to commercialize DSSCs due to (i) its scarcity and high cost [157], (ii) its dissolvability to form PtI_4 and H_2PtI_6 by-products in the presence of I_3^-/I^- redox electrolyte, which would deteriorate the long-term stability of the DSSC [158], and (iii) its relatively poor catalytic activity toward the cobalt redox couples [159]. Thus, it is highly desired to develop low-cost materials with high catalytic performance. In this regard, various materials including carbonaceous [160–163], conducting polymers [164, 165], transition metals [166, 167], and their composites [168–170] have been investigated and reviewed [157, 171]. However, most of these materials are synthesized with toxic reagents and/or by complex synthesis processes that limit their large-scale manufacturability. Recently, waste-derived nanomaterials have been investigated as CEs for DSSCs to reduce the production cost of DSSCs and the environmental impact of waste materials. The recent progress of utilizing these nanomaterials as catalysts for DSSCs is reviewed in the following discussion. Jiang et al. [172] prepared a highly ordered mesoporous carbon material by direct carbonization of bamboo and oak wood under argon atmosphere. The obtained porous carbon showed good catalytic activity and high conductivity (Fig. 19a–b). The power conversion efficiency (PCE) of the assembled DSSC with the bamboo and oak wood electrodes was 4.53 and 7.98%, respectively. Furthermore, a carbonized sucrose-coated eggshell membrane (CSEM) with hierarchically porous microstructure has been prepared by immersing the eggshell membrane in 40% sucrose solution followed by pre carbonization and carbonization at 800 °C [173]. The CSEM showed comparable catalytic ability as the thermally decomposed Pt CE toward the I_3^-/I^- redox couple. Moreover, the assembled DSSC with CSEM displayed PCE of 6.71%, which is better than the devices assembled with a Pt, and the prepared porous carbon was 6.63 and 6.71%, respectively.

Recently, Cha et al. prepared quince leaf-derived porous carbon using fallen quince leaves by alkali treatment and pyrolysis at 700–900 °C [174]. The as-prepared carbon showed high electrocatalytic activity toward I_3^-/I^-. Accordingly, the assembled DSSCs showed PCE of 5.52%, which is comparable to the Pt-based CE (6.56%). Xu et al. utilized twenty kinds of plant biowastes in three groups as raw materials to prepare porous carbon materials by a one-step pyrolysis process [175]. These groups are: (i) eleven kinds of woods (weeping willow, phoenix, camphor, Chinese fir, maple, peach, poplar, cypress, tea oil camellia, orange, and chinaberry), (ii) seven leaves (pine needles, camphor, palm, maple, poplar, Chinese fir, and red after-wood), and (iii) two papers (filter paper and facial tissue). The individually prepared porous carbons were utilized as CEs in DSSCs.

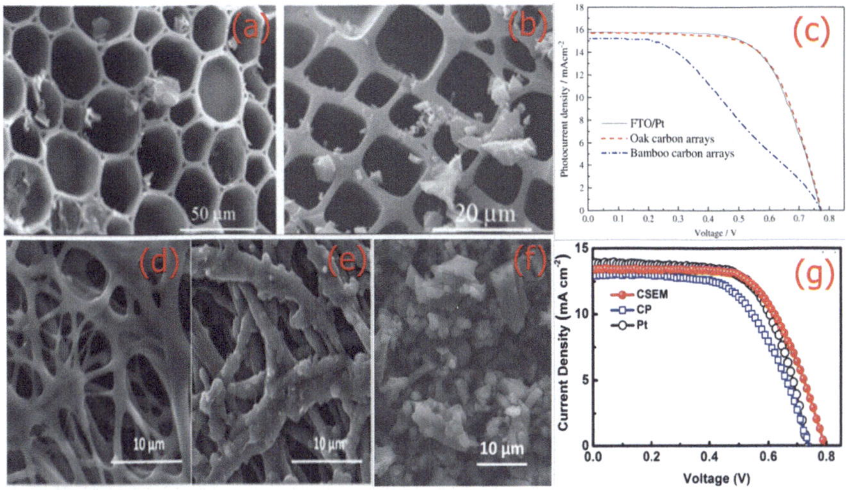

Fig. 19 SEM images of **a** bamboo mesoporous carbon, **b** oak mesoporous carbon arrays, and **c** J–V curves of DSSCs with different CEs. Adapted from Ref. [172] with permission, Copyright 2017, Elsevier. **d** SEM images of an eggshell membrane, **e** CSEM, **f** CSEM after grinding with a mortar and pestle, and **g** J–V curves of DSSCs with various CEs. Adapted from Ref. [173] with permission, Copyright 2017, Wiley-VCH

The highest PCE values of the DSSCs with CEs prepared from the porous carbon derived from the first and the second groups are 1.91 and 1.85%, respectively, while that of the assembled devices with the CEs prepared by the porous carbon derived from the third group is 4.72%, which is about 75% of that of the device with the Pt CE. Recently, Wang et al. used pomelo peel to prepare porous carbon by microwave pyrolysis, two-step activation, and hydrothermal carbonization combining chemical activation methods [176]. The as-prepared carbon with a two-step activation carbonization process showed a smaller size with an excellent catalytic performance toward the reduction of I_3^-. The PCE of the related DSSC device was 6.94%, very close to that obtained by Pt CE (6.71%).

Rice husk (RH) is one of the highly abundant wastes and represents about 20% of the total weight of rice. Currently, attention has been drawn to the use of RH as a sustainable material in advanced technology applications [177]. Currently, converting RH into porous carbons for DSSCs applications displayed much attention. For instance, Wang et al. [178] have prepared RH-derived porous carbon by pyrolysis. The sensitized carbon displayed hierarchical porous structure with large pore sizes. Moreover, the coated RH-derived carbon on FTO glass substrate as CE for DSCs with I_3^-/I^- redox couples displayed higher catalytic ability toward reduction of I_3^- than the standard Pt CE. The assembled DSSC displayed a PCE of 6.32%, close to that of the Pt CE (6.69%). The high content of silica in RH can improve the total catalytic activity. Ahmad et al. [179] extracted nano-silicon with activated carbons from RH by a one-step chemical activation method to prepare

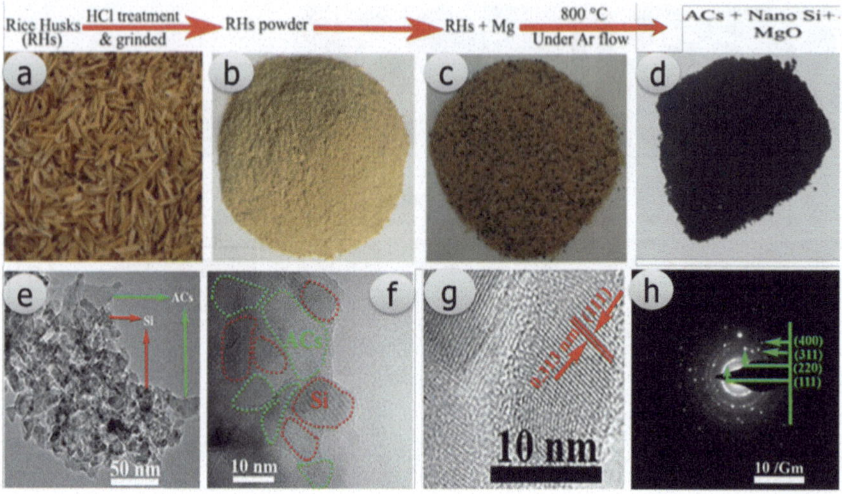

Fig. 20 **a–d** Flowchart for extraction of nano-Si@ACs from rice husk. **a** Photograph of RH after threshing, **b** RH after grinding, **c** RH/Mg mixture, and **d** after thermal decomposition and magnesiothermic reduction processes. **e**, **f** TEM image of nano-Si@ACs composite, **g** HRTEM image of nano-Si, respectively, and **i** FFT pattern of nano-Si extracted from RH simultaneously. Adapted from Ref. [179] with permission, Copyright 2017, Nature

nano-Si@ACs as catalyst for DSSCs as described in Fig. 20. The as-prepared nano-Si@ACs displayed a high surface area with a porous structure. Additionally, the nano-Si@ ACs displayed high reduction ability toward I_3^-/I^- redox couple and the related DSSC displayed a PCE of 8.01%, which is higher than that the device with Pt CE (7.20%).

5 Conclusions

Increasing the consumption of fossil fuel is a major challenge for our environment; thus, intensive efforts have been made to develop sustainable conversion and storage energy systems. Electrochemical-based techniques such as water splitting, fuel cell, and dye-sensitized solar cells displayed a potential ability to offer sustainable energy for future generations. However, the high cost and low abundance of noble metals are the bottlenecks facing their practical applications. Accordingly, significant research has been done to find cheap, abundant, and eco-friendly materials with high electrical conductivity and electrocatalytic performance. On the other hand, the waste materials (biowaste and electronic wastes) are considered significant challenges facing our environment. These unwanted materials have many valuable metals; thus, the disposal of these materials without management is considered not only an environmental problem, but also a non-responsible way to lose our resources. Therefore, an intensive research work could reuse the waste-recovered nanomaterials in the electrocatalytic applications. Usually, bio-waste materials are recovered via pyrolysis process at appropriate temperature atmosphere, while the metallurgy (hydrometallurgy) is utilized to recover the valuable metals from the electronics wastes followed by thermal treatment. Some of the recovered nanomaterials can be utilized as extracted without modifications, while the others were used to support materials to enhance the catalytic performance of the guest materials. Most of the waste-recovered nanomaterials displayed a promising catalytic performance toward water-splitting reactions (HER, OER), ORR in both fuel cells and metal–air batteries, and in DSSCs applications. The catalytic activity and the overall performance of their related devices were comparable or even better than those with benchmark nobel metals. Due to the high catalytic activity, simple preparation routes, high availability with low cost, and eco-friendly, the waste recovered nanomaterials have the potential to replace the noble metal-based electrocatalysts which will promote the commercialization of their related electrochemical devices.

6 Future Perspectives

Further studies are still needed to overcome the challenges of the waste-recovered nanomaterials. Until now, only a few types of nanomaterials have been recovered, and most of them are from biomass. However, there are numerous waste stocks available as other valuable sources in our societies with high quantities such as mine wastes and electronic waste materials other than LiBs. Therefore, future research should be directed to recover nanomaterial from these waste materials considering the commercialization issues-related electrochemical devices. Moreover, it can be noticed that most of the studies are focusing on utilizing the waste-recovered nanomaterials in electrochemical water splitting and metal–air batteries. However, there are few studies focused on DSSCs applications. Therefore, more research efforts should be directed to make DSSCs available in the market at low cost.

References

1. International Energy Outlook (2018) https://www.eia.gov/outlooks/ieo/
2. Annual Energy Outlook (2019) www.eia.gov/aeo
3. Gong J, Sumathy K, Qiao Q, Zhou Z (2017) Review on dye-sensitized solar cells (DSSCs): advanced techniques and research trends. Renew Sustai Energy Rev 68:234–246
4. Smalley RE (2003) Our energy challenge. Public lecture presented at low library. Columbia University 23
5. REN21, Renewables 2018 Global Status Report. REN21 Secretariat, Paris

6. Directive (EU) 2018/2001 of the European Parliament and of the Council of 11 December 2018 on the promotion of the use of energy from renewable sources. https://eur-lex.europa.eu/eli/dir/2018/2001/2018-12-21
7. Downes CA, Marinescu SC (2017) Electrocatalytic metal-organic frameworks for energy applications. Chemsuschem 10(22):4374–4392
8. Dutta T, Kim K-H, Deep A, Szulejko JE, Vellingiri K, Kumar S, Kwon EE, Yun S-T (2018) Recovery of nanomaterials from battery and electronic wastes: a new paradigm of environmental waste management. Renew Sustain Energy Rev 82:3694–3704
9. Bennett JA, Wilson K, Lee AF (2016) Catalytic applications of waste derived materials. J Mater Chem A 4(10):3617–3637
10. Sadh PK, Duhan S, Duhan JS (2018) Agro-industrial wastes and their utilization using solid state fermentation: a review. Bioresour Bioprocess 5:1
11. Habeeb OA, Ramesh K, Ali GAM, Yunus RM (2017) Application of response surface methodology for optimization of palm kernel shell activated carbon preparation factors for removal of H_2S from industrial wastewater. Jurnal Teknologi 79(7):1–10
12. Hegde G, Abdul Manaf SA, Kumar A, Ali GAM, Chong KF, Ngaini Z, Sharma KV (2015) Biowaste sago bark based catalyst free carbon nanospheres: waste to wealth approach. ACS Sustain Chem Eng 5(9):2247–2253
13. Ali GAM, Divyashree A, Supriya S, Chong KF, Ethiraj AS, Reddy M, Algarni H, Hegde G (2017) Carbon nanospheres derived from Lablab purpureus for high performance supercapacitor electrodes: a green approach. Dalton Trans 46(40):14034–14044
14. Ali GAM, Tan LL, Jose R, Yusoff MM, Chong KF (2014) Electrochemical performance studies of MnO_2 nanoflowers recovered from spent battery. Mater Res Bull 60:5–9
15. Ali GAM, Yusoff MM, Feng CK (2015) Electrochemical properties of electrodeposited MnO_2 nanoparticles. Adv Mater Res 1113:550–553
16. Ali GAM, Abdul Manaf SA, Kumar A, Chong KF, Hegde G (2014) High performance supercapacitor using catalysis free porous carbon nanoparticles. J Phys D-Appl Phys 47 (49):495307–495313
17. Ali GAM, Yusoff MM, Algarni H, Chong KF (2018) One-step electrosynthesis of MnO_2/rGO nanocomposite and its enhanced electrochemical performance. Ceram Int 44(7):7799–7807
18. Ali GAM, Supriya S, Chong KF, Shaaban ER, Algarni H, Maiyalagan T, Hegde G (2019) Superior supercapacitance behavior of oxygen self-doped carbon nanospheres: a conversion of Allium cepa peel to energy storage system. Biomass Conv Bioref
19. Ali GAM, Manaf SAA, Divyashree A, Chong KF, Hegde G (2016) Superior supercapacitive performance in porous nanocarbons. J Energy Chem 25(4):734–739
20. Ali GAM, Habeeb OA, Algarni H, Chong KF (2018) CaO impregnated highly porous honeycomb activated carbon from agriculture waste: symmetrical supercapacitor study. J Mater Sci 54:683–692
21. Habeeb OA, Ramesh K, Ali GAM, Yunus RM (2017) Experimental design technique on removal of hydrogen sulfide using CaO-eggshells dispersed onto palm kernel shell activated carbon: experiment, optimization, equilibrium and kinetic studies. J Wuhan University Technol-Mater Sci Ed 32(2):305–320
22. Ali GAM, Yusoff MM, Shaaban ER, Chong KF (2017) High performance MnO_2 nanoflower supercapacitor electrode by electrochemical recycling of spent batteries. Ceram Int 43:8440–8448
23. Habeeb OA, Ramesh K, Ali GAM, Yunus RM (2017) Low-cost and eco-friendly activated carbon from modified palm kernel shell for hydrogen sulfide removal from wastewater: adsorption and kinetic studies. Desalin Water Treat 84:205–214
24. Kirubakaran AJ, Shailendra Nema RK (2009) A review on fuel cell technologies and power electronic interface. Renew Sustain Energy Rev 13(9):2430–2440
25. Brian CH, Steele AH (2001) Materials for fuel-cell technologies. Nature 414(6861):345–352
26. Yang M-Q, Wang J, Wu H, Ho GW (2018) Noble metal-free nanocatalysts with vacancies for electrochemical water splitting. Small 14:1703323

27. Trasatti S (1999) Water electrolysis: who first? J Electroanal Chem 476:90–91
28. Jiao L, Zhou YX, Jiang HL (2016) Metal-organic framework-based CoP/reduced graphene oxide: high-performance bifunctional electrocatalyst for overall water splitting. Chem Rev 7 (3):1690–1695
29. Ewan B, Allen R (2005) A figure of merit assessment of the routes to hydrogen. Int J Hydrogen Energy 30(8):809–819
30. Cheng Y, Jiang SP (2015) Advances in electrocatalysts for oxygen evolution reaction of water electrolysis-from metal oxides to carbon nanotubes. Prog Nat Sci Mater Int 25(6):545–553
31. Ganguly P, Harb M, Cao Z, Cavallo L, Breen A, Dervin S, Dionysiou DD, Pillai SC (2019) 2D nanomaterials for photocatalytic hydrogen production. ACS Energy Lett 4:1687–1709
32. Han L, Dong S, Wang E (2016) Transition-metal (Co, Ni, and Fe)-based electrocatalysts for the water oxidation reaction. Adv Mater 28(42):9266–9291
33. Anantharaj SE, Rao S, Sakthikumar K, Karthick K, Mishra S, Kundu Subrata (2016) Recent trends and perspectives in electrochemical water splitting with an emphasis on sulfide, selenide, and phosphide catalysts of Fe Co, and Ni: a review. ACS Catalysis 6(12):8069–8097
34. Khan MA, Zhao H, Zou W, Chen Z, Cao W, Fang J, Xu J, Zhang L, Zhang J (2018) Recent progresses in electrocatalysts for water electrolysis. Electrochem Energy Rev 1(4):483–530
35. Shi Y, Zhang B (2016) Recent advances in transition metal phosphide nanomaterials: synthesis and applications in hydrogen evolution reaction. Chem Soc Rev 45(6):1529–1541
36. Lee J-S, Tai Kim S, Cao R, Choi N-S, Liu M, Lee KT, Cho J (2011) Metal-air batteries with high energy density: Li–Air versus Zn–Air. Adv Energy Mater 1(1):34–50
37. Drouet S, Creus J, Collière V, Amiens C, García-Antón J, Sala X, Philippot K (2017) A porous Ru nanomaterial as an efficient electrocatalyst for the hydrogen evolution reaction under acidic and neutral conditions. Chem Commun 53(85):11713–11716
38. Chen Z, Cummins D, Reinecke BN, Clark E, Sunkara MK, Jaramillo TF (2011) Core-shell MoO_3-MoS_2 nanowires for hydrogen evolution: a functional design for electrocatalytic materials. Nano Lett 11(10):4168–4175
39. Zheng X, Xu J, Yan K, Wang H, Wang Z, Yang S (2014) Space-confined growth of MoS_2 nanosheets within graphite: the layered hybrid of MoS_2 and graphene as an active catalyst for hydrogen evolution reaction. Chem Mater 26(7):2344–2353
40. Xue N, Diao P (2017) Composite of few-layered MoS_2 grown on carbon black: tuning the ratio of terminal to total sulfur in MoS_2 for hydrogen evolution reaction. J Phys Chem C 121 (27):14413–14425
41. Hemamala I, Karunadasa EM, Sun Yujie, Majda Marcin, Long Jeffrey R, Chang Christopher J (2012) A molecular MoS_2 edge site mimic for catalytic hydrogen generation. Science 335(6069):698–702
42. Benck JD, Hellstern TR, Kibsgaard J, Chakthranont P, Jaramillo TF (2014) Catalyzing the hydrogen evolution reaction (HER) with molybdenum sulfide nanomaterials. ACS Catalysis 4(11):3957–3971
43. Wu Y, Liu X, Han D, Song X, Shi L, Song Y, Niu S, Xie Y, Cai J, Wu S, Kang J, Zhou J, Chen Z, Zheng X, Xiao X, Wang G (2018) Electron density modulation of $NiCo_2S_4$ nanowires by nitrogen incorporation for highly efficient hydrogen evolution catalysis. Nat Commun 9(1):1425
44. Zhu D, Liu J, Wang L, Du Y, Zheng Y, Davey K, Qiao S-Z (2019) A 2D metal–organic framework/$Ni(OH)_2$ heterostructure for an enhanced oxygen evolution reaction. Nanoscale 11(8):3599–3605
45. Abdalla S, Al-Marzouki F, Obaid A (2017) High-efficient and low-cost catalyst for hydrogen evolution reaction: nickel phosphide nano-spheres. J Renew Sustain Energy 9(2):023104
46. Zhang R, Wang X, Yu S, Wen T, Zhu X, Yang F, Sun X, Wang X, Hu W (2017) Ternary $NiCo_2Px$ nanowires as pH-universal electrocatalysts for highly efficient hydrogen evolution reaction. Adv Mater 29(9):1605502

47. Cai Z, Wu A, Yan H, Xiao Y, Chen C, Tian C, Wang L, Wang R, Fu H (2018) Hierarchical whisker-on-sheet NiCoP with adjustable surface structure for efficient hydrogen evolution reaction. Nanoscale 10(16):7619–7629
48. Du C, Yang L, Yang F, Cheng G, Luo W (2017) Nest-like NiCoP for highly efficient overall water splitting. ACS Catalysis 7(6):4131–4137
49. Che Q, Bai N, Li Q, Chen X, Tan Y, Xu X (2018) One-step electrodeposition of a hierarchically structured S-doped NiCo film as a highly-efficient electrocatalyst for the hydrogen evolution reaction. Nanoscale 10(32):15238–15248
50. J. O. Bockris AKNRaMG-A (2000) Modern electrochemistry 2A. Fundamentals of electrodics. Kluwer Academic, New York
51. Suen NT, Hung SF, Quan Q, Zhang N, Xu YJ, Chen HM (2017) Electrocatalysis for the oxygen evolution reaction: recent development and future perspectives. Chem Soc Rev 46 (2):337–365
52. Guidelli R, Compton RG, Feliu JM, Gileadi E, Lipkowski J, Schmickler W, Trasatti S (2014) Defining the transfer coefficient in electrochemistry: an assessment (IUPAC technical report). Pure Appl Chem 86(2):245–258
53. Dincer I, Acar C (2015) Review and evaluation of hydrogen production methods for better sustainability. Int J Hydrogen Energy 40(34):11094–11111
54. Zheng Y, Jiao Y, Jaroniec M, Qiao SZ (2015) Advancing the electrochemistry of the hydrogen-evolution reaction through combining experiment and theory. Angew Chem Int Ed 54(1):52–65
55. Zheng Y, Jiao Y, Vasileff A, Qiao S-Z (2018) The hydrogen evolution reaction in alkaline solution: from theory, single crystal models, to practical electrocatalysts. Angew Chem Int Ed 57(26):7568–7579
56. Zhang Y-Y, Zhang X, Wu Z-Y, Zhang B-B, Zhang Y, Jiang W-J, Yang Y-G, Kong Q-H, Hu J-S (2019) Fe/P dual doping boosts the activity and durability of CoS_2 polycrystalline nanowires for hydrogen evolution. J Mater Chem A 7(10):5195–5200
57. Kibsgaard J, Tsai C, Chan K, Benck JD, Nørskov JK, Abild-Pedersen F, Jaramillo TF (2015) Designing an improved transition metal phosphide catalyst for hydrogen evolution using experimental and theoretical trends. Energy Environ Sci 8(10):3022–3029
58. Li T, Jin H, Liang Z, Huang L, Lu Y, Yu H, Hu Z, Wu J, Xia BY, Feng G, Zhou J (2018) Synthesis of single crystalline two-dimensional transition-metal phosphides via a salt-templating method. Nanoscale 10(15):6844–6849
59. Mir RA, Pandey OP (2020) An ecofriendly route to synthesize C–Mo_2C and C/N–Mo_2C utilizing waste polyethene for efficient hydrogen evolution reaction (HER) activity and high performance capacitors. Sustain Energy Fuels 4:655–669
60. Zhang Z, Yang S, Li H, Zan Y, Li X, Zhu Y, Dou M, Wang F (2019) Sustainable carbonaceous materials derived from biomass as metal-free electrocatalysts. Adv Mater 31 (13):e1805718
61. Corma A, de la Torre O, Renz M, Villandier N (2011) Production of high-quality diesel from biomass waste products. Angew Chem Int Ed 50(10):2375–2378
62. Christopher O, Tuck EP, Horváth István T, Sheldon Roger A, Poliakoff Martyn (2012) Valorization of biomass: deriving more value from waste. Science 337(10):695–699
63. Liu X, Zhang M, Yu D, Li T, Wan M, Zhu H, Du M, Yao J (2016) Functional materials from nature: honeycomb-like carbon nanosheets derived from silk cocoon as excellent electrocatalysts for hydrogen evolution reaction. Electrochim Acta 215:223–230
64. Zheng Y, Jiao Y, Zhu Y, Li LH, Han Y, Chen Y, Du A, Jaroniec M, Qiao SZ (2014) Hydrogen evolution by a metal-free electrocatalyst. Nat Commun 5:3783
65. Yao Zheng YJ, Li LH, X T, C Ying, Jaroniec M, Qiao SZ (2014) Toward design of synergistically active carbon-based catalysts for electrocatalytic hydrogen evolution. ACS Nano 8(5):5290–5296
66. Dai L, Xue Y, Qu L, Choi HJ, Baek JB (2015) Metal-free catalysts for oxygen reduction reaction. Chem Rev 115(11):4823–4892

67. Prabu N, Saravanan RSA, Kesavan T, Maduraiveeran G, Sasidharan M (2019) An efficient palm waste derived hierarchical porous carbon for electrocatalytic hydrogen evolution reaction. Carbon 152:188–197
68. Yan Q, Yang X, Wei T, Zhou C, Wu W, Zeng L, Zhu R, Cheng K, Ye KK, Zhu K, Yan J, Cao D, Wang G (2020) Porous beta-Mo_2C nanoparticle clusters supported on walnut shell powders derived carbon matrix for hydrogen evolution reaction. J Colloid Interface Sci 563:104–111
69. Kumar A, Chaudhary DK, Parvin S, Bhattacharyya S (2018) High performance duckweed-derived carbon support to anchor NiFe electrocatalysts for efficient solar energy driven water splitting. J Mater Chem A 6(39):18948–18959
70. Ng JWD, García-Melchor M, Bajdich M, Chakthranont P, Kirk C, Vojvodic A, Jaramillo TF (2016) Gold-supported cerium-doped NiOx catalysts for water oxidation. Nat Energy 1:16053
71. Shi Q, Fu S, Zhu C, Song J, Du D, Lin Y (2019) Metal–organic frameworks-based catalysts for electrochemical oxygen evolution. Mater Horizons 6:684–702
72. Katsounaros I, Cherevko S, Zeradjanin AR, Mayrhofer KJ (2014) Oxygen electrochemistry as a cornerstone for sustainable energy conversion. Angew Chem Int Ed 53(1):102–121
73. Frydendal R, Paoli EA, Knudsen BP, Wickman B, Malacrida P, Stephens IEL, Chorkendorff I (2014) Benchmarking the stability of oxygen evolution reaction catalysts: the importance of monitoring mass losses. ChemElectroChem 1(12):2075–2081
74. Kotz R, Lewerenz HJ, Stucki S (1983) XPS studies of oxygen evolution on Ru and RuO_2 anodes. J Electrochem Soc 130:825–829
75. Lyu F, Wang Q, Choi SM, Yin Y (2019) Noble-metal-free electrocatalysts for oxygen evolution. Small 15(1):1804201
76. Liang Y, Li Y, Wang H, Zhou J, Wang J, Regier T, Dai H (2011) Co_3O_4 nanocrystals on graphene as a synergistic catalyst for oxygen reduction reaction. Nat Mater 10:780
77. Zhang L, Xiao J, Wang H, Shao M (2017) Carbon-based electrocatalysts for hydrogen and oxygen evolution reactions. ACS Catalysis 7(11):7855–7865
78. Cheng Y, Liao F, Shen W, Liu L, Jiang B, Li Y, Shao M (2017) Carbon cloth supported cobalt phosphide as multifunctional catalysts for efficient overall water splitting and zinc–air batteries. Nanoscale 9(47):18977–18982
79. Song F, Bai L, Moysiadou A, Lee S, Hu C, Liardet L, Hu X (2018) Transition metal oxides as electrocatalysts for the oxygen evolution reaction in alkaline solutions: an application-inspired renaissance. J Am Chem Soc 140(25):7748–7759
80. Xue Y, Ren Z, Xie Y, Du S, Wu J, Meng H, Fu H (2017) CoSex nanocrystalline-dotted CoCo layered double hydroxide nanosheets: a synergetic engineering process for enhanced electrocatalytic water oxidation. Nanoscale 9(42):16256–16263
81. Zhang R, Tang C, Kong R, Du G, Asiri AM, Chen L, Sun X (2017) Al-doped CoP nanoarray: a durable water-splitting electrocatalyst with superhigh activity. Nanoscale 9 (14):4793–4800
82. Devi MM, Ojha KN, Ganguli AK, Jha M (2018) Transformation of waste tin-plated steel to iron nanosheets and their application in generation of oxygen. Int J Environ Sci Technol 16 (7):3669–3678
83. Jothi VR, Bose R, Rajan H, Jung C, Yi SC (2018) Harvesting electronic waste for the development of highly efficient eco-design electrodes for electrocatalytic water splitting. Adv Energy Mater 8(34):1802615
84. Babar P, Lokhande A, Karade V, Pawar B, Gang MG, Pawar S, Kim JH (2019) Towards highly efficient and low-cost oxygen evolution reaction electrocatalysts: an effective method of electronic waste management by utilizing waste Cu cable wires. J Colloid Interface Sci 537:43–49
85. Natarajan S, Anantharaj S, Tayade RJ, Bajaj HC, Kundu S (2017) Recovered spinel $MnCo_2O_4$ from spent lithium-ion batteries for enhanced electrocatalytic oxygen evolution in alkaline medium. Dalton Trans 46(41):14382–14392

86. Aboelazm EA, Ali GAM, Chong KF (2018) Cobalt oxide supercapacitor electrode recovered from spent lithium-ion battery. Chem Adv Mater 3:67–74
87. Aboelazm EAA, Ali GAM, Algarni H, Yin H, Zhong YL, Chong KF (2018) Magnetic electrodeposition of the hierarchical cobalt oxide nanostructure from spent lithium-ion batteries: its application as a supercapacitor electrode. J Phys Chem C 122(23):12200–12206
88. Li J, He X, Zeng X (2017) Designing and examining e-waste recycling process: methodology and case studies. Environ Technol 38(6):652–660
89. Chen N, Qi J, Du X, Wang Y, Zhang W, Wang Y, Lu Y, Wang S (2016) Recycled $LiCoO_2$ in spent lithium-ion battery as an oxygen evolution electrocatalyst. RSC Adv 6 (105):103541–103545
90. Yang Y, Yang H, Cao H, Wang Z, Liu C, Sun Y, Zhao H, Zhang Y, Sun Z (2019) Direct preparation of efficient catalyst for oxygen evolution reaction and high-purity Li_2CO_3 from spent Li $Ni_{0.5}$ $Mn_{0.3}Co_{0.2}O_2$ batteries. J Cleaner Prod 236:117576
91. Pegoretti VCB, Dixini PVM, Magnago L, Rocha AKS, Lelis MFF, Freitas MBJG (2019) High-temperature (HT) $LiCoO_2$ recycled from spent lithium ion batteries as catalyst for oxygen evolution reaction. Mater Res Bull 110:97–101
92. Wanwan Mei XY, Li L, Tong Y, Lei Y, Li P, Zheng Z (2020) Rational electrochemical recycling of spent $LiFePO_4$ and $LiCoO_2$ batteries to Fe_2O_3/CoPi photoanode for water oxidation. ACS Sustain Chem Eng 8(9):3606–3616
93. Amiinu IS, Zhang J, Kou Z, Liu X, Asare OK, Zhou H, Cheng K, Zhang H, Mai L, Pan M, Mu S (2016) Self-organized 3D porous graphene dual-doped with biomass-sponsored nitrogen and sulfur for oxygen reduction and evolution. ACS Appl Mater Interfaces 8 (43):29408–29418
94. Guan C, Liu X, Elshahawy AM, Zhang H, Wu H, Pennycook SJ, Wang J (2017) Metal–organic framework derived hollow CoS_2 nanotube arrays: an efficient bifunctional electrocatalyst for overall water splitting. Nanoscale Horizons 2(6):342–348
95. Ji L, Wang J, Teng X, Meyer TJ, Chen Z (2019) CoP nanoframes as bifunctional electrocatalysts for efficient overall water splitting. ACS Catalysis 10(1):412–419
96. Deng S, Zhang K, Xie D, Zhang Y, Zhang Y, Wang Y, Wu J, Wang X, Fan HJ, Xia X, Tu J (2019) High-index-faceted Ni_3S_2 branch arrays as bifunctional electrocatalysts for efficient water splitting. Nano-Micro Lett 11:12
97. Tiwari JN, Dang NK, Sultan S, Thangavel P, Jeong HY, Kim KS (2020) Multi-heteroatom-doped carbon from waste-yeast biomass for sustained water splitting. Nat Sustain 3:556–563
98. Seh ZW, Kibsgaard J, Dickens CF, Chorkendorff I, Nørskov JK, Jaramillo TF (2017) Combining theory and experiment in electrocatalysis: insights into materials design. Science 355(6321):eaad4998
99. Morozan A, Jousselme B, Palacin S (2011) Low-platinum and platinum-free catalysts for the oxygen reduction reaction at fuel cell cathodes. Energy Environ Sci 4(4):1238–1254
100. Mahmood A, Guo W, Tabassum H, Zou R (2016) Metal-organic framework-based nanomaterials for electrocatalysis. Adv Energy Mater 6(17):1600423
101. Yeager E (1986) Dioxygen electrocatalysis: mechanisms in relation to catalyst structure. J Mol Catal 38:5–25
102. Zhang J (2008) PEM fuel cell electrocatalysts and catalyst layers. Springer Verlag, London
103. Song Z, Cheng N, Lushington A, Sun X (2016) Recent progress on MOF-derived nanomaterials as advanced electrocatalysts in fuel cells. Catalysts 6(8):116
104. Zhang Y, Deng L, Hu H, Qiao Y, Yuan H, Chen D, Chang M, Wei H (2020) Pomelo peel-derived, N-doped biochar microspheres as an efficient and durable metal-free ORR catalyst in microbial fuel cells. Sustain Energy Fuels 4:1642–1653
105. Sui S, Wang X, Zhou X, Su Y, Riffat S, Liu C-j (2017) A comprehensive review of Pt electrocatalysts for the oxygen reduction reaction: nanostructure, activity, mechanism and carbon support in PEM fuel cells. J Mater Chem A 5(5):1808–1825

106. Ren Q, Wang H, Lu X-F, Tong Y-X, Li G-R (2018) Recent progress on MOF-derived heteroatom-doped carbon-based electrocatalysts for oxygen reduction reaction. Adv Sci 5 (3):1700515
107. Lim B, Jiang M, Camargo PHC, Cho EC, Tao J, Lu X, Zhu Y, Xia Y (2009) Pd-Pt bimetallic nanodendrites with high activity for oxygen reduction. Science 324(5932):1302–1305
108. Dai S, Chou J-P, Wang K-W, Hsu Y-Y, Hu A, Pan X, Chen T-Y (2019) Platinum-trimer decorated cobalt-palladium core-shell nanocatalyst with promising performance for oxygen reduction reaction. Nat Commun 10(1):440
109. Huang C, Li C, Shi G (2012) Graphene based catalysts. Energy Environ Sci 5(10):8848–8868
110. Qu L, Liu Y, Baek J-B, Dai L (2010) Nitrogen-doped graphene as efficient metal-free electrocatalyst for oxygen reduction in fuel cells. ACS Nano 4(3):1321–1326
111. Gong K, Du F, Xia Z, Durstock M, Dai L (2009) Nitrogen-doped carbon nanotube arrays with high electrocatalytic activity for oxygen reduction. Science 323(5915):760–764
112. Qu K, Zheng Y, Dai S, Qiao SZ (2015) Polydopamine–graphene oxide derived mesoporous carbon nanosheets for enhanced oxygen reduction. Nanoscale 7(29):12598–12605
113. Deng L, Yuan H, Cai X, Ruan Y, Zhou S, Chen Y, Yuan Y (2016) Honeycomb-like hierarchical carbon derived from livestock sewage sludge as oxygen reduction reaction catalysts in microbial fuel cells. Int J Hydrogen Energy 41(47):22328–22336
114. Liu Y, Ruan J, Sang S, Zhou Z, Wu Q (2016) Iron and nitrogen co-doped carbon derived from soybeans as efficient electro-catalysts for the oxygen reduction reaction. Electrochim Acta 215:388–397
115. Li C, Sun F, Lin Y (2018) Refining cocoon to prepare (N, S, and Fe) ternary-doped porous carbon aerogel as efficient catalyst for the oxygen reduction reaction in alkaline medium. J Power Sources 384:48–57
116. Guo C, Hu R, Liao W, Li Z, Sun L, Shi D, Li Y, Chen C (2017) Protein-enriched fish "biowaste" converted to three-dimensional porous carbon nano-network for advanced oxygen reduction electrocatalysis. Electrochim Acta 236:228–238
117. Kaur P, Verma G, Sekhon SS (2019) Biomass derived hierarchical porous carbon materials as oxygen reduction reaction electrocatalysts in fuel cells. Prog Mater Sci 102:1–71
118. Lihua Zhou PF, Cai X, Zhou S, Yuan Y (2016) Naturally derived carbon nanofibers as sustainable electrocatalysts for microbial energy harvesting: a new application of spider silk. Appl Catal B 188:31–38
119. Li S, Xu R, Wang H, Brett DJ, Ji S, Pollet BG, Wang R (2017) Ultra-high surface area andmesoporous N-doped carbon derived from sheep bones with high electrocatalytic performance toward the oxygen reduction reaction. J Solid State Electrochem 21:2947–2954
120. Yongxi Zan ZZ, Liu H, Dou M, Wang F (2017) Nitrogen and phosphorus co-doped hierarchically porous carbons derived from cattle bones as efficient metal-free electrocatalysts for the oxygen reduction reaction. J Mater Chem A 5:24329–24334
121. Chaudhari KN, Song MY, Yu JS (2014) Transforming hair into heteroatom-doped carbon with high surface area. Small 10(13):2625–2636
122. Yuan W, Feng Y, Xie A, Zhang X, Huang F, Li S, Zhang X, Shen Y (2016) Nitrogen-doped nanoporous carbon derived from waste pomelo peel as a metal-free electrocatalyst for the oxygen reduction reaction. Nanoscale 8(16):8704–8711
123. Ma M, You S, Wang W, Liu G, Qi D, Chen X, Qu J, Ren N (2016) Biomass-derived porous Fe_3C/tungsten carbide/graphitic carbon nanocomposite for efficient electrocatalysis of oxygen reduction. ACS Appl Mater Interfaces 8(47):32307–32316
124. Wang Y, Zhu M, Wang G, Dai B, Yu F, Tian Z, Guo X (2017) Enhanced oxygen reduction reaction by in situ anchoring Fe(2)N nanoparticles on nitrogen-doped pomelo peel-derived carbon. Nanomaterials (Basel) 7(11)
125. Li M, Zhang H, Xiao T, Wang S, Zhang B, Chen D, Su M, Tang J (2018) Low-cost biochar derived from corncob as oxygen reduction catalyst in air cathode microbial fuel cells. Electrochim Acta 283(1):780–788

126. Huang Z-F, Wang J, Peng Y, Jung C-Y, Fisher A, Wang X (2017) Design of efficient bifunctional oxygen reduction/evolution electrocatalyst: recent advances and perspectives. Adv Energy Mater 7(23):1700544
127. Lu G, Li Z, Fan W, Wang M, Yang S, Li J, Chang Z, Sun H, Liang S, Liu Z (2019) Sponge-like N-doped carbon materials with Co-based nanoparticles derived from biomass as highly efficient electrocatalysts for the oxygen reduction reaction in alkaline media. RSC Adv 9(9):4843–4848
128. Liu Y, Su M, Li D, Li S, Li X, Zhao J, Liu F (2020) Soybean straw biomass-derived Fe–N co-doped porous carbon as an efficient electrocatalyst for oxygen reduction in both alkaline and acidic media. RSC Adv 10(12):6763–6771
129. Qiao Liu YD, Zhao Q, Pan F, Zhang B, Zhang J (2014) A direct synthesis of nitrogen-doped carbon nanosheets with high surface area and excellent oxygen reduction performance. Langmuir 30(27):8238–8245
130. Xie S, Huang S, Wei W, Yang X, Liu Y, Lu X, Tong Y (2015) Chitosan waste-derived Co and N Co-doped carbon electrocatalyst for efficient oxygen reduction reaction. ChemElectroChem 2(11):1806–1812
131. Li Y, Lu J (2017) Metal-air batteries: will they be the future electrochemical energy storage device of choice? ACS Energy Lett 2(6):1370–1377
132. Thalji MR, Ali GAM, Algarni H, Chong KF (2019) Al^{3+} ion intercalation pseudocapacitance study of $W_{18}O_{49}$ nanostructure. J Power Sources 438:227028
133. Ali GAM, Wahba OAG, Hassan AM, Fouad OA, Chong KF (2015) Calcium-based nanosized mixed metal oxides for supercapacitor application. Ceram Int 41(6):8230–8234
134. Ali GAM, Fouad OA, Makhlouf SA, Yusoff MM, Chong KF (2014) Co_3O_4/SiO_2 nanocomposites for supercapacitor application. J Solid State Electrochem 18(9):2505–2512
135. Ali GAM, Makhlouf SA, Yusoff MM, Chong KF (2015) Structural and electrochemical characteristics of graphene nanosheets as supercapacitor electrodes. Rev Adv Mater Sci 40 (1):35–43
136. Fu J, Cano ZP, Park MG, Yu A, Fowler M, Chen Z (2017) Electrically rechargeable Zinc-air batteries: progress, challenges, and perspectives. Adv Mater 29(7):1604685
137. Lars Öjefors LC (1978) An iron—air vehicle battery. J Power Sources 2(3):287–296
138. Zhang J, Zhou Q, Tang Y, Zhang L, Li Y (2019) Zinc-air batteries: are they ready for prime time? Chem Rev 10(39):8924–8929
139. Wang M, Lei X, Hu L, Zhang P, Hu H, Fang J (2017) High-performance waste biomass-derived microporous carbon electrocatalyst with a towel-like surface for alkaline metal/air batteries. Electrochim Acta 250:384–392
140. Lei X, Wang M, Lai Y, Hu L, Wang H, Fang Z, Li J, Fang J (2017) Nitrogen-doped micropore-dominant carbon derived from waste pine cone as a promising metal-free electrocatalyst for aqueous zinc/air batteries. J Power Sources 365:76–82
141. Ma Z, Wang K, Qiu Y, Liu X, Cao C, Feng Y, Hu P (2018) Nitrogen and sulfur co-doped porous carbon derived from bio-waste as a promising electrocatalyst for zinc-air battery. Energy 143:43–55
142. Choudhary R, Patra S, Madhuri R, Sharma PK (2017) Cow dung derived PdNPs@WO3 porous carbon nanodiscs as trifunctional catalysts for design of zinc-air batteries and overall water splitting. ACS Sustain Chem Eng 5(11):9735–9748
143. Yang L, Zeng X, Wang D, Cao D (2018) Biomass-derived FeNi alloy and nitrogen-codoped porous carbons as highly efficient oxygen reduction and evolution bifunctional electrocatalysts for rechargeable Zn-air battery. Energy Storage Mater 12:277–283
144. Wei J, Zhao S, Ji L, Zhou T, Miao Y, Scott K, Li D, Yang J, Wu X (2018) Reuse of Ni-Co-Mn oxides from spent Li-ion batteries to prepare bifunctional air electrodes. Resour Conserv Recycl 129:135–142
145. Smil V (2006) Energy at the crossroad, organisation for economic co-operation and development. Paris 12
146. Anders Hagfeldt GB, Sun L, Kloo L, Pettersson H (2010) Dye-sensitized solar cells. Chem Rev 110:6595–6663

147. Grätzel M (2001) Photoelectrochemical cells. Nature 414:338–344
148. Chapin DM, Fuller CS, Pearson GL (1954) A new silicon p-n junction photocell for converting solar radiation into electrical power. J Appl Phys 25(5):676–677
149. Eser RWBaE (1997) Polycrystalline thin film solar cells: present status and future potential. Annu Rev Mater Res 27:625–653
150. Shockley W, Queisser HJ (1961) Detailed balance limit of efficiency of p-n junction solar cells. J Appl Phys 32(3):510–519
151. Brain O'Regan MG (1991) Alow- cost, high-effeciency solarcell based on dye-sensitized colloidal TiO_2 films. Nature 353:737–740
152. Carey GH, Abdelhady AL, Ning Z, Thon SM, Bakr OM, Sargent EH (2015) Colloidal quantum dot solar cells. Chem Rev 115(23):12732–12763
153. Akihiro Kojima KT, Shirai Y, Miyasaka T (2009) Organometal halide perovskites as visible-light sensitizers for photovoltaic cells. J Am Chem Soc 131:6050–6051
154. Ito S, Chen P, Comte P, Nazeeruddin MK, Liska P, Péchy P, Grätzel M (2007) Fabrication of screen-printing pastes from TiO_2 powders for dye-sensitised solar cells. Prog Photovoltaics Res Appl 15(7):603–612
155. Murakami TN, Grätzel M (2008) Counter electrodes for DSC: application of functional materials as catalysts. Inorg Chim Acta 361(3):572–580
156. Wu J, Li Y, Tang Q, Yue G, Lin J, Huang M, Meng L (2014) Bifacial dye-sensitized solar cells: a strategy to enhance overall efficiency based on transparent polyaniline electrode. Sci Rep 4:4028
157. Thomas S, Deepak TG, Anjusree GS, Arun TA, Nair SV, Nair AS (2014) A review on counter electrode materials in dye-sensitized solar cells. J Mater Chem A 2(13):4474–4490
158. Olsen E, Hagen G, Eric Lindquist S (2000) Dissolution of platinum in methoxy propionitrile containing LiI/I_2. Sol Energy Mater Sol Cells 63(3):267–273
159. Mathew S, Yella A, Gao P, Humphry-Baker R, Curchod BFE, Ashari-Astani N, Tavernelli I, Rothlisberger U, Nazeeruddin MK, Grätzel M (2014) Dye-sensitized solar cells with 13% efficiency achieved through the molecular engineering of porphyrin sensitizers. Nat Chem 6:242
160. Takurou N, Murakami SI, Qing Wang Md, Nazeeruddin Khaja, Bessho Takeru, Cesar Ilkay, Liska Paul, Humphry-Baker Robin, Comte Pascal, Péchy Péter, Grätzel Michael (2006) Highly efficient dye-sensitized solar cells based on carbon black counter electrodes. J Electrochem Soc 153:2255–2261
161. Wu C-S, Chang T-W, Teng H, Lee Y-L (2016) High performance carbon black counter electrodes for dye-sensitized solar cells. Energy 115:513–518
162. Xu X, Huang D, Cao K, Kang M, Zakeeruddin SM, Gratzel M (2013) Electrochemically reduced graphene oxide multilayer films as efficient counter electrode for dye-sensitized solar cells. Sci Rep 3:1489
163. Lodermeyer F, Costa RD, Casillas R, Kohler FTU, Wasserscheid P, Prato M, Guldi DM (2015) Carbon nanohorn-based electrolyte for dye-sensitized solar cells. Energy Environ Sci 8(1):241–246
164. Gong F, Xu X, Zhou G, Wang ZS (2013) Enhanced charge transportation in a polypyrrole counter electrode via incorporation of reduced graphene oxide sheets for dye-sensitized solar cells. Phys Chem Chem Phys 15(2):546–552
165. Heo SY, Koh JK, Kim JK, Lee CS, Kim JH (2014) Three-dimensional conducting polymer films for Pt-free counter electrodes in quasi-solid-state dye-sensitized solar cells. Electrochim Acta 137:34–40
166. Kung CW, Chen HW, Lin CY, Huang KC, Vittal R, Ho KC (2012) CoS acicular nanorod arrays for the counter electrode of an efficient dye-sensitized solar cell. ACS Nano 6:7016–7025
167. Chun-Ting Li Y-LT, Ho Kuo-Chuan (2016) Earth abundant silicon composites as the electrocatalytic counter electrodes for dye-sensitized solar cells. ACS Appl Mater Interfaces 8(11):7037–7046

168. Shengjie Peng LT, Liang J, Mhaisalkar SG, Ramakrishna S (2012) Polypyrrole nanorod networks/carbon nanoparticles composite counter electrodes for high-efficiency dye-sensitized solar cells. ACS Appl Mater Interfaces 4(1):397–404
169. Dong F, Guo Y, Xu P, Yin X, Li Y, He M (2017) Hydrothermal growth of MoS_2/Co_3S_4 composites as efficient Pt-free counter electrodes for dye-sensitized solar cells. Sci China Mater 60(4):295–303
170. Ahmed ASA, Xiang W, Gu A, Hu X, Saana IA, Zhao X (2018) Carbon black/silicon nitride nanocomposites as high-efficiency counter electrodes for dye-sensitized solar cells. New J Chem 42:11715–11723
171. Wu M, Ma T (2012) Platinum-free catalysts as counter electrodes in dye-sensitized solar cells. Chemsuschem 5(8):1343–1357
172. Jiang QW, Li GR, Wang F, Gao XP (2010) Highly ordered mesoporous carbon arrays from natural wood materials as counter electrode for dye-sensitized solar cells. Electrochem Commun 12(7):924–927
173. Wang C-L, Liao J-Y, Chung S-H, Manthiram A (2015) Carbonized eggshell membranes as a natural and abundant counter electrode for efficient dye-sensitized solar cells. Adv Energy Mater 5(6):1401524
174. Cha SM, Nagaraju G, Sekhar SC, Bharat LK, Yu JS (2018) Fallen leaves derived honeycomb-like porous carbon as a metal-free and low-cost counter electrode for dye-sensitized solar cells with excellent tri-iodide reduction. J Colloid Interface Sci 513:843–851
175. Xu S, Liu C, Wiezorek J (2018) 20 renewable biowastes derived carbon materials as green counter electrodes for dye-sensitized solar cells. Mater Chem Phys 204:294–304
176. Wang C, Yun S, Xu H, Wang Z, Han F, Zhang Y, Si Y, Sun M (2020) Dual functional application of pomelo peel-derived bio-based carbon with controllable morphologies: an efficient catalyst for triiodide reduction and accelerant for anaerobic digestion. Ceram Int 46 (3):3292–3303
177. Shen Y (2017) Rice husk silica derived nanomaterials for sustainable applications. Renew Sustain Energy Rev 80:453–466
178. Wang G, Wang D, Kuang S, Xing W, Zhuo S (2014) Hierarchical porous carbon derived from rice husk as a low-cost counter electrode of dye-sensitized solar cells. Renewable Energy 63:708–714
179. Ahmad WB, Yang Z, Khan J, Jing W, Jiang F, Chu L, Liu N, Li L, Gao Y (2016) Extraction of nano-silicon with activated carbons simultaneously from rice husk and their synergistic catalytic effect in counter electrodes of dye-sensitized solar cells. Sci Rep 6:39314

Agriculture Waste Recycling Technologies

Recycling of Nanosilica Powder from Bamboo Leaves and Rice Husks for Forensic Applications

Nik Fakhuruddin Nik Hassan, Cik Norhazrin Che Hamzah, Revathi Rajan, and Yusmazura Zakaria

Abstract Fingermarks are of the commonly found evidence at crime scenes or on items submitted about a crime. However, most of these fingermarks are latent and cannot be seen with the naked eye. Powdering is the most common method employed to develop latent fingermark on non-porous surfaces. However, existing commercially available fingermark powders suffer several limitations, such as health problems. Hence, the objective of this study was of nanosilica powders synthesis from agricultural wastes, i.e. bamboo leaves (BL) and rice husks (RH), as a green approach and later applied for fingermark development. To obtain highly purified eco-friendly silica powder, acid leaching of RH and BL was carried out to remove impurities and metallic elements. Thermal combustion of BL and RH under controlled conditions had produced silica ash, and the addition of ash into sodium hydroxide produced sodium silicate solution. The addition of acetone as a polar solvent into sodium silicate before precipitation with acetic acid yielded a spherical form of nanosilica. The yield percentage of nanosilica from RH (12.16%) was higher than that of nanosilica from BL (6.9%). The characterization of synthesized nanosilica was carried out using FESEM, EDX, and ATR-FTIR spectroscopy. FESEM analysis of the nanosilica produced was spherical. The EDX elemental spectra showed significant silicon elements and oxygen in the BL and RH nanosilica. FTIR analysis showed predominant absorbance peaks at 1057 and 1060 cm^{-1} corresponding to siloxane bonds. The synthesized nanosilica powders were further applied to visualize latent fingermarks on various substrates. The powdering technique using nanosilica powders yielded good quality and clarity of

N. F. Nik Hassan (✉) · C. N. Che Hamzah · R. Rajan
Forensic Science Programme, School of Health Sciences, Universiti Sains Malaysia, Health Campus, 16150 Kubang Kerian, Kelantan, Malaysia
e-mail: nikf@usm.my

R. Rajan
Forensic Science Programme, Department of Biotechnology, Faculty of Applied Sciences, UCSI University, Cheras, Kuala Lumpur, Malaysia

Y. Zakaria
Biomedicine Programme, School of Health Sciences, Universiti Sains Malaysia, Health Campus, 16150 Kubang Kerian, Kelantan, Malaysia

A. S. H. Makhlouf and G. A. M. Ali (eds.), *Waste Recycling Technologies for Nanomaterials Manufacturing*, Topics in Mining, Metallurgy and Materials Engineering, https://doi.org/10.1007/978-3-030-68031-2_11

developed fingermark on most of the tested surfaces as compared to commercially available white fingerprint powder. In conclusion, nanosilica powders were successfully synthesized from agricultural wastes and applied in the field of forensic science as latent fingermark detection material for the powdering technique.

Keywords Forensic science · Fingermarks · Nanosilica · Bamboo leaves · Rice husks

List of Abbreviations

ATR	Attenuated total reflectance
BL	Bamboo leave
BLA	Bamboo leave ash
BLP	Bamboo leave powder
EDX	Energy dispersive x-ray
FESEM	Field emission scanning electron microscope
FTIR	Fourier transform infrared spectroscopy
ICP-MS	Inductively coupled plasma-mass spectrometry
NSP	Nanosilica powder
RH	Rice husk
RHA	Rice husk ash
RHP	Rice husk powder

1 Introduction

Fingermark is an important and commonly found physical evidence in forensic investigation [1, 2]. It is formed by the impression of friction skin of the fingers on any material touched by the fingers [3]. The persistence and uniqueness of fingermark make it one of the most important physical evidence in criminal investigation [4]. Persistence of fingermark refers to unchanged of fingermarks from the time of formation during pregnancy until decomposition after death. When the body grows, fingermark will become larger, but the specific characteristics will remain unchanged. When a person gets older, the appearance of fingermark will decline but classification and identification are still not affected. Fingermarks are also unique to an individual. For example, identical twins have different fingermarks [5].

Fingermark can provide information about the person who had deposited the fingermark found at the crime scene [6]. Latent, patent, and plastic fingermark are three types of fingermarks that can be encountered. Haan [7] reported that the latent fingermarks are the most common form of fingermark evidence encountered at a crime scene and give the most problem during the development process. Latent fingermark cannot be seen with the naked eye, and some enhancement methods are needed to make it visible before collection and comparison [8].

The pattern of the fingermark can be grouped into four classes according to the Henry-Galton classification system, namely arches, whorls, loops, and compounds [9, 10]. By studying the ridge pattern closely, it can be observed that the parallel ridge flow is interrupted by irregularities at specific intervals. These irregularities are called minutiae, and there are seven types, namely ridge termination, bifurcations, hooks or spurs, lake, island, independent ridge, and crossovers [11, 12]. These provide further identification of an individual in forensic investigation.

The fingermark consists of a mixture of substances excreted from three glands, which are eccrine, apocrine, and sebaceous glands [13–15]. In latent fingermark residues, eccrine secretions are water-soluble components [16, 17] such as amino acids, proteins, urea, uric acid, lactic acid, sugars, creatinine, and choline, and the inorganic constituents are chlorides, metal ions, sulphates, phosphates, ammonia as well as water. Sebaceous secretions are non-water-soluble components such as glycerides, fatty acids, wax esters, squalene, sterol esters, and sterols [18–20]. Fingermark residues also comprise contaminants such as food residue, dust, cosmetics, and drug traces or drug metabolites [21–24].

The powdering method is the first and common development method applied for visualization of latent fingermark [2, 7, 25]. There are few types of powders used for this purpose, such as black powder [26, 27], white powder [28], magnetic and metallic powders [29, 30], and luminescent powders [31, 32]. However, the exposure to these powders during application can cause health problem [33]. For example, overexposure to fingermark powders can cause eye irritation, skin irritation, and inhalation irritation [34]. Therefore, some alternative steps must be taken to reduce the risk of health problems to the person handling the fingermark powders.

This study focused on the synthesis of nanosilica powders (NSPs) from agricultural wastes, namely BL and RH and their effectiveness for the visualization of latent fingermark on various types of surfaces. Agricultural wastes like BL and RH are rich in silica content, and nanosilica synthesized from these wastes can yield highly pure silica [35]. The synthesis was widely studied recently due to the low cost of production, safe to use, and eco-friendly products. Hence, this study was carried out to synthesize fingermark powders from natural products, which should be safe for the person handling these powders.

Nanoparticles are particles with a size between 1 and 100 nm, but these particles are further classified according to diameter [36]. Ultrafine particles are between 1 and 100 nm, fine particles are between the size of 100 and 2500 nm, and coarse particles are between 2500 and 10,000 nm. In the forensic field, nanoparticles are much smaller than particles of commercial fingermark powders used for fingermark detection, which are 1–10 μm in size [37]. Therefore, it will give better development of fingermark than commercially available fingermark powders.

NSPs were reported to be used as fingermark powders for the detection of latent fingermark with sufficient clarity [38]. These NSPs exist in the amorphous state and have a high surface area, small size, has the versatility and ability to modify their surfaces [39]. The surface modification versatility of the powders makes them able to selectively detect specific compounds in the fingermark residue [40]. Due to these features, they can produce a better quality of developed fingermark when applied as a means for the detection of latent fingermark. Another research used dye-doped nanophosphors of dimension 300–500 nm for fingermark development and reported positive outcomes for extremely dry fingermarks [41].

$ZnO–SiO_2$ had demonstrated precise ridge details [38]. On the contrary, the commercial white fingermark powder did not correctly adhere to the fingermarks ridges and diminished the patterns. It adhered to the whole surface area because of the sticky nature of its constituents, namely titanium dioxide (TiO_2). The $ZnO–SiO_2$ nanopowder produced a distinct image of latent fingermarks on the tested surface with lower background interference, depicting excellent clarity of ridge details that are of paramount importance in criminal investigation [1]. $ZnO–SiO_2$ is more effective as compared to the commercially white powder for developing latent fingermarks, where an excellent third-level ridge detail was utilizing this synthesized powder. To date, there is limited research reported on the application of nanosilica from agricultural wastes for the detection and visualization of latent fingermarks. Revathi et al. [42] demonstrated that nanosilica powder from rice husk was successfully synthesized for the development of latent fingermarks. Some research proved the potential application of combination nanosilica with other mixtures like $ZnO–SiO_2$ and synthetic silica like tetraethyl orthosilicate (TEOS) for the development of latent fingermarks [43].

Zinc oxide (ZnO) is employed as a nanopowder for fingermark detection due to its adhesive properties and interaction with lipids and proteins [44]. Sebaceous fingermark consists of the non-water-soluble compound, and latent fingermarks can be visualized after the enhancement with a mixture of $ZnO–SiO_2$ nanopowder due to the non-covalent or hydrophobic interaction between the powders and the residue [45]. $ZnO–SiO_2$ nanopowder is white. Hence, it will produce a perfect contrast when used on the dark-colored surface. Manufacturing of pure silica is energy-intensive when conventional raw materials are utilized [46, 47]. The most common NSPs precursors are silicon alkoxides or silicates, which in turn are synthesized from raw material like sand through the smelting method [48, 49]. This process requires high energy, high temperature, high pressure, and also strong acidity [50].

Synthesis of nanoparticles can be categorized into two most basic methods that are top-down approach or the bottom-up approach. The purity of the final product and particle size distribution is ultimately governed by the nature of the manufacturing process and any integral purification steps involved. The top-down or the physical method of nanoparticle creation is the fine division of the bulk element into their respective nanosized counterpart [51].

The bottom-up approach is a more commonly adapted chemical route of nanoparticle synthesis. In this approach, nucleation of the specific atom acts as template for crystal or particle growth. The subsequent or simultaneous aggregation of the atoms forms nanoparticles of the required size [52–55]. The basis of the wet chemical reduction in the formation of an aqueous solution containing the element of interest and reducing the ions into atoms using any reduction agents [56]. Sometimes, a catalyst or stabilizing agents accompany the process to aid in the formation of specific shape and size distribution of such nanoparticles via electrostatic repulsion or stearic stabilization [57]. The six routes of NSPs synthesis from synthetic precursors and waste products are vapour-phase reaction, thermal decomposition, chemical digestion, precipitation, sol-gel technique, and fungus-mediated transformation [58–83]. A newly synthesized nanomaterial may be characterized using analytical instruments to understand its intrinsic structure and properties [84]. New nanomaterial may be characterized by its morphological, structural, and optical features, as well as particle size and surface area analysis [85–94].

Plants have a natural silica synthesizing system that converts water-soluble silicic acid seeped from the ground into amorphous silica by precipitation and polymerization [95–99]. Agricultural wastes such as rice husk (RH), bamboo leave (BL), sugarcane bagasse, and groundnut shell are the natural sources that contain a large amount of silica content [80, 100–102]. The waste materials are burnt during harvesting and produced ash that is rich in carbon and silica contents [103, 104]. These types of NSPs are safe to use and can reduce health problems as they are synthesized from agricultural wastes [105]. The low-cost production of nanosilica from agricultural, natural resources makes it to become widely used today in many fields [35, 106, 107]. According to Chakraverty et al. [108], the combustion of rice husk resulted in rice husk ash (RHA), and at low cost, this raw material is utilized to produce silica powder.

Umemura and Takenaka [109] reported that bamboo is a plentiful natural resource in the tropical and temperate regions. Bamboo is a plant that absorbs and accumulates high amounts of silicon from the soil [110]. The silica content in the bamboo increases from roots to leaves [111]. Various studies have demonstrated various methods to obtain nanosilica from BL [71, 110], and these studies showed some similarities in the purification treatment of bamboo leaves. The purification treatment is carried out in two steps: leaching using an acid solution to remove metallic impurities and sintering to reduce the carbon content of the leached samples. Vaibhav et al. [100] reported that the acid leaching of BL could increase the content of pure silica by removing soluble elemental impurities. Similarly, Wang et al. [112] had also indicated in their study that acid leaching of BL could remove alkaline impurities by preventing the formation of glassy state silicate.

FESEM analysis had shown that BL ash samples had a specific arrangement and shape [100]. Sintering of BL in the furnace at 900 °C produced agglomerated cluster particles, but the size and shape of these particles were altered after the acid leaching process. According to a study by Wang et al. [113], the FESEM images indicated the presence of nanosilica in the bamboo leaves. The elemental distribution observed using EDX illustrated that the Si was arranged uniformly within their organic components.

The exploitation of RH for commercial purposes is being extensively explored to synthesis nanomaterial at a much lower cost. The extraction of cellulose and silica from RHA has been widely researched to produce silica using the low-cost method from a sustainable source [114]. This also reduces the ecological impact of accumulating RH and prevent air pollution caused by the burning of this waste material [115]. Houston [116] reported that the RH was tough, resistant to degradation, and contained a high content of ash. Primary components of RH are hydrated silica and lignin, cellulose, and hemicellulose [117]. Evaporation and polymerization of Si on the outer surface of plants make it concentrated to form a cellulose–silica membrane [118]. The characterization of silica present in RH by FESEM shows that the silica is localized in the epidermis layer and also in the spaces between the epidermal cells [119]. Silica in rice plants is present in inorganic linkages and bonded strongly to the organic components which cannot dissolve in alkali and can resist very high temperature [47].

After the calcination process in the furnace, the organic compound in the RH will be decomposed and form rice husk ash (RHA). RHA is known as slightly impure silica that contains a small amount of metallic impurities such as calcium, potassium, sodium, and others [120]. RHA is categorized into low-carbon grey ash, high-carbon char, and white ash. Various factors are responsible for determining the quality of the RHA, such as temperature, time of ashing, and soaking and heating rate [121]. The thermal treatment of RH is needed to remove the carbon content in RH. According to Della et al. [122], the carbon content in RHA had decreased to 0.14% after thermal treatment for 6 h in the furnace.

The previous study reported a few processes for preparing silica from RH, which includes direct combustion, combustion after pretreatment, hydrothermal method, reaction with sodium carbonate (Na_2CO_3), and reaction with sodium hydroxide (NaOH) [118]. Nakata et al. [123] reported that RH could produce silica via direct combustion without the pretreatment process. The combustion temperature and combustion instruments are two factors that can affect the quality of silica produced. The RHA formed at combustion temperature below 800 °C is in an amorphous form, where the diameter of the ash particles obtained is in the average diameter of 20 μm. At combustion temperature above 900 °C, the RHA formed is in cristobalite and tridymite.

Combustion after the pretreatment process is related to the use of acid in pretreatment prior to combustion. HCl is the common acid used for pretreatment of the RH sample [124]. The acid leaching using 1 M HCl can remove the metallic impurities present in the RH sample [108]. This process did not affect the amorphic of the silica produced, and it yielded white colour RHA of high purity. The acid leaching of RH using HCl solution before combustion at 600 °C produced 99.5% pure silica with a high specific area (approximately 260 m^2 g^{-1}) [125]. The high specific area of silica has high reaction activity, and this will enhance the quality of the silica produced.

The reaction temperature in the hydrothermal method is lower as compared to the combustion method, and hence, amorphous silica can be retained in RH. According to Wu et al. [126], silica with high purity and surface area silica can be obtained by utilizing the appropriate ratio of RH, concentrated HNO_3, and water and proceed with optimized reaction time and temperature of 3 h at 160–180 °C, respectively. Concentrated HNO_3 and longer retention time do not influence the reaction result, but the addition of a high amount of water will cause incomplete oxidation of RH.

Reaction with Na_2CO_3 refers to the process, where the RH is first carbonated, and then, the produced RHA will react with Na_2CO_3 solution to obtain pure silica.

The yield and purity of silica are affected by lower and higher temperatures, respectively, which will affect the yield and quality of the silica. If the Na_2CO_3 concentration is more than 15%, then the silica yield is more than 90%. The best combination is the reaction temperature around 600 °C and reaction time more than 3.5 h [127].

In the method of the reaction with NaOH, the RHA will be dissolved in NaOH to produce sodium silicate (Na_2SiO_3). Then, H_2SO_4 is added drop by drop to the Na_2SiO_3 solution to produce silica [128]. When observed under FESEM, a sample of RHA without acid leaching showed the fibre-like appearance while acid treatment RHA sample showed that the particle form is in NSPs [100].

Noushad et al. [129] had demonstrated that NSPs could be obtained in the spherical form from rice husk by the addition of specific solvents. The use of a solvent affects the particles of nanosilica and their shape. The particles are in irregular shape if no solvent or non-polar solvent is utilized. On the contrary, if polar solvents are used, then the silica particles become well rounded and spherical. The uses of two different polar solvents do not show any differences concerning the particle size of the silica. The silica particles are solvated in polar solvents, and the surface charge will be increased and result in a spherical shape of the silica particles.

Other studies had reported that the nanosilica could also be synthesized from RH by precipitation method [130]. In the precipitation method, a few types of acid are used to precipitate the silica. H_2SO_4 is the common acid used as a precipitating agent. The acids are added drop by drop into the Na_2SiO_3 until the pH of the solution reaches the range 7.5–8.5. The precipitation method is generally employed to produce stable nanosilica. The degree of agglomeration of the silica particles obtained from RH can be controlled by the use of different acids and solvents in the precipitation process. Noushad et al. [129] reported that the agglomeration was observed to be higher when nitric acid was utilized for precipitation in conjugation with polar solvents such as ethanol and 2-propanol. On the contrary, when non-polar solvents such as diethyl ether and toluene were employed, the agglomeration was observed to be higher if orthophosphoric acid was added before precipitation.

2 Methodology

2.1 Materials and Reagents

BL (*Bamboosa vulgaris*) was obtained from the Golf Course of USM Health Campus, Kubang Kerian Kelantan. RH from local white rice type was obtained from Kilang Beras Bernas Peringat, Kota Bharu Kelantan. Sodium hydroxide (NaOH) pellets and 37% hydrochloric acid (HCl) were purchased from Merck (Germany); acetone (CH_3COCH_3) and glacial acetic acid (CH_3COOH) were purchased from Sigma-Aldrich (Germany). Distilled water and deionized water were used when applicable. Sulphuric acid (H_2SO_4, 95–98%) reagent grade was purchased from Sigma-Aldrich (Singapore). Methanol, ethanol, and isopropanol were also purchased from Sigma-Aldrich (USA).

2.2 Synthesis of Nanosilica from Bamboo Leave and Rice Husk

There were three main steps involved in the synthesis of nanosilica, namely (a) washing and acid treatment, (b) thermal treatment in the furnace, and (c) extraction of silica in NaOH solution.

2.3 Washing and Acid Treatment

BL and RH were washed thoroughly with tap water and then soaked for one hour with normal tap water and rinsed. After one hour, the samples were again soaked for another hour in distilled water and rinsed to remove contaminants. The samples were then laid out onto the aluminium tray and dried in the air oven for 24 h at 100 °C. The samples were then cut into smaller sizes and then ground using a blender. The blended samples were then weighed, and the mass before acid leaching was recorded. Samples were then heated with an HCl solution on a hot plate at the ratio 5 g of bamboo leave powder (BLP) or rice husk owder (RHP) to 100 mL of 1 M HCl solution for approximately 2 h stirring using a magnetic stir. Samples were then cooled, decanted, and washed with warm distilled water. The acid-treated BL and RH samples were then transferred onto an aluminium tray and dried in the oven at 100 °C for 24 h. The samples were heated on a hotplate for 2 h until charring to release the volatile gases.

2.4 Thermal Treatment

The charred acid-treated BL and RH samples undergone heat treatment to acquire the ashes. Samples were burnt in a Carbolite CWF 1300 (United Kingdom) at 700 ° C for 5 h. The ashing steps resulted in the formation of white ashes and were designated as bamboo leaves ashes (BLA) and rice husk ashes (RHA). The weight of BLA and RHA was recorded.

2.5 Extraction of Silica

A sample of 0.5 g BLA or RHA was dissolved in 100 mL of 1 M NaOH solution. The solution was heated and stirred continuously for 2 h before filtering to remove the residue. The resultant viscous, transparent, and colourless solution was Na_2SiO_3. It was cooled down at room temperature and kept in the oven at 1000 °C for 24 h.

2.6 Synthesis of Nanosilica

After extraction of silica in NaOH, the preparation of nanosilica was followed by optimization methods. Four conditions were optimized, which were solvents addition, different polar solvents used, volumes of solvent added into Na_2SiO_3, and different drying processes. The optimized method was used for the mass production of nanosilica from BL and RH. 40 mL acetone was added into the Na_2SiO_3 solution before the precipitation process. Then, 5 M CH_3COOH was added drop by drop into the Na_2SiO_3 solution under constant stirring until it reached pH 8. This step was carried in the sonicator. The formed silica precipitate was magnetically stirred for 45 min on a hot plate. The silica precipitate was then centrifuged at 3124 RCF for 5 min using a Universal 32R CTF15-009, Hettich Zentrifugen (Germany). The supernatant was discarded and then washed with hot deionized water three times to make it alkali free. Lastly, the clean silica precipitate was left overnight in the freeze dryer (Inshinbase Freeze Drier, Korea EDE Netherland) for the drying process until white NSPs were formed. The weight of NSPs was recorded, and the percentage yield of nanosilica from BL and RH was calculated.

2.7 Characterization of Nanosilica Synthesized from Bamboo Leave and Rice Husk

The original BL and RH, acid-treated BL and RH, BLA and RHA, and nanosilica particle morphology (size and shape) were characterized using a Quanta FEG 450 FESEM (Czech Republic). The elemental composition of nanosilica powder synthesized from BL and RH was characterized using EDX (Czech Republic). G3281A 7700 Series ICP-MS from Agilent Technologies (Boblingan, Germany) was used for elemental determination present in original BL and RH and acid-treated BL and RH. BRUKER-TENSOR 27 ATR-FTIR Spectrophotometer (USA) was used to observe the peak of SiO_2 stretching and bending vibrations in NSPs samples from BL and RH.

2.8 *Development of Latent Fingermarks from Bamboo Leave and Rice Husk*

2.8.1 Materials and Surfaces

Eight types of non-porous surfaces/substrates (car plate number, plastic file, black glass, and black painted car metal, black tiles, ceramic bowl, CD disc, and magnetic strip bank card) and two semi-porous surfaces (glossy paper and black tape) were chosen as substrates for latent fingermark deposition. All the surfaces (except ceramic bowl, black tape, CD disc, and magnetic strip bank card) were cut to 4 cm × 4 cm size. Most of the surfaces were black to increase the contrast of white powder with a dark background. Commercial white fingermark powder (SIRCHIE, USA) and squirrel brushes (SIRCHIE, USA) were employed for the development of latent fingermarks on tested substrates.

2.8.2 Depletion Studies of Split Fingermarks

The depletion studies were carried out to minimize the potential variability in fingermark composition from the same donor. The donor was asked to wash her hands with soap and water before the experiment. The nose and face areas were rubbed using dried right thumb before deposition onto the substrates to obtain the sebaceous secretion type of latent fingermarks. All substrates were initially cleaned with acetone. Each piece of the substrates was arranged side by side. Then, five fingermarks were deposited in a depletion series on the four corners (quarters) of a substrate continuously using the same finger to form quartered split fingermarks in the centre part of the substrate. Then, all the fingermark samples on a substrate were left at room temperature for about 1 h. The arranged pieces of the substrate were separated. For depletion 1, the first corner fingermark was developed accordingly

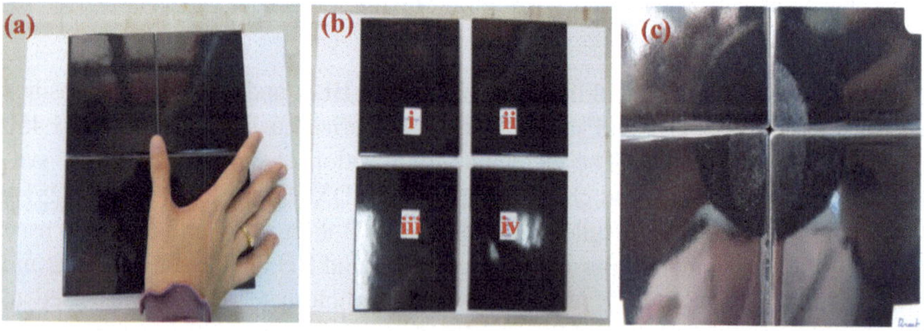

Fig. 1 Deposition of fingermarks on the substrate (**a**), development of latent fingermarks using three powders (**b**), and the developed fingermarks for each corner of the substrate were recombined and photographed with a bridge camera (**c**)

using commercial fingerprint white powder (i), the second corner fingermark was left undeveloped (ii), the third corner mark was developed with RH nanosilica powder (iii), and the fourth corner mark was developed with BL nanosilica powder (iv). Then, the same procedure was repeated for depletion 2 until depletion 5. The pieces of the substrate were recombined for comparison purposes to determine the effectiveness of each powder. All developed fingermarks were photographed using a bridge camera (Powershot SX60 HS Canon, Japan) to obtain the images. This method was repeated in triplicates for all the six substrates. Figure 1a–c depict the procedures in depletion studies.

3 Results and Discussion

3.1 Optimization Methods for the Synthesis of Nanosilica

The optimization was carried out to find the optimal design to synthesize nanosilica from agricultural wastes. In this optimization method, several parameters were tested, namely effect on solvent addition, different polar solvents used, different volumes of acetone added into Na_2SiO_3, and different drying processes of precipitated nanosilica. The results suggested that the addition of 40 mL acetone followed by precipitating with 5 M CH_3COOH produced a smooth spherical form of nanosilica. The combination of this solvent and acid was utilized in the mass production method of nanosilica from both BL and RH.

3.1.1 Images of Bamboo Leave and Rice Husk in Different Conditions

Figure 2a, b show the colours of BL and acid-treated BL which changed from original green leaves colour to yellow-brown. The colour of RH and acid-treated RH (after leaching with 1 M HCl) had changed to brown (Fig. 3b) from the original yellow colour of RH (Fig. 3a) after charring and leaching processes. The images of BLA and RHA prepared after calcining in the furnace at 700 °C are shown in Figs. 2c and 3c. Figures 2d and 3d depict the amorphous nanosilica of high purity from BL and RH after precipitation with acid. The results were in agreement with a previous study [121]. Rapid heating of RH would result in black particles as the oxidation of carbon did not occur prior to surface melting. The combustion temperature at 700 °C was ideal as RHA formed would show higher impurity and crystallization at lower (500 °C) and higher (1000 °C) temperatures, respectively. However, it was postulated that the specific surface area had decreased with increasing temperature.

Metallic ingredients that are present in BL and RH were also thought to affect the quality of silica produced. Potassium, for example, can cause surface melting, amorphous silica crystallization, and fixation of carbon in RHA [121]. The decrease

Fig. 2 Original BL (**a**), acid-treated BL (**b**), BLA with a combustion temperature of 700 °C (**c**), and nanosilica powder of BL (**d**)

in surface area of silica can occur due to the strong interaction between the metallic ions and silica. Therefore, treatment of BL and RH with an acidic solution, HCl solution, was carried out to remove impurities and to recover silica with high purity. This leaching step prevents glassy state silicate from forming while the organic compound is removed by burning [113]. The chemical reaction between acid and metals had led to the leaching out of metals from the acidic solution during filtration. The colour changes during digestion from green to yellow for BL and yellow to brown for RH were observed in this study.

After RH is acid treated using HCl, the cellulose in RH is leached out, and the blackening of carbohydrate occurs as the oxygen is removed. The proteins are also degraded to amino acids. It has been previously reported that the impurities in RH were substantially removed by acid treatment [119]. Acid-treated BL and RH with 1 M HCl was indicated to be the best chemical treatments for the removal of metallic ingredients to produce a high surface of silica [121].

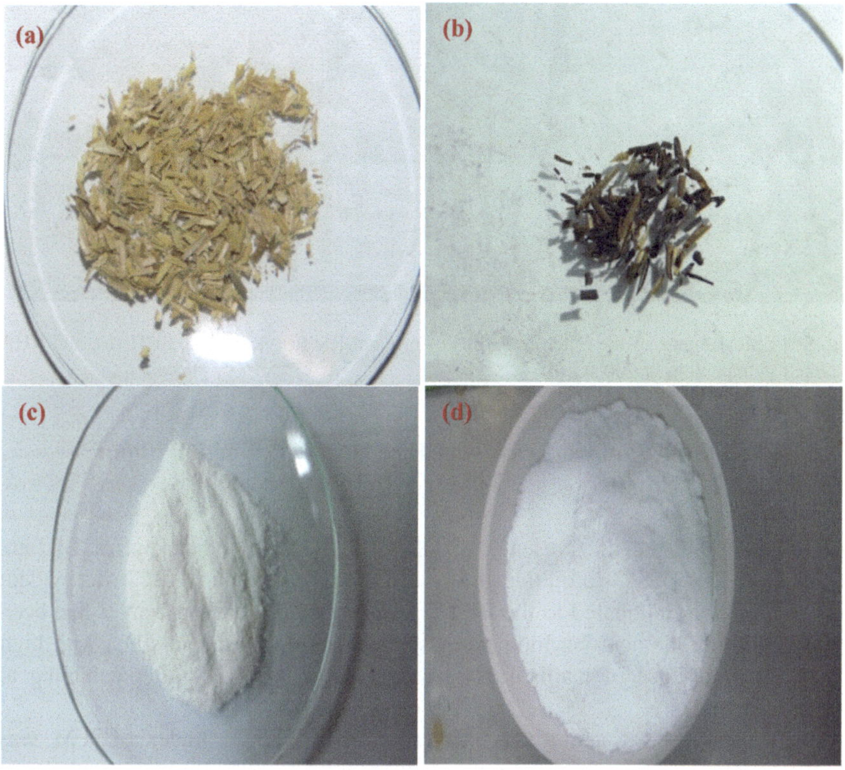

Fig. 3 Original RH (**a**), acid-treated RH (**b**), RHA with combustion temperature of 700 °C (**c**), and nanosilica powder of RH (**d**)

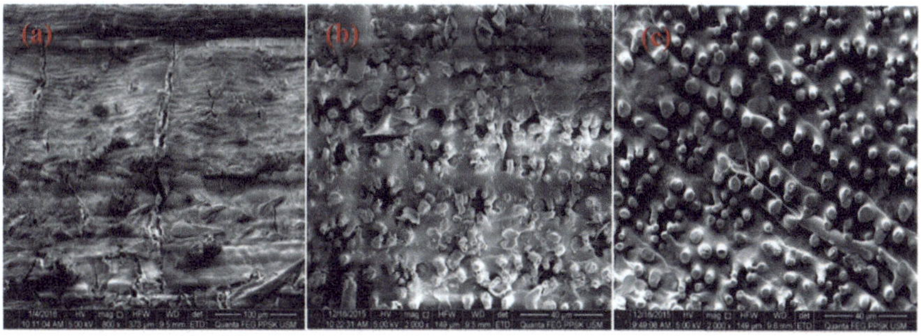

Fig. 4 FESEM images of the outer epidermis of original BL (**a**), acid-treated BL (**b**), and BLA obtained by calcining at 700 °C (**c**)

Fig. 5 FESEM image of the outer epidermis of original RH (**a**), acid-treated RH (**b**), and RHA obtained by calcining at 700 °C (**c**)

3.1.2 FESEM of Bamboo Leave and Rice Husk (with and Without Acid Leaching)

Figure 4a, b indicate the presence of silica on the outer epidermis of BL and acid-treated BL. The silica on the surface of BL treated with 1 M HCl was evident and prominent as compared to the BL because the impurities from leaves had been removed during the acid leaching process. This process produced silica with high purity. The presence of nanosilica was observed in the BLA after calcining at 700 °C (Fig. 4c).

The appearance of acid-treated RH (Fig. 5b) observed under FESEM was similar to RH (Fig. 5a). The silica was distributed in the rice husk, particularly on the outer epidermis, protuberances, and trichomes. The presence of silica can be seen more clearly in RHA after calcining at 700 °C (Fig. 5c).

3.1.3 Yield Percentage of Nanosilica from Bamboo Leave and Rice Husk

The means of yield percentage of nanosilica from BL and RH from three replicates were 6.9% (w/w) and 12.16% (w/w), respectively. This demonstrates that the yield percentage of nanosilica from BL was lower than that of RH. The yield of silica is strongly dependent on the type of acid used for acid leaching as well as the concentration of NaOH. According to Selvakumar et al. [131], the yield of silica was higher (85%) when HCl was used for acid leaching and at 1 M concentration of NaOH. RH has a higher mean yield percentage of silica than BL because they contain more silica than BL, and the portion of silica present in the RH is not dissolved in alkali and resistant to high temperatures during the thermal treatment process in the furnace [118].

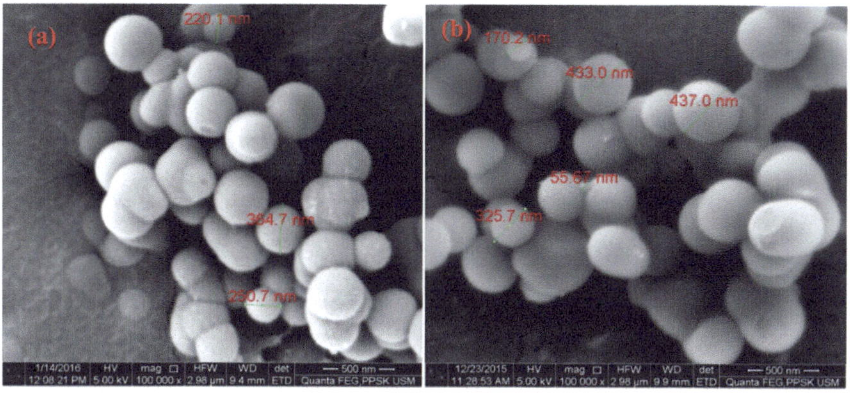

Fig. 6 FESEM image of BL nanosilica powders (**a**) and RH nanosilica powders (**b**) (100000 × magnification)

3.1.4 Nanosilica Synthesized from Bamboo Leave and Rice Husk

In this study, precipitation process was used as a method to synthesize nanosilica from BL and RH. The addition of polar solvent before precipitation with acid has successfully produced the spherical form of nanosilica from BL and RH with an average size of 50–450 nm (Fig. 6a, b). The use of different types of solvents affected the shape of the synthesized nanosilica. The shape of nanosilica particles was observed to be irregular if no solvent or non-polar solvents were utilized. On the contrary, if polar solvents were utilized, then the nanosilica particles produced were spherical [129]. It was evident in this study that the use of polar solvent, i.e. acetone, had successfully produced spherical shaped nanosilica particles. The polar solvents may have solvated the nanosilica particles, hence enhancing the surface charge that resulted in the spherical shape of nanosilica particles [132]. The addition of 5 M acetic acid, CH_3COOH as a precipitating agent during precipitation process, had lowered the agglomeration of the silica particles synthesized from BL and RH.

3.1.5 EDX Analysis of Bamboo Leave and Rice Husk Nanosilica

The silicon content of the samples was estimated using EDX spectroscopy. The EDX elemental spectrum of BL nanosilica indicates significant silicon elements and oxygen and sodium (Na) impurity (Fig. 7a). Carbon (C) and Aurum (Gold) peaks were also observed in the spectrum. The EDX elemental spectrum of RH contains major elements of silicon and oxygen and carbon (C) and Aurum (Gold) (Fig. 7b). No trace elements or any impurities were recorded in EDX elemental spectrum in RH samples, indicating complete removal of impurities during the acid leaching process. The presence of carbon peak was due to carbon tape used to attach the nanosilica samples to the sample holder, while the gold peak was

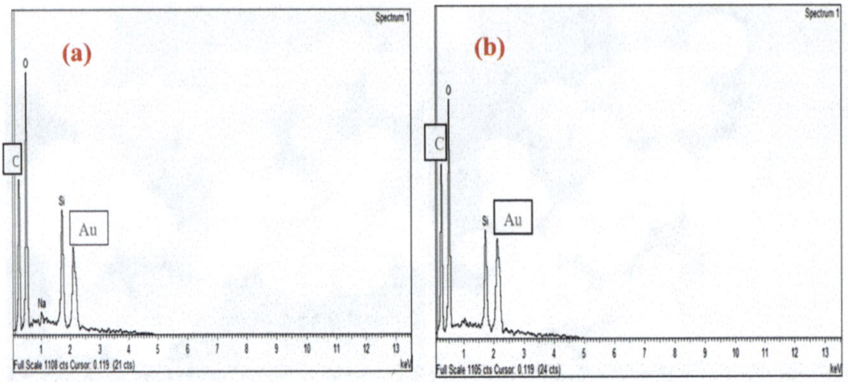

Fig. 7 EDX elemental spectra of BL nanosilica (**a**) and RH nanosilica (**b**)

Table 1 Percentage of elements present in BL and RH nanosilica

Element	BL nanosilica		RH nanosilica	
	Weight (%)	Atomic (%)	Weight (%)	Atomic (%)
O K	53.30	66.48	55.91	69.00
Na K	2.18	1.89	–	–
Si K	44.52	31.63	44.09	31.00
Total	100.00		100.00	

attributable to gold used to coat the nanosilica samples before the analysis. Table 1 shows the percentage of elements present in BL nanosilica comprising of 55.30% of oxygen and 44.52% of silicon elements as well as 2.18% of sodium element. On the other hand, RH nanosilica consists mainly of 55.91% of oxygen and 44.09% of

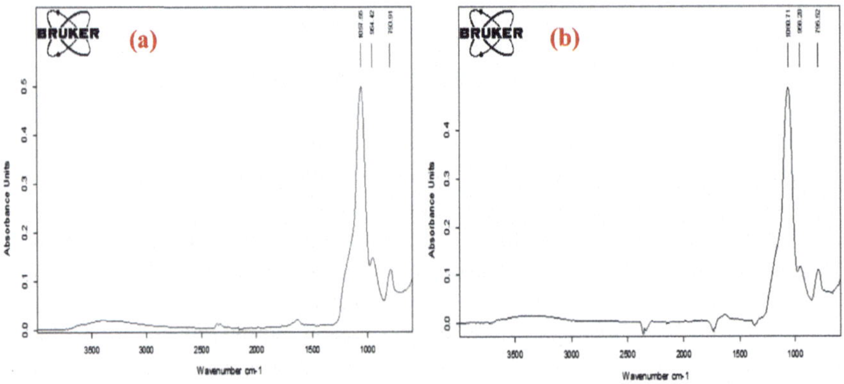

Fig. 8 FTIR spectra of BL nanosilica (**a**) and RH nanosilica (**b**)

silicon as major elements. The EDX analysis indicates that nanosilica was successfully synthesized from both BL and RH.

3.1.6 FTIR Analysis of Bamboo Leave and Rice Husk Nanosilica

The major chemical groups present in nanosilica of BL and RH were identified from the FTIR spectra, as shown in Fig. 8a, b. The broad peak around 2800 and 3750 cm^{-1} was assigned to silanol and hydroxyl groups. No peak was observed between 2800 and 3000 cm^{-1} indicating the absence of organic components in the nanosilica following the combustion and extraction processes, as previously reported by Rafiee et al. [121]. The predominant absorbance peaks at 1057 cm^{-1} in BL spectrum and 1060 cm^{-1} in RH spectrum were assigned to siloxane bonds (Si–O–Si).

3.1.7 ICP-MS Analysis of Bamboo Leave and Rice Husk Without and with Acid Leaching

ICP-MS analysis was carried out for elemental determination in the samples. Figure 9 shows the bar chart of elemental compositions in BL without and with acid leaching using ICP-MS. Twenty-five elements were successfully detected. It is noted that the concentrations of all elements present in the acid-leached sample were below that of the present in a sample without acid leaching except for two elements: Be and Th due to incomplete removed during the acid leaching process. The results indicate that the acid leaching process has removed most of the impurities and metallic elements present in the samples.

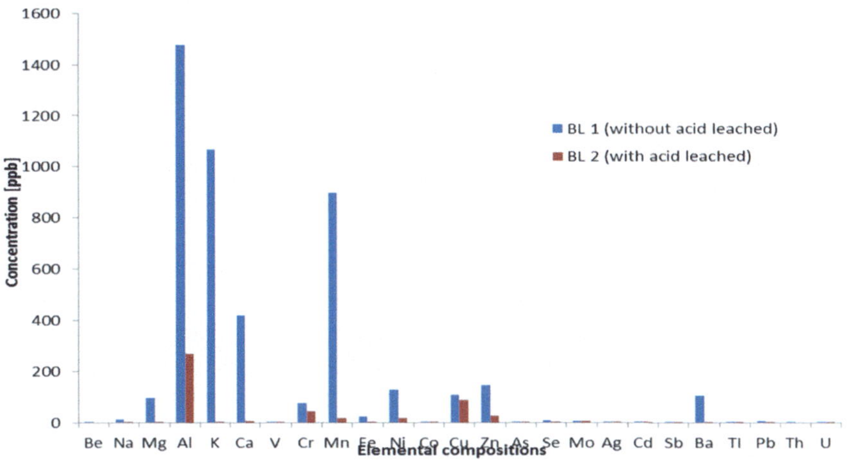

Fig. 9 Elemental compositions of untreated and acid-treated BL samples analyzed using ICP-MS

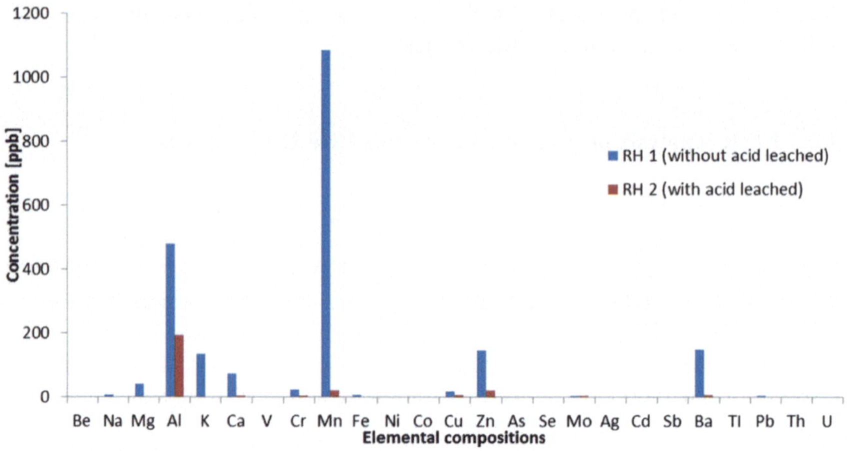

Fig. 10 Elemental compositions of untreated and acid-treated RH samples analyzed using ICP-MS

Figure 10 shows the bar chart of elemental compositions in RH without and with acid leaching using ICP-MS. Twenty-two elements were detected, and the concentrations of all elements present in the acid-leached sample were below that of the present in a sample without acid leaching except for Be elements. Similarly to BL samples, the acid leaching process has shown to remove the impurities and metallic elements that present in RH samples.

3.2 Development of Fresh Latent Fingermarks Using Nanosilica

Depletion series studies were carried out in this research to determine the relative sensitivity of the tested BL and RH nanosilica powders for developing latent fingermarks on the specific surfaces. In depletion studies, a series of fingermarks were deposited by the contact of the same finger with the surface. The method for the development of latent fingermarks is sensitive if the depletion series progressively continues to develop fingermarks [133].

Split fingermarks study was performed to minimize the potential impact of the variability in fingermark composition from the same donor. This method allows a direct comparison between each part of fingermarks as it eliminates uncontrolled variables such as deposition pressure if individual fingermark is deposited. All samples for depletion series were carried out in triplicates. Figure 11a–e represent the images of split developed fingermarks in depletion series on black painted car metal plate. All tested powders developed good quality fingermarks in depletions 1 until 5. It is noted that the quality of the developed fingermarks degraded from

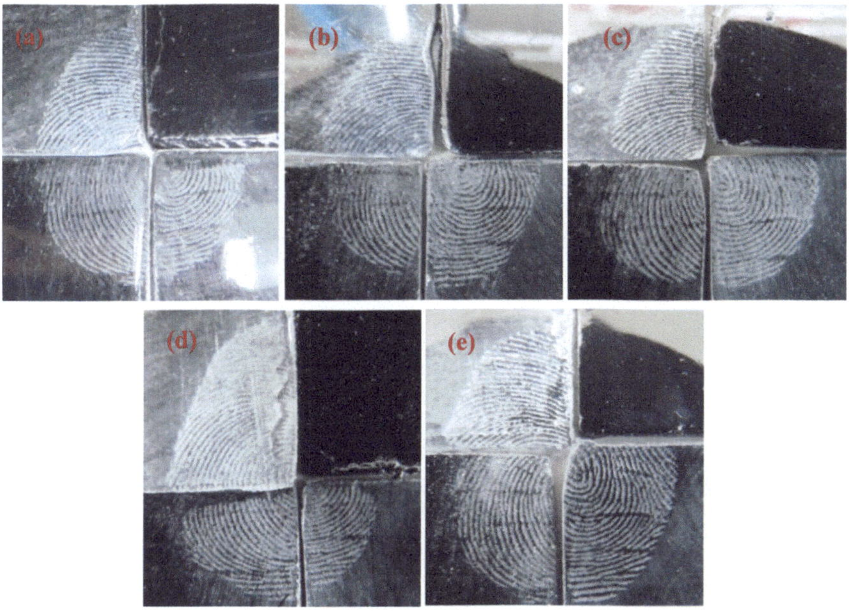

Fig. 11 Developed fingermarks on black painted car metal plate for depletion 1 (**a**), depletion 2 (**b**), depletion 3 (**c**), depletion 4 (**d**), and depletion 5 (**e**)

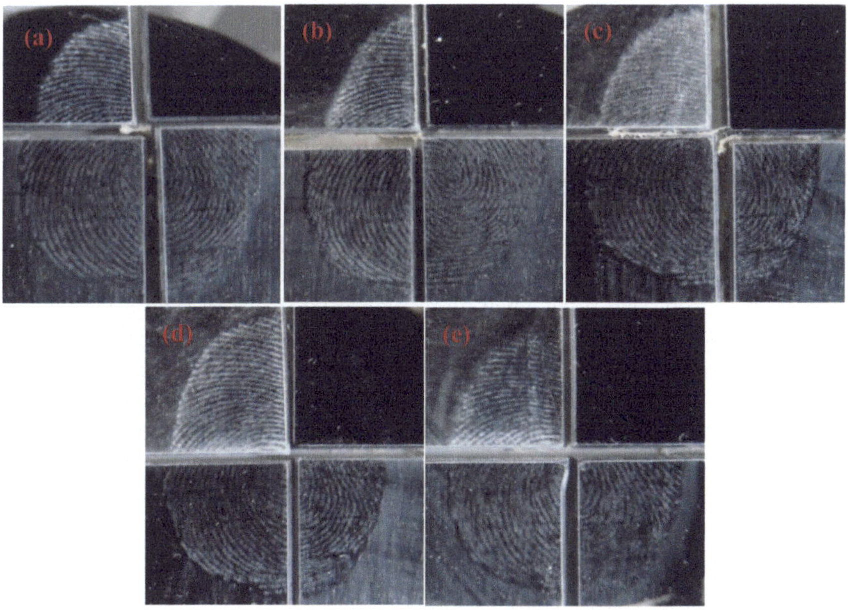

Fig. 12 Developed fingermarks on car plate number for depletion 1 (**a**), depletion 2 (**b**), depletion 3 (**c**), depletion 4 (**d**), and depletion 5 (**e**)

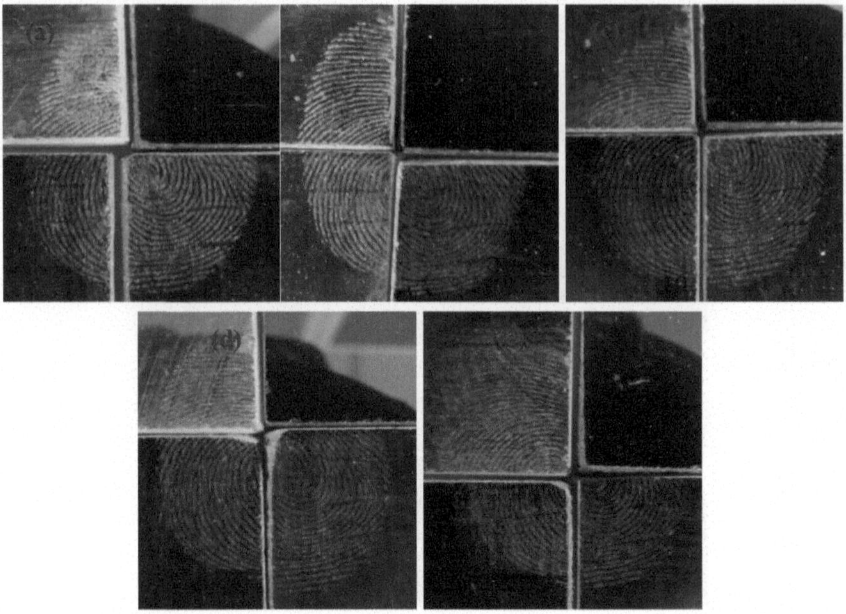

Fig. 13 Developed fingermarks on black glass for depletion 1 (**a**), depletion 2 (**b**), depletion 3 (**c**), depletion 4 (**d**), and depletion 5 (**e**)

depletion 1 to depletion 5 due to the decrease of fingermark residue on the surface of the substrate. The loss of ridge detail, however, was not prominent.

Figure 12a–e illustrates the images of the depletion series of split developed fingermarks on plastic car plate numbers. Similarly to the development on painted car metal, all three powders produced good and clear ridge detail of fingermarks in depletions 1 until 4, but for depletion 5, the fingermark developed appeared as smudges.

The images of the depletion series of split developed fingermarks on the black glass are shown in Fig. 13a–e. White powder, RH nanosilica, and BL nanosilica powders produced clear ridge detail of developed fingermarks in all depletions.

Figure 14a–e illustrates the images of the depletion series of split developed fingermarks on a black tile. All tested powders successfully developed excellent clarity of fingermarks, of which quality had decreased from depletions 1–5.

Figure 15a–e shows the images of the split depletion series of developed fingermarks on plastic file. NSPs developed good quality fingermarks in comparison with commercially white fingermark powder.

Figure 16a–e illustrates the images of the depletion series of split developed fingermarks on glossy paper. It is noted that fingermarks of poor quality were developed using all three powder for all depletions. Faint ridge detail was observed but hardly identified. This was due to the less amount of residue left on the surface that led to a low amount of powders adhered to it. Glossy paper is a semi-porous

Fig. 14 Developed fingermarks on black tiles for depletion 1 (**a**), depletion 2 (**b**), depletion 3 (**c**), depletion 4 (**d**), and depletion 5 (**e**)

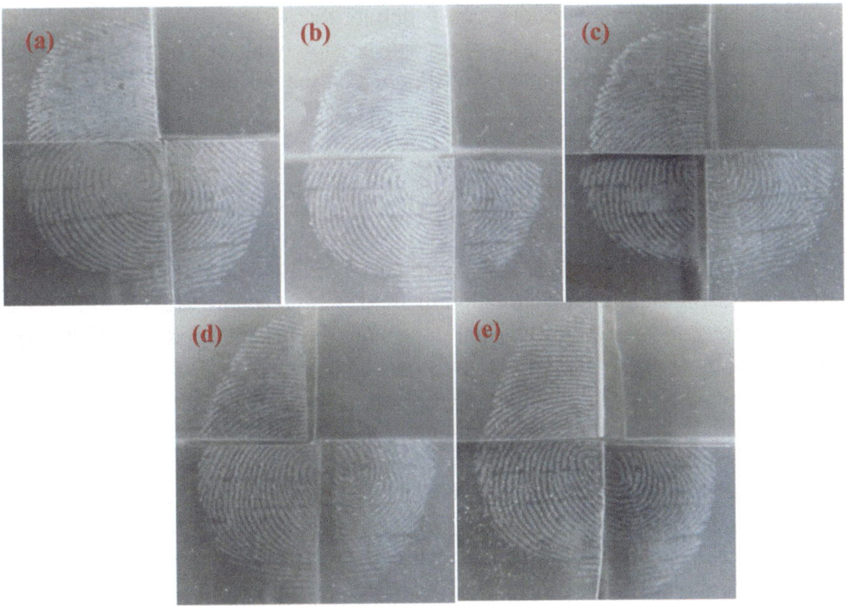

Fig. 15 Developed fingermarks on plastic files for depletion 1 (**a**), depletion 2 (**b**), depletion 3 (**c**), depletion 4 (**d**), and depletion 5 (**e**)

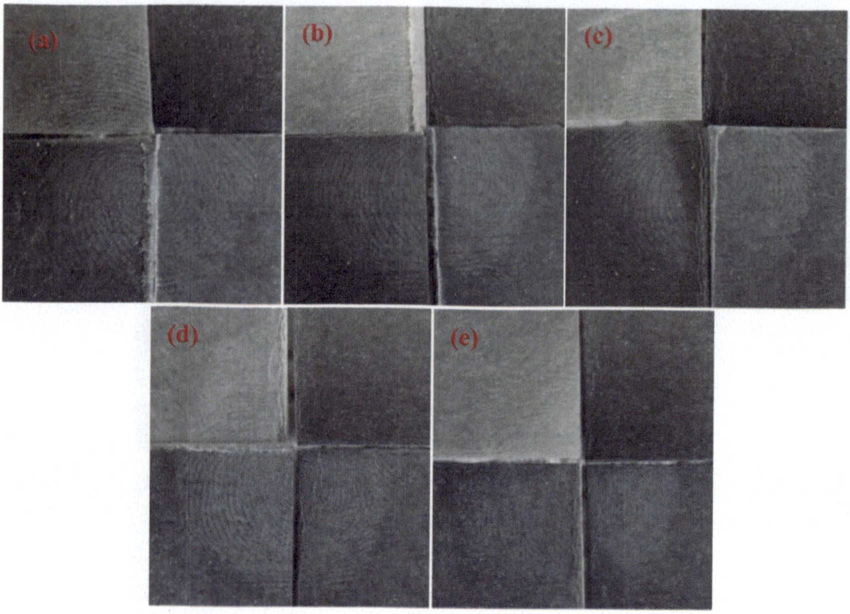

Fig. 16 Developed fingermarks on glossy paper for depletion 1 (**a**), depletion 2 (**b**), depletion 3 (**c**), depletion 4 (**d**), and depletion 5 (**e**)

surface; hence, fingermark residue can penetrate the surface after the deposition. The water-soluble deposit in the residue will be slowly absorbed after deposition while the water-insoluble residue retains longer on the surface.

4 Conclusions

Fine BL and RH nanosilica particles were successfully synthesized from agricultural wastes using a simple precipitation method. Acid leaching of BL and RH was performed to remove impurities and to obtain highly purified silica powder. Thermal treatment of BL and RH samples by calcining at 700 °C for 5 h in furnace yielded BLA and RHA containing high-purity silica. Amorphous nanosilica in spherical form was obtained with the addition of acetone prior to precipitation process with acetic acid. Nanosilica produced from BL and RH was successfully characterized using FESEM-EDX and FTIR spectroscopy. The FESEM images showed that the spherical form of nanosilica was synthesized from both samples with an average size of 50–450 nm. A high percentage of nanosilica powders were extracted from both wastes. ICP-MS analysis demonstrated result that the elemental composition in acid-leached samples was lower than that of without acid leaching.

This indicates that the acid leaching process has removed the impurities and metallic elements present in the samples.

Depletion series of split fingermarks studies were carried out to minimize the potential impact of the variability in fingermark composition from the same donor. The quality of developed fingermarks using three powders (white powder, BL, and RH nanosilica powders) decreased from depletions 1 to depletion 5 due to the decrease in fingermark residue. In conclusion, nanosilica synthesized from BL and RH has great potential as effective and low-cost fingermark powder for developing latent fingermarks in forensic investigation.

5 Future Perspectives

Studies on the synthesis of nanosilica from agricultural wastes should be explored due to the wide applications of nanosilica using cost-effective and eco-friendly products. Alternatively, other agricultural waste products that have been reported to be high in silica contents such as groundnut shell and sugarcane bagasse can be used as sources to synthesize nanosilica for fingermark detection materials.

In this study, the synthesis of nanosilica was carried out by precipitation method. It is thus recommended that other methods shall be conducted, such as the sol–gel method. This method is a well-known technique for the synthesis of silica with uniform, small particle sizes, and varied morphologies. The method involves the transition of a system from liquid sol into a solid gel phase, and it can be divided into five steps, which are forming a solution, gelation, ageing, drying, and densification.

The application of a variety of fluorescent dyes such as basic red 28, basic yellow 40, methylene blue, and rhodamine B should be studied by incorporating them in order to produce luminescent nanosilica. Luminescent nanosilica can also be functionalized chemically to recognize specifically the fingermark residue and produce a luminescent image of developed fingermark.

Other fingermark secretions such as eccrine and normal/natural secretions can be employed during the deposition of latent fingermark in order to test the effectiveness of the development using nanosilica on a wide variety of secretions in fingermark deposit.

References

1. Daluz HM (2015) Fundamentals of fingerprint analysis, 1st ed. CRC Press, Boca Raton FL
2. Houck MM (2016) Forensic fingerprints, 1st ed. Elsevier, London
3. Hawthorne MR (2009) Fingerprints: analysis and understanding, 6 ed. CRC Press, Boca Raton, pp 142–142
4. Saferstein R, Graf SL (2001) Evaluation of a reflected ultraviolet imaging system for fingerprint detection. J Forensic Ident 51(4):385–393

5. Stigler SM (1995) Galton and identification by fingerprints. Genetics 140(3):857–860
6. Tilstone WJ, Savage KA, Clark LA (2006) Forensic science, an encyclopedia of history, method and techniques, 1st ed. ABC-CLIO Corporate, Santa Barbara, pp 307–307
7. Haan PV-D (2006) Physics and fingerprints. Contemp Phys 47(4):209–230
8. Exline DL, Wallace C, Roux C, Lennard C, Nelson MP, Treado PJ (2003) Forensic applications of chemical imaging: Latent fingerprint detection using visible absorption and luminescence. J Forensic Sci 48(5):1047–1053
9. Lee HC, Gaensslen RE (2001) Advance in fingerprint technology, 2nd ed. CRC Press, Boca Raton, pp 105–175
10. Caplan RM (1990) How fingerprints came into use for personal identification. J Am Acad Dermatol 23(1):109–114
11. Neumann C, Champod C, Yoo M, Genessay T, Langenburg G (2015) Quantifying the weight of fingerprint evidence through the spatial relationship, directions and types of minutiae observed on fingermarks. Forensic Sci Int 248:154–171
12. Gutiérrez E, Galera V, Martínez JM, Alonso C (2007) Biological variability of the minutiae in the fingerprints of a sample of the Spanish population. Forensic Sci Int 172(2-3):98–105
13. Asano KG, Bayne CK, Horsman KM, Buchanan MV (2002) Chemical composition of fingerprints for gender determinantion. J Forensic Sci 47(4):805–807
14. Jelly R, Patton ELT, Lennard C, Lewis SW, Lim KF (2009) The detection of latent fingermarks on porous surfaces using amino acid sensitive reagents: a review. Anal Chim Acta 652(1-2):128–142
15. Frick AA, Fritz P, Lewis SW (2015) Chemistry of print residue, 2nd ed., vol 2. Elsevier, pp 351–356
16. Carey FA, Giuliano RM (2011) Organic chemistry, 8th ed. Mc-Graw Hill, New York
17. Mathews CK, Appling DR, Anthony-Cahill SJ, van Holde KE (2013) Biochemistry, 4th ed. Pearson, Ontario, Canada
18. Champod C, Lennard C, Margot P, Stoilovic M (2004) Fingerprints and other ridge skin impressions, 1 ed. CRC Press, Boca Raton, pp 1–10
19. Girod A, Ramotowski R, Weyermann C (2012) Composition of fingermark residue: a qualitative and quantitative review. Forensic Sci Int 223(1-3):10–24
20. Weyermann C, Roux C, Champod C (2011) Initial results on the composition of fingerprints and its evolution as a function of time by GC/MS analysis. J Forensic Sci 56(1):102–108
21. Ramotowski R (2001) Composition of latent print residue.CRC Press, Boca Raton, pp 63–104
22. West MJ, Went MJ (2008) The spectroscopic detection of exogenous material in fingerprints after development with powders and recovery with adhesive lifters. Forensic Sci Int 174 (1):1–5
23. Croxton RS, Baron MG, Butler D, Kent T, Sears VG (2010) Variation in amino acid and lipid composition of latent fingerprints. Forensic Sci Int 199(1-3):93–102
24. Groeneveld G, de Puit M, Bleay S, Bradshaw R, Francese S (2015) Detection and mapping of illicit drugs and their metabolites in fingermarks by MALDI MS and compatibility with forensic techniques. Sci Rep 5:11716
25. Jones BJ, Reynolds AJ, Richardson M, Sears VG (2010) Nano-scale composition of commercial white powders for development of latent fingerprints on adhesives. Sci Justice 50(3):150–155
26. Sodhi GS, Kaur J (2005) The forgotten Indian pioneers of fingerprint science. Curr Sci 88 (1):185–191
27. Goode GC, Morris JR (1983) Latent fingerprints: a review of their origin, composition and methods for detection. Forensic Sci Int 25(2):1–137
28. Friesen JB (2015) Forensic chemistry: the revelation of latent fingerprints. J Chem Educ 92 (3):497–504
29. James JD, Wilshire B (1993) New powder metallurgy techniques for fingerprint detection. Key Eng Mater 86–87:11–16

30. Gürbüz S, Özmen Monkul B, Ipeksaç T, Gürtekin Seden M, Erol M (2015) A systematic study to understand the effects of particle size distribution of magnetic fingerprint powders on surfaces with various porosities. J Forensic Sci 60(3):727–736
31. Connatser RM, Prokes SM, Glembocki OJ, Schuler RL, Gardner CW, Lewis SA, Lewis LA (2010) Toward surface-enhanced raman imaging of latent fingerprints. J Forensic Sci 55 (6):1462–1470
32. Chadwick S, Maynard P, Kirkbride P, Lennard C, McDonagh A, Spindler X, Roux C (2012) Styryl dye coated metal oxide powders for the detection of latent fingermarks on non-porous surfaces. Forensic Sci Int 219:208–214
33. Ashish BNK (2015) Efficacy of Robin® powder blue for latent fingerprint development on various surfaces. Egyptian J Forensic Sci 5(4):166–173
34. Shah SNA, Shah Z, Hussain M, Khan M (2017) Hazardous effects of titanium dioxide nanoparticles in ecosystem. Bioinorg Chem Appl 2017:1–12
35. Mor S, Manchanda CK, Kansal SK, Ravindra K (2017) Nanosilica extraction from processed agricultural residue using green technology. J Clean Prod 143:1284–1290
36. Horikoshi S, Serpone N (2013) Introduction to nanoparticles, pp 1–24
37. Wilshire B (1996) Advance in fingerprint detection. Endeavour 20(1):12–15
38. Arshad A, Farrukh MA, Ph D, Ali S (2015) Development of latent fingermarks on various surfaces using ZnO–SiO2. J Forensic Sci 60(5):1182–1187
39. Mishra P, Chakraverty A, Banerjee HD (1985) Production and purification of silicon by calcium reduction of rice-husk white ash. Journal of Materials Science 20:4387–4391
40. Becue A, Cantu AA (2012) Fingermark detection using nanoparticles. CRC Press, Boca Raton, pp 307–79
41. Reip A, Withnall R, Silver J (2010) A study of silica coated phosphors for latent fingerprint visualisation. AIP Conf Proc 1267:531–533
42. Rajan R, Zakaria Y, Shamsuddin S, Nik Hassan NF (2019) Fluorescent variant of silica nanoparticle powder synthesised from rice husk for latent fingerprint development. Egyptian J Forensic Sci 9(1):50
43. Moret S, Bécue A, Champod C (2014) Nanoparticles for fingermark detection: an insight into the reaction mechanism. Nanotechnology 25(42):425502
44. Bouvy C, Marine W, Su B-L (2007) ZnO/mesoporous silica nanocomposites prepared by the reverse micelle and the colloidal methods: photoluminescent properties and quantum size effect. Chem Phys Lett 438:67–71
45. Skelton R (2011) A survey of the Forensic Sciences. Lulu.com, Raleigh, pp 173–182
46. Mittal D (1997) Silica from ash. Resonance 2(7):64–66
47. Kalapathy U, Proctor A, Shultz J (2000) A simple method for production of pure silica from rice hull ash. Biores Technol 73:257–262
48. Jal PK, Sudarshan M, Saha A, Patel S, Mishra BK (2004) Synthesis and characterization of nanosilica prepared by precipitation method. Colloids Surf A Physicochem Eng Asp 240:173–178
49. Shen Y, Zhao P, Shao Q (2014) Porous silica and carbon derived materials from rice husk pyrolysis char. Microporous Mesoporous Mater 188:46–76
50. Bansal V, Ahmad A, Sastry M (2006) Fungus-mediated biotransformation of amorphous silica in rice husk to nanocrystalline silica. J Am Chem Soc 128(43):14059–14066
51. Balasooriya ER, Jayasinghe CD, Jayawardena UA, Ruwanthika RWD, De Silva RM, Udagama PV (2017) Honey mediated green synthesis of nanoparticles: new era of safe nanotechnology. J Nanomater 2017:1–10
52. Merza KS, Al-Attabi HD, Abbas ZM, Yusr HA (2012) Comparative study on methods for preparation of gold nanoparticles. Green Sustain Chem 2:26–28
53. Polte J (2015) Fundamental growth principles of colloidal metal nanoparticles—a new perspective. CrystEngComm 17:6809–6830
54. Jana NR, Gearheart L, Murphy CJ (2001) Seed-mediated growth approach for shape-controlled synthesis of spheroidal and rod-like gold nanoparticles using a surfactant template. Adv Mater 13(18):1389–1393

55. Rahman IA, Padavettan V (2012) Synthesis of silica nanoparticles by sol-gel: size-dependent properties, surface modification, and applications in silica-polymer nanocomposites—a review. J Nanomater 2012:132424
56. Sharma V, Park K, Srinivasarao M (2009) Colloidal dispersion of gold nanorods: Historical background, optical properties, seed-mediated synthesis, shape separation and self-assembly. Mater Sci Eng R: Rep 65:1–38
57. Abid JP, Wark AW, Brevet PF, Girault HH (2002) Preparation of silver nanoparticles in solution from a silver salt by laser irradiation. ChemComm:792–793
58. Shekar S, Sander M, Riehl RC, Smith AJ, Braumann A, Kraft M (2012) Modelling the flame synthesis of silica nanoparticles from tetraethoxysilane. Chem Eng Sci 70:54–66
59. Adam F, Chew T-S, Andas J (2011) A simple template-free sol–gel synthesis of spherical nanosilica from agricultural biomass. J Sol-Gel Sci Technol 59:580–583
60. Hanafi S (1980) Surface properties of silicas produced by thermal treatment of rice-husk ash. Thermochim Acta 37:137–143
61. James J, Rao MS (1986) Silica from rice husk through thermal decomposition. Thermochim Acta 97:329–336
62. Coradin T, Boissiere M, Livage J (2006) Sol-gel chemistry in medicinal science. Curr Med Chem 13(1):99–108
63. Milea CA and Bogatu C (2011) The influence of parameters in silica sol-gel process. Bull Transilvania Univ Brasov Eng Sci 4(1):59–66
64. Beganskiene A, Sirutkaitis V, Kurtinaitiene M, Juskenas R, Kareiva A (2004) FTIR, TEM and MNR investigations of Stober silica nanoprticles. Mater Sci 10(4):287–290
65. Estella J, Echeverria JC, Laguna M, Garrido JJ (2007) Effects of aging and drying conditions on the structural and textural properties of silica gels. Microporous Mesoporous Mater 102:274–282
66. Mohanta K, Kumar D, Parkash O (2012) Properties and industrial applications of rice husk: a review. Int J Emerg Technol Adv Eng 2(10):86–90
67. Rajanna SK, Kumar D, Vinjamur M, Mukhopadhyay M (2015) Silica aerogel microparticles from rice husk ash for drug delivery. Ind Eng Chem Res 54:949–956
68. Tabatabaei S, Shukohfar A, Aghababazadeh R, Mirhabibi A (2006) Experimental study of the synthesis and characterisation of silica nanoparticles via the sol-gel method. J Phys: Conf Ser 26:371–374
69. Park SK, Kim KD, Kim HT (2002) Preparation of silica nanoparticles: determination of the optimal synthesis conditions for small and uniform particles. Colloids Surf A Physicochem Eng Asp 197:7–17
70. Zaky RR, Hessien MM, El-Midany AA, Khedr MH, Abdel-Aal EA, El-Barawy KA (2008) Preparation of silica nanoparticles from semi-burned rice straw ash. Powder Technol 185 (1):31–35
71. Kow K-W, Yusoff R, Aziz ARA, Abdullah EC (2014) From bamboo leaf to aerogel: preparation of water glass as a precursor. J Non-Crystalline Solids 386:76–84
72. Schramm LL (2014) Emulsions, foams, suspensions, and aerosols: microscience and applications, 2nd ed. Wiley, Weinheim
73. Hessien MM, Rashad MM, Zaky RR, Abdel-Aal EA, El-Barawy KA (2009) Controlling the synthesis conditions for silica nanosphere from semi-burned rice straw. Mater Sci Eng, B 162:14–21
74. Kongmanklang C, Rangsriwatananon K (2014) Hydrothermal synthesis of high crystalline silicalite from rice husk ash. J Spectro 2015:1–5
75. Ghorbani F, Sanati AM, Maleki M (2015) Production of silica nanoparticles from rice husk as agricultural waste by environmental friendly technique. Environ Stud Persian Gulf 2 (1):56–65
76. Jembere AL, Fanta SW (2017) Studies on the synthesis of silica powder from rice husk ash as reinforcement filler in rubber tire tread part: Replacement of commercial precipitated silica. Int J Mater Sci Appl 6(1):37–44

77. Tungkananurak K, Kerdsiri S, Jadsadapattarakul D, Burns DT (2007) Semi-micro preparation and characterization of mesoporous silica microspheres from rice husk sodium silicate using a non-ionic surfactant as a template: application in normal phase HPLC columns. Microchim Acta 159:217–222
78. Stanley R, Nesaraj A (2014) Effect of surfactants on the wet chemical synthesis of silica nanoparticles. J Eng Appl Sci 12(1):9–21
79. Nittaya T, Apinon N (2008) Synthesis and characterization of nanosilica from rice husk ash prepared by precipitation method. J Nat Sci 7(1):59–65
80. Premaratne WAPJ, Priyadarshana WMGI, Gunawardena SHP, De Alwis AAP (2013) Synthesis of nanosilica from paddy husk ash and their surface functionalization. J Sci Univ Kelaniya 8:33–48
81. Pineda T, Soares CHL, Hotza D, Casas-Botero AE, Ramírez-Carmona M, Torres-Taborda M (2012) Extracellular synthesis of silica oxide particles by Fusarium oxysporum from rice husk ash. Mater Sci Forum 727–728:1153–1157
82. Fouad OA, Makhlouf SA, Ali GAM, El-Sayed AY (2011) Cobalt/silica nanocomposite via thermal calcination-reduction of gel precursors. Mater Chem Phys 128(1–2):70–76
83. Ali GAM, Fouad OA, Makhlouf SA, Yusoff MM, Chong KF (2014) Co_3O_4/SiO_2 nanocomposites for supercapacitor application. J Solid State Electrochem 18(9):2505–2512
84. Murdock RC, Braydich-Stolle L, Schrand AM, Schlager JJ, Hussain SM (2008) Characterization of nanomaterial dispersion in solution prior to in vitro exposure using dynamic light scattering technique. Toxicol Sci 101(2):239–253
85. Zhou W, Apkarian RP, Wang ZL, Joy D (2007) Fundamentals of scanning electron microscopy. In: Zhou W, Wang ZL (eds) Scanning microscopy for nanotechnology. Springer, New York, pp 1–40
86. Egerton RF (2016) Physical principles of electron microscopy: an introduction to TEM, SEM, and AEM, 2nd ed. Springer, New York, pp 1–196
87. Als-Nielsen J, McMorrow D (2001) Elements of modern X-ray physics, 2nd ed. Wiley, Hoboken
88. Schneider R (2011) Energy-Dispersive X-Ray Spectroscopy (EDXS). In: Friedbacher G, Bubert H (eds). Wiley, Weinheim
89. Eckert M (2012) Max von Laue and the discovery of X-ray diffraction in 1912. Ann Phys 524(5):83–85
90. Khan I, Saeed K, Khan I (2017) Nanoparticles: properties, applications and toxicities. Arab J Chem: 1–24
91. Fouad OA, Ali GAM, El-Erian MAI, Makhlouf SA (2012) Humidity sensing properties of cobalt oxide/silica nanocomposites prepared via sol-gel and related routes. Nano 7(5):1250038
92. Ali GAM, Fouad OA, Makhlouf SA (2013) Structural, optical and electrical properties of sol-gel prepared mesoporous Co3O4/SiO2 nanocomposites. J Alloy Compd 579:606–611
93. Abdel Ghafar HH, Ali GAM, Fouad OA, Makhlouf SA (2015) Enhancement of adsorption efficiency of methylene blue on Co_3O_4/SiO_2 nanocomposite. Desalin Water Treat 53 (11):2980–2989
94. Ali GAM, Fouad Osama A, Makhlouf Salah A (2016) Electrical properties of cobalt oxide/ silica nanocomposites obtained by sol-gel technique. Am J Eng Appl Sci 9(1):12–16
95. Yoshida S, Onishi Y, Kitagishi K (1959) Soil science and plant nutrition the chemical nature of silicon in rice plant. Soil Plant Food 5(1):23–27
96. Lu PHY (2012) Highly pure amorphous silica nano-disks from rice straw. Powder Technol 225:149–155
97. Habeeb OA, Ramesh K, Ali GAM, Yunus RM (2017) Experimental design technique on removal of hydrogen sulfide using CaO-eggshells dispersed onto palm kernel shell activated carbon: experiment, optimization, equilibrium and kinetic studies. J Wuhan Univ Technol-Mater Sci Ed 32(2):305–320
98. Habeeb OA, Ramesh K, Ali GAM, Yunus RM (2017) Low-cost and eco-friendly activated carbon from modified palm kernel shell for hydrogen sulfide removal from wastewater: adsorption and kinetic studies. Desalin Water Treat 84:205–214

99. Ali GAM, Habeeb OA, Algarni H, Chong KF (2018) CaO impregnated highly porous honeycomb activated carbon from agriculture waste: symmetrical supercapacitor study. J Mater Sci 54:683–692
100. Vaibhav V, Vijayalakshmi U, Roopan SM, Mohana Roopan S (2015) Agricultural waste as a source for the production of silica nanoparticles. Spectrochim Acta Part A Mol Biomol Spectrosc 139:515–520
101. Sivasubramanian SKS (2015) Synthesis and characterisation of silica nano particles from coconut shell. Int J Pharma Bio Sci 6(1):530–536
102. Aminullah, Rohaeti E, Irzaman (2015) Reduction of high purity silicon from bamboo leaf as basic material in development of sensors manufacture in satellite technology. Proc Environ Sci 24:308–316
103. Hegde G, Abdul Manaf SA, Kumar A, Ali GAM, Chong KF, Ngaini Z, Sharma KV (2015) Biowaste sago bark based catalyst free carbon nanospheres: waste to wealth approach. ACS Sustain Chem Eng 5(9):2247–2253
104. Ali GAM, Manaf SAA, Kumar A, Chong KF, Hegde G (2016) Superior supercapacitive performance in porous nanocarbons. J Energy Chem 25(4):734–739
105. Napierska D, Thomassen LCJ, Lison D, Ja Martens, Hoet PH (2010) The nanosilica hazard: another variable entity. Part Fibre Toxicol 7(1):39
106. Kamath SR, Proctor A (1998) Silica gel from rice hull ash: preparation and characterization. Cereal Chem 75(4):484–487
107. Liu N, Huo K, McDowell MT, Zhao J, Cui Y (2013) Rice husks as a sustainable source of nanostructured silicon for high performance Li-ion battery anodes. Sci Rep 3(1919):1–7
108. Chakraverty A, Mishra P, Banerjee HD (1988) Investigation of combustion of raw and acid-leached rice husk for production of pure amorphous white silica. J Mater Sci 23(1):21–24
109. Umemura M, Takenaka C (2014) Biological cycle of silicon in moso bamboo (Phyllostachys pubescens) forests in central Japan. Ecol Res 29(3):501–510
110. Meng-Hao D, Jin-Ping Z, Jing-Wen W, Li-Song H, Wei-Fang S (2012) A study on the extraction technology of bioactive silicon from bamboo leaves. Electr J Environ Agric Food Chem 11(1):23–31
111. Currie HA, Perry CC (2007) Silica in plants: biological, biochemical and chemical studies. Ann Bot 100(7):1383–1389
112. Wang W, Martin JC, Fan X, Han A, Luo Z, Sun L (2012) Silica nanoparticles and frameworks from rice husk biomass. ACS Appl Mater Interfaces 4:977–981
113. Wang L, Gao B, Peng C, Peng X, Fu J, Chu PK, Huo K (2015) Bamboo leaf derived ultrafine Si nanoparticles and Si/C nanocomposites for high-performance Li-ion battery anodes. Nanoscale 7(33):13840–13847
114. Marin DC, Vecchio A, Ludueña LN, Fasce D, Alvarez VA, Stefani PM (2015) Revalorization of rice husk waste as a source of cellulose and silica. Fibers Polym 16 (2):285–293
115. Carmona VB, Oliveira RM, Silva WTL, Mattoso LHC, Marconcini JM (2013) Nanosilica from rice husk: extraction and characterization. Ind Crops Prod 43(1):291–296
116. Houston DF (1972) Rice: chemistry and technology. American Association of Cereal Chemists. Monograph Series. American Association of Cereal Chemists, St Paul, pp 517–517
117. Park B-D, Wi SG, Lee KH, Singh AP, Yoon T-H, Kim YS (2003) Characterization of anatomical features and silica distribution in rice husk using microscopic and micro-analytical techniques. Biomass Bioenergy 25:319–327
118. Sun L, Gong K (2001) Silicon-based materials from rice husks and their applications. Ind Eng Chem Res 40:5861–5877
119. Krishnarao RV, Godkhindi MM (1992) Distribution of silica in rice husks and its effect on the formation of silicon carbide. Ceram Int 18(4):243–249
120. Abu Bakar BHPR, Abdulaziz H (2012) Malaysian rice husk ash—improving the durability and corrosion resistance of concrete: pre-review. Concr Res Lett 1(1):6–13

121. Rafiee E, Shahebrahimi S, Feyzi M, Shaterzadeh M (2012) Optimization of synthesis and characterization of nanosilica produced from rice husk (a common waste material). Int Nano Lett 2(1):29
122. Della VP, Kühn I, Hotza D (2002) Rice husk ash as an alternate source for active silica production. Mater Lett 57(4):818–821
123. Nakata Y, Suzuki M, Okutani T, Akiyama T (1989) Preparation and properties of SiO_2 from rice Hulls. J Ceram Soc Jpn 97(8):842–849
124. Yalçin N, Sevinç V (2001) Studies on silica obtained from rice husk. Ceram Int 27(2):219–224
125. Real C, Alcalá MD, Criado JM (1996) Preparation of silica from rice husks. J Am Ceram Soc 79(8):2012–2016
126. Wu X (1996) Studies on the hydrothermal conditions of the extraction of high-purity silica from rice husks. Bull Chin Ceram Soc 15(4):36
127. Yu XW, Xu GH, Zhou YY, Zhao GP, Shang SN (1996) A new technology pf producing high quality white carbon black with rice husk, rice straw and wheat straw. Bull Chin Ceram Soc 15(3):48
128. Lu FY, Lu AJ (1994) Study on preparation of precipitated silica from ammonium hydrogen. J Chem Fertiliser Ind 21(4):53
129. Mohammed NIA, Adam H, Dasmawati M, Abdul RI (2012) A simple method of obtaining spherical nanosilica from rice husk. Int J Adv Sci Eng Inf Technol 2:141–143
130. Thuadaij N, Nuntiya A (2008) Preparation of nanosilica powder from rice husk ash by precipitation method. Chiang Mai J Sci 35(1):206–211
131. Selvakumar KV, Umesh A, Ezhilkumar P, Gayatri S, Vinith P, Vignesh V (2014) Extraction of silica from burnt paddy husk. Int J ChemTech Res 6(9):4455–4459
132. Ibrahim IAM, Zikry AAF, Sharaf MA (2010) Preparation of spherical silica nanoparticles: Stober silica. J Am Sci 6(11):985–989
133. Fieldhouse S (2011) Consistency and reproducibility in fingermark deposition. Forensic Sci Int 201(1):96–100

Recycling of Nanosilica from Agricultural, Electronic, and Industrial Wastes for Wastewater Treatment

Tarek A. Seaf El-Nasr, Hassanien Gomaa, Mohammed Y. Emran, Mohamed M. Motawea, and Abdel-Rahman A. M. Ismail

Abstract Water pollutants are detrimental to human life. High doses of organic or inorganic toxins in drinking water may be responsible for various diseases such as kidney maladies, nervous order disturbances. Water pollution can be arising from urban activities such as industrialization and mining or naturalistic actions such as biological conversions and geological denudation of earth. Treating water and wastewater containing toxicants via an economical and straightforward process is crucial for sustaining humanbeing healthcare and the diverse ecosystems. Currently, controlled removal of hazardous species in water sources, either organic or inorganic, using secure and operative methods still represented great defiance. Therefore, the development of cost-effective and rapid techniques to remove pollutants from the contaminated water was reported. This chapter outlines the latest studies for the preparation of nanosilica (NS) from agricultural, electronic, and industrial waste for the removal of various toxins from polluted water through adsorption strategy. The waste-derived NS can be modified through physical or chemical processes to get highly efficient and economic features for environmental usage. Removal of organic and inorganic pollutant species from polluted water sources using efficient and cost-effective waste-derived NS was ascertained. The NS obtained from natural and urban waste through interrogation of the recently published data, without generation of secondary waste-contaminated water, leads to reduce the risk of water pollution for tomorrow. Furthermore, factors that may promote the performance of waste-derived NS toward the removal of pollutants are summarized, e.g., pH, contact time, temperature, particle size, surface activity, and porosity.

Keywords Organic pollutants · Inorganic pollutants · Wastewater · Agriculture waste · Industrial waste · Electronic waste · Waste-derived nanosilica

T. A. Seaf El-Nasr
Faculty of Science, Department of Chemistry, Jouf University, Sakaka P.O. Box 2014 Aljouf, Saudi Arabia

T. A. Seaf El-Nasr · H. Gomaa (✉) · M. Y. Emran · M. M. Motawea · A.-R.A. M. Ismail
Faculty of Science, Department of Chemistry, Al-Azhar University, Assiut 71524, Egypt
e-mail: gomaa.h1989@gmail.com; h.gomaa@azhar.edu.eg

A. S. H. Makhlouf and G. A. M. Ali (eds.), *Waste Recycling Technologies for Nanomaterials Manufacturing*, Topics in Mining, Metallurgy and Materials Engineering, https://doi.org/10.1007/978-3-030-68031-2_12

List of Abbreviations

CTAB	Cetyl Trimethyl Ammonium Bromide
e-waste	Electronic waste
LCD	Liquid crystal display
NS	Nanosilica
RHA	Rice husk ash
WHO	World Health Organization

1 Introduction

Water is an extraordinarily important issue for the existence of earthly life. It is a source of life and imperative, but in the non-attendance of immaculate potable water, a massive group of humans would be suffering globally. Due to the quadrupled global population in the twentieth century, the request for clean water has increased by seven times for various human uses [1–3]. Additionally, the rapid rate of industrialization, population growth, the sudden urbanization has contributed to significant water and soil pollution. The main sources of water pollution are toxic industrial wastewater and agricultural waste runoff [4]. Many hazardous substances, such as heavy and radioactive metals, medicines, pesticides, dyes, phenols, and others that contaminate water resources, are considered environmentally harmful [5–9]. The existence of toxins in water, whether organic or inorganic pollutants, is a significant concern as many of these contaminants are extremely poisonous and can stack in living creatures, causing extreme illnesses like cancer. The high concentration of such pollutants in water can also cause severe lung and kidney problems [10, 11]. Many studies have been reported to find solutions to this problem by using highly efficient and cost-effective materials.

Over recent years, the recycling of agrarian, manufacturing, and electronic waste has become a hotbed of research. Therefore, it is also advantageous that most of the waste has a higher silica content, apart from concerns about environmental issues resulting from the disposal of this solid waste. Finally, as a potential candidate for silica synthesis, it is excellent to recycle this waste. The silica nanoparticles have demonstrated infinite potential in various industries, with an estimated demand of 8.8 billion by 2020 [12, 13]. The NS may be found in the form of crystalline, amorphous, and gel. Nanosilica has been used for various applications such as optical fibers, coating, or electronic thin films. Furthermore, it is widely utilized in chromatography, catalysts, ceramics, etc. Because elevated-purity NS is costly to employ in different manufacturing purposes, scientists are thus seeking to recover high-level-purity NS from waste to minimize the NS recovery cost [14, 15].

Nanomaterials are potentially used in polluted-water remediation. The nanomaterial's surface morphology of inner and external outwards and surface area with pores nature/distribution can directly influence the processes of wastewater remediation of capacity, selectivity, uptake, and stability. Nanomaterials have unique mechanical properties, such as high strength and stability, even during long-term use. Moreover, nanomaterials have also unique chemical properties such as the abundance of surface-active sites, species of these active groups, and the extent to which highly produce performance and activity. Existing of active groups on the nanomaterial surface leads to chemical and physical changes, which normally generate positive effects to achieve the target goal [16, 17]. Nanomaterials of different morphologies such as nanorods, nanosphere, and nanotubes have been prepared for wastewater treatment. Nanomaterial synthesis is a costly process. Therefore, developing a cost-effective process for nanomaterials synthesis from agricultural, electronic, and industrial waste have been generated during the last decades [18–22].

The large nanoparticle surface effectively contributes to increasing the number of active sites of the nanomaterial surface, which enhances the potential functionality. In the development of water treatment technology, nanomaterials demonstrate several significant properties which could be helpful in the elimination of contaminants from polluted water. Some nanosized products have been applied for adsorption of toxins existing in wastewaters, including metal oxides, zeolites, carbon-based nanoparticle, nanoclays, nanocomposites, etc. [9, 23–25].

Nanosilica (SiO_2), as metal oxide, is an inorganic material composed in a three-dimensional network structure, together with a common bridge atom, leading to form a porous construction having a massive superficial area [26, 27]. The NS has outstanding physical and chemical characteristics, including non-tumescence in water, stable even at high temperatures, and not poisonous. Moreover, silica materials provide excellent interaction with a wide variety of inorganic and organic species through the surface existence of silanol (Si–OH) groups [28, 29]. This chapter provides an overview of the pollution source and types, recovery strategies of NS from agrarian, manufacturing, and electronic waste, removing of organic and inorganic pollutants from wastewater using NS. The factors affecting the waste removal efficiency was also discussed in terms of NS's surface area and porosity and the performance of NS material during the adsorption process.

2 Sources of Water Pollution

Although water is the source of life on Earth, millions of people around the world do not have access to clean and fresh water due to the abundance of pollutants. A review by the World Health Organization (WHO) refers that around 2.6 billion persons have gained drinkable water resources until 2015. At the same time, 3.9 billion peoples will suffer from water deficiency areas by 2030. Global water demand doubles every 21 years due to rapid population growth and industrial activities [30]. So, the exploitation of a water security source is a global challenge for many countries, and many government organizations have given considerable attention to the development of cost-effective water treatment technology [31].

Fig. 1 Sources of water pollution, including point and non-point sources

Given the wide variety of sources polluting water, the sources of pollution can be split into two main sections, as shown in Fig. 1:

(i) Point source, which has a direct and specific source for pollution, including all liquid wastes from factories and refineries. The contaminants can reach the water from stored industrial water or underground crude oil storage system and other activities on the land surface when the sources of contamination above the water table [32].
(ii) Non-point source (also called widespread sources of contamination) in which pollutants are generated from various resources such as runoff from agrarian fields, acid rain, radioactive waste, oil spillage, and civilian regions. This type of pollution is so difficult to be controlled because there are sundry and non-identifiable routes by which toxins get in the surface-water or groundwater systems and then reach to the ecosystems [33–35].

Moreover, water can be contaminated by some chemical substances which found naturally in rocks and soils which act as natural resources of pollution; these pollutants include arsenic (As), iron (Fe), manganese (Mn), sulfates (SO_4^{2-}), chlorides (Cl^-), fluorides (F^-), etc. [33–36]. Now, we need to protect freshwater resources more than ever before through developing new water resources to meet the world's needs of healthy water, leading to the finding of innovative, cost-effective, and new technologies/protocols and made a vigorous effort to serve humankind by scientists.

2.1 Organic Pollutants

Dyes, fertilizers, pesticides, organo-halides, pharmaceutical ingredients, phenols, and surfactants, as listed in Table 1, are the main categories of organic contaminants in water. These products reached a hazardous level as the high demand for industrial production and the expansion of human activities. The organic pollutant removal gained significant interest due to various reasons as follows: (a) wide range applications in different industries, (b) long-period stability, and (c) high bio-gathering, leading to significant impacts on human health and different ecosystems. In the last decades, numerous researchers are intensifying their efforts to find solutions to the organic pollutants, which is known as its inability to decompose environmentally, as it causes serious damage to the health of humans. Organic pollutants leak into water mostly by humans because most of them are human-made chemicals using pesticides and residues for different chemical processes as well because of other industrial chemicals and cause many environmental problems because it lasts a long time. The greatest danger is the presence of these pollutants in the soil and water in their access to the biological food chain through several different methods. Natural (e.g., terpenoids, sugars, alkaloids, etc.) and synthetic organic pollutants are two categories of organic contaminants that can pollute the freshwater system. Laboratory-synthetic organic pollutants (e.g., dyes, phenols, etc.) are fabricated using various chemicals, which posteriorly drainage into the freshwater resources [37–41].

Persistent organic contaminates attract much attention because of the environmental damage it causes in the long run, while the emerging organic pollutants are a significant source of organic contamination for freshwater because it includes many

Table 1 Some hazardous effects of organic pollutants

Organic pollutants	Hazardous effect	References
Ciprofloxacin sulfamethoxazole	Endocrine systems damage	[44]
Direct dyes	Carcinogenic	[45]
Chlorobenzene	Nervous systems defect, carcinogenic	[46]
Toluene	Eyes excitement, harms of liver and kidney	[47]
Nitrobenzene	Nervous systems defect, carcinogenic	[48]
Benzene	Defect in the heart and brain functions	[49]
Phenol	Cardiovascular disease, carcinogenic	[50]
Metoprolol	Cardio and neuro-defects in watery creatures	[51]
Acid dyes	Skin excitement, respiratory insufficiency, eyes burning	[52]
Basic dyes	Carcinogenic, vomiting, skin irritation	[53]
4-Chlorophenol	Skin excitement, respiratory insufficiency, eyes burning	[54]
2-naphthol	Human blood system trouble and kidney illnesses	[55]

compounds such as pharma products, cosmetics, insecticides, animal medications, manufacturing by-products, food-industries waste, developed laboratory material, etc. [42]. Some old persistent organic contaminants (such as hexachlorocyclohexanes, dichlorodiphenyltrichloroethane, and polychlorinated biphenyls) were still repeatedly detected in the ecosystems, despite they were restricted and prohibited for many years. Regarding the concentration of the emerging organic pollutants (such as polybrominated diphenyl ethers, perfluorooctane sulfonate, and polycyclic aromatic hydrocarbons) will increase in the ecosystems with enhancing the industrial and population growth [43].

2.2 *Inorganic Pollutants*

Inorganic species such as weighty metal ions present a significant function in the essential physiological operations in living creatures, from microorganism to humans. Some of these metals, especially the poisonous weighty metals, have no role in the biological operations and are found in minimal quantities predominantly. These poisonous weighty metals such as iron (Fe), lead (Pb), cadmium (Cd), chromium (Cr), zinc (Zn), mercury (Hg), selenium (Se), nickel (Ni), aluminum (Al), copper (Cu), uranium (U), arsenic (As) at a high-level concentration are usually harmful to ecological systems [56, 57]. The major sources of inorganic contaminations are from electronic and untreated industrial waste, farming waste, sewage water, and erosion of rocks; therefore, treating these wastes is a global challenge. Not only soil pollution created from heavy metal as a result of water pollution but produced food can also be influenced where the growth of different crops in toxic metal-polluted regions leads to a mutation in the plant metabolism and biological functions, leading to metal aggregation, minimize biomass production, and lowering in plant outgrowth. Heavy metals are accumulation gradually from plant to animals and finally in human vital body organs [58–60]. People who are exposed to large amounts of toxic heavy metals may be infected with a wide variety of diseases such as hematic, kidney toxicity, cardiac issues, osteoporosis, cancer, etc. Due to the wide impact of heavy metals on human health, many countries have set standards for the permissible limit of heavy metals in food and water to avoid harmful effects. Many organizations, the most important of which is WHO, have issued periodic reports on the danger of these pollutants (i.e., heavy metals) to human and animal life, as explained in Table 2.

3 Waste as a Secondary Source of Nanosilica

3.1 *Agriculture Waste*

Agricultural waste can be defined as a waste generated from the diverse agrarian processes and farming actions, according to the United Nations. Agricultural wastes

Table 2 Hazardous effect of some inorganic pollutant species

Inorganic pollutant species	Permissible level (ppm)	Natural and industrial sources	Some of the health effect	References
Aluminum	0.2	Recycling of aluminum scrap and Bauxite	Neurotoxicity and Alzheimer's disease	[61]
Antimony	0.005	Alloys, paint pigments, rubber compounding. Semiconductor devices, sulfide ore mining	Lung and heart diseases	[62]
Arsenic	0.01	Geogenic/natural processes, smelting operations, volcanic action, mineral ore	liver tumors, cancer	[63]
Barium	0.3	The petroleum industry, steel industry, production of semiconductors, barite, and witherite	Cardiovascular disease risk and cerebral thrombosis.	[64]
Cadmium	0.003	Batteries waste, electronic waste, paint sludge, incinerations, anthropogenic sources.	chronic lung disease, dyspnoea, cancer	[65]
Chromium	0.05	Industrial coolants, chromium salts manufacturing, volcanic dust, and gases.	Nasal cancer, kidney and liver defects, eye excitement	[66]
Copper	2	Electroplating, mining, copper industries	nose and eyes excitement, headache	[67]
Iron	0.3	Steel manufacturing, iron mining	cardiac defects, diabetes, and cancer	[68]
Lead	0.01	Lead batteries, paints, electronic waste, volcanic products	blood pressure, anemia, cancer, kidney defects	[69]
Manganese	0.5	production of dry battery cells, matches, fireworks, and volcanic activity.	growth lag, sexual weakness, eye darkness, fever	[70]
Mercury	0.001	Chlor-alkali plants, geologic deposits of mercury	lack of appetite, headache, palsy	[71]
Nickel	0.02	Electroplating, thermal power plant, nickel–cadmium battery industries, soil and dust volcanoes	lung defects and cancer of lungs and intestinal	[65]
Zinc	3	Electroplating, surface water, soil, and rock.	Brain fever, vomiting, skin excitement	[72]

(continued)

Table 2 (continued)

Inorganic pollutant species	Permissible level (ppm)	Natural and industrial sources	Some of the health effect	References
Cobalt	0.002	Sewage sludge, phosphate fertilizers, processing of cobalt alloys, erosion of rocks	asthma, vomiting, cardiac problems, thyroid defect	[73, 74]
Uranium	0.002	Nuclear reactor wastes, phosphate rock	carcinogenic	[75]
Thorium	0.0001	Nuclear reactor wastes, and monazite	Cancer of lung and colorectal, liver diseases	[76]

are produced daily, and the amount of waste is created in the production of major crops such as rice, wheat, sugarcane, maize, and others. Agricultural waste is a primary source of environmental pollution, mainly for soil and various water sources [77–80]. On the other hand, there are economical and effective types of absorbents, called bio-adsorbent, which is derived from natural sources such as agricultural wastes. The bio-adsorbents were used in water purification as natural absorbers for many pollutants, whether organic or inorganic species. Despite this, agricultural wastes are an inexpensive raw material; moreover, reusing it can lead to solving many problems such as reduce the accumulated waste and protect the ecosystems. Bio-adsorbent such as NS that is extracted from various agricultural waste was characterized by easy technique to extract, needs easy operations, superior adsorption capacity, eclectic adsorption of pollutant species, economic, easy availability, superior mechanical strength, and chemically stable and easy regenerated for multiple repeated using. There is scarcely researched agricultural wastes as adsorbents at present, so it is of great prominence to consider the enforcement of agricultural wastes as an alternative source of efficient bio-adsorbents [81]. Agricultural waste is considered a promising, unexpansive, and sustainable source of nanomaterials, which have an extensive application now, waste-derived NS (extra pure) can be prepared from agriculture waste as a low-cost material such as wheat chaff, rice peel, rice chaff, bagasse, corn-cobs. Figure 2 shows the graphical clarification of NS recovered from various biosources such as bamboo culm, sugar bagasse, and rice peel and the chemical composition of rice husk as a single example for agricultural waste. Moreover, Table 3 shows various agriculture residues used to produce the NS as bioadsorbent.

3.2 *Electronic Waste*

Electronic waste (e-waste) includes all items of electric and electronic devices and can be defined as the items that disposed of by owners as trash without re-use in various areas and conditions around the world [95]. One of the most wastes

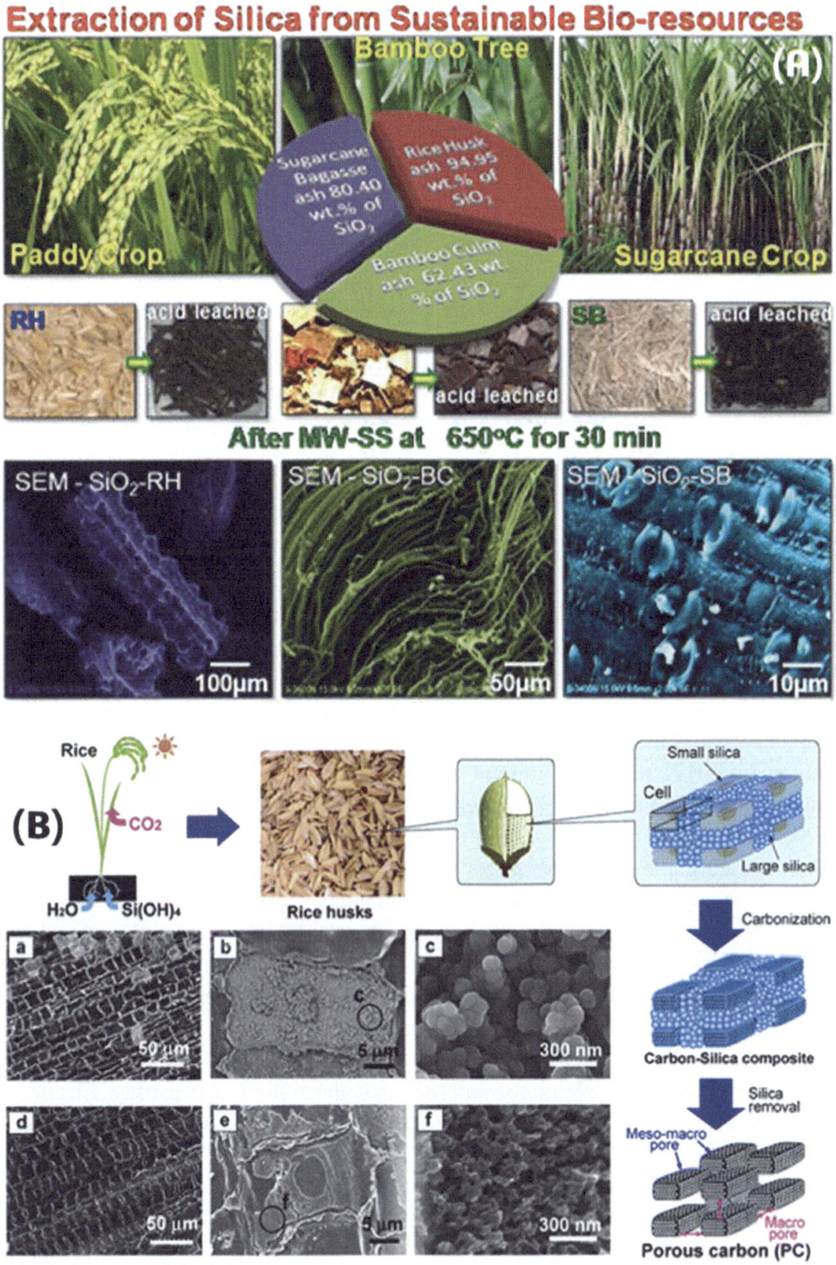

Fig. 2 Graphical clarification of **a** NS recovered from various biosources such as bamboo culm, sugar bagasse, and rice peel. Adapted with permission from Ref. [82], Copyright 2020, American Chemical Society, and **b** the chemical composition of rice husk as a single example for agricultural waste. Adapted with permission from Ref. [83], Copyright 2020, Royal Society of Chemistry

Table 3 Agriculture residues used for NS production

Silica source	Average SiO_2 particle size	Products	References
RHA	0.5–0.7 μm	Amorphous silica	[84]
Sugarcane waste ash	20 nm	SiO_2 NPs	[85]
Paddy straw	15–20 nm	NS	[86]
Olive stone	15–68 nm	Crystalline silica	[87]
Pinecone	37 nm	Silica nanoparticles	[88]
Wheat straw	75–320 nm	Amorphous hydrated silica	[89]
Rice black straw liquor	50–100 nm	Lignin-modified silica nanoparticles	[90]
Coconut husk ash	–	Crystalline silica and amorphous silica	[91]
Rice husk ash	10–15 nm	Amorphous silica	[92]
Bottom ash of sugar industry	–	Amorphous silica	[93]
Horsetail (Equisetum arvense)	9–32 nm	Mesoporous silica	[94]

expanding rapidly in the globe is electronic waste. Currently, 3.6 billion people from 7.4 billion people in the world are on the Internet, and in the next few years, this number will grow. 44.7 Mt of e-waste is produced over the world in 2016 [95, 96]. This is due to increased acceleration in the use of mobiles, screens, printers, computers, batteries, and other electronic devices. According to the e-waste global monitor in 2017, the total amount of e-waste will achieve 55.2 Mt by 2021. Figure 3 shows the total global growth in the past years and the expected increase by 2021 and top 10 nations by the quantity of e-waste produced in 2016 according to the literature [95–97]. The major danger in this subject is that 20% (around 8.9 Mt) only of e-waste in 2016 was documented and duly recycled, while 80% (around 35.8 Mt) was not authenticated. In advanced nations, 4% (around 1.7 Mt) of e-waste may be discarded into the rest of the waste. In contrast, the destiny of 76% (around 34.1 Mt) of e-waste is still obscure, where these quantities can be overthrown, traded, or recycled under bad circumstances [98, 99]. The rapid growth of e-waste presents a major warning to the ecosystem and humans due to the presence of many toxic elements/materials and the pollution it causes. Electronic wastes have various material which has harmful effects on human life. Therefore, the reprocessing of e-waste leads to the recovery of valuable metals, plastics, and other components, which in turn leads to safeguarding the ecosystems from the risks of these pollutants. Recycling e-waste is a major economic concern as a secondary source for many valuable materials such as NS that can be reused in many applications [98, 99]. Table 4 displays a different NS that can be extracted from various types of e-waste with recovery efficiency.

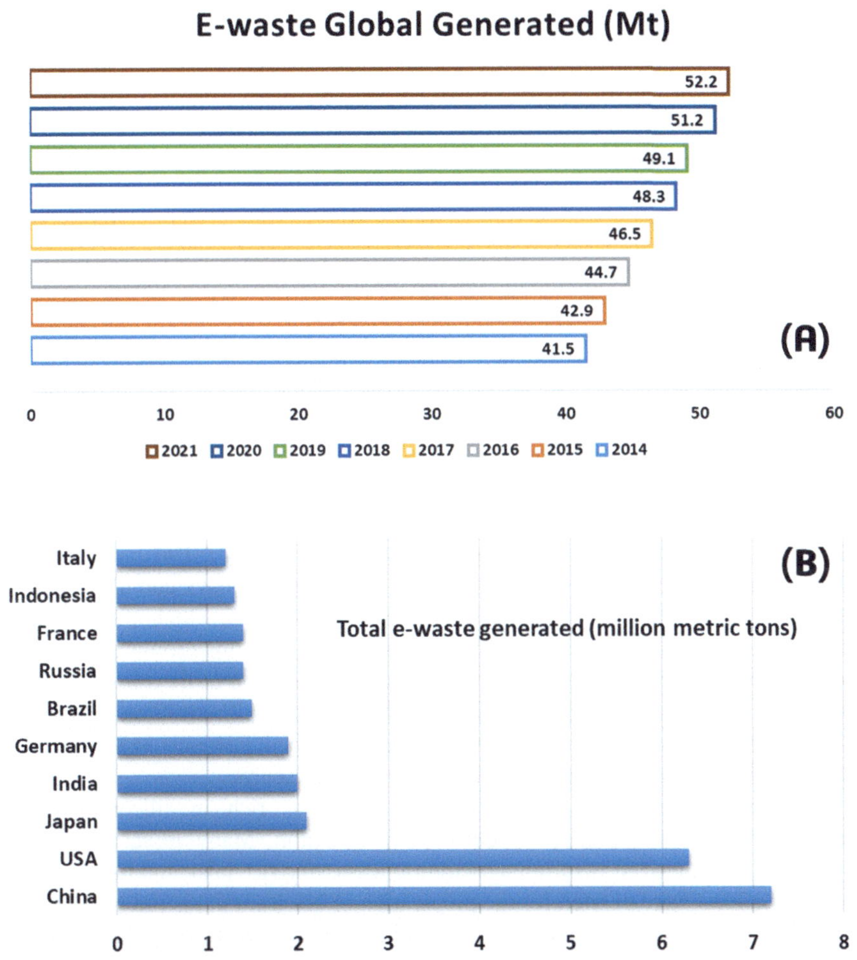

Fig. 3 **a** E-waste global generated from 2014 till 2021. **b** Top 10 nations by the quantity of e-waste produced in 2016 [97]

3.3 Industrial Waste

The industrial wastes, also called consumables or by-products, are utilized seldom for any additional objectives. The by-product kind makes it to be simply accessible and very cost-effective also. These by-products, for example, coal fly ash and paper-mill solid wastes (boiler ash, lime sludge, fly ash, and primary sludge), presently present a range of disposal troubles due to a huge quantity and their poisonousness behave [105, 106]. Using this waste as a low-cost advantage to produce nanomaterials such as NS will offer a two-fold benefit of ecological protection and valuable material reusing. The number of industrial by-products

Table 4 NS and silicon recovered from e-waste

Waste type	Target material	Recovery (%)	References
Printed circuit board	Mesoporous silica	99	[100]
Electronic packing resin waste	Mesoporous silica	80	[101]
Photonic and semiconductor industries waste	Mesoporous silica	100	[102]
Solar grade silicon waste	Silicon particle	99	[103]
Polyvinyl waste	Nanostructure silicon	High purity	[104]

(or waste) can be partially lowered through using some of its components as low-cost adsorbents; these adsorbents can be developed to reduce wastewater pollution at an affordable cost. Industrial wastes derived NS has got adsorption/removal capability and can be used for removing inorganic and organic pollutants from wastewater. A massive quantity of fly ash waste is being generated globally as the by-product of coal-burning from power-generation manufactories, and the discarding of this ash is a significant ecological challenge [107]. The coal waste fly ash contains mainly of alumina silica with a few amounts of heavy and light metals. Industrial coal fly ash can be used to prepare the valuable NS, which is an efficient way to generate novel nanosized materials from fly ash, and this promotes reducing the environmental contamination caused by the accumulation of fly ash waste. The coal-fired power plant generates an annual release of up to 750 million tons of fly ash, but the rate is just below 25%. SiO_2's main component is 40–60 wt% in carbon fly ash, making it a big, cheap, and easy to use silica source.

Furthermore, desilicated polluted water, which is rich in silicon, can be employed as a raw source for porous NS preparation. Many researchers have successfully synthesized NS from agricultural waste like rice husk and bagasse ash, but a few efforts were exerted for the preparation of NS from industrial waste [108, 109]. Figure 4 shows the potential applications of fly ash, bearing in mind the natural benefits of this manufacturing mineral.

4 The Strategy of Synthesis of Nanosilica from Different Solid Wastes

Researchers have made considerable efforts over recent decades to improve the feasibility of industrializing NS and successfully developed goods using various solid waste as a low-cost alternative source. Here, we will briefly review the commonly used strategies to prepare the NS from agricultural, electronic, and industrial waste to use it as a highly efficient adsorbent to remove the organic and inorganic contaminants from polluted water.

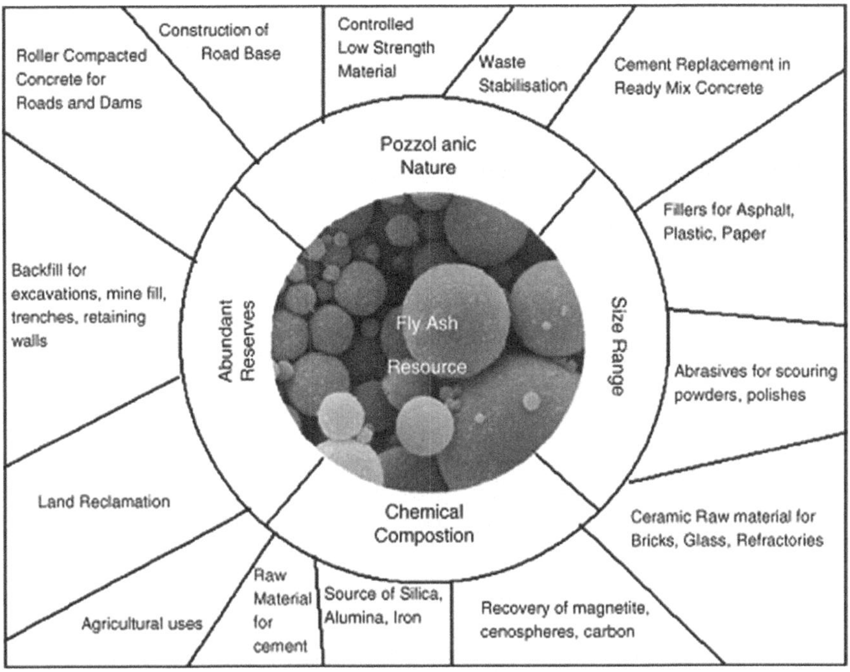

Fig. 4 Potential applications of industrial fly ash, such as preparation of NS. Adapted with permission from Ref. [110], Copyright 2006, Elsevier

4.1 Nanosilica Recovered from Agricultural Waste

NS is a key raw material for various industries, such as ceramics. It has a significant number of uses in various areas of the ceramics industry as raw material. Quartz is a crystalline silica form that is primarily used in industrial applications. The use of amorphous silica recently slowly increased because of specific properties distinct from crystalline silica, i.e., increased amorphous silica reactivity [111]. Fused silica is costly, because it is an industrial commodity, as the principal source of amorphous silica in the industrial. There are two major methods to production the waste-derived NS from different types of agro-waste, such as rice husk, wheat straw, sugarcane bagasse, olive stone. These methods include chemical, thermal and biogenic methods. Scheme 1 shows the schematic chart of crystalline and amorphous NS extraction from agro-waste through chemical, thermal, and biogenic methodologies [111].

In the thermal method, tubular reactor, muffle furnace, cyclonic furnace rotary kiln, step-grate oven, and fluidized vessel are utilized for recovery of the NS from agriculture waste, in thermal treatment method. This method can be carried out in both domestic and industrial operations in open fireplaces or different kinds of boilers. Oxygen in the biomass serves as an oxidizing agent during a burning

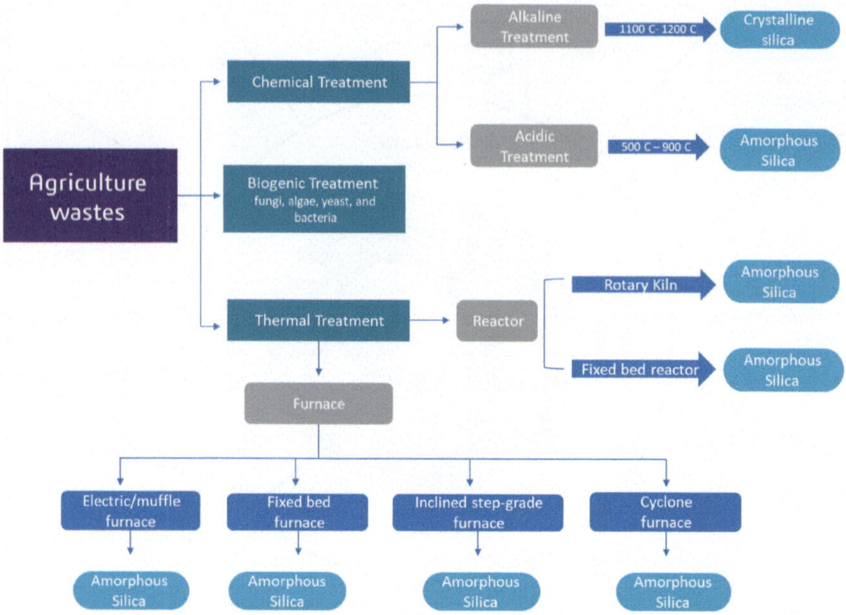

Scheme 1 Extraction of crystalline and amorphous NS from agricultural waste through chemical, biogenic, and thermal methodologies

oxidizing reaction that is exothermic. Biomass is oxidized during burning, producing a combustion product (ash) and heat energy. The major shortcomings of the thermal technique are long period reaction, hot-spot product growth, shortage of air flowing for entire burning and oxidation of carbon-based materials, etc. [112]. Figure 5 displays the production of NS from biomass power plant fly ash. As a result, there is a growing demand for high-purity silica. Combustion-derived NS comprises less than 95 wt% with other components that consist of various alkaline oxides and impurities (without chemical treatment using acids or alkaline).

The chemical method done by two major operations includes the alkaline and acidic extraction. Pure and high amount silica particles are produced by this method; the advantage of this method is that we can control easy to the size, shape, with high purity. But, the long time of the reaction and high cost of the chemical reactions which are used make this method more expensive. The NS content can be increased, however, to more than 99 wt% with appropriate acid or alkali treatment of agricultural waste ash. Consequently, several researchers have taken various chemical routes to obtain highly pure NS from rice husk, wheat straw, sugarcane bagasse, coconut husk, olive stone, banana peel ash, bean pod ash, Bambusa vulgaris leaves, palm kernel shell ash, etc. Recently, NS extraction by thermal/chemical treatment processes with low environmental impacts has now become possible. Briefly, in two stages, they recovered NSS from agro-waste: (i) agro-waste conversion into ash; and then (ii) NSS formation [114, 115]. Figure 6 displays how the production of NS from

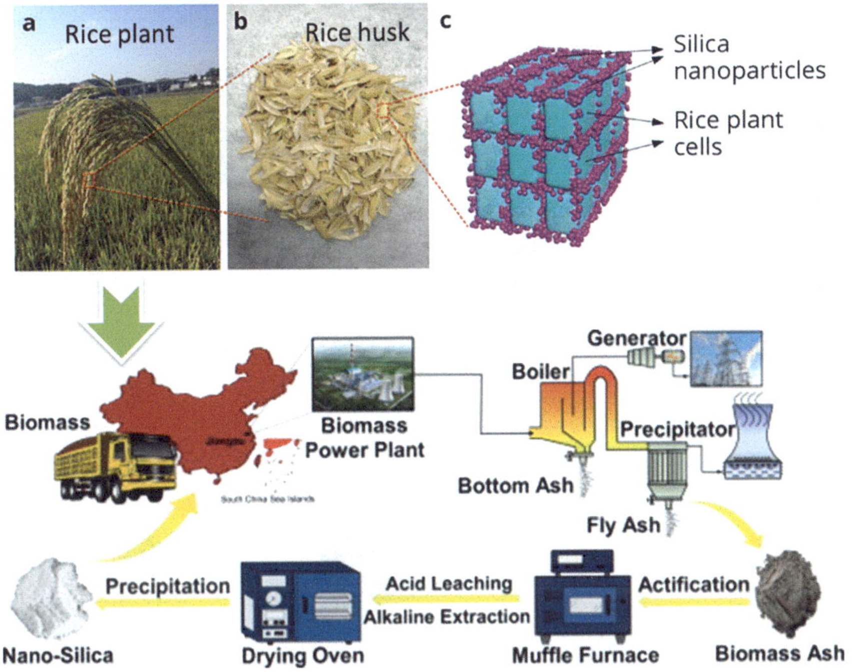

Fig. 5 Production of NS from agricultural waste through thermal approach. Adapted with permission from Ref. [113], Copyright 2020, Elsevier

agricultural waste through the chemical approach [116]. In general, NS can be extracted from different types of agro-waste as follows: (i) washing the waste to remove any impurities and dust; (ii) treating with HCl at 70 °C to remove minerals; (iii) after filtration, washing with distilled water, the agro-waste was burned at high temperature to obtain ash; (iv) waste ash was ground; (v) the waste ash powder was calcined at 700 °C until getting the white as-made product; (vi) dissolving the white product in NaOH to get sodium silicate; (vii) precipitation the nanosized SiO_2 using H_2SO_4; (viii) the as-made white gel product of NS was washed several times and then dried at 50 °C for overnight [92, 117].

Moreover, NS can be extracted from agricultural waste using microorganisms, called the biogenic synthesis process. Recently, the availability, low costs, and eco-friendly nature of NS have given silica a significant interest in the material sciences and biomedicine fields. Several living organisms, such as mollusks, sponges, and protozoa, have developed to produce biogenic silica at a fantastic rate of gigatons per year [118]. The bioprocesses appreciate of microorganisms, such as fungi, algae, yeast, and bacteria, as a rich source of numerous enzymes with expanding manufacturing ability. These microorganisms display a variety of metabolic capability and produce enzymes that can reveal the ability of the uncooperative lignocellulosic of plant cell wall formations. Therefore, the cracking of carbon structures of natural

Fig. 6 Production of NS from agricultural waste through chemical approach. Adapted with permission from Ref. [116], Copyright 2020, MDPI

materials is a method to get their desires of biogenic components essential for plant growth [119, 120]. The usage of fungi as biocatalysts in the production of NS from agro-waste has numerous features more than other microorganisms. These characteristics include simple management and control, simply removed via filtration processes, effortless of biomass handling, and cost-effective.

4.2 Nanosilica Recovered from Electronic Waste

Electronics manufacturing is single of the most rapidly developing industrial sectors in the globe. E-waste, however, has become a big issue of pollution. Estimates of the global semiconductor industry's sales in 2009 exceeded USD 226 billion, which is expected to grow 13% annually. But, the disposal of e-waste, which results in the manufacture of large amounts of electronic products, has become a major environmental issue. The global annual output of e-waste is estimated at approximately 20–50 million tons, including old computers, smartphones, TVs, stereos, radios, etc. E-waste is generally made up of over a thousand different component-based materials [121]. These materials can be divided into metals and non-metals categories. Materials for electronic packaging are typically a combination of epoxy, phenol, silica, and others. Due to its distinct and desirable properties, epoxy polymeric composite offers a broad range of industrial applications such as the packing of electronics; these properties include superior thermal stability, low shrinkage, excellent chemical resistance, and strong rigidity [122]. Approximately, 80 wt% of silica is used for the chemical composition of resin waste packaging for mass production of NS. The use of recycled e-waste, therefore, offers advantages for the disposal and recuperation of valuable materials.

In general, NS can be extracted from electronic waste as follows: (i) grinding the parts that expected to contain silica to a fine powder; (ii) remove the metallic impurities through refluxing the obtained powder with HCl solution; (iii) washing with deionized water for many times to remove the acidity, then drying for overnight at 100 °C; (iv) dissolving the dried powder in 4.0 M NaOH solution under constant stirring for 6 h at boiling temperature to get sodium silicate; (v) centrifugation and filtration several times to obtain a colorless, transparent solution; (vi) mixing with constant stirring of the sodium silicate solution with a surfactant to get porosity in the final product; (vii) adding of H_2SO_4 solution until the gel formed as a result of silica formation; (viii) hydrothermal heat using stainless-steel autoclave for 2 days; (ix) filtration, washing, drying, then calcination to get the final white product of NS [103].

One of the NS-recovery approaches from e-wastes, such as liquid crystal display LCD, can be explained as follows: (i) washing the waste silica powder by deionized water; (ii) then adding HF solution, CTAB dropwise, NH_4OH solution (0.2 CTAB: 1 SiO_2: 12 NH_4OH: 10 HF: 120 H_2O); (iii) stirring for 8 h at the room temperature; (iv) filtration and washing with deionized water, then drying for overnight at 110 ° C; (v) calcination at 550 °C for 6 h to remove CTAB. The obtained mesoporous SiO_2 nanoparticles from thin-film-transistor LCD had an average diameter of 100 nm [123].

4.3 Nanosilica Recovered from Industrial Waste

Due to its high emissions, low availability and serious pollution, industrial solid waste, such as silica fume, coal fly ash, paper mill solid wastes posed serious threats to social and environmental stability. Therefore, solid waste disposal has been a significant issue in the past few years through recycling processes to exploit its valuable components [124]. Synthesis procedures of NS from industrial solid wastes have two features comparing to commercial NS that synthesized from pure chemicals: (i) waste-derived NS is cheaper than traditional; and (ii) using industrial solid wastes as NS sources lead to by-products recycling. There are two main challenges to produce NS form industrial waste. Firstly, the synthesis parameters of industrial waste to generate pure, stable NS for different potential uses such as wastewater treatment sustain a great focusing. Secondly, the removal of the non-silica phase and waste to increase the amount of NS generated and boost its production, and economic performance is still a challenge.

Substantial efforts were made to extract and develop porous SiO_2 nanomaterials from fly ashes with desirable textural properties to boost their environmental efficiency, as follows: (i) extraction of Si from fly ashes, through calcination at 800 °C under air circumstances to eliminate organic ingredients, called desilication process; (ii) ground and sieved to obtain fine powder product; (iii) mixing with a solution of NaOH in the Teflon-lined vessel at 110 °C for with continuous stirring; (iv) filtration to get the sodium silicate solution; (v) mixing of the obtained solution with surfactant in the Teflon-lined reactor under gases atmosphere at 80 °C under stirring until the pH ranges to 10.5–11.5; (vi) the synthetic NS was extracted by liquidation, rinsing, and desiccation at 105 °C; (vii) calcination at 550 °C for 4 h to get the porous NS final product [109, 125]. Furthermore, silica aerogels from industrial fly ashes can be prepared as follow: (i) synthesis of Na-silicate from fly ash waste by interaction hydrothermally with NaOH; (ii) getting of porous SiO_2-aerogels from the gained Na-silicate solution using H_2SO_4 solution; (iii) solvent exchange using anhydrous ethanol, surface adjustment and desiccation of wet-gels at 120 °C for 2 h at normal pressure and air circumstances [126]. There is another route to extract NS from fly ashes as follows: (i) mixing (in a planetary ball mill) and heated (using a muffle furnace) of fly ashes with $CaCO_3$ for 2 h at 1200 °C; (ii) acid leaching using HCl solution (1 g:10 mL); (iii) filtration, washing by distilled water and drying to obtain the solid product of porous NS [108]. Figure 7 shows the schematic design of NS synthesis from fly ash in general.

5 Treatment of Wastewater Using Nanosilica

In recent decades, there is considerable attention from scientists in the nanotechnology field. The term "nano" has become frequent in many ads for various products, as people use it to indicate the singular features of these products. It also draws the attention of researchers and governments around the world. For example, total investments in the nanotechnology field in the USA amounted to about three and a half billion dollars [128]. Wastewater treatment is one of the potentials filed to sustain the water resources and reduce the human health risk. Water contamination is considered one of the most critical environmental problems facing the world due to the high environmental risks that it causes.

Additionally, it constitutes a threat to water sources due to being polluted by inorganic and organic pollutants [129]. With the increase in the population, the world will suffer from a severe lack of basic water resources, which underscores the urgent need for new efficient technologies to recycle the discharged wastewater. There are many commonly used methods used in treating wastewater, such as membrane filtration methods used in the treatment of wastewater such as ion exchange [130], membrane filtration [130], reverse osmosis [131], chemical precipitation [132], photocatalysis [133], coagulation-flocculation [134], and

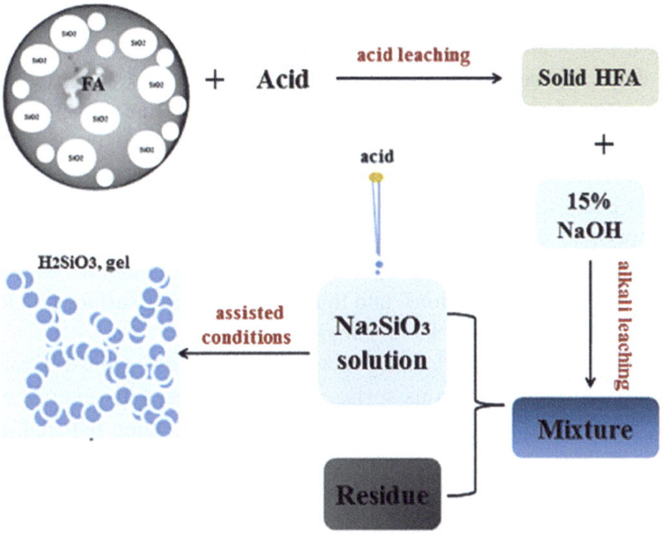

Fig. 7 Synthesis of NS from industrial fly ash via a new methodology. Adapted with permission from Ref. [127], Copyright 2017, Elsevier

adsorption [135]. However, all of these methods have corrosion problems and toxic sludge formation. Also, their efficiency in removing pollutants is not as much as required [136]. Even though adsorption is the most promising of these methods, its traditional sorbents still have low efficiency and sorption capacity [137].

Nanotechnology has proven to be the most efficient process for treating wastewater. This is due to the ability of nanomaterials to remove pollutants with a high degree of efficiency. The second question that needs to be asked is "What are the characteristics that made nanomaterials more efficient than other materials in this field?" The answer is that nanomaterials have some unique properties, which make them more advantageous than other materials, such as i) the higher surface area, which increases with decreasing the particle size. This property leads to higher efficiencies and faster adsorption rates. Moreover, by this property, we can apply nanomaterials in many fields, for example, catalysis and adsorbents; ii) low cost, catalytic potential, and high sorption capacity; iii) its surface modification can be achieved by various functional groups to enhance its performance and increase its applications [138, 139]. The focus of this review is on nano-adsorbents, more accurately, treating wastewater by adsorption onto nanomaterials. Nano-adsorbents can be prepared using atoms of some elements such as (silica, activated carbon, clay materials, and metal oxides) on the surface of nanoparticles. These atoms have high chemical activity and sorption capacity. To make sure that an adsorbent is ideal, some important characteristics must be met: (1) higher surface area and higher sorption capacity; (2) pollutants can be easily separated from its surface; (3) be recyclable to avoid environmental pollution; (4) having a large number of the active site to be able to adsorb various pollutants, nano-adsorbents has proved in many studies that it meets all of these properties [140, 141].

One of the common examples of nano-adsorbent that is used for wastewater treatment is waste-derived nanosilica. Silica (SiO_2) is present in nature as sand or quartz. It has a 3D network structure, where the silica atom is attached to four oxygen atoms by four covalent bonds. Nanosilica is believed to be a promising choice for the decontamination of various pollutants due to its unique chemical and physical properties, such as the powerful mechanical strength, large surface area, easy modification, low toxicity, high biocompatibility, thermal solidity which facilitates their access to metal ions, and thus their deep bulk surface due to its large pores, even if the reaction rate is very high [142, 143].

Furthermore, it has a lower loading of various functional groups because of the existence of (Si–OH) groups on its surface, which enhances its removal efficiency. As a result, the dispersion capacity of its nanoparticles is being reduced, as well as its applications have been largely limited. Therefore, researchers are aiming to resolve this problem by using various surface modification approaches to enhance its adsorption capacity by increasing its surface affinity for organic substances. Silica nanoparticles can be adapted by using various functional groups, for instance, NH_2 functionalized mesoporous silica, which possesses a high surface. The mechanism wherein how this group works lies in its ability to form coordination bonds with the adsorbate (such as organic and inorganic pollutants), thereby improving the effectiveness of removing these pollutants [144, 145].

6 Effect of Nanosilica's Surface Area and Porosity on the Wastewater Treatment Efficiency

Two major factors are responsible for enhancing the removal efficiency of the adsorbent materials (surface area and porosity, see Fig. 8). Firstly, there is a powerful relation between surface and particle size, as the surface area of a nanoparticle increases as its size decreases, since the smaller sized particles have a more significant surface expanse to volume ratio than the bigger fragments. Larger surface area ensures more external active sites for surface processes such as adsorption, which enhances the removal efficiency of pollutants. The use of mesoporous materials has been very popular in recent years owing to its outstanding properties, which enhances its applications such as adsorption, medical, sensors, energy, and catalysis. The unique properties of mesoporous materials are found in their extremely high surface area which provides a large number of outer active sites, its large pore volumes which enhances the process of loading of adsorbed species, and its tunable pore sizes which facilitate the entry of atoms, ions, and large molecules into materials bulk which in turn increase the activity of its surface area, leading to boost the adsorption capacity and removal efficiency of adsorbents (NSS) toward the organic and inorganic pollutants. Various conventional physical and chemical methods are accessible for preparing mesoporous materials. Among them, the template method is believed to be a promising choice in the recent years due to its effectiveness in preparing nanomaterials. The template method is not affected by the conditions of preparation, in addition to its capability to command the particle structure, morphology, and nanomaterials size. Furthermore, the template method is easy and implementation operation. The template method is subdivided into two methods, soft and hard template. There are some main differences between soft and hard templates such as the easier building and removal of a soft template than hard template since the formation of a soft template occurs within the reaction, while a hard template is prepared before the reaction [146, 147]. Figure 8 shows a brief scheme to explain the use of extracted NS in the water treatment field and the trapping of inorganic pollutants inside the meso-grooves of the waste-derived NS absorbent.

7 Effect of Nanosilica's Morphology on the Treatment Behavior

Besides surface area and porosity, the morphology of nano-adsorbents is also one of the most important factors that have a significant impact on the adsorption performance. This can be illustrated by looking at the relationship between the structural morphology of the adsorbents and their physical and chemical properties. Many studies have shown that the difference in adsorbent morphology changes their physical and chemical properties, which in turn changes their adsorption mechanism and delivers different performances. For example, cerium dioxide (CeO_2) was prepared in three different morphologies (nanorods, nano-octahedrons, and nano-cubes). The prepared materials in the various morphological structure of diameter, length, and interplanar spacing lead to present variable chemical properties. Moreover, the difference in the crystallinity led to different structural defects (inverse relationship) and grain sizes. Subsequently, by studying the adsorption isotherm of these materials, the results support the variation in the adsorption property as changing the morphology and crystallinity. The adsorption capacity of nanorods (71.5 mg g^{-1}) is higher than octahedrons (28.3 mg g^{-1}) and nano-cubes (7.0 mg g^{-1}). All the above emphasizes the importance of the adsorbent morphology and its vital role in adsorption efficiency [148, 149].

As we speak of the importance of the morphology, the concept of the "Hierarchy" should be highlighted. The term "Hierarchy" comes from the Greek word (hierarchy), which generally describes arranging of something like objects, values, and concepts. The concept of hierarchy is found in various systems, such as social in which the vertices are divided into major groups, and groups are subdivided into smaller groups, and so on, into multiple scales. It is also found in technical systems such as networks of electricity that are hierarchical in every city. Lightning and nerves also show a hierarchical structure for easy transport and

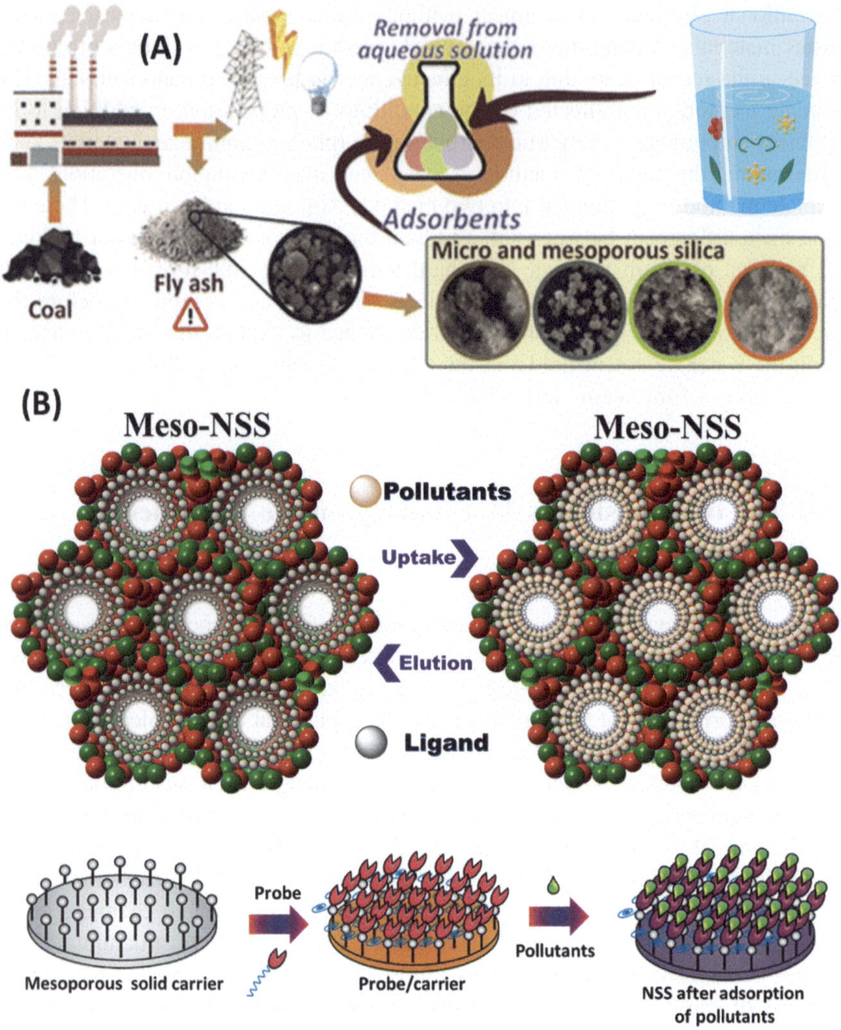

Fig. 8 **a** Brief scheme to show the use of extracted NS in the water treatment field. Adapted with permission from Ref. [126], Copyright 2020, American Chemical Society, **b** mesoporous NS contains grooves in mesoscale which can act as a trapper for the target pollutant species

efficient energy and information delivery. From a scientific point of view, hierarchical materials can be classified into three types: hierarchical porous materials, structural materials, and smart materials. Hierarchical porous materials provide two or more types of pores, such as meso with macropores. A hierarchical material must meet two requirements; firstly, its structural elements must contain further than one length scale; secondly, the function of each of these structural elements must be very distinct but complementary. Hierarchical materials are more advantageous than other materials owing to their unique properties such as the higher compressive strength, the ability of hierarchical materials to eliminate a wide range of pollutants of different sizes. Hierarchical materials have an excessive deal of attentiveness to use it in many fields such as environmental technology, the utilization of carbon dioxide (CO_2), and storage and conversion of energy. As a result, various strategies have been used in the synthesis of hierarchical material, such as templating strategy (soft and hard), self-formation, bioinspired strategy, and biomimetic strategy [150, 151].

8 Inorganic Pollutants Adsorption Using Nanosilica

Removal from wastewater of inorganic contaminants, including heavy metals, is currently one of the most important environmental issues, because of its significant threats to human health and the environment. The universal call encouraged scientists to find operative strategies to remedy the contaminated water and to get clean drinking water, thereby protecting the ecosystems. Different strategies for eliminating inorganic compounds from various water sources have been contrived, including precipitation using chemicals, filtration through membranes, ion exchange, electrochemical methods, and so on. Because of the disadvantages of these strategies, for example (a) the generation of large volumes of relatively low-density sludge, (b) precipitation of a specific metal at optimum pH may let of another metal to dissolve at the same pH condition, (c) using of complexing agents may prevent the formation of some metals hydroxides, (d) formation of toxic fumes at acidic conditions, (e) precipitation of colloidal precipitates, (f) costly expensive, (g) low efficiency, and (h) handling intricate, the adsorption process is considered the most convenient method in both laboratory and large scale-up applications. Treating wastewater, due to civilian and manufacturing activity, depends on the controlling of suitable strategy, materials, and techniques to have a positive impact on the water goodness by reducing contaminants [152–154].

Elimination of inorganic pollutants from contaminated water is an important issue because of the deleterious effects of these toxins. Various techniques, such as adsorption processing, were used to remove these ions from watery inorganic-contaminated waters. The adsorption mechanism is based on the mass transfer from the solution bulk (i.e., wastewater) to the solid form, known as adsorbent, while the target ions that trapped at the surface are known as "adsorbent." Due to low operating cost, high efficacy of toxic metals removal, high-quality treated water provided by the treatment of water with adsorption technology, efficient use in a wide pH range, the re-use of adsorbents through elution/desorption processes. The adsorption method depends on the pH value of the interaction medium, adsorbent dosage, stirring time, metal concentration, the area and porosity of the adsorbent, the temperature, and flow rate through batch contact phase and fixed-bed column techniques. Mainly, pH plays a significant role in raising and decreasing the efficiency of inorganic pollutants removal, which may possess a great impact on the active sites of adsorbent surface, selectivity, and speed of the adsorption process. Adsorption technique has recently grown in various industries into one of the main wastewater treatment technology. Through this process, chemical or physical binding can be occurred to associate the adsorbent with toxic metals. Thus, the adsorption process can be divided into two categories, physisorption and chemisorption conditional, on the type of interaction between target species and used adsorbent [155, 156].

Metal oxide nanoparticles like NS extracted from different types of waste have emerged in recent years as an enabling medium for the elimination of inorganic toxins from polluted water. Thus, owing to their steadiness and the easy of toxins removal behavior, there is a need for these nanostructured adsorbents to use in the wastewater treatment field. An adsorbent should preferably have ample binding sites to properly adsorb inorganic contaminants. Carbon, chitosan, clay, etc., are considered the main conventional adsorbents for the removal of inorganic toxins. Many drawbacks such as poor adsorption power, lack of practical tuning, low selectivity, and sensitivity and difficult reusability could be present in these adsorbents. At the same time, new nanosorbents derived from waste with good properties for water decontamination are used to address these drawbacks [157].

The special properties NS, such as large surface areas regulated surface-active interaction sites, and porosity (micro-, meso-, macro-porous), enforced its ability as a highly efficient adsorbent for inorganics removal from water. NS also reveals a non-poisonous and environmentally safe sorbent. Change NSS surfaces by grafting additional functional groups such as $-NH_2$, –COOH, –SH, phosphonate, poly (ethylene glycol), octadecyl, carboxylate, octadecyl groups, etc., have been improved adsorption, strong adsorption rate, and enhance the sensitivity and selectivity behavior. The 3D porous network of NS offers a strong mechanical force and a wide surface expanse, good thermal solidity, and insoluble in water. These specific features of NS allow the inorganic ions to easily and fast reach the inner surface-active sites. For the trapping of inorganic contaminants from polluted water, the key surface characteristics of NSS are, for example, surface expanse, functional units, surface charge, particle size, and pores-based permeability. Because of its high cost, NS use is minimal. Therefore, the use of NS for commercial rubbish-water remediation is not feasible economically. To address this trouble, large-amount waste materials such as agricultural, manufacturing, and electronic waste can provide a substitutional source of an effective adsorbent such as NS for the treatment of polluted water, which will minimize the total cost of the wastewater remediation process. Figure 9 shows the synthesis of NS from agricultural, industrial fly ash, and electronic wastes and adsorption of inorganic pollutants from wastewater [158, 159].

NSS structure plays a major part in its adsorption chemistry behavior, through the interior and exterior surface. The arrangement of interlinked SiO_4 tetrahedral leads to form siloxane (Si–O–Si group) and silanol (Si–OH group) on the exterior surface. The silanols can be split into free silanols groups, where the Si atom at the exterior surface has three bonds in deep structure, while the fourth bond is assigned to the single –OH group called bridged silanols. Hydrogen bonds can be formed

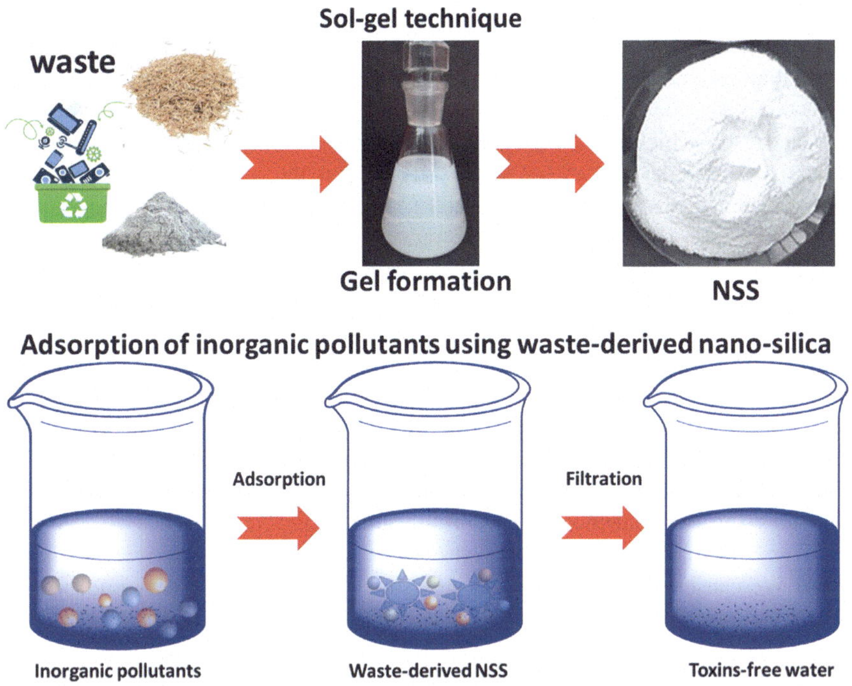

Fig. 9 Schematic design of NS fabrication from agricultural, industrial fly ash, and electronic wastes and adsorption of inorganic pollutants from wastewater

because of silanols–OH groups affinity. Furthermore, geminal silanols are composed of two hydroxyl groups bound to the same Si atom, which are too nearby to form hydrogen bonds. The increased acidity of the SiO_2 surface grants it an elevated level of chemical activity, which ensures that functional silanes can react with other organically binding agents to obtain a novel NS with great features. Surface amendment of silica includes all processes contributing to improvements in the surface's chemical composition, leading to remove target species selectively. Surface modifications may be made either by the physical process (thermal or hydrothermal), which changes the ratio of silanolto siloxane at the NS surface or by the chemical method resulting in a modification of the chemical properties of the surface of NS. The adjustment greatly affects the adsorption characteristics. The chemisorption of silica-surfaced chelating groups provides high efficiency, sensitivity, and selectivity during the inorganics adsorption process [160, 161].

The two main tracks for modulation of nanomaterials by covalent bonding of organic chelating agents and NS are co-condensation and grafting processes, which are thoroughly studied in a lot of published papers. The technique of grafting after synthesis was used for incorporating the pre-synthesized porous NS wall into a wide variety of active groups. The features of grafting-prepared functionalized NS are determined by their structural characteristics and chemical structure.

Co-condensation, in which organosilane and silica precursors are condensed together, is another major way for incorporating functionality. Co-condensing shows many advantages, including uniform functional group spreading and quicker fabrication. Also, large numbers of functional groups inside/outside the porous NS surface can be achieved through the co-condensation strategy [162].

The low-cost, large-scale processes depend on the adsorption strategy. The adsorbent must preserve its structure and morphology with high efficiency during repeated cycles of adsorption-elution processes. Numerous specialists detailed reusing of the spent adsorbents they utilized predominantly by elution with HCl and HNO_3 aqueous solutions or treating by EDTA as a complex operator bonded to the target ions. Treating spent adsorbent using acidic eluent leads to convert the surface-active sites to protonated sites, which in turn provides a rise in the positive surface charge. Repeated use of the adsorbent material may cause to lose its structural form, leading to a reduction in the adsorption capability, due to losing a lot of functional, active sites. In some cases, adsorption efficiency enhances after the first adsorption-elution cycle, because the eluent agent may help to remove any impurities and can contribute to exhibit the active sites [163, 164].

Diverse factors may impact the adsorption process either positively or negatively. We can emphasize the factors that should be in interest during adsorption studies for inorganic pollutants removal as follows:

(i) pH-solution: the surface of NS adsorbent can exhibit of various ionic functional, active sites with the change of pH values; this surly be altered because of pH variations in the system. Thus, an electrostatic mechanism of interaction between the NS adsorbent and inorganic contaminates will directly regulate the adsorption process.

(ii) Initial target ions concentration: the number of target inorganic species can have a favorable impact on the NS adsorbent adsorbing effectiveness. This conception may be explained mainly in the external active sites of the porous NS surface by the adsorption process where the initial concentration of target species is low, where migration to active interior sites deeper in the porous nano-adsorbent at using high initial concentrations of target ions. Besides, relevant information can also be obtained about the maximum adsorption capacity of NS adsorbent, and the theoretical adsorption isothermic models can be interpreted more [165].

(iii) Dosage of adsorbent: the dose of NS adsorbent is a significant factor that can directly affect an adsorbent's removal effectiveness while the various targets are removed. The increased adsorption effectiveness may refer to the boosted availability of interior and active exterior locations on the NS's surface [166].

(iv) Stirring time: the stirring time changes are a paramount factor to be investigated in the inorganics removal operation as from this study, initial rate information can be obtained to an adsorption process equilibrium. During stirring time studies, the theoretical adsorption kinetic models can be studied more thoroughly. Stirring time has a positive effect as it allows targets to reach the largest possible number of active sites. On the other hand, it may have a negative effect, because prolonging the stirring time may lead to breaking the bonds between target ions and active sites of NS [167].

(v) The temperature of the adsorption process medium: the temperature in the inorganics removal process is a key essential factor in providing detailed information on the kinetic adsorption behavior. Furthermore, temperature studies can provide worthy information on thermodynamic factors, for instance, ΔG (Gibbs energy), ΔH (enthalpy), and ΔS (entropy), to explain the adsorption process mechanism [168].

9 Organic Pollutants Adsorption Using Nanosilica

Nowadays, contamination of water by various organic pollutants such as phenols, dyes, polymers is one of the foremost challenges the world faces today since enormous volumes of these substances are utilized in many industries. Serious health problems are caused by discharging the organic pollutants into the water, which calls attention to removing them before discharging into the environment [169]. The importance of study these organic pollutants is due to their (a) useful applications, and consequent disposal to terrestrial lands and aquatic bodies, (b) longer perseverance, (c) high resistance, and (d) major influences on the ecosystems and human healthiness. As a result, the treatment of water and the removal of organic pollutants are gaining a great deal of attention from researchers to make it valuable for both households and agrarian utilization. There is considerable attention to the remediation of wastewater, especially the use of economical and environmentally friendly methods. Speaking of this field, the technological deployment in the removal of organic pollutants and the interaction between these materials and materials used for its removal has become significant [170]. pH is one of the most significant factors influencing the efficiency of adsorbents. This effect varies depending on the type of adsorbate (i.e., organic pollutants); for example: if the adsorbate is positively charged, the NS adsorbent active sites must be negative. While in the case of negatively charged adsorbate, the NS adsorbent active sites must be positively charged. The pH value will be directly proportional to the adsorbed amount of adsorbate. This can be illustrated as follow: In the case of low pH, the presence of H^+ will compete for the positive pollutants species on the active sites of the adsorbent, which in turn reduce the adsorbed amount of organic pollutants; while at high value of pH, the positive pollutants species will not be in any competition. This tendency led to an increase in adsorption capacity. Furthermore, at high pH value, the presence of OH^- ions will compete for the negatively charged adsorbate. Generally, the charge of NS adsorbent active sites can be controlled by the pH value [171].

Increasing the adsorbent dosage enhances the removal efficiency by increases the number of active sites, which in turn accelerates the adsorption process of organic pollutants [172]. Another factor that has a major impact on the adsorption efficiency is the initial pH of the dye solution due to its impact on the adsorbent surface charge and binding sites [173]. In addition to that increasing the concentration of the adsorbate (organic compound) also has a significant impact on the adsorption capability of clay minerals. Finally, the adsorption capability is also affected by the ionic strength of suspension; as a result, we can determine the adsorption capacity of bentonite clays for organic compounds through its ionic strength [174]. On the other hand, the adsorption of organic compounds on illicit clays is also affected by some factors such as the existence of electrolytes in the suspension and the increase of NaCl concentration in suspension, which in turn reduces the adsorption efficiency of tetracycline [175]. Particle size is a significant factor in the application of this field. The addition of appropriate additive agents such as alcohol, amine, inorganic bases, and inorganic salts, to control the particle size, is a promising choice. These agents can change the condensation of NS precursor, in addition to its ability to increase the reaction rate, leading to formation of particles with smaller size. In addition, there are many factors influence on the adsorption efficiency such as textural characteristics, the adsorbent surface functionality, and the form of interactions posed between the NS as adsorbent and the dye molecules like hydrogen-bonding, ion-exchange, coordination, electrostatic interactions, and acid-base interactions [176].

The surface of NS is known to have a negatively charged surface that is susceptible to trapping/removal of electron-deficient species; this is because the silica surface is always stacked with groups of hydroxy and ethereal linkages. The binding nature between adsorbate and silica can be illustrated by the adsorption isotherm. During the adsorption of aromatic compounds, it found that the hydrogen connection with silanol (–OH group) includes the pi-cloud. The adsorption of cationic and non-ionic surfactants can be done involving hydrogen bonding. Occasionally, a polar portion of the surfactants additionally plays a role in the trapping/removal process. The adsorption mechanism of a cationic surfactant or polyethylene glycol can be affected by alkali treatment; this change is due to differences in the composition of NS or modified NS surface. It was indicated that the powerful interaction with the NS surface occurs through hydrogen bonds onto the silanol group surface. Based on the above, the surface interaction between the nano-adsorbents and pollutants is a very important factor that has a major impact on the removal efficiency of noxious waste from water. The functionalization of the surface of the nanoparticles can further enhance these interactions, which in turn improves the removal efficiency. The presence of silanol groups on the surface of NS improves its reactivity, so we can use this feature to graft different organic functionalities on the surface of these particles by using alkoxysilane grafting chemistry [177, 178].

Many literature reports on the functional method of alkoxysilane grafting indicated the facility of this method and explained how to use it on silica nanoparticles to convey the required functionality. The surface interactions between the dye molecule and nanomaterial can be improved by the grafted functionalities. We can control the reversibility of these surface interactions by varying some conditions like temperature, the pH of the solution, etc. Getting the desorption of dyes from the particle surface to generate the adsorption sites is the main object of this process. It is a very difficult process because of the strong surface interactions between dyes and silica nanoparticles (silicon dioxide). Polymer-grafted silica nanoparticles are also considered to be a promising choice for the elimination of dyes and further impurities from water. The existence of functional groups on the backbone of polymer chains branches grafted on nanoparticles surface plays the main part in increasing the ability of particles to be complexed toward pollutants. Moreover, the polymer shell around nanoparticles provides the protection of the surface of the silica from exposure to acid or base-based dye solutions in water. The colloidal stability and contact time of particles with polluted water can be enhanced by the polymer grafting [179, 180].

10 Conclusion

Agricultural, electronic, and industrial by-products or wastes are inexpensive and plentiful NS precursors that could be complementary NS resources. Consequently, surveys in the latest years have registered the manufacture of NS from these wastes. The recycling route could effectively switch unusable trashes into high-value SiO_2-based nanomaterials, leading to resolving the difficulty of waste dumping; it will be beneficial in terms of environmentally and economically. Moreover, proper treatment of wastewater containing organic and inorganic pollutants has become one of the significant issues in the avoidance and management of water contamination. Therefore, waste-derived NS finds a wide range of applications in wastewater treatment as an efficient adsorbent.

11 Future Prospective

This chapter demonstrated that the agrarian, electronic, and industrial wastes could be applied as little-cost primal materials for the manufacturing of NS with high purity to eliminate the organic and inorganic toxins from polluted water, as well as showed the preparation strategies of NS from these wastes. Nevertheless, this kind of NS source will provide pluses, not only as an economical primal material, but also provide a solution additionally to reduce the garbage. Waste-derived NS can be utilized in different applications such as electronics and optical materials, solar cell productions, storage media, constructions and building materials, medicine, drug delivery, molecular separation, catalysis, glass, and ceramics industries.

References

1. Vardhan KH, Kumar PS, Panda RC (2019) A review on heavy metal pollution, toxicity and remedial measures: current trends and future perspectives. J Mol Liq 290:111197
2. Khanzada NK, Farid MU, Kharraz JA, Choi J, Tang CY, Nghiem LD, Jang A, An AK (2020) Removal of organic micropollutants using advanced membrane-based water and wastewater treatment: a review. J Membr Sci 598:117672
3. Gomaa H, Shenashen MA, Yamaguchi H, Alamoudi AS, Abdelmottaleb M, Cheira MF, El-Naser TAS, El-Safty SA (2018) Highly-efficient removal of AsV, Pb^{2+}, Fe^{3+}, and Al^{3+} pollutants from water using hierarchical, microscopic TiO_2 and $TiOF_2$ adsorbents through batch and fixed-bed columnar techniques. J Clean Prod 182:910–925
4. Essien EA, Kavaz D (2019) Effective and reusable nano-silica synthesized from barley and wheat grass for the removal of nickel from agricultural wastewater. Environ Sci Pollut Res 26(25):25802–25813
5. Carolin CF, Kumar PS, Saravanan A, Joshiba GJ, Naushad M (2017) Efficient techniques for the removal of toxic heavy metals from aquatic environment: a review. J Environ Chem Eng 5(3):2782–2799
6. Sadeek SA, Negm NA, Hefni HH, Wahab MMA (2015) Metal adsorption by agricultural biosorbents: adsorption isotherm, kinetic and biosorbents chemical structures. Int J Biol Macromol 81:400–409
7. Padilla-Ortega E, Leyva-Ramos R, Flores-Cano J (2013) Binary adsorption of heavy metals from aqueous solution onto natural clays. Chem Eng J 225:535–546
8. Ethiraj AS, Uttam P, KV, Chong KF, Ali GAM (2020) Photocatalytic performance of a novel semiconductor nanocatalyst: copper doped nickel oxide for phenol degradation. Mater Chem Phys 242:122520
9. Gupta VK, Agarwal S, Sadegh H, Ali GAM, Bharti AK, Hamdy AS (2017) Facile route synthesis of novel graphene oxide-β-cyclodextrin nanocomposite and its application as adsorbent for removal of toxic bisphenol A from the aqueous phase. J Mol Liq 237:466–472
10. Singhon R, Husson J, Knorr M, Lakard B, Euvrard M (2012) Adsorption of Ni (II) ions on colloidal hybrid organic–inorganic silica composites. Colloids Surf, B 93:1–7
11. Gomaa H, Khalifa H, Selim M, Shenashen M, Kawada S, Alamoudi AS, Azzam A, Alhamid A, El-Safty S (2017) Selective, photoenhanced trapping/detrapping of arsenate anions using mesoporous blobfish head TiO_2 monoliths. ACS Sustain Chem Eng 5 (11):10826–10839
12. Osman NS, Sapawe N (2019) Waste material as an alternative source of silica precursor in silica nanoparticle synthesis—a review. Mater Today: Proc 19:1267–1272
13. Singh SP, Endley N (2020) Fabrication of nano-silica from agricultural residue and their application. In: Nanomaterials for agriculture and forestry applications. Elsevier, pp 107–134
14. Sapawe N, Osman NS, Zakaria MZ, Fikry SASSM, Aris MAM (2018) Synthesis of green silica from agricultural waste by sol-gel method. Mater Today: Proc 5(10):21861–21866
15. Mo J, Yang Q, Zhang N, Zhang W, Zheng Y, Zhang Z (2018) A review on agro-industrial waste (AIW) derived adsorbents for water and wastewater treatment. J Environ Manage 227:395–405
16. Gupta SS, Bhattacharyya KG (2011) Kinetics of adsorption of metal ions on inorganic materials: a review. Adv Coll Interface Sci 162(1–2):39–58
17. Sandoval OGM, Orozco AEL, Valenzuela S, Trujillo GCD (2019) Modified amorphous silica from a geothermal central as a metal adsorption agent for the regeneration of wastewater. Water Resour Ind 21:100105
18. Ali GAM, Habeeb OA, Algarni H, Chong KF (2018) CaO impregnated highly porous honeycomb activated carbon from agriculture waste: symmetrical supercapacitor study. J Mater Sci 54:683–692

19. Habeeb OA, Ramesh K, Ali GAM, Yunus RM (2017) Application of response surface methodology for optimization of palm kernel shell activated carbon preparation factors for removal of H2S from industrial wastewater. Jurnal Teknologi 79(7):1–10
20. Habeeb OA, Ramesh K, Ali GAM, Yunus RM (2017) Low-cost and eco-friendly activated carbon from modified palm kernel shell for hydrogen sulfide removal from wastewater: adsorption and kinetic studies. Desalin Water Treat 84:205–214
21. Habeeb OA, Ramesh K, Ali GAM, Yunus RM (2017) Experimental design technique on removal of hydrogen sulfide using CaO-eggshells dispersed onto palm kernel shell activated carbon: experiment, optimization, equilibrium and kinetic studies. J Wuhan Univ Technol-Mater Sci Ed 32(2):305–320
22. Habeeb OA, Ramesh K, Ali GAM, Yunus RM (2017) Isothermal modelling based experimental study of dissolved hydrogen sulfide adsorption from waste water using eggshell based activated carbon. Malaysian J Anal Sci 21(2):334–345
23. Gomaa H, El-Safty S, Shenashen M, Kawada S, Yamaguchi H, Abdelmottaleb M, Cheira M (2018) Three-dimensional, vertical platelets of ZnO carriers for selective extraction of cobalt ions from waste printed circuit boards. ACS Sustain Chem Eng 6(11):13813–13825
24. Kumar R, Chawla J (2014) Removal of cadmium ion from water/wastewater by nano-metal oxides: a review. Water Qual Exposure Health 5(4):215–226
25. Seyed Arabi SM, Lalehloo RS, Olyai MRTB, Ali GAM, Sadegh H (2019) Removal of congo red azo dye from aqueous solution by ZnO nanoparticles loaded on multiwall carbon nanotubes. Physica E 106:150–155
26. Abdel Ghafar HH, Ali GAM, Fouad OA, Makhlouf SA (2015) Enhancement of adsorption efficiency of methylene blue on Co_3O_4/SiO_2 nanocomposite. Desalin Water Treat 53 (11):2980–2989
27. Ali GAM, Barhoum A, Gupta VK, Nada AA, El-Maghrabi H, Kanthasamy R, Shaaban ER, Algarni H, Chong KF (2020) High surface area mesoporous silica for hydrogen sulfide effective removal. Curr Nanosci 16(2):226–234
28. Vunain E, Mishra A, Mamba B (2016) Dendrimers, mesoporous silicas and chitosan-based nanosorbents for the removal of heavy-metal ions: a review. Int J Biol Macromol 86:570–586
29. Poole CF (2015) High-performance precoated stationary phases. In: Poole CF (ed) Instrumental thin-layer chromatography. Elsevier, Boston, pp 31–51
30. Gitis V, Hankins N (2018) Water treatment chemicals: trends and challenges. J Water Process Eng 25:34–38
31. Hodges BC, Cates EL, Kim J-H (2018) Challenges and prospects of advanced oxidation water treatment processes using catalytic nanomaterials. Nat Nanotechnol 13(8):642–650
32. Speight JG (2020) Natural water remediation: chemistry and technology. Butterworth-Heinemann
33. Arévalo R, Pinedo S, Ballesteros E (2007) Changes in the composition and structure of Mediterranean rocky-shore communities following a gradient of nutrient enrichment: descriptive study and test of proposed methods to assess water quality regarding macroalgae. Mar Pollut Bull 55(1–6):104–113
34. Arnt TTF (1965) Device for collecting flotsam, especially waste oil spillage on a water surface. Google Patents
35. Manivasakam N (2005) Physico-chemical examination of water sewage and industrial effluents. Pragati Prakashan, India
36. Sadegh H, Ali GAM, Makhlouf ASH, Chong KF, Alharbi NS, Agarwal S, Gupta VK (2018) MWCNTs-Fe_3O_4 nanocomposite for Hg(II) high adsorption efficiency. J Mol Liq 258:345–353
37. Gupta K, Gupta D, Khatri OP (2019) Graphene-like porous carbon nanostructure from Bengal gram bean husk and its application for fast and efficient adsorption of organic dyes. Appl Surf Sci 476:647–657
38. Schwarzenbach RP, Escher BI, Fenner K, Hofstetter TB, Johnson CA, Von Gunten U, Wehrli B (2006) The challenge of micropollutants in aquatic systems. Science 313 (5790):1072–1077

39. Shannon MA, Bohn PW, Elimelech M, Georgiadis JG, Marinas BJ, Mayes AM (2010) Science and technology for water purification in the coming decades. In: Nanoscience and technology: a collection of reviews from nature Journals. World Scientific, pp 337–346
40. Richardson SD, Kimura SY (2016) Water analysis: emerging contaminants and current issues. Anal Chem 88(1):546–582
41. Giahi M, Pathania D, Agarwal S, Ali GAM, Chong KF, Gupta VK (2019) Preparation of Mg-doped TiO_2 nanoparticles for photocatalytic degradation of some organic pollutants. Stud Univ Babes-Bolyai, Chem 64(1):7–18
42. Lapworth D, Baran N, Stuart M, Ward R (2012) Emerging organic contaminants in groundwater: a review of sources, fate and occurrence. Environ Pollut 163:287–303
43. Kim K-H, Parker D, Li F, Tang K-T (2012) Sampling and analysis of odorants and volatile organic compounds in air, water, and soil. J Anal Methods Chem 1(1):308013
44. Lin D, Chen K, Chan EW-C, Chen S (2015) Increasing prevalence of ciprofloxacin-resistant food-borne Salmonella strains harboring multiple PMQR elements but not target gene mutations. Sci Rep 5(1):1–8
45. Abdel-Ghani NT, El-Chaghaby GA, Helal FS (2015) Individual and competitive adsorption of phenol and nickel onto multiwalled carbon nanotubes. J Adv Res 6(3):405–415
46. Zhang K, Sun P, Faye MCA, Zhang Y (2018) Characterization of biochar derived from rice husks and its potential in chlorobenzene degradation. Carbon 130:730–740
47. Sahin A, Ozer V, Akkaya F, Kaplan S (2018) Successful resuscitation of prolonged cardiac arrest occurring in association with 'skunk'and toluene toxicity. Turkish J Emergency Med 18(4):172–175
48. Qin Q, Xu Y (2016) Enhanced nitrobenzene adsorption in aqueous solution by surface silylated MCM-41. Microporous Mesoporous Mater 232:143–150
49. Stähelin PM, Valério A, Ulson S, da Silva A, Valle JAB, de Souza AAU (2018) Benzene and toluene removal from synthetic automotive gasoline by mono and bicomponent adsorption process. Fuel 231:45–52
50. Gioxari A, Kogiannou D, Kalogeropoulos N, Kaliora A (2016) Phenolic compounds: bioavailability and health effects. In: Encyclopedia of food and health. Amsterdam: Academic Press
51. Soares RO, Torres PH, da Silva ML, Pascutti PG (2016) Unraveling HIV protease flaps dynamics by constant pH molecular dynamics simulations. J Struct Biol 195(2):216–226
52. Zhao F, Repo E, Yin D, Meng Y, Jafari S, Sillanpää M (2015) EDTA-cross-linked β-cyclodextrin: an environmentally friendly bifunctional adsorbent for simultaneous adsorption of metals and cationic dyes. Environ Sci Technol 49(17):10570–10580
53. Zhu W, Liu L, Liao Q, Chen X, Qian Z, Shen J, Liang J, Yao J (2016) Functionalization of cellulose with hyperbranched polyethylenimine for selective dye adsorption and separation. Cellulose 23(6):3785–3797
54. Uwizeyimana H, Wang M, Chen W, Khan K (2017) The eco-toxic effects of pesticide and heavy metal mixtures towards earthworms in soil. Environ Toxicol Pharmacol 55:20–29
55. Li Y, Meas A, Shan S, Yang R, Gai X (2016) Production and optimization of bamboo hydrochars for adsorption of Congo red and 2-naphthol. Biores Technol 207:379–386
56. Thekkudan VN, Vaidyanathan VK, Ponnusamy SK, Charles C, Sundar S, Vishnu D, Anbalagan S, Vaithyanathan VK, Subramanian S (2016) Review on nanoadsorbents: a solution for heavy metal removal from wastewater. IET Nanobiotechnol 11(3):213–224
57. Lee SP, Ali GAM, Algarni H, Chong KF (2019) Flake size-dependent adsorption of graphene oxide aerogel. J Mol Liq 277:175–180
58. Bode-Aluko CA, Pereao O, Ndayambaje G, Petrik L (2017) Adsorption of toxic metals on modified polyacrylonitrile nanofibres: a review. Water Air Soil Pollut 228(1):35
59. Manirethan V, Raval K, Balakrishnan RM (2020) Adsorptive removal of trivalent and pentavalent arsenic from aqueous solutions using iron and copper impregnated melanin extracted from the marine bacterium Pseudomonas stutzeri. Environ Pollut 257:113576

60. Singh J, Sharma M, Basu S (2018) Heavy metal ions adsorption and photodegradation of remazol black XP by iron oxide/silica monoliths: kinetic and equilibrium modelling. Adv Powder Technol 29(9):2268–2279
61. Krewski D, Yokel RA, Nieboer E, Borchelt D, Cohen J, Harry J, Kacew S, Lindsay J, Mahfouz AM, Rondeau V (2007) Human health risk assessment for aluminium, aluminium oxide, and aluminium hydroxide. J Toxicol Environ Health, Part B 10(S1):1–269
62. Cooper RG, Harrison AP (2009) The exposure to and health effects of antimony. Indian J Occup Environ Med 13(1):3
63. Chung J-Y, Yu S-D, Hong Y-S (2014) Environmental source of arsenic exposure. J Preventive Med Public Health 47(5):253
64. Kravchenko J, Darrah TH, Miller RK, Lyerly HK, Vengosh A (2014) A review of the health impacts of barium from natural and anthropogenic exposure. Environ Geochem Health 36 (4):797–814
65. Heffron J, Marhefke M, Mayer BK (2016) Removal of trace metal contaminants from potable water by electrocoagulation. Sci Rep 6:28478
66. Samaratunga SS, Nishimoto J, Tabata M (2008) Extraction of chromium (VI) by salting-out with a homogeneous, mixed solvent of water and 2-propanol: a laboratory study. Environ Sci Pollut Res 15(1):27
67. Li M, Liu Z, Wang S, Calatayud DG, Zhu W-H, James TD, Wang L, Mao B, Xiao H-N (2018) Fluorescence detection and removal of copper from water using a biobased and biodegradable 2D soft material. Chem Commun 54(2):184–187
68. Sayed A, Othman IMM, Hamam M, Gomaa H, Gadallah MI, Mostfa MA, Ali HRH, Emran MY, Abdel-Hakim M, Mahross MH (2021) A novel fluorescent sensor for fast and highly selective turn-off detection of Fe3+ in water and pharmaceutical samples using synthesized azopyrazole-benzenesulfonamide derivative. J Mol Struct 1225:129175
69. La DD, Thi HPN, Nguyen TA, Bhosale SV (2017) Effective removal of Pb (II) using a graphene@ ternary oxides composite as an adsorbent in aqueous media. New J Chem 41 (23):14627–14634
70. Tobiason JE, Bazilio A, Goodwill J, Mai X, Nguyen C (2016) Manganese removal from drinking water sources. Curr Pollut Rep 2(3):168–177
71. Ahmed S, Brockgreitens J, Xu K, Abbas A (2017) A nanoselenium sponge for instantaneous mercury removal to undetectable levels. Adv Func Mater 27(17):1606572
72. Hasan M, Begum L, Hosain S, Poddar P, Chowdhury A, Ali F (2017) Study on heavy metals (Zinc and Lead) in drinking water of Tannery area, adjacent areas and outside village areas. J Environ Anal Toxicol 7(2):1–3
73. Tizro S, Baseri H (2017) Removal of cobalt ions from contaminated water using magnetite based nanocomposites: effects of various parameters on the removal efficiency. J Water Environ Nanotechnol 2(3):174–185
74. Gomaa H, Shenashen M, Yamaguchi H, Alamoudi A, El-Safty S (2018) Extraction and recovery of Co^{2+} ions from spent lithium-ion batteries using hierarchical mesosponge γ-Al_2O_3 monolith extractors. Green Chem 20(8):1841–1857
75. Abdien HG, Cheira MF, Abd-Elraheem MA, Saef El-Naser TA, Zidan IH (2016) Extraction and pre-concentration of uranium using activated carbon impregnated trioctyl phosphine oxide. Elixir Appl Chem 100:43462
76. Shiraishi K, Igarashi Y, Yamamoto M, Nakajima T, Los I, Zelensky A, Buzinny M (1994) Concentrations of thorium and uranium in freshwater samples collected in the former USSR. J Radioanal Nucl Chem 185(1):157–165
77. Agamuthu P (2009) Challenges and opportunities in agro-waste management: an Asian perspective. In: Inaugural meeting of first regional 3R forum in Asia
78. Nagendran R (2011) Agricultural waste and pollution. In: Waste. Elsevier
79. Ali GAM, Divyashree A, Supriya S, Chong KF, Ethiraj AS, Reddy MV, Algarni H, Hegde G (2017) Carbon nanospheres derived from Lablab purpureus for high performance supercapacitor electrodes: a green approach. Dalton Trans 46(40):14034–14044

80. Hegde G, Abdul Manaf SA, Kumar A, Ali GAM, Chong KF, Ngaini Z, Sharma KV (2015) Biowaste sago bark based catalyst free carbon nanospheres: waste to wealth approach. ACS Sustain Chem Eng 5(9):2247–2253
81. Dai Y, Sun Q, Wang W, Lu L, Liu M, Li J, Yang S, Sun Y, Zhang K, Xu J (2018) Utilizations of agricultural waste as adsorbent for the removal of contaminants: a review. Chemosphere 211:235–253
82. Praneetha S, Murugan AV (2015) Development of sustainable rapid microwave assisted process for extracting nanoporous Si from earth abundant agricultural residues and their carbon-based nanohybrids for lithium energy storage. ACS Sustain Chem Eng 3(2):224–236
83. Tabata S, Iida H, Horie T, Yamada S (2010) Hierarchical porous carbon from cellassemblies of rice husk for in vivo applicationsElectronic supplementary information (ESI) available: the experimental procedure and materials characterization data. MedChemComm 1(2):136. https://doi.org/10.1039/c0md00011f
84. Bakar RA, Yahya R, Gan SN (2016) Production of high purity amorphous silica from rice husk. Proc Chem 19:189–195
85. Rovani S, Santos JJ, Corio P, Fungaro DA (2018) Highly pure silica nanoparticles with high adsorption capacity obtained from sugarcane waste ash. ACS Omega 3(3):2618–2627
86. Kauldhar BS, Yadav SK (2018) Turning waste to wealth: a direct process for recovery of nano-silica and lignin from paddy straw agro-waste. J Clean Prod 194:158–166
87. Naddaf M, Kafa H, Ghanem I (2020) Extraction and characterization of nano-silica from olive stones. Silicon 12(1):185–192
88. Assefi M, Davar F, Hadadzadeh H (2015) Green synthesis of nanosilica by thermal decomposition of pine cones and pine needles. Adv Powder Technol 26(6):1583–1589
89. Chen H, Wang F, Zhang C, Shi Y, Jin G, Yuan S (2010) Preparation of nano-silica materials: the concept from wheat straw. J Non-Cryst Solids 356(50–51):2781–2785
90. Zhang X, Zhao Z, Ran G, Liu Y, Liu S, Zhou B, Wang Z (2013) Synthesis of lignin-modified silica nanoparticles from black liquor of rice straw pulping. Powder Technol 246:664–668
91. Anuar MF, Fen YW, Zaid MHM, Matori KA, Khaidir REM (2018) Synthesis and structural properties of coconut husk as potential silica source. Results Phys 11:1–4
92. Mor S, Manchanda CK, Kansal SK, Ravindra K (2017) Nanosilica extraction from processed agricultural residue using green technology. J Clean Prod 143:1284–1290
93. Amin N-u, Khattak S, Noor S, Ferroze I (2016) Synthesis and characterization of silica from bottom ash of sugar industry. J Clean Prod 117:207–211
94. Mohtasham N, Gholizadeh M (2020) Nano silica extracted from horsetail plant as a natural silica support for the synthesis of H3PW12O40 immobilized on aminated magnetic nanoparticles (Fe3O4@ SiO2-EP-NH-HPA): a novel and efficient heterogeneous nanocatalyst for the green one-pot synthesis of pyrano [2, 3-c] pyrazole derivatives. Research on Chemical Intermediates
95. Baldé C, Forti V, Gray V, Kuehr R, Stegmann P (2017) The Global E-waste Monitor–2017, United Nations University (UNU), International Telecommunication Union (ITU) & International Solid Waste Association (ISWA), Bonn/Geneva/Vienna. Electronic Version:978–992
96. Kumar A, Holuszko M, Espinosa DCR (2017) E-waste: an overview on generation, collection, legislation and recycling practices. Resour Conserv Recycl 122:32–42
97. Khan A, Asiri AM (2019) E-waste recycling and management: present scenarios and environmental issues, vol 33. Springer
98. Baxter J, Lyng K-A, Askham C, Hanssen OJ (2016) High-quality collection and disposal of WEEE: environmental impacts and resultant issues. Waste Manag 57:17–26
99. Awasthi AK, Li J, Koh L, Ogunseitan OA (2019) Circular economy and electronic waste. Nat Electron 2(3):86–89
100. Bazargan A, Bwegendaho D, Barford J, McKay G (2014) Printed circuit board waste as a source for high purity porous silica. Sep Purif Technol 136:88–93

101. Liou T-H (2011) A green route to preparation of MCM-41 silicas with well-ordered mesostructure controlled in acidic and alkaline environments. Chem Eng J 171(3):1458–1468
102. Lin L-Y, Bai H (2013) Efficient method for recycling silica materials from waste powder of the photonic industry. Environ Sci Technol 47(9):4636–4643
103. Liu Y, Kong J, Zhuang Y, Xing P, Yin H, Luo X (2019) Recycling high purity silicon from solar grade silicon cutting slurry waste by carbothermic reduction in the electric arc furnace. J Clean Prod 224:709–718
104. Woditsch P, Koch W (2002) Solar grade silicon feedstock supply for PV industry. Sol Energy Mater Sol Cells 72(1-4):11–26
105. Wang B, Zhou Y, Li L, Wang Y (2018) Preparation of amidoxime-functionalized mesoporous silica nanospheres (ami-MSN) from coal fly ash for the removal of U (VI). Sci Total Environ 626:219–227
106. Vashistha P, Singh S, Dutt D, Kumar V (2019) Sustainable utilization of paper mill solid wastes via synthesis of nano silica for production of belite based clinker. J Clean Prod 224:557–565
107. Zhang S, Ravi S, Lee Y-R, Ahn J-W, Ahn W-S (2019) Fly ash-derived mesoporous silica foams for CO_2 capture and aqueous Nd^{3+} adsorption. J Ind Eng Chem 72:241–249
108. Li C-C, Qiao X-C (2016) A new approach to prepare mesoporous silica using coal fly ash. Chem Eng J 302:388–394
109. Yan F, Jiang J, Tian S, Liu Z, Shi J, Li K, Chen X, Xu Y (2016) A green and facile synthesis of ordered mesoporous nanosilica using coal fly ash. ACS Sustain Chem Eng 4(9):4654–4661
110. Wang S, Wu H (2006) Environmental-benign utilisation of fly ash as low-cost adsorbents. J Hazard Mater 136(3):482–501
111. Hossain SS, Mathur L, Roy P (2018) Rice husk/rice husk ash as an alternative source of silica in ceramics: a review. J Asian Ceram Societies 6(4):299–313
112. Chandrasekar G, You K-S, Ahn J-W, Ahn W-S (2008) Synthesis of hexagonal and cubic mesoporous silica using power plant bottom ash. Microporous Mesoporous Mater 111(1–3):455–462
113. Liang G, Li Y, Yang C, Zi C, Zhang Y, Hu X, Zhao W (2020) Production of biosilica nanoparticles from biomass power plant fly ash. Waste Manag 105:8–17
114. Imoisili PE, Ukoba KO, Jen T-C (2019) Synthesis and characterization of amorphous mesoporous silica from palm kernel shell ash. Boletín de la Sociedad Española de Cerámica y Vidrio
115. Falk G, Shinhe G, Teixeira L, Moraes E, de Oliveira AN (2019) Synthesis of silica nanoparticles from sugarcane bagasse ash and nano-silicon via magnesiothermic reactions. Ceram Int 45(17):21618–21624
116. Alam MM, Hossain MA, Hossain MD, Johir M, Hossen J, Rahman MS, Zhou JL, Hasan A, Karmakar AK, Ahmed MB (2020) The potentiality of rice husk-derived activated carbon: from synthesis to application. Processes 8(2):203
117. Alves RH, Reis TVDS, Rovani S, Fungaro DA (2017) Green synthesis and characterization of biosilica produced from sugarcane waste ash. J Chem:6129035
118. Athinarayanan J, Periasamy VS, Alhazmi M, Alatiah KA, Alshatwi AA (2015) Synthesis of biogenic silica nanoparticles from rice husks for biomedical applications. Ceram Int 41 (1):275–281
119. Pieła A, Żymańczyk-Duda E, Brzezińska-Rodak M, Duda M, Grzesiak J, Saeid A, Mironiuk M, Klimek-Ochab M (2020) Biogenic synthesis of silica nanoparticles from corn cobs husks: dependence of the productivity on the method of raw material processing. Bioorg Chem:103773
120. Ahmad-Alyosef H, Uhlig H, Münster T, Kloess G, Einicke W-D, Gläser R, Enke D (2014) Biogenic silica from rice husk ash—sustainable sources for the synthesis of value added silica. Chem Eng Trans 37:667–672
121. Dutta T, Kim K-H, Deep A, Szulejko JE, Vellingiri K, Kumar S, Kwon EE, Yun S-T (2018) Recovery of nanomaterials from battery and electronic wastes: a new paradigm of environmental waste management. Renew Sustain Energy Rev 82:3694–3704

122. Rizal S, Fizree H, Hossain MS, Gopakumar DA, Ni ECW, Khalil HA (2020) The role of silica-containing agro-industrial waste as reinforcement on physicochemical and thermal properties of polymer composites. Heliyon 6(3):e03550
123. Lin Y-C, Lin L-Y, Gao M-Y, Fang Y-P (2013) Mesoporous silica nanoparticles synthesized from liquid crystal display manufacturing extracts as a potential candidate for a drug delivery carrier: evaluation of their safety and biocompatibility. Int J Nanomed 8:3833
124. Wang Y, Chen J, Lei X, Ren Y, Wu J (2018) Preparation of high silica microporous zeolite SSZ-13 using solid waste silica fume as silica source. Adv Powder Technol 29(5):1112–1118
125. Wang B, Zhou Y, Li L, Xu H, Sun Y, Du Y, Wang Y (2018) In situ synthesis of TiO_2-doped mesoporous silica from coal fly ash for the photocatalytic degradation of dyes. Ind Eng Chem Res 57(46):15632–15637
126. de Oliveira FF, Moura KO, Costa LS, Vidal CB, Loiola AR, do Nascimento RF (2020) Reactive adsorption of parabens on synthesized micro-and mesoporous silica from coal fly ash: pH effect on the modification process. ACS Omega 5(7):3346–3357
127. Gao M, Ma Q, Lin Q, Chang J, Ma H (2017) A novel approach to extract SiO_2 from fly ash and its considerable adsorption properties. Mater Des 116:666–675
128. Yang H, Xia Y (2007) Bionanotechnology: enabling biomedical research with nanomaterials. Adv Mater 19(20):3085–3087
129. Fu F, Wang Q (2011) Removal of heavy metal ions from wastewaters: a review. J Environ Manage 92(3):407–418
130. Zhang X, Ye C, Pi K, Huang J, Xia M, Gerson AR (2019) Sustainable treatment of desulfurization wastewater by ion exchange and bipolar membrane electrodialysis hybrid technology. Sep Purif Technol 211:330–339
131. Chan B, Dudeney A (2008) Reverse osmosis removal of arsenic residues from bioleaching of refractory gold concentrates. Miner Eng 21(4):272–278
132. Reyes-Serrano A, López-Alejo JE, Hernández-Cortázar MA, Elizalde I (2020) Removing contaminants from tannery wastewater by chemical precipitation using CaO and $Ca(OH)_2$. Chin J Chem Eng 28(4):1107–1111
133. Yu M, Wang J, Tang L, Feng C, Liu H, Zhang H, Peng B, Chen Z, Xie Q (2020) Intimate coupling of photocatalysis and biodegradation for wastewater treatment: mechanisms, recent advances and environmental applications. Water Res:115673
134. Meng X, Khoso SA, Jiang F, Zhang Y, Yue T, Gao J, Lin S, Liu R, Gao Z, Chen P (2020) Removal of chemical oxygen demand and ammonia nitrogen from lead smelting wastewater with high salts content using electrochemical oxidation combined with coagulation–flocculation treatment. Sep Purif Technol 235:116233
135. Yu K, Sun C, Zhang B, Hassan M, He Y (2019) Size-dependent adsorption of antibiotics onto nanoparticles in a field-scale wastewater treatment plant. Environ Pollut 248:1079–1087
136. Ray PZ, Shipley HJ (2015) Inorganic nano-adsorbents for the removal of heavy metals and arsenic: a review. RSC Adv 5(38):29885–29907
137. Lakherwal D (2014) Adsorption of heavy metals: a review. Int J Environ Res Develop 4 (1):41–48
138. Kolahalam LA, Viswanath IK, Diwakar BS, Govindh B, Reddy V, Murthy Y (2019) Review on nanomaterials: synthesis and applications. Mater Today: Proc 18:2182–2190
139. Sadegh H, Shahryari GR, Masjedi A, Mahmoodi Z, Kazemi M (2016) A review on carbon nanotubes adsorbents for the removal of pollutants from aqueous solutions. Int J Nano Dimension 7(2):109–120
140. Sadegh H, Ali GAM, Gupta VK, Makhlouf ASH, Shahryari-ghoshekandi R, Nadagouda MN, Sillanpää M, Megiel E (2017) The role of nanomaterials as effective adsorbents and their applications in wastewater treatment. J Nanostruct Chem 7(1):1–14
141. Anjum M, Miandad R, Waqas M, Gehany F, Barakat M (2019) Remediation of wastewater using various nano-materials. Arabian J Chem 12(8):4897–4919

142. Gao Z, Qiao M, Tan M, Peng H, Ding L (2020) Surface functionalization of mesoporous silica nanoparticles with pyronine derivative for selective detection of hydrogen sulfide in aqueous solution. Colloids Surf, A 586:124194
143. Thakkar A, Raval A, Chandra S, Shah M, Sircar A (2019) A comprehensive review of the application of nano-silica in oil well cementing. Petroleum
144. Hao S, Verlotta A, Aprea P, Pepe F, Caputo D, Zhu W (2016) Optimal synthesis of amino-functionalized mesoporous silicas for the adsorption of heavy metal ions. Microporous Mesoporous Mater 236:250–259
145. Ma X-k, Lee N-H, Oh H-J, Kim J-W, Rhee C-K, Park K-S, Kim S-J (2010) Surface modification and characterization of highly dispersed silica nanoparticles by a cationic surfactant. Colloids Surf, A 358(1–3):172–176
146. Holban AM, Grumezescu AM, Andronescu E (2016) Inorganic nanoarchitectonics designed for drug delivery and anti-infective surfaces. In: Surface chemistry of nanobiomaterials. Elsevier, pp 301–327
147. Xie Y, Kocaefe D, Chen C, Kocaefe Y (2016) Review of research on template methods in preparation of nanomaterials. J Nanomater
148. Wang M, Wen B, Fan B, Zhang H (2020) Study on adsorption mechanism of silicate adsorbents with different morphologies and pore structures towards formaldehyde in water. Colloids Surf, A:124887
149. Kang D, Yu X, Ge M (2017) Morphology-dependent properties and adsorption performance of CeO_2 for fluoride removal. Chem Eng J 330:36–43
150. Schwieger W, Machoke AG, Weissenberger T, Inayat A, Selvam T, Klumpp M, Inayat A (2016) Hierarchy concepts: classification and preparation strategies for zeolite containing materials with hierarchical porosity. Chem Soc Rev 45(12):3353–3376
151. Clauset A, Moore C, Newman ME (2008) Hierarchical structure and the prediction of missing links in networks. Nature 453(7191):98–101
152. Lee Z, Zaini M, Tang S (2019) Porous nanomaterials for heavy metal removal. Springer, Cham, Switzerland
153. Sandoval OGM, Trujillo GCD, Orozco AEL (2018) Amorphous silica waste from a geothermal central as an adsorption agent of heavy metal ions for the regeneration of industrial pre-treated wastewater. Water Resour Ind 20:15–22
154. Abdullah N, Yusof N, Lau W, Jaafar J, Ismail A (2019) Recent trends of heavy metal removal from water/wastewater by membrane technologies. J Ind Eng Chem 76:17–38
155. Gomaa H, Shenashen MA, Elbaz A, Yamaguchi H, Abdelmottaleb M, El-Safty SA (2021) Mesoscopic engineering materials for visual detection and selective removal of copper ions from drinking and waste water sources. J Hazard Mater 406:124314
156. Azimi A, Azari A, Rezakazemi M, Ansarpour M (2017) Removal of heavy metals from industrial wastewaters: a review. ChemBioEng Rev 4(1):37–59
157. Wadhawan S, Jain A, Nayyar J, Mehta SK (2020) Role of nanomaterials as adsorbents in heavy metal ion removal from waste water: a review. J Water Process Eng 33:101038
158. Omotunde I, Okoronkwo A, Oluwashina O (2018) Derived and thiourea-functionalized silica for cadmium removal: isotherm, kinetic and thermodynamic studies. Appl Water Sci 8 (1):21
159. Ahmad N, Sereshti H, Mousazadeh M, Nodeh HR, Kamboh MA, Mohamad S (2019) New magnetic silica-based hybrid organic-inorganic nanocomposite for the removal of lead (II) and nickel (II) ions from aqueous solutions. Mater Chem Phys 226:73–81
160. El-Nahhal IM, El-Ashgar NM (2007) A review on polysiloxane-immobilized ligand systems: synthesis, characterization and applications. J Organomet Chem 692(14):2861–2886
161. Shahata MM (2016) Adsorption of some heavy metal ions by used different immobilized substances on silica gel. Arabian J Chem 9(6):755–763
162. Da'na E (2017) Adsorption of heavy metals on functionalized-mesoporous silica: a review. Microporous Mesoporous Mater 247:145–157

163. Mureseanu M, Reiss A, Stefanescu I, David E, Parvulescu V, Renard G, Hulea V (2008) Modified SBA-15 mesoporous silica for heavy metal ions remediation. Chemosphere 73 (9):1499–1504
164. Goubert-Renaudin S, Etienne M, Sp B, Meyer M, Denat F, Lebeau B, Walcarius A (2009) Factors affecting copper (II) binding to multiarmed cyclam-grafted mesoporous silica in aqueous solution. Langmuir 25(17):9804–9813
165. Costa JAS, de Jesus RA, da Silva CMP, Romão LPC (2017) Efficient adsorption of a mixture of polycyclic aromatic hydrocarbons (PAHs) by Si–MCM–41 mesoporous molecular sieve. Powder Technol 308:434–441
166. Costa JA, Garcia AC, Santos DO, Sarmento VH, de Mesquita ME, Romao LP (2015) Applications of inorganic–organic mesoporous materials constructed by self-assembly processes for removal of benzo [k] fluoranthene and benzo [b] fluoranthene. J Sol-Gel Sci Technol 75(3):495–507
167. Costa JAS, de Jesus RA, Santos DO, Mano JF, Romao LP, Paranhos CM (2020) Recent progresses in the adsorption of organic, inorganic, and gas compounds by MCM-41-based mesoporous materials. Microporous Mesoporous Mater 291:109698
168. Awad AM, Shaikh SM, Jalab R, Gulied MH, Nasser MS, Benamor A, Adham S (2019) Adsorption of organic pollutants by natural and modified clays: a comprehensive review. Sep Purif Technol:115719
169. Mahmoudi F, Amini MM (2020) Confined crystallization of microporous metal-organic framework within mesoporous silica with enhanced hydrostability: Ultrafast removal of organic dyes from aqueous solutions by MIL-68 (Al)@ SBA-15 composite. J Water Process Eng 35:101227
170. Gusain R, Gupta K, Joshi P, Khatri OP (2019) Adsorptive removal and photocatalytic degradation of organic pollutants using metal oxides and their composites: a comprehensive review. Adv Colloid Interface Sci:102009
171. Tahervand S, Jalali M (2017) Sorption and desorption of potentially toxic metals (Cd, Cu, Ni and Zn) by soil amended with bentonite, calcite and zeolite as a function of pH. J Geochem Explor 181:148–159
172. Naguib DM, Badawy NM (2020) Phenol removal from wastewater using waste products. J Environ Chem Eng 8(1):103592
173. Wang H, Gao H, Chen M, Xu X, Wang X, Pan C, Gao J (2016) Microwave-assisted synthesis of reduced graphene oxide/titania nanocomposites as an adsorbent for methylene blue adsorption. Appl Surf Sci 360:840–848
174. Zhang L, Zhu M, Guo L, Li L, Shi J (2009) Bilirubin adsorption property of mesoporous silica and amine-grafted mesoporous silica. Nano-Micro Lett 1(1):14–18
175. Okiel K, El-Sayed M, El-Kady MY (2011) Treatment of oil–water emulsions by adsorption onto activated carbon, bentonite and deposited carbon. Egyptian J Pet 20(2):9–15
176. Awad AM, Jalab R, Benamor A, Naser MS, Ba-Abbad MM, El-Naas M, Mohammad AW (2020) Adsorption of organic pollutants by nanomaterial-based adsorbents: an overview. J Mol Liq:112335
177. Zhang Y-y, Liu Q, Yang C, Wu S-c, Cheng J-h (2019) Magnetic aluminum-based metal organic framework as a novel magnetic adsorbent for the effective removal of minocycline from aqueous solutions. Environ Pollut 255:113226
178. Parida SK, Dash S, Patel S, Mishra B (2006) Adsorption of organic molecules on silica surface. Adv Coll Interface Sci 121(1–3):77–110
179. Jadhav SA, Garud HB, Patil AH, Patil GD, Patil CR, Dongale TD, Patil PS (2019) Recent advancements in silica nanoparticles based technologies for removal of dyes from water. Colloid Interface Sci Commun 30:100181
180. Arica TA, Kuman M, Gercel O, Ayas E (2019) Poly (dopamine) grafted bio-silica composite with tetraethylenepentamine ligands for enhanced adsorption of pollutants. Chem Eng Res Des 141:317–327

Extraction of Silica and Lignin-Based Nanocomposite Materials from Agricultural Waste for Wastewater Treatment Using Photocatalysis Technique

Radwa A. El-Salamony and Asmaa M. El Shafey

Abstract Nowadays, agricultural waste has become increasingly alarming, as it can cause considerable environmental impact. However, it can be utilized for many applications, such as energy production, chemical recovery, and dye adsorptions. Henceforth, rice husk (RH) and ash (RHA) have been used as substitutes to produce silica. Porous and high surface area silica can be widely used in adsorption, separation, thermal insulation, and catalysis. Therefore, it is of great importance to find an economical way to produce nanosilica. Besides, lignin has been underused among all the fractions of biomass because of its complicated structure. Apart from cellulose and hemicellulose, lignin has a significant fraction of biomass. In comparison to other natural adsorbents, lignin residues contain agricultural and wood residues and can be extracted during the precipitation process from black liquor. In addition, lignin residues have higher bio-adsorption capacity and affinity. On the other hand, the industrial activities in developed countries are generating pollutants as wastes into the water stream. The photocatalysis technique is considered as one of the most sustainable approaches, especially for the treatment of water from pollutants. However, its usage has some limitations, such as its high bandgap energy for semiconductor materials, the limitation of photoresponse under sunlight, the short lifetime of active e^-/h^+ radicals and the difficulty of separation from water. To overcome these challenges, many techniques have been used and adopted: metal or non-metallic doping, surface modification, using high surface area support as silica, alumina, lignin and carbon materials. This chapter aims to summarize the recent progress in extracted silica and lignin from agricultural waste and their applications as heterogeneous photocatalysts for wastewater treatment.

Keywords Photocatalysis · Agricultural waste · Silica nanoparticles · Lignin · Wastewater treatment

---

R. A. El-Salamony (✉)
Egyptian Petroleum Research Institute, Nasr City, Cairo 11727, Egypt
e-mail: radwa2005@hotmail.com

A. M. El Shafey
Faculty of Science and Arts, King Khalid University, Sarat Ebida, Abha, Saudi Arabia

A. S. H. Makhlouf and G. A. M. Ali (eds.), *Waste Recycling Technologies for Nanomaterials Manufacturing*, Topics in Mining, Metallurgy and Materials Engineering, https://doi.org/10.1007/978-3-030-68031-2_13

List of Abbreviations

AFM	Atomic force microscopic
CTAB	Cetyle trimethyl ammonium bromide
PEG	Poly ethylene glycol
RHA	Rice husk ash
RH	Rice husk
TEM	Transmission electron microscope
TEOS	Tetra ethyl ortho-silicate
UV	Ultra violet
AFM	Atomic force microscopic

1 Introduction

Rice husk is a derivative in the agricultural manufacturing [1]. China and India are the main countries producing rice husk. It is isolated from rice grains through the milling progress, due to its low nutrition value. The separated rice husk is usually being used as animal feed, burning fuel and other applications after pyrolysis [2–4]. Burning rice husk under a controlled temperature produces ash with a composition of 60% amorphous silica, 10-40% carbon, and an insignificant amount of minerals [5]. Rice husk ash can also be used in the construction material [3, 6], as the adsorbent for hazardous pollutants removal [7–9] as well as the production of silica composite or activated carbon [10, 11].

Silicon dioxide (SiO_2), or commonly known as silica, is one of the valuable inorganic compounds with high reusability. Silicon is accounted for approximately 25.7% of earth's crust weight and considered as the most abundant element in the earth's crust after oxygen [12]. Silica forms quartz, sand, or flint, which exist in gel, crystalline, or amorphous forms. Nowadays, most silica is produced from quartz or sand by smelting quartz sand with sodium carbonate at 1300 °C [13]. However, it is a costly process [14]. Silica is widely used in many industrially significant products such as electronics, ceramic, pharmaceutics, detergents, adhesives and polymer materials. Due to progressive properties like small-diameter particles, high surface area, ultrafine, silica powders have many high-tech applications, such as thixotropic agents, thermal insulators, composite fillers, solar cell, etc. [15–19]. Some of the natural herbal silica source (an agricultural waste or industrial by-product) makes it extractable at a lower temperature and hence provides a low energy method as an alternative to the current high-energy methods of silica extraction [20].

On the other hand, the vascular plants in the woody stems of arborescent angiosperms (hardwoods) contain a cell wall component, which is called lignin [21]. The thing that distinguishes lignin content is the variation of its content between 15 and 40%. Lignin acts as a fundamental role in managing water transport through the cell wall as it functions as a water sealant in the stems. It also offers protection for plants by inhibiting biological attack by penetration of hampering enzyme. Lignin has numerous advantages as it acquires the woody stems' rigidity and impact resistance plus its permanent glue, which could bond the cells together in the woody stems. The hardwoods contain 18–25% lignin content, but the soft one contains 25–35%, and the cellulose content is between 41 and 45% [22].

In comparison to other natural adsorbents, lignin residues contain agricultural and wood residues and can be extracted during the precipitation process from black liquor; they have higher bio-adsorption capacity and affinity. Lignin has several advantages based on its structure as it possesses a hydroxyl group that plays a fundamental role in its interaction with water. The first idea about lignin from many centuries ago was that it was a very cheap by-product without any commercial value and was treated as waste material. The industrial revolution catalysed the usage of lignin in polymers synthesis such as polyester, polyethers and polystyrene derivatives due to its ability to recover a wide range of chemicals, as discussed by Hatakeyama et al. [23]. Lignin had also been utilized as integrated material in several industrial applications. Natural lignin provides several advantages than synthetic ones, such as biodegradability, availability in various industrial wastes, eco-friendly, low cost, as well as stabilizing, antimicrobial and antioxidant properties. The applications of lignin vary according to its extraction process level as it determines the physicochemical properties. Various functional groups can be found in softwood, hardwood and kraft lignin, which determine their properties according to the reactivity and functionality.

Besides, silica and lignin are extracted from renewable resources such as biomass, and they will form low-cost and eco-friendly material. Both of them can be utilized as photocatalyst support for its potential application in contaminant photodegradation. The interaction between the components in hybrid photocatalysts acts as an essential role in detecting photocatalytic performance. The application of silica and lignin composite material as photocatalyst support provides uniform particle distribution that affects the properties of the photocatalyst. However, several limitations, such as high bandgap energy of semiconductors, low photoresponse under visible light region and recombination of electron-hole, have to be overcome before the utilization of silica and lignin composite material in a commercial product.

2 Preparation of Silica from Rice Husk Waste

Sol–gel synthesis is one of the most common methods of converting ash into silica gel [24–28]. In this method, silica is synthesized from ash through simultaneous hydrolysis and condensation reaction where a sol of sodium silicate, silicon alkoxide or halide gels are converted into a polymeric network of gel [29]. The synthesis of silica using this method leads to silica precipitation under certain conditions like the restriction of gel growth that involves coagulation and precipitation step during its preparation [30].

2.1 *Strong Acid Leaching Treatment Method*

Leaching procedure is an effective form of silica extraction [31, 32]. In the typical method, the first step, silica is dissected as sodium silicate (Na_2SiO_3) from ash using sodium hydroxide (NaOH). In the second step, silica is precipitated with sulphuric acid from sodium silicate [33, 34]; according to Eqs. (1) and (2):

$$SiO_2 + 2NaOH \rightarrow Na_2SiO_3 + H_2O \quad (1)$$

$$Na_2SiO_3 + H_2SO_4 \rightarrow SiO_2 + Na_2SO_4 + H_2O \quad (2)$$

This process will eliminate any impurities in the agricultural wastes [35, 36]. Many investigations have been conducted on strong acid leaching of rice husk to produce silica such as HCl [37–39], HNO_3 [40], H_2SO_4 [41]. Boiling before thermal treatment is proven to be effective in removing most metallic impurities substantially and producing entirely white ash-silica in a high specific surface area [42].

2.1.1 Porous Silica

Porous silica with high surface areas can be synthesized from rice husk char within 10 h using polyethylene glycol (PEG, molecular weight = 20,000) as a template [43]. The preparation method mainly includes the production of sodium silicate, then precipitation by ortho-phosphoric acid and calcination. The textural properties of silica are affected by the amount of used PEG where the PEG can be detached from the composites by calcination. Changing the PEG dosage from 100 to 176 mg could produce porous silica with specific surface areas ranging from 709 to 936 $m^2\ g^{-1}$.

2.1.2 Silica Aerogel

The nano-porous silica aerogel can be prepared from RHA by the sol–gel method [44]. A small amount of TEOS is added before the gelation of hydrosol, followed by washing with ethanol and finally drying the pre-treated silica gel at atmospheric pressure for 10 h to yield the silica aerogel [25, 45, 46]. Silica aerogel can also be produced by the same process using water as the pore fluid without TEOS doped [47]. Adding TEOS increased mainly the texture properties of the prepared silica aerogel and decreased its density [26–28, 46]. It has a surface area of 315 $m^2\ g^{-1}$, an average pore size of 9.8 nm and a pore volume of 0.78 $cm^3\ g^{-1}$.

2.1.3 Spheroid Silica

Nanosized amorphous silica was prepared by thermally treated Vietnamese rice husk at 600 °C following the sol-gel method [48]. The effect of adding CTAB surfactant during the preparation method on the texture properties of SiO_2 had been reported. The results indicated that the dispersed and uniform particle size were produced in the increasing of CTAB concentration from 0.5 to 2.0 wt%; but, above this concentration, the particle aggregation occurred. Figure 1 presents the TEM images of NP-SiO_2 prepared using different concentrations of CTAB. Also, a pure spheroid amorphous silica with a particle size of 10–50 nm can be extracted at 81% yield by using a hydrothermal extraction process with nitric acid in the autoclave at different temperatures and recovery time [49]. The extraction temperature of 160 °C for 2 h gives the maximum yield of silica.

Four different chemicals were used in the preparation of porous silica from RH: H_2SO_4, HCl, oxalic acid ($C_2H_2O_4$), an ionic liquid (1-butyl-3-methylimidazolium hydrogen sulphate) [50]. The chemical composition and texture properties of the obtained silica were analysed. Both sulphuric-acid and the ionic liquid treatment showed significantly increased in ash content, silica purity and surface value. The image of the prepared silica was shown in Fig. 2.

2.1.4 Nanodisks Silica

Pure amorphous circular nanodisks silica with S_{BET} of 509.5 m^2 g^{-1} was derived from the rice straw ash using a frozen process with 90.8% yield [51]. Initially, silica gel was prepared using base dissolution and acid precipitation process [52, 53]. Then, the silica-gel was rapidly frozen using N_2 liquid and dried to eliminate water. Figure 3 shows TEM and AFM images of the amorphous silica, and it confirmed the formation of rounded nanodisks with an average diameter and thickness of 172 and 3.09 nm, respectively. This nanodisks silica has a high prospect to be applied as molecular sieves, adsorbents, and catalyst supports.

Many factors determine the silica content of agricultural waste like the type of crops, soil content of crops, maturation, etc. For example, sugarcane bagasse ash, corn cob, and bamboo leaf ash can yield up to 50, 66, and 76% silica, respectively [54–56]. A study achieved on bamboo *Gigantochloa scortechinii* species culm revealed that the highest silica extracted from a matured bamboo (5.5 years) rather than a younger one [57]. Sapawe et al. [58] reported that the silica extracted from the waste in the order of decreasing as sugarcane bagasse ash > bamboo leaf > corncob > bamboo culm > banana peel > cigarette ash.

2.2 *Organic Acid Leaching Treatment Method*

There are many problems with using strong acid in the extraction of silica as a dangerous chemical. They are high cost where their storage requires the corrosion resistance container, and there is a negative environmental impact during disposal of used acid. An organic acid is an alternate chemical in the leaching process because it has a low level of risk compared to the strong acid.

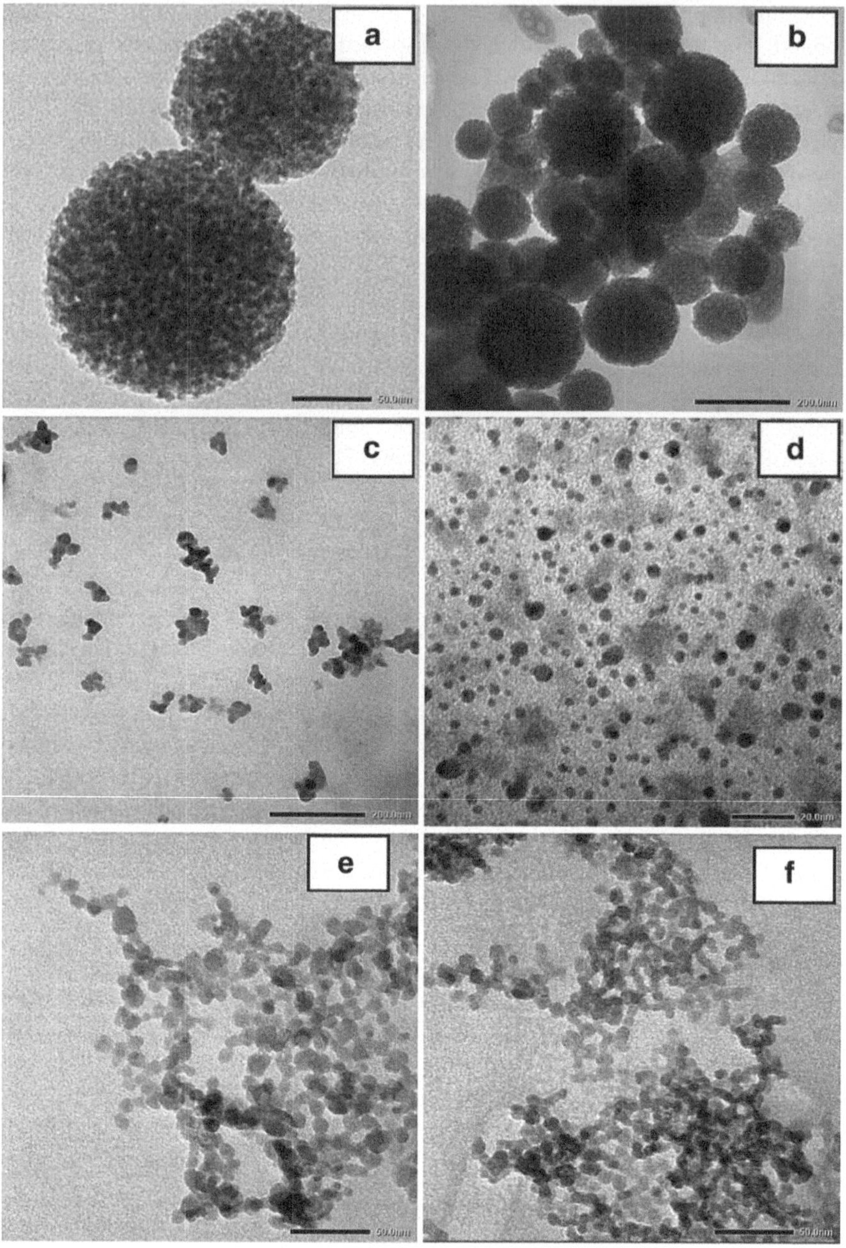

Fig. 1 TEM images of silica nanoparticles prepared from CTAB. 0.5 (**a**), 1.0 (**b**), 1.5 (**c**), 2.0 (**d**), 2.5 (**e**), and 3.0 wt% (**f**). Adapted with permission from Ref. [48], Copyright 2020, Springer

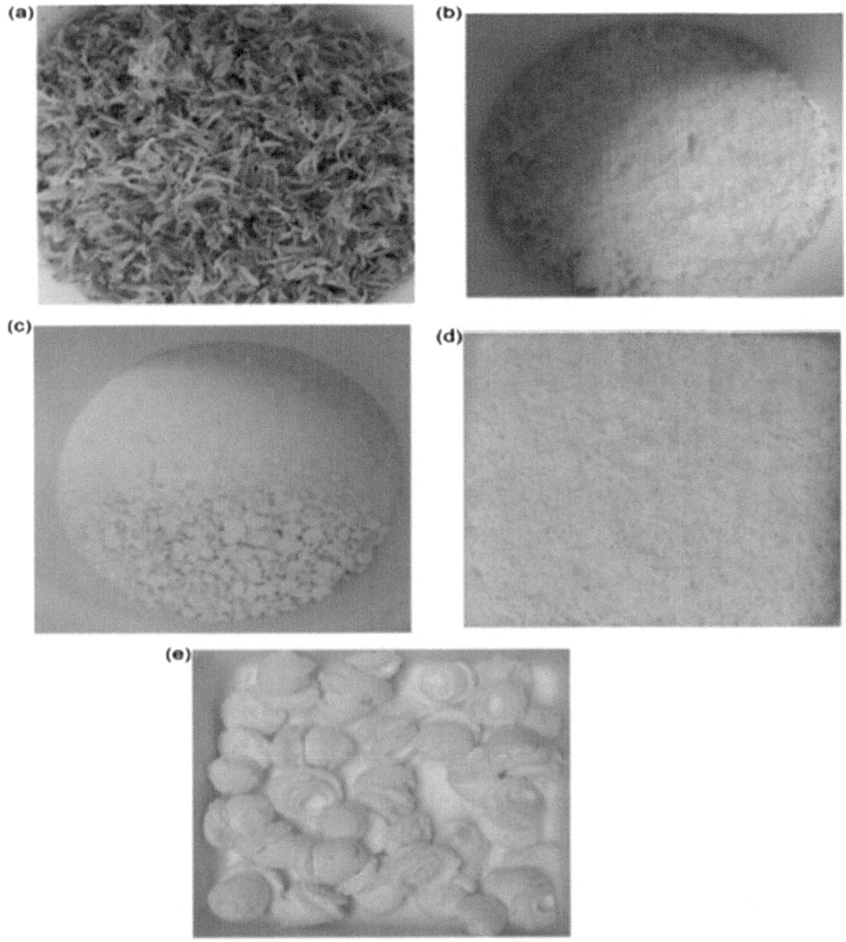

Fig. 2 The silica prepared with (**a**) no chemical treatment, (**b**) H_2SO_4, (**c**) HCl, (**d**) oxalic acid, and (**e**) ionic liquid treatments. Adapted with permission from Ref. [50], Copyright 2017, Elsevier

The citric acid solution showed a good result in the preparation of high-purity amorphous silica with 99.5–99.77 mass% from rice husks [59]. GC–MS analysis confirmed the hydrolysis reaction during the leaching process, whereas the delaminating step occurred by the chelate reaction between carboxyl groups and metallic impurities contained in husks to remove them as metal complexes. A green and cost-effective method with K_2CO_3 at a lower temperature used for the commercial production of silica and activated carbon simultaneously from rice husk ash [60]. Na_2CO_3 was used in the same method as the reagent for silica extraction, and the precipitator was waste gas [61]. Silica yield reached 72.52% impregnation ratio, and it is affected by the concentration of reagent and reaction time. The reagent for the extraction can be reused by adding a small amount of Na_2CO_3 to its initial

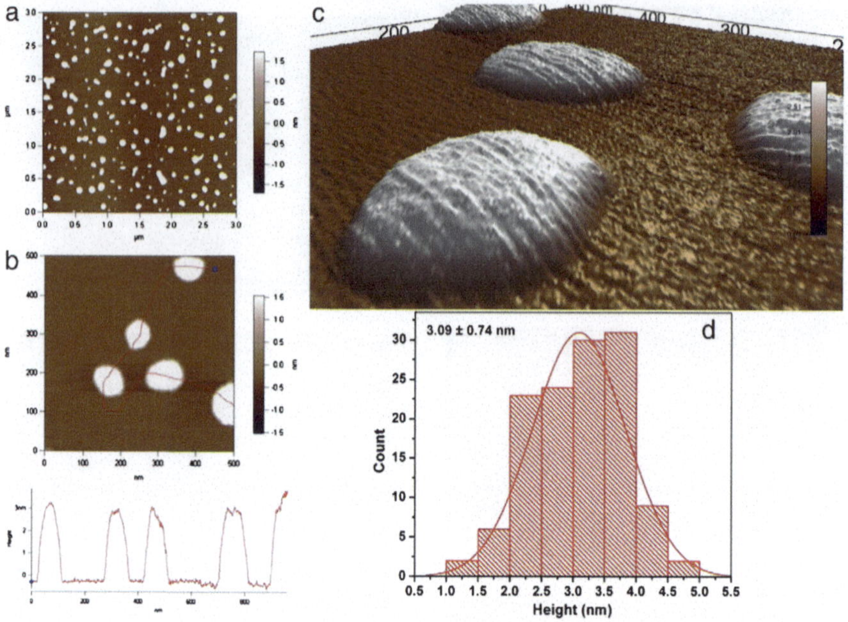

Fig. 3 AFM images show the topography of silica: **a** height image, **b** section analysis of height (thickness) and **c** 3D simulation of silica nanodisks with different resolutions. The height distribution **d** of silica nanodisks were determined from over 150 particles in AFM height images. Adapted with permission from Ref. [51], Copyright 2012, Elsevier

concentration. Ratep [62] prepared the glass silica from Na_2O and Li_2O reagents, and silica was extracted from waste content. As silica makes up 97% of the rice husk, increasing the heat treatment time can effectively increase the purity of silica.

3 Photocatalytic Activity of RHA-Silica

The extraction of silica from natural resources such as biomass provides a cheap alternative for the silica-based composite material with enhanced photocatalytic performance. The photophysical and photochemical mechanisms of the process should be understood to minimize the photocatalytic hindrance. Photoactive material and light source play vital roles in the photocatalytic performance. The huge gap between valence and conduction band in semiconductors can hinder the redox reaction, as well as high-energy UV radiation is required for the photochemical reaction, which are regarded as the major obstacles for the photocatalytic transformation upscaling [63–67]. To prepare silica-based catalyst, the sol–gel technique has been recommended due to the ease of homogeneous mixing between both metal and support precursors [68]. Many scholars have described synthetic routes for the preparation of photocatalysts based on rice husk silica and used them in water treatment. Silica–tin nanotubes (RHA-10Sn10Ti) and titania-ceria incorporated silica catalyst (RHS-50Ti10Ce) were synthesized via a simple sol-gel method using nitric oxide [69, 70]. The photoactivity was estimated on the degradation of methylene blue dye under UV-irradiation. A 20 mg of RHA-10Sn10Ti successfully removed 99% of 20 mL of MB solution (12 mg L^{-1}) in 60 min irradiation time. However, RHS-50Ti10Ce removed 99.5% of MB in 210 min using 0.25 g of catalyst in 400 ml of dye (80 mg L^{-1}). Also, as a low cost and green RHSi-3% Fe catalyst was produced through a modified alkali extraction process for heterogeneous Fenton-like degradation of Rhodamine B [71]. At a pH of 3.0, 100% of decolourization was completed in 10 min. TiO_2/SiO_2 composites were synthesized using $TiCl_4$ and RHA-SiO_2, and their photoactivity was examined in the degradation of terephthalic acid under UV-C light [72]. The role of hydrolysis temperature was examined, and the composite synthesized at 95 °C was found to be the most active catalyst than that hydrolysis at 60 °C. Photocatalytic composite materials (RHA)-TiO_2/ZnO was prepared using a ball mill and shaped into the desired size (3.5 cm × 3.5 cm × 0.5 cm) to investigate the photodegradation of real textile composites using sunlight [73]. Rice straw ash containing titania, RSA-TiO_2 (TiO_2, 50%) was prepared by the mechano-mixing method in disc-shaped forms and applied in photodegradation of methylene blue dye under visible light [74]. The photodegradation of 25 ppm MB at pH 5.85 was rapidly increased with the irradiation time up to 60 min, after which the rate of reaction became almost constant. The RSA-binder disc was found to be high reusability with acceptable results even after 4 cycles of usage.

4 Preparation of Lignin

4.1 Preparation of Lignin from Wood

Wood treatment with nitric acid and alkaline solutions was the reason for lignin discovery by Payen in 1838 [75] as the treatment produced two fractions; one of them was insoluble that was called cellulose, and the other one was incrusted and later named by Schulze as lignin [76]. Lignin results from enzyme-initiated dehydrogenation polymerization of three precursors: trans-sinapyl, trans-coniferyl and trans-P-coumaryl [77]. The number of methoxyl groups on the aromatic ring is the tool to distinguish between the 3 precursors. There are two types of lignin, protolignin and lignin. Protolignin refers to lignin associated with cells, while lignin is designated for those extracted from wood. Lignin possesses exceptional stability bonds, which include bi-phenyl C-C linkages between an aliphatic and aromatic carbon as well as hydrolysis-resistant ether links. They are all produced from enzyme-initiated polymerization and render lignin to be resistant on degradationt, but the α-aryl ether bond is the only bond in lignin that causes weak and hydrolyzable linkages. The final structure is determined according to the types of plant-based precursors and their relative used amount. All lignin have similar base structure as the functional groups in the molecule are recorded per phenyl propane unit, a C_6-C_3 unit, and C-atoms within each propane unit are counted due to a common notification to eliminate molecular weight influence. Because of various lignins can coexist within the same plant, the lignin types can be classified into two main classes, namely Quaiacyllignins and Guaiacyl-syringyllignins.

A large amount of lignin present in woody stems between the fibres of cellulose and its content varies with different plant species. Various types of lignin, such as bagasse lignin, wheat straw and softwood lignin are used for wastewater treatments nowadays [78]. Lignin can be extracted from paper and pulp industry. Lignin is considered the second crucial agricultural by-product after cellulose. Large amounts of lignin can be obtained from the biorefinery process and hydrated polymers. Outstanding achievements can be attained by modifications in lignin [79]. As lignin is one of the main parts of lignocellulosic biomass, so lignocellulosic fractionation operation is aimed to breakdown the cell wall structure to isolate cellulose, hemicellulose and lignin. The biomass structure can be altered by pre-treating lignocellulosic. The pre-treating of lignocellulosic also increases the amenability of cellulose and hemicellulose to hydrolytic enzymes and more useful for the pulp fibre manufacture which requires various modification depending on the final use of product [80, 81]: (1) hemicellulose degradation product minimization, (2) inhibition of ethanol fermentation by limiting by-product formation, (3) minimizing environmental leverages (air and water contaminations) and energy/water use reduction, (4) lowering operation costs, (5) minimization of the amount of the chemical needed for pre-treatments, and (6) raw materials and products flexibility process.

Pre-treatment is required to process biomass feedstock for lignin production. Pre-treatment processes can be categorized into physical, chemical and biological methods. Various methods are often combined in order to raise the yield. Physical operations involve stone grinding, refiner mechanical pulping and fiberization [81]. The crystallinity and particle size of lignocellulosic material determine the energy requirements for physical pretreatments, which determines its suitability for the full-scale process [82]. To alter the lignocellulosic biomass structure and chemical composition, biological treatment is aimed to use ligninolytic microorganisms and bacteria to increase the enzyme digestion of modified biomass. However, it requires a huge amount of space, and its delicate growth conditions make it to be less attractive commercially [83]. For chemical pretreatment, numerous chemicals such as formic acid, acetic acid, hydrogen peroxide, sodium hydroxide or distilled water are utilized for lignin extraction [84, 85].

4.2 *Preparation of Lignin from Rice Husk*

4.2.1 Alkaline Hydrogen Peroxide Method

This process starts with purchasing of rice husk from a local source (rice milling factory) and then smashed into a mesh size of 40–60 followed by a drying process for 6 h at 60 °C. Hydrogen peroxide, sodium hydroxide and mineralized water were obtained from an authorized chemical company. According to Chesson-Datta [86, 87] method, lignin and cellulose were produced and analysed according to the following steps: 1 g of dried sample was added into demineralized water at 100 °C and heated for 1 h, followed by filtration and washing by 300 mL of demineralized water. The filtered residue was dried in an oven until it reached a constant weight. The dried residue was then mixed with 150 mL 1 N H_2SO_4, followed by heating for 1 h at 100 °C. The mixture was then filtered and washed by 300 mL demineralized water, followed by a drying process. The resulted residue was immersed with H_2SO_4 for 4 h, followed by the addition of 150 mL of 1 N H_2SO_4. It was then refluxed in an oil bath for 1 h, followed by demineralized water washing (400 mL). The mixture was then heated at 105 °C until constant weight and collected as ash. An alkaline hydrogen peroxide solution was used for lignin separation as 20 g of rice husk ash was put in the 500 mL flask followed by 120 ml demineralized water addition involving 1% H_2O_2 (volume/weight ratio of 1:6) followed by addition of NaOH (2 N, pH = 9). The extraction process was performed at 100 °C by utilizing an oil bath for 3 h. The filter paper was used to separate the solid from the solution [88].

4.2.2 Chemical Pretreatment by Microwave Irradiation for Delignification

Rice husk was purchased from a local source and crushed to a mesh size of 1 mm. To have high value and accurate lignin extraction, the rice husk was subjected to the soxhlet extraction overnight by utilizing the removal procedure of NREL extractives [89, 90]. The rice husk was left in the airtight container, and 2% sulphuric acid, 3.5% sodium hydroxide, 3.5% sodium carbonate and 5% hydrogen peroxide were prepared. 10 mL of each solution was mixed with 1 g of rice husk for 48 h. Microwave irradiation was utilized for heating for 5 min, and the steps were repeated [91].

4.2.3 Reflux Conditions (Organosolv Lignin)

Deep washing of rice husk with n-hexane and ethanol in soxhlet extractor is performed in order to eliminate different extraction residues such as wax, proteins and lipids from the cell wall [92]. Sulfuric acid was used as a catalyst with different concentrations from 0.5 to 3.0 V/V %, and 100 mL of ethanol/water mixture (using a concentration of ethanol 25–99%) was prepared. 5 g of rice husk was added into the ethanolic solution with sulfuric acid and left in reflux for 24 h. The solution was filtered after reflux in order to eliminate the husk. The filtrate was evaporated, and the extracted lignin was obtained as a solid powder. Water was added into the lignin, and it was ready for characterization, as shown in Fig. 4.

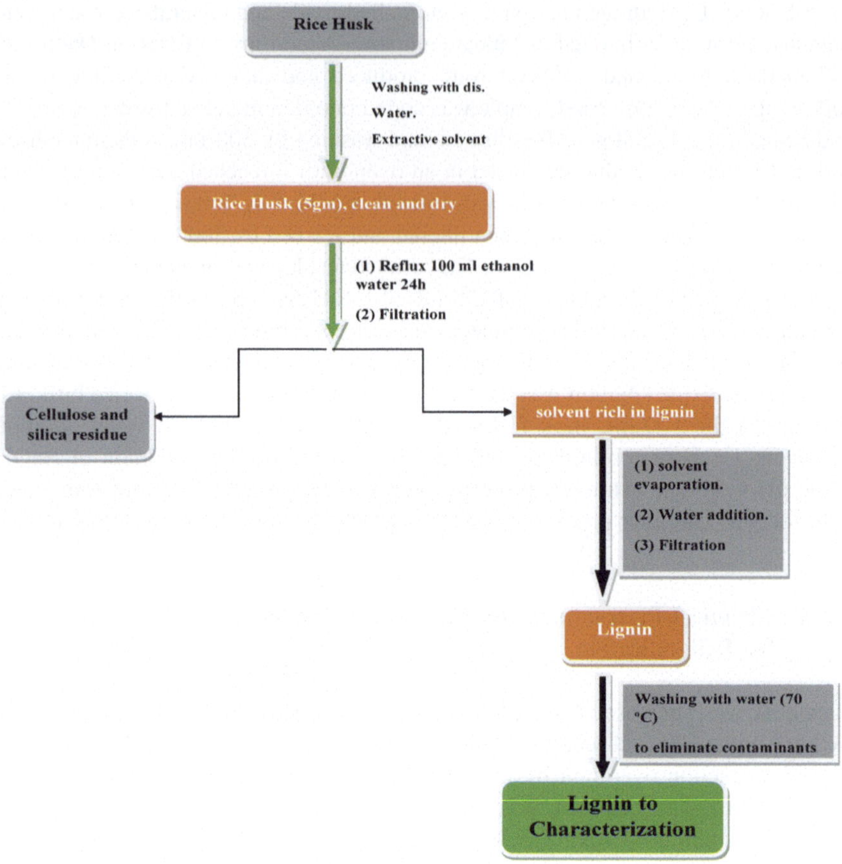

Fig. 4 Lignin extraction from rice husk. Adapted with permission from Ref. [90], Copyright 2017, Science Signpost Publishing

5 Types of Lignin According to the Preparation Method

5.1 Kraft Lignin

Kraft lignin is known as softwood lignin. Rigidity, substitution and cross-linkages origin are the tool to determine numerous types of lignin. The prevalent type in softwood lignin in lignin G (guaiacyl) structure and cross-linked with various units [78, 93] as it is manufactured in kraft (sulphate) cooking process. Kraft cooking offers 85% of lignin manufacture all over the world. It requires desolvation of 90–95% of the wood in sodium hydroxide and sodium sulphide solution [94].

5.2 Hard Lignin

The predominant structure is the syringe unit and guaiacyl that aid in preserving the linear lignin structure [78, 93], whereas the S (syringy) and G (guaiacyl) structures are the controllers in the hardwood.

5.3 Lignin Alkali

Lignin alkali is known as soda lignin as the soda pulping process is the cooking material resulted from wood pulp manufacture during the pulping process utilizing sodium hydroxide. The main by-product of this process is considered as the fundamental constituents of black liquor, which is called by soda lignin. Soda lignin can be prepared from different resources, but they have similar characterizations such as (1) low molecular weight [95], (2) glass transition temperature and (3) phenolic hydroxyl content.

In comparison to kraft lignin and lignosulphonates, lignin alkali results from the cooking process of soda anthraquinone, which is considered the cooking process without sulphur. Soda lignin can be dispersed into a low molecular fraction, and this has a reflection on its physical characterization.

5.4 Lignosulphonates

It results from the sulphate cooking process proceeds a wood delignification by HSO_3^- and SO_3^-ions. It possesses high molecular weight soluble water [77]. Lignosulphonates include a large number of charged groups, so it is considered water-soluble anionic polyelectrolyte, so the resulted in sulfonated lignin can be degraded.

5.5 Organosolv Lignin

In this pulping process, we use many solvents such as acetic acid, formic acid, ethanol and peroxy-organic acid as it is separated through solubilization [96, 97]. Sulphur-free technologies are the only one preferred for lignin revaluation among all chemical pretreatments.

6 Photocatalytic Composite Based on Lignin

These researches discussed lignin structure, chemical, biosynthetic properties and degradation mechanisms. The essential methods of lignin transformation into valuable chemicals and lignin involve:

(1) Many valuable and efficient surfactants and flocculants can be synthesized from lignin by managing its chemical reaction preparation methods and degradation mechanisms, which are utilized in numerous domains such as environment-related industries that involve oil, coal, agriculture and light. Light degradation has offered cheap phenolic- and furan-type micromolecular chemical raw materials.
(2) The competition between polymer matrix and lignin to execute high dispersion limits can be promoted by lignin chemical structure alterations and blending compatibility.

The lignin microstructure altered materials can be regulated to get the most utilization of lignin in consolidation, oxidation impedance, flame slow down, absorption of UV light and nucleating crystallization, etc. Lignin degradation releases a category of materials with catechol, which is identical to dopamine-rendering bio-based adhesives [98], by controlling lignin viscosity properties. Bio-based material with reasonable prices can be synthesized from furan-type macromolecules [99]. Urea-formaldehyde resin, phenolic resin and lignin-altered rubbers are all considered examples of lignin-altered material that can be used in several practical implementations. The adhesives advancement is dependent on lignin-altered urea-formaldehyde and phenolic resin and is influenced by the liberation of aldehyde resin matrix nature of constraints. Otherwise, lignin-altered polyurethane can be utilized in the foaming materials domain [100, 101]. Lignocellulosic fibres and nanofibres can be synthesized by spinning or electrospinning that will be transformed into carbon fibres for many technology implementations [102]. Lignin-altered materials offer applications in many domains such as electrospinning film [103], self-assembled film [104] and hydrogel [105]. All the applications are dependent on lignin structural characteristics, and lignin is also used for boosting adsorption. Other techniques are based on lignin physico-chemical properties by chemical reactions or physical blending, and the altered materials feature to have a wide pH scale, solvent, temperature response smart materials [103]. Viruses [106] and tumours [107] remediation have been discovered by utilizing lignin for biological activities. Lignin-altered materials can be used for technological implementations of the film [108], nanotube [109], and other uses such as drug and gene vector, which have been revealed. The complicated chemical structure of lignin has boosted the advance of its high-value utilization in the material domain [110–112], as indicated in Fig. 5.

Fundamental advantage can be attained by depositing a photoactive material within the lignin-based supports voids, especially if the photoactive component is an oriented or assembled model. Photocatalysts formed on lignin semiconductors

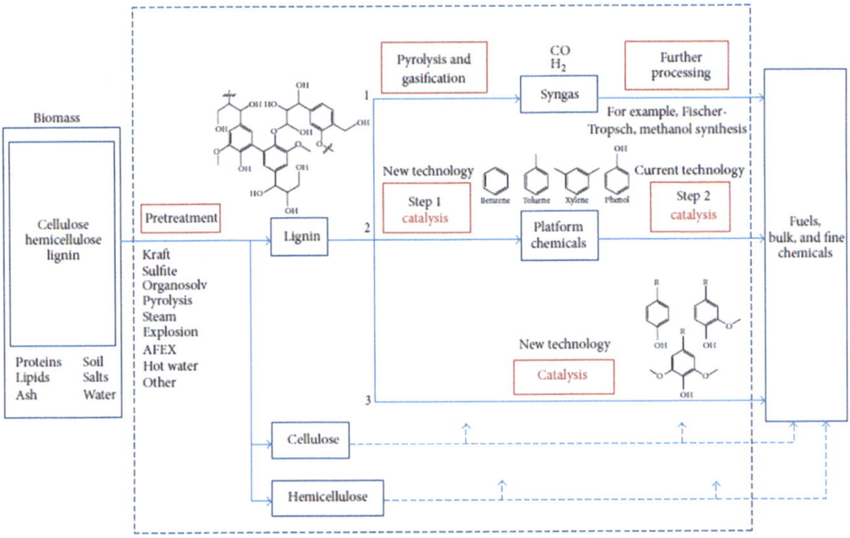

Fig. 5 Lignocellulosic biorefinery scheme with particular emphasis on the lignin stream. Adapted with permission from Ref. [113], Copyright 2010, Hindawi Publishing Corporation

composites have great potential for water contamination treatment and have acquired more interest [114, 115]. The lignin utilization as photocatalyst support requires research interest in the semiconductor-support interaction and contamination degradation process. The incorporation of lignin in semiconductors such as ZnO, TiO_2 or CuO enhances the pollutant degradation in comparison to pure metal oxide [114–116]. Various ways can be done to achieve for this purpose [66]; for example, lignin-TiO_2 composites were synthesized through dry and wet milling methods [66, 117]. Lignin-based carbon-ZnO nanocomposites can be prepared using alkali lignin and zinc nitrate, which is considered a low-cost and eco-friendly method by utilizing industrial alkali-lignin [110, 118]. CuO–ZnO nanocomposites can be prepared by utilizing sodium lignosulfonate and zinc carbonate [119]. An exceptional feature of this method is reactants isolation by washing and filtration from final production. CuO particles are attained by sodium hydroxide, and copper nitrate with aminated lignin as Wang et al. [120] discussed a CuO-nanocatalyst by using aminated lignin through solid-phase technique. Lignin amine mesoporous zinc oxide hybrid catalyst possesses characterized high sunlight photocatalytic activity and inserting amine groups to the lignin through amination reduction enhances flocculation and surface activity along with decolourization competence for wastewater remediation. Calcination temperature plays an essential part in detecting the texture properties and photocatalytic execution of the ZnO-lignin amine composites. Doping ZnO precursor with lignin amine participates to gain a smaller size and the increased specific surface area of ZnO particles by inhibiting the ZnO particles clustering. The photodegradation of methyl orange dye

(20 mg L^{-1}) using ZnO-LA composite calcined at 400 °C was reached 99.2 and 96.4% under solar radiation for 6 h and UV light irradiation for 1 h.

For the assembly of stable hybrid photocatalysts, lignin may be utilized as a template and plays an essential role in detecting photocatalytic execution [106]. Mesoporous TiO_2 nanoparticle preparation can be executed by utilizing $TiCl_4$ as a precursor and lignin as a template [44] and highly electronegative hydroxyl moieties on lignin surface improve an intense alliance in respect of electropositive metal ions. Porous ZnO-NP preparation utilizing zinc nitrate, sodium oxalate and lignin amine as an alkali template lead to lower coagulation and smaller size particles (15–44 nm) [118].

7 Conclusions

The current study is focused on the recent technology used in the extraction of silica and lignin nanoparticles from rice husk as agricultural waste. Many studies were successful in producing silica nanoparticles from rice husk waste by a simple leaching process. The method was started by heating rice husk waste into a basic solution, then treated with concentrated acid and dried as sodium silicate. In addition, the conversion of rice husk to ash is an effective method to eliminate the organic components in rice and produce high purity silica. However, there are safety concerns by using strong acid in the extraction of silica. An organic acid is an alternative chemical in the leaching process because it has a low level of risk as compared to the strong acid. Rice husk silica (RHS) is used in composites manufacture to enhance photocatalytic degradation of organic pollutants.

On the other hand, this chapter discusses lignin structure and preparation from wood and rice husk by different methods such as alkaline hydrogen peroxide method, chemical pretreatment by microwave irradiation for delignification, reflux conditions (Organosolv lignin), and the second method is the simplest. It also summarizes lignin types: kraft lignin, hard lignin, lignin alkali, lignosulphonates and organosolv lignin. The chapter also displayed lignin applications, which include several uses of lignin-based components in water treatment. Therefore, the combination of both production of silica or lignin from rice husk waste processes, and wastewater treatment in a one-pot reaction, would provide an attractive and sustainable approach for the valorization of agriculture residues and the decrease water pollution by integrating this process in the industrial factories.

8 Future Perspectives

A green and economical technique to manufacture high pure silica from rice husk waste is still needed. Indeed, silica isolated from ash still contains carbon material; though, this requires further treatment. Besides, many further studies on how to produce silica using organic acid are still required to encourage the industrial application.

On the practical scale, studies concerning lignin structure, managing methods and bioactivity mechanisms such as macromolecular degradation, products purification, lignin/polymer composites competition and several forms synthesis (films, fibres, and nanofibres, foams, hydrogels) must be encouraged to be utilized in numerous applications in the future. On the theoretical scale, studies on biomolecules altered polymer matrices should be expanded to control supermolecular micro-section mechanisms and molecular design protocols for hindering viruses and tumours in the future.

References

1. Pode R (2016) Potential applications of rice husk ash waste from rice husk biomass power plant. Renew Sustain Energy Rev 53:1468–1485
2. Alvarez J, Lopez G, Amutio M, Bilbao J, Olazar M (2014) Bio-oil production from rice husk fast pyrolysis in a conical spouted bed reactor. Fuel 128:162–169
3. Habeeb GA, Mahmud HB (2010) Study on properties of rice husk ash and its use as cement replacement material. Mater Res 13(2):185–190
4. He D, Ikeda-Ohno A, Boland DD, Waite TD (2013) Synthesis and characterization of antibacterial silver nanoparticle-impregnated rice husks and rice husk ash. Environ Sci Technol 47(10):5276–5284
5. Shen Y, Zhao P, Shao Q (2014) Porous silica and carbon derived materials from rice husk pyrolysis char. Microporous Mesoporous Mater 188:46–76
6. Shatat MR, Ali GAM, Gouda GA, Abdallah MH (2014) Rice husk ash for enhancing salts attack resistance of blended cement containing metakoalin. Can Chem Trans 2(3):274–285
7. Lakshmi UR, Srivastava VC, Mall ID, Lataye DH (2009) Rice husk ash as an effective adsorbent: evaluation of adsorptive characteristics for Indigo Carmine dye. J Environ Manage 90(2):710–720
8. Feng Q, Lin Q, Gong F, Sugita S, Shoya M (2004) Adsorption of lead and mercury by rice husk ash. J Colloid Interface Sci 278(1):1–8
9. Srivastava VC, Mall ID, Mishra IM (2008) Removal of cadmium(II) and zinc(II) metal ions from binary aqueous solution by rice husk ash. Colloids Surf, A 312(2–3):172–184
10. Sankar S, Sharma SK, Kaur N, Lee B, Kim DY, Lee S, Jung H (2016) Biogenerated silica nanoparticles synthesized from sticky, red, and brown rice husk ashes by a chemical method. Ceram Int 42(4):4875–4885
11. Le Van K, Luong Thi TT (2014) Activated carbon derived from rice husk by NaOH activation and its application in supercapacitor. Prog Nat Sci Mater Int 24(3):191–198
12. Chen H, Wang F, Zhang C, Shi Y, Jin G, Yuan S (2010) Preparation of nano-silica materials: the concept from wheat straw. J Non-Cryst Solids 356(50–51):2781–2785
13. Kalapathy U (2000) A simple method for production of pure silica from rice hull ash. Biores Technol 73(3):257–262
14. Chakraverty A, Mishra P, Banerjee HD (1988) Investigation of combustion of raw and acid-leached rice husk for production of pure amorphous white silica. J Mater Sci 23(1):21–24
15. Umeda J, Kondoh K (2008) High-purity amorphous silica originated in rice husks via carboxylic acid leaching process. J Mater Sci 43(22):7084–7090
16. Ali GAM, Fouad OA, Makhlouf SA, Yusoff MM, Chong KF (2016) Optical and electrochemical properties of Co_3O_4/SiO_2 nanocomposite. Adv Mater Res 1133:447–451

17. Abdel Ghafar HH, Ali GAM, Fouad OA, Makhlouf SA (2015) Enhancement of adsorption efficiency of methylene blue on Co_3O_4/SiO_2 nanocomposite. Desalin Water Treat 53 (11):2980–2989
18. Ali GAM, Fouad OA, Makhlouf SA (2013) Structural, optical and electrical properties of sol-gel prepared mesoporous Co_3O_4/SiO_2 nanocomposites. J Alloy Compd 579:606–611
19. Ali GAM, Fouad OA, Makhlouf SA, Yusoff MM, Chong KF (2014) Co_3O_4/SiO_2 nanocomposites for supercapacitor application. J Solid State Electrochem 18(9):2505–2512
20. Kalapathy U, Proctor A, Shultz J (2002) An improved method for production of silica from rice hull ash. Biores Technol 85(3):285–289
21. Erdtman H (1972) Lignins: occurrence, formation, structure and reactions. J Polym Sci, Part C: Polym Lett 10(3):228–230
22. Timell TE (1967) Recent progress in the chemistry of wood hemicelluloses. Wood Sci Technol 1(1):45–70
23. Hatakeyama H, Hatakeyama T (2009) Lignin structure, properties, and applications. Biopolymers. Springer, Berlin, Heidelberg, pp 1–63
24. Sun L, Gong K (2001) Silicon-based materials from rice husks and their applications. Ind Eng Chem Res 40(25):5861–5877
25. Li T, Wang T (2008) Preparation of silica aerogel from rice hull ash by drying at atmospheric pressure. Mater Chem Phys 112(2):398–401
26. Ali GAM, Fouad Osama A, Makhlouf Salah A (2016) Electrical properties of cobalt oxide/ silica nanocomposites obtained by sol-gel technique. Am J Eng Appl Sci 9(1):12–16
27. Ali GAM, Ahmed B, Vinod Kumar G, Amr Ahmed N, Heba E-M, Ramesh K, Essam Ramadan S, Hamed A, Kwok Feng C (2020) High surface area mesoporous silica for hydrogen sulfide effective removal. Curr Nanosci 16:226–234
28. Fouad OA, Ali GAM, El-Erian MAI, Makhlouf SA (2012) Humidity sensing properties of cobalt oxide/silica nanocomposites prepared via sol-gel and related routes. Nano 7(5):1250038
29. Roberts WO, Bergna H (2006) Colloidal silica: fundamentals and applications. CRC Taylor and Francis, Boca Raton
30. Jal PK, Sudarshan M, Saha A, Patel S, Mishra BK (2004) Synthesis and characterization of nanosilica prepared by precipitation method. Colloids Surf, A 240(1–3):173–178
31. Faizul CP, Abdullah C, Fazlul B (2012) Review of extraction of silica from agricultural wastes using acid leaching treatment. Adv Mater Res 626:997–1000
32. Kondoh K, Threrujirapapong T, Imai H, Umeda J, Fugetsu B (2009) Characteristics of powder metallurgy pure titanium matrix composite reinforced with multi-wall carbon nanotubes. Compos Sci Technol 69(7–8):1077–1081
33. Amutha K, Ravibaskar R, Sivakumar G (2010) Extraction, synthesis and characterization of nanosilica from rice husk ash. Int J Nanotechnol Appl 4(1):61–66
34. Umeda J, Imai H, Kondoh K (2009) Polysaccharide hydrolysis and metallic impurities removal behavior of rice husks in citric acid leaching treatment. Trans JWRI 38(2):13–18
35. Umeda J, Kondoh K (2010) High-purification of amorphous silica originated from rice husks by combination of polysaccharide hydrolysis and metallic impurities removal. Ind Crops Prod 32(3):539–544
36. Umeda J, Kondoh K (2008) Process optimization to prepare high-purity amorphous silica from rice husks via citric acid leaching treatment. Trans JWRI 37(1):13–17
37. Liou T-H (2004) Preparation and characterization of nano-structured silica from rice husk. Mater Sci Eng, A 364(1–2):313–323
38. Alshatwi AA, Athinarayanan J, Periasamy VS (2015) Biocompatibility assessment of rice husk-derived biogenic silica nanoparticles for biomedical applications. Mater Sci Eng, C 47:8–16
39. Yuvakkumar R, Elango V, Rajendran V, Kannan N (2012) High-purity nano silica powder from rice husk using a simple chemical method. J Exp Nanosci 9(3):272–281
40. Ghorbani F, Sanati AM, Maleki M (2015) Production of silica nanoparticles from rice husk as agricultural waste by environmental friendly technique. Environ Stud Persian Gulf 2 (1):56–65

41. Vaibhav V, Vijayalakshmi U, Roopan SM (2015) Agricultural waste as a source for the production of silica nanoparticles. Spectrochim Acta Part A Mol Biomol Spectrosc 139:515–520
42. Della VP, Kühn I, Hotza D (2002) Rice husk ash as an alternate source for active silica production. Mater Lett 57(4):818–821
43. Li D, Zhu X (2011) Short-period synthesis of high specific surface area silica from rice husk char. Mater Lett 65(11):1528–1530
44. Tadjarodi A, Haghverdi M, Mohammadi V (2012) Preparation and characterization of nano-porous silica aerogel from rice husk ash by drying at atmospheric pressure. Mater Res Bull 47(9):2584–2589
45. Mor S, Manchanda CK, Kansal SK, Ravindra K (2017) Nanosilica extraction from processed agricultural residue using green technology. J Clean Prod 143:1284–1290
46. Fouad OA, Makhlouf SA, Ali GAM, El-Sayed AY (2011) Cobalt/silica nanocomposite via thermal calcination-reduction of gel precursors. Mater Chem Phys 128(1–2):70–76
47. Liou T-H, Yang C-C (2011) Synthesis and surface characteristics of nanosilica produced from alkali-extracted rice husk ash. Mater Sci Eng, B 176(7):521–529
48. Le VH, Thuc CNH, Thuc HH (2013) Synthesis of silica nanoparticles from Vietnamese rice husk by sol–gel method. Nanoscale Res Lett 8(1):58
49. Tolba GMK, Barakat NAM, Bastaweesy AM, Ashour EA, Abdelmoez W, El-Newehy MH, Al-Deyab SS, Kim HY (2015) Effective and highly recyclable nanosilica produced from the rice husk for effective removal of organic dyes. J Ind Eng Chem 29:134–145
50. Lee JH, Kwon JH, Lee J-W, Lee H-S, Chang JH, Sang B-I (2017) Preparation of high purity silica originated from rice husks by chemically removing metallic impurities. J Ind Eng Chem 50:79–85
51. Lu P, Hsieh Y-L (2012) Highly pure amorphous silica nano-disks from rice straw. Powder Technol 225:149–155
52. Hessien MM, Rashad MM, Zaky RR, Abdel-Aal EA, El-Barawy KA (2009) Controlling the synthesis conditions for silica nanosphere from semi-burned rice straw. Mater Sci Eng, B 162(1):14–21
53. Sharifnasab H, Alamooti MY (2017) Preparation of silica powder from rice husk. Agric Eng Int:26–29
54. Rahman NA, Widhiana I, Juliastuti SR, Setyawan H (2015) Synthesis of mesoporous silica with controlled pore structure from bagasse ash as a silica source. Colloids Surf, A 476:1–7
55. Norsuraya S, Fazlena H, Norhasyimi R (2016) Sugarcane bagasse as a renewable source of silica to synthesize Santa Barbara amorphous-15 (SBA-15). Proc Eng 148:839–846
56. Madani Hosseini M, Shao Y, Whalen JK (2011) Biocement production from silicon-rich plant residues: perspectives and future potential in Canada. Biosys Eng 110(4):351–362
57. Hisham HN, Othman S, Rokiah H, Latif MA, Ani S, Tamizi MM (2006) Characterization of bamboo Gigantochloa scortechinii at different ages. J Trop For Sci:236–242
58. Sapawe N, Surayah Osman N, Zulkhairi Zakaria M, Amirul Shahab Syed Mohamad Fikry S, Amir Mat Aris M (2018) Synthesis of green silica from agricultural waste by sol-gel method. Mater Today: Proc 5(10):21861–21866
59. Greenwood NN, Earnshaw A (2012) Chemistry of the elements. Elsevier
60. Liu Y, Guo Y, Gao W, Wang Z, Ma Y, Wang Z (2012) Simultaneous preparation of silica and activated carbon from rice husk ash. J Clean Prod 32:204–209
61. An D, Guo Y, Zhu Y, Wang Z (2010) A green route to preparation of silica powders with rice husk ash and waste gas. Chem Eng J 162(2):509–514
62. Ratep A (2019) Preparation and characterization of silicate glasses from waste agriculture materials (rice husk and peanut peel). Silicon
63. Solehudin M, Sirimahachai U, Ali GAM, Chong KF, Wongnawa S (2020) One-pot synthesis of isotype heterojunction g-C_3N_4-MU photocatalyst for effective tetracycline hydrochloride antibiotic and reactive orange 16 dye removal. Adv Powder Technol 31(5):1891–1902

64. Sharifi A, Montazerghaem L, Naeimi A, Abhari AR, Vafaee M, Ali GAM, Sadegh H (2019) Investigation of photocatalytic behavior of modified ZnS:Mn/MWCNTs nanocomposite for organic pollutants effective photodegradation. J Environ Manage 247:624–632
65. Ethiraj AS, Uttam P, Varunkumar K, Chong KF, Ali GAM (2019) Photocatalytic performance of a novel semiconductor nanocatalyst: copper doped nickel oxide for phenol degradation. Mater Chem Phys 242:122520
66. Nair V, Dhar P, Vinu R (2016) Production of phenolics via photocatalysis of ball milled lignin–TiO_2 mixtures in aqueous suspension. RSC Adv 6(22):18204–18216
67. Giahi M, Pathania D, Agarwal S, Ali GAM, Chong KF, Gupta VK (2019) Preparation of Mg-doped TiO_2 nanoparticles for photocatalytic degradation of some organic pollutants. Stud Univ Babes-Bolyai, Chem 64(1):7–18
68. Lambert CK, Gonzalez RD (2001) The effect of pH and metal loading on the properties of Sol-Gel Rh/SiO_2. J Solid State Chem 158(2):154–161
69. Adam F, Appaturi JN, Khanam Z, Thankappan R, Nawi MAM (2013) Utilization of tin and titanium incorporated rice husk silica nanocomposite as photocatalyst and adsorbent for the removal of methylene blue in aqueous medium. Appl Surf Sci 264:718–726
70. Adam F, Muniandy L, Thankappan R (2013) Ceria and titania incorporated silica based catalyst prepared from rice husk: adsorption and photocatalytic studies of methylene blue. J Colloid Interface Sci 406:209–216
71. Gan PP, Li SFY (2013) Efficient removal of Rhodamine B using a rice hull-based silica supported iron catalyst by Fenton-like process. Chem Eng J 229:351–363
72. Banu Yener H, Helvacı ŞŞ (2015) Effect of synthesis temperature on the structural properties and photocatalytic activity of TiO_2/SiO_2 composites synthesized using rice husk ash as a SiO_2 source. Sep Purif Technol 140:84–93
73. Kabir MH, Kabir MF, Nigar F, Ahmed S, Mustafa AI, Ahsan M (2013) Preparation and characterization of rice husk ash (RHA)-TiO_2/ZnO composites and its application in treating effluents from textile industries. Bangladesh J Sci Indus Res 47(4):445–448
74. Hassan SA, El-Salamony RA (2014) Photocatalytic disc-shaped composite systems for removal of hazardous dyes in aqueous solutions. Can Chem Trans:57–71
75. Payen A (1838) Mémoire sur la composition du tissu propre des plantes et du ligneux. Comptes rendus 7:1052–1056
76. Schulze F (1857) Beitrage zur Kenntniss des lignins. Chem Zentr 21:321–325
77. Fan J, Zhan H (2008) Optimization of synthesis of spherical lignosulphonate resin and its structure characterization. Chin J Chem Eng 16(3):407–410
78. Thakur VK, Thakur MK (2014) Recent trends in hydrogels based on psyllium polysaccharide: a review. J Clean Prod 82:1–15
79. Huiquan C, Yunhuai G, Zhou F (1996) The situation and aspect of application for macromolecule (I)——cellulose, lignin and starch. Yunnan Chem Technol 1:01
80. Sannigrahi P, Pu Y, Ragauskas A (2010) Cellulosic biorefineries—unleashing lignin opportunities. Curr Opin Environ Sustain 2(5–6):383–393
81. Fengel D, Wegener G (2011) Wood: chemistry, ultrastructure, reactions, Walter de Gruyter
82. Brodeur G, Yau E, Badal K, Collier J, Ramachandran KB, Ramakrishnan S (2011) Chemical and physicochemical pretreatment of lignocellulosic biomass: a review. Enzyme Res 2011:1–17
83. Zheng Y, Pan Z, Zhang R (2009) Overview of biomass pretreatment for cellulosic ethanol production. Int J Agric Biol Eng 2(3):51–68
84. Mankar S, Chaudhari A, Soni I (2012) Lignin in phenol-formaldehyde adhesives. Int J Knowl Eng 3(1):116–118
85. Watkins D, Nuruddin M, Hosur M, Tcherbi-Narteh A, Jeelani S (2015) Extraction and characterization of lignin from different biomass resources. J Mater Res Technol 4(1):26–32
86. Maryana R, Jatmiko T, Prasetyo D, Rizal W, Suwanto A, Praharasti A, Indrianingsih A, Anwar M, Rizaluddin A (2019) Evaluation of high purity cellulose production from pretreated various agricultural biomass wastes. In: IOP Conference Series: Earth and Environmental Science, IOP Publishing

87. Maryana R, Jatmiko TH, Prasetyo DJ, Rizal WA, Suwanto A, Praharasti AS, Indrianingsih AW, Anwar M, Rizaluddin AT (2019) Evaluation of high purity cellulose production from pretreated various agricultural biomass wastes. IOP Conf Ser: Earth Environ Sci 251:012001
88. Ma'ruf A, Pramudono B, Aryanti N (2017) Lignin isolation process from rice husk by alkaline hydrogen peroxide: lignin and silica extracted, Author(s)
89. Sluiter A, Ruiz R, Scarlata C, Sluiter J, Templeton D (2005) Determination of extractives in biomass. Laboratory Analytical Procedure (LAP) 1617
90. da Rosa MP, Beck PH, Müller DG, Moreira JB, da Silva JS, Durigon AMM (2017) Extraction of organosolv lignin from rice husk under reflux conditions. Biol Chem Res:87–98
91. Laghari S, Tunio M, Laghari A, Laghari A, Ali A (2018) Delignification of rice husk by microwave assisted chemical pretreatment. Eng Technol Appl Sci Rese 8(3):3084–3087
92. Browning BL (1967) Methods of wood chemistry, vols I and II. Wiley, New York
93. Doherty WOS, Mousavioun P, Fellows CM (2011) Value-adding to cellulosic ethanol: lignin polymers. Ind Crops Prod 33(2):259–276
94. Chakar FS, Ragauskas AJ (2004) Review of current and future softwood kraft lignin process chemistry. Ind Crops Prod 20(2):131–141
95. Botros SH, Eid MAM, Nageeb ZA (2005) Thermal stability and dielectric relaxation of natural rubber/soda lignin and natural rubber/thiolignin composites. J Appl Polym Sci 99 (5):2504–2511
96. Xu F, Sun J-X, Sun R, Fowler P, Baird MS (2006) Comparative study of organosolv lignins from wheat straw. Ind Crops Prod 23(2):180–193
97. Naseer A, Jamshaid A, Hamid A, Muhammad N, Ghauri M, Iqbal J, Rafiq S, Khuram S, Shah NS (2019) Lignin and lignin based materials for the removal of heavy metals from waste water—an overview. Z Phys Chem 233(3):315–345
98. Zhang W, Ma Y, Xu Y, Wang C, Chu F (2013) Lignocellulosic ethanol residue-based lignin–phenol–formaldehyde resin adhesive. Int J Adhes Adhes 40:11–18
99. Zeng C, Seino H, Ren J, Hatanaka K, Yoshie N (2013) Bio-Based furan polymers with self-healing ability. Macromolecules 46(5):1794–1802
100. Jeong H, Park J, Kim S, Lee J, Ahn N (2013) Compressive viscoelastic properties of softwood kraft lignin-based flexible polyurethane foams. Fibers Polym 14(8):1301–1310
101. Pan X, Saddler JN (2013) Effect of replacing polyol by organosolv and kraft lignin on the property and structure of rigid polyurethane foam. Biotechnol Biofuels 6(1):12
102. Wang X-S, Okazaki N, Suzuki T, Funaoka M (2003) Effect of calcium on the catalysis of nickel in the production of crystallized carbon from lignocresol for electromagnetic shielding. Chem Lett 32(1):42–43
103. Dallmeyer I, Chowdhury S, Kadla JF (2013) Preparation and characterization of kraft lignin-based moisture-responsive films with reversible shape-change capability. Biomacromol 14(7):2354–2363
104. Pereira AA, Martins GF, Antunes PA, Conrrado R, Pasquini D, Job AE, Curvelo AAS, Ferreira M, Riul A, Constantino CJL (2007) Lignin from sugar cane bagasse: extraction, fabrication of nanostructured films, and application. Langmuir 23(12):6652–6659
105. Kim YS, Kadla JF (2010) Preparation of a thermoresponsive lignin-based biomaterial through atom transfer radical polymerization. Biomacromol 11(4):981–988
106. Lopez BSG, Yamamoto M, Utsumi K, Aratsu C, Sakagami H (2009) A clinical pilot study of lignin—ascorbic acid combination treatment of herpes simplex virus. In vivo 23(6):1011–1016
107. Brück TB, Brück DW (2011) Oxidative metabolism of the anti-cancer agent mitoxantrone by horseradish, lacto-and lignin peroxidase. Biochimie 93(2):217–226
108. Raschip IE, Hitruc EG, Oprea AM, Popescu M-C, Vasile C (2011) In vitro evaluation of the mixed xanthan/lignin hydrogels as vanillin carriers. J Mol Struct 1003(1–3):67–74
109. Ten E, Ling C, Wang Y, Srivastava A, Dempere LA, Vermerris W (2014) Lignin nanotubes as vehicles for gene delivery into human cells. Biomacromol 15(1):327–338

110. Kumar A, Anushree Kumar J, Bhaskar T (2020) Utilization of lignin: a sustainable and eco-friendly approach. J Energy Inst 93(1):235–271
111. Zakzeski J, Bruijnincx PCA, Jongerius AL, Weckhuysen BM (2010) The catalytic valorization of lignin for the production of renewable chemicals. Chem Rev 110(6):3552–3599
112. Chen X, Kuo D-H, Lu D, Hou Y, Kuo Y-R (2016) Synthesis and photocatalytic activity of mesoporous TiO_2 nanoparticle using biological renewable resource of un-modified lignin as a template. Microporous Mesoporous Mater 223:145–151
113. Awungacha Lekelefac C, Busse N, Herrenbauer M, Czermak P (2015) Photocatalytic based degradation processes of lignin derivatives. Int J Photoenergy 2015:1–18
114. Wang H, Qiu X, Zhong R, Fu F, Qian Y, Yang D (2017) One-pot in-situ preparation of a lignin-based carbon/ZnO nanocomposite with excellent photocatalytic performance. Mater Chem Phys 199:193–202
115. Miao T-T, Guo Y-R, Pan Q-J (2013) The SL-assisted synthesis of hierarchical ZnO nanostructures and their enhanced photocatalytic activity. J Nanopart Res 15(6)
116. Zhou J, Zhang M, Zhu Y (2015) Photocatalytic enhancement of hybrid C_3N_4/TiO_2 prepared via ball milling method. Phys Chem Chem Phys 17(5):3647–3652
117. Altavilla C (2016) Upconverting nanomaterials: perspectives, synthesis, and applications, CRC Press
118. Wang H, Qiu X, Liu W, Yang D (2017) Facile preparation of well-combined lignin-based carbon/ZnO hybrid composite with excellent photocatalytic activity. Appl Surf Sci 426:206–216
119. Wu F, Wang X, Hu S, Hao C, Gao H, Zhou S (2017) Solid-state preparation of CuO/ZnO nanocomposites for functional supercapacitor electrodes and photocatalysts with enhanced photocatalytic properties. Int J Hydrogen Energy 42(51):30098–30108
120. Wang X, Wu F, Duan Y, Wang Y, Hao C, Ge C (2016) Lignin-assisted solid-phase synthesis of nano-CuO for a photocatalyst with excellent catalytic activity and high performance supercapacitor electrodes. RSC Adv 6(70):65644–65653

Recovery of Nanomaterials from Agricultural and Industrial Wastes for Water Treatment Applications

Enas Amdeha

Abstract Water is the origin of life, even though millions around the world agonize from the shortage of clean water. On the other hand, over two billion tons of solid waste are produced annually and represent the major source for wastewater, with an increase in numbers over the next few decades. According to the UN's Sustainable Development Goals, materials science research has to focus on recovering methods and proper waste managing. This includes two major issues, wastewater treatment and solid wastes, through one process, i.e. removing waste by waste. In this chapter, a comprehensive list of cheap materials prepared from diverse types of agricultural and industrial wastes and their performance towards the reduction of numerous aquatic pollutants has been summarized. This chapter provides an essential perception towards the use of solid wastes as materials/precursor materials for the preparation of nanomaterials to be used as adsorbents, supports and/or photocatalysts for water treatment. Furthermore, the challenges for upcoming studies of adsorbents and photocatalysts derived from waste were explored.

Keywords Adsorption · Agricultural wastes · Industrial wastes · Nanomaterials · Organic pollutants · Photocatalysis · Water treatment

List of Abbreviations

4-NP	4-nitrophenol
AC	Activated carbon
BFD	Blast furnace flue dust
CES	Calcined egg shell
COD	Chemical oxygen demand
CR	Congo red

E. Amdeha (✉)
Process Design and Development Department, Egyptian Petroleum Research Institute, Cairo 11727, Egypt
e-mail: enas_amdeha@yahoo.com

A. S. H. Makhlouf and G. A. M. Ali (eds.), *Waste Recycling Technologies for Nanomaterials Manufacturing*, Topics in Mining, Metallurgy and Materials Engineering, https://doi.org/10.1007/978-3-030-68031-2_14

CTAB	Cetyl trimethyl ammonium bromide
DTA-TGA	Differential thermal and thermogravimetric analysis
EB	Evans blue
EBT	Eriochrome black t
EDS	Energy dispersive spectroscopy
EDTA	Ethylene diamine tetraacetic acid
EDXRF	Energy dispersive X-ray fluorescence
fcc	Face-centred cubic
FID	Flame ionization detection
FTIR	Fourier Transform Infra-Red spectroscopy
GC	Gas chromatography
GCMS	Gas chromatography-mass spectroscopy
HRTEM	High-resolution transmission electron microscopy
MB	Methylene blue
MO	Methyl orange
MR	Methylene red
MWCNTs	Multi-wall carbon nanotubes
NCMs	Nano-composite materials
NMs	Nano-materials
NPs	Nano-particles
ODOE	Optimal Design of Experiment
Rh 6G	Rhodamine 6G
RhB	Rhodamine B
RSM	Response surface methodology
SBH	Sodium boro-hydride
SCB	Sugar cane bagasse
SDS	Sodium dodecyl sulfate
SY	Sunset yellow
TEM	Transmission electron microscope
TZ	Tartrazine
UV–Vis	UV–visible
VOCs	Volatile organic compounds
VSM	Vibrating sample magnetometer
XRD	X-ray diffraction

1 Introduction

Environmental pollution in the twenty-first century is marked as one of the most significant issues. Although water is the source of life, millions around the world are struggling from the lack of healthy and clean drinking water as it has been contaminated by numerous organic and inorganic contaminants existing in the manufacturing waste, resulting in about 75% of all diseases in developing countries, especially for females and kids [1–3]. Waste materials, including domestic, agricultural, and industrial contaminants, negatively affect people's health and the environment [4–6]. It produces millions of tons of agricultural waste year after year. Taking into consideration the economic growth and environmental sustainability, it is enormously essential to valorize these wastes. In the modern period of the Industrial Revolution, the enormous emphasis is on green technology, including zero or less waste production, the recycling of resources, and the use of biological/ recyclable raw materials [7–11]. Resource recycling is, among others, a highly significant method by which valuable materials can be recovered from any form of waste material. This not merely successfully decreases the quantity of waste produced but similarly reduces the environment's resource problem [12]. Proper management of waste is a very significant issue as one considers people increasing and the resultant rise in consumption [13]. Therefore, it is essential to highlight waste from the agro-industry, such as rice husk, sugar cane bagasse, bamboo, and palm kernel shell. These wastes have a common component of natural polymers such as cellulose, hemicellulose, lignin, and small quantities of inorganic substances, e.g. silica [14]. The production of solid wastes globally is more than two billion tonnes yearly with a continuous increase in the future. According to UN's Sustainable Development Goals (SDGs) for a peaceful and flourishing world, materials science research focuses on recovering techniques and proper waste controlling as essential to addressing the worldwide waste catastrophe. The converting of low-valued waste materials, especially agricultural and industrial sources, into high-valued nanomaterials/nanocomposite materials (NMs/NCMs) to be used in different environmental applications in water treatment could solve two major issues, wastewater treatment and solid wastes, through one process, i.e. removing waste by waste [15, 16].

Nanomaterials possess stunningly engineered structures, e.g. high surface area, the distinctive character of electron conduction, and other surface-active positions. There are different forms of nanomaterials, such as nanotubes, nanowires, nanofibers, and nanoparticles. These nanomaterials and their composites with supports such as polymers, silica, zeolites, and carbonaceous materials are used as adsorbents, catalysts, or support for pollutant removal [17–21].

Water pollution comes from various sources. Hazardous pollutants in wastewater include many biological and chemical pollutants. Chemical pollutants are divided into two major parts: (a) organic pollutants, e.g. dyes, pesticides, insecticides, pharmaceutical compounds, etc., and (b) inorganic pollutants, e.g. heavy metals. These pollutants cause a decrease in water quality [22–24]. Many dye molecules have an aromatic structure, making them harmful for people and marine life as these dyes are stable, hard to be degraded, and carcinogenic [25]. In addition, prevent solar light penetration, hinder the photosynthetic response, and affect the marine lifecycle [26, 27]. Heavy metals, mercury (Hg), cadmium (Cd), lead (Pb), and arsenic (As), etc., are among the toxic inorganic pollutants that may harmfully affect the living things because they are carcinogens even at low concentration [28]. Also, they cause some health problems such as proteinuria, anaemia, lung and skin cancer, and neuropathy [29–32]. They are produced from several industrial activities, e.g. metal plating, metallurgical, leather treating, batteries, and electronic devices industries [33, 34]. Hence, it is important to eliminate or reduce these pollutants to the acceptable concentration before releasing for a sustainable green environment. Among various available water treatment technologies, e.g. adsorption, photooxidation, chemical-oxidation, advanced oxidation processes, filtration, coagulation, etc. [35–39], adsorption and photocatalysis are superior because of suitability, simplicity of setup, and design. The adsorption method is one of the best water treatment techniques that can remove various pollutants. However, heterogeneous photocatalytic oxidation processes are promising methods that can completely degrade the organic pollutants without the release of any toxic products [40, 41]. Therefore, adsorption and photocatalytic degradation are more applicable to water pollution control. The photocatalysis theory (Fig. 1) is based on using a suitable light irradiation source to stimuli a photocatalyst to produce free radicals clusters responsible for the redox reaction [42]. This chapter discusses the importance of using waste as raw materials to prepare NMs/NCMs for the adsorption and photocatalytic degradation of dyes of toxic materials, dyes, and heavy metals, in the water phase. This chapter lists different types of wastes used as resources for producing NMs/NCMs and discusses their applications towards pollutants removal. The chapter also discusses the challenges and possible further research directions.

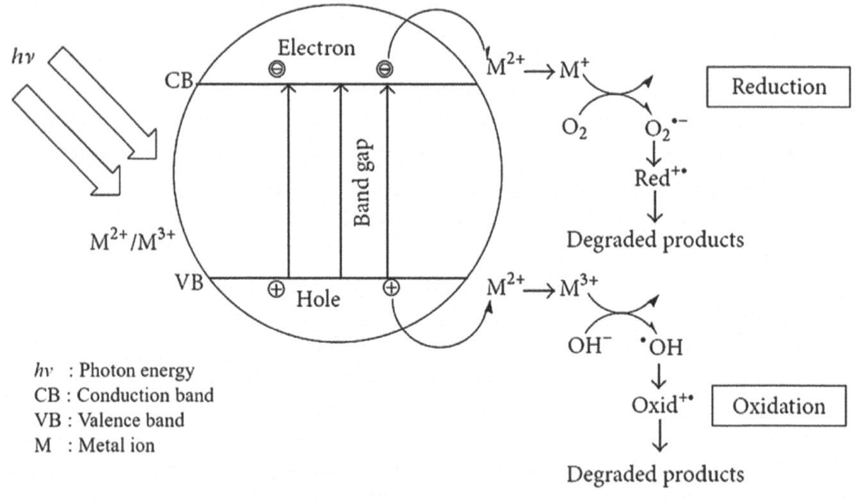

Fig. 1 Main photocatalytic processes over semiconductor photocatalyst. Adapted with permission from Ref. [43], Copyright 2015, Hindawi Publishing Corporation

2 Water Pollutants

2.1 Dyes as Organic Pollutants

The water colour is widely regarded as an indication of its consistency, such that in aqueous systems, very low quantities of certain pollutant dyes (even <1 ppm) are relatively visible and undesirable [44, 45]. Dyes are broadly applied in the fabric industry, coatings, newspapers, beauty products, and similar industries [46–49]. Synthetic dyes are an important group of organic contaminants that are disposed of from their industries as wastewater. As they possess a complicated aromatic structure that hard to be degraded, they will threaten the survival of the aquatic life with deficient of the required sunlight in coloured water [50, 51]. In addition, inflammation, eye burns, heart pulse rate abnormally, and cytotoxicity are some of the damaging impacts of dye pollution on humans [52, 53]. Dye production rates are about 700–105 metric tons a year, consisting of more than 100,000 different dyes. Around 10–15% of effluent dyes are disposed of in the environment [44, 50, 52, 54]. Therefore, searching for a green, sustainable, and cost-effective technology for water treatment is of great importance. The existing techniques for wastewater treatment involve photocatalysis degradation [55, 56], nanofiltration [57], ozonation [58], anodic oxidation, and electrocoagulation [59]. Among these techniques, adsorption and photocatalysis technologies are suitable for dye removal and degradation, respectively, because of their pros of effectiveness, affordability, and ease of the process.

2.2 Heavy Metals as Inorganic Pollutants

Other dangerous water pollutants are heavy metals because of their toxicity and other side cons [60–64]. The utmost familiar heavy metal pollutants are Hg, Cd, Pb, Cu, Ni, Cr, etc. (Table 1). There are many traditional techniques for heavy metal ions removals from wastewater, such as chemical precipitation, reduction, electrochemical methods, ion exchange, adsorption, floatation [65–69]. Adsorption is the best favourable technique as it is simple, inexpensive, effective, and easy to recycle [70].

Table 1 List of some heavy metals and their effects

Metal ion	Effects
Cadmium	Hypertension, migraines renal disorders
Copper	Nausea, vomiting, and hypertension
Nickel	Causes nausea, cyanosis, oedema, dermatitis
Lead	Carcinogenic causes abnormalities in neonatal, kidney problems, etc.
Chromium	Ulcers on the skin, exceeding nausea, coma, and also, death
Mercury	Chromosomal aberrations, visual and phycological impairment

3 Agricultural Waste-Based Materials

3.1 Activated Carbon from Wastes

Due to the ease of recycling, availability, no or low-cost, and excellent adsorption performance of activated carbon (AC), it is commonly used as adsorbent [11, 71–74]. Though the adsorbent cost determines the process' cost, AC is much recommended for dyes and heavy metals removal [75]. Thus, the usage of cheap and profuse source materials with a high carbon yield has been focused on by many researchers in the previous periods. Shokry et al. [76] used Maghara Egyptian coal as a useful carbon source for the production of nanoactivated carbon (NAC). The modified NAC was successfully used for the Methylene Blue (MB) dye adsorption [77]. The presence of −OH groups on the surface of NAC due to NaOH activation enhances the adsorption ability of NAC for the cationic MB by establishing H-bond between −OH of NACs and the MB molecules [78]. This adsorption process followed the Langmuir behaviour demonstrating monolayer coverage of MB on NAC homogenous active sites with 28.09 mg g^{-1} as a maximum monolayer adsorption capacity. Diluted HCl is a favourable eluent for recycling of NAC up to 10 times with excellent stability as the efficiency showed slightly decline between cycles 1 and 10 (98.7–96.2%) (Fig. 2).

El-Salamony et al. [79] prepared AC from the waste of sugar cane bagasse (SCB) using chemical carbonization method at low temperature. The prepared AC is used as a support for nano-TiO_2 via a mechano-mixing method using a different loading of TiO_2 (10, 20, 50, and 70 wt%). It was found that the 20 TiO_2–AC exhibiting the best performance of MB dye degradation, which is 96% after 3 h under visible light irradiation due to the synergistic effect between adsorption and photocatalytic degradation (Fig. 3).

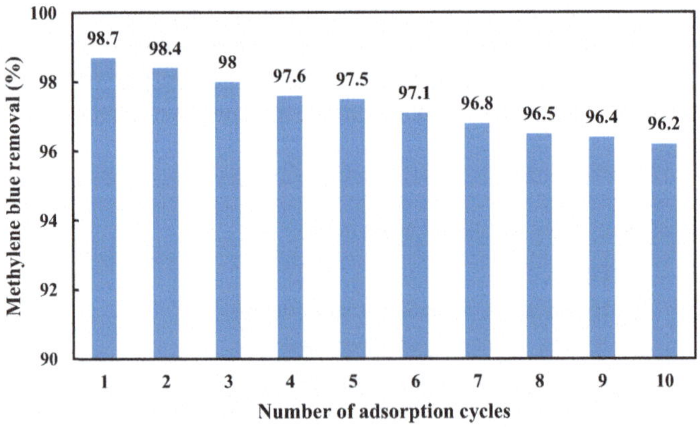

Fig. 2 Reusability of NAC (dose = 10 g L^{-1}, MB conc. = 50 mg L^{-1}, time = 90 min). Adapted with permission from Ref. [76], Copyright 2019, Elsevier

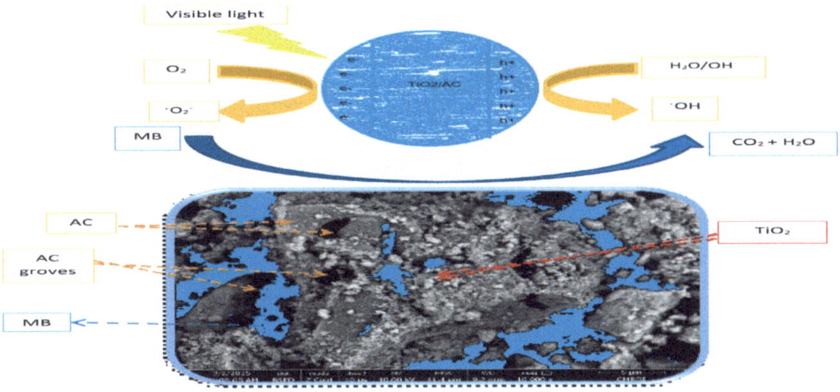

Fig. 3 Graphical scheme for the photocatalytic degradation of MB over TiO_2–AC. Adapted with permission from Ref. [79], Copyright 2017, Taylor & Francis

Sruthi et al. [80] prepared AC from coconut shells Nanosized (AC-NP) and tamarind seeds Nanosized (AT-NP). The prepared AC-NP and AT-NP were used to remove nickel extracted from battery waste. AC-NP have proved to be more efficient (87%) compared to AT-NP (74%). Dieme et al. [81] used Senegal agricultural wastes such as millet stalks, cashew-shells, and rice husks for the preparation of AC and Fe-loaded AC. The prepared nanocomposites were used for the removal of arsenic (As). The Fe-loaded millet stalks AC has (7926 µg g^{-1}) as a maximum adsorption capacity, making it a promising adsorbent for the removal of heavy metals, especially arsenate.

3.2 *Rice Husk-Based Materials*

Rosa et al. [14] produced cellulose and lignin-based natural adsorbent from rice husk waste (RHW) for the removal of MB dye. The prepared adsorbent has a surface area of 14.771 m^2 g^{-1} and provided excellent thermal stability, and the functional groups of lignin were identified after treatment. The maximum equilibrium adsorption capacity of MB on RHW was 1350 mg g^{-1} after 30 s. This adsorption behaviour followed the pseudo-first-order with R^2 of 0.99, and the isotherm followed Freundlich and Sips models.

By microwave and steam activation, Yang et al. [82] prepared a magnetic cobalt–iron modified porous carbon derived from rice straw for the removal of elemental Hg. The pore structure of the porous carbon is enhanced by the microwave activation; moreover, the diffusion of Co oxides and Fe oxides is promoted by the ultrasound-assisted impregnation. The effective primary species for high Hg^0 removal are cobalt oxides and iron oxides. XPS reveals that elemental Hg^0 can be converted by the chemisorption oxygen and the lattice oxygen (originated from the Co^{3+}/Co^{2+} and Fe^{3+}/Fe^{2+}) on the modified porous carbon (Fig. 4).

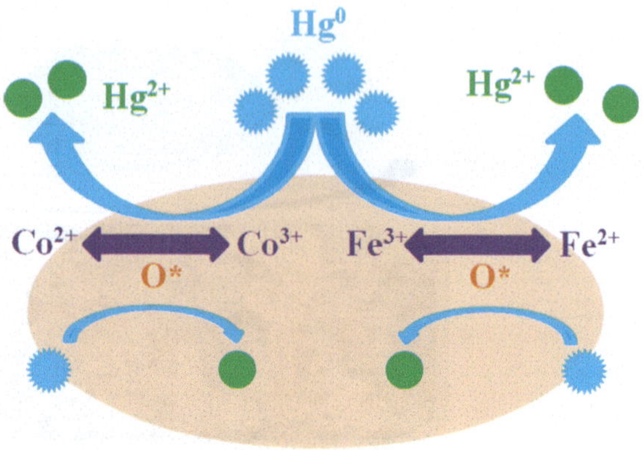

Fig. 4 Chemisorption oxygen and the lattice oxygen on the modified porous carbon. Adapted with permission from Ref. [82], Copyright 2020, Elsevier

3.3 *Peels-Based Materials*

As the waste of fruit peels is generally disposed of as agricultural wastes, Tavker and Sharma [41] isolated cellulose from some seasonal fruit (pomegranate, orange, banana, and sweet lime). This isolated cellulose is used for the preparation of cellulose/MoS_2 by the in situ hydrothermal approach. Cellulose/MoS_2 has an excellent photocatalytic degradation efficiency towards Rhodamine B (RhB) (96%). The presence of cellulose as support with MoS_2 delays the recombination of photo-generated charge carriers. This catalyst is a promising one as it can degrade about 87% of RhB after the fifth cycle (Fig. 5).

Narasaiah and Mandal [83] used vacant cotton boll peel waste extract, type of phytochemicals, as a reducing agent for Pd^{+2} to Pd^0 nanoparticles (Pd NPs). The stability of Pd NPs was reported by surface capping. The prepared Pd NPs exhibited high efficiency for some dyes; congo red (CR), methyl orange (MO), sunset yellow (SY), and tartrazine (TZ); (95–97% within 14 min) using $NaBH_4$ as electron donors, where Pd NPs act as a carrier that transferred the electrons from $NaBH_4$ to the pollutants (Fig. 6).

Karnan and Selvakumar [84] used the rambutan extract (Nephelium lappaceum L.) peel for the preparation of crystalline ZnO nanoparticles that have 3.32 eV

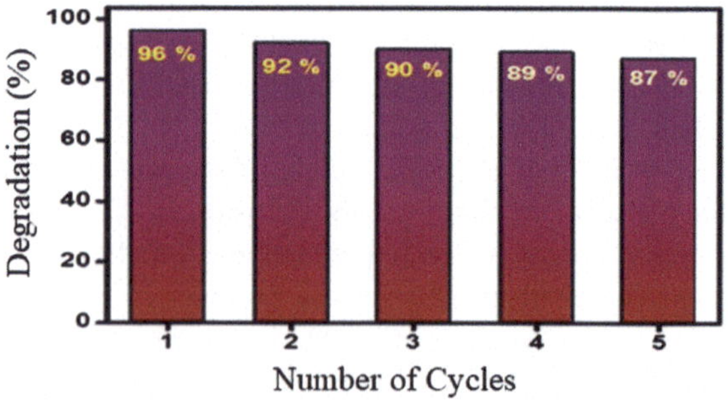

Fig. 5 Recyclability of Cel/MoS_2 nanostructure. Adapted with permission from Ref. [41], Copyright 2020, Elsevier

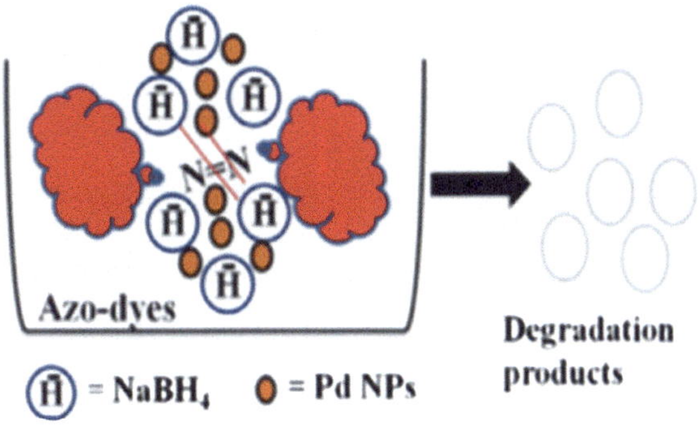

Fig. 6 Probable mechanism of dye reduction using Pd NPs. Adapted with permission from Ref. [83] Copyright 2020, Elsevier

bandgap. The prepared ZnO has a surface area of 69 m^2 g^{-1} with particle size 25–40 nm. ZnO showed enhanced photocatalytic degradation of MO under UV light (83.99% after 2 h) and mineralized the dye into less toxic products, as confirmed by the measurement of the chemical oxygen demand (COD) values.

3.4 Miscellaneous Agricultural Waste-Based Materials

According to the World Health Organization (WHO), clean water should not have <0.003 mg L^{-1} of Cd [85]. As the effluents of battery industries have Cd as a hazardous pollutant, Basu et al. [86] studied the adsorption and mechanism of Cd on lentil's husk (LH) as agricultural waste. The surface area and pore volume of the LH were 6.16 m^2 g^{-1} and 0.0053 cm^3 g^{-1}, respectively. The equilibrium adsorption capacity was 107.31 mg g^{-1} at pH 5 within 1 h. This process followed the pseudo-second-order kinetics and Langmuir model, which confirming the uniform distribution of Cd ions on the surface of LH. In the dual experiment, in which Cd and Pb were used as pollutants, the LH demonstrated a favourite in the adsorption of Pb more than Cd.

SEM images characterized the grainy morphology of LH with the high surface area for Cd adsorption (Fig. 7a) that had become irregular after Cd binding (Fig. 7b, c). EDAX spectra showed signals of C, O, Si, P, S, Ca in the LH (Fig. 7d), while the Ca signal was replaced by Cd in the metal-loaded LH indicates the involvement of the ion exchange mechanism (Fig. 7e). From many eluents, HNO_3, HCl, ethylenediaminetetraacetic acid (EDTA), H_2SO_4, and Na_2CO_3 that were studied for the desorption ability to regenerate the LH, HNO_3 exhibited the best performance, desorbing ~96.0% of the Cd. After five successive cycles, a small decline in the LH efficiency occurred due to the disruption of cell integrity by treatment with HNO_3 during desorption or physico-chemical damage of biomass due to shaking [87]. By using real effluents containing Cd (initial concentrations are 5.76 and 6.50 mg L^{-1}), the final Cd concentrations at pH 5 using LH were 0.19 and 1.54 mg L^{-1} which is well below 2 mg L^{-1} [86].

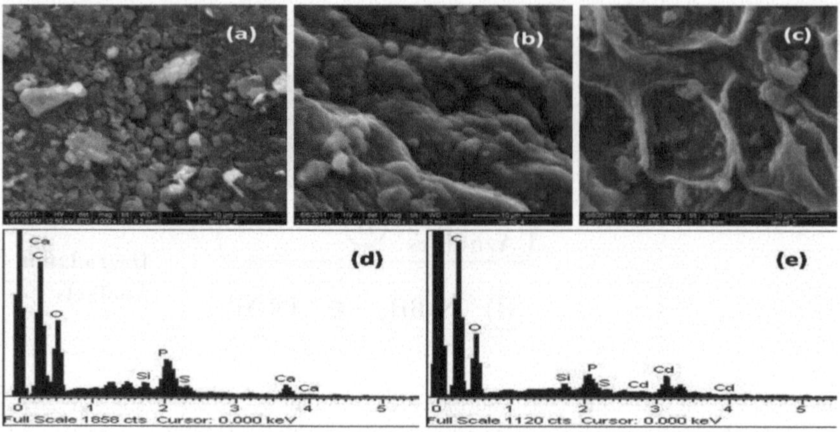

Fig. 7 SEM images of LH (**a**); Cd-loaded LH (**b**); Cd-loaded LH at higher magnification (**c**); EDAX spectra of LH (**d**); Cd-loaded LH (**e**). Adapted with permission from Ref. [86], Copyright 2017, Elsevier

Cellulose is a highly favourable raw material presented at low cost for the preparation of many functional materials [88]. Banana fibre is a natural, biodegradable, and eco-friendly fibre, composed of cellulose, hemicellulose, and lignin [3]. Rani et al. [3] extracted cellulose nanocrystals (CNCs) from banana fibre, then prepared a novel composite of (CNCs)-graft-n butyl acrylate copolymer for the lead removal. This composite has a high adsorption efficiency towards the removal of Pb under these conditions (pH 5, 360 min, dose = 40 g L^{-1}, and Pb (II) conc. = 125 mg L^{-1}). This behaviour is confirmed by the SEM images as the adsorbent was seemed to be heterogeneous and covered with clusters. This process followed the pseudo-second-order kinetics and Freundlich model. The prepared composite can be recycled three times by HCl then treated with EDTA.

With 159 million tonnes of tomatoes produced each year globally, Yargıç et al. [34] studied the Cu(II) ion adsorption by HCl treated tomato factory waste. The treated waste exhibited a high removal efficiency of 92.08% under the optimum conditions; pH 8, dose = 4 g L^{-1}, Cu (II) conc. = 50 mg L^{-1}. This process has an exothermic nature and followed Langmuir isotherm and pseudo-second-order, exhibiting that the rate-limiting step that controlled the sorption process may be chemisorption.

Shorea robusta (common name Sal), Indian forest trees, produce a huge quantity of Sal seeds as main jungle by-products. These seeds are beneficial in many industries as they have high-fat content [89]. The residue Sal after extraction of oil (deoiled seed cake) DOC remains under-utilized or useless in deficiency of any important use and thus existing at a low cost [90]. Singhal and Gupta [91] synthesized Ag nanoparticles (AgNPs) from Sal DOC, an Indian forest-based industrial waste, as a natural and cheap substrate. The covering of AgNPs with Sal DOC extract biomolecules and the stability of AgNPs is confirmed by Fourier transform

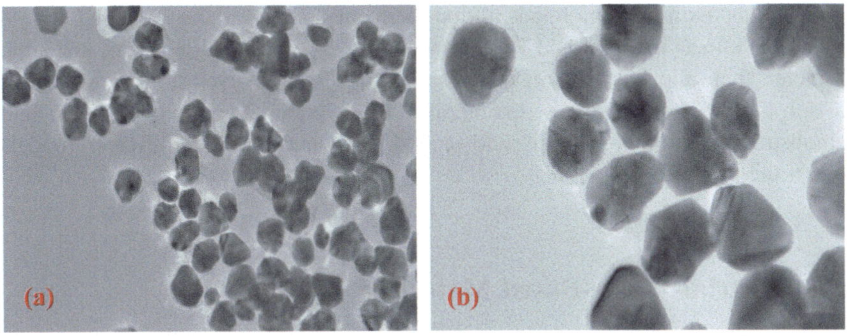

Fig. 8 TEM image of synthesized AgNPs (**a**) and capping of AgNPs with Sal DOCE biomolecules (**b**). Adapted with permission from Ref. [91], Copyright 2018, Elsevier

Table 2 Degradation of a mixture of five dyes and their recycling behaviour after 3 h. (Dyes + 1 ml AgNPs (1589 ppm) + 1 ml $NaBH_4$ (100 mM))

Catalyst	10 ppm/dye				25 ppm/dye				References
	1st	2nd	3rd	4th	1st	2nd	3rd	4th	
AgNPs	94.23	90.61	90.04	61.08	95.20	84.92	71.05	59.96	[91]
XRWS-AgNPs	96.04	95.82	24.79	–	94.35	46.13	17.25	–	[12]

infrared spectroscopy (FTIR) and zeta potential of −33.2, respectively. Figure 8a showed the transmission electron microscope (TEM) images of AgNPs. The AgNPs (35–70 nm) have polygonal/irregular shapes. The inconsistency of nanoparticles' shape and size in biological synthesis is common and also described by others [92]. The existence of Sal DOCE's organic molecules as a capping agent appeared as a separate bright layer surrounded the AgNPs (Fig. 8b) [93]. The prepared AgNPs have high catalytic activity towards the degradation of the tested dyes, congo red (CR), methyl orange (MO), methylene blue (MB), evans blue (EB), and eriochrome black T (EBT), independently in addition to the mixture of these dyes (Table 2). AgNPs were also assessed for the treatment of real industrial coloured wastewater. The AgNPs recycled up to four times for dye degradation.

Global attention is paid to the use of renewable and functional clusters of nano lignin as a bio-additive in manufacturing materials. Sohni et al. [94] used waste from palm kernel shells to prepare chitosan nanocomposite based on novel lignin nanoparticles (Chit/LNP) for the removal of MB. (Chit/LNP) recorded a maximum adsorption capacity of 74.07 mg g^{-1} following the Langmuir model and pseudo-second-order kinetics as a result of both nano-size and LNP combined synergistic effects that improved dye uptake. This adsorption process is spontaneous and has an endothermic nature and can be recycled up to four times, with high efficiency of 89%. Allafchian et al. [95] used a cress seed mucilage CSM-based magnetic iron oxide nanoparticles (MIONs) biosorbent for MB removal that exhibited high adsorption performance (44.6 mg g^{-1}) due to the synergistic effects of CSM and Fe_3O_4 nanoparticles, along with a strong magnetic character that allowed the regeneration of up to five cycles. The nature of bio-sorption was spontaneous, exothermic, followed by Langmuir isotherm and demonstrated that both the pseudo-second and intra-particle diffusion models govern the process.

4 Industrial Waste-Based Materials

4.1 Eggshells-Based Materials

The food industry uses chicken eggs, and the eggshells are thrown away, accounting for hundreds of thousands of tonnes worldwide. This eggshell waste is an essential raw material with a high percentage of $CaCO_3$ that can be used in various sectors, including food production, handicrafts or fertilizers [73, 74, 96–98]. $CaCO_3$ may be used as an adsorbent for heavy metals and neutralizing acids [99, 100]. Eggshell also developed as an efficient adsorbent due to the presence of many pores. Moreover, this waste has fascinating features for use as an environmentally friendly catalyst in many applications, but there is a further need to extend the use of this low-cost green catalyst [96].

Abdel-Khalek et al. [101] investigated the removal of dye (MB and CR tested as model pollutants) using a low-cost eggshell matrix (eggshell + membrane). The surface morphology of the eggshells has a porous structure and is composed of different sizes with irregular particles, as shown in the SEM images (Fig. 9a). The adsorption of cationic MB and anionic CR molecules on the positively charged surface of the eggshells is explained by the electrical double layer and the electrostatic attraction mechanisms, respectively. This process is spontaneous, has exothermic nature, and followed second-order-rate and Freundlich isotherm exhibiting multilayer adsorption. After the adsorption, there is a noticeable change

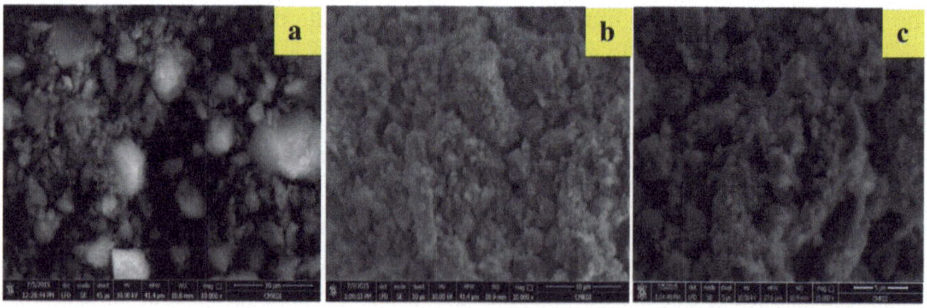

Fig. 9 SEM images of the untreated eggshell (**a**) adsorbed MB on eggshell at low magnification (**b**) and high magnification (**c**). Adapted with permission from Ref. [101], Copyright 2017, Elsevier

in the eggshell textural structure (Fig. 9b, c). By treating 1000 mg g^{-1} of dye concentration, the obtained maximum adsorption capacities were 94.9 mg g^{-1} (MB) and 49.5 mg g^{-1} (CR).

Balaz et al. [102] pulverized eggshell (ES) and its membrane (ESM) by ball-mill. These ES and ESM were used to remove heavy metal ions. Milling was found to be useful only in the case of ES. After milling, no change in IR spectra occurred, confirming no chemical changes. The maximum adsorption capacities for Ag (I) were 52.9 mg g^{-1} (ESM) and 55.7 mg g^{-1} (ES), and the efficiency order for the tested heavy metals was Ag(I) > Cd(II) > Zn(II) confirming the selective nature of ESM. This process followed the Freundlich and Langmuir for Ag (I) and Cd (II), respectively.

Singh et al. [103] supported TiO_2 nanoparticles on the nano eggshell waste (TS-ES) and used it for photocatalytic degradation of dyes. As the amount of eggshell support affecting the degradation efficiency, different loading concentrations have been studied. It was found that 1:1 ratio (photocatalyst: support) experienced the highest degradation efficiency for the mixture of MB and Rh 6G (Rhodamine 6G). Compared to the bare TiO_2, the prepared nanocomposite (TS-ES) exhibits enhanced photocatalytic degradation efficiency in solar irradiation due to the synergistic effect of TiO_2 nanoparticles and high surface area offered by nano eggshell as support.

Nasrollahzadeh et al. [104] used the extract of Orchis mascula L's leaves as a stabilizer and reducing agent for the preparation of Cu/eggshell, Fe_3O_4/eggshell, and Cu/Fe_3O_4/eggshell nanocomposites. These composites have been used to reduce 4-nitrophenol (4-NP) and some dyes (MO, CR, MB, and RhB). The Cu and Fe_3O_4 NPs (about 17 nm) are evenly spread on the eggshell surface due to the highly porous nature of eggshell particles that increase the connection space [105–110]. The eggshell porous is almost coated with Cu and Fe_3O_4 NPs, causing changes in the surface roughness (Fig. 10a). The Cu/Fe_3O_4/eggshell can be recycled up to seven cycles with the same performance and without any change in the size and shape of the catalyst (Fig. 10b).

The plausible mechanism for the degradation of the tested dyes with Cu/eggshell nanocomposite is presented in Fig. 11.

Seyahmazegi et al. [111] used low-cost calcined eggshell (CES) to prepare multiwall carbon nanotubes MWCNTs/CES hydrothermally for the adsorption of CR. Compared to CES (58.14 mg g^{-1}) and untreated eggshell (NES) (5.76 mg g^{-1}), MWCNTs/CES had the highest adsorption capacity (136.99 mg g^{-1}). This adsorption process was spontaneous, has an exothermic nature, and followed the pseudo-second-order kinetic and Langmuir model. The possible interactions between CR and MWCNTs/CES are suggested (Fig. 12). As the used ratio (MWCNTs: CES) is low, this catalyst is a favourable one from an industrial point of view.

The use of groundwater for the daily uses has some limitations on the quality of water, as the dissolved heavy metals are present in groundwater in addition to a high level of acidity. Setiawan et al. [32] used NaClO and HCl to develop nanoporous eggshells using eggshells of different birds for the adsorption purposes of

Fig. 10 FESEM of Cu/Fe_3O_4/eggshell nanocomposite (**a**) and the recovered sample (**b**). Adapted with permission from Ref. [104], Copyright 2016, Elsevier

heavy metals and neutralization of acid present in the water. The obtained result revealed that the yield of duck eggshell (56.23%) is the best among chicken (48.28%) and quail (37.75%) eggshell (Fig. 13).

4.2 Electronic Waste-Based Materials

Electronic devices have emerged as an essential element of modern lifestyle. The demand for electronic products keeps increasing at a rapid rate, and so does the electronic waste (e-waste) [112]. One of the most challenging issues the world has faced is what to do with discarded electronic devices or e-waste. Consequently,

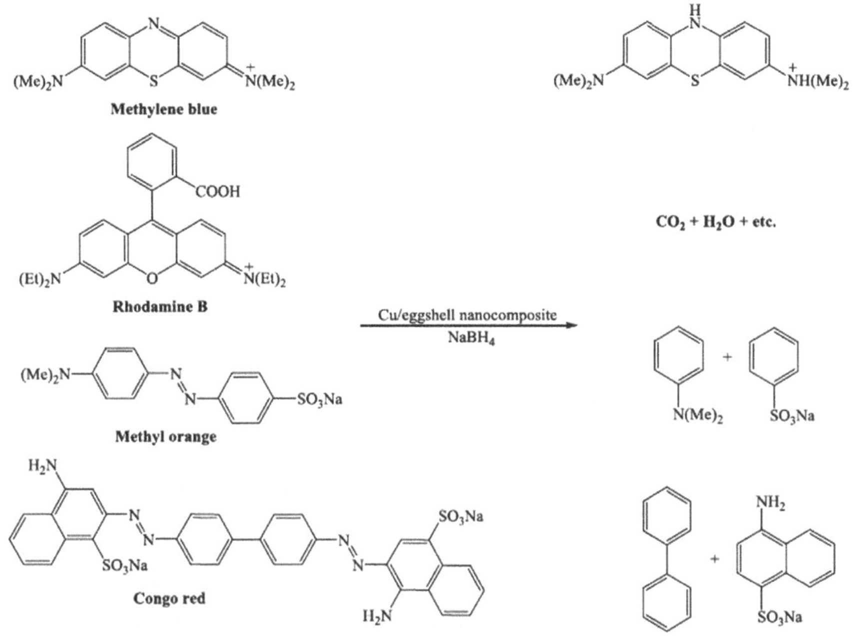

Fig. 11 Plausible mechanism for the degradation of the dye with Cu/eggshell nanocomposite. Adapted with permission from Ref. [104], Copyright 2016, Elsevier

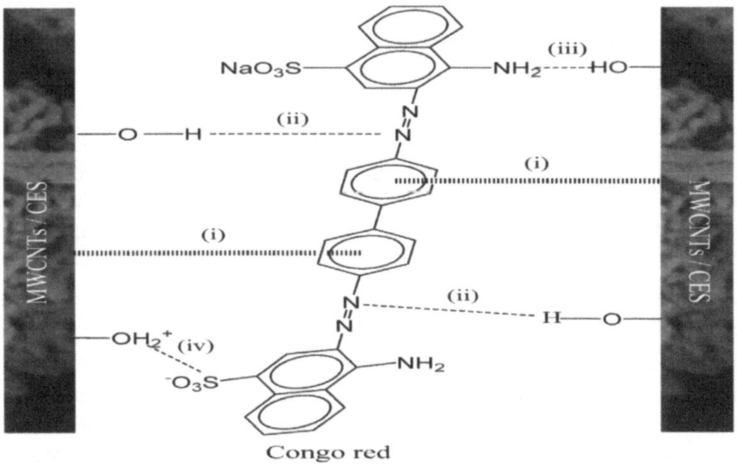

Fig. 12 Suggested interactions between CR and MWCNTs/CES: (i) π-π interaction, (ii) H-bonding between the OH group of MWCNTs/CES and electronegative residues in the CR, (iii) H-bonding between OH group of MWCNTs/CES and aromatic residue in the CR and (iv) ionic interaction between them. Adapted with permission from Ref. [111], Copyright 2016, Elsevier

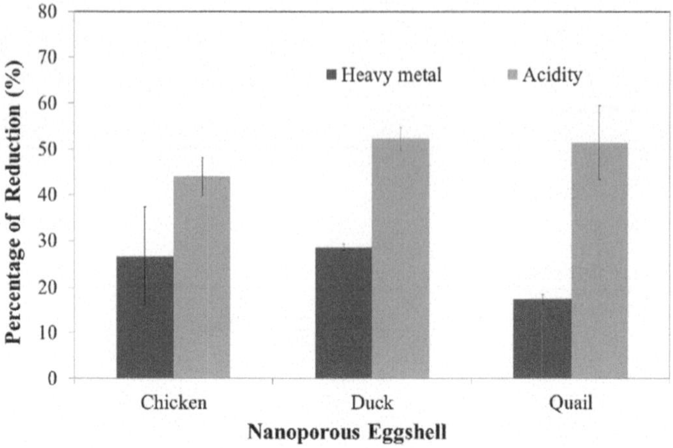

Fig. 13 Heavy metal removal and acidity neutralization over nanoporous eggshells from different sources. Adapted with permission from Ref. [32], Copyright 2018, Elsevier

a new global approach is needed to manage the volume and flow of e-wastes, including carefully planned and regulated recycling programs [113]. Recycling not only reduces pressure on conventional resources but also mitigates the environmental impact associated with e-waste disposal. Direct transformation of e-wastes to value-added materials helps to conserve resources and, at the same time, prevents the environmental impacts of conventional disposal [5, 114].

Nekouei et al. [115] have worked on the improvement of the recycling of metallic e-wastes by transforming them into a high value-added material. In their study, a spent solution from e-waste processing was transformed into t-SnO_2 nanoparticles (NPs) quantum dots (QDs). The NPs have a high surface area of 241 $m^2\ g^{-1}$, and each NP (≈3.9 nm) contains two crystals (≈2.6 nm), which indicates that these NPs can be considered as quantum dots (QDs). The produced t-SnO_2 NPs successfully removed about 92% MB dye under 450 min UV irradiation from simulated industrial wastewater. According to semiconductor theories, photocatalytic activity is driven after the excitation of the QDs by UV light and the production of effective factors in the charge separation efficiency. A short-distance bulk diffusion by the few-nanometre diameter of the QDs makes it much easier for photo-excited charge carriers to move that extend the lifetime of charge carrier and hence improve the quantum yield and photodegradation efficiency [116]. Regarding the electron–hole theory, a comprehensive mechanism for the MB degradation using SnO_2 is [117]:

$$SnO_2 + h\nu \rightarrow e^-(SnO_2) + h^+(SnO_2) \tag{1}$$

$$e^-(SnO_2) + O_2 \rightarrow SnO_2 + O_2^- \tag{2}$$

$$O_2^{-} + H_2O \rightarrow HO_2^{\bullet} + OH^{-} \quad (3)$$

$$HO_2^{\bullet} + H_2O \rightarrow H_2O_2 + OH^{\bullet} \quad (4)$$

$$H_2O_2 \rightarrow 2OH^{\bullet} \quad (5)$$

$$OH^{\bullet} + MB \rightarrow CO_2 + H_2O \quad (6)$$

$$h^{+}(SnO_2) + MB \rightarrow CO_2 + H_2O \quad (7)$$

The printed circuit boards (PCBs) are commonly used as the principal elements of most electronic devices. Huge volumes of e-waste produce a large volume of waste PCBs annually [118]. PCBs include many dangerous products such as brominated flame retardant (BFR) and heavy metals and also valuable materials such as Cu metal and precious metals. PCBs, in particular, contain around 20 wt% of Cu as metallic constituents [119]. Therefore, the reuse of these Cu resources is significant.

Xiu and Zhang [120] introduced waste PCBs into an electrokinetic (EK) (Fig. 14) system with nano-TiO_2 suspension as catholyte after the pretreatment by supercritical water oxidation (SCWO). The EK technology is simple and easy to operate on a large scale as it is efficient and clean for recovering Cu nanocomposites from PCBs wastes. The particle size of Cu_2O was about 40 nm. The prepared Cu_2O/TiO_2 (6 h EK) completely degraded MB dye (10 mg g^{-1} MB concentration, within 60 min). Compared to commercial P25, Cu_2O/TiO_2 has high efficiency due to the presence of Cu_2O on the TiO_2 that increases the electron transfer to oxygen rate and generates a high quantity of holes for the dye degradation [120].

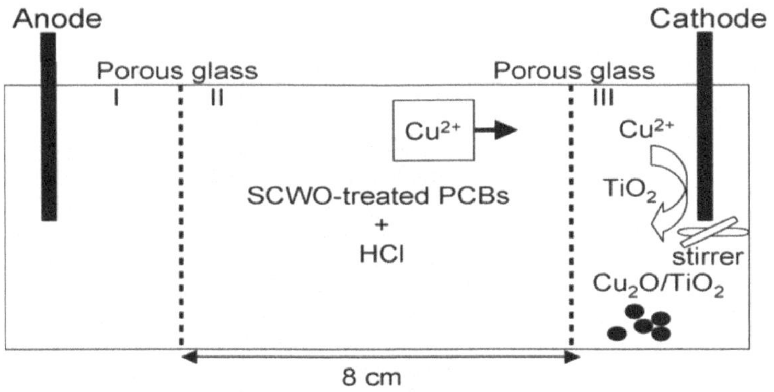

Fig. 14 Schematic drawing of the EK setup. Adapted with permission from Ref. [120], Copyright 2009, Elsevier

4.3 Blast Furnace Dust-Based Materials

One of the major problematic manufacturing variables in the engineering of iron and steel is the blast furnace dust (BFD) waste. The key constituents are iron oxides, carbon, and small quantities of refractory oxides, such as Si, Ca, and Mg [121]. BFD is environmentally harmful because of the large amounts produced every year, and the presence of alkalis and some heavy metals. For such reasons, the development of environmentally and economically viable destinations for BFD waste is of significant importance. Direct recycling of BFD is complicated due to the presence of some pollutants, e.g. Zn, Pb, and alkali metals may cause some problems in the blast furnace [122]. Several types of research have studied the environmental uses of wastes of steel and iron as adsorbents for dyes [123–125], pesticides [126], and phenols [127, 128]. Amorim et al. [129] used BFD for the degradation of (RR195) dye by photo-Fenton processes (Fig. 15).

As maghemite, haematite, and magnetite are present as Fe sources in the BFD waste, the dissolution produces Fe^{3+} more than Fe^{2+} in the Fenton reaction. The BFD morphology (Fig. 16a) exhibited different sizes (200–800 μm) of irregular particles, the largest containing C (mostly) and Ca, Si, Al, Fe, and O (traces) (Fig. 16b), and some small spherical particles (1–17 μm). The clear spherical particles have a prevalence of Fe, O, Si, and Al. The rate of RR195 degradation increased by the use of BFD as it achieves complete removal with a few minutes, suggesting that the BFD acts as Fe provider and also plays a role fasting the reaction kinetics exhibiting a promising behaviour in the photo-Fenton-like process. Compared to the traditional Fenton process (with H_2SO_4), the use of BFD as a catalyst is favoured due to some factors: (i) the catalyst is plentiful and cheap; (ii) the liquid–solid separation is simple; (iii) the catalyst can be recycled up to three times with more than 85% efficiency; and (iv) the degradation rate is enhanced [129].

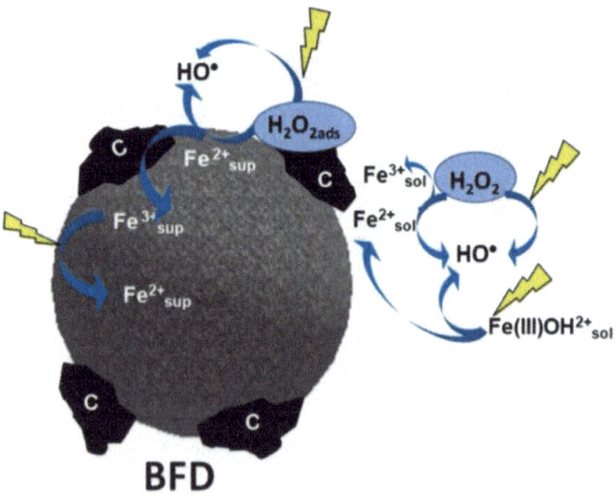

Fig. 15 Photodegradation of (RR195) dye by BFD using photo-Fenton processes. Adapted with permission from Ref. [129], Copyright 2013, Elsevier

Fig. 16 SEM images (**a**) and the corresponding nergy-dispersive spectroscopy (EDS) (**b**) for BFD. Adapted with permission from Ref. [129], Copyright 2013, Elsevier

Wu et al. [130] used solvolysis and hydrothermal processes for the preparation of metals (Al, Zn, Ti)-doped haematite from blast furnace dust. The Fe and all metals (Al, Zn, and Ti) are recycled from BFD as raw materials with no chemical agents. Under simulated sunlight irradiation (Fig. 17), the BFD-derived haematite exhibited more efficiency towards MB degradation (72.2%) than that of pristine one (62.5%) as a result of the co-doping effects of the BFD-derived metals, Al^{3+}, Zn^{2+}, and Ti^{4+}, to α-Fe_2O_3.

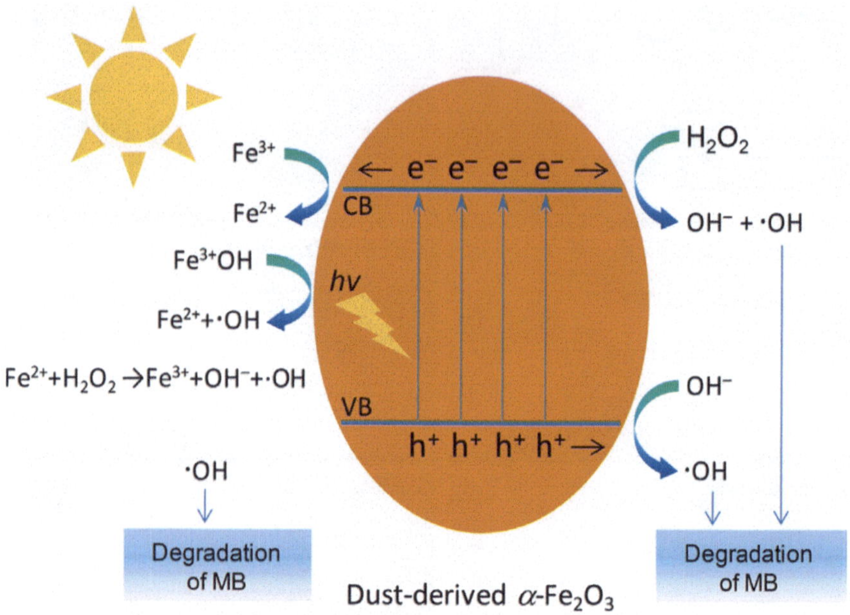

Fig. 17 Schematic diagram representing the MB photodegradation using the BFD-derived α - Fe_2O_3. Adapted with permission from Ref. [130], Copyright 2016, Elsevier

4.4 Miscellaneous Industrial Wastes-Based Materials

Photographic and X-ray films use an immense amount of silver (Ag) as silver halides (AgX) [131]. AgX's sensitivity to light for the processing of photographs is used in radiography and photography [132]. Used X-ray films produce a significant amount of Ag that can be recovered. In the absence of any substantial use, these films are thrown away after their use and thus constitute a high source of waste generation. The recovery of Ag from this waste will reduce the related environmental impact. Ag has been recognized for centuries as a valuable and unique metal and is widely used in many industries, such as the electronics, photographic and chemical industries [133–136]. Due to the high cost and precious nature of this valuable metal, it would be of considerable importance, if it was extracted from the waste and recycled for other essential uses. Used X-ray scanning film and MRI scanning provide a decent path for Ag recovery [137].

To complete their previous work [91], Singhal and Gupta [12] synthesized AgNPs from Ag metal recovered from X-ray sheets waste, in the form of Ag nitrate ($AgNO_3$), and used Sal deoiled seed cake (DOC) as a reducing-capping agent. The distinctive yellowish-brown colour appeared after 30 min of interaction between $AgNO_3$ and Sal DOC. The X-ray waste synthesized (XRWS) AgNPs have been confirmed by the presence of a peak at 485–495 in UV-visible spectrophotometer. XRWS-AgNPs were polycrystalline and had face-centred cubic (fcc) lattice. TEM exhibited that the majority of XRWS-AgNPs is polygonal in shape (size range 30–150 nm) with some flower-like aggregates (Fig. 18a). Also, the manifestation of a sharp peak at 3 keV in energy-dispersive X-ray fluorescence (EDXRF), a specific to Ag element, confirmed the formation of Ag nanoparticles (Fig. 18b), while the others (8 keV and 9 keV) represented the Cu in the grid used for TEM [138]. The XRWS-AgNPs displayed an excellent performance towards the degradation of five azo dyes (CR, MO, MB, EB, and EBT) individually as well as a mixture of these dyes following first- and second-order rate kinetic and can be recovered twice with excellent efficiency (Table 2). The synthesis of AgNPs from waste broadens the use of the Ag in wastewater treatment applications due to the low-cost process.

The toner ink includes nanoparticles, which are polymers [139] consisting of magnetic oxides of Fe, Cu, Si, and Mn [140]. The particle size of toner ink was in the range of ~8–16 μm or higher. They can penetrate the lungs and negatively affecting the health due to their small size. Residual material known as waste old ink was thrown away after the use of toner ink and may further contribute to the release of volatile organic compounds (VOCs) as well as the discharge of heavy metals that could affect the environment and human health [141]. Nevertheless, about 70% of ink cartridges still need an effective strategy for safe drainage or recyclability. Given that the quantity of thrown away waste ink is significant, it is strongly desirable to adopt a responsible strategy for its use and to convert it into the more beneficial material. For this reason, Saini et al. [142] isolated and synthesized functionalized iron oxide nano-carbons (f-FeO-NC) from a freely abundant waste C black precursor, black toner ink, as an active photocatalytic nanomaterial for CR removal (Fig. 19).

The f-FeO-NC is similar to the flakes shape (Fig. 20a, b) and has two types of interplanar distance, 0.21 and 0.27 nm (Fig. 20c) corresponding to the diffraction planes (1 1 0) and (1 0 4) of Fe metal in f-FeO-NC, respectively. f-FeO-NC attains the highest photocatalytic activity (~99%) of CR under 150 min of sunlight irradiation, compared to the dark (~15%). This reaction followed pseudo-first-order kinetics.

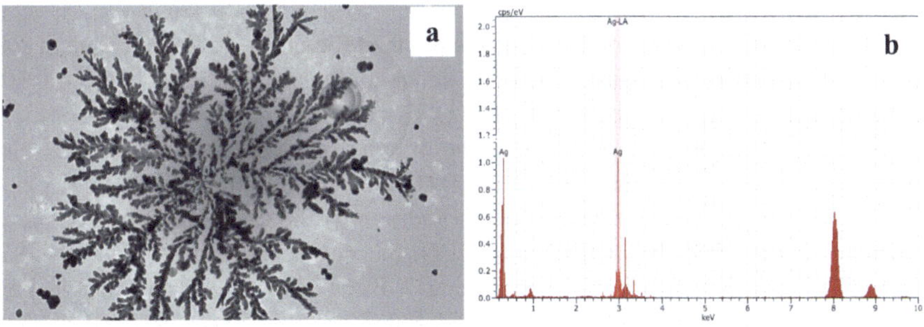

Fig. 18 TEM micrographs of a flower-like aggregate of XRWS-AgNPs (**a**) and EDXRF of XRWS-AgNPs (**b**). Adapted with permission from Ref. [12], Copyright 2019, Elsevier

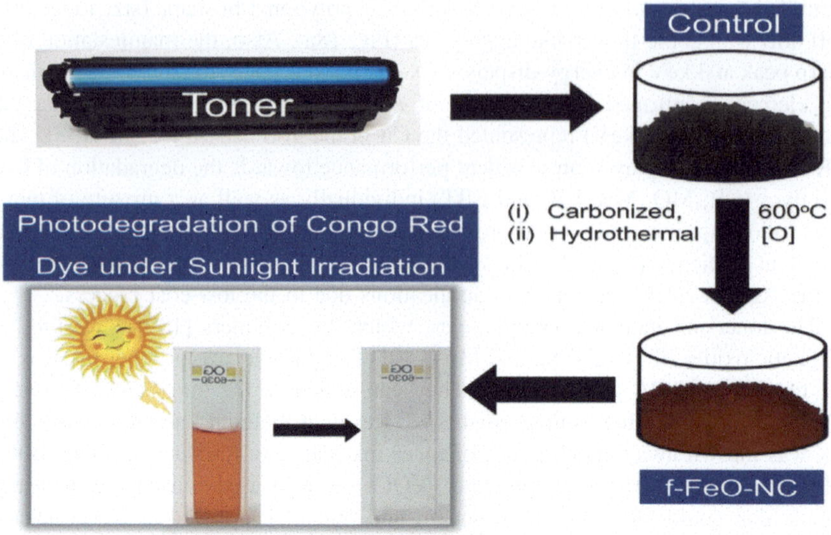

Fig. 19 Isolation of f-FeO-NC from waste black toner ink and the usage for degradation of CR under sunlight irradiation. Adapted with permission from Ref. [142], Copyright 2019, Elsevier

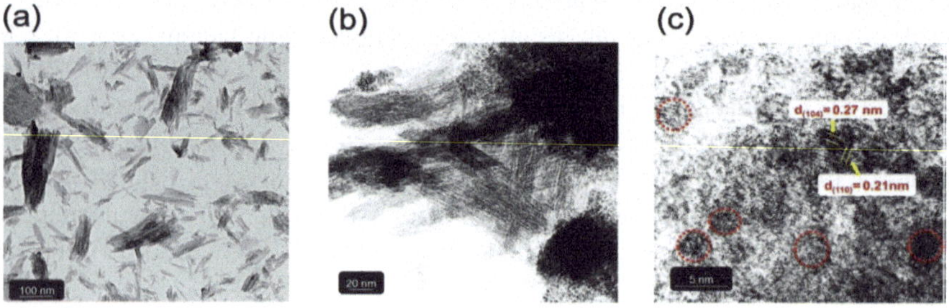

Fig. 20 f-FeO-NC (**a**) and (**b**) TEM and (**c**) High-resolution TEM. Adapted with permission from Ref. [142], Copyright 2019, Elsevier

A trap experiment was conducted to understand the conceivable mechanism for the CR photodegradation using f-FeO-NC under sunlight irradiation. Photocatalysts usually generate reactive species such as –OH$^{\bullet}$, superoxide, and holes [143], which responsible for the photodegradation of the organic compounds. For this, different scavengers such as t-BuOH (for –OH$^{\bullet}$), BZQ (for superoxide radical), and Na_2-EDTA (for hole scavenger) were used in addition to the experiment without any scavenger [144]. With Na_2-EDTA and BZQ, the performance decreased to ~80 and ~50%, respectively, but in the case of t-BuOH, no significant change has occurred. So, the superoxide radicals and holes are the dominant species that control the CR degradation in which the conceivable mechanism is proposed, as shown in Fig. 21.

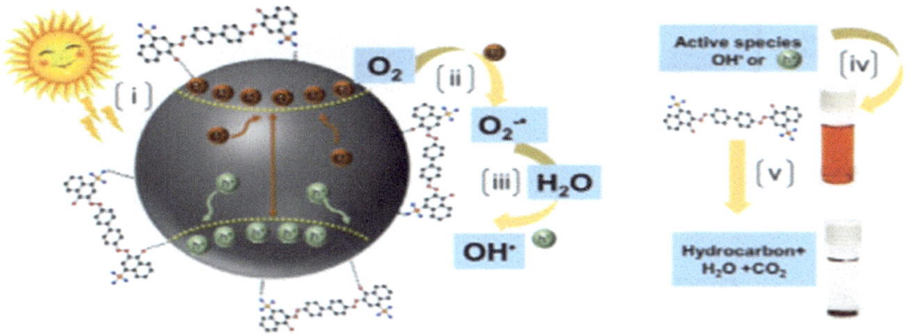

(i) f-FeO-NC + hυ → $e^- + h^+$; (ii) $e^- + O_2 \rightarrow O_2^{-\cdot}$

(iii) $O_2^{-\cdot} + H_2O \rightarrow OH^{\cdot}$

(iv) & (v) $OH^{\cdot} / h^+$ + CR → Hydrocarbon

Fig. 21 Possible mechanism of CR dye degradation. Adapted with permission from Ref. [142], Copyright 2019, Elsevier

Ductile cast iron production expanded sharply due to its high strength, its impact on ductility, and its superior ductility to other types of steel due to the improvement in the graphite structure through crystallization from lamellar to a globular form. The graphite structure alteration relates to the application of Mg in pure or alloy shapes [145–147]. The core wire process is a basic graphite structure improvement process that is based on the injection of core wire into molten cast iron. Mg reaction with the molten Fe is highly challenging due to the reduced Mg boiling point, which results in a lower adsorption rate and dramatic discharge of MgO fume [148]. The generated dust is accumulated by a filtration system that prevents air pollution and generates a solid waste that needs special management to reduce the risk of land pollution and respiratory illness. The obtained MgO waste management is of high significance from environmental and financial terms. It requires safe recovery and recycling as a resource through an environmentally friendly method that encourages the production of other beneficial materials. Pourrahim et al. [148] extracted nanoporous MgO from ductile Fe industry waste, about 88% MgO, by precipitation associated with combined surfactants for the treatment of reactive dyes (Fig. 22). The obtained adsorption capacity was 1 g g^{-1} due to the porous adsorbent structure.

Singh et al. [16] have reused the solid waste sludge produced through the wastewater electrochemical (EC) treatment in the form of nanocomposite materials (NCMs) (Fig. 23). Prepared iron oxide, a hexagonal Fe_2O_3 type phase with trevorite ($NiFe_2O_4$)-type cubic phase, NCMs exhibited a high efficiency towards the methylene red (MR) dye degradation through the wet per-oxidation due to nanocrystalline and highly efficient catalytic properties.

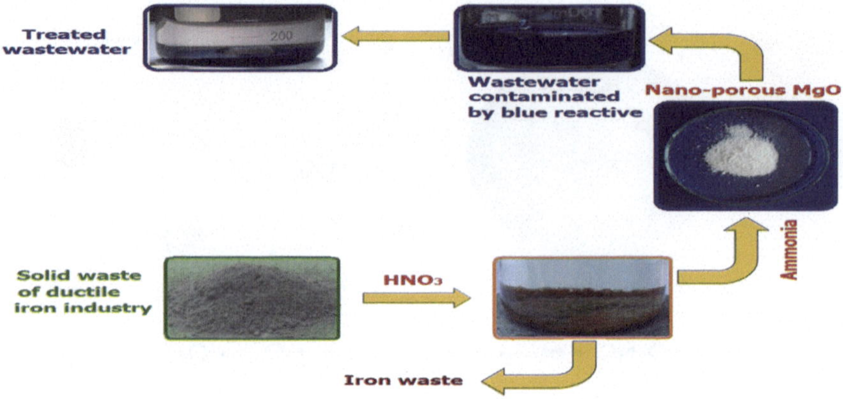

Fig. 22 Extraction of MgO from the ductile cast Fe industry waste. Adapted with permission from Ref. [148], Copyright 2020, Elsevier

Fig. 23 Recycled iron oxide (NCMs) from electrochemical solid waste. Adapted with permission from Ref. [16], Copyright 2016, Elsevier

5 Conclusion

This chapter summarizes the various studies on the preparation and characterization of waste-based nanomaterials and their performance towards the removal of wastewater pollutants. Agricultural and industrial wastes are considered to be major wastes. Activated carbon is prepared from different wastes and has demonstrated good performance towards the removal of dyes and heavy metals. Also, the use of fruit peels for pollutant removal is studied as these wastes are generally disposed of as agricultural wastes. The food industry uses chicken eggs, and the eggshells are thrown away, accounting for thousands of tonnes worldwide. This eggshell waste is a significant source of $CaCO_3$ that can be used in a variety of applications, especially for environmental use. Moreover, as technology updated continuously, the use of electronic devices has increased and hence the electronic waste. Making good use of this waste is of great importance, and global attention is paid to have zero waste by the removal of waste by waste.

6 Future Perspectives

To achieve high performance, fast rate, and low treatment cost, further studies on the properties and the internal structure of the wastes (agricultural/industrial) itself and the prepared materials are demanded. To the best of our knowledge, there is a lack of articles focused on the recycling of ink toner and blast furnace dust into high-value nanomaterials for water treatment applications. However, the recyclability of the prepared waste-based catalyst was studied, much more dedication work, and further investigations are needed to expand this research area. Increasing the recycling numbers improves the performance of wastes-based nanomaterials in large-scale wastewater treatment applications. Moreover, the desorbing and regeneration of the adsorbents and reusing it is a promising trend. To attain a green environment and achieve the SDGs, proper waste disposal is key to continue to use and recycle the wastes correctly.

References

1. Chojnacka K (2010) Biosorption and bioaccumulation—the prospects for practical applications. Elsevier, pp 299–307
2. Bhatnagar A, Sillanpää M (2010) Utilization of agro-industrial and municipal waste materials as potential adsorbents for water treatment—a review. Chem Eng J 157(2–3):277–296
3. Rani K, Gomathi T, Vijayalakshmi K, Saranya M, Sudha PN (2019) Banana fiber Cellulose Nano Crystals grafted with butyl acrylate for heavy metal lead (II) removal. Int J Biol Macromol 131:461–472
4. Ali GAM, Tan LL, Jose R, Yusoff MM, Chong KF (2014) Electrochemical performance studies of MnO_2 nanoflowers recovered from spent battery. Mater Res Bull 60:5–9
5. Ali GAM, Yusoff MM, Shaaban ER, Chong KF (2017) High performance MnO_2 nanoflower supercapacitor electrode by electrochemical recycling of spent batteries. Ceram Int 43:8440–8448
6. Aboelazm EAA, Ali GAM, Chong KF (2018) Cobalt oxide supercapacitor electrode recovered from spent lithium-ion battery. Chem Adv Mater 3:67–74
7. Sethurajan M, van Hullebusch ED, Nancharaiah YV (2018) Biotechnology in the management and resource recovery from metal bearing solid wastes: Recent advances. J Environ Manage 211:138–153

8. Jawoor SS, Kumbar M, Patil SA (2018) Green synthesis of biologically active transition metal nanoparticles containing novel Schiff base via catalyst free hydrothermal reaction: Structural, biological and morphology study. Appl Organomet Chem 32(5):e4322–e4322
9. Ali GAM, Manaf SAA, Kumar A, Chong KF, Hegde G (2016) Superior supercapacitive performance in porous nanocarbons. J Energy Chem 25(4):734–739
10. Ali GAM, Supriya S, Chong KF, Shaaban ER, Algarni H, Maiyalagan T, Hegde G (2019) Superior supercapacitance behavior of oxygen self-doped carbon nanospheres: a conversion of Allium cepa peel to energy storage system. Biomass Conv Bioref. https://doi.org/10.1007/s13399-019-00520-3
11. Habeeb OA, Ramesh K, Ali GAM, Yunus RM (2017) Low-cost and eco-friendly activated carbon from modified palm kernel shell for hydrogen sulfide removal from wastewater: adsorption and kinetic studies. Desalin Water Treat 84:205–214
12. Singhal A, Gupta A (2019) Sustainable synthesis of silver nanoparticles using exposed X-ray sheets and forest-industrial waste biomass: assessment of kinetic and catalytic properties for degradation of toxic dyes mixture. J Environ Manage 247:698–711
13. Giusti L (2009) A review of waste management practices and their impact on human health. Pergamon, pp 2227–2239
14. Da Rosa MP, Igansi AV, Lütke SF, Sant'anna Cadaval TR, Do Santos ACR, De Oliveira Lopes Inacio AP, De Almeida Pinto LA, Beck PH (2019) A new approach to convert rice husk waste in a quick and efficient adsorbent to remove cationic dye from water. J Environ Chem Eng 7(6):103504
15. Samaddar P, Ok YS, Kim KH, Kwon EE, Tsang DCW (2018) Synthesis of nanomaterials from various wastes and their new age applications. Elsevier, pp 1190–1209
16. Singh S, Srivastava VC, Mandal TK, Mall ID, Lo SL (2016) Synthesis and application of green mixed-metal oxide nanocomposite materials from solid waste for dye degradation. J Environ Manage 181:146–156
17. Kreyling WG, Semmler-Behnke M, Chaudhry Q (2010) A complementary definition of nanomaterial. Nano Today 5(3):165–168
18. Qu X, Alvarez PJJ, Li Q (2013) Applications of nanotechnology in water and wastewater treatment. Water Res 47(12):3931–3946
19. Barhoum A, Shalan AE, El-Hout SI, Ali GAM, Abdelbasir SM, Abu Serea ES, Ibrahim AH, Pal K (2019) A broad family of carbon nanomaterials: classification, properties, synthesis, and emerging applications. In: Barhoum A, Bechelany M, Makhlouf A (eds) Handbook of nanofibers. Springer, Cham, pp 1–40
20. Sadegh H, Ali GAM, Abbasi Z, Nadagoud MN (2017) Adsorption of ammonium ions onto multi-walled carbon nanotubes. Stud Univ Babes-Bolyai, Chem 62(2):233–245
21. Sadegh H, Ali GAM, Gupta VK, Makhlouf ASH, Shahryari-ghoshekandi R, Nadagouda MN, Sillanpää M, Megiel E (2017) The role of nanomaterials as effective adsorbents and their applications in wastewater treatment. J Nanostruct Chem 7(1):1–14
22. Goyal A, Srivastava VC, Kushwaha JP (2017) Treatment of highly acidic wastewater containing high energetic compounds using dimensionally stable anode. Chem Eng J 325:289–299
23. Klatte S, Schaefer HC, Hempel M (2017) Pharmaceuticals in the environment—a short review on options to minimize the exposure of humans, animals and ecosystems. Sustain Chem Pharm 5:61–66
24. Omwoma S, Lalah JO, Kueppers S, Wang Y, Lenoir D, Schramm KW (2017) Technological tools for sustainable development in developing countries: the example of Africa, a review. Elsevier, pp 67–81
25. Zeng QF, Fu J, Zhou Y, Shi YT, Zhu HL (2009) Photooxidation degradation of Reactive brilliant red K-2BP in aqueous solution by ultraviolet radiation/sodium hypochlorite. Clean Soil, Air, Water 37(7):574–580
26. Luo P, Zhao Y, Zhang B, Liu J, Yang Y, Liu J (2010) Study on the adsorption of Neutral Red from aqueous solution onto halloysite nanotubes. Water Res 44(5):1489–1497

27. Cai Z, Sun Y, Liu W, Pan F, Sun P, Fu J (2017) An overview of nanomaterials applied for removing dyes from wastewater. Environ Sci Pollut Res 24(19):15882–15904
28. Ong S-A, Toorisaka E, Hirata M, Hano T (2013) Comparative study on kinetic adsorption of Cu(II), Cd(II) and Ni(II) ions from aqueous solutions using activated sludge and dried sludge. Appl Water Sci 3(1):321–325
29. Chen L, Zhou S, Shi Y, Wang C, Li B, Li Y, Wu S (2018) Heavy metals in food crops, soil, and water in the Lihe River Watershed of the Taihu Region and their potential health risks when ingested. Sci Total Environ 615:141–149
30. Ghaderpoori M, Kamarehie B, Jafari A, Ghaderpoury A, Karami M (2018) Heavy metals analysis and quality assessment in drinking water—Khorramabad city. Iran. Data in Brief 16:685–692
31. Jaishankar M, Tseten T, Anbalagan N, Mathew BB, Beeregowda KN (2014) Toxicity, mechanism and health effects of some heavy metals, pp 60–72
32. Setiawan BD, Rizqi O, Brilianti NF, Wasito H (2018) Nanoporous of waste avian eggshell to reduce heavy metal and acidity in water. Sustain Chem Pharm 10:163–167
33. Soetaredjo FE, Kurniawan A, Ki OL, Ismadji S (2013) Incorporation of selectivity factor in modeling binary component adsorption isotherms for heavy metals-biomass system. Chem Eng J 219:137–148
34. Yargiç AS, Yarbay Şahin RZ, Özbay N, Önal E (2015) Assessment of toxic copper(II) biosorption from aqueous solution by chemically-treated tomato waste. J Clean Prod 88:152–159
35. Yuan R, Ramjaun SN, Wang Z, Liu J (2012) Photocatalytic degradation and chlorination of azo dye in saline wastewater: kinetics and AOX formation. Chem Eng J 192:171–178
36. Alventosa-deLara E, Barredo-Damas S, Alcaina-Miranda MI, Iborra-Clar MI (2012) Ultrafiltration technology with a ceramic membrane for reactive dye removal: optimization of membrane performance. J Hazard Mater 209–210:492–500
37. Cai Z, Gong Y, Liu W, Fu J, O'Reilly SE, Hao X, Zhao D (2016) A surface tension based method for measuring oil dispersant concentration in seawater. Mar Pollut Bull 109(1):49–54
38. Amdeha E, El-Salamony RA, Al-Sabagh AM (2019) Enhancing the photocatalytic activity of Ga_2O_3–TiO_2 nanocomposites using sonication amplitudes for the degradation of Rhodamine B dye. Appl Organomet Chem 34(2). https://doi.org/10.1002/aoc.5336
39. Amdeha E, Mohamed RS (2021) A green synthesized recyclable ZnO/MIL-101(Fe) for Rhodamine B dye removal via adsorption and photo-degradation under UV and visible light irradiation. Environ Technol (UK) 42(6):842–859
40. Bargozideh S, Tasviri M (2018) Construction of a novel BiSI/MoS_2 nanocomposite with enhanced visible light driven photocatalytic performance. New J Chem 42(22):18236–18241
41. Tavker N, Sharma M (2020) Designing of waste fruit peels extracted cellulose supported molybdenum sulfide nanostructures for photocatalytic degradation of RhB dye and industrial effluent. J Environ Manage 255:109906
42. Wu Mh, Li L, Liu N, Wang Dj, Xue Yc, Tang L (2018) Molybdenum disulfide (MoS_2) as a co-catalyst for photocatalytic degradation of organic contaminants: a review. Inst Chem Eng:40–58
43. Mahlambi MM, Ngila CJ, Mamba BB (2015) Recent developments in environmental photocatalytic degradation of organic pollutants: the case of titanium dioxide nanoparticles—a review. J Nanomater 2015:1–29
44. Rafatullah M, Sulaiman O, Hashim R, Ahmad A (2010) Adsorption of methylene blue on low-cost adsorbents: a review. Elsevier, pp 70–80
45. Singh RP, Singh PK, Gupta R, Singh RL (2019) Treatment and recycling of wastewater from textile industry. Springer, Singapore, pp 225–266
46. Seyed Arabi SM, Lalehloo RS, Olyai MRTB, Ali GAM, Sadegh H (2019) Removal of congo red azo dye from aqueous solution by ZnO nanoparticles loaded on multiwall carbon nanotubes. Phys E 106:150–155

47. Solehudin M, Sirimahachai U, Ali GAM, Chong KF, Wongnawa S (2020) One-pot synthesis of isotype heterojunction g-C_3N_4-MU photocatalyst for effective tetracycline hydrochloride antibiotic and reactive orange 16 dye removal. Adv Powder Technol. 31(5):1891–1902
48. Abdel Ghafar HH, Ali GAM, Fouad OA, Makhlouf SA (2015) Enhancement of adsorption efficiency of methylene blue on Co_3O_4/SiO_2 nanocomposite. Desalin Water Treat 53 (11):2980–2989
49. Agarwal S, Sadegh H, Majid M, Makhlouf ASH, Ali GAM, Memar AOH, Shahryari-ghoshekandi R, Tyagi I, Gupta VK (2016) Efficient removal of toxic bromothymol blue and methylene blue from wastewater by polyvinyl alcohol. J Mol Liq 218:191–197
50. Aksu Z (2005) Application of biosorption for the removal of organic pollutants: a review. Elsevier, pp 997–1026
51. El-Bindary AA, El-Sonbati AZ, Al-Sarawy AA, Mohamed KS, Farid MA (2015) Removal of hazardous azopyrazole dye from an aqueous solution using rice straw as a waste adsorbent: kinetic, equilibrium and thermodynamic studies. Spectrochim Acta Part A: Mol Biomol Spectrosc 136(PC):1842–1849
52. Zhou L, Huang J, He B, Zhang F, Li H (2014) Peach gum for efficient removal of methylene blue and methyl violet dyes from aqueous solution. Carbohyd Polym 101(1):574–581
53. Crini G (2006) Non-conventional low-cost adsorbents for dye removal: a review. Elsevier, pp 1061–1085
54. Chowdhury S, Mishra R, Saha P, Kushwaha P (2011) Adsorption thermodynamics, kinetics and isosteric heat of adsorption of malachite green onto chemically modified rice husk. Desalination 265(1–3):159–168
55. Hamad H, Bailón-García E, Morales-Torres S, Carrasco-Marín F, Pérez-Cadenas AF, Maldonado-Hódar FJ (2018) Physicochemical properties of new cellulose-TiO_2 composites for the removal of water pollutants: developing specific interactions and performances by cellulose functionalization. J Environ Chem Eng 6(4):5032–5041
56. Hamad H, Bailón-García E, Maldonado-Hódar FJ, Pérez-Cadenas AF, Carrasco-Marín F, Morales-Torres S (2019) Synthesis of Ti_xO_y nanocrystals in mild synthesis conditions for the degradation of pollutants under solar light. Appl Catal B 241:385–392
57. Chakraborty S, De S, Basu JK, DasGupta S (2005) Treatment of a textile effluent: application of a combination method involving adsorption and nanofiltration. Desalination 174(1):73–85
58. El Hassani K, Kalnina D, Turks M, Beakou BH, Anouar A (2019) Enhanced degradation of an azo dye by catalytic ozonation over Ni-containing layered double hydroxide nanocatalyst. Sep Purif Technol 210:764–774
59. El-Ashtoukhy ESZ, Amin NK, Abd El-Latif MM, Bassyouni DG, Hamad HA (2017) New insights into the anodic oxidation and electrocoagulation using a self-gas stirred reactor: a comparative study for synthetic C.I Reactive Violet 2 wastewater. J Clean Prod 167:432–446
60. Badawy NA, El-Bayaa AA, Amdeha E (2015) Removal of Cu (II), Ni (II), and Co (II) ions from aqueous solution using low cost adsorbent: Isotherms, kinetics and equilibrium aspects. J Chem Pharm Res 7(5):589–598
61. Valipour M, Singh VP (2016) Global experiences on wastewater irrigation: challenges and prospects. Springer, pp 289–327
62. Lee SP, Ali GAM, Algarni H, Chong KF (2019) Flake size-dependent adsorption of graphene oxide aerogel. J Mol Liq 277:175–180
63. Shayegan H, Ali GAM, Safarifard V (2020) Amide-functionalized metal-organic framework for high efficiency and fast removal of Pb(II) from aqueous solution. J Inorg Organomet Polym Mater. 30:3170–3178
64. Shayegan H, Ali GAM, Safarifard V (2020) Recent progress in the removal of heavy metal ions from water using metal-organic frameworks. ChemistrySelect 5(1):124–146
65. Vijayaraghavan K, Balasubramanian R (2015) Is biosorption suitable for decontamination of metal-bearing wastewaters? A critical review on the state-of-the-art of biosorption processes and future directions. Academic Press, pp 283–296

66. Al-Zoubi H, Ibrahim KA, Abu-Sbeih KA (2015) Removal of heavy metals from wastewater by economical polymeric collectors using dissolved air flotation process. J Water Process Eng 8:19–27
67. Misra RK, Jain SK, Khatri PK (2011) Iminodiacetic acid functionalized cation exchange resin for adsorptive removal of Cr(VI), Cd(II), Ni(II) and Pb(II) from their aqueous solutions. J Hazard Mater 185(2–3):1508–1512
68. Huang Y, Wu D, Wang X, Huang W, Lawless D, Feng X (2016) Removal of heavy metals from water using polyvinylamine by polymer-enhanced ultrafiltration and flocculation. Sep Purif Technol 158:124–136
69. Fu F, Wang Q (2011) Removal of heavy metal ions from wastewaters: a review. Academic Press, pp 407–418
70. Ajitha P, Vijayalakshmi K, Saranya M, Gomathi T, Rani K, Sudha PN, Sukumaran A (2017) Removal of toxic heavy metal lead (II) using chitosan oligosaccharide-graft-maleic anhydride/polyvinyl alcohol/silk fibroin composite. Int J Biol Macromol 104:1469–1482
71. Habeeb OA, Ramesh K, Ali GAM, Yunus RM (2017) Optimization of activated carbon synthesis using response surface methodology to enhance H_2S removal from refinery wastewater. J Chem Eng Ind Biotechnol 1(1):1–17
72. Habeeb OA, Ramesh K, Ali GAM, Yunus RM, Thanusha TK, Olalere OA (2016) Modeling and optimization for H_2S adsorption from wastewater using coconut shell based activated carbon. Aust J Basic Appl Sci 10(17):136–147
73. Habeeb OA, Ramesh K, Ali GAM, Yunus RM, Olalere OA (2017) Kinetic, isotherm and equilibrium study of adsorption of hydrogen sulfide from wastewater using modified eggshells. IIUM Eng J 18(1):13–25
74. Habeeb OA, Ramesh K, Ali GAM, Yunus RM (2017) Isothermal modelling based experimental study of dissolved hydrogen sulfide adsorption from waste water using eggshell based activated carbon. Malaysian J Anal Sci 21(2):334–345
75. Duman O, Tunç S, Gürkan Polat T (2015) Adsorptive removal of triarylmethane dye (Basic Red 9) from aqueous solution by sepiolite as effective and low-cost adsorbent. Microporous Mesoporous Mater 210:176–184
76. Shokry H, Elkady M, Hamad H (2019) Nano activated carbon from industrial mine coal as adsorbents for removal of dye from simulated textile wastewater: Operational parameters and mechanism study. J Mater Res Technol 8(5):4477–4488
77. Skouteris G, Saroj D, Melidis P, Hai FI, Ouki S (2015) The effect of activated carbon addition on membrane bioreactor processes for wastewater treatment and reclamation—a critical review. Elsevier, pp 399–410
78. Maneerung T, Liew J, Dai Y, Kawi S, Chong C, Wang CH (2016) Activated carbon derived from carbon residue from biomass gasification and its application for dye adsorption: kinetics, isotherms and thermodynamic studies. Biores Technol 200:350–359
79. El-Salamony RA, Amdeha E, Ghoneim SA, Badawy NA, Salem KM, Al-Sabagh AM (2017) Titania modified activated carbon prepared from sugarcane bagasse: adsorption and photocatalytic degradation of methylene blue under visible light irradiation. Environ Technol (UK) 38(24):3122–3136
80. Sruthi T, Ashok J, Nissy MS, Rao CK, Vangalapati M (2019) Extraction and removal of nickel from battery waste, using nano sized activated carbon of coconut shell and tamarind seed powders in a column. Mater Today: Proc 18(7):5240–5246
81. Dieme MM, Villot A, Gerente C, Andres Y, Diop SN, Diawara CK (2017) Sustainable conversion of agriculture wastes into activated carbons: energy balance and arsenic removal from water. Environ Technol (UK) 38(3):353–360
82. Yang W, Chen H, Han X, Ding S, Shan Y, Liu Y (2020) Preparation of magnetic Co-Fe modified porous carbon from agricultural wastes by microwave and steam activation for mercury removal. J Hazard Mater 381:120981
83. Narasaiah BP, Mandal BK (2020) Remediation of azo-dyes based toxicity by agro-waste cotton boll peels mediated palladium nanoparticles. J Saudi Chem Soc 24(2):267–281

84. Karnan T, Selvakumar SAS (2016) Biosynthesis of ZnO nanoparticles using rambutan (Nephelium lappaceum L.) peel extract and their photocatalytic activity on methyl orange dye. J Mol Struct 1125:358–365
85. Pillai SS, Deepa B, Abraham E, Girija N, Geetha P, Jacob L, Koshy M (2013) Biosorption of Cd(II) from aqueous solution using xanthated nano banana cellulose: equilibrium and kinetic studies. Ecotoxicol Environ Saf 98:352–360
86. Basu M, Guha AK, Ray L (2017) Adsorption behavior of cadmium on husk of lentil. Process Saf Environ Prot 106:11–22
87. Saeed A, Iqbal M (2003) Bioremoval of cadmium from aqueous solution by black gram husk (Cicer arientinum). Water Res 37(14):3472–3480
88. Lin N, Dufresne A (2014) Nanocellulose in biomedicine: current status and future prospect. Eur Polymer J 59:302–325
89. Singh S, Gupta A (2014) Comparative fermentation studies on amylase production by Aspergillus flavus TF-8 using Sal (Shorea robusta) deoiled cake as natural substrate: characterization for potential application in detergency. Ind Crops Prod 57:158–165
90. Mahdi BA, Bhattacharya A, Gupta A (2012) Enhanced lipase production from Aeromonas sp. S1 using Sal deoiled seed cake as novel natural substrate for potential application in dairy wastewater treatment. J Chem Technol Biotechnol 87(3):418–426
91. Singhal A, Gupta A (2018) Efficient utilization of Sal deoiled seed cake (DOC) as reducing agent in synthesis of silver nanoparticles: Application in treatment of dye containing wastewater and harnessing reusability potential for cost-effectiveness. J Mol Liq 268:691–699
92. Ahmad N, Sharma S, Alam MK, Singh VN, Shamsi SF, Mehta BR, Fatma A (2010) Rapid synthesis of silver nanoparticles using dried medicinal plant of basil. Colloids Surf, B 81 (1):81–86
93. Muthukrishnan SBS, Muthukumar MSM, Rao Mv SKT (2015) Catalytic degradation of organic dyes using synthesized silver nanoparticles: a green approach. J Biorem Biodegrad 06(05):1–9
94. Sohni S, Hashim R, Nidaullah H, Lamaming J, Sulaiman O (2019) Chitosan/nano-lignin based composite as a new sorbent for enhanced removal of dye pollution from aqueous solutions. Int J Biol Macromol 132:1304–1317
95. Allafchian A, Mousavi ZS, Hosseini SS (2019) Application of cress seed musilage magnetic nanocomposites for removal of methylene blue dye from water. Int J Biol Macromol 136:199–208
96. Laca A, Laca A, Díaz M (2017) Eggshell waste as catalyst: a review. Academic Press, pp 351–359
97. Habeeb OA, Ramesh K, Ali GAM, Yunus RM (2017) Experimental design technique on removal of hydrogen sulfide using CaO-eggshells dispersed onto palm kernel shell activated carbon: experiment, optimization, equilibrium and kinetic studies. J Wuhan Univ Technol-Mater Sci Ed 32(2):305–320
98. Ali GAM, Habeeb OA, Algarni H, Chong KF (2018) CaO impregnated highly porous honeycomb activated carbon from agriculture waste: symmetrical supercapacitor study. J Mater Sci 54:683–692
99. Hammarstrom JM, Sibrell PL, Belkin HE (2003) Characterization of limestone reacted with acid-mine drainage in a pulsed limestone bed treatment system at the Friendship Hill National Historical Site, Pennsylvania, USA. Appl Geochem 18(11):1705–1721
100. Suzuki S (2002) Black tea adsorption on calcium carbonate: a new application to chalk powder for brown powder materials. Colloids Surf, A 202(1):81–91
101. Abdel-Khalek MA, Abdel Rahman MK, Francis AA (2017) Exploring the adsorption behavior of cationic and anionic dyes on industrial waste shells of egg. J Environ Chem Eng 5(1):319–327
102. Baláž M, Ficeriová J, Briančin J (2016) Influence of milling on the adsorption ability of eggshell waste. Chemosphere 146:458–471

103. Singh R, Kumari P, Chavan PD, Datta S, Dutta S (2017) Synthesis of solvothermal derived TiO_2 nanocrystals supported on ground nano egg shell waste and its utilization for the photocatalytic dye degradation. Opt Mater 73:377–383
104. Nasrollahzadeh M, Sajadi SM, Hatamifard A (2016) Waste chicken eggshell as a natural valuable resource and environmentally benign support for biosynthesis of catalytically active Cu/eggshell, Fe_3O_4/eggshell and Cu/Fe_3O_4/eggshell nanocomposites. Appl Catal B 191:209–227
105. Gao Y, Xu C (2012) Synthesis of dimethyl carbonate over waste eggshell catalyst. Catal Today 190(1):107–111
106. Viriya-empikul N, Krasae P, Puttasawat B, Yoosuk B, Chollacoop N, Faungnawakij K (2010) Waste shells of mollusk and egg as biodiesel production catalysts. Biores Technol 101(10):3765–3767
107. Liao Dexiang D, Zheng W, Li X, Yang Q, Yue X, Guo L, Zeng G (2010) Removal of lead (II) from aqueous solutions using carbonate hydroxyapatite extracted from eggshell waste. J Hazard Mater 177(1–3):126–130
108. Fan TX, Chow SK, Zhang D (2009) Biomorphic mineralization: from biology to materials. Pergamon, pp 542–659
109. Yoo S, Hsieh JS, Zou P, Kokoszka J (2009) Utilization of calcium carbonate particles from eggshell waste as coating pigments for ink-jet printing paper. Biores Technol 100(24):6416–6421
110. Wei Z, Xu C, Li B (2009) Application of waste eggshell as low-cost solid catalyst for biodiesel production. Biores Technol 100(11):2883–2885
111. Seyahmazegi EN, Mohammad-Rezaei R, Razmi H (2016) Multiwall carbon nanotubes decorated on calcined eggshell waste as a novel nano-sorbent: application for anionic dye Congo red removal. Chem Eng Res Des 109:824–834
112. Nekouei RK, Pahlevani F, Rajarao R, Golmohammadzadeh R, Sahajwalla V (2018) Direct transformation of waste printed circuit boards to nano-structured powders through mechanical alloying. Mater Des 141:26–36
113. Wang Z, Zhang B, Guan D (2016) Take responsibility for electronic-waste disposal. Nature Publishing, pp 23–25
114. Aboelazm EAA, Ali GAM, Algarni H, Yin H, Zhong YL, Chong KF (2018) Magnetic electrodeposition of the hierarchical cobalt oxide nanostructure from spent lithium-ion batteries: its application as a supercapacitor electrode. J Phys Chem C 122(23):12200–12206
115. Khayyam Nekouei R, Pahlevani F, Mayyas M, Maroufi S, Sahajwalla V (2019) Direct transformation of waste printed circuit boards into high surface area t-SnO_2 for photocatalytic dye degradation. J Environ Chem Eng 7(3):103133
116. Zhang P, Wang L, Zhang X, Hu J, Shao G (2014) Three dimensional porous networks of ultra-long electrospun SnO_2 nanotubes with high photocatalytic performance. Nano-Micro Lett 7(1):86–95
117. Kim SP, Choi MY, Choi HC (2016) Photocatalytic activity of SnO_2 nanoparticles in methylene blue degradation. Mater Res Bull 74:85–89
118. Goosey M, Kellner R (2002) A scoping study end-of-life printed circuit boards, pp 1–44
119. Li J, Lu H, Guo J, Xu Z, Zhou Y (2007) Recycle technology for recovering resources and products from waste printed circuit boards. Environ Sci Technol 41(6):1995–2000
120. Xiu FR, Zhang FS (2009) Preparation of nano-Cu_2O/TiO_2 photocatalyst from waste printed circuit boards by electrokinetic process. J Hazard Mater 172(2–3):1458–1463
121. Das B, Prakash S, Reddy PSR, Misra VN (2007) An overview of utilization of slag and sludge from steel industries. Resour Conserv Recycl 50(1):40–57
122. Jaafar I, Griffiths AJ, Hopkins AC, Steer JM, Griffiths MH, Sapsford DJ (2011) An evaluation of chlorination for the removal of zinc from steelmaking dusts. Miner Eng 24 (9):1028–1030
123. Bhatnagar A, Jain AK (2005) A comparative adsorption study with different industrial wastes as adsorbents for the removal of cationic dyes from water. J Colloid Interface Sci 281 (1):49–55

124. Gupta VK, Suhas S (2009) Application of low-cost adsorbents for dye removal—a review. Academic Press, pp 2313–2342
125. Jain AK, Gupta VK, Bhatnagar A, Suhas S (2003) Utilization of industrial waste products as adsorbents for the removal of dyes. J Hazard Mater 101(1):31–42
126. Gupta VK, Ali I, Suhas S, Saini VK (2006) Adsorption of 2,4-D and carbofuran pesticides using fertilizer and steel industry wastes. J Colloid Interface Sci 299(2):556–563
127. Bhatnagar A (2007) Removal of bromophenols from water using industrial wastes as low cost adsorbents. J Hazard Mater 139(1):93–102
128. Jain AK, Suhas S, Bhatnagar A (2002) Methylphenols removal from water by low-cost adsorbents. J Colloid Interface Sci 251(1):39–45
129. Amorim CC, Leão MMD, Moreira RFPM, Fabris JD, Henriques AB (2013) Performance of blast furnace waste for azo dye degradation through photo-fenton-like processes. Chem Eng J 224(1):59–66
130. Wu ZJ, Wang LC, Gao ZF, Liu WM, Wu XR (2016) Recycling blast furnace dust into metals (Al, Zn and Ti)-doped hematite with enhanced photocatalytic activity. J Environ Chem Eng 4(1):341–345
131. Malinowski J (1960) Synthesis of silver halides for photographic research. J Photographic Sci 8(2):69–71
132. Atta AM, El-Mahdy GA, Al-Hodan HA, Ezzat AO (2016) Application of stabilized monodisperse antimicrobial silver nanoparticles produced from X-ray and photographic films as corrosion inhibitor for carbon steel in aqueous acidic solution. Int J Electrochem Sci 11(9):8082–8095
133. Shankar S, Prasad R, Selvakannan P, Jaiswal L, Laxman R (2015) Green synthesis of silver nanoribbons from waste X-ray films using alkaline protease. Mater Express 5(2):165–170
134. Bhargava S, Uma V (2019) Rapid extraction of Cu(II) heavy metal from industrial waste water by using silver nanoparticles anchored with novel Schiff base. Sep Sci Technol (Philadelphia) 54(7):1182–1193
135. Jena SK (2018) Application of silver nanoparticles for water treatment. J Adv Nanobiotechnol 2(4):1–11
136. Pappas S, Turaga U, Kumar N, Ramkumar S, Kendall RJ (2017) Effect of concentration of silver nanoparticles on the uptake of silver from silver nanoparticles in soil, pp 80–90
137. Amira HAA, Eida MAK (2016) Recovery of silver from used X-ray film using alkaline protease from Bacillus subtilis sub sp. subtilis. Afr J Biotechnol 15(26):1413–1416
138. Edison TNJI, Atchudan R, Sethuraman MG, Lee YR (2016) Reductive-degradation of carcinogenic azo dyes using Anacardium occidentale testa derived silver nanoparticles. J Photochem Photobiol, B 162:604–610
139. Jujun R, Jia L, Zhenming X (2013) Improvements of the recovery line of waste toner cartridges on environmental and safety performances. Environ Sci Technol 47(12):6457–6462
140. Gaikwad V, Kumar U, Pahlevani F, Piadasa A, Sahajwalla V (2017) Thermal transformation of waste toner powder into a value-added ferrous resource. ACS Sustain Chem Eng 5 (12):11543–11550
141. Yordanova D, Angelova S, Dombalov I (2014) Utilisation options for waste toner powder, pp 140–144
142. Saini D, Aggarwal R, Anand SR, Sonkar SK (2019) Sunlight induced photodegradation of toxic azo dye by self-doped iron oxide nano-carbon from waste printer ink. Sol Energy 193:65–73
143. Ramírez-Aparicio J, Sánchez-Martínez A, Ramírez-Bon R (2016) Photodecolorization of rhodamine under sunlight irradiation driven by chabazite. Sol Energy 129:45–53
144. Nezar S, Laoufi NA (2018) Electron acceptors effect on photocatalytic degradation of metformin under sunlight irradiation. Sol Energy 164:267–275
145. Hsu CH, Lin KT (2011) A study on microstructure and toughness of copper alloyed and austempered ductile irons. Mater Sci Eng, A 528(18):5706–5712

146. Chaengkham P, Srichandr P (2011) Continuously cast ductile iron: Processing, structures, and properties. J Mater Process Technol 211(8):1372–1378
147. Labrecque C (1998) Review ductile iron: fifty years of continuous development. Can Metall Q 37(5):343–378
148. Pourrahim S, Salem A, Salem S, Tavangar R (2020) Application of solid waste of ductile cast iron industry for treatment of wastewater contaminated by reactive blue dye via appropriate nano-porous magnesium oxide. Environ Pollut 256:113454

Carbon Nanomaterials Synthesis-Based Recycling

Mohamed F. Sanad

Abstract Nowadays, the concept of recycling becomes one of the most general research topics, which for most of the researchers, involves the weekly ritual of placing waste resources such as cans and cardboard into oversizing and placing them outside the houses. Carbon-based nanomaterials possess unique physical and chemical characteristics that make them attractive to use in different directions. For this type of recycling strategies to be more practical, they must be relatively simple and highly resourceful. Carbon nanomaterial's nucleation is a challenging process, as it requires temperature, renewable sources, and specific types of catalysts. Recently, reported recycling activities include hydrocarbon-rich organic and polymeric material waste as the primary source. In this chapter, discuss the synthesis of carbon nanomaterials using pyrolysis systems in detail. Furthermore, the chapter explores recent step-ups made in the context of this direction.

Keywords Carbon nanomaterials · Graphene · Recycling · Pyrolysis · CNFs

List of Abbreviations

3D	Three dimensional
BET	Brunauer–Emmett–Teller
BJH	Barrett–Joyner–Halenda
CNFs	Carbon nanofibers
CNTs	Carbon nanotubes
DTA	Differential thermal analysis
DFT	Density functional theory

M. F. Sanad (✉)
Basic Science Departments, Modern Academy for Engineering and Technology, Maadi, Egypt
e-mail: mfsanad@miners.utep.edu

Basic Science Department, British University in Egypt, El-Sherouk City, Misr-Ismailia Road, Cairo 11837, Egypt

Chemistry Department, Faculty of Science, Ain-Shams University, Abbasia, Cairo, Egypt

University of Texas at El Paso, 500 West University Avenue, El Paso, TX 79968, USA

A. S. H. Makhlouf and G. A. M. Ali (eds.), *Waste Recycling Technologies for Nanomaterials Manufacturing*, Topics in Mining, Metallurgy and Materials Engineering, https://doi.org/10.1007/978-3-030-68031-2_15

FESEM	Field emission scanning electron microscope
FTIR	Fourier transform infrared spectroscopy
LPCVD	Low pressure chemical vapor deposition
MWCNTs	Multi–walled carbon nanotubes
rGO	Reduced graphene oxide
SWCNTs	Single–walled carbon nanotubes
SWCNHs	Single–walled carbon nanohorns
TEM	Transmission electron microscope
TGA	Thermogravimetric analysis
UV–Vis	Ultraviolet–visible spectrophotometry
USEPA	United states environmental protection agency
XRD	X–ray diffraction

1 Introduction

Carbon nanomaterials have revolutionized all key directions of science from drug targeting to the nuclear industry [1–5]. The modern chemical synthesis of carbon nanomaterials is very critical for large product manipulation and industrial-level [5–11]. For disabling this problem, researchers need to look for new resources of nanomaterials which can satisfy the industrial demand [11–19]. Carbon nanohybrids first reported in 1958 and assembled in 1991 [19–25], when the synthesized multi-walled carbon nanotubes (MWCNTs) and single-walled carbon nanotubes (SWCNTs) first prepared using an arc discharge methodology [26]. Since then, a new direction of knowledge in carbon materials science has emerged. But the question here, why carbon nanomaterial? [27–36]. Given the high value of recycling of manufactured products [37], there is strong attention among the researchers in recovering certain products for the reuse of precursors or nanomaterials directly. Carbon nanomaterials are set to become very vital for the future because they are light and highly stable. Carbon nanomaterials signify high importance in a wide variety of derivatives, such as a wide gap of beneficial compounds such as fullerenes, 2D graphene, and carbon nanotubes. Multi-walled carbon nanotubes possess high importance due to their brilliant applications in aerospace and energy [38, 39]. However, due to the deficiency in the information which reported for recycled carbon nanomaterials, it may be declared as hazardous waste material, which causes a loss of valuable sources. Nanomaterials could exist in the waste recycling lines, which might include purely manufactured carbon nanotubes and liquid suspensions [35]. Matters contaminated with nanoparticles such as wipes, solid matrices, ceramics present everywhere, and highly available with integrated carbon nanomaterials [40, 41]. The potential to collect and separate carbon nanoparticles and nanocomposites from the waste stream to reuse them has also been explored widely [41–47]. The major part of the pre-tested processes is the standard separation technique, such as pyrolysis and solvent evaporation, with high energy demand [48–51]. Some particular persecutions should be taken into consideration to ensure that nanomaterials will never hurt human health and nature. Therefore, nanocomposites that are dangerous, toxic, or highly reactive should be eliminated [52–55]. Researchers highly recommend taking some particular persecutions before design recycling plans [11, 36, 56–58]. Hence, it is essential to recognize the characteristics of specific wastes before developing effective disposal systems. Safety precautions determine the disposal systems required for the handling of nanomaterials must be based on current knowledge and take into consideration. Different parameters should control the process, liable on the type of material, electrical, chemical, and physical operating of nanotechnology-containing waste is possible deactivation waste mixtures. The defect in general regulations related to the application and disposal of nano-waste, besides the recycling of nano-waste-containing fragments, is very limited to specific levels. One of the most critical limitations of those processes is the size, and nanoparticles are difficult to monitor; due to its tiny size, it can diffuse in water systems, causing harm to human health and the environment. Also, legislation is significant to organize the scale of waste-containing carbon nanomaterials in the mega-markets and the use of them after separation. Where possible, recycling of nano-waste is the most desirable target. Governments must adjust assessments and monitoring the manufacturers [59–61]. In summary, manufacturers should make extensive health impact research investigations before placing the products on the markets. Nano-designers must also explore whether these manufacturing systems could represent a risk to the environment. From the other point of view, some products should not only be allowed into the marketplace. Additionally, manufacturers producing such waste as a by-product of their industrial systems should prove that their nanoparticles are non-hazardous to human health. Recently, the disposal project, funded by the European Union, targeted to recycle MWCNT created during production to turn them into other novel plastic composites among nanoscale. Polymers with conductive properties require particular and highly expensive fillers, so the projects on studying how to recycle carbon nanomaterials production from these polymers waste should include pads with cheap constituents. Researchers reported some injected plastic material of the reused conductive polymers to demonstrate this. Additionally, recycling carbon nanotube may not be as simple as with plastics, and this pushes the potential directions to be limited by the quality of the final product.

In this chapter, we discuss the recycling of carbonic composites and nanomaterials, specifically carbon nanotubes, graphite, carbon nanofibers (CNFs), and reduced graphene (rGO) by different pyrolysis systems. Furthermore, the chapter explores recent improvements through the context of this reuse system.

2 Recycling of Carbonic Nanomaterials Using Various Pyrolysis Systems

Pyrolysis method is considered one of the thermochemical-based reactions in which heat converts matter into tiny fragments in the absence of air. Previously, it was applied to produce significant petroleum fractions from heavy crude oil. Pyrolysis is the most prevalent used method in removing organic waste.

Generally, the pyrolysis of organo-waste produces volatile fragments and leaves a solid residue filled with carbon. High-level pyrolysis, which leads to the waste of carbon as it usually has another name like carbonization. Pyrolysis is also considered the first step in the processes of combustion; besides, another outstanding example of this method is the charring of wood. Nowadays, this system has gained considerable attention for its use in the conversion of wastes into valuable carbon products. Several research directions have shown that relevant organic products can be separated from feeds such as waste tires using pyrolysis [62–65]. The production of advanced materials such as carbon-based nanoparticles with high yield using the pyrolysis system as can be seen in Fig. 1 [66–70].

As mentioned before, carbon nanomaterials can be multilayers graphite, monolayer graphene [71], or cylindrical carbon nanotube, as can be observed in Fig. 2 [72–74]. MWCNT has been synthesized using some plastic disposal. Plastic disposal, especially, the types made of a polymeric hydrocarbon containing compounds are commonly found in soil, water, and factories waste. Researchers have recommended using this plastic polymeric waste as a carbonaceous precursor to making carbon nanocompounds like carbon tubes with both types single-walled and multi-walled [75–77]. Assembly of MWCNTs from polymers is achievable through

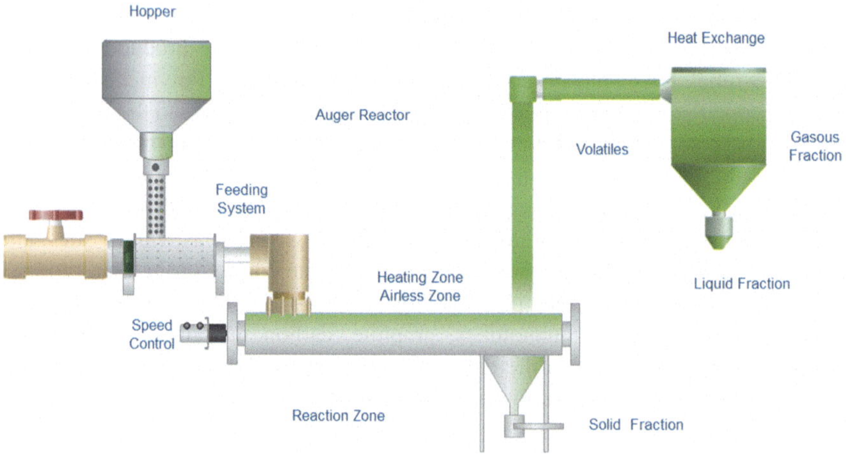

Fig. 1 Schematic representation of the auger reactor for the pyrolysis of waste to carbon nanomaterials

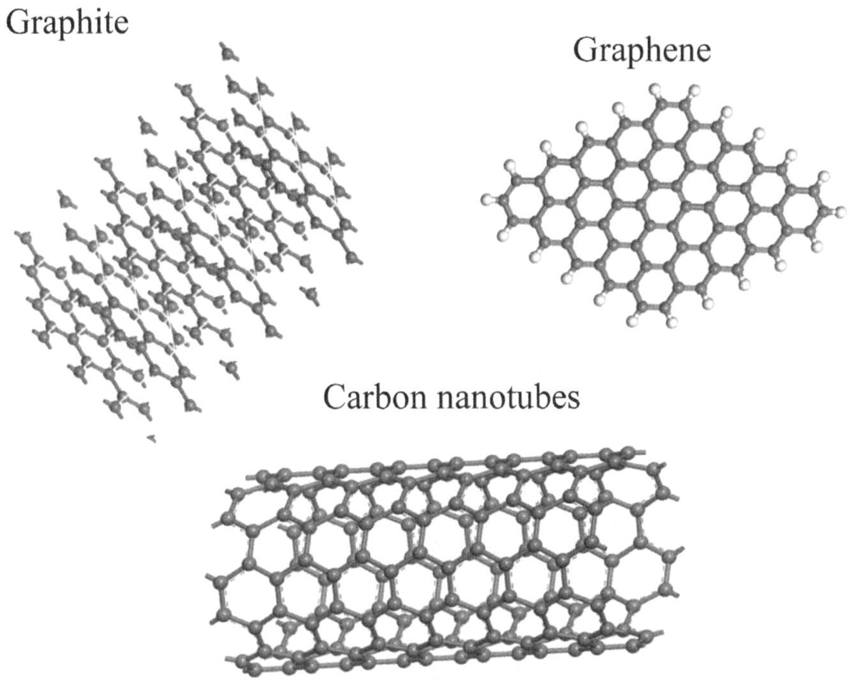

Fig. 2 Schematic representation of different types of valuable carbon nanomaterials

pyrolysis and thermal treatment [59, 62]. The possibility of turning all the products of pyrolysis into nanomaterials is very efficient, and this was reported recently. However, the industrial design of such reactions needs further study and more knowledge about the required precautions. This process shows an excellent output regarding the manufacturing of valuable resources from waste components [75, 76].

Potential research work that has been reported recently has shown promising results in the application of wood disposal to generate nanomaterials and products with a higher valuation [78–81]. Wood waste mostly contains cellulose, which in turn can be applied to produce carbonaceous components in the shape of carbon nanostructure. The reaction environment is a crucial factor here, which makes it feasible to make carbon nanocompounds and nanotubes in various geometries. As an example, graphitic carbon nanocomposites with many types of morphologies can be designed by the hydro-thermal handling of cellulose at high temperatures [82–84]. Hollow carbonic microstructures can also be assembled from cellulose by high-temperature pyrolysis (CO_2 laser, ~2200 °C) [82–87]. However, those kinds of reactions often generate specific quantities of pollutants, particularly in the form of exhausts.

Furthermore, the production cost of the nanomaterials is relatively very high. There is a study that explores a pyrolysis reactor that has excellent potential to produce carbon nanofibers in a very cheap way with a low amount of pollutants.

This reactor uses a starting material of waste wood to create carbon with low emissions of dioxide, unlike other resources of processes [64, 82, 88, 89]. Furthermore, the process does not apply toxic gases whose production is not usually environmentally friendly. Unlike the commercial methods utilized to produce carbon nanomaterials, this system is still in a developing spotlight and is typically a two-step method. In the beginning steps of the reaction, precursor's garbage is created by fixed-bed pyrolysis, while in the second round, these products are settled over catalysts regularly nickel or cobalt. Additionally, these materials are accumulated over the support such as silica and alumina; the idea is to permit enough dispersion of the content of the catalyst particles to create few active locators during the nucleation form. The wide gap of thoughts has analytically proved that those active sites should be only among the nanoscale range. Additionally, a study carried out by other researchers utilizes that metallic iron-based motivators dedicate the highest activity for MWCNT nucleation associated with other metals catalysts. Another example of the low-density-based type of polyethylene (LDPE) polymer plastic was applied as a source in this type of pyrolysis. Other research aligned multi-walled carbon nanomaterials were manufactured in single-step pyrolysis and chemical ousting process [90–93].

2.1 Fixed-Bed Class of Pyrolysis Using Water Vapor

The hydrous designation is the pyrolysis division of reactions based on water vapor. Multiple research work had been established a remarkable rise in the production of carbon nanomaterials when the water was applied [94–96]. The basic idea of pyrolysis, as mentioned before, is accomplished to produce precursors over the surface of the catalyst in the second level of this mechanism; however, a vapor is introduced in the process [97–100]. Adding water vapor may be pre-set either by direct injection simultaneously with transporting using bubbler before the pyrolysis sector. The mode behind this is the recognized exceptional nucleation of carbon nanomaterials through the injection of water vapors to the reaction. In this reaction, ethylene was settled over iron-based catalysts, while the well-ordered quantity of water vapors was also inoculated at the same time [101–104]. Water effect research was conducted by some researchers using acetylene decomposition over the same condition (iron catalyst) in the chemical vapor deposition chamber. It has been reported that the presence of water significantly induced carbon material growth; moreover, there is an apparent effect of water on the annealing step. Also, it has been founded that the presence of hindered Ostwald ripening is usually based on the water where the larger particles grow over the smaller ones until later are executed [101–105]. Many examinations have been conveyed to inspect the effect of steam introduction to the chemical vapor deposition reaction. The bed pyrolysis of polymers in nitrogen (N_2) gas has been carried out at temperature near 500 °C, and the created gaseous stream was then passed over a specific type of catalyst at a sophisticated temperature. It was reported that filamentous carbon deposits reduced by increasing the vapor releasing rate. Low-density polyethylene had the highest potent carbon yield, while other polymers like polypropylene and polystyrene had given the highest returns when the excess amount was used. Therefore, it was reviewed that the influx of water vapor has important role in the reduction of the recycling of carbon nanomaterials. But in the case of hydrogen delivering, it displays the flexibility of process in terms of the carbon outputs. Graphitic structured germination has been reported by water introduction in pyrolysis/CVD reaction in which coconut coir dust was utilized as feed. Introducing water is well proven for chemical vapor ousting reaction; additional studies are needed for its effect in two-step pyrolysis based on the CVD manner [66, 106, 107].

2.2 *Fixed-Bed Pyrolysis Using Microwave*

Using microwave in pyrolysis is like using the thermal pyrolysis, where heat is produced by microwave. In some cases, this method is preferred to convert complete waste to carbon nanomaterial at one time [108–110]. Materials like CNFs and CNTs can be obtained from renewable sources using microwave radiation. However, it may be a limited process due to a lot of challenges like the selection of the raw waste and specific catalyst. Also, gas resources can be a problem for large-scale production [89, 111–114]. However, there is the research of high purity multi-walled carbon nanotube production using a pyrolysis-based microwave. In other studies, gumwood was applied as a precursor, and silicon carbide has been added as a microwave support catalyst. The described mechanism is quite different from normal thermal pyrolysis. It is assumed that limited hotspots are created as a result of microwave operating limitations, which have much more formidable temperatures [88].

2.3 *Chemical Vapor Deposition*

Recycling of valuable materials from GaN influences researchers to obtain carbon nanomaterials [115–118]. Leaching-based kinetics of gallium precious waste is investigated, and the process is adjusted. The gallium precious waste dust has been characterized by XRD and other structural analysis followed by digestion. Different acids are used to find out the best for the leaching method of the gallium and indium separation [119, 120]. Concentrated hydrochloric acid is relatively better, having reasonably better leaching efficiency. Various leaching systems parameters, like the effect of medium, temperature, and concentration of the used catalyst, also have been investigated [121].

The process is customarily utilized in the semiconductor industry to produce thin films, as shown in Fig. 3 [122, 123]. In conventional chemical vapor ousting or deposition, the wafer or the substrate is revealed to one or more light origins, which

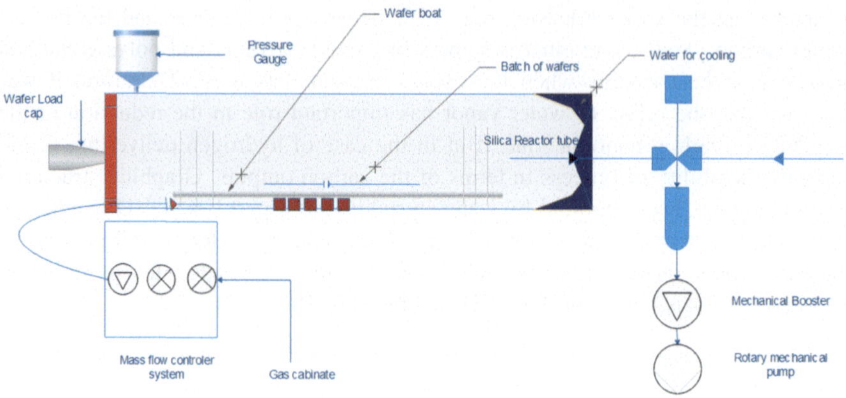

Fig. 3 Schematic representation of the chemical vapor deposition system

disintegrates on the substrate surface to provide the desired material. Microfabrication retreats widely use chemical vapor deposition to transform materials into different types of forms. Chemical vapor deposition is practiced in a variety of formats. Depositions processes generally are diffrent in the way by which chemical reactions are initiated into the following: Atmospheric pressure, when the response performed under atmospheric pressure and ultrahigh vacuum chemical vapor deposition at a slight pressure, is typically below 10−8 torr. The major categories of chemical vapor deposition present today are classified only by the vapor, either LPCVD or UHVCVD. A new type of chemical vapor deposition is the plasma-assisted CVD; the method employs plasma to determine the chemical reaction rates of the precursors. This technique allows removal at lower temperatures flows, which is often crucial in the recycling of semiconductors [119, 124, 125].

3 Resources for Carbon Materials Recycling

3.1 Reformation Using Sawdust

The development of new carbon and carbon derivative products related to rapid industrialization is considered as a unique platform for removing new pollutants that are increasingly affecting the environment. Sawdust, as a resource of carbon nanomaterial, was prepared using successive steps, starting from washing with DI water [126, 127]. Currently, the most convenient recycling way used to recover carbon fiber is performed under high thermal energy. Solvolysis usually displays a solvent to solubilize the precursor. So far, the commercialization of solvolysis has been through batch-based reactor processes. However, inline processing is finally in the works. Vamsidhar Patlollae developed a new way to reuse CNF from reinforced plastic test coupons during a project on self-healing polymers [126, 127].

3.2 Multi-hierarchical Carbonic Materials as Representative Recycling of Waste

In seaside fields, household trash, particularly the seafood consumption from our homes, is ubiquitous. Massive amounts of family trash were created every day [127]. Rice bran can be a useful reference for carbon nanocomposites with multiple complex chemical compositions using heat carbonization. Stimulated by these investigations, we believe the extraordinary microstructure of the crab crust, which makes it reasonable to develop into multi-hierarchical carbon elements [127–130].

3.3 Carbon Nanospheres from Trash Tires Pyrolysis Overtop Ferrocene Synergist

Tires scrap can be a hopeful carbon root for the release of carbonic nanomaterials because of their low value and great abundance [131]. Carbon nanomaterials may be designed from tire waste via pyrolysis over the specific type of catalyst at high temperatures.

A lot of characterizations can be applied to characterize the products. BET measurements also have been used to describe the total pore volume. The environmental strains of the carbon materials manufactured from waste tire powder were assessed by the life cycle assessment (LCA) way, as represented in Fig. 4 [70, 132].

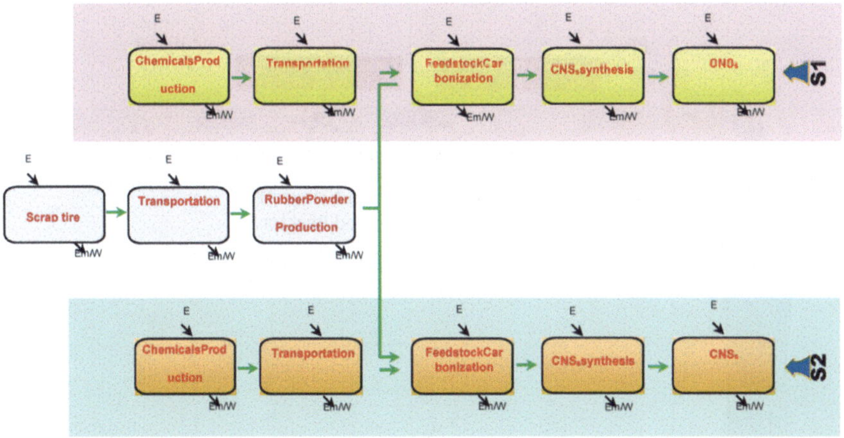

Fig. 4 Schematic diagram of carbon nanomaterials production from the waste tire. Adapted from Ref. [133] Copyright 2020, Elsevier

3.4 *Reuse of Rubbish Rubber Particles by Mechano-Chemical Alteration*

Recently, it has been reported that it is possible to obtain rubber powder, which is originally carbon nanomaterial via mechano-chemically using twin-screw extruder at medium temperature. The synthesized rubber has been characterized to extend the process to the manipulation area. A comparison between the modified and unmodified method of treating material show that the content of oxygenated collections upon the surface of the mechano-chemically restrained rubber powder was more critical for modified one than that for unmodified one. The implemented temperature is a crucial factor and restricting parameter here to constrain the conversion manner. The original temperature and the maximum weight loss rate of qualified rubber powder were decreased after adjustment. These results revealed that the surface activity of the modified rubber was improved while the crosslinking period degree was decreased under the effect of the chemical alteration effect. The rubber analyzer examination results reveal that the processability of modified rubber powder was enhanced, as well as the hysteresis loss was decreased after conversion [134].

3.5 *Catalytic Reformation of Solid Plastics to Precious Carbon Nanotubes*

The recycling systems of plastic are an enormous challenge among all recycling aspects. Conventional gasification of a hard type of polymeric pieces of vinyl has been extensively recorded [135]. The produced carbon materials have been recharacterized by scanning different systems to check purity and structure, as

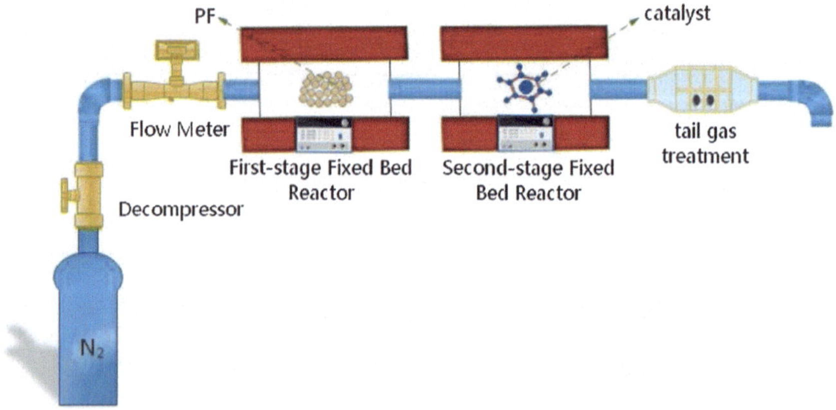

Fig. 5 Schematic diagram of the experimental system of CNT production from hard plastics. Adapted from Ref. [137] Copyright 2020, Elsevier

shown in Fig. 5. Different catalysts have been used to yield of the article, where Ni catalysts are about 25% with purity is 94%. Compared with Fe, nanocatalysts have higher yield and purity, which reached 35% with 98% purity [136].

3.6 *Chemical Reuse and Recycle of Carbon Fibers Reinforced Epoxy Resin*

The carbon nanomaterial, expressly strands strengthened epoxy mastics, was obtained in the air at a higher pressure and temperature with microstructure in fiber form, as presented in Fig. 6 [137].

3.6.1 Honeycomb Activated Carbon Producer from Agriculture Waste

Carbon-based nanomaterials have been prepared from different types of agriculture waste [138–143]. Activated carbon has been prepared from the palm kernel shell, not only pure also with metal oxide impregnation. The prepared material is measured as supercapacitor electrodes by testing their electrochemical properties. Finally, this calcium suspension has been utilized for the AC impregnation. All trials have been done at 30 °C under controlled temperature and blending for 2 h. Finally, the impregnated substances were heated at 800 °C for two h in an attempt to get the CaO/ACPKS [144].

3.6.2 Green Approach for Carbon Nanospheres Production

Published articles show that Purpureus seeds were recovered to produce carbon nanomaterials only by drying and heating. The drained Lablab Purpureus

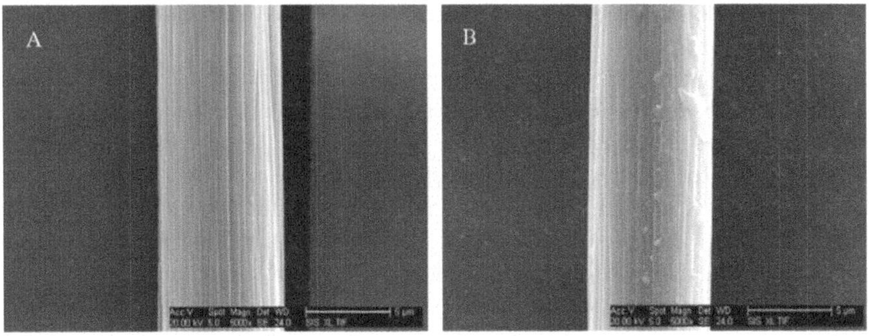

Fig. 6 SEM micrographs of the surface of the carbon fibers before (**a**) and after (**b**) treatment. Adapted from Ref. [137] Copyright 2010, Elsevier

(LP) seeds were initially grounded to a granular powder operating a grinder at a medium speed close to 12 E3 rpm. The material presented appeared as a fine powder obtained after grinding and sieved to get a consistent particle size around 60 μm. The finely ground dress sample obtained after sieving was independently pyrolyzed under nitrogen-controlled conditions atmosphere with a flow rate of 150 mL cm^{-3} at different ramping temperatures and heating rate of 10 °C min^{-1} using a quartz tube furnace. To achieve a higher degree of purity, silica pyrolysis has been veiled using natural lignocellulosic materials. Then, samples were removed using 2.5 M sodium hydroxide aqueous mixture followed by rinsing with double deionized water [140]. Initially, a weight loss is close to ~8% at specific temperatures as a result of the water molecules desorption. After that, in the second staging, a weight loss of roughly about ~61% is reached in the temperature range of 150–390 °C. This thermal degradation ends at a temperature of ~400 °C; therefore, any higher than that can be analyzed at the critical level of synthesizing carbon nanospheres. Henceforth, we can see a gradual weight loss in the precursor up to a temperature of 1200 °C and in the whole reaction. The EDX spectrum exploits the presence of a plentiful amount of carbon content, which reached 61.1 wt% in comparison with oxygen (40 wt%) along with a small ratio of other elements. Fourier transform infrared (FTIR) analysis is applied to the study of the functional group, which is already presented in the sample. FTIR spectrum employs the active bonds in the precursor, which related to the synthesis of carbon nanomaterials after pyrolysis. The broad peak at 3432 cm^{-1} determines a hydrogen-bonded –O–H stretching vibration of α–cellulose and also affirms the presence of compounds with alcoholic groups like polyphenols.

Also, the presence of peaks at 2926 and 2864 cm^{-1} shows C–H stretching, which is linked to lignocellulosic elements. The band observed at 1649 cm-1 also represents the C = O stretching vibration, which is conjugated carbonyl of lignin. The group at 1418 cm^{-1} dedicate the sweet-smelling skeletal combined with C–H deforming and stretching in lignin. Also, another group at 1028 cm^{-1} is due to the C–O vibrational stretching of primary flavonoids [140].

3.6.3 Synthesis of Carbon Nanospheres by Pyrolysis from Biowaste Sago Bark

Sago bark is the waste usually defined as the solid powder produced from the sago starch processing manufacturing. Mainly because of its lignin content, sago bark can be applied for the sustainable production of carbon nanospheres. Thus, it has been reported that sago bark is employed as a precursor for the synthesis of carbon nanomaterials by a simple environmentally pyrolysis system. This system produced carbon nanomaterials with highly distributed particles of carbon among nanoscale sizes. TGA analysis has also been done to investigate the effect of temperature over the material to set the optimum temperature. The scanning electron microscope, together with EDX analysis, dispenses a high content of carbon in the sago bark, as displayed in Fig. 7a and b. It also confers the stability of the carbon nanomaterials

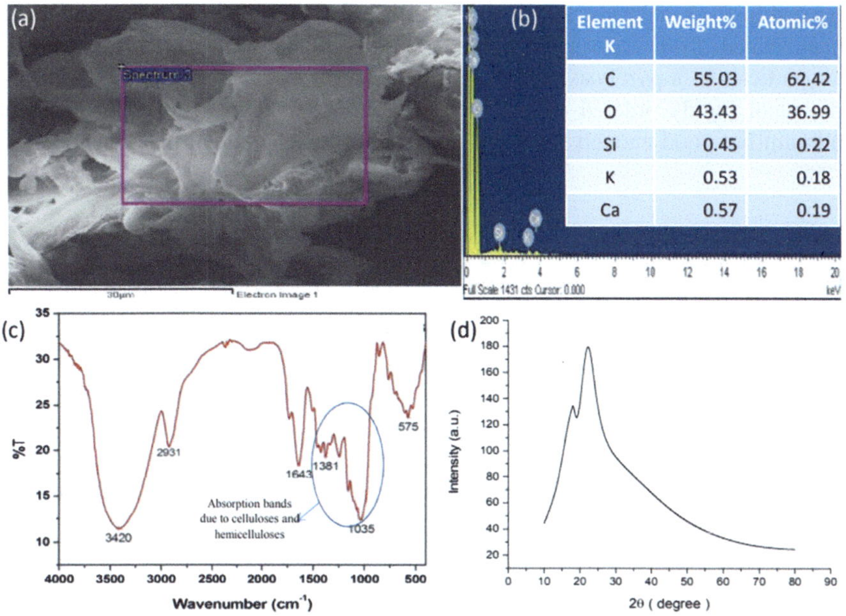

Fig. 7 (**a**) SEM image, (**b**) EDX spectrum, which confirmed the purity of the material with other silica compounds, (**c**) the FT-R spectrum of the prepared material, and (**d**) XRD pattern of the as-synthesized material. Adapted from Ref. [139] Copyright 2015, American Chemical Society

after pyrolysis at high temperature by seeing the recorded stability also over a low-temperature range exploiting complete carbonization. FTIR analysis was an essential analysis tool in elucidating the functional groups of the sago bark. According to Sun et al., the cellulose part of the sago bark is composed of xylose and glucose as the major constituents of the isolated hemicelluloses besides other sugars with noticeable amounts of galactose. The intense vibration group, which can be seen around 1643 cm^{-1}, can be related to the carbonyl group (C = O), which gives a view about the high carbon content included in sago bark. Besides this, the other absorption bands due to structural multifariousness of celluloses and hemi-celluloses, which include C–OH, C–H, –C–C–O functional groups, can be seen in the region between 2931 and 1500 to 900 cm^{-1}. Other bands that appeared at 1160 and 1035 cm^{-1} recommend the presence of arabinose residues together with α-glucan of hemicelluloses. The various absorption peak at 575 cm^{-1} might result due to the presence of some particles related to oxides, which agglomerated in sago bark, as shown in Fig. 7c.

According to the results are incompatible with the literature review [139], the fibrous scraps with coarse behavior may behave as an incomplete template for the integration of the nanospheres. Also, the XRD spectrum of sago bark exploited two peaks, which are attributed to the microcrystalline character of the cellulose. The first peak at $2\theta = 17.74°$ corresponding to (1 1 0) and the other one at $2\theta = 22.09°$ corresponding to (1 2 0) plane, respectively, as shown in Fig. 7d. Thus, as it is revealed in EDX analysis, FTIR, and XRD studies, high content of carbon can be amalgamated from sago bark. These considerations motivate us to fabricate nanoparticles at a pyrolysis temperature of 500 °C, which is related to the development of highly ordered carbon materials, and these temperatures are not that optimum but good enough to show the possibility of the parent materials.

3.6.4 Nanocarbons Developed Utilizing Biowaste Oil Palm Sheets as a Precursor

Given that leaves of the oil palm used as waste material for the design of porous carbon nanomaterials. The material dried in an oven at a temperature a little higher than 110 °C for two days to eradicate all the precipitation. Then, the sample was creased vigorously with a high cycle grinder and further sieved to the particle size of around 60 μm. Carbon nanotubes have been designed by the mono-step catalyst for the pyrolysis system in a tube furnace at higher than 500 °C under a controlled nitrogen atmosphere with continuous rate flow of 150 mL cm^{-3} for more than 1 h at a steaming rate of 10 °C min^{-1} accompanied by reducing to ambient conditions. Sodium hydroxide with a concentration of 2.5 M is applied to eliminate silica, which is classified as a pollutant here. The following method is optimized to design samples for electrochemical studies [145]. The prepared samples are well characterized using electron microscopy, XRD, FTIR, and Raman, as shown in Fig. 8. According to the characterization results, XRD pattern the as prepared carbon nanotubes at 700 °C from this waste show peaks at 2θ of 27°, 43.55°, 49.45°, and 59.10° that refer to graphite carbon nanomaterials (96-901-2231), as shown in Fig. 8j. Besides, the peak at $2\theta = 21°$ is near to the reflection card for carbon (ICDD card 96-901-4005). 26.85° peak is because of the great crystalline cellulose matters which are formed due to the hemicelluloses and celluloses of the precursor. The peaks at 44.35°, 50.25°, and 59.10° deed the graphitic nature of the given carbon materials. Carbon materials purchased from the construction using this precursor developed with graphitic structure. Furthermore, 68° displays this peak corresponds to the (2 2 0) plane. It can also be remarked that the intensity of this peak reduced in comparison with other peaks, which confirms the reduction in the crystallinity. FTIR spectrum reveals a band at 1365 cm^{-1} is recognized as the disorder-induced characteristic of graphite together with the other band, which develops at 1610 cm^{-1}. The Raman band between 2700 and 2900 cm^{-1} matches to the overtone of the D band is known as 2D; also, the ID/IG ratio is 0.903. To examine the morphology of the achieved carbon nanomaterials, as shown in Fig. 8 [145].

Field emission scanning electron microscope and transmitting electron microscope have been performed, as shown in Fig. 8, and dedicate the spherical shape without any impurities and irregular shapes in the carbon materials. The average particle size confirmed using TEM, which found to be around 20–40 nm which is shown in Fig. 8c. Homogeneity of particles was judged using the TEM image. According to the reported histogram, it exploits an average particle size distribution, which would be ultimate for the electrochemical report as it smooths the ion diffusion between the

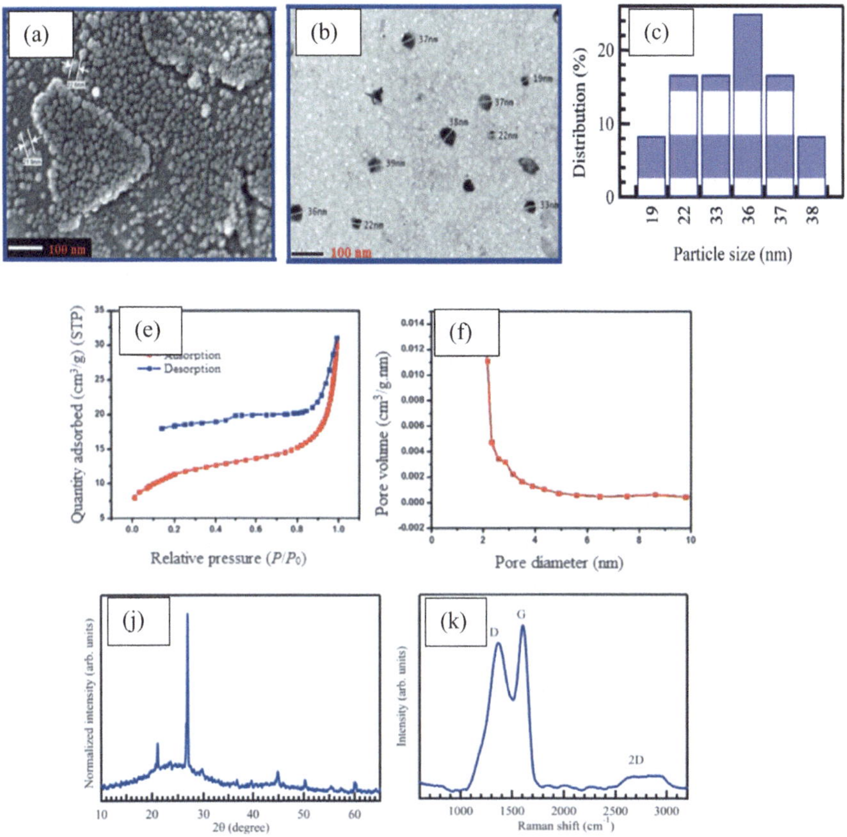

Fig. 8 (**a**) FESEM, (**b**) TEM pictures, and (**c**) particle size distribution for Polycarbon nanomaterials. (**e**) Nitrogen adsorption-desorption isotherms for porous carbon nanomaterials and (**f**) pore volume with pore diameter, (**j**) XRD pattern, and (**k**) Raman spectrum for PCNs. Adapted from Ref. [145] Copyright 2016, Elsevier

particles. The surface area and pore width were measured using the nitrogen adsorption-desorption system using BET with de-gassing at 200 °C for 12 h. The results devote that carbon materials having a surface area close to 37.3 and 22 m^2 g^{-1} of t-plot micropore values. Also, the carbon materials exploited a micropore percentage close to 56.4% and a pore radius of 0.99 nm, as shown in Fig. 8f [145].

3.6.5 Exchange of Allium Cepa Peels to Energy Storage Arrangement-Based Carbon Nanospheres

Mesoporous carbon nanospheres can recycle from biowaste, *Allium cepa* peels, which are identified as an onion or dry peels utilizing the catalyst-free pyrolysis approach. The synthesis process comprises an unusable bio-precursor that agglomerate in millions of tons per annum. After pyrolysis at 800, 900, and 1000 ° C, the carbon nanomaterial is directly applied for super capacitance investigation without further activation processes. These research studies support carbon nanomaterials obtained from *Allium cepa* wastes to be used as promising materials for the supercapacitor direction [58]. Typically, the process of building onion peel waste is applied here as a raw material for the integration of carbon nanomaterials. The waste material was collected from the southern side of great India, washed thoroughly with water, and subsequently dried. The adequately dried waste was restricted into a fine powder using ~62 μm sieve, which was then exposed to pyrolysis at three constant temperatures, viz.: 800, 900, and 1000 °C under a controlled atmosphere.

4 Conclusions

In summary, upcycling material waste into high-value carbon nanomaterials is a requirement and sustainable solution with a promising output as it allows for the transfer of post-consumer products to more value-added products and can help to eliminate the burden of solid wastes on the universe. It is a highly multidisciplinary collaboration between the sciences to reach the best results. Partnerships among fundamental science, engineering, and business management can improve this recycling system to be more feasible and applicable. Knowledge and experience from science and industry on existing facilities, such as those for recycling, gasification, and pyrolysis, as well as those of chemical vapor deposition, can be applied to improve the efficiency of recycling.

5 Future Perspectives

This chapter looks ahead, beyond the projected large-scale market penetration of carbon nanomaterial-based applications, to the time when the spent waste will be ready for final recycling. It explores different types of working systems for the recycling of nanomaterials, particularly carbon. The national economy is growing attention as a potential route for our society to enhance prosperity while reducing demands on restricted raw materials and minimizing adverse externalities. This challenge requires advanced collaboration mechanisms particulary, the intersection of plastics and packaging in a specific way. How can collaboration along with the global recycling management and after-use value chain, as well as with governments, achieve systemic change to overcome stalemates in today's recycling economy to move to a more circular model. It is also crucial to modify the pyrolysis system to be more selective and more efficient in the mass production level.

References

1. Zhang J, Terrones M, Park CR, Mukherjee R, Monthioux M, Koratkar N, Kim YS, Hurt R, Frackowiak E, Enoki T (2016) Carbon science in 2016: Status, challenges and perspectives. Carbon 98(70):708–732
2. Medenbach L, Escher I, Köwitsch N, Armbrüster M, Zedler L, Dietzek B, Adelhelm P (2018) Sulfur spillover on carbon materials and possible impacts on metal–sulfur batteries. Angew Chem Int Ed 57(41):13666–13670
3. Schmies H, Hornberger E, Anke B, Jurzinsky T, Nong HN, Dionigi F, Kühl S, Drnec J, Lerch M, Cremers C, Strasser P (2018) Impact of Carbon Support Functionalization on the Electrochemical Stability of Pt Fuel Cell Catalysts. Chem Mater 30(20):7287–7295
4. Ghawanmeh AA, Ali GAM, Algarni H, Sarkar SM, Chong KF (2019) Graphene oxide-based hydrogels as a nanocarrier for anticancer drug delivery. Nano Res 12(5):973–990
5. Yang Y, Ramirez C, Wang X, Guo Z, Tokranov A, Zhao R, Szlufarska I, Lou J, Sheldon BW (2017) Impact of carbon nanotube defects on fracture mechanisms in ceramic nanocomposites. Carbon 115:402–408
6. Wei J, Wang G, Chen F, Bai M, Liang Y, Wang H, Zhao D, Zhao Y (2018) Sol-Gel Synthesis of Metal-Phenolic Coordination Spheres and Their Derived Carbon Composites. Angew Chem Int Ed 57(31):9838–9843
7. Wang Q, Wang Y, Duan B, Zhang M (2016) Modified Sol-Gel Synthesis of Carbon Nanotubes Supported Titania Composites with Enhanced Visible Light Induced Photocatalytic Activity. J Nanomater 2016:1–6
8. Habibi S and Jamshidi M (2019) Sol–gel synthesis of carbon-doped TiO2 nanoparticles based on microcrystalline cellulose for efficient photocatalytic degradation of methylene blue under visible light. Environmental Technology:1–15
9. Li D, Zang J, Zhang J, Ao K, Wang Q, Dong Q, Wei Q (2016) Sol-Gel Synthesis of Carbon Xerogel-ZnO Composite for Detection of Catechol. Materials 9(4):282
10. Sivakumar M, Sakthivel M, Chen S-M, Veerakumar P, Liu S-B (2017) Sol-Gel Synthesis of Carbon-Coated $LaCoO_3$for Effective Electrocatalytic Oxidation of Salicylic Acid. ChemElectroChem 4(4):935–940
11. Habeeb OA, Ramesh K, Ali GAM, Yunus RM (2017) Low-cost and eco-friendly activated carbon from modified palm kernel shell for hydrogen sulfide removal from wastewater: adsorption and kinetic studies. Desalination Water Treat 84:205–214
12. Yang B, Zhang M, Lu Z, Luo J, Song S, Zhang Q (2018) From Poly(p-phenylene terephthalamide) Broken Paper: High-Performance Aramid Nanofibers and Their Application in Electrical Insulating Nanomaterials with Enhanced Properties. ACS Sustain Chem & Eng 6(7):8954–8963
13. Zambrzycki M, Fraczek-Szczypta A (2018) Conductive hybrid polymer composites based on recycled carbon fibres and carbon nanofillers. J Mater Sci 53(10):7403–7416
14. Li J, Zhou Q, Liu Y, Lei M (2017) Recyclable nanoscale zero-valent iron-based magnetic polydopamine coated nanomaterials for the adsorption and removal of phenanthrene and anthracene. Sci Technol Adv Mater 18(1):3–16
15. Chen Z, Xie Y, Huang W, Qin C, Yu A, Lai G (2019) Exonuclease-assisted target recycling for ultrasensitive electrochemical detection of microRNA at vertically aligned carbon nanotubes. Nanoscale 11(23):11262–11269
16. Zhu G, Chen T, Hu Y, Ma L, Chen R, Lv H, Wang Y, Liang J, Li X, Yan C, Zhu H, Liu H, Tie Z, Jin Z, Liu J (2017) Recycling PM2.5 carbon nanoparticles generated by diesel vehicles for supercapacitors and oxygen reduction reaction. Nano Energy 33:229–237
17. Boonruksa P, Bello D, Zhang J, Isaacs JA, Mead JL, Woskie SR (2016) Exposures to nanoparticles and fibers during injection molding and recycling of carbon nanotube reinforced polycarbonate composites. J Eposure Sci Environ Epidemiol 27(4):379–390

18. Sanad MF, Abdellatif SO, Ghali HA (2019) Enhancing the performance of photovoltaic operating under harsh conditions using carbon-nanotube thermoelectric harvesters. J Mater Sci: Mater Electron 30(22):20029–20036
19. Meneses Brassea BP, Sanad MF, Blazer DS, Pourmiri S, and El-Gendy AA (2020) Functionalized superparamagnetic Fe@ Au Core/shell nanoparticles and their feasibility for magnetic hyperthermia. Bulletin of the American Physical Society
20. Aacharya R and Chhipa H (2020) Nanocarbon fertilizers: Implications of carbon nanomaterials in sustainable agriculture production, in Carbon Nanomaterials for Agri-Food and Environmental Applications. Elsevier, pp 297–321
21. Lee WJ, Lim J, and Kim SO (2017) Carbon Nanomaterials: Nitrogen Dopants in Carbon Nanomaterials: Defects or a New Opportunity? (Small Methods 1–2/2017). Small Methods 1(1-2)
22. Kim M, Jang J, Cha C (2017) Carbon nanomaterials as versatile platforms for theranostic applications. Drug Discov Today 22(9):1430–1437
23. Abdel Messih MF, Shalan AE, Sanad MF, Ahmed MA (2019) Facile approach to prepare ZnO@SiO2 nanomaterials for photocatalytic degradation of some organic pollutant models. J Mater Sci: Mater Electron 30(15):14291–14299
24. Sanad MF, Shalan AE, Bazid SM, Abu Serea ES, Hashem EM, Nabih S, Ahsan MA (2019) A graphene gold nanocomposite-based 5-FU drug and the enhancement of the MCF-7 cell line treatment. RSC Advances 9(53):31021–31029
25. Nabih S, Shalan AE, Serea ESA, Goda MA, Sanad MF (2019) Photocatalytic performance of TiO2@SiO2 nanocomposites for the treatment of different organic dyes. J Mater Sci: Mater Electron 30(10):9623–9633
26. Iijima S (1991) Helical microtubules of graphitic carbon. Nature 354(6348):56–58
27. Frank S (1998) Carbon Nanotube Quantum Resistors. Science 280(5370):1744–1746
28. Maazinejad B, Mohammadnia O, Ali GAM, Makhlouf ASH, Nadagouda MN, Sillanpää M, Asiri AM, Agarwal S, Gupta VK, Sadegh H (2020) Taguchi L9 (34) orthogonal array study based on methylene blue removal by single-walled carbon nanotubes-amine: Adsorption optimization using the experimental design method, kinetics, equilibrium and thermodynamics. J Mol Liq 298:112001
29. Jiang K, Li Q, Fan S (2002) Spinning continuous carbon nanotube yarns. Nature 419 (6909):801
30. Ali GAM, Lih Teo EY, Aboelazm EAA, Sadegh H, Memar AOH, Shahryari-Ghoshekandi R, Chong KF (2017) Capacitive performance of cysteamine functionalized carbon nanotubes. Mater Chem Phys 197:100–104
31. Javey A, Guo J, Wang Q, Lundstrom M, Dai H (2003) Ballistic carbon nanotube field-effect transistors. Nature 424(6949):654–657
32. Seyed Arabi SM, Lalehloo RS, Olyai MRTB, Ali GAM, Sadegh H (2019) Removal of congo red azo dye from aqueous solution by ZnO nanoparticles loaded on multiwall carbon nanotubes. Physica E 106:150–155
33. Britto PJ, Santhanam KSV, Ajayan PM (1996) Carbon nanotube electrode for oxidation of dopamine. Bioelectrochem Bioenerg 41(1):121–125
34. Niu C, Sichel EK, Hoch R, Moy D, Tennent H (1997) High power electrochemical capacitors based on carbon nanotube electrodes. Appl Phys Lett 70(11):1480–1482
35. McEuen PL, Fuhrer MS, Hongkun P (2002) Single-walled carbon nanotube electronics. IEEE Trans Nanotechnol 1(1):78–85
36. Ali GAM, Yusoff MM, Algarni H, Chong KF (2018) One-step electrosynthesis of MnO_2/rGO nanocomposite and its enhanced electrochemical performance. Ceram Int 44(7):7799–7807
37. Shulman VL (2019) Tire Recycling. In: Waste. Elsevier, pp 489–515
38. Liao H, Shi Y, Liu X, Shen N, Deng Q (2019) A non-probabilistic model of carbon footprints in remanufacture under multiple uncertainties. J Clean Prod 211:1127–1140
39. Rodygin KS, Vikenteva YA, Ananikov VP (2019) Calcium-Based Sustainable Chemical Technologies for Total Carbon Recycling. Chemsuschem 12(8):1483–1516

40. Yu J, Chi C, Zhu B, Qiao K, Cai X, Cheng Y, Yan S (2020) High adsorptivity and recycling performance activated carbon fibers for Cu(II) adsorption. Sci Total Environ 700:134412
41. Ali GAM, Manaf SABA, Kumar A, Chong KF, Hegde G (2014) High performance supercapacitor using catalysis free porous carbon nanoparticles. J Phys D Appl Phys 47 (49):495307
42. Santo KP, Vishnyakov A, Brun Y, Neimark AV (2019) Critical Conditions of Adhesion and Separation of Functionalized Nanoparticles on Polymer Grafted Substrates. J Phys Chem C 123(26):16091–16106
43. Gaete J, Molina L, Alfaro I, Yañez J, Valenzuela F, Basualto C (2019) Recovery and separation of rhenium and molybdenum from aqueous solutions that simulate mine waters using magnetite nanoparticles functionalized with amine-derivative groups. Miner Eng 136:66–76
44. Yang J, Loh XJ, Tan BH, Li Z (2018) pH-Responsive Poly(dimethylsiloxane) Copolymer Decorated Magnetic Nanoparticles for Remotely Controlled Oil-in-Water Nanoemulsion Separation. Macromol Rapid Commun 40(5):1800013
45. Tang S-Y, Mitchell DRG, Zhao Q, Yuan D, Yun G, Zhang Y, Qiao R, Lin Y, Dickey MD, Li W (2019) Phase Separation in Liquid Metal Nanoparticles. Matter 1(1):192–204
46. Couzon N, Maillard M, Chassagneux F, Brioude A, Bois L (2019) Photoelectrochemical Behavior of Silver Nanoparticles inside Mesoporous Titania: Plasmon-Induced Charge Separation Effect. Langmuir 35(7):2517–2526
47. Lee J-J, Jeong KJ, Kohane DS (2019) Magnetic separation using nanoparticles, Google Patents
48. Barry PH, de Moor JM, Giovannelli D, Schrenk M, Hummer DR, Lopez T, Pratt CA, Segura YA, Battaglia A, Beaudry P, Bini G, Cascante M, d'Errico G, di Carlo M, Fattorini D, Fullerton K, Gazel E, González G, Halldórsson SA, Iacovino K, Ilanko T, Kulongoski JT, Manini E, Martínez M, Miller H, Nakagawa M, Ono S, Patwardhan S, Ramírez CJ, Regoli F, Smedile F, Turner S, Vetriani C, Yücel M, Ballentine CJ, Fischer TP, Hilton DR, Lloyd KG (2019) Forearc carbon sink reduces long-term volatile recycling into the mantle. Nature 568(7753):487–492
49. Ross MB, De Luna P, Li Y, Dinh C-T, Kim D, Yang P, Sargent EH (2019) Designing materials for electrochemical carbon dioxide recycling. Nature Catalysis 2(8):648–658
50. Oie T, Shimada K (2020) Cleaning solution and cleaning method for material comprising carbon-incorporated silicon oxide for use in recycling wafer, Google Patents
51. Zhu J-H, Chen P-y, Su M-n, Pei C, Xing F (2019) Recycling of carbon fibre reinforced plastics by electrically driven heterogeneous catalytic degradation of epoxy resin. Green Chem 21(7):1635–1647
52. Ameh T, Sayes CM (2019) The potential exposure and hazards of copper nanoparticles: A review. Environ Toxicol Pharmacol 71:103220
53. Jahangirian H, Kalantari K, Izadiyan Z, Rafiee-Moghaddam R, Shameli K, Webster TJ (2019) A review of small molecules and drug delivery applications using gold and iron nanoparticles. Int J Nanomed 14:1633–1657
54. Flores-López LZ, Espinoza-Gómez H, Somanathan R (2018) Silver nanoparticles: Electron transfer, reactive oxygen species, oxidative stress, beneficial and toxicological effects. Mini review. J Appl Toxicol 39(1):16–26
55. Turan NB, Erkan HS, Engin GO, Bilgili MS (2019) Nanoparticles in the aquatic environment: Usage, properties, transformation and toxicity—A review. Process Saf Environ Prot 130:238–249
56. Ali GAM, Tan LL, Jose R, Yusoff MM, Chong KF (2014) Electrochemical performance studies of MnO_2 nanoflowers recovered from spent battery. Mater Res Bull 60:5–9
57. Ali GAM, Yusoff MM, Shaaban ER, Chong KF (2017) High performance MnO_2 nanoflower supercapacitor electrode by electrochemical recycling of spent batteries. Ceram Int 43:8440–8448

58. Ali GAM, Supriya S, Chong KF, Shaaban ER, Algarni H, Maiyalagan T, Hegde G (2019) Superior supercapacitance behavior of oxygen self-doped carbon nanospheres: a conversion of Allium cepa peel to energy storage system. Biomass Conv Bioref https://doi.org/10.1007/s13399-019-00520-3
59. Udayana C, Kumara S (2020) Private and Government Partnership Yayasan Lengis Hijau and Denpasar City Government—Recycling used-cooking-oil into Bio-diesel for Gensets and School Buses, Center for Open Science
60. Tan Y, Guo C (2019) Research on Two-Way Logistics Operation with Uncertain Recycling Quality in Government Multi-Policy Environment. Sustainability 11(3):882
61. Zand F, Yaghoubi S, Sadjadi SJ (2019) Impacts of government direct limitation on pricing, greening activities and recycling management in an online to offline closed loop supply chain. J Clean Prod 215:1327–1340
62. Veksha A, Yin K, Moo JGS, Oh W-D, Ahamed A, Chen WQ, Weerachanchai P, Giannis A, Lisak G (2020) Processing of flexible plastic packaging waste into pyrolysis oil and multi-walled carbon nanotubes for electrocatalytic oxygen reduction. J Hazard Mater 387:121256
63. Huang X, Yamasaki K, Kudo S, Sperry J, Hayashi J-i (2020) Influence of ionic liquid type on porous carbon formation during the ionothermal pyrolysis of cellulose. J Anal Appl Pyrol 145:104728
64. Sriplai N, Mongkolthanaruk W, Eichhorn SJ, Pinitsoontorn S (2019) Magnetic bacterial cellulose and carbon nanofiber aerogel by simple immersion and pyrolysis. J Mater Sci 55 (9):4113–4126
65. Shi K, Yan J, Menéndez JA, Luo X, Yang G, Chen Y, Lester E, Wu T (2020) Production of H2-Rich Syngas From Lignocellulosic Biomass Using Microwave-Assisted Pyrolysis Coupled With Activated Carbon Enabled Reforming. Frontiers in Chemistry 8
66. George D, Garcia A, Pham Q, Perez MR, Deng J, Nguyen MT, Zhou T, Martinez-Chapa SO, Won Y, Liu C, Lo RC, Ragan R, Madou M (2020) Fabrication of patterned graphitized carbon wires using low voltage near-field electrospinning, pyrolysis, electrodeposition, and chemical vapor deposition. Microsystems & Nanoengineering 6(1)
67. Casazza AA, Spennati E, Converti A, Busca G (2020) Production of carbon-based biofuels by pyrolysis of exhausted Arthrospira platensis biomass after protein or lipid recovery. Fuel Process Technol 201:106336
68. Haig J, Ascough PL, Wurster CM, Bird MI (2020) A rapid throughput technique to isolate pyrogenic carbon by hydrogen pyrolysis for stable isotope and radiocarbon analysis. Rapid Communications in Mass Spectrometry 34(10)
69. Ota M, Arai K, Kuribara A, Deguchi Y, Onozuka K (2020) Preparation of Silicon Nanoparticles by Pyrolysis of Organosilicon Compounds inside the Porous Carbon. J Mater Sci Chem Eng 08(02):1–6
70. Heidari A, Younesi H (2020) Synthesis, characterization and life cycle assessment of carbon nanospheres from waste tires pyrolysis over ferrocene catalyst. J Environ Chem Eng 8 (2):103669
71. Lai S-M, Chu Y-L, Chiu YT, Chang M-C, Hsieh T-Y, Hsieh M-H (2020) Effect of pyrolysis carbon black from waste tires on the properties of styrene–butadiene rubber compounds. Polymers and Polymer Composites:096739112090288
72. Shrivastava P, Alam SN, Panda D, Sahoo SK, Maity T, Biswas K (2020) Effect of addition of multiwalled carbon nanotube/graphite nanoplatelets hybrid on the mechanical properties of aluminium. Diam Relat Mater 104:107715
73. Dul S, Ecco LG, Pegoretti A, Fambri L (2020) Graphene/Carbon Nanotube Hybrid Nanocomposites: Effect of Compression Molding and Fused Filament Fabrication on Properties. Polymers 12(1):101
74. Datsyuk V, Trotsenko S, Trakakis G, Boden A, Vyzas-Asimakopoulos K, Parthenios J, Galiotis C, Reich S, Papagelis K (2020) Thermal properties enhancement of epoxy resins by incorporating polybenzimidazole nanofibers filled with graphene and carbon nanotubes as reinforcing material. Polym Testing 82:106317

75. Manzoor J, Sharma M, Sofi IR, Dar AA (2020) Plastic Waste Environmental and Human Health Impacts, in Handbook of Research on Environmental and Human Health Impacts of Plastic Pollution, IGI Global. pp 29–37
76. Biswal T, Kar PK (2020) Plastic Pollution and Its Effect on the Environment, in Handbook of Research on Environmental and Human Health Impacts of Plastic Pollution, IGI Global. pp 1–28
77. Khayyam Nekouei R, Maroufi S, Assefi M, Pahlevani F, Sahajwalla V (2020) Thermal Isolation of a Clean Alloy from Waste Slag and Polymeric Residue of Electronic Waste. Processes 8(1):53
78. Bonora R, Boaretti C, Campea L, Roso M, Martucci A, Modesti M, Lorenzetti A (2020) Combined AOPs for Formaldehyde Degradation Using Heterogeneous Nanostructured Catalysts. Nanomaterials 10(1):148
79. Kuehr S, Meisterjahn B, Schröder N, Knopf B, Völker D, Schwirn K, Schlechtriem C (2020) Testing the bioaccumulation of manufactured nanomaterials in the freshwater bivalve Corbicula fluminea using a new test method. Environmental Science: Nano 7(2):535–553
80. Gusain R, Kumar N, Ray SS (2020) Recent advances in carbon nanomaterial-based adsorbents for water purification. Coord Chem Rev 405:213111
81. Abdel Maksoud MIA, Elgarahy AM, Farrell C, AaH A-M, Rooney DW, Osman AI (2020) Insight on water remediation application using magnetic nanomaterials and biosorbents. Coord Chem Rev 403:213096
82. Yi R, Wang R, Duan J, Fang Z, Li H, Chen Z, Zhou A, Sun Y (2020) Rational design of hierarchically porous $NiCo_2O_4$ and Bi_2O_3 nanostructure: Anchored on 3D nitrogen doped carbonized melamine foam for flexible asymmetric supercapacitor. Electrochim Acta 338:135845
83. Li L, Zhang H, Cheng A, Zhong W, Li Z, Li Z (2020) Ultrasmall metal oxide nanocrystals embedded in nitrogen-doped carbon networks based on one-step pyrolysis of bi-functional metallo-organic molecules for high-performance lithium-ion batteries. Electrochim Acta 331:135430
84. Xu F, Qiu Y, Han H, Jiang G, Zhao R, Zhang E, Li H, Wang H, Kaskel S (2020) Manipulation of carbon framework from the microporous to nonporous via a mechanical-assisted treatment for structure-oriented energy storage. Carbon 159:140–148
85. Liu Y, Liu Q, Wang L, Yang X, Yang W, Zheng J, Hou H (2020) Advanced Supercapacitors Based on Porous Hollow Carbon Nanofiber Electrodes with High Specific Capacitance and Large Energy Density. ACS Appl Mater Interfaces 12(4):4777–4786
86. Dong Y, Jiang X, Mo J, Zhou Y, Zhou J (2020) Hollow CuO nanoparticles in carbon microspheres prepared from cellulose-cuprammonium solution as anode materials for Li-ion batteries. Chem Eng J 381:122614
87. Shen M, Gao K, Xiang F, Wang B, Dai L, Zheng L, Baker F, Duan C, Zhang Y, Sun S, Ni Y (2020) Nanocellulose-assisted synthesis of ultrafine Co nanoparticles-loaded bimodal micro-mesoporous N-rich carbon as bifunctional oxygen electrode for Zn-air batteries. J Power Sources 450:227640
88. Jiang Q, Liu M, Shao C, Li X, Liu H, Li X, Liu Y (2020) Nitrogen doping polyvinylpyrrolidone-based carbon nanofibers via pyrolysis of g-C3N4 with tunable chemical states and capacitive energy storage. Electrochim Acta 330:135212
89. Ellison CR, Hoff R, Mărculescu C, Boldor D (2020) Investigation of microwave-assisted pyrolysis of biomass with char in a rectangular waveguide applicator with built-in phase-shifting. Appl Energy 259:114217
90. Ye Q, Amirkhanian S, Li J, Chen Z (2020) Effects of Waste Polyethylene on the Rheological Properties of Asphalt Binder. J Test Eval 48(3):20190220
91. Kamali AR (2020) Molten Salt Conversion of Plastics into Highly Conductive Carbon Nanostructures. In: Green Production of Carbon Nanomaterials in Molten Salts and Applications. Springer Singapore, pp 109–140

92. Laycock B, Wang X, Liu R-F, Annamalai PK, Cork J, Derstine C, Mills M, McFarland EW (2020) Pyrolysis of brominated polyethylene as an alternative carbon fibre precursor. Polym Degrad Stab 172:109057
93. Ali A, Baheti V, Khan MZ, Ashraf M, Militky J (2019) Development of electrically conductive composites based on recycled resources. J TextE Inst 111(1):16–25
94. Cheng L, Gu J, Wang Y, Zhang J, Yuan H, Chen Y (2020) Polyethylene high-pressure pyrolysis: Better product distribution and process mechanism analysis. Chem Eng J 385:123866
95. Tanui JK, Kioni PN, Mirre T, Nowitzki M, Karuri NW (2020) The influence of particle packing density on wood combustion in a fixed bed under oxy-fuel conditions. Energy 194:116863
96. Yu J, Xu J, Li Z, He W, Huang J, Xu J, Li G (2019) Upgrading pyrolytic carbon-blacks (CBp) from end-of-life tires: Characteristics and modification methodologies. Frontiers of Environmental Science & Engineering 14(2)
97. Yan M, Zhang S, Wibowo H, Grisdanurak N, Cai Y, Zhou X, Kanchanatip E (2020) Biochar and pyrolytic gas properties from pyrolysis of simulated municipal solid waste (SMSW) under pyrolytic gas atmosphere. Waste Disposal & Sustainable Energy 2(1):37–46
98. Tokmurzin D, Kuspangaliyeva B, Aimbetov B, Abylkhani B, Inglezakis V, Anthony EJ, Sarbassov Y (2020) Characterization of solid char produced from pyrolysis of the organic fraction of municipal solid waste, high volatile coal and their blends. Energy 191:116562
99. Quesada L, Calero M, Martín-Lara MÁ, Pérez A, Blázquez G (2020) Production of an Alternative Fuel by Pyrolysis of Plastic Wastes Mixtures. Energy Fuels 34(2):1781–1790
100. Arun J, Gopinath KP (2019) Chemical Recycling of Electronic-Waste for Clean Fuel Production, in E-waste Recycling and Management. Springer International Publishing. pp 111–126
101. Al-Haj Ibrahim H (2020) Introductory Chapter: Pyrolysis, in Recent Advances in Pyrolysis, IntechOpen
102. Zhong R, Xu J, Hui D, Bhosale SS, Hong R (2019) Pyrolytic preparation and modification of carbon black recovered from waste tyres. Waste Manage Res 38(1):35–43
103. Solis M, Silveira S (2020) Technologies for chemical recycling of household plastics—A technical review and TRL assessment. Waste Manag 105:128–138
104. Salehi E, Shafie M (2020) Adsorptive removal of acetaldehyde from water using strong anionic resins pretreated with bisulfite: An efficient method for spent process water recycling in petrochemical industry. J Water Process Eng 33:101025
105. Park K-B, Jeong Y-S, Guzelciftci B, Kim J-S (2020) Two-stage pyrolysis of polystyrene: Pyrolysis oil as a source of fuels or benzene, toluene, ethylbenzene, and xylenes. Appl Energy 259:114240
106. Sedlovets DM, Knyazev MA, Trofimov OV, Koveshnikov SV (2020) Graphene-like film CVD on e-beam exposed SiO2/Si by the pyrolysis of different oxygen-containing precursors. Fullerenes, Nanotubes, Carbon Nanostruct 28(4):309–312
107. Zhang Y, Wang Q, Ramachandran CS (2020) Synthesis of carbon nanotube reinforced aluminum composite powder (CNT-Al) by polymer pyrolysis chemical vapor deposition (PP-CVD) coupled high energy ball milling (HEBM) process. Diam Relat Mater 104:107748
108. Ghorbannezhad P, Park S, Onwudili JA (2020) Co-pyrolysis of biomass and plastic waste over zeolite- and sodium-based catalysts for enhanced yields of hydrocarbon products. Waste Manag 102:909–918
109. Zubair Yahaya A, Rao Somalu M, Muchtar A, Anwar Sulaiman S, Ramli Wan Daud W (2020) Effects of temperature on the chemical composition of tars produced from the gasification of coconut and palm kernel shells using downdraft fixed-bed reactor. Fuel 265:116910
110. Hanif MU, Zwawi M, Capareda SC, Iqbal H, Algarni M, Felemban BF, Bahadar A, Waqas A (2019) Influence of Pyrolysis Temperature on Product Distribution and Characteristics of Anaerobic Sludge. Energies 13(1):79

111. Chen W, Fang Y, Li K, Chen Z, Xia M, Gong M, Chen Y, Yang H, Tu X, Chen H (2020) Bamboo wastes catalytic pyrolysis with N-doped biochar catalyst for phenols products. Appl Energy 260:114242
112. Li Z, Zhong Z, Zhang B, Wang W, Seufitelli GVS, Resende FLP (2020) Effect of alkali-treated HZSM-5 zeolite on the production of aromatic hydrocarbons from microwave assisted catalytic fast pyrolysis (MACFP) of rice husk. Sci Total Environ 703:134605
113. Suriapparao DV, Vinu R, Shukla A, Haldar S (2020) Effective deoxygenation for the production of liquid biofuels via microwave assisted co-pyrolysis of agro residues and waste plastics combined with catalytic upgradation. Biores Technol 302:122775
114. Jiang L, Wang Y, Dai L, Yu Z, Wu Q, Zhao Y, Liu Y, Ruan R, Ke L, Peng Y, Xia D, Jiang L (2020) Integrating pyrolysis and ex-situ catalytic reforming by microwave heating to produce hydrocarbon-rich bio-oil from soybean soapstock. Biores Technol 302:122843
115. Teah HY, Sato T, Namiki K, Asaka M, Feng K, Noda S (2020) Life Cycle Greenhouse Gas Emissions of Long and Pure Carbon Nanotubes Synthesized via On-Substrate and Fluidized-Bed Chemical Vapor Deposition. ACS Sustainable Chemistry & Engineering 8 (4):1730–1740
116. Zheng X, Huo X, Zhang Y, Wang Q, Zhang Y, Xu X (2019) Cardiovascular endothelial inflammation by chronic coexposure to lead (Pb) and polycyclic aromatic hydrocarbons from preschool children in an e-waste recycling area. Environ Pollut 246:587–596
117. Xu P, Pan S, Yongkang Q, Liu J (2019) Deposition prevention plate and method for manufacturing the same, method for recycling material, and thin film deposition apparatus, Google Patents
118. Altay MC, Eroglu S (2019) Chemical vapor deposition of Ge thin films from solid GeO2 and C2H5OH. Thin Solid Films 677:22–27
119. Zhong Y, Chen Y, Liu K, Cai Y (2019) Method of Purifying and Recycling Normal-pressure Waste Hydrogen by Full Temperature Range Pressure Swing Adsorption (FTrPSA) in Manufacturing Process of Semiconductor, Google Patents
120. Choi M, Byun Y, Yang J (2019) A Study on Deport Maintenance Technology for Recycling Observation Window of the K1A1 Tank Commander's Primary Thermal Sight. J Soc Korea Ind Syst Eng 42(3):89–94
121. Kwon SJ, Seo HK, Ahn S, Lee TW (2019) Value-Added Recycling of Inexpensive Carbon Sources to Graphene and Carbon Nanotubes. Adv Sustain Syst 3(1):1800016
122. Wei W, Zhang Z, You G, Shan Y, Xu Z (2019) Preparation of recyclable MoO3 nanosheets for visible-light driven photocatalytic reduction of Cr(vi). RSC Advances 9(49):28768–28774
123. Hedayati A, Barnett C, Swan G, Orbaek White A (2019) Chemical Recycling of Consumer-Grade Black Plastic into Electrically Conductive Carbon Nanotubes. C 5(2):32
124. Kato T, Ishiwatari Y, Furukawa M, Yamaguchi H, Igarashi S (2012) Homogeneous Liquid-Liquid Extraction Method in order to Recycle Indium from Liquid Crystal Display, in Proceedings of the 14th Asia Pacific Confederation of Chemical Engineering Congress. Research Publishing Services
125. Guo Y, Shen P-C, Su C, Lu A-Y, Hempel M, Han Y, Ji Q, Lin Y, Shi E, McVay E, Dou L, Muller DA, Palacios T, Li J, Ling X, Kong J (2019) Additive manufacturing of patterned 2D semiconductor through recyclable masked growth. Proc Natl Acad Sci 116(9):3437–3442
126. Ahsan MA, Islam MT, Hernandez C, Kim H, Lin Y, Curry ML, Gardea-Torresdey J, Noveron JC (2018) Adsorptive Removal of Sulfamethoxazole and Bisphenol A from Contaminated Water using Functionalized Carbonaceous Material Derived from Tea Leaves. J Environ Chem Eng 6(4):4215–4225
127. Ahsan MA, Jabbari V, Islam MT, Kim H, Hernandez-Viezcas JA, Lin Y, Díaz-Moreno CA, Lopez J, Gardea-Torresdey J, Noveron JC (2018) Green synthesis of a highly efficient biosorbent for organic, pharmaceutical, and heavy metal pollutants removal: Engineering surface chemistry of polymeric biomass of spent coffee waste. J Water Process Eng 25:309–319

128. Yu J, Lu W, Pei S, Gong K, Wang L, Meng L, Huang Y, Smith JP, Booksh KS, Li Q, Byun J-H, Oh Y, Yan Y, Chou T-W (2016) Omnidirectionally Stretchable High-Performance Supercapacitor Based on Isotropic Buckled Carbon Nanotube Films. ACS Nano 10 (5):5204–5211
129. Ling Z, Wang Z, Zhang M, Yu C, Wang G, Dong Y, Liu S, Wang Y, Qiu J (2016) Sustainable Synthesis: Sustainable Synthesis and Assembly of Biomass-Derived B/N Co-Doped Carbon Nanosheets with Ultrahigh Aspect Ratio for High-Performance Supercapacitors.Adv Funct Mater 26(1):1
130. Hou J, Jiang K, Wei R, Tahir M, Wu X, Shen M, Wang X, Cao C (2017) Popcorn-Derived Porous Carbon Flakes with an Ultrahigh Specific Surface Area for Superior Performance Supercapacitors. ACS Appl Mater Interfaces 9(36):30626–30634
131. Wang X (2019) A Refined Assessment Methodology for Wastewater Treatment Alternatives, in Springer Theses. Springer Singapore, pp 57–77
132. Zeynalov EB, Friedrich JF, Tagiyev DB, Huseynov AB, Magerramova MY, Abdurehmanova NA (2018) Review on nanostructures from catalytic pyrolysis of gas and liquid carbon sources. Materials Testing 60(7–8):783–793
133. Wang Z, Shen D, Wu C, Gu S (2018) State-of-the-art on the production and application of carbon nanomaterials from biomass. Green Chem 20(22):5031–5057
134. Liu H, Wang X, Jia D (2020) Recycling of waste rubber powder by mechano-chemical modification. J Clean Prod 245:118716
135. Veerappan G, Ramasamy E, Gowreeswari B (2019) Economical and Highly Efficient Non-Metal Counter Electrode Materials for Stable Dye-Sensitized Solar Cells. In: Dye-Sensitized Solar Cells. Elsevier, pp 397–435
136. Sajdak MM (2019) Optimization Frameworks in Resource Management and Process Engineering. In: *Plastics to Energy*. Elsevier, pp 425–442
137. Bai Y, Wang Z, Feng L (2010) Chemical recycling of carbon fibers reinforced epoxy resin composites in oxygen in supercritical water. Mater Des 31(2):999–1002
138. Habeeb OA, Ali GAM (2017) Application of Response Surface Methodology for Optimization of Palm Kernel Shell Activated Carbon Preparation Factors for Removal of H_2S from Industrial Wastewater. Jurnal Teknologi (Sciences and Engineering) 79(7):1–10
139. Hegde G, Abdul Manaf SA, Kumar A, Ali GAM, Chong KF, Ngaini Z, Sharma KV (2015) Biowaste Sago Bark Based Catalyst Free Carbon Nanospheres: Waste to Wealth Approach. ACS Sustainable Chem Eng 3(9):2247–2253
140. Ali GAM, Divyashree A, Supriya S, Chong KF, Ethiraj AS, Reddy MV, Algarni H, Hegde G (2017) Carbon nanospheres derived from Lablab purpureus for high performance supercapacitor electrodes: a green approach. Dalton Trans 46(40):14034–14044
141. Habeeb OA, Ramesh K, Ali GAM (2017) Experimental design technique on removal of hydrogen sulfide using CaO-eggshells dispersed onto palm kernel shell activated carbon: Experiment, optimization, equilibrium and kinetic studies. J Wuhan Univ Technol Mater Sci Ed 32(2):305–320
142. Habeeb OA, Ramesh K, Ali GAM, Yunus RM, Olalere OA (2017) Kinetic, Isotherm and Equilibrium Study of Adsorption of Hydrogen Sulfide From Wastewater Using Modified Eggshells. IIUM Eng J 18(1):13–25
143. Habeeb OA, Kanthasamy R, Ali GAM, bin Mohd R (2017) Optimization of Activated Carbon Synthesis Using Response Surface Methodology to Enhance H_2S Removal From Refinery Wastewater. J Chem Eng Ind Biotechnol 1:17
144. Ali GAM, Habeeb OA, Algarni H, Chong KF (2018) CaO impregnated highly porous honeycomb activated carbon from agriculture waste: symmetrical supercapacitor study. J Mater Sci 54:683–692
145. Ali GAM, Manaf SAA, Divyashree A, Chong KF, Hegde G (2016) Superior supercapacitive performance in porous nanocarbons. J Energy Chem 25(4):734–739

Recent Trends of Recycled Carbon-Based Nanomaterials and Their Applications

M. Abd Elkodous, Gharieb S. El-Sayyad, Mohamed Gobara, and Ahmed I. El-Batal

Abstract "There's Plenty of Room at the Bottom: An Invitation to Enter a New Field of Physics" said Richard Feynman in 1959, this lecture opened the way to the new field of science which we know today as nanotechnology. Materials' manipulation at a very small size, ranges from 1 to 100 nm (nanoworld or the nano-edge) is well-known as nanotechnology. Since then, a lot of investigations and research were devoted by many researchers around the globe to keep an eye on the different properties and behavior of nanomaterials. Materials with at least one nanoscale dimension are called nanomaterials that have outstanding features compared to their bulk counterparts. These exceptional characteristics are due to the relatively-high surface area and the relatively-large surface atoms compared to those in the inner mass. Thus, nanomaterials have attractive chemical, physical, electronic, physiological, and optical properties. In this chapter, we are covering the historical overview and origin of nanomaterials to their recent applications. In addition, types and applications of recycled carbon-based nanomaterials as an example have also been discussed.

Authors: M. Abd Elkodousa and Gharieb S. El-Sayyad are equally contibuted.

M. Abd Elkodous (✉)
Department of Electrical and Electronic Information Engineering,
Toyohashi University of Technology, Toyohashi, Aichi 441-8580, Japan
e-mail: mohamed.hamada.abdlekodous.xi@tut.jp

M. Abd Elkodous
Center for Nanotechnology (CNT), School of Engineering and Applied Sciences,
Nile University, Sheikh Zayed, Giza 16453, Egypt

G. S. El-Sayyad (✉) · A. I. El-Batal
Drug Microbiology Lab, Drug Radiation Research Department, National Center for Radiation Research and Technology (NCRRT), Egyptian Atomic Energy Authority (EAEA),
Cairo, Egypt
e-mail: Gharieb.Elsayyad@eaea.org.eg

G. S. El-Sayyad · M. Gobara
Chemical Engineering Department, Military Technical College (MTC),
Egyptian Armed Forces, Cairo, Egypt

A. S. H. Makhlouf and G. A. M. Ali (eds.), *Waste Recycling Technologies for Nanomaterials Manufacturing*, Topics in Mining, Metallurgy and Materials Engineering, https://doi.org/10.1007/978-3-030-68031-2_16

Keywords Nanomaterials · Recycling · Carbon nanomaterials · Banana fibers · Argania spinosa seeds · Corn grains · Sugarcane fibers · Oil palm shells

List of Abbreviations

C60	Fullerenes
CDs	Carbon dots
CNFs	Carbon nanofibers
CNHs	Carbon nanohorns
CNMs	Carbon nanomaterials
CNOs	Carbon nanoonions
CNPs	Carbon nanoparticles
CNTs	Carbon nanotubes
DDGS	Distiller's dried grains
DNA	Deoxyribonucleic acid
FCNDs	Fluorescent carbon nanodots
FESEM	Field emission scanning electron microscope
GO	Graphene oxide
HRTM	High resolution transmission microscope
NMs	Nanomaterials
NPs	Nanoparticles
QDs	Quantum dots
SEM	Scanning electron microscope

1 Introduction

1.1 Overview of Nanomaterials

In general, nanomaterials (NMs) have gained much attention, interest, and prominence as a result of their controllable physicochemical properties such as light absorption, electrical conductivity, and melting point, leading to superior properties compared with their bulk counterparts [1–4]. Although there is no one internationally-accepted definition of NMs, they can be commonly-known as the materials with one nanoscale dimension, which ranges from 1 to 100 nm [4, 5]. Currently, NMs are considered the cornerstones of the broad area of nanoscience and technology [4, 6]. NMs are attractive as a result of their promising and versatile uses in medicine, electronics, energy, environment, biology, industry, and other fields [7–17]. There are two reasons behind the novel properties of NMs, which are not common in their bulk materials; the first is the relatively-higher surface area, and the second is the quantum effects, which play an essential role in tuning materials' properties at the nanoscale [18–22].

1.2 Origin of Nanomaterials

Nanomaterials were not recently-originated; they have been formed and existed for billions of years since the big bang [23]. There are three possible sources for NMs. Firstly, natural sources such as combustion, ocean spray, the decay of radioactive gases such as radon, volcanic ash, and fires of forests, the developed NMs from these sources and called natural NMs [24–26]. Secondly, incidental NMs, which were unintentionally-formed by human activities like iron oxy/hydroxide, oxygen-deficient TiO_2, and smoke particles which were formed as a result of using fires by the ancient humans [24, 27–29]. Thirdly, engineered NMs, which are considered the most recent-class of NMs, these NMs were prepared and designed for particular applications and can be found in many commercial products, including cosmetics, sun blockers, clothing, sporting goods, and other various daily items [30–34].

2 Recycled Nanomaterials

There is a significant increase in household consumer products that contain nanomaterials as a result of the desirable properties of NMs [28, 35–39]. Thus, many strategies have been designed to extract NMs from waste; for example, gold nanoparticles (NPs) and inorganic quantum dots (QDs) have been extracted from the computer and television displays. In addition, carbon dots (CDs – carbon NPs with a diameter of less than 10 nm) have been obtained from heating small pieces of plastic bags with hydrogen peroxide [40]. Interestingly, silicon carbide NPs have been created from heating discarded CDs with sand in the furnace [41]. Magnetic NPs can be isolated from wastewater using external magnets [42, 43]. Another method investigated wasted glass (silica) for the formation of silicon NMs which are essential for energy storage applications (batteries) [44]. Finally, food wastes known as biomass can be very important sources of NMs such as banana fibers, sugarcane, and oil palm, which can be used to produce carbon NMs [45–49].

3 Classification of Recycled Nanomaterials

Biomass is a low cost, easily-accessible, publicly-administered, environmentally-beneficial, and renewable carbon source [50–52]. Understanding the right biomass disposal is an essential challenge because it is a byproduct and/or bio-waste of farming, manufacturing, and forestation [53]. Carbon nanomaterials (CNMs), like graphene, carbon nanotubes (CNTs), and carbon nanofibers (CNFs), developed from fossil fuels are the common carbon substances today due to their exceptional features and widespread purposes [54]. It is well-established that carbon substances

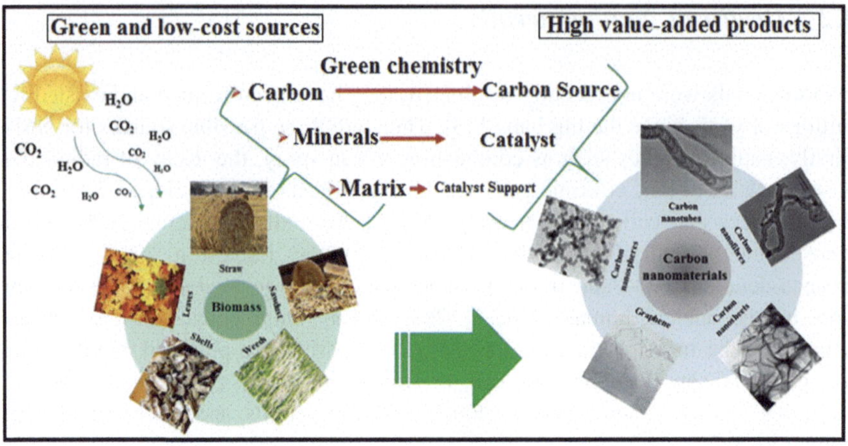

Fig. 1 Biomass-based CNMs Production. Adapted from Ref. [54] Copyright 2018, American Chemical Society

perform critical functions in supporting social improvement and technological processes [55]. Diamond, graphite, carbon fibers, and activated carbon are widely-used in everyday uses, leading to high progress in the industry by supplying power and fostering environmental technology [56].

Recent advancements in the construction of CNMs (e.g., CNFs, CNTs, and graphene) from biomass [54, 57] is shown in Fig. 1. Several carbon-based biomass methods for the construction of CNMs were widely-performed, such as pyrolytic conversion (thermo-chemical activation, catalytic graphitization, and catalytic graphitization), combustion, cyclic oxidation, and mechanical activation [54, 58]. Care is also required to investigate the employed biomass type as a cost-effective, sustainable catalyst or a co-catalyst to obtain improved CNM preparations. Biomass-based CNMs employ mentis that were used in many applications [22, 54, 59, 60].

3.1 Carbon Nanomaterials from Banana Fibers

Since nanotechnology evolution as a separate discipline, many nanomaterials' classes were prepared, which are probably-applicable for various purposes that range from healthcare stocks to high-performance building supplies [61–63]. Banana (Moses) is a family of the Musaceae group [64]. Interestingly, banana is the world's fourth common and major food product after rice, grain, and corn. It is supporting the economies of several developed nations [65]. The production of banana reached 6,179,190 tons in 2014 (Indonesia) [66]. 20% of banana crops turn into waste that is not generally-utilized [67]. Unlike banana leaves and tuber, the stalks and skin midrib are normally-disposed and just utilized on a short-range like farming media usage [68], fertilizer composition, and additional construction in injury medications. The carbon content in the terminal of the banana stalk is extremely-essential, which equals 86.99% [69]. An alternative application of banana waste is a natural substance in the development of activated carbon plates for supercapacitor cell purposes. Numerous investigations on natural substances such as banana stem for different applications, including supercapacitor plates, were reported [47, 70–73].

Carbon substances were extracted from banana fibers by treating the fibers with chemical materials like $ZnCl_2$ and KOH to enlarge the surface area and to improve their electrochemical behavior [47]. The usage of these substances was facilitated by a neutral electrolyte. The obtained surface area of carbon materials was up-and-coming due to the efficient pore formations [47]. SEM microscopy and N_2 adsorption/desorption analysis were used to investigate the surface structural and surface arear properties of the obtained carbon substances, respectively [74]. The surface area of the 15% $ZnCl_2$-treated materials was determined to be 1100 m^2 g^{-1} [47]. The electrochemical characteristics of as-prepared and treated carbon samples were assessed by using galvanostatic charge—discharge and cyclic voltammetry analyses [47]. Carbon materials possessing large surface area were very promising in the electrochemical applications in neutral electrolyte media [75].

Several applications in nano-pharmacology and nano-biomedicine were generated by examining the CNMs as intelligent distribution systems and carriers as drug vehicles [76–78]. CNMs can be used with a suitable dosage of medicines or as additional vital materials for effective distribution to a particular target position inside the cell [79, 80]. Although the corresponding systems were extensively-used in plant operations, nanotechnology in plant biology and sustainable crops (agriculture) did no raise serious concerns within the scientific community. Yet, the effects of CNMs on seedling germination and growth were slightly less studied and less known compared to similar investigations on animals [81]. Research papers on the impact of CNMs on different varieties of plants have been newly-developed [82–84]. Notwithstanding these applications, the impacts of the CNMs on morphological, anatomical, molecular, morphological, biochemical, and physiological methods and their tools in various plants are not yet completely-recognized, and a call for full investigations is a must [85].

Although the main purpose of nanomaterials' application in plants is to improve agricultural fertility, several research investigations have suggested significant issues about the possible antagonistic results of NPs on human fitness and the ecosystem [86]. First, articles have explained the novel NMs' characteristics and identified their promising applications as roots sprout, germination enhancers, herbicides, pesticides, fertilizer vehicles, DNA, and phytohormones to other cells of the shoot [87]. Although more investigation is promising, it is important to explain the influence of NPs' features on specific plants [88]. Systematic experimental investigations about CNMs' uptake and quantity in many sections of the plants are still in the first trials. Given that, many researchers selected the specific literature with a new strategy on the field of plant nano-technology; specifically uptake, translocation, and CNMs' quantity in plant's trunk and their role in controlling the

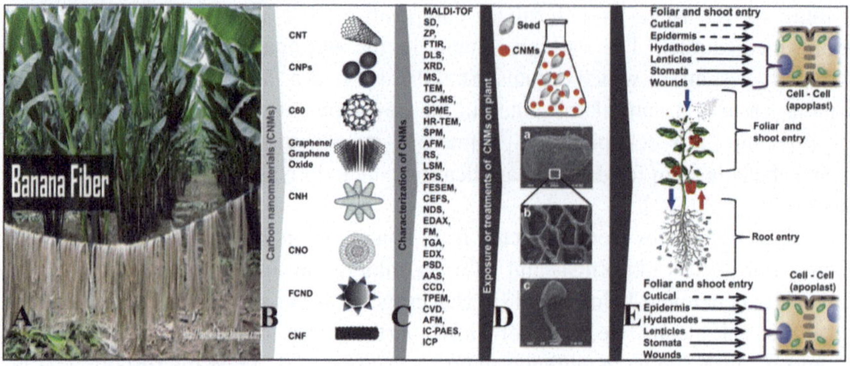

Fig. 2 Schematic diagram showing: (**a**) Banana fibers and the synthesis of different CNMs, (**b**) different kinds of CNMs (C60: fullerenes, CNTs: carbon nanotubes, CNHs: carbon nanohorns, GO: graphene oxide, CNPs: carbon nanoparticles, FCNDs: fluorescent carbon nanodots, CNFs: carbon nanofibers, CNOs: carbon nanoonions), (**c**) Different CNMs' characterization techniques. (**d–e**): CNMs effect (carbon nanomaterials) on seed germination of Digitalis and growth, uptake, bioaccumulation, and CNMs' transport. Adapted from Ref. [89] Copyright 2019, Elsevier

plant germination and plant's reply to potential toxicity after the construction of CNMs from banana fibers as shown in Fig. 2.

3.2 *Carbon Nanomaterials from* Argania Spinosa *Seeds*

Porous carbon materials like activated carbon are of outstanding value for their applications in supercapacitor plates because they can be easily-synthesized from low-cost biomass debris. Therefore, the activated carbon was fabricated from rice and other biomass-derived materials [90], waste coffee beans [91], pistachio husks and firewood [92], corn grains [93], sugarcane bagasse [94], sunflower seed shells [95], and cassava peel waste [96].

The argan tree (*Argania spinosa)* is an endemic class found in the Southwest area of Morocco; its seeds and roots are utilized to provide oil for ornamental applications [97]. The treatment of argan oil generates enormous amounts of biomass residuals that essentially-utilized in warming [98].

A large amount of activated carbon was developed by KOH-treatment of argan seed husks [99]. The synthesized activated carbon possessed high surface area and porosity partially-attributed to added oxygen and nitrogen functionalities [100]. Graphene, carbon aerogel, activated carbon, carbon nanotubes, and others with a great surface area, and versatile applications were collected after the utilization of *Argania spinosa* [99], as displayed in Fig. 3.

The electrochemical potential of the prepared activated carbon by KOH-treated *Argania spinosa* seeds was reported within the written articles on different activated carbon, explaining that *Argania spinosa* seeds are attractive biomass for the

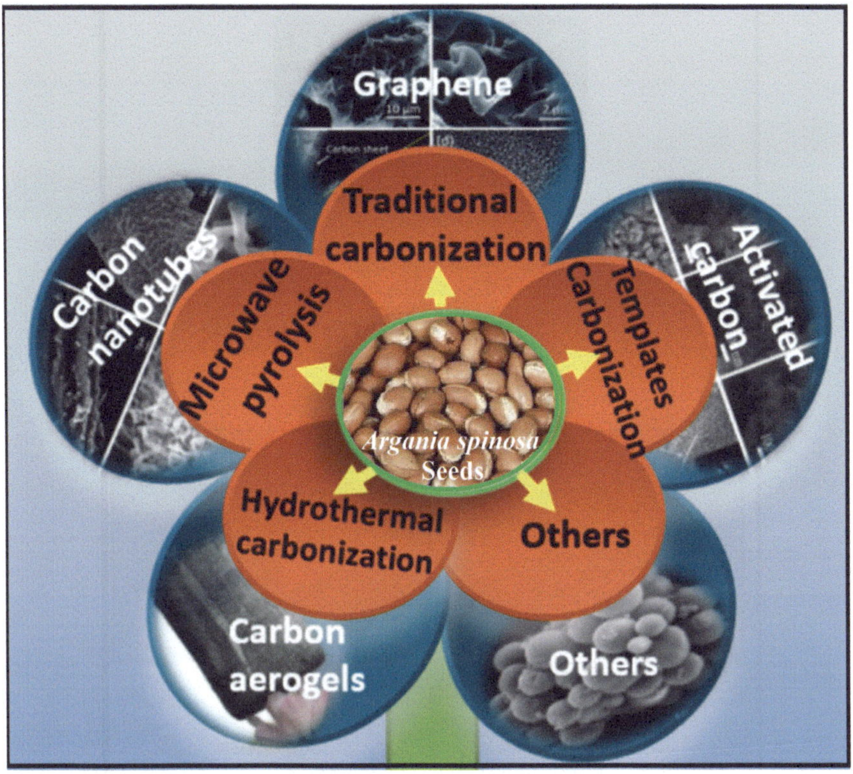

Fig. 3 Synthesis of activated carbon, graphene, carbon nanotubes, carbon aerogel, and others by the usage of *Argania spinosa* seeds

development of acceptable activated carbon which may be used in supercapacitor purposes. The most significant capacitance obtained was about 350 F g^{-1} at 12 mA g^{-1} with a 94% retention capacitance at 1 A g^{-1}, because of the high surface area, suitable and well-formed micro- and/or mesopore structure, and the content of nitrogen in the activated carbon.

3.3 Carbon Nanomaterials from Corn Grains, Sugarcane Fibers, and Oil Palm Shells

The conventional sugar generation linked with the developing creation of ignitable ethanol, obtained from sugarcane manufacturing to be an example of the common primary parts of Brazilian economics [101]. The Brazilian operators of ethanol and sugar developed nearly 650 million tons (sugarcane) in 2011 that produced about 143 million tons (bagasse) [102]. A unique procedure to develop NMs utilizing the gases produced through the sugarcane (bagasse) pyrolysis was used [103].

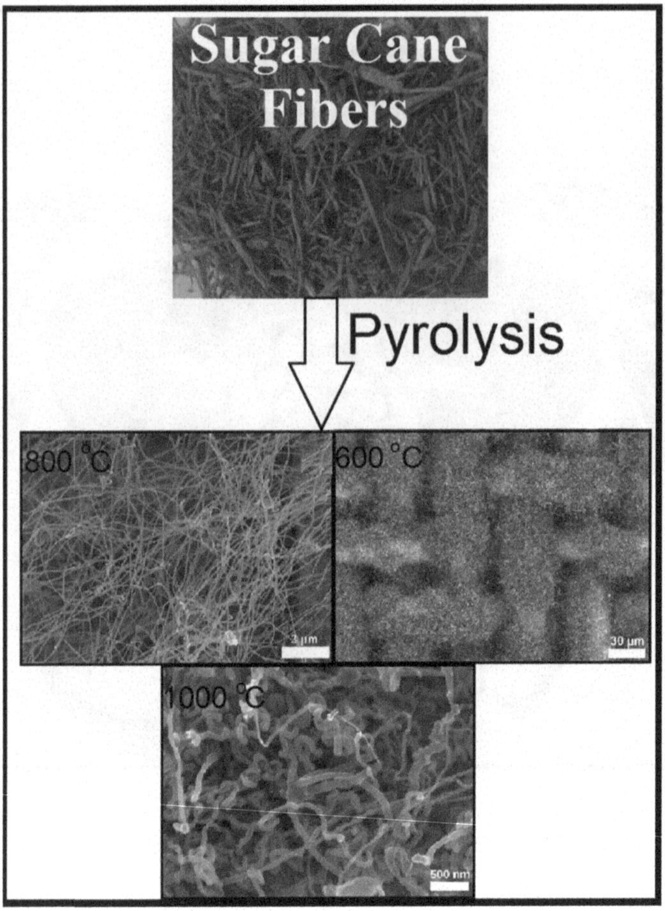

Fig. 4 SEM images of different CNMs synthesized using sugarcane fibers by the pyrolysis at different temperatures

A new method to create CNMs utilizing the gases generated through the pyrolysis of different wastes (corn and bagasse) was founded in the Laboratory of Northeastern University in Boston, USA [104]. In this chapter, we have considered the catalyst substrates formed among the pyrolysis vapors of the sugarcane (bagasse) to decide if NMs were developed and further to explain the design of such substances as shown in Fig. 4.

Nanotechnology fostered a business of about 15 trillion dollars in 2011, although 350 billion dollars granted only to NMs [105]. Certain elements with exceptional engineering, thermal and electrical properties [106]. In 2012, the world production of CNMs was about 3550 tons and predicted to grow at a composite yearly increase percentage of 35.5%. Developing businesses for CNMs (like nanofibers and nanotubes) aimed at reducing their generation charges [107]. To overcome nanomaterials' price, it is necessary to decrease the costs correlated with the acquisition of raw materials [108]. By using sugarcane (bagasse) as raw-elements for CNMs production, the cost may be reduced [109].

Corn grain-based ethanol generation in North America was quickly-developing, from around 20 billion liters in 2005 to 40 billion liters in 2010 [110]. Most of the manufactures can transform corn grains to ethanol essentially by two methods: dry grinding or wet milling. During wet milling, seeds of corn are grinded into fundamental ingredients (germ, fiber, and starch), which end in a different product such as waters and beneficial co-products [111]. The dry-grind design includes six principal steps consisting of grinding, cooking, and liquefaction as the first main steps, followed by saccharification, fermentation, and parting process [112]. The products of dry grinding involve distillers condensed grains (DDGS), ethanol fuel, and carbon dioxide [113]. The dry-grinded corn is qualified for more than 75% of corn-based ethanol generation due to more cost-effectiveness and high yields of ethanol [114].

After changing the starch division of corn with accepted ferments and some catalysts to create CO_2 and ethanol, ethanol is separated through the distillation process and the residual debris is evaporated, providing DDGS [115]. Now, DDGS is used as cattle supplies because of its great contents of protein. They also have a large amount of fibers that restrict their applications to ruminant nutrition, which might produce health issues in mammals [116].

The production of DDGS should be developed simultaneously with ethanol generation, DDGS was manufactured in the USA around 32.5 million tons in 2010 [113]. Therefore, it is better to notice different benefits of DDGS for the stock market and consequently to support reducing the cost of ethanol manufacturing [117]. DDGS can be utilized for energy production because of their power content of about 30 MJ kg^{-1}, which exceeded the power of lignite fuels (15–25 MJ kg^{-1}) and resembled that of bituminous fuels (35 MJ kg^{-1}) [118].

Waste-to-power biotechnology was applied to decrease residues' quantity and to produce energy outdoors utilizing clean non-renewable sources [119]. Biomass consumption in DDGS and sugarcane bagasse can be turned into valuable kinds of power handling different sorts of technologies like biochemical/biological, thermochemical and/or mechanical removal [120].

The chemical–thermal technology remains the common modern technology; this involves the gasification, pyrolysis, liquefaction, and combustion methods [121]. In recent research, the pyrolysis method was utilized, and the excess biomass remained disintegrated through thermal processing under an oxygen-free atmosphere (i.e., N_2) [122]. The cost-effective breakdown of biomass by warm processing to generate power has been extensively-examined [123]. A method where regular biomass debris are pyrolyzed to produce probably beneficial vaporous effluents (including CO, hydrocarbons, and H_2) while, at the same time, creating CNMs, like CNTs, on catalytic materials [124], is shown in Fig. 5.

Activated carbon can be developed through a high fraction of raw-substances, particularly from the agro-industrial wastes like coconut husks [125], almond shells [126], hazelnut shells [127, 128], cherry gems [129], eucalyptus [130], apricot

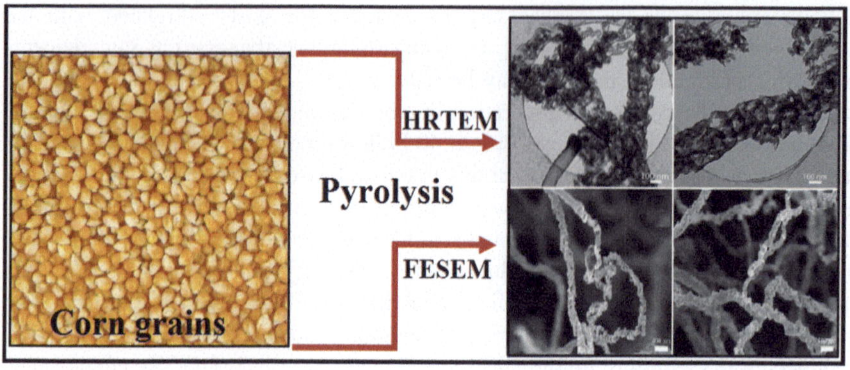

Fig. 5 HRTM and FESEM images of different CNMs synthesized using corn grains by the pyrolysis process

grains, nuts, grape grains [129], peach and olive masses [131], bagasse of sugarcane [132], palm kernel shell [133, 134], oil palm leaves [49, 135], palm kernel shell [133, 134, 136–138], and oil palm blocks [139].

Oil palm production was developed by the production of large amounts of wastes at farm areas, oil plants, and factories, as presented in Fig. 6. It was reported that processing presented periodically around 2.50 million tons of palm mesocarp tissue, 1.45 million tons of oil palm husks, and 4.15 million tons of dry fruit remain as by-products [140].

It is a mixed benefit over the oil palm industry if the husk excess toilet can be converted into beneficial and relevant outcomes. It is expected that around 1150 kg (palm nuts) can be produced per the hectare of the oil palm [141]. The analyzed study data of the content of ash and the produced carbon in the palm husk noted proper by-products during the generation of activated carbon substances [142, 143]. Earlier investigations revealed that the exterior surface of the produced activated carbon developed from the oil palm shells was about 955 $m^2 g^{-1}$ [144].

However, no synthetic active education on oil palm husks has been performed. To improve the surface area of the synthesized carbon, CO and ZnCl were utilized as chemical catalysts [145].

The effect of porosity and surface area of the produced activated carbon on the strength of the produced zinc suspension over the starting materials must be considered [146]. 10 gm of the dried oil palm husks were combined with 100 cm^3 of 2–40% ZnCl solutions (w/w).

It should be noted that the ratio between the starting materials and the solution must be regularly-maintained. Impregnation was performed at around 75°C under a water bath until the excess water disappeared. The specimen was next concentrated in an oven near 125°C for about 12 h. The specimen of the loaded shells was formerly-placed in a stainless steel core reactor (50 mm width and 155 mm length) [147].

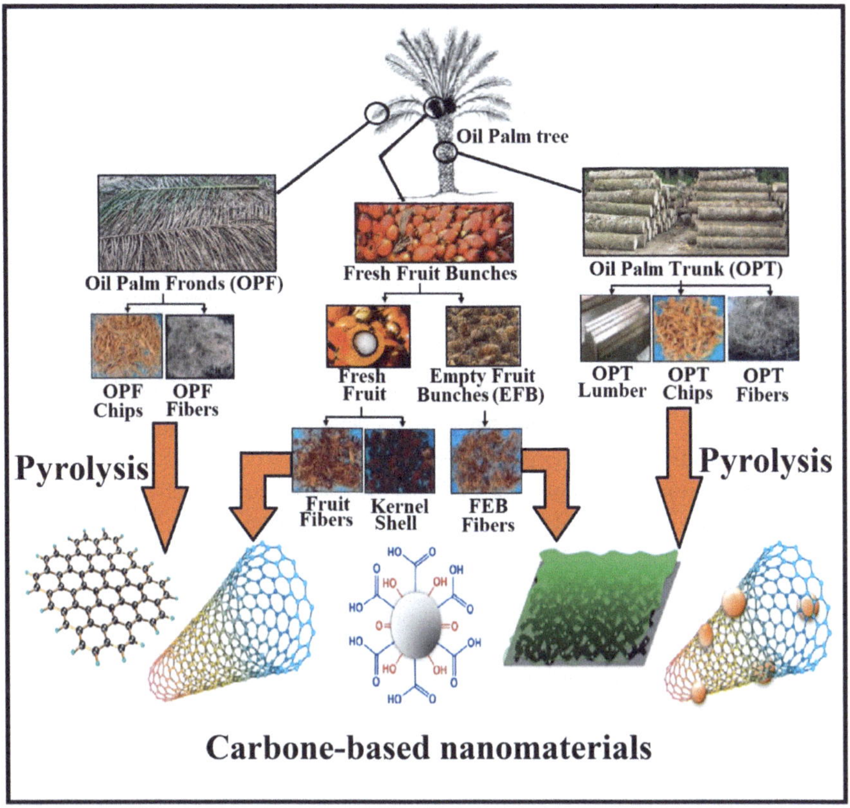

Fig. 6 Usage of palm oil tree by-products for the production of carbon nanomaterials (CNMs). Adapted from Ref. [149] Copyright 2018, Intechopen

The mentioned reactor remained placed in a designed electric furnace. The temperature of pyrolysis was fixed at 550°C supplemented with an inactive stream of nitrogen gas for 4 h. Finally, CO_2 gas passed through the sample for 60 min for producing carbon nanomaterials (CNMs) [148].

4 Applications of the Recycled Nanomaterials

Since the 1980s, avast agreement of examination has been conveyed on nanoparticles and their applications. The nanoparticles have a broad variety of purposes like materials science, chemistry, biochemistry, electronics, and pharmacy. The small particle sizes, ranging from 1 to 100 nm, give unusual mechanical, thermal, optical, and chemical characteristics to certain substances. This small size not only offers unique opportunities to decrease the amount of materials used but also enhances their efficiency. The progressive development in synthesis approaches allows governing the shape, size, stability, and dispersibility of the prepared nanoparticles [150]. The reactivity of the nanomaterials varies according to size, shape, degree of aggregation, and the surface function of these particles [151]. However, nanotechnology has many challenges in the coming few years comprising: costs, scaling-up production, and safety. Safety concerns would govern the spreading of nanomaterials applications awaiting to understand the biological activity of these materials [152]. Nanomaterials can penetrate the epidermis into the body forming free radicals that may oxidize DNA and proteins [153, 154]. Recently, a new branch of science appeared, "nanotoxicology" concerning the negative effects of the nanomaterials on biological systems [152, 155]. Many approaches were applied to control the toxicity of some nanomaterials, such as surface modification [17, 156]. Recycling the nanomaterials would also decrease their adverse environmental impacts. Compared with other techniques of recycling nanoparticles, immerge a magnetic property would offer special advantages in transporting, detection, and handling beside the cost-effectiveness of these active small size materials [157].

In addition, the applications of magnetic nanoparticles extended to cover biomedicine, data storage, magnetic resonance imaging, and catalysis [158–160]. The main advantage of using magnetically-prepared catalysts is that it can be easily-collected and separated using an external magnet. This approach not only supporting economic and eco-friendly requirements but also evades the loss of catalyst taking place in traditional centrifugation and filtration techniques [161].

Magnetic nanoparticles have many applications, including adsorption and separation of pollutants from aqueous solutions. The magnetic nanoabsorbents have the merit of both fast and high capacity adsorption process [162, 163]. Surface treatment of the magnetic nanoparticles with functional groups can offer both purification and disinfection of water during wastewater treatment [164–166].

Carbon nanotubes, CNTs, are one of the most proposing candidates for water treatment. They are chemically-stable adsorbents material that can possess functional groups on their surface, increasing both sensitivity and selectivity toward heavy metals pollutants [167]. Magnetic CNTs have been used for water treatment [168, 169]. They, in combination with layered double hydroxides, were used to adsorb oil from the water where efficiency of 99% can be achieved [169]. In addition, CNTs modified with natural polymers (chitosan) can adsorb endocrine-disrupting compounds from aqueous solution [170]. Recently, CNTs have been used in membranes for water purification due to their large surface area and low-pressure drop during water transport [171, 172]. They can enhance many properties of the membrane, such as strength, disinfection, rejection, and permeability [173, 174].

5 Conclusions

In this chapter, we have revealed the origin, characteristics, various types of nanomaterials, and their multiple applications and usage in daily products such as sun blockers, sporting goods, and clothing. In addition, recycled nanomaterials from the biomass have been analyzed. Among recycled nanomaterials, recycled carbon-based nanomaterials like carbon nanotubes (CNTs), carbon nanofibers (CNFs), and graphene, along with their different applications, have been discussed in detail. Moreover, we have revealed the different origins and resources of carbon-based nanomaterials, such as banana fibers, *Argania spinosa* seeds, sugarcane, oil palm shells, and corn grains. Besides, optimized synthetic routes and physical, chemical, thermal, and mechanical characteristics of the prepared materials have also been analyzed. Finally, multiple applications of the recycled carbon-based nanomaterials in many disciplines as energy storage (supercapacitors), environment (wastewater treatment), nanomedicine (drug delivery systems), and industry (catalysts) have been extensively-covered. This chapter highlights the attractive properties and great potential of recycled nanomaterials and carbon-based nanomaterials in particular for many useful applications.

6 Future Perspectives

Despite the promising and versatile applications of CNMs, concerns about any possible toxicology and impact on living organisms, are affecting their applications. The physicochemical characteristics of CNMs play a significant role in their toxicological behavior. Size of these materials may allow cell nucleus insertion and reaction with the genetic material. Additionally, the initiation of free radicals inside living cells may oxidize the biological macromolecules like proteins, carbohydrates, and genetic DNA. However, different factors can control this toxicity, including the dose of CNMs, route of manipulation and particle size. Currently, more investigations are required for scaling-up the production of CNMs from laboratory to industry. In addition, cost-effective, easily-manipulated and more effective routes for the preparation of CNMs and to ensure their biosafety are critical challenges.

References

1. Wong CW, Chan YS, Jeevanandam J, Pal K, Bechelany M, Abd Elkodous M, El-Sayyad GS (2020) Response Surface Methodology Optimization of Mono-dispersed MgO Nanoparticles Fabricated by Ultrasonic-Assisted Sol-Gel Method for Outstanding Antimicrobial and Antibiofilm Activities. J Cluster Sci 31:367–389
2. Govindasamy G, Pal K, Abd Elkodous M, El-Sayyad GS, Gautam K, Murugasan P (2019) Growth dynamics of CBD-assisted CuS nanostructured thin-film: optical, dielectric and novel switchable device applications. J Mater Sci: Mater Electron 30(17):16463–16477

3. Abd Elkodous M, El-Sayyad GS, Abdel Maksoud MIA, Abdelrahman IY, Mosallam FM, Gobara M, El-Batal AI (2019) Fabrication of ultra-pure anisotropic zinc oxide nanoparticles via simple and cost-effective route: implications for UTI and EAC medications. Biol Trace Elem Res 196:297–317
4. Fouad OA, Makhlouf SA, Ali GAM, El-Sayed AY (2011) Cobalt/silica nanocomposite via thermal calcination-reduction of gel precursors. Mater Chem Phys 128(1):70–76
5. Jeevanandam J, Barhoum A, Chan YS, Dufresne A, Danquah MK (2018) Review on nanoparticles and nanostructured materials: history, sources, toxicity and regulations. Beilstein J Nanotechnol 9(1):1050–1074
6. Ali GAM, Fouad OA, Makhlouf SA (2013) Structural, optical and electrical properties of sol-gel prepared mesoporous Co_3O_4/SiO_2 nanocomposites. J Alloy Compd 579:606–611
7. Nathan A, Ahnood A, Cole MT, Lee S, Suzuki Y, Hiralal P, Bonaccorso F, Hasan T, Garcia-Gancedo L, Dyadyusha A (2012) Flexible electronics: the next ubiquitous platform. Proceedings of the IEEE 100(Special Centennial Issue):1486–1517
8. Xu P, Zeng GM, Huang DL, Feng CL, Hu S, Zhao MH, Lai C, Wei Z, Huang C, Xie GX (2012) Use of iron oxide nanomaterials in wastewater treatment: a review. Sci Total Environ 424:1–10
9. Charitidis CA, Georgiou P, Koklioti MA, Trompeta A-F, Markakis V (2014) Manufacturing nanomaterials: from research to industry. Manuf Rev 1:11–29
10. De Marchi L, Pretti C, Gabriel B, Marques PA, Freitas R, Neto V (2018) An overview of graphene materials: Properties, applications and toxicity on aquatic environments. Sci Total Environ 631:1440–1456
11. Aneggi E, de Leitenburg C, Boaro M, Fornasiero P, Trovarelli A (2020) Catalytic applications of cerium dioxide. In: Cerium Oxide (CeO_2): Synthesis, Properties and Applications. Elsevier, pp 45–108
12. Elkodous MA, Hassaan A, Ghoneim A, Abdeen Z (2018) C-dots dispersed macro-mesoporous TiO_2 photocatalyst for effective waste water treatment. Charact Appl Nanomater 1(2):1–9
13. Asiya SI, Pal K, El-Sayyad GS, Elkodous MA, Demetriades C, Kralj S, Thomas S (2019) Reliable optoelectronic switchable device implementation by CdS nanowires conjugated bent-core liquid crystal matrix. Org Electron 82:105592
14. Jeevanandam J, Sundaramurthy A, Sharma V, Murugan C, Pal K, Abdel Kodous MH, Danquah MK (2020) Chapter 4—Sustainability of One-Dimensional Nanostructures: Fabrication and Industrial Applications. In: G Szekely, A Livingston (eds) Sustainable Nanoscale Engineering. Elsevier, pp 83–113
15. Elkhenany H, Abd Elkodous M, Ghoneim NI, Ahmed TA, Ahmed SM, Mohamed IK, El-Badri N (2020) Comparison of different uncoated and starch-coated superparamagnetic iron oxide nanoparticles: Implications for stem cell tracking. Int J Biol Macromol 143:763–774
16. Fouad OA, Ali GAM, El-erian MAI, Makhlouf SA (2012) Humidity sensing properties of cobalt oxide/silica nanocomposites prepared aia sol–gel and related routes. NANO 07 (05):1250038
17. Ghawanmeh AA, Ali GAM, Algarni H, Sarkar SM, Chong KF (2019) Graphene oxide-based hydrogels as a nanocarrier for anticancer drug delivery. Nano Res 12(5):973–990
18. Kohle FF, Hinckley JA, Li S, Dhawan N, Katt WP, Erstling JA, Werner-Zwanziger U, Zwanziger J, Cerione RA, Wiesner UB (2019) Amorphous Quantum Nanomaterials. Adv Mater 31(5):1806993
19. Liu P, Williams JR, Cha JJ (2019) Topological nanomaterials. Nature Reviews Materials:1
20. Varghese RJ, Parani S, Thomas S, Oluwafemi OS, Wu J (2019) Introduction to nanomaterials: synthesis and applications. In: Nanomaterials for Solar Cell Applications. Elsevier, pp 75–95
21. Lee SP, Ali GAM, Algarni H, Chong KF (2019) Flake size-dependent adsorption of graphene oxide aerogel. J Mol Liq 277:175–180

22. Hegde G, Abdul Manaf SA, Kumar A, Ali GAM, Chong KF, Ngaini Z, Sharma KV (2015) Biowaste Sago Bark Based Catalyst Free Carbon Nanospheres: Waste to Wealth Approach. ACS Sustainable Chem Eng 3(9):2247–2253
23. Hochella MF, Mogk DW, Ranville J, Allen IC, Luther GW, Marr LC, McGrail BP, Murayama M, Qafoku NP, Rosso KM, Sahai N, Schroeder PA, Vikesland P, Westerhoff P, Yang Y (2019) Natural, incidental, and engineered nanomaterials and their impacts on the Earth system. Science 363(6434):eaau8299
24. Hochella MF, Mogk DW, Ranville J, Allen IC, Luther GW, Marr LC, McGrail BP, Murayama M, Qafoku NP, Rosso KM (2019) Natural, incidental, and engineered nanomaterials and their impacts on the Earth system. Science 363(6434):eaau8299
25. Damiati S, Haslam C, Awan S, Kodzius R, Sleytr U, Schuster B (2019) Natural Nanomaterials as Building Blocks for Designing Biosensors
26. Xie M, Xu M, Chen X, Li Y (2019) Recent progress of supercritical carbon dioxide in producing natural nanomaterials. Mini Rev Med Chem 19(6):465–476
27. Madkour LH (2019) Where are Nanomaterials (Nms) Found? In: *Nanoelectronic Materials*. Springer, pp 91–100
28. Radwan IMO (2019) Fate of Nanomaterials in the Environment: Effects of Particle Size, Capping agent and Surface Cleaning Products on the Stability of Silver Nanomaterials. In: Colloidal Consumer Products. University of Cincinnati
29. Ghio A, Soukup J, Madden M, Berntsen J, Hays M, Gonzalez D, Paulson S (2019) Fulvic Acid Participates in the Biological Effects of Cigarette and Wood Smoke Particles. In: A50. Molecular mechanisms of environmental and occupational lung diseases. American Thoracic Society. p A1823
30. Abd Elkodous M, El-Sayyad GS, Abdelrahman IY, El-Bastawisy HS, Mohamed AE, Mosallam FM, Nasser HA, Gobara M, Baraka A, Elsayed MA, El-Batal AI (2019) Therapeutic and diagnostic potential of nanomaterials for enhanced biomedical applications. Colloids Surf, B 180:411–428
31. Pal K, Sajjadifar S, Abd Elkodous M, Alli YA, Gomes F, Jeevanandam J, Thomas S, Sigov A (2019) Soft, Self-Assembly Liquid Crystalline Nanocomposite for Superior Switching. Electron Mater Lett 15(1):84–101
32. Thirugnanasambandan T, Pal K, Elkodous MA, Prasath H, Kulasekarapandian K, Ayeshamariam A, Jeevanandam J (2018) Aggrandize efficiency of ultra-thin silicon solar cell via topical clustering of silver nanoparticles. Nano-Structures & Nano-Objects 16:224–233
33. Abd Elkodous M, El-Sayyad GS, Nasser HA, Elshamy AA, Morsi M, Abdelrahman IY, Kodous AS, Mosallam FM, Gobara M, El-Batal AI (2019) Engineered Nanomaterials as Potential Candidates for HIV Treatment: Between Opportunities and Challenges. J Cluster Sci 30(3):531–540
34. Pal K, Elkodous MA, Mohan MLNM (2018) CdS nanowires encapsulated liquid crystal in-plane switching of LCD device. J Mater Sci: Mater Electron 29(12):10301–10310
35. Khan ST, Malik A (2019) Engineered nanomaterials for water decontamination and purification: From lab to products. J Hazard Mater 363:295–308
36. Ali GAM, Tan LL, Jose R, Yusoff MM, Chong KF (2014) Electrochemical performance studies of MnO_2 nanoflowers recovered from spent battery. Mater Res Bull 60:5–9
37. Ali GAM, Yusoff MM, Shaaban ER, Chong KF (2017) High performance MnO_2 nanoflower supercapacitor electrode by electrochemical recycling of spent batteries. Ceram Int 43:8440–8448
38. Aboelazm EAA, Ali GAM, Algarni H, Yin H, Zhong YL, Chong KF (2018) Magnetic Electrodeposition of the Hierarchical Cobalt Oxide Nanostructure from Spent Lithium-Ion Batteries: Its Application as a Supercapacitor Electrode. J Phys Chem C 122(23):12200–12206
39. Ali GAM, Yusoff MM, Algarni H, Chong KF (2018) One-step electrosynthesis of MnO_2/rGO nanocomposite and its enhanced electrochemical performance. Ceram Int 44(7):7799–7807

40. Hu Y, Yang J, Tian J, Jia L, Yu J-S (2014) Green and size-controllable synthesis of photoluminescent carbon nanoparticles from waste plastic bags. RSC Advances 4 (88):47169–47176
41. Rajarao R, Ferreira R, Sadi SHF, Khanna R, Sahajwalla V (2014) Synthesis of silicon carbide nanoparticles by using electronic waste as a carbon source. Mater Lett 120:65–68
42. Myakonkaya O, Hu Z, Nazar MF, Eastoe J (2010) Recycling functional colloids and nanoparticles. Chemistry–A Eur J 16(39):11784–11790
43. Myakonkaya O, Guibert C, Eastoe J, Grillo I (2010) Recovery of nanoparticles made easy. Langmuir 26(6):3794–3797
44. Li C, Liu C, Wang W, Mutlu Z, Bell J, Ahmed K, Ye R, Ozkan M, Ozkan CS (2017) Silicon derived from glass bottles as anode materials for lithium ion full cell batteries. Scientific reports 7(1):917
45. Sugumaran P, Susan VP, Ravichandran P, Seshadri S (2012) Production and characterization of activated carbon from banana empty fruit bunch and Delonix regia fruit pod. J Sustain Energy & Environ 3(3):125–132
46. Van Thuan T, Quynh BTP, Nguyen TD, Bach LG (2017) Response surface methodology approach for optimization of Cu^{2+}, Ni^{2+} and Pb^{2+} adsorption using KOH-activated carbon from banana peel. Surf Interfaces 6:209–217
47. Subramanian V, Luo C, Stephan AM, Nahm KS, Thomas S, Wei B (2007) Supercapacitors from Activated Carbon Derived from Banana Fibers. J Phys Chem C 111(20):7527–7531
48. Ali GAM, Abdul Manaf SA, Kumar A, Chong KF, Hegde G (2014) High performance supercapacitor using catalysis free porous carbon nanoparticles. J Phys D-Appl Phys 47 (49):495307–495313
49. Ali GAM, Manaf SAA, Chong KF, Hegde, G (2016) Superior supercapacitive performance in porous nanocarbons. J Energy Chem 25(4):734–739
50. Bello A, Manyala N, Barzegar F, Khaleed AA, Momodu DY, Dangbegnon JK (2016) Renewable pine cone biomass derived carbon materials for supercapacitor application. Rsc Advances 6(3):1800–1809
51. Ali GAM, Habeeb OA, Algarni H, Chong KF (2018) CaO impregnated highly porous honeycomb activated carbon from agriculture waste: symmetrical supercapacitor study. J Mater Sci 54:683–692
52. Habeeb OA, Ramesh K, Ali GAM, Yunus RM, Olalere OA (2017) Kinetic, Isotherm and Equilibrium Study of Adsorption of Hydrogen Sulfide From Wastewater Using Modified Eggshells. IIUM Eng J 18(1):13–25
53. Blaschek HP, Ezeji TC, Scheffran J (2010) Biofuels from agricultural wastes and byproducts: an introduction. Biofuels from agricultural wastes and byproducts. Wiley, Ames:3–10
54. Wang Z, Shen D, Wu C, Gu S (2018) State-of-the-art on the production and application of carbon nanomaterials from biomass. Green Chem 20(22):5031–5057
55. Scarlat N, Dallemand J-F, Monforti-Ferrario F, Nita V (2015) The role of biomass and bioenergy in a future bioeconomy: Policies and facts. Environ Dev 15:3–34
56. Derbyshire F, Jagtoyen M, Andrews R, Rao A, Martin-Gullon I, Grulke EA (2001) Carbon materials in environmental applications. Chemistry and physics of carbon:1–66
57. Gadipelli S, Guo ZX (2015) Graphene-based materials: Synthesis and gas sorption, storage and separation. Prog Mater Sci 69:1–60
58. Cornejo A (2013) The Thermo-catalytic Decomposition of Methane for Economical and Emission-free Hydrogen Production. University of Western Australia
59. Fernandes E, Pires R, Reis R (2017) Cork biomass biocomposites: Lightweight and sustainable materials. In: Lignocellulosic Fibre and Biomass-Based Composite Materials. Elsevier, pp 365–385
60. Hegde G, Abdul Manaf SA, Kumar A, Ali GAM, Chong KF, Ngaini Z, Sharma KV (2015) Biowaste sago bark based catalyst free carbon nanospheres: waste to wealth approach. ACS Sustain Chem & Eng 5(9):2247–2253

61. Endes C, Camarero-Espinosa S, Mueller S, Foster E, Petri-Fink A, Rothen-Rutishauser B, Weder C, Clift M (2016) A critical review of the current knowledge regarding the biological impact of nanocellulose. J Nanobiotechnology 14(1):78
62. Guo S, Dong S (2011) Graphene nanosheet: synthesis, molecular engineering, thin film, hybrids, and energy and analytical applications. Chem Soc Rev 40(5):2644–2672
63. Joshi M, Bhattacharyya A (2011) Nanotechnology—a new route to high-performance functional textiles. Textile Progress 43(3):155–233
64. Ampomah EO (2009) Efficacy of Four Botanicals and Two Chemical Fungicides In The Control of Crown Rot Disease of Banana (Musa Spp AAAA) Cv. Medium Cavendish
65. Weis AJ, Weis T (2007) The global food economy: The battle for the future of farming. Zed Books
66. Ilham IF (2018) Efektivitas Pemberian Kompos Daun Sirih Hutan (Piper aduncum L.) dan Pupuk Kandang Sapi Terhadap Pertumbuhan Bibit Pisang FHIA-17
67. Ku Hamid KH (2013) A preliminary study of banana stem juice as a plant-based coagulant for treatment of spent coolant wastewater. J Chem 2013(165057). https://doi.org/10.1155/2013/165057
68. Robinson JC, Saúco VG (2010) Bananas and plantains, vol 19. Cabi
69. Taer E, Kurniawan P, Taslim R, Agustino A, Afrianda A (2018) Carbon electrode based on durian shell: effects concentration of chemical activator agent (Potassium hydroxide).J Phys: Conf Ser. IOP Publishing
70. Noremberg BS, Silva RM, Paniz OG, Alano JH, Gonçalves MR, Wolke SI, Labidi J, Valentini A, Carreño NL (2017) From banana stem to conductive paper: A capacitive electrode and gas sensor. SensS Actuators B: Chem 240:459–467
71. Madhu R, Veeramani V, Chen S-M (2014) Heteroatom-enriched and renewable banana-stem-derived porous carbon for the electrochemical determination of nitrite in various water samples. Sci Rep 4:4679
72. Taer E, Susanti Y, Awitdrus, Sugianto, Taslim R, Setiadi R, Bahri S, Agustino, Dewi P, Kurniasih B (2018) The effect of CO2 activation temperature on the physical and electrochemical properties of activated carbon monolith from banana stem waste. In: AIP Conference Proceedings. AIP Publishing LLC
73. Ghosh S, Santhosh R, Jeniffer S, Raghavan V, Jacob G, Nanaji K, Kollu P, Jeong SK, Grace AN (2019) Natural biomass derived hard carbon and activated carbons as electrochemical supercapacitor electrodes. Sci Rep 9(1):1–15
74. Tan YH, Davis JA, Fujikawa K, Ganesh NV, Demchenko AV, Stine KJ (2012) Surface area and pore size characteristics of nanoporous gold subjected to thermal, mechanical, or surface modification studied using gas adsorption isotherms, cyclic voltammetry, thermogravimetric analysis, and scanning electron microscopy. J Mater Chem 22(14):6733–6745
75. Calvo E, Lufrano F, Staiti P, Brigandì A, Arenillas A, Menéndez J (2013) Optimizing the electrochemical performance of aqueous symmetric supercapacitors based on an activated carbon xerogel. J Power Sources 241:776–782
76. Stirling DA (2018) Nanotechnology Applications. In: The Nanotechnology Revolution, Pan Stanford, pp 281–434
77. Tibbals HF (2010) Medical nanotechnology and nanomedicine. CRC Press
78. Mohajeri M, Behnam B, Sahebkar A (2019) Biomedical applications of carbon nanomaterials: drug and gene delivery potentials. J Cell Physiol 234(1):298–319
79. Gupta H, Bhandari D, Sharma A (2009) Recent trends in oral drug delivery: a review. Recent Pat Drug Deliv Formul 3(2):162–173
80. Sahoo SK, Misra R, Parveen S (2017) Nanoparticles: a boon to drug delivery, therapeutics, diagnostics and imaging. In: Nanomedicine in cancer. Pan Stanford, pp 73–124
81. Saxena T, Rumyantsev S, Dutta P, Shur M (2014) CdS based novel photo-impedance light sensor. Semicond Sci Technol 29(2):025002
82. Nair R, Mohamed MS, Gao W, Maekawa T, Yoshida Y, Ajayan PM, Kumar DS (2012) Effect of carbon nanomaterials on the germination and growth of rice plants. J Nanosci Nanotechnol 12(3):2212–2220

83. Lin S, Reppert J, Hu Q, Hudson JS, Reid ML, Ratnikova TA, Rao AM, Luo H, Ke PC (2009) Uptake, translocation, and transmission of carbon nanomaterials in rice plants. Small 5(10):1128–1132
84. Lahiani MH, Dervishi E, Chen J, Nima Z, Gaume A, Biris AS, Khodakovskaya MV (2013) Impact of carbon nanotube exposure to seeds of valuable crops. ACS Appl Mater Interfaces 5(16):7965–7973
85. Verma SK, Das AK, Patel MK, Shah A, Kumar V, Gantait S (2018) Engineered nanomaterials for plant growth and development: A perspective analysis. Sci Total Environ 630:1413–1435
86. Servin A, Elmer W, Mukherjee A, De la Torre-Roche R, Hamdi H, White JC, Bindraban P, Dimkpa C (2015) A review of the use of engineered nanomaterials to suppress plant disease and enhance crop yield. J Nanopart Res 17(2):92
87. Prajitha N, Athira S, Mohanan P (2019) Bio-interactions and risks of engineered nanoparticles. Environ Res 172:98–108
88. Lahiani MH, Dervishi E, Ivanov I, Chen J, Khodakovskaya M (2016) Comparative study of plant responses to carbon-based nanomaterials with different morphologies. Nanotechnology 27(26):265102
89. Verma SK, Das AK, Gantait S, Kumar V, Gurel E (2019) Applications of carbon nanomaterials in the plant system: A perspective view on the pros and cons. Sci Total Environ 667:485–499
90. Guo Y, Qi J, Jiang Y, Yang S, Wang Z, Xu H (2003) Performance of electrical double layer capacitors with porous carbons derived from rice husk. Mater Chem Phys 80(3):704–709
91. Abioye AM, Ani FN (2015) Recent development in the production of activated carbon electrodes from agricultural waste biomass for supercapacitors: a review. Renew Sustain Energy Rev 52:1282–1293
92. Tan I, Ahmad A, Hameed B (2008) Optimization of preparation conditions for activated carbons from coconut husk using response surface methodology. Chem Eng J 137(3):462–470
93. Balathanigaimani M, Shim W-G, Lee M-J, Kim C, Lee J-W, Moon H (2008) Highly porous electrodes from novel corn grains-based activated carbons for electrical double layer capacitors. Electrochem Commun 10(6):868–871
94. Rufford TE, Hulicova-Jurcakova D, Khosla K, Zhu Z, Lu GQ (2010) Microstructure and electrochemical double-layer capacitance of carbon electrodes prepared by zinc chloride activation of sugar cane bagasse. J Power Sources 195(3):912–918
95. Li X, Xing W, Zhuo S, Zhou J, Li F, Qiao S-Z, Lu G-Q (2011) Preparation of capacitor's electrode from sunflower seed shell. Biores Technol 102(2):1118–1123
96. Ismanto AE, Wang S, Soetaredjo FE, Ismadji S (2010) Preparation of capacitor's electrode from cassava peel waste. Biores Technol 101(10):3534–3540
97. El Mousadik A, Petit R (1996) High level of genetic differentiation for allelic richness among populations of the argan tree [Argania spinosa (L.) Skeels] endemic to Morocco. Theor Appl Genet 92(7):832–839
98. Charrouf Z, Guillaume D (2008) Argan oil: Occurrence, composition and impact on human health. Eur J Lipid Sci Technol 110(7):632–636
99. Elmouwahidi A, Zapata-Benabithe Z, Carrasco-Marín F, Moreno-Castilla C (2012) Activated carbons from KOH-activation of argan (Argania spinosa) seed shells as supercapacitor electrodes. Biores Technol 111:185–190
100. Jain A, Balasubramanian R, Srinivasan M (2016) Hydrothermal conversion of biomass waste to activated carbon with high porosity: A review. Chem Eng J 283:789–805
101. Quintero J, Montoya M, Sánchez OJ, Giraldo O, Cardona C (2008) Fuel ethanol production from sugarcane and corn: comparative analysis for a Colombian case. Energy 33(3):385–399
102. Teixeira SR, Souza A, Peña AFV, Lima R, Miguel ÁG (2011) Use of charcoal and partially pirolysed biomaterial in fly ash to produce briquettes: sugarcane bagasse. Alternative Fuel:346

103. de Rezende Pinho A, de Almeida MBB, Mendes FL, Ximenes VL (2014) Production of lignocellulosic gasoline using fast pyrolysis of biomass and a conventional refining scheme. Pure Appl Chem 86(5):859–865
104. Zhuo C, Alves JO, Tenorio JA, Levendis YA (2012) Synthesis of carbon nanomaterials through up-cycling agricultural and municipal solid wastes. Ind Eng Chem Res 51(7):2922–2930
105. Newberry D, Uldrich J (2010) The next big thing is really small: How nanotechnology will change the future of your business. Random House
106. Sharma VP, Sharma U, Chattopadhyay M, Shukla V (2018) Advance applications of nanomaterials: a review. Mater Today: Proc 5(2):6376–6380
107. Vanderroost M, Ragaert P, Devlieghere F, De Meulenaer B (2014) Intelligent food packaging: The next generation. Trends Food Sci Technol 39(1):47–62
108. Adeleye AS, Conway JR, Garner K, Huang Y, Su Y, Keller AA (2016) Engineered nanomaterials for water treatment and remediation: Costs, benefits, and applicability. Chem Eng J 286:640–662
109. Kumar R, Singh RK, Singh DP (2016) Natural and waste hydrocarbon precursors for the synthesis of carbon based nanomaterials: graphene and CNTs. Renew Sustain Energy Rev 58:976–1006
110. Awika JM (2011) Major cereal grains production and use around the world. In: Advances in Cereal Science: Implications to Food Processing and Health Promotion. ACS Publications. Chapter 1:1–13. https://doi.org/10.1021/bk-2011-1089.ch001
111. Wang H, Wang T, Johnson LA (2017) Corn degerming ethanol fermentation processes, Google Patents
112. Bothast R, Schlicher M (2005) Biotechnological processes for conversion of corn into ethanol. Appl Microbiol Biotechnol 67(1):19–25
113. Rosentrater KA, Ileleji K, Johnston DB (2012) Manufacturing of fuel ethanol and distillers grains-current and evolving processes. Distillers grains: Production, properties, and utilization. CRC Press, Boca Raton, FL:73–102
114. Solomon BD, Barnes JR, Halvorsen KE (2007) Grain and cellulosic ethanol: History, economics, and energy policy. Biomass Bioenerg 31(6):416–425
115. McAloon A, Taylor F, Yee W, Ibsen K, Wooley R (2000) Determining the cost of producing ethanol from corn starch and lignocellulosic feedstocks. National Renewable Energy Lab, Golden, CO, USA
116. Rausch KD, Belyea RL (2006) The future of coproducts from corn processing. Appl Biochem Biotechnol 128(1):47–86
117. Gardner B (2007) Fuel ethanol subsidies and farm price support. J Agric Food Ind Organ 5 (2):1–22
118. Broer KM, Peterson C (2019) Gasification. Thermochemical Processing of Biomass: Conversion into Fuels, Chemicals and Power:85–123
119. Herbert GJ, Krishnan AU (2016) Quantifying environmental performance of biomass energy. Renew Sustain Energy Rev 59:292–308
120. Kumar M, Bhadrecha P, Pirzadah TB, Malik B, Verma A, Kumar V, Prasad R, Pachouri U, Rehman RU (2015) Power reservoirs of jumble-based biomass in Asia. In: Agricultural Biomass Based Potential Materials. Springer, pp 455–470
121. Zhang L, Xu CC, Champagne P (2010) Overview of recent advances in thermo-chemical conversion of biomass. Energy Convers Manag 51(5):969–982
122. Paisley MA (2004) Biomass gasification system and method. Google Patents
123. Claassen P, Van Lier J, Contreras AL, Van Niel E, Sijtsma L, Stams A, De Vries S, Weusthuis R (1999) Utilisation of biomass for the supply of energy carriers. Appl Microbiol Biotechnol 52(6):741–755
124. Van Zwieten L, Kimber S, Morris S, Chan K, Downie A, Rust J, Joseph S, Cowie A (2010) Effects of biochar from slow pyrolysis of papermill waste on agronomic performance and soil fertility. Plant Soil 327(1–2):235–246

125. Pandolfo A, Amini-Amoli M, Killingley J (1994) Activated carbons prepared from shells of different coconut varieties. Carbon (New York, NY) 32(5):1015–1019
126. Khalfaoui B, Meniai AH, Borja R (1995) Removal of copper from industrial wastewater by raw charcoal obtained from reeds. J Chem Technol Biotechnol 64(2):153–156
127. Balci S, Doğu T, Yücel H (1994) Characterization of activated carbon produced from almond shell and hazelnut shell. J Chem Technol & Biotechnol: Int Res Process, Environ Clean Technol 60(4):419–426
128. Smuthkochorn A, Katunyoo N, Kaewtrakulchai N, Atong D, Soongprasit K, Eiad-ua A (2019) Nanoporous carbon from Cattial leaves for carbon dioxide capture. Mater Today: Proc 17:1240–1248
129. Ayea TT, Aye SW, Win MM, Htay HH, Win PP (2019) Preparation and characterization of activated carbon from coconut shell. Prep Charact Act Carbon Coconut Shell 19(1):18
130. Hu Z, Srinivasan M (1999) Preparation of high-surface-area activated carbons from coconut shell. Microporous Mesoporous Mater 27(1):11–18
131. Tung MT, Luong VD, Trang PM, Van Tuyen L, Bich Thuy HT, Wu NL (2018) The synthesis and characterization of high purity mixed microporous/mesoporous activated carbon from rice husk. Vietnam J Chem 56(6):684–688
132. Zhang J, Zhao X, Li M, Lu H (2018) High-quality and low-cost three-dimensional graphene from graphite flakes via carbocation-induced interlayer oxygen release. Nanoscale 10 (37):17638–17646
133. Habeeb OA, Ramesh K, Ali GAM, RbM Yunus (2017) Low-cost and eco-friendly activated carbon from modified palm kernel shell for hydrogen sulfide removal from wastewater: adsorption and kinetic studies. Desalination Water Treat 84:205–214
134. Habeeb OA, Ramesh K, Ali GAM (2017) Experimental design technique on removal of hydrogen sulfide using CaO-eggshells dispersed onto palm kernel shell activated carbon: Experiment, optimization, equilibrium and kinetic studies. J Wuhan Univ Technol Mater Sci Ed 32(2):305–320
135. Ali GAM, Manaf SABA, Kumar A, Chong KF, Hegde G (2014) High performance supercapacitor using catalysis free porous carbon nanoparticles. J Phys D Appl Phys 47 (49):495307
136. Habeeb OA, Kanthasamy R, Ali GAM, bin Mohd R (2017) Optimization of Activated Carbon Synthesis Using Response Surface Methodology to Enhance H_2S Removal From Refinery Wastewater. J Chem Eng Ind Biotechnol V1(1):17
137. Habeeb OA, Ramesh K, Ali GAM, Yunus RBM (2017) Isothermal modelling based experimental study of dissolved hydrogen sulfide adsorption from waste water using eggshell based activated carbon. Malays J Anal Sci 21(2):334–345
138. Habeeb OA, Ali GAM (2017) Application of Response Surface Methodology for Optimization of Palm Kernel Shell Activated Carbon Preparation Factors for Removal of H2S from Industrial Wastewater. J Teknol (Sci Eng) 79(7):1–10
139. Nasir S, Hussein MZ, Zainal Z, Yusof NA (2018) Carbon-based nanomaterials/allotropes: A glimpse of their synthesis, properties and some applications. Materials 11(2):295
140. Maji IK, Saari MY, Habibullah MS, Utit C (2017) Measuring the economic impacts of recent oil price shocks on oil-dependent economy: evidence from Malaysia. Policy Studies 38(4):375–391
141. Sivasankari R, Kumaran P, Normanbhay S, Shamsuddin AH. Preliminary experimental results of Sewage Sludge (SS) Co-digestion with Palm Oil Mill Effluent (POME) for Enhanced Biogas Production in Laboratory Scale Anaerobic Digester. In: IOP Conference Series: Earth and Environmental Science. IOP Publishing
142. Bansal R, Donnet J, Stoeckli F (1988) Active Carbon. Marcel Dekker, New York
143. Shaikh F, Jadakar S, Kamat R, Pathan H (2018) Activated carbon/transition metal oxides thin films for supercapacitors. Mat Res Found 26:1–30. https://doi.org/10.21741/9781945291579-3
144. Abnisa F, Daud WMAW (2014) A review on co-pyrolysis of biomass: an optional technique to obtain a high-grade pyrolysis oil. Energy Convers Manag 87:71–85

145. Kwiatkowski M, Broniek E (2017) An analysis of the porous structure of activated carbons obtained from hazelnut shells by various physical and chemical methods of activation. Colloids Surf, A 529:443–453
146. Ioannidou O, Zabaniotou A (2007) Agricultural residues as precursors for activated carbon production—a review. Renew Sustain Energy Rev 11(9):1966–2005
147. Hussein M, Tarmizi RH, Zainal Z, Ibrahim R, Badri M (1996) Preparation and characterization of active carbons from oil palm shells. Carbon (New York, NY) 34 (11):1447–1449
148. Hidayu A, Muda N (2016) Preparation and characterization of impregnated activated carbon from palm kernel shell and coconut shell for CO_2 capture. Procedia Engineering 148:106–113
149. Dungani R, Aditiawati P, Aprilia S, Yuniarti K, Karliati T, Suwandhi I, Sumardi I (2018) Biomaterial from oil palm waste: properties, characterization and applications. Palm Oil 31. https://doi.org/10.5772/intechopen.76412
150. Zhou K, Zhou X, Liu J, Huang Z (2020) Application of magnetic nanoparticles in petroleum industry: a review. J Pet Sci Eng 106943
151. Jain A, Ranjan S, Dasgupta N, Ramalingam CJ (2018) Nanomaterials in food and agriculture: an overview on their safety concerns and regulatory issues. Crit Rev Food Sci Nutr 58(2):297–317
152. Amin YM, Hawas AM, El-Batal A, Elsayed SHHE (2015) Evaluation of acute and subchronic toxicity of silver nanoparticles in normal and irradiated animals. Br J Pharmacol Toxicol 6(2):22–38
153. Oberdörster G, Oberdörster E, Oberdörster JJ (2005) Nanotoxicology: an emerging discipline evolving from studies of ultrafine particles. Environ Health Perspect 113 (7):823–839
154. Jiang X, Wu Y, Gray P, Zheng J, Cao G, Zhang H, Zhang X, Boudreau M, Croley TR, Chen CJN (2018) Influence of gastrointestinal environment on free radical generation of silver nanoparticles and implications for their cytotoxicity. NanoImpact 10:144–152
155. Aillon KL, Xie Y, El-Gendy N, Berkland CJ, Forrest ML (2009) Effects of nanomaterial physicochemical properties on in vivo toxicity. Adv Drug Deliv Rev 61(6):457–466
156. Choi AO, Cho SJ, Desbarats J, Lovrić J, Maysinger D (2007) Quantum dot-induced cell death involves Fas upregulation and lipid peroxidation in human neuroblastoma cells. J Nanobiotechnology 5(1):1
157. Amiri M, Eskandari K, Salavati-Niasari M (2019) Magnetically retrievable ferrite nanoparticles in the catalysis application. Adv Colloid Interface Sci
158. Gupta AK, Gupta M (2005) Synthesis and surface engineering of iron oxide nanoparticles for biomedical applications. Biomater 26(18):3995–4021
159. Mornet S, Vasseur S, Grasset F, Veverka P, Goglio G, Demourgues A, Portier J, Pollert E, Duguet E (2006) Magnetic nanoparticle design for medical applications. Prog Solid State Chem 34(2–4):237–247
160. Hyeon T (2003) Chemical synthesis of magnetic nanoparticles. Chem Commun 8:927–934
161. Lu AH, EeL Salabas, Schüth F (2007) Magnetic nanoparticles: synthesis, protection, functionalization, and application. Angew Chem Int Ed 46(8):1222–1244
162. Liu J, Zhao Z, Feng H, Cui F (2012) One-pot synthesis of Ag–Fe 3 O 4 nanocomposites in the absence of additional reductant and its potent antibacterial properties. J Mater Chem 22 (28):13891–13894
163. Yang B, Wei Y, Liu Q, Luo Y, Qiu S, Shi Z (2019) Polyvinylpyrrolidone functionalized magnetic graphene-based composites for highly efficient removal of lead from wastewater. Colloids Surf, A 582:123927
164. Reddy DHK and Yun Y-S (2016) Spinel ferrite magnetic adsorbents: alternative future materials for water purification? Coord Chem Rev 315:90–111
165. Tang SC, Lo IM (2013) Magnetic nanoparticles: essential factors for sustainable environmental applications. Water Res 47(8):2613–2632

166. Sadegh H, Ali GAM, Makhlouf ASH, Chong KF, Alharbi NS, Agarwal S, Gupta VK (2018) MWCNTs-Fe_3O_4 nanocomposite for Hg (II) high adsorption efficiency. J Mol Liq 258:345–353
167. Fiyadh SS, AlSaadi MA, Jaafar WZB, AlOmar MK, Fayaed SS, Mohd NSB, Hin LS, El-Shafie A (2019) Review on heavy metal adsorption processes by carbon nanotubes. Journal of Cleaner Production
168. Verma B, Balomajumder C (2019) Surface modification of one-dimensional Carbon Nanotubes: A review for the management of heavy metals in wastewater. Environmental Technology & Innovation:100596
169. Zhang B, Huang K, Wang Q, Li G, Wu T, Li Y (2020) Highly efficient treatment of oily wastewater using magnetic carbon nanotubes/layered double hydroxides composites. Colloids Surf, A 586:124187
170. Mohammadi AA, Dehghani MH, Mesdaghinia A, Yaghmaian K, Es' haghi Z (2019) Adsorptive removal of endocrine disrupting compounds from aqueous solutions using magnetic multi-wall carbon nanotubes modified with chitosan biopolymer based on response surface methodology: Functionalization, kinetics, and isotherms studies. Int J Biol Macromol 155:1019–1029
171. Ihsanullah (2019) Carbon nanotube membranes for water purification: Developments, challenges, and prospects for the future. Sep Purif Technol 209:307–337
172. Lee J, Jeong S, Liu Z (2016) Progress and challenges of carbon nanotube membrane in water treatment. CritAl Rev Environ Sci Technol 46(11–12):999–1046
173. Rizzuto C, Pugliese G, Bahattab MA, Aljlil SA, Drioli E, Tocci E (2018) Multiwalled carbon nanotube membranes for water purification. Sep Purif Technol 193:378–385
174. Li S, Liao G, Liu Z, Pan Y, Wu Q, Weng Y, Zhang X, Yang Z, Tsui OK (2014) Enhanced water flux in vertically aligned carbon nanotube arrays and polyethersulfone composite membranes. J Mater Chem A 2(31):12171–12176

Heteroatoms Doped Porous Carbon Nanostructures Recovered from Agriculture Waste for Energy Conversion and Storage

Diab Khalafallah, Mingjia Zhi, and Zhanglian Hong

Abstract Biomass-derived porous carbons (BPCs) represents one of the most diverse classes of materials with exceptional properties such as high specific surface area, wide availability, biodegradability, low cost, and tunable porous features. A broad range of new carbon materials for suitable applications including water purification, catalyst supports and electrodes for electrochemical capacitors, sensing, and fuel cells have been developed. This not only increased the economic benefits and sustainability of chemical industry but also minimized the environmental impacts. The wide application of various energy technologies for specific purposes is mainly reliant on the design of electrode materials, particularly carbon electrodes. In this chapter, recent developments and breakthroughs of BPCs are presented. Characteristics controlling mechanisms behind their performance, especially pore structure and surface functionality, are discussed, which will direct the rational design of BPCs for practical use. In addition, the progress on application of these materials as electrodes for electrochemical devices such as fuel cells, CO_2 capture, water splitting, and lithium-ion batteries, is summarized.

Keywords BPCs · Heteroatoms doping · Porous structured materials · Energy conversion and storage

D. Khalafallah (✉) · M. Zhi (✉) · Z. Hong (✉)
State Key Laboratory of Silicon Material, School of Materials Science and Engineering, Zhejiang University, 38 Zheda Road, Hangzhou 310027, China
e-mail: diab@zju.edu.cn

M. Zhi
e-mail: mingjia_zhi@zju.edu.cn

Z. Hong
e-mail: hong_zhanglian@zju.edu.cn

D. Khalafallah
Mechanical Design and Materials Department, Faculty of Energy Engineering, Aswan University, P.O. Box 81521, Aswan, Egypt

A. S. H. Makhlouf and G. A. M. Ali (eds.), *Waste Recycling Technologies for Nanomaterials Manufacturing*, Topics in Mining, Metallurgy and Materials Engineering, https://doi.org/10.1007/978-3-030-68031-2_17

List of Abbreviations

BPCs	Biomass-derived porous carbons
CV	Cyclic voltammetry
DFT	Density functional theory
ESCs	Electrochemical supercapacitors
HER	Hydrogen evolution reaction
HTC	Hydrothermal carbonization
LiBs	lithium-ion batteries
LSV	Linear sweep voltammetry
OER	Oxygen evolution reaction
ORR	Oxygen reduction reaction
SEM	Scanning electron microscopy
TEM	Transmission electron microscopy

1 Introduction

Challenges aroused from the growing environmental pollution guided the employment of agricultural biomass as sustainable carbon-rich precursors for constructing carbonaceous materials [1–3]. BPCs collected from fungi, corn grain, lignocellulosic materials, rice waste, banana, fish scales, starch, and celtuce leaves represent one of the most important groups of materials with distinct properties such as high specific surface area, high availability, biodegradability, low cost, and tunable porous features [4]. Owing to their unique characteristics, BPCs have received an extensive attention in energy and environmental fields, and hence, exploring highly efficient carbonaceous materials from renewable biomass resources is a critical research issue and features the concepts of green chemistry. During recent years, many advances have been made in carbon technology to improve the existing techniques as well as adopt new processing techniques [5–8].

Several synthetic techniques and post-modification methods were reported for the access of macro/meso/microporous carbons from biomass precursors for efficient electrochemical applications. Accordingly, the electrical and physicochemical properties of BPCs are highly reliant on the synthetic strategies, which enhancing their performance for energy technologies [9–12]. Prior to the exploration of renewable sources, large amounts of carbon materials have been fabricated from classical fossil fuels under relatively severe conditions (i.e., chemical vapor deposition and direct pyrolysis of organic precursors), but the accessibility of those adopted procedures was limited [13]. As a simple straightforward and eco-friendly approach, hydrothermal carbonization (HTC) of biomass resources carried out at mild temperatures below 300 °C in the presence of water has displayed a high potential to tailor functional carbon materials with modifiable final products. Besides HTC, other approaches, such as ionothermal carbonization and molten salt carbonization, were introduced as effective complementary tactics [14–16]. Introducing organic functionalities or heteroatom such as sulfur (S), phosphorus (P), nitrogen (N), boron (B), and oxygen (O) onto porous carbon surfaces can be considered as an effective tool to promote the carbon properties and provide attractive features for economic use in energy conversion and storage technologies (i.e., supercapacitors, fuel cell, batteries, CO_2 capture, etc.) [17–27]. Since the carbon itself is a conducting material while heteroatom dopants provide a redox-like behavior, leading to great enhancements in overall kinetics and charge storage capability of electrodes [28–35]. Post-synthetic reactions are typically implemented to immobilize metal functionalities onto the surface of mesoporous complexes. Additionally, grafting different moieties on the surface during condensation with surface functional groups has been widely used in recent years. Integrating the carbon production and biomass waste disposal into a biomass refinery would surprisingly diminish the cost of carbon, which is one of the major handicaps to the widespread investigation of biomass [36–39]. The top-down routes are incapable of optimizing the material properties including engineered architecture, surface chemistry, and porosity for specified applications. Hence, developing tunable methodologies are desirable to control the surface functionality of interfacial properties synergistically.

In response to sustainability issues, BPCs can play an interesting role in the future applications and implementations due to the ingenuity of integrated devices for advanced uses. From a fundamental and comprehensive studies viewpoint, BPCs are very interesting for electrochemical energy generation. Numerous promising works based on porous carbon composites were reported recently for rechargeable batteries and supercapacitors [40–51]. Particularly, conductive carbon supports/substrates can offer many nucleation sites for the growth and crystallization of catalytic species, enabling a full utilization of active materials, reinforced electric conductivity, and as well as an extensive electroactive area [52–55]. Most essentially, the chemically stable carbon matrices alleviate the induced stress during chemical reactions, guaranteeing a prolonged cycle life of electrode [56]. The interactions between the carbon surface functional groups and catalytically active guest atoms favor their intimate integration, implying a unique structural integrity and a low interfacial resistance [57–64].

In this chapter, we substantiate the sustainable aspects of various BPCs-based materials, the nature of their solid sources, and synthetic approaches. After a general discussion of strengths toward practical applications, a critical overview of the recent advancement of functionalizing the hierarchically structured compounds is presented.

2 Synthetic Strategies of Carbon from Biomass Precursors

Biomass, which often denotes plants or plant-based complexes, can be grown through CO_2-assisted biological photosynthesis approach under sunlight. The current statistics indicate that the worldwide biomass production per year is close to 104.9 petagrams; however, an enormous amount of waste biomass (i.e., agricultural residues and forest by-products) have been directly burned, leading to a serious air pollution and weak energy utilization efficiency [65, 66]. Naturally formed species are used as precursors for a broad range of porous materials with favourable functionalities of living configurations bank on their hierarchical architectures. The conventional synthetic methodology is based on direct pyrolysis and chemical activation process. The thermochemical conversion methodologies are necessary to prepare carbonaceous materials and eject other unwanted elements. The properties of resultant carbons strongly correlate with the employed synthesis strategies.

2.1 Hydrothermal Carbonization

Similar to traditional coal formation, HTC is a viable strategy to convert sustainable alternatives to functional carbon materials in an aqueous solution under relatively low temperatures [14]. The biomass precursors can be either raw plant materials or segregated, carbohydrates. Furthermore, copious agricultural/fruits wastes and forest by-products can be employed as raw products. The low-temperature HTC method can be used for functionalizing carbonaceous materials through simple chemistries and polymerization reactions [67–70]. The yield, surface structure, and induced surface functional groups of BPCs hinge on the carbonizing temperature, reaction time and media. The raw materials can decompose into minimal monomers, followed by polymerization and aromatization reactions. Then, a high-temperature activation procedure is necessary to tune further the surface chemistry, porosity, and graphitization degree of hydrothermally carbonized products [71, 72]. HTC process is advantageous concerning other strategies in terms of simple technique and instrumentations, minimized toxicological effect, high atom economy, and low energy consumed [73]. This approach offers a substantial possibility to introduce additional materials and functionalities into carbon frameworks during hydrothermal process through the availability of water-soluble functional species. Remarkably, the content of incorporated heteroatoms into carbon structure during HTC remains unchanged even after high-temperature pyrolysis because the heteroatoms are covalently doped and maintained as a part of the materials. Furthermore, an appropriate graphitization degree can be attained within the as-prepared carbons.

2.2 Pyrolysis Method

The pyrolysis has been attempted as a reliable method to prepare different carbonaceous materials with the biomass as the main source. Pyrolysis of biomass is regarded as a thermochemical reaction, in which the raw materials are mostly transformed into gases, liquid oil and finally biochar with a full oxygen denial or a very slight oxygen stream [74, 75]. Pyrolysis methods are divided into chemical pyrolysis, self-activation, and gas-assisted pyrolysis. Gas-assisted pyrolysis is generally carried out between 600 and 1200 °C in the presence of O_2, steam, or CO_2. Air activation is conducted at low temperatures below 500 °C and generates carbon with relatively low surface areas [76–78]. The inherent morphologies of some precursors can be maintained during air activation, and oxygen functional groups are created onto the carbon surface. Porous carbon can be fabricated from rice husk biomass by a simple pyrolysis process. The rice husk was treated at 700 ° C under N_2 atmosphere and mixed with NaOH granules at 900 °C [79]. Besides, porous carbons were prepared by pyrolysis methods from plant leaves, grass, and gynostemia. In addition to plants, waste can be employed as precursors to obtain porous carbons by pyrolysis, including litchi exocarp, waste frying oil, peanut shell, watermelon peel, and peanut skin [80]. Different parameters are influencing the biomass pyrolysis, which can be addressed as follow: One related to properties of raw materials such as biomass particle size, types of biomass, and their characteristics. The other is inextricably associated with the reaction conditions (i.e., pyrolysis temperature and time) [81–83]. However, constructing BPCs with an excellent porosity and graphitic structure is not an easy task owing to their complex composition and strong chemical bonding. Pyrolysis temperature showed a considerable effect on the porous nature and graphitic structure of biomass. Increasing the reaction temperature with a certain range can significantly affect both surface area and micropore numbers, which is favourable for attaining an excellent porous material. The graphitic structure of biomass can gradually be improved with the increment of pyrolysis temperature, resulting in a more orderly arrangement and an enhanced graphitization degree [84–88]. Thus, the formation of hierarchically graphitic carbon with a macro/meso/microporous structure is promoted with elevating pyrolysis temperature. As the graphitization features of product improved, pyrrolic N can be converted into pyridinic N and then converted into graphitic N in the final material [89, 90]. According to reports, the formation procedure of BPCs with a well-developed pore structure and graphitic layers should be precisely addressed to direct further expansion of synthesis by pyrolysis in the future.

2.3 Microwave Method

Consideration should be taken into time and energy-saving issues. Compared to a traditional heating system where energy is supplied by conductive or convective devices, microwave irradiation is an innovative conceptual approach in the field of chemical synthesis due to its fast heating property, shortened time, and energy saving. The microwave-assisted hydrothermal method was reported for the synthesis of carbon spheres through tuning the glucose-based solution at a very short time compared to the classical hydrothermal carbonization process (6–24 h) [91]. Besides, there have been several reports on the employment of microwave irradiation to prepare porous carbon materials [1–3]. The final product structure and characteristic depend on the starting raw materials and can be tuned under certain reaction conditions. A variety of biomass resources such as pine needles, pinecone hulls, and cornstalks were applied as carbon precursors. Loose textures of biomass compounds usually produce porous carbons with an improved surface area compared to compact-textured resources. Therefore, there is no doubt that microwave strategy displays a promising potential for synthesizing porous carbon complexes.

2.4 *Template-Directed Synthesis*

Activation methodologies derived carbon materials show a narrow pore size distribution in the supermicropore range (between 0.7 and 2 nm), which is not sufficient for the rapid diffusion of electrolyte ions [92]. To this end, the template method is a robust strategy to optimize the porosity features of BPCs for specific uses. The template simply acts like a strong skeleton around which other carbonaceous plateforms are created. It guides the formation of pores and enhances the structural ordering level. In the case of hard templating or nanocasting tactics, the resulting carbon materials reversely repeat the original surface engineering of templates. This method accomplishes via a multi-step process: (i) assembling of templates with a well-defined porosity, (ii) depositing the carbon source into the as-prepared template, (iii) carbonizing and linking the carbon raw materials to form a strong organic–inorganic complex, and (iv) elimination of templates to construct a porous framework. The formed carbonaceous material into the pores of host template turns to a stable and uniform carbon matrix; meanwhile, the vacancies occupied by the template are converted into open pores. Several compounds have been fabricated, such as carbon nanosheets, porous carbon monoliths, peanut-like carbon, and ordered mesoporous carbon [2]. The hard template can be applied to create numerous mesopores in the carbon matrix. In this context, nanocrystalline cellulose with a chiral nematic-like structure, long-range orientation, and high specific surface area could be prepared through silica films-assisted hydrolysis reaction (Fig. 1) [93]. In addition, glucose-derived carbon was obtained by the dual templating method, in which the combined role of silica and ice templating with physical activation induced the formation of macro/meso/micropores within the resulting product [94].

Essentially, hard-templating derived structured porous carbons reveal significant merits; however, there are some inherent drawbacks associated with the tendency of pore structure to collapse. Also, the removal of host template entails a

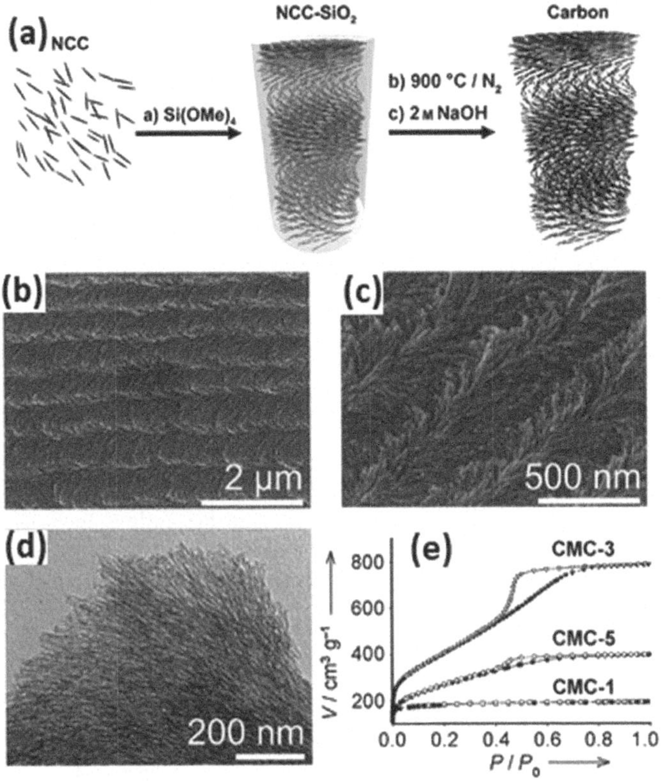

Fig. 1 **a** Representative diagram for the formation of chiral nematic mesoporous carbon. **b** and **c** SEM images and **d** TEM image of the CMC-3 sample. **e** N_2 adsorption–desorption loops of the as-fabricated product. Reproduced with permission from Ref. [11], Copyright 2017, Royal Society of Chemistry

resource-consuming process, a long time, and requires toxic reagents, which arousing arguments related to environmental impact. Soft-templating can be a convenient and green alternative route to the hard template method. The collapse of the pore structure can be reduced because the molecular species stabilize and protect the pore's carbonization treatment at elevated temperatures. The self-assembly of organic–organic complexes via a soft-template route demonstrates a good potential as an effective pathway to synthesize carbons materials with controlled pore structures. Block copolymers, particularly, Pluronic® F127, are the typical growth-directing surfactants in soft-template. Under optimized concentration of F127, micelles would be generated in the reaction solution, and their hydrophobic cores assembled by the PPO chains of F127 represent the source of open pores. Carbon materials derived from highly branched lignin with well-defined mesopores were successfully obtained with the help of F127 [95, 96].

However, only a few works reported the formation of porous carbon by soft-templating protocols; with the emergence of new and advanced methodologies, it is currently in the developing stage. One of the major disadvantages is the intrinsic instability of surfactant species at elevated temperatures and thereby, carbonization reaction with soft templates is hindered at higher temperatures. This might be attributed to the weak interactions between these templates and the used carbon precursors. In sharp contrast, hard templates afford a strong interfacial interaction during grafting reaction. In recent years, melamine sulfate was introduced to reinforce the thermal stability of soft templates. The complexes could maintain their ordered pore structure even after carbonizing at 900 °C [97]. Thus, selecting the proper soft template is of great importance due to different hydrogen-bonding capabilities of block copolymers-carbohydrates complexes. To clarify this matter, sugarcane bagasse was adopted as a robust scaffold for synthesizing hierarchically porous carbon products with a meso/microporous nature [98]. To achieve this goal, F127 reagent was used as a structure-directing salt and phenol-formaldehyde as a carbon starting material. Compared to the absent bulk structure of underlying scaffolds, a monolithic porous carbon material could be grown. It was believed that the sugarcane bagasse surface hydroxyl groups interacted with the formaldehyde resins could further strengthen the thermal stability of monolithic product.

2.5 *Ionothermal Carbonization*

Ionothermal approach refers to the reaction performed in ionic liquids (ILs). ILs that mainly developed for battery applications as molten electrolytes below 100 °C have attracted an interesting attention as benign solvents in the synthesis of BPCs (Fig. 2) [99]. Beyond their originally designed conception, these solvents play an important role in various fields such as chemical synthesis, CO_2 capture, and catalysis due to their excellent thermal stability, suppressed solvent volatility, and reasonable solubility of biomass. As an analog of hydrothermal synthesis, ionothermal process is widely investigated for the fabrication of metal nanoparticles (i.e., Ru, Pd, Ir, etc.), organic-inorganic compounds (i.e., metal-organic frameworks), and inorganic materials (i.e., zeolite, TiO_2, SiO_2, etc.) with unique structures and properties [100, 101]. Such a self-templated technique did not require additional substrates or additives, and the porosity can be optimized by their anions size. But the widespread practical utilization is obstructed mainly by the high cost of those ionic liquids.

The pyrolysis of ILs with cross-linkable anions or cations represents a straightforward strategy to produce N-rich porous carbons. Recent works evidenced the activity of ILs in solubilizing biomass, for example, commercial 1-butyl-3-methylimidazolium chloride ([Bmim]Cl) can allow the dissolution of more cellulose because of its ability to generate hydrogen bonds with the solutes to facilitate the solubility of carbohydrates and their underlying acidity function can accelerate the breakage of the existed hydrogen-bonding network. It is appropriate

Fig. 2 A typical illustration of ionothermal synthesis of BPCs. Reproduced with permission from Ref. [1], Copyright 2016, Royal Society of Chemistry

to incorporate a molten component as an in situ catalyst into an ionothermal reaction to meet specific requirements [100–102]. In this regard, iron (Fe)-containing ILs-based solvent was used to convert carbohydrate precursors into carbon materials, and the porosity of product was stabilized by Fe species since the [Bmim][Cl] alone can provide only nanopores. In addition, the catalytic effect of Fe-assisted ILs offers a more obvious aromatic region. The ionothermal treatment of biomass can enhance the chemical and physical interactions between solvent and reagents, leading to the formation of carbon materials [99]. The chemical reactions involved in ionothermal synthesis are very complicated. Different carbon resources result in specific intermediates and thereby control the final product structure and morphology. It is accepted that the reaction mechanism of ionothermal process obeys a similar pathway of dehydration/polymerization, polycondensation based on the hydroxymethylfurfural and some organic acids as well as furfural [2]. The acids would decrease the pH and stimulate the in situ dehydration of carbohydrates. The products obtained from polymerization–polycondensation stages of intermediates possess an apparent aromatic character. More interesting, the N doped property of ionothermal synthesis is an additional grant for constructing functional carbon materials in comparison with hydrothermal-derived carbon materials. In which water molecules generate a high pressure at elevated temperatures, the ionothermal process is more comfortable to handle even with uncapped vessels. Ionothermal-derived carbons exhibit slightly higher yields with larger surface areas. Although ILs can be possibly recovered after the reaction and used without further purifying, their high price stills a significant challenge toward their wide utilization in industry. Therefore, ionothermal treatment presents a unique complementary protocol for the rational design of functional carbons using biomass raw materials.

Motivated by their structure-directing feature and unique capability on dissolving biomass, we can ask if ILs can be proposed as solvents for transforming renewable biomass into useful carbons via ionothermal strategy.

3 Activation Processes

Direct pyrolysis of biomass precursors often results in poorer porous structures and lower specific surface areas, which are essential key parameters for electrochemical utilization. Activation is considered as an effective method to adjust the porous structure and increase the surface area of carbon materials. Based on activation mechanism, the activation process can be classified into two main categories: chemical activation and thermal/physical activation (Table 1).

3.1 Chemical Activation

The chemical activation of carbons is usually obtained after direct heating of precursors at a range from 450 to 800 °C in presence of chemical reagents such as potassium hydroxide (KOH), zinc chloride ($ZnCl_2$), sodium hydroxide (NaOH), boric acid (H_3BO_3), phosphoric acid (H_3PO_4), and potassium bicarbonate ($KHCO_3$). Several types of porous structures within the carbon skeleton can be established with these chemical activating agents (Table 1). The chemical activation mechanism is complex and unconfirmed, but several reaction mechanisms were proposed according to reaction parameters, chemical activating reagent, and carbon resource. As a well-known type of chemical activation salt, the use of KOH species usually associated with the formation of K_2CO_3, K_2O, H_2O, CO, and CO_2 as reaction products, and the resulting carbon matrices were etched into abundant meso/micropores with a high specific surface area [115, 116]. The *in situ* produced water vapor in the process can facilitate the carbon gasification and improve the porosity, while the intermediate metallic component (i.e., K) might intercalate into the carbonaceous frameworks and extend the lattices, forming a highly porous architecture. In brief, KOH dehydrates into K_2O at high temperature, and H_2 releases as a result of interaction between carbon scaffold and water molecules (reactions **1** and 2) [117]. The evolution of CO and CO_2 gasses occurs between 500 and 650 °C (reactions 3 and 4). The reaction of CO_2 and K_2O produces K_2CO_3, which is further reduced to K or decomposed into K_2O by carbon (reactions 5–8) [117, 118].

Table 1 Related structural parameters of BPCs produced from a different plant-, fruit-, microorganism-, and animal-based precursors

Biomass resource	Activation method	Final product	Specific surface area (m^2g^{-1})	Pore volume (m^3g^{-1})	Surface morphology	References
Bamboo char	K_2FeO_4 at 800 °C for 2 h	Porous graphitic biomass carbon	1732	0.97	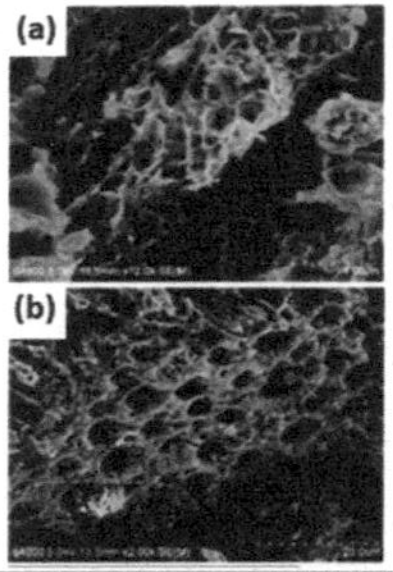	[103]
Bacillus subtilis	KOH at 800 °C for 2 h	Heteroatom doped carbon	1578	1.092	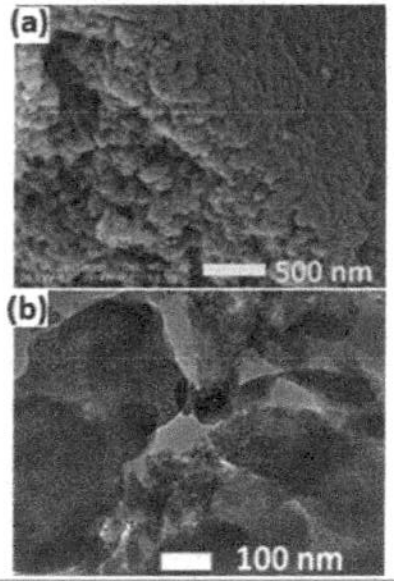	[104]

(continued)

Table 1 (continued)

Biomass resource	Activation method	Final product	Specific surface area (m^2g^{-1})	Pore volume (m^3g^{-1})	Surface morphology	References
Chicken eggshell membrane	Atmospheric air at 300 °C for 2 h	Carbonized eggshell membrane	221.2	0.13	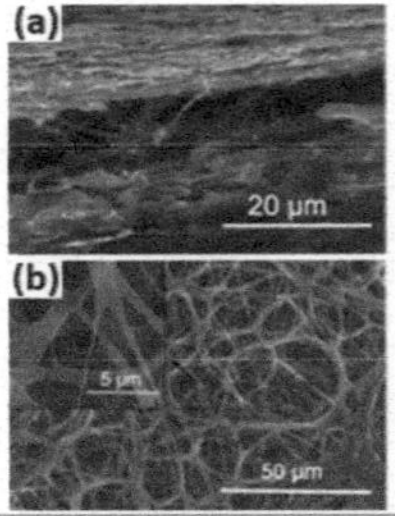	[105]
Auricularia	KOH-assisted one-pot carbonization and activation	Porous graphene-like carbon	1103	0.54	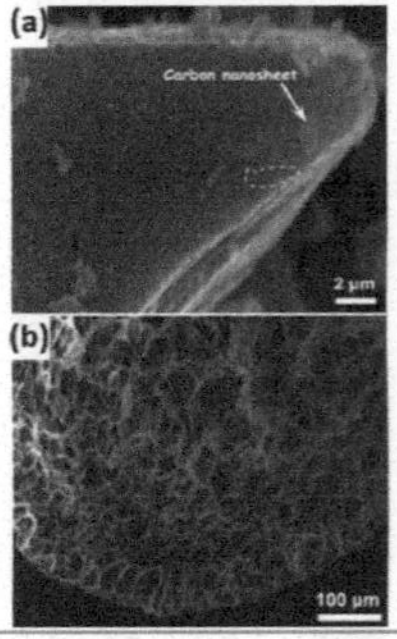	[42]

(continued)

Table 1 (continued)

Biomass resource	Activation method	Final product	Specific surface area (m^2g^{-1})	Pore volume (m^3g^{-1})	Surface morphology	References
Dry elm samara	KOH at 700 °C for 2 h	Porous carbon nanosheets	1947	1.33	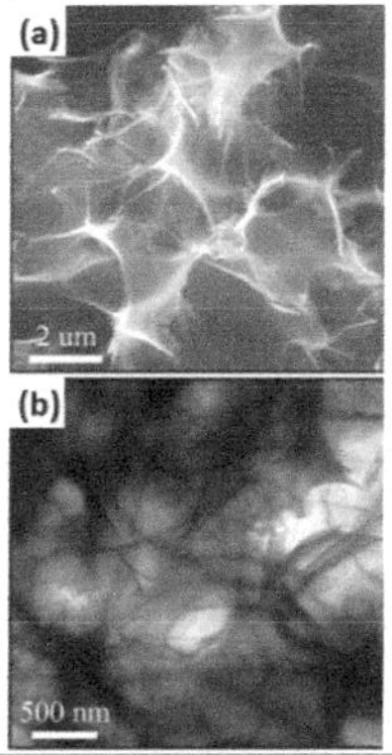	[106]
Human hair	KOH at 800 °C for 2 h	Heteroatom doped porous carbon flakes	1306	0.9	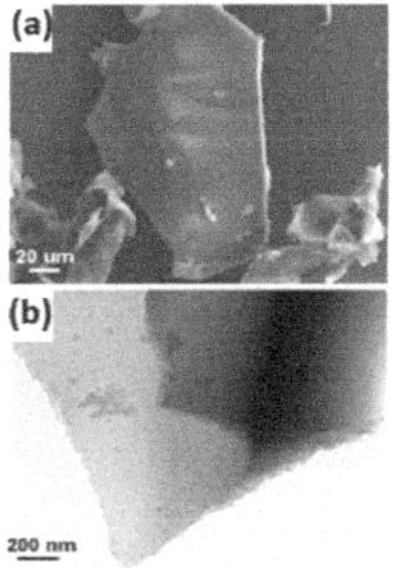	[40]

(continued)

Table 1 (continued)

Biomass resource	Activation method	Final product	Specific surface area (m^2g^{-1})	Pore volume (m^3g^{-1})	Surface morphology	References
Oil tea shell	$ZnCl_2$-assisted one-pot carbonization and activation	Activated carbon	2851	2.68	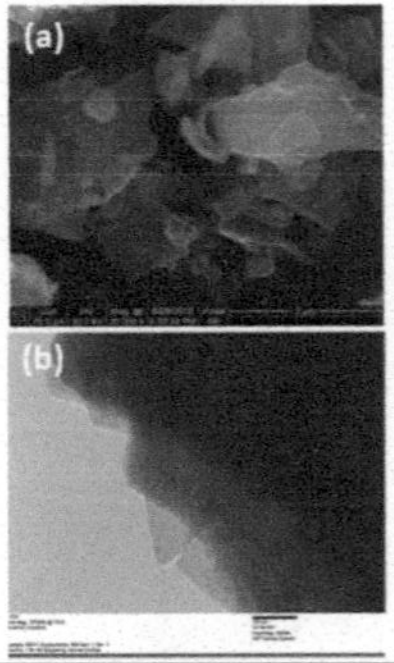	[107]
Fig fruit (inner part)	KOH-assisted one-pot carbonization and activation 900 °C for 2 h	Highly porous foam-like carbon	2337	1.005	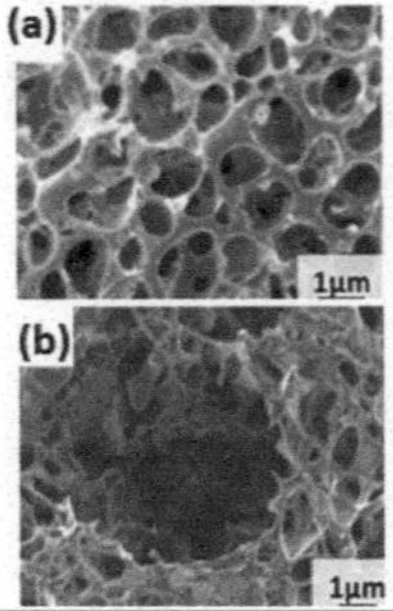	[108]

(continued)

Table 1 (continued)

Biomass resource	Activation method	Final product	Specific surface area (m^2g^{-1})	Pore volume (m^3g^{-1})	Surface morphology	References
Orange peel	KOH at 600 °C for 1 h	3D nanoporous carbon	2160	0.779	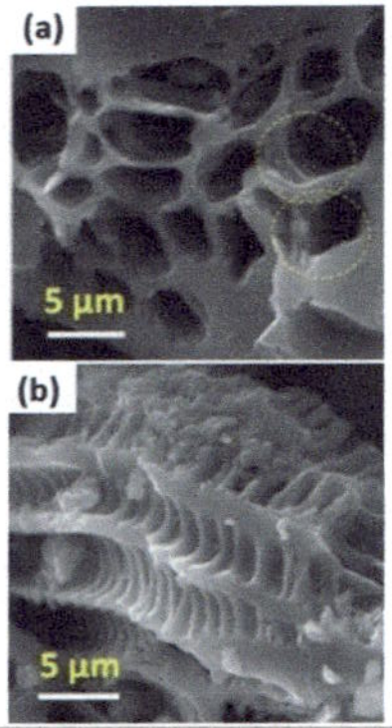	[109]
Shiitake mushroom	KOH at 800 °C for 2 h	Hierarchically porous activated carbon	2988	1.76	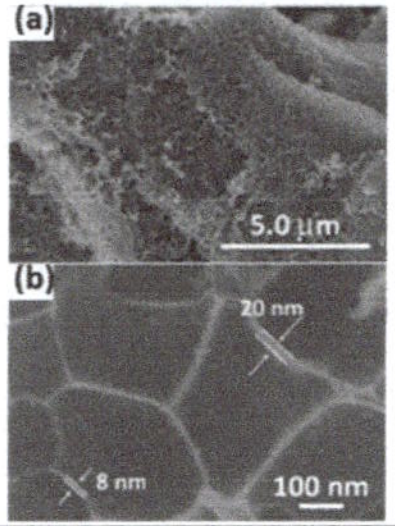	[110]

(continued)

Table 1 (continued)

Biomass resource	Activation method	Final product	Specific surface area (m^2g^{-1})	Pore volume (m^3g^{-1})	Surface morphology	References
Kombucha	KOH at 700 °C for 1 h	Hierarchically porous carbon	917	0.41	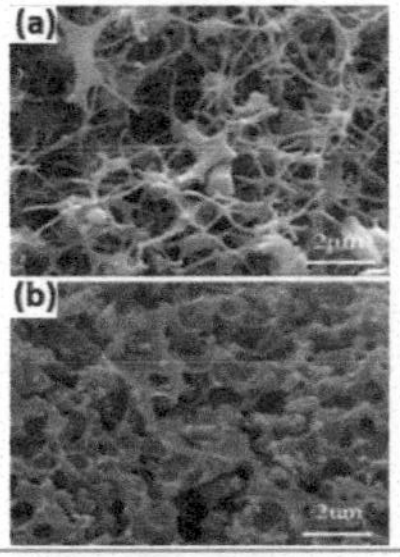	[111]
Orange peel	KOH at 800 °C for 2 h	Interconnected hollow-structured carbon	1391	0.72	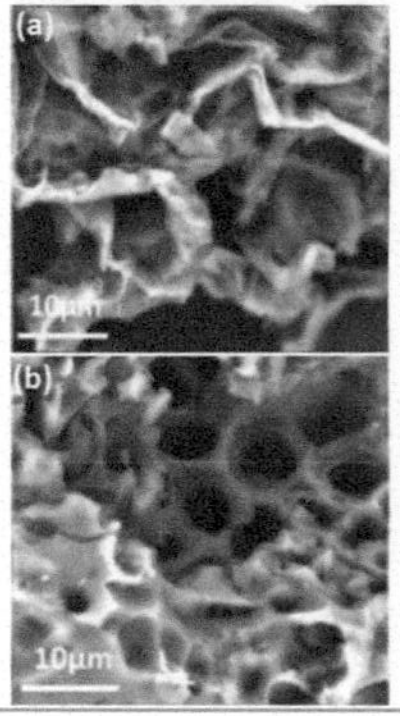	[41]

(continued)

Table 1 (continued)

Biomass resource	Activation method	Final product	Specific surface area (m^2g^{-1})	Pore volume (m^3g^{-1})	Surface morphology	References
Chicken egg white	KOH at 900 °C for 3 h	Egg white-derived activated carbon	3250	1.97	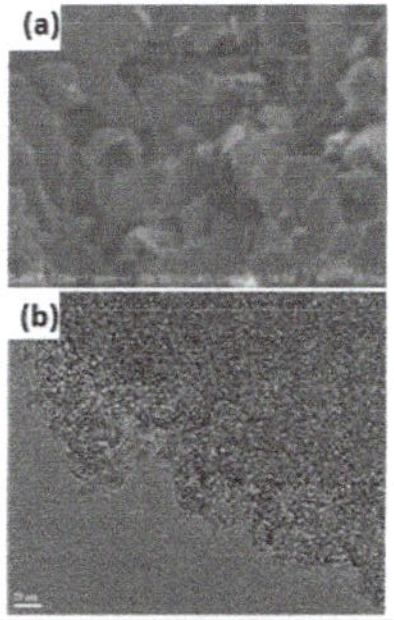	[112]
Poplar catkin	$ZnCl_2$ at 800 °C for 2 h	N and O dual doped carbon	1462.5	1.31		[28]

(continued)

Table 1 (continued)

Biomass resource	Activation method	Final product	Specific surface area (m^2g^{-1})	Pore volume (m^3g^{-1})	Surface morphology	References
Willow catkin	KOH at 400 °C for 3 h & 850 °C for 1 h, and then air at 300 °C for 1 h	N, S-co-doped porous carbon nanosheet	1533	0.92	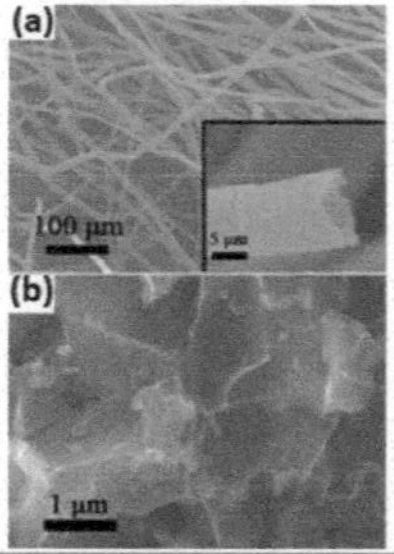	[113]
Wheat flour	KOH at 800 °C for 2 h	Hierarchically porous N doped carbon	1294	–	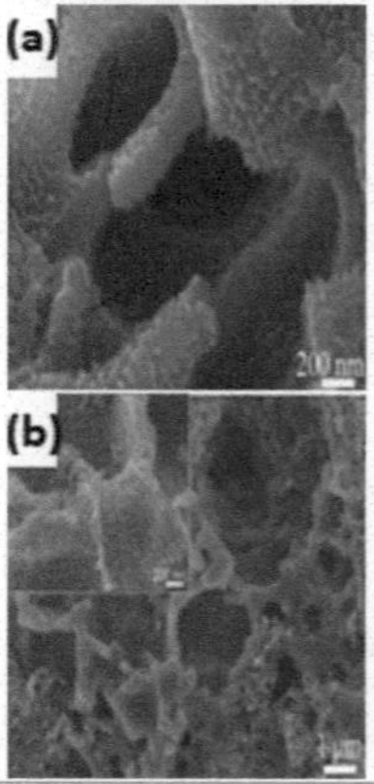	[114]

$$2KOH \rightarrow K_2O + H_2O \tag{1}$$

$$H_2O + C \rightarrow H_2 + CO \tag{2}$$

$$H_2O + CO \rightarrow H_2 + CO_2 \tag{3}$$

$$K_2O + CO_2 \rightarrow K_2CO_3 \tag{4}$$

$$K_2CO_3 \rightarrow CO_2 + K_2O \tag{5}$$

$$C + CO_2 \rightarrow 2CO \tag{6}$$

$$K_2CO_3 + 2C \rightarrow 3CO + 2K \tag{7}$$

$$C + K_2O \rightarrow CO + 2K \tag{8}$$

Various BPCs with different architectures and pore texture have been developed from renewable biomass resources such as plants, animal organs, and waste materials using KOH activation. Most important, the porous structure can be significantly tuned through controlling the activation conditions (i.e., activation temperature/time, and the ratio of precursor/activation agent). The higher activating temperature creates defects in the molecular structure and induces the formation of dangling vacancies. These dangling bonds engender surface-free radicals and generate various surface functionalities, particularly, oxygen-based functional groups [119]. Such functionalities may faradically store the electric charge and subject to electrochemically redox reactions. Besides, hydrophilic functionalities such as –OH, –COOH, and –C = O can be produced which not only boost the wettability of carbonaceous electrode materials but also improve the capacitive performance [120]. $ZnCl_2$ activation possesses a similar effect in increasing the surface area because of its Lewis acid nature. Under optimized activation conditions, $ZnCl_2$ can facilitate the aromatic condensation reactions of biomass precursors at a higher temperature and enhance the dehydration process at a lower temperature. This activation salt was used to prepare porous graphitic carbons from different starting materials like chitosan, coffee beans, coconut shell, sugar cane bagasse, rice husk, peanut shell, etc. [121–125]. But the high price of $ZnCl_2$ reagent and corrosive property of the released hydrogen chloride during pyrolysis treatment at higher temperatures are two major side effects. Activating natural biomass with H_3PO_4 agent generates P-containing groups into the carbon frameworks and micropores. The grafted P-containing functional groups are beneficial for fine-tuning the stability of electrochemical supercapacitors (ESCs) electrodes operated with a wide potential window [126, 127]. A more effective and green chemical activating method based on potassium bicarbonate was recently proposed to synthesize carbon materials from hydrochars with a large surface area and a high pore volume [128]. The obtained surface area was greater or comparable to that of

KOH activation. The low cost, less corrosive property, and improved yield of this environmentally friendly technique fit the industrial requirement.

3.2 *Physical Activation*

Porous carbons can be formed by the physical activation of biomass precursors with a bulk density and a high yield in presence of oxidizing gasifying agents in the temperature range from 800 to 1000 °C. Chemical activation of carbons suffers from the corrosiveness of chemical salts and demanding an appropriate washing of the final product, which may eliminate the chemical agents. CO_2 and H_2O vapor are the fundamental processes of physical activation [129]. Activating carbon materials with CO_2/H_2O involves the C–CO_2/H_2O reaction, leading to carbon atoms removal and improving the porous architecture. Impressively, the CO_2 atmosphere-assisted carbon activation exhibited a good capability to increase the surface area and pore volume of the as-prepared products. On the other hand, the H_2O activation is largely restricted on the surface of carbon because the diffusion rate of H_2O is much slower than the reaction rate of C–H_2O [130]. Therefore, the CO_2 activation produces an expanded surface area and much microporosity in carbon texture than steam activation, and these properties are strongly correlated with the activation temperature, gas flow rate, and holding period. Compared to physical activation, chemical activation has several advantages including lower activation temperature, smaller reaction time, larger surface area, and tuned microporosity. Moreover, physical activation shows a poor capability to optimize the electronic structure and surface chemistry of carbons, which limits its extensive use to some extent.

3.3 *Self-Activation*

Self-activation of BPCs requires no additional activating reagents, which minimizes the cost and simplifies the process. This technique achieves through two mechanisms: one investigates the emitted gases during the pyrolysis treatment of biomass precursors rather than activating gas or chemicals to self-activating the resulting carbon. The other uses the inorganic species (K-containing component) that already existing inside the biomass to *in situ* activates carbon during the calcination process [131, 132]. The self-activation of carbon is impacted by the generated gases, and the activation degree can be influenced by tailoring the gas flow rates, yielding a controllable surface area up to 2000 $m^2\ g^{-1}$. Thus, compared to classical activation strategies, the self-activating approach represents an eco-friendly and more cost-effective approach.

4 Heteroatom Doped Porous Carbon Matrix

Surface functionalization of porous carbon can be implemented to further fine-tune the binding with guest atoms and optimize the interfacial and bulk characteristics of materials, suggesting advanced applications and opportunities. Strategies to modulate the surface of porous carbon have come into being with specific applications in mind. In general, functionalization of the surface can be achieved via various ways including in situ and post-synthesis processes to embed different chemical functionalities preferable in electrochemical systems such as catalysis. The incorporation of heteroatoms into carbon lattice can promote the electrical, semiconducting, and mechanical properties of carbon materials [57, 59–61]. These heteroatoms may improve the conductivity, impact the wettability, and consequently enlarge the electroactive surface area. Depositing electron-rich atoms into carbon materials has its advantages for sustainable technologies. Recent studies indicated that the doped heteroatoms and surface functional groups are beneficial to enhance performances of the carbon electrodes.

The doping of N atoms into porous carbon skeletons would enable a shift of the Fermi level to the conducting band, which is a key for supercapacitor applications. The collaborative effect of N/O-containing functional groups was visible in modulating capacitance. The chemisorption features of reactive carbon surface can be modified via boron atoms incorporation. The phosphorus atoms stabilize the oxygenated groups during gas-phase electrocatalysis and chemical charging, resulting in an improved stability and selectivity. Overall, incorporating heteroatoms become blissful when biomass has been employed as resources for constructing carbon materials with a desirable porous structure, unique electronic feature, and partial graphitization [133–136].

The embedment of heteroatoms into carbon matrices can be accomplished via the following pathways:

i. Low-temperature treatment of heteroatoms-containing precursors (i.e., thiourea, sulfur powder, sodium hypophosphite, ammonium hydrogen phosphate, or red phosphorus, boric acid, melamine-based polymer, etc.), and carbon materials at different weight ratios.
ii. Low-temperature treatment of carbon materials with the effective flow of N_2 gas, H_2/N_2 and NH_3/O_2 mixtures.

The locations of N, B, S, P-centers into carbon lattice of porous carbon scaffolds directly contribute to the efficiency of materials for catalysis, charge storage, and gas adsorption applications. For example, hierarchical carbons prepared from carbonizing N-containing salts and BPCs afford N centers, which lack during the carbonization process [127].

NH_3-assisted heat treatment of magnesium hydroxide templating provides a unique N doping into carbon frameworks and enhances the surface area with the generation of various N groups [22]. This happens due to the interaction of two carbon atoms with the edge-N atoms which prohibits the growth of carbon lattice, leading to the formation of micropores. The establishment of micropores can be ascribed to the released hydrogens which etchs the carbons and accelerates the carbon atoms removal from the lattice. The synthesis of phosphate-functionalized hierarchical porous carbons (HPCs) as supercapacitor electrodes is also interesting owing to the expanding of operational voltage range with aqueous electrolytes and subsequent enhancement in the obtained energy can be assured.

The combined doping of N and B into BPCs with a hierarchical porous structure is also accessible via the self-assembly heat treatment in presence of N and B-containing sources. The N, B co-doped carbon materials with multi-length-connected frameworks revealed graphite-like architectures and could intensively tuned the electrical conductivity. The enhancement in electron mobility of carbon materials was probably assigned to the B and N substitution and stabilization of π electrons [53]. These changes might reinforce the C–O bonds and deteriorate the C–C bonds. The generated pores were suitable for facile diffusion of electrolyte ions throughout diminishing the molecular diffusion limitations. Moreover, introducing oxygen functional groups is another tactic to modulate the surface electron properties of carbon materials (reactions **9** and 10). These functional groups with electron-acceptor properties can improve the wettability and boost ion accessibility.

$$-\mathrm{COOH} \leftrightarrow \mathrm{H}^{+} + \mathrm{e}^{-} + -\mathrm{COO} \tag{9}$$

$$> \mathrm{C{-}OH} \leftrightarrow \; > \mathrm{C} = \mathrm{O} + \mathrm{H}^{+} + \mathrm{e}^{-} \tag{10}$$

O doped carbons are usually achieved by selecting O-enriched biomaterials, acid treatment, via KOH activation and air activation [58, 59]. Condensation of proteins at high temperatures offers plenty of O groups onto carbon surface. The redox reaction between the O functional groups and electrolyte ions enables an exceptional pseudocapacitance. However, the irreversibility of reactions between the oxygenated groups and electrolyte ions may cause a high rate of self-discharging, electrolyte dissociation, and weak cycling durability, while the carboxyl groups may prevent the organic electrolyte ions transportation into the inner pores and reduce electric conductivity, which in turn decline the rate capability and increase the internal resistance.

S species onto carbon surface modify the wettability and act as electroactive sites for redox reactions. The electrons on the *p*-orbital of S intertwine with the *p*-orbital of graphite sp^2 hybridization, which promotes the electron delocalization among carbon backbones and thereby enhances the electronic structure [49, 137]. The robust interfacial adsorption between S species and electrolyte ions allows the full utilization of porous carbon materials. It is notable that, contrary to the non-doped carbon, the S-functionalized carbon surface showed a smaller charge transfer resistance owing to the improved polarity [137].

From the above discussion, introducing heteroatoms into localized graphitic carbon matrices can sufficiently improve the conductivity of ions for a wide range of applications. Notably, multiple heteroatoms doping represents comparatively a new route to greatly improve the electronic structure for electrochemical energy conversion and storage applications. Such a pathway depends on the choice of solvents and molecular design of monomers to effectively incorporate heteroatoms into carbon skeletons during condensation reaction. For example, O, N, and S co-doped porous carbon was successfully prepared from ant powder via simultaneous carbonization and activation reactions (Fig. 3a) [138]. The *in situ* doping of N and S moieties was achieved by fatty acids and proteins. The combined roles of S, N, and O dopants contributed to a high capacitance. Moreover, N, O, and P-co-doped soybean dreg-derived carbon material was prepared with a tuned conductivity and an enhanced charge storage capability. As an asymmetric supercapacitor electrode material, the integrated device worked at 1.9 V in Na_2SO_4 electrolyte and delivered a high specific capacitance and an excellent energy density (Fig. 3b and c) [139]. Nevertheless, controlling the electrochemical accessibilities and porosities of heteroatoms enriched BPCs still need to be improved.

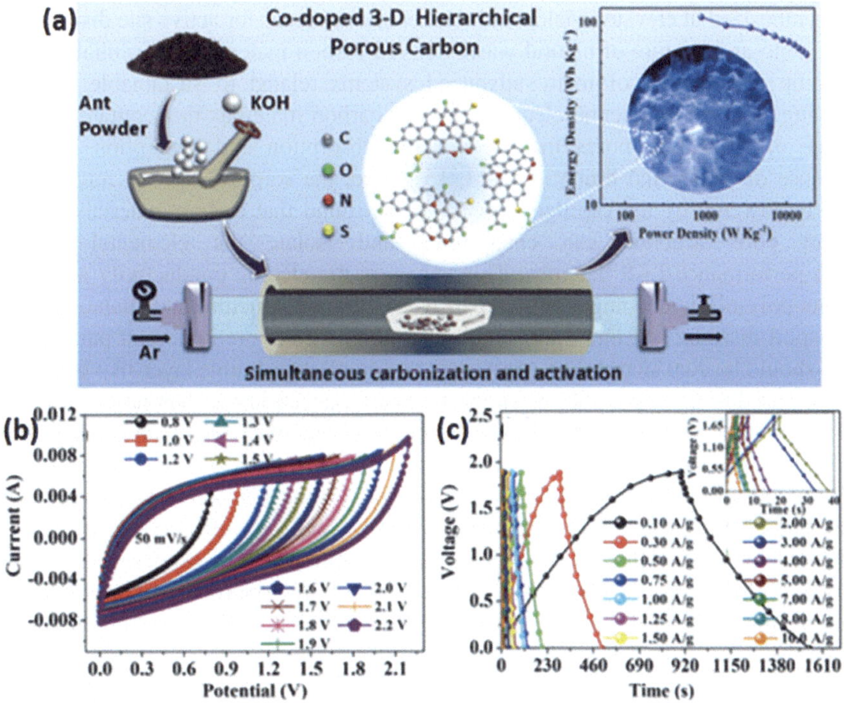

Fig. 3 **a** Schematic configuration for the synthesis of ant powder-derived carbon. **b** CV curves collected with different potential windows and **c** galvanostatic charge–discharge curves of N, O, P-co-doped soybean dreg-derived carbon optimized with different current densities. Reproduced with permission from Ref. [60], Copyright 2019, Royal Society of Chemistry

5 Synergistic Effect of Macro/Meso/Micropores for Applications

Currently, the consumption of traditional fossil fuels such as natural gas, oil, and coal has accelerated dramatically due to population explosion and worldwide economic expansion. Furthermore, emissions due to the combustion of these non-renewable resources and greenhouse gas have caused serious environmental problems. Therefore, implementing green, renewable, and efficient systems for energy conversion, as well as new energy storage is of a great importance. Fuel cells, supercapacitors, lithium-ion batteries (LiBs), and solar cells are among the most promising candidates [9–11]. In addition, considering the cost issue, high-performance devices should have specific features and be rationally established with abundant and renewable materials, which hold the key to fundamental advances in energy harvesting. The applicability of BPCs has been examined in several fields such as metal ion removal, methane storage, and heterogeneous catalysis. HPCs derived from natural biomass with well-defined pore topologies and sizes can guarantee shorter diffusion pathways with a minimized resistance to mass transportation [9]. Compared to conventional carbons, these porous materials can be synthesized at elevated yields with higher surface areas for active site dispersion. The nano-architecture of natural waste-derived carbon materials is essential for the specific requirement of many advanced systems related to sustainable energy, sensing, and environmental cleaning. Porous carbon materials that contain a large range of meso/micropores may reduce the absorption and desorption kinetics because of diffusional limitations [14, 15]. Also, the sorption rates in micropores can be lowered by the capillary forces. It was found that carbon materials with a smart pore structure can encapsulate and isolate the elemental S for high-performance Li-S batteries. This enhances the electric conductivity and prohibits polysulfide decomposition. Apart from the pore structure and shape, the ion transport resistance of inner pores and diffusion distances are important parameters and should be considered. The construction of an electric double layer-like behavior is supposed to be impossible when the solvated ions are larger than pore sizes, but in ascertained conditions, the electric double layer can exist even with smaller pore diameters than solvated ions by using organic electrolytes. For biomass-assisted HPCs, electrolyte ions can be smoothly delivered through macro/meso/micropores to surfaces [2, 3]. The promising energy and power performances of HPCs demonstrated that the micropores could be effectively investigated for charge storage. However, the decreased capacitances of microporous and mesoporous carbon materials at the higher discharging current densities might be originated from the solute diffusion process.

5.1 CO_2 Storage Materials

The increasing global warming is a crucial environmental challenge. The high emission rate of greenhouse gases such as CO and CO_2 is a key factor contributing to this serious problem. Carbon capturing and storing is an anticipated approach to diminish the emissions of industrial resources. The use of porous compounds to uptake CO_2 from effluent gases is an efficient strategy because the regeneration of adsorbents requires less energy [140, 141]. The captured CO_2 can be further utilized to generate fuels or as a feedstock to form liquid fuels. Carbon capturing in the shape of CO_2 represents a coherent extension of biomass conversion into fuels and solar-to-fuel. The reversible adsorption-desorption is very identical to the green leaf that can uptake CO_2 in sunlight time and release in dark. A material with functional pores would serve as an artificial leaf that capable of absorbing and releasing CO_2 into atmospheric air under wet and dry conditions, respectively [142]. Hierarchical porous carbons are attractive alternatives to conventional adsorbents because of their low cost, excellent hydrophobicity, good chemical resistance, and high availability [143]. The hydrophobic cores of carbon materials can modify their capability and efficiency for CO_2 capturing and make them more selective for CO_2 over N_2 gas. Moreover, CO_2 adsorption kinetics strongly correlates to micropore diffusion, and the obtained analyses matched well with the Freundlich mechanistic model. These materials exhibited a high adsorption capacity with an excellent reusability without a significant loss in efficiency owing to the strength between the porous carbon and CO_2 induced electrostatic interactions [144]. The existence of narrow micropores is very important to attain a high adsorption capability (Fig. 4) [143].

A suitable pore structure is usually the first consideration in the construction and preparation of porous adsorbents for a selective gas separation. Taking into account the factors influencing the capture efficiency (i.e., adsorption capacity, selectivity, and kinetics), the following formation principles of sorbents should be highly appreciated: (i) abundant microporosity with appropriate geometries identical to CO_2 molecules and well-distributed surface functionalities are appealing aspects to improve the CO_2 uptake capacity, (ii) for establishing a highly interconnected hierarchical pore system, the quantity and mutually interconnected modes of macro/meso/micropores should be considered to allow better adsorption-diffusion dynamics of CO_2. Thus, the primary focus is to prepare carbons with a multi-scale, highly interconnected pore structure, and good mechanical properties for actual practical use. N doped porous carbons with the controlled meso/microporous structures are highly desirable for achieving rapid adsorption–desorption kinetics. It is expected that N-functionalities in biochar will be beneficial for enhancing the CO_2 adsorption properties which are non-porous. These materials open a new door for the fabrication of effective CO_2 adsorbents-based carbonaceous materials. Accordingly, N doped porous carbon was developed from coconut shells in the presence of urea as a N source (Fig. 5a) [145]. The as-synthesized material revealed a high CO_2 uptake ascribed to the high microporosity and doped N. It was demonstrated that the pyrrole- and pyridine-N groups contributed significantly to

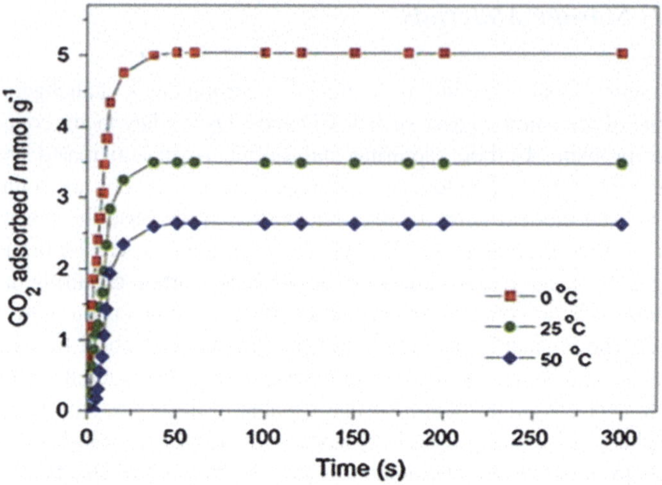

Fig. 4 Adsorption kinetics of CO_2 at different temperatures through a microporous carbon material. Reproduced with permission from Ref. [143], Copyright 2015, Elsevier

the CO_2 capture efficiency than quaternary- and pyridine-N species. In addition, Arundo donax biomass and chitosan-derived N-functionalized activated carbons with $ZnCl_2$ activation displayed an exceptionally high CO_2 capture (Fig. 5b) [146]. The polarity of N sites promoted the interactions between the N doped activated porous carbon and CO_2 molecules.

Functionalizing BPCs with suitable metallic species encapsulation is another route to improve the CO_2 adsorption capacity owing to the increased number of active sites. Porous carbon prepared from a pine cone and KOH activation containing metal and nitrogen dopants with large micropores exhibited an impressive CO_2 adsorption (Fig. 6a) [147]. However, the CO_2 adsorption property of activated porous carbons was varied according to the activation temperature and impregnation conditions. Furthermore, KOH-assisted porous carbon activation derived from abundant London plane leaves with bimetallic well-dispersed magnesium (Mg) and calcium (Ca) components showed an improved CO_2 adsorption capacity [148]. The excellent performance of the composite was mainly assigned to the higher microporosity degree and larger surface area (Fig. 6b).

Alternative raw materials should be explored to provide a cost-effective and an economical pathway to fulfill the projected demands. The use of BPCs in industries and related sectors appears to be a comparable option. Therefore, the exploration of the next generation of functionalized BPCs through integrating with organic and inorganic nanostructures is necessary. Combining BPCs-based materials with inexpensive metal oxides, metal nitrides, and metal chalcogenides is a prospect for establishing composite materials with interesting properties that might improve the overall efficiency of these cost-effective materials and open numerous opportunities and possibilities for real application.

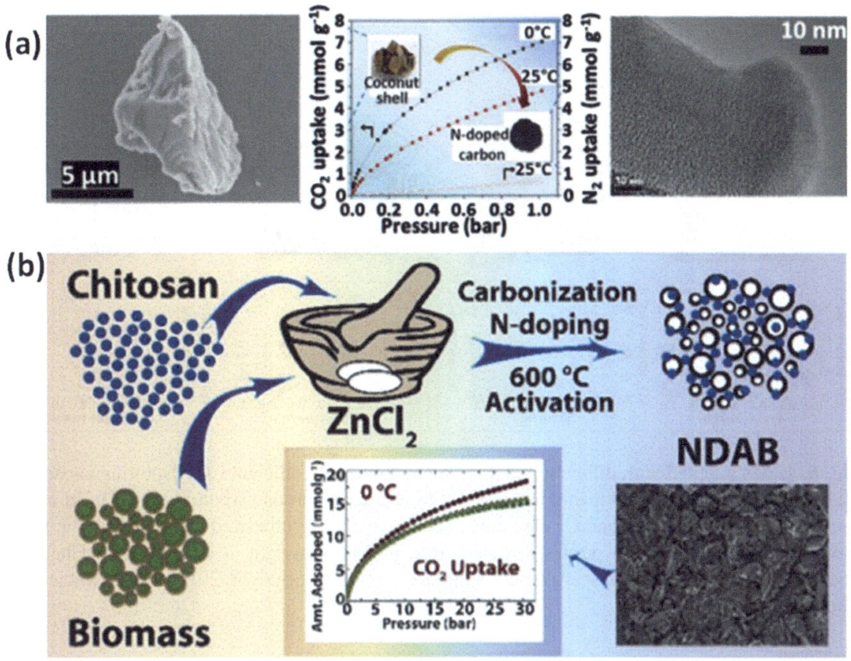

Fig. 5 **a** N doped porous carbons synthesized from coconut shell and and their activity for CO_2 capturing under various conditions of pressure and temperature. **b** Synthesis of Arundo donax/chitosan-assisted N doped carbon and the corresponding CO_2 adsorption activity [**a** Reproduced with permission from Ref. [145], Copyright 2016, American Chemical Society. **b** Reproduced with permission from Ref. [146], Copyright 2017, Royal Society of Chemistry]

5.2 *Fuel Cells and Electrocatalysis*

Fuel cells have been emerged to be one of the most potential technologies for commercial and industrial power generation because of their high efficiency and longer operating times compared to battery systems. They can directly produce an electrical energy from a chemical energy without a combustion process. Despite the huge research efforts offered to enhance the efficiency and durability of fuel cell devices, their broad-scale commercialization is still problematic. The high price and less abundance of widely employed platinum (Pt)-based catalysts are one the drawbacks. Besides, these electrocatalysts suffer from poisoning by CO by-product, which may block the active surfaces and deactivate the catalyst. In the year 2014, the EU founded a critical list comprising 54 precursors based on their possible threats and economic value in which Pt was in the top risk family [149]. Thereby, it is essential to design a cheaper Pt-free catalyst with a comparable electrochemical activity and higher stability. One possible way to replace or even reduce the loading

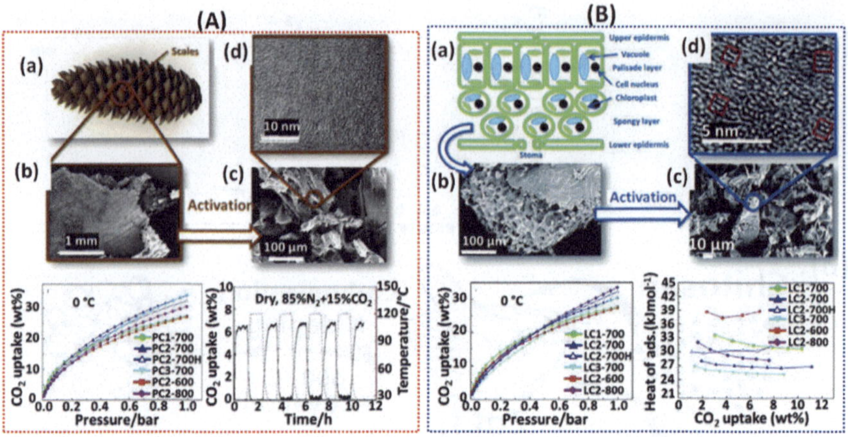

Fig. 6 BPCs adopted with different metallic functionalities for efficient CO_2 capturing capacities. **a** High level of N and Ca dopants integrated into activated porous carbon obtained from a pine cone. **b** Well-dispersed Mg and Ca elements in porous carbon synthesized from the London plane leaves. [**a** Reproduced with permission from Ref. [147], Copyright 2016, American Chemical Society. **b** Reproduced with permission from Ref. [148], Copyright 2015, Royal Society of Chemistry]

of noble metals in electrocatalysis is to apply heteroatom doped carbons and/or transition metal-based catalysts.

ORR is a kinetically slow process that takes place at the cathode part of fuel cells and involves multi-electron and multi-step reactions, which constitutes the efficiency and cost of fuel cells [150, 151]. Generally speaking, the cathode ORR electrocatalyst material is the most critical factor affecting the practical application of such devices. At present, the most widely employed catalysts with an excellent electrocatalytic activity for ORR are Pt-based catalysts, but their high price, unimpressive durability, weak tolerance to fuel cross-over, and poisoning resistance are greatly hindering the economic value of fuel cells. With the growing scales of global warming and environmental pollution, the demand for the huge utilization of fuel cells has become more necessary. It is well known that earth-abundant biomass precursors with self-functionalized species can be easily harvested from nature. BPCs have been examined as precious metal-free electrocatalysts for ORR with a high electrocatalytic activity (Table 2).

Doping N heteroatom into the hierarchical architectures of BPCs is essential for ORR catalysis. The N functionalities can significantly modulate the ORR electrocatalysis process, mainly in these aspects:

Table 2 Performances of some BPCs for ORR electrocatalysis

Biomass	Onset potential	Half potential	Achieved stability	References
Pomelo peel	–	0.86 V *vs.* RHE	92.5% retention after 10,000 s	[152]
Eclipta prostrata	0.25 V *vs.* Hg/HgO	–	94% retention after 18,000 s	[153]
Catkin	−98 mV *vs.* Ag/AgCl	−194 mV *vs.* Ag/AgCl	88.9% retention after 20,000 s	[154]
Soybean	0.910 V *vs.* RHE	0.821 V *vs.* RHE	91.22% retention after 7200 s	[155]
Kelp	0.90 V *vs.* RHE	0.74 V *vs.* RHE	More than 95% retention over 9 h	[156]
Water lettuce	–	0.887 V *vs.* RHE	91% retention after 30,000 s	[157]
Taro stem	–	0.87 V *vs.* Hg/HgO	Around 96.5% retention after 20,000 s	[158]
Goat skin trimming	–	50 mV *vs.* Ag/AgCl	About 90% retention after 6 h	[159]
Reed stalk	–	0.99 V *vs.* RHE	94.8% retention after 20,000 s	[160]
Amaranth	0.27 V *vs.* Hg/HgO	–	A well-defined cathodic peak after 1000 cycles	[161]
Chlorella	–	0.87 V *vs.* RHE	96% retention after 50,000 s	[162]
White beech mushroom	0.11 V *vs.* Hg/HgO	–	Outstanding long-term stability	[163]

i. Introducing nitrogen N species facilitate the breakage of electric impartiality of carbon surface and re-stabilize the charge intensity.
ii. N functionalization will induce defects in carbon structure and stimulate the catalytic performance.
iii. N dopants motivate the electron delocalization of adjoining carbon atoms, leading to a high positive charge and thus boosting the adsorption and electroreduction of O_2 onto carbon surface.
iv. N doping modifies the electric conductivity and efficiently promote the ORR process.

Accordingly, N-rich goat skin wastes underwent a simple pyrolysis process to construct carbon-like nano-onions for ORR catalysis in an alkaline electrolyte with an excellent electrocatalytic performances exceeded those of conventional Pt/C catalyst [159]. The pyridinic N species played a significant role as active sites for ORR. The high surface area and hollow graphitic structure contributed positively to better ORR activity. Moreover, N doped porous graphitic carbon for ORR electrocatalysis was synthesized from amaranth through a one-step direct pyrolysis route [161]. Results confirmed that the collaborative roles of graphitic N and pyridinic N contributed significantly to ORR process and improved the overpotential as the core active site participated in the whole process. Also, the catalytic activity was greatly affected by the tuned electrical conductivity and large specific surface area.

More importantly, the ORR performance showed a significant correlation with the pore size distribution of BPCs. The changes in vertical coordinates of pore size distribution and their impact on the catalytic activity of BPCs toward ORR electrocatalysis were well-verified. In addition, the density of active sites and mass transfer could be influenced by the vertical coordinates of pore size distribution that reflect the effect of average number and volume of generated pores. In this context, meso/microporous carbon with N/O dopants optimized from white beech mushroom via hydrothermal carbonization and high-temperature annealing process was tested for ORR electrocatalysis [163]. The vertical coordinate's change of pore size profiles was realized only by controlling the pore volume of the as-obtained material. The study illustrated that the catalyst with a pore volume of 0.45 cm^3 g^{-1} possessed the highest electrocatalytic performance and was approximately similar to that of Pt/C catalyst (Fig. 7a and b). The novel catalyst exhibited better long-term cycling performance, superior methanol tolerance, and CO resistance compared to Pt/C catalyst (Fig. 7c–f). The ORR process takes place in the liquid/gas/solid boundaries of catalytic materials and involves the transfer of both electrons/protons, O_2, and water inflow or outflow. Thus, HPCs with functional open architectures are preferable for the ease penetration and fast transport of ions and O_2 molecules, to promote the ORR activity.

In general, the focus on BPCs as efficient ORR catalysts or co-catalysts has entered deepwater areas concerning porous structure control, electronic structure fine-tuning, and heteroatom functionalization. Many important ideas, valuable opinions, and strategies have been born. However, the following issues should be solved in the future.

i. The activity and stability of these materials in acidic environment are still insufficient. Thus, the changes in ORR activity under various pH ranges need to be carefully analyzed.
ii. The specific stability mechanism should be further addressed despite the good durability of these catalysts.
iii. The origin, composition, and availability of catalytically active sites for ORR should be much investigated. DFT approach and effective in situ characterization instruments are needed to illustrate the catalytic role of each active site and the construction process of the ORR active center.

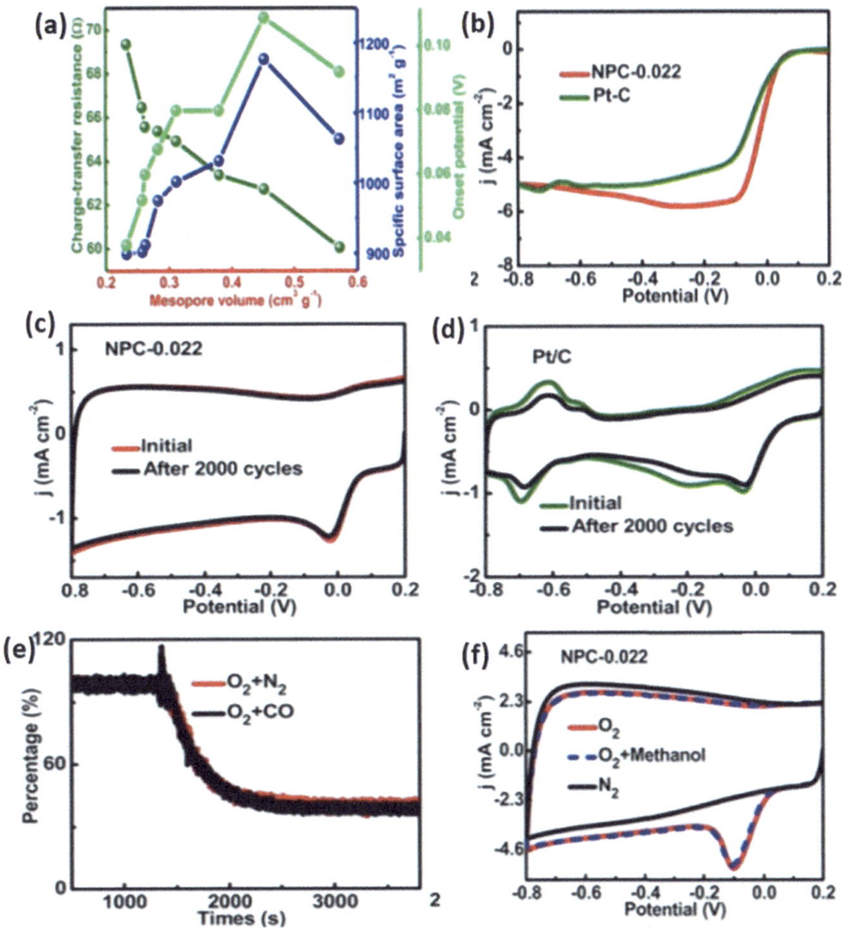

Fig. 7 **a** Effect of mesopore volume, specific surface area, and charge transfer resistance on ORR activity of N, O co-doped meso/microporous carbon optimized from white beech mushroom (NPC-0.022). **b** LSV plots of NPC-0.022 and commercial Pt/C catalyst. **c** Cycling stability test of NPC-0.022 and **d** Pt/C analyzed by CV curves over 2000 consecutive cycles. **e** CO tolerance test and **f** methanol tolerance of the integrated NPC-0.022 catalyst. Reproduced with permission from Ref. [163], Copyright 2019, Elsevier

5.3 *Water Splitting*

Molecular hydrogen (H_2) is an effective energy carrier. It can be formed from natural gas through methane reforming, but the resulting CO_2 aggravates the greenhouse effect. The overall water electrocatalysis generates both H_2 and O_2. An effective catalyst for HER is desirable to decrease overpotential and increase the faradic efficiency. It is worth mentioning that Pt-based electrocatalysts showed a higher efficiency for HER catalysis with a near-zero overpotential. On the other hand, the OER is an anodic process, established in water catalysis; however, the associated sluggish kinetics represents a crucial parameter determining the overall water electrocatalysis process. Ru, Pt, and their alloys with the rare availability and high price provide the best performance for OER, although their long-term stability is still inadequate. Over recent years, numerous studies have been offered to minimize the usage of expensive noble metals but maintained a high activity by constructing specific catalyst supports. BPCs with a robust chemical stability in both alkaline and acidic electrolytes can render an industrially viable chance. For an instant, a sustainable alternative substrate was collected from turfgrass with a competitive cost compared to commercially available activated carbon [164]. Coupling inexpensive metallic components with BPCs revealed a promising electrochemical performance. Co_3O_4 decorated blood-derived carbon was applied for the catalysis of ORR based on the surface reactions between $Co^{2+}/Co^{3+}/Co^{4+}$ species of the corresponding catalyst (Fig. 8a). The metallic species demonstrated an intimate connection with the carbonaceous frameworks and provided a synergistic effect, which resulted in an excellent activity and a high long-term stability (Fig. 8b and c) [165].

BPCs are often prepared to support metallic compounds to modify the electronic structure and thereby promote the catalytic performances for both HER and OER processes. A very few attempts indicated that heteroatoms doped nanocarbons displayed HER activity, but their performances are much lower than those of metal-based electrocatalysts. The synthesis of BPCs incorporated non-noble metal complex systems is an attractive approach to substitute noble metal based catalysts. The strong interfacial coupling between encapsulated metal nanoparticles and graphitic carbon layers enables a higher chemical stability over a wide pH domain.

5.4 Lithium-ion batteries

The battery manufacturing has become a potential spot as a new superior member of global green energy storage exploration [166–168]. These batteries have been widely applied in several portable electronic instruments, aerospace, electric cars, and other fields. As secondary devices, LiBs represent one of the most promising comprehensive energy storage systems because of their high specific capacity, outstanding energy density, large operational voltage windows, and low rate of self-discharge [169, 170]. The electrode material is a fundamental element that significantly influences the overall performance and application value of these devices. Graphite-based compounds have been applied as negative electrodes for LiBs, in which the chemical energy converted into an electrical energy throughout the deintercalation of li ions from the cathode and subsequently interact with the graphite at the anode side during charging reaction. Meanwhile, the discharging reaction follows the reverse scenario. However, the unsatisfied rate performance of graphite anodes due to the growth of metallic li dendrite onto their surfaces at high

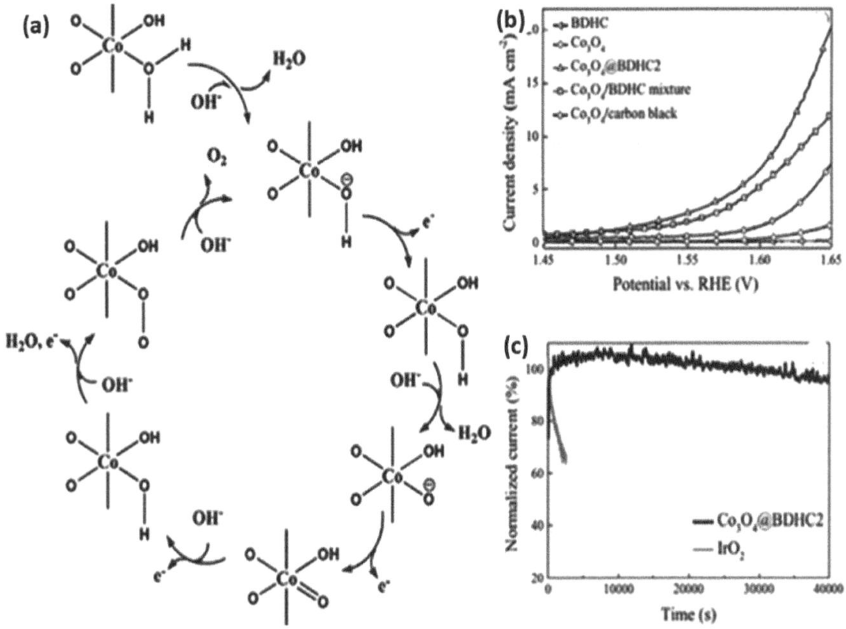

Fig. 8 **a** The OER mechanism over Co_3O_4 electrode, **b** LSV curves of the as-prepared catalysts in N_2-saturated 0.1 M KOH with 10 mV s^{-1} sweep rate and 1600 rpm. **c** Long-term durability response of catalysts. Reproduced with permission from Ref. [1], Copyright 2016 Royal Society of Chemistry

rates may limit their opportunities to be commercially popularized [171, 172]. The employment of disordered carbon enabled an enhanced capacity and allowed more li to intercalate, but the rate of charge was low for approaching the practical use. On one hand, silicon materials with a relatively high theoretical capacity are suffering from capacity critical attenuation because of serious volume expansion during charging and discharging processes. Furthermore, transition metals oxides (i.e., NiO, MnO, Co_3O_4 or Fe_3O_4) and their hybrids have been proposed as anode materials for LiBs to replace graphite [173–175]. Unfortunately, their commercial usage is still limited due to their humble electronic conductivity and extreme volumetric changes during the insertion/extraction processes of li ions. Thus, further development of LiBs makes great demands on exploring new electrode materials with a high energy density, Coulombic efficiency, and cycle performance. Nowadays, researchers try to develop more sustainable materials. Porous frameworks of carbon nanomaterials are promising alternative electrodes in energy storage devices. In this regard, BPCs with tunable features might serve as suitable materials for anodes in LiBs. The biomass raw materials are naturally occurring, and therefore, no tedious policies need to be fulfilled for materials crystallization

Table 3 Performances of selected BPCs as electrode materials for LiBs

Biomass	Capacity	Rate capability	Cyclic performance	References
Agaric	1465.1/894.5 mA h g^{-1} discharge/ charge capacities at the first cycle	218.3 mA h g^{-1} discharge capacity at 10 A g^{-1}	Almost no decay after 1000 cycles	[47]
Fish scale	500 and 480 mA h g^{-1} at reversible capacities 75 mA g^{-1}	224.7 and 232.5 mA h g^{-1} discharge capacities at 2000 mA g^{-1}, respectively	About 80% capacity retention	[48]
Soybean	Storage capacity of 360 mA h g^{-1} in the 100th at 0.5 C	About 190 mA h g^{-1} and 125 mA h g^{-1} capacities at 8 C and 16 C, respectively	310 mA h g^{-1} in the 500th at 2 C and 275 mA h g^{-1} at 4 C after 1000 cycles	[46]
Shell of broad bean	845.2 mA h g^{-1} discharge capacity at 0.5 C	—	261.5 mA h g −1 discharge capacity after 100 cycles	[49]
Natural silk	1865 mA h g^{-1} specific capacity at 0.1 A g^{-1}	212 mA h g^{-1} reversible capacity at at 100 C	92% capacity retention after 10,000 cycles at 2 A g^{-1}	[43]
Peanut dregs	1288 mA h g^{-1} irreversible capacity at the initial cycle	731 mA h g^{-1} reversible capacity at 100 mA g^{-1}	286 mA h g^{-1} capacity at 1000 mA g^{-1} over 1000 cycles	[44]
Microalgae	445 mA h g^{-1} and 370 mA h g^{-1} specific charge capacities at 1 C, respectively	445 and 370 mA h g^{-1} charge capacities at 0.1 C and 1 C, respectively	95% capacity retention after 500 cycles at 1 C	[50]
Degreasing cotton	1298.1/860 mA h g^{-1} discharge/ charge capacities at the first cycle	950 mA h g^{-1} after 100 cycles at 1C and 850 mA h g^{-1} after 200 cycles at a 2 C rate	1070 mA h g^{-1} reversible capacity of after 430 cycles at 0.2 C	[83]

and growth, which itself is an economic benefit. Such carbon nanostructures are capable of providing short pathways for fast transport of li-ions and better accessibility of electrolytes. BPCs with a fast electron transport can buffer the large volume changes and restrain the large concentration of active nanoparticles. Hence, as far as we know, these hierarchical frameworks have been employed as anode materials in LiBs (Table 3).

In general, good anode materials should fulfill these important characteristics:

i. High surface area and abundant active sites, thus allowing larger contact regions at electrode/electrolyte interfaces, tuning the li^+ adsorption and promoting the charge transfer.
ii. An appropriate hierarchical porous nature. Mesopores and micropores enable convenient channels for the rapid and facile ions transportation as well as ensure better mass transport.
iii. Optimized surface functionality and higher graphitized structures further enhance the electrochemical activity, electric conductivity, and create defects.

A highly ordered porous structure of wheat stalk was developed by hydrothermal process and graphitization treatment (Fig. 9a) [176]. The interconnected graphitic carbon nanosheets could provide more accessible sites for storing li-ion and modified the mobility of electrons (Fig. 9b). As anode material, the electrode displayed improved energy storage characteristics with an interesting capacity, good rate capability, and cycling performance benefiting from its excellent graphitic structure and slight voltage hysteresis (Fig. 9c–e). Furthermore, a scalable activation-graphitization strategy was adopted to synthesis N-functionalized HPCs with a high degree of graphitization from natural silk [43]. The LiBs assembled with the developed ordered nanosheets as negative electrode material exhibited a high capacity compared to the benchmark graphite electrode.

Some raw biomass resources with a specific nanostructure may be destroyed under the rigid thermal treatment. Worse still, the usage of harsh activating reagents to fabricate BPCs would corrode the products during calcination process. Based on the above, formic acid was used to eliminate the hemicellulose of rice husk (Fig. 10a and b) to produce fibrous porous carbon under hydrothermal treatment and carbonization reaction in an inert atmosphere. Thereafter, the carbonized yield was chemically treated with ammonium hydrogen difluoride (NH_4HF_2) to obtain the fibrous carbon network's final product [177]. As illustrated in Fig. 10c, the product obtained from direct pyrolysis of rice husk treated with formic acid showed a compact morphology, while the fibrous product displayed hierarchical porous network-like structures with well-defined pores (Fig. 10d). The fibrous porous-based integrated anode for LiBs exhibited an improved electrochemical performance in terms of discharge and reversible capacities with a good cycling capability (Fig. 10e and f). The unique macro/mesoporous networks and improved electronic structure contributed to the superior activity.

At the same time, it is also very important to control the heteroatom doping of BPCs to establish LiBs with greater efficiencies. Recently, human hair was used as a precursor to prepare porous graphitic carbon microtubes with heteroatom co-doping and dominant pores via the help of nickel nitrate hydroxide activation (Fig. 11a) [45]. As a negative electrode material for LiBs, the graphitic carbon achieved a high capacity at the higher current density with a good cycling performance (Fig. 11b and c). Hierarchical porous matrices offered the following functionalities: (i) a wide and an effective contact at electrode/electrolyte interfaces

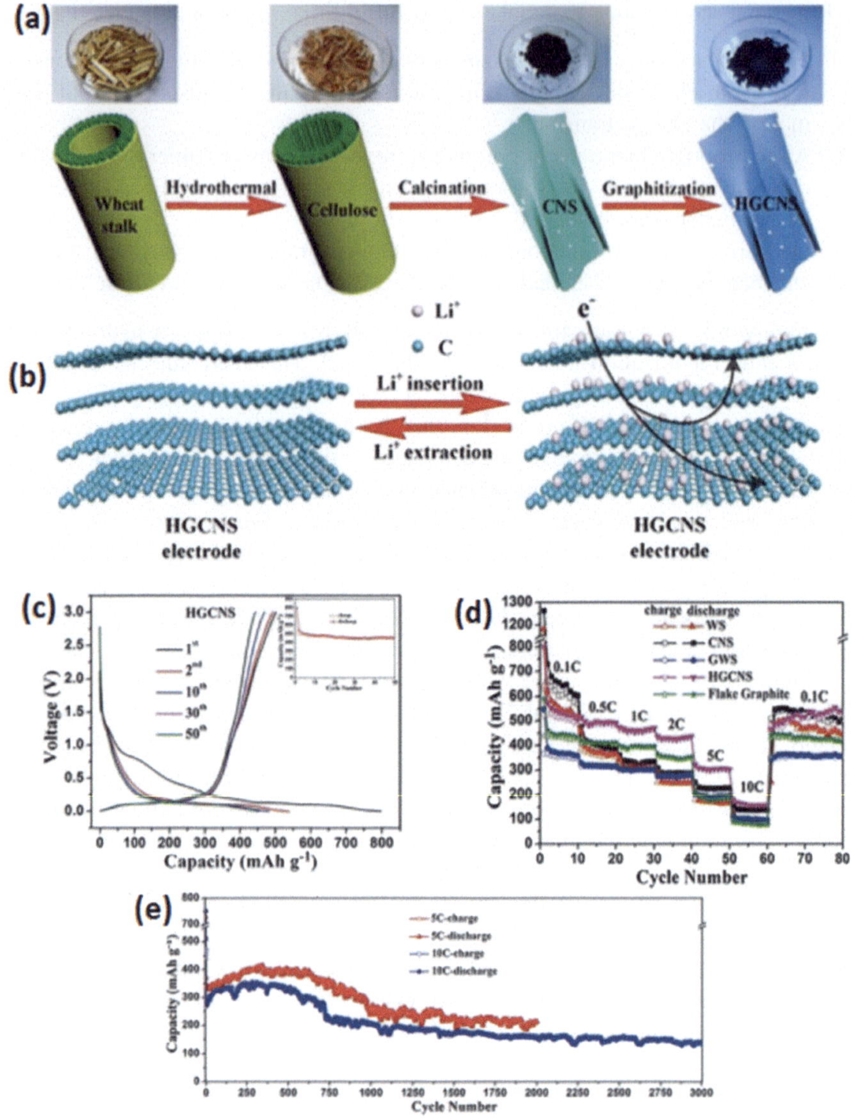

Fig. 9 **a** Schematic configuration for the synthesis of ordered porous carbon with hierarchical graphitic structure (HGCNS) adopted from the wheat stalk. **b** Schematic illustration for the electron mobility and li^+ storage scenario into the HGCNS electrode. **c** Galvanostatic discharge/charge plots, **d** rate capability under different current densities, and **e** cycle performance of the electrode. Reproduced with permission from Ref. [4], Copyright 2020, Royal Society of Chemistry

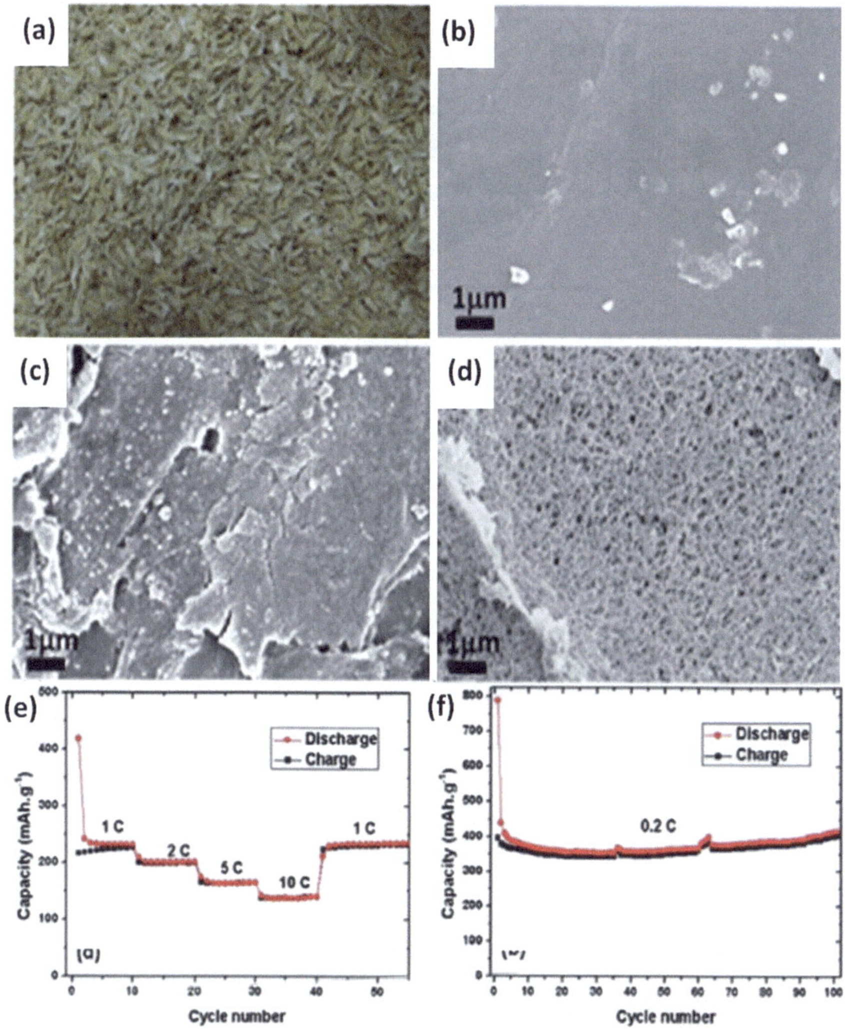

Fig. 10 **a** An optical image and **b** SEM image of pure rice husk. **c** SEM image of rice husk after being subjected to formic acid treatment and **d** SEM image of rice husk with a combined treatment of formic acid-NH_4HF_2. **e** Rate performance obtained with various current densities and **f** capacity retention performance of the proposed anode versus the cycle number. Reproduced with permission from Ref. [177], Copyright 2013, Royal Society of Chemistry

enabling more fast redox reactions and modulated capacity, (ii) improved cycle stability with the favorable intercalation–deintercalation of ions during reactions, implying a better utilization of the active electrode material, and (iii) a rapid electron transfer caused by the tuned conductivity, heteroatoms doping and defects, and thereby guaranteeing a fast Faraday reaction.

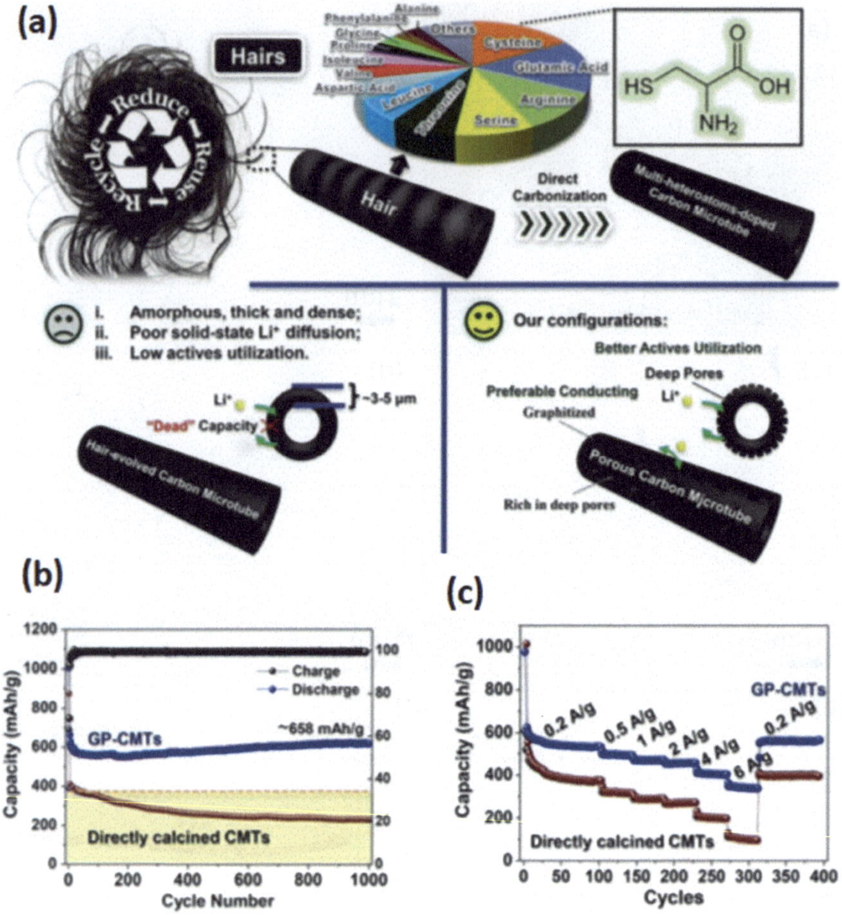

Fig. 11 **a** Schematic diagram elucidating the investigation of human hair-derived functionalized porous graphitic carbon microtubes for LiBs applications. **b** Galvanostatic cyclic behavior and **c** programmed cyclic records of electrode measured with a half-cell system. Reprinted with permission from ref. [4], Copyright 2020, Royal Society of Chemistry

6 Conclusions, Challenges and Future Prospectives

As is known to all, the enormous amounts of renewable biomass resources and well-established fabrication strategies underlie the fantastic porous structure and physicochemical properties of BPCs for addressing the most critical environmental issues and energy problem. Valuable BPCs could be assembled through varieties of synthetic processes to satisfy special uses. The most advanced application of BPCs in electrochemical energy conversion and storage devices, namely the utilization in fuel cells, water electrocatalysis, CO_2 capture, and LiBs have been elaborated in this chapter. The related works and analyses have offered nano-science in the formation of BPCs with a condensed emphasis on the green energy concepts to greatly expand their future development and applications. However, there are still some challenges that impede their broad investigation.

Thus, it is very significant to be aware of the handicaps to further spread the commercial benefit of BPCs, as follow:

1. The high synthesis cost is still a great obstacle against their applications. The dependence on energy is quite high since pyrolysis at elevated temperatures is an indispensable process. Therefore, developing a low-energy route is a hot topic of current research. Highly efficient chemical activators and self-activation systems may be the newly sought.
2. The appropriate selection of biomass raw material and formation method identifies the effective construction of HPCs. Accordingly, screening out feedstocks with various characteristics, corresponding production methods and conditions to establish a systematic route is desirable in future research. Suitable raw resources with the right constituents/ingredients and controlled processing parameters can lead to the precise design of BPCs with tailored features for advanced applications.
3. It is urgently to strengthen the applications biomass-derived HPCs since their current potential uses are largely focused on electrochemical energy storage devices. Many new applications should be discovered in the future such as photocatalysis, HER electrocatalysis, and gas adsorbents.
4. Although, the major advantages of BPCs over porous graphitic carbon materials including very low cost and wide availability, it is desirable to enhance the performance and conduct *in-depth* research on the factors influencing the efficiency improvement for supercapacitors, LiBs, and fuel cells because the performance of BPCs is still a bit behind that of well-known graphitic carbon materials obtained from other fine resources (i.e., graphene and carbon nanotubes).
5. To date, few works were introduced for BPCs-based electrodes in Li-S batteries and other new energy storage systems such as Mg (Al, Mn)-ion batteries. Benefiting from the robust structure, good surface/interface chemistry, these green carbonaceous materials can be readily used for newly emerging energy storage technologies throughout the customized control of their chemical and physical characteristics.

Hence, it is very promising to develop scalable and high-performance carbon materials from crude biomass with simplified and effective approaches to achieve widespread practical applications in the future.

References

1. Deng J, Li M, Wang Y (2016) Biomass-derived carbon: synthesis and applications in energy storage and conversion. Green Chem 18:4824–4854
2. De S, Balu AM, van der Waal JC, Luque R (2015) Biomass-derived porous carbon materials: synthesis and catalytic applications. ChemCatChem 7:1608–1629
3. Barhoum A, Shalan AE, El-Hout SI, Ali GAM, Abdelbasir SM, Abu Serea ES, Ibrahim AH, Pal K (2019) A broad family of carbon nanomaterials: classification, properties, synthesis, and emerging applications, Handbook of nanofibers. Springer International Publishing, New York City, USA, pp 1–40
4. Chen Q, Tan X, Liu Y, Liu S, Li M, Gu Y, Zhang P, Ye S, Yang Z, Yang Y (2020) Biomass-derived porous graphitic carbon materials for energy and environmental applications. J. Mater Chem A 8:5773–5811
5. Varma RS (2019) Biomass-derived renewable carbonaceous materials for sustainable chemical and environmental applications. ACS Sustainable Chem Eng 7:6458–6470
6. Shen X, Shamshina JL, Berton P, Gurau G, Rogers RD (2016) Hydrogels based on cellulose and chitin: fabrication, properties, and applications. Green Chem 18:53–75
7. Visakh PM, Thomas S (2010) Preparation of bionanomaterials and their polymer nanocomposites from waste and biomass. waste and biomass valorization. Waste Biomass Valorization 1:121–134
8. Ali GAM, Divyashree A, Supriya S, Chong KF, Ethiraj AS, Reddy M, Algarni H, Hegde G (2017) Carbon nanospheres derived from lablab purpureus for high performance supercapacitor electrodes: a green approach. Dalton Trans 46:14034–14044
9. Titirici MM, White RJ, Brun N, Budarin VL, Su D, Monte FD, Clark JH, MacLachlan MJ (2015) Sustainable carbon materials. Chem Soc Rev 44:250–290
10. Liu WJ, Jiang H, Yu HQ (2019) Emerging applications of biochar-based materials for energy storage and conversion. Energy Environ Sci 12:1751–1779
11. Wang J, Nie P, Ding B, Hao X, Dong S, Dou H, Zhang X (2017) Biomass derived carbon for energy storage devices. J Mater Chem A 5:2411–2428
12. Ali GAM, Abdul Manaf SA, Kumar A, Chong KF, Hegde G (2014) High performance supercapacitor using catalysis free porous carbon nanoparticles. J Phys D 47:495307–495313
13. Eftekhari A, Jafarkhani P, Moztarzadeh F (2006) Synthesis of carbon nanotubes from solid carbon sources by direct microwave irradiation. Carbon 44:1343–1345
14. White RJ, Budarin V, Luque R, Clark JH, Macquarrie DJ (2009) Tuneable porous carbonaceous materials from renewable resources. Chem Soc Rev 38:3401–3418
15. Hu B, Wang K, Wu L, Yu SH, Antonietti M, Titirici MM (2010) Engineering carbon materials from the hydrothermal carbonization process of biomass. Adv Mater 22:813–828
16. Falco C, Caballero FP, Babonneau F, Gervais C, Laurent G, Titirici MM, Baccile N (2011) Hydrothermal carbon from biomass: structural differences between hydrothermal and pyrolyzed carbons via ^{13}C solid state NMR. Langmuir 27:14460–14471
17. Khalafallah D, Quan X, Ouyang C, Zhi M, Hong Z (2021) Heteroatoms doped porous carbon derived from waste potato peel for supercapacitors. Renew Energ 170:60–71
18. Zhao Y (2015) Oxygen-rich hierarchical porous carbon derived from artemia cyst shells with superior electrochemical performance. ACS Appl Mater Interfaces 7:1132–1139
19. Zhang D, Zheng L, Ma Y, Lei L, Li Q, Li Y, Luo H, Feng H, Hao Y (2014) Synthesis of nitrogen and sulfur-codoped 3D cubic-ordered mesoporous carbon with superior performance in supercapacitors. ACS Appl Mater Interfaces 6:2657–2665
20. Hao E, Liu W, Liu S, Zhang Y, Wang H, Chen S, Cheng F, Zhao S, Yang H (2017) Rich sulfur doped porous carbon materials derived from ginkgo leaves for multiple electrochemical energy storage device. J Mater Chem A 5:2204–2214
21. Lahaye J, Nanse G, Bagreev A, Strelko V (1999) Porous structure and surface chemistry of nitrogen containing carbons from polymers. Carbon 37:585–590

22. Wang DW, Li F, Yin LC, Lu X, Chen ZG, Gentle IR, Lu GQM, Cheng HM (2012) Nitrogen-doped carbon monolith for alkaline supercapacitors and understanding nitrogen-induced redox transitions. Chem Eur J 18:5345–5351
23. Hassen D, Shenashen MA, El-Safty AR, Elmarakbi A, El-Safty SA (2018) Anisotropic N-Graphene-diffused Co_3O_4 nanocrystals with dense upper-zone top-on-plane exposure facets as effective ORR electrocatalysts. Sci Rep 8:3740
24. Xing W, Liu C, Zhou Z, Zhang L, Zhou J, Zhuo S, Yan Z, Gao H, Wang G, Qiao SZ (2012) Superior CO_2 uptake of N-doped activated carbon through hydrogen-bonding interaction. Energy Environ Sci 5:7323–7327
25. Han C, Wang S, Wang J, Li M, Deng J, Li H, Wang Y (2014) Controlled synthesis of sustainable N-doped hollow core-mesoporous shell carbonaceous nanospheres from biomass. Nano Res. 7:1809–1819
26. Liu Q, Duan Y, Zhao Q, Pan F, Zhang B, Zhang J (2014) Direct synthesis of nitrogen-doped carbon nanosheets with high surface area and excellent oxygen reduction performance. Langmuir 30:8238–8245
27. Chaudhari NK, Song MY, Yu JS (2014) Heteroatom-doped highly porous carbon from human urine. Sci. Rep. 4:5221
28. Gao S, Li X, Li L, Wei X (2017) A versatile biomass derived carbon material for oxygen reduction reaction, supercapacitors and oil/water separation. Nano Energy 33:334–342
29. Zhu L, Shen F, Smith RL, Yan L, Li L, Qi X (2017) Black liquor-derived porous carbons from rice straw for high-performance supercapacitors. Chem Eng J 316:770–777
30. Hou J, Jiang K, Tahir M, Wu X, Idrees F, Shen M, Cao C (2017) Tunable porous structure of carbon nanosheets derived from puffed rice for high energy density supercapacitors. J Power Sources 371:148–155
31. Hou J, Jiang K, Wei R, Tahir M, Wu X, Shen M, Wang X, Cao C (2017) Popcorn-derived porous carbon flakes with an ultrahigh specific surface area for superior performance supercapacitors. ACS Appl Mater Interfaces 9:30626–30634
32. Peng C, Yan X, Wang R, Lang J, Ou Y, Xue Q (2013) Promising activated carbons derived from waste tea-leaves and their application in high performance supercapacitors electrodes. Electrochim Acta 87:401–408
33. Cao J, Zhu C, Aoki Y, Habazaki AH (2018) Starch-derived hierarchical porous carbon with controlled porosity for high performance supercapacitors. ACS Sustainable Chem. Eng. 6:7292–7303
34. Huang J, Liang Y, Hu H, Liu S, Cai Y, Dong H, Zheng M, Xiao Y, Liu Y (2017) Ultrahigh-surface-area hierarchical porous carbon from chitosan: acetic acid mediated efficient synthesis and its application in superior supercapacitors. J. Mater. Chem. A 5:24775–24781
35. Boyjoo Y, Cheng Y, Zhong H, Tian H, Pan J, Pareek VK (2017) From waste Coca Cola® to activated carbons with impressive capabilities for CO_2 adsorption and supercapacitors. Carbon 116:490–499
36. Ling Z, Wang Z, Zhang M, Yu C, Wang G, Dong Y, Liu S, Wang Y, Qiu J (2016) Sustainable synthesis and assembly of biomass-derived B/N co-doped carbon nanosheets with ultrahigh aspect ratio for high-performance supercapacitors. Adv Funct Mater 26:111–119
37. Zhou Y, Ren J, Xia L, Wu H, Xie F, Zheng Q, Xu C, Lin D (2017) Nitrogen-doped hierarchical porous carbon framework derived from waste pig nails for high-performance supercapacitors. ChemElectroChem. 4:3181–3187
38. Fan P, Ren J, Pang K, Cheng Y, Wu X, Zhang Z, Ren J, Huang W, Song R (2018) Cellulose-solvent-assisted, one-step pyrolysis to fabricate heteroatoms-doped porous carbons for electrode materials of supercapacitors. ACS Sustainable Chem. Eng. 6:7715–7724
39. Ali GAM, Habeeb OA, Algarni H, Chong KF (2018) Cao impregnated highly porous honeycomb activated carbon from agriculture waste: symmetrical supercapacitor study. J Mater Sci 54:683–692

40. Qian W, Sun F, Xu Y, Qiu L, Liu C, Wang S, Yan F (2014) Human hair-derived carbon flakes for electrochemical supercapacitors. Energy Environ Sci 7:379–386
41. Ranaweera CK, Kahol PK, Ghimire M, Mishra SR, Gupta RK (2017) Orange-peel-derived carbon: designing sustainable and high-performance supercapacitor electrodes. J. Carbon Res. 3:25
42. Long C, Chen X, Jiang L, Zhi L, Fan Z (2015) Porous layer-stacking carbon derived from in-built template in biomass for high volumetric performance supercapacitors. Nano Energy. 12:141–151
43. Hou J, Cao C, Idrees F, Ma X (2015) Hierarchical porous nitrogen-doped carbon nanosheets derived from silk for ultrahigh-capacity battery anodes and supercapacitors. ACS Nano 9:2556–2564
44. Yuan G, Li H, Hu H, Xie Y, Xiao Y, Dong H, Liang Y, Liu Y, Zheng M (2019) Microstructure engineering towards porous carbon materials derived from one biowaste precursor for multiple energy storage applications. Electrochim Acta 326:134974
45. Zhu J, Liu S, Liu Y, Meng T, Ma L, Zhang H, Kuang M, Jiang J (2018) Graphitic, porous, and multiheteroatom codoped carbon microtubes made from hair waste: a superb and sustained anode substitute for li-ion batteries. ACS Sustain. Chem. Eng. 6:13662–13669
46. Zhang D, Wang G, Xu L, Lian J, Bao J, Zhao Y, Qiu J, Li H (2018) Defect-rich N-doped porous carbon derived from soybean for high rate lithium-ion batteries. Appl Surf Sci 451:298–305
47. Ali GAM, Supriya S, Chong KF, Shaaban ER, Algarni H, Maiyalagan T, Gurumurthy H (2019) Superior supercapacitance behavior of oxygen self-doped carbon nanospheres: a conversion of Allium cepa peel to energy storage system. Biomass Conv Bioref. https://doi.org/10.1007/s13399-019-00520-3
48. Selvamani V, Ravikumar R, Suryanarayanan V, Velayutham D, Gopukumar S (2015) Fish scale derived nitrogen doped hierarchical porous carbon—a high rate performing anode for lithium ion cell. Electrochimca Acta 182:1–10
49. Xu G, Han J, Ding B, Nie P, Pan J, Dou H, Li H, Zhang X (2015) Biomass-derived porous carbon materials with sulfur and nitrogen dual-doping for energy storage. Green Chem 17:1668–1674
50. Wu F, Huang R, Mu D, Wu B (2016) Chen Y (2016) Controlled synthesis of graphitic carbon-encapsulated α-Fe_2O_3 nanocomposite via low-temperature catalytic graphitization of biomass and its lithium storage property. Electrochim Acta 187:508–516
51. Ali GAM, Manaf SAA, Divyashree A, Chong KF, Hegde G (2016) Superior supercapacitive performance in porous nanocarbons. J. Energy Chem. 25:734–739
52. Wei J, Zhou D, Sun Z, Deng Y, Xia Y, Zhao DY (2013) A controllable synthesis of rich nitrogen-doped ordered mesoporous carbon for CO2 capture and supercapacitors. Adv Funct Mater 23:2322–2328
53. Guo DC, Mi J, Hao GP, Dong W, Xiong G, Li WC, Lu AH (2013) Ionic liquid $C_{16}mimBF_4$ assisted synthesis of poly(benzoxazine-*co*-resol)-based hierarchically porous carbons with superior performance in supercapacitors. Energy Environ Sci 6:652–659
54. Vinu A, Terrones M, Golberg D, Hishita S, Ariga K, Mori T (2005) Synthesis of mesoporous bn and bcn exhibiting large surface areas via templating methods. Chem Mater 17:5887–5890
55. Linares N, Serrano E, Rico M, Balu AM, Losada E, Luque R, Garc.a-Mart.nez J (2011) Incorporation of chemical functionalities in the framework of mesoporous silica. Chem Commun 47:9024–9035
56. Zhang LL, Zhao XS (2009) Carbon-based materials as supercapacitor electrodes. Chem Soc Rev 38:2520–2531
57. Zhang LL, Li HH, Shi YH, Fan CY, Wu XL, Wang HF, Sun HZ, Zhang JP (2016) A novel layered sedimentary rocks structure of the oxygen-enriched carbon for ultrahigh-rate-performance supercapacitors. ACS Appl Mater Interfaces 8:4233–4241

58. Ye JS, Liu X, Cui HF, Zhang WD, Sheu FS, Lim TM (2005) Electrochemical oxidation of multi-walled carbon nanotubes and its application to electrochemical double layer capacitors. Electrochem Commun 7:249–255
59. Lota G, Tyczkowski J, Kapica R, Lota K, Frackowiak E (2010) Carbon materials modified by plasma treatment as electrodes for supercapacitors. J Power Sources 195:7535–7539
60. Lyu L, Seong KD, Ko D, Choi J, Lee C, Hwang T, Cho Y, Jin X, Zhang W, Pang H, Piao Y (2019) Recent development of biomass-derived carbons and composites as electrode materials for supercapacitors. Mater. Chem. Front. 3:2543–2570
61. Wang J, Shen L, Nie P, Yun X, Xu Y, Dou H, Zhang X (2015) N-doped carbon foam based three-dimensional electrode architectures and asymmetric supercapacitors. J. Mater. Chem. A 3:2853–2860
62. Jia L, Kang L, Xiang G, Yao B, Huo K, Cheng Y, Cheng X, Chen D, Bo W, Sun W, Ding D, Liu M, Huang L (2015) Oxygen-and nitrogen-enriched 3D porous carbon for supercapacitors of high volumetric capacity. ACS Appl Mater Interfaces 7:24622–24628
63. Fuertes AB, Sevilla M (2015) Superior capacitive performance of hydrochar-based porous carbons in aqueous electrolytes. Chemsuschem 8:1049–1057
64. Yun YS, Cho SY, Shim J, Kim BH, Chang SJ, Baek SJ, Huh YS, Tak Y, Park YW, Park S (2013) Microporous carbon nanoplates from regenerated silk proteins for supercapacitors. Adv Mater 25:1993–1998
65. Hegde G, Abdul Manaf SA, Kumar A, Ali GAM, Chong KF, Ngaini Z, Sharma KV (2015) Biowaste sago bark based catalyst free carbon nanospheres: Waste to wealth approach, ACS Sustain. Chem Eng 5:2247–2253
66. Wang F, Wu X, Yuan X, Liu Z, Zhang Y, Fu L, Zhu Y, Zhou Q, Wu Y, Huang W (2017) Latest advances in supercapacitors: from new electrode materials to novel device designs. Chem Soc Rev 46:6816–6854
67. Hirose T, Fujino T, Fan TX, Endo H, Okabe T, Yoshimura M (2002) Effect of carbonization temperature on the structural changes of woodceramics impregnated with liquefied wood. Carbon 40:761–765
68. Salvador F, Sanchez-Montero MJ, Izquierdo C (2007) C/H_2O reaction under supercritical conditions and their repercussions in the preparation of activated carbon. J Phys Chem C 111:14011–14020
69. Wang Q, Li H, Chen LQ (2001) Huang XJ (2001) Monodispersed hard carbon spherules with uniform nanopores. Carbon 39:2211–2214
70. Sun XM, Li YD (2004) Colloidal carbon spheres and their core/shell structures with noble-metal nanoparticles. Angew Chem Int Ed 43:597–601
71. Rabenau A (1985) The Role of hydrothermal synthesis in preparative chemistry. Angew. Chem. Int. Ed. Engl. 24:1026–1040
72. Hacncl MW (1992) Recent progress in coal structure research. Fuel 71:1211–1223
73. Titirici M-M, White RJ, Falco C, Sevilla M (2012) Black perspectives for a green future: hydrothermal carbons for environment protection and energy storage. Energy Environ Sci 5:6796–6822
74. Kan T, Strezov V, Evans TJ (2016) Lignocellulosic biomass pyrolysis: A review of product properties and effects of pyrolysis parameters. Renew. Sust. Energ. Rev. 57:1126–1140
75. Habeeb OA, Ramesh K, Ali GAM, Yunus RM (2017) Experimental design technique on removal of hydrogen sulfide using cao-eggshells dispersed onto palm kernel shell activated carbon: experiment, optimization, equilibrium and kinetic studies. J Wuhan Univ Technol Mater Sci Ed 32:305–320
76. Du S, Valla JA, Bollas GM (2013) Characteristics and origin of char and coke from fast and slow, catalytic and thermal pyrolysis of biomass and relevant model compounds. Green Chem 15:3214–3229
77. Babu BV (2008) Biomass pyrolysis: a state-of-the-art review. Biofuel, Bioprod Biorefin. 2:393–414
78. Wang S, Dai G, Yang H, Luo Z (2017) Lignocellulosic biomass pyrolysis mechanism: A State-of-the-art review. Prog Energ Combust. 62:33–86

79. Wang Z, Yu J, Zhang X, Li N, Liu B, Li Y, Wang Y, Wang W, Li Y, Zhang L, Dissanayake S, Suib SL, Sun L (2016) Large-scale and controllable synthesis of graphene quantum dots from rice husk biomass: a comprehensive utilization strategy. ACS Appl Mater Interfaces 8:1434
80. Meng W, Bai X, Wang B, Liu Z, Lu S, Yang B (2019) Biomass-derived carbon dots and their applications. Energy Environ. Mater. 2:172–192
81. Manyà JJ (2012) Pyrolysis for biochar purposes: A review to establish current knowledge gaps and research needs. Environ Sci Technol 46:7939–7954
82. Wang Z, Shen D, Wu C, Gu S (2018) State-of-the-art on the production and application of carbon nanomaterials from biomass. Green Chem 20:5031–5057
83. Ru H, Bai N, Xiang K, Zhou W, Chen H, Zhao XS (2016) Porous carbons derived from microalgae with enhanced electrochemical performance for lithium-ion batteries. Electrochim Acta 194:10–16
84. Smith AJ, MacDonald MJ, Ellis LD, Obrovac MN, Dahn JR (2012) A small angle X-ray scattering and electrochemical study of the decomposition of wood during pyrolysis. Carbon 50:3717–3723
85. Antal MJ, Grønli M (2003) The art, science, and technology of charcoal production. Ind Eng Chem Res 42:1619–1640
86. Wang RZ, Huang DL, Liu YG, Zhang C, Lai C, Zeng GM, Cheng M, Gong XM, Wan J, Luo H (2018) Investigating the adsorption behavior and the relative distribution of Cd^{2+} sorption mechanisms on biochars by different feedstock. Bioresour Technol 261:265–271
87. Zhou L, Liu Y, Liu S, Yin Y, Zeng G, Tan X, Hu X, Hu X, Jiang L, Ding Y, Liu S, Huang X (2016) Investigation of the adsorption-reduction mechanisms of hexavalent chromium by ramie biochars of different pyrolytic temperatures. Bioresour Technol 218:351–359
88. Liu WJ, Jiang H, Yu HQ (2015) Development of biochar-based functional materials: toward a sustainable platform carbon material. Chem Rev 115:12251–12285
89. Fromm O, Heckmann A, Rodehorst UC, Frerichs J, Becker D, Winter M, Placke T (2018) Carbons from biomass precursors as anode materials for lithium ion batteries: New insights into carbonization and graphitization behavior and into their correlation to electrochemical performance. Carbon 128:147–163
90. Yu J, Tang L, Pang Y, Zeng G, Feng H, Zou J, Wang J, Feng C, Zhu X, Ouyang X, Tan J (2020) Hierarchical porous biochar from shrimp shell for persulfate activation: A two-electron transfer path and key impact factors. Appl Catal B 260:118160
91. Jung A, Han S, Kim T, Cho WJ, Lee KH (2013) Synthesis of high carbon content microspheres using 2-step microwave carbonization, and the influence of nitrogen doping on catalytic activity. Carbon 60:307–316
92. Knox JH, Kaur B, Millward GR, Chromatogr J (1986) Structure and performance of porous graphitic carbon in liquid chromatography. J Chromatogr A 352:3–25
93. Shopsowitz KE, Hamad WY, MacLachlan MJ (2011) Chiral nematic mesoporous carbon derived from nanocrystalline cellulose. Angew Chem Int Ed 50:10991–10995
94. Estevez L, Dua R, Bhandari N, Ramanujapuram A, Wang P, Giannelis EP (2013) A facile approach for the synthesis of monolithic hierarchical porous carbons – high performance materials for amine based CO_2 capture and supercapacitor electrode. Energy Environ Sci 6:1785–1790
95. Kubo S, White RJ, Yoshizawa N, Antonietti M, Titirici MM (2011) Ordered Carbohydrate-Derived Porous Carbons. Chem Mater 23:4882–4885
96. Saha D, Li Y, Bi Z, Chen J, Keum JK, Hensley DK, Grappe HA, Meyer HM, Dai S, Paranthaman MP, Naskar AK (2014) Studies on supercapacitor electrode material from activated lignin-derived mesoporous carbon. Langmuir 30:900–910
97. Wang S, Han C, Wang J, Deng J, Zhu M, Yao J, Li H, Wang Y (2014) Controlled synthesis of ordered mesoporous carbohydrate-derived carbons with flower-like structure and N-doping by self-transformation. Chem Mater 26:6872–6877

98. Huang CH, Doong RA (2012) Sugarcane bagasse as the scaffold for mass production of hierarchically porous carbon monoliths by surface self-assembly. Microporous Mesoporous Mater 147:47–52
99. Zhang P, Gong Y, Lv Y, Guo Y, Wang Y, Wang C, Li H (2012) Ionic liquids with metal chelate anions. Chem Commun 48:2334–2336
100. Lee JS, Mayes RT, Luo H, Dai S (2010) Ionothermal carbonization of sugars in a protic ionic liquid under ambient conditions. Carbon 48:3364–3368
101. Xie ZL, White RJ, Weber J, Taubert A, Titirici MM (2011) Hierarchical porous carbonaceous materials *via* ionothermal carbonization of carbohydrates. J Mater Chem 21:7434–7442
102. Zhang P, Gong Y, Wei Z, Wang J, Zhang Z, Li H, Dai S, Wang Y (2014) Updating biomass into functional carbon material in ionothermal manner. ACS Appl Mater Interfaces 6:12515–12522
103. Gong Y, Li D, Luo C, Fu Q, Pan C (2017) Highly porous graphitic biomass carbon as advanced electrode materials for supercapacitors. Green Chem 19:4132–4140
104. Zhu H, Yin J, Wang X, Wang H, Yang X (2013) Microorganism-derived heteroatom-doped carbon materials for oxygen reduction and supercapacitors. Adv Funct Mater 23:1305–1312
105. Li Z, Zhang L, Amirkhiz BS, Tan X, Xu Z, Wang H, Olsen BC, Holt CMB (2012) Mitlin, D. Carbonized chicken eggshell membranes with 3D architectures as high-performance electrode materials for supercapacitors. Adv. Energy Mater. 2:431–437
106. Chen C, Yu D, Zhao G, Du B, Tang W, Sun L, Sun Y, Besenbacher F, Yu M (2016) Three-dimensional scaffolding framework of porous carbon nanosheets derived from plant wastes for high-performance supercapacitors. Nano Energy 27:377–389
107. Zhou M, Gomez J, Li B, Jiang YB, Deng S (2017) Oil tea shell derived porous carbon with an extremely large specific surface area and modification with MnO_2 for high-performance supercapacitor electrodes. Appl. Mater. Today 7:47–54
108. Ba H, Wang W, Pronkin S, Romero T, Baaziz W, Nguyen-Dinh L, Chu W, Ersen O, Pham-Huu C (2018) Biosourced foam-like activated carbon materials as high-performance supercapacitors. Adv. Sustain. Syst. 2:1700123
109. Subramani K, Sudhan N, Karnan M, Sathish M (2017) Orange peel derived activated carbon for fabrication of high-energy and high-rate supercapacitors. ChemistrySelect 2:11384–11392
110. Cheng P, Gao S, Zang P, Yang X, Bai Y, Xu H, Liu Z, Lei Z (2015) Hierarchically porous carbon by activation of shiitake mushroom for capacitive energy storage. Carbon 93:315–324
111. Dai C, Wan J, Geng W, Song S, Ma F, Shao J (2017) KOH direct treatment of kombucha and in situ activation to prepare hierarchical porous carbon for high-performance supercapacitor electrodes. J Solid State Electrochem 21:2929–2938
112. Li B, Dai F, Xiao Q, Yang L, Shen J, Zhang C, Cai M (2016) Activated carbon from biomass transfer for high-energy density lithium-ion supercapacitors. Adv. Energy Mater. 6:1600802
113. Li Y, Wang G, Wei T, Fan Z, Yan P (2016) Nitrogen and sulfur co-doped porous carbon nanosheets derived from willow catkin for supercapacitors. Nano Energy 19:165–175
114. Yu P, Zhang Z, Zheng L, Teng F, Hu L, Fang X (2016) A novel sustainable flour derived hierarchical nitrogen-doped porous carbon/polyaniline electrode for advanced asymmetric supercapacitors. Adv. Energy Mater. 6:160111
115. Yan J, Wang Q, Wei T, Fan Z (2014) Supercapacitors: recent advances in design and fabrication of electrochemical supercapacitors with high energy densities. Adv. Energy Mater. 4:1300816
116. Cheng P, Li T, Yu H, Zhi L, Liu Z, Lei Z (2016) Biomass-derived carbon fiber aerogel as a binder-free electrode for high-rate supercapacitors. J Phys Chem C 120:2079–2086
117. Wang J, Kaskel S (2012) KOH activation of carbon-based materials for energy storage. J Mater Chem 22:23710–23725
118. Wang H, Gao Q, Hu J (2009) High hydrogen storage capacity of porous carbons prepared by using activated carbon. J Am Chem Soc 131:7016–7022

119. Lozano-Castello D, Calo JM, Cazorla-Amoros D, Linares-Solano A (2007) Carbon activation with KOH as explored by temperature programmed techniques and the effects of hydrogen. Carbon 45:2529–2536
120. Sahu V, Shekhar S, Sharma RK, Singh G (2015) Ultrahigh performance supercapacitor from lacey reduced graphene oxide nanoribbons. ACS Appl Mater Interfaces 7:3110–3116
121. Hayashi JI, Kazehaya A, Muroyama K, Watkinson AP (2000) Preparation of activated carbon from lignin by chemical activation. Carbon 38:1873–1878
122. Sun L, Fu Y, Tian C, Yang Y, Wang L, Yin J, Ma J, Wang R, Fu H (2014) Isolated Boron and Nitrogen Sites on Porous Graphitic Carbon Synthesized from Nitrogen-Containing Chitosan for Supercapacitors. Chemsuschem 7:1637–1646
123. Rufford TE, Hulicova-Jurcakova D, Zhu Z, Lu GQ (2008) Nanoporous carbon electrode from waste coffee beans for high performance supercapacitors. Electrochem Commun 10:1594–1597
124. Sun L, Tian C, Li M, Meng X, Wang L, Wang R, Yin J, Fu H (2013) From coconut shell to porous graphene-like nanosheets for high-power supercapacitors. J. Mater. Chem. A 1:6462–6470
125. Mozammel HM, Masahiro O, Bhattacharya SC (2002) Activated charcoal from coconut shell using $ZnCl_2$ activation. Biomass Bioenergy 22:397–400
126. Guangzhen Z, Yanjiang L, Guang Z, Junyou S, Ting L, Likun P (2019) Biomass-based N, P, and S self-doped porous carbon for high-performance supercapacitors. ACS Sustainable Chem. Eng. 7:12052–12060
127. Patel MA, Luo F, Khoshi MR, Rabie E, Zhang Q, Flach CR, Mendelsohn R, Garfunkel E, Szostak M, He H (2016) P–doped porous carbon as metal free catalysts for selective aerobic oxidation with an unexpected mechanism. ACS Nano 10:2305–2315
128. Sevilla M, Fuertes AB (2016) A green approach to high-performance supercapacitor electrodes: the chemical activation of hydrochar with potassium bicarbonate. Chemsuschem 9:880–1888
129. Wei J, Iglesia E (2004) Isotopic and kinetic assessment of the mechanism of reactions of CH_4 with CO_2 or H_2O to form synthesis gas and carbon on nickel catalysts. J Catal 224:370–383
130. Yang K, Peng J, Xia H, Zhang L, Srinivasakannan C, Guo S (2010) Textural characteristics of activated carbon by single step CO_2 activation from coconut shells. J. Taiwan Inst. Chem. E. 41:367–372
131. Bommier C, Xu R, Wang W, Wang X, Wen D, Lu J, Ji X (2015) Self-activation of cellulose: A new preparation methodology for activated carbon electrodes in electrochemical capacitors. Nano Energy 13:709–717
132. Raymundo-Pinero E, Cadek M, Beguin F (2009) Tuning carbon materials for supercapacitors by direct pyrolysis of seaweeds. Adv Funct Mater 19:1032–1039
133. Panchakarla LS, Subrahmanyam KS, Saha SK, Achutharao G, Krishnamurthy HR, Waghmare UV, Rao CNR (2009) Synthesis, structure, and properties of boron- and nitrogen-doped graphene. Adv Mater 21:4726–4730
134. Wu ZS, Winter A, Chen L, Sun Y, Turchanin A, Feng X, Mullen K (2012) three-dimensional nitrogen and boron co-doped graphene for high-performance all-solid-state supercapacitors. Adv Mater 24:5130–5135
135. Xu G, Ding B, Pan J, Han J, Nie P, Zhu Y, Sheng Q, Dou H (2015) Porous nitrogen and phosphorus co-doped carbon nanofiber networks for high performance electrical double layer capacitors. J. Mater. Chem. A 3:23268–23273
136. Huang C, Sun T, Hulicova-Jurcakova D (2013) Wide electrochemical window of supercapacitors from coffee bean-derived phosphorus-rich carbons. Chemsuschem 6:2330–2339
137. El-Sawy AM, Mosa IM, Su D, Guild CJ, Khalid S, Joesten R, Rusling JF, Suib SL (2016) Controlling the active sites of sulfur-doped carbon nanotube–graphene nanolobes for highly efficient oxygen evolution and reduction catalysis. Adv. Energy Mater. 6:1501966

138. Zhao G, Chen C, Yu D, Sun L, Yang C, Zhang H, Sun Y, Besenbacher F, Yu M (2018) One-step production of O-N-S co-doped three-dimensional hierarchical porous carbons for high-performance supercapacitors. Nano Energy 47:547–555
139. Deng Y, Ji Y, Wu H, Chen F (2019) Enhanced electrochemical performance and high voltage window for supercapacitor based on multi-heteroatom modified porous carbon materials. Chem Commun 55:1486–1489
140. Scott V, Gilfillan S, Markusson N, Chalmers H, Haszeldine RS (2013) Last chance for carbon capture and storage. Nat Clim Chang. 3:105–111
141. Haszeldine RS (2009) Carbon sequestration. Science 325:1644–1645
142. Wang T, Lackner KS, Wright A (2011) Moisture Swing Sorbent for Carbon Dioxide Capture from Ambient Air. Environ Sci Technol 45:6670–6675
143. Parshetti GK, Chowdhury S, Balasubramanian R (2015) Biomass derived low-cost microporous adsorbents for efficient CO_2 capture. Fuel 148:246–254
144. Sevilla M, Fuertes AB (2011) Sustainable porous carbons with a superior performance for CO_2 capture. Energy Environ Sci 4:1765–1771
145. Chen J, Yang J, Hu G, Hu X, Li Z, Shen S, Radosz M, Fan M (2016) Enhanced CO_2 capture capacity of nitrogen-doped biomass-derived porous carbons. ACS Sustain. Chem. Eng. 4:1439–1445
146. Singh G, Kim IY, Lakhi KS, Joseph S, Srivastava P, Naidu R, Vinu A (2017) Heteroatom functionalized activated porous biocarbons and their excellent performance for CO_2 capture at high pressure. J Mater Chem 5:21196–21204
147. Zhu B, Shang C, Guo Z (2016) Naturally nitrogen and calcium-doped nanoporous carbon from pine cone with superior CO_2 capture capacities. ACS Sustain. Chem. Eng. 4:1050–1057
148. Zhu B, Qiu K, Shang C, Guo Z (2015) Naturally derived porous carbon with selective metal-and/or nitrogen-doping for efficient CO_2 capture and oxygen reduction. J. Mater. Chem. A 3:5212–5222
149. European academies science advisory council (2016) priorities for critical materials for a circular economy. German National Academy of Sciences Leopoldina, Halle
150. Khalafallah D, Alothman OY, Fouad H, Khalil KA (2018) Nitrogen and carbon functionalized cobalt phosphide as efficient non-precious electrocatalysts for oxygen reduction reaction electrocatalysis in alkaline environment. Electroanal Chem 809:96–104
151. Khalafallah D, Akhtar N, Alothman OY, Fouad H, Khalil KA (2017) Self-assembled dopamine nanolayers wrapped carbon nanotubes as carbon-carbon bi-functional nanocatalyst for highly efficient oxygen reduction reaction and antiviral drug monitoring. Solid State Sci 71:51–60
152. Wang N, Li T, Song Y, Liu J, Wang F (2018) Metal-free nitrogen doped porous carbons derived from pomelo peel treated by hypersaline environments for oxygen reduction reaction. Carbon 130:692–700
153. Gao S, Wei X, Liu H, Geng K, Wang H, Moehwald H, Shchukin D (2015) Transformation of worst weed into N-, S-, and P-tridoped carbon nanorings as metal-free electrocatalysts for the oxygen reduction reaction. J. Mater. Chem. A 3:23376–23384
154. Li M, Xiong Y, Liu X, Han C, Zhang Y, Bo X, Guo L (2015) Iron and nitrogen co-doped carbon nanotube@hollow carbon fibers derived from plant biomass as efficient catalysts for the oxygen reduction reaction. J. Mater. Chem. A 3:9658–9667
155. Liu Y, Ruan J, Sang S, Zhou Z, Wu Q (2016) Iron and nitrogen co-doped carbon derived from soybeans as efficient electro-catalysts for the oxygen reduction reaction. Electrochim Acta 215:388–397
156. Wu D, Zhu C, Shi Y, Jing H, Hu J, Song X, Si D, Liang S, Hao C (2019) Biomass-derived multilayer-graphene-encapsulated cobalt nanoparticles as efficient electrocatalyst for versatile renewable energy applications. ACS Sustainable Chem. Eng. 7:1137–1145
157. Liu l, Zeng G, Chen J, Bi L, Dai L, Wen Z (2018) N-doped porous carbon nanosheets as pH-universal ORR electrocatalyst in various fuel cell devices. Nano Energy 49:393–402

158. He D, Zhao W, Li P, Liu Z, Wu H, Liu L, Han K, Liu L, Wan Q, Butt FK, Qu X (2019) Bifunctional biomass-derived 3D nitrogen-doped porous carbon for oxygen reduction reaction and solid-state supercapacitor. Appl Surf Sci 465:303–312
159. Chatterjee K, Ashokkumar M, Gullapalli H, Gong Y, Vajtai R, Thanikaivelan P, Ajayan PM (2018) Nitrogen-rich carbon nano-onions for oxygen reduction reaction. Carbon 130:645–651
160. Wang G, Deng Y, Yu J, Zheng L, Du L, Song H, Liao S (2017) From chlorella to nestlike framework constructed with doped carbon nanotubes: a biomass-derived, high-performance, bifunctional oxygen reduction/evolution catalyst. ACS Appl Mater Interfaces 9:32168–32178
161. Gao S, Geng K, Liu H, Wei X, Zhang M, Wang P, Wang J (2015) Transforming organic-rich amaranthus waste into nitrogen-doped carbon with superior performance of the oxygen reduction reaction. Energy Environ Sci 8:221–229
162. Wei Q, Yang X, Zhang G, Wang D, Zuin L, Banham D, Yang L, Ye S, Wang Y, Mohamedi M, Sun S (2018) An active and robust Si-Fe/N/C catalyst derived from waste reed for oxygen reduction. Appl. Catal. B: Environ. 237:85–93
163. Zhao Y, Li X, Jia X, Gao S (2019) Why and how to tailor the vertical coordinate of pore size distribution to construct ORR-active carbon materials? Nano Energy 58:384–391
164. Kalyani P, Anitha A, Darchen A (2013) Activated carbon from grass – A green alternative catalyst support for water electrolysis. Int J Hydrogen Energy 38:10364–10372
165. Zhang C, Antonietti M, Fellinger TP (2014) Blood ties: Co_3O_4 decorated blood derived carbon as a superior bifunctional electrocatalyst. Adv Funct Mater 24:7655–7665
166. Larcher D, Tarascon JM (2014) Towards greener and more sustainable batteries for electrical energy storage. Nat. Chem. 7:19–29
167. Winter M, Barnett B, Xu K (2018) Before Li ion batteries. Chem Rev 118:11433–11456
168. Li M, Lu J, Chen Z, Amine K (2018) 30 Years of lithium-ion batteries. Adv Mater 30:1800561
169. Armand M, Tarascon JM (2008) Building better batteries. Nature 451:652–657
170. Tarascon JM, Armand M (2001) Issues and challenges facing rechargeable lithium batteries. Nature 414:359–367
171. Wang JG, Jin D, Liu H, Zhang C, Zhou R, Shen C, Xie K, Wei B (2016) All-manganese-based Li-ion batteries with high rate capability and ultralong cycle life. Nano Energy 22:24–532
172. Xia Y, Xiao Z, Dou X, Huang H, Lu X, Yan R, Gan Y, Zhu W, Tu J, Zhang W, Tao X (2013) Green and facile fabrication of hollow porous MnO/C microspheres from microalgaes for lithium-ion batteries. ACS Nano 7:7083–7092
173. Zang J, Ye J, Qian H, Lin Y, Zhang X, Zheng M, Dong Q (2018) Hollow carbon sphere with open pore encapsulated MnO_2 nanosheets as high-performance anode materials for lithium ion batteries. Electrochim Acta 260:783–788
174. Eftekhari A (2017) Low voltage anode materials for lithium-ion batteries. Energy Storage Mater. 7:157–180
175. Taberna PL, Mitra S, Poizot P, Simon P, Tarascon JM (2006) High rate capabilities Fe_3O_4-based Cu nano-architectured electrodes for lithium-ion battery applications. Nat Mater 5:567–573
176. Zhou X, Chen F, Bai T, Long B, Liao Q, Ren Y, Yang J (2016) Interconnected highly graphitic carbon nanosheets derived from wheat stalk as high performance anode materials for lithium ion batteries. Green Chem 18:2078–2088
177. Wang L, Schnepp Z, Titirici MM (2013) Rice husk-derived carbon anodes for lithium ion batteries. Rice husk-derived carbon anodes for lithium ion batteries. J. Mater. Chem. A 1:5269–5273

Recycled Activated Carbon-Based Materials for the Removal of Organic Pollutants from Wastewater

Seyedehmaryam Moosavi, Chin Wei Lai, Omid Akbarzadeh, and Mohd Rafie Johan

Abstract Wastewater treatment has been drawing more and more attention due to increasing water pollution. The most common sources of water pollution are heavy metals, dyes, plastics, and foods from industries. Animal, agriculture, and farm wastes are other important causes of water pollution. Several adsorbents have been tested for wastewater treatment. Among these, activated carbon (AC) is the best and the most effective adsorbent for a specific class of pollutants because of its abundant starting materials availability, high surface area, surface reactivity, adsorption efficiency, and porosity structure. Although commercial AC has outstanding potential in the industry of water treatment, the usage is limited because of the high cost. The high initial cost and expensive regeneration of AC motivate the search for low cost, disposable ACs from conventional wastes materials to remove dyes, volatile organic compounds, heavy metals, and organic pollutants. The main goal of this chapter is to compare and list the advantages and disadvantages and methods of AC preparation from wastes materials as adsorbents and their application in water treatment to remove pollutants.

Keywords Recycling · Activated carbon · Wastewater treatment · Pollutant removal · Adsorption

List of Abbreviations

AC	Activated carbon
AOP	Advanced oxidation processes
BET	Brunauer–emmett–teller
BOD	Biological oxygen demand
C_0	Initial concentration
C_e	Equilibrium concentration
CMS	Carbon molecular sieves
CNS	Cashew nutshell

S. Moosavi (✉) · C. W. Lai · O. Akbarzadeh · M. R. Johan
Nanotechnology & Catalysis Research Centre (NANOCAT), Institute for Advanced Studies (IAS), University for Malaya (UM), Level 3, Block A, 50603 Kuala Lumpur, Malaysia
e-mail: m.moosavi1987@gmail.com

A. S. H. Makhlouf and G. A. M. Ali (eds.), *Waste Recycling Technologies for Nanomaterials Manufacturing*, Topics in Mining, Metallurgy and Materials Engineering, https://doi.org/10.1007/978-3-030-68031-2_18

DW	Distilled water
FESEM	Field emission scanning electron microscope
FTIR	Fourier transform infrared spectroscopy
GAC	Granular Activated carbon
m	Mass
MB	Methylene blue
ms	Mass of adsorbent
NOM	Natural organic matter
PAC	Powdered Activated carbon
q_t	Concentration at t time
RR 24	Reactive red 24
TEM	Transmission electron microscope
TGA	Thermogravimetric analysis
USEPA	United States Environmental Protection Agency
UV–Vis	Ultraviolet–Visible Spectrophotometry
XRD	X–Ray Diffraction

1 Introduction

1.1 Types of Pollutants

There are lots of substances in treated and untreated wastewater that are toxic and very harmful for humans, plants, and animals. These pollutants are very persistent and have negative impacts on the environment. The primary wastewater contaminants are organic pollutants, nitrogen, phosphorus, hydrocarbons, heavy metals, microbes, and endocrine disruptors are the major contaminants in wastewater that lead to adverse effects on both human health and the environment. These pollutants cause the propagation of pathogenic organisms like viruses protozoa, fungi, and bacteria. Therefore, it is imperative to treat the wastewater before discharging into the rivers, etc.

1.2 Water and Wastewater Treatment

Water is essential for human beings' natural resources. Although access to safe and clean water is vital for living organisms, ecosystems, and human beings, substantial amounts of inadequately treated wastewater from domestic, industry, and agriculture are still discharged into clean bodies of water. Consequently, this leads to water contamination with direct and indirect implications for human beings and natural ecosystems. Furthermore, most of the wastewater flows into the oceans and causes severe damage to the environment in many ways [1]. One of the most dangerous organic pollutants is dyes, and 15% are let out into the environment [2, 3]. Annually, dyes production was about 10,000 tons in the world wild, which is due to their usage in lots of industries like food, textiles, cosmetics, plastic, paper, pharmaceutical, and inks [3, 4]. Among all these industries, the textile industry had the highest damage to the environment (various types of dyes). Annually, around 280,000 tons of dyes were dispersed to the environment from the textile industry, especially in the marine environment [5]. Moreover, some types of dyes have toxic compounds, which can cause severe damage to the ecosystem and health. In the past decades, water usage is increased due to industrialization, agricultural expansion, and urbanization. Besides, the over usage of various toxic chemicals caused the pollution of soil, vegetables, and also water bodies fouling to a greater extent.

Pollutions of water are a very persistent and serious problem and have evolved gradually. The time for recognition of these kinds of problems and essential applications for prevention measures takes a longer time. There were lots of complaints and report about stinking water and industrial waste disposal within overcrowded cities. Consumption of organic pollutants in large quantities by industries and their discharge into the environment made them a serious issue, which needs to be managed [5]. So, it is vital to reduce the pollutant to acceptable international concentrations before being discharged to the environment. As the organic pollutant removal is difficult, the World Health Organization (WHO) recommended the numbers of processes for the reduction of pollutant concentration to the limits. These processes are very effective for organic pollutants removal and are including adsorption [6–8], chemical coagulation [9], filtration [10], and photodegradation [11–15]. However, their disadvantage is the production of secondary waste.

The methods of wastewater treatment are broadly classified into chemical, physical, and biological processes. The selection of any of the mentioned methods is based on the pollutant concentration in the solution and the treatment cost. The cost factors restrict most of the methods since they are in demand for considerable financial input, which makes pollution control less favourable. Adsorption (among all types of treatment) is confirmed to be very effective and easy to perform [16–19].

1.3 Industrial Wastewater Treatment

According to the industrial process, industrials plants have unique methods, and the water after treatment is discharged in the sewage system. There are three general methods for wastewater treatment, including chemical, biological, and physical methods [20].

1.3.1 Chemical Methods

Chemical techniques such as irradiation, coagulation, precipitation, ozonation, chemical adsorption, and oxidation are used for the separation of toxic dyes and metals [21, 22]. These processes are effortful and rapid, and no adsorbent weight loss is reported because of regeneration. The main disadvantages of the mentioned methods are their limited versatility, capability, high cost, and energy consumption. These methods cause secondary pollution problem and produce a large volume of sludge. Chemical adsorption (chemisorption) is the formation of the chemical bond between the adsorbent and adsorbate that is a monolayer. Chemisorption is an irreversible method and takes place at high temperature [23]. Nowadays, advanced oxidation processes (AOP) becomes one of the significant technologies in the water treatment field, especially for biodegradable pollutants. AOP is used for the degradation and destruction of organic compounds by generating hydroxyl radicals [24]. The most studied AOPs are the Fenton reaction process using H_2O_2 in the presence of the catalyst [25, 26]. The unique advantages of the Fenton oxidation process are high easy operation, degradation efficiency, mild reaction conditions, low cost, and environmentally friendly materials [27].

1.3.2 Biological Methods

Biological methods are the most attractive alternative for physical and chemical methods [28]. This method includes pollutants microbial degradation, which is a prevalent technique for the treatment of wastewater. However, some technical constraints restricted their applications. These methods are sensitive towards variants and need a large land area. Therefore, they are not suitable for design and operation [29]. Using activated sludge for biological methods can be very effective for reducing biological oxygen demand (BOD) but not sufficient for removing colour, due to the prolonged oxidation rate [21].

1.3.3 Physical Methods

Physical treatment techniques are the most popular treatment types, which consist of adsorption and membrane filtration. Membrane filtration is appointed as any kind of wastewater filtering and cleaning [30]. Membrane filtration can reject various pollutants simultaneously. This method offers some advantages like a smaller plant footprint, and also the usage of chemicals is very little or even no required. However, their lifetime is limited compare to other methods, as membrane fouling occurs very quickly. Membrane fouling and sludge production are the main drawbacks of membrane technology since it causes permeate flux decline and changes in selectivity. Other disadvantages of membrane technology are high capital cost, frequent pore-clogging, and high cost of periodic replacement [31].

Membrane filtration is commonly used for metallic contaminants removal in wastewater [32].

Adsorption is the most used and most famous methods for water treatment, especially using cheap adsorbents, easy regeneration, and no pre-treatment step required [33]. Adsorption is an easy and effective method which has relatively free from by-products. The advantage of this method is high dye removal percentage; however, it has some problems associated with adsorbent disposal and posts contamination by used adsorbents. Adsorption is confirmed as a preferable method in many applications due to its availability, simple design, and operation [34–36]. Many researchers have studied the development of cheap adsorbents with high efficiency for dye removal [37, 38].

2 Activated Carbon

As early as 1550 BC, powdered charcoal has been reported in an Egyptian papyrus and used for the medicinal purpose from prehistoric times [39]. The adsorption properties of charcoal were discovered by Karl Wilhelm Scheele, which is a Swedish chemist, in 1773 for the first time. In 1785, Lovitz, which is a Russian academician, observed the effectiveness of using charcoal in removing the colour from the solution. In 1794, scientists from England attempted to purify cane sugar using wood char. In 1900 and 1901, two processes of activation were discovered, and cellulosic precursor was used to producing activated carbon (AC) as patents. AC production on a commercial basis has followed these patents as the basic scheme up to now. In 1909, the powdered AC (industrial scale) was prepared for the first time. In 1911, a novel type of AC (peat AC) was synthesized using steam. In 1915, during World War I, Germany developed granular AC to protect its soldiers against Cl_2 gas using filtering medium in gas masks [40]. In the early twentieth century AC has been specified as a popular adsorbent in treating the wastewater. Although conventional materials are essential in the AC industry, they can still be replaced.

The high initial cost and expensive regeneration of AC motivate the search for low cost, disposable ACs from other feedstocks [41]. In the last decade, countless studies and rapid communications related to various adsorbents development have been reported because of their importance, quality, and cost-effectiveness for wastewater treatment [42]. AC, with a porous structure and large surface area, is including of wide range of amorphous carbonaceous materials. Figure 1a presents the crystallites and graphite structure of AC. The AC high adsorption capacity of many organic compounds has been widely studied and proven by many studies [43]. AC is available as granular (GAC) and powdered AC (PAC) [44]. Figure 1b shows the GAC and PAC.

Rao et al. [45] perceived that this imperious need creates a growing interest in deriving AC from abundant and cheap waste materials. Specific waste and natural materials can be used as a cost-effective adsorbent to decontaminate domestic or

Fig. 1 (**a**) AC graphite structure (**b**) Granular and powdered AC

industrial water pollutions. The ease of activation, processing cost, purity, low degradation during storage, stability in quality is an essential factor for the selection of starting materials [46]. Considerable amounts of various agricultural residues are stored worldwide as environment solid pollutants [47]. Correspondingly, different studies have been conducted for producing AC from some solid agriculture wastes (macadamia nutshell [48], bagasse [49], coffee residue [50], orange peel [51], banana peel [52]). Sud et al. [53] quoted various other agricultural wastes like sawdust, rice husk, black gram husk, waste tea leaves, and bark of trees. Figure 2 presents the dye removal efficiency of adsorbents from some of the polysaccharides on three types of dyes after 12 h [43].

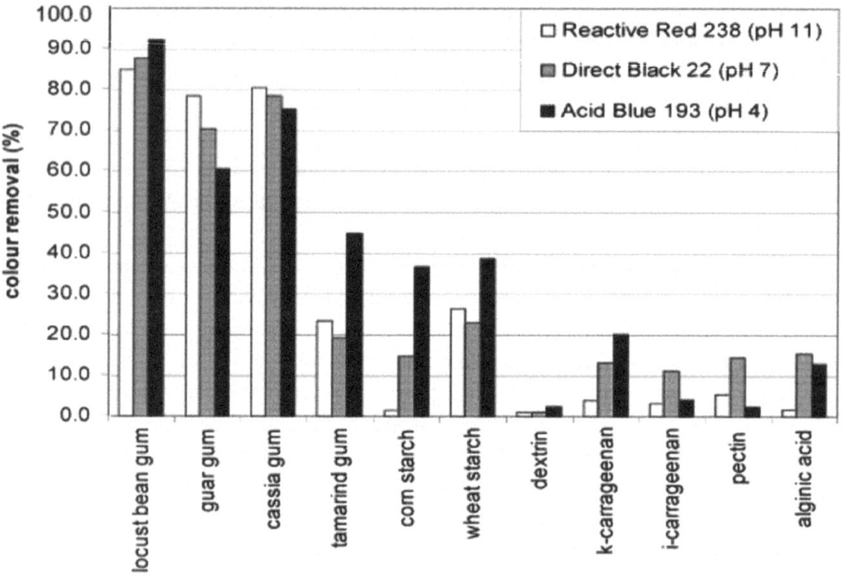

Fig. 2 Dye removal percentage of different adsorbents. Adapted with permission from Ref. [43] Copyright 2013, American Chemical Society

These low-cost materials are desirable; Sun et al. [54] stated that because lignite was less aromatic in structure, its produced surface area and micropore volume of AC were lower. Colella et al. [55] studied the adsorption isotherms of a wood-based (Chemviron carbon) and lignite-based AC for chlorinated phenols removal. Their result indicated that wood-based AC had a more vital adsorption ability compared to the lignite-based one. Besides, the presence of additional inorganic solutes had almost no effect on adsorption. Ahmaruzzaman [56] mentioned that bark is a rich forest residue and rice husk is the abundant by-product from the industry of rice milling with about more than 100 million tonnes.

The following criteria should be considered concerning the selection of materials: (1) low cost and easy availability, (2) low mineral and high carbonaceous content, (3) the easy process of activation (e.g., calcined coke is arduous to activate while wood char is relatively easy), (4) low degradation during storage, and (5) high adsorption capacity.

The presence of O and H also has remarkable effects on AC's surface characterizes and adsorption behaviours [57]. It can be produced from the original raw material directly through the activation method or via post-treatment after preparation. Although several carbons are produced free of O from phenol formaldehyde or other polymers, the O surface group will be formed afterwards by O chemisorption. The O in the surface group is responsible for the properties of AC [58]. The functionality of the O groups is influenced by the carbonaceous temperatures, surface area, particle size, and ash content. The addition of O groups increases the hydrophilicity and enhances the surface charge density. Surface O groups (carboxylic and carbonyl) on AC are capable of H-bonding with aromatic compounds with a functional group, such as phenol. These O groups are located at the entrance of the C pores, especially carboxylic groups, which are highly accessible, hydrophilic, and polar, which indicated that they could also attract water molecules through H-bonds between them. Water is preferential and can easily have H-bonding with the O groups; it will form water clusters to hinder the adsorption of the large phenol molecules [59]. According to this mechanism, water adsorption is dominant on the surface of O groups through the H-bonding effect compared with hydrophobic phenol compounds. AC contains amorphous carbonaceous materials that are introduced as high porous and large surface areas. The AC surface area is dependent on the activation process and used raw material and can be in the range of 100–1000 $m^2\ g^{-1}$ or even higher. Surface oxidation is also a critical factor in enhancing surface activity. Surface chemistry is one of the most crucial properties that dictate the adsorption mechanism and capacity towards different contaminants. C consists of hexagonal graphite sheet-like structures with carbons attached to each other covalently. On treatment, several C atoms are replaced by other atoms, such as O, S, P, and N, based on the method of treatment [60]. The different sizes and shapes of micro-, meso-, and macropores (Fig. 3) make the structure of AC complex and heterogeneous. Currently, various types of available AC can be used as adsorbents or catalysts. Thus, AC is known as a very useful and common adsorbent material in the industrial treatment of wastewater.

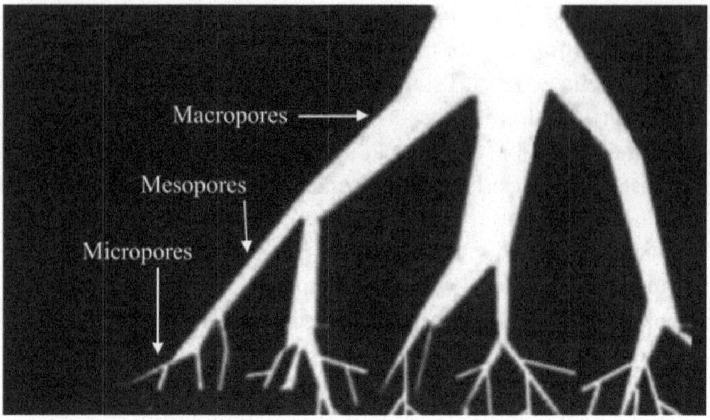

Fig. 3 Internal structure of activated carbon adsorbent. Adapted with permission from Ref. [61] Copyright 2006, Elsevier

3 Preparation of Activated Carbon

AC usually produced using all polymeric compounds via two steps, namely carbonization and activation [62]. Figure 4 presents a flowsheet of two-step AC preparation. A preliminary process of carbonization (pyrolysis) for the transformation of cellulose structures into a carbonaceous material is necessary for organic raw materials, such as peat, wood, coconut shells, or sawdust. Pyrolysis is typically carried out in the condition of inert gas in the temperature range of 450–900 °C. The number of O and H functional groups of cellulose structure can be removed using dehydrating chemicals under pyrolytic conditions at increased temperatures, which lead to cellulose structure destruction wherein the carbon skeleton is left.

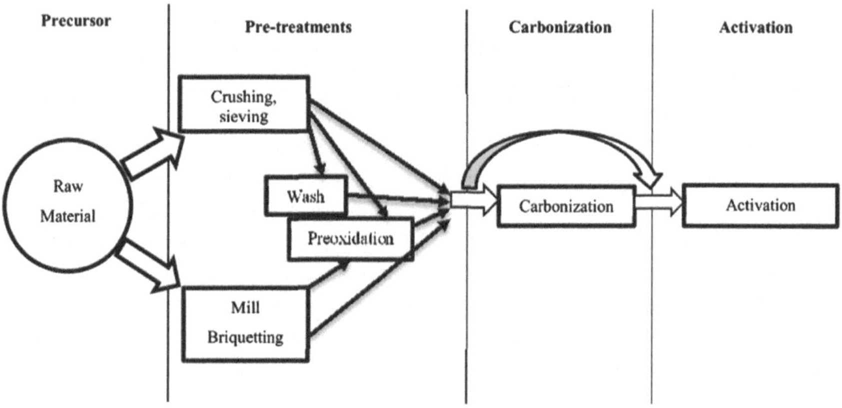

Fig. 4 Production of AC flowsheet. Adapted with permission from Ref. [61] Copyright 2006, Elsevier

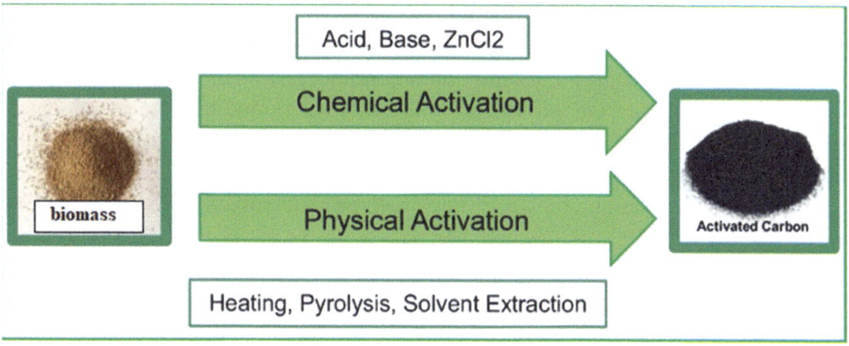

Fig. 5 Methods of activation for AC preparation. Adapted with permission from Ref. [68] Copyright 2019, Springer

N or He was used to assist the breakdown of the raw materials, and the pores can be achieved through an activation process. Illan-Gomez [63] proved that the produced AC possesses a broader pore size distribution after treatment at 700 °C than 500 °C. The time of the pyrolysis process does not influence microporosity formation. The porosity and passage development of the AC with the large surface area is achieved through the activation step via the oxidation of the charred residue. In practice, the final AC porosity is measured by the extent of activation.

Two methods of activation, including chemical and physical activation, are usually used to remove the structure's most reactive carbon atoms (Fig. 5). These processes cause an increment of AC porosity, surface area, and adsorption efficiency. Thus, the operation processes costs are decreased [64, 65]. The first step of physical activation is pyrolysis (carbonization) of the carbonaceous precursor (at 600–900 °C), and the second step is the activation process using hot gases. The high activation temperature in the range from 800 to 1000 °C is needed to maintain a high reaction rate and ensure the burn-off level. The activating agent has an evident impact on the properties of C and activation degree. Fu et al. [66] reported the preparation of black liquor lignin AC with steam (physical activation method). The optimized temperature and time carbonizations are 450 °C and 1 h, which are reasonable economically. Mui et al. [67] studied the waste tires AC and demonstrated that CO_2–AC displayed a small pore volume and sizeable external surface area.

The chemical activation process is including, first washing the raw materials, drying, crushing them into suitable sizes, and then impregnation with the chemical activant. The concentration of activant is set required activant mass ratio to the raw material; then, it is stirred until homogeneity is achieved. The mixture is carbonized and activated in a range of 400–800 °C in an inert atmosphere. After that, the mixture is washed using 0.1 M HCl solution and DW to remove residual activant, and the pH is close to 7, then dried and stored. Optimum operating conditions like carbonization, activation, their temperature, time, and activant ratio allow for the production of carbon with specific application properties [69]. Some researchers reported that the chemical activation process needs lower temperatures and soaking time in comparison with the physical activation process. Therefore, chemical activation needs less energy consumption and production cost; however, it requires residual treatment from the washings. Some studies reported that the increment of the activant precursor ratio to the optimum point caused an increase of adsorption capacity increases.

In chemical activation, biowaste raw materials are mixed with an activating agent, like hydroxides (i.e., NaOH, KOH), acids (e.g., H_3PO_4 [70]), or other chemicals such as $ZnCl_2$ [71] and K_2CO_3 [72]. Ehrburger et al. [73] reported that in the chemical activation process, minerals like SiO_2, FeS, Al_2O_3 will consume the alkaline activating agents like KOH or K_2CO_3, and thereby inhibit the formation of activated sites and porosity. Labus [74] used the activation method using KOH and reported that it is effective in producing AC with high surface area (1800 m^2 g^{-1}) and porous structure. The production of super AC (with KOH) with the highest surface area (3362 m^2 g^{-1}) was obtained [75, 76]. It was concluded that the ratio of impregnation, activation temperature (800 °C), and time were essential factors on the surface area improvement.

According to the work of Illan-Gomez [63], the chemical activation method is much better than the physical activation for coal-based AC production. Chemical activation will lower the inherent mineral content of the coal and produce low ash content AC. However, for commercialized AC, the physical activation is used, thus far because of its cost-effective production [77].

4 Effects of Carbonization Temperature on Activated Carbon

Ahmadpour and Do [48] and Wigmans [78] indicated that the pore volume evolution is highly correlated with carbonization temperature. KOH as the activating agent (at a high temp. of 800 °C) caused large pore volume and surface area. Meanwhile, low temperature (500–600 °C) is preferred when $ZnCl_2$ is used as the chemical agent. Teng [76] who measured the production of waste tires AC confirmed that maximum pore volume is achieved in the presence of KOH at 800 °C. Further increase in temperature caused pore volume decrement (Table 1). This

Table 1 Pyrolysis temperatures impact on the waste tires AC surface properties [79]

Pyrolysis temp. (°C)	Carbon yield (%)	BET surface area (m^2 g^{-1})	Pore volume (m^3 g^{-1})	Average pore diameter (nm)
600	26	116	0.22	7.7
700	16	474	0.38	3.2
800	12	411	0.57	5.5
900	11	306	0.45	5.9

phenomenon can be due to the breakdown of crosslinks in the C matrix because of severe thermal stress. The breakdown will result in the rearrangement of the carbonaceous matters and the collapse of pores. Another potential reason is the destruction of pore structures during the gasification process because of the high temperature. It is reported that high pyrolysis temperature reduces C yield but increases surface area and pore volume [79]. These trends can be primarily attributed to C gasification to form and release CO_2. Considerable gasification occurred at high temperature, thereby causing the extensive breakdown of pore walls, thereby enlarging the surface area and pore volume. This result revealed that the extent of C removal by gasification-dominated pore development. This tendency is similar to that of physical activation. Saka [80] reported that temperature increment from 300 to 600 °C caused burnt C, the release of volatile matters during carbonization and thus, the increment of weight loss ($\sim$87 wt%).

4.1 Effect of Carbonization Time on Activated Carbon

Carbonization time has a considerable efficiency on the AC properties. Ahmadpour and Do [48] studied the characteristics of AC made from bituminous coal in KOH and $ZnCl_2$ series and observed that prolonged time is recommended for these series, especially KOH, because the weight loss, micropore volume, the BET surface area, and benzene adsorption can be boosted. Saka [80] studied the effect of time on the properties of AC. With increased time from 15 to 30 min at the 300–600 °C, the BET, and adsorption efficiency, significantly increased and maximized (1289 and 1209 mg g^{-1} at 30 min). However, activation time increments from 30 to 60 min resulted in the enlargement or even collapse of several pores, and the AC BET surface area decreased.

4.2 Activated Carbon Physical and Chemical Properties

Adsorption effectiveness and performance are other critical screening criteria in choosing feedstock in addition to cost-effectiveness, C content, and availability. AC is frequently tailored physically and chemically for various adsorptive needs. The surface's chemical nature and porous structure are essential for AC adsorption. The AC's structure and distribution of pore size are caused by the access of the adsorbates to the AC and can be tailored for the aim of selectivity. According to the IUPAC, the AC pore size distribution includes micropores of ($\leq$2 nm), macropores of $\geq$50 nm), and mesopores of (2–50 nm) [81–83]. Given that the relationship between the pore size and surface area are inverse, mesopores and macropores have a small surface area because of their pore size. Thus, most of the adsorption occurs in micropores. Figure 6 shows fox nutshell AC FESEM images before and after chemical activation.

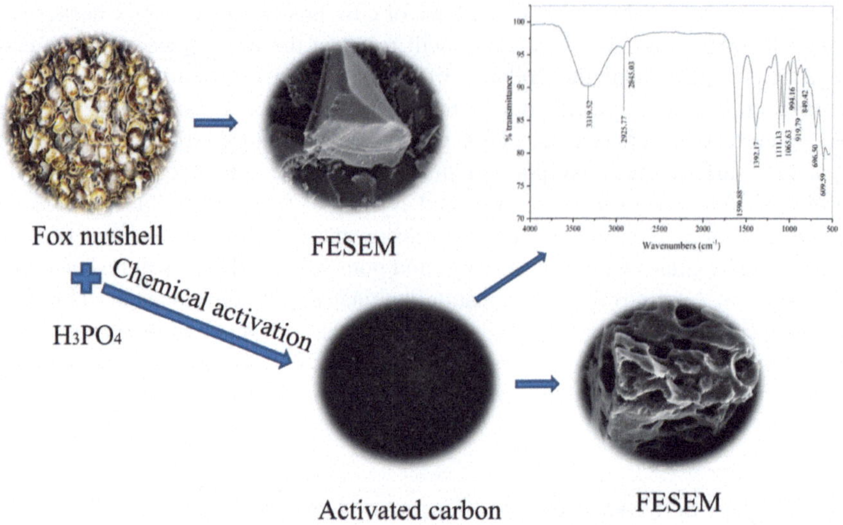

Fig. 6 FESEM images of fox nutshell AC before and after chemical activation. Adapted with permission from Ref. [84] Copyright 2016, Elsevier

5 Improving the Physical and Chemical Properties of Activated Carbon

AC can be altered in shape to meet different needs. Forms like granules, pellets, powders, fibres, and cloth are considered as flexible options commonly. Powdered AC is for direct addition, and granular AC is often used in adsorption columns. AC adsorption is a mass transfer process, which implies that interaction between the carbon surfaces with the adsorbates is decisive. Since the distribution of pore size and structure are responsible for the access of the adsorbates to the AC, they can be tailored for the aim of selectivity. For example, 0.4 nm carbon molecular sieves (CMS) can adsorb the disc-like benzene molecule, whereas zeolite with the same size fails to adsorb benzene [85]. Benzene has difficulty in entering the zeolite pores because of the incompatible shapes. In comparison, the pore shape of the CMS allows the benzene molecules to penetrate and occupy the pores.

The adsorption characteristics of AC are studied broadly by the carbon chemical nature on its surface. Rodriguez-Reinoso [85] mentioned that the importance of chemical nature was recognized is in the middle 1970s since the physical properties of the adsorbents cannot sufficiently explain some adsorptive behaviour. Awing to many studies of the carbon surface chemistry, people gained a much better understanding of the carbon adsorption behaviours. Acid–base characteristics and functional groups were found to contribute to the adsorption properties. The surface of the AC is modified using various treatments according to different objective adsorbates. That means that the carbon structure is altered to meet the different requirements; for instance, the pH of the surface is found to be increased with a larger extent of activation. A functional group like the aromatic ring was also indicated to have the potential to affect the adsorption property.

6 Adsorption

Adsorption is the substance concentration changes at its interface concerning the concentration of the bulk part of the system. The basic adsorption definition is presented in Fig. 7. In water treatment, adsorption is the process that the adsorbate contained in the water is transported into the adsorbent (in solid) porous structure and adsorbed onto its inner surface by chemical or physical reaction [86]. AC adsorption occurs in several steps. The process begins as adsorbate(s) bulk transfer from the water onto the surface film surrounding the carbon. The following step is film diffusion transport, where the adsorbate(s) spreads out from the surface film to the entrance of the pores. Next, internal pore transfer occurs where the adsorbates transit through the carbon pores via pore diffusion into the AC pores. However, intraparticle diffusion can become the limiting step depending on adsorbates size because the pores within AC can vary from about 4 angstroms to greater than 500 angstroms. More specifically, in the narrower pores (e.g., <20 angstroms), larger adsorbates are slower to diffuse because the size of the pores and the adsorbate are similar in size. Additionally, pore diffusion may be restricted in the presence of background organic compounds due to some pores being blocked by more extensive natural organic matter (NOM) molecules [87]. With pore restriction may be overcome with the proper selection of the PAC and/or GAC. The last step in

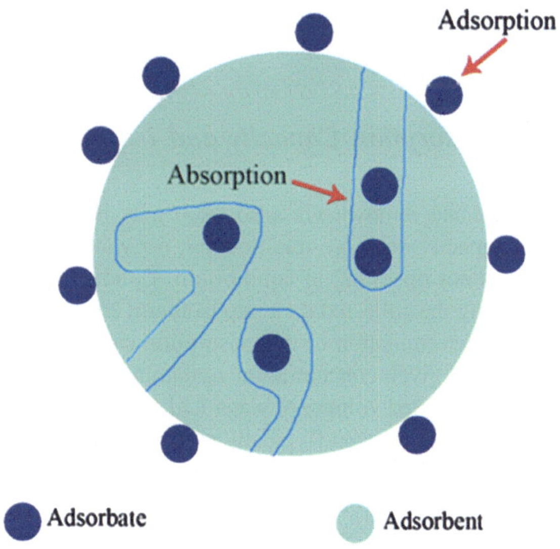

Fig. 7 Illustration of the absorption and adsorption processes

adsorption is the physical or chemical bonding of the contaminant to the AC surface or its functional groups.

For porous materials, separation of the adsorption from the absorption process is not possible; therefore, applying the term sorption can account for the two processes. While absorption phenomena occur within the interior of the sorbent, adsorption is a surface-based phenomenon. Adsorption is a contributing phenomenon for wastewater quality improvement, which has fluid behaviour in a force field extorted by the solid surface.

Three main interactions that are present in aqueous solution adsorption processes are adsorbate–aqueous interaction, adsorbate–surface interaction, and aqueous–surface interaction. Adsorbate–aqueous interaction is appointed with the adsorbate solubility, while the adsorbate–surface interaction and aqueous–surface interaction are related to surface chemistry. Depending on the strength of adsorbate–surface interaction, adsorption is classified into chemical and physical adsorption [88]. Physical adsorption is van der Waals interactions between adsorbent and adsorbate including field gradient-quadrupole, field-dipole, and polarization. Van der Waals interactions are directly related to the polarity and size of the compounds. The bigger and the more nonpolar compounds tend to have stronger interactions with the adsorbent [89]. The adsorbate affinity onto the adsorbent in aqueous solution is depended on (a) adsorbate solution medium properties like pH, temperature, the organic or inorganic compounds competition, and the ionic strength, and (b) adsorbent properties such as functional group, structure, surface properties, and charge density [90]. It is believed that physisorption is mainly meant by which AC works to adsorb most contaminants from water. Chemical adsorption occurs when the forces between the adsorbate and adsorbents are electrostatic chemical or covalent bonds. In general, the attraction will have higher bond energy, and monolayer adsorption is present because of the new bonds [89]. Unlike physisorption, chemisorption is irreversible without excess energy to reverse the process. Figure 7 presents the absorption and adsorption processes.

6.1 Adsorption Capacity and Isotherms Contaminant

Contaminant removal via adsorption may be controlled by its adsorption equilibrium capacity or, more traditionally, for water treatment by kinetics since PAC or GAC is not operating at equilibrium. Conducting experimental adsorption studies and fitting the collected data into isotherm equations develops adsorption isotherms. For the determination of AC adsorption capacity, the experimental studies involve exposing a given concentration contaminant (the adsorbate) to series of containers, each with a fixed volume of water and each containing a different mass of AC. Once equilibrium is reached (typically no more than 24 h) work, the remained amount of the pollutant in the water is measured, and the amount of adsorbed pollutants onto AC is calculated from the equation [91, 92]:

$$q = \frac{x}{m} = \frac{(C_0 - C_e) \times V}{m_s} \tag{1}$$

where q_e is the mass of adsorbed adsorbate (x) per unit mass of adsorbent (m) at equilibrium (mg contaminant/g AC), C_0 is the adsorbate initial concentration (mg L^{-1}), C_e is the final equilibrium of adsorbate concentration after adsorption (mg L^{-1}), V is the water volume in the reactor (L), and ms is the mass of adsorbent (g). The q_e values and the experimental isotherm data calculated from the equation can be used to determine adsorption isotherms based on equations developed by Brunauer, Emmet, and Teller (BET isotherm), Freundlich, and Langmuir. Historically, the Freundlich and Langmuir isotherms are the best to describe the AC adsorption capacity onto the adsorbate in equilibrium using bulk solution concentration.

6.2 *Kinetic of Adsorption*

In the process of adsorption, understating the kinetics and the involved mechanisms is crucial for controlling of process. Since it helps identify the factors information that affects adsorbent and adsorbate interaction and the reaction rate, two factors that have an impact on the adsorption kinetics are the reactions of sorption and the mass transfer process [8, 93]. The mass transfer steps are controlled by three factors including (1) adsorbent's physical form and intrinsic structure such as crystallinity, particle size, and molecular weight, (2) the intrinsic characteristics of the dye solution, and (3) and finally, the conditions of the adsorption process (e.g., temperature, pH and primary concentration). Many types of research have been developed by various kinetic models to study the adsorption mechanism on the adsorbent beads surface that match the experimental data. This information is employed to develop the system designs for waterwaste treatment applications [94].

6.2.1 Pseudo-First-Order

For the pseudo-first-order model, it is assumed that the change of solution uptake rate with time has an inverse relationship with changes of saturated concentration and the solid uptake value in time t [95]. Pseudo-first-order can be defined as:

$$\log(q_e - q_t) = \log q_e - \frac{k_1 t}{2.303} \tag{2}$$

where k_1 (min^{-1}) is the constant rate of pseudo-first-order and q_e and q_t are the adsorption capacity (mg g^{-1}) of MB onto the magnetic beads at equilibrium and at t (min), respectively.

6.2.2 Pseudo-Second Order

The pseudo-second-order model explained the velocity dependence on the adsorption capacity in the solid phase but no dependence on the adsorbed substance concentration:

$$\frac{t}{q_t} = \frac{1}{k_2 q_e^2} + \frac{t}{q_e} \tag{3}$$

where k_2 (g mg^{-1} min^{-1}) is the rate constant of pseudo-second-order.

6.2.3 Intraparticle Diffusion

Generally, the solid–liquid adsorption system involves intra-particle diffusion, film diffusion, and mass transfer. However, mass action is neglected in adsorption kinetic studies because of the rapid nature of the process. Therefore, the kinetic process is always determined with either intra-particle diffusion or liquid film diffusion [96]. For intraparticle diffusion plots, there are three steps to describe the adsorbate mechanism diffused on the adsorbent particles. First, the adsorbate is transported from the solution on the adsorbent outer surface layer by molecular diffusion as external film diffusion. Second, the adsorbate transport from the surface particle of adsorbent into interior levels as internal diffusion. Third, the solute particles adsorption from active sites into the pores interior surface (Fig. 8) [97].

The intraparticle diffusion and can be expressed as:

$$q_t = k_i t^{\frac{1}{2}} \tag{4}$$

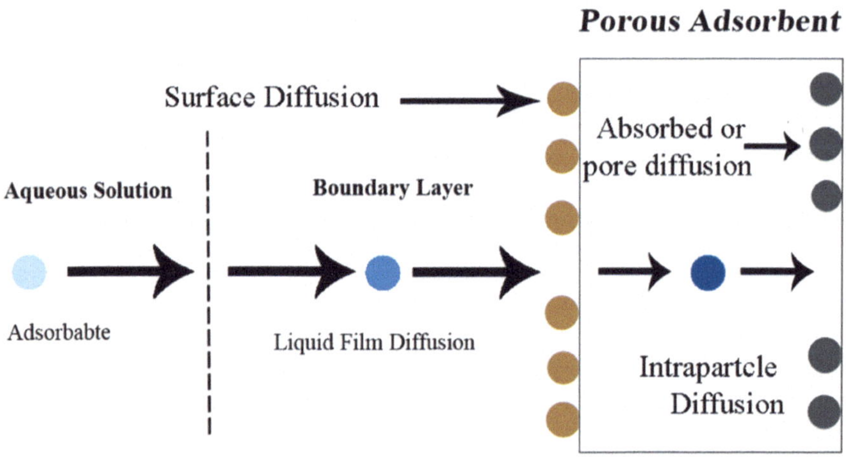

Fig. 8 Steps of adsorption

where q_t is the adsorption capacity at t (min), and k_i (mg g^{-1} $min^{-1/2}$) is the constant rate of intraparticle diffusion which obtained from the linear plot slop of q_t versus $t^{1/2}$.

6.3 Contaminant Removal of Activated Carbon Adsorbents from Aqueous Solutions

Many adsorbents are used in selective adsorption studies, for instance, AC, biological materials, synthetic resins, and mineral oxides [49]. Among these, AC is the best and most famous adsorbent for a particular class of pollutants. Gang et al. [98] reported the arsenic adsorption capability of iron-impregnated chitosan granulated adsorbent from water. The adsorption behaviour was examined by column and batch studies. They studied the efficiency of initial concentration and pH were investigated, and the study resulted that the highest adsorption capacity of 6.48 mg g^{-1} was obtained at pH 8, and initial As(III) concentration of 1007 μg L^{-1}. The methylene blue (MB) dye removal from aqueous solution with cashew nutshell (CNS) was studied by Kumar et al. [99]. Different parameters like adsorbent dosage, pH, and temperature were determined. The study suggested that the increment of pH from 2 to 10 increased the dye adsorption efficiency due to the negatively charged surface of CNS at higher pH that improves the positively charged dye adsorption through electrostatic forces.

Mohan et al. [49] investigated the zinc and cadmium adsorption using bagasse AC. The study reported that the adsorption fitted the Langmuir and Freundlich models. Several factors, like significant diffusion coefficients, entropy, mass transfer, and activation energy, were measured, and it was reported that the adsorption at low and high concentrations occurred through a film diffusion mechanism. Kadirvelu et al. [100] analysed the adsorption of the toxic metal like Hg (II), Pb (II), Cu (II), Cd (II), and Ni (II), onto coir pith ACs. The results suggested that in the pH ranges of 4–5, maximum adsorption was observed. This could be because of $M(OH)_2$ and $M(OH)^+$ formation from partial hydrolysis of metal ions. At pH 2–3, the competition between M^{2+} and H^+ ions for exchangeable ion sites led to lower metal ions removal.

Sulphuric acid activation of sunflower oil cake ACs for MB adsorption was carried out by Karagoz et al. [101]. AC with an impregnation ratio of 0.85 was measured to have the highest specific surface area in comparison with those with impregnation ratio of 0 and 1.9. The highest adsorption was reached at pH 6 where the AC surface was charged negatively. From the kinetics study, the maximum adsorption was achieved at 900 min for AC with an impregnation ratio of 0 and 0.85. However, 1440 min was the required time for AC with an impregnation ratio of 1.9 to achieve maximum adsorption. The observed phenomena could be because of the longer time needed for the dyes molecules to reach the boundary layer, diffuse into the fist, surface, and then into the pores of the adsorbent. The studies also reported that the adsorption process fitted Langmuir isotherm and pseudo-second-order kinetics.

Table 2 Adsorption capacity of AC from different biomass materials for some dyes

AC from Biomass material	Pollutant	Adsorption capacity (mg g^{-1})	References
Coconut bunch waste AC	MB	70.92	[102]
Tea waste AC	MB Acid Blue 25	85.16 203.34	[103] [104]
Bamboo cane powder AC	Lanasyn orange Lanasyn grey	2600 3040	[105]
Bamboo-based AC	Basic blue Reactive Black 5 MB	143.2 489.9 454.2	[106–108]
Cashew nutshell AC	MB	476	[109]
Posidonia oceanica AC	MB	285.7 2681.9	[110, 111]
Tobacco AC	MB	1264.51 35.7	[112, 113]
Oil palm AC	MB Janus green B Methylene blue dye	90.9 67 103.02	[114–116]

Similar adsorption studies were also performed on AC from Jerusalem artichoke stalk-based, pinecone, and pineapple waste biomass [117, 118]. Although many studies have been done on the natural and recycled materials as adsorbents, issues such as the availability of these materials that differ from one region to another region have developed the interest to study synthetic adsorbents. Table 2 presents the AC adsorption capacity.

7 Activated Carbon Recycling and Reactivation

Recycling is an essential factor in AC development. The recycling and reactivation of saturated AC are crucial for reducing the operational costs and wastes leftover [119–123]. Restoring the activity of the exhausted AC has been highly attractive because of several reasons. The cost of AC production is high in general. The cost of using reactivated carbon is sometimes much cheaper than that of the original AC because of the relative ease of the operating process [124–127]. After the AC adsorbs the contaminants and reaches its adsorption capacity, it turns into the noxious solid waste that requires proper treatment or disposal. Incineration is undesirable because AC is not recycled, and several air pollutants with high toxicities might be emitted during the incineration process [128]. Similarly, the discharge of the depleted AC via landfill will also add pollutants to the environment. Table 3 presents the pollutant removal of AC after regeneration. Regeneration of adsorbent is used to restore the used adsorbent adsorptive capacity. Figure 9 presents the general AC recycling description. The methods of regeneration are based on either adsorbate decomposition or desorption. The former can be achieved by

Table 3 Pollutant removal of activated carbon after the regeneration process

Adsorbent	Pollutant	Adsorption efficiency	References
Sludge-based AC	MB RR 24	4th cycles = 85% removal efficiency 4th cycles = 38.2% removal efficiency	[134]
Fe_3O_4/AC AC	MB MB	1st cycle = 185.7 mg/ g^{-1}, 5th cycles = 56.4 mg g^{-1} 1st cycle = 123.6 mg g^{-1}, 5th cycles = 52.2 mg g^{-1}	[135]
AC	Red Yellow	3rd cycles = 52% removal efficiency 3rd cycles = 28% removal efficiency	[136]
AC	Crystal violet	– Regeneration time = 15 min, Approximately 100% efficiency over several cycles. – Regeneration Time = 5 min, 80% efficiency after 1 cycle.	[137]
AC	Acid orange	5th cycles = 90% removal efficiency	[138]
AC	Blue R Blue G	4th cycles = 96% removal efficiency 4th cycles = 98% removal efficiency	[139]
AC	Reactive black 5	2nd cycles = 50% removal efficiency	[140]

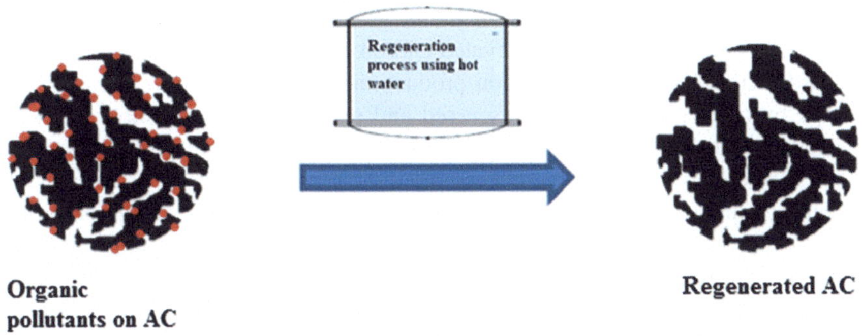

Fig. 9 General description of activated carbon recycling. Adapted with permission from Ref. [131] Copyright 2018, Elsevier

decreasing the pressure, temperature increment, or replacing the adsorbate. The latter can be achieved by thermal, microwave, and chemical processes. Selecting the appropriate method of regeneration is dependent on the contaminants nature, cost of regeneration, and its conditions of the process. The most used method for regeneration is using steam or inert atmosphere for thermal regeneration [129, 130].

The advantages of thermal regeneration are its simple and economically attractive process, and its disadvantages are including the production of hazardous by-products, significant carbon loss, and time-consuming process. The thermal regeneration method needs a high amount of energy to increase the temperature. Other disadvantages of reusing are the reduction of adsorbent's capacity after cycle repetition, the formation of carbon in the recycling/reusing process, and the abrasion and carbon friction and collapsing during consecutive heating and cooling processes that lead to transforming carbon portion to dust.

An alternative recycling/reusability method is chemical regeneration, although it is not as effective as thermal treatment. The advantages of the chemical regeneration method are its simple process, no loss of carbon, rapid heating, and adequate capacity for recovery of the adsorbates [132]. Microwave regeneration is another carbon regeneration method, which seems to be finding favour [133]. The advantages of this method are short treatment time, low energy consumption, high restoration of carbon porosity, more rapid heating, shorter regeneration time, better performance of the carbon adsorption, less damage to the carbon, larger surface area, and the porosity of the carbon remains unaltered.

8 Conclusion

The present chapter book was focused on activated carbon synthesis using biowaste materials for organic pollutants removal. Specific waste and cheap materials are appropriate for activated carbon production to decontaminate industrial or domestic wastewater treatment. The criteria, fictional groups, structure, and chemical and physical properties of biowaste activated carbon are discussed and reviewed. The standard methods of activated carbon production and their effect on removal percentage by other researchers are discussed and reviewed. Carbonization and activation time and temperature have the highest effect on synthesized activated carbon. The activated carbon adsorption process and the effect of various types of activated carbon on removal efficiency are discussed. The organic pollutant adsorption of the biowaste activated carbon, including materials, properties, and adsorption details, is reviewed. The regeneration efficiency of activated carbon and methods of regeneration is explained in detail. Hence, activated carbon from cheap and abundant biowaste can be used as an efficient adsorbent in industry. Nevertheless, the application of biomaterials activated carbon still is limited and requires more attention to the processes and properties. Activated carbon adsorbents with high removal efficiency and reusability in the organic pollutant removal of wastewater treatment will be successfully commercialized for large-scale removal of emerging water pollutants to benefit the society.

9 Future Perspectives

Synthesis of effective activated carbon derived from various other abundant agricultural waste resources like jackfruit peel, tobacco stalk, bagasse, moringa seed, and peach seeds may also be considered for the removal of water pollutants. Other activating agents, such as p-toluenesulfonic acid, can be tested as activating agents for activated carbon preparation.

References

1. Palaniappan M, Gleick PH, Allen L, Cohen MJ, Christian-Smith J, Smith C, Ross N (2010) *Clearing the waters: a focus on water quality solutions*
2. Attia AA, Girgis BS, Fathy NA (2008) Removal of methylene blue by carbons derived from peach stones by H_3PO_4 activation: batch and column studies. Dyes Pigm 76(1):282–289
3. Dotto GL, Vieira ML, Pinto LA (2012) Kinetics and mechanism of tartrazine adsorption onto chitin and chitosan. Ind Eng Chem Res 51(19):6862–6868
4. Aşçı YS (2017) Removal of textile dye mixtures by using modified Mg–Al–Cl layered double hydroxide (LDH). J Dispersion Sci Technol 38(7):923–929
5. Idan IJ, Jamil SNABM, Abdullah LC, Choong TSY (2017) Removal of reactive anionic dyes from binary solutions by adsorption onto quaternized kenaf core fiber. International Journal of Chemical Engineering 2017
6. Walker G, Weatherley L (2000) Prediction of bisolute dye adsorption isotherms on activated carbon. Process Saf Environ Prot 78(3):219–223
7. Maazinejad B, Mohammadnia O, Ali GAM, Makhlouf ASH, Nadagouda MN, Sillanpää M, Asiri AM, Agarwal S, Gupta VK, Sadegh H (2020) Taguchi L9 (34) orthogonal array study based on methylene blue removal by single-walled carbon nanotubes-amine: Adsorption optimization using the experimental design method, kinetics, equilibrium and thermodynamics. J Mol Liq 298:112001
8. Sadegh H, Ali GAM, Makhlouf ASH, Chong KF, Alharbi NS, Agarwal S, Gupta VK (2018) MWCNTs-Fe_3O_4 nanocomposite for Hg(II) high adsorption efficiency. J Mol Liq 258: 345–353
9. Vandevivere PC, Bianchi R, Verstraete W (1998) Treatment and reuse of wastewater from the textile wet-processing industry: Review of emerging technologies. J Chem Technol & Biotechnol: Int Res Process, Environ Clean Technol 72(4):289–302
10. Mohan D, Singh KP, Singh G, Kumar K (2002) Removal of dyes from wastewater using flyash, a low-cost adsorbent. Ind Eng Chem Res 41(15):3688–3695
11. Chu W, Ma C-W (2000) Quantitative prediction of direct and indirect dye ozonation kinetics. Water Res 34(12):3153–3160
12. Solehudin M, Sirimahachai U, Ali GAM, Chong KF, Wongnawa S (2020) One-pot synthesis of isotype heterojunction g-C_3N_4-MU photocatalyst for effective tetracycline hydrochloride antibiotic and reactive orange 16 dye removal. Adv Powder Technol 32(5):1891–1902
13. Sharifi A, Montazerghaem L, Naeimi A, Abhari AR, Vafaee M, Ali GAM, Sadegh H (2019) Investigation of photocatalytic behavior of modified ZnS:Mn/MWCNTs nanocomposite for organic pollutants effective photodegradation. J Environ Manage 247:624–632
14. Giahi M, Pathania D, Agarwal S, Ali GAM, Chong KF, Gupta VK (2019) Preparation of Mg-doped TiO_2 nanoparticles for photocatalytic degradation of some organic pollutants. Stud Univ Babes-Bolyai, Chem 64(1):7–18
15. Ethiraj AS, Uttam P, K V, Chong KF, Ali GAM (2019) Photocatalytic performance of a novel semiconductor nanocatalyst: Copper doped nickel oxide for phenol degradation. Mater Chem Phys 242:122520
16. Seyed Arabi SM, Lalehloo RS, Olyai MRTB, Ali GAM, Sadegh H (2019) Removal of congo red azo dye from aqueous solution by ZnO nanoparticles loaded on multiwall carbon nanotubes. Physica E 106:150–155
17. Sadegh H, Ali GA, Nia HJ, Mahmoodi Z (2019) Nanomaterial Surface Modifications for Enhancement of the Pollutant Adsorption From Wastewater. In: Nanotechnology Applications in Environmental Engineering, p 143
18. Hamidreza S, Gomaa AMA (2019) Potential Applications of Nanomaterials in Wastewater Treatment: Nanoadsorbents Performance. In: Athar H, Sirajuddin A (eds) Advanced Treatment Techniques for Industrial Wastewater. IGI Global, Hershey, PA, USA, pp 51–61
19. Sadegh GAMA H, Abbasi Z, Nadagoud MN (2017) Adsorption of Ammonium Ions onto Multi-Walled Carbon Nanotubes. Stud Univ Babes-Bolyai, Chem 62(2):233–245

20. Abed HO, Kanthasamy R, Ali GAM, Sethupathi S, Yunus RBM (2017) Hydrogen sulfide emission sources, regulations, and removal techniques: a review. Rev Chem Eng 34(6):837–854
21. Blackburn RS (2004) Natural polysaccharides and their interactions with dye molecules: applications in effluent treatment. Environ Sci Technol 38(18):4905–4909
22. Sadegh H, Ali GA, Gupta VK, Makhlouf ASH, Shahryari-ghoshekandi R, Nadagouda MN, Sillanpää M, Megiel E (2017) The role of nanomaterials as effective adsorbents and their applications in wastewater treatment. J Nanostructure Chem 7(1):1–14
23. Webb PA (2003) Introduction to chemical adsorption analytical techniques and their applications to catalysis. Micromeritics Instrument Corp., Technical Publications, Chennai
24. Mourao P, Carrott P, Carrott MR (2006) Application of different equations to adsorption isotherms of phenolic compounds on activated carbons prepared from cork. Carbon 44 (12):2422–2429
25. Neamţu M, Catrinescu C, Kettrup A (2004) Effect of dealumination of iron (III)—exchanged Y zeolites on oxidation of Reactive Yellow 84 azo dye in the presence of hydrogen peroxide. Appl Catal B 51(3):149–157
26. Sajab MS, Chia CH, Chan CH, Zakaria S, Kaco H, Chook SW, Chin SX (2016) Bifunctional graphene oxide–cellulose nanofibril aerogel loaded with Fe (III) for the removal of cationic dye via simultaneous adsorption and Fenton oxidation. RSC advances 6(24):19819–19825
27. Hu X, Liu B, Deng Y, Chen H, Luo S, Sun C, Yang P, Yang S (2011) Adsorption and heterogeneous Fenton degradation of 17α-methyltestosterone on nano Fe_3O_4/MWCNTs in aqueous solution. Appl Catal B 107(3–4):274–283
28. Keskinkan O, Goksu M, Yuceer A, Basibuyuk M, Forster C (2003) Heavy metal adsorption characteristics of a submerged aquatic plant (Myriophyllum spicatum). Process Biochem 39 (2):179–183
29. Bhattacharyya KG, Sarma A (2003) Adsorption characteristics of the dye, Brilliant Green, on Neem leaf powder. Dyes Pigm 57(3):211–222
30. Zahoor M (2014) Magnetic adsorbent used in combination with ultrafiltration membrane for the removal of surfactants from water. Desalination and Water Treatment 52(16–18):3104–3114
31. Cho D-W, Jeon B-H, Chon C-M, Schwartz FW, Jeong Y, Song H (2015) Magnetic chitosan composite for adsorption of cationic and anionic dyes in aqueous solution. J Ind Eng Chem 28:60–66
32. Kurniawan TA, Chan GY, Lo W-H, Babel S (2006) Physico–chemical treatment techniques for wastewater laden with heavy metals. Chem Eng J 118(1–2):83–98
33. Wang S, Boyjoo Y, Choueib A, Zhu ZH (2005) Removal of dyes from aqueous solution using fly ash and red mud. Water Res 39(1):129–138
34. Sanghi R, Bhattacharya B (2002) Review on decolorisation of aqueous dye solutions by low cost adsorbents. Color Technol 118(5):256–269
35. Meshko V, Markovska L, Mincheva M, Rodrigues AE (2001) Adsorption of basic dyes on granular acivated carbon and natural zeolite. Water Res 35(14):3357–3366
36. Bulut Y, Aydın H (2006) A kinetics and thermodynamics study of methylene blue adsorption on wheat shells. Desalination 194(1–3):259–267
37. Fan L, Luo C, Sun M, Li X, Lu F, Qiu H (2012) Preparation of novel magnetic chitosan/ graphene oxide composite as effective adsorbents toward methylene blue. Biores Technol 114:703–706
38. Kyzas GZ, Lazaridis NK, Mitropoulos AC (2012) Removal of dyes from aqueous solutions with untreated coffee residues as potential low-cost adsorbents: Equilibrium, reuse and thermodynamic approach. Chem Eng J 189–190:148–159
39. Hassler JW (1974) Purification with activated carbon; industrial, commercial, environmental. Chemical Publishing, New York
40. Yehaskel A (1978) Activated carbon. Manufacture and Regeneration. Noyes Data Corporation, Park Ridge, NJ United States

41. Moosavi S, Gan S, Zakaria SJCC, Technology (2019) Functionalized cellulose beads with activated carbon Fe3O4/CoFe2O4 for cationic dye removal 53(7–8):815-825
42. Wong S, Ngadi N, Inuwa IM, Hassan O (2018) Recent advances in applications of activated carbon from biowaste for wastewater treatment: A short review. J Clean Prod 175:361–375
43. Suginta W, Khunkaewla P, Schulte A (2013) Electrochemical Biosensor Applications of Polysaccharides Chitin and Chitosan. Chem Rev 113(7):5458–5479
44. Moosavi S, Lai CW, Gan S, Zamiri G, Akbarzadeh Pivehzhani O, Johan MRJAO (2020) Application of Efficient Magnetic Particles and Activated Carbon for Dye Removal from Wastewater
45. Madhavarao M, Ramesh A, Purnachandrarao G, Seshaiah K (2006) Removal of copper and cadmium from the aqueous solutions by activated carbon derived from Ceiba pentandra hulls. J Hazard Mater 129(1–3):123–129
46. Girgis BS, El-Hendawy A-NA (2002) Porosity development in activated carbons obtained from date pits under chemical activation with phosphoric acid. Microporous Mesoporous Mater 52(2):105–117
47. Fiol N, Villaescusa I, Martínez M, Miralles N, Poch J, Serarols J (2006) Sorption of Pb(II), Ni(II), Cu(II) and Cd(II) from aqueous solution by olive stone waste. Sep Purif Technol 50 (1):132–140
48. Ahmadpour A, Do DD (1997) The preparation of activated carbon from macadamia nutshell by chemical activation. Carbon 35(12):1723–1732
49. Mohan D, Singh KP (2002) Single- and multi-component adsorption of cadmium and zinc using activated carbon derived from bagasse—an agricultural waste. Water Res 36(9):2304–2318
50. Boudrahem F, Aissani-Benissad F, Aït-Amar H (2009) Batch sorption dynamics and equilibrium for the removal of lead ions from aqueous phase using activated carbon developed from coffee residue activated with zinc chloride. J Environ Manage 90(10):3031–3039
51. Foo KY, Hameed BH (2012) Preparation, characterization and evaluation of adsorptive properties of orange peel based activated carbon via microwave induced K_2CO_3 activation. Biores Technol 104:679–686
52. Ma J, Huang D, Zou J, Li L, Kong Y, Komarneni S (2014) Adsorption of methylene blue and Orange II pollutants on activated carbon prepared from banana peel. J Porous Mater 22 (2):301–311
53. Sud D, Mahajan G, Kaur M (2008) Agricultural waste material as potential adsorbent for sequestering heavy metal ions from aqueous solutions – A review. Biores Technol 99 (14):6017–6027
54. Sun J, Hippo EJ, Marsh H, O'Brien WS, Crelling JC (1997) Activated carbon produced from an Illinois Basin coal. Carbon 35(3):341–352
55. Colella LS, Armenante PM, Kafkewitz D, Allen SJ, Balasundaram V (1998) Adsorption Isotherms for Chlorinated Phenols on Activated Carbons. J Chem Eng Data 43(4):573–579
56. Ahmaruzzaman M (2008) Adsorption of phenolic compounds on low-cost adsorbents: A review. Adv Coll Interface Sci 143(1–2):48–67
57. Liang X, Chi J, Yang Z (2018) The influence of the functional group on activated carbon for acetone adsorption property by molecular simulation study. Microporous Mesoporous Mater 262:77–88
58. Ahumada E, Lizama H, Orellana F, Suárez C, Huidobro A, Sepúlveda-Escribano A, Rodríguez-Reinoso F (2002) Catalytic oxidation of Fe(II) by activated carbon in the presence of oxygen. Carbon 40(15):2827–2834
59. Gupta VK, Agarwal S, Sadegh H, Ali GAM, Bharti AK, Hamdy AS (2017) Facile route synthesis of novel graphene oxide-β-cyclodextrin nanocomposite and its application as adsorbent for removal of toxic bisphenol A from the aqueous phase. J Mol Liq 237:466–472
60. Do DD (1996) A model for surface diffusion of ethane and propane in activated carbon. Chem Eng Sci 51(17):4145–4158

61. Bandosz TJ (2006) Desulfurization on activated carbons, in Interface Science and Technology, Elsevier, pp 231–292
62. Pui WK, Yusoff R, Aroua MK (2019) A review on activated carbon adsorption for volatile organic compounds (VOCs). Rev Chem Eng 35(5):649–668
63. Illán-Gómez M, Garcia-Garcia A, Salinas-Martinez de Lecea C, Linares-Solano A (1996) Activated carbons from Spanish coals. 2. Chemical activation. Energy & Fuels 10(5):1108–1114
64. Jiménez V, Sánchez P, Romero A (2017) Materials for activated carbon fiber synthesis, in Activated Carbon Fiber and Textiles. Elsevier, pp 21–38
65. Merino C, Soto P, Vilaplana-Ortego E, Gomez de Salazar JM, Pico F, Rojo JM (2005) Carbon nanofibres and activated carbon nanofibres as electrodes in supercapacitors. Carbon 43(3):551–557
66. Fu K, Yue Q, Gao B, Sun Y, Zhu L (2013) Preparation, characterization and application of lignin-based activated carbon from black liquor lignin by steam activation. Chem Eng J 228:1074–1082
67. Mui ELK, Ko DCK, McKay G (2004) Production of active carbons from waste tyres—a review. Carbon 42(14):2789–2805
68. Saleem J, Shahid UB, Hijab M, Mackey H, McKay G (2019) Production and applications of activated carbons as adsorbents from olive stones. Biomass Conversion and Biorefinery 9 (4):775–802
69. Foo KY, Hameed BH (2012) A rapid regeneration of methylene blue dye-loaded activated carbons with microwave heating. J Anal Appl Pyrol 98:123–128
70. Shamsuddin MS, Yusoff NRN, Sulaiman MA (2016) Synthesis and Characterization of Activated Carbon Produced from Kenaf Core Fiber Using H_3PO_4 Activation. Procedia Chemistry 19:558–565
71. Zyoud A, Nassar HNI, El-Hamouz A, Hilal HS (2015) Solid olive waste in environmental cleanup: Enhanced nitrite ion removal by $ZnCl_2$-activated carbon. J Environ Manage 152:27–35
72. Mestre AS, Bexiga AS, Proença M, Andrade M, Pinto ML, Matos I, Fonseca IM, Carvalho AP (2011) Activated carbons from sisal waste by chemical activation with K_2CO_3: Kinetics of paracetamol and ibuprofen removal from aqueous solution. Biores Technol 102 (17):8253–8260
73. Ehrburger P, Addoun A, Addoun F, Donnet J-B (1986) Carbonization of coals in the presence of alkaline hydroxides and carbonates: Formation of activated carbons. Fuel 65 (10):1447–1449
74. Labus K, Gryglewicz S, Machnikowski J (2014) Granular KOH-activated carbons from coal-based cokes and their CO_2 adsorption capacity. Fuel 118:9–15
75. Li S, Han K, Li J, Li M, Lu C (2017) Preparation and characterization of super activated carbon produced from gulfweed by KOH activation. Microporous Mesoporous Mater 243:291–300
76. Teng H, Lin Y-C, Hsu L-Y (2000) Production of Activated Carbons from Pyrolysis of Waste Tires Impregnated with Potassium Hydroxide. J Air Waste Manag Assoc 50(11):1940–1946
77. Park S-J, Kim K-D (2001) Influence of activation temperature on adsorption characteristics of activated carbon fiber composites. Carbon 39(11):1741–1746
78. Wigmans T (1989) Industrial aspects of production and use of activated carbons. Carbon 27 (1):13–22
79. Hsieh C-T, Teng H (2000) Influence of mesopore volume and adsorbate size on adsorption capacities of activated carbons in aqueous solutions. Carbon 38(6):863–869
80. Saka C (2012) BET, TG–DTG, FT-IR, SEM, iodine number analysis and preparation of activated carbon from acorn shell by chemical activation with $ZnCl_2$. J Anal Appl Pyrol 95:21–24
81. Lastoskie C, Gubbins KE, Quirke N (1993) Pore size distribution analysis of microporous carbons: a density functional theory approach. J Phys Chem 97(18):4786–4796

82. Divyashree A, Hegde G (2015) Activated carbon nanospheres derived from bio-waste materials for supercapacitor applications—a review. RSC Advances 5(107):88339–88352
83. Lakshmi SD, Avti PK, Hegde G (2018) Activated carbon nanoparticles from biowaste as new generation antimicrobial agents: A review. Nano-Structures & Nano-Objects 16:306–321
84. Kumar A, Jena HM (2016) Preparation and characterization of high surface area activated carbon from Fox nut (Euryale ferox) shell by chemical activation with H_3PO_4. Results in Physics 6:651–658
85. Rodríguez-reinoso F (1998) The role of carbon materials in heterogeneous catalysis. Carbon 36(3):159–175
86. Toth J (2002) Adsorption theory, modeling and analysis. CRC Press, New York
87. Crittenden JC, Reddy PS, Arora H, Trynoski J, Hand DW, Perram DL, Summers RS (1991) Predicting GAC Performance With Rapid Small-Scale Column Tests. J-Am Water Work Assoc 83(1):77–87
88. Worch E (2012) Adsorption Technology in Water Treatment. Walter de Gruyter
89. Crittenden JC, Trussell RR, Hand DW, Howe KJ, Tchobanoglous G (2012) MWH's Water Treatment. Wiley
90. Taty-Costodes VC, Fauduet H, Porte C, Ho Y-S (2005) Removal of lead (II) ions from synthetic and real effluents using immobilized Pinus sylvestris sawdust: Adsorption on a fixed-bed column. J Hazard Mater 123(1–3):135–144
91. Agarwal S, Sadegh H, Majid M, Makhlouf ASH, Ali GAM, Memar AOH, Shahryari-ghoshekandi R, Tyagi I, Gupta VK (2016) Efficient removal of toxic bromothymol blue and methylene blue from wastewater by polyvinyl alcohol. J Mol Liq 218:191–197
92. Abdel Ghafar HH, Ali GAM, Fouad OA, Makhlouf SA (2015) Enhancement of adsorption efficiency of methylene blue on Co_3O_4/SiO_2 nanocomposite. Desalination Water Treat 53 (11):2980–2989
93. Wei C-C, Pathiraja IK, Fabry E, Schafer K, Schimp N, Hu T-P, Norcio LP (2018) Removal of Acid Yellow 25 from Aqueous Solution by Chitin Prepared from Waste Snow Crab Legs. J Encapsulation Adsorpt Sci 08(03):139–155
94. Liu Y, Wang W, Jin Y, Wang A (2011) Adsorption Behavior of Methylene Blue from Aqueous Solution by the Hydrogel Composites Based on Attapulgite. Sep Sci Technol 46 (5):858–868
95. Reddy DHK and Yun Y-S (2016) Spinel ferrite magnetic adsorbents: Alternative future materials for water purification? Coord Chem Rev 315:90–111
96. Oubagaranadin JUK, Sathyamurthy N, Murthy ZJJoHM (2007) Evaluation of Fuller's earth for the adsorption of mercury from aqueous solutions: a comparative study with activated carbon. 142(1–2):165–174
97. Wu F-C, Tseng R-L, Juang R-S (2009) Initial behavior of intraparticle diffusion model used in the description of adsorption kinetics. Chem Eng J 153(1–3):1–8
98. Gang DD, Deng B, Lin L (2010) As(III) removal using an iron-impregnated chitosan sorbent. J Hazard Mater 182(1–3):156–161
99. Kumar PS, Ramalingam S, Sathishkumar K (2010) Removal of methylene blue dye from aqueous solution by activated carbon prepared from cashew nut shell as a new low-cost adsorbent. Korean J Chem Eng 28(1):149–155
100. Kadirvelu K (2001) Removal of heavy metals from industrial wastewaters by adsorption onto activated carbon prepared from an agricultural solid waste. Biores Technol 76(1):63–65
101. Karagoz S, Tay T, Ucar S, Erdem M (2008) Activated carbons from waste biomass by sulfuric acid activation and their use on methylene blue adsorption. Biores Technol 99 (14):6214–6222
102. Song J, Zou W, Bian Y, Su F, Han R (2011) Adsorption characteristics of methylene blue by peanut husk in batch and column modes. Desalination 265(1–3):119–125
103. Uddin MT, Islam MA, Mahmud S, Rukanuzzaman M (2009) Adsorptive removal of methylene blue by tea waste. J Hazard Mater 164(1):53–60

104. Auta M, Hameed BH (2011) Preparation of waste tea activated carbon using potassium acetate as an activating agent for adsorption of Acid Blue 25 dye. Chem Eng J 171(2):502–509
105. Pradhananga R, Adhikari L, Shrestha R, Adhikari M, Rajbhandari R, Ariga K, Shrestha L (2017) Wool Carpet Dye Adsorption on Nanoporous Carbon Materials Derived from Agro-Product. C 3(4):12
106. Liu Q-S, Zheng T, Li N, Wang P, Abulikemu G (2010) Modification of bamboo-based activated carbon using microwave radiation and its effects on the adsorption of methylene blue. Appl Surf Sci 256(10):3309–3315
107. AkankshaKalra HP, Hui C, Mackey H, Ansari T, Saleem J, McKay G (2019) Adsorption of dyes from water on to bamboo-based activated carbon-error analysis method for accurate isotherm parameter determination. J Water Sci Eng 1(1):1–11
108. Hameed B, Din A, Ahmad A (2007) Adsorption of methylene blue onto bamboo-based activated carbon: Kinetics and equilibrium studies. J Hazard Mater 141(3):819–825
109. Spagnoli AA, Giannakoudakis DA, Bashkova S (2017) Adsorption of methylene blue on cashew nut shell based carbons activated with zinc chloride: The role of surface and structural parameters. J Mol Liq 229:465–471
110. Dural MU, Cavas L, Papageorgiou SK, Katsaros FK (2011) Methylene blue adsorption on activated carbon prepared from Posidonia oceanica (L.) dead leaves: Kinetics and equilibrium studies. Chemical Engineering Journal 168(1):77–85
111. Elmorsi RR, El-Wakeel ST, Shehab El-Dein WA, Lotfy HR, Rashwan WE, Nagah M, Shaaban SA, Sayed Ahmed SA, El-Sherif IY, Abou-El-Sherbini KS (2019) Adsorption of Methylene Blue and Pb^{2+} by using acid-activated Posidonia oceanica waste. Scientific Reports 9(1)
112. Mudyawabikwa B, Mungondori HH, Tichagwa L, Katwire DM (2017) Methylene blue removal using a low-cost activated carbon adsorbent from tobacco stems: kinetic and equilibrium studies. Water Sci Technol 75(10):2390–2402
113. Ghosh RK, Reddy DD (2013) Tobacco Stem Ash as an Adsorbent for Removal of Methylene Blue from Aqueous Solution: Equilibrium, Kinetics, and Mechanism of Adsorption. Water, Air, & Soil Pollution 224(6)
114. Ahmad A, Loh M, Aziz J (2007) Preparation and characterization of activated carbon from oil palm wood and its evaluation on Methylene blue adsorption. Dyes Pigm 75(2):263–272
115. Leng Chew T, Husni H (2019) Oil palm frond for the adsorption of Janus Green dye. Materials Today: Proceedings 16:1766–1771
116. Setiabudi HD, Jusoh R, Suhaimi SFRM, Masrur SF (2016) Adsorption of methylene blue onto oil palm (Elaeis guineensis) leaves: Process optimization, isotherm, kinetics and thermodynamic studies. J Taiwan Inst Chem Eng 63:363–370
117. Yu L, Luo Y-m (2014) The adsorption mechanism of anionic and cationic dyes by Jerusalem artichoke stalk-based mesoporous activated carbon. J Environ Chem Eng 2(1):220–229
118. Mahamad MN, Zaini MAA, Zakaria ZA (2015) Preparation and characterization of activated carbon from pineapple waste biomass for dye removal. Int Biodeterior Biodegradation 102:274–280
119. El Gamal M, Mousa HA, El-Naas MH, Zacharia R, Judd S (2018) Bio-regeneration of activated carbon: A comprehensive review. Sep Purif Technol 197:345–359
120. Ali GAM, Habeeb OA, Algarni H, Chong KF (2018) CaO impregnated highly porous honeycomb activated carbon from agriculture waste: symmetrical supercapacitor study. J Mater Sci 54:683–692
121. Habeeb OA, Ramesh K, Ali GAM, Yunus RM, Olalere OA (2017) Kinetic, Isotherm and Equilibrium Study of Adsorption of Hydrogen Sulfide From Wastewater Using Modified Eggshells. IIUM Eng J 18(1):13–25
122. Habeeb OA, Ramesh K, Ali GAM, Yunus RM (2017) Application of response surface methodology for optimization of palm kernel shell activated carbon preparation factors for removal of H_2S from industrial wastewater. J Teknologi 79(7):1–10

123. Habeeb OA, Ramesh K, Ali GAM, Yunus RM (2017) Low-cost and eco-friendly activated carbon from modified palm kernel shell for hydrogen sulfide removal from wastewater: adsorption and kinetic studies. Desalination Water Treat 84:205–214
124. Habeeb OA, Ramesh K, Ali GAM, Yunus RM (2017) Experimental design technique on removal of hydrogen sulfide using CaO-eggshells dispersed onto palm kernel shell activated carbon: Experiment, optimization, equilibrium and kinetic studies. J Wuhan UniTechnol Mater Sci Ed 32(2):305–320
125. Habeeb OA, Ramesh K, Ali GAM, Yunus RM (2017) Isothermal modelling based experimental study of dissolved hydrogen sulfide adsorption from waste water using eggshell based activated carbon. Malays J Anal Sci 21(2):334–345
126. Habeeb OA, Ramesh K, Ali GAM, Yunus RM (2017) Optimization of activated carbon synthesis using response surface methodology to enhance H_2S removal from refinery wastewater. J Chem Eng Ind Biotechnol 1(1):1–17
127. Habeeb OA, Ramesh K, Ali GAM, Yunus RM, Thanusha TK, Olalere OA (2016) Modeling and optimization for H_2S adsorption from wastewater using coconut shell based activated carbon. Aust J Basic Appl Sci 10(17):136–147
128. Toledo LC, Silva ACB, Augusti R, Lago RM (2003) Application of Fenton's reagent to regenerate activated carbon saturated with organochloro compounds. Chemosphere 50 (8):1049–1054
129. Ania CO, Parra JB, Menéndez JA, Pis JJ (2005) Effect of microwave and conventional regeneration on the microporous and mesoporous network and on the adsorptive capacity of activated carbons. Microporous Mesoporous Mater 85(1–2):7–15
130. Bonjour J, Chalfen J-B, Meunier F (2002) Temperature Swing Adsorption Process with Indirect Cooling and Heating. Ind Eng Chem Res 41(23):5802–5811
131. Sühnholz S, Kopinke F-D, Weiner B (2018) Hydrothermal treatment for regeneration of activated carbon loaded with organic micropollutants. Sci Total Environ 644:854–861
132. Yuen FK, Hameed BH (2009) Recent developments in the preparation and regeneration of activated carbons by microwaves. Adv Coll Interface Sci 149(1–2):19–27
133. Xin-hui D, Srinivasakannan C, Jin-sheng L (2014) Process optimization of thermal regeneration of spent coal based activated carbon using steam and application to methylene blue dye adsorption. J Taiwan Inst Chem Eng 45(4):1618–1627
134. Li W-H, Yue Q-Y, Gao B-Y, Ma Z-H, Li Y-J, Zhao H-X (2011) Preparation and utilization of sludge-based activated carbon for the adsorption of dyes from aqueous solutions. Chem Eng J 171(1):320–327
135. Wang F, Zhang J, Jia D (2019) Facile synthesis of shell-core structured Fe_3O_4@ACS as recyclable magnetic adsorbent for methylene blue removal. J Dispersion Sci Technol 40 (12):1736–1743
136. Lu P-J, Lin H-C, Yu W-T, Chern J-M (2011) Chemical regeneration of activated carbon used for dye adsorption. J Taiwan Inst Chem Eng 42(2):305–311
137. Brown NW, Roberts EPL, Garforth AA, Dryfe RAW (2004) Electrochemical regeneration of a carbon-based adsorbent loaded with crystal violet dye. Electrochim Acta 49(20):3269–3281
138. Zhou MH, Lei LC (2006) Electrochemical regeneration of activated carbon loaded with p-nitrophenol in a fluidized electrochemical reactor. Electrochim Acta 51(21):4489–4496
139. Shende RV, Mahajani VV (2002) Wet oxidative regeneration of activated carbon loaded with reactive dye. Waste Manag 22(1):73–83
140. Saroyan HS, Giannakoudakis DA, Sarafidis CS, Lazaridis NK, Deliyanni EA (2017) Effective impregnation for the preparation of magnetic mesoporous carbon: application to dye adsorption. J Chem Technol Biotechnol 92(8):1899–1911

Rice Husk-Derived Nanomaterials for Potential Applications

Shimaa Hosny Ali, Mohammed Y. Emran, and Hassanien Gomaa

Abstract Agricultural waste reuse has been gained much interest in various environmental and industrial aspects. The agricultural waste is a rich source of various nanomaterials such as nanosilica (NS), nanocarbon (NC), and nanozeolite (NZ). Through the rice planting to the final product, 20% husk was left as a by-product from the total weight, so-called rice husk (RH). The economic consideration of RH refers to the rich source of NS, NC, and NZ, where RH contents are 70–85% organic matter and inorganic residues (20–25%). Also, rice husk ash (RHA) contents are 60% silica and 10–40% carbon as well as another mineral's composition. Therefore, there are many approached for producing and extracting NS and NC from RH and RHA using different effective physical and chemical recycling methodologies. The extracted NS from RH and RHA was successfully used for the development and production of NZ. The production of NS, NC, and NZ from RH and RHA is a potential approach for reducing the environmental effect and raising the economic values with highly effective materials in various applications. The valuable usage of NS, NC, and NZ is very high, which can be used in various industrial and environmental aspects such as electronics, ceramic, catalyst support, adsorbents, ion exchangers, water treatment, antimicrobial products, biosensing, and biomedical applications. In this chapter, the various extraction and synthesis methodologies for the controlled formation of NS, NC, and NZ from RH and RHA were reported. Besides, the potential usage and sustainable applications of these materials were successfully presented.

S. H. Ali
Department of Chemistry, Faculty of Science, New Valley University, New Valley, Egypt

M. Y. Emran (✉) · H. Gomaa (✉)
Department of Chemistry, Faculty of Science, Al-Azhar University-Assiut Branch, Assiut, Egypt
e-mail: mohamedyossef@azhar.edu.eg; emranm527@gmail.com

H. Gomaa
e-mail: h.gomaa@azhar.edu.eg; gomaa.h1989@gmail.com

A. S. H. Makhlouf and G. A. M. Ali (eds.), *Waste Recycling Technologies for Nanomaterials Manufacturing*, Topics in Mining, Metallurgy and Materials Engineering, https://doi.org/10.1007/978-3-030-68031-2_19

Keywords Agricultural waste · Rice husk · Recycling · Nanosilica · Nanozeolite · Nanocarbon · Potential applications

List of Abbreviations

0D	Zero-dimensional
1D	One-dimensional
2D	Two-dimensional
3D	Three-dimensional
AFM	Atomic force microscopy
Ag NPs	Silver nanoparticles
Au NPs	Gold nanoparticles
AWM	Agricultural waste management
BRHA	Black rice husk ash
CNTs	Carbon nanotubes
CVD	Chemical vapor deposition
DI water	Deionized water
EDX	Energy-dispersive X-ray spectroscopy
FTIR	Fourier transform infrared
GO	Graphene oxide
GQDs	Graphene quantum dots
GR	Graphene
HR-TEM	High-resolution transmission electron microscopy
ICP-MS	Inductively coupled plasma mass spectrometry
LiBs	Lithium-ion batteries
MP	Magnetic particle
MWCNTs	Multi-walled carbon nanotubes
NC	Nanocarbon
NPs	Nanoparticles
NS	Nanosilica
NZ	Nanozeolite
PAA	Polyacrylic acid
PDMS	Polydimethylsiloxane
QD	Quantum dot
RH	Rice husk
RHA	Rice husk ash
SEM	Scanning electron microscope
SWCNTs	Single-walled carbon nanotubes
TEM	Transmission electron microscopy
WRHA	White rice husk ash
XRD	X-ray diffraction
ZSM-5	Zeolite Socony Mobil–5

1 Introduction

The waste generated from agricultural activities has been known as agricultural waste [1]. The agricultural wastes (agro-wastes) are produced from various agriculture activities and industries such as animal excreta and cadavers, food-treating waste, crop waste, and harmful and poisonous agricultural wastes of pesticides, insecticides, and herbicides [1, 2]. However, the agricultural wastes contain various organic and mineral sources that can be extracted and used for a variety of applications like feedstock for energy production and remedy of hazardous materials [3]. In recent years, the agricultural wastes are increased as increasing the world population due to the increase of agricultural production (i.e., livestock wastes, residues from crops, and agro-industrial wastes). The huge quantity of agricultural waste ($\sim$998 million tons) causes significant environmental problems [4]. In addition, organic waste amounts in any farm represent 80% of the total solid waste, where the amount can be expressed to 5.27 kg/day/1000 kg live weight [5].

Recently, agricultural waste management (AWM) restrains high policymaker concerns for green agriculture and sustainable development [6–8]. To control the AWM for potential reuse, and to reduce the environmental hazard of pollution of air, water, and land resources, a lot of approaches and considerations were presented. Moreover, the controlling of AWM needs more effort in various aspects, such as using technology and inducements, and a change in philosophy and attitudes. The designed system of AWM is fixed to control and utilize the agricultural by-products to maintain/improve the environmental quality and sustainability of water, plant, soil, and energy resources [3]. Scheme 1 shows the AWM flowchart, which contains the essential functions of production, treatment, collection, transfer, storage, and utilization, as presented in scheme 1.

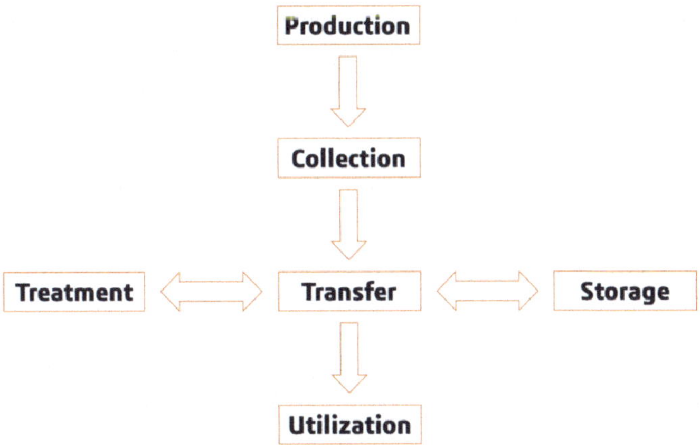

Scheme 1 The flowchart of agricultural waste management functions

The AWM can be used in various potential uses and significant applications such as fertilizer application, anaerobic digestion, animal feed, adsorbents, heat source, direct combustion, and production of nanomaterial [3]. Rice regards, as one of the top, consumed food universally [9]. The production of rice paddy is about ~502.09 million metric tons in June 2020, and this number would be expected to increase regularly due to the predictable requirement of the world population [10], while initiatives toward privatization have been implemented to increase the cultivated area and boost rice production in recent years. The world still faces challenges in agricultural production, especially this season as of ongoing COVID-19 pandemic. Industry and the media have reported shortages of fertilizers and pesticides and fuel shortages, increasing the possibility of limited production this year.

Rice husk (RH) is an agriculture waste of industrial rice treating and represents the 20 wt% of rice waste weight. Moreover, the organic substances represent 70–80% of RH contents. The composition of RH is mainly composed of lignin, cellulose, SiO_2, alkalis, and other hint elements (20–30%) [11]. On the other hand, rice husk ash (RHA) usually produces from burning of RH and is regarded as the second by-product of the rice industry. RHA is roughly about 25% of the original husk weight and induces environmental pollutions and disposal problems [12, 13]. By sampling RH and RHA production from the rice industry, one thousand kg of paddy cereals produces about 200 kg of RH, and about 50 kg of RHA is produced after burning. High production of amorphous silica can be extracted from RHA (85–95% of RHA weight). The high amorphous silica contents from the RHA have raised the broad applicability and high economic values in several activities such as ceramics, construction, chemicals, and electronics, among others [14]. Furthermore, RH and RHA play a significant character in nanomaterial synthesis, as shown in Fig. 1a. Figure 2b and c shows different applications of RH and RHA.

The nanomaterials are widely synthesized and employed for several applications due to the numerous interesting physical and chemical characteristics such as high quantum yield, high surface expanse, and novel morphological construction [17–27]. These interesting features make them proper candidates in numerous areas of science and nanotechnology for a wide collection of applications [28–36]. Various methods and approaches were designed and established for synthesis of nanomaterials such as pyrolysis, vapor-phase reaction, plant extracts, chemical vapor synthesis and laser evaporation, solgel treatment, hydrothermal treatment, and solution precipitation method [27, 37–44]. Some of these approaches require high energy techniques and convoluted methodologies [45, 46]. In recent years, the removal of nanomaterials from natural biological resources like microorganisms, plants, and by-products and waste from agriculture has gotten a great interest [47]. This method offers several benefits over traditional manners including flexibility, economical, a higher safety margin, and reducing environmental pollution. Particularly, the preparing of nanomaterials from agriculture waste, especially nanosilica, and production of nanocarbon is regarded as a renewable source for solving environmental degradation problems and economic substitutes for industrial feedstock [48].

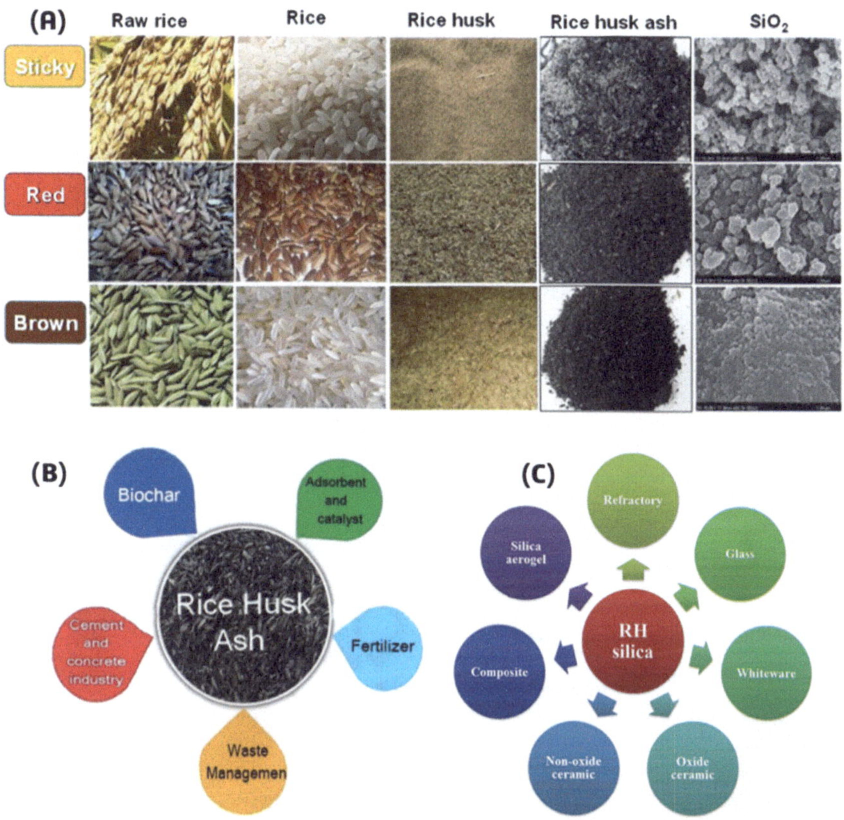

Fig. 1 **a** Photo- and scanning electron microscope (SEM) images of the rice plant, rice, rice husk, rice husk ash, and synthesized nanosilica from sticky, red and brown rice. Adapted with permission from Ref. [15], Copyright 2016, Elsevier. **b** The various rice husk ash applications in many valuable fields. Adapted with permission from Ref. [16], Copyright 2019, Elsevier. **c** The various applications of silica derived from rice husk in multi-industries such as ceramic industry. Adapted with permission from Ref. [9], Copyright 2018, CC BY license, Taylor & Francis

Silica is one of the plentiful and commonly used materials in numerous applications like thixotropic agents, thermal insulator, and composite fillings [49]. Particularly, the silica nanoparticles have been used in biosensor design, drug delivery, cell labeling, imaging, and separation [26, 39, 40, 50–53]. The production of silica nanoparticles was extensively produced from RH/RHA, and other agricultural by-products and wastes [17, 48, 54]. For example, the preparing of nanosilica from RH, coconut shell, and wheat straw was approached [55, 56].

Carbon-based nanomaterials attained great interest in numerous uses such as energy, sensor/biosensors, catalysis, water treatment, and other potential applications. There are many types of carbon-based materials involving fullerene, graphene, nanodiamonds, nanospheres, nanotubes, and activated carbon [21, 22, 30,

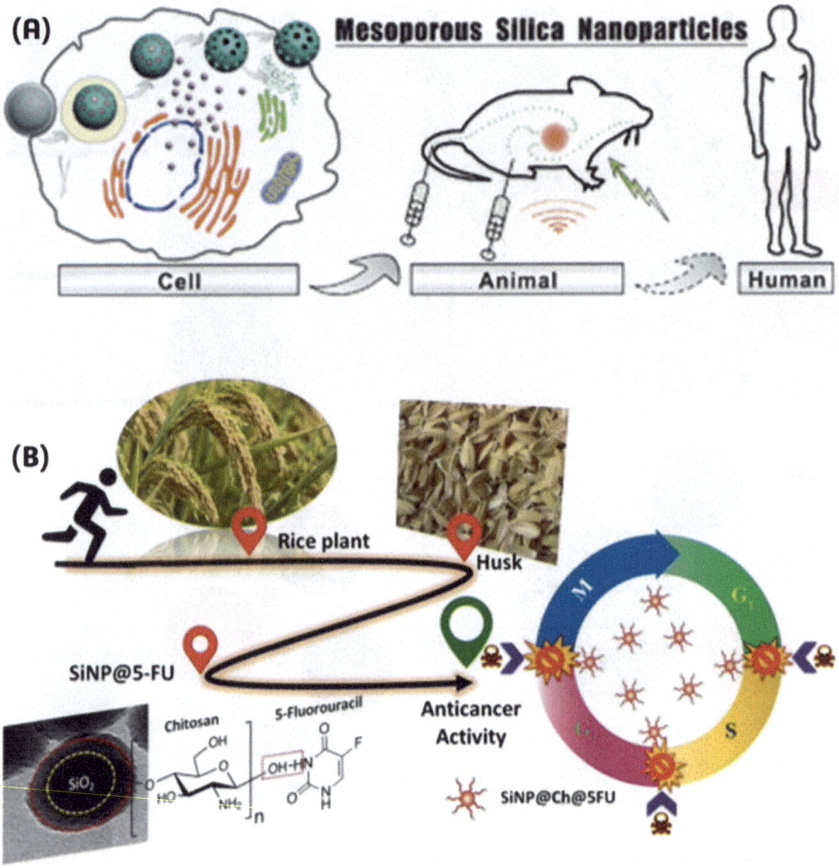

Fig. 2 **a** Valuable and medical application of nanosilica as drug delivery-based materials with high biocompatibility and low cytotoxicity. Adapted with permission from Ref. [56], Copyright 2011, RSC. **b** The using of nanosilica derived from rice husk modified by chitosan for loading of 5-fluorouracil as an anticancer treatment drug. Adapted with permission from Ref. [150], Copyright 2020, Elsevier

31, 57]. The carbon-based material possesses potential physical and chemical properties that are extensively used for several uses [58, 59]. Typically, the carbon-based materials can be categorized in accordance with the morphological structures of zero-dimensional (0D) such as fullerenes and carbon dots, one-dimensional (1D) such as carbon nanotubes (CNTs), two-dimensional (2D) such as graphene, and three-dimensional (3D) such as graphite [60, 61]. One of the carbon-based materials is CNTs that possesses significant properties involving high chemical and mechanical stability, and fascinating electrical conductivity. Moreover, graphene is one of the 2D carbon materials that possess fascinating properties of highly transmitted light and graphene edges [62]. Till now, there are many attempts for productions and applications of carbon-based nanomaterials such as CNTs, and graphene such as electronics, sensors, strengthening of composite materials, and storage devices [19, 20, 22, 63]. Recently, the green synthesis of carbon-based materials using the waste product has been achieved great consideration. RH contents are rich in cellulose and lignin, which make it a useful and abundant source of carbon nanomaterials [64]. Within this chapter, we illustrate the synthesis of carbon nanomaterials from RH.

Zeolites are microporous crystal materials with complicated pores and grooves [65]. Moreover, zeolites are employed as ion exchangers, catalysts, and adsorbents [66, 67]. Several kinds of nanometer-sized zeolites, involving zeolite A, and faujasite-X and -Y have been prepared by the hydrothermal method [68, 69]. RH acts as the silica sources for preparing of zeolite A- [70], β- [71], and -ZSM-5 [72]. Alternatively, to previously used pure chemical sources, RH is a practical source of silica, as it is economy, less selective, and very active [73]. Initially, RH is completely burnt to create ash, turning the husk's organic siliceous material into white ash called RHA. Therefore, using a suitable alkali solution, amorphous silica can be extracted from RHA and further used for synthesis zeolites [74].

This chapter seeks to outline recent advancements in the utilization of agriculture waste for the source and synthesis of various valuable nanomaterials and their significant uses. Specifically, we present the use of RH and RHA for control production of nanomaterials such as NS, NC, and NZ. Nowadays, these nanomaterials participate in various applications. Besides, the valuable utilization of these nanomaterials in several areas such as catalysis, electronic industries, ceramic production, biosensor, and energy, and in high technologies of semiconductors, smart chips, and so on has been produced. Therefore, synthesis of these nanomaterials from agriculture wastes such as RH and RHA induces many benefits, including low cost, renewable sources, and high environmental impact. Moreover, various potential applications of derived nanomaterials from RH and RHA are discussed.

2 Rice Husk and Rice Husk Ash

2.1 Rice Husk Composition

RH is the paddy cereal's outermost layer, which is isolated from the rice cereals during the grinding operations [75]. The chemical components of RH are varied from one sample to another due to the kind of paddy, crop time, climate, and geographical circumstances [76]. The RH is a deficient nourishment substance and seldom employed as animal nutrition. Moreover, the RH usually is used for various aspects such as fertilizer additives, as a fuel for cooking, and in landfill applications or for paving. Other RH uses involve partition boards, production of biochar which is significant material for the improvement of the soil and generating bioenergy, an alternate source of renewable energy. Nevertheless, only a small amount of RH is currently used for the production and composting of silica.

RH is composed of 24.3% hemicelluloses, 19.2% lignin, 34.4% cellulose, 18.85% ash, and 3.25% trace elements. Hemicelluloses are utilized as a recourse of activated carbon, xylose, and silicon dioxide [77]. RH comprises major elementary components as 37.05% for carbon, 8.80% for hydrogen, 11.06% for nitrogen, 9.01% for silicon, and 35.03% for oxygen [11]. The husk contains silica from 17 to 25% [78, 79]. RHs have bulk density of 96–100 kg m^{-3}, hardness (Mohr's scale) 5–6, 22.29% for ash, 31–37% for O, 0.23–0.32% for N, 0.04–0.08% for S, and 4–5% for H [80]. The composition of RH depends on several factors such as a variety of rice, type of fertilizer used, chemistry of soil, and even the geographical location of the production [81].

2.2 *Rice Husk Ash Composition*

RHA is produced from RH after incineration and represents 17–20% weight of RH, which is lightweight and bulky. The RHA is a porous material with an approximate density of 180–200 kg m^{-3}. RHA is classified into white rice husk ash (WRHA), and black rice husk ash (BRHA) depends on the combustion process conditions [82]. The controlled pyrolysis of RH under air produces WRHA. Typically, WRHA contains pure SiO_2 (>95%) in a hydrated amorphous shape with high porosity and reactive surface [83]. Meanwhile, the BRHA is formed from controlled combustion of RH in the inert atmosphere such as nitrogen, leading to produce carbon and silica products [84]. The existence of a high quantity of SiO_2 renders it an appropriate material for manufacturing uses. The chemical composition of RHA is 89% SiO_2, 18.24% C, 1% CaO, 1.20% Al_2O_3, 1.22% K_2O, and 1.28% Fe_2O_3 [85]. Moreover, trace elements of Na, Cu, K, Mg, Ca, Mn, Fe, and Zn are founded [86]. The RHA properties of color, activity, impurity contents, and SiO_2 contents are affected by burning conditions (e.g., temperature and duration time), heating rate, burning equipments, acid/alkaline leaching pretreatment, source and geographical position, variety of crops, and the type of fertilizer used [87–97]. Table 1 illustrates the RHA chemical composition under various analytical studies. The changes in the composition of RHA primarily differ according to the kind of soil, climate, agricultural management, and other parameters.

3 Synthesis and Application of Nanosilica from Rice Husk and Rice Husk Ash-Based Resources

Synthesis of the nanosilica (NS) industrial scale usually uses sodium silicate as the principal silicone component. In this process, they need large amounts of energy for the production of Na_2SiO_3 by melting the silica sand and Na_2CO_3 at 1573 K [98]. Thus, producing of NS from silicone-containing biomass such as RH and RHA is

Table 1 Chemical composition of rice husk ash

SiO_2	Al_2O_3	Fe_2O_3	K_2O	CaO	MgO	Na_2O	P_2O_5	References
96.71	0.09	0.01	0.69	–	–	–	0.23	[87]
88.32	0.46	0.67	2.91	0.67	.44	–	–	[88]
94.36	0.26	0.23	0.65	1.56	0.86	0.39	–	[89]
94.6	0.3	0.3	1.3	0.4	0.3	0.2	0.3	[90]
97.89	0.02	0.16	0.18	0.27	0.09	0.18	0.13	[91]
99.14	0.03	0.03	0.12	0.16	0.08	0.06	0.29	[92]
95.55	0.13	0.03	0.05	0.56	0.09	–	0.45	[93]
99.58	0.17	0.03	0.02	0.04	0.02	–	0.11	[94]
92.81	–	0.312	1.021	0.417	0.212	2.658	1.071	[95]
86.0	5.12	1.12	1.82	1.26	0.48	0.05	0.48	[96]
87.4	0.4	0.3	3.39	0.9	0.6	0.04	–	[97]

highly needed for establishing enormous economic values. The RH is one of the biomasses that provides effective heat and riches in biological oils with lignocellulose [99]. Besides, the high RH's global annual production is high, with approximately 100 million tons that increase the potentiality as a source of energy and a source of various valuable materials [100]. Consequently, various approaches were preceded for investigating economic, environmentally friendly, and easy ways to extract high-purity silica from RH/RHA. There are two conventional techniques for RH silica preparation (combustion and chemical method). The combustion technique is performed at high temperatures for the burning of the rice husk to eliminate carbon and organic material. The chemical technique with appropriate acid or alkaline was used for RH/RHA treatment to extract silica through sodium silicate [101].

3.1 *Synthesis of Nanosilica*

Biomass-based silica growth was formed based on thermal and chemical techniques. Scheme 2 describes the various techniques used for the manufacturing of NS from agricultural wastes.

3.1.1 Thermal Techniques

The thermal techniques include the use of different types of furnaces such as muffle furnace, fluidized bed reactor, fixed bed furnace, and other thermal practices. There are many drawbacks of thermal technology, such as the need for more reaction time, the formation hot spots, incomplete oxidation of carbon due to the absence of free-flowing air, and others [102]. NS is extracted from RH/RHA on a laboratory

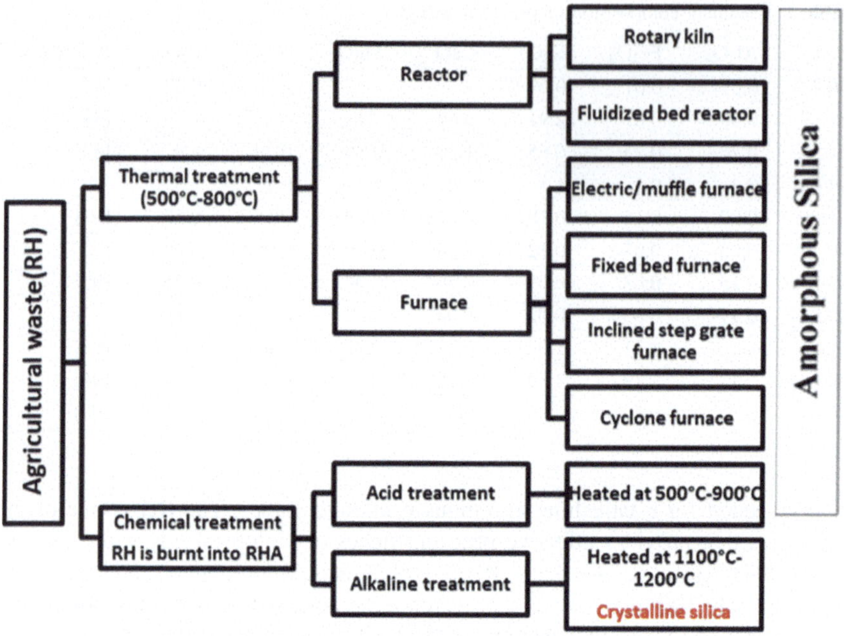

Scheme 2 Several techniques employed for generating nanosilica from farming wastes

scale by an electric/muffle furnace. The greatest drawback of applying this technique is the time consuming and a lower rate of production. Moreover, the NS from RH was achieved using a fixed bed furnace. The amorphous nanosilica was prepared using fixed bed furnace [103]. In this technique, RH raw has been treated by acid and pyrolysis at 600–1200 °C, and at 1000 °C the amorphous silica changed to crystalline form. Hamad successfully produced silica from RHA in a muffle furnace and fixed bed reactor at 500–1150 °C [104]. In addition, a fluidized bed reactor was used for decreasing the energy combustions and increasing high yield from the incineration process. Huang et al. extracted NS from WRHA using fluidized bed reactor [105]. The amorphous silica can be produced at varying temperatures and speeds in fluidized bed bubbling pilot plants [92]. Genieva et al. produced silica material from RH by rice-milling phase [106].

The major product from agricultural waste was produced in N_2 atmosphere using a fluidized bed reactor. Luan and Chou extracted silica from RH in an adapted fluidized bed vessel during the pilot flame [107]. Consequently, the resulted silica product exhibited high activity. The benefits of fluidized bed reactor are including uniform temperature distribution, fast rate of reaction, efficiency in carbon conversion, low operating temperature, and high combustion strength. The construction of this furnace involves a feeding module, chamber for combustion, and ash precipitation chamber. This furnace is usually used to produce RHA from RH. The drawback of using this method is poor production of RHA and high yield of non-combustible carbon content. In this method, the RH is supplied from the top of the vessel, while air streams from the underside [108]. The cyclonic furnace is another furnace and acts as an incineration method for the production of RHA from RH [109]. The advantages of this method are conducted with the production of lower carbon contents. A rotating kiln is a pyro-treating system employed in the continuous procedure to elevate the materials to a high temperature (calcination) and increase the production of RHA. Sugita presented a patent for producing active RHA using a rotary kiln. In this method, an upstream rotary kiln carbonized RH that warmed at 300–400 °C by electrical heaters, flames, or other heat supplies. Carbonized RH is provided in a rotating furnace and carbonization at 600 °C. The drawback of this method is conducted with extra fuel needed to prevent RHA from being crystallized, long reaction period time, and high energy consumption. Table 2 shows the properties of thermal furnaces that are utilized for the extraction of NS from RH/RHA.

3.1.2 Chemical Method

The high silica quantities with high purity have been extracted from RHA using chemical techniques such as alkaline and acid extraction. The chemical technique provides high purity and quantity of NS as compared to the NS produced from combustion. Meanwhile, this technique exhibits a high cost due to the relatively long reaction time (24–48 h) and follows various steps using various chemical sorts. Before chemical treatment, RH has been incinerated to produce RHA, and then the RHA is chemically treated for production of NS. The chemical technique of acid or alkaline RH/RHA treatment increased the SiO_2 contents to over 99%. Hence, many researchers have used various chemical methods to extract higher content of NS.

Table 2 Properties of thermal furnace utilized for the extraction of nanosilica from rice husk ash

Thermal furnace	Properties
Electric/muffle furnace	This technology needs a long time and a lower rate of production
Fluidized bed reactor	Fluidized bed reactor has several advantages as combustion strength is high, low operating temperature, simple process, fast start-up, and easier extraction of ash [110]
Inclined step grate furnace	Construction and operation are uncomplicated, but the combustion is ineffectual, and ash isolation caused smoke and sparkle [108]
Rotary kiln	Carbonizes RH by turning of RH into RHA This process generates WRHA [111]
Cyclone furnace	Soponronnarit et al. demonstrate a cyclone furnace to boost the furnace efficacy of 16% by increasing the airflow (90%) [112]

Alkaline Extraction

The extraction of amorphous silica from agriculture waste such as RH and RHA using alkaline and acid neutralization is regarded as an effective and easy technique. Scheme 3 shows the alkali extraction technique of NS from RH [113]. Acid treatment of RHA by HCl at 75 °C in a water bath for 4 h was performed to get rid of the metallic impurities. The precipice was filtered and washed by distilled water until the pH reaches 7 and was then dried at 110 °C for 12 h. After that, the filtrate was treated with NaOH to obtain Na_2SiO_3 solution that was continuously stirred for 1 h at 90 °C. The solution of Na_2SiO_3 was treated with ethanol, and deionized water (DI water) was added under continuous stirring for 10 min. The H_3PO_4 (3 M) titration was proceeded for the mixture until gel formation. The yellow gel was centrifuged and washed by DI water to eliminate excess Na_2SiO_3 and Na_3PO_4, and then calcinated to produce NS.

$$SiO_{2(s)} + 2NaOH_{(l)} \rightarrow Na_2SiO_{3(l)} + H_2O_{(l)} \quad (1)$$

$$3Na_2SiO_{3(l)} + 2H_3PO_{4(l)} \rightarrow SiO_{2(s)} + 2Na_3PO_{4(l)} + 3H_2O_{(l)} \quad (2)$$

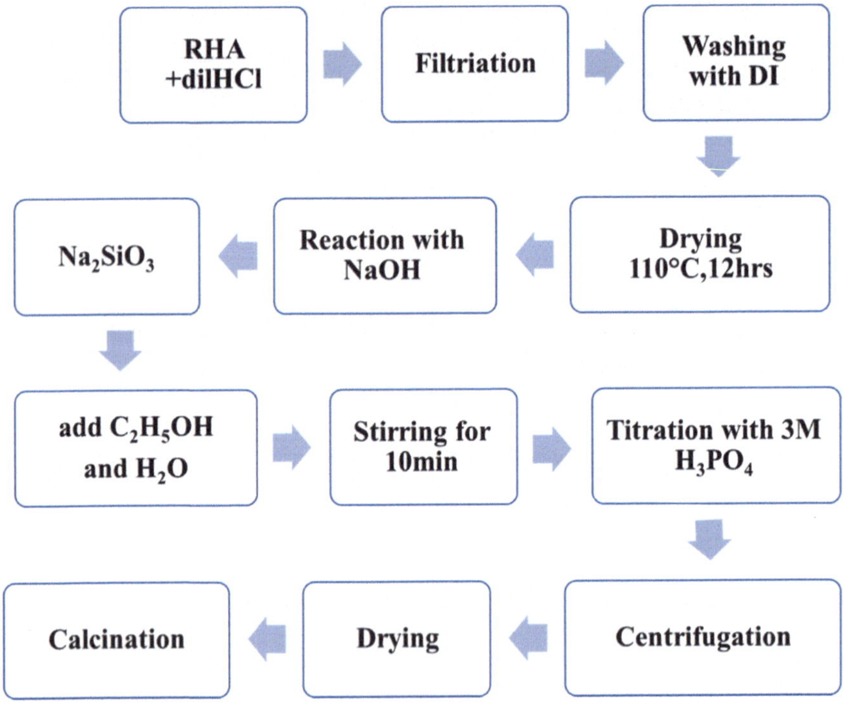

Scheme 3 Extraction of nanosilica from rice husk ash

NS powder was prepared from RH in the high surface area using an alkaline extraction technique (NaOH) [114]. The prepared NS has been investigated by several techniques (Fourier transform infrared (FTIR), X-ray diffraction (XRD), SEM, and transmission electron microscopy (TEM)). The purity of NS was established using EDX assessment, and it was over 95% SiO_2 in the obtained sample. Paranhos prepared NS using the same procedure with some adjustments [115]. NS is produced from RH following the two steps: (i) transformation of RH to RHA and (ii) production of amorphous SiO_2. Concentrated solutions of H_2SO_4 and H_3PO_4 were used to precipitate nano-SiO_2 from Na_2SiO_3 solution. The NS was derived from RHA using the alkali extraction method [56]. The optimized parameters were studied, such as the concentration of acid and alkaline, pH of gelation process, time of aging, and temperature of aging. NS with various surface areas and particle size has resulted in the effects of different acids. Song et al. applied the Taguchi method for silica production from RH. The method depends on the following of two steps: (a) involving the transformation of RH into RHA and (b) the production of NS from RHA [116].

To conclude the 1st step, RH was completely rinsed and leached to eliminate the impurities accompanied using 820 μm sieve mesh. Then, the samples were washed by DI water, then stirred continuously for 1 h, and after that dried at 70 °C to eliminate the sticky dust followed by leaching at 90 °C by 1 M HCl for 2 h. The leaching samples were continually rinsed by DI water and annealed at 700 °C for 2 h. The dried sample was denoted as RHA. The 2nd step is the production of NS from RHA. RHA was treated by 1.5 M NaOH and then heated at 90 °C for 1 h. The obtained solution of Na_2SiO_3 was diluted by DI water to 1 M concentration and titrated with 1 M HCl to reach the neutral pH. The solution was stirred until the gel formation, then dried for seven days at 70 °C for aging, and accompanied by centrifugation at 10^4 rpm. Rehman et al. prepared nanosilica using the RHA silica source. Silica nanoparticles were extracted from RH using NaOH alkaline solgel method [117]. Thuc and Thuc's technique synthesized nanosilica using the RHA. The RH was thermally treated at optimum conditions of 600 °C for 4 h to produce the RHA [118]. The NaOH solution was used to produce Na_2SiO_3 solution and then deposit the NS by addition of H_2SO_4 (pH = 4) in the water/butanol solution. There are many developing methods for extraction of nanosilica from RHA using NaOH solution (alkaline extract) and then treated by the acid solution of H_2SO_4 [119, 120]. The extracted SiO_2 nanoparticles were illustrated using various physical methods such as SEM, EDX, XRD, and FTIR [121]. Therefore, the alkaline extraction from RHA was performed to produce various SiO_2 nanoparticles with high purity which are used in different applications.

Acid Extraction

NS was produced based on the acid treatment of RHA. Sankar et al. treated the RHA by acid solution after the formation of RHA [15]. The RHA was annealed at 700 °C in a muffle furnace (with a rising rate of 5 °C min^{-1} for 2 h) under atmospheric pressure. After this, white-obtained mesoporous SiO_2 nanopowder was obtained. Carmona et al. prepared amorphous and white nanosilica from RH using lower acid solution [122]. Mahmud et al. employed HCl for acid leaching and thermal combustion at 700 °C to produce NS with high purity (greater than 99%) and high surface area [123]. Rafiee and Shahebrahimi synthesized very pure amorphous SiO_2 of 98.8%, nanoscaled particles, high surface domain, and an amorphous shape of 99.9%. The RH was treated by acid, followed by thermal combustion. TEM image showed the average NS size is 6 nm [124].

It was found that NS can be extracted by two major techniques of thermal and chemical methods. These techniques have been widely adopted and developed (Scheme 4). A fluidized bed reactor generated NS with high purity of 92–96% via thermal approaches at 800:950 °C for 4:8 h [125]. While the alkaline-modified chemical technique appeared promising properties, it generated NS with high purity of 95% at 110 °C (mild condition) [126]. The low-cost method in the preparation of NS using green technology was documented by Mor et al. [48]. The RHA was initially treated with NaOH and left at 100 °C for 2 h in an autoclave to prepare the viscous mixture and then dilute with DI-H_2O for phase split. The filtrate was treated by to form the precipitate, then washed by DI-H_2O for a number of periods, and after that desiccated at 50 °C in an oven to produce high pure NS of 99%.

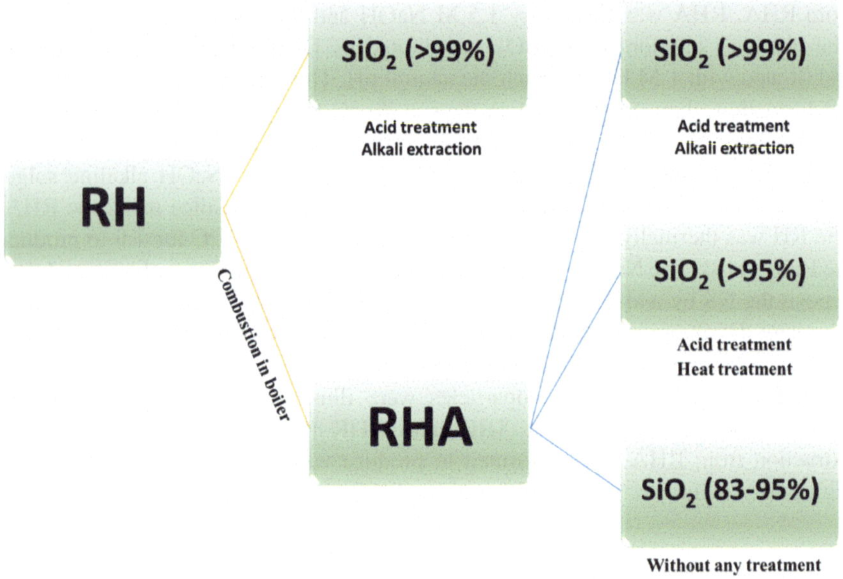

Scheme 4 The various methods to extract nanosilica from rice husk and rice husk ash

3.2 Potential Applications of Nanosilica

NS is regarded as one of the potential inorganic materials that broadly applied in numerous functions such as drug delivery, optical imaging agents, water remediation, rare earth element extraction, and so on [127]. NS is not toxic and highly biocompatible. The NS has been potentially used, due to its intrinsic features of macroscopic shape, chemical functionality, and mesoporous structure (pore sizes between 2 and 50 nm).

NS structure has three unique functional positions, namely: pore walls, pore entrance, and particles' interior/exterior surfaces. NS particles of surface functionalization, high specific area, and mesoporous network featured a specific and unique surface [128]. Besides, there are many approached for adding more beneficial characteristics to NS by adding a non-siliceous group to the pore walls that not partially block the mesopores [129–131]. Therefore, this chapter enclosed the importance of NS in a variety of areas of usage such as medicine, the agricultural field, and ecological bioremediation.

3.2.1 Biomedical Applications of Nanosilica

Bioimaging and Biosensing

Many techniques are available for the modification of NS to make them effective in medicinal chemistry [132–134]. In this context, there are many approached for surface modifications and further use in medicinal purposes like drug transport and bioimaging. The affinity ligands binding to receptors over-expressed in cancer cells can activate the surface of nano-SiO_2. Additionally, the development is designed to improve the permeability and retention impact to improve the effectiveness of cancer treatment [135]. The porous construction of NS has been regarded as the ideal scaffolds in drug delivery that induced the ability to load the proteins for biosensing applications [136]. Moreover, the NS surface modifications enhance the sustainable and regulated access of loaded drugs or biomolecules to specific positions, which make it important candidates for the treatment of cancer [137]. Selvan et al. explained the functional and multifunctional synthesis of NS for application in bioimaging and biosensing. Nanoparticles (NPs) are functionalized for biological applications to involve magnetic particles (MP), quantum dots (QD), and noble metal NPs [138]. The purpose of preparing these functional systems was optical detection and sensing applications, such as ultrasensitive protein detection. Thus, in the field of bioimaging, the efficacy of silica coating in imaging living cells has been demonstrated. Cancer cell targeting is widely employed as a simple method to enhance the attraction of NS to tumor cells through receptors–ligand interaction, leading to increasing the cellular uptake and drug delivery system. The grafted N-folate-3-aminopropyl with NS was internalized into HeLa cells [139].

Therefore, the NS has been widely used for the loading of various active materials such as drugs, ligands, and so on for bioimaging, biosensing, and drug delivery applications.

Drug Delivery Systems

For using the NS in the drug delivery applications, there are many reports approached in this field. Kempen et al. used NS for stem cell therapy. The insulin-like growth factor was used as a pro-survival agent model to establish its role in cardiac stem cell therapy [140]. The ultrasonic signal enabled the intracardiac delivery of fibrous tissue and enhanced the growth factor delivery, leading to increase the cell survival after an injury from ischemia and transplantation of cells [141]. The ability of NS for encapsulating various antigens' size and composition was demonstrated. The encapsulation process was accomplished by adjusting the NS pore size and/or surface functionalization to obtain the required strength in the antigen particle system in addition to the required kinetic profile [142]. The commercial anticancer drug such as doxorubicin hydrochloride was encapsulated at solgel-derived NS/modified NS [143]. However, attracting interest in this field was reported with several experimental drugs such as betulinic acid [144] or curcumin [145]. Such experiments demonstrated the anticancer drug delivery activity that contained strong internalization to the cells without using of an identifiable fragment on NS surface. Besides, many approaches have been implemented using NS-based materials modified with metallodrugs such as titanocene [146, 147] or organotin(IV) compounds [148]. The metallodrug-modified materials, therefore, do not act as a standard drug delivery system. In fact, these substances do not need to be released as a metallodrug to become carcinogenic. Therefore, the cytotoxic activity of these materials does not require a triggering release mechanism. Also, the incorporation of metallodrugs into NS enables the hydrophobic active species use for studying of anticancer activity. The NS system was used for the in vitro model of monastrol targeted delivery into the cancer cells (Fig. 2) [149]. Additionally, NS is a photo-absorber material that converts the absorbed optical energy into heat to treat cancer.

3.2.2 Application of Nanosilica in the Agricultural Field

NS exhibits great potential applications in agriculture field as of the potential chemical and physical properties [151]. NS was widely used as nanopesticides, nanoherbicides, and nanofertilizers. In addition, the NS was used as a carrier for nucleotide, protein, and other chemical delivery into plants. A lot of reports show that mesoporous NS increases the efficiency and durability of commercial pesticides. In the case of fertilizer delivery, the NS incorporated with the organic fertilizers improves the plant productivity and induces the role of NS as fertilizer delivery [152]. Mesoporous NS with a specific pore size of 2-10 nm acted as an effective transport vector for urea, B, and N-based fertilizers, as summarized in Table 3 [152, 153]. Thus, NS performs a potential carrier of fertilizers for use in growing specific crops by delivering various herbicides and fertilizers to plants (see Fig. 3).

3.2.3 Use of Nanosilica in Environmental Remediation

NS has been widely used in environmental remediation of the hazardous compound to reach the water. The NS was employed as an adsorbent/trapper for the weighty metal's elimination from polluted plants. Contaminated plants were joined to electrically charged NS and analyzed by using ICP-MS. To remove boron, mesoporous NS functionalized with cellulose acetate matrix to covalently bind with boron [159]. NS can be employed to eliminate lead ions with high efficiency, due to the enormous surface expanse and the active surface of NS that negatively charged at the localized surface. Therefore, a companion with the positively charged Pb ions occurs. Nuclear sites are suffering from lack of radioactive components. Therefore, NS was used for elimination of radioactive species from radioactive wastes [160].

3.2.4 Use of Nanosilica in Water Decontamination

The water purification from hazardous elements and compounds has a great interest due to the large and excessive population [161]. The NS is regarded as one of the highly used materials due to high surface area, porous construction, controlled morphological design, and active doping with various metals. There are various approaches for the preparation and design of active adsorbent. From that, the NS was coated with a thin film of polydimethylsiloxane (PDMS) using chemical vapor deposition. These PDMS-coated nano-SiO_2 are used in the oil/water mixture for selective oil gelation, enabling rapid separation, and elimination of oil from the water [161]. The silver nanoparticles (Ag NPs) were loaded and connected to the SiO_2 particles to reduce agglomeration among the Ag NPs and release of the particles in water ecosystems [162]. The NS was employed for the adsorption of

Table 3 Using of nanosilica as fertilizers in agriculture management

Composition	Herbicide/fertilizer	Benefits	References
NS	Farmyard droppings and NPK nourishments	The NPK nourishments were considerably delivered and improved the growth traits	[157]
Si-NP	Encapsulated urease	Enhancement of the instability of urease by adsorption of urease at the surface of Si-NP	[158]
Si-NP	Encapsulated urease	The slow release of entrapped urea in soil than in water	[152]

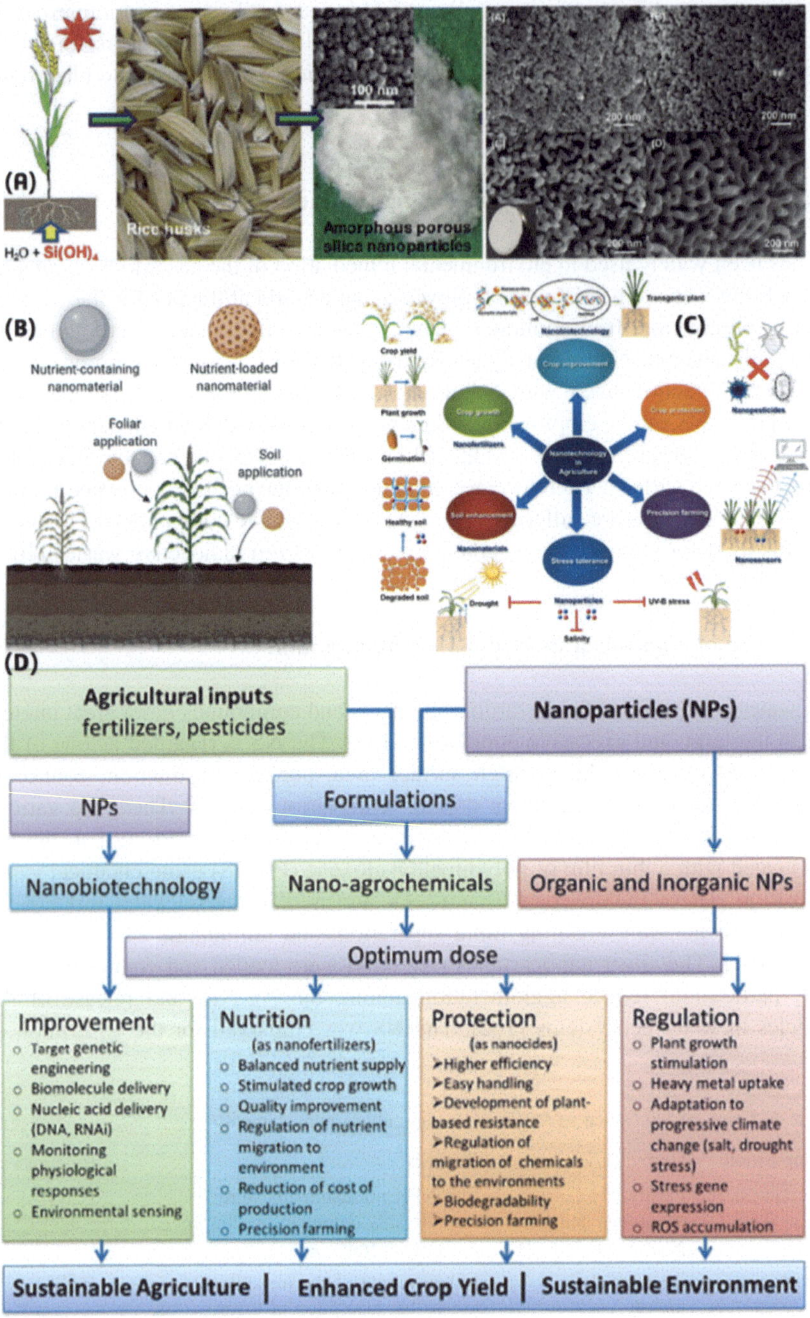
(A)
H_2O + $Si(OH)_4$
Rice husks
100 nm
Amorphous porous silica nanoparticles
200 nm
(B)
Nutrient-containing nanomaterial
Nutrient-loaded nanomaterial
Foliar application
Soil application
(C)
(D)
Agricultural inputs
fertilizers, pesticides
Nanoparticles (NPs)
NPs
Formulations
Nanobiotechnology
Nano-agrochemicals
Organic and Inorganic NPs
Optimum dose
Improvement
Target genetic engineering
Biomolecule delivery
Nucleic acid delivery (DNA, RNAi)
Monitoring physiological responses
Environmental sensing
Nutrition
(as nanofertilizers)
Balanced nutrient supply
Stimulated crop growth
Quality improvement
Regulation of nutrient migration to environment
Reduction of cost of production
Precision farming
Protection
(as nanocides)
Higher efficiency
Easy handling
Development of plant-based resistance
Regulation of migration of chemicals to the environments
Biodegradability
Precision farming
Regulation
Plant growth stimulation
Heavy metal uptake
Adaptation to progressive climate change (salt, drought stress)
Stress gene expression
ROS accumulation
Sustainable Agriculture | Enhanced Crop Yield | Sustainable Environment

◂**Fig. 3** **a** FE-SEM images of nanosilica extracted from rice husk for the potential applications in agriculture fields. Adapted with permission from Ref. [154], Copyright 2012, ACS. **b** Nutrient-based nanotechnology may be employed as fertilizers either directly on the soil or on the plant surface. Adapted with permission from Ref. [155], Copyright 2019, RSC. **c** and **d** Simplified survey of possible nanomaterial applications in prospective farming development. Adapted with permission from Ref. [156], Copyright 2019, MDPI

toxic dyes like methylene blue and methyl red [163]. Moreover, the NS was used as photocatalyst support for the disintegration of various organic pollutants. The NS was decorated by photoactive materials such as Ag and Au, and showed highly reactive photocatalyst for degradation of methyl red dyes [164].

3.2.5 Application of Nanosilica in Solar Cells

The use of nanostructured materials in solar power engineering is presently one of the highly encouraging ways of reducing solar cell costs and increasing their efficiency [165]. A variety of experimental and theoretical studies has been performed to utilize new silicon-based solar cells with valuable characteristics. Gribov et al. demonstrated a film consisting of NS. The designed solar cell efficiency was improved by about 12% [165]. Arunmetha et al. produced the NS from quartz and used it in hybrid solar cell applications. The NS was widely used in portable electronic devices and other valuable applications [166]. The silicon was produced from silicate (extracted from RHA) with a high purity of ≈99.99% [167]. The produced silicone using the RHA as the row source decreases the high cost of the silicone manufacture from sands and minimizes the ecological danger impact of agriculture waste.

3.2.6 Application of Nanosilica in Batteries

High-purity and crystalline structures of silicone are used in battery material preparation (Fig. 4). The conventional silicon production process is highly expensive [75]. Therefore, various research focusing and development are focusing on commercial production of highly pure silicon with low cost and high surface-to-volume [168]. As we described earlier in the extraction of NS from RH and RHA, a lot of approaches were developed for the production of silicon with high purity. The silicone was produced as a reduction of NS using the carbothermal reduction process at elevated temperature [167]. Moreover, the porous nanosilicon was produced by magnesiothermic reduction of sand as the source of silica. 3D network of nanosilicon (8–10 nm) was produced after magnesiothermic reduction in the presence of NaCl as effective heat scavenger. The designed electrodes based on carbon-coated nanosilicon exhibited high capacity (1024 mA h g^{-1} at 2 A g^{-1}) and long-term stability (1000 cycles) [169]. Moreover, the nanosilica as lithium-ion batteries (LiBs) anodes has proceeded high potential as compared with silica

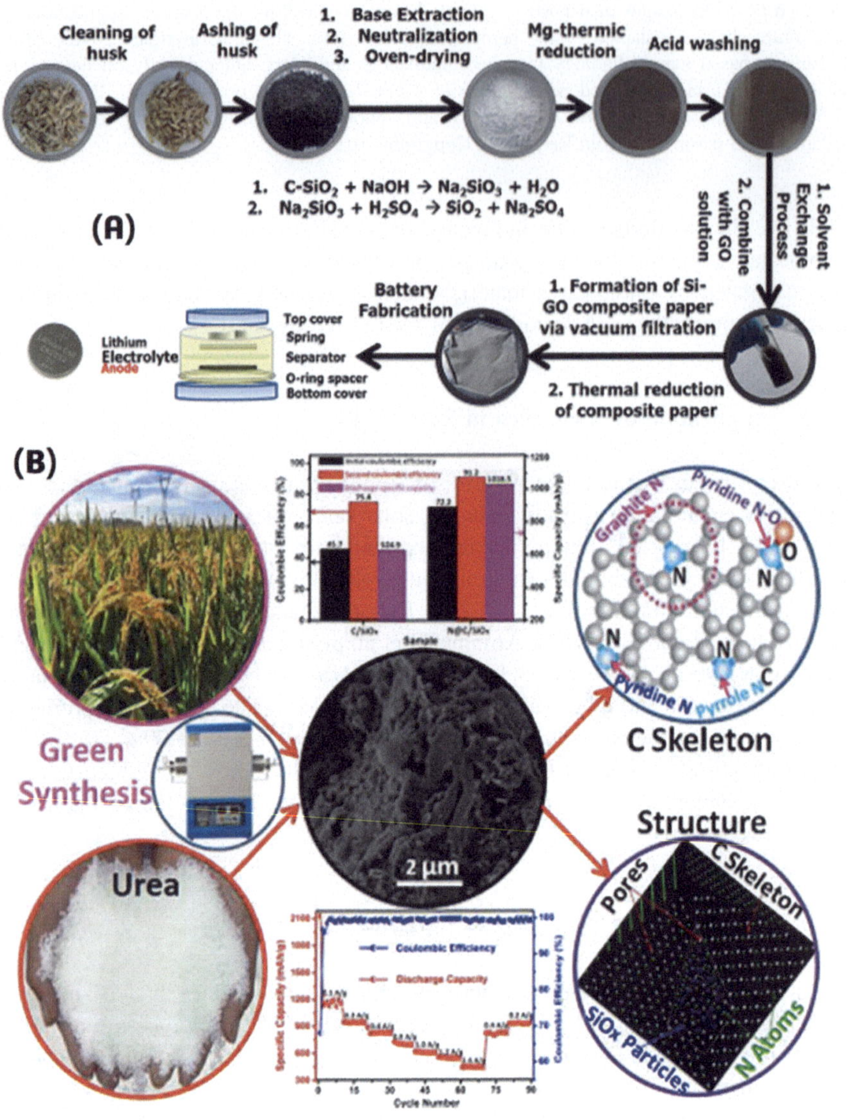

Fig. 4 **a** Synthesis of Si nanoparticles from rice husk-derived SiO_2 for battery fabrication. Adapted with permission from Ref. [171], Copyright 2014, RSC. **b** Porous N-doped carbon/SiO_x composite synthesis process from rice husk to produce Li-ion battery anode. Adapted with permission from Ref. [172], Copyright 2020, Elsevier

composites [170]. Therefore, the potential user of silicon materials as anode materials for LiBs production is potentially needed in high demands and increases the potential applications of silicone products in energy storage field.

4 Nanocarbon from Rice Husk

Carbon is widely abundant element that is capable of polymerizing at the atomic scale, producing lengthy carbon chains [173]. There are many types of carbon-based materials called "allotropes" or "allotropic modifications" of a particular chemical element such as diamond and graphite. The carbon-based nanomaterials are categorized depending on their geometrical arrangement and construction (Fig. 5) [174]. Carbon-based materials in nanostructure are present in various shapes such as tube, and the tubes can be divided into various structures of CNTs and MWCNTs. The NC-based material is widely employed in numerous technical purposes such as the manufacture of conductive plastics, micro- and nanoelectronics, gas storage, antifouling paints, gas biosensors, textile industry with specific property, batteries, and others [175]. Recently, the production of nanocarbon (NC) from the waste

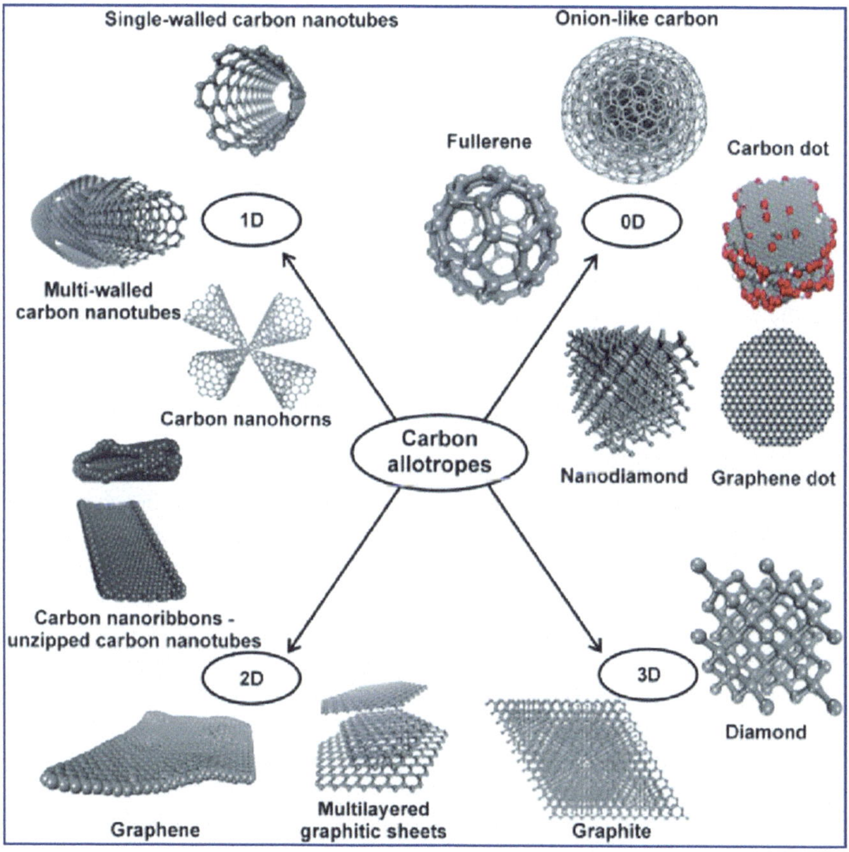

Fig. 5 Carbon materials-based classification based on their dimensionality. Adapted with permission from Ref. [177], Copyright 2015, American Chemical Society

product has received great attention. The RH contains cellulose and lignin richly, which is considered to be good and economical source of NCs [176].

4.1 Activated Carbon

AC is the oldest known adsorbent that is widely used in the removal of inorganics [178, 179] and organics [180] pollutants. AC shows a potential property of high surface vicinity, great mechanical and thermal solidity, and low acid/base character. Bituminous coal [181], wood [182], coconut shell [183], peat [184], petroleum, pitch [185], and polymers [186] are commonly used as a precursor for AC preparation. Despite the prolific use of AC as adsorbent, there are some limitations of adsorbent cost production and regeneration [187]. Adsorbent cost may be classified according to (i) the availability that includes (a) agro-waste such as peat, rice, and lignite, (b) food industry and management of sludge, rice husk, slag, fly ash, red mud, etc., and (c) chemically synthesized products, and (ii) the nature such as (d) inorganic and (e) organic [188]. RH is typically utilized as an energy resource of low value [189]. The production of AC from RH added high value for RH-based resources for water and wastewater treatment. Basically, the AC is produced from two main processes: a) heat treatment of raw materials under inert gas flow and b) chemical treatment of the produced carbon using KOH, NaOH, $ZnCl_2$, K_2CO_3, and H_3PO_4. In addition, the AC activation has been proceeded physically using CO_2, steam, or air. However, the AC using chemical method is broadly utilized due to minimal power needed and high production [75]. Therefore, we will discuss typically the preparation of activated carbons from RH as raw material.

4.1.1 Methods of Preparing Activated Carbon from Rice Husk

There were many approaches for the preparation of AC from RH based on various activation processes. The AC from RH was produced as presented in scheme 5, in which, 20 mesh size RH was weighed and then, cleaned, washed, and heated/carbonized for various time intervals to around 600 °C. The RH was activated chemically in the existence of NaOH, $ZnCl_2$, and H_3PO_4 at 60 °C for approximately 1 h. After that, the activated carbon was produced from activated RH at 900 °C [190]. AC was produced from RHs by following these steps: (a) air carbonization of RH at 300 °C, (b) then silica nanoparticles were extracted using NaOH solution, (c) after that the AC chemically activated using NaOH, $ZnCl_2$, and KOH, and (d) complete carbonization proceeded at 900 °C under N_2 gas flow [191]. Le Van and Luong Thi [192] prepared AC from RH by following these steps: (i) the RH washing for removing of grime and further impurities and drying in a dryer at 110 °C for overnight, (ii) carbonization at 400 °C for 1.5 h under N_2 stream atmosphere (300 mL min^{-1}), (iii) chemical treatment by NaOH and then drying for 12 h at 120 °C, and (iv) heat treatment at a high temperature of 400 °C

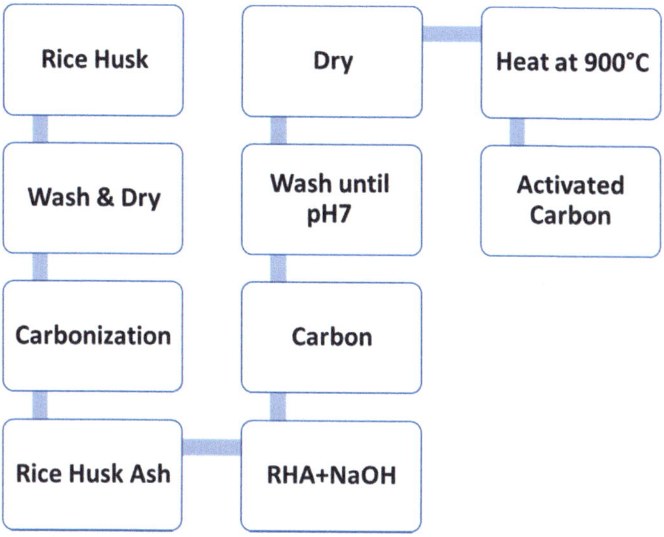

Scheme 5 Activated carbon process production from rice husk

for 20 min and then raising the temperature to reach 650–800 °C under the N_2 stream for 1 h. The produced AC product was powdered and treated by 0.1 M HCl solution and then rinsed by hot distilled H_2O several times to reach the neutral pH (6.6–7.0). All these methods used various steps to produce activated carbon, and they mainly depend on natural sources such as RHs which provide the sustainable process for environmental managements.

4.1.2 Preparation of Carbon Nanotube from Rice Husk

There are many techniques for making carbon-based materials from agriculture waste, especially RH. In this approach, Asnawi et al. used a microwave oven to produce carbon nanostructures, specifically CNTs, from RH (scheme 6). The economical and fast methods are based on the catalytic production of CNTs from RH after various washing and drying processes [57]. Fathy et al. synthesized CNTs by carbonization of pre-treated RHs using chemical vapor deposition. The synthesis of CNTs was performed by double proceedings: (i) hydrothermal remediation and (ii) chemical vapor deposition. This approach was discussed in some attempts to generate CNTs bundles by (i) hydrothermally treated rice stubble bolstered either by ferrocene–nickel catalysts or by ferrocene and then (ii) chemical vapor deposition of camphor supported either by ferrocene–nickel catalysts or by ferrocene. The CNTs morphology was investigated using TEM and SEM, while the thermal stability and electronic properties were evaluated using TGA and Raman spectroscopy techniques, respectively [193].

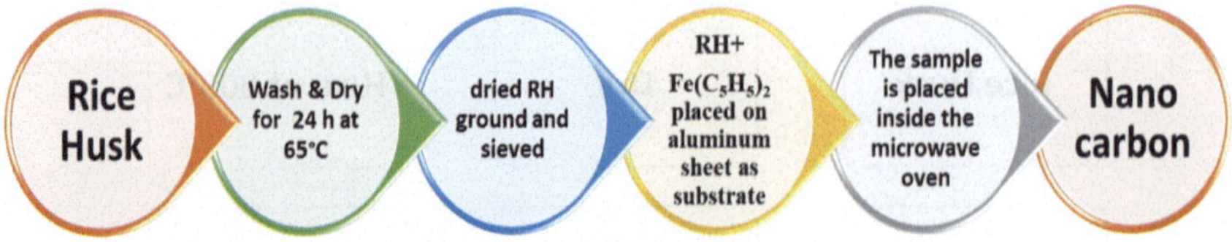

Scheme 6 Synthesis of nanocarbon from rice husk

4.1.3 Preparation of Graphene from Rice Husk

Graphene is one of the carbon-based materials that is composed of dimensional (2D) array with sp^2-hybridized carbon atoms. The GR properties are significant in various fields because of the excellent physical, chemical, optical, electronic, and mechanical characteristics [194, 195]. Numerous methods have been established to synthesize GR with different desired sizes, purity, and efflorescence of the specific product [196]. The synthesis of GR is divided into two main categories, (a) topmost-down and (b) bottom-topmost styles (Table 4). For the topmost-down approach, there are two commonly used methods which are the mechanical peeling of graphite and the chemical exfoliation of graphene oxide (chemical method), whereas, for the bottom-topmost approach, the most used methods are chemical vapor deposition (CVD) and epitaxial thermal augmentation of graphene on scaffold.

The GR was produced from RH using KOH at 800 °C with an impregnation ratio of 1:2 [197]. The explained methodology uses RH as a carbon source. RH acted as a carbon source, and KOH acted as an activating agent for GR production. The GR layers were prepared after four successive steps: (a) pre-carbonization, (b) desilication in 1 M NaOH solution, (c) chemical activation, and exfoliation of the (d) carbonized rice husk. In addition, GR nanosheet production from brown RH was produced by carbonization of RH and then activated by KOH activation to create sustainable supercapacitor electrodes. The GR nanosheets displayed ultra-slim crumpled-silk-cameo-wave sheetlike morphology with large specific surface area (1225 m^2 g^{-1}) and porous construction as presented in Fig. 6a. The graphene quantum dots (GQDs) were synthesized from RH biomass with high yield (ca. 15% wt) [198]. The various characterization techniques such as HR-TEM, atomic force microscopy (AFM), and Raman spectroscopy were used for establishing the size, morphology, and structure of the GQDs-derived RH. Singh, Bahadur, and Pal synthesized GR layers using RHA and KOH [199]. RHA acted as carbon supplier for GR production and as a defensive barrier against parent RH and KOH mixture oxidation (Fig. 6b). Therefore, rice by-products of RH and RHA acted as a potential source of GR that increases the economic and environmental use of agriculture waste products.

Table 4 Various approaches for the synthesis of graphene [197]

Type	Method	Figure/illustration	Advantages	Disadvantages
Top-down approach (from graphite)	Mechanical exfoliation	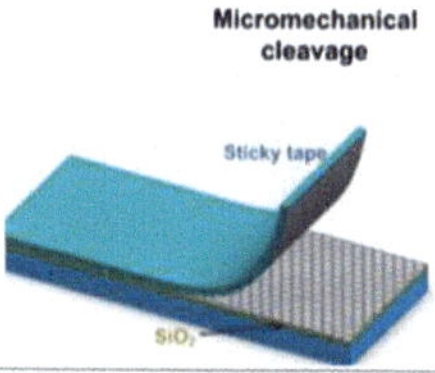	Fewer defects	Neither scalable nor capable for mass production
	Chemical method	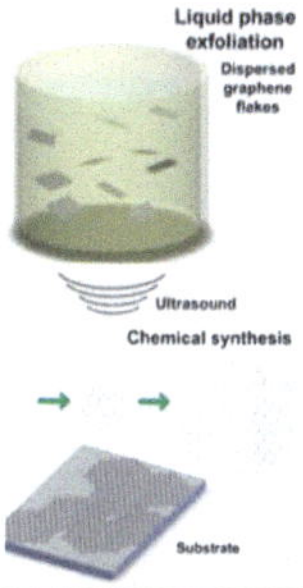	Cost-effective and suitable for mass production	Utilize many toxic chemicals throughout the synthesis process
Bottom-up approach (from carbon)	Chemical vapor deposition (CVD)	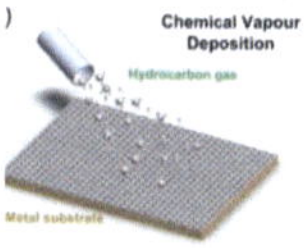	Compatible with the current complementary metal–oxide–semiconductor (CMOS) technology due to large area and high-quality graphene produced	Expensive and involves complex transfer process
	Epitaxial growth	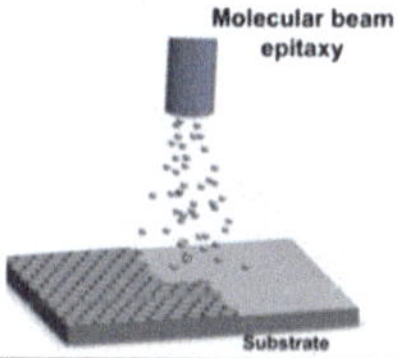	No defects for every single graphene island	Discontinuous

4.2 *Potential Applications of Nanocarbon*

NCs are broadly employed in several functions due to the fascinating physical and chemical characters. In 2013, several thousand tons of CNTs was produced [201]. NCs-based materials were used in various applications such as ceramic by direct expanding CNTs in a cement template [202]. On the other hand, CNTs inter in the manufacture of soft turbine blade materials and aquatic turbines [173]. Besides, the potential use in the fields of automotive industry [203], aviation [204], and sports equipment is widely used [201]. Moreover, NCs were potentially used in medical applications of drug and gene delivery [205], and cosmetics [206]. Another NC

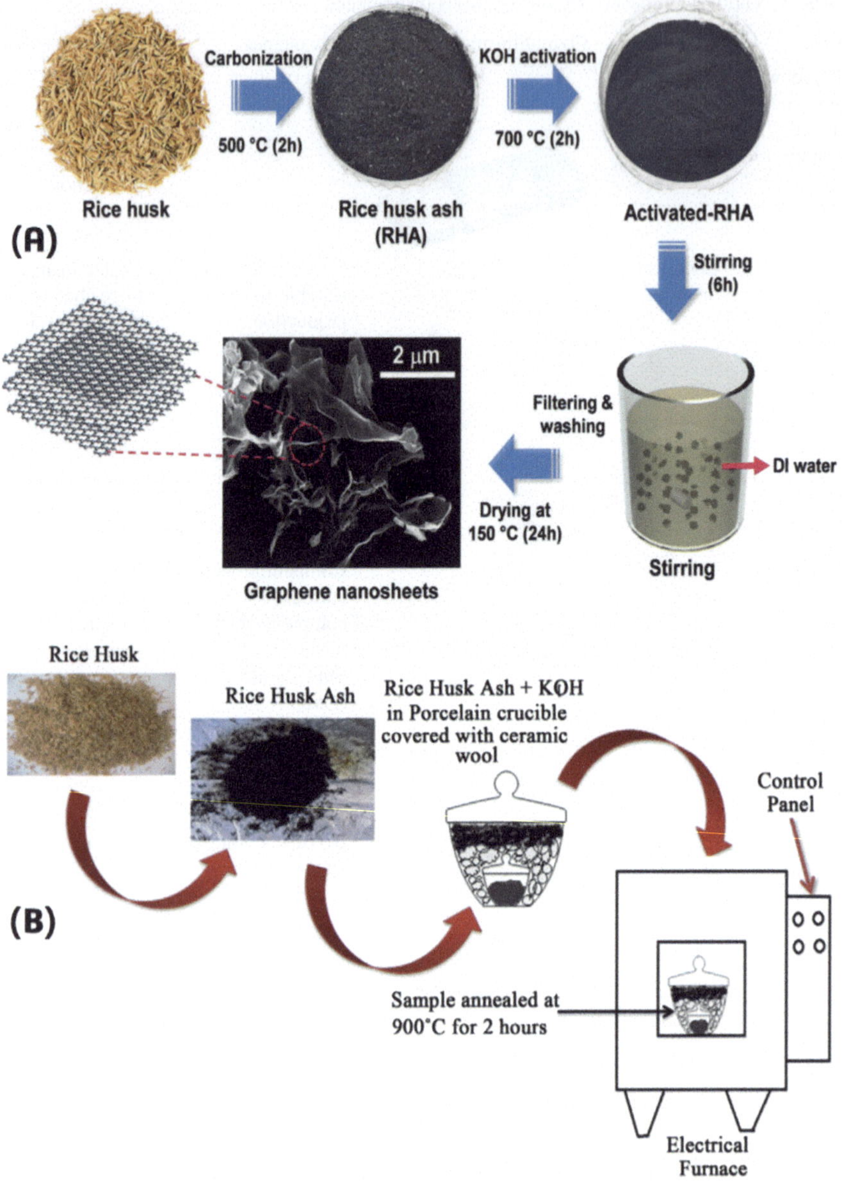

Fig. 6 **a** Synthesis of graphene nanosheets from the brown rice husks. Adapted with permission from Ref. [200], Copyright 2017, Royal Society of Chemistry. **b** Experiential setup for the synthesis of rice husk ash-derived graphene. Adapted with permission from Ref. [199], Copyright 2017, CC BY license, Scientific Research

form is graphene that is used in various fields of various biochemical sensors, electronics, solar cells, and others [207]. Moreover, the carbon-based nanomaterials have been potentially used in various environmental and agricultural applications that are summarized in Fig. 7.

4.2.1 Ecological Uses of Nanocarbon

Environmental pollution is one of the world's main. To improve environmental sustainability strategies for the remediation of pollutants, it is necessary to produce innovative approached for producing a potential conventional method. In this context, carbon-based nanomaterials make a significant contribution. NC materials exhibit large surface area, surface functionalization, and porous construction leading to the potential application as sorbents for the elimination of organic and inorganic toxins. From NC materials, AC has been commonly utilized as an adsorbent for traditional wastewater remediation. The utilization of nanoscaled carbon material in this context has a promising potential approach for enhancing the wastewater treatment with numerous examples in various kinds of the literature [208, 209]. The various hazardous materials, such as micro-cysteines (cyanobacterial toxins) [210], lead [211], and copper ions [212], were removed based on adsorption process at the surface of CNTs. Moreover, MWCNTs have been utilized for adsorption of antibiotics [213], herbicides [214], nitrogen, and phosphorus in wastewater [215]. By the way, both fullerenes and CNTs are potentially used to remove various organic pollutants [216], such as agricultural insecticide of lindane

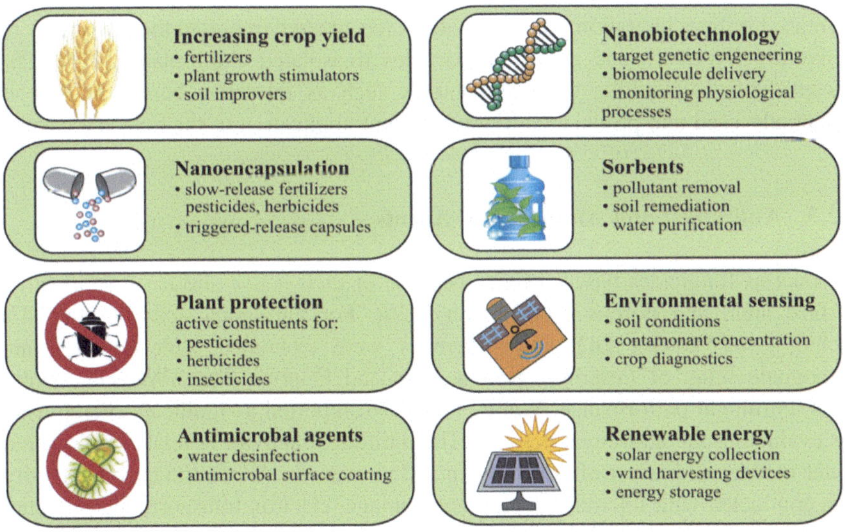

Fig. 7 Summarized nanoscaled carbon material applications in environmental and farming areas. Adapted with permission from Ref. [173], Copyright 2016, CC BY license, Springer Nature

[217] and insistent polychlorinated biphenyls [218]. The major benefits of NCs are huge surface vicinity, superior mechanical and thermal solidity, aromatic compound affinity, and high potential antibacterial property [219]. Various contaminants were desorbed at NCs [220]. The CNTs were used as a disinfectant and antimicrobial resistance after coating with antimicrobial. The hybrid nanoparticles of silver-(Ag)-coated CNTs acted as antimicrobial materials. The designed materials can be used in biomedical appliances and antibacterial management techniques [221]. Al-Hakami et al. termed a water disinfection process depending on microwave-dependent interactions between functionalized CNTs. The CNTs functionalized with the aliphatic alcohol-1-octadecanol ($C_{18}H_{38}O$) proceeded fabulous antimicrobial features [222]. Therefore, NCs and their composites are considered potential materials for various environmental remediation such as water treatment from potential organic and inorganic hazardous, and water disinfectants, and so-on.

4.2.2 Nanoencapsulation and Intelligent Delivery Methods

Clever delivery systems are an encouraging agrochemical targeting strategy having several potential benefits. Encapsulated agrochemicals have improved stability and safety against degradation to decrease the number of the used agrochemicals and enhance their performance [223]. Sarlak et al. reported fungicides encapsulated with citric acid in MWCNTs. The fungicide toxicity was decreased, and the efficiency to antifungal activity was increased. Graphene oxide films can be encapsulated with slow-release fertilizers [224]. Soil-portable nourishments such as potassium nitrate were encapsulated in graphene oxide as host materials that increase fertilizer liberation process [224]. A recent study reported the possibility of delivering SWCNTs and ceria nanoparticles to isolated chloroplasts [225]. The encapsulation of various variable contents such as fertilizers based on NCs is potentially used and produced a highly efficient application to the delivery targets.

4.2.3 Antifungal and Antibacterial Agents

NCs act as fungicides-based activity because of their active surface functionality, surface area, and porous construction [226]. For antifungal activity, the NCs including MWCNTs, GO, and fullerenes were examined with double plant pathogenic fungi of Fusarium graminearum and F. poae. The SWCNTs exhibit great antifungal performance. Wang et al. [226] reported a significant prerequisite for evaluating the antifungal activity. The antifungal activity is related to lowered water content and arrest of development. Moreover, the GO antimicrobial activity was conducted with microbial membrane damage, electron transport, and potential membrane disturbance [227]. Moreover, the GO exhibited a good antibacterial activity, where GO sheets can be wrapped bacterial cells and effectively separate bacteria from their environment. The longer GO sheets displayed a better antibacterial performance compared to small GO sheets [228]. These studies provide the potential use of NCs for antifungal and antibacterial applications.

4.2.4 Medical Applications of Nanocarbon

There are various types of NCs, including CNT, GO, and GQD that are used for biomedical applications as shown in Fig. 8. CNTs are widely used in biomedical applications. CNTs acted as a host to a broad assortment of diagnostic and therapeutic agents of medicines, inoculations, genomes, and antibodies. Moreover, CNTs are expressed as a brilliant tool for medicine delivery usage without body metabolism. Subsequently, CNTs have been used for biosensor diagnosis, tissue regeneration, drug and pollutant extraction, and enantiomer chiral drug separation. Furthermore, CNTs have recently been identified as a promising antioxidant [229]. The CNTs were used as drug delivery-based materials and can be resumed briefly as follows; the specific drug was immobilized at the surface of CNTs, and then CNTs drug capsule was entered to the animal/human body by classical mode (oral, injection), and another mode such as a controlled external magnet to the target organ. The targeted drug encapsulated on the CNTs reached the cells, and the contents were spilt into the cells [230]. In general, functionalized CNTs could transport interesting molecules through the nuclear membrane and/or cytoplasmic membrane. Therefore, the drug@CNT verifies safer and more effective drug delivery. Moreover, CNTs-immobilized specific drug was used as cancer therapy to treat tumors [231]. The antitumor immunotherapy was used for local antitumor hyperthermia therapy as a cancer treatment strategy [232].

GR is potentially used for drug delivery due to its interesting properties of easy functionalization. GR surface can be functionalized to facilitate the immobilization of many biological segments as drug delivery materials [233]. GR and GO act as nanocarriers for effective gene and drug delivery. The GR and GO were used as a drug carrier of anticancer drugs for charging of pirfenidone in the medication of subarachnoid hemorrhage. GR and GO were used to prevent the undesired release of drugs into the bloodstream during drug transport, which effectively transport antitumor drugs to tumor cells/tissues [234]. The DOX was loaded at reduced GO–Au nanorods for producing of photothermal treatment and chemotherapy-based capsules. The drug was released by exposure to near-infrared photothermal as heating influence and acidic nature of the cancer microenvironment [235]. Cheon et al. designed chemo- and photothermal treatment of brain cancers based on loading of DOX at the surface of Au NPs/GO [236]. The gold nanoparticles (Au NPs) induce the absorption peaks, and at specific laser power and wavelength, the area under treatment was exposed for heat and photoeffect. Chen group designed a drug delivery protocol based on loading of DOX at the treated surface of Au NPs–chitosan–g-poly (N-isopropyl acrylamide) folic acid–GO. The thermosensitive hydrogel was used for breast tumor treatment [237]. Dual chemotherapeutic drugs of lipid bilayers (lipo-GNS) modified with cancer-aiming protein were loaded into

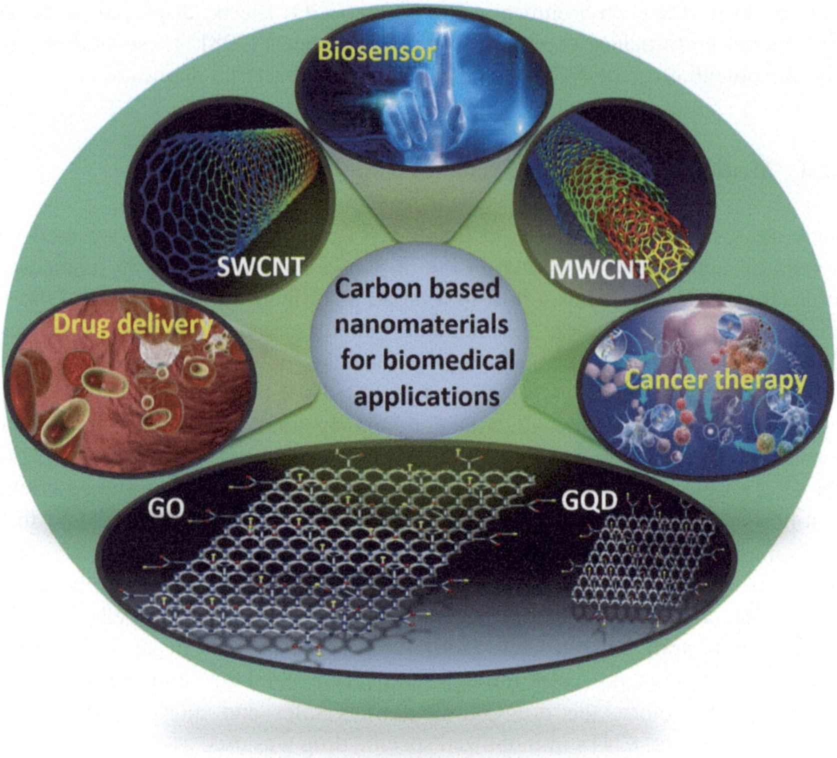

Fig. 8 Possible uses of nanoscaled carbon materials in biomedical areas. Adapted with permission from Ref. [233], Copyright 2018, CC BY license, PMC

sponge-like GR nanosheet (graphene nanosponge) [238]. The fabricated lipo-GNS accumulated in the tumor sites. Hence, effective control of the xenograft cancers in 16 days was assigned. Mesoporous silica-coated polydopamine-functionalized reduced GO was modified by hyaluronic acid and DOX for acting as a drug capsule [239]. The pH-dependent and close infrared-triggered DOX release made the designed drug delivery as an effective chemo-photothermal agent.

4.2.5 Application of Nanocarbon in Water Purification

The NCs have been widely utilized as adsorbents for contaminated water remediation. The CNTs showed excellent adsorption abilities and superior adsorption efficacy for many forms of pollutants, such as dichlorobenzene [240], ethylbenzene [241], Zn^{2+} [242], Pb^{2+}, Cu^{2+}, and Cd^{2+} [243], as well as dyes [244]. Recently, MWCNTs and SWCNTs have been used as adsorbents for water purification [242]. CNTs, rGO, GO, and other NCs doped/dressed with other metal oxides to improve the adsorption, mechanical, optical, and electrical features were established [245]. The functionalization of NCs by various groups of oxygen, nitrogen, sulfur, and phosphorous enhances their dispersibility, enhances the specific surface vicinity, and increases the selectivity and adsorption capability. For example, CNTs supported by magnetic iron oxide have been used for removal of chromium ions, GO-TiO_2 was used for removing Cd^{2+}, Pb^{2+}, and Zn^{2+} from wastewater, and GO-TiO_2 and rGO-TiO_2 hybrid materials efficiently remove the Cr^{5+} from the aqueous solutions with high efficiency [246–248]. The polyacrylic acid (PAA)/Fe_3O_4/GO was efficiently used for removal of Pb^{2+}, Cd^{2+}, and Cu^{2+}, while the Fe_2O_3/rGO acted as adsorbent for elimination of the As^{5+} from watery solutions [249, 250]. Therefore, the NCs and its composites were widely used for potential removing of various hazards from water such as heavy metals, organic compounds by adsorption or degradation.

5 Nanozeolite

Microporous crystals of aluminosilicates known as zeolites can be broadly employed as adsorbents, ion-exchangers, catalysts, etc. [66]. Natural zeolites are often produced in closed alkaline and salty water systems at low temperatures. Chemically synthesized zeolite from aluminosilicate solutions was produced using hydrothermal procedures in the existence of surfactant templates to form various types of NZ, including faujasite-X and -Y, NaA, ZSM-5, and silicalite-1 (72). RH was used as the source of silica for producing of A-, beta-, and ZSM-5 zeolites. The production of silica (the source of NZ) from RH can be produced by concerting of the organo-siliceous raw material of the rice shell into white ash RH by burning, and then silica would be extracted from white ash using alkali solution (69). The methods of nanozeolite (NZ) production from RH waste are summarized as follows.

5.1 Preparation of Nanozeolite from Risk Husk

The NZ (NaA) was synthesized using SiO_2 recovered from RH as a supplier of SiO_2 by the hydrothermal method [65]. In this method, the amorphous SiO_2 with 87.988 wt% NS was recovered from RHA using an appropriate alkali solution. The sizes of NaA crystals varying from 50 to 120 nm were produced without adding any organic additives. NZ (sodalite) was prepared using extracted silica from RH without organic additives by a hydrothermal method (scheme 7) [251]. In addition, ZSM-5 and analcime zeolites were prepared by a hydrothermal technique using SiO_2 recovered from RH via Ghasemi et al. This method provides the significant economic value of low cost for zeolite synthesis from local biowastes and without an extra source of alumina [252]. Various factors of time, crystallization temperature, and SiO_2/Al_2O_3 ratios on the properties of zeolites have been investigated.

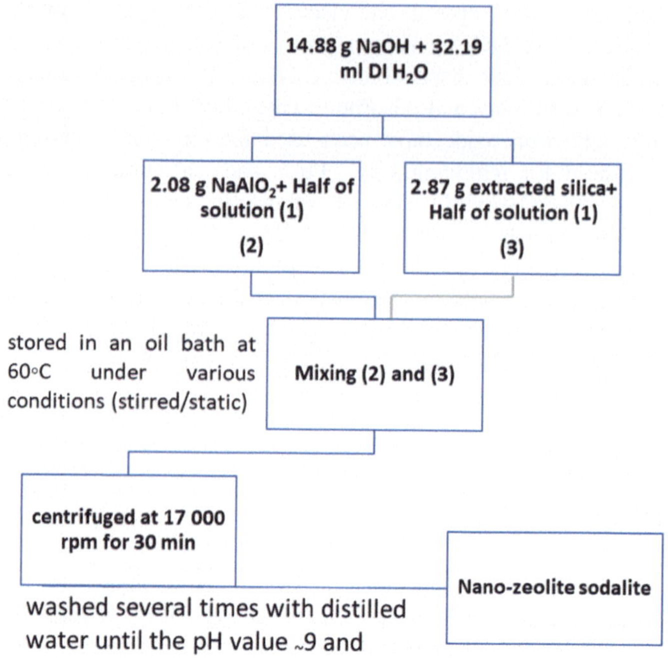

Scheme 7 The schematic diagram of typical synthesis stages of nanozeolite sodalite

The degree of crystallinity of analcime and ZSM-5 was 95.86 and 89.56%, respectively. By using silica extracted from RHA, Tan et al. synthesized both zeolites NaA and NaY using hydrothermal approach [253]. The zeolite NaY (faujasite) was effectively manufactured with commercial chemical seed. Therefore, the RHA is used as the source of preparing NZ. The NZ is composed of silica and alumina to form NZ, and the silica source is the RHA, which lowers the fee of the final product of NZ and increases the economic values of agriculture waste.

5.2 *Potential Applications of Nanozeolite*

Figure 9 shows the various applications that utilize zeolite such as environmental protection, construction, agriculture, medicine/hygiene, gas treatment, and petrochemistry. The zeolite applications are usually associated with their porosity and specific surface space. Zeolite adsorption properties are controlled depending on the capability of adsorbate molecules to penetrate zeolite voids (pore diameter ranged from 0.4 nm up to 1.3 nm). NZs are used as catalysts in various industries such as petrochemical industry [254], as adsorbents for air-, soil-, and water-cleansing, removing radioactive poisons, adsorption of refrigeration [255, 256], and detergent industry [257].

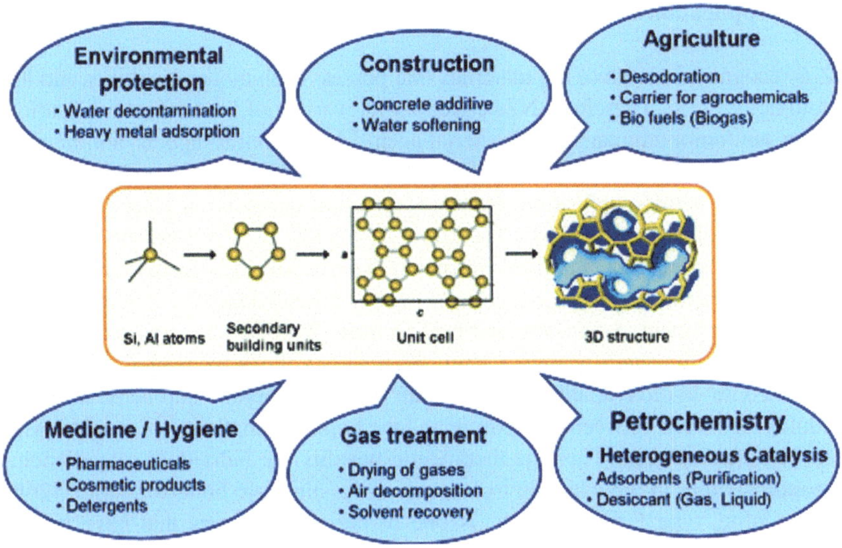

Fig. 9 Nanozeolite usage in various fields such as medicine/hygiene, gas treatment, petrochemistry, agriculture, construction, and environmental protection. Adapted with permission from Ref. [258], Copyright 2018, Royal Society of Chemistry

5.2.1 Usage of Nanozeolite in Water Remediation

Natural zeolites have been used in wastewater treatment for a long time. Wastewater from various resources contains potentially hazardous materials such as weighty metals of Zn, Fe, Hg, Cd, As, Cr, Cu, Mn, Pb, Cd, and so on that acted as a significant environmental threat. Natural zeolites have been exploited for the removal of various weighty metals. The removal process was broadly researched via several techniques, including chemical precipitation, membrane filtration, ion-exchange, coagulation–flocculation, flotation, adsorption, and electrochemical methods [259]. Different natural zeolites showed a good ion-exchange capability to get rid of cations. For increasing capacity and selectivity, there are many approached for modification of natural zeolites such as acid treatment and functionalization of surfactant [260]. The natural zeolites were used for concurrently extracting ammonia and humic acid. The results showed that zeolite displayed the best performing at pH near to that of natural waters for the simultaneous removal of ammonia and humic acid [261]. The usage of natural and modified zeolites for the elimination of Fe and Mn ions from groundwater specimens has been further investigated, and the removal efficiency was in the range of 22–90% and 61–100% for Fe and Mn, respectively [262].

5.2.2 Application of Nanozeolite in Biomedical

NZ is recognized as one of the materials that possess high biocompatibility and low cytotoxicity [263]. Therefore, NZs cover a wide variety of biomedical applications such as antitumor adjuvants, antibacterial agents, MRI contrast agents, antidiarrheal agents, bone formation, development of Alzheimer's disease, hemodialytic, drug delivery, and dental application. Bone regeneration depends on osteoblast proliferation and differentiation [264]. Therefore, NZ can be used as implants to replace conventional materials. Zeolites' intrinsic features of 3D microstructures and unique topography lead to form a network with nanoscaled pores and be suitable for cell bone attachment, expansion, and proliferation [264]. Zeolite–hydroxyapatite is coated on a composite of stainless steel and titanium alloys for renewing medicine [265]. Zeolite–hydroxyapatite properties are super-hydrophilic and acted as elastic modulus mismatch between the coating and the bone. It can, therefore, enable faster recovery after surgery. This has therapeutic benefits for individuals experiencing osteoporosis due to the capability of zeolite A to stimulate bone formation [266]. Clinoptilolite, type F and type W, is one of the zeolite forms that have a great potential for application of hemodialysis in ammonia ion-exchange systems. The zeolites have been used as antitumor adjuvants [267]. NZ was used as an adjuvant in cancer treatment [268]. The oral dose of natural clinoptilolite to mice and dogs with different types of tumors leads to reduce the tumor size and improve the survival time and overall health. Therefore, zeolites play a key role in various applications that produce highly economical and environmental impact.

6 Conclusion

The economic efficiency of the agriculture wastes such as by-products and wastes of rice industry such as RH and RHA is discussed. The potential use of these wastes has been conducted with the production of potential materials including NS, NC, and NZs. The agricultural wastes are regarded as the rich source of various nanomaterials such as NS, NZ, and NC. RH comprises 70–85% organic matter and inorganic residues (20–25%), while RHA contains 60% SiO_2, 10–40% carbon, and other ingredients. Therefore, RH and RHA are regarded as a potential source of valuable and widely used materials of NS, NZ, and NCs. In precise, we concluded and expressed the various methods that are used for the potential synthesis of NS, NZ, and NCs from RH and RHA. There are many approaches used for increasing the potential values of the agriculture wastes of RH and RHA. The potential and widely used materials in various applications in industrial and environmental aspects such as electronics, ceramic, catalyst support, adsorbents, ion exchangers, water treatment, antimicrobial products, biosensing, and biomedical applications were discussed. Therefore, the valuable uses of various agriculture-derived materials of NS, NZ, and NCs were discussed in detail. Moreover, the potential applications in various aspects were expressed. The increase of agriculture waste value environmentally and industry would be deadly developed for various potential uses in human health products such as cosmetics, a medical application such as drug delivery, implanting growth such as fertilizer delivery, and other agriculture activities. In addition, the produced materials play a key role in water treatment such as adsorbents, membranes, and ion exchangers due to high surface area, and porous construction with the easy functionalized surface. Besides, the conducted ceramic materials and composite composition attain a great potential application in bone recovery, acid and alkali-resistive ceramics, biosensing applications for detection of various biomolecules and biomarkers, and adherent surface as antibacterial, antifungal activity. In conclusion, the waste management and potential use of the produced materials attained a great environmental and useful industrial impacts for sustainable resources and decreasing waste hazardous influence.

7 Future Prospective

The fast growth of population within the huge building and using nanotechnology have been destroying the ecological system. Therefore, there are many global calls for environmental protection from human activities. In this point of view, the conversion of wastes such as agriculture waste into valuable products covers both potentially needed and interesting ways of clean environmental impact and high tech needed. The global warming increased as the production of various gases as direct combustion of agriculture wastes such as rice by-products in air. RH and RHA burning in a suitable furnace convey the source of energy that can be used for various aspects and can be converted to available products based on nanosilica. The huge production of high-tech production products depends on the use of metal/metal oxides and composite materials that are present in RH/RHA, and potentially used in battery, transistors, alloys, and other products. So, using the agriculture waste of rice by-products leads to produce a highly economical and sustainable source for developing the high-tech industry. Moreover, there are too many industries depending on the nanosilica, such as detergents, cosmetics, medicine, analytical devices, and so on. Therefore, a low-cost source of these products can be achieved by using a highly economic row source. Furthermore, NC and NZ are potentially used in various aspects such as water treatment, catalytic reactions, sensors, medicine such as drug delivery, and so on. Availability of these resources with high purity, intrinsic features, functionalized surface, porous construction, and high surface area increases the potential application of these materials. The RH/RHA can be treated to produce highly efficient materials with novel characteristics. Production of these materials with high yield, from low-cost source, and high purity leads to establishing a great approach in various environmental and industrial uses.

References

1. Zhang Z, Gonzalez AM, Davies EGR, Liu Y (2012) Agricultural Wastes. Water Environ Res 84(10):1386–1406
2. Ali GAM, Divyashree A, Supriya S, Chong KF, Ethiraj AS, Reddy MV, Algarni H, Hegde G (2017) Carbon nanospheres derived from Lablab purpureus for high performance supercapacitor electrodes: a green approach. Dalton Trans 46(40):14034–14044
3. Obi FO, Ugwuishiwu BO, Nwakaire JN (2016) Agricultural Waste Concept, Generation, Utilization and Management. Niger J Technol 35(4):957
4. Agamuthu P (2009) Challenges and opportunities in agro-waste management: An Asian perspective. in Inaugural meeting of first regional 3R forum in Asia
5. Hadiya UB, Javed F, Jadeja KY (2018) Approach towards enhancing sanitation in godavadi village through sustainable solid waste management. Int Res J Eng Technol 5(4):2067–2072
6. Hai HT, Tuyet NTA (2010) Benefits of the 3R approach for agricultural waste management (AWM) in Vietnam: Under the framework of joint project on Asia Resource Circulation Research. Institute for Global Environmental Strategies
7. Hegde G, Abdul Manaf SA, Kumar A, Ali GAM, Chong KF, Ngaini Z, Sharma KV (2015) Biowaste sago bark based catalyst free carbon nanospheres: waste to wealth approach. ACS Sustain Chem & Eng 5(9):2247–2253
8. Ali GAM, Abdul Manaf SA, Kumar A, Chong KF, Hegde G (2014) High performance supercapacitor using catalysis free porous carbon nanoparticles. J Phys D Appl Phys 47 (49):495307–495313
9. Hossain SS, Mathur L, Roy PK (2018) Rice husk/rice husk ash as an alternative source of silica in ceramics: A review. J Asian Ceram Soc 6(4):299–313
10. Lokare K (2019) Rising from the Ashes: Renewable Silica from Rice Husk Ash
11. Sarangi M, Bhattacharyya S, Behera R (2009) Effect of temperature on morphology and phase transformations of nano-crystalline silica obtained from rice husk. Phase TransitS 82 (5):377–386
12. Kishore R, Bhikshma V, Prakash PJ (2011) Study on strength characteristics of high strength rice husk ash concrete. Procedia Engineering 14:2666–2672
13. Shatat MR, Ali GAM, Gouda GA, Abdallah MH (2014) Rice husk ash for enhancing salts attack resistance of blended cement containing metakoalin. Can Chem Trans 2(3):274–285
14. Foletto EL, Hoffmann R, Hoffmann RS, Portugal UL Jr, Jahn SL (2005) Aplicabilidade das cinzas da casca de arroz. Quim Nova 28(6):1055–1060
15. Sankar S, Sharma SK, Kaur N, Lee B, Kim DY, Lee S, Jung H (2016) Biogenerated silica nanoparticles synthesized from sticky, red, and brown rice husk ashes by a chemical method. Ceram Int 42(4):4875–4885
16. Moayedi H, Aghel B, Nguyen H, Rashid ASA (2019) Applications of rice husk ash as green and sustainable biomass. Journal of Cleaner Production:117851
17. Naddaf M, Kafa H, Ghanem I (2019) Extraction and Characterization of Nano-Silica from Olive Stones. Silicon 12(1):185–192
18. Emran MY, El-Safty SA, Shenashen MA, Minowa T (2019) A well-thought-out sensory protocol for screening of oxygen reactive species released from cancer cells. Sensors and Actuators B: Chemical 284:456–467
19. Emran MY, Khalifa H, Gomaa H, Shenashen MA, Akhtar N, Mekawy M, Faheem A, El-Safty SA (2017) Hierarchical CN doped NiO with dual-head echinop flowers for ultrasensitive monitoring of epinephrine in human blood serum. Microchim Acta 184 (11):4553–4562
20. Emran MY, Mekawy M, Akhtar N, Shenashen MA, El-Sewify IM, Faheem A, El-Safty SA (2018) Broccoli-shaped biosensor hierarchy for electrochemical screening of noradrenaline in living cells. Biosens Bioelectron 100:122–131

21. Emran MY, Shenashen MA, Abdelwahab AA, Abdelmottaleb M, El-Safty SA (2018) Facile synthesis of microporous sulfur-doped carbon spheres as electrodes for ultrasensitive detection of ascorbic acid in food and pharmaceutical products. New J Chem 42(7):5037–5044
22. Emran MY, Shenashen MA, Abdelwahab AA, Abdelmottaleb M, Khairy M, El-Safty SA (2018) Nanohexagonal Fe 2 O 3 electrode for one-step selective monitoring of dopamine and uric acid in biological samples. Electrocatalysis 9(4):514–525
23. Gomaa H, Khalifa H, Selim M, Shenashen M, Kawada S, Alamoudi AS, Azzam A, Alhamid A, El-Safty S (2017) Selective, photoenhanced trapping/detrapping of arsenate anions using mesoporous blobfish head TiO2 monoliths. ACS Sustain Chem & Eng 5 (11):10826–10839
24. Thalji MR, Ali GAM, Algarni H, Chong KF (2019) Al^{3+} ion intercalation pseudocapacitance study of $W_{18}O_{49}$ nanostructure. J Power Sources 438:227028
25. Ali GAM, Habeeb OA, Algarni H, Chong KF (2018) CaO impregnated highly porous honeycomb activated carbon from agriculture waste: symmetrical supercapacitor study. J Mater Sci 54:683–692
26. Fouad OA, Makhlouf SA, Ali GAM, El-Sayed AY (2011) Cobalt/silica nanocomposite via thermal calcination-reduction of gel precursors. Mater Chem Phys 128(1–2):70–76
27. Ali GAM, Fouad OA, Makhlouf SA (2013) Structural, optical and electrical properties of sol-gel prepared mesoporous Co_3O_4/SiO_2 nanocomposites. J Alloy Compd 579:606–611
28. Emran MY, Shenashen MA, Abdelwahab AA, Khalifa H, Mekawy M, Akhtar N, Abdelmottaleb M, El-Safty SA (2018) Design of hierarchical electrocatalytic mediator for one step, selective screening of biomolecules in biological fluid samples. J Appl Electrochem 48(5):529–542
29. Emran MY, Shenashen MA, Mekawy M, Azzam AM, Akhtar N, Gomaa H, Selim MM, Faheem A, El-Safty SA (2018) Ultrasensitive in-vitro monitoring of monoamine neurotransmitters from dopaminergic cells. SensS Actuators B: Chem 259:114–124
30. Emran MY, Shenashen MA, Morita H, El-Safty SA (2018) One-step selective screening of bioactive molecules in living cells using sulfur-doped microporous carbon. Biosens Bioelectron 109:237–245
31. Emran MY, Shenashen MA, Morita H, El-Safty SA (2018) 3D-ridge stocked layers of nitrogen-doped mesoporous carbon nanosheets for ultrasensitive monitoring of dopamine released from PC12 cells under K + stimulation. Adv Healthc Mater 7(16):1701459
32. Gomaa H, Shenashen M, Yamaguchi H, Alamoudi A, El-Safty S (2018) Extraction and recovery of Co 2 + ions from spent lithium-ion batteries using hierarchical mesosponge γ-Al 2 O 3 monolith extractors. Green Chem 20(8):1841–1857
33. Barhoum A, Shalan AE, El-Hout SI, Ali GAM, Abdelbasir SM, Abu Serea ES, Ibrahim AH, Pal K (2019) A Broad Family of Carbon Nanomaterials: Classification, Properties, Synthesis, and Emerging Applications. In: Barhoum A, Bechelany M, Makhlouf A (eds) Handbook of Nanofibers. Springer International Publishing, Cham, pp 1–40
34. Ali GAM, Lih Teo EY, Aboelazm EAA, Sadegh H, Memar AOH, Shahryari-Ghoshekandi R, Chong KF (2017) Capacitive performance of cysteamine functionalized carbon nanotubes. Mater Chem Phys 197:100–104
35. Ali GAM, Tan LL, Jose R, Yusoff MM, Chong KF (2014) Electrochemical performance studies of MnO_2 nanoflowers recovered from spent battery. Mater Res Bull 60:5–9
36. Abdel Ghafar HH, Ali GAM, Fouad OA, Makhlouf SA (2015) Enhancement of adsorption efficiency of methylene blue on Co_3O_4/SiO_2 nanocomposite. Desalination Water Treat 53 (11):2980–2989
37. Pokropivny V, Lohmus R, Hussainova I, Pokropivny A, Vlassov S (2007) Introduction to nanomaterials and nanotechnology. Tartu University Press Ukraine
38. Gomaa H, Shenashen MA, Yamaguchi H, Alamoudi AS, Abdelmottaleb M, Cheira MF, El-Naser TAS, El-Safty SA (2018) Highly-efficient removal of AsV, Pb^{2+}, Fe^{3+}, and Al^{3+} pollutants from water using hierarchical, microscopic TiO_2 and $TiOF_2$ adsorbents through batch and fixed-bed columnar techniques. J Clean Prod 182:910–925

39. Ali GAM, Fouad OA, Makhlouf SA (2016) Electrical properties of cobalt oxide/silica nanocomposites obtained by sol-gel technique. Am J Eng Appl Sci 9(1):12–16
40. Fouad OA, Ali GAM, El-Erian MAI, Makhlouf SA (2012) Humidity sensing properties of cobalt oxide/silica nanocomposites prepared via sol-gel and related routes. Nano 7(5):1250038
41. Ali GAM, Yusoff MM, Shaaban ER, Chong KF (2017) High performance MnO_2 nanoflower supercapacitor electrode by electrochemical recycling of spent batteries. Ceram Int 43:8440–8448
42. Ali GAM, Thalji MR, Soh WC, Algarni H, Chong KF (2020) One-step electrochemical synthesis of MoS_2/graphene composite for supercapacitor application. J Solid State Electrochem 24(1):25–34
43. Ali GAM, Yusoff MM, Algarni H, Chong KF (2018) One-step electrosynthesis of MnO_2/rGO nanocomposite and its enhanced electrochemical performance. Ceram Int 44(7):7799–7807
44. Ali GAM, Yusoff MM, Ng YH, Lim NH, Chong KF (2015) Potentiostatic and galvanostatic electrodeposition of MnO_2 for supercapacitors application: A comparison study. Curr Appl Phys 15(10):1143–1147
45. Tolba GM, Barakat NA, Bastaweesy A, Ashour E, Abdelmoez W, El-Newehy MH, Al-Deyab SS, Kim HY (2015) Effective and highly recyclable nanosilica produced from the rice husk for effective removal of organic dyes. J Ind Eng Chem 29:134–145
46. Gomaa H, El-Safty S, Shenashen M, Kawada S, Yamaguchi H, Abdelmottaleb M, Cheira M (2018) Three-dimensional, vertical platelets of ZnO carriers for selective extraction of cobalt ions from waste printed circuit boards. ACS Sustain Chem & Eng 6(11):13813–13825
47. Griffin S, Masood MI, Nasim MJ, Sarfraz M, Ebokaiwe AP, Schäfer K-H, Keck CM, Jacob C (2018) Natural nanoparticles: A particular matter inspired by nature. Antioxidants 7(1):3
48. Mor S, Manchanda CK, Kansal SK, Ravindra K (2017) Nanosilica extraction from processed agricultural residue using green technology. J Clean Prod 143:1284–1290
49. Pek YS, Wan AC, Shekaran A, Zhuo L, Ying JY (2008) A thixotropic nanocomposite gel for three-dimensional cell culture. Nat Nanotechnol 3(11):671
50. Tang L, Cheng J (2013) Nonporous silica nanoparticles for nanomedicine application. Nano Today 8(3):290–312
51. Singh P, Srivastava S, Chakrabarti P, Singh SK (2017) Nanosilica based electrochemical biosensor: A novel approach for the detection of platelet-derived microparticles. SensS Actuators B: Chem 240:322–329
52. Tao Z (2014) Mesoporous silica-based nanodevices for biological applications. RSC Adv 4 (36):18961–18980
53. Ali GAM, Barhoum A, Gupta VK, Nada AA, El-Maghrabi H, Kanthasamy R, Shaaban ER, Algarni H, Chong KF (2020) High surface area mesoporous silica for hydrogen sulfide effective removal. Curr Nanosci 16(2):226–234
54. Pouroutzidou GK, Theodorou GS, Kontonasaki E, Tsamesidis I, Pantaleo A, Patsiaoura D, Papadopoulou L, Rhoades J, Likotrafiti E, Lioutas CB (2019) Effect of ethanol/TEOS ratios and amount of ammonia on the properties of copper-doped calcium silicate nanoceramics. J Mater Sci Mater Med 30(9):98
55. Chen H, Wang F, Zhang C, Shi Y, Jin G, Yuan S (2010) Preparation of nano-silica materials: The concept from wheat straw. J Non-Cryst Solids 356(50–51):2781–2785
56. Liou T-H, Yang C-C (2011) Synthesis and surface characteristics of nanosilica produced from alkali-extracted rice husk ash. Mater Sci Eng, B 176(7):521–529
57. Asnawi M, Azhari S, Hamidon MN, Ismail I, Helina I (2018) Synthesis of Carbon Nanomaterials from Rice Husk via Microwave Oven. J Nanomater 2018:1–5
58. Greil P (2015) Perspectives of nano-carbon based engineering materials. Adv Eng Mater 17 (2):124–137
59. Zhang H, Zhang X, Sun X, Ma Y (2013) Shape-controlled synthesis of nanocarbons through direct conversion of carbon dioxide. Scientific reports 3:3534
60. Villarreal CC, Pham T, Ramnani P, Mulchandani A (2017) Carbon allotropes as sensors for environmental monitoring. Current Opinion in Electrochemistry 3(1):106–113

61. Jiang J-W, Leng J, Li J, Guo Z, Chang T, Guo X, Zhang T (2017) Twin graphene: A novel two-dimensional semiconducting carbon allotrope. Carbon 118:370–375
62. Wang Z, Ogata H, Morimoto S, Ortiz-Medina J, Fujishige M, Takeuchi K, Muramatsu H, Hayashi T, Terrones M, Hashimoto Y, Endo M (2015) Nanocarbons from rice husk by microwave plasma irradiation: From graphene and carbon nanotubes to graphenated carbon nanotube hybrids. Carbon 94:479–484
63. Akhtar N, Emran MY, Shenashen MA, Khalifa H, Osaka T, Faheem A, Homma T, Kawarada H, El-Safty SA (2017) Fabrication of photo-electrochemical biosensors for ultrasensitive screening of mono-bioactive molecules: the effect of geometrical structures and crystal surfaces. J Mater Chem B 5(39):7985–7996
64. Kure N, Hamidon MN, Azhari S, Mamat N, Yusoff H, Isa B, Yunusa Z (2017) Simple microwave-assisted synthesis of carbon nanotubes using polyethylene as carbon precursor. Journal of Nanomaterials 2017
65. Ghasemi Z, Younesi H (2011) Preparation and Characterization of Nanozeolite NaA from Rice Husk at Room Temperature without Organic Additives. J Nanomater 2011:1–8
66. Breck DW (1974) Zeolite Molecular Sieves. Wiley, New York
67. Groenen E, Kortbeek A, Mackay M, Sudmeijer O (1986) Double-ring silicate anions in tetraalkyl-ammonium hydroxide/silicate solutions; their possible role in the synthesis of silicon-rich zeolites. Zeolites 6(5):403–411
68. Xu M, Cheng M, Bao X (2000) Growth of ultrafine zeolite Y crystals on metakaolin microspheres. Chem Commun 19:1873–1874
69. Wang H, Wang Z, Yan Y (2000) Colloidal suspensions of template-removed zeolite nanocrystalsElectronic supplementary information (ESI) available: Suppl. Figs. 1–3: TGA curves, XRD patterns and IR spectra. See http://www.rsc.org/suppdata/cc/b0/b006518h Chemical Communications (23):2333–2334
70. Nur H (2001) Direct synthesis of NaA zeolite from rice husk and carbonaceous rice husk ash. Indones J Agric Sci 1:40–45
71. Prasetyoko D, Ramli Z, Endud S, Hamdan H, Sulikowski B (2006) Conversion of rice husk ash to zeolite beta. Waste Manag 26(10):1173–1179
72. Rawtani AV, Rao MS, Gokhale K (1989) Synthesis of ZSM-5 zeolite using silica from rice-husk ash. Ind Eng Chem Res 28(9):1411–1414
73. Ramli Z (1995) Rhenium-impregnated zeolites: synthesis, characterization and modification ascatalysts in the metathesis of alkanes. Universiti Teknologi, Malaysia
74. Bhagiyalakshmi M, Anuradha R, Palanichamy M, Jang HT (2010) Dexterous template-free synthesis of ferrisilicate with MFI morphology using rice husk ash. J Non-Cryst Solids 356 (23–24):1204–1209
75. Pode R (2016) Potential applications of rice husk ash waste from rice husk biomass power plant. Renew Sustain Energy Rev 53:1468–1485
76. Chandrasekhar S, Satyanarayana K, Pramada P, Raghavan P, Gupta T (2003) Review processing, properties and applications of reactive silica from rice husk—an overview. J Mater Sci 38(15):3159–3168
77. Babaso PN, Sharanagouda H (2017) Rice Husk and Its Applications: Review. Int J Curr Microbiol Appl Sci 6(10):1144–1156
78. Conradt R, Pimkhaokham P, Leela-Adisorn U (1992) Nano-structured silica from rice husk. J Non-Cryst Solids 145:75–79
79. Real C, Alcala MD, Criado JM (1996) Preparation of silica from rice husks. J Am Ceram Soc 79(8):2012–2016
80. Muthadhi A, Anitha R, Kothandaraman S (2007) Rice husk ash-Properties and its uses: a review. Journal of the Institution of Engineers. India. Civil Engineering Division 88(5):50–56
81. Muntohar AS (2004) Utilization of uncontrolled burnt rice husk ash in soil improvement. Civ Eng Dimens 4(2):100–105
82. Ugheoke IB, Mamat O (2012) A critical assessment and new research directions of rice husk silica processing methods and properties. Maejo Int J Sci Technol 6(3):430

83. Vlaev L, Markovska I, Lyubchev L (2003) Non-isothermal kinetics of pyrolysis of rice husk. Thermochim Acta 406(1–2):1–7
84. Ghaly A, Mansaray K (1999) Comparative study on the thermal degradation of rice husks in various atmospheres. Energy Sources 21(10):867–881
85. Mohamed R, Mkhalid I, Barakat M (2015) Rice husk ash as a renewable source for the production of zeolite NaY and its characterization. Arab J Chem 8(1):48–53
86. Zou Y, Yang T (2019) Rice Husk, Rice Husk Ash and Their Applications, in Rice Bran and Rice Bran Oil Chemistry, Processing and Utilization, pp 207–246
87. Bakar RA, Yahya R, Gan SN (2016) Production of high purity amorphous silica from rice husk. Procedia Chemistry 19:189–195
88. Bhardwaj A, Hossain S, Majhi MR (2017) Preparation and characterization of clay bonded high strength silica refractory by utilizing agriculture waste. Boletín de la Sociedad Española de Cerámica y Vidrio 56(6):256–262
89. Della VP, Kühn I, Hotza D (2002) Rice husk ash as an alternate source for active silica production. Mater Lett 57(4):818–821
90. Fernandes IJ, Calheiro D, Kieling AG, Moraes CA, Rocha TL, Brehm FA, Modolo RC (2016) Characterization of rice husk ash produced using different biomass combustion techniques for energy. Fuel 165:351–359
91. Ghorbani F, Sanati AM, Maleki M (2015) Production of silica nanoparticles from rice husk as agricultural waste by environmental friendly technique. Environmental Studies of Persian Gulf 2(1):56–65
92. Gomes GMF, Philipssen C, Bard EK, Dalla Zen L, de Souza G (2016) Rice husk bubbling fluidized bed combustion for amorphous silica synthesis. J Environ Chem Eng 4(2):2278–2290
93. Habeeb GA, Mahmud HB (2010) Study on properties of rice husk ash and its use as cement replacement material. Materials research 13(2):185–190
94. Ma Z, Ye J, Zhao C, Zhang Q (2015) Gasification of rice husk in a downdraft gasifier: the effect of equivalence ratio on the gasification performance, properties, and utilization analysis of byproducts of char and tar. BioResources 10(2):2888–2902
95. Nehdi M, Duquette J, El Damatty A (2003) Performance of rice husk ash produced using a new technology as a mineral admixture in concrete. Cem Concr Res 33(8):1203–1210
96. Umeda J, Kondoh K (2008) High-purity amorphous silica originated in rice husks via carboxylic acid leaching process. J Mater Sci 43(22):7084–7090
97. Vayghan AG, Khaloo A, Rajabipour F (2013) The effects of a hydrochloric acid pre-treatment on the physicochemical properties and pozzolanic performance of rice husk ash. Cement Concr Compos 39:131–140
98. Affandi S, Setyawan H, Winardi S, Purwanto A, Balgis R (2009) A facile method for production of high-purity silica xerogels from bagasse ash. Adv Powder Technol 20(5):468–472
99. Tang Z, Zhang Y, Guo Q (2010) Catalytic hydrocracking of pyrolytic lignin to liquid fuel in supercritical ethanol. Ind Eng Chem Res 49(5):2040–2046
100. Yakovlev V, Yeletsky P, Lebedev MY, Ermakov DY, Parmon V (2007) Preparation and investigation of nanostructured carbonaceous composites from the high-ash biomass. Chem Eng J 134(1–3):246–255
101. Sun L, Gong K (2001) Silicon-based materials from rice husks and their applications. Ind Eng Chem Res 40(25):5861–5877
102. Patel KG, Shettigar RR, Misra NM (2017) Recent Advance in Silica Production Technologies from Agricultural Waste Stream. J Adv Agric Technol 4(3)
103. Yang H, Liu B, Chen Y, Li B, Chen H (2015) Influence of inherent silicon and metals in rice husk on the char properties and associated silica structure. Energy Fuels 29(11):7327–7334
104. Hamad M (1981) Combustion of rice hulls in a static bed. Energy Agric 1:311–315
105. Huang S, Jing S, Wang J, Wang Z, Jin Y (2001) Silica white obtained from rice husk in a fluidized bed. Powder Technol 117(3):232–238
106. Genieva S, Turmanova S, Dimitrova A, Vlaev L (2008) Characterization of rice husks and the products of its thermal degradation in air or nitrogen atmosphere. J Therm Anal Calorim 93(2):387–396

107. Luan TC, Chou TC (1990) Recovery of silica from the gasification of rice husks/coal in the presence of a pilot flame in a modified fluidized bed. Ind Eng Chem Res 29(9):1922–1927
108. Bautista E, Aldas R, Gagelonia E (1996) Rice hull furnaces for paddy drying: The Philippine rice research institute's experience. In: ACIAR PROCEEDINGS
109. Singh R, Maheshwari R, Ojha T (1980) Development of a husk fired furnace. J Agric Eng Res 25(2):109–120
110. Rozainee M, Ngo S, Salema AA, Tan K (2008) Fluidized bed combustion of rice husk to produce amorphous siliceous ash. Energy Sustain Dev 12(1):33–42
111. Sugita S (1994) Method of producing active rice husk ash, Google Patents
112. Soponronnarit S, Swasdisevi T, Wetchacama S, Shujinda A, Srisawat B (2007) Cyclonic rice husk furnace and its application on paddy drying. Int Energy J 1(2)
113. Zulkifli NSC, Ab Rahman I, Mohamad D, Husein A (2013) A green sol–gel route for the synthesis of structurally controlled silica particles from rice husk for dental composite filler. Ceram Int 39(4):4559–4567
114. Hassan A, Abdelghny A, Elhadidy H, Youssef A (2014) Synthesis and characterization of high surface area nanosilica from rice husk ash by surfactant-free sol–gel method. J Sol-Gel Sci Technol 69(3):465–472
115. Zulfiqar U, Subhani T, Husain SW (2015) Towards tunable size of silica particles from rice husk. J Non-Cryst Solids 429:61–69
116. Song S, Cho H-B, Kim HT (2018) Surfactant-free synthesis of high surface area silica nanoparticles derived from rice husks by employing the Taguchi approach. J Ind Eng Chem 61:281–287
117. Rehman MSU, Umer MA, Rashid N, Kim I, Han J-I (2013) Sono-assisted sulfuric acid process for economical recovery of fermentable sugars and mesoporous pure silica from rice straw. Ind Crops Prod 49:705–711
118. Thuc CNH, Thuc HH (2013) Synthesis of silica nanoparticles from Vietnamese rice husk by sol–gel method. Nanoscale Res Lett 8(1):1–10
119. Awizar DA, Othman NK, Jalar A, Daud AR, Rahman IA, Al-Hardan N (2013) Nanosilicate extraction from rice husk ash as green corrosion inhibitor. Int J Electrochem Sci 8(2):1759–1769
120. Haq IU, Akhtar K, Malik A (2014) Effect of experimental variables on the extraction of silica from the rice husk ash. J Chem Soc Pak 36(3):382
121. Coorey R (2017) Application of sago pith waste and nanosilica from rice husk ash as hybrid bio-nanofiller composite for food plastic packaging. Ukr Food J 6(4):618–631
122. Carmona V, Oliveira R, Silva W, Mattoso L, Marconcini J (2013) Nanosilica from rice husk: extraction and characterization. Ind Crops Prod 43:291–296
123. Mahmud A, Megat-Yusoff P, Ahmad F, Farezzuan AA (2016) Acid leaching as efficient chemical treatment for rice husk in production of amorphous silica nanoparticles. ARPN J Eng Appl Sci 11(22):13384–13388
124. Rafiee E, Shahebrahimi S (2012) Nano silica with high surface area from rice husk as a support for 12-tungstophosphoric acid: an efficient nano catalyst in some organic reactions. Chin J Catal 33(7–8):1326–1333
125. Soltani N, Bahrami A, Pech-Canul M, González L (2015) Review on the physicochemical treatments of rice husk for production of advanced materials. Chem Eng J 264:899–935
126. Adam F, Chew T-S, Andas J (2011) A simple template-free sol–gel synthesis of spherical nanosilica from agricultural biomass. J Sol-Gel Sci Technol 59(3):580–583
127. Duan R, Li C, Liu S, Liu Z, Li Y, Yuan Y, Hu X (2016) Determination of adenine based on the fluorescence recovery of the L-Tryptophan–Cu2 + complex. Spectrochim Acta Part A Mol Biomol Spectrosc 152:272–277
128. Slowing II, Vivero-Escoto JL, Trewyn BG, Lin VS-Y (2010) Mesoporous silica nanoparticles: structural design and applications. J Mater Chem 20(37):7924–7937
129. Balaji T, El-Safty SA, Matsunaga H, Hanaoka T, Mizukami F (2006) Optical sensors based on nanostructured cage materials for the detection of toxic metal ions. Angew Chem Int Ed 45(43):7202–7208

130. Awual MR, Ismael M, Yaita T, El-Safty SA, Shiwaku H, Okamoto Y, Suzuki S (2013) Trace copper (II) ions detection and removal from water using novel ligand modified composite adsorbent. Chem Eng J 222:67–76
131. Awual MR, Yaita T, El-Safty SA, Shiwaku H, Suzuki S, Okamoto Y (2013) Copper (II) ions capturing from water using ligand modified a new type mesoporous adsorbent. Chem Eng J 221:322–330
132. Ji X, Wang H, Song B, Chu B, He Y (2018) Silicon nanomaterials for biosensing and bioimaging analysis. Frontiers in chemistry 6:38
133. Jafari S, Derakhshankhah H, Alaei L, Fattahi A, Varnamkhasti BS, Saboury AA (2019) Mesoporous silica nanoparticles for therapeutic/diagnostic applications. Biomed Pharmacother 109:1100–1111
134. Gubala V, Giovannini G, Kunc F, Monopoli MP, Moore CJ (2020) Dye-doped silica nanoparticles: synthesis, surface chemistry and bioapplications. Cancer Nanotechnology 11 (1):1–43
135. Ma M, Zheng S, Chen H, Yao M, Zhang K, Jia X, Mou J, Xu H, Wu R, Shi J (2014) A combined "RAFT" and "Graft From" polymerization strategy for surface modification of mesoporous silica nanoparticles: towards enhanced tumor accumulation and cancer therapy efficacy. J Mater Chem B 2(35):5828–5836
136. El-Safty S, Shahat A, Awual MR, Mekawy M (2011) Large three-dimensional mesocage pores tailoring silica nanotubes as membrane filters: nanofiltration and permeation flux of proteins. J Mater Chem 21(15):5593–5603
137. Owens GJ, Singh RK, Foroutan F, Alqaysi M, Han C-M, Mahapatra C, Kim H-W, Knowles JC (2016) Sol–gel based materials for biomedical applications. Prog Mater Sci 77:1–79
138. Selvan ST, Tan TTY, Yi DK, Jana NR (2010) Functional and multifunctional nanoparticles for bioimaging and biosensing. Langmuir 26(14):11631–11641
139. Slowing I, Trewyn BG, Lin VS-Y (2006) Effect of surface functionalization of MCM-41-type mesoporous silica nanoparticles on the endocytosis by human cancer cells. J Am Chem Soc 128(46):14792–14793
140. Kempen PJ, Greasley S, Parker KA, Campbell JL, Chang H-Y, Jones JR, Sinclair R, Gambhir SS, Jokerst JV (2015) Theranostic mesoporous silica nanoparticles biodegrade after pro-survival drug delivery and ultrasound/magnetic resonance imaging of stem cells. Theranostics 5(6):631
141. Davis ME, Hsieh PC, Takahashi T, Song Q, Zhang S, Kamm RD, Grodzinsky AJ, Anversa P, Lee RT (2006) Local myocardial insulin-like growth factor 1 (IGF-1) delivery with biotinylated peptide nanofibers improves cell therapy for myocardial infarction. Proc Natl Acad Sci 103(21):8155–8160
142. Kupferschmidt N, Qazi KR, Kemi C, Vallhov H, Garcia-Bennett AE, Gabrielsson S, Scheynius A (2014) Mesoporous silica particles potentiate antigen-specific T-cell responses. Nanomedicine 9(12):1835–1846
143. Prokopowicz M, Żegliński J, Gandhi A, Sawicki W, Tofail SA (2012) Bioactive silica-based drug delivery systems containing doxorubicin hydrochloride: In vitro studies. Colloids Surf B 93:249–259
144. Sánchez-Muñoz S, Gómez-Ruiz S, Pérez-Quintanilla D, Morante-Zarcero S, Sierra I, Prashar S, Paschke R, Kaluđerović GN (2012) Preliminary study of the anticancer applications of mesoporous materials functionalized with the natural product betulinic acid. Chem Med Chem 7(4):670–679
145. Bollu VS, Barui AK, Mondal SK, Prashar S, Fajardo M, Briones D, Rodríguez-Diéguez A, Patra CR, Gómez-Ruiz S (2016) Curcumin-loaded silica-based mesoporous materials: Synthesis, characterization and cytotoxic properties against cancer cells. Mater Sci Eng, C 63:393–410
146. Kaluđerović GN, Pérez-Quintanilla D, Žižak Ž, Juranić ZD, Gómez-Ruiz S (2010) Improvement of cytotoxicity of titanocene-functionalized mesoporous materials by the increase of the titanium content. Dalton Trans 39(10):2597–2608

147. García-Peñas A, Gómez-Ruiz S, Pérez-Quintanilla D, Paschke R, Sierra I, Prashar S, del Hierro I, Kaluđerović GN (2012) Study of the cytotoxicity and particle action in human cancer cells of titanocene-functionalized materials with potential application against tumors. J Inorg Biochem 106(1):100–110
148. Bulatović MZ, Maksimović-Ivanić D, Bensing C, Gómez-Ruiz S, Steinborn D, Schmidt H, Mojić M, Korać A, Golić I, Pérez-Quintanilla D (2014) Organotin (IV)-loaded mesoporous silica as a biocompatible strategy in cancer treatment. Angew Chem Int Ed 53(23):5982–5987
149. Zheng Y, Fahrenholtz CD, Hackett CL, Ding S, Day CS, Dhall R, Marrs GS, Gross MD, Singh R, Bierbach U (2017) Large-Pore Functionalized Mesoporous Silica Nanoparticles as Drug Delivery Vector for a Highly Cytotoxic Hybrid Platinum–Acridine Anticancer Agent. Chemistry–A Eur J 23(14):3386–3397
150. Durgalakshmi D, Rishvanth R, Rakkesh RA, Bargavi P, Balakumar S, Aruna P, Ganesan S (2020) Chitosan mediated 5-Fluorouracil functionalized silica nanoparticle from rice husk for anticancer activity. Int J Biol Macromol
151. Abdel-Haliem ME, Hegazy HS, Hassan NS, Naguib DM (2017) Effect of silica ions and nano silica on rice plants under salinity stress. Ecol Eng 99:282–289
152. Wanyika H, Gatebe E, Kioni P, Tang Z, Gao Y (2012) Mesoporous silica nanoparticles carrier for urea: potential applications in agrochemical delivery systems. J Nanosci Nanotechnol 12(3):2221–2228
153. Torney F, Trewyn BG, Lin VS-Y, Wang K (2007) Mesoporous silica nanoparticles deliver DNA and chemicals into plants. Nat Nanotechnol 2(5):295–300
154. Wang W, Martin JC, Fan X, Han A, Luo Z, Sun L (2012) Silica Nanoparticles and Frameworks from Rice Husk Biomass. ACS Appl Mater Interfaces 4(2):977–981
155. Kopittke PM, Lombi E, Wang P, Schjoerring JK, Husted S (2019) Nanomaterials as fertilizers for improving plant mineral nutrition and environmental outcomes. Environmental Science: Nano 6(12):3513–3524
156. Shang Y, Hasan M, Ahammed GJ, Li M, Yin H, Zhou J (2019) Applications of nanotechnology in plant growth and crop protection: a review. Molecules 24(14):2558
157. Janmohammadi M, Amanzadeh T, Sabaghnia N, Ion V (2016) Effect of nano-silicon foliar application on safflower growth under organic and inorganic fertilizer regimes. Botanica Lithuanica 22(1):53–64
158. Hossain K-Z, Monreal CM, Sayari A (2008) Adsorption of urease on PE-MCM-41 and its catalytic effect on hydrolysis of urea. Colloids Surf B 62(1):42–50
159. Mattos BD, Rojas OJ, Magalhães WL (2017) Biogenic silica nanoparticles loaded with neem bark extract as green, slow-release biocide. J Clean Prod 142:4206–4213
160. Sibag M, Choi B-G, Suh C, Lee KH, Lee JW, Maeng SK, Cho J (2015) Inhibition of total oxygen uptake by silica nanoparticles in activated sludge. J Hazard Mater 283:841–846
161. (a) Cho YK, Park EJ, Kim YD (2014) Removal of oil by gelation using hydrophobic silica nanoparticles. J Ind Eng Chem 20(4):1231–1235. (b) Gomaa H, Shenashen MA, ElbazA, Yamaguchi H, Abdelmottaleb M, El-Safty SA (2021) Mesoscopic engineering materials for visual detection and selective removal of copper ions from drinking and waste watersources. J Hazard Mater 406:124314
162. Park S, Ko Y-S, Jung H, Lee C, Woo K, Ko G (2018) Disinfection of waterborne viruses using silver nanoparticle-decorated silica hybrid composites in water environments. Sci Total Environ 625:477–485
163. Karim A, Jalil A, Triwahyono S, Sidik S, Kamarudin N, Jusoh R, Jusoh N, Hameed B (2012) Amino modified mesostructured silica nanoparticles for efficient adsorption of methylene blue. J Colloid Interface Sci 386(1):307–314
164. Jeelani PG, Mulay P, Venkat R, Ramalingam C (2019) Multifaceted Application of Silica Nanoparticles. A Review. Silicon:1–18
165. Gribov B, Zinov'ev K, Kalashnik O, Gerasimenko N, Smirnov D, Sukhanov V, Kononov N, Dorofeev S (2017) Production of Silicon Nanoparticles for Use in Solar Cells. Semiconductors 51(13):1675–1680

166. Arunmetha S, Vinoth M, Srither SR, Karthik A, Sridharpanday M, Suriyaprabha R, Manivasakan P, Rajendran V (2017) Study on Production of Silicon Nanoparticles from Quartz Sand for Hybrid Solar Cell Applications. J Electron Mater 47(1):493–502
167. Marchal JC, Krug DJ III, McDonnell P, Sun K, Laine RM (2015) A low cost, low energy route to solar grade silicon from rice hull ash (RHA), a sustainable source. Green Chem 17 (7):3931–3940
168. Acharya H, Datta S, Banerjee H, Basu S (1982) Low-temperature preparation of polycrystalline silicon from silicon tetrachloride. Mater Lett 1(2):64–66
169. Favors Z, Wang W, Bay HH, Mutlu Z, Ahmed K, Liu C, Ozkan M, Ozkan CS (2014) Scalable synthesis of nano-silicon from beach sand for long cycle life Li-ion batteries. Scientific reports 4:5623
170. Shen Y (2017) Rice husk silica-derived nanomaterials for battery applications: a literature review. J Agric Food Chem 65(5):995–1004
171. Wong DP, Suriyaprabha R, Yuvakumar R, Rajendran V, Chen Y-T, Hwang B-J, Chen L-C, Chen K-H (2014) Binder-free rice husk-based silicon–graphene composite as energy efficient Li-ion battery anodes. J Mater Chem A 2(33):13437–13441
172. Cui J, Zhang H, Liu Y, Li S, He W, Hu J, Sun J (2020) Facile, economical and environment-friendly synthesis process of porous N-doped carbon/SiOx composite from rice husks as high-property anode for Li-ion batteries. Electrochim Acta 334:135619
173. (a) Zaytseva O, Neumann G (2016) Carbon nanomaterials: production, impact on plant development, agricultural and environmental applications. Chem Biol Technol Agric 3:17. (b) Abdien HG, Cheira MF, Abd-Elraheem MA, Saef El-Naser TA, Zidan IH (2016) Extraction and pre-concentration of uranium using activated carbon impregnated tri-octylphosphine oxide. Elixir Appl. Chem. 100:43462
174. Rauti R, Musto M, Bosi S, Prato M, Ballerini L (2019) Properties and behavior of carbon nanomaterials when interfacing neuronal cells: How far have we come? Carbon 143:430–446
175. Khare R (2005) Carbon nanotube based composites-a review. J Miner Mater Charact Eng 4 (01):31
176. Ummah H, Suriamihardja D, Selintung M, Wahab A (2015) Analysis of chemical composition of rice husk used as absorber plates sea water into clean water. ARPN J Eng Appl Sci 10(14):6046–6050
177. Georgakilas V, Perman JA, Tucek J, Zboril R (2015) Broad family of carbon nanoallotropes: classification, chemistry, and applications of fullerenes, carbon dots, nanotubes, graphene, nanodiamonds, and combined superstructures. Chem Rev 115(11):4744–4822
178. Gupta V, Rastogi A (2008) Biosorption of lead from aqueous solutions by green algae Spirogyra species: kinetics and equilibrium studies. J Hazard Mater 152(1):407–414
179. Gupta V, Srivastava S, Mohan D, Sharma S (1998) Design parameters for fixed bed reactors of activated carbon developed from fertilizer waste for the removal of some heavy metal ions. Waste Manag 17(8):517–522
180. Gupta VK, Jain C, Ali I, Chandra S, Agarwal S (2002) Removal of lindane and malathion from wastewater using bagasse fly ash—a sugar industry waste. Water Res 36(10):2483–2490
181. Zhang C, Long D, Xing B, Qiao W, Zhang R, Zhan L, Liang X, Ling L (2008) The superior electrochemical performance of oxygen-rich activated carbons prepared from bituminous coal. Electrochem Commun 10(11):1809–1811
182. Sahu J, Acharya J, Meikap B (2010) Optimization of production conditions for activated carbons from Tamarind wood by zinc chloride using response surface methodology. Biores Technol 101(6):1974–1982
183. Yang K, Peng J, Srinivasakannan C, Zhang L, Xia H, Duan X (2010) Preparation of high surface area activated carbon from coconut shells using microwave heating. Biores Technol 101(15):6163–6169
184. Donald J, Xu CC, Hashimoto H, Byambajav E, Ohtsuka Y (2010) Novel carbon-based Ni/Fe catalysts derived from peat for hot gas ammonia decomposition in an inert helium atmosphere. Appl Catal A 375(1):124–133

185. Ma J, Tan J, Du X, Li R (2010) Effects of preparation parameters on the textural features of a granular zeolite/activated carbon composite material synthesized from elutrilithe and pitch. Microporous Mesoporous Mater 132(3):458–463
186. Seredych M, Deliyanni E, Bandosz TJ (2010) Role of microporosity and surface chemistry in adsorption of 4, 6-dimethyldibenzothiophene on polymer-derived activated carbons. Fuel 89(7):1499–1507
187. Yuen FK, Hameed B (2009) Recent developments in the preparation and regeneration of activated carbons by microwaves. Adv Coll Interface Sci 149(1–2):19–27
188. Gupta V (2009) Application of low-cost adsorbents for dye removal–a review. J Environ Manage 90(8):2313–2342
189. Guo Y, Rockstraw DA (2007) Activated carbons prepared from rice hull by one-step phosphoric acid activation. Microporous Mesoporous Mater 100(1–3):12–19
190. Ghosh R (2013) A Review Study on Precipitated Silica and Activated Carbon from Rice Husk. J Chem Eng & Process Technol 4(04)
191. Shrestha LK, Thapa M, Shrestha RG, Maji S, Pradhananga RR, Ariga K (2019) Rice husk-derived high surface area nanoporous carbon materials with excellent iodine and methylene blue adsorption properties. C—Journal Carbon Res 5(1):10
192. Le Van K, Luong Thi TT (2014) Activated carbon derived from rice husk by NaOH activation and its application in supercapacitor. Prog Nat Sci: Mater Int 24(3):191–198
193. Fathy NA (2017) Carbon nanotubes synthesis using carbonization of pretreated rice straw through chemical vapor deposition of camphor. RSC Adv 7(45):28535–28541
194. Geng D, Wang H, Yu G (2015) Graphene single crystals: size and morphology engineering. Adv Mater 27(18):2821–2837
195. Novoselov KS, Fal V, Colombo L, Gellert P, Schwab M, Kim K (2012) A roadmap for graphene. Nature 490(7419):192–200
196. Bhuyan MSA, Uddin MN, Islam MM, Bipasha FA, Hossain SS (2016) Synthesis of graphene. International. Nano Lett 6(2):65–83
197. Ismail MS, Yusof N, Yusop MZM, Fauzi A (2019) Synthesis and characterization of graphene derived from rice husks. Malays J Fundam Appl Sci 15(4):516–521
198. Wang Z, Yu J, Zhang X, Li N, Liu B, Li Y, Wang Y, Wang W, Li Y, Zhang L, Dissanayake S, Suib SL, Sun L (2016) Large-Scale and Controllable Synthesis of Graphene Quantum Dots from Rice Husk Biomass: A Comprehensive Utilization Strategy. ACS Appl Mater Interfaces 8(2):1434–1439
199. Singh P, Bahadur J, Pal K (2017) One-Step One Chemical Synthesis Process of Graphene from Rice Husk for Energy Storage Applications. Graphene 06(03):61–71
200. Sankar S, Lee H, Jung H, Kim A, Ahmed ATA, Inamdar AI, Kim H, Lee S, Im H, Kim DY (2017) Ultrathin graphene nanosheets derived from rice husks for sustainable supercapacitor electrodes. New J Chem 41(22):13792–13797
201. De Volder MF, Tawfick SH, Baughman RH, Hart AJ (2013) Carbon nanotubes: present and future commercial applications. science 339(6119):535–539
202. Nasibulin AG, Shandakov SD, Nasibulina LI, Cwirzen A, Mudimela PR, Habermehl-Cwirzen K, Grishin DA, Gavrilov YV, Malm JE, Tapper U (2009) A novel cement-based hybrid material. New J Phys 11(2):023013
203. Peddini S, Bosnyak C, Henderson N, Ellison CJ, Paul DR (2014) Nanocomposites from styrene-butadiene rubber (SBR) and multiwall carbon nanotubes (MWCNT) part 1: Morphology and rheology. Polymer 55(1):258–270
204. Luinge H (2011) Nano-modified materials in aviation: Carbon nanotubes for lighter airplane outer shells. Konstruktion
205. Uritu CM, Varganici CD, Ursu L, Coroaba A, Nicolescu A, Dascalu AI, Peptanariu D, Stan D, Constantinescu CA, Simion V (2015) Hybrid fullerene conjugates as vectors for DNA cell-delivery. J Mater Chem B 3(12):2433–2446
206. Bergeson LL, Cole MF (2012) Fullerenes used in skin creams. Nanotech L & Bus 9:114
207. Choi W, Lahiri I, Seelaboyina R, Kang YS (2010) Synthesis of graphene and its applications: a review. Crit Rev Solid State Mater Sci 35(1):52–71

208. Das R, Ali ME, Hamid SBA, Ramakrishna S, Chowdhury ZZ (2014) Carbon nanotube membranes for water purification: a bright future in water desalination. Desalination 336:97–109
209. Khan NA, Khan SU, Ahmed S, Farooqi IH, Dhingra A, Hussain A, Changani F (2019) Applications of Nanotechnology in Water and Wastewater Treatment: A Review. Asian J Water Environ Pollut 16(4):81–86
210. Yan H, Gong A, He H, Zhou J, Wei Y, Lv L (2006) Adsorption of microcystins by carbon nanotubes. Chemosphere 62(1):142–148
211. Li Y-H, Wang S, Wei J, Zhang X, Xu C, Luan Z, Wu D, Wei B (2002) Lead adsorption on carbon nanotubes. Chem Phys Lett 357(3–4):263–266
212. Dichiara AB, Webber MR, Gorman WR, Rogers RE (2015) Removal of copper ions from aqueous solutions via adsorption on carbon nanocomposites. ACS Appl Mater Interfaces 7 (28):15674–15680
213. Zhang L, Song X, Liu X, Yang L, Pan F, Lv J (2011) Studies on the removal of tetracycline by multi-walled carbon nanotubes. Chem Eng J 178:26–33
214. Deng J, Shao Y, Gao N, Deng Y, Tan C, Zhou S, Hu X (2012) Multiwalled carbon nanotubes as adsorbents for removal of herbicide diuron from aqueous solution. Chem Eng J 193:339–347
215. Zheng X, Su Y, Chen Y, Wei Y, Li M, Huang H (2014) The effects of carbon nanotubes on nitrogen and phosphorus removal from real wastewater in the activated sludge system. RSC Adv 4(86):45953–45959
216. Kraus GA, Seidl PR, and Stankiewicz A, *Green materials for sustainable water remediation and treatment*. 2013: Royal Society of Chemistry
217. Srivastava M, Abhilash P, Singh N (2011) Remediation of lindane using engineered nanoparticles. J Biomed Nanotechnol 7(1):172–174
218. Wang L, Fortner JD, Hou L, Zhang C, Kan AT, Tomson MB, Chen W (2013) Contaminant-mobilizing capability of fullerene nanoparticles (nC60): Effect of solvent-exchange process in nC60 formation. Environ Toxicol Chem 32(2):329–336
219. Yoo J, Ozawa H, Fujigaya T, Nakashima N (2011) Evaluation of affinity of molecules for carbon nanotubes. Nanoscale 3(6):2517–2522
220. Wang H, Ma H, Zheng W, An D, Na C (2014) Multifunctional and recollectable carbon nanotube ponytails for water purification. ACS Appl Mater Interfaces 6(12):9426–9434
221. Jung JH, Hwang GB, Lee JE, Bae GN (2011) Preparation of airborne Ag/CNT hybrid nanoparticles using an aerosol process and their application to antimicrobial air filtration. Langmuir 27(16):10256–10264
222. Al-Hakami SM, Khalil AB, Laoui T, and Atieh MA (2013) Fast disinfection of Escherichia coli bacteria using carbon nanotubes interaction with microwave radiation. Bioinorganic chemistry and applications 2013
223. González-Melendi P, Fernández-Pacheco R, Coronado MJ, Corredor E, Testillano P, Risueño MC, Marquina C, Ibarra MR, Rubiales D, Pérez-de-Luque A (2008) Nanoparticles as smart treatment-delivery systems in plants: assessment of different techniques of microscopy for their visualization in plant tissues. Ann Bot 101(1):187–195
224. Zhang M, Gao B, Chen J, Li Y, Creamer AE, Chen H (2014) Slow-release fertilizer encapsulated by graphene oxide films. Chem Eng J 255:107–113
225. Giraldo JP, Landry MP, Faltermeier SM, McNicholas TP, Iverson NM, Boghossian AA, Reuel NF, Hilmer AJ, Sen F, Brew JA (2014) Plant nanobionics approach to augment photosynthesis and biochemical sensing. Nat Mater 13(4):400–408
226. Wang X, Liu X, Chen J, Han H, Yuan Z (2014) Evaluation and mechanism of antifungal effects of carbon nanomaterials in controlling plant fungal pathogen. Carbon 68:798–806
227. Liu S, Hu M, Zeng TH, Wu R, Jiang R, Wei J, Wang L, Kong J, Chen Y (2012) Lateral dimension-dependent antibacterial activity of graphene oxide sheets. Langmuir 28 (33):12364–12372
228. Akhavan O, Ghaderi E (2010) Toxicity of graphene and graphene oxide nanowalls against bacteria. ACS Nano 4(10):5731–5736

229. He H, Pham-Huy LA, Dramou P, Xiao D, Zuo P, Pham-Huy C (2013) Carbon nanotubes: applications in pharmacy and medicine. Biomed Res Int 2013:578290
230. Singh B, Baburao C, Pispati V, Pathipati H, Muthy N, Prassana S, and Rathode BG (2012) Carbon nanotubes. A novel drug delivery system. International Journal of Research in Pharmacy and Chemistry 2(2):523–532
231. Al-Jamal KT, Nerl H, Müller KH, Ali-Boucetta H, Li S, Haynes PD, Jinschek JR, Prato M, Bianco A, Kostarelos K (2011) Cellular uptake mechanisms of functionalised multi-walled carbon nanotubes by 3D electron tomography imaging. Nanoscale 3(6):2627–2635
232. Elhissi A, Ahmed W, Hassan IU, Dhanak V, and D'Emanuele A (2012) Carbon nanotubes in cancer therapy and drug delivery. Journal of drug delivery 2012
233. Maiti D, Tong X, Mou X, Yang K (2018) Carbon-Based Nanomaterials for Biomedical Applications: A Recent Study. Front Pharmacol 9:1401
234. Gera Y (2017) Applications of Graphene in Medicine and Electronics: A Review. IOSR Journal of Applied Physics 09(04):19–28
235. Song J, Yang X, Jacobson O, Lin L, Huang P, Niu G, Ma Q, Chen X (2015) Sequential drug release and enhanced photothermal and photoacoustic effect of hybrid reduced graphene oxide-loaded ultrasmall gold nanorod vesicles for cancer therapy. ACS Nano 9(9):9199–9209
236. Cheon YA, Bae JH, Chung BG (2016) Reduced graphene oxide nanosheet for chemo-photothermal therapy. Langmuir 32(11):2731–2736
237. Fong YT, Chen C-H, Chen J-P (2017) Intratumoral delivery of doxorubicin on folate-conjugated graphene oxide by in-situ forming thermo-sensitive hydrogel for breast cancer therapy. Nanomaterials 7(11):388
238. Su S, Wang J, Vargas E, Wei J, Martínez-Zaguilán R, Sennoune SR, Pantoya ML, Wang S, Chaudhuri J, Qiu J (2016) Porphyrin immobilized nanographene oxide for enhanced and targeted photothermal therapy of brain cancer. ACS Biomaterials Science & Engineering 2(8):1357–1366
239. Shao L, Zhang R, Lu J, Zhao C, Deng X, Wu Y (2017) Mesoporous silica coated polydopamine functionalized reduced graphene oxide for synergistic targeted chemo-photothermal therapy. ACS Appl Mater Interfaces 9(2):1226–1236
240. Peng X, Li Y, Luan Z, Di Z, Wang H, Tian B, Jia Z (2003) Adsorption of 1, 2-dichlorobenzene from water to carbon nanotubes. Chem Phys Lett 376(1–2):154–158
241. Lu C, Su F, Hu S (2008) Surface modification of carbon nanotubes for enhancing BTEX adsorption from aqueous solutions. Appl Surf Sci 254(21):7035–7041
242. Cho H-H, Wepasnick K, Smith BA, Bangash FK, Fairbrother DH, Ball WP (2010) Sorption of aqueous Zn [II] and Cd [II] by multiwall carbon nanotubes: the relative roles of oxygen-containing functional groups and graphenic carbon. Langmuir 26(2):967–981
243. Li Y-H, Ding J, Luan Z, Di Z, Zhu Y, Xu C, Wu D, Wei B (2003) Competitive adsorption of Pb2 + , Cu2 + and Cd2 + ions from aqueous solutions by multiwalled carbon nanotubes. Carbon 41(14):2787–2792
244. Madrakian T, Afkhami A, Ahmadi M, Bagheri H (2011) Removal of some cationic dyes from aqueous solutions using magnetic-modified multi-walled carbon nanotubes. J Hazard Mater 196:109–114
245. Ray PZ, Shipley HJ (2015) Inorganic nano-adsorbents for the removal of heavy metals and arsenic: a review. RSC Adv 5(38):29885–29907
246. Zhao Y, Zhao D, Chen C, Wang X (2013) Enhanced photo-reduction and removal of Cr (VI) on reduced graphene oxide decorated with TiO2 nanoparticles. J Colloid Interface Sci 405:211–217
247. Tan L, Wang Y, Liu Q, Wang J, Jing X, Liu L, Liu J, Song D (2015) Enhanced adsorption of uranium (VI) using a three-dimensional layered double hydroxide/graphene hybrid material. Chem Eng J 259:752–760
248. Gupta V, Rastogi A (2008) Equilibrium and kinetic modelling of cadmium (II) biosorption by nonliving algal biomass Oedogonium sp. from aqueous phase. J Hazard Mater 153(1–2):759–766

249. Li L, Zhou G, Weng Z, Shan X-Y, Li F, Cheng H-M (2014) Monolithic Fe2O3/graphene hybrid for highly efficient lithium storage and arsenic removal. Carbon 67:500–507
250. Zhang W, Shi X, Zhang Y, Gu W, Li B, Xian Y (2013) Synthesis of water-soluble magnetic graphene nanocomposites for recyclable removal of heavy metal ions. J Mater Chem A 1 (5):1745–1753
251. Ghasemi Z, Younesi H, Kazemian H (2011) Synthesis of nanozeolite sodalite from rice husk ash without organic additives. Can J Chem Eng 89(3):601–608
252. Sari ZGLV, Younesi H, Kazemian H (2014) Synthesis of nanosized ZSM-5 zeolite using extracted silica from rice husk without adding any alumina source. Appl Nanosci 5(6):737–745
253. Tan W-C, Yap S-Y, Matsumoto A, Othman R, Yeoh F-Y (2011) Synthesis and characterization of zeolites NaA and NaY from rice husk ash. Adsorption 17(5):863–868
254. Song Z, Huang Y, Xu WL, Wang L, Bao Y, Li S, Yu M (2015) Continuously adjustable, molecular-sieving "gate" on 5A zeolite for distinguishing small organic molecules by size. Sci Rep 5:13981
255. Molla A, Ioannou Z, Mollas S, Skoufogianni E, Dimirkou A (2017) Removal of chromium from soils cultivated with maize (Zea mays) after the addition of natural minerals as soil amendments. Bull Environ Contam Toxicol 98(3):347–352
256. Voronina A, Blinova M, Semenishchev V, Gupta D (2015) Returning land contaminated as a result of radiation accidents to farming use. J Environ Radioact 144:103–112
257. Fruijtier-Pölloth C (2009) The safety of synthetic zeolites used in detergents. Arch Toxicol 83(1):23–35
258. Bacakova L, Vandrovcova M, Kopova I, Jirka I (2018) Applications of zeolites in biotechnology and medicine – a review. Biomater Sci 6(5):974–989
259. Fu F, Wang Q (2011) Removal of heavy metal ions from wastewaters: a review. J Environ Manage 92(3):407–418
260. Wang S, Peng Y (2010) Natural zeolites as effective adsorbents in water and wastewater treatment. Chem Eng J 156(1):11–24
261. Moussavi G, Talebi S, Farrokhi M, Sabouti RM (2011) The investigation of mechanism, kinetic and isotherm of ammonia and humic acid co-adsorption onto natural zeolite. Chem Eng J 171(3):1159–1169
262. Inglezakis V, Doula M, Aggelatou V, Zorpas A (2010) Removal of iron and manganese from underground water by use of natural minerals in batch mode treatment. Desalination Water Treat 18(1–3):341–346
263. Derakhshankhah H, Jafari S, Sarvari S, Barzegari E, Moakedi F, Ghorbani M, Shiri Varnamkhasti B, Jaymand M, Izadi Z, Tayebi L (2020) Biomedical Applications of Zeolitic Nanoparticles, with an Emphasis on Medical Interventions. Int J Nanomedicine 15:363–386
264. Bedi RS, Zanello LP, Yan Y (2009) Osteoconductive and osteoinductive properties of zeolite MFI coatings on titanium alloys. Adv Func Mater 19(24):3856–3861
265. Bedi RS, Chow G, Wang J, Zanello L, Yan YS (2012) Bioactive Materials for Regenerative Medicine: Zeolite-Hydroxyapatite Bone Mimetic Coatings. Adv Eng Mater 14(3):200–206
266. Cefali EA, Nolan JC, McConnell WR, Walters DL (1995) Pharmacokinetic study of zeolite A, sodium aluminosilicate, magnesium silicate, and aluminum hydroxide in dogs. Pharm Res 12(2):270–274
267. Pavelic K, Katic M, Sverko V, Marotti T, Bosnjak B, Balog T, Stojkovic R, Radacic M, Colic M, Poljak-Blazi M (2002) Immunostimulatory effect of natural clinoptilolite as a possible mechanism of its antimetastatic ability. J Cancer Res Clin Oncol 128(1):37–44
268. Colic M, Pavelic K (2000) Molecular mechanisms of anticancer activity of natural dietetic products. J Mol Med 78(6):333–336

Recycle Strategies to Deal with Metal Nanomaterials by Using Aquatic Plants Through Phytoremediation Technique

Jyoti Mehta, Moharana Choudhury, Arghya Chakravorty, Rehab A. Rayan, Neeta Laxman Lala, and Andrews Grace Nirmala

Abstract An expanding need for nanotechnology in different enterprises may cause a vast situation scattering of nanoparticles in the coming years. The most widely recognized recuperation technique utilized so far includes using magnets to isolate iron-containing nanoparticles from complex blends, including wastewater. A few strategies have additionally been produced for the extraction, partition, and re-utilization of costly gold nanoparticles from various fluids. Pollution of multiple contaminants similar to metal nanoparticles (MNPs), Cu, Ni, Zn, Cd Ag, Pb, etc. exists well known to cause toxicity on the aquatic ecosystem. Macrophytes *like Trapa spp., Lemna spp., Eichhornia spp., Vallisneria spp., and Pistia spp.,* etc., will be used to remove the MNPs from the contaminated water in an eco-friendly and cost-effective way. Phytoremediation has been effectively actualized in various

J. Mehta (✉)
Department of Environmental Sciences, Central University of Jharkhand,
Brambe, Ranchi 835205, Jharkhand, India
e-mail: jyotimehta300@gmail.com

M. Choudhury
Voice of Environment (VoE), Guwahati 781034, Assam, India
e-mail: moharanasil@gmail.com

A. Chakravorty
School of Bio Sciences and Technology, Vellore Institute of Technology,
cVellore 632014, India
e-mail: arghya.kgp.AC@gmail.com

R. A. Rayan
Department of Epidemiology, High Institute of Public Health, Alexandria University,
Alexandria 21526, Egypt
e-mail: rayanr@alexu.edu.eg

N. L. Lala
Voice of Environment (VoE), Guwahati 781034, India
e-mail: neet_25@yahoo.com

A. G. Nirmala
Centre for Nanotechnology Research, Vellore Institute of Technology,
Vellore 632014, India
e-mail: anirmalagladys@gmail.com

A. S. H. Makhlouf and G. A. M. Ali (eds.), *Waste Recycling Technologies for Nanomaterials Manufacturing*, Topics in Mining, Metallurgy and Materials Engineering, https://doi.org/10.1007/978-3-030-68031-2_20

areas, including military destinations, agrarian fields, present-day units, mine dumps, sludge, and common wastewater treatment plants, by productive limit concerning expelling different natural and inorganic toxins through procedures, for example, extraction, debasement, or adjustment. Aquatic macrophytes speak to a diverse gathering of plants with a significant probability of expulsion/corruption into an assortment of pollutants, together with overwhelming metals, inorganic/natural poisons, radiogenic wastes, and explosives. The current examination highlights aquatic plants' work through phytoremediation progressions utilizing reasonable gathering regardless of presence free-swimming, underwater, or developing. Understanding the top capacities of sea-going macrophytes their relevance for more extensive utilization in phytoremediation innovations with developed swamps is underlined.

Keywords Aquatic plants · Toxicity · Phytoremediation · Nanomaterials · Recycling · Water

List of Abbreviations

CNTs	Carbon nanotubes
EDTA	Ethylene tetra acetic acid
ISO	International organization for standardization
MNPs	Metal nanoparticles
NMs	Nanoscale metals

1 Introduction

In the field of production and manufacturing, nanotechnology gives its vast involvement [1–3] results in their far-reaching with the new discharges into the atmosphere, particularly in water. The daily of metal nanoparticles in our day-to-day life reflect adverse effects in our environment. The food chain directly connects with water because water is the direct source for survival in nature, and these metal nanoparticles disturb it rapidly. The pH level of water is not right, and it is beneficial for other purposes to use as irrigation, drinking, cooking, etc., due to the presence of the metal nanoparticles, the water is not healthy, and day-to-day the aquatic bodies get disturbed and imbalanced. The use of metal nanoparticles in cosmetics, pesticides, fertilizers, paint, and other products causes an adverse impact on the water ecosystem, so we have to solve these problems naturally. As the ecological dangers of metal nanoparticles use today are not explained [4, 5], the potential strategies for their removal from the water condition, to keep up the nature of standard water.

It is grasped that water plants can rapidly hold water with different toxins [6–8], so they are commonly used in phytoremediation [9–11]. Phytoremediation presents from moves using green plants to expel risky designed blends from land and water proficient, wind, and soil. The upsides of phytoremediation are a high remediation level that is not undermined to the physical and designed strategies, ease, normal security, the chance of further extraction of contaminants from the green mass of plants [12, 13], and checking the cleaning framework. There are positive encounters and focused systems for phytoremediation of the mill vaults run utilizing area and water capable macrophytes for water decontaminating in world practices [14].

The water plants have high remediation potential for macronutrients taking into account their general immediate turn of events, development, and high biomass improvement. Utilizing these plants are very cost-effective, and it is also eco-friendly. These all are biodegradable and readily available in many places. Having sponge body structure made them hold the water into them and make them survive in the natural environment, water lettuce (*Pistia stratiotes*) and Water Duckweed (*Lemna minor*) Hyacinth (*Eichhornia crassipes*) alongside other macrophytes are visible as nanometals collector plants used for remediation from substantial metallic contaminated water. They absorb carbon from the atmosphere also.

The macrophytes like, *Leptospermum laevigatum, P. stratiotes, Sphaerotilus natans* are increasingly obstructive to nanomaterials activity as far as the photosynthetic framework's accuracy in this way has a higher potential for phytoremediation. Developed wetlands alongside water plants were broadly applied worldwide for wastewater treatment [15, 16]. The decision water plant species for developing powerful metal nanoparticle is a noteworthy issue to improve phytoremediation [17, 18].

Phytoremediation origination is a practical and eco-friendly revolution for irresistible metal as well as nanometal expulsion from aquatic environments. Aquatic macrophytes are reliable devices to expel awesome nanomaterial from aquatic forms and a great deal of consideration worldwide. The living and dead aquatic plants go about as a bio purification gadget used for overpowering metals together in ordinary and built wetlands. Present-day floods and fortifying treated city wastewater can be improved through macrophytes, and sorted out biomass might be reused.

2 Phytoremediation

The biological remediation technique, shown in Fig. 1, is an environmentally sustainable means for recovering heavy metals from polluted soil that involves phytoremediation, bioremediation, and a combination of both approaches. Phytoremediation is a rising technique using green flora to clear the ecosystem from

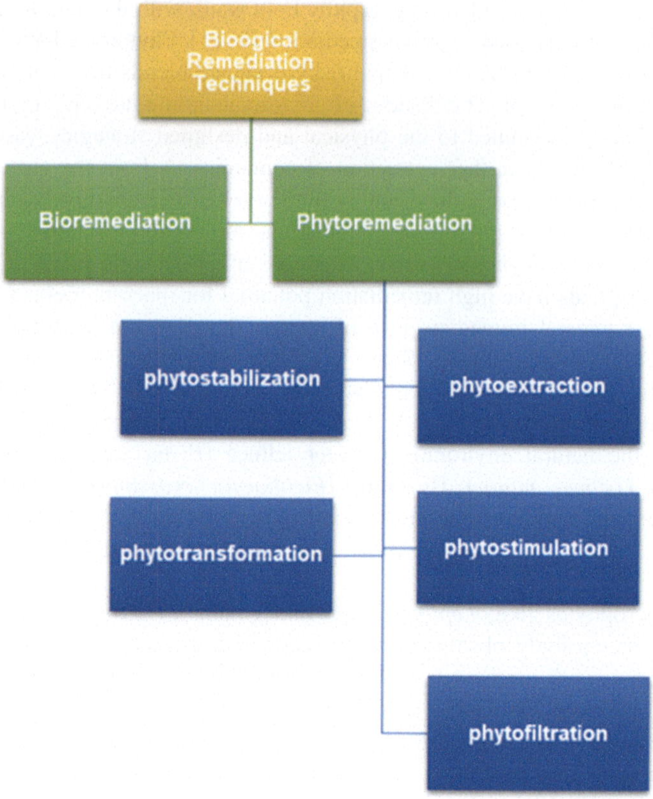

Fig. 1 Different biological remediation approaches for heavy metal-contaminated soil and water

pollutants and has been regarded as a non-invasive and affordable solution to the traditional reclamation procedures. There are several sorts of phytoremediation, as shown in Fig. 1. Phytoextraction is regarded to remediate toxic, heavy metal-polluted soils [1].

Reclaiming soil can retrieve an eco-friendly solution [2]. Phytoremediation, also known as green remediation, botano-remediation or agro-remediation, is a sustainable, green, and economical technique in relation to traditional methods, where rapidly growing plants can detoxify pollutants in water and soil [3]. In 1983, the phytoremediation principle was initiated, and yet, it is at the examination level. Phytoremediation is the optimum technique to work with mild-to-moderate metal-polluted soils and could be used along with other conventional techniques to eliminate pollutants. It is effective whenever plants could uptake and store a high concentration of metal pollutants [4].

2.1 Types of Phytoremediation

2.1.1 Phytostabilization

Phytostabilization referred to as photo-restoration is a flora-derived deactivation technique to work on metal-polluted by minimizing bioavailability and mobility of heavy minerals, hence reducing leakage and access to underground water and nutrients cycle, accordingly [4, 5]. In phytostabilization, metal pollutants' chemical and physical immobilization via uptake by roots and attachment with various soil refinements [5]. Phosphate fertilizers, organic material, clay minerals, and biosolids are highly effective soil refinements in immobilizing heavy metals. Plants minimize contracting with or the migration of pollutants, the leakage of water, and soil erosion, and [6]. Phytostabilization is not ongoing but a managing technique since, after all, metal pollutants remain in the soil [7].

2.1.2 Phytostimulation

Phytostimulation, referred to as rhizodegradation, is the decay of the rhizosphere's natural contaminants augmented by germ action [8]. The rhizosphere is soil mass, almost 1 mm close to the root, and is impacted by root action [9]. In the rhizosphere, microbial action is augmented by various means: The plants supply the ecosystem to optimize microbial species, and the roots oxygenate the rhizosphere for aerobic changes; the root supply organic carbon; the root leaks, which involve carbohydrates and amino acids to enhance native microbes and the mycorrhizae fungi degrade elements which the bacteria cannot degrade [10].

2.1.3 Phytotransformation

Phytotransformation, known as phytodegradation, breaks down organic elements by plants enzymes or metabolic procedures without microbes; therefore, plants are the biospheres green liver. In phytovolatilization, a process confined to organic elements Se or Hg heavy metals, plant transpiration occurs where plants consume soils pollutants, turn them into gaseous elements, and then liberate them into the air [11].

2.1.4 Phytofiltration

In phytofiltration, the plants' roots are used to remediate less polluted surface and groundwater and wastewater [8]. Initially, plants are provided with polluted water to adapt and then relocate to a polluted area for reclamation. The roots are harvested after saturation. Phytofiltration could be a blastofiltration via seedlings, a rhizofiltration via plant roots, or caulofiltration via plant shoots [12]. Pollutants are absorbed, adsorbed, or deposited, and leakage into underground water is minimized in such a process. Root leaks modify the rhizosphere's pH, leading to metals precipitation on the plant roots [13]. In rhizofiltration, both earthly and marine rapidly growing plants could be used to retrieve several heavy metals. Generally, earthly flora is utilized because the roots are lengthier and fibrous [14].

2.1.5 Phytoextraction

In phytoextraction, also called biomining or phytomining, rapidly flourishing flora eliminates heavy minerals from the soil and water [15]. Phytoextraction involves ongoing or natural and chemically driven phytoextraction [16]. As demonstrated in Fig. 2, ongoing phytoextraction eliminates heavy metals by the roots network and then shifted to the above-the-ground plant parts [17]. Harvested plant biomass can generate biogas and or get combusted. Combusted plant biomass can restore metals discarded in deserted lands or retained in stones. Such a technique is the best to minimize the concentration of metal pollutants in the soil via plant roots and shoots with no implications on the soil qualities where harvestable plant tissues can recover metals [18]. Ongoing harvesting and cropping can reduce pollutants level in the soil [19]. Moreover, such an evolving sustainable technique would be cost-effective compared to the current reclamation approaches [20].

The utilized plants for phytoextraction should be fast-growing, of great biomass and wide roots network, and adapting and reserving great levels of heavy metals [21]. In ongoing phytoextraction, natural hyperaccumulators plants are utilized at metalliferous locations [22]. Phytoextraction includes sorbing a certain portion of metals at the surface of the root, uptaking of bioavailable metals via the cellular membrane, immobilizing few metals portions through roots into vesicles, uptaking moving metals by the xylem, and translocating minerals from the roots to the stems and leaves of the plants [23].

2.2 *Pros and Cons of Phytoremediation*

Phytoremediation is a cost-effective technique to remediate heavy metals-polluted soil in relation to traditional physicochemical approaches [24]. Following phytoextraction, the harvested plant's biomass can generate bioenergy [25]. Phytoremediation is a promising clean-up technique; however, it has some drawbacks such as consuming a long time, the slowly growing plant species that could accumulate metals, and the low bioavailability of heavy metals [26].

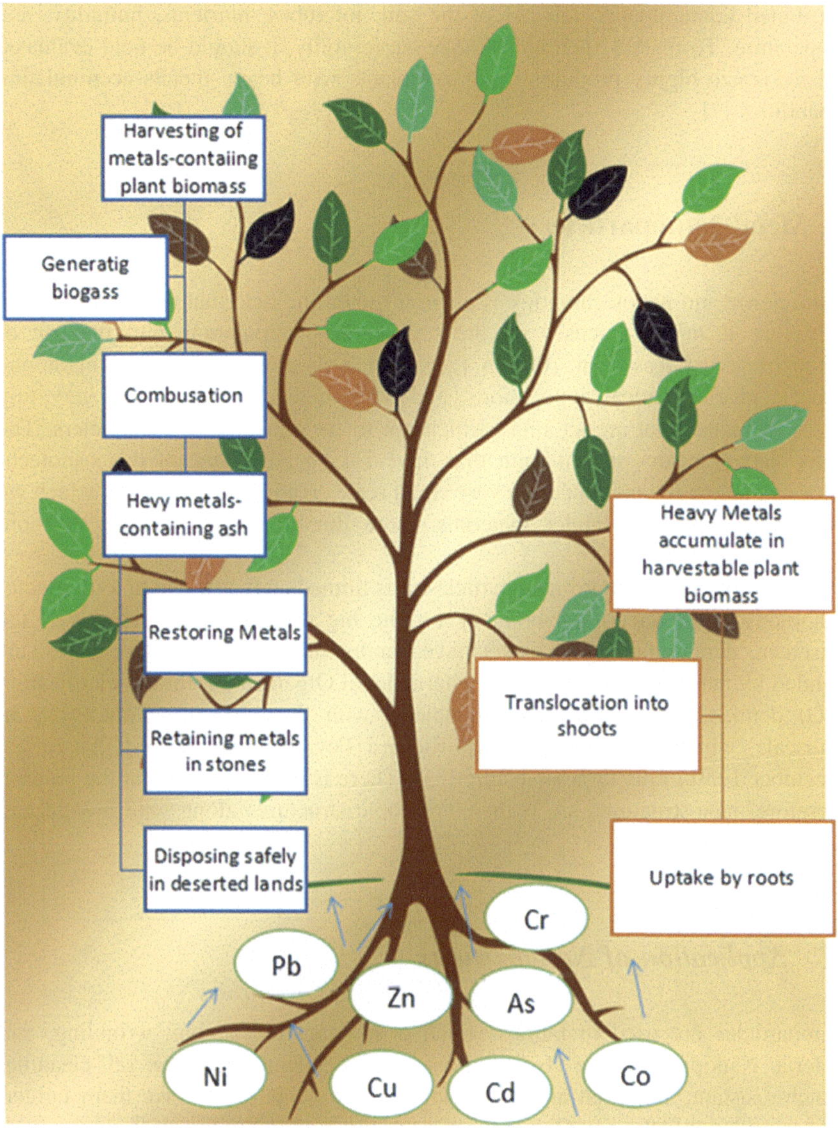

Fig. 2 Procedure of phytoextraction

3 The Future of Phytoremediation

Phytoremediation is an evolving discipline that needs more investigations. Findings could vary according to greenhouse and field conditions [27]. Phytoremediation procedure at the field is impacted by various parameters like the inconsistently distributed contaminants, the pH of the soil, microbes, nutrients, humidity, and temperature. To market such a technique successfully, it should be field-evaluated and recognize highly producing biomass plants with heavy metals accumulating capabilities [7].

4 Metal Nanoparticles

Nanoscience and nanotechnology is a current scientific field that tends to dominate daily life. It mainly focuses on the formation, manipulation, and the use of nanoparticles of less than 100 nm [28], which is used significantly in atomic physics, pharmaceutical applications, and other fields [29]. They are very tiny pieces of matter that are so small which are to be measured in nanometers. The prefix ‘nano’ means one-billionth that tiny! Taking advantage of this nanotechnology, it tailors the materials at a very small scale creating the properties which we are interested in like lighter materials, more durable, more reactive, or more conductive.

When you hear the word nanoparticles, it is immediately referred to as computer technology or engineering laboratory for the big electronics toy. However, the sunscreen, deodorant, tennis racket have nanoparticles; we are likely to be surrounded by nanoparticles every day. International Organization for Standardization (ISO) defines nanoparticles as nanoobjects with three external dimensions in nanoscale, which is 1–100 nm [30]. Richard Feynman first started this on 29 December 1959 in his famous lecture [31]. There is plenty of room at the bottom. Therefore, nanostructures are both microscopic structures along with atomic and molecular structures.

4.1 Application of Nanoparticles

Nanoparticles are used in fabrics which is resistant to staining, wrinkling, and bacteria. Nanoparticles are also used as water repellent, anti-glare, self-cleaning, scratch-resistant, and even anti-microbial, used in vehicles to make them lighter, cleaner, and more efficient. One of the most useful places is in the electronics world. Nanoparticles help reduce the power consumption of electronics, decrease the weight and thickness of electronic screens, and increase the computer’s speed. The wide use of metal nanoparticles in our daily life is shown in Fig. 3. Nanoparticles are used in health products and medical treatments as well. Products like deodorant, sunscreen, and cosmetics employ nanoparticles. While household products are degreasers, stain removers, air filters, purifiers, and paint-resist stains and dirt on walls.

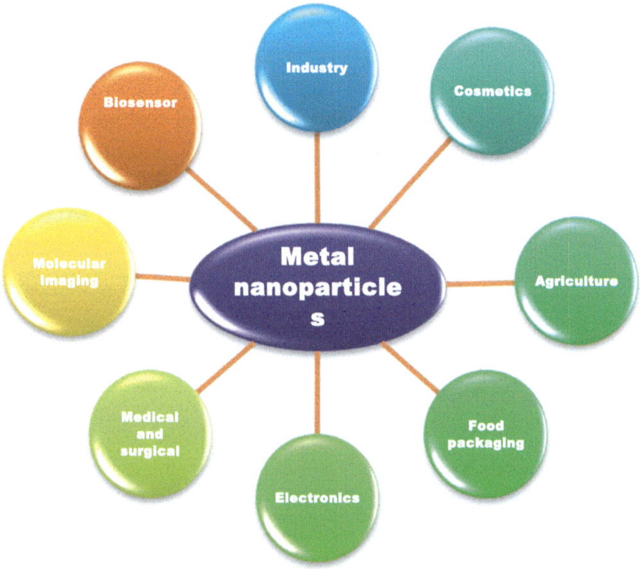

Fig. 3 Wide use of metal nanoparticles in our daily life

4.2 *Different Types of Nanoparticles*

Nanoparticles are categorized into multiple categories ranging in diameter, shape, physicochemical properties. There are ceramic nanoparticles, carbon-containing nanoparticles, metal nanoparticles, polymeric nanoparticles, nanoparticles centered on lipids, semiconductor nanoparticles. Carbon-based nanoparticles consist mainly of two carbon nanotubes (CNTs) and fullerenes, mainly used for structural reinforcement as they are 100 times stronger than steel [32]. Ceramics nanoparticles are inorganic, carbide, carbonate, oxide, and phosphate solids, used in photocatalysis, color degradation, drug delivery, and imaging [33]. Metal nanoparticles are formed from metal precursors. These nanoparticles were used in research areas such as biomolecular detection and imaging and environmental and bioanalytical applications [34]. Semiconductor nanoparticles have metal and non-metal properties. These particles show different properties when tuned, mainly used in electronics, photocatalysis, photo-optics, and water splitting applications [35]. Polymeric nanoparticles are carbon nanoparticles, used in drug delivery and diagnosis [36]. Lipid nanoparticles are spherical, with a diameter of 10–100 nm. The heart is composed of lipid and lipophilic molecules. Such nanoparticles are used in the biomedical field as drug carriers and cancer treatment transmission and RNA release [37].

4.3 Strategies Used to Synthesize Nanoparticles

Physical and chemical methods produce nanoparticles. Some of the commonly used are ion-sputtering, solvothermal synthesis, and sol-gel technique. There are two approaches to synthesize nanoparticle, namely (a) bottom-up approach and (b) top-down approach. The bottom-up approach is a process that builds toward larger and more complex systems at a molecular level and maintaining precise control of molecular structure [38]. While in the top-down approach, the bulk material is converted to fine particles.

4.4 Synthesis of Nanoparticles

There are various methods to synthesize nanoparticles, which are neutral to pH, low cost, and environmentally friendly. Sol-gel processing [39] is a wet-chemical technique where a chemical solution or colloidal particles is used to produce an integrated network forming a gel whereby alkoxides and metal chlorides are the main precursors used. Solvothermal synthesis [40] is similar to the hydrothermal route producing chemical compounds. It allows precise control over the size, shape distribution, and crystallinity of the metal oxide nanoparticles or nanostructure products. Chemical reduction methods [41] are cost-effective and widely available for mass production. Sodium borohydride, sodium citrate, hydrazine hydrate are some of the reducing agents used in which ionic salts undergo a reduction process in an appropriate medium in the presence of a surfactant used. Laser ablation [42] is a technique for removing materials from a solid surface. The absorbed laser energy and evaporates involves material heated at low laser flux. Inert gas condensation [43] is an ultra-high vacuum chamber filled with helium or argon gas at a typical pressure of a few 100 Pa's where different metals are evaporated in a crucible inside. As a result, the interatomic collisions with gas atoms within the chamber evaporate metal atoms losing their kinetic energy resulting in condensation in the form of small crystals accumulated on the liquid nitrogen-filled cold finger. Synthesis using bio-organisms [44] is compatible with green chemistry principles, which are environmentally friendly, non-toxic, and safe reagents mainly involved in green synthesis of the nanoparticle.

Biosynthesis of nanoparticles states that plants are known to possess various therapeutic compounds, which are traditional medicine known since ancient times. Plants are said to be advantageous as it is safe to handle, easily available and contain a broad range of biomolecules such as alkaloids, terpenoids, phenols, flavonoids, tannins, quinones, etc. known to mediate the synthesis of nanoparticles.

A nanoparticle synthesized by plants is faster and more stable compared to other cases [45]. Platinum nanoparticles are synthesized using leaf extract of *Diopyros kaki,* which is possible due to the biomolecules present in the extract but not an enzyme-mediated process [46]. Iron and silver nanoparticles were studied using the bran extract of Sorghum. Biological synthesis of gold nanoparticles using *Nycthanthes arbortristis* ethanolic flower extract was evaluated, resulting in spherical shaped gold nanoparticles with a size of 19.8 nm [47]. Selenium/protein composites using *Capsicum annuum* L. extract resulted in an increased concentration of *C. annuum* leaf extract with low pH increases the selenium/protein composites size and thickness of the shell [48]. Aqueous leaves extract of the *Sorbus aucuparia* was used as a reducing agent for the synthesis of silver and gold nanoparticles. Seaweed *Padina tetrastromatica* leaf extract results in the formation of silver nanoparticles, which is confirmed by analytical techniques.

5 Obtrusive Aquatic Plants Utilized in Phytoremediation

Macrophytes are maritime plants, found in or near water in eminent and skimming or brought downstage. They take after vascular plants specific with cells (tracheids), which are used to nimbly minerals and water with the help of its basic establishments. The major stage in macrophytes is the sporophyte, which has two approaches of chromosomes for every cell (i.e., diploid). Freshwater macrophytes in this manner portrayed are routinely collected liable to living things: making plants, shimmering fragile leaf plants, brought down plants, and coasting free plants [49, 50].

They need to create from seed, from vegetative propagates, or by development from neighboring masses. On the off chance that a macrophyte masses set up from seed or vegetative propagules, the water speed must permit these units to settle at the site. They need low energy in water for continuance at the site. They found that 10–20 weeks were required for macrophytes to set up >1% spread. Subsequently, a long interflow period is required to permit vegetation to make the ideal biomass. Truly, even scarcely any floods in a stream inside a year will diminish macrophyte wealth, separated and streams without flood aggravations [50, 51].

The effect of light on stream macrophyte deftly and riches can be plenteous ordinary changes in macrophyte biomass in calm streams are related to a mix of changes in light and temperature among summer and winter. For improvement, they require inorganic carbon and enhancements. The roots, similarly as leaves, expected an enormous activity in macrophytes to alter the carbon and enhancements [52]. Macrophytes offer refuge to edge and other land and water proficient animals. They comparably produce oxygen and feed some fish and normal life [49, 53].

5.1 Varieties of Macrophytes

Submerged macrophytes contain a couple of pteridophytes (e.g., the *Isoetes*), various greeneries, and charophytes (stonewort green improvement *Chara, Nitella)*, as like different angiosperms. Emergent macrophytes advancements occur in brought down soils, from where the water table is about 0.5 m underneath the earth. They are routinely rhizomatous perennials (e.g., *Glyceria, Eleocharis, Typha,* and *Phragmites).* Free-coasting macrophytes are an especially changed pack that is non-rooted to the base; in any case, lives free in the water. Shifting in structure and domain, they are long plants with sprout, floating and skimming leaves taking everything together around made lowered roots (e.g., *Eichhornia, Trapa, Hydrocharis)* to minute surface drifting or brought down plants with not very many or no roots (e.g., *Lemnaceae, Azolla, Salvinia).* Regenerative organs are going over the water surface (e.g., land and water proficient Utricularia) yet inconsistently brought down (e.g., *Ceratophyllum*). Aquatic macrophytes are related to the bedrock. Floating leaved macrophytes occur on water brought down buildup profundities at the water from about 0.5–3 m, having long (e.g., *the waterlilies* Nuphar and Nymphea versatile or short (e.g., *Brasenia, Potamogeton natans*) petioles and the skimming leaves are accessible on them [54, 55].

6 Role of Different Aquatic Macrophytes in Metal Nanoparticle Removal

6.1 Role of Water Hyacinth—(Eichhornia crassipis)

Water hyacinth is an aquatic macrophyte; it shows both tropical nature and subtropical nature. There are a total of seven species of water hyacinth; among all of them, *Eichhornia crassipis* is very common, develops quick and profoundly open-minded to contamination, and also utilized in the management of wastewater because of the high ingestion limit of overwhelming metals [56, 57]. The specialty of *E. crassipes* is high creation of biomass at appropriate climatic conditions; because of this characteristic feature, it is having a potent arsenic evacuation limit than the other aquatic macrophytes [58]. Water hyacinth shows a high development rate and enormous vegetative generation, throughout the year in every season and through its root or shoots framework. The water hyacinth plant absorbs different toxic metals like lead, nickel, manganese, zinc, copper, and cadmium [59, 60]. *E. crassipes* may be the best way to eliminate heavy metals. Water hyacinth was used for wastewater management to improve the quality of water. It does so by lowering the organic and inorganic nutrient levels [60, 61]. Irfan et al., Irfan treated water for one month at different concentrations of chromium (Cr) and copper (Cu) by using *E. crassipes*. The water hyacinth successfully eliminated dangerous metals without even any indications of impacting anything like this. The macrophyte was efficiently removed Cr and Cu after 30 days of the experiment; 80.94 and 95.5% accordingly [62].

6.2 *Role of Mosquito Fern*—(Azolla caroliniana) *and Mustard Green*—(Brassica juncea)

Azolla is remarkably efficient in the accumulation of toxic metals and can take away pollutants from water [63]. The majority of phytoremediations studied in *Zea mays, Helianthus annulus,* and *Brassica juncea* between 1995 and 2009. Many investigations have shown that *B. juncea* has capability highly effective in remediation of soil and water by accumulating hazardous cadmium (Cd) metal. The efficiency of zinc removal is high because of the increased production of biomass. In all three species, the removal efficiency of zinc, copper, and lead has been shown to be high in zinc and copper removal in its footsteps as compared to the other three species, which is *Brassica carinata, Brassica oleracea,* and *B. juncea*; and zinc and plumage have nearly continuously been reported [64].

6.3 *Role of Water Lettuce*—(Pistia stratiotes) *and Duckweeds*—(Lemnoideae)

Pistia stratiotes is a fast-growing aquatic macrophyte with large biomass. Due to the widespread root system, it shows extrinsic effectiveness in metal removal. Dead *P. stratiotes* are found to be highly proficient and inexpensive alternatives to the removal from the industrial waste of diluted heavy metals such as cadmium and lead. During 1-month water treatment, the *P. stratoites* removed chromium (Cr) at 77.3% and copper (Cu) at four separate concentrations of these heavy metals [60, 62, 65].

Duckweed is an aquatic plant that floats freely. In many aquatic conditions, it grows rapidly. Optimum plant growth temperature is between 5 and 35 °C, and a common pH range is between 3.5 and 10.5 [62]. These macrophytes are usually found in ponds and wetlands. These macrophytes (*Lemna spp.*) show a high capacity to exclude hazardous metals from the stagnant water. *L. minor* develops well enough on pH range 6–9 and absorbs up to 90% water-soluble lead, whereas high nitrate and ammonia levels inhibit its growth [66].

6.4 *Role of Hydrilla*—(Hydrilla verticillata) *and Duckweed*—(Spirodela intermedia)

Hydrilla verticillata, a thick layer of the water body, is an aquatic weed. Every part of this plant takes part in the pollutant removal process. Denny and Wilkins have been reported that shoots are more capable in the accumulation of toxic metals instead of the roots. It absorbed 98% water-soluble lead within 7 days of the experiment [67, 68]. *Spirodela intermedia* is a floating aquatic plant that is able to accumulate water-soluble cadmium, lead, and chromium from the surrounding, even under various weather conditions, showing a high growth rate. It prevents the development of algae across the whole water surface, even restricts the sunlight penetration and, eventually, the process of photosynthesis [69, 70].

6.5 *Role of Giant Bulrush*—(Schoenoplectus californicus)*; Ricciaceae*—(Ricciocarpus natans)*; Hydrocharitaceae*—(Vallisneria spiralis)

The giant bulrush is geographically wide ranging. It is just a vascular plant growing underneath water flow all along Asian, American, and African continents. It absorbs sediments to full uptake the nutrients and minerals through its root system. The plant is extremely tolerant of higher metal concentrations in waterways [71]. *Ricciocarpus natans* is an aquatic macrophyte that can absorb elements directly from the water stream [72]. The plant body lack seeds, leaves, roots, and vascular tissues. The research compared three different aquatic plant bodies *P. stratiotes, E. crassipes*, and *S. polyrrhiza* to test their heavy metal removal performance, while *E. crassipes* were observed with more proficiency than *P. stratiotes* and *S. polyrrhiza*. The outer intake of ethylene tetra acetic acid (EDTA) in *S. polyrrhiza* plant demonstrated greater metal intake like arsenic (III) and arsenic (V) [73].

A 21-day experiment on *Vallisneria spiralis* was conducted to test its ability to absorb Cu and Cd at dissimilar concentrations in a ready pan with high absorption of sediments in the shoots and roots by reducing chlorophyll. The positive association was found between soil and plant metals and/or between water/plant metals. *S. natans* demonstrated a great ability to accumulate copper (Cu), zinc (Zn), and Iron (Fe) from variable concentrations at different time duration [74].

7 Metal Nanoparticle Recycling and Removal Through Different Types of Phytoremediation

Phytoextraction has achieved attention worldwide over the last 20 years. It is also coined as phytoaccumulation, photoabsorption, and phytosequestration. Phytoextraction includes a root system that absorbs soil or water pollutants and accumulates them in aboveground biomass like plant shoots. Phytoabsorption is clearly observed in *B. juncea* and *T. caerulescens* plants [75, 76]. The diagrammatic presentation of different natural phytoremediation techniques and the different physiological activities of plant body during the operation are shown in Fig. 4.

Using roots, the plants acquire pollutants and accumulate it in the fibrous roots or carry them into other parts of the upper portion of the plant. A plant can continue this cycle until removed. Tiny numbers of toxins linger in the soil after disposal,

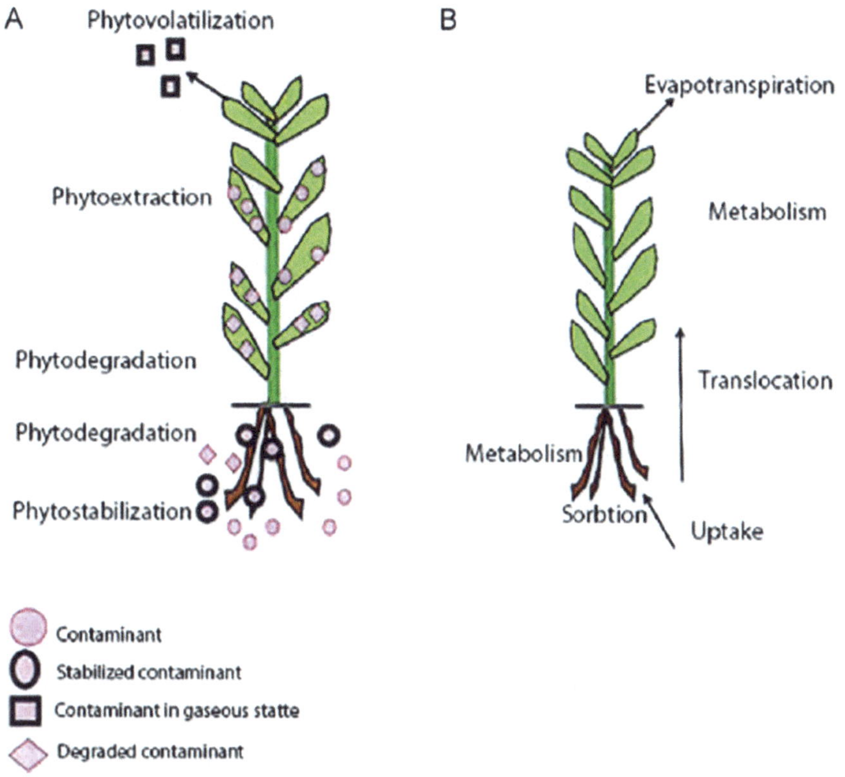

Fig. 4 **a** Diagrammatic presentation of different natural phytoremediation techniques; **b** different physiological activity of plant body during the operation. Adapted with permission from Ref. [102] Copyright 2016, Elsevier

and the growth and elimination process must be performed periodically across a variety of crops to get significant washing. Following this, the treated soil will sustain many macrophytes. The critical time for removal depends on the form and volume of metal contamination, the duration of the increasing cycle, and the usefulness of plant exclusion [76].

The technique is ideal for treating large areas of the field, and pollution rates in these areas are small to moderate because the growth of macrophytes cannot be maintained in highly contaminated soils. Metals from the soil would be absorbable and assimilated by the root system. But, both processes, such as high metal accumulation and high biomass formation, result in optimum metal removal [76]. Figure 5 shows the diagrammatic presentation of phytoextraction.

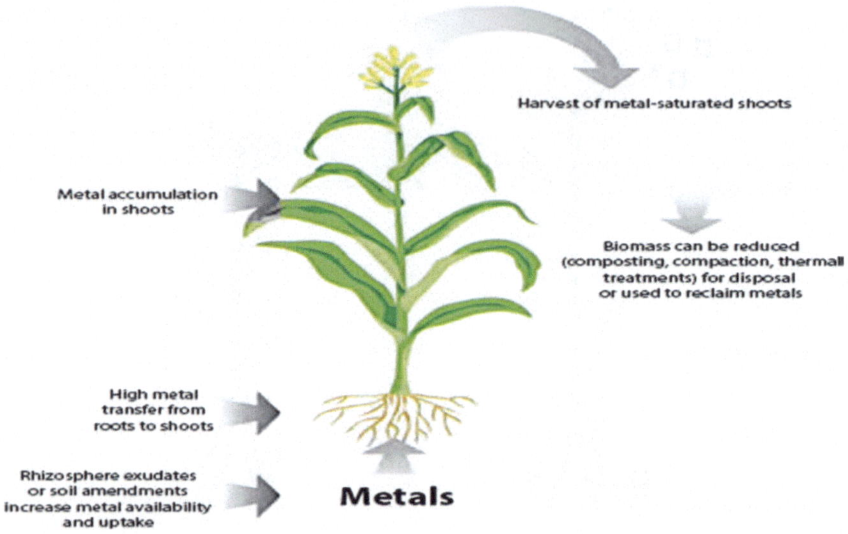

Fig. 5 Diagrammatic presentation of phytoextraction. Adapted with permission from Ref. [102] Copyright 2016, Elsevier

7.1 Mechanism of Phytostabilization

Phytoimmobilization is another name of phytostabilization. It refers to the use of different plants to contain toxins in contaminated soils [60]. It is a plant-based remediation process. It helps to stabilize pollutants and inhibits mobilization and liquid–erosion contact. It provides hydraulic control, reducing pollutant upstanding mobility. This reduces pollutant movement externally or chemically through root absorption [77, 78]. This strategy is used to prevent environmental pollutant proliferation and release, avoiding their movement and entry into the food and water system [79]. Phytostabilization's main objective is soil pollutant recovery. In this method, toxins are not contained in plant tissues, and less readily viable, resulting in less damage to plants, humans, and wildlife. *A. tenuis* and *F. rubra* are commercially available to treat heavy metal (eg. zinc, lead and copper) contaminated soils.

This method reduces pollutant flow and migration to surface water. This approach is also used to recover the plant lifecycle at places where normal planting miscarries to stay alive due to the high metal content in surface soils or surface material physical resistance. Metal-tolerant plants are used to defend habitats in damaged areas. There will be less chance of transferring pollutants through soil erosion and uncovered soil. This is used for Zn, Hg, As, Cd, Cu, and Cr [80]. The diagrammatic presentation of phytostabilization is shown in Fig. 6. Phytostabilization is cheap, eco-friendly, easy to install, or to use and provides visual benefit to many other remediation technologies [78]. This is most effective in areas with inorganic compounds that have smooth soils but is suitable for treating a

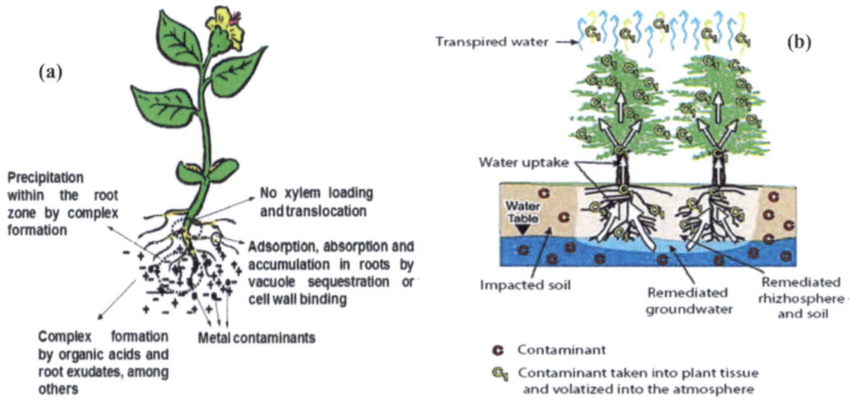

Fig. 6 Diagrammatic presentations of phytostabilization. Adapted with permission from Ref. [102] Copyright 2016, Elsevier

wide range of places where there are large areas of surface pollution. Phytostabilization is not practicable in highly polluted conditions, as plant growth and survival are unlikely [77].

7.2 *Mechanism of Rhizofiltration*

Rhizofiltration is the process of collecting and precipitating root hairs hold in hazardous metals from effluents. Water hyacinth, duckweed, and pennywort are widely utilized for rhizofiltration process [81].

In the root system, plant roots metabolize different chemicals that establish biogeochemical environments that allow toxins to precipitate into the roots or water sources. If the plant roots are polluted with toxins, either the stems or whole plants are deposited [82].

7.3 *Mechanism of Phytotransformation*

Phytotransformation is the transformation of soil and water contaminants that are either biological or mineral contaminants. Phytotransformation refers to petrochemical areas and other dumping fields, e.g., industrial products, ammunition waste, chlorinated solvents, diesel spills, landfill leachates [83].

7.4 Mechanism of Phytovolatilization

Phytovolatization is the mechanism by which plants receive contaminants from their environments and then spread the pollutants. Transpiration is the mechanism in which water travels from the base roots to the upward portion of the plant, which evaporates in the pores of the herb. Cultivated tobacco, swamp lily, spring wheat, mouse-ear cress, water hyssop, and white clover are widely used plants for phytovolatilization technique. Phytovolatization is the cycle where a plant takes a pollutant and initiates transpiration. Therefore, the plant emits a contaminant or transformed form of a contaminant into the environment. Another method that was renamed phytodegradation is a phytoremediation cycle related to phytovolatilization [83–85].

Destinations that utilize this system of phytovolatilization may not require a lot of oversight after the ranch of these plants. This remediation technique has the extra points of interest, for instance, these locales are less upset, fewer odds of disintegration, and plants utilized in this procedure need not be discarded. Phytovolatilization would not be appropriate for places that are close to the exceptionally populated locales or at some different spots with particular climate designs that embrace the speedy settlement of precarious mixes [86, 87].

8 Recycling of Metal Nanoparticles

Given the high estimation of extraordinarily produced nanoscale metals (NMs), manufacturers and consumers have a keen interest in reusing or recycling these NMs. Nonetheless, they could be articulated or classified as hazardous waste due to a lack of knowledge on NMs, which may incite missing sensitive materials [88]. NMs could be presented below in the waste stream reuse process [89], as shown in Fig. 7.

However, the prospect of isolating nanoparticles from the squander medium to reuse them was studied [90]. Most of the trials are ordinary detachment procedures with high vitality requests, similar to centrifugation or dissolvable vanishing. For

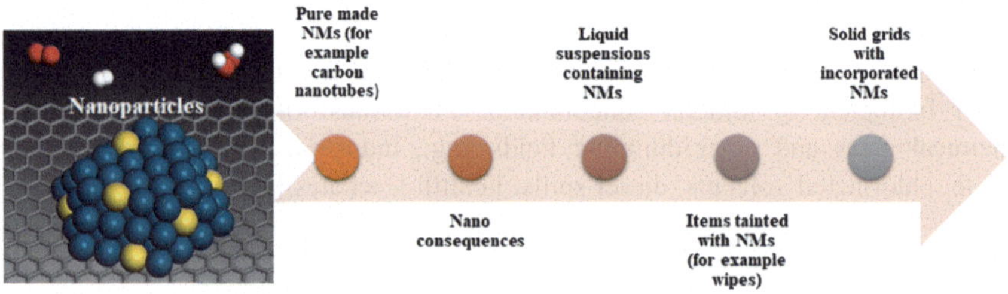

Fig. 7 Schematic diagram by representing steps of recycling

example, the use of alluring environments, pH and the rmoresponsive materials, subnuclear antisolvents, or nanostructured colloidal solvents provide persuasive and capable methodologies for reusing nanoparticles without the need for specific costs, time use, or imperativeness [91]. However, a call for reusing NMs is required to further focus on their natural recyclability properties, such as dry, mechanical, and material properties. A discussion on how the architecture could enhance the disposal of certain products in an accurate way would better understand appropriate reuse and reuse decisions. Today, a number of innovative reactions for reusing NMs have been applied, i.e., the use of alluringly recoverable support for the immobilization of gold nanoparticle impulses, which ensures that the catalyst section is quick, perfect, fast and efficient at the completion of the treatment cycle [92]. To address the issue of reuse and mull over the acts of OECD part countries on nanomaterial prosperity, the OECD Chemicals Committee hosts developed a Working Get-together on Manufactured Nanomaterials.

9 Nanoparticle Waste Treatment

Various creators have also portrayed the effects of anthropogenic pollution and sullying on humans. For shellfish Daphnia magna, the typical research species [93–96], hardly circulated reports inquiring about the effect on the maritime environment, essentially pushed to C60, nanotubes, and titanium dioxide. There are various new requests for NMs that need the right reactions: how to evaluate prologue to NMs, what are the potential prosperity effects of NMs, what are the potential biological effects of NMs, how well would we have the option to review the threats from NMs, what do we need to know regardless of everything. Responses to these questions provide a credible overview of the sensible hypothesis presented in 2006 by the SCENIHR Scientific Advisory Collection on Current and Newly Found Health Dangers [96, 97]. Balanced advice (after free gathering) on the reasonableness of current approaches to tackle the possible dangers of engineered and external impact of nanotechnology [97]. The SCENIHR shuts down the NMs that have exceptional (eco-) toxicological properties than the substances in mass structure and that their perils should be checked inferior to the situation. There are three broad origins of waste from nanomaterials in the packaging of division and recovery thoughts of safe nanomaterials, with categorization similar to customary waste [98]. A schematic representation of nanoparticle waste treatment is shown in Fig. 8.

In 2011, the European Commission launched topical analyses of nanomaterials and the EU's biological system, a study circulated in accordance with the EU's Waste Policy Coherence [99]. Customer involvement may help keep an eye on information needs about things lifecycle necessities during the early periods of nanoproduct design. Nanoproduct tests may be implemented to display fluctuating degrees of end-of-thing life processes and knowledge similar to government reusing delivery centers. Non-manufacturers should communicate with consumers during

Fig. 8 Schematic representation of nanoparticle waste treatment

the lifecycle of order and get some information about their interactions, including the potential for abuse, whether deliberate or accidental.

The course suggests that nanosquander should be labeled hazardous waste and as a foundation, double sacked, enclosed in an inflexible impermeable container, and naturally wrapped in a solid framework [100]. In the key case, the nanoobject capture and recommendation will be viewed as a general prerequisite for the administrators appropriate nanowaste. Where air dispersion of nanoobjects possess translocation of nanoobjects to organic structures out of reach the result, similarly as declaration within and integrating workplaces for investigation and nanomanufacturing; For example, parameters such as nanoobject scale, game plan, and dissolvability would need evaluation for any time the nanoproduct lifecycle is ignored before the program evacuation procedures can be considered in standard operating procedure (SOP) [101].

10 Removal and Reusing of Items Containing Nanotechnology

Removal of nanomaterials and items in which they are containing ought to be performed with specific consideration to guarantee that nanomaterials can possibly represent a danger to human wellbeing, and the earth is not discharged. Nanomaterials that are hazardous, poisonous, or synthetically receptive ought to be killed. Where conceivable, nanowaste ought to be reused [103–109].

Nanowaste can be the outcome or side effect of modern or business forms. Due to the expansive scope of existing nanomaterials, a single methodology for removal will not get the job done for all classes of nanomaterials. Subsequently, it is imperative to comprehend the properties of explicit nanowastes before creating powerful removal rehearses. The created security measures and removal systems essential for dealing with the nanosquander must be in light of current information and take into account existing enactment. The removal techniques must guarantee that the waste is deactivated by its perilous properties. Contingent upon the sort of material, warm, concoction, or physical preparation of nanotechnology-containing waste is conceivable deactivation arrangements [110–119].

11 Conclusion

The use of MNPs is not a big issue, but the toxicity which is created by this cause adverses impact on the whole ecosystem. So, the recycling of MNPs with various new techniques will help to make eco-friendly and cost-effective. They can use it in recreation with new products in different faces, and it can be used in the pharmaceutical, agriculture, killing of bacteria, and water purification.

To save the water bodies' ecosystem and maintain the green zone, the aquatic plants are the best option to maintain the stability of the water bodies. These plants are also easily available everywhere; no extra care is required for its survival. These plants are used as toxicity removal of MNPs from water due to root and spongy structure. These plants act as a natural filter which used to maintain the water pollutant free. These aquatic plants provide food for aquatic organisms for survival, and they also absorb the carbon from the atmosphere. So, with the MNPs recycle and toxicity removal by plants together helps us to male the green environment.

12 Future Prospective

The absence of exacting arrangements and guidelines identified with the utilization and removal of nanotechnology, not with standing the reusing of nanomaterial-containing items, are basic issues. Nanowaste is famously hard to contain and screen; because of its little size, it can spread in water frameworks or become airborne, making hurt human wellbeing and nature. Enactment is required so as to direct the offer of items containing nanomaterials in the commercial center and their further removal after use. Where conceivable, reusing of nanomaterials is the most attractive result.

Governments must execute appraisals, guidelines, and observing measures for nanotechnology producers. Preceding putting nanomaterial-put together items with respect to the advertising, broad ecological and wellbeing sway considered must be played out; these must incorporate investigations identified with the poisonousness what's more, synthetic reactivity of any new nanomaterial. From this, sheltered removal and reusing systems can be set up.

Nanomaterial makers (or a free body, or EPA) should likewise decide if these substances or producing procedures could represent a hazard to general wellbeing or the earth. Items should possibly be permitted into the commercial center if there is no hazard, or if the hazard can be controlled through defensive measures. Concentrated modern nanowaste ought to be weakened and deactivated preceding removal. Furthermore, organizations creating such waste as a side effect of their mechanical tasks must be required to demonstrate EPA that their nanowaste is non-risky to the condition and to human wellbeing. Recently created nanomaterials must not be discharged to the market without fitting removal systems. Recently created nanowaste removal systems must be analyzed and endorsed by government organizations dependent on undisputed proof gave by the case dwelling association. To give adequate proof, the association may do tests itself or allude to existing logical systems and claims.

Acknowledgements Authors are thankful to Mr. Deepak Kumar; Voice of Environment (VoE) and Centre for Nanotechnology Research, VIT Vellore, for encouraging them during this project work to complete successfully.

References

1. Pérez-de-Luque A, Rubiales D (2009) Nanotechnology for parasitic plant control. Pest Manag Sci 65:540–545. https://doi.org/10.1002/ps.1732
2. Roco MC (2003) Nanotechnology: convergence with modern biology and medicine. Curr Opin Biotechnol 14:337–346. https://doi.org/10.1016/s0958-1669(03)00068-5
3. Saharan V (2011) Advances in nanobiotechnology for agriculture. In: Current topics in biotechnology & microbiology. Lap Lambert Academic Publishing AG & CO. KG, Dudweller Landstr, Germany, pp 156
4. Handy RD, Shaw BJ (2007) Toxic effects of nanoparticles and nanomaterials: implications for public health, risk assessment and the public perception of nanotechnology. Health Risk Soc 9(2):125–144
5. Owen R, Handy RD (2007) Formulating the problems for environmental risk assessment of nanomaterials. Environ Sci Technol 41:582–588
6. Murithi G, Onindo CO, Muthakia GK (2012) Kinetic and equilibrium study for the sorption of Pb(II) ions from aqueous phase by water hyacinth (Eichhornia crassipes). Bull Chem Soc Ethiopia 26(2):181–193
7. Parra LM, Torres G, Arenas AD, Sánchez E, Rodríguez K (2012) Phytoremediation of low levels of heavy metals using duckweed (Lemna minor). Abiotic stress responses in plants: metabolism, productivity and sustainability. Springer, New York, pp 451–463
8. Alvarado S, Guédez M, Lué-Merú MP, Nelson G, Alvaro A, Jesús AC et al (2008) Arsenic removal from waters by bioremediation with the aquatic plants water hyacinth (Eichhornia crassipes) and lesser duckweed (Lemna minor). Bioresour Technol 99:8436–8440
9. Prasad M, Malec P, Waloszek V (2001) Physiological responses of Lemna trisulca L. (duckweed) to cadmium and copper bioaccumulation. Plant Sci 161:881–889
10. Taghiganji M, Khosravi M, Rakhshaee R (2012) Phytoremediation potential of aquatic macrophyte Azolla. Ambio 41:122–137
11. Bennicelli R, Stezpniewska Z, Banach A, Szajnocha K, Ostrowski J (2004) The ability of Azolla caroliniana to remove heavy metals (Hg(II), Cr(III), Cr(VI)) from municipal waste water. Chemosphere 55:141–146

12. Smadar E, Benny C, Tel-Or E, Lorena V, Antonio C, Aharon G (2011) Removal of silver and lead ions from water wastes using Azolla filiculoides, an aquatic plant, which adsorbs and reduces the ions into the corresponding metallic nanoparticles under microwave radiation in 5 min. Water Air Soil Pollut 218:365–370
13. Leblebici Z, Aksoy A (2011) Growth and lead accumulation capacity of Lemna minor and Spirodela polyrhiza (Lemnaceae): interactions with nutrient enrichment. Water Air Soil Pollut 214:175–184
14. Rahman MA, Hasegawa H (2011) Aquatic arsenic: phytoremediation using floating macrophytes. Chemosphere 83:633–646
15. Sasmaza A, Obek E (2012) The accumulation of silver and gold in Lemna gibba exposed to secondary effluents. Chem Erde-Geochem 72:149–152
16. Gorito AM, Ribeiro AR, Almeida CMR, Silva AMT (2017) A review on the application of constructed wetlands for the removal of priority substances and contaminants of emerging concern listed in recently launched EU legislation. Environ Pollut 227:428–443. https://doi.org/10.1016/j.envpol.2017.04.060
17. Mesa J, Mateos-Naranjo E, Caviedes MA et al (2015) Scouting contaminated estuaries: heavy metal resistant and plant growth promoting rhizobacteria in the native metal rhizoaccumulator Spartina maritima. Mar Pollut Bull 90:150–159. https://doi.org/10.1016/j.marpolbul.2014.11.002
18. Galal TM, Eid EM, Dakhil MA, Hassan LM (2017) Bioaccumulation and rhizofiltration potential of Pistia stratiotes L. for mitigating water pollution in the Egyptian wetlands. Int J Phytorem 20:440–447. https://doi.org/10.1080/15226514.2017.1365343
19. Ashraf S, Ali Q, Zahir ZA et al (2019) Phytoremediation: environmentally sustainable way for reclamation of heavy metal polluted soils. Ecotoxicol Environ Saf 174:714–727. https://doi.org/10.1016/j.ecoenv.2019.02.068
20. Martin TA, Ruby MV (2004) Review of in situ remediation technologies for lead, zinc, and cadmium in soil. Remediat J 14:35–53. https://doi.org/10.1002/rem.20011
21. Mahar A, Wang P, Ali A et al (2016) Challenges and opportunities in the phytoremediation of heavy metals contaminated soils: a review. Ecotoxicol Environ Saf 126:111–121. https://doi.org/10.1016/j.ecoenv.2015.12.023
22. Khalid S, Shahid M, Niazi NK et al (2017) A comparison of technologies for remediation of heavy metal contaminated soils. J Geochem Explor 182:247–268. https://doi.org/10.1016/j.gexplo.2016.11.021
23. Singh S (2012) Phytoremediation: a sustainable alternative for environmental challenges. https://www.semanticscholar.org/paper/Phytoremediation%3A-A-Sustainable-Alternative-for-Singh/037d1b0ab68659be168ad436f139451562550681. Accessed 28 Mar 2020
24. Akhtar MS, Chali B, Azam T (2013) Bioremediation of arsenic and lead by plants and microbes from contaminated soil. Res Plant Sci 1:68–73. https://doi.org/10.12691/plant-1-3-4
25. Vangronsveld J, Herzig R, Weyens N et al (2009) Phytoremediation of contaminated soils and groundwater: lessons from the field. Environ Sci Pollut Res 16:765–794. https://doi.org/10.1007/s11356-009-0213-6
26. Mukhopadhyay S, Maiti SK (2010) Phytoremediation of metal enriched mine waste: a review. Am-Eurasian J Agric Amp Environ Sci 9:560–575
27. Pilon-Smits E (2005) Phytoremediation. Annu Rev Plant Biol 56:15–39. https://doi.org/10.1146/annurev.arplant.56.032604.144214
28. Ji P, Sun T, Song Y et al (2011) Strategies for enhancing the phytoremediation of cadmium-contaminated agricultural soils by Solanum nigrum L. Environ Pollut 159:762–768. https://doi.org/10.1016/j.envpol.2010.11.029
29. Mobasser S, Firoozi AA (2016) Review of nanotechnology applications in science and engineering. J Civil Eng Urban 6(4):84–93
30. Anandalakshmi K, Venugobal J, Ramasamy V (2015) Characterization of silver nanoparticles by green synthesis method using Pedalium murex leaf extract and their antibacterial activity. Appl Nanosci 6:399–408. https://doi.org/10.1007/s13204-015-0449-z

31. Boverhof DR, Bramante CM, Butala JH, Clancy SF, Lafranconi M, West J, Gordon SC (2015) Comparative assessment of nanomaterial definitions and safety evaluation considerations. Regul Toxicol Pharmacol 73(1):137–150
32. Feynman R (1960) There's plenty of room at the bottom. An invitation to enter a new field of science, lecture, annual meeting of the American Physical Society, California Institute of Technology, December 29, 1959. Caltech Eng Sci 23:22–36
33. Maiti D, Tong X, Mou X, Yang K (2019) Carbon-based nanomaterials for biomedical applications: a recent study. Front Pharmacol 9. https://doi.org/10.3389/fphar.2018.01401
34. Sorescu M, Diamandescu L, Sanns A et al (2007) Synthesis and characterization of ceramic nanoparticles system based on anatase-doped hematite. Res Lett Mater Sci 2007:1–5. https://doi.org/10.1155/2007/48962
35. Chokkareddy R, Redhi GG (2018) Green synthesis of metal nanoparticles and its reaction mechanisms. Green Metal Nanoparticles, 113–139
36. Matyushkin LB, Ryzhov OA, Aleksandrova OA, Moshnikov VA (2016) Synthesis of metal and semiconductor nanoparticles in a flow of immiscible liquids. Semiconductors 50 (6):844–847
37. Bennet D, Kim S (2014) Polymer nanoparticles for smart drug delivery. Application of nanotechnology in drug delivery, 257–310, Chapter 8
38. García-Pinel B, Porras-Alcalá C, Ortega-Rodríguez A, Sarabia F, Prados J, Melguizo C, López-Romero JM (2019) Lipid-based nanoparticles: application and recent advances in cancer treatment. Nanomaterials 9(4):638
39. Prathna TC, Mathew L, Chandrasekaran N et al (2010) Biomimetic synthesis of nanoparticles: science, technology and applicability. In: Biomimetics learning from nature. InTech
40. Ahlawat DS, Kumari R, Rachna Yadav I (2014) Synthesis and characterization of sol-gel prepared silver nanoparticles. Int J Nanosci 13:1450004. https://doi.org/10.1142/s0219581x14500045
41. Li J, Wu Q, Wu J (2015) Synthesis of nanoparticles via solvothermal and hydrothermal methods. In: Handbook of nanoparticles. Springer, Berlin, pp 1–28
42. Manikam VR, Cheong KY, Razak KA (2011) Chemical reduction methods for synthesizing Ag and Al nanoparticles and their respective nanoalloys. Mater Sci Eng B 176:187–203. https://doi.org/10.1016/j.mseb.2010.11.006
43. Reza Sadrolhosseini A, Adzir Mahdi M, Alizadeh F, Abdul Rashid S (2019) Laser ablation technique for synthesis of metal nanoparticle in liquid. In: Laser technology and its applications. IntechOpen
44. Suryanarayana C, Prabhu B (2007) Synthesis of nanostructured materials by inert-gas condensation methods. In: Nanostructured materials. Elsevier, Netherlands, pp 47–90
45. Singh P, Kim Y-J, Zhang D, Yang D-C (2016) Biological synthesis of nanoparticles from plants and microorganisms. Trends Biotechnol 34:588–599. https://doi.org/10.1016/j.tibtech.2016.02.006
46. Iravani S, Zolfaghari B (2013) Green synthesis of silver nanoparticles usingpinus eldaricabark extract. Biomed Res Int 2013:1–5. https://doi.org/10.1155/2013/639725
47. Song JY, Kwon E-Y, Kim BS (2009) Biological synthesis of platinum nanoparticles using Diopyros kaki leaf extract. Bioprocess Biosyst Eng 33:159–164. https://doi.org/10.1007/s00449-009-0373-2
48. Das RK, Gogoi N, Bora U (2011) Green synthesis of gold nanoparticles using Nyctanthes arbortristis flower extract. Bioprocess Biosyst Eng 34:615–619. https://doi.org/10.1007/s00449-010-0510-y
49. Li S, Shen Y, Xie A et al (2007) Rapid, room-temperature synthesis of amorphous selenium/protein composites using Capsicum annuum Lextract. Nanotechnol 18:405101. https://doi.org/10.1088/0957-4484/18/40/405101
50. Riis T, Sand-Jensen K (2006) Dispersal of plant fragments in small streams. Freshw Biol 51:274–286. https://doi.org/10.1111/j.1365-2427.2005.01496.x

51. Hansen LP, Sargent TJ (1980) Formulating and estimating dynamic linear rational expectations models. J Econ Dyn Control 2:7–46
52. Carr KL (1997) US Patent 5,616,268. US Patent and Trademark Office, Washington, DC
53. Haury J, Aïdara LG (1999). Macrophyte cover and standing crop in the River Scorff and its tributaries (Brittany, northwestern France): scale, patterns and process. In: Biology, ecology and management of aquatic plants. Springer, Dordrecht, pp 109–115
54. Kennamer JD, Honnold J, Bradford J, Hendricks M (2000) Differences in disclosure of sexuality among African American and White gay/bisexual men: implications for HIV/AIDS prevention. AIDS Educ Prev 12(6):519
55. Riis L, Bellotti AC, Castaño O (2003) In field damage of high and low cyanogenic cassava due to a generalist insect herbivore Cyrtomenus bergi (Hemiptera: Cydnidae). J Econ Entomol 96(6):1915–1921
56. Akhtar, ABT, Yasar A, Ali R, Irfan R (2017) Phytoremediation using aquatic macrophytes. In: Phytoremediation. Springer, Cham, pp 259–276
57. Ebel A, Memmesheimer M, Jakobs HJ, Feldmann H (2007) Advanced air pollution models and their application to risk and impact assessment. In: Ebel A, Davitashvili T (eds) Air, water and soil quality modelling for risk and impact assessment. Springer, Dordrecht, pp 83–92
58. Fang YY, Yang XE, Chang HQ, Pu PM, Ding XF, Rengel Z (2007) Phytoremediation of nitrogen-polluted water using water hyacinth. J Plant Nutr 30(11):1753–1765
59. Mishra S, Sharma S, Vasudevan P (2008) Comparative effect of biofertilizers on fodder production and quality in guinea grass (*Panicum maximum Jacq.*). J Sci Food Agric 88 (9):1667–1673
60. Giraldo E, Garzon A (2002) The potential for water hyacinth to improve the quality of bogota river water in the muna reservoir: comparison with the performance of waste stabilization ponds. Water Sci Technol 42:103–110
61. Singh D, Tiwari A, Gupta R (2012) Phytoremediation of lead from wastewater using aquatic plants. J Agric Technol 8(1):1–11
62. David GK, Blondeau M, Schiltz S, Penel A, Lewit-Bentley (2003) YodA from *Escherichia coli* is a metal-binding, lipocalin-like protein. J Biol Chem 278:43728–43735
63. Irfan R (2012) Removal efficiency of toxic metals by aquatic macrophytes *Eichhornia crassipes* and *Pistia stratiotes*
64. Rai PK, Tripathi BD (2009) Comparative assessment of *Azolla pinnata* and *Vallisneria spiralis* in Hg removal from G.B. Pant Sagar of Singrauli Industrial region, India. Environ Monit Assess 148:75–84
65. Gorelova SV, Frontasyeva MV (2017) The use of higher plants in biomonitoring and environmental bioremediation. In: Phytoremediation. Springer, Cham, pp 103–155
66. Miretzky P, Saralegui A, Fernandez Cirelli A (2004) Aquatic macrophytes potential for the simultaneous removal of heavy metals (Buenos Aires, Argentina). Chemosphere 57:997–1005
67. Chong Y, Hu H, Qian Y (2003) Effects of inorganic nitrogen compounds and pH on the growth of duckweed. J Environ Sci 24:35–40
68. Denny H, Wilkins D (1987) Zinc tolerance in Betula spp. II. Microanalytical studies of zinc uptake into root tissues. New Physician 106:525–534
69. Gallardo T, Maria, Benson F, Robert, Martin F (1999) Lead accumulation by three aquatic plants. Symposia papers presented before the division of environmental chemistry. Am Chem Soc 39(2):46–47
70. Cardwell AJ, Hawker DW, Greenway M (2002) Metal accumulation in aquatic macrophytes from Southeast Queensland, Australia. Chemosphere 48:653–663
71. Hammouda O, Gaber A, Abdel-Raouf N (1995) Microalgae and wastewater treatment. Ecotoxicol Environ Saf 31(3):205–210
72. Arreghini S, Cabo LD, Iorio AFD (2006) Phytoremediation of two types of sediment contaminated with Zn by *Schoenoplectus americanus*. Int J Phytoremediation 8:223–232

73. Samecka-Cymerman A, Kempers J (1996) Bioaccumulation of heavy metals by aquatic macrophytes around Wrocław, Poland. Ecotoxicol Environ Saf 43:242–247
74. Mishra VK, Tripathi BD (2008) Concurrent removal and accumulation of heavy metals by the three aquatic macrophytes. Bioresour Technol 99(15):7091–7097
75. da Gomes MAC, Hauser-Davis RA, de Souza AN, Vitória AP (2016) Metal phytoremediation: general strategies, genetically modified plants and applications in metal nanoparticle contamination. Ecotoxicol Environ Saf 134:133–147. https://doi.org/10.1016/j.ecoenv.2016.08.024
76. Rafati M, Khorasani N, Moattar F, Shirvany A, Moraghebi F, Hosseinzadeh S (2011) Phytoremediation potential of *Populus alba* and *Morus alba* for cadmium, chromuim and nickel absorption from polluted soil. Int J Environ Res 5(4):961–970
77. Blaylock MJ, Huang JW (2000) Phytoextraction of metals. Phytoremediation of toxic metals: using plants to clean-up the environment. Wiley, New York, pp 53–70
78. Berti WR, Cunningham SD (2000) Phytostabilization of metals. Phytoremediation of toxic metals: using plants to clean-up the environment. Wiley, New York, pp 71–88
79. Schnoor JL (2000) Phytostabilization of metals using hybrid poplar trees. Phytoremediation of toxic metals: using plants to clean-up the environment. Wiley, New York, pp 133–150
80. Erakhrumen Agbontalor A (2007) Review phytoremediation: an environmentally sound technology for pollution prevention, control and remediation in developing countries. Educ Res Rev 2(7):151–156
81. Etim EE (2012) Review: phytoremediation and its mechanisms. Int J Environ Bioenergy 2 (3):120–136
82. Rawat K, Fulekar MH, Pathak B (2012) Rhizofiltration: a green technology for remediation of heavy metals. Int J Innov in Biosci 2(4):193–199
83. Ensley BD (2000) Rationale for the use of phytoremediation. Phytoremediation of toxic metals: using plants to clean-up the environment. Wiley, New York, pp 205–210
84. Liu P, Qiu GL, Shang LH (2007) Phytoremediation of mercury contaminated soil: a review. Chinese J Ecol 6:27
85. Banuelos GS, Lin ZQ, Wu L, Terry N (2002) Phytoremediation of selenium-contaminated soils and waters: fundamentals and future prospects. Rev Environ Health 17(4):291–306
86. Rugh C, Bizily SP, Meagher RB (2000) Phytoreduction of environmental mercury pollution. Phytoremediation of toxic metals: using plants to clean-up the environment. Wiley, New York, pp 151–170
87. Padmavathiamma PK, Li LY (2007) Phytoremediation technology: hyperaccumulation metals in plants. Water Air Soil Pollut 184:105–126
88. Chakravorty A, Rather GA, Ali A et al (2020) Nano approach. In: Ethnopharmacological investigation of Indian spices. IGI Global, pp 205–241
89. Myakonkaya O, Hu Z, Nazar MF, Eastoe J (2010) Recycling functional colloids and nanoparticles. Chem Eur J 16:11784–11790. https://doi.org/10.1002/chem.201000942
90. SCENIHR (2010) Opinion on the scientific basis for the definition of the term "nanomaterial". Scientific committee on emerging and newly identified health risks (SCENIHR), European Commission Brussels, Belgium
91. Chang C-H, Paul BK, Remcho VT et al (2008) Synthesis and post-processing of nanomaterials using microreaction technology. J Nanopart Res 10:965–980. https://doi.org/10.1007/s11051-007-9355-y
92. EU (2011) Commission recommendation of 18 October 2011 on the definition of nanomaterial (2011/696/EU). Official J L 275:38–40
93. Oliveira RL, Kiyohara PK, Rossi LM (2010) High performance magnetic separation of gold nanoparticles for catalyticoxidation of alcohols. Green Chem 12:144–149. https://doi.org/10.1039/b916825g
94. Basegio T, Haas C, Pokorny A et al (2006) Production of materials with alumina and ashes from incineration of chromium tanned leather shavings: environmental and technical aspects. J Hazard Mater 137:1156–1164. https://doi.org/10.1016/j.jhazmat.2006.03.059

95. Maximova N, Dahl O (2007) A set up of a modern analytical laboratory for wastewaters from pulp and paper industry. Chem Soc Rev 36:1323. https://doi.org/10.1039/b515226g
96. Ellis JB (2006) Pharmaceutical and personal care products (PPCPs) in urban receiving waters. Environ Pollut 144:184–189. https://doi.org/10.1016/j.envpol.2005.12.018
97. Baun A, Hartmann NB, Grieger K, Kusk KO (2008) Ecotoxicity of engineered nanoparticles to aquatic invertebrates: a brief review and recommendations for future toxicity testing. Ecotoxicology 17:387–395. https://doi.org/10.1007/s10646-008-0208-y
98. SCENIHR (2006) Modified opinion (after public consultation) on the appropriateness of existing methodologies to assess the potential risks associated with engineered and adventitious products of nanotechnologies
99. Internet page. http://www.nanotechia.org/sectors/recycling-waste
100. Mudgal S, Monier V, Long LV, André N, Anderson G (2011, August 11) European Commission (DG ENV)—Study on coherence of waste legislation—Final report
101. Edwards J, Harford A, Wright P, Priestly B (2007) Current OHS best practices for the Australian nanotechnology industry: a position paper by the NanoSafe Australia Network. J Occup Health Saf Aust New Zeal 23(4):315. http://mams.rmit.edu.au/72nuxiavskpg.pdf
102. Rashidi O, Ramya R, Baharuddin ZM, Hashim KSH, Yaman M (2015) Response of *Lemna minor* and *Salninia natans* as phyroremediation agents towards Fe, Cu and Zn toxicities via in vivo model system. Jurnal Teknologi 77(30):101–109
103. Yadav R, Arora P, Kumar S, Chaudhury A (2010) Perspectives for genetic engineering of poplars for enhanced phytoremediation abilities. Ecotoxicology 19:1574–1588. https://doi.org/10.1007/s10646-010-0543-7
104. Krämer U (2018) The plants that suck up metal. Ger Res 40:18–23. https://doi.org/10.1002/germ.201870307
105. Javed MT, Tanwir K, Akram MS et al (2019) Chapter 20—phytoremediation of cadmium-polluted water/sediment by aquatic macrophytes: role of plant-induced pH changes. In: Hasanuzzaman M, Prasad MNV, Fujita M (eds) Cadmium toxicity and tolerance in plants. Academic Press, pp 495–529
106. Dhanwal P, Kumar A, Dudeja S et al (2017) Recent advances in phytoremediation technology. In: Kumar R, Sharma AK, Ahluwalia SS (eds) Advances in environmental biotechnology. Springer, Singapore, pp 227–241
107. Pajević S, Borišev M, Nikolić N et al (2016) Phytoextraction of heavy metals by fast-growing trees: a review. In: Phytoremediation. Springer, Berlin, pp 29–64
108. Ghosh M, Singh SP (2005) A review on phytoremediation of heavy metals and utilization of its by products
109. Jadia CD, Fulekar M (2008) Phytoremediation: the application of vermicompost to remove zinc, cadmium, copper, nickel and lead by sunflower plant
110. Garbisu C, Alkorta I (2001) Phytoextraction: a cost-effective plant-based technology for the removal of metals from the environment. Bioresour Technol 77:229–236. https://doi.org/10.1016/S0960-8524(00)00108-5
111. Vandenhove H, Hees MV, Winckel SV (2001) Feasibility of phytoextraction to clean up low-level uranium-contaminated soil. Int J Phytoremediation 3:301–320. https://doi.org/10.1080/15226510108500061
112. Wan X, Lei M, Chen T (2016) Cost–benefit calculation of phytoremediation technology for heavy-metal-contaminated soil. Sci Total Environ 563–564:796–802. https://doi.org/10.1016/j.scitotenv.2015.12.080
113. Sytar O, Brestic M, Taran N, Zivcak M (2016) Chapter 14—plants used for biomonitoring and phytoremediation of trace elements in soil and water. In: Ahmad P (ed) Plant metal interaction. Elsevier, Netherlands, pp 361–384
114. Usman K, Al-Ghouti MA, Abu-Dieyeh MH (2018) Phytoremediation: halophytes as promising heavy metal hyperaccumulators. Heavy Met. https://doi.org/10.5772/intechopen.73879

115. Adrees M, Ali S, Rizwan M et al (2015) Mechanisms of silicon-mediated alleviation of heavy metal toxicity in plants: a review. Ecotoxicol Environ Saf 119:186–197. https://doi.org/10.1016/j.ecoenv.2015.05.011
116. Rakhshaee R, Giahi M, Pourahmad A (2009) Studying effect of cell wall's carboxyl–carboxylate ratio change of Lemna minor to remove heavy metals from aqueous solution. J Hazard Mater 163:165–173. https://doi.org/10.1016/j.jhazmat.2008.06.074
117. Tian Y, Zhang H (2016) Producing biogas from agricultural residues generated during phytoremediation process: possibility, threshold, and challenges. Int J Green Energy 13:1556–1563. https://doi.org/10.1080/15435075.2016.1206017
118. Ramamurthy AS, Memarian R (2012) Phytoremediation of mixed soil contaminants. Water Air Soil Pollut 223:511–518. https://doi.org/10.1007/s11270-011-0878-6
119. MacCormack TJ, Goss GG (2008) Identifying and predicting biological risks associated with manufactured nanoparticles in aquatic ecosystems. J Ind Ecol 12:286–296. https://doi.org/10.1111/j.1530-9290.2008.00041.x

Advanced Waste Recycling Technologies for Manufacturing of Nanomaterials for Green Energy Applications

Tahany Mahmoud, Mohamed A. Sayed, A. A. Ragab, and Eslam A. Mohamed

Abstract The effect of waste accumulation can be enormously violent for several societies in developing countries. In the overworld, this problem becomes more difficult as there are no obvious specific strategies for actual solid waste management that causes severe environmental hazards. This chapter discusses the recent advances of wastes recycling environmentally friendly technologies to provide economic value toward reducing the high cost and additional ways for nanomaterial production in the petroleum field. It shows how carbon nanostructures can be formed from different waste materials such as waste natural oil, plastic wastes, heavy oil residue, waste engine oil, deoiled asphalt, and scrap tire that have the potential to cause incredible environmental damage in the form of water, air, and land pollution. Particular sorts of this wastes can be reused. However, most of them are left in landfill sites; waste recycling approach has a tremendous economic value besides to their environmental impact: for example, waste reduction, resource conservation, energy conservation, reduction gas emissions from the greenhouse, and reducing the extent of pollution in air and water sources. The chapter is consisting of three parts. Part one is describing carbon nanostructures such as carbon nanotubes, fibers, porous carbon, and microspheres that can be produced from different waste materials. The second part deals with a review of waste materials in the petroleum field that has the probable to cause incredible environmental damage in the form of water, air, and land pollution. Finally, the third part will discuss the multidisciplinary green approach toward the acquisition of high-value carbon-based nanomaterials as a natural precursor by using waste materials.

Keywords Activated carbon · Waste · Green energy recycling technologies · Nanomaterials manufacturing

---

T. Mahmoud (✉) · A. A. Ragab · E. A. Mohamed (✉)
Petroleum Application Department, Egyptian Petroleum Research Institute, Cairo, Egypt
e-mail: tahany.mahmoud.a@gmail.com

E. A. Mohamed
e-mail: eslamazmy60@yahoo.com

M. A. Sayed
Refining Department, Egyptian Petroleum Research Institute, Cairo, Egypt

A. S. H. Makhlouf and G. A. M. Ali (eds.), *Waste Recycling Technologies for Nanomaterials Manufacturing*, Topics in Mining, Metallurgy and Materials Engineering, https://doi.org/10.1007/978-3-030-68031-2_21

List of Abbreviations

AC	Activated carbon
LOI	Ignition loss
PCBs	Printed circuit boards
WCOs	Waste cooking oils

1 Introduction

There is a general awareness that nanotechnologies will have an important influence on emerging clean and green technologies with significant environmental benefits. The best models are the usage of nanotechnology in areas extending from water treatment to hydrogen applications and energy breakthroughs. Renewable energy applications probably are the scopes where nanotechnology will create its first large-scale commercial revolutions. Green nanotechnology can proactively affect the designing of nanomaterials and products by removing or diminishing pollution from the manufacture of the nanomaterials, taking a life cycle style to nanoproducts to assesses and alleviate where environmental effects might take place in the product series, design toxicity out of nanomaterials and using nanomaterials to treat or remediate present environmental problems. Regardless, the noticeable fields of utilization nanomaterials in the extents of solar biofuels, cells, and fuel cells, green nanotechnology applications may include a clean production method, like manufacturing nanoparticles with sunlight or the reusing of waste materials from industry into nanomaterials, such as the production of nanotubes from diesel soot. An extensive scope of techniques involving hybrid, biological, physical, and chemical techniques was utilized to produce nanoparticles (Fig. 1). Commercial activated carbon (AC) is the desired adsorbent for micropollutants elimination from the aqueous phase and other numerous benefits such as energy storage, battery recycling, and energy conversion; nevertheless, its common usage is limited owing to high related costs [1–4]. To reduce management costs, efforts have been ready to find low-cost precursors of unconventional AC, such as waste materials. Several types of research describe the usage of waste materials for the synthesis of AC, but these investigations are controlled with each kind of wastes, synthesis methods, or applications in different fields [4–7]. The existing chapter illustrates and assesses literature devoted both to the manufacturing of AC via recycling several kinds of waste materials and to its usage in green energy. Traditional (from the wood industry and agriculture) and unconventional (from the industrial activities and municipal) wastes can be recycled to prepare AC, which can be practical in numerous processes of treatment.

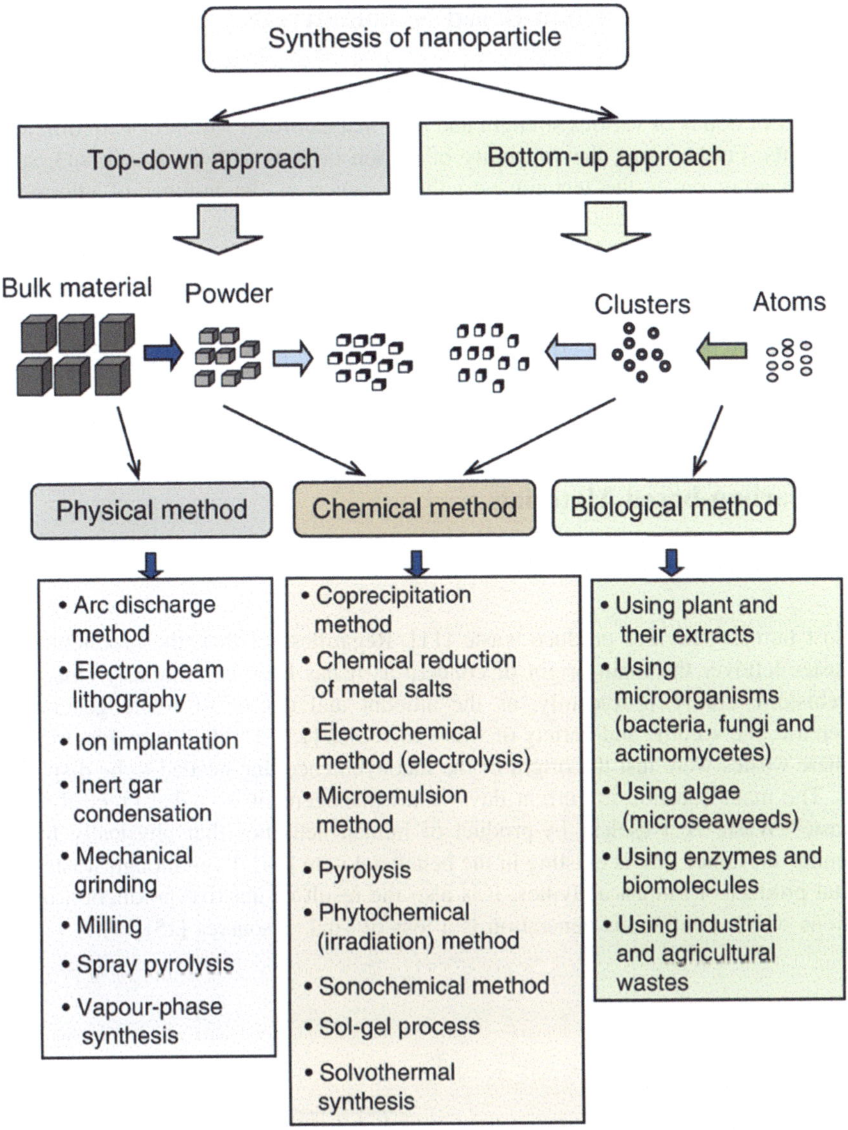

Fig. 1 Various methodologies and strategies for preparing nanoparticles. Adapted with permission from Ref. [8] Copyright 2014, Hindawi Publishing Corporation

2 Carbon and Carbon-Based Nanomaterials

Carbon is the most diverse element in the periodic table [9], due to the various number of bonds of various strength and type that can form with it or with different elements. Furthermore, the capability of carbon orbitals to hybridize in sp3, sp2, and sp array cover the technique to the presence of the number of allotropes (Fig. 2).

Until now, the natural three arising allotropes of carbon (graphite, diamond, and amorphous carbon), have been combined by further ones arising from preparation methods (for example, carbon nanotubes, graphene, nanodiamonds, carbon nanohorns, fullerenes (Fig. 3)) [10].

3 Waste Materials as Carbon Sources to produce Carbon-based Materials

3.1 The Meaning of Waste

Most human activities produce waste [11]. Regardless of that, the production of wastes leftover the main origin of concern as it has been continuously since the prehistoric era [12]. Recently, as the amount and degree of waste generation increase, the volume and variety of waste increases [13]. Unlike the prehistoric era where wastes were just the origin of the inconvenience that needed to be disposed of. The main question in current day wastes management is—what precisely is a waste? Waste is a useless by-product of human activities that physically has a similar substance that is existing in the beneficial item [14]. Even though waste is a vital product of human activities, it is also the result of inactive production operations whose continuous generation is a loss of vital resources [15].

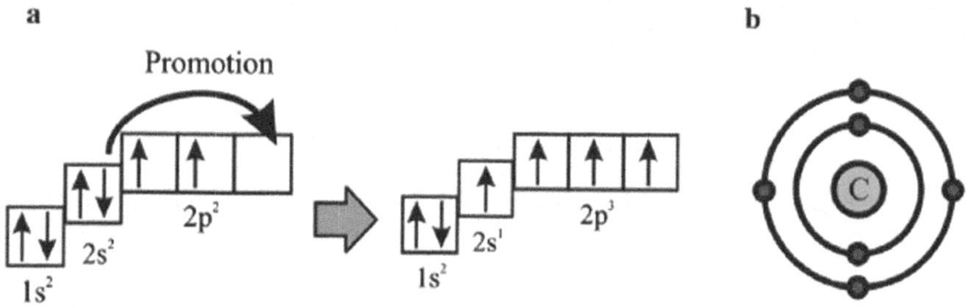

Fig. 2 Structure of a carbon molecule and carbon-based nanoparticles **a** electronic design of a carbon molecule after and before the promotion of one s-electron; **b** schematic portrayal of a carbon molecule structure with two-electron orbitals around the six electrons and nucleus distributed on them. Adapted with permission from Ref. [10] Copyright 2007, Royal Society of Chemistry

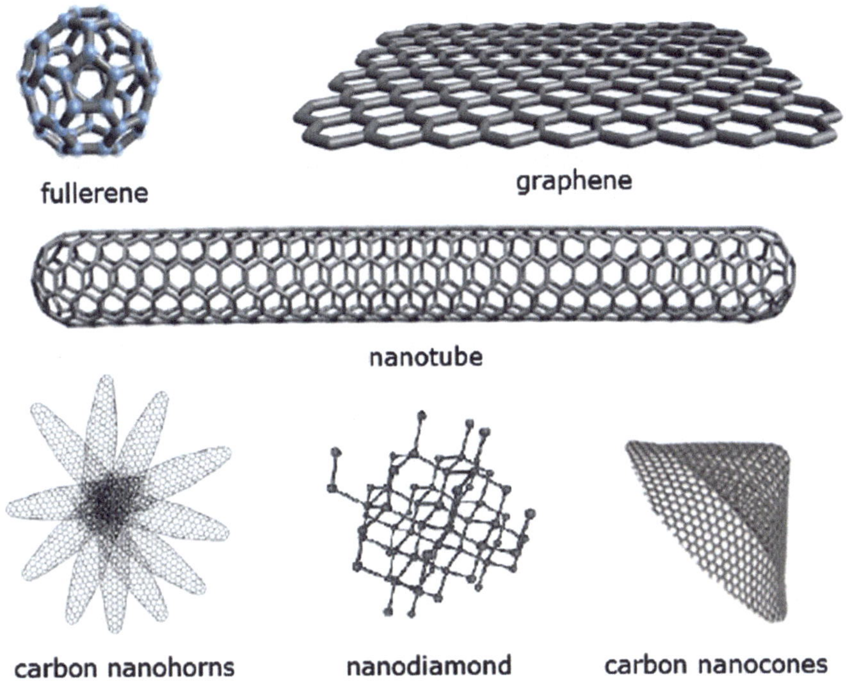

Fig. 3 Carbon allotropes derived from preparation methods. Adapted with permission from Ref. [10] Copyright 2007, Royal Society of Chemistry

3.2 *Classification and Types of Waste*

Waste arises in various forms, and its characterization can be stated in different forms. Some general characteristics used in the classification of waste include the physical properties, physical states, reusable and biodegradable potentials, the degree of environmental impact, and source of production [16, 17]. Wastes can be categorized according to their sources into agricultural waste, domestic and industrial waste, demolition and construction waste, and commercial waste. Likewise, waste can be characterized dependent on their environmental effect on hazardous waste and non-hazardous waste. According to the physical properties, waste can be categorized approximately to three major types; these are liquid, solid, and gaseous waste [14].

3.3 *Solid Waste*

Solid waste is useless or undesirable solid materials created by human activities (industrial, residential, and commercial activities in a known area) that are in a solid or semisolid form [18]. It might be classified by its source (industrial, local, agricultural, commercial) [19, 20], as per its substance (glass, plastic, natural materials, paper, metal, etc.), or as indicated by its hazardous potential (flammable, infectious, poisonous, non-toxic, radioactive, and so on). The expression municipal solid waste is generally supposed to involve all of the waste arises in a community with the exemption of waste created by treatment plants, municipal services, agricultural, and industrial processes [21–23].

3.4 *Liquid Waste*

Liquid waste is a main problem on the world, owing to about 71% (surface of the earth) is contain water. As indicated by the environmental protection agency, liquid waste is defined as any waste material that passes the meaning of a liquid. On the other hand, liquid waste can be defined as any shape of liquid residue that is dangerous for the environment or people. It can be massive or sludgy, or even simply liquid, such as with lab waste. Commonly, this waste shape comes from cars, restaurants, homes, any facility that involve washing instrument or industrial buildings which used tank-clearing processes or laboratories. Now, it is established in septic tanks, grease traps, surfactants, prescribed waste, wash-waters, photographic waste, oily water, or the more exemplary clinical or medical wastes, solvents, resins, paint, inks and dyes, pesticides, and chemical and laboratory wastes.

4 Environmental and Health Impacts of Waste

Waste that is not completely managed, particularly excreta and other solid and liquid waste collected from households and the community, is an international concern in expressions of environmental pollution, economic sustainability and public inclusion [24, 25], which need incorporated appreciation and holistic approaches for its solution [26]. Waste management has appeared as a main environmental hazard for cities in developing countries around the world. In an investigation showed by UNDP in (1997), more than one hundred and fifty mayors over the world classify solid waste disposal issues their second most imperious urban challenges exceed only via unemployment and surveyed by urban poverty [27]. In developing countries, a significant percentage of urban waste is dumped both on the streets and street sides, non-supported dump sites and inopen locales that destructively influence kind ecological disposition or in the waterways evacuation system [28]. Actually, wastes pose several dangers to community health and harmfully affect fauna and flora in addition to the environment mainly when it is not applicably collected and disposed of. Unrestrained disposal produces severe heavy metals contamination taking place in the water, plants, and soil [29]. Plastic waste is another cause of ill health when burnt or recycled inefficiently.

Waste, therefore, needs to have an effective management system in place, beginning from the source of generation. Wastes that end up in water bodies affect all life forms standing in water. Also, they could cause injury to the animals which drink from this contaminated water. Dangerous chemicals that dumped into the soil (impurities) could hurt plants when they hold the pollutants. If persons eat animals and plants which have been in interaction with contaminated soils, there may be a bad influence on their health. For example, chemicals and sludge waste from leather tanning heavy metals, which when disposed of in soil or water, are absorbed by plants. The bioaccumulation of these heavy metals in animals and humans is known to cause health impacts. Leachate from used batteries discarded in the environment also has damaging impact upon exposure. Air pollution owing to the burning of wastes could make breathing complications and some opposite health effects as impurities are absorbed to some parts in the body from the lungs. Leachate from unscientifically disposed waste on land forms a very damaging blend of chemicals that might produce from dangerous materials inflowing groundwater, soil, or surface water. The increasing amount of waste and lack of an adequate and efficient system of waste management in villages contributes to the overall health and hygiene problems in these areas. We are quite familiar with the frequent outbreak of diseases such as malaria, diarrhea, dysentery, cholera, etc., especially among young children. Therefore, the most urgent need for waste management comes from the point of view of health.

From an environmental context, the large amount of agricultural waste is a resource that can enrich the soil. However, if left unattended and mixed with other kinds of waste, this can pollute the environment. A clean environment is essential for healthy living. The decaying waste, including dung, can contaminate the soil, water, and air, causing nitrate pollution of soil and water and the release of methane and hydrogen sulfide into the atmosphere, depending on the type of waste whether chemical, physical or biological, and the degree of pollution will vary. Some of the pollutants may be poisonous or toxic and may affect the lives of the people and animals. Some of the pollutants may be absorbed by the crop plants, and edible portions may become unfit for consumption, while others may pollute the air or water.

5 Waste Management

A great volume of organic matter is produced from dairy farms and animal shelters, agricultural activities, manufacturing activities, and agro-based trading in rural areas. This appreciated resource can be used by composting/transforming it into a value-added end product called manure. The main objective of composting organic waste should not be to get rid of solid organic waste but to produce high-quality fertilizer to feed our soil nutrients-organic-hungry materials. Plastic waste recycling also has major prospects for resource preservation and greenhouse. Gas emissions decrease, for example, the conversion of plastic waste into diesel fuel. This resource protection objective is extremely significant for most of the local and national and governments, where plastic waste recycling also has major prospects for resource preservation and greenhouse gas emissions decrease, for example, the conversion of plastic waste into diesel fuel. This resource conservation goal is extremely significant for a large portion of the local and national governments, where fast economic and industrialization progress is putting much force on natural resources.

5.1 *Importance of Waste Management*

Safe disposal of waste can lead to:

- **Health benefits** from safe disposal of waste that pollutes the environment.
- **Economic benefits** and environmental benefits through recycling/reuse of products that would have been rejected as waste.
- **Aesthetic benefits** from a clean environment.

5.2 *Solid Waste Management*

5.2.1 Principal Phases of Solid Waste Management

The activities involved in solid waste management from the point of generation to the final disposal can be accumulated into six main stages [18]:

- Waste identification.
- Handling, separation, and capacity at the source.
- Collection.
- Transport or transfer.
- Processing and transformation.
- Disposal.

In any case, a suitable management of solid waste is accomplished through the progress and implementation of variation of tools for example enforcement procedures, legislations, recycling, and ability of waste recovery, just as the existence of infra-structures and facilities for safe treatment, dealing with and removal of such waste [22].

5.2.2 Maintainable Technique for Solid Waste Management

It must believe an incorporated waste management framework which can be characterized as the decision and use of right strategies, projects, and technologies to attain explicit waste management goals which concur with the regional conditions and requirements [22].

Sustainable Methodology for Solid Waste Management

It found four essential management alternatives:

(A) **Source reduction**

It is the most significant factor of any dynamic management system. Reduction includes any activity that decreases the toxicity or volume of waste preceding removal or treatment [23].

(B) **Waste recovery**

Recycling of materials concern the recovery of specific waste kinds to be utilized in new products, as well as the alteration of specific types of waste into energy.

(C) **Waste treatment**

When waste cannot be stopped or limited through recycling, then, there is a requirement for systems planned to reduce waste toxicity or volume by utilizing treatment advances before removal [21]. Treatment techniques are chosen depend on the proportion, form, and type of waste materials and include biological and thermal treatment.

(D) **Landfill disposal**

Waste management processes cannot totally exclude the requirement for landfills. Up till now, waste removal in landfills stay the smallest required methodology. However, it found two fundamental landfill forms for suitable waste removal (Sanitary/secure landfills and Controlled dumps) [24].

5.3 *Liquid Waste Management*

It is the main class of waste management and cannot be easily hand-picked up and removed from an environment. It spreads out and without difficulty contaminates different sources of liquid. This form of waste can additionally soak into groundwater and soil which effect on the plants, the animals, and the humans in the area of the pollution.

6 Green Approach Toward the Acquisition of Carbon-Based Nanomaterial

Conventional sources for carbon materials production are toluene, ethane, acetylene, methane, and one more hydrocarbon depend on petroleum product. These sources have numerous hindrances, for example, expensive and restricted accessibility. It found numerous researches on alternative precursors for carbon materials production utilizing waste materials as carbon sources such as utilizing natural bio-hydrocarbon sources for example: turpentine [30], eucalyptus [31], neem [32], palm [2, 4, 33, 34], sesame oil, olive, and corn [35, 36]. These sources give an environmentally methodology and less expensive but less success because it beneficial in human life. Consequently, the survey on other powerful sources was ready utilizing waste materials. This chapter show a survey on some kinds of waste materials, for example, printed circuit boards, waste pyrolysis oil, waste cooking and engine oil, glycerol, waste normal oil, chicken fat, plastic wastes and scrap tire elastic as beginning material for production of carbon-based nanomaterials. A concise clarification of their preparation technique is additionally illustrated. Likewise, aromatic hydrocarbon that contain rich carbon amount such (heavy oil residue and asphalt) can be utilized as carbon sources to create carbon-based nanomaterials.

6.1 Activated Carbon-Supported Materials

The demand on AC is increasing in many applications. Because of absence of the main basic materials, such hard coal, coconut shells or wood in numerous countries different biomass matters were used for production of AC. Waste biomass like nut shells, straw matters, coffee grounds, olive stones what's more, spent grain is changed over thermally in two stages. First the biomass go through the process of pyrolysis at temp. from 500 to 600 °C in presence of N_2 gas. The liquid and gaseous pyrolysis items can be utilized powerfully for electricity creation. Second, treating char (solid residue) in an activation procedure at temp. from 800 to 1000 ° C in steam air so as to improve the char surface area which was investigated by standard BET technique. The expansion of surface area relies upon the form of biomass and on the initiation factors.

6.1.1 Origin and Source of Activated Carbon

Firstly, AC has been utilized as adsorbent by Egyptians as purifying agent and medicinal purposes. Karl Wilhelm (1773) used the charcoal to adsorb the gases. Following a couple of years, using AC in sugar industry as color removal agent. In the twentieth century, in Germany, the main AC unit was established. It was employed as color removal/sugar purifying agent. The production of AC from plants was used as color removal agent for pharmaceutical and food items in 1900. Now, AC assumes a dynamic job in several applications for example: medication, purification, fuel storage, air filtration, hydrogen storage, metal decaffeination and extraction, and water sewage treatment.

6.1.2 Activated Carbon Preparation

AC has been mostly used for energy storage, air cleaning, and water treatment mainly for the expulsion of harmful metal particles ions from its properties of pore textural such large specific surface area, average pore width, the volume of micropore and resources of the outside functional groups (active) that have oxygen [37]. AC was prepared from several carbonaceous materials, for example, fruit and vegetable wastes, agriculture wastes, electronic and biological wastes, plastic wastes, and so on, by the carbonization procedure after that chemical or physical activation [38–40].

Activated Carbon from Agricultural Wastes

AC, which prepared from different waste materials, appears to be economical. The sources of AC that depend on waste can be classified into two classifications: human-made and natural wastes. Firstly, the natural wastes are utilized natural materials like agriculture items, woods, natural products, seeds, and vegetables. Second, the human-made wastes made by human activities contain plastic, industrial, and electronic wastes.

Agricultural waste is undesirable materials produced absolutely from the outcomes of several farming tasks, which identified with the rising of harvests. Crop residue production is abundant where farming is rehearsed, and its creation is more than 3107 Mt/a for 25 varieties of legumes and 17 varieties of cereals and 3758 (Mt/a) for 27 food crops at the commercial level [41]. A lot of agricultural wastes, for example, corn cobs, rice husk, groundnut shell, tamarind seeds, etc., have been produced daily in several agricultural processes. In light of information from 227 countries, worldwide production of agricultural deposits for oats was assessed to be about 3.7 × 1012 t/a [41]. Waste management and removal expenses of agricultural waste materials could be reduced by using this material as the resource materials in various fields, for example, wastewater treatment, medical, engineering, and so on. These wastes are biodegradable and can be transformed into AC [42]. The plenty of agricultural waste produced worldwide stays a critical environmental issue, and AC preparation suggests the possibility of utilizing it as a renewable carbon source to produce AC (highly porous) (Fig. 4).

Shrestha et al. [43] used the agricultural waste seed to produce AC from Lapsi in India. They found higher adsorption of AC obtained by using Lapsi seed and made chemical activation then carbonization for Ni(II) and Pb(II) removal. Kadirvelu et al. synthesized AC using many solid wastes obtained from agricultural, for example, silk cotton hull, maize cob, and sago waste for adsorption Ni (II) from aqueous solution [44]. Similarly, Taer et al. [45] have prepared the AC electrode using coconut husk via a mix of physical and chemical activation techniques. The electrochemical properties of the electrode presented a high specific capacitance of 184 F g^{-1}. Furthermore, olive stone waste is a crude material utilized for AC manufacture, which is the superior material because it is plentiful and low-cost, as

Fig. 4 Sketch shows the preparation of activated carbon from agricultural waste materials

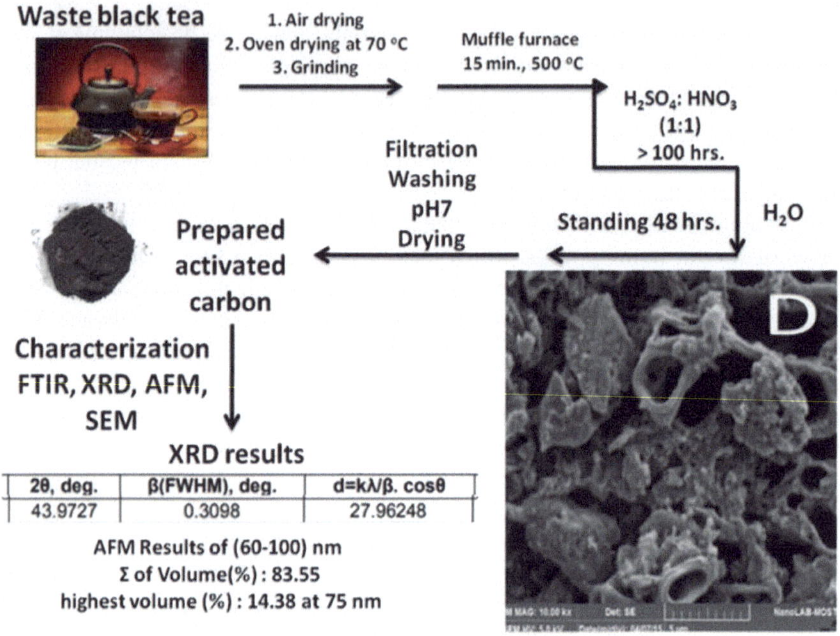

Fig. 5 Activated carbon preparation steps from waste cooked black tea. Adapted with permission from Ref. [48], Copyright 2018, ISRES Publishing

they are wealthy in carbon atoms (40–45 wt%). Also, Al-slaibi and collaborators have enhanced the techniques of the microwave for preparing AC from olive stone and applied it in the removal of Cd(II) [46], Chand and collaborators used a wheat straw to prepare AC by using the carbonization procedure and applied it in removal Cr(VI) from aqueous solution [47]. Hammud et al. [48] have prepared AC from waste cooked black tea and applied it as Chemical Dyes—Filter, as shown in Fig. 5.

Activated Carbon from Biological Wastes

The biological wastes found in nature are capable for self-replecation and may produce the effects according to organisms. The significant biological wastes incorporate cow-like skin acquired from leather industries, buffing dust, and so on, which may cause itself a grave problem until it recycled. Therefore, the preparation of AC from this method can be used in environmental remediation [49]. Lopez-Anton and collaborators have investigated the AC of bio-wastes obtained from the process of vegetable tanning and applied in Hg(II) adsorption in the process of oxy-combustion. The great extent of Hg oxidizes to Hg(II) below AC is significant to think about the economic mood, to decrease Hg(II) outflows through coal combustion procedure, and removal of 60% of Hg(II) was obtained [50].

Activated Carbon from Fruit Wastes

Fruit wastes arise during the fruit production processes when sorting and selecting are done. The progress of organic products may deliver two kinds of waste, such as the liquid waste of juice and the solid waste of stones, seeds, and so on. Several solid fruit wastes such as grape seeds, orange strips, palm shells, banana strips, pomegranate strips, medlar seeds, and so on are used to prepare AC for toxic removal from water [51, 52]. Now, the regard for palm shell has enhanced as it tends to be a unique hotspot to prepare excellent AC [53]. Likewise, Awwad and collaborators have synthesized AC from date seeds via chemical and physical activation [54]. Also, solid waste such as medlar seed waste that has a large potential for the synthesis of AC. The medlar is an apple formed organic product with a ruddy tinged shading that has various sizes with weight, 10–80 g and diameter, 1.5–3 cm. Medlar has been utilized as jam and jelly, which has been financially detected by the food industry. What is more, it also used for the treatment of bladder, kidney stones, and diuretic [55].

Activated Carbon from Plastic Wastes

Waste plastic materials is a manufactured organic amorphous solid materials which come about because of oil and gaseous petrol. Several waste plastic materials as elastic tires, bottles of polyethylene terephthalate (PET), polystyrene tires, and so on are transformed into AC and used for heavy metal removal from water. PET from bottle wastes were used to prepare AC through chemical activation (using KOH) and physical activation (using steam under limited condition). Similarly, Mendoza and collaborates have prepared AC from polyethylene terephthalate bottle wastes for Fe(III) removal [56].

Activated Carbon from Electronic Wastes

E-waste means disposed of electronic equipments. The source of e-wastes are cell phones, waste PCs, office electronic, and entertainment equipments, fridges, and TV devices which are destined for recycling. A minimal cost AC material obtained from several e-wastes, for example, printed circuit boards (PCBs) and disposal of printed track sheets have been used for cleaning water by removal of heavy metal [57]. The PCBs are central elements in the electronic items. The greatest eco-friendly innovations used for PCBs reusing split the PCBs into a metallic and non-metallic powder [58]. Xu and collaborators have concentrated on reusing the e-waste material to improve the AC for removal Cd(II) from wastewater. The most extreme take-up limit of the freshly determined PCB material for Cd(II) has found to be 2.1 mol g^{-1} [59]. The huge disposal of PCBs universal prompts the environmental dangers for a satisfactory solution. Hadi and collaborators have prepared AC from PCBs and applied it as an adsorbent for Cu(II), Zn(II), and Pb(II) removal [60] (Fig. 6).

Activated Carbon from Vegetable Wastes

It is generally contained several vegetable issues and can be decayed by microorganisms. Now, many vegetable wastes, for example, putrescible vegetables, pumpkin stem waste, soybean hulls, and so forth, have been used for the preparation of AC to remove harmful metal from water. Mostly, wastes are burnt in the atmosphere to obtain the capacity of the quick expulsion of water and soil contamination in the earth [61]. AC can be synthesized from different kinds of

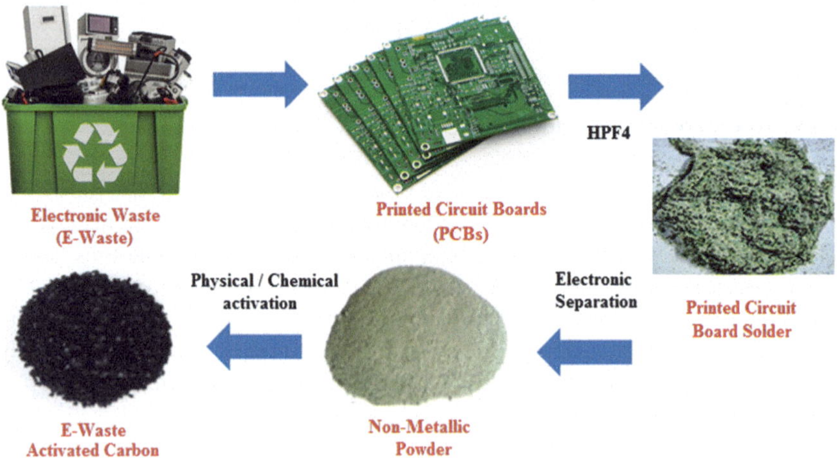

Fig. 6 Sketch shows the preparation of activated carbon from electronic waste materials

Fig. 7 The sketch shows the preparation of activated carbon from vegetable waste materials

vegetable waste materials as carrot remains. Carrot residue is made out of cellulose and lignin with the ability of removal to tie metal cations because of the existing functional groups such as phenolic and carboxylic gatherings [62]. The obtained carrot residue was dried for the time being at 60 °C, and the resulted AC was adjusted chemically for the utilization of the adsorption process (Fig. 7).

6.2 *Using Vegetable Wastes to Prepare AC and Their Application for Fabrication of Biodiesel from Waste Cooking Oils*

The growing production of WCOs from domestic as well as sources of manufacturing is an increasing problem everywhere in the world. The remainder has routinely discharged down the drain, consequential make big complications for wastewater management vegetation, or is combined with the nutrition sequence to animal nurturing, therefore, becoming a possible reason for personal medical issues [63]. There are numerous end-uses for this waste, as the manufacture of detergents or of energy by anaerobic breakdown, thermal cracking [64], and additional newly the manufacture of biofuel. The biofuel is unpolluted, naturally, non-toxic plus environmental fuel [65]. As well, biofuel was described as has excellent physicochemical properties such as high fire and flash point, low density (low viscosity), and remarkable octan number. Furthermore, biofuel may be utilized in the current engines devoid of adjustments [66]. There are numerous procedures to create biofuels using waste cooking oil as a source, for example, transesterification reaction [67], catalytic cracking (CC) [68], and pyrolysis [69, 70].

Consequently, catalytic cracking methods have broadly utilized in the conversion of waste cooking oil to eco-friendly gasoline. Several catalysts have been utilized in the CC method of waste cooking oils, for example, AC. AC has broadly utilized as a catalyst because its shapeless form (amorphus) which get there to create an extremely surface area and interior pore building. AC frequently equipped from biological substance which surrounds high carbon, so farmed waste has remarkable excellent as its low price [71, 72].

6.2.1 Activated Carbon Preparation

Processing of Peach Seeds

Sadeek et al. [73] prepared AC from peach seeds. Peach seeds were cleansed with condensed water numerous times then drying it by exposed it to warm temperature at 40 °C for 5 h and pulverizes by pull grinder to get average diameter particles around 0.2–0.4 mm. In order to the elimination of fats, amino acids, and contaminations in the beginning substance, the particles were emptied in flask contains a solution of 10% $ZnCl_2$ (10 g $ZnCl_2$/100 mL distilled water). Then, the matrix warming to 90 °C around two hours. Followed washed by condensed water and saturated solution of zinc chloride for one day, and lastly, cleansed numerous rounds with purified water till pure solution.

6.2.2 Conversion of Preserved Mixture to Activated Carbon

The gained treated matrix of peach seeds was converted to AC in muffle heater in five steps according to temperature ramps. The first two rises stayed at 150 as well as 250 °C for one and a half hours; then, two rises stayed at 350 and 450 °C for 60 min, but the most recent ramp was at 800 °C for ten min [73, 74]. The gotten AC has washed away with sufficiently condensed water till Cl^- ion free solution was gotten and tested by silver nitrite. The gotten AC was labeled as PAC.

6.2.3 Activated Carbon Doped by Transition Metals

AC (30 g) was cured by manganese sulfate 500 ml and a mixture of ferric chloride (500 ppm by weight) in the flask at warming around 60 °C for one day. Then the intermediate was left toward losing it is heat to ambient heat degree, then wash away and poured off the cured AC was achieved for numerous rounds. Lastly, the AC doped with metals was cleared and dehydrated at 50 °C around one day and labeled as Mn-PAC and Fe-PAC [73].

6.2.4 Waste Cooking Oil Cracking by Prepared Catalyst

Specification of Waste Cooking Oils

The waste cooking oils (WCOs) are used in the cracking process was described according to ASTM. The specification of the waste cooking oils show extraordinary acid value, 2.00 (mg KOH/g), kinematic viscosity, at 40 °C (34 mm^2 s^{-1}), oxidation constancy (18), iodine value (55), saponification significance (200), and its fatty acid composition were: myristic acid (0.5%), palmitic acid (21%), stearic acid (5%), oleic acid (53%), linoleic acid (13.9), eicosanoic acid (0.2). This information indicated the exceptional causticity and incredible unsaturation degree in the used oil [73].

Biofuel Physical Specification

The prepared catalysts Fe.PAC, Mn.PAC, and PAC were utilized in several ratio percentages as catalysts in the CC operation of WCOs to get biofuels. The biofuels were specification for their fuel feature counting: viscosity, density at 40 °C, fire point, cetane number, flash point, pour point, carbon residue, sulfur content, and ash or residue. The lowermost density (ρ) of the created biofuels were 0.8654, 0.8478, and 0.8112 g cm^{-3} by utilizing 3.0% from PAC, Mn.PAC and Fe.PAC catalysts ratio, separately. At that point, the PAC impetus made biofuel with the most extreme thickness at 0.8654 g cm^{-3} [73]. The lowermost SO_x predictable in the gained biofuels from the CC operation was 0.001, 0.0006, as well as 0.0009 wt% utilizing 5% of PAC, Mn.PAC and Fe.PAC promoters, correspondingly [74–80].

The Mechanism of Catalytic Cracking of Waste Cooking Oil

The catalytic cracking (CC) reaction mechanism of waste cooking oil utilizing the prepared AC as well as its improved types, maybe explained depending on the recorded values as of the representative features of the gained biofuel. BET characterization indicated the occurrence of different kinds of the holes at the surface of the prepared catalysts; this pores different in pore volume and pore radius, the higher pore volume corresponding to macropores, the middle pore volume are mesopores, and small pore volume related to micropores, as shown in Fig. 8. The sorts of responses that existed in the various pores of the promoters are comparable, i.e., CC process. Be that as it may, the difference among the pores through the CC process is the size of the reactants. High hole volume may steward an extensive variability of particles that deviate from one another in their chain reactant length. If there should arise an occurrence of slighter pores (sub-micropores, micropores, and mesopores), the introduced atoms are much slighter in volume, for example, shorter chain lengths. The properties of the reactants in the pores depict the properties of the items. In micropores and mesopores, the reactants are littler in volume, and

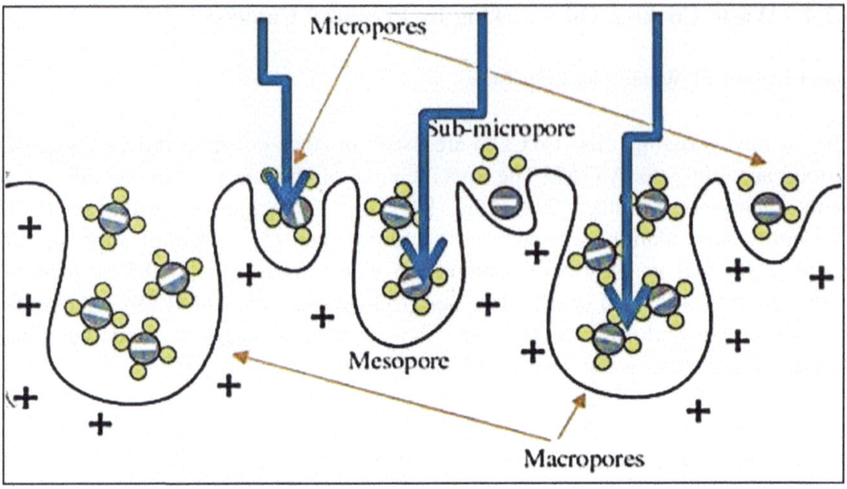

Fig. 8 Various pore structures existing in activated carbon structure. Adapted with permission from Ref. [73], Copyright 2020, Elsevier

along these lines, the created particles are unsurprising to be light volume hydrocarbons [81–86]. Be that as it may, in macropores, the reactants are prolonged chain and massive volume particles (triglycerides) that yield extraordinary hydrocarbon chains as items from CC process. It is observed that the attendance of Fe, as well as Mn particle in the prepared AC network, reduced the pore volumes if compared with PAC. Values exposed extremely existing of high pore volume (macropores) in the PAC than Fe.PAC and Mn.PAC promoters. Thus, the action that happened to utilize the doped PAC has happened with small molecules, and the CC yields are predicted to be short-chain hydrocarbons instead of the long-chain products [87, 88].

6.3 *Activated Carbon from Petroleum Residue*

In Alberta, Canada, millions of huge amounts of oiled sand coke are produced each year as a result of the progress of bitumen. Owing to its high carbon content, petroleum sandblasting coke can be a proper indicator for preparing AC [89].

6.3.1 Using Oil Sands Coke to Prepare Activated Carbon

In north-eastern, the Athabasca oil sands region Alberta, Canada contains about 174 billion barrels of bitumen [90]. When mining, the bitumen is improved to synthetic crude oil with either deferred or liquid thermal cooking techniques [91]. During oil production, large amounts of dry and wet waste are produced. Great volumes of wastewater extracted from bitumen are warehoused nearby the waste ponds. More than 4108 cubic meters of water affected by the process are stored in the waste pans [92]. The main dry waste that is produced during cooking and liquid coke operations is delayed. Syncrude Canada Ltd. produces more than 5 million tons of coke annually [91]. Syncrude liquid coke is now stored in place (more than 30 million tons); Suncor, on the other hand, is burned in small amounts to raise steam and electricity [93]. Emissions of nitrogen oxides, sulfur, and different polluting reduce the usage of late coke as fuel. It is estimated that a total of one billion cubic meters of coke is produced through the life of the oil sands processes in Alberta [91]. There is a necessity for management and safe disposal. The use of adsorbents while removing organic acids from wastewater is an emerging technique for oil sands production [90]. Nevertheless, the associated costs for the oil sands industry with obtaining great amounts of adsorbents and preparing them to treat the waste ponds are considered a significant investment cost. The promptly accessible crude carbon source (around 80–85% by weight) [94] for the companies of oil sands is their stored coke. Whereas, petroleum coal was usually used as an indication for the improvement of a stimulant carbon, for example [95, 96], few investigations showed activated oil sands coke for wastewater cure using KOH in chemical activation [97]. However, this method is favored owing to the shorter activation time desirable to obtain products with higher surface area, Panfilo and collaborators [98] confirm that only fifty percent of the KOH utilized for activation process was recoverable. Physical activation has the benefit of usage presented activation agents, for example, steam.

Be that as it may, physical activation has not been fine investigated on fluid coke and oil sands hindered. They showed physical activation of oil sands liquid coke utilizing an initial temperature of 850 °C for 6 h and steam of CO_2. The results indicated a microporous volume (0.244 cm^3 g^{-1}) and a surface area of BET of (319 m^2 g^{-1}) [99]. Shawwa and collaborators utilized steam and CO_2 through activation of delayed coke for 4 h at 850 °C to obtain AC with a methylene blue significance of (100 mg g^{-1}) for handling wastewater of mash mill, though additional description of the material was not finalized [100]. Stavropoulos and collaborators premeditated a higher total carbon produce for physically activated petroleum coke among carbon materials (like charcoal, wood, used tires, lignite and, etc.) foremost to lower total outlay prices to the client (in view of cost-effective assessment of an AC manufacture plant) [99]. These researches show that the physical activation of overdue and fluid oil sands coke can produce adsorbents with higher surface area valuable for industry of the oil sands; nevertheless, thoroughly, the adsorbents characterization cannot seem to be finished, and the preeminent physical activation considerations investigation essential for high surface area adsorbents production out of fluid and delayed coke still cannot seem to be discovered. Researches have likewise yet characterized activated fluid and delayed coke comparison.

6.3.2 Using Asphalt and Heavy Oil Fly Ash to Prepare Activated Carbon

Fly ash is the solid residue after burning heavy oil and coal. In general, fly ash from coal contains non-combustible materials—mainly aluminum, silicon oxides, calcium, iron, and Mg in charcoal—as well as a little quantity of carbon which residues after partial combustion. Coal fly ash can be utilized in concrete as a cement substitute according to its chemical structure. One of the most significant characteristics of fly ash, especially as an indication of its appropriateness for use in concrete, is the ignition loss (LOI), a quantity of unburned carbon residual in fly ash. Fly ash that has a high percentage of LOI (carbon content with high percent) is formed at low temperatures' combustion necessary to operate low nitrogen oxide burners, not compatible to utilize as a substitute for cement. Therefore, fly ash with high carbon content is usually disposed of in landfills. Likewise, heavy oil fly ash has a comparatively high content of heavy metals, especially nickel oxide (about 1% NiO) and vanadium oxide (usually 3% V_2O_5). Applying systems with low temperatures to meet stringent NO_x specifications give high levels of carbon remaining in the heavy ash fly ash. Currently, ash fly from heavy oil is utilized as fuel for cement kilns or is disposed of in landfills that is harmful to the environment. The search for heavy metal recovery (Vanadium, Nickel) in the oil fly ash by filtration with an acid solution takes a significant consideration. The remaining carbon residue after filtering can be used as an introduction to AC. Nevertheless, more research is necessary for modern technologies for more efficiently using high-carbon ashes. Asphalt is a layer as material got from a waste of petrol. Asphalt is generally utilized in construction industries such as coatings and insulators, lanes, and road paving seals. It is likewise utilized in building materials and thermal applications as a binder. High carbon content materials, for example, asphalt, which utilized as precursors of AC (Fig. 9).

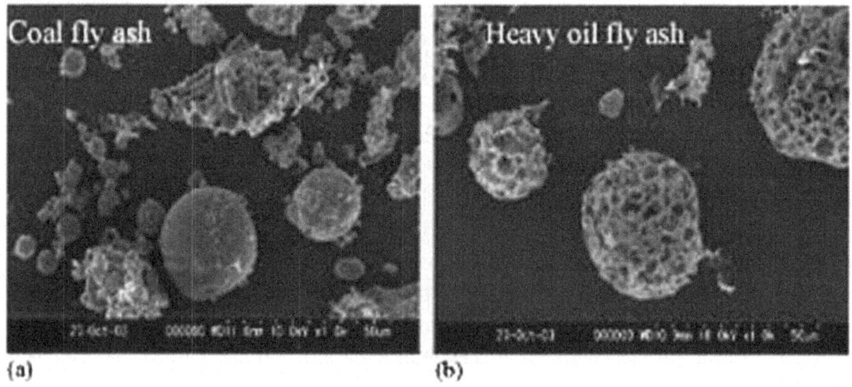

Fig. 9 SEM images of **a** fly ash of coal and **b** fly ash of heavy oil. Adapted with permission from Ref. [100], Copyright 2020, Elsevier

There are literature reviews on the preparation of AC from petroleum remains. Generally, the product of coke from pyrolysis (solid carbon) of heat-like materials is low if the samples are not pre-treated before carbonization. The step of activation is essential to produce high carbon–carbon atoms. Studies of converting several carbon precursors to active carbon have presented that pre-oxidation shows a significant part in forming O_2 groups that decrease the material's thermoplastic properties. Molecular oxygen has been reported to be able to correlate with polycyclic hydrocarbons on the field. Through the O_2 treatment of the layer, functional groups containing oxygen are designed, and the interlinked reactions among the functional groups containing the oxygen result in transitional stability in the volatile pitch molecules to the temperatures in which carbon–carbon bonds can be made in pyrolysis. Some researchers also researched synthesizing zeolite from the fly ash of coal. In addition, heavy ash is fly ash as a source of AC. Its chief component is carbon. Though this probability has not yet been fully realized, it is observed that both heavy oil and coal ash are made up of large micrometer pellets. In particular, heavy ash is formed by carbon cenospheres, that is, hollow particles or spongy. Activated carbonates with two-dimensional porous arrangements, wherein gaps among fly ash pellets work as small pores and asphalt coating layer as an indicator of fine pores if asphalt is utilized as a carbon source in the synthesis of AC by fly ash (Fig. 10). Also, asphalt can play as a binder for the particles of fly ash. Additionally, it is believed that the starter of asphalt in pores of pellets' fly ash, then carbonization will cause growth in fine pores within the large pores of pellets' fly ash [100]. Benjamin et al. prepared high-quality activated porous carbons and carbon nanotube membranes by using asphaltenes as a waste of the petroleum industry, as shown in Fig. 11 [101].

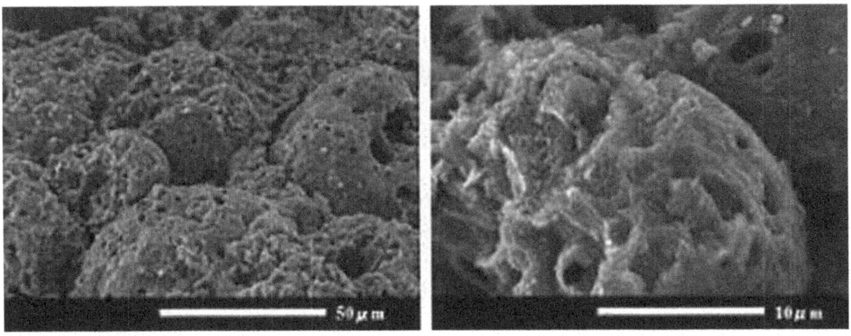

Fig. 10 SEM images of the asphalt–heavy oil fly ash derived carbon. Adapted with permission from Ref. [100], Copyright 2020, Elsevier

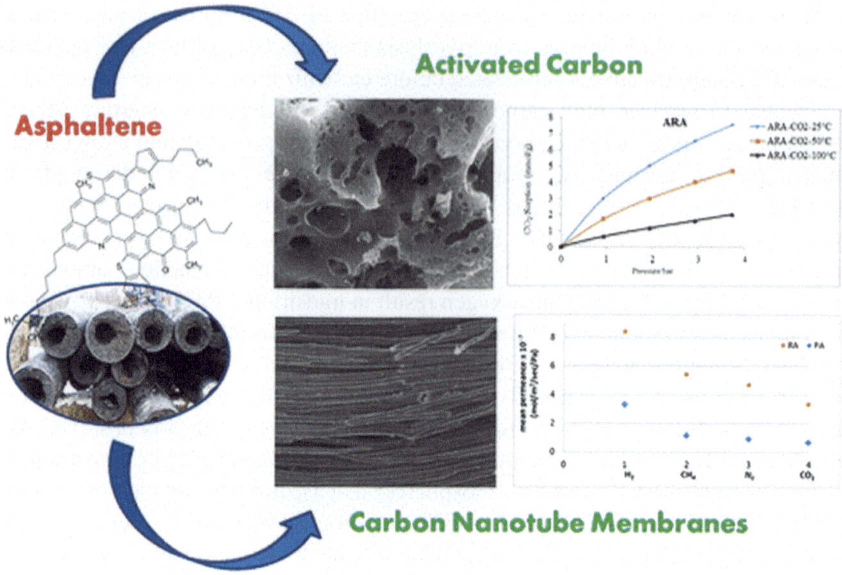

Fig. 11 Asphaltene-derived AC and carbon nanotube membranes for CO_2 separation. Adapted with permission from Ref. [101], Copyright 2018, American Chemical Society

6.3.3 Using Spent Lubricating Oil to Prepare Activated Carbon

Industrial lubricants and spent cars are one of the main hazardous waste produced in all the world. Improper disposal of this waste will have a negative effect on the environment. It is expected that pollution of one million gallons of drinking water caused by one gallon of utilized oil, covering an area of 8 acres of surface water. Oil's layers' spills on the outside of the water can obstruct the sun's rays and prevent photosynthesis. It stops the recovery of liquefied oxygen and kills fish and different organisms. This pollution can influence the environment on the health of people, plants, and animals. Thusly, many attempts have been made to properly recycle and dispose the spent oil. These strategies incorporate biodegradation, refining, land application, burning, and profound good removal. Recycling is a complicated advance contrasted with assembling another variant of Lubricant [102–104]. In addition, burning, land use, or removal of profound wells can cause air contamination or defile groundwater resources. Jordan, just as neighboring nations, suffers from a lack of drinking water. Accordingly, it must not just stop water resources from spent oil spills, yet in addition, treat wastewater for reuse.

Wastewater treatment of minerals and poisonous organic materials is professionally achieved by the adsorption process using a filter of AC. Poisons' removal by AC since adsorption is a procedure that reduces the grouping of poisonous metals to a sub-part per million. Commercial AC can be delivered from various kinds of crude materials, such as coal, wood, nutshells, and fruit's stones. Hence the adsorption capacity of AC against specific target adsorbate is strongly affected by the raw material type and treatment technology as well. The lubricants used in this research consist of 3 main sets of oils: crankcase oil, transmission fluids, and gear oils. The characterization of the oil is shown in Table 1. The considered samples exhibited enough quantity of ash as a product of corrosion of the interior surface of the machines or the introduction of air dust during combustion. The oil samples were processed using sulfuric and nitric acids.

They were activated by adding sulfuric or a blend of nitric and sulfuric acid to oil tests at ambient temperature. In the first case, no impact on the blend was observed until the blend was warmed to 100 °C where the chemical reaction occurred. In the subsequent case, HNO_3 was continuously added to the blend at ambient temperature, which prompted a fast increment in the temperature of the solution to 150 °C and arrived in one moment because of an exothermic response. In the two cases, the arrangements were further warmed, while carbon dioxide, carbon monoxide, nitrous oxide, and sulfur (observed in the subsequent system) were discharged outside the solution. Upon activation with these acids, a few sulfonation and nitrogen responses happened on the outside of the oil shale tests (surface), resulting in the production of AC with large surface functional groups. Chemical analysis of the produced alternating current with the consumer lubricant oil appears in Table 2. Unmistakably, the oil is activated with acids, then washing with water and treated with NaOH that wash off every heavy metal.

Table 1 Characterization of the used lubricating oil

Test	Results	Method
Specific gravity at 20 °C	0.91	ASTM D 1298
Flash point	422 K	ASTM D 93–94
Viscosity at 38 °C	107 cp	ASTM D 445–74
Ash content	1.3 wt%	ASTM D 582
Asphaltene content	5.9 wt%	ASTM D 582

Table 2 Elemental analysis of used lubricating oil and AC produced

Element	Lubricating oil used	AC produced
Sulfur, wt%	0.4	–
Iron, ppm	1358	11 (mg Kg^{-1})
Lead, ppm	1320	–
Calcium, ppm	2364	–
Zinc, ppm	1182	–
Magnesium, ppm	607	–
Copper, ppm	17	–
Barium, ppm	11	–
Aluminum, ppm	76	2 (mg Kg^{-1})
Chromium, ppm	3.7	–

Small quantities of barium for Fe and Al were observed on the carbon surface. Iron and aluminum can be present in the AC owing to the passivation reaction between these elements and the acids. Furthermore, the reaction with hydroxide will precipitate, such as $Al(OH)_3$ and Fe $(OH)_3$ [105, 106] .

7 Conclusions

Nanomaterials, containing different forms of carbon-based nanomaterials like carbon nanotubes, graphene, carbon particles, and activated carbon, are readiness for large-scale industrial production to obtain widespread interest and application areas such as sensors, supercapacitors, bio-medical, and gas storage uses to request the usage of heavy hydrocarbons and organic chemicals to achieve the requirement.

Conventional precursors for the production of carbon-based nanostructure are acetylene, methane, toluene, ethane, and more hydrocarbon created from fossil fuels; they have minimal availability, produce pollution and will not be existing in nearby future. In this state, numerous waste materials can be utilized for carbon nanostructures manufacture via various techniques. The utilization of waste materials as a precursor of carbon sources will not only find a plentiful and inexpensive source for the manufacture of carbon materials but also as an unconventional use of waste material, which is dangerous to the environment.

The activated carbon prepared from the waste materials appears to be profitable; various waste materials were used in the synthesis of AC, such as: agriculture wastes (Lapsi seed, olive stone, sago waste, agro-waste barley straw, rice straw, wheat straw, coconut tree sawdust, maize cob and silk cotton hull), biological wastes (buffing dust, bovine skin got from leather industries), fruit wastes (the liquid waste of juice and its wash waters. and solid waste (stones, peel skin, orange peels, seeds,, banana peels, pomegranate peels, medlar seeds, palm shells, grape seeds), plastic wastes (bottles of polyethylene terephthalate, polystyrene tires, rubber tires), electronic wastes (mobile phones, entertainment device electronics, discarded computers, refrigerators, office electronic equipment, printed circuit boards, and television sets), vegetable wastes (putrescible vegetables, soybean hulls, pumpkin stem waste, and carrot residue), and petroleum waste (oil sands coke, oily fly ash and spent lubricating oil).

As a case study, peach seeds were utilized for the synthesis of AC doped by transition metals in its framework as a heterogeneous catalyst, which used in the catalytic cracking of west cocking oil into biofuel. The prepared catalyst showed excellent physical and chemical properties, and the obtained biofuel also has excellent properties. The work here is still extensively open for future research to get high-value nanomaterial from inexpensive sources and less difficult techniques.

8 Future Perspectives

Today, government, industry, and specialists advance waste recycling technology as a solution for some problems. Waste recycling technology has been introduced as an answer to the issue to provide of creating nations over the years. In the previous decades, waste recycling technology has been tightened as an instrument to give treatment to illnesses all through the world, yet just incompletely in the created countries. Presently, waste recycling technology has high innovation to encourage elective, sustainable sources as the basis to decrease worldwide NO_X and SO_X gas impressions and an earth-wide temperature boost (greenhouse effect). In any case, we may need to understand that if waste recycling technology can help provide these issues, it is not the main arrangement. It is only one of the advancements and approaches that expected to avoid the issues. In the interpretation of the asset shortage and vitality reliance, the nation is confronting, squanders the executives are completely important to accomplish and keep up a reasonable society, particularly in the field of sustainable power source where the feed-in tariff is among the best in the world. Environmental matters appreciate unique consideration for years and the area profits by administrative help, including different motivating force estimates, for example, the feed-in tariff or the New-Town programs. The consciousness of the populace in regards to these inquiries being high, i.e., on account of refinement estimates beginning from youth, it is in light of a legitimate concern for organizations and politic to encourage this segment and ceaselessly execute new natural measures, until to now, consideration has essentially been paid to the third R (reuse), and the plan thinks about that new measures ought to be taken to cultivate the reduction of waste and reuse of items to diminish the general amount of waste, in this way decreasing the requirement for reusing. Among the endeavors and activities, the plan refers to cutting food misfortune, growing the reuse advertise, instructing the purchasers to change their way of life or requesting that the business segments give dependable and spare items. Another significant test is that a lessening in the amount of waste is not out of the ordinary, thinking about the different measures targeting decreasing waste age in any case. The present limit of existing waste treatment offices is, in this manner, adequate to cover the interest. Under these conditions, a downtrend of capital speculation by makers and decrease sought after from the open division that used to help the waste and reusing market were as of late watched. The natural business advertises anticipated to move from traditional waste removal to new vitality frameworks, for example, those for vitality sparing, sustainable power source, and earth inviting items.

References

1. Ali GAM, Habeeb OA, Algarni H, Chong KF (2018) CaO impregnated highly porous honeycomb activated carbon from agriculture waste: symmetrical supercapacitor study. J Mater Sci 54:683–692
2. Habeeb OA, Ramesh K, Ali GAM, Yunus RM (2017) Application of response surface methodology for optimization of palm kernel shell activated carbon preparation factors for removal of H_2S from industrial wastewater. J Teknologi 79(7):1–10
3. Habeeb OA, Ramesh K, Ali GAM, Yunus RM (2017) Low-cost and eco-friendly activated carbon from modified palm kernel shell for hydrogen sulfide removal from wastewater: adsorption and kinetic studies. Desalin Water Treat 84:205–214
4. Habeeb OA, Ramesh K, Ali GAM, Yunus RM (2017) Experimental design technique on removal of hydrogen sulfide using CaO-eggshells dispersed onto palm kernel shell activated carbon: experiment, optimization, equilibrium and kinetic studies. J Wuhan Univ Technol-Mater Sci Ed 32(2):305–320
5. Habeeb OA, Ramesh K, Ali GAM, Yunus RM (2017) Isothermal modelling based experimental study of dissolved hydrogen sulfide adsorption from waste water using eggshell based activated carbon. Malaysian J Anal Sci 21(2):334–345
6. Habeeb OA, Ramesh K, Ali GAM, Yunus RM (2017) Optimization of activated carbon synthesis using response surface methodology to enhance H_2S removal from refinery wastewater. J Chem Eng Ind Biotechnol 1(1):1–17
7. Habeeb OA, Ramesh K, Ali GAM, Yunus RM, Thanusha TK, Olalere OA (2016) Modeling and optimization for H_2S adsorption from wastewater using coconut shell based activated carbon. Aust J Basic Appl Sci 10(17):136–147
8. Patra JK, Baek K-H (2014) Green nanobiotechnology: factors affecting synthesis and characterization techniques. J Nanomater 2014:1–12
9. Silva GA (2008) Nanotechnology approaches to crossing the blood-brain barrier and drug delivery to the CNS. BMC Neurosci 9(S3)
10. Pantarotto D, Briand J-P, Prato M, Bianco A (2004) Translocation of bioactive peptides across cell membranes by carbon nanotubes Electronic supplementary information (ESI) available: details of the synthesis and characterization, cell culture, TEM, epifluorescence and confocal microscopy images of CNTs 1, 2 and fluorescein. http://www.rsc.org/suppdata/cc/b3/b311254c. Chemical Communications (1):16
11. Brunner PH, Rechberger H (2015) Waste to energy—key element for sustainable waste management. Waste Manag 37:3–12
12. Chandler AJ, Eighmy T, Hartlen J, Hjelmar O, Kosson D, Sawell S, van der Sloot HA, Vehlow J (1990) Treatise on municipal solid waste incinerator residues. Contaminated soil '90. Springer, Netherlands, pp 1309–1310
13. Vergara SE, Tchobanoglous G (2012) Municipal solid waste and the environment: a global perspective. Annu Rev Environ Resour 37(1):277–309
14. White P, Franke M, Hindle P (1995) A lifecycle inventory of solid waste. Integrated solid waste management: a lifecycle inventory. Springer, US, pp 37–62
15. Cheremisinoff NP (2003) Municipal solid waste. In: Handbook of solid waste management and waste minimization technologies. Elsevier, Netherlands, pp 34–95
16. Demirbas A (2011) Waste management, waste resource facilities and waste conversion processes. Energy Convers Manag 52(2):1280–1287
17. Dixon N, Jones DRV (2005) Engineering properties of municipal solid waste. Geotext Geomembr 23(3):205–233
18. Tchobanoglous G, Theisen H, Vigil S (1993) Integrated solid waste management: engineering principles and management issues. McGraw-Hill
19. Ali GAM, Divyashree A, Supriya S, Chong KF, Ethiraj AS, Reddy MV, Algarni H, Hegde G (2017) Carbon nanospheres derived from Lablab purpureus for high performance supercapacitor electrodes: a green approach. Dalton Trans 46(40):14034–14044

20. Hegde G, Abdul Manaf SA, Kumar A, Ali GAM, Chong KF, Ngaini Z, Sharma KV (2015) Biowaste sago bark based catalyst free carbon nanospheres: waste to wealth approach. ACS Sustain Chem Eng 5(9):2247–2253
21. Kreith F (1999) Handbook of solid waste management
22. Ali GAM, Supriya S, Chong KF, Shaaban ER, Algarni H, Maiyalagan T, Hegde G (2019) Superior supercapacitance behavior of oxygen self-doped carbon nanospheres: a conversion of Allium cepa peel to energy storage system. Biomass Conv Bioref. https://doi.org/10.1007/s13399-019-00520-3
23. Ali GAM, Manaf SAA, A D, Chong KF, Hegde G (2016) Superior supercapacitive performance in porous nanocarbons. J Energy Chem 25(4):734–739
24. Gupta N, Yadav KK, Kumar V (2015) A review on current status of municipal solid waste management in India. J Environ Sci 37:206–217
25. de Souza Vitorino, Melaré A, Montenegro González S, Faceli K, Casadei V (2017) Technologies and decision support systems to aid solid-waste management: a systematic review. Waste Manag 59:567–584
26. Bing X, Bloemhof JM, Ramos TRP, Barbosa-Povoa AP, Wong CY, van der Vorst JGAJ (2016) Research challenges in municipal solid waste logistics management. Waste Manag 48:584–592
27. Agagu O (2009) Threats to the Nigerian environment: a call for positive action. Being a paper delivered at the, 3–11
28. Njeru J (2006) The urban political ecology of plastic bag waste problem in Nairobi, Kenya. Geoforum 37(6):1046–1058
29. Vongdala N, Tran H-D, Xuan T, Teschke R, Khanh T (2018) Heavy metal accumulation in water, soil, and plants of municipal solid waste landfill in Vientiane, Laos. Int J Environ Res Public Health 16(1):22
30. Afre RA, Soga T, Jimbo T, Kumar M, Ando Y, Sharon M, Somani PR, Umeno M (2006) Carbon nanotubes by spray pyrolysis of turpentine oil at different temperatures and their studies. Microporous Mesoporous Mater 96(1–3):184–190
31. Ghosh P, Afre RA, Soga T, Jimbo T (2007) A simple method of producing single-walled carbon nanotubes from a natural precursor: eucalyptus oil. Mater Lett 61(17):3768–3770
32. Kumar R, Tiwari R, Srivastava O (2011) Scalable synthesis of aligned carbon nanotubes bundles using green natural precursor: neem oil. Nanoscale Res Lett 6(1):92
33. Suriani AB, Azira AA, Nik SF, Md Nor R, Rusop M (2009) Synthesis of vertically aligned carbon nanotubes using natural palm oil as carbon precursor. Mater Lett 63(30):2704–2706
34. Ali GAM, Abdul Manaf SA, Kumar A, Chong KF, Hegde G (2014) High performance supercapacitor using catalysis free porous carbon nanoparticles. J Phys D Appl Phys 47 (49):495307–495313
35. Azmina MS, Suriani AB, Salina M, Azira AA, Dalila AR, Asli NA, Rosly J, Nor RM, Rusop M (2012) Variety of bio-hydrocarbon precursors for the synthesis of carbon nanotubes. Nano Hybrids 2:43–63
36. Dalila AR, Suriani AB, Rosmi MS, Rosazley R, Rosli J, Rusop M (2013) Carbon nanotubes: a brief outlook on history, synthesis methods and various bio-hydrocarbon sources. Adv Mater Res 832:792–797
37. Ayranci E, Duman O (2009) In-situ UV-visible spectroscopic study on the adsorption of some dyes onto activated carbon cloth. Sep Sci Technol 44(15):3735–3752
38. Abioye AM, Ani FN (2015) Recent development in the production of activated carbon electrodes from agricultural waste biomass for supercapacitors: a review. Renew Sustain Energy Rev 52:1282–1293
39. Skodras G, Diamantopoulou I, Zabaniotou A, Stavropoulos G, Sakellaropoulos GP (2007) Enhanced mercury adsorption in activated carbons from biomass materials and waste tires. Fuel Process Technol 88(8):749–758
40. Shehzad A, Bashir MJK, Sethupathi S, Lim J-W (2015) An overview of heavily polluted landfill leachate treatment using food waste as an alternative and renewable source of activated carbon. Process Saf Environ Prot 98:309–318

41. Bentsen NS, Felby C, Thorsen BJ (2014) Agricultural residue production and potentials for energy and materials services. Prog Energy Combust Sci 40:59–73
42. Nabais JMV, Laginhas C, Carrott MMLR, Carrott PJM, Amorós JEC, Gisbert AVN (2013) Surface and porous characterisation of activated carbons made from a novel biomass precursor, the esparto grass. Appl Surf Sci 265:919–924
43. Shrestha RM, Varga I, Bajtai J, Varga M (2013) Design of surface functionalization of waste material originated charcoals by an optimized chemical carbonization for the purpose of heavy metal removal from industrial waste waters. Microchem J 108:224–232
44. Kadirvelu K, Kavipriya M, Karthika C, Radhika M, Vennilamani N, Pattabhi S (2003) Utilization of various agricultural wastes for activated carbon preparation and application for the removal of dyes and metal ions from aqueous solutions. Biores Technol 87(1):129–132
45. Taer E (2018) Activated carbon electrode made from coconut husk waste for supercapacitor application. Int J Electrochem Sci, 12072–12084
46. Alslaibi TM, Abustan I, Ahmad MA, Foul AA (2013) Cadmium removal from aqueous solution using microwaved olive stone activated carbon. J Environ Chem Eng 1(3):589–599
47. Chand R, Watari T, Inoue K, Torikai T, Yada M (2009) Evaluation of wheat straw and barley straw carbon for Cr(VI) adsorption. Sep Purif Technol 65(3):331–336
48. Hammud KK, Aboob RM, Hassan MS, Zanad DE, Khalīl MH (2018) Preparation of activated carbon from waste cooked tea for using as chemical dyes-filter. Eurasia Proc Sci Technol Eng Math 4:15–20
49. Gil RR, Ruiz B, Lozano MS, Martín MJ, Fuente E (2014) VOCs removal by adsorption onto activated carbons from biocollagenic wastes of vegetable tanning. Chem Eng J 245:80–88
50. Lopez-Anton MA, Gil RR, Fuente E, Díaz-Somoano M, Martínez-Tarazona MR, Ruiz B (2015) Activated carbons from biocollagenic wastes of the leather industry for mercury capture in oxy-combustion. Fuel 142:227–234
51. Guo J, Lua A (2000) Preparation and characterization of adsorbents from oil palm fruit solid wastes
52. Issabayeva G, Aroua MK, Sulaiman NMN (2006) Removal of lead from aqueous solutions on palm shell activated carbon. Biores Technol 97(18):2350–2355
53. Daud W (2004) Comparison on pore development of activated carbon produced from palm shell and coconut shell. Biores Technol 93(1):63–69
54. Awwad NS, El-Zahhar AA, Fouda AM, Ibrahium HA (2013) Removal of heavy metal ions from ground and surface water samples using carbons derived from date pits. J Environ Chem Eng 1(3):416–423
55. Gruz J, Ayaz FA, Torun H, Strnad M (2011) Phenolic acid content and radical scavenging activity of extracts from medlar (Mespilus germanica L.) fruit at different stages of ripening. Food Chem 124(1):271–277
56. Mendoza-Carrasco R, Cuerda-Correa EM, Alexandre-Franco MF, Fernández-González C, Gómez-Serrano V (2016) Preparation of high-quality activated carbon from polyethyleneterephthalate (PET) bottle waste. Its use in the removal of pollutants in aqueous solution. J Environ Manage 181:522–535
57. Xiang D, Mou P, Wang J, Duan G, Zhang HC (2006) Printed circuit board recycling process and its environmental impact assessment. Int J Adv Manuf Technol 34(9–10):1030–1036
58. Wong MH, Wu SC, Deng WJ, Yu XZ, Luo Q, Leung AOW, Wong CSC, Luksemburg WJ, Wong AS (2007) Export of toxic chemicals—a review of the case of uncontrolled electronic-waste recycling. Environ Pollut 149(2):131–140
59. Xu M, Hadi P, Chen G, McKay G (2014) Removal of cadmium ions from wastewater using innovative electronic waste-derived material. J Hazard Mater 273:118–123
60. Hadi P, Barford J, McKay G (2014) Selective toxic metal uptake using an e-waste-based novel sorbent–Single, binary and ternary systems. J Environ Chem Eng 2(1):332–339
61. Nahil MA, Williams PT (2010) Activated carbons from acrylic textile waste. J Anal Appl Pyrol 89(1):51–59

62. Dhiman P, Naushad M, Batoo KM, Kumar A, Sharma G, Ghfar AA, Kumar G, Singh M (2017) Nano Fe x Zn 1 − x O as a tuneable and efficient photocatalyst for solar powered degradation of bisphenol A from aqueous environment. J Clean Prod 165:1542–1556
63. Zhang Y, Dubé MA, McLean DD, Kates M (2003) Biodiesel production from waste cooking oil: 2. Economic assessment and sensitivity analysis. Bioresour Technol 90(3):229–240
64. Nam LTH, Vinh TQ, Loan NTT, Tho VDS, Yang X-Y, Su B-L (2011) Preparation of bio-fuels by catalytic cracking reaction of vegetable oil sludge. Fuel 90(3):1069–1075
65. Ajala OE, Aberuagba F, Odetoye TE, Ajala AM (2015) Biodiesel: sustainable energy replacement to petroleum-based diesel fuel—a review. ChemBioEng Rev 2(3):145–156
66. Zhang Q, Wang T, Xu Y, Zhang Q, Ma L (2014) Production of liquid alkanes by controlling reactivity of sorbitol hydrogenation with a Ni/HZSM-5 catalyst in water. Energy Convers Manag 77:262–268
67. Negm NA, Rabie AM, Mohammed EA (2018) Molecular interaction of heterogeneous catalyst in catalytic cracking process of vegetable oils: chromatographic and biofuel performance investigation. Appl Catal B 239:36–45
68. Negm NA, Betiha MA, Alhumaimess MS, Hassan HMA, Rabie AM (2019) Clean transesterification process for biodiesel production using heterogeneous polymer-heteropoly acid nanocatalyst. J Clean Prod 238:117854
69. Rabie AM, Mohammed EA, Negm NA (2018) Feasibility of modified bentonite as acidic heterogeneous catalyst in low temperature catalytic cracking process of biofuel production from nonedible vegetable oils. J Mol Liq 254:260–266
70. Baskar G, Soumiya S (2016) Production of biodiesel from castor oil using iron (II) doped zinc oxide nanocatalyst. Renew Energy 98:101–107
71. Nekouei S, Nekouei F, Canselier JP (2016) Solid-phase extraction for simultaneous separation and preconcentration of Fe(III) and Zn(II) traces using three chelatants and Ramelak bark-derived activated carbon as a new bio-sorbent. Sep Sci Technol 52(5):824–833
72. Negm NA, Sayed GH, Yehia FZ, Habib OI, Mohamed EA (2017) Biodiesel production from one-step heterogeneous catalyzed process of Castor oil and Jatropha oil using novel sulphonated phenyl silane montmorillonite catalyst. J Mol Liq 234:157–163
73. Sadeek SA, Mohammed EA, Shaban M, Abou Kana MTH, Negm NA (2020) Synthesis, characterization and catalytic performances of activated carbon-doped transition metals during biofuel production from waste cooking oils. J Mol Liq 306:112749
74. Joshi S, Adhikari M, Pradhananga RR (2013) Adsorption of fluoride ion onto zirconyl-impregnated activated carbon prepared from lapsi seed stone. J Nepal Chem Soc 30:13–23
75. International A (2010) Standard test method for sulfur in petroleum and petroleum products by energy dispersive X-ray fluorescence spectrometry. ASTM International
76. ASTM (1996) Standard test method for pour point of petroleum products. Annual book of ASTM standards
77. D-97 AS (2006) Standard test method for pour point of petroleum products. ASTM International West Conshohocken, PA
78. Standard A (2007) D5773: Standard test method for cloud point of petroleum products (constant cooling rate method). ASTM, West Conshohocken, PA (USA)
79. ASTM Committee D-2 on Petroleum Products LF and Lubricants (2015) Standard test method for kinematic viscosity of transparent and opaque liquids (and calculation of dynamic viscosity). ASTM International
80. International A (2011) Standard test method for density, relative density, and API gravity of liquids by digital density meter. ASTM International
81. Senra M, McCartney SN, Soh L (2019) The effect of bio-derived additives on fatty acid methyl esters for improved biodiesel cold flow properties. Fuel 242:719–727
82. Alslaibi TM, Abustan I, Ahmad MA, Foul AA (2013) A review: production of activated carbon from agricultural by-products via conventional and microwave heating. J Chem Technol Biotechnol 88(7):1183–1190

83. Negm NA, Shaalan MA, Barouty GSE, Mohamed MY (2016) Preparation and evaluation of biodiesel from Egyptian castor oil from semi-treated industrial wastewater. J Taiwan Inst Chem Eng 63:151–156
84. Negm NA, Zahran MK, Abd Elshafy MR, Aiad IA (2018) Transformation of Jatropha oil to biofuel using transition metal salts as heterogeneous catalysts. J Mol Liq 256:16–21
85. Ibrahim MM, Agblevor FA, El-Zawawy WK (2010) Isolation and characterization of cellulose and lignin from steam-exploded lignocellulosic biomass. BioResources 5(1):397–418
86. Sun K, Lu J, Ma L, Han Y, Fu Z, Ding J (2015) A comparative study on the catalytic performance of different types of zeolites for biodiesel production. Fuel 158:848–854
87. Giuliano Albo PA, Lago S, Wolf H, Pagel R, Glen N, Clerck M, Ballereau P (2017) Density, viscosity and specific heat capacity of diesel blends with rapeseed and soybean oil methyl ester. Biomass Bioenerg 96:87–95
88. Negm NA, Sayed GH, Habib OI, Yehia FZ, Mohamed EA (2017) Heterogeneous catalytic transformation of vegetable oils into biodiesel in one-step reaction using super acidic sulfonated modified mica catalyst. J Mol Liq 237:38–45
89. Small CC, Hashisho Z, Ulrich AC (2012) Preparation and characterization of activated carbon from oil sands coke. Fuel 92(1):69–76
90. Allen EW (2008) Process water treatment in Canada's oil sands industry: I. Target pollutants and treatment objectives. J Environ Eng Sci 7(2):123–138
91. Fedorak P, Coy D (2006) Oil sands cokes affect microbial activities. Fuel 85(12–13):1642–1651
92. Paslawski JC, Headley JV, Hill GA, Nemati M (2008) Biodegradation kinetics of trans-4-methyl-1-cyclohexane carboxylic acid. Biodegradation 20(1):125–133
93. Furimsky E (1998) Gasification of oil sand coke: review. Fuel Process Technol 56(3):263–290
94. Karanfil T, Kilduff JE (1999) Role of granular activated carbon surface chemistry on the adsorption of organic compounds. 1. Priority Pollutants. Environ Sci Technol 33(18):3217–3224
95. Chang C-F, Chang C-Y, Tsai W-T (2000) Effects of burn-off and activation temperature on preparation of activated carbon from corn cob agrowaste by CO2 and steam. J Colloid Interface Sci 232(1):45–49
96. Kawano T, Kubota M, Onyango MS, Watanabe F, Matsuda H (2008) Preparation of activated carbon from petroleum coke by KOH chemical activation for adsorption heat pump. Appl Therm Eng 28(8–9):865–871
97. Shawwa AR, Smith DW, Sego DC (2001) Color and chlorinated organics removal from pulp mills wastewater using activated petroleum coke. Water Res 35(3):745–749
98. DiPanfilo R, Egiebor NO (1996) Activated carbon production from synthetic crude coke. Fuel Process Technol 46(3):157–169
99. Stavropoulos GG, Zabaniotou AA (2009) Minimizing activated carbons production cost. Fuel Process Technol 90(7–8):952–957
100. Azhar Uddin M, Shinozaki Y, Furusawa N, Yamada T, Yamaji Y, Sasaoka E (2007) Preparation of activated carbon from asphalt and heavy oil fly ash and coal fly ash by pyrolysis. J Anal Appl Pyrol 78(2):337–342
101. Kueh B, Kapsi M, Veziri CM, Athanasekou C, Pilatos G, Reddy KSK, Raj A, Karanikolos GN (2018) Asphaltene-derived activated carbon and carbon nanotube membranes for CO2 separation. Energy Fuels 32(11):11718–11730
102. Harrison C (1994) The engineering aspects of a used oil recycling project. Waste Manag 14 (3–4):231–235
103. Ali MF, Rahman F, Hamdan AJ (1996) Techno-economic evaluation of waste lube oil rerefining. Int J Prod Econ 42(3):263–273
104. Gañan J, González-García CM, González JF, Sabio E, Macías-García A, Díaz-Díez MA (2004) Preparation of activated carbons from bituminous coal pitches. Appl Surf Sci 238(1–4):347–354

105. Shawabkeh R, Khlaifat A, Khashman O, Tarawneh S (2006) Synthesis of activated carbon from spent lubricating oil and application for adsorption of cadmium and lead ions from aqueous solution. NATO security through science series. Springer, Netherlands, pp 195–200
106. Luechinger NA, Grass RN, Athanassiou EK, Stark WJ (2010) Bottom-up fabrication of metal/metal nanocomposites from nanoparticles of immiscible metals. Chem Mater 22 (1):155–160

Nanoformulated Materials from Citrus Wastes

Radwa Mahmoud Azmy

Abstract Citrus peels are a rich source of essential oils (EOs); these oils have efficient antioxidant and insecticidal properties. However, using these EOs is restricted because of technical obstacles such as their poor solubility in water and rapid rate of vaporization. Nanotechnology enables the formulation of the EOs into promising, efficient nanomaterials. These materials can be used in different fields such as water treatment, production of eco-friendly insecticides, and the food industry. Also, the citrus peels are a great source of nanocellulose, which is considered as a promising material used in water treatment and composite industry. This chapter sheds light on the employment of nanotechnology in the production of different nanomaterials from the agricultural wastes of citrus crops.

Keywords Citrus waste · Essential oils · Waste management · Recycling · Nanocellulose · Nanoemulsion · Eco-friendly insecticides · Food preservation · Antimicrobial

List of Abbreviations

EOs	Essential oils
C	Citrus
LC	Lethal concentration
NE	Nanoemulsion
NEs	Nanoemulsions
NPs	Nanoparticles

R. M. Azmy (✉)
Entomology Department, Faculty of Science, Ain Shams University, Cairo, Egypt
e-mail: radwa.azmy@yahoo.com; radwa.azmy@sci.asu.edu.eg

A. S. H. Makhlouf and G. A. M. Ali (eds.), *Waste Recycling Technologies for Nanomaterials Manufacturing*, Topics in Mining, Metallurgy and Materials Engineering, https://doi.org/10.1007/978-3-030-68031-2_22

1 Introduction

Agricultural wastes can be defined as the wastes produced through the different farming activities causing environmental pollution. The actual need for sustainable materials is directing the research focus towards the green biodegradable materials. Citrus fruits belong to Rutaceae and comprise around 140 genera and 1300 species, including *Citrus sinensis* (orange), *C. aurantifulia* (lime), *C. limonum* (lemon), and *C. reticulata* (mandarin). Citrus fruits are very common cultivated fruits throughout the world with increasing production every year [1, 2]. After consumption, about 40–60% of the citrus fruit is considered wastes [3]. Annually, about 110–120 million tons of wastes are produced worldwide from industries of citrus fruits producing great challenges of waste management [4]. For example, the industry of orange juice production leads to a significant quantity of liquid and solid wastes, up to 8–20 million tons every year [5]. Citrus wastes do not have commercial importance, and the accumulation of these wastes generates a serious problem to the environment, such as many leachates, heavy odour, moreover, attracting flies and rats [6]. However, these residues comprise rich materials (such as essential oils (EOs), insoluble and soluble carbohydrates, pectin, and cellulose); these components can be the base of many industrial practices [7–10]. Lately, several attempts have been made to develop new procedures to produce valuable materials from these wastes motivated by economic and environmental concerns [11–17].

Nanotechnology is growing rapidly and interacting with many other scientific fields creating new innovative applications. Green nanotechnology is a fruitful multidisciplinary field in the agricultural sector. The applications of the green nanotechnology attracted attention, especially in the formulation of new nanoinsecticides. A nanoinsecticide is a formulation that comprises components in the size of the nanometer range with unique characteristics [18].

This work aims to highlight the production of nanomaterials from citrus wastes such as the formulation of nanoinsecticides from EOs extracted from citrus wastes (peels of the fruits). In addition, this chapter discusses the extraction of

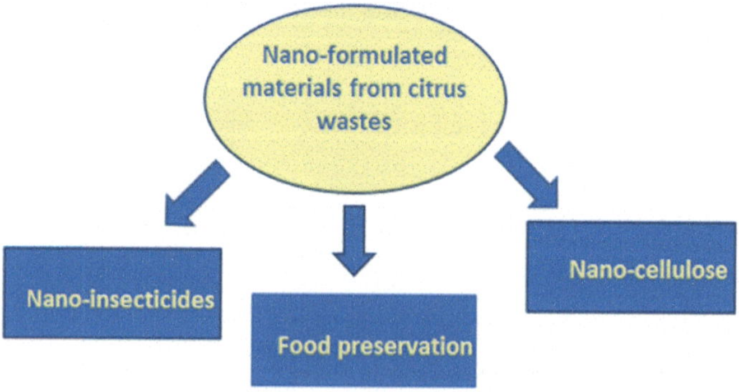

Fig. 1 Main fields for the recycled nanomaterials from citrus wastes

nanocellulose from citrus peels and its use in water treatment and different composite materials. Besides, this chapter enumerates the use of nanomaterials from citrus wastes in food products. These different nanomaterials present promising approaches to overcome pollution problems through the recycling of agricultural wastes and also present effective eco-friendly products. The main areas discussed in this chapter are shown in Fig. 1.

2 Nanoinsecticides Formulated from Citrus Essential Oils

Nanopesticide is a term used to describe a variety of formulations that include elements in the size range of nanometer, such as nanoparticles (NPs), nanocapsules, and nanoemulsions (NEs). The purpose of the nanoformulations is to increase the poor solubility and guarantee a constant slow release of the active ingredients. Citrus EOs are considered as by-products in the industry of juice; such EOs have been evaluated as insecticides to control several insect pests [19, 20]. EOs thought to be promising in insecticides industries. However, the use of EOs is often limited due to several reasons, such as high volatility and low water solubility [21]. The nanotechnology can solve the obstacles of EOs application as insecticides through the production of novel delivery systems. A promising approach to overcome those obstacles is to incorporate the EOs into the formulation of NEs, and the nanoencapsulation pokes to get NPs coating the EOs [20].

These new nanoformulations enhance the efficacy of EOs because of the greater surface area, the sustained release, the generation of systemic activity due to the smaller size of particles and the higher mobility. The nanoformulations also enable the avoiding of the organic solvents used in the application of conventional pesticides [22].

2.1 Nanoemulsions of Essential Oils

A nanoemulsion (NE) is an emulsion that contains tiny particles ranging in size from 10 to 100 nm [23, 24]. The small size of the particles related to the wavelength of light causes the NEs to tend to be slightly turbid or transparent. NEs can be formulated from the following major components:

i. The aqueous phase: This phase primarily consists of water.
ii. The oil phase: This phase includes various non-polar components such as EOs. The nature, stability, and properties of the NE are dependent on the physicochemical properties of the oil phase, such as solubility in water, chemical stability, polarity, density, viscosity, and interfacial tension [21].

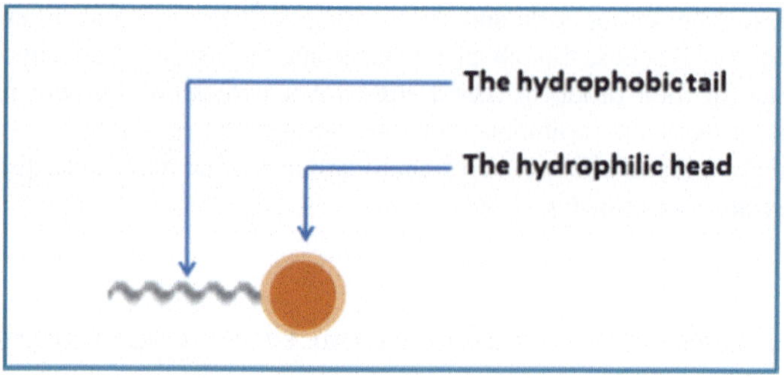

Fig. 2 Schematic representation of the emulsifier molecule

iii. The emulsifiers: It is a surface-active molecule that can adsorb to surfaces of droplets; it facilitates the partition of the oil droplets and prevents them from aggregation. The emulsifier molecule consists of a hydrophobic tail that faces the oil phase and a hydrophilic head that faces the aqueous phase, as shown schematically in Fig. 2.

The emulsifier molecules on the surface of the Nes particles act as a shell surrounding a core of lipophilic material. The selection of suitable emulsifier is very crucial for the appropriate design of the NE [25, 26]. The nature of emulsifier has a crucial impact on the kind of homogenization mechanism used to formulate the NE. The stability of the NE under the environmental pressures such as the pH, the ionic strength, cooling, heating or storage for a long time depends on the kind of the emulsifier used [27]. The emulsifier helps to prevent the system breakdown through various mechanisms, such as gravitational separation, droplet flocculation, and phase separation, as shown in Fig. 3.

2.1.1 Formulation of the Nanoemulsion

Nanoemulsions can be formulated through different approaches classified into low-energy and high-energy approaches according to the underlying principle [21, 28].

Low-Energy Approaches

The formulation of nanoemulsions using the low-energy approaches depends on the formation of oil droplets spontaneously in mixtures of water, oil, and emulsifier when their environment is changed [27].

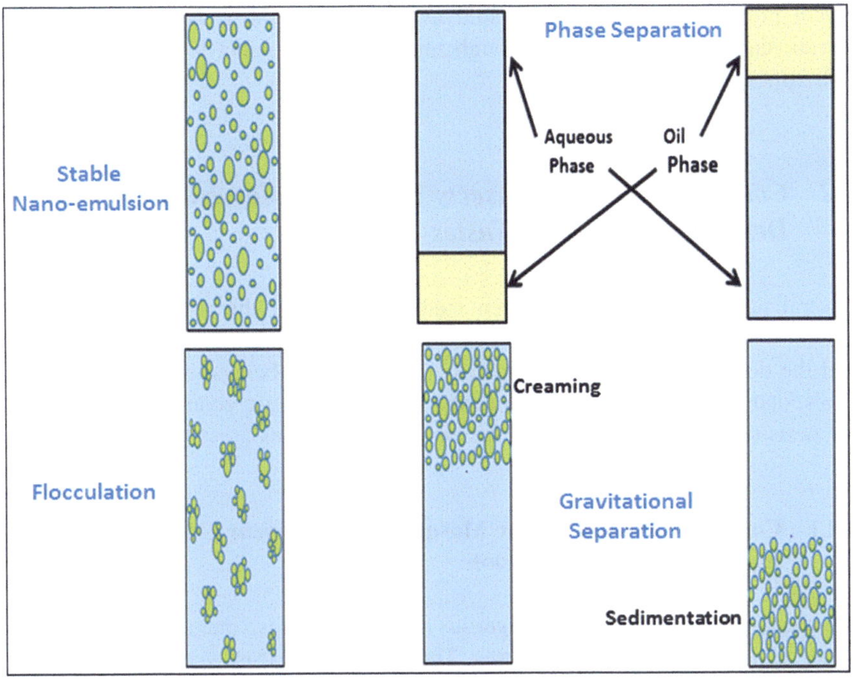

Fig. 3 Schematic representation of the stable nanoemulsion and breakdown of the system through various mechanisms, such as flocculation, gravitational separation, and phase separation. Adapted with permission from Ref. [27] Copyright 2017, Taylor and Francis Group

High-Energy Approaches

The formulation of NEs by high-energy approaches is done by using mechanical devices able to cause powerful forces that can disrupt the aqueous and oil phases into tiny oil droplets. These approaches include micro-fluidizers, sonication methods, and high-pressure valve homogenizers [28, 29].

2.1.2 Preparation of Essential Oils Nanoparticles

To prepare encapsulated NPs of EOs, polyethylene glycol is heated at 65 °C till it melts. Afterwards, EOs are mixed with polyethylene glycol; the mixture then has to be stirred heavily for 30 min to guarantee the dispersion of the EOs in the polyethylene glycol. Then, cooling of the mixture for 2 h at −4 °C to form the NPs spontaneously, then grinding in a refrigerated mortar (0 °C) and sieving by a sieve mesh 230. The powders have to be kept in sealed polyethylene pouches then stored at 27 ± 2 °C in a desiccator comprising calcium chloride to avoid absorption of moisture [22]. EOs in solid controlled-release NPs coated by polyethylene glycol prevent degradation and rapid evaporation of the EOs and enhance the insecticidal activities of them through ingestion and contact with the insect pest [22, 30].

2.2 Control of Harmful Insects Using Nanoinsecticides Derived from Citrus Wastes

Several nanoinsecticides have been studied to control the medically necessary insects which transmit dangerous diseases such as the disease-vector mosquitoes and the domestic cockroaches which live near human beings in the houses, hospitals, and restaurants. In addition to the insects that cause economic loss such as the pests of the stored grains and the pests of field plants.

2.2.1 Control of Disease-Vector Mosquito *Culex pipiens* Using Citrus Essential Oils Nanoemulsion

Mosquitoes transmit several dangerous human diseases, including malaria, rift valley fever, and encephalitis viruses. The control of disease-vector mosquitoes is facing ecological and economic challenges regarding the environmental consideration, besides the expansion of resistance of the mosquito species against the conventional chemical insecticides [31, 32]. There is a critical necessity for novel eco-friendly control tactics; the use of botanical-based insecticides is auspicious because they are more eco-friendly, while the synthetic insecticides pollute the environment and harm the non-target organisms.

The EOs of *C. sinensis* (Rutaceae) contain many constituents such as monoterpenes, limonene, pinene, octanal, and terpinolene [31]. Nanotechnology makes it possible to overcome the obstacles of EOs application as insecticides, such as poor water solubility and vaporization of volatile compounds. The formulation of NEs from EOs conserves the biological activity of the EOs; this formulation could be used effectively to control the aquatic larval stage of *Culex pipiens* [33].

Azmy et al. prepared NE from EOs extracted from peels of *C. sinensis* (agricultural wastes) by the high-energy ultrasonication method. This study contributed to nanobiotechnology presenting an effective larvicide against the disease-vector mosquito *C. pipiens with* LC_{50} equal to 27.4 ppm; the droplet size distribution of the NE is shown in Fig. 4; the mean droplet diameter of the nanoemulsion was calculated to be 78.8 ± 14.2 nm [26, 33]. The stabilization of the NE is due to the surfactant, as it acts as a mechanical barrier that prevents the accumulation of the NE droplets [34]. Larvicidal activity of the NE may be due to the primary component (limonene), which is reported to have insecticidal properties [34].

The high efficacy of the NE may be a result of the tiny size of droplets of the NE, which increases the surface area and facilitates the penetration of the active

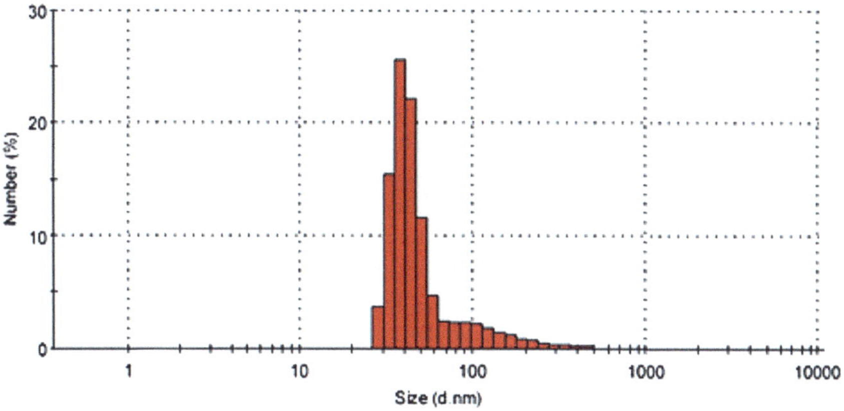

Fig. 4 Droplet size distribution of the nanoemulsion droplets. Adapted with permission from Ref. [33]

ingredients in the EOs into the mosquito larvae. The botanical molecules can interact with larval body enzymes and hormones, bind to cellular membranes, and thus interfere with the biochemical pathways of the mosquito larvae [35, 36]. In addition, NEs of EOs have various advantages, including long shelf life and biodegradation [37].

2.2.2 Control of the German Cockroach Pest

Blattella germanica is a cockroach pest commonly found in urban environments as houses, hospitals, restaurants, and food production facilities [38]. This insect is a vector of several pathogens, such as bacteria, protozoa, viruses, and helminths [39, 40]. The development of resistance in the cockroach populations against synthetic insecticides as organophosphates, organochlorines, pyrethroid, and carbamates, and the concern about the environment demands new and safe control approaches [41–45]. Bio-insecticides made of EOs can be used as an alternative method for pest management [46–49]. The application of EOs is progressively considered for control programs because they are generally less toxic to human beings and the environment than the synthetic neurotoxic insecticides [50]. However, some obstacles related to EOs poor water solubility, volatility, lack of persistence, and a tendency to oxidation should be overcome before using them as an alternative control system [50]. Nanoformulation of the EOs could be the solution to the EOs limitations, and it also provides a controlled release of EOs.

Nanoformulations of EOs are more soluble, with higher surface area, smaller particle size, and lower toxicity because no organic solvents which are used as in conventional pesticides and their formulations [20]. Thus, this formulation is safe to be used in domestic places such as houses, schools, and restaurants, where the cockroach pest exists. González et al. used polyethylene glycol as a coating material for EOs extracted from fruit peels (agricultural waste) of bergamot, *Citrus reticulate*, EOs to control *B. germanica* [37]. Polyethylene glycol has a full scale of solubility and non-interference with enzymatic activities of the living organisms and the natural excretion from them [51]. This nanoformulation of the EOs causes a remarkable rise in the contact toxicity due to the persistent release of the terpenes in the EOs and improves the contact activity against the first instar and adults of *B. germanica*. González et al. assumed that the polyethylene glycol stabilizes the EOs and decreases the volatility of the terpenes. They noticed that the major components of EOs in the nanoformulation did not change chemically through the time of storage, so no hydrolytic or oxidized derivatives were found from the original compound. This is an indication of no breakdown of the active constituents [37]. The nanoformulation enables better penetration of EOs constituents into the insect tissues [52]. Faster penetration may occur by contact with the cuticle of the insect or through penetration of the digestive tract [52, 53]. The nanoformulation droplets exhibit a large surface area; they can increase the exposure time of the biologically active ingredients of the EOs to the insect tissues.

2.2.3 Control of Stored Grains Pests

Rhizopertha dominica is a common serious pest of stored grains such as rice, wheat and corn. At the same time, *Tribolium castaneum* is a common secondary pest in stored grains. Control of such pests depends on synthetic insecticides as pyrethroids, organophosphates, and fumigants [54, 55]. These insecticides are cost-effective, but cause problems like environmental pollution and resistant behaviour besides the negative effects on human health and the other organisms [56–58] . EOs display efficient repellent and toxic effects against several stored product insect pests [59–61]. Despite these promising effects, EOs have problems in the application concerning volatility, stability, and sustainability [37].

González et al. studied the incorporation of *C. reticulate* (bergamot) EOs in stable controlled-release NPs against both *R. dominica* and *T. castaneum*. They used polyethylene glycol as a carrier; it prevents fast evaporation and enhances the stability and insecticidal activities of these EOs [22]. The designed solid nanoformulation of the EOs is favourable for the control of pests in the stored grain, as it does not affect the humidity, which is a significant factor regarding the seed's quality [22]. Insect cuticle is produced by the epidermal cells, and it covers the whole body of the insect. The cuticle extends in the foregut, hindgut, and the tracheal system. It consists of various layers: wax, epicuticle, then exocuticle, and endocuticle. Exocuticle and endocuticular layers consist of crystalline chitin nanofibres in a matrix of polyphenols, protein, water, and minor quantities of lipids [62].

It is suggested that constituents of the EOs alone (non-polar or minimally polar) diffuse vertically and/or horizontally in the insect cuticle [22]. The horizontal diffusing occurs when these constituents reach the tracheal system; then, they carry on moving to the other tissues in the insect and finally reach their sites of action. At the same time, the vertical diffusing occurs when the EOs constituents enter through the tegument towards the epidermis and then enter into the organism [63, 64].

Terpenes of the EOs studied in González et al. work was polar; thus, the external layer of wax could facilitate the horizontal diffusion of this terpene to the detriment of their vertical diffusion through the hydrophilic endocuticle. In contrast, the diffusion of EOs NPs could occur through both the vertical and horizontal diffusion because the NPs had a matrix of polyethylene glycol 6000 with its amphiphilic nature (has both hydrophilic and hydrophobic properties); it is soluble in water and some polar organic solvents. This different uptake pathway between the EOs only and the NPs of EOs may be related to the improvement of the toxicological activity against the insect. Moreover, the nanoformulation has a large surface area resulting in more adhesiveness of bergamot EOs coated with polyethylene glycol to the tissues of the insect, extending the exposure time and contact with the active ingredients [22]. Giunti et al., 2019 developed NE containing sweet orange (*C. sinensis*) EOs and evaluated its repellence activity against two insect pests of stored grains: *Tribolium confusum* and *Cryptolestes ferrugineus*. The developed nanoformulation showed acute toxicity against both insects when tested as cold aerosol and fumigant. The NE was effective in repelling and controlling the target pests. Cold aerosol treatments with nanoformulations of EOs are promising alternatives for the sanitation of warehouses, production areas, and machinery [65].

2.2.4 Control of Tomato Crop Pest

The tomato crop has a great economic significance worldwide; it is threatened by the tomato borer *Tuta absoluta* [66]. This insect has a great reproductive potential reaching 13 generations in the year [67]. The larvae feed into the plant leaves, fruits, and stems, causing a serious loss in tomato yields. The high rate of growth and severe damage caused by this tomato crop pest forced the farmers to raise the times of insecticides application [67]. This misuse resulted in the resistance towards these insecticides [68] and also caused a negative impact on the non-target living organisms, such as the pollinators and natural enemies [69, 70]. As a result of these negative impacts, alternative safe insecticides are needed. Among the various plant-derived active materials, EOs showed efficacy in controlling insect pests [71]. Despite the promising properties of EOs, there are some drawbacks that can affect their application like environmental degradation, poor solubility in water, volatility, and tendency to oxidize [50].

The encapsulation of EOs into NPs with polyethylene glycol as a coating material could solve these problems and improve their efficacy due to the small size of the particles [22]. This formulation improves the solubility of the EOs in water and controls the release of the active ingredients [72]. Campolo et al. formulated polyethylene glycol NPs containing EOs of different citrus peels (lemon, mandarin, and sweet orange) to control these tomato borer (*T. absoluta*) larvae through contact and ingestion route. The results showed higher insecticidal activity of NPs on larvae through ingestion; NPs of the mandarin EOs had the highest efficacy through the ingestion toxicity [20]. NPs have more mobility than the bulk materials; this property enables the fast penetration of the active constituents into the insect tissues [73, 74]. In addition, NPs can release the active ingredients of the EOs gradually at the site of action [75].

3 Application of Nanomaterials of Citrus Wastes in the Food Industry

Currently, the preservation of foodstuff products is the most vital concern in the food industry [76]. Processed food products are exposed to contamination with microbes and spoilage by the enzymatic activity. As a result of cutting the food product, the inside nutrients are exposed to the enzymes and microorganisms that reduce the shelf life [77].

The chemical materials used to inhibit some enzymes and decrease the growth rate of microbes in the food products may have toxicological effects in the long run [78, 79]. So, new alternatives for these chemical treatments should be developed through new technologies [80]. The use of EOs is considered as an efficient alternative. EOs are botanical products which consist of a combination of constituents as terpenoids, terpenes, and aromatic phenol-derived constituents with high antioxidant and antimicrobial efficiency. EOs showed promising use for food application because of its ability to prolong the shelf lifetime of processed food by preventing lipid oxidation [67, 81, 82]. The antimicrobial impact of EOs is related to the dissolving of the bacterial cytoplasmic membrane [83]; some studies reported the ability of EOs to inhibit *Staphylococcus aureus* 10 and *Bacillus cereus* [84].

Although EOs thought to be promising in food industries, the use of EOs is often limited as a result of many factors such as rapid volatility, strong odour, and low solubility in water. On the other hand, it is challenging to mix oily compounds in aqueous products due to the chemical and physical instability when applying in the food systems [83]. Thus, the formation of new innovative delivery methods to enhance the effectiveness of EOs is a demand in the foodstuff industries. A promising approach to overcome such obstacles is to incorporate the EOs in the formulation of NEs. The NE is composed of two immiscible phases, the EOs, and the aqueous phase, with droplets of size from 10 to 100 nm [85]. NEs are stable emulsions in which the surfactant is used to stabilize EOs and water phases by decreasing their surface tension [21]. The application of NEs in beverages and drinks is growing due to the small droplet size and transparency or only slight turbidly in the product [86]. Citrus EOs are generally recognized as safe by the United States food and drug administration [87]; they can be used with edible coatings and films to provide protection without significant impact on the sensory properties of the food (Fig. 5).

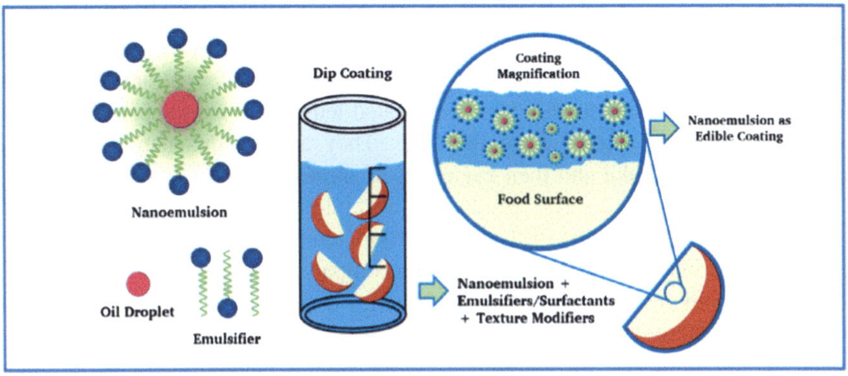

Fig. 5 Schematic demonstration of using nanoemulsion as food coatings. Adapted with permission from Ref. [88] Copyright 2018, MDPI

3.1 *Preservation of Fish Products*

Wu et al. designed NE coating from EOs extracted from citrus peels to preserve silvery pomfret fish (*Pampus argenteus*); this nanoemulsion coating was formed through the incorporation of the EOs to chitosan coating (biopolymer-based edible coating) onto the fish surface. They studied the preservation effect of the NE coating compared to the coating with conventional emulsions; the NE coating showed potential antimicrobial activities and acted as a barrier to oxidative reaction and gas exchange during the refrigerated storage process. The NE coating was efficient in preventing the growth of microorganisms and the changes in the product chemistry during the storage. As a result, the NE coating extended the shelf life of the fish product from 12 to 16 days [85]. This work showed that the NE coating containing citrus EOs could be used efficiently in seafood preservation.

Severino et al. and Donsì et al. used coating based on chitosan containing NE of EOs extracted from the citrus agricultural wastes, mandarin peels, on green beans. Their experiments were associated with several non-thermal treatments against *Listeria innocua*; the results revealed a promising application of this type of NE in the food industry [86, 88].

3.2 *Increasing the Shelf Life of the Cake*

Citrus wastes principally the peels are a rich source of dietary fibres and antioxidants; they can be used in processing healthy food products. The shelf life of foodstuff products is limited; manufacturers of foods need to prevent the oxidation process for the sake of shelf life prolongation of such products. Therefore, antioxidants became essential food additives [89].

Synthetic antioxidants may cause toxic or carcinogenic effects. So, natural resources gained attention, especially, by-products of several juice factories [85]. The extract of dried peels of lemon and orange could be potential antioxidants in foodstuff products. Mahmoud et al. encapsulated lemon and orange extract studied its efficiency on the shelf life of the cake. They applied this nano formulated antioxidants on the cake and then evaluated its influence on the cake sensory and stability properties [90]. The results showed strong antioxidant activity and extension of the storage time of the cake at room temperature. The influence of the nanoencapsulation was estimated and then compared with butylhydroxytoluene, which is a common synthetic antioxidant. There was no notable change in the taste, colour, odour, texture foodstuff industries whole acceptability of the cake samples. Encapsulation empowers the entrapment of active ingredients inside a carrier, then transport, and discharges these active ingredients through a controlled system [86].

3.3 Processed Cheese Supplemented with Nanoliposomes

Processed cheese is widely consumed and has a high nutritional value, but it requires efficient preservation during the long shelf life [90]. Phenolic compounds are known for their antimicrobial and antioxidant activity; the use of EOs containing phenolic compounds improves the nutritional value and reduces the spoilage rate of the processed cheese [91]. These compounds can be extracted from natural sources, such as citrus fruits, and the consumers are more concerned with the compounds from natural resources. However, the interaction between the proteins in milk and the phenolic compounds causes the loss of phenolic compounds activity and reduces the nutritional value of the cheese [92].

El-Messery et al. prepared nanoliposomes as encapsulation technique of mandarin peel extract to be added to the processed cheese as an alternative to bioactive phenolic compounds. These nanoliposomes can help in the conservation of phenolic compounds from the mandarin peel and protect these phenolic compounds from interaction with milk proteins [93]. They found that the encapsulation efficiency of the nanoliposomes during the cold storage of the processed cheese was stable for three months. In addition, they noticed no change in the phenolic content, the physical and chemical properties of the cheese samples. This study reveals that nanoliposomes of EOs is a promising technique to produce processed cheese with a high content of phenolics and prevent the phenolics from interaction and retain their efficiency.

3.4 Mechanism of Antimicrobial Activity of the Essential Oils Nanoformulations

Hydrophobicity is the important characteristic of EOs, and the lipophilic nature of their constituents permits them to interact with the fatty acids of the microbial cell membrane [94]. It is well known that gram-positive bacteria are susceptible to EOs in comparison to gram-negative bacteria [95]. This may be due that gram-negative bacteria have complex and rigid membrane rich in lipopolysaccharide that limits the diffusion of hydrophobic constituents through it. On the other hand, gram-positive is surrounded by a thick peptidoglycan wall. It is not sufficiently dense to resist the small antimicrobial molecules which facilitate access to the cellular membrane [96, 97]. Numerous reports support that the bioactive constituents in EOs might attach to the cell surface and penetrate the phospholipid bilayer of the cell membrane. Therefore, their accumulation disturbs the structure of the cell membrane and influences the cell metabolism leading to cell death, as shown in Fig. 6 [98, 99]. The antimicrobial efficiency of EOs can be enriched by encapsulating with several nanomaterials such as liposomes, NEs, and polymeric NPs. The nanoencapsulation of the EOs control the release of the active constituents, decrease the volatility, and protect it from the environmental pressure [100].

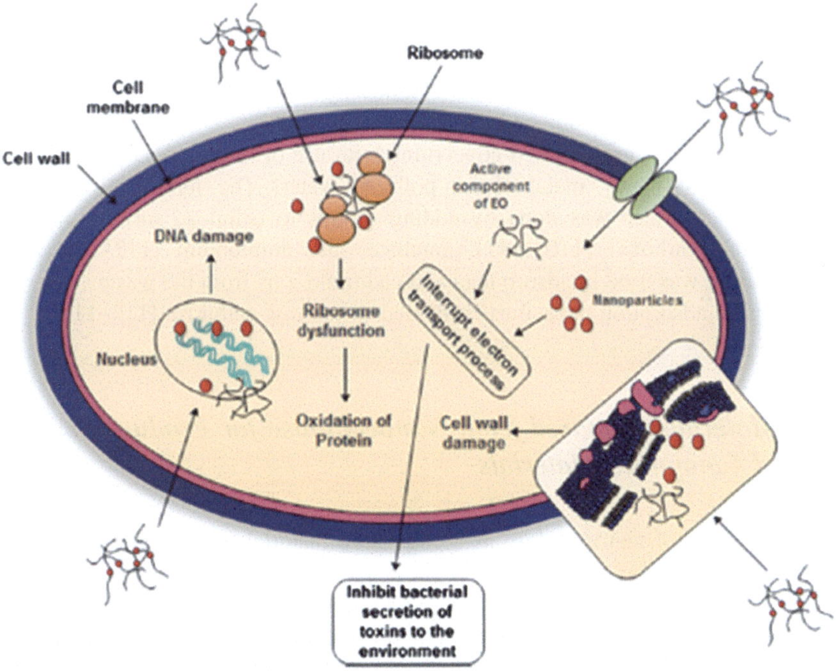

Fig. 6 Schematic demonstration of possible nanoencapsulated EOs mechanism of actions. Adapted with permission from Ref. [94] Copyright 2017, Elsevier

4 Nanocellulose Derived from Citrus Wastes

Cellulose, in the form of nanostructures, is known as nanocellulose; it is one of the most noticeable green materials. Nanocellulose gained increasing interests due to its attractive characteristics such as biocompatibility, nontoxicity, abundance, and renewability [101]. Several applications of nanocellulose attracted attention, such as the water treatment to remove pollutants and the manufacturing of high-performance composites. Owing to the low cost and energy, the extraction of nanocellulose from the agricultural wastes is a smart alternative for waste treatment [102].

4.1 Water Treatment Using Nanocellulose Derived from Citrus Wastes

Nonhazardous adsorption materials like nanocellulose are an important source for the elimination of the water pollutants without a dangerous effect on human health and the environment. Nanocellulose can be produced from agricultural wastes through different methods; bacterial, chemical, and physical, depending on the fibre content [103–105]. Several studies reported citrus peels as one of the best sources of nanocellulose [102, 106, 107]. Nanocellulose affords an alternative to the conventional adsorbent materials as zeolite or activated carbon. The use of nanocellulose in water treatment was effectively reported for the removal of heavy metal and organic pollutants [108]. High adsorption achieved by the nanocelluloses is due to its several reactive groups and large surface areas.

Several studies were reported concerning with use of modified nanocellulose in the removal of heavy metals from polluted water. The modification of the nanocellulose surface was done by adding groups to cellulose surfaces such as amine [109] carboxyl [110, 111], xanthate, and ammonium [112]. Modified nanocellulose was used to adsorb lead and cadmium ions from the wastewater with much higher adsorption capacities compared to the raw cellulose [113–115].

4.2 Materials Prepared from Nanocellulose for Production of Composite Materials

The application of nanocellulose as a composite material brought attention to the brilliant properties of nanocellulose. These properties include lightweight, high strength, and biodegradability [12, 15], in addition to its high rigidity as supporting material, which makes it ideally used in composite materials [27]. Such biomaterial has high applicability in numerous industries such as plastics and polymer composites, films, gels, foams, cosmetics, and implant material, concrete, coatings, screens. Nanocelluloses have a much larger surface area than cellulose, so it is an eco-friendly source for paper production with considerable quality. Nanocellulose can be used in high strength and lightweight material for electronics, coatings, and latex paints [116]. Moreover, nanocellulose can be applied in drug delivery as a biodegradable tissue scaffold and the production of filter paper used in the treatment of water and oil recovery [117].

5 Conclusion

This chapter summarizes the use of nanotechnology to convert citrus wastes into innovative nanomaterials through different formulations such as NEs and nanocapsules. These materials have excellent properties such as biodegradability and high efficiency and can be used in various fields. The nanoinsecticides can be applied to control disease-vector mosquitoes, the domestic cockroach pest, the stored gain pests, and the plant cop pests as a tomato. Also, these innovative nanomaterials can be used in the food industry to prolong the shelf life of many products, such as processed cheese, cake, and fish products. Finally, the chapter mentioned the application of the nanomaterials of the citrus waste in water treatment and their great application in various composite materials. However, there is still a significant gap in knowledge about the whole lifecycles of the nanomaterials in general. It is crucial to assess the consequences of using nanomaterials on the environment. Along with the determination of the optimum dose and safe limits and the interactions with the food matrices, other packaging materials, and antimicrobial compounds.

6 Future Perspectives

Despite reducing pollution through recycling of the citrus wastes and the production of a new type of promising materials, more studies are mandatory to estimate the effect of these new nanomaterials on the living organisms, especially the human beings and the environment, including the water sources and the soil. Environmental issues have to be taken into consideration when applying the nanomaterials in different products such as coatings, latex paints, plastics, foams, and cosmetics because these products can be frequently touched and handled by people. The excellent penetration power of NPs due to their tiny size requires excellent attention. Generally, NPs can enter organisms during inhalation or ingestion and can accumulate within the body to various tissues and organs due to long exposure. Moreover, the use of nanomaterials in packaging and food processing may cause a new collection of risks. Consumers are noticed to be accepting of nanomaterials in the out packaging more than in food processing. There is a need for safety assessments with an increasing number of food products containing nanomaterials. For example, studying the migration of NPs from food packaging into the foodstuff and the problem of accumulation of these particles in the living organisms. In addition, the various reaction factors such as pressure, pH, temperature, and time have to be evaluated as it may affect the nature of the properties of the nanomaterials.

References

1. Okwu DE, Emenike IN (2006) Evaluation of the phytonutrients and vitamin contents of Citrus fruits. Int J Mol Med Adv Sci 2(1):1–6
2. de Castro MDL (2014) Towards a comprehensive exploitation of agrofood residues: olive tree–olive oil as example. Comptes Rendus Chim 17(3):252–260
3. Martín M, Siles J, El Bari H, Chica A, Martin A (2010) Orange peel: organic waste or energetic resource. Technol Syst Differ Manure Org Waste Treat Opt 1:25–28
4. Sharma K, Mahato N, Cho MH, Lee YR (2017) Converting citrus wastes into value-added products: economic and environmently friendly approaches. Nutrition 34:29–46
5. Awan AT, Tsukamoto J, Tasic L (2013) Orange waste as a biomass for 2G-ethanol production using low cost enzymes and co-culture fermentation. Rsc Adv 3(47):25071–25078
6. Lin J, Zuo J, Gan L, Li P, Liu F, Wang K, Chen L, Gan H. (2011) Effects of mixture ratio on anaerobic co-digestion with fruit and vegetable waste and food waste of China. J Environ Sci 23(8):1403–1408
7. Sharma K, Mahato N, Lee YR (2019) Extraction, characterization and biological activity of citrus flavonoids. Rev Chem Eng 35(2):265–284
8. Mahato N, Sharma K, Koteswararao R, Sinha M, Baral E, Cho MH (2019) Citrus essential oils: extraction, authentication and application in food preservation. Crit Rev Food Sci Nutr 59(4):611–625
9. Mahato N, Sharma K, Sinha M, Cho MH (2018) Citrus waste derived nutra-/pharmaceuticals for health benefits: current trends and future perspectives. J Funct Foods 40:307–316
10. Mahato N, Sharma K, Nabybaccus F, Cho MH (2016) Citrus waste reuse for health benefits and pharma-/neutraceutical applications. Era's J Med Res 3(1):20–32
11. Su H, Tan F, Xu Y (2016) Enhancement of biogas and methanization of citrus waste via biodegradation pretreatment and subsequent optimized fermentation. Fuel 181:843–851
12. Negro V, Mancini G, Ruggeri B, Fino D (2016) Citrus waste as feedstock for bio-based products recovery: review on limonene case study and energy valorization. Bioresour Technol 214:806–815
13. Mamma D, Christakopoulos P (2014) Biotechnological potential of citrus peels. Springer, Dordrecht, pp 273-291
14. Zema DA, Calabrò PS, Folino A, Tamburino V, Zappia G, Zimbone SM (2018) Valorisation of citrus processing waste: a review. Waste Manag 80:252–273
15. Fagbohungbe MO, Herbert BMJ, Hurst L, Li H, Usmani SQ, Semple KT (2016) Impact of biochar on the anaerobic digestion of citrus peel waste. Bioresour Technol 216:142–149
16. Kah M, Beulke S, Tiede K, Hofmann T (2013) Nanopesticides: state of knowledge, environmental fate, and exposure modeling. Crit Rev Environ Sci Technol
17. Mahato N, Sinha M, Sharma K, Koteswararao R, Cho MH (2019) Modern extraction and purification techniques for obtaining high purity food-grade bioactive compounds and value-added co-products from citrus wastes. Foods 8(11):523
18. Campolo O, Malacrinò A, Zappalà L, Laudani F, Chiera E, Serra D, Russo M, Palmeri V. (2014) Fumigant bioactivity of five Citrus essential oils against Tribolium confusum. Phytoparasitica 42(2):223–233
19. Song J-E, Kim J-M, Lee N-H, Yang J-Y, Lee H-S (2016) Acaricidal and insecticidal activities of essential oils against a stored-food mite and stored-grain insects. J Food Prot 79 (1):174–178

20. Campolo O, Cherif A, Ricupero M, Siscaro G, Grissa-Lebdi K, Russo A, Cucci LM, Di Pietro P, Satriano C, Desneux N, Biondi A (2017) Citrus peel essential oil nanoformulations to control the tomato borer, Tuta absoluta: chemical properties and biological activity. Sci Rep 7(1):1–10
21. McClements DJ (2015) Food emulsions: principles, practices, and techniques. CRC press
22. González JOW, Gutiérrez MM, Ferrero AA, Band BF (2014) Essential oils nanoformulations for stored-product pest control–characterization and biological properties. Chemosphere 100:130–138
23. Mason TG, Wilking JN, Meleson K, Chang CB, Graves SM (2006) Nanoemulsions: formation, structure, and physical properties. J Phys: Condens Matter 18:635–666. https://doi.org/10.1088/0953-8984/18/41/r01
24. Tadros T, Izquierdo P, Esquena J, Solans C (2004) Formation and stability of nano-emulsions. Adv Colloid Interface Sci 108:303–318
25. Anton N, Gayet P, Benoit J-P, Saulnier P (2007) Nano-emulsions and nanocapsules by the PIT method: an investigation on the role of the temperature cycling on the emulsion phase inversion. Int J Pharm 344(1–2):44–52
26. Kralova I, Sjöblom J (2009) Surfactants used in food industry: a review. J Dispers Sci Technol 30(9):1363–1383
27. McClements DJ, Rao J (2011) Food-grade nanoemulsions: formulation, fabrication, properties, performance, biological fate, and potential toxicity. Crit Rev Food Sci Nutr 51 (4):285–330
28. Leong TSH, Wooster TJ, Kentish SE, Ashokkumar M (2009) Minimising oil droplet size using ultrasonic emulsification. Ultrason Sonochem 16(6):721–727
29. Gutiérrez JM, González C, Maestro A, Solè I, Pey CM, Nolla J (2008) Nano-emulsions: new applications and optimization of their preparation. Curr Opin Colloid Interface Sci 13 514 (4):245–251
30. Velikov KP, Pelan E (2008) Colloidal delivery systems for micronutrients and nutraceuticals. Soft Matter 4(10):1964–1980
31. Akono PN, Dongmo PM, Tonga C, Kouotou S, Kekeunou S, Magne GT, Lehman LG, Menut C (2015) Larvicidal activity of essential oils from pericarps of ripe Citrus fruits cultivated in Cameroon on pyrethroids sensitive and resistant strains of Anopheles gambiae Giles, 1902. J Entomol Zool Stud 3:334–339
32. Anjali CH, Khan SS, Margulis-Goshen K, Magdassi S, Mukherjee A, Chandrasekaran N (2010) Formulation of water-dispersible nanopermethrin for larvicidal applications. Ecotoxicol Environ Saf 73(8):1932–1936
33. Azmy RM, El Gohary EGE, Mahmoud DM, Salem DAM, Abdou MA, Salama MS (2019) Assessment of larvicidal activity of nanoemulsion from Citrus sinensis essential oil on Culex pipiens L. (Diptera: Culicidae). Egypt J Aquat Biol Fish 23(3):61–67
34. Saad MMG (2013) Chemical composition and biological activities of four Citrus essential oils. J Plant Prot Pathol 4(9):767–780
35. Fallatah SA (2010) Histopathological effects of fenugreek (Trigonella foenumgraceum) extracts on the larvae of the mosquito Culex quinquefasciatus. J Arab Soc Med Res 5 (2):123–130
36. Mishra P, Tyagi BK, Chandrasekaran N, Mukherjee A (2018) Biological nanopesticides: a greener approach towards the mosquito vector control. Environ Sci Pollut Res 25 (11):10151–10163
37. González JOW, Stefanazzi N, Murray AP, Ferrero AA, Band BF (2004) Novel nanoinsecticides based on essential oils to control the German cockroach. J Pest Sci 88 (2):393–404, 2015
38. Fotedar R, Shriniwas UB, Verma A (1991) Cockroaches (Blattella germanica) as carriers of microorganisms of medical importance in hospitals. Epidemiol Infect 107(1):181–187
39. Pai H-H, Ko YC, Chen ER (2003) Cockroaches (Periplaneta americana and Blattella germanica) as potential mechanical disseminators of Entamoeba histolytica. Acta Trop 87 (3):355–359

40. Cochran DG (1995) Insecticide resistance. Understanding and controlling the German cockroach. Oxford University Press, New York, pp 171–192
41. Hemingway J, Small GJ, Monro AG (1993) Possible mechanisms of organophosphorus and carbamate insecticide resistance in German cockroaches (Dictyoptera: Blattelidae) from different geographical areas. J Econ Entomol 86(6):1623–1630
42. Valles SM, Yu SJ (1996) Detection and biochemical characterization of insecticide resistance in the German cockroach (Dictyoptera: Blattellidae). J Econ Entomol 89(1):21–26
43. Wei Y, Appel AG, Moar WJ, Liu N (2001) Pyrethroid resistance and cross-resistance in the German cockroach, Blattella germanica (L). Pest Manag Sci Former Pestic Sci 57(11):1055–1059
44. Casida JE, Durkin KA (2013) Neuroactive insecticides: targets, selectivity, resistance, and secondary effects. Annu Rev Entomol 58:99–117
45. Köhler H-R, Triebskorn R (2013) Wildlife ecotoxicology of pesticides: can we track effects to the population level and beyond? Science (80-) 341(6147):759–765
46. Tripathi AK, Upadhyay S, Bhuiyan M, Bhattacharya PR (2009) A review on prospects of essential oils as biopesticide in insect-pest management. J Pharmacogn Phyther 1(5):52–63
47. Talbert R, Wall R (2012) Toxicity of essential and non-essential oils against the chewing louse, Bovicola (Werneckiella) ocellatus. Res Vet Sci 93(2):831–835
48. Athanassiou CG, Kavallieratos NG, Evergetis E, Katsoula A-M, Haroutounian SA (2013) Insecticidal efficacy of silica gel with Juniperus oxycedrus ssp. oxycedrus (Pinales: Cupressaceae) essential oil against Sitophilus oryzae (Coleoptera: Curculionidae) and Tribolium confusum (Coleoptera: Tenebrionidae). J Econ Entomol 106(4):1902–1910
49. Isman MB, Miresmailli S, Machial C (2011) Commercial opportunities for pesticides based on plant essential oils in agriculture, industry and consumer products. Phytochem Rev 10 (2):197–204
50. Moretti MDL, Sanna-Passino G, Demontis S, Bazzoni E (2002) Essential oil formulations useful as a new tool for insect pest control. AAPS PharmSciTech 3(2):64–74
51. Danprasert K, Kumar R, Ming H, Gupta P, Shakil NA, Prasad AK, Parmar VS, Kumar J, Samuelson LA, Watterson AC. (2003) Synthesis of novel poly (ethylene glycol) based amphiphilic polymers. Eur Polym J 39(10):1983–1990
52. Nel AE, Mädler L, Velegol D, Xia T, Hoek EM, Somasundaran P, Klaessig F, Castranova V, Thompson M. (2009) Understanding biophysicochemical interactions at the nano–biointerface. Nat Mater 8(7):543–557
53. Margulis-Goshen K, Magdassi S (2013) Nanotechnology: an advanced approach to the development of potent insecticides. In: Advanced Technologies for Managing Insect Pests. Springer, Berlin, pp 295–314
54. Saroukolai AT, Moharramipour S, Meshkatalsadat MH (2004) Insecticidal properties of Thymus persicus essential oil against Tribolium castaneum and Sitophilus oryzae. J Pest Sci 83(1):3–8, 2010
55. Edde PA (2012) A review of the biology and control of Rhyzopertha dominica (F.) the lesser grain borer. J Stored Prod Res 48:1–18
56. Kljajić P, Perić I (2006) Susceptibility to contact insecticides of granary weevil Sitophilus granarius (L.) (Coleoptera: Curculionidae) originating from different locations in the former Yugoslavia. J Stored Prod Res 42(2):149–161
57. Islam MS, Hasan MM, Lei C, Mucha-Pelzer T, Mewis I, Ulrichs C (2004) Direct and admixture toxicity of diatomaceous earth and monoterpenoids against the storage pests Callosobruchusmaculatus (F.) and Sitophilus oryzae (L.). J Pest Sci 83(2):105–112, 2010
58. Pimentel MAG, LRD'A F, Guedes RNC, Sousa AH, Tótola MR (2009) Phosphine resistance in Brazilian populations of Sitophilus zeamais motschulsky (Coleoptera: Curculionidae). J Stored Prod Res 45(1):71–74
59. Ali A, Ahmad F, Biondi A, Wang Y, Desneux N (2004) Potential for using Datura alba leaf extracts against two major stored grain pests, the khapra beetle Trogoderma granarium and the rice weevil Sitophillus oryzae. J Pest Sci 85(3):359–366, 2012

60. Isman MB (2006) Botanical insecticides, deterrents, and repellents in modern agriculture and an increasingly regulated world. Annu Rev Entomol 51:45–66
61. Regnault-Roger C (2013) Essential oils in insect control. Nat Prod Berlin, Ger. Springer, Berlin, pp 4087–4102
62. Vincent JFV, Wegst UGK (2004) Design and mechanical properties of insect cuticle. Arthropod Struct Dev 33(3):187–199
63. Sfara V, Zerba EN, Alzogaray RA (2009) Fumigant insecticidal activity and repellent effect of five essential oils and seven monoterpenes on first-instar nymphs of Rhodnius prolixus. J Med Entomol 46(3):511–515
64. Tarelli G, Zerba EN, Alzogaray RA (2009) Toxicity to vapor exposure and topical application of essential oils and monoterpenes on Musca domestica (Diptera: Muscidae). J Econ Entomol 102(3):1383–1388
65. Giunti G, Palermo D, Laudani F, Algeri GM, Campolo O, Palmeri V (2019) Repellence and acute toxicity of a nano-emulsion of sweet orange essential oil toward two major stored grain insect pests. Ind Crops Prod 142:111869
66. Desneux N, Luna MG, Guillemaud T, Urbaneja A (2011) The invasive South American tomato pinworm, Tuta absoluta, continues to spread in Afro-Eurasia and beyond: the new threat to tomato world production. J Pest Sci 84(4):403–408
67. Garzia GT, Siscaro G, Biondi A, Zappalà L (2012) Tuta absoluta, a South American pest of tomato now in the EPPO region: biology, distribution and damage. EPPO Bull 42(2):205–210
68. Haddi K, Berger M, Bielza P, Cifuentes D, Field LM, Gorman K, Rapisarda C, Williamson MS, Bass C. (2012) Identification of mutations associated with pyrethroid resistance in the voltage-gated sodium channel of the tomato leaf miner (Tuta absoluta). Insect Biochem Mol Biol 42(7):506–513
69. Biondi A, Zappalà L, Stark JD, Desneux N (2013) Do biopesticides affect the demographic traits of a parasitoid wasp and its biocontrol services through sublethal effects? PLoS One 8 (9):e76548
70. Abbes K, Biondi A, Kurtulus A, Ricupero M, Russo A, Siscaro G, Chermiti B, Zappalà L (2015) Combined non-target effects of insecticide and high temperature on the parasitoid Bracon nigricans. PLoS One 18;10(9):e0138411
71. Cardiet G, Fuzeau B, Barreau C, Fleurat-Lessard F (2004) Contact and fumigant toxicity of some essential oil constituents against a grain insect pest Sitophilus oryzae and two fungi, Aspergillus westerdijkiae and Fusarium graminearum. J Pest Sci 85(3):351–358
72. Sasson Y, Levy-Ruso G, Toledano O, Ishaaya I (2007) Nanosuspensions: emerging novel agrochemical formulations. In: Insecticides Design Using Advanced Technologies. Springer, Berlin, pp 1–39
73. Gogos A, Knauer K, Bucheli TD (2012) Nanomaterials in plant protection and fertilization: current state, foreseen applications, and research priorities. J Agric Food Chem 60 (39):9781–9792
74. de Oliveira JL, Campos EVR, Bakshi M, Abhilash PC, Fraceto LF (2014) Application of nanotechnology for the encapsulation of botanical insecticides for sustainable agriculture: prospects and promises. Biotechnol Adv 32(8):1550–1561
75. Rasooli I (2007) Food preservation–a biopreservative approach. Food 1(2):111–136
76. Madden JM (1992) Microbial pathogens in fresh produce–the regulatory perspective. J Food Prot 55(10):821–823
77. Madden JM (1992) Microbial pathogens in fresh produce–The regulatory perspective. J Food Prot 55(10):821–823
78. Beuchat LR (1998) World Health Organization. Surface decontamination of fruits and vegetables eaten raw: a review.
79. Hansen FT, Sørensen JL, Giese H, Sondergaard TE, Frandsen RJN (2012) Quick guide to polyketide synthase and nonribosomal synthetase genes in Fusarium. Int J Food Microbiol 155(3):128–136
80. Giatrakou V, Ntzimani A, Savvaidis IN (2010) Effect of chitosan and thyme oil on a ready to cook chicken product. Food Microbiol 27(1):132–136

81. López-Malo A, Alzamora SM, Guerrero S (2000) Natural antimicrobials from plants. In: Minimally processed fruits and vegetables. Fundamentals aspects and applications. Aspen Publishers, Gaithersburg, MD, p 237
82. Ghosh V, Mukherjee A, Chandrasekaran N (2014) Eugenol-loaded antimicrobial nanoemulsion preserves fruit juice against, microbial spoilage. Colloids Surfaces B Biointerfaces 114:392–397
83. Prakash A, Baskaran R, Paramasivam N, Vadivel V (2018) Essential oil based nanoemulsions to improve the microbial quality of minimally processed fruits and vegetables: a review. Food Res Int 111:509–523
84. Chang Y, McLandsborough L, McClements DJ (2012) Physical properties and antimicrobial efficacy of thyme oil nanoemulsions: influence of ripening inhibitors. J Agric Food Chem 60 (48):12056–12063
85. Wu C, Wang L, Hu Y, Chen S, Liu D, Ye X (2016) Edible coating from citrus essential oil-loaded nanoemulsions: physicochemical characterization and preservation performance. RSC Adv 6(25):20892–20900
86. Severino R, Vu KD, Donsì F, Salmieri S, Ferrari G, Lacroix M (2014) Antibacterial and physical effects of modified chitosan based-coating containing nanoemulsion of mandarin essential oil and three non-thermal treatments against Listeria innocua in green beans. Int J Food Microbiol 191:82–88
87. Persico P, Ambrogi V, Carfagna C, Cerruti P, Ferrocino I, Mauriello G (2009) Nanocomposite polymer films containing carvacrol for antimicrobial active packaging. Polym Eng Sci 49(7):1447–1455
88. Zambrano-Zaragoza ML, González-Reza R, Mendoza-Muñoz N, Miranda-Linares V, Bernal-Couoh TF, Mendoza-Elvira S, Quintanar-Guerrero D (2018) Nanosystems in edible coatings: a novel strategy for food preservation. Int J Mol Sci 19:705
89. Donsì F, Marchese E, Maresca P, Pataro G, Vu KD, Salmieri S, Lacroix M, Ferrari G (2015) Green beans preservation by combination of a modified chitosan based-coating containing nanoemulsion of mandarin essential oil with high pressure or pulsed light processing. Postharvest Biol Technol 106:21–32
90. Buňka F, Hrabě J, Kráčmar S (2004) The effect of sterilisation on amino acid contents in processed cheese. Int Dairy J 14(9):829–831
91. Martillanes S, Rocha-Pimienta J, Cabrera-Bañegil M, Martín-Vertedor D, Delgado-Adámez J (2017) Application of phenolic compounds for food preservation: food additive and active packaging. Phenolic Compd Act London, UK IntechOpen, pp 39–58
92. Mehanna NS, Hassan ZMR, El-Din HMF, Ali AA-E, Amarowicz R, El-Messery TM (2014) Effect of interaction phenolic compounds with milk proteins on cell line. Food Nutr Sci 5 (22):2130
93. El-Messery TM, El-Said MM, Farahat ESA (2019) Production of functional processed cheese supplemented with nanoliposomes of mandarin peel extract. Sciences (New York) 22 (5):247–256
94. Rai M, Paralikar P, Jogee P, Agarkar G, Ingle AP, Derita M, Zacchino S (2017) Synergistic antimicrobial potential of essential oils in combination with nanoparticles: emerging trends and future perspectives. Int J Pharm 519(1–2):67–78
95. Azhdarzadeh F, Hojjati M (2016) Chemical composition and antimicrobial activity of leaf, ripe and unripe peel of bitter orange (Citrus aurantium) essential oils. Nutr Food Sci Res 3 (1):43–50
96. Holley RA, Patel D (2005) Improvement in shelf-life and safety of perishable foods by plant essential oils and smoke antimicrobials. Food Microbiol 22(4):273–292
97. Hyldgaard M, Mygind T, Meyer RL (2012) Essential oils in food preservation: mode of action, synergies, and interactions with food matrix components. Front Microbiol 3:12
98. Bajpai VK, Sharma A, Baek K-H (2013) Antibacterial mode of action of Cudrania tricuspidata fruit essential oil, affecting membrane permeability and surface characteristics of food-borne pathogens. Food Control 32(2):582–590

99. Lv F, Liang H, Yuan Q, Li C (2011) In vitro antimicrobial effects and mechanism of action of selected plant essential oil combinations against four food-related microorganisms. Food Res Int 44(9):3057–3064
100. Bilia AR, Guccione C, Isacchi B, Righeschi C, Firenzuoli F, Bergonzi MC (2014) Essential oils loaded in nanosystems: a developing strategy for a successful therapeutic approach. Evidence-based complement. Altern Med 2014
101. Trache D, Tarchoun AF, Derradji M, Hamidon TS, Masruchin N, Brosse N, Hussin MH (2020) Nanocellulose: from fundamentals to advanced applications. Front Chem 8
102. Mariño M, Lopes da Silva L, Durán N, Tasic L (2015) Enhanced materials from nature: nanocellulose from citrus waste. Molecules 20(4):5908–5923
103. Kardam A, Raj KR, Srivastava S, Srivastava MM (2014) Nanocellulose fibers for biosorption of cadmium, nickel, and lead ions from aqueous solution. Clean Technol Environ Policy 16(2):385–393
104. Moon RJ, Martini A, Nairn J, Simonsen J, Youngblood J (2011) Cellulose anomaterials review: structure, properties and nanocomposites. Chem Soc Rev 40(7):3941–3994
105. Aspler JJ, Bouchard J, Hamad W, Berry R, Beck S, Drolet F, Zou X. (2013) Review of nanocellulosic products and their applications. Biopolym Nanocomposites Process Prop Appl 461–508
106. Thygesen LG, Hidayat BJ, Johansen KS, Felby C (2011) Role of supramolecular cellulose structures in enzymatic hydrolysis of plant cell walls. J Ind Microbiol Biotechnol 38(8):975–983
107. Zain M, Fazelin N, Yusop M, Ahmad I (2015) Preparation and characterization of cellulose and nanocellulose from pomelo (Citrus grandis) albedo. Nutr Food Sci
108. Kalia S, Boufi S, Celli A, Kango S (2014) Nanofibrillated cellulose: surface modification and potential applications. Colloid Polym Sci 292(1):5–31
109. Shivakumar Singh P, Vidyasagar GM (2014) Biosynthesis, characterization, and antidermatophytic activity of silver nanoparticles using Raamphal plant (Annona reticulata) aqueous leaves extract. Indian J Mater Sci 2014
110. Donia AM, Atia AA, Abouzayed FI (2012) Preparation and characterization of nano-magnetic cellulose with fast kinetic properties towards the adsorption of some metal ions. Chem Eng J 191:22–30
111. Yu X, Tong S, Ge M, Wu L, Zuo J, Cao C, Song W (2013) Adsorption of heavy metal ions from aqueous solution by carboxylated cellulose nanocrystals. J Environ Sci 25(5):933–943
112. Pillai SS, Deepa B, Abraham E, Girija N, Geetha P, Jacob L, Koshy M. (2013) Biosorption of Cd (II) from aqueous solution using xanthated nano banana cellulose: equilibrium and kinetic studies. Ecotoxicol Environ Saf 98:352–360
113. Stephen M, Catherine N, Brenda M, Andrew K, Leslie P, Corrine G (2011) Oxolane-2, 5-dione modified electrospun cellulose nanofibers for heavy metals adsorption. J Hazard Mater 192(2):922–927
114. Boufi S (2014) Nanofibrillated cellulose: sustainable nanofiller with broad potentials use. In: Biomass and Bioenergy. Springer, Berlin, pp 267–305
115. Hokkanen S, Repo E, Sillanpää M (2013) Removal of heavy metals from aqueous solutions by succinic anhydride modified mercerized nanocellulose. Chem Eng J 223:40–47
116. Mandal A, Chakrabarty D (2011) Isolation of nanocellulose from waste sugarcane bagasse (SCB) and its characterization. Carbohydr Polym 86(3):1291–1299
117. Tsukamoto J, Durán N, Tasic L (2013) Nanocellulose and bioethanol production from orange waste using isolated microorganisms. J Braz Chem Soc 24(9):1537–1543
118. Mahmoud KF, Ibrahim MA, Mervat E-D, Shaaban HA, Kamil MM, Hegazy NA (2016) Nano-encapsulation efficiency of lemon and orange peels extracts on cake shelf life. Am J Food Technol 11(3):63–75

Bottom-Up Approach Through Microbial Green Biosynthesis of Nanoparticles from Waste

Rania Azouz

Abstract The small invisible nanofactories of bacteria, fungi, and microalgae represent a green, cheap, and easy biogenic method to build nanosized metal particles in a bottom-up approach. Using waste as the source of metal ions helps environmental remediation and increases waste value by recovering rare, precious, and essential metals. The recovery happens in an eco-friendly way after adjustment of growth conditions and salts concentrations. Models for the waste solutions and real wastewaters were used as the source of platinum group metals and rare metals to be recovered by microorganisms that can process and resist high heavy metals concentrations. The produced nanoparticles are highly efficient candidates for catalytic applications and other potential applications. The high concentration of gold in solid electronic waste scrap is the target for cyanogenic bacteria to solubilize gold, refined by another bacterium in the form of nanogold that can be reused in electronic devices. Microbial biosynthesis of nanoparticles has advantages that far exceed that of other methods, though it is difficult to control the shape, size, and size distribution of nanoparticles produced by this method, and it has low production to be used commercially. Further production enhancement and control over the produced nanoparticles are expected.

Keywords Green chemistry · Microorganisms · Biosynthesis · Nanoparticles · Waste recycling

List of Abbreviations

EDS	Energy-dispersive X-ray spectroscopy
CD-RW	Re-writable compact discs
DVD-RW	Re-writable digital versatile discs
IC	Intracellular

R. Azouz (✉)
Clinical Microbiology Unit, Clinical and Chemical Pathology Department, Faculty of Medicine, Beni Suef University, Beni Suef, Egypt
e-mail: rania_azouz@med.bsu.edu.eg

R. Azouz
Medical Administration, Beni Suef University, Beni Suef, Egypt

A. S. H. Makhlouf and G. A. M. Ali (eds.), *Waste Recycling Technologies for Nanomaterials Manufacturing*, Topics in Mining, Metallurgy and Materials Engineering, https://doi.org/10.1007/978-3-030-68031-2_23

MIC	Minimal inhibitory concentrations
MW	Microwave
MYE	Malate yeast extract
NADH	Nicotinamide adenine dinucleotide
NADPH	Nicotinamide adenine dinucleotide phosphate
NPs	Nanoparticles
PGM	Platinum group metals
PIXE	Proton-induced X-ray emission
QCM	Quartz crystal microbalance
rpm	run per minute
SEM	Scanning electron microscope
TEM	Transmission electron microscope
US	United States
UV	Ultra-violet

1 Introduction

Several free-living bacteria, yeasts, molds, microalgae, and macroalgae are successfully used to produce nanoparticles. Bacteria are the most prominent microorganisms used in nanoparticles' biosynthesis due to their remarkable ability to reduce heavy metal ions and produce well-characterized nanoparticles extracellularly and/or intracellularly [1]. Bacteria are 0.2–2 um in diameter and 2–8 um in length, and nanoparticles are particles with at least one dimension 10–100 nm [2]. This living factory stands for a renewable, clean, cheap, and facile method for fabricating nanoparticles by adjusting growth conditions, shaking speed, and metal salts concentrations [3–5]. *Pseudomonas stutzeri* was the first to be used for this purpose when Joerger et al. in 2000 used *P. stutzeri* AG259 to synthesize Ag nanoparticles with size less than 200 nm [6]. Later on Au, Ag, Zn, and Fe nanoparticles have been successfully produced [7], using other bacteria [8–10].

Industrial wastewaters hold high concentrations of platinum group metals (PGM) (i.e., ruthenium, rhodium, palladium, osmium, iridium, and platinum) [11]. Recovery of those metals from waste is a motivating approach because of PGM's scarceness in the earth's crust, increased demand for them, and difficulty of their extraction from ores as they usually exist in low concentrations [12]. Using microbes for this purpose has the advantage of low cost and green approach [13].

Solid electronic waste far exceeds in amount other types of solid waste. It is expected to reach 52 million tons by 2021 [14] due to the increasing demand for electronic devices and some devices' short lifespan. In some countries, the lifetime of mobile phones is less than two years [15]. Recovery of precious metals and base metals from electronic waste is a promising approach as it is estimated that gold contained in one ton of computers electronic scrap, for example, is more than that recovered from 17 tons of gold ores [16], and the concentration of copper is 40 times more than that in its ores [17]. The conventional methods for gold recovery are either high energy-consuming (pyrometallurgy) or causing chemical pollution (hydrometallurgy) due to the use of concentrated chemicals (e.g., concentrated nitric and hydrochloric acids) [18]. The biologic method of recovery avoids all disadvantages of conventional methods. The amount of gold recovered by bacteria from electronic waste scrap can be enhanced by several methods [17].

2 Green Chemistry and Its Basic Principles

Green chemistry is an increasingly adopted approach that aims to augment the benefit and efficiency of chemical reactions and minimize waste production in an eco-friendly way that prioritizes human health and environmental protection. It is hard for any chemical reaction to be entirely green, though Anastas and Warner described 12 principles to be applied whenever possible [19]. They are presented in Table 1.

Biologic synthesis of nanoparticles applies well to all principles of green chemistry. The sustainable feedstocks of the used living microorganisms or plant extracts are the source of enzymes and other biomolecules that construct and stabilize nanoparticles without the need for hazardous chemicals, high energy sources, or waste management. The produced particles have physical and chemical properties that suit their function much more efficiently than those produced by conventional methods [20].

3 Different Approaches for the Production of Nanoparticles

The fabrication of nanoparticles depends on two basic approaches. The two methods are described below.

3.1 Top-Down Approach

One of the two approaches calls for fragmentation of the bulk of material to small and even smaller particles till the desired size is reached using external energy force as thermal energy, pressure, light, or kinetic force. This approach is termed the top-down approach. Due to cavities and rough surfaces of the produced particles through this approach, it is difficult to get the admirable crystallography. However, this approach has the advantage of timely large-scale production. This is well

Table 1 Twelve principles of green chemistry

Principle	Explanation
To prevent is better than to cure	It is better to plan to prevent or minimize waste generation than to manage or clean up waste, either gaseous, solid, solution, or noise
Atom economy	Maximize the number of atoms from reagents incorporated in the final product to increase the reaction efficiency
Nonhazardous chemical synthesis	Synthetic methods and chemical reactions have to be as safe as possible considering the hazards of all incorporated chemicals, including waste
Designing safer chemicals	Design chemicals to be less toxic, safer and more effective in their anticipated function
Safer solvents and auxiliaries	The safe green solvent should be noninflammable, nontoxic, nonvolatile, not mutagenic, commonly available, cheap, and degradable
Energy economy	Ambient temperature and pressure are preferred to using heat, cooling, pressure, or vacuum pressure conditions
Renewable feeds tocks	Use renewable raw materials (e.g., plant and microbial sources) rather than limited petrochemical compounds
Reduce derivatives	Avoid or minimize derivatives (e.g., blocking, protecting, or modifying groups) to avoid adding more reagents or producing additional waste
Catalysis	Using catalysts decreases the time of reaction, declines energy needed, reduces waste, and increases specificity
Degradable waste	Design chemical products that do not persist in the environment or produce toxic degradations when disintegrated
Real-time waste	Advanced techniques allow monitoring of chemical reactions as soon as they take place to prevent the production or release of any hazardous waste
Accident prevention	The use of inherently safer compounds minimizes the risk of chemical accidents like fire, explosion, corrosion, or release of toxic substances

represented in the ball milling method, which depends on grinding microparticles in balls mill where kinetic energy is transferred to the microparticles powder to get the desired nanoscale particles. This method has a drawback of significant surface defects that badly affects the surface reaction properties of the produced nanoparticles [21]. Spray pyrolysis is another method that requires high pressure and temperature to force the fluid of metal salts through a narrow opening to be burned on the other side where nanoparticles can be recovered from the solid ash. Lithography is also a method of a top-down approach in which electron beam, X-ray, UV light, or electrostatic power are used to cut nanoscale units through solid bulk material [21, 22].

3.2 *Bottom-Up Approach*

The other approach, the bottom-up approach, stands for using metal atoms or liquid metal ions reduction to propagate, nucleate, and build nanoparticles. This approach has the main gain of producing smaller homogenous particles with the excellent shape of crystals and perfect surface configurations [23–27]. Common methods based on this approach are discussed below.

3.2.1 Chemical Reduction Method

It is the traditional method for noble metal nanoparticle synthesis in which metal salt ions are reduced using organic or inorganic reducing agents (e.g., sodium borohydride $NaBH_4$, hydrazine, hydrogen, or alcohols) in a suitable medium with added stabilizing molecules. This method has a drawback of using hazardous and toxic chemicals. It also has an impurity issue due to chemical residues on the produced nanocrystals [28, 29].

3.2.2 Electrochemical Reduction Method

It was first described in 1994. The metal precursor at the anode is dissolved in a supporting electrolyte solution, and the produced metal ionic salts are reduced at the cathode. The size of the produced particles is controlled through a change in current density [30–34].

3.2.3 Microwave Method

It is a rapid and straightforward method that uses regular heating of microwave irradiations to solve ionic metal salts, surfactants, and reducing agents. The shape and size of crystals can be adjusted by changing reaction factors [35].

3.2.4 Reverse Micelle Method

It is a simple method of reaction in ambient air without the need for heat or expensive reagents. This method is used to fabricate metal oxide nanoparticles inside reverse micelles. Surfactants are dissolved in organic solvents to which aqueous metal source is added to produce reverse micelles inside which nanoparticles are formed [36].

3.2.5 Laser Ablation

It is a simple method that produces a well-controlled size of nano-products. The laser is applied to a reliable target flooded with a solution of surfactant and metal salts, which become reduced to produce nanoparticles [37].

3.2.6 Green Biological Method

It involves using the reducing ability of bacteria, archaea, fungi, algae, and plant extracts to reduce metal ions to produce, nucleate, form, and stabilize nanoparticles. It avoids the hazards of traditional chemical methods as the reducing and stabilizing agents are safe enzymes and biomolecules (proteins, peptides, polysaccharides, and secondary metabolites) synthesized by the living microorganisms or plants, so they are reproducible, and there is no waste to be cleaned. Besides, there is no need for high energy sources. Nanoparticles produced by this method have comparatively small size and excellent physicochemical properties [38].

4 Microorganisms Used for Nanoparticles Synthesis

Members of all domains of the three-domain system of classification were used for nanoparticle synthesis. Environmental bacteria, rapidly growing molds, and algae were successfully used for this target. Bacteria are the most commonly used because of their ubiquitous existence, high ability to adapt to extreme growth conditions, and nonexpensive requirements for growth. However, the low production limits their use on an industrial scale. Genetic manipulation of bacterial strains to scale up the production for industrial purposes is relatively more accessible than other microorganisms. Gram-negative bacteria are broadly used for NPs biosynthesis; nevertheless, gram-positive bacteria [39, 40] are also used for biosynthesis, especially *actinomycetes,* because of their ability to synthesize various bioactive protein compounds [41, 42].

Fungi are characterized by their large surface area of hyphae, which enables higher production than bacteria. Also, synthesis is mostly extracellular, so it is easier to purify and collect NPs. However, genetic manipulation to scale up production is challenging in fungi [43].

Algae are becoming more popular in NPs synthesis. They are photoautotrophs that do not have roots, stem, or leaves of plants. Green algae are mostly microscopic, while brown algae are macroscopic algae that are usually present in coastal waters. Red algae are multicellular and unicellular algae that live in deeper areas of the ocean, and a few of them produce toxins [44]. Diatoms are unicellular aquatic eukaryotic microalgae. Their characteristic cell walls of the micro and nanoscale pattern of silica are proposed for several technological practices [45]. Algae are

Table 2 Microorganisms used for nanoparticles synthesis

Microorganisms	Bacteria	Fungi	MicroAlgae
Classification	Domain: Bacteria; Kingdom: Not assigned	Domain: Eukarya; Kingdom: Fungi	Domain: Eukarya; Kingdom: Plantae
Advantages	High growth rate Adaptable to harsh conditions Relative monodispersity Easy genetic modification	Large surface area Bulky production Mostly extracellular synthesis	Rapid production Medical applications Good stability Nonpathogenic (few produce toxins)
Disadvantages	Potential pathogens Low production Purification in IC synthesis	Potential pathogens Difficult genetic manipulation	Relative polydispersity
Examples	*Kocuria flava* for CuNPs [47] *Desulfovibrio desulfuricans* and *E. coli* for Pd, Rh, and PtNPs [48] *Cupriavidus metallidurans* for AuNPs [49]	*Cladosporium oxysporum* for AuNPs [50] *Fusarium oxysporum* for AgNPs [51] *Saccharomyces cerevisiae* for PdNPs [52]	*Chlorella vulgaris* for PdNPs [53] *Tetraselmis kochinensis* for AuNPs [54] *Diatom frustules* for AgNPs [55]

characterized for their rapid production and stability of the produced NPs, which are frequently used for biomedical applications [46] (Table 2).

5 Mechanism of Microbial Synthesis of Nanoparticles

Microbial synthesis of metal NPs is a measure for detoxifying soluble toxic metal ions to be reduced and precipitated as solid nontoxic nanoscale particles either extracellularly and/or intracellularly. Microbial oxidoreductase enzymes use cofactors of the reduced form of nicotinamide adenine dinucleotide (NADH) and the reduced form of nicotinamide adenine dinucleotide phosphate (NADPH) (e.g., NADH-dependent nitrate reductase, NADPH-dependent sulfite reductase, and cysteine desulfhydrase) for electron transfer. Biomolecules as peptides, proteins, and polysaccharides are used for biosorption, growth, nucleation, and stabilization of NPs [56]. It was supposed that not only reductase enzymes but also electron shuttle are required for the reduction of metal ions. Species of *Fusarium* mold that contain both the reductase enzymes and anthraquinone can form NPs, while *Fusarium moniliform* that contains nitrate reductase failed to synthesize NPs due to the absence of the electron shuttle anthraquinone [57].

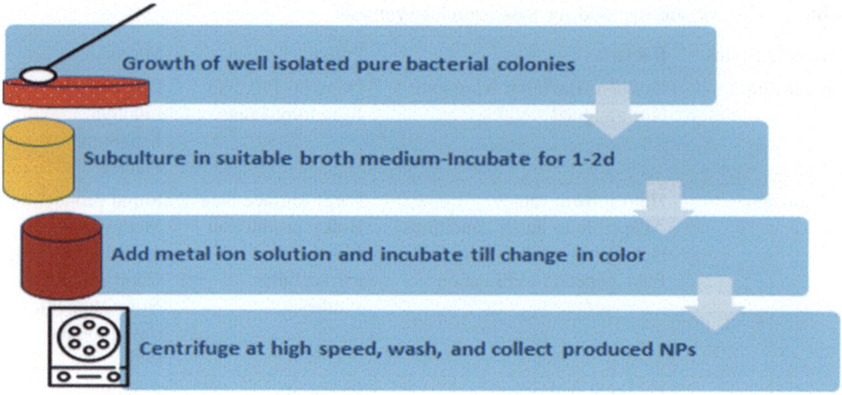

Fig. 1 Operating steps for extracellular bacterial production of nanoparticles

Many microorganisms can synthesize nanoscale particles extracellularly and/or intracellularly. The intracellular process involves the interaction of the metal ions with charged portions of the cell wall for the ion transport system and then enzymes of the cell wall perform the reduction process. Extracellular synthesis is achieved through secreted reductases that implement the bioreduction of metal ionic salts to consistent NPs [58].

For the production of NPs synthesized extracellularly, the pure isolated bacterium is subcultured in a suitable broth medium with adjusted pH and incubated for 24–84 h in a rotating shaker under optimum temperature and oxygenation. The inoculated broth is centrifuged, and the supernatant is incubated with added sterile metal salt solution until the color of the culture media changes denoting NPs synthesis (Fig. 1). Different speed centrifugations are implemented, followed by high-speed centrifugation and washing with water, methanol, or ethanol to finally collect NPs at the bottom of the centrifuge tube [59]. Extracellular synthesis is preferred over the intracellular synthesis of nanoparticles as it does not involve extra steps to disrupt the cell wall of bacteria to get the NPs released and collected [58].

In the case of intracellular production of NPs, the isolated microorganism is subcultured in a suitable broth culture medium and incubated for 1–2 days under optimum growth conditions. Centrifugation and wash with sterile water are followed to collect the biomass, which is then dissolved in sterile water and sterilized metal salt is added. The reaction mixture is incubated until its color changes indicating the formation of NPs, which are released after the breakdown of the cell wall through repeated ultrasonication, washing, and centrifugation [59].

6 Reaction Parameters Affecting the Biogenic Synthesis of Nanoparticles

Growth conditions (pH, temperature, and shaking conditions), salinity, and metal ions concentrations can affect the size, shape, dispersity, localization, aggregation, and efficiency of the produced NPs. A study for the effect of growth temperature on AgNPs synthesis using *Morganella psychrotolerans* reported that at 25 °C hexagonal, spherical, and triangular NPs were fabricated, while at 20 °C, only spherical 2–5 nm particles were formed. On decreasing temperature to 15 °C, spherical particles in addition to nanoplates were gained. When the temperature was further reduced to 4 °C, larger 10–70 nm spherical NPs were produced, and the plate form markedly outnumbered the spherical form [60].

A recent study described the effect of microwave (MW) irradiation for gram-negative bacteria *Desulfovibrio desulfuricans* and *E. coli* before the addition of Pd(II) to be reduced to Pd^0 nanoparticles. The produced NPs after MW energy application showed narrower size variation and more intracellular localization compared to less monodispersity and more surface distribution in control untreated cells. In a related study comparing the catalytic power of PdNps produced by MW treated cells to untreated cells, the biocatalyst produced by treated cells showed a 50% increase in the reaction rate and double selectivity toward the desired product [61].

The effect of metal ion concentration, temperature, and pH was assessed in a study using the bacterium *Arthrobacter* sp. for the synthesis of AgNPs. Cubic particles of size 9–72 nm were formed by adding low concentration (1 mM) of silver nitrate, at 70 °C and pH 7–8. Aggregation of AgNps resulted when the concentration of silver nitrate was increased to 3 mM at 70 °C and pH 7–8. Increasing temperature from 70 to 90 °C was accompanied by more rapid synthesis. Below pH 5 and above pH 8, no AgNPs were produced [62].

Using shaking conditions during the biologic reduction of Au ions by *Trichothecium* sp. fungal biomass resulted in intracellular localization of the produced NPs while under stable conditions, NPs were produced extracellularly. This may be due to the prevention of the release of reducing enzymes and proteins on shaking conditions [63].

7 Microorganisms Used for the Synthesis of Nanoparticles from Wastewaters

Effluents of mining metals, electrochemical plating, and electrical converters production have mixed metals content with high concentrations of PGM. Several studies used models for wastewaters resembling pH and salinity of real waste to assess biosorption, recovery, and reduction of precious metals, PGM, and rare metals from waste. Few studies used real wastewaters with or without dilution or pH adjustment [13].

7.1 Cupriavidus metallidurans *for Pd Nanoparticles Synthesis*

Cupriavidus metallidurans is a facultative anaerobic gram-negative bacterium that inhabits soil and water contaminated with heavy metals (e.g., mining sites, near factories, and sewage plants) as they have great ability to resist high concentrations of numerous metal ions including Cu^{+}, Cu^{2+}, Cd^{2+}, Ag^{+}, Au^{+}, Au^{3+}, SeO_3^{2-}, SeO_4^{2-}, CrO_4^{2-}. They have about 24 gene clusters coding for metal detoxification carried mainly on two megaplasmids [64]. The full taxonomy for *C. metallidurans* is Domain; Bacteria, Phylum; *Proteobacteria*, Class; *Betaproteobacteria*, Order; *Burkholderiales*, Family; *Burkholderiaceae*, Genus; *Cupriavidus*, species; *C. metallidurans* [65]. Studies that describe the synthesis of PdNPs by *C. metallidurans* necessitate the presence of hydrogen, formate, or glutaraldehyde for adsorption, reduction and formation of PdNPs by the bacterial cells (Fig. 2) [66].

Palladium (Pd) is a lustrous white precious metal that is more expensive than gold. The produced metal is less than the global demand. It is an essential component of catalytic converters that turn the toxic emissions (carbon monoxide and nitrogen dioxide) of motor vehicles into carbon dioxide, nitrogen, and water vapor [67]. The alloy of palladium and gold is one of the alloys used commonly in white gold jewelry as the original white color of palladium does not require polishing by rhodium plating. Palladium is also used in dentistry, electronics, solar energy-producing units, and chemical applications. It is produced as a secondary product during the extraction of nickel and platinum. It comes mainly from Russia, South Africa, the USA, and Canada [68]. Bacterial synthesis of Pd NPs seems an excellent way to decrease the gap between production and demand for Pd. Several bacteria (e.g., C. *metallidurans*, *E. coli*, and *D. desulfuricans*) can reduce Pd^{II} to Pd^{0}intracellularly and/or extracellularly (Fig. 3) [61].

Real undiluted industrial wastewater was used in a valuable study as a source of Pd^{II} to be reduced by *C. metallidurans* DS M2839 and *Cupriavidus necator* DS M428 separately, to Pd^{0} nanocrystals that were instantly used as active catalyst for organic reactions. The active hydrogenases of the selected strains used in the study were suspected of inducing nucleation to reduce Pd^{II} using hydrogen as the electron donor [69–71]. The acidic leachate containing a mixture of heavy metals was used undiluted and without alteration of pH. Desalting and degassing of the waste solution were accomplished before initiating the process of biorecovery. After 24 h incubation, the deep orange color of the waste solution changed to be colorless, denoting NPs synthesis. TEM images of pure cultures of *C. necator* and *C. metallidurans* showed nanoparticles precipitated on the bacterial cell surface after using H_2 as the electron donor, which ensued to be essential for reduction PdNPs

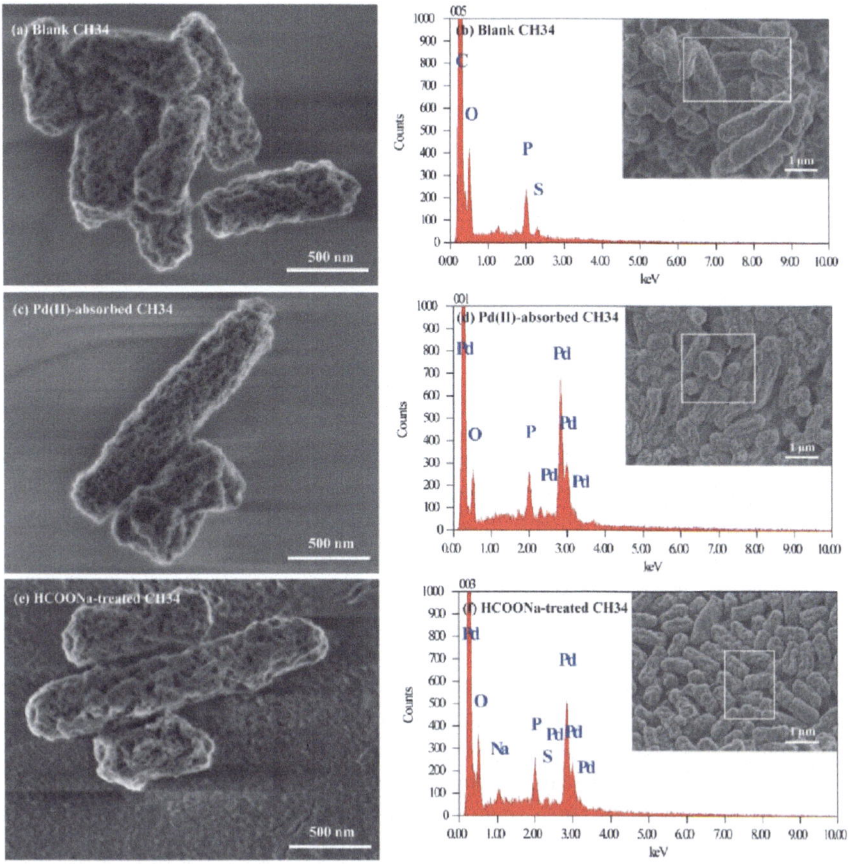

Fig. 2 Secondary electron SEM micrographs and EDS analysis for *Cupriavidus metallidurans* CH34 cells without formate or glutaraldehyde (**a**, **b**), with formate (**c**, **d**) and with glutaraldehyde (**e**, **f**) showing EDS confirmed uptake of Pd by bacterial cells only in the presence of formate or glutaraldehyde. Adapted with permission from Ref. [66], Copyright 2020, Elsevier

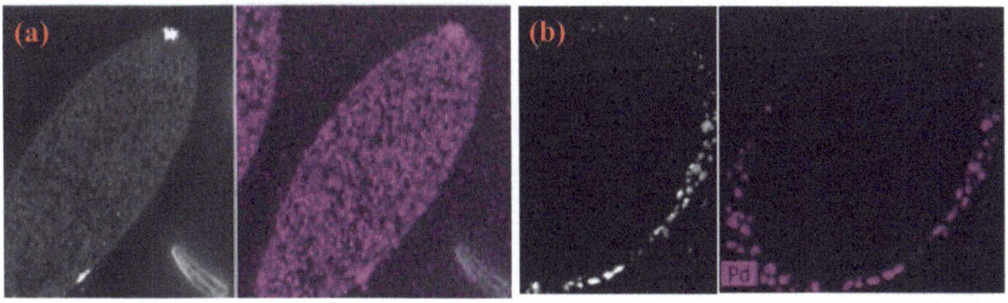

Fig. 3 **a** Elemental mapping showing the distribution of Pd NPs mainly intracellularly in the cytoplasm of *E. coli MC4100* **b** Elemental mapping showing the distribution of Pd NPs in cell surface layers of *D. desulfuricans* (MDPI open access). Adapted with permission from Ref. [61] Copyright 2019, MPDI

synthesis. Energy-dispersive X-ray analysis demonstrated that metal ions rather than Pd^0 were precipitated on the bacterial surface. However, Pd^0 nanoparticles are the main components denoting more efficient recovery than other heavy metals. The produced nanoparticles were used effectively as a catalyst in two useful organic reactions [72].

7.2 Desulfovibrio desulfuricans *and Pd Nanoparticles Synthesis*

Desulfovibrio desulfuricans is an obligate anaerobic gram-negative bacterium showing curved rods or spirilloid morphology [73]. It is classified as Bacteria (Domain), *Proteobacteria* (phylum), *Deltaproteobacteria* (Class), *Desulfovibrionales* (Order); *Desulfovibrionaceae* (Family); *Desulfovibrio* (Genera); *D. desulfuricans* (Species). *Desulfovibrio* genus is the best example of sulfur-reducing bacterial genera. It inhabits anaerobic soil, marine and estuarine sediments and the intestinal tract of animals and humans. It reduces sulfur or sulfate using organic sources like ethanol, lactate, or fatty acids as the electron donors resulting in the production of H_2S, which combines with iron developing black precipitate that is responsible for the black color of sediments [44].

In one of the few studies using real industrial waste, *D. desulfuricans* ATCC 29577 was added to waste solution after dilution and supplementation with additional Pd^{II}, in the presence of H_2 or formate as an electron donor. After 48 h incubation, the yellowish-orange color of the solution disappeared with the formation of a solid black precipitate, indicating a reduction of metal ions and nanoparticle synthesis endorsed by the hydrogenase activity of *D. desulfuricans*. Composition analysis of the black precipitate by Proton-Induced X-ray Emission (PIXE) appeared to be mostly Pd^0 nanoparticles [48]. Like previous studies, the size of biosynthesized Pd0 nanocrystals was half produced by chemical methods [74, 75]. The produced bio-Pd^0 nanocrystals were successfully used directly without additional steps as a catalyst to reduce the hazardous carcinogenic Cr^{VI} to the less harmful Cr^{III} helping environmental remediation of Cr^{VI} [48].

7.3 Rhodopseudomonas palustris *and Recovery of Ruthenium*

Rhodopseudomonas palustris inhabits versatile environments (wet soil, springs, freshwater puddles, and saltwater coastal sediments) owing to its extremely diverse metabolic activities (phototrophic, autotrophic, heterotrophic, chemotrophic, and organotrophic). It can produce hydrogen efficiently throughout the degradation of organic waste [76]. The full classification is Bacteria (Domain); *Proteobacteria*

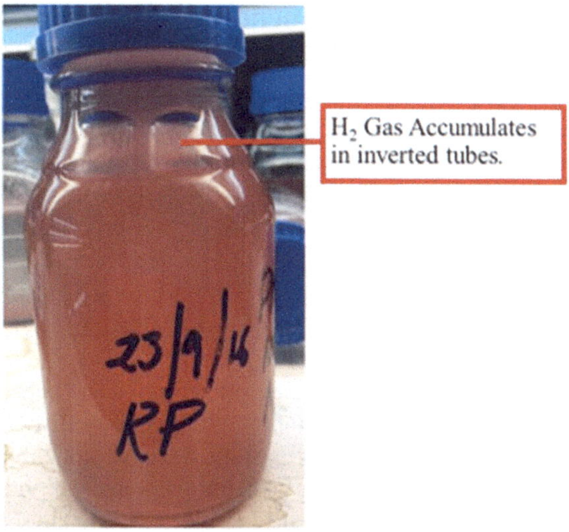

Fig. 4 Pure culture of Rhodopseudomonas palustris (CGA009) grown anaerobically in photosynthetic medium (PM) containing palm oil which is decomposed by the growing bacteria with production of hydrogen. Adapted with permission from Ref. [81] Copyright 2020, Elsevier

(Phylum); *Alphaproteobacteria* (Class); *Rizobiales* (Order); *Bradyrhizobiacae* (Family); *Rheudopeudomonas* (Genus); *R. palustris* (Species). It is facultative photosynthetic facultative anaerobic gram-negative rods belonging to the photosynthetic purple nonsulfur bacteria. In depleted oxygen conditions, it develops its characteristic purple color as it synthesizes absorbing light pigment for photosynthesis (Fig. 4). It can modulate photosynthesis according to the amount of light available. In weak light conditions, it increases the production of pigment to harvest more light [77].

Ruthenium (Ru) is one of the PGM. It is known for its relatively lower price than other metals of PGM, so it offers an excellent alternative to gold and rhodium coating to get a durable and chemically stable finish [78]. It is used as a hardener in alloys with other metals. It has biomedical diagnostic applications as it is used to measure calcitonin, ferritin, and cyclosporine level in blood [79].

In a trial to recover Ru from waste effluents with an economical, eco-friendly method, two strains, *Rheudopseudomonas palustris* AV33 and *R. palustris* SC0 were incubated with two different (acidic 1.8 pH and alkaline 8.7 pH) undiluted waste effluents of Ru-plating containing Cu, Zn, and Ni in addition to Ru at 65 °C (similar to the temperature used in electrochemical plating process). The Ru recovered was 42–72% of Ru amount adsorbed by the bacterial biomass, as the amount differs according to the type of waste used. Acid pretreatment of waste to decrease contaminants seems to increase Ru biosorption and recovery [80].

7.4 Pseudomonas mendocina *for Reduction of Tellurium*

Pseudomonas mendocina is a gram-negative rod-shaped flagellated aerobic bacterium that was first isolated in 1969 from environmental samples collected from surface soil and water in Mendoza province in Argentina. It was first thought to be *P. stutzeri*, though its inability to use maltose or starch as a carbon source indicated a novel species [82]. It is classified as Bacteria (Domain), *Proteobacteria* (Phylum), *Gammaproteobacteria* (Class); *Pseudomonadales* (Order); *Pseudomonadacae* (Family); *Pseudomonas* (genus); *P. mendocina* (Species). It carries resistance genes to different heavy metals, and it is assumed to be sharing in environmental bioremediation [83].

Tellurium (Te) lies in the same group in the periodic table with oxygen, sulfur, and selenium. Its concentration in the earth's crust (1part per billion) is lower than gold and platinum [84]. It exists in the environment in several forms, including the organic dimethyl telluride and the inorganic telluride (Te^{II}, Te^{2-}). The insoluble elemental tellurium Te^0 is the least toxic, while the soluble inorganic tellurite (Te^{IV}, TeO_3^{2-}) and tellurate (Te^{VI}, TeO_4^{2-}) are highly toxic in low concentrations [85]. Tellurium has been traditionally used to form alloys with steel and copper. The semiconducting property of telluride metal is useful in existing and potential applications as in solar photovoltaic cells, fluorescent labels in biomedical diagnostics, and thermoelectric supplies in cooling systems. Tellurium suboxides have useful applications in re-writable compact discs (CD-RW), re-writable digital versatile discs (DVD-RW), and memory chips. The increased use of tellurium has a drawback of environmental contamination with its toxic forms [84]. Bacterial biosynthesis of tellurium was studied using different species of the *Pseudomonas*genus [86, 87], and nanorods of the produced tellurium were characterized using a scanning electron microscope (SEM) and a transmission electron microscope (TEM) after treatment of the precipitated particles of tellurium (Fig. 5e, f) [87].

Pseudomonas mendocina MCM B-180 strain was studied for its ability to reduce tellurite to elemental tellurium to be applied for waste effluents containing tellurite. The study was performed using different initial concentrations of tellurite under variable temperatures and pH in the presence of sucrose and diammonium hydrogen phosphate as a source for carbon and nitrogen to compensate for the scarcity of organic matter in waste effluents. Efficient reduction (about 99%) of tellurium was sustained over a wide range of temperatures (2–45 °C) and pH (5.5–8.5 pH) and with different initial tellurite concentrations (10–100 mg L^{-1}) [86].

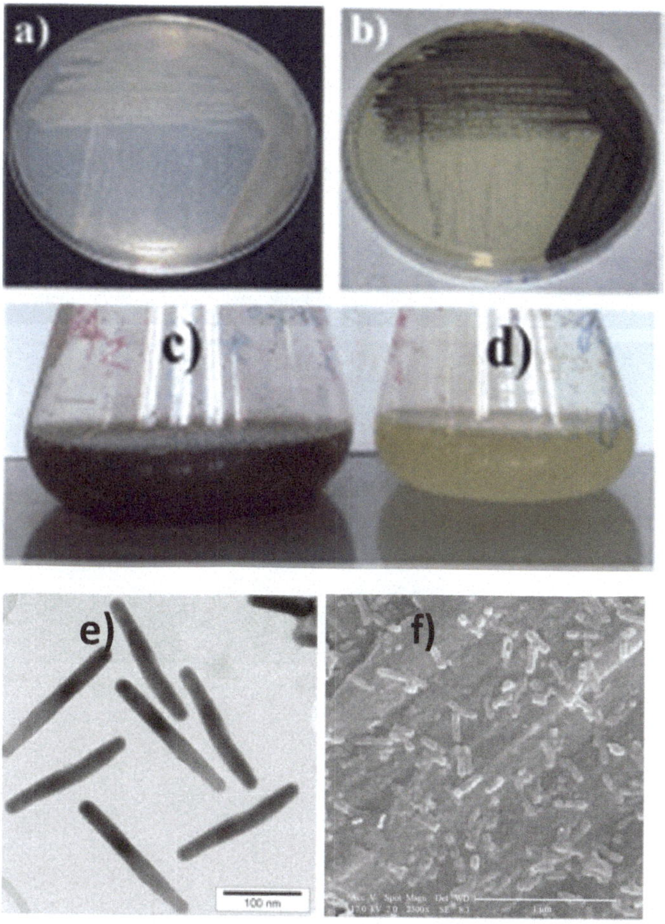

Fig. 5 *P. pseudoalcaligenes* Te strain grown on nutrient agar plates **a** in the absence and **b** in the presence of Te^{4+}, Fluid culture of *P. pseudoalcaligenes* Te strain grown **c** with and **d** without Te^{4+}. TEM image (**e**) and SEM image (**f**) of Tenanorods synthesized by *P. pseudoalcaligenes* Te strain following 24 h incubation with Te ions at 30 °C. Adapted with permission from Ref. [87] Copyright 2015, American Chemical Society

7.5 *Raotella sp and Echirechia sp for Te Nanorods Production*

The genera of Raotella and *Echirechia* both belong to *the Enterabacteriacae* family descending from *Gammaproteobacteria* class and *proteobacteria* phylum. *Raotella* genus was named after the French microbiologist Raoul. Out of the four species belonging to this genus, three species (*R. planticola*, *R. ornithinolytica*, and *R. terrigena*) were reclassified in 2001 from *klebsiella* genus to the new genus

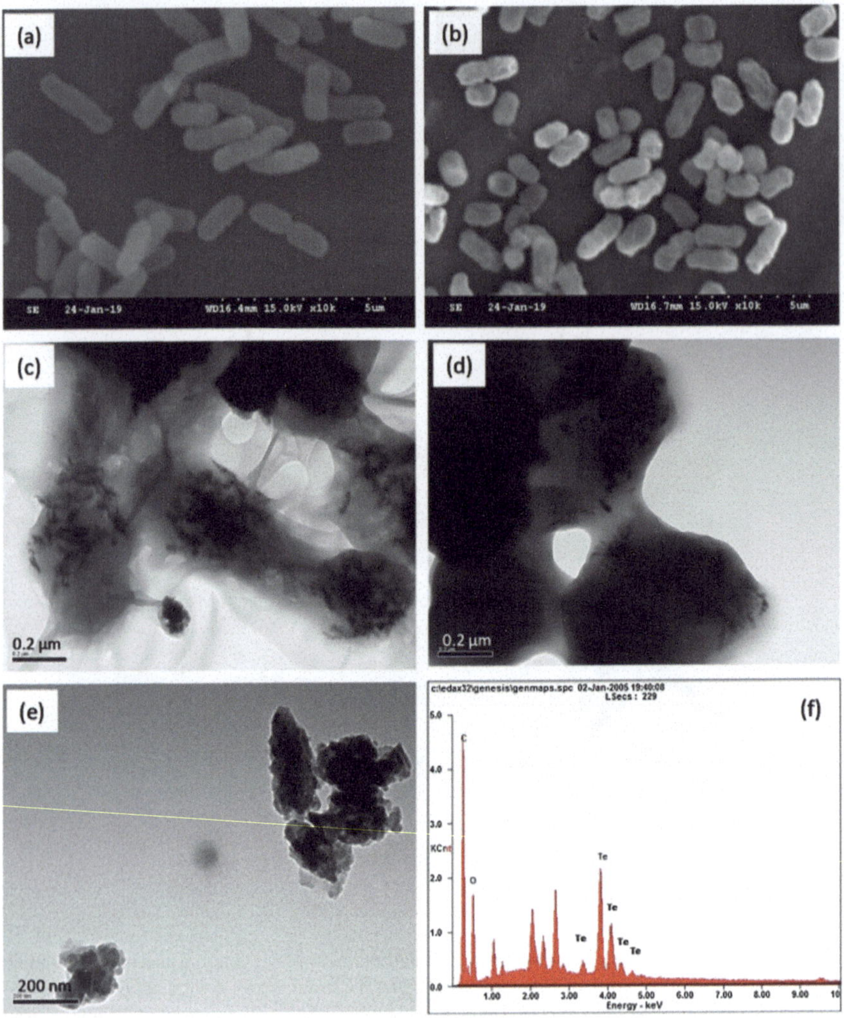

Fig. 6 SEM images of 100.000 × magnification for **a** *Raotella* sp WAY strain and **b** *Escherichia* sp. WYS strain. TEM images of intracellular elemental Te nanoparticles **c** in WAY strain cells and **d** in WYS strain cells. **e** TEM image of the extracellular aggregation of elemental Te nanoparticles. **f** EDS spectrum of the elemental Te nanoparticles. Adapted with permission from Ref. [89] Copyright 2019, Elsevier

Raotella. It is a gram-negative, rod-shaped, encapsulated, nonmotile, aerobic bacterium that usually lives freely related to soil, water, and plants [88].

In a recent study, two bacterial strains, *Raotella* sp. WAY strain and *Escherichia* sp. WYS showed high resistance to tellurite indicated by the highest minimal inhibitory concentrations (MIC) among bacterial strains isolated from municipal wastewater. The two strains were incubated separately with tellurite under wide ranges of temperature, pH, and salinity to identify the best conditions for tellurite reduction. Tellurite in a concentration of 0.5 mM was added to each bacterium's pure culture in a nutrient broth medium and incubated at different temperatures and pH with shaking at 180 rpm, till reduction of tellurite to elemental tellurium. Results showed that both bacterial strains could accomplish tellurite reduction at temperatures of 20–37 °C, initial pH of 5–9, and salt concentration lower than 5%. Intracellular and extracellular nanorods of tellurium were detected and characterized using SEM (Fig. 6c–e). The black precipitate was confirmed to be elemental tellurium with Energy-dispersive X-ray spectroscopy (EDS) showing the Te signal spectrum (Fig. 6f). The reduction efficiency of *Raotella* sp. WAY strain was about twofold that of *Escherichia* sp. WYS strain. The study designated that both bacterial strains (*Raotella* sp. more effective than *Escherichia* sp.) can be used to remediate the highly toxic tellurite Te^{IV} to the least toxic tellurium Te^{0} nanorods that have many useful industrial applications [89].

8 Microorganisms Used for the Synthesis of Nanoparticles from Solid Waste

Metals represent about 30% of electronic waste scrap and are mostly copper, a fair amount of base metals (aluminum, zinc, iron, and lead), and lower precious metals (silver, gold). The remaining constituents of electronic scrap are ceramics, plastics, and recalcitrant oxides [17]. It is estimated that only 20% of electronic waste is recycled, and the remaining is a potential cause of environmental pollution if incinerated or used in landfills [90].

8.1 Chromobacterium violaceum *and* Delftia acidovorans *for Gold Recovery*

Chromobacterium violaceum is a gram-negative rod-shaped facultative anaerobic bacterium whose total taxa are: Bacteria (Domain); *Proteobacteria* (Phylum); *Betaproteobacteria* (Class); *Neisseriaceae* (Family); *Chromobacterium* (Genus); *C. violaceum* (Species). It inhabits the soil and freshwater in tropical and subtropical areas [91]. It produces a purple pigment violacein [92] (Fig. 7a) that has antimicrobial properties [91].

Delftia acidovorans is a rod-shaped gram-negative, aerobic, nonspore-forming bacterium that is classified as Bacteria (Domain); *Proteobacteria* (Phylum); *Betaproteobacteria* (Class); *Burkholderiales* (Order); *Comamonadaceae* (Family); *Delftia* (Genus); *D. acidovorans* (Species). It is usually isolated from soil and is associated with gold nuggets [93].

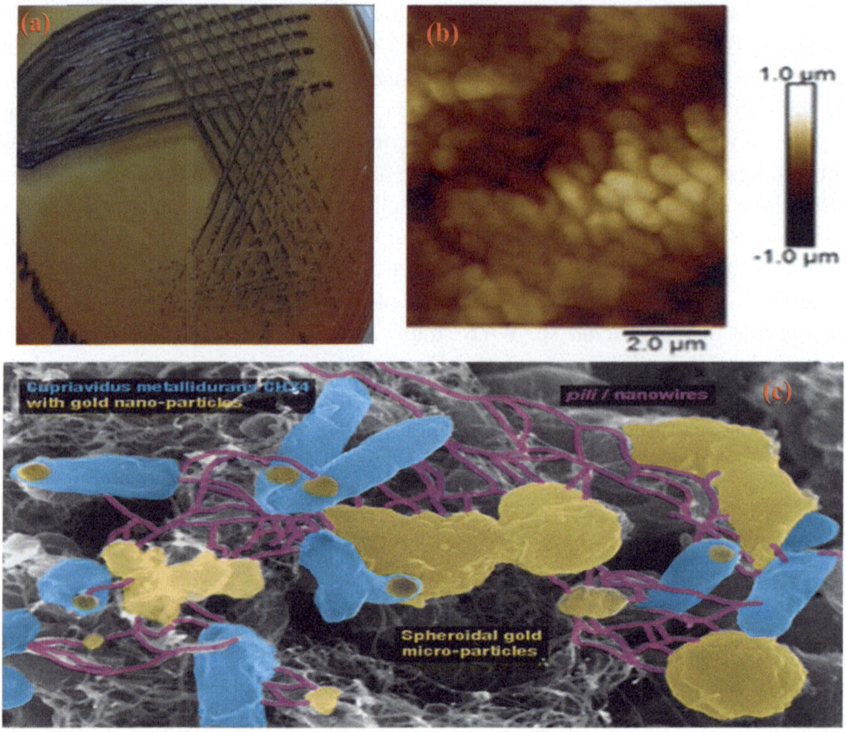

Fig. 7 **a** Violet pigmentation of the *Chromobacterium violaceum* isolated colonies on MacConkey agar **b** Images of Atomic Force Microscope 10 × 10 um^2 scans of *D. acidovorans* grown on QCM surface (Adapted with permission from Ref. [98] Copyright 2019, American Chemical Society) **c** Secondary electron micrographs of *C. metallidurans* cells with intracellular gold nanoparticles and extracellular gold aggregates connected to cells by nanowires. Adapted with permission from Ref. [101] Copyright 2013, American Chemical Society

Gold (Au) is usually used in electronic devices because of its extreme resistance to corrosion and good electric conductivity. The amount of gold in electronic scrap is 10 g–10 kg for each ton [94], which far exceeds its concentrations in natural ores (0.5–13.7 g per ton gold rocks) [95]. Bioleaching and bioremediation terms are increasingly used. The former means extracting metals from ores or waste, while the later means getting rid of heavy metals contaminants. Both use microorganisms to extract or get rid of metals [96]. Alkaline bioleaching is called for the use of the cyanogenic bacterium *C. violaceum* to liquefy gold, which is then harvested using chemical methods or another bacterium *D. acidovorans*, which produces an extracellular peptide that forms a complex with gold [17].

Chromobacterium violaceum produces cyanide from glycine using the enzyme hydrogen cyanide synthase encoded by *hcnABC* operon under quorum control, which restricts cyanide production to 20 mg cyanide per liter of bacterial culture containing 10^6 colony-forming units. Metabolic engineering for strains of *C.*

violaceum eluded quorum control using exogenous promoters resulting in 68% more cyanide production and more gold (more than twofold) recovery compared to strains of the wild type. It is assumed that further improvement can be achieved through metabolic engineering [94]. Other trials to enhance cyanide production comprised pretreatment of waste to remove copper, which competes with gold for cyanide, and the use of mutated *C. violaceum* that can survive higher pH to decrease the loss of cyanide, which tends to form volatile hydrogen cyanide at lower pH [17].

Delftia acidovorans is used as gold refiner from the cyanide containing solution. It produces extracellular peptide (delftibactin) that forms a complex with gold to be precipitated in spherical and octahedral nanocrystals [90, 97].

A Japanese study recently developed a method for real-time quantification and monitoring of gold production from Au^{3+} ions by *D. acidovorans* biofilm grown on gold-coated quartz crystal microbalance (QCM) that was used as a surface-sensitive transducer (Fig. 7b). Variations in the resonance frequency of the oscillating quartz designate for additional mass to the QCM, while the change in energy dissipation denotes the degree of intimate contact with the QCM surface. This method represents an affordable eco-friendly method for monitoring the production of gold nanoparticles [98].

Another unique bacterium that can be used for gold recovery is *C. metallidurans* (Fig. 7c). It can rapidly adsorb gold complexes from solutions in less than one minute, seemingly in a pH-dependent way. It is armed by multiple gene clusters and an operon for gold induced genes that protect against gold toxicity through reduction, efflux, or probably adding methyl to gold complexes [99]. It can be used for the complete regaining of gold from low-grade sources as mining effluents and contaminated soil [100].

9 Applications and Advantages of Nanoparticles Produced by Microbes Using Waste

Because of the small size, large surface area, excellent physicochemical properties, and stability of biogenically produced NPs, they are excellent catalysts that can be used in different catalytic applications as environmental remediation of hazardous substances to be reduced to less harmful forms and degradation of dyes used in many industries as they are usually toxic and suffer low-efficiency removal by conventional methods [102]. The biogenic method is also applied for the recovery and recycling of precious and rare metals with low concentrations in the earth's crust.

As mentioned before, Pd NPs are synthesized by two *Cupriavidus* spp. using industrial waste effluents were used efficiently as catalysts in important organic reactions [72]. Another study described the use of Pd NPs biorecovered by *D. desulfuricans* and *E. coli*, from wastewaters to reduce the mutagenic Cr^{VI} to the less harmful Cr^{III}. The same study compared the biosynthesized catalyst to a chemically produced one and showed much more efficiency of the biosynthesized Pd catalyst besides the sustainable production of the biocatalyst by the living bacteria [48].

The rare element Te biosynthesized from waste effluents can be reused in many industrial applications that benefit its unique properties [84]. An Indian study developed a method for recovery and reduction of telluride from solar photovoltaic cells waste scrappings after treatment with nitric acid to solubilize silver, cadmium, and telluride. Removal of silver and cadmium was accomplished through biosorbent columns, as reported in a previous study [103]. Following adjustment of pH and addition of organic matter, the remaining telluride was reduced efficiently in 12 h using *P. mendocina* MCM B-180 strain [86].

Ru metal which is usually used in the electroplating industry, was biorecovered efficiently from the real waste effluents of electroplating plants to be reused in the same plating industry [80].

Gold and copper recovered by bacteria from solid electronic waste that contain much more gold and copper than their ores can be reused in electric circuits of computers and other electronic devices. This biogenic method for recovery protects the environment from the polluting chemical method and saves the energy of the pyrogenic methods that are conventionally used to recover gold and copper [17].

10 Limitations of Microbial Biologic Method for Nanoparticles Synthesis

Despite the various advantages of the green method of microbial synthesis of NPs, it has some limitations. The low yield of the produced NPs is the main obstacle that hinders using this method on a large production scale. This can be solved by metabolic genetic engineering for microbial strains used in biosynthesis [43]. Recombinant *E. coli* ABLE C strains were designed to overexpress glutathione synthetase or gamma-glutamlcysteine synthetase through the transformation of plasmids carrying the amplified genes encoding the enzymes to be overexpressed. When the recombinant cells were exposed to cadmium sulfide CdS, and sodium sulfide, the product of CdS nanocrystals was 2.5 times more than that of the wild type [104]. Also, recombinant *E. coli* strains have been organized to acquire genes for biosynthesis of about sixty different nanosized materials [43]. A recent study compared the adsorption of Pd and Pt from wastewater by a genetically modified *E. coli* BL21 (EC20) to wild strains of *E. coli* BL21 and *Providencia vermicola. E. coli* was modified to express EC20 protein to increase the sulfhydryl sites (functional groups in Pd^{II} adsorption) on its surface [105]. The highest adsorption of Pd (97.5%) and Pt (100%) from wastewater was by the modified *E. coli* BL21 (EC20). The reuse of the three bacterial species for Pd adsorption was possible with the highest efficiency (93%) for *E. coli* (EC20) [106].

Difficult control of the size, shape, and size distribution of the produced particles is a big hindrance as their control depends on numerous factors like pH, temperature, light, nutrients, oxygenation, salts concentrations, mixing speed, and harvesting time. Another limitation is the tedious purification steps, mainly when NPs are synthesized intracellularly. Besides, the physicochemical methods to release NPs from cells could interfere with the structure of NPs and may cause aggregation of NPs [107].

The mechanisms of microbial biosynthesis of nanoparticles are not fully revealed and described. Not all enzymes and proteins involved in the production and stabilization of NPs are identified. The form of the biocatalyst used in the process of production may have some restrains. Using culture supernatant or microbial extract to increase the reaction rate may affect the stability of the produced particles. Expensive cofactors like NADH or NADPH have to be added when using crude or purified enzymes instead of the whole cells to avoid extra purification steps. In whole cells, there is no need to add cofactors as they are reproducible and sustainable by the living cells [108].

11 Conclusions

Microbial biosynthesis of nanosized particles is a green, low cost, and easy method. Bacteria, molds, yeasts, and microalgae can reduce metal ion salts to nucleate and build NPs. Microbial species, especially bacterial species that live in metal contaminated areas, have resistant genes to various metals, so they are suitable candidates for the recovery and synthesis of nanoparticles of precious and rare metals present in the waste. *C. metallidurans*, *P. mendocina*, *Raotella spp.*, *D. desulfuricans* and *R. palustris* are good examples of bacteria used for PGM and rare metals recovery from industrial waste effluents. This helps decrease the gap between increased demand and low production of these metals. The cyanogenic bacteria *C. violaceum* that can solubilize gold in electronic waste scrap enable the recovery of gold by other bacteria as *D. acidovarans*. The benefits of the microbial biosynthesized NPs exceed that of other methods. However, polydispersity, the need for purification steps, and the low yield hinder the wide use of this method in the industry. Genetically, modified species enable several folds to increase in production.

12 Future Perspectives

The desirable properties of the biosynthesized NPs motivated comprehensive studies for new and expanding uses in the biosensor, catalytic, and bioremediation applications. The exact mechanisms for biosynthesis are not fully revealed and described. Future thorough studies are needed to uncover the mechanistic details of synthesis and the particular role of all enzymes and proteins involved in adsorption, reduction, and stabilization of NPs. This will enable better control of the shape, size, and size variations of the produced crystalline NPs. To scale up production, metabolic engineering methods were able to modify microbial strains to achieve several folds increase in the produced NPs compared to wild strains. Comparative proteome studies of different microbial species can define additional targets for metabolic engineering, and further enhancement of the production is expected. The commercial-scale production will augment the need for more consistent methods that can characterize the produced nanoscale materials from several aspects, especially size, size variations, and chemistry of the surface ligands, which can affect the potential applications. A standardized protocol involving a combination of characterization techniques is anticipated.

References

1. Verma ML, Sharma S, Dhiman K, Jana AK (2019) Microbial production of nanoparticles: mechanisms and applications. In: Microbial nanobionics. Springer, Berlin, pp 159–176
2. Pantidos N, Horsfall LE (2014) Biological synthesis of metallic nanoparticles by bacteria, fungi and plants. J Nanomed Nanotechnol 5(5):1
3. Darroudi M, Ahmad MB, Abdullah AH, Ibrahim NA (2011) Green synthesis and characterization of gelatin-based and sugar-reduced silver nanoparticles. Int J Nanomed 6:569
4. Prakash N, Soni N (2011) Factors affecting the geometry of silver nanoparticles synthesis in Chrysosporium tropicum and Fusarium oxysporum. Am J Nanotechnol 2(1):112–121
5. Vanaja M, Rajeshkumar S, Paulkumar K, Gnanajobitha G, Malarkodi C, Annadurai G (2013) Kinetic study on green synthesis of silver nanoparticles using Coleus aromaticus leaf extract. Adv Appl Sci Res 4(3):50–55
6. Joerger R, Klaus T, Granqvist CG (2000) Biologically produced silver–carbon composite materials for optically functional thin-film coatings. Adv Mater 12(6):407–409
7. Marooufpour N, Alizadeh M, Hatami M, Lajayer BA (2019) Biological synthesis of nanoparticles by different groups of bacteria. In: Microbial nanobionics. Springer, Berlin, pp 63–85
8. Mohanpuria P, Rana NK, Yadav SK (2008) Biosynthesis of nanoparticles: technological concepts and future applications. J Nanopart Res 10(3):507–517
9. Iravani S, Korbekandi H, Mirmohammadi SV, Zolfaghari B (2014) Synthesis of silver nanoparticles: chemical, physical and biological methods. Res Pharm Sci 9(6):385
10. Prasad NVKVT, Subba Rao Kambala V, Naidu R (2011) A critical review on biogenic silver nanoparticles and their antimicrobial activity. Curr Nanosci 7(4):531–544
11. Dobson RS, Burgess JE (2007) Biological treatment of precious metal refinery wastewater: a review. Miner Eng 20(6):519–532
12. Bernardis FL, Grant RA, Sherrington DC (2005) A review of methods of separation of the platinum-group metals through their chloro-complexes. React Funct Polym 65(3):205–217
13. Pat-Espadas AM, Cervantes FJ (2018) Microbial recovery of metallic nanoparticles from industrial wastes and their environmental applications. J Chem Technol Biotechnol 93 (11):3091–3112
14. Baldé C, Forti V, Gray V, Kuehr R, Stegmann P (2017) The global E-waste Monitor–2017, United Nations University (UNU), International Telecommunication Union (ITU) & International Solid Waste Association (ISWA), Bonn/Geneva/Vienna. ISBN Electronic Version, pp 978–992
15. Natarajan G, Ting Y-P (2014) Pretreatment of e-waste and mutation of alkali-tolerant cyanogenic bacteria promote gold biorecovery. Biores Technol 152:80–85

16. Bleiwas DI, Kelly TD (2001) Obsolete computers, "gold mine", or high-tech trash? resource recovery from recycling. US Geological Survey Denver, USA
17. Das S, Natarajan G, Ting Y-P (2017) Bio-extraction of precious metals from urban solid waste. In: AIP conference proceedings. AIP Publishing LLC
18. Ashiq A, Kulkarni J, Vithanage M (2019) Chapter 10—Hydrometallurgical recovery of metals from e-waste. In: Prasad MNV, Vithanage M (eds) Electronic waste management and treatment technology. Butterworth-Heinemann, pp 225–246
19. Anastas PT, Warner JC (1998) Principles of green chemistry. Green chemistry: theory and practice, pp 29–56
20. Parveen K, Banse V, Ledwani L (2016) Green synthesis of nanoparticles: their advantages and disadvantages. In: AIP conference proceedings. AIP Publishing LLC
21. Pareek V, Bhargava A, Gupta R, Jain N, Panwar J (2017) Synthesis and applications of noble metal nanoparticles: a review. Adv Sci Eng Med 9(7):527–544
22. Luechinger NA, Grass RN, Athanassiou EK, Stark WJ (2010) Bottom-up fabrication of metal/metal nanocomposites from nanoparticles of immiscible metals. Chem Mater 22 (1):155–160
23. Cao YC, Jin R, Mirkin CA (2002) Nanoparticles with Raman spectroscopic fingerprints for DNA and RNA detection. Science 297(5586):1536–1540
24. Fouad OA, Makhlouf SA, Ali GAM, El-Sayed AY (2011) Cobalt/silica nanocomposite via thermal calcination-reduction of gel precursors. Mater Chem Phys 128(1–2):70–76
25. Fouad OA, Ali GAM, El-Erian MAI, Makhlouf SA (2012) Humidity sensing properties of cobalt oxide/silica nanocomposites prepared via sol-gel and related routes. Nano 7 (5):1250038
26. Ali GAM, Fouad OA, Makhlouf SA (2013) Structural, optical and electrical properties of sol-gel prepared mesoporous Co3O4/SiO2 nanocomposites. J Alloy Compd 579:606–611
27. Ali GAM, Fouad OA, Makhlouf SA, Yusoff MM, Chong KF (2014) Co_3O_4/SiO_2 nanocomposites for supercapacitor application. J Solid State Electrochem 18(9):2505–2512
28. Zhang X-F, Liu Z-G, Shen W, Gurunathan S (2016) Silver nanoparticles: synthesis, characterization, properties, applications, and therapeutic approaches. Int J Mol Sci 17 (9):1534
29. Ali GAM, Makhlouf SA, Yusoff MM, Chong KF (2015) Structural and electrochemical characteristics of graphene nanosheets as supercapacitor electrodes. Rev Adv Mater Sci 40 (1):35–43
30. Reetz MT, Helbig W (1994) Size-selective synthesis of nanostructured transition metal clusters. J Am Chem Soc 116(16):7401–7402
31. Ali GAM, Tan LL, Jose R, Yusoff MM, Chong KF (2014) Electrochemical performance studies of MnO_2 nanoflowers recovered from spent battery. Mater Res Bull 60:5–9
32. Ali GAM, Yusoff MM, Feng CK (2015) Electrochemical properties of electrodeposited MnO_2 nanoparticles. Adv Mater Res 1113:550–553
33. Ali GAM, Yusoff MM, Shaaban ER, Chong KF (2017) High performance MnO_2 nanoflower supercapacitor electrode by electrochemical recycling of spent batteries. Ceram Int 43:8440–8448
34. Ali GAM, Yusoff MM, Ng YH, Lim NH, Chong KF (2015) Potentiostatic and galvanostatic electrodeposition of MnO_2 for supercapacitors application: a comparison study. Curr Appl Phys 15(10):1143–1147
35. Li AY, Kaushik M, Li C-J, Moores A (2016) Microwave-assisted synthesis of magnetic carboxymethyl cellulose-embedded Ag–Fe_3O_4 nanocatalysts for selective carbonyl hydrogenation. ACS Sustain Chem Eng 4(3):965–973
36. Noh J, Yi M, Hwang S, Im KM, Yu T, Kim J (2016) A facile synthesis of rutile-rich titanium oxide nanoparticles using reverse micelle method and their photocatalytic applications. J Ind Eng Chem 33:369–373
37. Mafuné F, Kohno J-Y, Takeda Y, Kondow T, Sawabe H (2000) Formation and size control of silver nanoparticles by laser ablation in aqueous solution. J Phys Chem B 104(39):9111–9117

38. Gour A, Jain NK (2019) Advances in green synthesis of nanoparticles. Artif Cells Nanomed Biotechnol 47(1):844–851
39. Ahmad V, Jamal QMS, Shukla AK, Alam J, Imran A, Abaza UM (2017) Bacilli as biological nano-factories intended for synthesis of silver nanoparticles and its application in human welfare. J Cluster Sci 28(4):1775–1802
40. Korbekandi H, Iravani S, Abbasi S (2012) Optimization of biological synthesis of silver nanoparticles using Lactobacillus casei subsp. casei. J Chem Technol Biotechnol 87(7):932–937
41. Karthik L, Kumar G, Kirthi AV, Rahuman AA, Bhaskara Rao KV (2014) Streptomyces sp. LK3 mediated synthesis of silver nanoparticles and its biomedical application. Bioprocess Biosyst Eng 37(2):261–267
42. Buszewski B, Railean-Plugaru V, Pomastowski P, Rafińska K, Szultka-Mlynska M, Golinska P, Wypij M, Laskowski D, Dahm H (2018) Antimicrobial activity of biosilver nanoparticles produced by a novel Streptacidiphilus durhamensis strain. J Microbiol Immunol Infect 51(1):45–54
43. Iravani S, Varma RS (2019) Biofactories: engineered nanoparticles via genetically engineered organisms. Green Chem 21(17):4583–4603
44. Tortora G, Berdell R, Funke L (2013) The prokaryotes: domains bacteria and archaea—the eukaryotes: fungi, algae, protozoa, and helminths. Microbiology: an introduction, 11th edn. Pearson, Yorkshire, pp 299–348
45. Pytlik N, Brunner E (2018) Diatoms as potential "green" nanocomposite and nanoparticle synthesizers: challenges, prospects, and future materials applications. MRS Commun 8 (2):322–331
46. Gahlawat G, Choudhury AR (2019) A review on the biosynthesis of metal and metal salt nanoparticles by microbes. RSC Adv 9(23):12944–12967
47. Kaur H, Dolma K, Kaur N, Malhotra A, Kumar N, Dixit P, Sharma D, Mayilraj S, Choudhury AR (2015) Marine microbe as nano-factories for copper biomineralization. Biotechnol Bioprocess Eng 20(1):51–57
48. Mabbett AN, Sanyahumbi D, Yong P, Macaskie LE (2006) Biorecovered precious metals from industrial wastes: single-step conversion of a mixed metal liquid waste to a bioinorganic catalyst with environmental application. Environ Sci Technol 40(3):1015–1021
49. Montero-Silva F, Durán N, Seeger M (2017) Synthesis of extracellular gold nanoparticles using Cupriavidus metallidurans CH34 cells. IET Nanobiotechnol 12(1):40–46
50. Bhargava A, Jain N, Khan MA, Pareek V, Dilip RV, Panwar J (2016) Utilizing metal tolerance potential of soil fungus for efficient synthesis of gold nanoparticles with superior catalytic activity for degradation of rhodamine B. J Environ Manage 183:22–32
51. Rajput S, Werezuk R, Lange RM, McDermott MT (2016) Fungal isolate optimized for biogenesis of silver nanoparticles with enhanced colloidal stability. Langmuir 32(34):8688–8697
52. Sriramulu M, Sumathi S (2018) Biosynthesis of palladium nanoparticles using Saccharomyces cerevisiae extract and its photocatalytic degradation behaviour. Adv Nat Sci Nanosci Nanotechnol 9(2):025018
53. Arsiya F, Sayadi MH, Sobhani S (2017) Green synthesis of palladium nanoparticles using Chlorella vulgaris. Mater Lett 186:113–115
54. Senapati S, Syed A, Moeez S, Kumar A, Ahmad A (2012) Intracellular synthesis of gold nanoparticles using alga Tetraselmis kochinensis. Mater Lett 79:116–118
55. Chetia L, Kalita D, Ahmed GA (2017) Synthesis of Ag nanoparticles using diatom cells for ammonia sensing. Sens Bio-Sens Res 16:55–61
56. Grasso G, Zane D, Dragone R (2020) Microbial nanotechnology: challenges and prospects for green biocatalytic synthesis of nanoscale materials for sensoristic and biomedical applications. Nanomaterials 10(1):11
57. Ali J, Ali N, Wang L, Waseem H, Pan G (2019) Revisiting the mechanistic pathways for bacterial mediated synthesis of noble metal nanoparticles. J Microbiol Methods 159:18–25

58. Ovais M, Khalil AT, Ayaz M, Ahmad I, Nethi SK, Mukherjee S (2018) Biosynthesis of metal nanoparticles via microbial enzymes: a mechanistic approach. Int J Mol Sci 19 (12):4100
59. Singh P, Kim Y-J, Zhang D, Yang D-C (2016) Biological synthesis of nanoparticles from plants and microorganisms. Trends Biotechnol 34(7):588–599
60. Ramanathan R, O'Mullane AP, Parikh RY, Smooker PM, Bhargava SK, Bansal V (2011) Bacterial kinetics-controlled shape-directed biosynthesis of silver nanoplates using Morganella psychrotolerans. Langmuir 27(2):714–719
61. Gomez-Bolivar J, Mikheenko IP, Macaskie LE, Merroun ML (2019) Characterization of palladium nanoparticles produced by healthy and microwave-injured cells of Desulfovibrio desulfuricans and Escherichia coli. Nanomaterials 9(6):857
62. Yumei L, Yamei L, Qiang L, Jie B (2017) Rapid biosynthesis of silver nanoparticles based on flocculation and reduction of an exopolysaccharide from arthrobacter sp. B4: its antimicrobial activity and phytotoxicity. J Nanomater 2017
63. Ahmad A, Senapati S, Khan MI, Kumar R, Sastry M (2005) Extra-/intracellular biosynthesis of gold nanoparticles by an alkalotolerant fungus, Trichothecium sp. J Biomed Nanotechnol 1(1):47–53
64. Van Houdt R, Provoost A, Van Assche A, Leys N, Lievens B, Mijnendonckx K, Monsieurs P (2018) Cupriavidus metallidurans strains with different mobilomes and from distinct environments have comparable phenomes. Genes 9(10):507
65. Janssen PJ, Van Houdt R, Moors H, Monsieurs P, Morin N, Michaux A, Benotmane MA, Leys N, Vallaeys T, Lapidus A (2010) The complete genome sequence of Cupriavidus metallidurans strain CH34, a master survivalist in harsh and anthropogenic environments. PLoS One 5(5)
66. Tan L, Jones T, Poitras J, Xie J, Liu X, Southam G (2020) Biochemical synthesis of palladium nanoparticles: the influence of chemical fixatives used in electron microscopy on nanoparticle formation and catalytic performance. J Hazard Mater, 122945
67. Balasubramanian R, He J, Wang LK (2009) 15 Control, management, and treatment of metal emissions from motor vehicles. Heavy metals in the environment, 475
68. Wollaston WH. Palladium (II). History 7440:05-3
69. Vignais P, Colbeau A (2004) Molecular biology of microbial hydrogenases. Curr Issues Mol Biol 6(2):159–188
70. Heinekey DM (2009) Hydrogenase enzymes: recent structural studies and active site models. J Organomet Chem 694(17):2671–2680
71. Ludwig M, Cracknell JA, Vincent KA, Armstrong FA, Lenz O (2009) Oxygen-tolerant H2 oxidation by membrane-bound [NiFe] hydrogenases of Ralstonia species coping with low level H2 in air. J Biol Chem 284(1):465–477
72. Gauthier D, Søbjerg LS, Jensen KM, Lindhardt AT, Bunge M, Finster K, Meyer RL, Skrydstrup T (2010) Environmentally benign recovery and reactivation of palladium from industrial waste by using gram-negative bacteria. Chemsuschem 3(9):1036–1039
73. Amonette JE, Russell CK, Carosino KA, Robinson NL, Ho JT (2003) Toxicity of Al to Desulfovibrio desulfuricans. Appl Environ Microbiol 69(7):4057–4066
74. Yong P, Rowson NA, Farr JPG, Harris IR, Macaskie LE (2002) Bioreduction and biocrystallization of palladium by Desulfovibrio desulfuricans NCIMB 8307. Biotechnol Bioeng 80(4):369–379
75. Lloyd JR, Yong P, Macaskie LE (1998) Enzymatic recovery of elemental palladium by using sulfate-reducing bacteria. Appl Environ Microbiol 64(11):4607–4609
76. Vincenzini M, Balloni W, Mannelli D, Florenzano G (1981) A bioreactor for continuous treatment of waste waters with immobilized cells of photosynthetic bacteria. Experientia 37 (7):710–712
77. Adessi A, Spini G, Presta L, Mengoni A, Viti C, Giovannetti L, Fani R, De Philippis R (2016) Draft genome sequence and overview of the purple non sulfur bacterium Rhodopseudomonas palustris 42OL. Stand Genom Sci 11(1):24

78. Rao CRK, Trivedi DC (2005) Chemical and electrochemical depositions of platinum group metals and their applications. Coord Chem Rev 249(5):613–631
79. Sahu AK, Dash DK, Mishra K, Mishra SP, Yadav R, Kashyap P (2018) Properties and applications of ruthenium. In: Noble and precious metals-properties, nanoscale effects and applications. IntechOpen
80. Colica G, Caparrotta S, De Philippis R (2012) Selective biosorption and recovery of Ruthenium from industrial effluents with Rhodopseudomonas palustris strains. Appl Microbiol Biotechnol 95(2):381–387
81. Phongjarus N, Suvaphat C, Srichai N, Ritchie RJ (2018) Photoheterotrophy of photosynthetic bacteria (Rhodopseudomonas palustris) growing on oil palm and soybean cooking oils. Environ Technol Innov 10:290–304
82. Palleroni NJ, Doudoroff M, Stanier RY, Solanes R, Mandel M (1970) Taxonomy of the aerobic pseudomonads: the properties of the Pseudomonas stutzeri group. Microbiology 60 (2):215–231
83. Ramos-González M-I, Ben-Bassat A, Campos M-J, Ramos JL (2003) Genetic engineering of a highly solvent-tolerant Pseudomonas putida strain for biotransformation of toluene to p-hydroxybenzoate. Appl Environ Microbiol 69(9):5120–5127
84. Chivers T, Laitinen RS (2015) Tellurium: a maverick among the chalcogens. Chem Soc Rev 44(7):1725–1739
85. Kim D-H, Kanaly RA, Hur H-G (2012) Biological accumulation of tellurium nanorod structures via reduction of tellurite by Shewanella oneidensis MR-1. Biores Technol 125:127–131
86. Rajwade J, Paknikar K (2003) Bioreduction of tellurite to elemental tellurium by Pseudomonas mendocina MCM B-180 and its practical application. Hydrometallurgy 71 (1–2):243–248
87. Forootanfar H, Amirpour-Rostami S, Jafari M, Forootanfar A, Yousefizadeh Z, Shakibaie M (2015) Microbial-assisted synthesis and evaluation the cytotoxic effect of tellurium nanorods. Mater Sci Eng, C 49:183–189
88. Sękowska A (2019) The many faces of Raoultella spp. Advances in Hygiene & Experimental Medicine/Postepy Higieny i Medycyny Doswiadczalnej 73
89. Nguyen VK, Choi W, Ha Y, Gu Y, Lee C, Park J, Jang G, Shin C, Cho S (2019) Microbial tellurite reduction and production of elemental tellurium nanoparticles by novel bacteria isolated from wastewater. J Ind Eng Chem 78:246–256
90. Kwok R (2019) Inner workings: how bacteria could help recycle electronic waste. Proc Natl Acad Sci 116(3):711–713
91. Batista JH, da Silva Neto JF (2017) Chromobacterium violaceum pathogenicity: updates and insights from genome sequencing of novel Chromobacterium species. Front Microbiol 8:2213
92. Ahmad WA, Venil CK, Aruldass CA (2015) Production of violacein by chromobacterium violaceum grown in liquid pineapple waste: current scenario. In: Liong M-T (ed) Beneficial microorganisms in agriculture, aquaculture and other areas. Springer, Cham, pp 45–58
93. Khalifa AYZ, AlMalki M (2019) Polyphasic characterization of Delftia acidovorans ESM-1, a facultative methylotrophic bacterium isolated from rhizosphere of Eruca sativa. Saudi J Biol Sci 26(6):1262–1267
94. Tay SB, Natarajan G, Rahim MNBA, Tan HT, Chung MCM, Ting YP, Yew WS (2013) Enhancing gold recovery from electronic waste via lixiviant metabolic engineering in Chromobacterium violaceum. Sci Rep 3(1):2236
95. Korte F, Spiteller M, Coulston F (2000) The cyanide leaching gold recovery process is a nonsustainable technology with unacceptable impacts on ecosystems and humans: the disaster in Romania. Elsevier, Netherlands
96. Zhuang W-Q, Fitts JP, Ajo-Franklin CM, Maes S, Alvarez-Cohen L, Hennebel T (2015) Recovery of critical metals using biometallurgy. Curr Opin Biotechnol 33:327–335
97. Shuster J, Reith F (2018) Reflecting on gold geomicrobiology research: thoughts and considerations for future endeavors. Minerals 8(9):401

98. Funari R, Ripa R, Söderström B, Skoglund U, Shen AQ (2019) Detecting gold biomineralization by delftia acidovorans biofilms on a quartz crystal microbalance. ACS Sens 4(11):3023–3033
99. Reith F, Brugger J, Zammit CM, Nies DH, Southam G (2013) Geobiological cycling of gold: from fundamental process understanding to exploration solutions. Minerals 3(4):367–394
100. Rana S, Mishra P, Za Wahid, Thakur S, Pant D, Singh L (2020) Microbe-mediated sustainable bio-recovery of gold from low-grade precious solid waste: A microbiological overview. J Environ Sci 89:47–64
101. Fairbrother L, Etschmann B, Brugger J, Shapter J, Southam G, Reith F (2013) Biomineralization of gold in biofilms of Cupriavidus metallidurans. Environ Sci Technol 47(6):2628–2635
102. Fang X, Wang Y, Wang Z, Jiang Z, Dong M (2019) Microorganism assisted synthesized nanoparticles for catalytic applications. Energies 12(1):190
103. Paknikar K, Rajwade J, Pethkar A, Goyal D, Bilurkar P, Mate N (1996) An integrated chemical-microbiological approach for the disposal of waste thin film cadmium telluride photovoltaic modules. MRS Online Proceedings Library Archive 447
104. Chen YL, Tuan HY, Tien CW, Lo WH, Liang HC, Hu YC (2009) Augmented biosynthesis of cadmium sulfide nanoparticles by genetically engineered Escherichia coli. Biotechnol Prog 25(5):1260–1266
105. Tan L, Xiao Y, Cui H, Xu H, Xu M, Wu H, Dong H, Qiu G, Liu X, Xie J (2018) Influence of sulfhydryl sites on palladium (II) adsorption by displaying EC20 on the surface of Escherichia coli. J Chem Technol Biotechnol 93(12):3569–3581
106. Tan L, Wu H, Cui H, Xu H, Xu M, Xiao Y, Qiu G, Liu X, Dong H, Xie J (2020) Selective adsorption of palladium and platinum from secondary wastewater using Escherichia coli BL21 and Providencia vermicola. Bioprocess Biosyst Eng 43(10):1885–1897
107. Iravani S (2014) Bacteria in nanoparticle synthesis: current status and future prospects. International scholarly research notices 2014
108. Korbekandi H, Iravani S, Abbasi S (2009) Production of nanoparticles using organisms. Crit Rev Biotechnol 29(4):279–306

Plastic and Polymeric Waste Recycling Technologies

Recycling the Plastic Wastes to Carbon Nanotubes

Atika Alhanish and Gomaa A. M. Ali

Abstract This chapter introduces the reader to utilizing plastic wastes as a precursor for the fabrication of carbon nanotubes and the efforts done for this purpose. In addition, it provides a brief introduction to the topic, and an overview of the fundamental concepts of carbon nanotubes, including structure, types, and growth mechanism, is given. The conventional methods of fabricating carbon nanotubes are discussed. Moreover, it describes the methods used to convert plastic waste to carbon nanotubes in detail, while also highlighting the factors affecting each process's efficiency and the recent progress in this regard.

Keywords Recycling · Plastic Wastes · Carbon Nanotubes

List of Abbreviations

AAO	Anodic aluminum oxide
BCNTs	Bamboo carbon nanotubes
BR	Polybutadiene rubber
CCVD	Ceramic boat in a horizontal quartz tube
CNTs	Carbon nanotubes
CO	Carbon monoxide
CO CVD	Carbon monoxide CVD
CoAc	Cobalt acetate
CVD	Chemical vapor deposition
HDPE	High-density polyethylene
HiPCO	High-pressure catalytic decomposition of carbon Monoxide
LDPE	Low-density polyethylene
MA-PP	Maleated polypropylene

A. Alhanish (✉)
Chemical Engineering Department, Faculty of Petroleum and Natural Gas Engineering, University of Zawia, Zawia, Libya
e-mail: atika.alhanish@zu.edu.ly

G. A. M. Ali
Chemistry Department, Faculty of Science, Al–Azhar University, Assiut 71524, Egypt
e-mail: gomaasanad@azhar.edu.eg

A. S. H. Makhlouf and G. A. M. Ali (eds.), *Waste Recycling Technologies for Nanomaterials Manufacturing*, Topics in Mining, Metallurgy and Materials Engineering, https://doi.org/10.1007/978-3-030-68031-2_24

MWNT	Multiwalled nanotube
NR	Natural rubber
OMMT	Modified montmorillonite
PE	Polyethylene
PECVD	Plasma-enhanced CVD
PET	Polyethylene terephthalate
PF	Phenolic formaldehyde resin
PP	Polypropylene
PS	Polystyrene
PVC	Polyvinyl chloride
SBR	Styrene-butadiene rubber
SWNT	Single-walled nanotube
VLS	Vapor–liquid-solid

1 Introduction

Plastics (polymers) are widely used in modern life, as we interact with them every day. They are versatile, durable, lightweight, and cheap with tunable properties, and plastics applications are expected to increase as newly developed types of polymers are discovered consistently to meet today's requirements [1]. However, plastics have a large carbon print, which raised concerns about their impact on the ecosystem in the form of water, air, and land, as also their wastes dramatically accumulated. It is well known that most of the plastics degrade in hundreds of years in normal conditions; thus, their disposal is a global issue [2]. The threat of plastic waste increased public awareness of the importance of recycling those materials, although a few of them reaching the recycling facilities [3]. The recycling of plastic waste is restricted by screening, separation methods followed by cleaning and hot melting, which increases the cost of recycling [4]. On the other hand, landfill and incineration have been used as solutions to overcome this problem [5]. Still, decreasing the space of lands for landfilling and the costs associated with incineration besides the emission of hazard contaminants in flue gas makes them quite stringent [6].

Although plastic conveys the privileges of containing high carbon content and high energy content, which paved the way to be considered a carbon source for the production of carbon nanomaterials [7]. It has been revealed that plastic waste could be recycled under suitable conditions for producing carbon nanotubes (CNTs) of different forms and qualities [8]. The concept of recycling plastic waste to CNTs was first reported in 1997 [9, 10]. Subsequently, numerous efforts explored different types of plastic waste, conversion processes, various growth conditions, and different CNTs quality and quantity had been reported. Since they have been discovered, CNTs have extended our ability to design devices such as wires, molecular probes, pumps, among others [11]. They are small molecules with numerous potentials, and they gained attention in most domains of science and engineering [12]. However, high production costs limit their usage.

Currently, the commercial production of CNTs consumes massive amounts of premium feedstocks such as carbon monoxide (CO), methane (CH_4), hydrogen (H_2), and ethylene(–C=C). Thus, the need for cost-effective and eco-friendly materials is an urgent demand. Hence, using plastic waste as a precursor for synthesizing CNTs is considered a sustainable and economically favorable solution to promote the production of CNTs on a large scale and save the earth from plastic waste accumulation [13].

2 Fundamental Concepts of Carbon Nanotubes

2.1 Overview of Carbon Nanotubes

CNTs have been one of the emerging materials for multiple decades; their existence was observed in 1952 [14], but at that time, they had not been synthesized in macroscopic amounts where no fabrication process for this purpose was known. However, the history of CNTs discovery is a matter of argument [15]. At the beginning of 1999s, the cutting edge of synthesizing and observing CNTs was done by Iijima [16, 17]. Subsequently, various routes had been developed. Structurally, single CNTs have tubular structures that possess a large surface area made of sp^2 bonded carbon atoms with a diameter in the range of nanometer and length in micrometers. CNTs are allotropes that belong to fullerene family. Based on CNTs structure, they are generally categorized into (1) single-walled nanotube (SWNT), (2) multiwalled nanotube (MWNT), (3) polymerized SWNT, (4) nanotours, and (5) nanobuds (Fig. 1).

The main forms of CNTs are SWCNTs and MWCNTs. SWCNTs are composed of a wrapped sheet of graphene according to a specific atomic arrangement of carbon atoms within it known as chirality [14, 18–20]. The way that the graphene sheet is wrapped is represented by integers of the chiral vector (n, m). Based on the chiral vector, SWCNTs are divided into two standard forms: zigzag and armchair, besides a chiral's non-standard form. This chirality is responsible for SWCNTs' unique property that they can behave as a semiconductor in some forms or as metal in others [21].

On the other hand, MWCNTs comprised of a rolling up of multiple graphene sheets fitted one inside the other known as Russian doll structure or as a single sheet rolls itself like a parchment into multiple wall tubes know as parchment structure [14, 22, 23]. The number of walls ranging from 2, i.e., double-wall CNTs, to less than a hundred [21]. However, thin MWCNTs about (3–6) are mostly favorable than thick MWCNTs [24]. The MWCNTs also possess unique properties with a relatively high transition temperature [25–28]. The main differences between SWCNTs and MWCNTs are presented in Table 1.

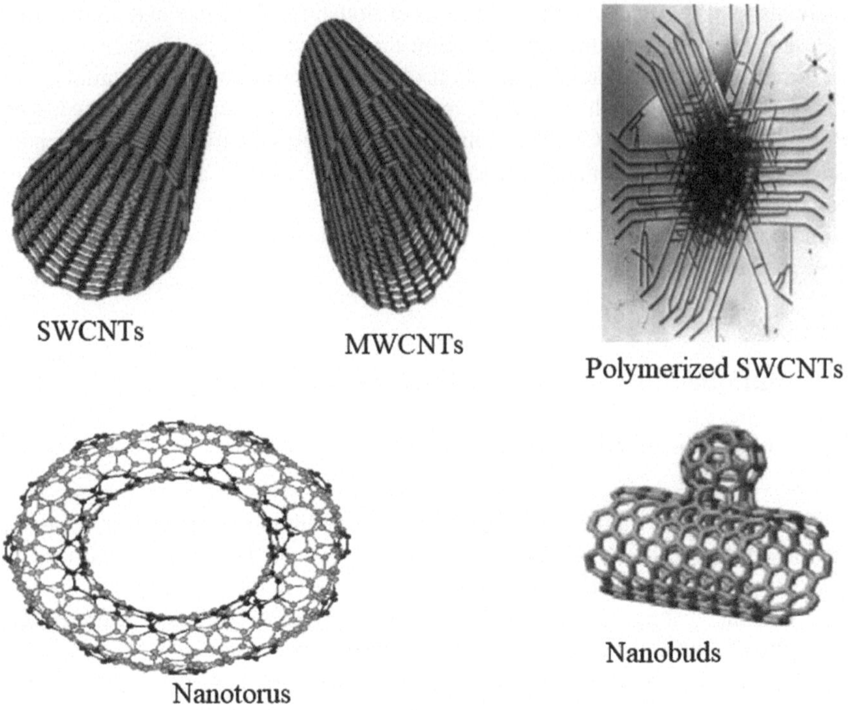

Fig. 1 Types of CNTs. Adapted from Ref. [11] Copyright 2011, Elsevier

Table 1 Differences between SWCNTs and MWCNTs [29]

SWCNTs	MWCNTs
Single-layer of graphene Catalyst is required for the synthesis	Multiple layers of graphene It can be produced without a catalyst
Bulk synthesis is difficult, as it requires proper control overgrowth and atmospheric conditions	Bulk synthesis is easy
Not fully dispersed, and form bundled structures. Resistivity is usually in the range of 10^{-4}–10^{-3} Ω m	Homogeneously dispersed with no apparent bundled formation. Resistivity is usually in the range of 1.8×10^{-5}–6.1×10^{-5} Ω m
Purity is poor. Typical SWCNTs content in as-prepared samples by chemical vapor deposition (CVD) method is about 30–50 wt% However, high purity of up to 80% has been reported by using the arc discharge synthesis method	Purity is high Typical MWCNTs content in as-prepared samples by CVD method is about 35–90 wt %
A chance of defect is more during functionalization	A chance of defect is less, especially when synthesized by the arc discharge method
Characterization and evaluation are easy It can be easily twisted and are more pliable	It has a very complex structure. It cannot be easily twisted

2.2 *Growth of Carbon Nanotubes*

To control CNTs properties and quality, understanding the growth mechanism is vital [30]. The growth mechanism of CNTs depends on the synthesized method. For instance, MWCNTs grow without a catalyst in both laser ablation and arc discharge methods. In contrast, a catalyst is vital for increasing MWCNTs in the chemical vapor deposition method (CVD) [11]. Various mechanisms have been reported, but no standard mechanism is accepted to demonstrate the growth behavior [30]. Vapor–liquid-solid (VLS) mechanism is a model widely used to illustrate the growth mechanism of CNTs. However, this mechanism is still contentious and under development [30]. Thus, the growth mechanism still requires more work to reveal the variety of observed features [11]; while different mechanisms could be involved during the formation of CNTs [29].

2.3 *Progress in Carbon Nanotubes*

The diversity in CNTs structures such as a number of walls, diameter size, chiral angle, and length made them a versatile material escorted by superior properties that conventional materials cannot rival; paving the way for unlimited applications; consequently, CNTs could be used in the global industry in different areas [11, 31]. For instance, NASA's significant investment recently developed novel CNTs based on composites for spacecraft body parts [25]. Despite the massive progress in CNTs research, the application is still restricted because some of the superlative properties of CNTs were predicted just for ideal atomic arrangement [24]. Also, the field is still struggling to produce well-defined CNTs on an industrial scale at a viable cost for some applications, especially in the case of SWCNTs [25]. Hence, efforts are carried out consistently to overcome these challenges; as one of the recent struggles, a research team at the University of Liege has invented two approaches that had been used industrially via Nanocyl SA Co. to produce pure CNTs with consistent quality at a competitive cost [31].

3 Conventional Pathways for Synthesizing Carbon Nanotubes

Synthesizing of CNTs has been a global burning issue for about two decades [32]. The first synthesis of CNTs was done serendipity by arc discharge [16]; nowadays, various techniques are available. Today's most common methods are arc discharge, combustion/flame, CVD, and laser ablation [33]. The mutual objective is providing energy to the carbon precursor to formulate CNTs. In arc discharge, current is used as an energy source while heat is supplied from a furnace in CVD, and a laser with high intensity is used in the laser ablation method [11].

Several parameters affect the type and the properties of CNTs produced, such as heat source used, type of carbon precursor, and reaction conditions [34]. In general, those techniques are requiring the use of high temperatures; thus, they have been replaced by CVD techniques with low temperatures (~800 °C) [14]. Generally, CVD with a low temperature in the range of (600–900 °C) is used to produce MWCNTs, while the high-temperature technique (900–1200 °C) is still used to produce SWCNTs [11]. Laser ablation and arc discharge are mainly used to produce SWCNTs despite the expensive equipment required and the small quantities produced with high impurities [11].

For large-scale production, CVD has been preferred for its simplicity, controllable process, ability to produce high aligned CNTs, and the existence of a variety of techniques [34]. The grown CNTs in this method occur over the surface of the catalyst as a result of the decomposition of carbon precursor, and different forms of CNTs could be produced, such as thin or thick films, powder, straight or coiled, aligned or entangled, or even a desired architecture [11]. Several techniques had been developed over the years based on the CVD method (Fig. 2); five of them considered as promising routes: high-pressure catalytic decomposition of carbon monoxide (HiPCO), methane CVD, plasma-enhanced CVD (PECVD), alcohol CVD, and carbon monoxide CVD (CO CVD) [29].

The drawback of CVD, as well as flame synthesis, is the requirement of injection of expensive high flammable chemicals, such as H_2 to promote the reduction needed to produce CNTs [35]; besides consuming a large amount of electric power on an industrial scale, in contrast, combustion does not require supplying electricity [13]. In comparison, CNTs produced from laser ablation and arc discharge only in powdered form and the tubes tangled into bundles. Those techniques were the first

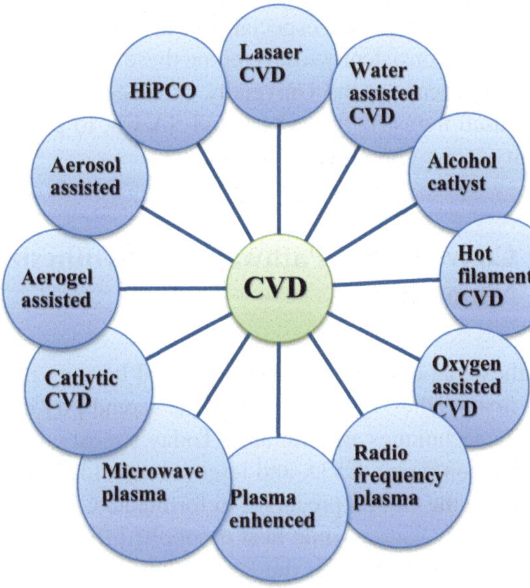

Fig. 2 Illustration of techniques based on CVD for CNTs production. Adapted from Ref. [34] Copyright 2017, Islamic Azad University

Table 2 Basic details of the common methods [29]

Method	Arc discharge	Laser ablation	CVD
Process	Connect two graphite rods to a power supply; place them a few millimeters apart. At 100 amps, carbon vaporizes and forms a hot plasma	Blast graphite with intense laser pulses; use the laser pulses rather than electricity to generate carbon gas from which the CNTs form; try various conditions until hit on one that produces prodigious amounts of SWCNTs	Place substrate in oven, heat to high temperature, and slowly add a carbon-bearing gas such as methane. As gas decomposes, it frees up carbon atoms, which recombine in the form of CNTs
Conditions	Low-pressure inert gas (He)	Ar or N_2 gas at 500 Torr	High temperatures within 500 to 1000 °C at atmospheric pressure
Typical yield	30–90%	Up to 70%	20–100%
SWCNT	Short tubes with diameters with an individual diameter (1–2 nm)	Long bundles of tubes (5–20 μm)	Long tubes with diameters ranging from 0.6 to 4 nm
MWCNT	Short tubes with inner the diameter of 1–3 nm and the outer diameter of approximately 10 nm	Not very high interest in this route	Long tubes with a diameter ranging from 10 to 240 nm
Carbon source	Pure graphite	Graphite	Fossil-based hydrocarbon and botanical hydrocarbon
Cost	High	High	Low
Advantages	It can quickly produce SWNT, MWCNTs. SWCNTs have few structural defects; MWCNTs without catalyst, not too expensive, open-air synthesis possible	Good quality, higher yield, and narrower distribution of SWNT than arc discharge	Easiest to scale up to industrial production; long length, simple process, SWNT diameter controllable, and quite pure
Disadvantages	Tubes tend to be short with random sizes and directions; often needs much purification	Costly technique because it requires expensive lasers a high-power requirement	Often riddled with defects

methods allowing the synthesis of SWCNTs in relatively large amounts (in grams). The common feature between both ways is the condensation of hot gaseous carbon produced from evaporation solid carbon. In general, both techniques require vacuum conditions and a large amount of energy; they pose difficulties for large production [29]. Table 2 summarizes the most popular methods used: laser ablation, arc-discharge, and CVD.

Other methods used for synthesizing CNTs include: decomposition Sic, torsion of graphene layers, dipping graphite in cold water, heat treatment of polymer, electrolysis, pyrolysis, among others, and some of them are non-standard techniques [34].

Most of these methods produce CNTs in powder form that contain a small amount of CNTs with impurities such as amorphous carbon, nanocrystalline graphite, and some metal fractions from the catalyst, which will affect CNTs properties; thus, a purification step is required, which will raise the cost of the production processes. Therefore, there is a necessity to adopt eco-friendly synthesis methods utilizing renewable, natural, and cheaper materials [32].

4 Synthesis Techniques of Carbon Nanotubes from Plastic Waste

To mitigate the threat of increasing plastic waste, numerous efforts had been made to investigate the best ways to recycle them in terms of sustainability. The current research trend is investigating the ability to utilize plastic waste as a precursor for synthesizing nanomaterials as a low-cost route. Several strategies have been reported to synthesize CNTs from plastic waste; they differ in the type of plastic waste, conversion route, type and concentration of catalyst used, growth conditions, quantity and quality of CNTs produced. Among them, tremendous efforts investigated heat treatment (pyrolysis or carbonization) under various conditions such as the presence or absence of catalysts in the inert or oxidative atmosphere under high pressure or atmospheric pressure producing gases oils for further treatment [36]. These conversion processes are the most used to recycle plastic waste to valuable products such as CNTs, where life cycle assessment studies found that pyrolysis of plastic waste was a good environmental option in terms of global warming [37]. Thus, this chapter will focus on recycling plastic wastes into CNTs and related approaches. Those processes' main objective is to break carbon chains of polymer to small carbon precursors via thermal degradation or pyrolysis in different types of reactors under varying conditions [8].

Generally, most of the carbons in the plastic waste converted to carbon material during heating through aromatization were the rest of the carbons, and non-carbon atoms in the plastic such as O, N, H, and Cl are converted to gaseous such as CO_2, CO, H_2O, NH_3, HCL, and CH_4. Direct pyrolysis (carbonization) could be conducted for plastics containing oxygen in their structure like polyethylene terephthalate (PET) at high temperature and in the presence of inert gas or air to prevent the pyrolyzed product from ignition. In contrast, for plastics without oxygen in their structure, such as polyolefines, a pre-oxidation/pre-chemical treatment is required before pyrolyzing them. In general, catalytic pyrolysis is considered the most common approach used to recycle plastic waste into CNTs through dissolution and precipitation mechanisms. The catalyst promotes the dehydrogenation of plastic waste completely to produce high carbon content [38]. Besides being used in one step, this technique is often used in two steps and is still consistently developed [39].

The various methods are categorized based on the primary reaction steps, and some recent efforts will be covered in detail. Also, the role of different catalysts in various approaches will be discussed.

4.1 One-Step Processes

In this category, the formation of CNTs occurs upon the generation of carbon precursors derived from plastic waste. The typical approach for this category is as following: plastic waste mixed with catalysts. Then, heat is applied in the reaction chamber to pyrolyze the plastic; subsequently, carbon precursor (in gas or liquid phase) is produced, which plays the role of carbon source for the growth of CNTs over the catalyst. Usually, treatment or purification step is required due to amorphous carbon and traces of catalyst in the collected powder of CNTs.

Generally, the conversion process of plastic waste to CNTs mechanism is sophisticated. It comprises different reaction steps of plastic waste cracking with high temperature, combustion/oxidation with high temperature, the formation of the thin film via CVD, and gas–solid chemical reaction through heterogeneous catalysts [13]. The dissociation-diffusion-precipitation mechanism is one of the most accepted explanation; it was widely used to explain the formation of CNTs from polyolefin degradation. In this model, plastic waste is dissociated by nanoparticles; subsequently, the produced carbon dissolves via dissociation reaction. Finally, the nanoparticle being supersaturated with carbon precipitation CNTs are formed [40]. However, this had been a long-standing issue due to the complexity of degradation products [41].

In general, various types of plastic wastes such as polypropylene (PP) [40, 42, 43], polystyrene (PS), polyethylene (PE) [44–46], low-density polyethylene (LDPE) [47], PET [48, 49], polyvinyl chloride (PVC) [50, 51], and tires [52, 53] among others have been used as carbon precursor for synthesizing CNTs through one-step and multistep processes.

Moreover, different types of catalyst have been examined, including zeolite [47], solid acids such as organically modified montmorillonite (OMMT) [43], combined catalysts [40], transition metals elements such as nickel (Ni) [46] and their compounds as cobalt acetate (CoAc) [3], nickel oxides (Ni_2O_3) [40], and ferrocene [44, 49]. In addition, heat for degradation is supplied from either electric furnace (autoclaves, fixed beds, fluidized beds, others) or combustion fuels. A variety of CNTs forms, quality, and quantities had been obtained at different conditions in this category. Table 3 summarized different efforts with different feedstocks and conditions with varied results using one-step processes.

In recent years, El-Essawy et al. [49] studied recycling PET waste to produce different carbon nanostructured materials via catalytic pyrolysis (catalytic thermal decomposition). Generally, this work studied the effect of different conditions on

Table 3 Different efforts using one-step processes for converting polymer waste to CNTs

Waste feed	Process description	Product specifications	References
PE	PE mixed with ferrocene and maleated polypropylene (MA-PP) in stainless steel autoclave placed in an electric furnace at 700 °C (pyrolysis), subsequently PE decomposed to form CNTs and some residual gases. The product was then cooled and purified	80 wt.% yield of straight and helical CNTs with a diameter ranging from 20 to 60 nm	[44]
PP	PE mixed with (OMMT, Ni supported on silica-alumina and MA-PP) in stirrer for 10 min. Then, the mixture placed in a crucible and heated with flame at 600 °C (combustion), leading to PE decomposition and CNTs formed. The product was then cooled and purified	MWCNTs with a hollow center and the diameter were about 33 nm; the yield increases with increasing content of the OMMT (Max. yield 41.16 wt.% at 10% OMMT)	[43]
PVA	PVA was dissolved in deionized water at 80 °C. Then, fly ash dispersed and sonicated for 5 min. The composite then shacked to remove bubbles. After cooling and drying, the sample placed in alumina crucible inside a quartz tube furnace (pyrolysis) at 500 °C for 10 min under N_2 atmosphere	CNTs ribbons with different shapes knotted and twisted, U- and spiral-shaped with width ranging (18–80 nm)	[50]
PVA	The process was carried out as almost the method mentioned above, but the fly ash catalyst was modified with NaOH at 85 °C for 8 h	45 wt.% yield of CNTs in the form of ropes, Y branch ropes, twisted ropes, and stacked cone sheet	[51]
PE/PP/PE individually and in mixture form	Polymer mixed with a catalyst (Fe, Ni) within stainless steel bubbling fluidized bed with quartz-rich sand or alumina particles (bed). In the N_2 atmosphere, different experiments were conducted, and the samples heated in the range from 450 to 850 °C (pyrolysis). Subsequently, PP decomposes at 700 °C, and CNTs obtained. The product then purified by sonication	MWCNTs with various yields and qualities for different starting polymers	[46]

decomposition of PET waste in an enclosed stainless-steel autoclave reactor at 800 °C under autogenic pressure, where different forms of carbon nanostructures obtained. One of their experiments was investigating the effect of heterogeneous oxidation on catalytic thermal decomposition products of PET waste in the presence of ferrocene as a catalyst. At first, PET waste was crashed with particle size in the range of (1–3 mm) and then placed in an autoclave in the center of the electric furnace at room temperature for 100 min. Secondly, the temperature in the furnace raised to 800 °C for 20–22 h followed by cooling.

A minor amount of MWCNTs was observed in TEM analysis in the sample of carbon powder formed by decomposition with 20 ml of 30% H_2O_2 as an oxidizing agent for 22 h. However, SWCNTs filled with fullerenes observed when using carbon powder with 15 ml of 30% H_2O_2 for 20 h. The authors explained this result occurs due to oxidation, which occurs in the edge of graphite flakes formed through PET's decomposition. Then, the H_2O_2 enters inner layers where H_2O_2 decomposes between layers causing the exfoliation of layers forming graphene sheets. Then, the individual sheet rolls around itself to reduce the surface energy forming MWCNTs. However, for better understanding, more investigation efforts, the interaction between H_2O_2 and graphene in this approach is needed.

In an earlier effort, Pol and Thiyagarajan [3] had recycled PP and HDPE into MWCNTs via catalytic pyrolysis in the presence of CoAc as a catalyst. The mixture was heated to 700 °C for 2 h in a closed autoclave under an inert or air atmosphere. Through the decomposition process (the bonds broke down at 600 °C), the carbon species (MWCNTs) self-assemble on the surface of the catalyst surface through solid–solid phase transformations, and part of graphite chemisorbed on to Co.

In general, carbon species produced via decomposition adhere in three places on the catalyst: (1) graphite flakes grow around catalyst, (2) additional flakes under the previous graphite flake forced to form a tubular structure which chemisorbed to the catalyst, and (3) carbon species are added to tubular CNTs [54]. These CNTs were used in the water purification application. Despite this, large-scale production using autoclave not suitable from the economic aspect was required.

In another study, Mishra et al. [42] synthesized high purity MWCNTs from PP waste by pyrolysis in CVD at 800 °C using Ni as a catalyst under H_2 and Ar atmosphere in one hour. When the concentration of plastic waste is too high, an overabundance of decomposition occurs, resulting in the formation of amorphous carbon. Thus, H_2 is vital to maintain catalyst activity for the growth of CNTs in CVD [42]. This approach's growth mechanism could be described as following: PP decomposes during interaction with catalyst surface inside the CVD system, and then the produced carbon diffuses onto the Ni surface. At some points, the Ni supersaturated with carbon, and growth of CNTs begin as perception on the Ni surface.

In this type of mechanism, the temperature gradient must occur. For instance, the decomposition of PP in the study of Mishra et al. started at about 400 °C, and carbon deposition on the catalyst's surface takes place in an exothermic reaction. After a while, carbon accumulated on the surface, forming nucleating sites, was CNTs growth, an endothermic reaction. The temperature gradient is created between the exothermic and endothermic reactions regions; the heat flow supports dissolved carbon atom. Also, the continuous dissolution of carbon atoms leads to deposition on a large surface area preventing catalyst poisoning [42].

CNTs with high purity like aspect ratio up to 1500, 30–50 nm in diameter and tens of microns in length could successfully be produced from PE waste using nickel dichloride ($NiCl_2$) as a cheap catalyst via pyrolysis under optimal conditions of 2 h of pyrolysis at 700 °C using a 0.75 wt.% of Ni/PE ratio. In this approach, a prescribed amount of $NiCl_2$ was dissolved in ethanol first, and then, the solution was subjected to a magnetic stirrer at room temperature. A powder of PE waste then mixed with the solution was initially desiccated at 50 °C in the oven overnight. Subsequently, the reactant precursor was placed in an alumina crucible and inserted into an alumina tube furnace, and then heated at a temperature between 500 and 800 °C before being held for 2 h in a flowing Ar atmosphere [46]. It was found that CNTs produced from this approach exhibit excellent adsorption capacity for methylene blue; furthermore, the CNTs could be recycled from the pollutant by a magnet, thanks to special magnetic properties of residual Ni catalyst.

Arc discharge also has been used to synthesize CNTs from plastic waste. A free catalyst and solvent approach was presented by Berkmans et al. [48]. In this study, PET waste was used as a carbon source via rotating cathode arc discharge. PET waste was first pyrolyzed in the N_2 atmosphere to form polymer char, subsequently subjected to an arc discharge. The temperature of the anode raised to about 300 °C during arc discharge generating plasma. Fine CNTs in the cathode are obtained with 364 nm in diameter and several micrometers in length, while longer and thinner CNTs about 95 nm in the anode. In this approach, the temperature has a significant impact on the quality of CNTs produced, where temperature difference in the cathode and the anode resulting in these differences in quality [54].

In a study conducted by Gong et al. [40], catalytic pyrolysis was explored for recycling PP to CNTs in the presence of Ni_2O_3 catalyst and activated carbon as co-catalyst. At first, PP was mixed with the preferred amount of Ni_2O_3 and activated carbon in mixer at 180 °C for 10 min. Then, the mixture was placed in a quartz tube reactor at different carbonization temperatures in the N_2 atmosphere; subsequently, CNTs are formed by carbonization with an outer diameter of 10–40 nm. In other work, Gong and his team studied catalytic pyrolysis of PP in the presence of Ni_2O_3 and CuCl as a combined catalyst. Different factors were examined in these studies, and a layer-by-layer assembling mechanism was proposed. The important findings from both studies can be summarized as follows:

(1) Using Ni_2O_3/activated carbon as a catalyst was synergistic on the recycling process of PP into CNTs;
(2) The presence of activated carbon promote the cracking of PP into light hydrocarbons and the dehydrogenation and aromatization of these hydrocarbons;
(3) Reducing the content of Ni_2O_3 lowered the yield of carbon;
(4) Also, replacing Ni_2O_3 by Fe_2O_3 and Co_2O_3 causes a decrease in the yield in this sequence: Ni > > Co > Fe; thus, Ni catalyst is more efficient in recycling PP into CNTs;

(5) The higher degree of graphitization for produced carbon is observed at higher carbonation temperature;
(6) The diameter of CNTs increased from 28.7–31.8 nm as the carbonation temperature increased from 720–920 °C;
(7) Combined Ni_2O_3 catalyst with halogens affected the degradation of PP, in the case of using CuCl the CNTs yield increases fast at low content of Cl;
(8) The ratio of combined catalysts had a vital effect on the morphology of produced carbon. At a ratio of ≤ 0.125 long fiber carbon obtained, while at a ration ranging from 0.25 to 0.5, short fibers occur, in ratio lower than 0.125 hollow CNTs obtained.

These findings show the important role of catalyst design in plastic waste recycling processes besides other crucial factors such as the type of plastic waste used, reaction temperature, and type of reactor. These factors synergistically play a vital role in the conversion processes that affect the shape, morphology, and structure of CNTs. It was also observed that high conversion yield could be achieved when plastic waste mixed with a catalyst and placed on a closed system such as crucible compared to open systems such as quartz tube. In a closed system, the reaction between polymer waste and catalyst stays longer, which could be the reason for higher conversion [13]. In addition, high temperatures, i.e., more than 700 °C lead to high yield, as shown in the studies mentioned above. Also, both the properties and the production of CNTs are mainly affected by the reactor type, where different types of reactors investigated in various works, and one such study investigated the impact of the main operating variables in fluidized bed reactor in batch and continuous mode on the composition and yield of the product as in the literature [55]. The role of the reactor type has been comprehensively discussed in this literature [6].

4.2 *Multistep Processes*

In this category, the formation of CNTs occurs after the generation of carbon precursors. Thermal decomposition (pyrolysis) occurs in the first step, then the gaseous product (carbon precursor) channeled to another reaction chamber to react with catalyst and form CNTs. The early reported efforts for investigating the conversion of plastic waste to CNTs applied this process [9, 10]. This process's privilege is the control of sub-processes involved in the conversion process individually [13]. Different approaches combine various steps developed, and multiple results were obtained; some of these attempts are summarized in Table 4.

One of the three-stage methods is mentioned in the table, showing a promising route on a scalable scale, as demonstrated by Zhuo and his co-workers [35]. In the mentioned work, the mechanism was simply converting the carbon in the polymer structure to hydrocarbons in the first step through pyrolysis or combustion. In the second step, hydrocarbon gases derived by pyrolysis were channeled to the catalytic chamber, where hydrocarbons eventually transformed into CNTs.

Table 4 Different efforts using multistep processes for converting polymer waste to CNTs

Feed waste	Process description	Product specifications	References
PE	In the first step, PE pyrolyzed in the chamber to produce a carbon precursor. In the second step, the carbon product passed over Ni catalyst in a quartz tube in the presence of He atmosphere at 4 atm and 420–450 °C in less than an hour	The yield of CNTs production was about at 3 × 10^{-3} g/cm^2 per h, crooked carbon nanotubes were obtained with diameters of 10–40 nm	[9]
Tire rubber	Catalyst powder with composition of (Co:Mg:Mn:Al;2.5:0.5:0.5:0.5) was dispersed in a ceramic boat in a horizontal quartz tube (CCVD) and heated to 650 °C in H_2 atmosphere for 1 h, where the scrap tire placed in the entrance of the tube at low temperature. Then, the rubber pushed to the heating zone about 400 °C and the atmosphere switched to N_2 for half hr. Subsequently, the system is left to cool	MWCNTs with hollow, short, and thick-wall structure	[56]
LDPE	LDPE was pyrolyzed in stainless steel reactor at 600 °C in N_2 atmosphere. In the second stage, the gaseous hydrocarbons channeled to a catalytic chamber (Ni, Co, and Fe, individually) at 800 °C to deposits carbon on the catalyst to form CNTs	Ni: CNTs and bamboo CNTs (BCNTs) with diameter 15–30 nm and length in few micrometers. Fe: MWCNTs and BCNTs with a larger diameter Co: narrow MWCNTs with diameter 5–20 nm and few micrometers in length. Cu: no CNTs observed	[57]
Polyamide (PA)	PA 6.6 was fed continuously to the first reactor and pyrolyzed at different temperatures (600–900 °C) in N_2 atmosphere. The pyrolysis gases (hydrocarbons and H_2) passed over a preferred amount of calcined catalyst (Fe/Al_2O_3) were placed in a crucible in a horizontal quartz reactor. BCNTs growing after 40 min over the catalyst surface	BCNTs (the best bamboo structure is obtained at 750 °C) with 20 nm in diameter and lengths in micrometer	[58]
HDPE/LDPE	Plastic pyrolyzed at 800 °C in porcelain boat in preheating pyrolyze section in the furnace (quartz tube) in N_2 atmosphere. In the second stage, pyrolyzed gases passed to venture (8 mm ID) were mixed with air or oxygen for 30 s	10% yield of MWCNTs (16 graphene layers and the distance between them about 0.36–0.38 nm) with diameters 15–84 nm and lengths of 1–5 μm	[35]

(continued)

Table 4 (continued)

Feed waste	Process description	Product specifications	References
	and light hydrocarbons generated (combustion stage). Combustion effluent was channeled through small screens of stainless steel (coated with Ni/Co) to catalyze the gases and produce CNTs at 750 °C in the third stage. The CNTs immersed in ethanol and then sonicated to remove substrates		
PET/Tires	PET/Tire, individually pyrolyzed in a furnace in the of 600–1000 ° C in N_2 atmosphere. Then, pyrolyzed gases channeled through mixing venture to synthesis reactor (furnace) at 1000 °C, were 15–19% of O_2 introduced in the venture and a ceramic filter placed between the two sections. In the third stage, combustion effluent passed over the catalyst in the quartz tube	Tire: Entangled CNTs with a diameter of 100 nm and about 40 μm PET: CNTs with a diameter of 50–200 nm and length of 20 μm	[53]
HDPE/PP/ Municipal plastic waste (MPW)	Plastics were pyrolyzed at 560–570 °C in a horizontal tubular reactor generating gaseous hydrocarbons. Then, the gases channeled into (CCVD) reactor, where CNTs formed at 700 °C in the presence of CoAc/FeAc based catalyst in 30 min	Bundle CNTs with ~50 μm in length and ~10 nm in diameter. The yield: 47.1% (waste HDPE), 48.1% (50% waste HDPE + 50% waste PP), and 60.7% (MPW)	[59]

In later work by almost similarly the same group [60], PP, PE, and PS were used in this method as carbon source; the pyrolyzed product of PP and PE was mixed and introduced to oxygen-containing gases or air in the combustion chamber while PS was not mixed oxygen-containing gases. All the samples were channeled to the synthesis chamber at 750 °C over stainless steel catalyst for 1 min. The results revealed different pyrolyzed products: light aromatic hydrocarbons from PS waste, while light aliphatic hydrocarbons from PE and PP wastes. Also, it was found that different products obtained after mixing pyrolyzed gases with O_2 such as H_2, CO, CO_2, and H_2O vapor; and under appropriate conditions CNTs growth. Moreover, H_2 and H_2O vapor have been used as cleaning agents successfully to keep the catalyst active. CNTs with length ranges from 3 to 6 μm were observed by SEM and TEM analysis in all samples. However, CNTs generated from PS has a smaller diameter than CNTs obtained from PP and PE.

In case of using more than two stages (Fig. 3), researches use a series of reactors such as furnaces, CVD or even tubular reactors separated by venture for gas mixing

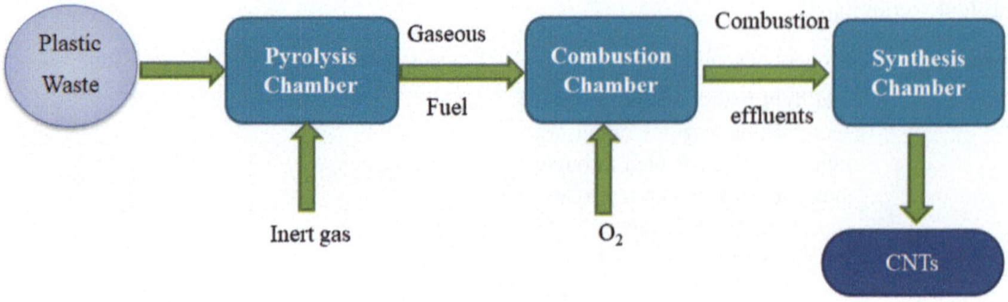

Fig. 3 Illustration of the three-stage process. Adapted from Ref. [60] Copyright 2011, Cambridge

where the pyrolyzed gas mixed with oxygen only in cases of using plastic with high carbon content such as PET and tire [61]. Although this method was also used for polyolefins, various CNTs were formed [35, 60]. However, due to the limited evolution in these conditions, a clarifying conclusion cannot be drawn.

Tremendous efforts for developing new technologies for recycling plastic wastes are sustained. Advanced thermochemical processes, including pyrolysis and gasification in a two-stage approach, have shown the most promise [62]. This technology has high energy efficiency and low emission of pollutants, where gasification of waste plastic generates H_2 syngas for fuel production, energy, and power. This process's efficiency can be enhanced by introducing catalysts in carbon conversion and H_2 production [63].

Several authors have explored using gasification for converting plastic wastes to valuable products, including CNTs, as a by-product. Recently, hydrous pyrolysis or steam gasification was widely studied for recycling plastic wastes into CNTs. In this method, steam injection is employed in catalytic pyrolysis processes; where pyrolysis of plastic waste to generate hydrocarbons in the first stage takes place, while they deposited over the catalyst in the second stage; although water injection either directly with a carrier gas or by transferring the gas through water bubbler before pyrolysis step [39]. One such study was carried out by [62], using LDPE, PP, and PS as carbon sources for synthesizing CNTs through two-stage pyrolysis (CVD) process. Gaseous hydrocarbons produced by pyrolysis at 600 °C in N_2 atmosphere, and then, the gases channeled were steam injected in the second stage over Ni/Al_2O_3 catalyst at 800 °C.

For all plastic wastes used, MWCNTs were produced with a diameter ranging from 10–20 nm and several microns in lengths. The results showed that more CNTs could produce from LDPE than either PP or PS, attributed to the formation of amorphous carbon from large hydrocarbons derived by pyrolysis of PS and PP. The study also investigated the impact of injection steam at different rates, and the results revealed increasing in carbon deposition and H_2 as the steam rate increased. Moreover, CNTs with high quality were obtained by controlling the steam rate. In contrast, another study found that the increase in the steam rate increases H_2 but decreases carbon deposition in the presence of Ni-Mn-Al catalyst [64]. Also, the later study reveals that at a low rate, the quality of CNTs improved. Thus, the catalyst used in this approach may have the main effect on the gasification step's impact on carbon deposition. On the other hand, another study confirms the findings of Acomb et al. [62] by using different plastic wastes. Zhang and Williams [65] produced CNTs and H_2 from waste tires through a two-stage catalytic pyrolysis/gasification process in the presence of Ni/Al_2O_3.

The results demonstrated that introducing steam to the processes enhanced CNTs and H_2 production. These results showed a significant enhancement in CNTs quantity and quality as well. In the same year, Wu et al. [63] applied hydrous pyrolysis for recycling PP as following: pyrolysis of PP in N_2 atmosphere waste at 500 °C inside steel crucible took place in the first stage, in the second stage, water is injected to provide steam for catalytic steam reforming of the pyrolyzed gases where the whole experiment took 40 min. The product was condensed using dry ice and cold air, where unreacted water and liquid oil were collected. This work did not discuss the product quality using this approach, rather discussing the catalysts' effect.

Recently, the production of CNTs through the pyrolysis/gasification process has been proposed using Fe and Ni-based catalysts [66]. Most of CNTs produced in this approach were with poor quality due to a granular catalyst such as Ni/Al_2O_3. A novel approach using template-based synthesis is to support the growth of CNTs with uniform structure through the hydrous pyrolysis process [67], by using anodic aluminum oxide (AAO) as a template catalyst, which is commonly used in conventional CNTs production. In this work, Ni-based AAO was used in recycling HDPE to CNTs through two-stage pyrolysis/gasification processes as the experiment conducted by [63] with little modifications.

In the first step, HDPE pyrolyzed at 500 °C generating hydrocarbon gases in the presence of N_2 gas. In the second step, the gases are channeled over a layer of conditioning catalyst (Ni/Al_2O_3) to modify the hydrocarbons gases before the growth of CNTs on the Ni/AAO catalyst at 700 °C, and water was injected in this step. It was found that using conditioning catalyst enhances CNTs production, where Ni/Al_2O_3 effectively cracked heavy hydrocarbons into small ones during gasification. In addition, increasing the amount of conditioning catalyst had improved the quality and yield of CNTs. However, it was reported that introducing steam generally decreases carbon deposition in this approach [38]. In the context, Nahil et al. study [64] found that most of the carbon deposits over the catalyst were reacted with steam producing H_2 at higher steam injection rates, and at lower steam injection, the quality of CNTs improved.

Recently, using three-stage catalytic pyrolysis/gasification (CVD) process successfully produced MWCNTs obtained from non-condensable gases derived from recycling different plastic packing containing PET with various percentage (11.8 and 27.5%) besides other plastics at 700 °C in the presence of Fe_2O_3 loaded on zeolite (7.5 g) and Ni loading on $CaCO_3$ catalyst (9.0 g), and CO_2 was used as gasifier agent. The study finds that various properties of MWCNTs depending on the plastic waste type were obtained and demonstrated that this approach effectively converted this kind of plastic wastes [68]. Most of the studies are performed using polyolefin wastes as a precursor for CNTs production in one-step and multistep processes. However, using multistep processes proved as a successful route for converting thermoset plastics into CNTs. In this context, a novel technique in the two-stage pyrolysis catalytic process through CVD was performed by Zhang et al. [69]. In this study, a range of reaction conditions has been explored to maximize the production of CNTs from different waste tires such as polybutadiene rubber (BR), styrene-butadiene rubber (SBR), and natural rubber (NR) in the presence of Ni/Al_2O_3 as a catalyst which was prepared by a wetness impregnation technique.

In the first stage, the rubber and tires' pyrolysis took place in a fixed bed reactor from ambient temperature to 600 °C and held at 600 °C for 20 min. The gaseous products are passed to the second fixed bed reactor over Ni/Al_2O_3 at 800 °C in N_2 atmosphere. N_2 continually purged the gaseous product to the condenser. The dry ice and water condensers performed the condensation of the liquid condenser. High rich gas contains H_2, Co and CH_4 were obtained beside carbon deposit containing amorphous carbon and hollow CNTs with a diameter between 5 and 10 nm and length of several microns. Generally, the char residue produced from the pyrolysis was 40 wt.% of waste car tires and 37 wt.% for waste truck tires; this difference is based on the tires' different formulation. The analysis also showed that the pyrolysis of truck tires generates high carbon deposition on the catalyst compared to car tires. Even more, the poor development of CNTs was obtained from waste car tires. In addition, attire: catalyst ratio 1:1, the maximum deposition of carbon from waste tires was found 14 wt.%, and for rubber samples, BR, SBR, and NR were 36, 40, and 36 wt.%, respectively.

The study found that changing the kinetic conditions in terms of tire, catalyst ratio in this approach could promote the yield of carbon deposited, including the amorphous, pyrolytic, filamentous, and CNTs on the catalyst to up to 14 wt.%.

In other work, Gou et al. [70] recycled phenolic formaldehyde resin (PF) by a two-stage pyrolysis catalytic process to produce CNTs in the presence of Ni and Fe as a catalyst, individually. A sample of PF was pyrolyzed at 500 °C in a fixed bed reactor; a gaseous product was generated. Then, the gaseous product was passed to a second fixed bed reactor over a catalyst at 800 °C in the presence of Ni atmosphere. CNTs with a diameter ranging from 15 to 20 nm and several microns in diameter were successfully produced, and the yield was about 24.07%. The purity of 93.13% is yielded when using Ni as a catalyst, while using Fe as catalyst CNTs yield was 34.39% with a purity of 97.45%.

The results showed a better performance of Fe catalyst in these conditions. These results are similar to the results obtained for recycling LDPE waste in a two-stage catalytic pyrolysis process by [57], where the yield and purity of CNTs produced by Fe supported on alumina (Fe/Al_2O_3) catalyst were 26 and 94.73%, respectively, compared to the yield and purity of CNTs produced by Ni supported on alumina (Ni/Al_2O_3) 12.2 and 47.37%, respectively. However, the calcination temperature may have a major effect in this case. In comparison, Ni shows higher efficiency than Fe in recycling PP to CNTs in a step process [40]. This result could be attributed to the combined catalyst's impact; thus, a definitive explanation cannot be made at this point; also, the study reveals that the CNTs yield using Ni_2O_3 and AC separately in two-steps was lower than using combined Ni_2O_3 and AC in one step.

In general, it was reported that using combined catalysts is a favorable strategy in order to control degradation of plastic wastes through a catalytic pyrolysis process, where high yield up to 65 wt.% with the presence of low catalyst contaminates about 5 wt.% in the carbon materials produced including CNTs [63]. Hence, the catalyst's design proved to be a key factor in controlling CNTs formation from recycled plastic wastes. Different works studied the effect of using various types of catalysts on the type of CNTs formed and the yield, where a few of these studies had been mentioned earlier, and recent efforts will be discussed.

Generally, catalysts used in catalytic pyrolysis processes (one-step or multistep) should have two vital functions: the ability to decompose plastics and grow CNTs [36]. Commonly, transition metal-based catalysts have been used in recycling plastic wastes into CNTs; due to their low cost and high catalytic activity compared to other metals. These catalysts are easily deactivated by encapsulating active sites in the catalyst's surface with amorphous carbon produced in the catalytic pyrolysis process [70]. Thus, various catalysts were developed to prohibit deactivation by amorphous carbon, i.e., soot, coke, or char.

One of the reported approaches found that CNTs can be produced from the amorphous carbon deposited over the catalyst by manipulating process conditions through recycling plastic waste processes [63]. However, one major concern in this approach is the possibility of CNTs encapsulation with catalyst particles, which increases the difficulty of CNTs recovery [71]. Using stainless steel was suggested to promote CNTs growth for various plastic wastes and minimize catalyst encapsulation. For instance, Alves et al. [53] used stainless steel mesh alloy (Fe:Cr:Ni) to recycle PET and tires. The study found that this catalyst's presence had prompted thermal dehydrogenation reaction and solid carbons yield, including CNTs and H_2. The study reported that $\sim$1000 °C stainless steel at high temperatures was useful for CNTs growth.

In recent work, stainless steel mesh loaded with Ni had been used in two-stage catalytic pyrolysis process of HDPE waste. Zhang et al. [71] investigated catalysts temperatures (700, 800, and 900 °C), and plastic to catalyst ratio (2:1 and 4:1) on the yield of pyrolyzed products, i.e., gas, liquid, and carbon products. The study reveals the deposition of carbon among other products produced during the process, i.e., gas and liquid hydrocarbons. The study reveals that CNTs quantity as well as quality affected by process conditions. For instance, the study demonstrated that increasing plastic to catalyst ratio decreases carbon deposition, whereas the yield of carbon obtained in this study was 32.5% at catalyst temperature of 700 °C was raised to 38% at 900 °C, and about 40% of the carbon deposited was MWCNTs with 10–20 nm in diameter and several microns in length. Also, the carbon formed at 700 °C is smaller than carbon filamentous formed at higher catalyst temperature when plastic to catalyst ratio was 2:1.

In the same context, Panahi et al. [72] investigated the influence of catalysts consisted of stainless steel-304, stainless steel-316, and stainless steel 316L wire cloth in the recycling of common waste plastic (PP, PE, PS, and PET). MWCNTs were produced through two stages of catalytic pyrolysis (quartz tube) process at 800 °C under N_2 atmosphere. In addition, the study investigated different treatment approaches for activating stainless steel for CNTs growth. Multistep pretreatment of stainless steel was employed to break the Cr oxide layer at the surface of the catalyst, which occurs due to the reaction of Cr with oxygen in the pyrolysis process. This treatment successfully activates the surface of stainless steel for CNTs growth.

It was demonstrated that the yield and quality of CNTs depending on the plastic waste type, stainless steel, and the pretreatment method used. Under the study conditions, PP gives the highest yield, followed by PE, PS, and finally, PET. With another novel strategy, Tripathi et al. [73] use stainless steel tube as a reactor and catalyst simultaneously. CVD made of stainless steel consisted of 45–60% Fe, 10–14% Ni, and 16–18% Cr and a few amounts of other transition metals. Waste centrifuge tubes are made of plastic recycled using the two-stage process. Under the Ar atmosphere, MWCNTs with high yield at 900 °C were obtained. In the second stage of this strategy, oxidation via static air was conducted, and after oxidation of Cr, Ni and Fe in the tube act as catalysts forming MWCNTs. The strategy with the non-oxidation step was conducted in other experiment sets, and carbon fibers were formed to demonstrate this strategy's effectiveness.

Another approach reported to prohibiting encapsulation is using H_2 atmosphere, where it enhances CNTs growth due to its cleaning effect that could prevent encapsulation of active sites in catalysts [67]. Experiments studying the influence of various variables related to the catalyst system used in recycling plastic waste were reported. For a metal-supported catalyst, the interaction is a key factor for determining the catalyst activity for CNTs growth. Acomb et al. [57] studied the impact of different metal supported on alumina catalysts, and the calcination temperature during the catalyst preparation on the recycling of LDPE into CNTs and H_2 was investigated. The study demonstrated the importance of interaction between the catalyst and the alumina in CNTs formation.

For instance, Fe and Ni showed good interactions, which were neither too strong nor too weak, forming a metal particle suitable for CNTs growth. At the same time, Co had a strong interaction, which caused metal particles' formation not suitable for CNTs growth. Thus, based on these results, the authors claim that Co is not a suitable catalyst for CNTs formation from plastic wastes [57]. In contrast, CNTs grow successfully over Co catalysts from recycled tires in the two-stage process [56]. Acomb et al. study reveals the important role of calcination temperature in this strategy, where at low calcination temperature, too large metal particles for CNTs growth produced, which cause the deactivation of catalyst due to the formation of amorphous carbon instead of CNTs. The low yield of CNTs was obtained at low calcination temperature because weak interaction occurs at low calcination temperature.

On the other hand, the amount of metals is mainly affecting the quality of CNTs. It was reported that the inner diameter of CNTs decreases when an excess amount of metals impregnated on the support surface [59]. Uncoated stainless steel and coated with a 4 nm layer of Ni and 4 nm layer of Co had been used as catalysts to investigate their effect on CNTs growth in the three-stage process [35]. The study found that CNTs grow in the presence of Ni layer more than in the presence of Co layer. However, the authors discuss the possibility of the influence of sample preparation on the results. A filter was introduced to deposition soot on the catalyst surface to prevent the deactivation of catalysts in the flame in three-stage processes. Zhuo et al. [35] use a filter made of high-temperature silicon carbide honeycomb barrier placed between the flame and the catalyst inside the quartz tube, where the filter has the efficiency of 97% to capture soot with particles larger than one micrometer. In another study, ceramic filters had been introduced [53].

Based on the literature, every catalyst reported being efficient in a particular system and inefficient in others. For example, using chlorinated compounds combined with AC had promoted the dehydrogenation of plastic wastes to produce low CNTs and more carbon fibers [36]. Solid acids are other catalysts used commonly in one stage process and are found to promote breaking the plastic chains through providing intermediate proton acidic sites [13]. It was reported that using Ni/Al_2O_3 was an active catalyst for the dual production of CNTs and H_2 from plastic waste [4]. For instance, Zhang et al. [52] reported that Ni/Al_2O_3 compared to Co/Al_2O_3, Cu/Al_2O_3, Fe/Al_2O_3, produces MWCNTs with high quality with the highest yield of H_2 from two-stage pyrolysis/gasification of waste tires.

In contrast, through the same reaction system (two-stage pyrolysis/gasification), LDPE waste is recycled into BCNTs with the best yield of both BCNTs and H_2 using Fe/Al_2O_3 to Ni, Co, and Cu based on alumina catalysts, respectively [57]. Using different catalyst combinations, Naihil et al. [64] studied using Ni-based on a ternary mixed oxide catalyst (Ni-Metal-Al) with different metals: Zn, Ca, Mg, Mn, or Ce. It was found that Ni-Mn-Al was the best catalyst system among the tested metals in producing MWCNTs and H_2 through the pyrolysis/gasification process of PP waste.

Moreover, in the recent effort, Aboul-Enein and Awadallah [47] investigated the catalytic activity of CoMo/MgO catalyst with different Co/Mo ratios (0.4, 1, 2.5, 6.5, and 14.5) for recycling LDPE waste into CNTs through simple two-step catalytic pyrolysis process in vertical and horizontal quartz reactors. The pyrolysis process was performed at 400 °C, and the growth of CNTs was conducted at 700 ° C. The study observed that using different Co/Mo ratios significantly affects the quality and the quantity of CNTs formed. The optimum yield of CNTs was about 1040 wt.% obtained at (6.5 CoMo/MgO) and the highest purity obtained by using catalysts with 6.5 and 14.5 wt.% of CoMo. On the other hand, it was found that by decreasing CoMo ratio from 14.5 to 0.4, the formation of CNTs with large diameter increases while CNTs with thinner diameter decreases. In addition, a catalyst with the highest ratio (14.5 CoMo/MgO) produces a portion of different CNTs forms, i.e., BCNTs, thin and thick MWCNTs with a hollow core.

Different important parameters related to the catalyst, such as the particle size, catalyst preparation method, and ratio of catalyst to plastic, have a crucial effect on the recycling process. All the approaches were conducted in the laboratory, and the cost of the method was not covered. The cost of a catalyst is an important issue for scaling up. A few efforts have investigated waste materials as catalysts in recycling plastic wastes such as red mud [74], and promising results were obtained [38].

Besides the importance of catalyst, another vital parameter influences the formation of CNTs from recycled plastic wastes is the gas atmosphere used through the pyrolysis process. N_2 is the most carrier gas used in recycling plastic waste through the pyrolysis process. In this manner, the rate of N_2 flow is crucial where the flow rate affects the residence time of the pyrolyzed gases over the catalyst and thus influences CNTs formation on the catalyst. Growth (catalyst) temperature is another key parameter that affects the quality and quantity of CNTs produced through catalytic pyrolysis. For instance, the pyrolysis of LDPE at 700–800 °C in the presence of Fe/Al_2O_3 produces CNTs with high quality and quantity, while at 900 °C less uniform CNTs are formed [75].

In another study, recycling PET into CNTs with a yield of 25% in the presence of molten salt at 550 °C [76]. While in recycling tire via pyrolysis/gasification process, the obtained results show that increases the temperature from 700 to 800 ° C promote the amorphous and filamentous carbon deposited on the catalyst from 6.6 to 13.3 wt.%, whereas at 900 °C, only filamentous carbon deposits with significant quantities where most of this carbon was MWCNTs [65]. Recently, Moo et al. [77] demonstrated the important role of synthesis temperature not in influencing the properties of CNTs obtained but also in the electrocatalytic activity through the two-stage catalytic pyrolysis process. Their results also revealed the vital role of the synthesis procedure on the electrocatalytic performance of CNTs obtained.

However, the applications of CNTs derived generally from carbon waste such as plastic wastes are limited because usually produced with low yield and low quality [54]. The causes cannot be restricted to some factors. However, the most observed through the literature are the presence of impurities in CNTs, usually containing amorphous carbon and catalyst residuals generated by both one-step and multistep approaches which is one of the factors affecting CNTs quality. Typically, the impurities represent 10 wt.% of the produced CNTs, and the lowest amount reported was 0.3 wt.% [13]. Thus, a treatment/purification step usually is required, which subsequently promotes the production cost. The low yield indirect pyrolysis under inert gas or air means that most of the carbon atoms of the plastic wastes such as polyolefins and PS are escaped in gases, which would lead to not only a huge waste of energy and resources but also serious air pollution. However, the carbon content also plays an important role in plastic wastes that contain low carbon content, such as PET (62.5 wt.%) and PVC (38.4 wt.%). Conversion to carbon materials with up to 20% yield could be achieved [36]. To the greatest extent, improving carbon recovery is the key to utilizing waste plastic to produce carbon materials. Despite, various CNTs derived from plastic wastes had been tested for different applications with promising results such as reinforcement in LDPE matrix [59, 63], adsorbent [46], electrocatalyst [77], and electrode [68].

Lastly, plastic waste recycling into CNTs depends on the conversion route, process condition, type of plastic waste, catalyst, and temperature. Thus, each approach's optimization is required to achieve the greatest recovery of CNTs from plastic wastes. The authors would also like to suggest several notable reviews that comprehensively cover detailed information about another aspect of recycling plastic wastes into CNTs [6, 8, 13, 37, 41].

5 Conclusions

This chapter presents various approaches for converting plastic wastes to CNTs based on the necessary reaction steps involved. Generally, the concept of utilizing polymer wastes as a precursor for CNTs production starts in 1997. However, the reviews covering this topic are limited, where each of them discussed the issue from a different aspect. Various techniques were used to convert different plastic wastes to CNTs using different feedstock, catalyst, reactor, and temperature were divergent results obtained. This chapter covered a wide range of studies emphasizing recent works and the role of catalyst as a critical factor in recycling plastic waste to CNTs. In addition, the process description, the product specification, and several mechanisms of some efforts are covered. The attempts are still in the laboratory stage, and further research for scaling up them is required.

6 Future Prospectives

Based on the available literature, some issues need more investigation, such as (1) catalyst performance depends on the process and plastic types used and has a significant effect on the specification of the product. Thus, investigating the use of cheap catalyst and the performance with optimum content is an essential demand for economic benefit was most of the used catalyst are expensive; (2) the growth parameters in terms of the reactor, reaction temperature, and catalyst type; and (3) fillers and/or additives in the waste plastic use must affect the CNTs produced, thus investigating certain types with different additives may help identify the proper conditions to be conducted for each type. In addition, it was found that thermoplastics and thermosets could be recycled to CNTs successfully by using multistep processes. Also, valuable by-products are obtained in most of the pyrolysis processes used for recycling plastic wastes such as H_2 with high content that could be used as recycling in CVD to minimize this approach's cost. In last, to expand the potential applications of these CNTs on an industrial scale, optimization of CNTs characteristics in terms of plastic waste used, conversion processes, catalyst type, and content, and kinetics of the decomposition must be investigated.

References

1. Tessnow-von Wysocki I, Le Billon P (2019) Plastics at sea: treaty design for a global solution to marine plastic pollution. Environ Sci Policy 100:94–104
2. Singh N, Hui D, Singh R, Ahuja IPS, Feo L, Fraternali F (2017) Recycling of plastic solid waste: a state of art review and future applications. Compos B Eng 115:409–422
3. Pol VG, Thiyagarajan P (2010) Remediating plastic waste into carbon nanotubes. J. Environ Monit 12(2):455–459

4. Okolie JA, Nanda S, Dalai AK, Berruti F, Kozinski JA (2020) A review on subcritical and supercritical water gasification of biogenic, polymeric and petroleum wastes to hydrogen-rich synthesis gas. Renew Sustain Energy Rev 119:109546
5. Williams PT (2006) Yield and composition of gases and oils/waxes from the feedstock recycling of waste plastic. In: Feedstock recycling and pyrolysis of waste plastics. John Wiley & Sons, Ltd., pp 285–313
6. Bazargan A, McKay G (2012) A review—synthesis of carbon nanotubes from plastic wastes. Chem Eng J 195–196:377–391
7. Czepirski L, Szczurowski J, Ba, ys M, aw, Makomaski G, Zieli, ski J, Ciesi, ska W, awa (2016) Novel carbonaceous nanomaterials from waste polymers. Curr Nanomaterials 1 (2):103–109
8. Mukherjee A, Debnath B, Ghosh SK (2018) Carbon nanotubes as a resourceful product derived from waste plastic—a review. In: Waste management and resource efficiency. Springer, Singapore, pp 915–934
9. Kukovitskii EF, Chernozatonskii LA, L'Vov SG, Mel'nik NN (1997) Carbon nanotubes of polyethylene. Chem Phys Lett 266(3–4):323–328
10. Kiselev NA, Sloan J, Zakharov DN, Kukovitskii EF, Hutchison JL, Hammer J, Kotosonov AS (1998) Carbon nanotubes from polyethylene precursors: structure and structural changes caused by thermal and chemical treatment revealed by HREM. Carbon 36(7–8):1149–1157
11. Purohit R, Purohit K, Rana S, Rana RS, Patel V (2014) Carbon nanotubes and their growth methods. Proc Mater Sci 6:716–728
12. Baughman RH (2002) Carbon nanotubes–the route toward applications. Science 297 (5582):787–792
13. Zhuo C, Levendis YA (2013) Upcycling waste plastics into carbon nanomaterials: a review. J Appl Polym Sci 131(4):39931
14. Prasek J, Drbohlavova J, Chomoucka J, Hubalek J, Jasek O, Adam V, Kizek R (2011) Methods for carbon nanotubes synthesis—review. J Mater Chem 21(40):15872
15. Ren Z, Lan Y, Wang Y (2012) Aligned carbon nanotubes: physics, concepts, fabrication and devices. Springer Science & Business Media, pp. 1–299
16. Iijima S (1991) Helical microtubules of graphitic carbon. Nature 354(6348):56–58
17. Iijima S, Ichihashi T (1993) Single-shell carbon nanotubes of 1-nm diameter. Nature 363 (6430):603–605
18. Sadegh H, Ali GAM, Abbasi Z, Nadagoud MN (2017) Adsorption of ammonium ions onto multi-walled carbon nanotubes. Stud Univ Babes-Bolyai Chem 62(2):233–245
19. Seyed Arabi SM, Lalehloo RS, Olyai MRTB, Ali GAM, Sadegh H (2019) Removal of congo red azo dye from aqueous solution by ZnO nanoparticles loaded on multiwall carbon nanotubes. Physica E 106:150–155
20. Ali GAM, Sadegh H, Yusoff MM, Chong KF (2019) Highly stable symmetric supercapacitor from cysteamine functionalized multi-walled carbon nanotubes operating in a wide potential window. Mater Today Proc 16:2273–2279
21. Nessim GD (2010) Properties, synthesis, and growth mechanisms of carbon nanotubes with special focus on thermal chemical vapor deposition. Nanoscale 2(8):1306
22. Ali GAM, Lih Teo EY, Aboelazm EAA, Sadegh H, Memar AOH, Shahryari-Ghoshekandi R, Chong KF (2017) Capacitive performance of cysteamine functionalized carbon nanotubes. Mater Chem Phys 197:100–104
23. Maazinejad B, Mohammadnia O, Ali GAM, Makhlouf ASH, Nadagouda MN, Sillanpää M, Asiri AM, Agarwal S, Gupta VK, Sadegh H (2020) Taguchi L9 (34) orthogonal array study based on methylene blue removal by single-walled carbon nanotubes-amine: adsorption optimization using the experimental design method, kinetics, equilibrium and thermodynamics. J Mol Liq 298:112001
24. Kumar M (2011) Carbon nanotube synthesis and growth mechanism. In: Yellampalli, S (eds.) Carbon nanotubes—synthesis, characterization, applications. IntechOpen, pp. 127–193
25. Saleh MH, Koller M (2019) Introductory chapter: carbon nanotubes. In: Saleh HE, El-Sheikh SMM (eds.) Perspective of carbon nanotubes. IntechOpen, pp. 3–9

26. Sadegh H, Ali GAM, Makhlouf ASH, Chong KF, Alharbi NS, Agarwal S, Gupta VK (2018) MWCNTs-Fe_3O_4 nanocomposite for Hg(II) high adsorption efficiency. J Mol Liq 258:345–353
27. Sharifi A, Montazerghaem L, Naeimi A, Abhari AR, Vafaee M, Ali GAM, Sadegh H (2019) Investigation of photocatalytic behavior of modified ZnS:Mn/MWCNTs nanocomposite for organic pollutants effective photodegradation. J Environ Manage 247:624–632
28. Sadegh H, Ali GAM, Agarwal S, Gupta VK (2019) Surface modification of MWCNTs with carboxylic-to-amine and their superb adsorption performance. Int J Environ Res 13(3):523–531
29. Saifuddin N, Raziah AZ, Junizah AR (2013) Carbon nanotubes: a review on structure and their interaction with proteins. J Chem 2013:1–18
30. Lu J, Miao J (2012) Growth mechanism of carbon nanotubes: a nano Czochralski model. Nanoscale Res Lett 7(1):356
31. Pirard SL, Douven S, Pirard J-P (2017) Large-scale industrial manufacturing of carbon nanotubes in a continuous inclined mobile-bed rotating reactor via the catalytic chemical vapor deposition process. Front Chem Sci Eng 11(2):280–289
32. Brinson BE, Gangoli VS, Kumar A, Hauge RH, Adams WW, Barron AR (2019) From newspaper substrate to nanotubes—analysis of carbonized soot grown on Kaolin sized newsprint. C. J Carbon Res 5(4):66
33. Venkataraman A, Amadi EV, Chen Y, Papadopoulos C (2019) Carbon nanotube assembly and integration for applications. Nanoscale Res Lett 14(1):220
34. Mirabootalebi SO, Akbari GH (2017) Methods for synthesis of carbon nanotubes—review. Int J Bio-Inorg Hybr Nanomater 6(2):49–57
35. Zhuo C, Hall B, Richter H, Levendis Y (2010) Synthesis of carbon nanotubes by sequential pyrolysis and combustion of polyethylene. Carbon 48(14):4024–4034
36. Chen S, Liu Z, Jiang S, Hou H (2020) Carbonization: a feasible route for reutilization of plastic wastes. Sci Total Environ 710:136250
37. Ahamed A, Veksha A, Yin K, Weerachanchai P, Giannis A, Lisak G (2020) Environmental impact assessment of converting flexible packaging plastic waste to pyrolysis oil and multi-walled carbon nanotubes. J Hazard Mater 390:121449
38. Sharma SS, Batra VS (2019) Production of hydrogen and carbon nanotubes via catalytic thermo-chemical conversion of plastic waste: review. J Chem Technol Biotechnol 95(1):11–19
39. Zahid MU, Pervaiz E, Hussain A, Shahzad MI, Niazi MBK (2018) Synthesis of carbon nanomaterials from different pyrolysis techniques: a review. Mater Res Express 5(5):052002
40. Gong J, Liu J, Wan D, Chen X, Wen X, Mijowska E, Jiang Z, Wang Y, Tang T (2012) Catalytic carbonization of polypropylene by the combined catalysis of activated carbon with Ni_2O_3 into carbon nanotubes and its mechanism. Appl Catal A 449:112–120
41. Gong J, Chen X, Tang T (2019) Recent progress in controlled carbonization of (waste) polymers. Prog Polym Sci 94:1–32
42. Mishra N, Das G, Ansaldo A, Genovese A, Malerba M, Povia M, Ricci D, Di Fabrizio E, Di Zitti E, Sharon M, Sharon M (2012) Pyrolysis of waste polypropylene for the synthesis of carbon nanotubes. J Anal Appl Pyrol 94:91–98
43. Tang T, Chen X, Meng X, Chen H, Ding Y (2005) Synthesis of multiwalled carbon nanotubes by catalytic combustion of polypropylene. Angew Chem Int Ed 44(10):1517–1520
44. Kong Q, Zhang J (2007) Synthesis of straight and helical carbon nanotubes from catalytic pyrolysis of polyethylene. Polym Degrad Stab 92(11):2005–2010
45. Zheng Y, Zhang H, Ge S, Song J, Wang J, Zhang S (2018) Synthesis of carbon nanotube arrays with high aspect ratio via ni-catalyzed pyrolysis of waste polyethylene. Nanomaterials 8(7):556
46. Arena U, Mastellone ML, Camino G, Boccaleri E (2006) An innovative process for mass production of multi-wall carbon nanotubes by means of low-cost pyrolysis of polyolefins. Polym Degrad Stab 91(4):763–768
47. Aboul-Enein AA, Awadallah AE (2019) Impact of Co/Mo ratio on the activity of CoMo/MgO catalyst for production of high-quality multi-walled carbon nanotubes from polyethylene waste. Mater Chem Phys 238:121879

48. Joseph Berkmans A, Jagannatham M, Priyanka S, Haridoss P (2014) Synthesis of branched, nano channeled, ultrafine and nano carbon tubes from PET wastes using the arc discharge method. Waste Manage 34(11):2139–2145
49. El Essawy NA, Konsowa AH, Elnouby M, Farag HA (2016) A novel one-step synthesis for carbon-based nanomaterials from polyethylene terephthalate (PET) bottles waste. J Air Waste Manag Assoc 67(3):358–370
50. Nath DCD, Sahajwalla V (2011) Application of fly ash as a catalyst for synthesis of carbon nanotube ribbons. J Hazard Mater 192(2):691–697
51. Nath DCD, Sahajwalla V (2012) Analysis of carbon nanotubes produced by pyrolysis of composite film of poly (vinyl alcohol) and modified fly ash. Mater Sci Appl 03(02):103–109
52. Zhang Y, Wu C, Nahil MA, Williams P (2015) Pyrolysis-catalytic reforming/gasification of waste tires for production of carbon nanotubes and hydrogen. Energy Fuels 29(5):3328–3334
53. Alves JO, Soares Tenório JA, Zhuo C, Levendis YA (2012) Use of Stainless Steel AISI 304 for catalytic synthesis of carbon nanomaterials from solid wastes. J Mater Res Technol 1 (3):128–133
54. Kwon S-J, Seo H-K, Ahn S, Lee T-W (2018) Value-added recycling of inexpensive carbon sources to graphene and carbon nanotubes. Adv Sustain Syst 3(1):1800016
55. Mastellone ML, Perugini F, Ponte M, Arena U (2002) Fluidized bed pyrolysis of a recycled polyethylene. Polym Degrad Stab 76(3):479–487
56. Yang W, Sun WJ, Chu W, Jiang CF, Wen J (2012) Synthesis of carbon nanotubes using scrap tyre rubber as carbon source. Chin Chem Lett 23(3):363–366
57. Acomb JC, Wu C, Williams PT (2016) The use of different metal catalysts for the simultaneous production of carbon nanotubes and hydrogen from pyrolysis of plastic feedstocks. Appl Catal B 180:497–510
58. Arnaiz N, Martin-Gullon I, Font R, Gomez-Rico MF (2018) Production of bamboo-type carbon nanotubes doped with nitrogen from polyamide pyrolysis gas. J Anal Appl Pyrol 130:52–61
59. Borsodi N, Szentes A, Miskolczi N, Wu C, Liu X (2016) Carbon nanotubes synthetized from gaseous products of waste polymer pyrolysis and their application. J Anal Appl Pyrol 120:304–313
60. Zhuo C, Hall B, Levendis Y, Richter H (2011) A novel technology for Green(er) Manufacturing of CNTs via recycling of waste plastics. In: MRS Proceedings 1317
61. Zhuo C, Alves JO, Tenorio JAS, Levendis YA (2012) Synthesis of carbon nanomaterials through up-cycling agricultural and municipal solid wastes. Ind Eng Chem Res 51(7):2922–2930
62. Acomb JC, Wu C, Williams PT (2014) Control of steam input to the pyrolysis-gasification of waste plastics for improved production of hydrogen or carbon nanotubes. Appl Catal B 147:571–584
63. Wu C, Nahil MA, Miskolczi N, Huang J, Williams PT (2016) Production and application of carbon nanotubes, as a co-product of hydrogen from the pyrolysis-catalytic reforming of waste plastic. Process Saf Environ Prot 103:107–114
64. Nahil MA, Wu C, Williams PT (2015) Influence of metal addition to Ni-based catalysts for the co-production of carbon nanotubes and hydrogen from the thermal processing of waste polypropylene. Fuel Process Technol 130:46–53
65. Zhang Y, Williams PT (2016) Carbon nanotubes and hydrogen production from the pyrolysis catalysis or catalytic-steam reforming of waste tyres. J Anal Appl Pyrol 122:490–501
66. Lopez G, Artetxe M, Amutio M, Alvarez J, Bilbao J, Olazar M (2018) Recent advances in the gasification of waste plastics. A critical overview. Renew Sustain Energy Rev 82:576–596
67. Liu X, Shen B, Yuan P, Patel D, Wu C (2017) Production of carbon nanotubes (CNTs) from thermochemical conversion of waste plastics using Ni/anodic aluminum oxide (AAO) template catalyst. Energy Proc 142:525–530
68. Veksha A, Yin K, Moo JGS, Oh W-D, Ahamed A, Chen WQ, Weerachanchai P, Giannis A, Lisak G (2020) Processing of flexible plastic packaging waste into pyrolysis oil and multi-walled carbon nanotubes for electrocatalytic oxygen reduction. J Hazard Mater 387:121256

69. Zhang Y, Wu C, Nahil MA, Williams P (2016) High-value resource recovery products from waste tyres. Proc Inst Civil Eng Waste Resource Manage 169(3):137–145
70. Gou X, Zhao D, Wu C (2020) Catalytic conversion of hard plastics to valuable carbon nanotubes. J Anal Appl Pyrol 145:104748
71. Zhang Y, Nahil MA, Wu C, Williams PT (2017) Pyrolysis–catalysis of waste plastic using a nickel–stainless-steel mesh catalyst for high-value carbon products. Environ Technol 38 (22):2889–2897
72. Panahi A, Wei Z, Song G, Levendis YA (2019) Influence of Stainless-Steel catalyst substrate type and pretreatment on growing carbon nanotubes from waste postconsumer plastics. Ind Eng Chem Res 58(8):3009–3023
73. Tripathi P, Durbach S, Coville N (2017) Synthesis of multi-walled carbon nanotubes from plastic waste using a stainless-Steel CVD reactor as catalyst. Nanomaterials 7(10):284
74. Adrados A, de Marco I, Caballero BM, López A, Laresgoiti MF, Torres A (2012) Pyrolysis of plastic packaging waste: a comparison of plastic residuals from material recovery facilities with simulated plastic waste. Waste Manage 32(5):826–832
75. Acomb JC, Wu C, Williams PT (2015) Effect of growth temperature and feedstock: catalyst ratio on the production of carbon nanotubes and hydrogen from the pyrolysis of waste plastics. J Anal Appl Pyrol 113:231–238
76. Zhang B, Song C, Liu C, Min J, Azadmanjiri J, Ni Y, Niu R, Gong J, Zhao Q, Tang T (2019) Molten salts promoting the "controlled carbonization" of waste polyesters into hierarchically porous carbon for high-performance solar steam evaporation. J Mater Chem A 7(40):22912–22923
77. Moo JGS, Veksha A, Oh W-D, Giannis A, Udayanga WDC, Lin S-X, Ge L, Lisak G (2019) Plastic derived carbon nanotubes for electrocatalytic oxygen reduction reaction: effects of plastic feedstock and synthesis temperature. Electrochem Commun 101:11–18

Conversion of Waste Cheap Petroleum Paraffinic Wax By-Products to Expensive Valuable Multiple Carbon Nanomaterials

Amr A. Nada, Fathi S. Soliman, Gomaa A. M. Ali, A. Hamdy, Hanaa Selim, Mohamed A. Elsayed, Mohamed E. Elmowafy, and Heba H. El-Maghrabi

Abstract The low atomic number of carbons, combined with the half-full shell of valence electrons and medium electronegativity, provide an important basis for strong covalent bonding to other carbon atoms and other elements. Moreover, this supports the wide differences of carbon-containing natural atoms, counting the atoms of life. Recent breakthroughs in carbon-based nanomaterials' science and technology use paraffinic waxes as a carbon source where it consists of not less than 18 carbon number per single paraffin crystal. Paraffinic waxes are considered a cheap by-product in the petroleum refinery, which is considered a source for nanocarbon synthesis, whether it is activated nanocarbon or carbon nanotubes and nanocarbon fibers. This chapter describes the separation of paraffinic petroleum wax, its purification, and characterization, besides that, the synthesis of nanocarbon and its evaluation.

Keywords Paraffin waxes · Carbon nanotubes · Activated carbon · Carbon fibers · Petroleum refinery

---

A. A. Nada (✉) · A. Hamdy · H. Selim
Department of Analysis and Evaluation, Egyptian Petroleum Research Institute, Cairo 11727, Egypt
e-mail: chem_amr@yahoo.com; amr.nada@epri.sci.eg

F. S. Soliman (✉) · H. H. El-Maghrabi (✉)
Department of Refining, Egyptian Petroleum Research Institute, Cairo 11727, Egypt
e-mail: hebachem@yahoo.com

G. A. M. Ali
Chemistry Department, Faculty of Science, Al-Azhar University, Assiut 71524, Egypt

M. A. Elsayed · M. E. Elmowafy
Chemical Engineering Department, Military Technical College, Cairo, Egypt

A. S. H. Makhlouf and G. A. M. Ali (eds.), *Waste Recycling Technologies for Nanomaterials Manufacturing*, Topics in Mining, Metallurgy and Materials Engineering, https://doi.org/10.1007/978-3-030-68031-2_25

List of Abbreviations

ACs	Activated carbons
CNTs	Carbon nanotubes
CVD	Chemical vapor deposition
EPA	Environmental protection agency
GO	Graphene oxide
MWCNTs	Multi wall carbon nanotubes
RGO	Reduced graphene oxide
SWCNTs	Single wall carbon nanotubes
TNT	Titanium nanotube
UV	Ultra-violet

1 Introduction

The general definition of crude waxes is the waxes found naturally in various crude oil fractions. They were initially considered by-products in the dewaxing of lubricating oils and gas oils, but today, they are valuable products for many industrial applications. The wax composition depends on the temperature at which it is crystallized, and the rise in molecular weight and/or boiling range makes it increasingly complex. The remains are referred to as oil or dewaxed oil after removal depending on the methane context. In contrast, natural gas is the main source of commercial H_2 and C production. Natural gas is an expensive option, and on the other hand, it is used by many commercial applications as an energy source. Therefore, a naturally occurring paraffin wax by-product was selected as a rich and cheap source of hydrocarbon. As a by-product of the dewaxing process during lubricating oil production, solid paraffin wax is produced generally [1–7].

Nanotechnology creation and growth induced a constant change in every field of modern science. To date, it has shifted the scientific community's outlook on catalysis and brought incredible changes in synthetic chemical processes [8–10]. Functioning catalytic supports are extremely useful in producing catalysts with fast recycling, a large area, and tunable mesoporous composition [10–15]. Several study papers sum up methods and techniques for metal recovery from spent Li-ion batteries, chloride pickle liquors from nickel-metal hydride batteries, and oxidic industrial by-products [16–24]. These materials have attracted considerable attention in catalysis due to the higher specialized surface area of nanomaterials compared to their bulk counterparts, yet separation and recovery from the reaction media are not a simple process. The magnetic nanocatalyst application has thus become a viable solution. Due to the supply's paramagnetic character, magnetically driven catalysts with an external magnet can be recovered easily and efficiently and then reused for further cycles. In many reactions with desired activities, magnet nanocatalysts were used. During catalyst preparation, their surface can work [25–32].

2 Petroleum Waxes

Waxes separated from oil are defined as the naturally occurring waxes in various crude oil fractions. Petroleum waxes are complex mixtures of hydrocarbons, including *n*-paraffin, branched-chain paraffins, and cycloparaffins in the range of C_{18}–C_{70}. The quality and quantity of waxes manufactured from crude oils depend on the crude source and the degree of refining to which it has been subjected before wax separation [7, 33–37].

Paraffin waxes command a good market because of their specific uses. The paraffin waxes are solid hydrocarbons at room temperature. Slack wax is a refinery term for the crude paraffin wax separated from base stocks' solvent dewaxing. Slack wax contains varying amounts of oil (ranging from 20 to 50 wt%) and must be removed to produce hard or finished waxes. If the slack wax is separated from residual oil fractions, the oil-bearing slack is frequently called petrolatum [7, 33, 38, 39].

Petroleum jelly is a general name applied to a slightly oiled crude microcrystalline wax. It is a semi-solid, jelly-like material. Petrolatum is obtained from a specific type of extra-large molecular weight petroleum distillates or residues. Ozokerite wax is a naturally occurring mineral wax, and it is also a microcrystalline wax. Ceresin is a microcrystalline wax; it is formerly given to the hard-white wax obtained from fully refined ozokerite. Petroleum ceresin is a similar microcrystalline wax but separated from petroleum. Ceresin and petroleum ceresins appear to have the same composition, structure, physical, and chemical properties [7].

3 Composition of Petroleum Waxes

Petroleum waxes are substances, which are solid at ordinary temperatures. Paraffin and microcrystalline waxes in their pure form consist of only solid saturated hydrocarbons. Petrolatum, in contrast to the other two waxes, contains both solid and liquid hydrocarbons. Petrolatum is semi-solid at normal temperatures and is quite soft as compared to the other two waxes. Paraffin wax is a solid and crystalline mixture of hydrocarbons; it is usually obtained in large crystals. It generally consists of normal paraffin ranging from C_{16}–C_{30} and may be higher. Proportions of slightly branched-chain paraffin ranging from C_{18}–C_{36} and naphthene's, especially alkyl-substituted derivatives of cyclopentane and cyclohexane; are also present [7, 36, 39–41]. The average molecular weight of these paraffin waxes is about 360–420. A paraffin wax (Fig. 1) melting at 53.5 °C showed a space lattice having C–C bond length of 1.52 Å, a C–C–C bond angle of 110 Å, a C–H bond length of 1.17 Å, and an H–C–H bond angle of 105 Å [42, 43].

Microcrystalline waxes are obtained from the vacuum residue. The source for the production of microcrystalline wax is petrolatum or bright stock.

Fig. 1 Paraffins wax

Microcrystalline waxes involve highly branched-chain paraffin, in contrast to the macrocrystalline, cycloparaffins, and small amounts of *n*-paraffins and alkylated aromatics. The actual chain length of the *n*-alkanes is approximately C34–C50. Long-chain, branched iso-alkanes predominantly contain chain lengths up to C70. The branched-chain structures of the composition C_nH_{2n+2} are found. Isoparaffin of monomethyl alkane, 2-methyl alkanes being found. As the methyl group's position moves farther from the end of the chain, the amount of the corresponding alkane becomes smaller [7, 36, 44]. The branched chains in the microcrystalline waxes are presented at random along the carbon chain meanwhile in paraffin wax, and they are located at the end of the chain.

The cycloalkanes, however, consist mainly of monocyclic systems. Monocyclopentyl, monocyclohexyl, dicyclohexyl paraffin, and polycyclic paraffin are also found. Some microcrystalline waxes are mainly composed of multiple-branched iso-paraffins and monocyclo paraffins. Moreover, nonhydrogenated microwaxes contain also mainly monocyclic and heterocyclic aromatic compounds. Microcrystalline waxes have higher molecular weights (600–800), densities, and refractive indices than paraffin waxes [7, 36, 37, 44].

Given that use of hydrocarbons, such as gasoline, kerosene, and diesel oil, waste waxes are still available, and that these fuels are common. Environmental security precisely and rigorously applied would decreased the greenhouse emissions by the limitation of used oil sources. Therefore, the new green energy solutions would be required to turn to convenient, safe, effective, and flexible energy carriers. Hydrogen isolated from paraffin suits these characteristics easily, where it can be produced and converted in reasonably high quality into electric energy with easy to store in large quantities [45–50]. By various techniques, carbon could be generated from catalytic decomposition, thermal plasma pyrolysis, thermolysis, the J $\times$ B arc-jet discharge method [47], or direct plasmolysis of methane and even from ethylene and acetylene [51], agricultural wastes, and also from polymers [48, 49, 52, 53].

Moreover, nanocarbon can be prepared from fuel storage in gas masks and insolvent recovery devices. Specific products, including cellulose raw materials such as timber, hemp, fruit core, and other agricultural waste, as a raw material of fuel such as coal, coke, coal bitumen, and polymer raw materials, can be used as raw materials, including waste from different types of bulky and plastic [54]. One of the simplest carbons in compounds could be paraffinic hydrocarbons separate and purified from waste petroleum waxes. Paraffin conversion technology is based on the C–H connection's catalytic cleavage to separate hydrogen and obtain carbon materials locally, usually CNTs [55–57]. The carbon generation's good interest source has been presented in recent papers [58, 59] called paraffin's. This is a wealthy type of carbon that can be separated and purified from waxes that generally appear as a by-product of the dewaxing stage during petroleum products production.

The principal kind of nanocarbon materials for catalytical applications, studied at present and available on a larger scale with chemical vapor deposition (CVD), remains multi-wall carbon nanotubes, or MWCNTs. The residual catalyst impurities in the MWCNTs obtained are limited, and the purity can be greater than 99.5%, which is enough for most applications. However, the results of catalytic applications can be altered by impurity sometimes. The industry in comparison with MWCNT is underdeveloped in terms of the production of single-wall CNTs (SWCNTs) instead (Fig. 2) [60, 61].

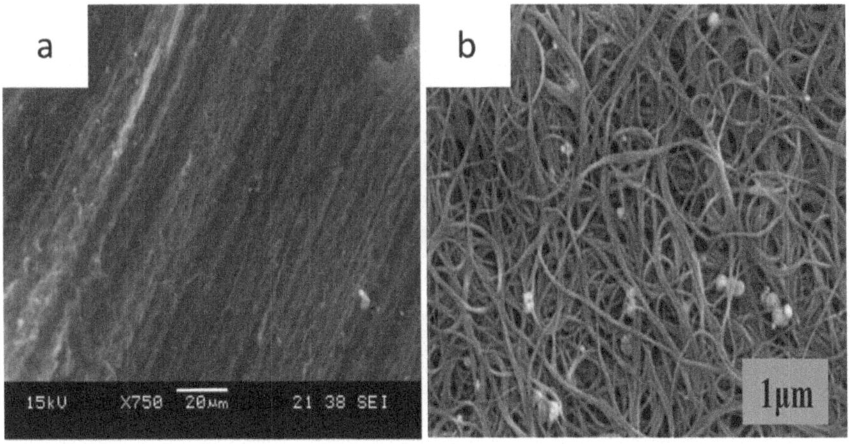

Fig. 2 SEM images of **a** MWCNTs obtained. **b** SWCNTs. Adapted from Refs. [60, 61], Copyright 2019, Elsevier

4 Types of Nanocarbons

4.1 Mesoporous Carbon

Mesoporous carbon (Fig. 3) can be prepared using hard and soft synthetic tempering methods [62]. The hard template method benefits from simplicity and consistency in the development of the OMC. However, it is complicated, time-consuming, and unfit for mass production with multiple steps required to produce structure matrices and subsequences of extraction in the silica template matrices' harsh chemical conditions. Soft tempering uses PEO-PPO-PEO resin with phenol and copolymer block to produce organic-organic copolymers and phenol resins with high OMCs [63–67].

4.2 Carbon Hierarchy

Carbon hierarchy is a combination of at least two nanocarbon (graph, carbon nanotubes or nanofiber, nanodiamonds, or polymers), which often has higher characteristics than those of the individual components [68]. It is essential to resolve the inherent drawbacks of mass transport in catalysts, balancing high surface area with acceptable usability and integrating various essential characteristics, like carbon nanotubes and nanofibers [69]. Electrocatalysis and energy systems (batteries or supercapacitors) need special requirements [70–78].

4.3 Activated Carbons

Activated carbons (ACs) are the oldest known adsorbents [79]. Activated carbons have a very porous structure of 500–2000 $m^2\ g^{-1}$. They have strong adsorption potential for various substances. They have been used to extract a wide range of contaminants, including organic and inorganic compounds, from the liquid or gas

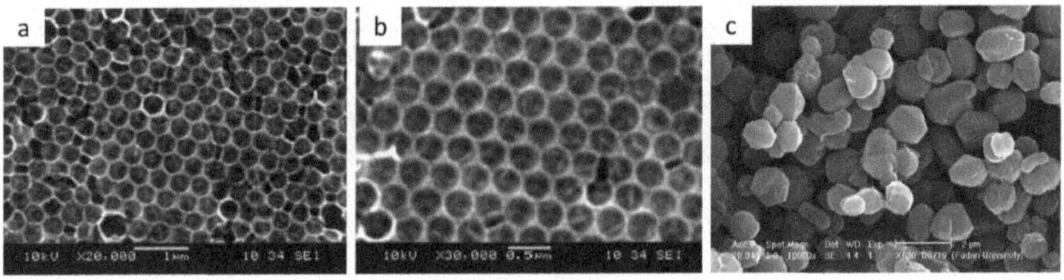

Fig. 3 SEM image of mesoporous carbon. Adapted from Ref. [66], Copyright 2012, Scientific. Net (**a**, **b**). Adapted from Ref. [67], Copyright 2007, American Chemical Society (**c**)

processes [80–92]. Activated carbon is available in two primary forms: powder-activated carbon and granular activated carbon in particular. The first type consists of fine particles of less than 0.2 mm of diameter with a large exterior surface area and a low resistance to diffusion.

Consequently, the adsorption rate is very high; however, the other consists of larger carbon particles with a diameter of approximately 5 mm and smaller visible surfaces than PAC. They are usually preferred to adsorb gasses and vapors and are used in fixed-bed filtration systems since they are better adapted to continuous touch [81, 93–98].

4.4 Graphene 2D Material

Graphene has been designed as a one-carbon layer structure of graphite that has been proposed to be termed graphene, related to the suffix -ene for polycyclic aromatic hydrocarbons, like naphthalene, anthracene, etc., and the prefix graph- from graphite [99–102]. Graphene has a two-dimensional carbon atom crystal network, and sp^2 orbital hybridization with 3σ and 1π bond of carbon atoms are violently packed in the atomic scale. In addition, the main structural start of other carbon allotropes (Fig. 4). Carbon atoms are bounded within the system by hard bonds into a honeycomb array formed of six-membered rings (Fig. 4), stacking of this layer on top of each other three-dimensional graphite crystal is formed. The atomic thickness in graphene is about 0.345 nm.

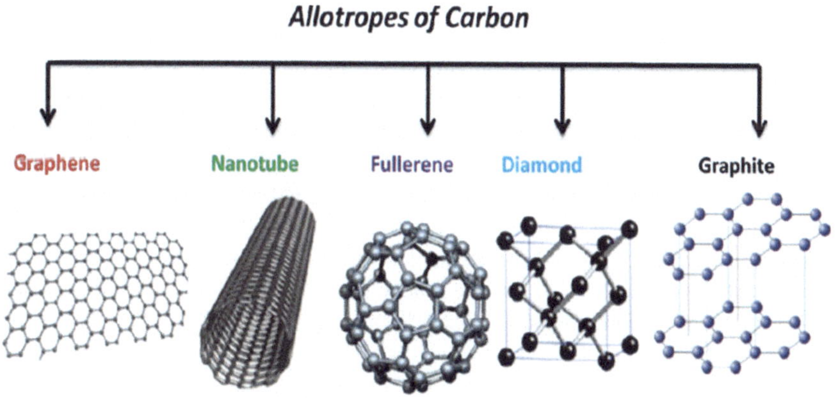

Fig. 4 Allotropes of carbon and their crystal structures. Adapted from Ref. [103], Copyright 2019, Springer

5 Application of Nanocarbon

The challenge of water remediation becomes larger as the paucity of water and population increases. Researchers create and elaborate the cheaper and faster remediation plants to combat this challenge, depending on a new technology with nanocarbon (Graphene).

5.1 Contaminant Adsorption on Graphene-Based Materials

Population growth, industrial activities, and agricultural intensification have led to a significant large pollution in the environment. All pollutants are very various, give an excellent concern for the environment and public health. As a result, there is a worldwide push to create significant innovations to capture pollutants from wastewater highly. Through all technologies, adsorption is a quick, low-cost, and easy way to remove pollutants from wastewater.

Adsorption is a method in which the contaminant (adsorbate) is removed via solid materials (adsorbent) through physical–chemical reactions [104, 105]. Here, we discuss graphene-based absorbents' application to uptake inorganic and organic pollutants from wastewater and identify the main absorption mechanisms.

Metal ions are major contaminants on undesirable wastewater and drinking water sources from human actions, like mining and industrial waste, pipes corrosion, and welded joints. Thus, interest increasing in lowering the concentration of polluted metals in wastewater. The permissible concentration of Cu and Pb in drinking water is 1.3 ppm and 15 ppb, respectively, in compliance with the US Environmental Protection Agency (EPA) [106].

Activated carbon (AC) was historically renowned for its ample adsorbent potential for a wide variety of pollutants [107]. In contrast, the widespread use of AC was limited by its high cost and hard regeneration. Graphene adsorbents have been sophisticated as an alternative to traditional adsorbents [108, 109]. Because of their scalable production, large surface areas, surface chemical tunable, and the presence of functional groups containing oxygen and non-corrosive properties, nanocomposites graphene were selected as novel adsorbents [110, 111].

Many researchers used GO as an excellent material for the uptake of metal ions in wastewater [109, 112–114]. Pristine graphene was made from a vacuum-fostered low-temperature exfoliation then heated to 500 and 700 °C (GNS500 and GNS700). GNS500 and GNS700 showed a high adsorption capacity of Pb(II) as opposed to pristine graphene and checked the role of carboxyl groups in Pb(II) adsorption mechanism [115]. El-Maghrabi et al. recorded 455 mg g^{-1} of adsorption potential to prepare UIV graphic ferberite using nuclear wastewater [116]. Many factors affect the adsorption capacity of GO, such as pH, ionic power, the number of GO groups with oxygen, and the presence of organic natural material [109, 113, 114, 117]. Firstly, the effect of ionic strength can be due to the race of metal ions with electrolytes (NaCl, KCl, and $NaClO_4$) on the GO surface [117]. Indeed, electrolytes have affected the hydrated particles' electrical double layer of metal ions, so variable the role of metal ions bond to the GO layers [117, 118]. Secondly, the influence of pH on metal ions' adsorption at low pH, the adsorption capacity decrescent [109, 113, 114, 117]. The most important way that influences metal ion adsorption behavior from aqueous wastewater solution via GO and GO material is their pH_{pzc} (pzc: point of zero charges), as confirmed in Fig. 5. With higher solution pH than pH_{pzc}, the surface of GO sheets is negatively charged due to the deprotonation of hydroxyl and carboxyl groups. In this case, the electrostatic interaction has been carried out between metal ions (+ve) and GO surface (−ve), leading to enhancement of adsorption capacity [109, 113, 117]. For pH less than pH_{pzc}, *charge repulsion* between metal ions and GO (positively charged). According to pH, ions may be produce hydroxide groups: $Me(OH)_3$, $Me(OH)_2$, and $Me(OH)^+$ [109, 113]. While the main adsorption mechanism is electrostatic interaction with oxide species and ions of metal, different interactions can occur. Huang et al. reported that the delocalized p-electrons in the graphene sp^2 system have been like Lewis bases donating electrons to metal ions [115].

5.1.1 Magnetic Graphene-Based Materials

Magnetic graphene-based materials are among the most interesting materials for the uptake of ions from wastewater with high sorption capacity [119–125]. The delocalized magnetic nanoparticles on GO block their accumulation, increasing their surface area [123]. In addition, the graphene composites with magnetic nanoparticles may be efficiently and rapidly separated from wastewater solutions by an external magnet and increasing binding sites for metal ions uptake via their high surface area [124]. Magnetic graphene material adsorption capacity, like GO, is affected by different in adsorbent amount, pH [121, 122, 124], contact time, [121,

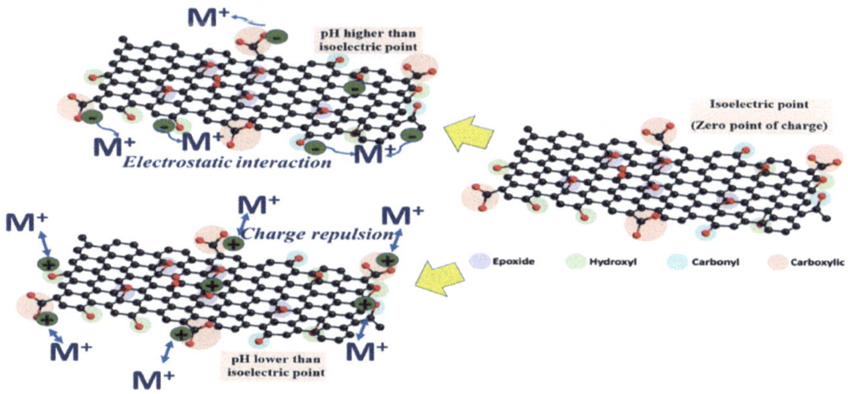

Fig. 5 Influence of pH on the adsorption of metal ions

122, 124] temperature, [123, 125] and presence of varying organic matter like fulvic acid [122].

5.1.2 Organic Molecules Graphene-Based Materials

Organic molecules graphene-based materials as novel adsorbents to capture metal ions from wastewater are ascribed to a synergistic impact by chelating metal ions with organic function group's graphene sheet surface. Grafting compound such as EDTA [126], chitosan [127], and polyacrylamide (PAM) [128]. The adsorption capacity of rGO for Pb(II) has been enhanced with PAM grafting from 500 to 1000 mg g^{-1}.

5.1.3 Thermo-Responsive Graphene-Based Materials

Thermo-responsive graphene-based materials as new adsorbents by graphene associated with non-covalent poly(*N*-isopropylacrylamide) [129] have been studied. At a low critical poly(*N*-isopropylacrylamide) temperature, nanocomposite material occurs at a low critical poly(*N*-isopropylacrylamide) (32 °C) temperature resulting in a nanocomposite material subject to rapid precipitation and aggregation, while at temperatures greater than 36 °C. It can then be resuspended when the temperature drops below 34 ℃ [129].

5.1.4 Anionic Toxics Capture

Anionic toxics capture from wastewater as phosphate (PO_4^-), perchlorate (ClO_4^-), and fluoride (F^-) [130–132]. Adsorption mechanism of the anion (e.g., F^-, Cl^-, and Br^-) was related to anion–*p* associations [132]. The interaction of anion–π is established on the attraction of the negative anion charge with the aromatic ring's electron-deficient ring on the graphene sheet [132]. Figure 6 displays the different ways of graphene-based adsorbents to capture metal ions pollutants from wastewater.

5.2 Photocatalysis Graphene-Based Materials

Heterogeneous photocatalysis is an efficient way for wastewater treatment because of its low cost and efficiency via the production of oxidative free radicals and species (like –OH, O_2^{-},˙ and H_2O_2) upon the presence of light energy [134, 135, 167]. They were described as having an electronic structure, including a filled valence band and a vacant conduction band [136]. When the semiconductor catalyst, introduced to light energy ($\lambda\upsilon$) that larger than the bandgap energy (Eg), The

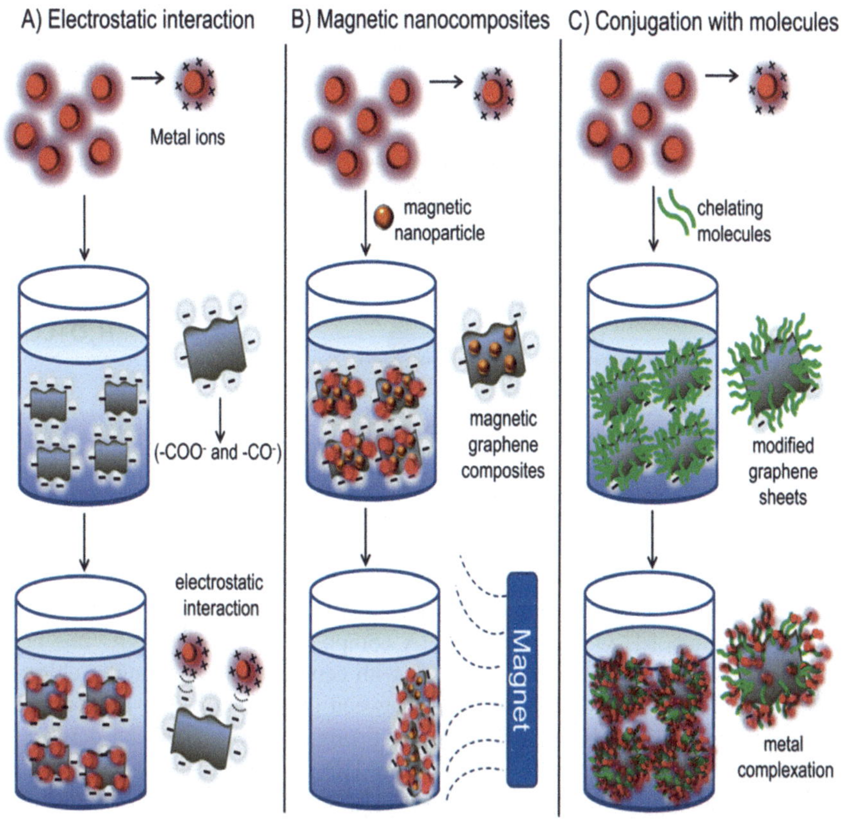

Fig. 6 Main ways of graphene-based adsorbents for the capture of metal ions pollutants from wastewater. Adapted from Ref. [133], Copyright 2015, RSC

pair electron of the valence band was excited at the leading ring, leaving a hole [135–137]. The electron/hole (e^-/h) pair in the semiconductor's surface helps form oxidative free radicals continues reactions made the organic molecules degradation and solar cells for energy production (Fig. 7).

Numerous researchers are interested in improving photocatalysts by enhancing efficiency [137] of water decomposition on TiO_2 under visible solar light, although TiO_2 is restricted via its absorption in the UV range. A novel nanosized photocatalysts via conjugation of TiO_2 nanoparticles, nanotubes, and other different morphologies with carbonaceous materials (CNTs, AC, and graphene) with different photocatalytic properties were developed to enhance the photocatalytic activity of TiO_2 [138–140, 168]. Graphene-based photocatalysts have high electron mobility, block the rapid recombination of electron–hole pairs, and improve photocatalytic activity [141–143]. Most processes used to fabricate photocatalysis of graphene nanocomposites are as follows: (1) Direct formation of nanomaterial on the GO sheet surface. A composite of TiO_2–GO has been fabricated by hydrolysis

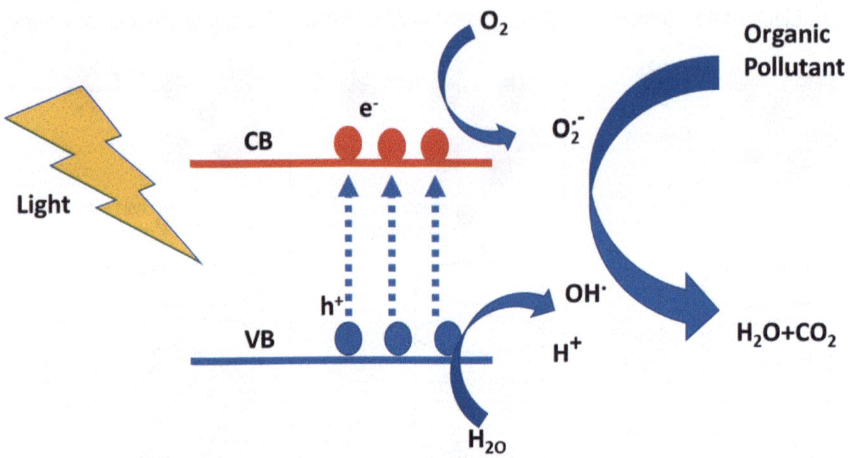

Fig. 7 Schematic of semiconductor excitation by bandgap illumination

of TiF_4 in an aqueous solution of GO at 60 °C for 24 h [144]. Liang et al. also reported cover GO layers with TiO_2 particles by hydrolyzing $Ti(BuO)_4$ at 80 °C in H_2SO_4 and ethanol-H_2O mixture [145]. A composite of Ag–AgX–GO has been obtained via reacting silver nitrate and GO with cetyltrimethylammonium chloride or bromide [146]. (2) The direct reaction between graphene or GO solutions with photoactive nanomaterials by sonication or stirring. A TiO2–graphene composite has been designed via sonication of GO with TiO_2 nanoparticles in ethanol, then using UV-irradiation to minimize GO layers [147]. Wang et al. also demonstrated that the $BiVO_4$–rGO nanocomposite preparation via electrostatic interactions of $BiVO_4$ positive charge and the GO sheet negative charge, then GO reduced by the hydrothermal method [140]. (30 hydrothermal process utilized to form nanocomposites on elevated temperature and pressure depends on different factors like the source of metal, pH, solvent, temperature, and time. A TiO2–graphene composite has been prepared via hydrothermal one-step to reduced GO sheet and P25 deposition on the graphene surface.

5.3 Photodegradation Mechanism on Graphene-Based Material

The graphene role improves the adsorptivity of pollutants and it modifies the light absorption of catalysts to be in visible range. Moreover, it enhances the charge transport/separation performance of P25-graphene composites. The rGO-TNT composites with different ratios of rGO were prepared for the malachite green dye. The ratio of 10% rGO on rGO-TNT has been given the largest photodegradation performance of malachite green [148]. The mechanism of degradation of organic coloring follows three main steps, as shown in Fig. 8. (1) Dye adsorption by p–p stacking interactions on the graphene surface. (2) Nanocomposite photoexcitation with UV or visible solar irradiation. Hence, the valence band electrons of the semiconductor nanoparticles are excited to the conduction band. These electrons were moved on the sp^2-hybridized system of the graphene layers. (3) The photoinduced electrons were traveled to O_2 to produce reactive oxygen species (ROS) used for organic dye degradation.

Graphene layer has been conjugated with different photocatalysts like bismuth vanadate ($BiVO_4$), silver orthophosphate (Ag_3PO_4), cadmium sulfide (CdS), and Ag/AgX (X=Br, Cl) nanomaterial to improve photodegradation of organic dyes [138, 140, 149, 150]. Xiong et al. reported the degradation of rhodamine B by using graphene-gold nanocomposite under visible light [151].

Composites of graphene photocatalysts have given high activity for the photodegradation of hydrocarbons. Such as, a photocatalyst of graphene-CdS has been synthesized via self-assembling GO layer (−ve) with CdS (+ve) for selective nitroaromatic compounds reduction [150, 152]. Many studies have found that pesticide, methanol, and endocrine (phenol, atrazine, and bisphenol) degradation by graphene photocatalysts has improved [145, 153, 154]. El-Maghrabi et al. reported photodegradation of phenol on TiO_2-nanotubes/graphene nanocomposites [155]. Zhang et al. reported an alternative mechanism of graphene role in the selective oxidation of alcohols and alkenes over graphene-ZnS nanocomposite [143]. Under visible light irradiation, ZnS cannot be photoexcited. Simultaneously, graphene layers acted as a photosensitizer by photoinduced electrons from the graphene itself and shuttled into the conductance band of ZnS to improve the photocatalytic performance of ZnS under visible light.

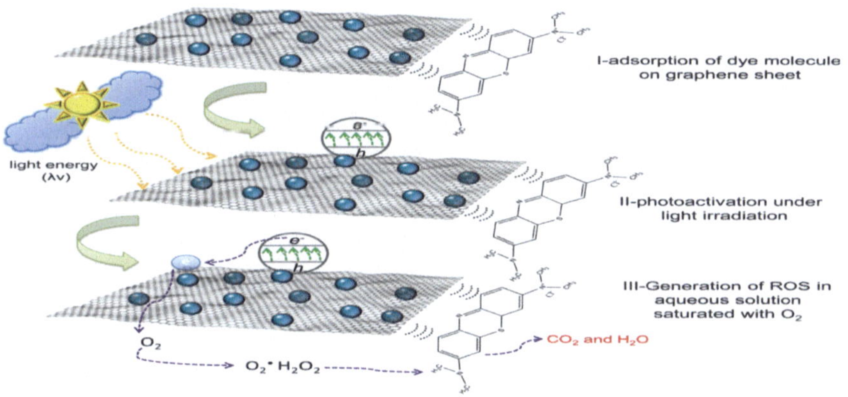

Fig. 8 Main three steps of methylene blue degradation on graphene nanocomposite. Adapted from Ref. [133], Copyright 2015, Royal Society of Chemistry

6 Conversion of Waste Paraffin to Carbon

The hydrocarbons decomposed via catalytic materials or without catalytic (as thermal and plasma). For decades, the methane thermal decomposition process has been used for the production of carbon black with hydrogen depending on catalyst types and temperature [156].

Catalysts tried to reduce thermal decomposition with hydrocarbons to the highest temperature. Common catalysts are mainly noble and mid-metal, including Ni, Fe, Pd, Co, etc., provided in pottery such as $A1_2O_3$ and SiO_2. Nickel continues to give all the catalysts tested the highest activity and is mostly used. The need for the decomposition of H_2 and carbon hydrocarbons as seen in many articles. Fuel efficiency, relative simplicity and compactness, renewable energy by-products, and reductions of CO_2 and CO are all the benefits of this method. The biggest challenge to the transition of catalytic technology from the laboratory to commercial scale is carbon accumulation on the catalyst sheet [157, 158].

Most of the studies were aimed at decomposing methane for stationary applications for CO_2-free H_2 growth. A powerful reactor must be configured to rely on the efficiency of a highly active catalyst. There are other proposals in the literature, and there are still several patents. Any of these are not thoroughly tested and are not suitable for onboard H_2 generation. Many kinds of reactors are tested, including tubular, fixed-bed, fluid surface, spout, and fluidized bed reactors, and the latter are considered the safest for a large operation. The bed of small catalyst particles in fluidized bed reactors functions as a well-mixed fluid body contributing to high heat and particles' mass conversion to water. However, further studies are needed for onboard applications for the use of fluidized bed reactors. Various problems must be tackled, including heat supply, catalyst removal, and regeneration [159, 160].

Several papers mentioned different H_2 and hydrocarbon methods of production [159, 161–164]. During thermal processes, partial combustion of the hydrocarbons and a water quenching were pyrolyzed in high temperatures to avoid a reverse reaction. Performance and efficiency were minimal. The inability to control carbon build-up continuously was another obstacle in the decomposition of hydrocarbons. The biggest downside of the process of thermal declination is the lack of energy due to carbon sequestration. Cracking will also be the favored choice for high H_2/C natural gas and other hydrocarbons. Moreover, the researcher recently concentrated on paraffin waxes as hydrocarbon which will act in the same trend as a source for carbon whether the carbon is activated type or in the form of carbon nanotube [165, 166].

7 Conclusions

Petroleum waxes are considered a valuable by-product separated from petroleum crude. Petroleum waxes easily refined to produce various fractions of pure paraffinic hydrocarbons. Paraffin waxes or paraffin hydrocarbon can be used as a cheap alternative source for methane or natural gas to synthesize multiple types of carbon, such as carbon nanotubes, graphene, and activated carbon. All of that, besides, of course, it acts as a reservoir for hydrogen production.

8 Future Perspectives

The production of carbon, hydrogen, methane and syngas via a low cost sources such as waxes, plastics and polymeric waste will be the upcoming scientific field.

References

1. Awadallah AE, Aboul-Enein AA, Aboul-Gheit AK (2014) Impact of group VI metals addition to Co/MgO catalyst for non-oxidative decomposition of methane into CO_x-free hydrogen and carbon nanotubes. Fuel 129:27–36
2. Awadallah AE, Aboul-Enein AA, Aboul-Gheit AK (2014) Effect of progressive Co loading on commercial Co–Mo/Al_2O_3 catalyst for natural gas decomposition to CO_x-free hydrogen production and carbon nanotubes. Energy Convers Manage 77:143–151
3. Li J, Dong L, Xiong L, Yang Y, Du Y, Zhao L, Wang H, Peng S (2016) High-loaded $NiCuSiO_2$ catalysts for methane decomposition to prepare hydrogen and carbon filaments. Int J Hydrogen Energy 41(28):12038–12048
4. Li Y, Li D, Wang G (2011) Methane decomposition to CO_x-free hydrogen and nano-carbon material on group 8–10 base metal catalysts: a review. Catal Today 162(1):1–48
5. Aboul-Enein AA, Soliman FS, Betiha MA (2019) Co-production of hydrogen and carbon nanomaterials using NiCu/SBA15 catalysts by pyrolysis of a wax by-product: effect of Ni–Cu loading on the catalytic activity. Int J Hydrogen Energy 44(59):31104–31120
6. Thomas R (2008) Process chemistry of lubricant base stocks: conventional base stock production. CRC Press, London
7. Mazee W (1973) Modern petroleum technology. Applied Science Publishers Ltd., on behalf of The Institute of Petroleum, Great Britain, p 782
8. Sharma R, Monga Y, Puri A (2014) Magnetically separable silica@Fe_3O_4 core–shell supported nano-structured copper (II) composites as a versatile catalyst for the reduction of nitroarenes in aqueous medium at room temperature. J Mol Catal A: Chem 393:84–95
9. Sharifi A, Montazerghaem L, Naeimi A, Abhari AR, Vafaee M, Ali GAM, Sadegh H (2019) Investigation of photocatalytic behavior of modified ZnS:Mn/MWCNTs nanocomposite for organic pollutants effective photodegradation. J Environ Manage 247:624–632
10. Ethiraj AS, Uttam P, Varunkumar K, Chong KF, Ali GAM (2019) Photocatalytic performance of a novel semiconductor nanocatalyst: copper doped nickel oxide for phenol degradation. Mater Chem Phys 242:122520
11. Mohapatra S, Rout SR, Das RK, Nayak S, Ghosh SK (2016) Highly hydrophilic luminescent magnetic mesoporous carbon nanospheres for controlled release of anticancer drug and multimodal imaging. Langmuir 32(6):1611–1620
12. Yue Q, Zhang Y, Wang C, Wang X, Sun Z, Hou X-F, Zhao D, Deng Y (2015) Magnetic yolk–shell mesoporous silica microspheres with supported Au nanoparticles as recyclable high-performance nanocatalysts. J Mater Chem A 3(8):4586–4594

13. Cui X, Zheng Y, Tian M, Dong Z (2017) Palladium nanoparticles supported on SiO_2@Fe_3O_4@m-MnO_2 mesoporous microspheres as a highly efficient and recyclable catalyst for hydrodechlorination of 2, 4-dichlorophenol and reduction of nitroaromatic compounds and organic dyes. Mol Catal 433:202–211
14. Giahi M, Pathania D, Agarwal S, Ali GAM, Chong KF, Gupta VK (2019) Preparation of Mg-doped TiO_2 nanoparticles for photocatalytic degradation of some organic pollutants. Stud Univ Babes-Bolyai Chem 64(1):7–18
15. Ali GAM, Wahba OAG, Hassan AM, Fouad OA, Chong KF (2015) Calcium-based nanosized mixed metal oxides for supercapacitor application. Ceram Int 41(6):8230–8234
16. Zhang X, Xie Y, Lin X, Li H, Cao H (2013) An overview on the processes and technologies for recycling cathodic active materials from spent lithium-ion batteries. J Mater Cycles Waste Manage 15(4):420–430
17. Zeng X, Li J, Singh N (2014) Recycling of spent lithium-ion battery: a critical review. Crit Rev Env Sci Technol 44(10):1129–1165
18. Ordoñez J, Gago EJ, Girard A (2016) Processes and technologies for the recycling and recovery of spent lithium-ion batteries. Renew Sustain Energy Rev 60:195–205
19. Wang W-Y, Yen CH, Lin J-L, Xu R-B (2019) Recovery of high-purity metallic cobalt from lithium nickel manganese cobalt oxide (NMC)-type Li-ion battery. J Mater Cycles Waste Manage 21(2):300–307
20. Shah K, Gupta K, Sengupta B (2017) Selective separation of copper and zinc from spent chloride brass pickle liquors using solvent extraction and metal recovery by precipitation-stripping. J Env Chem Eng 5(5):5260–5269
21. Cunha ML, Gahan CS, Menad N, Sandström Å (2008) Possibilities to use oxidic by-products for precipitation of Fe/As from leaching solutions for subsequent base metal recovery. Miner Eng 21(1):38–47
22. Kanamori T, Matsuda M, Miyake M (2009) Recovery of rare metal compounds from nickel–metal hydride battery waste and their application to CH_4 dry reforming catalyst. J Hazard Mater 169(1–3):240–245
23. Aboelazm EAA, Ali GAM, Chong KF (2018) Cobalt oxide supercapacitor electrode recovered from spent lithium-ion battery. Chem Adv Mater 3:67–74
24. Aboelazm EAA, Ali GAM, Algarni H, Yin H, Zhong YL, Chong KF (2018) Magnetic electrodeposition of the hierarchical cobalt oxide nanostructure from spent lithium-ion batteries: its application as a supercapacitor electrode. J Phys Chem C 122(23):12200–12206
25. Astruc D (2008) Transition-metal nanoparticles in catalysis: from historical background to the state-of-the art. Nanopart Catal 16:1–48
26. Somorjai GA, Frei H, Park JY (2009) Advancing the frontiers in nanocatalysis, biointerfaces, and renewable energy conversion by innovations of surface techniques. J Am Chem Soc 131(46):16589–16605
27. Lu AH, EeL S, Schüth F (2007) Magnetic nanoparticles: synthesis, protection, functionalization, and application. Angew Chem Int Ed 46(8):1222–1244
28. Polshettiwar V, Varma RS (2010) Green chemistry by nano-catalysis. Green Chem 12 (5):743–754
29. Shylesh S, Schuenemann V, Thiel WR (2010) Magnetically separable nanocatalysts: bridges between homogeneous and heterogeneous catalysis. Angew Chem Int Ed 49(20):3428–3459
30. Claesson EM, Mehendale NC, Gebbink RJK, van Koten G, Philipse AP (2007) Magnetic silica colloids for catalysis. J Magn Magn Mater 311(1):41–45
31. Yan J-M, Zhang X-B, Akita T, Haruta M, Xu Q (2010) One-step seeding growth of magnetically recyclable Au@ Co core–shell nanoparticles: highly efficient catalyst for hydrolytic dehydrogenation of ammonia borane. J Am Chem Soc 132(15):5326–5327
32. Torki M, Tangestaninejad S, Mirkhani V, Moghadam M, Mohammadpoor-Baltork I (2014) RuIII (OTf) Salophen CH_2–$NHSiO_2$–Fe: an efficient and magnetically recoverable catalyst for trimethylsilylation of alcohols and phenols with hexamethyldisilazane. Appl Organomet Chem 28(4):304–309
33. Prasad R (2000) Petroleum refining technology, Khanna

34. Gupta A, Severin D (1997) Characterization of petroleum waxes by high temperature gas chromatography-correlation with physical properties. Pet Sci Technol 15(9–10):943–957
35. Bennett H (1975) Industrial waxes. Chemical Pub. Co., New York
36. Letcher C (1984) Waxes. In: Encyclopedia of chemical technology, pp 466–481
37. Avilino S Jr (1994) Lubricant base oil and wax processing. Morcel Dekker Inc., New York, pp 17–36
38. Guthrie VB, Guthrie VB (1960) Petroleum products handbook, vol 33. McGraw-Hill, New York.
39. Concawe PW (1999) Related products. Report (99/110). Concawe, Brussels, p 1
40. Gottshall R, McCue C (1973) Criteria for quality of petroleum products. Applied Science Publishers Ltd., London (On behalf of the ….)
41. Freund M (1982) Paraffin products: properties, technologies, applications
42. Nakagawa H, Tsuge S, Itho T, Kimoto M (1983) Characterization of hydrocarbon waxes by gas–liquid chromatography with a high-resolution glass capillary column. J Chromatogr A 260:391–409
43. Vainshtein B, Pinsker Z (1950) Opredelenie polozheniya vodoroda v kristallicheskoi reshetke parafina. Dokl Akad Nauk SSSR 72(1):53–56
44. Meyer G (2009) Thermal properties of micro-crystalline waxes in dependence on the degree of deoiling. SOFW J 135(8):43–50
45. Yan X, Tian L, He M, Chen X (2015) Three-dimensional crystalline/amorphous Co/Co_3O_4 core/shell nanosheets as efficient electrocatalysts for the hydrogen evolution reaction. Nano Lett 15(9):6015–6021
46. Wang Y, Guo J, Wang T, Shao J, Wang D, Yang Y-W (2015) Mesoporous transition metal oxides for supercapacitors. Nanomaterials 5(4):1667–1689
47. Wang X, Hu A, Meng C, Wu C, Yang S, Hong X (2020) Recent advance in Co_3O_4 and Co_3O_4-containing electrode materials for high-performance supercapacitors. Molecules 25 (2):269
48. Yang Q, Lu Z, Sun X, Liu J (2013) Ultrathin Co_3O_4 nanosheet arrays with high supercapacitive performance. Scientific Reports 3:3537
49. Pal M, Rakshit R, Singh AK, Mandal K (2016) Ultra high supercapacitance of ultra small Co_3O_4 nanocubes. Energy 103:481–486
50. Cheng M, Duan S, Fan H, Su X, Cui Y, Wang R (2017) Core@shell CoO@Co_3O_4 nanocrystals assembling mesoporous microspheres for high performance asymmetric supercapacitors. Chem Eng J 327:100–108
51. Liu Z, Zhou W, Wang S, Du W, Zhang H, Ding C, Du Y, Zhu L (2019) Facile synthesis of homogeneous core-shell Co_3O_4 mesoporous nanospheres as high performance electrode materials for supercapacitor. J Alloy Compd 774:137–144
52. Zhang F, Yuan C, Lu X, Zhang L, Che Q, Zhang X (2012) Facile growth of mesoporous Co_3O_4 nanowire arrays on Ni foam for high performance electrochemical capacitors. J Power Sources 203:250–256
53. Li J, Zhang P, Zhao X, Chen L, Shen J, Li M, Ji B, Song L, Wu Y, Liu D (2019) Structure-controlled Co–Al layered double hydroxides/reduced graphene oxide nanomaterials based on solid-phase exfoliation technique for supercapacitors. J Colloid Interface Sci 549:236–245
54. Wang P, Zhou H, Meng C, Wang Z, Akhtar K, Yuan A (2019) Cyanometallic framework-derived hierarchical Co_3O_4–NiO/graphene foam as high-performance binder-free electrodes for supercapacitors. Chem Eng J 369:57–63
55. Babu IM, William JJ, Muralidharan G (2019) Ordered mesoporous Co_3O_4/CMC nanoflakes for superior cyclic life and ultra high energy density supercapacitor. Appl Surf Sci 480:371–383
56. Fan H, Quan L, Yuan M, Zhu S, Wang K, Zhong Y, Chang L, Shao H, Wang J, Zhang J, Cao C-n (2016) Thin Co_3O_4 nanosheet array on 3D porous graphene/nickel foam as a binder-free electrode for high-performance supercapacitors. Electrochim Acta 188:222–229

57. Chen H, Wang J, Liao F, Han X, Zhang Y, Xu C, Gao L (2019) Uniform and porous Mn-doped Co_3O_4 microspheres: Solvothermal synthesis and their superior supercapacitor performances. Ceram Int 45(9):11876–11882
58. Wang Y, Huo J (2020) C, N co-doped Co_3O_4 hollow spheres prepared via hydrothermal reaction for improved electrochemical activity. Int J Hydrogen Energy 45(1):1170–1176
59. Deng S, Xiao X, Chen G, Wang L, Wang Y (2016) Cd doped porous Co_3O_4 nanosheets as electrode material for high performance supercapacitor application. Electrochim Acta 196:316–327
60. Dlamini N, Mukaya HE, Van Zyl RL, Chen CT, Zeevaart RJ, Mbianda XY (2019) Synthesis, characterization, kinetic drug release and anticancer activity of bisphosphonates multi-walled carbon nanotube conjugates. Mater Sci Eng C 104:109967
61. Zhou Y, Liu Y, Zhou X, Gao Y, Gao C, Wang L (2019) High performance p-type organic thermoelectric materials based on metalloporphyrin/single-walled carbon nanotube composite films. J Power Sources 423:152–158
62. Li G, Chen M, Ouyang Y, Yao D, Lu L, Wang L, Xia X, Lei W, Chen S-M, Mandler D (2019) Manganese doped Co_3O_4 mesoporous nanoneedle array for long cycle-stable supercapacitors. Appl Surf Sci 469:941–950
63. Tao F, Zhao Y-Q, Zhang G-Q, Li H-L (2007) Electrochemical characterization on cobalt sulfide for electrochemical supercapacitors. Electrochem Commun 9(6):1282–1287
64. Yu Z, Tetard L, Zhai L, Thomas J (2015) Supercapacitor electrode materials: nanostructures from 0 to 3 dimensions. Energy Environ Sci 8(3):702–730
65. Qin Q, Chen L, Wei T, Liu X (2019) MoS_2/NiS yolk–shell microsphere-based electrodes for overall water splitting and asymmetric supercapacitor. Small 15(29):1803639
66. Wang CX, Sun CY, Yu CJ, Li YX, Zhao J, Huang DX, Song YB (2012) Fabrication of highly ordered carbon networks as catalyst supports for aerobic oxidation of glucose. Adv Mater Res 476–478:1186–1192
67. Zhang F, Gu D, Yu T, Zhang F, Xie S, Zhang L, Deng Y, Wan Y, Tu B, Zhao D (2007) Mesoporous carbon single-crystals from organic−organic self-assembly. J Am Chem Soc 129(25):7746–7747
68. Shinde SK, Ramesh S, Bathula C, Ghodake GS, Kim DY, Jagadale AD, Kadam AA, Waghmode DP, Sreekanth TVM, Kim HS, Nagajyothi PC, Yadav HM (2019) Novel approach to synthesize $NiCo_2S_4$ composite for high-performance supercapacitor application with different molar ratio of Ni and Co. Sci Rep 9(1):13717
69. Zhu T, Wang Z, Ding S, Chen JS, Lou XWD (2011) Hierarchical nickel sulfide hollow spheres for high performance supercapacitors. RSC Adv 1(3):397–400
70. Gaikar P, Pawar SP, Mane RS, Nuashad M, Shinde D (2016) Synthesis of nickel sulfide as a promising electrode material for pseudocapacitor application. RSC Adv 6(113):112589–112593
71. Hsu Y-K, Chen Y-C, Lin Y-G (2014) Synthesis of copper sulfide nanowire arrays for high-performance supercapacitors. Electrochim Acta 139(Suppl C):401–407
72. Zhang H, Yu X, Guo D, Qu B, Zhang M, Li Q, Wang T (2013) Synthesis of bacteria promoted reduced graphene oxide-nickel sulfide networks for advanced supercapacitors. ACS Appl Mater Interfaces 5(15):7335–7340
73. Hegde G, Abdul Manaf SA, Kumar A, Ali GAM, Chong KF, Ngaini Z, Sharma KV (2015) Biowaste sago bark based catalyst free carbon nanospheres: waste to wealth approach. ACS Sustain Chem Eng 5(9):2247–2253
74. Ali GAM, Supriya S, Chong KF, Shaaban ER, Algarni H, Maiyalagan T, Hegde G (2019) Superior supercapacitance behavior of oxygen self-doped carbon nanospheres: a conversion of Allium CEPA peel to energy storage system. Biomass Conv Bioref. https://doi.org/10.1007/s13399-019-00520-3
75. Ali GAM, Divyashree A, Supriya S, Chong KF, Ethiraj AS, Reddy MV, Algarni H, Hegde G (2017) Carbon nanospheres derived from Lablab purpureus for high performance supercapacitor electrodes: a green approach. Dalton Trans 46(40):14034–14044

76. Ali GAM, Manaf SAA, Divyashree A, Chong KF, Hegde G (2016) Superior supercapacitive performance in porous nanocarbons. J Energy Chem 25(4):734–739
77. Ali GAM, Habeeb OA, Algarni H, Chong KF (2018) CaO impregnated highly porous honeycomb activated carbon from agriculture waste: symmetrical supercapacitor study. J Mater Sci 54:683–692
78. Ali GAM, Abdul Manaf SA, Kumar A, Chong KF, Hegde G (2014) High performance supercapacitor using catalysis free porous carbon nanoparticles. J Phys D Appl Phys 47 (49):495307–495313
79. Zhang R, Lu C, Shi Z, Liu T, Zhai T, Zhou W (2019) Hexagonal phase NiS octahedrons co-modified by 0D-, 1D-, and 2D carbon materials for high-performance supercapacitor. Electrochim Acta 311:83–91
80. Chen W, Xia C, Alshareef HN (2014) One-step electrodeposited nickel cobalt sulfide nanosheet arrays for high-performance asymmetric supercapacitors. ACS Nano 8(9):9531–9541
81. Wang J, Dong S, Ding B, Wang Y, Hao X, Dou H, Xia Y, Zhang X (2017) Pseudocapacitive materials for electrochemical capacitors: from rational synthesis to capacitance optimization. Nat Sci Rev 4(1):71–90
82. Liu TC, Pell W, Conway B, Roberson S (1998) Behavior of molybdenum nitrides as materials for electrochemical capacitors comparison with ruthenium oxide. J Electrochem Soc 145(6):1882–1888
83. Choi D, Blomgren GE, Kumta PN (2006) Fast and reversible surface redox reaction in nanocrystalline vanadium nitride supercapacitors. Adv Mater 18(9):1178–1182
84. Balogun M-S, Qiu W, Wang W, Fang P, Lu X, Tong Y (2015) Recent advances in metal nitrides as high-performance electrode materials for energy storage devices. J Mater Chem A 3(4):1364–1387
85. Li L, Ge J, Wu F, Chen R, Chen S, Wu B (2010) Recovery of cobalt and lithium from spent lithium ion batteries using organic citric acid as leachant. J Hazard Mater 176(1):288–293
86. Saha S, Jana M, Khanra P, Samanta P, Koo H, Murmu NC, Kuila T (2015) Band gap engineering of boron nitride by graphene and its application as positive electrode material in asymmetric supercapacitor device. ACS Appl Mater Interfaces 7(26):14211–14222
87. Wang Y, Yang W, Yang J (2007) A Co–Al layered double hydroxides nanosheets thin-film electrode fabrication and electrochemical study. Electrochem Solid-State Lett 10(10):A233–A236
88. Prevot V, Forano C, Khenifi A, Ballarin B, Scavetta E, Mousty C (2011) A templated electrosynthesis of macroporous NiAl layered double hydroxides thin films. Chem Commun 47(6):1761–1763
89. Gao X, Liu X, Wu D, Qian B, Kou Z, Pan Z, Pang Y, Miao L, Wang J (2019) Significant role of Al in ternary layered double hydroxides for enhancing electrochemical performance of flexible asymmetric supercapacitor. Adv Func Mater 29(36):1903879
90. Gao X, Zhao Y, Dai K, Wang J, Zhang B, Shen X (2020) NiCoP nanowire@ NiCo-layered double hydroxides nanosheet heterostructure for flexible asymmetric supercapacitors. Chem Eng J 384:123373
91. Chen H, Hu L, Chen M, Yan Y, Wu L (2014) Nickel-Cobalt layered double hydroxide nanosheets for high-performance supercapacitor electrode materials. Adv Func Mater 24 (7):934–942
92. Guo XL, Liu XY, Hao XD, Zhu SJ, Dong F, Wen ZQ, Zhang YX (2016) Nickel-Manganese layered double hydroxide nanosheets supported on nickel foam for high-performance supercapacitor electrode materials. Electrochim Acta 194(Suppl C):179–186
93. Shao M, Ning F, Zhao Y, Zhao J, Wei M, Evans DG, Duan X (2012) Core–shell layered double hydroxide microspheres with tunable interior architecture for supercapacitors. Chem Mater 24(6):1192–1197
94. Zhang L, Wang J, Zhu J, Zhang X, San Hui K, Hui KN (2013) 3D porous layered double hydroxides grown on graphene as advanced electrochemical pseudocapacitor materials. J Mater Chem A 1(32):9046–9053

95. Habeeb OA, Ramesh K, Ali GAM, Yunus RM (2017) Low-cost and eco-friendly activated carbon from modified palm kernel shell for hydrogen sulfide removal from wastewater: adsorption and kinetic studies. Desalin Water Treat 84:205–214
96. Habeeb OA, Ramesh K, Ali GAM, Yunus RM (2017) Experimental design technique on removal of hydrogen sulfide using CaO-eggshells dispersed onto palm kernel shell activated carbon: experiment, optimization, equilibrium and kinetic studies. J Wuhan Univ Technol Mater Sci Ed 32(2):305–320
97. Habeeb OA, Ramesh K, Ali GAM, Yunus RM (2017) Application of response surface methodology for optimization of palm kernel shell activated carbon preparation factors for removal of H_2S from industrial wastewater. J Teknol 79(7):1–10
98. Habeeb OA, Ramesh K, Ali GAM, Yunus RM (2017) Isothermal modelling based experimental study of dissolved hydrogen sulfide adsorption from waste water using eggshell based activated carbon. Malays J Anal Sci 21(2):334–345
99. Marina PE, Ali GAM, See LM, Teo EYL, Ng E-P, Chong KF (2016) In situ growth of redox-active iron-centered nanoparticles on graphene sheets for specific capacitance enhancement. Arab J Chem. https://doi.org/10.1016/j.arabjc.2016.02.006
100. Ali GAM, Yusoff MM, Chong KF (2016) Graphene electrochemical production and its energy storage properties. ARPN J Eng Appl Sci 11(16):9712–9717
101. Gupta VK, Agarwal S, Sadegh H, Ali GAM, Bharti AK, Hamdy AS (2017) Facile route synthesis of novel graphene oxide-β-cyclodextrin nanocomposite and its application as adsorbent for removal of toxic bisphenol A from the aqueous phase. J Mol Liq 237:466–472
102. Ali GAM, Thalji MR, Soh WC, Algarni H, Chong KF (2020) One-step electrochemical synthesis of MoS_2/graphene composite for supercapacitor application. J Solid State Electrochem 24(1):25–34
103. Barhoum A, Shalan AE, El-Hout SI, Ali GAM, Abdelbasir SM, Abu Serea ES, Ibrahim AH, Pal K (2019) A broad family of carbon nanomaterials: classification, properties, synthesis, and emerging applications. In: Barhoum A, Bechelany M, Makhlouf A (eds) Handbook of nanofibers. Springer International Publishing, Cham, pp 1–40
104. Dąbrowski A (2001) Adsorption—from theory to practice. Adv Coll Interface Sci 93(1–3):135–224
105. Lee SP, Ali GAM, Algarni H, Chong KF (2019) Flake size-dependent adsorption of graphene oxide aerogel. J Mol Liq 277:175–180
106. DiGiulio D, Wilkin R, Miller C, Oberley G (2011) Investigation of groundwater contamination near Pavillion, Wyoming. US Environmental Protection Agency Draft Report. US EPA. https://www.epa.gov/region8/superfund/wy/pavillion/EPA_ReportOnPavillion_Dec-8-2011.pdf
107. Machida M, Mochimaru T, Tatsumoto H (2006) Lead (II) adsorption onto the graphene layer of carbonaceous materials in aqueous solution. Carbon 44(13):2681–2688
108. Li Y-H, Wang S, Luan Z, Ding J, Xu C, Wu D (2003) Adsorption of cadmium (II) from aqueous solution by surface oxidized carbon nanotubes. Carbon 41(5):1057–1062
109. Zhao H, Baker GA, Holmes SJO (2011) New eutectic ionic liquids for lipase activation and enzymatic preparation of biodiesel. Org Biomol Chem 9(6):1908–1916
110. Ren X, Chen C, Nagatsu M, Wang X (2011) Carbon nanotubes as adsorbents in environmental pollution management: a review. Chem Eng J 170(2–3):395–410
111. Sitko R, Zawisza B, Malicka E (2013) Graphene as a new sorbent in analytical chemistry. TrAC Trends Anal Chem 51:33–43
112. Wu W, Yang Y, Zhou H, Ye T, Huang Z, Liu R, Kuang Y (2013) Highly efficient removal of Cu (II) from aqueous solution by using graphene oxide. Water Air Soil Pollut 224(1):1372
113. Sitko R, Turek E, Zawisza B, Malicka E, Talik E, Heimann J, Gagor A, Feist B, Wrzalik R (2013) Adsorption of divalent metal ions from aqueous solutions using graphene oxide. Dalton Trans 42(16):5682–5689
114. Zhao G, Ren X, Gao X, Tan X, Li J, Chen C, Huang Y, Wang X (2011) Removal of Pb (II) ions from aqueous solutions on few-layered graphene oxide nanosheets. Dalton Trans 40 (41):10945–10952

115. Huang Z-H, Zheng X, Lv W, Wang M, Yang Q-H, Kang F (2011) Adsorption of lead (II) ions from aqueous solution on low-temperature exfoliated graphene nanosheets. Langmuir 27(12):7558–7562
116. El-Maghrabi HH, Abdelmaged SM, Nada AA, Zahran F, El-Wahab SA, Yahea D, Hussein G, Atrees M (2017) Magnetic graphene based nanocomposite for uranium scavenging. J Hazard Mater 322:370–379
117. Wang H, Yuan X, Wu Y, Huang H, Zeng G, Liu Y, Wang X, Lin N, Qi Y (2013) Adsorption characteristics and behaviors of graphene oxide for Zn (II) removal from aqueous solution. Appl Surf Sci 279:432–440
118. Zhao J, Wang Z, White JC, Xing B (2014) Graphene in the aquatic environment: adsorption, dispersion, toxicity and transformation. Environ Sci Technol 48(17):9995–10009
119. Yao Y, Yang Z, Zhang D, Peng W, Sun H, Wang S (2012) Magnetic $CoFe_2O_4$–graphene hybrids: facile synthesis, characterization, and catalytic properties. Ind Eng Chem Res 51 (17):6044–6051
120. Gollavelli G, Chang C-C, Ling Y-C (2013) Facile synthesis of smart magnetic graphene for safe drinking water: heavy metal removal and disinfection control. ACS Sustain Chem Eng 1 (5):462–472
121. Zhu J, Wei S, Gu H, Rapole SB, Wang Q, Luo Z, Haldolaarachchige N, Young DP, Guo Z (2011) One-pot synthesis of magnetic graphene nanocomposites decorated with core@ double-shell nanoparticles for fast chromium removal. Environ Sci Technol 46(2):977–985
122. Li J, Zhang S, Chen C, Zhao G, Yang X, Li J, Wang X (2012) Removal of Cu(II) and fulvic acid by graphene oxide nanosheets decorated with Fe_3O_4 nanoparticles. ACS Appl Mater Interfaces 4(9):4991–5000
123. Chandra V, Park J, Chun Y, Lee J, Hwang I, Kim K (2010) Water-dispersible magnetite-reduced graphene oxide composites for arsenic removal. ACS Nano 4(7):3979–3986
124. Liu M, Chen C, Hu J, Wu X, Wang X (2011) Synthesis of magnetite/graphene oxide composite and application for cobalt (II) removal. J Phys Chem C 115(51):25234–25240
125. WooáLee J, BináKim S (2011) Enhanced Cr (VI) removal using iron nanoparticle decorated graphene. Nanoscale 3(9):3583–3585
126. Madadrang CJ, Kim HY, Gao G, Wang N, Zhu J, Feng H, Gorring M, Kasner ML, Hou S (2012) Adsorption behavior of EDTA-graphene oxide for Pb (II) removal. ACS Appl Mater Interfaces 4(3):1186–1193
127. Liu L, Li C, Bao C, Jia Q, Xiao P, Liu X, Zhang Q (2012) Preparation and characterization of chitosan/graphene oxide composites for the adsorption of Au (III) and Pd (II). Talanta 93:350–357
128. Yang Y, Xie Y, Pang L, Li M, Song X, Wen J, Zhao H (2013) Preparation of reduced graphene oxide/poly (acrylamide) nanocomposite and its adsorption of Pb (II) and methylene blue. Langmuir 29(34):10727–10736
129. Wang L, Jiang L, Su D, Sun C, Chen M, Goh K, Chen Y (2014) Non-covalent synthesis of thermo-responsive graphene oxide–perylene bisimides-containing poly (*N*-isopropylacrylamide) hybrid for organic pigment removal. J Colloid Interface Sci 430:121–128
130. Vasudevan S, Lakshmi J (2012) The adsorption of phosphate by graphene from aqueous solution. RSC Adv 2(12):5234–5242
131. Zhang S, Shao Y, Liu J, Aksay IA, Lin Y (2011) Graphene–polypyrrole nanocomposite as a highly efficient and low cost electrically switched ion exchanger for removing ClO^{4-} from wastewater. ACS Appl Mater Interfaces 3(9):3633–3637
132. Shi G, Ding Y, Fang H (2012) Unexpectedly strong anion-π interactions on the graphene flakes. J Comput Chem 33(14):1328–1337
133. Perreault F, Fonseca de Faria A, Elimelech M (2015) Environmental applications of graphene-based nanomaterials. Chem Soc Rev 44(16):5861–5896
134. Chong MN, Jin B, Chow CW, Saint C (2010) Recent developments in photocatalytic water treatment technology: a review. Water Res 44(10):2997–3027

135. Ravelli D, Dondi D, Fagnoni M, Albini A (2009) Photocatalysis. A multi-faceted concept for green chemistry. Chem Soc Rev 38(7):1999–2011
136. Xiang Q, Yu J, Jaroniec M (2012) Graphene-based semiconductor photocatalysts. Chem Soc Rev 41(2):782–796
137. Leary R, Westwood A (2011) Carbonaceous nanomaterials for the enhancement of TiO_2 photocatalysis. Carbon 49(3):741–772
138. Khan Z, Chetia TR, Vardhaman AK, Barpuzary D, Sastri CV, Qureshi M (2012) Visible light assisted photocatalytic hydrogen generation and organic dye degradation by CdS–metal oxide hybrids in presence of graphene oxide. RSC Adv 2(32):12122–12128
139. Li B, Cao H (2011) ZnO@graphene composite with enhanced performance for the removal of dye from water. J Mater Chem 21(10):3346–3349
140. Wang Y, Wang W, Mao H, Lu Y, Lu J, Huang J, Ye Z, Lu B (2014) Electrostatic self-assembly of $BiVO_4$–reduced graphene oxide nanocomposites for highly efficient visible light photocatalytic activities. ACS Appl Mater Interfaces 6(15):12698–12706
141. Liu J, Bai H, Wang Y, Liu Z, Zhang X, Sun DD (2010) Self-assembling TiO_2 nanorods on large graphene oxide sheets at a two-phase interface and their anti-recombination in photocatalytic applications. Adv Func Mater 20(23):4175–4181
142. Zhang Y, Tang Z-R, Fu X, Xu Y-J (2011) Engineering the unique 2D mat of graphene to achieve graphene-TiO_2 nanocomposite for photocatalytic selective transformation: what advantage does graphene have over its forebear carbon nanotube? ACS Nano 5(9):7426–7435
143. Wang G, Zhang L, Zhang J (2012) A review of electrode materials for electrochemical supercapacitors. Chem Soc Rev 41(2):797–828
144. Lambert TN, Chavez CA, Hernandez-Sanchez B, Lu P, Bell NS, Ambrosini A, Friedman T, Boyle TJ, Wheeler DR, Huber DL (2009) Synthesis and characterization of titania−graphene nanocomposites. J Phys Chem C 113(46):19812–19823
145. Liang Y, Wang H, Casalongue HS, Chen Z, Dai H (2010) TiO_2 nanocrystals grown on graphene as advanced photocatalytic hybrid materials. Nano Res 3(10):701–705
146. Liu X, Pan L, Lv T, Lu T, Zhu G, Sun Z, Sun C (2011) Microwave-assisted synthesis of ZnO–graphene composite for photocatalytic reduction of Cr (VI). Catal Sci Technol 1 (7):1189–1193
147. Williams G, Seger B, Kamat PV (2008) TiO2-graphene nanocomposites. UV-assisted photocatalytic reduction of graphene oxide. ACS Nano 2(7):1487–1491
148. Perera S, Mariano R, Vu K, Nour N, Seitz O, Chabal Y, Balkus K Jr (2012) Hydrothermal synthesis of graphene-TiO_2 nanotube composites with enhanced photocatalytic activity. ACS Catal 2:949–956
149. Zhu M, Chen P, Liu M (2011) Graphene oxide enwrapped Ag/AgX (X=Br, Cl) nanocomposite as a highly efficient visible-light plasmonic photocatalyst. ACS Nano 5 (6):4529–4536
150. Chen Z, Liu S, Yang M-Q, Xu Y-J (2013) Synthesis of uniform CdS nanospheres/graphene hybrid nanocomposites and their application as visible light photocatalyst for selective reduction of nitro organics in water. ACS Appl Mater Interfaces 5(10):4309–4319
151. Xiong Z, Zhang LL, Ma J, Zhao X (2010) Photocatalytic degradation of dyes over graphene–gold nanocomposites under visible light irradiation. Chem Commun 46(33):6099–6101
152. Liu S, Chen Z, Zhang N, Tang Z-R, Xu Y-J (2013) An efficient self-assembly of CdS nanowires–reduced graphene oxide nanocomposites for selective reduction of nitro organics under visible light irradiation. J Phys Chem C 117(16):8251–8261
153. Zhai C, Zhu M, Bin D, Wang H, Du Y, Wang C, Yang P (2014) Visible-light-assisted electrocatalytic oxidation of methanol using reduced graphene oxide modified Pt nanoflowers-TiO_2 nanotube arrays. ACS Appl Mater Interfaces 6(20):17753–17761
154. Bhunia SK, Jana NR (2014) Reduced graphene oxide-silver nanoparticle composite as visible light photocatalyst for degradation of colorless endocrine disruptors. ACS Appl Mater Interfaces 6(22):20085–20092

155. El-Maghrabi HH, Nada EA, Soliman FS, Moustafa YM, Amin AE-S (2016) One pot environmental friendly nanocomposite synthesis of novel TiO_2-nanotubes on graphene sheets as effective photocatalyst. Egypt J Petrol 25(4):575–584
156. CDI (2016) Cobalt facts. Cited 15 July 2017. Available from: https://www.cobaltinstitute.org/
157. Statista (2016) World cobalt reserves as of 2016. Cited 13 Aug 2017. Available from: https://www.statista.com/statistics/264930/global-cobalt-reserves/
158. Cheang CY, Mohamed N (2016) Removal of cobalt from ammonium chloride solutions using a batch cell through an electrogenerative process. Sep Purif Technol 162:154–161
159. Pazik PM, Chmielewski T, Glass HJ, Kowalczuk PB (2016) World production and possible recovery of cobalt from the Kupferschiefer stratiform copper ore. E S Web Conf 8:01063
160. Crundwell FK (2011) Extractive metallurgy of nickel, cobalt and platinum group metals. Elsevier
161. Nelson E (2017) The hottest thing in the markets right now is an obscure metal mined in DR Congo. Batteries. Cited 15 Aug 2017. Available from: https://qz.com/969115/cobalt-is-the-unlikely-hero-of-financial-markets-right-now/
162. Randall T (2017) The electric-car boom is so real even oil companies say it's coming. Available from: https://www.bloomberg.com/news/articles/2017-04-25/electric-car-boom-seen-triggering-peak-oil-demand-in-2030s
163. Nitta N, Wu F, Lee JT, Yushin G (2015) Li-ion battery materials: present and future. Mater Today 18(5):252–264
164. Times F (2017) Glencore deals mining blow to African copper producers. Cited 12 Sept 2017. Available from: https://www.ft.com/content/3f1d9664-562f-11e5-a28b-50226830d644
165. Raugei M, Winfield P (2019) Prospective LCA of the production and EoL recycling of a novel type of Li-ion battery for electric vehicles. J Clean Product 213:926–932
166. van den Brink S, Kleijn R, Sprecher B, Tukker A (2020) Identifying supply risks by mapping the cobalt supply chain. Resour Conserv Recycl 156:104743
167. Nada AA, Tantawy HR, Elsayed MA, Bechelany M (2018) Elaboration of nano titania-magnetic reduced graphene oxide for degradation of tartrazine dye in aqueous solution. Solid State Sci 78:116–125
168. El-Maghrabi HH, Nada AA, Diab KR, Youssef AM, Hamdy A, Roualdesb S, Abd El-Wahab S (2018) Facile fabrication of $NiTiO_3$/graphene nanocomposites for photocatalytic hydrogen generation. J Photochem Photobio A: Chem 365:86–93

Recycling Polyethylene Terephthalate Waste to Magnetic Carbon/Iron Nanoadsorbent for Application in Adsorption of Diclofenac Using Statistical Experimental Design

Premanjali Rai and Kunwar P. Singh

Abstract A novel magnetic nanoadsorbent comprising carbon/iron composite was prepared from polyethylene terephthalate waste. The magnetic nanoadsorbent was characterized and applied in the adsorption of diclofenac from water. Batch adsorption experiments were conducted according to a three-factor three-level Box–Behnken design including temperature (°C), pH and adsorbent dose (g L^{-1}) as the process parameters. A polynomial regression model was used to predict and optimize the parameters for maximum adsorption capacity (mg g^{-1}) of the nanoadsorbent using response surface modeling. The magnetic nanoadsorbent exhibited a surface area of 288.88 m^2 g^{-1} and a saturation magnetization of 35.4 emu g^{-1}. Transmission electron microscopy of the nanoadsorbent depicted particle size range within 10–40 nm. The maximum adsorption capacity of the nanoadsorbent for diclofenac was 15.31 mg g^{-1} under optimized conditions of 42.65 °C, 5.74 pH and 1.04 g L^{-1} dose. High regression coefficient values ($R^2 = 0.987$) in the design experiments suggested considerable goodness of fit for the response surface model. Statistical analysis showed adsorption of diclofenac by the nanoadsorbent was significantly influenced by solution pH. FTIR analysis of the diclofenac loaded nanoadsorbent confirmed the adsorption of diclofenac by the emergence of new diagnostic peaks. The diclofenac loaded nanoadsorbent could be desorbed up to 69.88% by NaOH stripping, suggesting its reuse potential.

Keywords Polyethylene terephthalate · Activated carbon · Nanoadsorbent · Composite · Diclofenac · Adsorption · Optimization · Box–behnken design · Response surface modeling · Desorption

---

P. Rai (✉) · K. P. Singh
Environmental Chemistry Division, CSIR-Indian Institute of Toxicology Research, Mahatma Gandhi Marg, Post Box 80, Lucknow 226 001, India
e-mail: premanjalirai@yahoo.com

A. S. H. Makhlouf and G. A. M. Ali (eds.), *Waste Recycling Technologies for Nanomaterials Manufacturing*, Topics in Mining, Metallurgy and Materials Engineering, https://doi.org/10.1007/978-3-030-68031-2_26

Abbreviations

ANOVA	Analysis of variance
BBD	Box–Behnken design
DE	Desorption efficiency
EDX	Energy dispersive X-ray
FTIR	Fourier transform infrared
PC	Per cent contribution
PET	Polyethylene terephthalate
pH_{zpc}	Point of zero charge
RSM	Response surface modeling
RMSEP	Root mean square error of prediction
RSEP	Relative standard error of prediction
SEM	Scanning electron microscopy
TEM	Transmission electron microscopy
VSM	Vibrating sample magnetometer
XRD	X-ray diffraction

1 Introduction

The rationale behind identifying pharmaceuticals as contaminants of emerging concern emanates from their ubiquitous occurrence pattern, their persistent nature to resist removal in water/wastewater treatment plants and their eco-toxic effects exerted on the non-target organisms living in the aqua-terrestrial environment. Amongst all, diclofenac is one of the most frequently detected pharmaceutical compounds and is associated with the highest ecotoxicity within the group of non-steroidal anti-inflammatory drugs [1]. Aquatic ecotoxicity assays of diclofenac reveal the teratogenic and bioaccumulative potential of the drug from cytological alterations and other lethal effects [2, 3]. Pharmaceuticals have been considered as "contaminant of emerging concern" which was included in the previous Watch List of EU Decision 2015/495 [4] to monitor their presence and level in surface waters. Diclofenac is used for its analgesic and anti-inflammatory activity during rheumatoid arthritis, acute injury and dysmenorrhea in human and veterinary medicine. It is estimated that the global consumption volume of diclofenac is 940 tons per year [5]. In India, the manufacture of diclofenac for veterinary formulations was banned in the year 2006 after the drug was known to cause renal lesions in vultures which scavenged on diclofenac treated livestock bodies [6]. Nevertheless, this compound is actively consumed as an over-the-counter drug [7] and has been detected in the Indian rivers [8]. More recently, diclofenac levels have been detected at unsurpassed levels of up to 26.68 $\mu g\ L^{-1}$ in wastewaters of north India [9]. The widespread of diclofenac residues is evident by its presence in surface water [10] treated effluent [10], drinking water [11] and even in energy drinks [12] of different nations across the globe. Diclofenac demonstrates a low elimination rate of 21–40% in wastewater treatment plants [5] suggestively due to its low water-sludge distribution coefficient [13] and the presence of –Cl and –NH groups in its molecular structure which impairs the growth of sewage bacteria [10]. Hence, the need for an efficient diclofenac removal process which cost-prohibitive and avoids the formation of toxic by-products in water is continuously transpiring. Environmentally benign adsorption technology by the activated carbons fulfils the above-mentioned desirable characteristics. However, this application is often restricted by the high manufacturing costs of commercial activated carbons at around 700–5000 US$ per ton [14]. Therefore, low-cost adsorbent material, especially by recycled waste materials, has been suitably used for adsorption studies [15–17]. Adsorption susceptibility of diclofenac has been investigated by olive-waste cake activated carbon, commercial activated carbon, siliceous material, grape bagasse, biochar from orchard garden residues, moringa seeds, sugarcane bagasse, etc., as potential adsorbents in water [18–27]. PET is a redundant solid waste generated largely from its use in packaging of commodities. The amount of plastic waste along with PET commodities that is generated in India is voluminous. This is estimated to be around 5.6 million tons every year [28]. It also forms one of the most abundant plastics in the urban solid waste [29]. Globally, an estimated recycling rate of all plastics is only about 3.5% compared to 34% of paper, 22% of glass and 30% of metals. Man's behavioural propensity to the throw-away culture coupled with inequitable recovery has led to its increased accumulation as waste material in the environment. As this commodity is non-biodegradable under normal circumstances, vast scale littering of plastic wastes has led to the deterioration of environmental quality. The strewn plastics can create a toll on the quality of water, air and soil by simply choking drains or releasing carcinogenic compounds on burning. On the other hand, plastics might generate toxic leachates which seep into groundwater, thereby contaminating potential drinking water sources. The leachates contain human endocrine-disrupting compounds such as the phthalates which can be responsible for causing several diseases including cancer and physical infirmities. India experiences high temperatures (>40 °C) during the summer months which may induce leaching of phthalates and other stabilizers from the plastic wastes. Such leachates continue to contaminate the groundwater which eventually reaches human beings from drinking water. Plastic waste also finds its way into the marine environment where cases of plastic debris ingestion by the marine animals [30] have been reported. Despite various conjectures and regulations on their use, their vast scale littering in the surroundings continues and is a very common eye sore in the Indian landscape. Hence, the metamorphosis of such obnoxious blackballed material into a novel artefact will definitely lessen public outcry. Besides providing environmental respite, the strategy of converting wastes into functional materials may reduce the logistics on waste management in any country worldwide Hence, due to the environmental discord posed by PET wastes and its high carbon content [31], PET has been chosen as the precursor to the magnetic carbon/iron composite in this study. On the other hand, magnetism actuated removal process assists in the rapid, wholesome, non-invasive collection and reuse of the magnetic particle functionalized adsorbent. However, multifactor experiments in adsorption require a large number of experiments by varying each process parameter while keeping the rest constant, in order to establish the adsorbent-adsorbate interactions and optimization of process parameters. This limitation can be overcome by statistically designed experiments and mathematical modeling of the dataset. The main aim of such design is to build models, evaluate the independent and relative effects of the process parameters involved in building the model and optimize the process conditions from a nominal number of experimental trials. In this study, the optimization of the process parameters, namely temperature, solution pH and adsorbent dose for maximum adsorption of diclofenac by a newly developed magnetic carbon/iron nanoadsorbent derived from polyethylene terephthalate (PET) wastes has been reported.

In a nutshell, the main aim of this chapter is to study the application of low-cost magnetic carbon recycled from PET waste in the optimized adsorption of diclofenac from water using a three-factor Box–Behnken design (BBD) of experiments and response surface modeling (RSM). The details on structural and chemical characteristics of the developed nanoadsorbent were obtained through relevant characterization studies. Sorption isotherm, reaction kinetics and desorption studies have been conducted to gain insight into the nature of the sorption process. Fourier transform infrared (FTIR) analysis of the magnetic carbon adsorbent after diclofenac adsorption has been reported for better comprehension of the sorption mechanism.

2 Materials and Methods

2.1 Chemicals and Reagents

All chemicals used in this study were of analytical reagent grade. Pure standards of diclofenac were procured from Sigma-Aldrich (USA). Other reagents such as the anhydrous ferric chloride ($FeCl_3$), sodium hydroxide (NaOH), sodium borohydride ($NaBH_4$) and ethyl alcohol (C_2H_5OH) were purchased from Merck (India) and SD Fine-Chem Limited (India). All chemical solutions were prepared in Milli-Q water (Millipore, Milford, MA, USA).

2.2 Recycling PET Wastes into Magnetic Carbon

The detailed route of synthesis for the preparation of magnetic carbon from waste PET commodities has been given in our earlier published study [32]. Pyrolysis of the waste PET pellets was conducted in nitrogen (N_2) gas at a flow rate of 200 cm^3 min^{-1} under ramp conditions at 5 °C min^{-1} until 450 °C and held as such for 1 h. This step was conducted to maximize the evolution of volatile components. After that, the temperature was elevated to 725 °C with a hold time of 2 h. The char obtained was cooled and prepared for physical activation. For activation, an initial flow of N_2 at 10 cm^3 min^{-1} was given until a temperature of 925 °C was obtained and held as such for 1 h. The flow was later switched to that of carbon dioxide (CO_2) with a soaking time of 2 h. The activated carbon was again cooled in N_2 gas and then stored in desiccators. For the preparation of the magnetic nanoadsorbent, iron oxides were intercalated with the carbon matrix by a wet impregnation method. For this, a $FeCl_3$ solution (0.5 M) was prepared in a mixture solvent comprising C_2H_5OH and modified Milli-Q water (60:100 v/v). The activated carbon obtained after pyrolysis and activation of PET waste was added into the $FeCl_3$ solution under a continuous stirring condition by a magnetic stirrer for 2 h. 3 M NaOH was added dropwise to raise the pH above the pH_{zpc} of activated carbon previously determined as 7. This ensured metal-carboxyl coordination between the negatively charged carboxyl groups on the activated carbon and positively charged ferric hydroxide sol. Subsequently, $NaBH_4$ (1 M) was added to the suspension as an external reducing agent in order to obtain a black sol comprising a composite of iron oxides impregnated on the carbon matrix. The iron impregnated carbon was filtered and washed thoroughly with ethanol to be dried at 80 °C for 30 min. Upon drying, the magnetic carbon/iron composite was seen to be attracted to an external magnet even from outside the walls of the container.

2.3 Characterization

The BET surface area (S_{BET}) and pore volume of the magnetic carbon/iron nanoadsorbent and the activated carbon were measured by N_2 physisorption experiments conducted on the adsorption analyser Quantachrome Autosorb 1C (USA) at −196 °C. The pore volume distribution was estimated from the gas adsorbed at relative pressure P/Po of 0.99. X-ray diffraction (XRD) patterns were obtained from a powder X-ray diffraction system on JSO-Debyeflex 2002 diffractometer (Germany) using Cu–K_α radiation at a wavelength of 1.54 Å within 2θ range of 5° − 90°. The average crystallite size (d) was measured by the Debye Scherrer equation, $d = \frac{K\lambda}{\beta \cos \theta}$ [33, 34] here parameters K, λ, θ and beta (β) are the constants (close to unity), X-ray wavelength, Bragg angle and the full width at half maximum measured in radians, respectively. The surface morphology of the magnetic carbon was scanned by scanning electron microscopy (SEM) on a surface electron microscope (LeO 430, England) at 20 kV. The shape and dimension of the iron oxide nanoparticles were found by transmission electron microscopy (TEM) performed on Technai G^2 transmission electron microscope (Netherlands) at 80 kV. The functionality of the magnetic carbon groups was interpreted from their infrared spectra within a spectral range of 4000–450 cm^{-1} by Perkin Elmer FTIR RX1 spectrophotometer (USA). The magnetic susceptibility of the magnetic adsorbent was measured by vibrating sample magnetometer by EV7 VSM (USA) at room temperature. The pH at which there is zero charge (pH_{zpc}) on the adsorbent surface was determined using the pH drift method [35] while measuring the pH by a calibrated 744 Metrohm pH meter (Switzerland).

2.4 Adsorption Experiment

A standard stock solution of diclofenac at 1000 mg L^{-1}, prepared in Milli-Q water was appropriately diluted to obtain working solutions of the desired concentration. A typical protocol of adsorption experiment comprised the addition of the adsorbent at a varying dose (0.4–1.6 g L^{-1}) to the diclofenac solutions (25 ml) adjusted to varying pH (5–10). The mixtures were stirred in a temperature-regulated water bath shaker (YSI-417 Yorco, India) at a fixed temperature within a range of 20–50 °C. After reaching equilibrium, the residual diclofenac concentration was measured spectrophotometrically by a Perkin Elmer UV–Vis absorbance spectrophotometer (absorbance accuracy of ± 0.004) at the maximum absorbance wavelength of diclofenac (270 nm). The adsorbent's adsorption capacity at equilibrium, expressed as q_e (mg g^{-1}) was determined by the following mass balance equation (Eq. 1) [16, 36]:

$$q_e = \frac{C_0 - C_e}{w} V \quad (1)$$

where C_0 is the initial diclofenac concentration (mg L^{-1}), C_e is the equilibrium diclofenac concentration (mg L^{-1}), V is the volume of the solution (L) and w is the mass of the adsorbent (g).

2.5 Box–Behnken Design and Response Surface Modeling

Geometrically, Box–Behnken design (BBD) is considered as a free, spherical and rotatable quadratic design with the experimental points lying on the central point and midpoints of the edges of the sphere [37]. BBD as a second-order design, apart from reducing the number of experiments, avoids the combinations of variables at extreme cases and provides orthogonal blocking for estimation of the parameters independent of the block effects [38]. The number of experiments in BBD is defined as $N = 2\,k\,(k - 1) + n_c$ (where N is the number of experimental runs, k pertains to the number of parameters and n_c is the number of runs at the centre) [38]. A three-factor BBD including temperature (ºC), pH and adsorbent dose (g L^{-1}) at constant diclofenac concentration (60 mg L^{-1}) yielded $N = 15$ at $n_c = 3$. The BBD matrix for the present study in actual and coded values of the three process parameters at three levels is given in Table 1.

The actual factors (uncoded) were transformed into their coded values, i.e. −1 (low), 0 (central point or middle) and +1 (high) by using Eq. (2):

Table 1 Process variables and their limits in actual (A) and coded (C) values

Variable (X_i)	Units	Limits			Step change (Δx_i)
		−1 (C)	0 (C)	+1 (C)	
Temperature	°C	20 (A)	35 (A)	50 (A)	15
pH	–	5 (A)	7.5 (A)	10 (A)	2.5
Dose	g L^{-1}	0.4 (A)	1.0 (A)	1.6 (A)	0.6

$$x_i = \frac{X_i - X_0}{\Delta X_i} \tag{2}$$

where x_i is the dimensionless coded value of the ith independent variable, X_i is the uncoded value of the ith independent variable, X_0 is the uncoded value of the ith independent variable at the centre point and ΔX_i is the step change in variable X_i. A quadratic model was built using the process variables and analysing the response (y) as the adsorption capacity (mg g^{-1}) of the magnetic carbon/iron composite. This empirical model is known as a response surface model (RSM) and is expressed as the second-order polynomial equation written as [17, 39–41]:

$$y = \beta_0 + \sum_{i=1}^{n} \beta_i X_i + \sum_{i=1}^{n} \beta_{ii} X_i^2 + \sum_{i=1}^{n} \sum_{j=1}^{n} \beta_{ij} X_i X_j \tag{3}$$

where y is the predicted response, X_i's are the independent input variables and the parameters β_0, β_i, β_{ii} and β_{ij} are the model constant, linear, quadratic and interaction coefficients, respectively. The coefficients were estimated by multiple regression method and used for predicting the response. Model fitting, statistical evaluation, plotting of 3D graphs and optimization were done by Design-Expert-9 Stat Ease (Minneapolis, USA).

2.6 Model Adequacy and Process Optimization

The adequacy of the selected model for prediction of the response variable was tested by analysis of variance (ANOVA). The statistical tools studied to testify the same were the Fisher variation ratio (F-value), probability value (p-value), the sum of squares (SS), the mean sum of squares (MSS), students t distribution (t-value), determination coefficient (R^2), adjusted determination coefficient (R^2_{adj}) and chi-square test (χ^2). For process optimization, the desirability function approach was applied. The optimization procedure in desirability function involves two steps: (1) finding the levels of the independent process parameters that simultaneously deliver the most desired predicted response (dependent variable) and (2) maximize the overall desirability concerning controllable factors [39, 42]. The approach is to convert and assign the response, $y_i(x)$, a number between 0 to 1 by a desirability function, $d_i(y_i)$, such that $d_i(y_i) = 0$ for an undesirable value of y_i and $d_i(y_i) = 1$ for completely desirable response. The desirability scores for individual predicted values of each response are combined to obtain an overall desirability function, D, as the geometric mean of the d_i values:

$$D = (d_1 \times d_2 \cdots d_n)^{1/n} \tag{4}$$

where n is the number of responses in the measure. In the present study, the optimization of process parameters for maximum adsorption capacity (mg g^{-1}) was achieved using the following desirability function [39]:

$$d_i(y_i) = \begin{cases} 0 & \text{if} \quad y_i(x) < y_{i,\min} \\ \left(\frac{y_i(x) - y_{i,\min}}{y_{i,\max} - y_{i,\min}}\right)^s & \text{if} \quad y_{i,\min} \leq \hat{Y}_i(x) \leq y_{i,\max} \\ 1 & \text{if} \quad y_i(x) > y_{i,\max} \end{cases} \tag{5}$$

where $y_{i,\min}$ and $y_{i,\max}$ are the lowest and highest acceptable limits for the response y_i. The value of weight (s) was kept at its default value of 1. For optimization, the input process variables were set to "in range" amongst all the input goals while the response was set to "maximize" goal. The goals were combined to obtain an overall desirability function, and the optimization algorithm in Design-Expert software seeks to maximize this function by starting at random points and proceeding to a maximum via the steepest slope.

3 Results and Discussion

3.1 Characterization

The surface parameters and elemental composition of the developed adsorbents (activated carbon (AC) and the magnetic counterpart i.e., the magnetic carbon/iron composite (MAC)) have been summarized in Table 2. The S_{BET} of the magnetic carbon/iron composite was 288.8 m^2 g^{-1} compared to 659.6 m^2 g^{-1} for the raw activated carbon. The decrease in surface area is attributed to the presence of the particles of iron oxide scattered along with the pores and surface of the activated carbon. The bulk intercalation of iron oxides with the carbon matrix is testified from the significant decrease in total pore volume (V_{tot}) of the activated carbon from 0.36 to 0.17 cm^3 g^{-1} after magnetic modification. The N_2 adsorption–desorption isotherms of both the samples have been given in Fig. 1. As per the classification of adsorption isotherms by the International Union of Pure and Applied Chemistry (IUPAC) [43], the observed isotherm curves of the activated carbon and its magnetic counterpart depicts the dominant presence of the micropores (Type I). The pore volume distribution curve has been provided in Fig. 1b. The effective pore diameter of about 1.4 nm contributed the maximum to the total pore volume.

Table 2 Physicochemical properties of AC and MAC

	S_{BET} (m^2 g^{-1})	V_{tot} [a] (cm^3 g^{-1})	V_{micro} [b] (cm^3 g^{-1})	V_{meso} [c] (cm^3 g^{-1})	pH_{pc}	C	H	N	O[d]	Fe[e]
AC	659.6	0.36	0.30	0.04	7	77.22	2.02	1.37	19.39	–
MAC	288.8	0.17	0.14	0.03	6.8	50.26	1.34	0.25	23.85	24.3

[a]Total pore volume at $P/P_0 \approx 0.99$
[b]Calculated by density functional theory (DFT) method
[c]Calculated by Barrett-Joyner-Halenda (BJH) method
[d]Calculated by difference
[e]Analysed by EDAX

Table 3 Three-factor BBD matrix and the corresponding experimental and predicted adsorption capacities of MAC (mg g^{-1})

Exp. No.	*T* (°C)		pH		*D* (g^{1-1})		Ads cap (mg g^{-1})	
	A	C	A	C	A	C	Exp	Pred
1	20	–1	5	–1	1	0	9.31	9.23
2	20	–1	10	1	1	0	4.23	4.86
3	50	1	5	–1	1	0	15.89	15.26
4	50	1	10	1	1	0	5.19	5.27
5	20	–1	7.5	0	0.4	–1	8.95	8.39
6	20	–1	7.5	0	1.6	1	6.77	6.78
7	50	1	7.5	0	0.4	–1	9.96	9.95
8	50	1	7.5	0	1.6	1	11.11	11.67
9	35	0	5	–1	0.4	–1	11.99	12.62
10	35	0	5	–1	1.6	1	12.42	12.49
11	35	0	10	1	0.4	–1	5.32	5.25
12	35	0	10	1	1.6	1	6.12	5.49
13	35	0	7.5	0	1	0	14.71	14.35
14	35	0	7.5	0	1	0	13.79	14.35
15	35	0	7.5	0	1	0	14.55	14.35

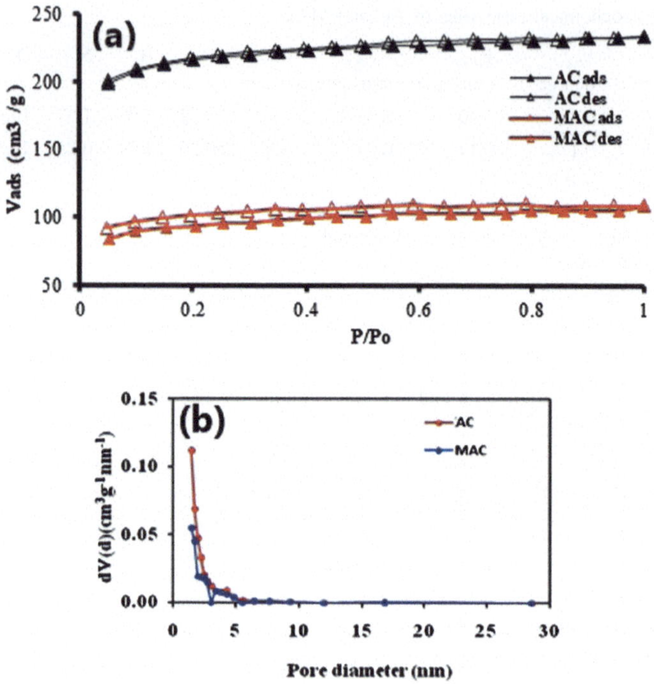

Fig. 1 Nitrogen physisorption isotherms of PET-derived AC and MAC at 77 K (**a**) and their pore size distribution (**b**)

The CHN element analysis gave a carbon content of 77.22% in PET-derived activated carbon and is close to the carbon content of 80% reported for KOH activated PET [44]. The EDX peak analysis of the magnetic carbon composite provided an approximate Fe content in wt% as 24%. The point of zero charge established by the pH drift method was found to be 6.8 for the magnetic carbon. The XRD diffraction patterns have been shown in Fig. 2. As it can be seen, there is a broad diffraction band at 2θ within 20º–30º, and also a comparatively weak intensity band at $2\theta \sim 43^{\circ}$ in the carbon sample are ascribed to reflections 002 and 101, respectively [31, 45]. The presence of these bands denotes that the structure of the activated carbon intermediates between graphitic (crystalline) and amorphous state in a random layer lattice [31]. In the XRD spectra of the magnetic carbon, the bands mentioned above completely disappeared and distinct peaks were found at 29.72º, 35.24º, 42.80º, 56.68º and 62.58º, corresponding to the bands of Fe_3O_4 at reflections 220, 311, 400, 511 and 440, respectively [46]. The average Scherrer crystallite size of iron oxide nanoparticles on the magnetic carbon was found to be 19.2 nm using $2\theta \sim 35.24^{\circ}$ in the diffraction pattern.

The SEM images (Fig. 3) depict the change in morphology of activated carbon after magnetic modification. The PET-derived activated carbon exhibits a smooth texture with nanosized perforations uniformly developed throughout the surface

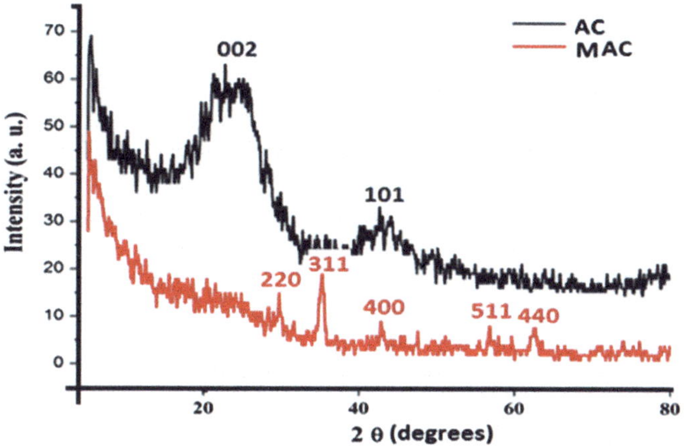

Fig. 2 XRD patterns of AC and MAC

(Fig. 3a, b). At the same time, clusters of iron oxides can be seen to be randomly dispersed on the surface of the magnetic carbon/iron composite delivering a rugged and coarse morphology (Fig. 3c, d). TEM analysis of the composite (Fig. 4) revealed spherical shape morphology with particle size ranging from 10 to 40 nm.

The VSM measurement on the magnetic composite gave saturation magnetization of 35.4 emu g^{-1}, remanence and coercivity values of 1.98 emu g^{-1} and 24 Oesterd, respectively indicating ferromagnetic characteristics of the nanoadsorbent (Fig. 5). The FTIR analysis of both the samples showed the copious presence of different functional groups (Fig. 6). Firstly, the broadband around 3400 cm^{-1} arises from the hydrogen-bonded O–H stretching vibration in hydroxyl groups of phenols, alcohols, carboxylic acids or physisorbed water. An intense band positioned at 1630 cm^{-1} is attributed to the stretched vibration of the C=C bond present in the aromatic structure of the activated carbon matrix. Another band at 1217 cm^{-1} is due to the stretched vibrations of epoxy (C–O) group while the intense band positioned at 770 cm^{-1} is associated with the out of plane deformation by O–H bending [47]. At the lower frequency region (650–450 cm^{-1}), the peak at 613 cm^{-1} arises on account of the stretched vibrations of Fe–O bond from Fe_3O_4 in the magnetic carbon [48].

3.2 *Box–Behnken Design and Response Surface Modeling*

The experimental values for the adsorption capacity of the magnetic nanoadsorbent as obtained from the quadratic design matrix (BBD) have been given in Table 3. The adsorption capacity of the developed nanoadsorbent ranged from 4.23 to 15.89 mg g^{-1} corresponding to a removal percentage of 7.05–24.82% at a fixed

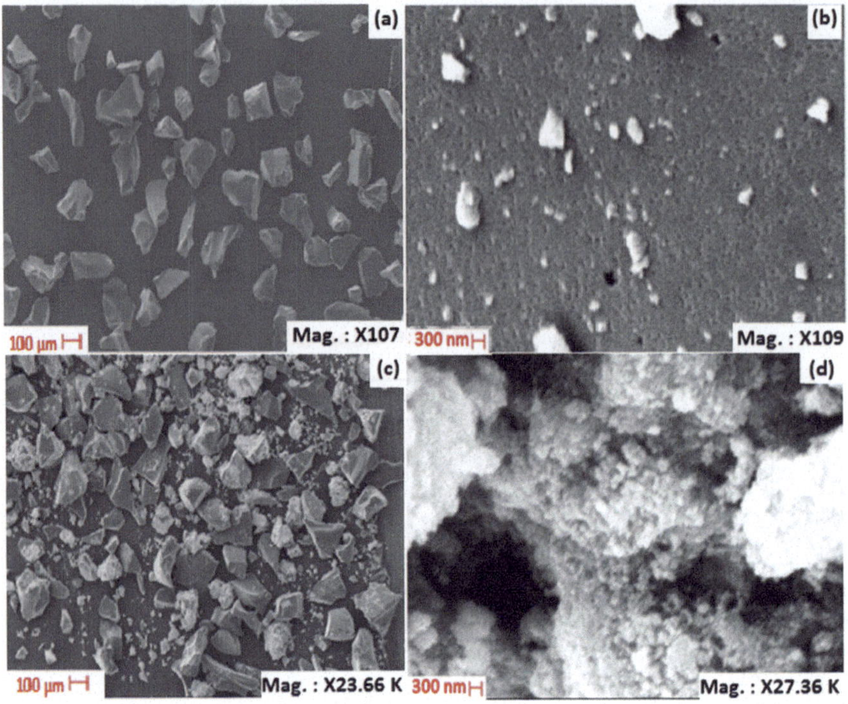

Fig. 3 SEM micrographs of AC particles (**a**), AC surface (**b**), MAC particles (**c**), MAC surface (**d**)

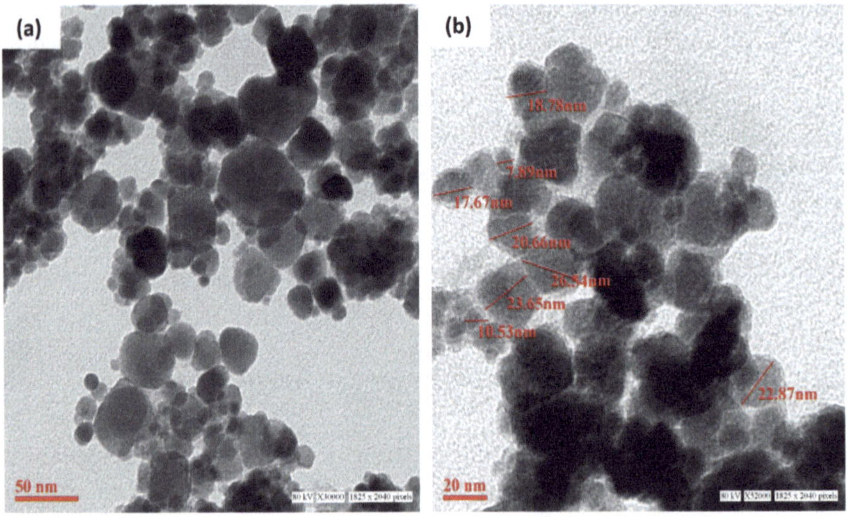

Fig. 4 TEM images of MAC at magnification × 30,000 (**a**), magnification × 52,000 (**b**)

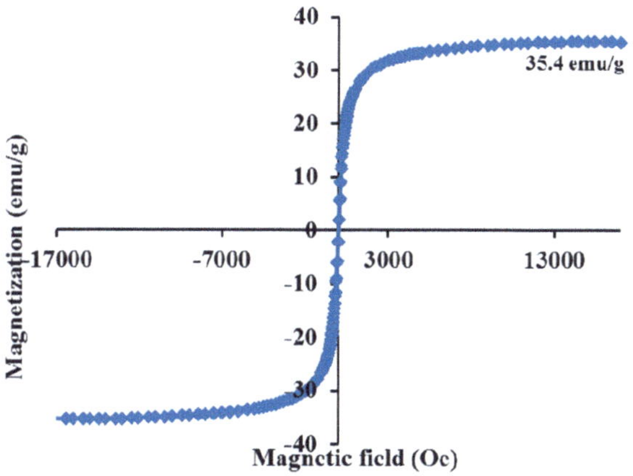

Fig. 5 VSM plot of MAC showing saturation magnetization value of 35.4 emu g^{-1}

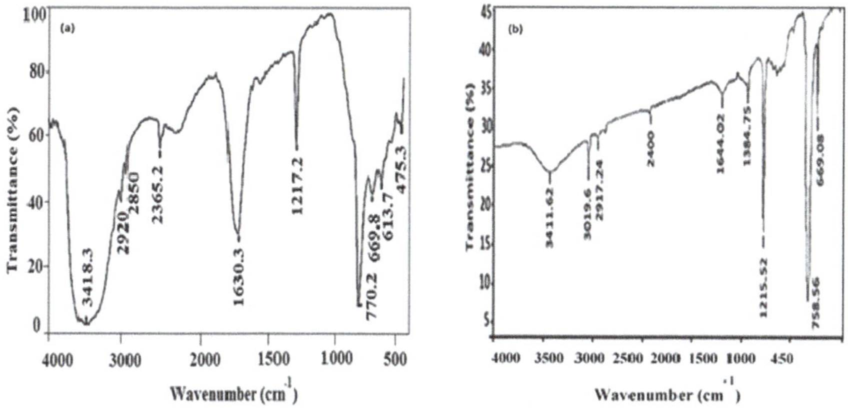

Fig. 6 FTIR spectra of MAC before DIC adsorption (**a**), MAC after DIC adsorption (**b**)

diclofenac concentration of 60 mg L^{-1}. In RSM, a polynomial regression was performed on the dataset, and the best-fitted equation in terms of the coded values of the process parameters was obtained (Eq. 6):

$$Y\left(\text{mg g}^{-1}\right) = 14.35 + 1.61(X_T) - 3.59\left(X_{pH}\right) + 0.025(X_D) - 2.73\left(X_T^2\right) - 2.97\left(X_{pH}^2\right) - 2.42\left(X_D^2\right) - 1.40\left(X_T * X_{pH}\right) + 0.83(X_T * X_D) + 0.092\left(X_{pH} * X_D\right) \quad (6)$$

The model (Eq. 6) was used to predict the adsorption capacity of the magnetic nanoadsorbent, and the adequacy of the fitted model was evaluated by ANOVA (Table 4). In the chosen quadratic model, a high F-value of 42.19 and p-value

Table 4 Analysis of variance (ANOVA) of the response surface quadratic model used in prediction of diclofenac adsorption by MAC

Source	Coeff	SS	DF	MSS	F-value	p-value	t-value	PC	Remarks
Model	14.35	205.68	9	22.85	42.19	0.0003	68.33	–	Significant
X_T	1.61	20.77	1	20.77	38.35	0.0016	12.38	9.59	Significant
X_{pH}	−3.59	103.32	1	103.32	190.76	< 0.0001	−27.61	47.73	Significant
X_D	0.025	0.000	1	0.000	0.000	0.9272	0.19	0	
X^2T	−2.73	27.52	1	27.52	50.81	0.0008	−14.36	12.71	Significant
X^2pH	−2.97	32.46	1	32.46	59.93	0.0006	−15.63	14.99	Significant
X^2D	−2.42	21.67	1	21.67	40.01	0.0015	−12.73	10.01	Significant
$X_T * X_{pH}$	−1.40	7.90	1	7.90	14.58	0.0124	−7.56	3.64	Significant
$X_T * X_D$	0.83	2.77	1	2.77	5.12	0.0731	4.48	1.27	
$X_{pH} * X_D$	0.092	0.034	1	0.034	0.063	0.8115	0.49	0.015	
Residual		2.71	5	1.10					
Lack of fit		2.22	3	1.82	3.07	0.2553			Not significant
Pure error		0.48	2	0.015					

(0.0003) less than 0.05 at 95% confidence were obtained. The F-value is the ratio between the mean square of the model and the residual error [49] and a considerable value of F with low p-value suggests the significance of the chosen model. The calculated F-value was also higher than the critical F-value ($F_{0.05,df,(n-df+1)}$) = 4.77 for 9 degrees of freedom (df), indicating that the model is valid for navigation of design space. The lack of fit was observed to be insignificant ($p = 0.2553$), which otherwise generally indicates misfit of the selected model. The robustness in the model prediction of diclofenac adsorption by the nanoadsorbent was further testified by its good precision value which was obtained as 17.32. Adequate precision is a measure of signal to noise ratio, and a ratio greater than 4 is desirable to conclude the presence of an adequate signal in the model [50]. The R^2 value represents the ratio of the explained variation to the total variation of the response [51]. The R^2 for the chosen model was 0.987, which indicates only 1.3% of the total variation was not satisfactorily explained by the model. The R_{adj}^2 statistic helps in correcting the number of unnecessary parameters which may otherwise erroneously increase the R^2 in the model. The R_{adj}^2 value obtained herein was 0.963 and is close to unity which nullifies the need for further model reduction. The predictive ability of the model was also evaluated by the root mean square error of prediction (RMSEP) and relative standard error of prediction (RSEP):

$$\mathrm{RMSEP} = \sqrt{\frac{\sum_{i=1}^{N}\left(y_{\mathrm{pred},i} - y_{\mathrm{meas},i}\right)^2}{N}} \tag{7}$$

$$\mathrm{RSEP} = \sqrt{\frac{\sum_{i=1}^{N}\left(y_{\mathrm{pred},i} - y_{\mathrm{meas},i}\right)^2}{\sum_{i=1}^{N}\left(y_{\mathrm{meas},i}\right)^2}} \times 100 \tag{8}$$

where $y_{\mathrm{pred},i}$ and $y_{\mathrm{meas},i}$ represent the model-predicted and measured values of the response variable and N represents the number of experimental observations. Low values of RMSEP (0.43) and RSEP (3.98) were obtained for the selected model, which indicates model adequacy in predicting the adsorption capacity of the nanoadsorbent. The chi-square (χ^2) test was also calculated to ensure the fitness of the selected model.

$$\chi^2 = \sum_{i=1}^{N}\frac{\left(y_{\mathrm{meas},i} - y_{\mathrm{pred},i}\right)^2}{y_{\mathrm{pred},i}} \tag{9}$$

The calculated chi-square value ($\chi_{\mathrm{cal}}^2 = 0.31$) was found to be less than the critical value ($\chi_{\mathrm{crit}}^2 = 23.69$), suggesting a satisfactory model fit with no significant difference between the observed and predicted responses. Hence, the model was found to be statistically significant and best suited for the prediction of diclofenac adsorption by the developed nanoadsorbent. The correlation plot (Fig. 7a) between the experimental and model-predicted adsorption capacities depicted that the points lie close to each other, thus, confirming model stability and good prediction ability of the model.

The significance of the model components was tested by the Student's *t*-test and the *p*-value. The *t* statistic is a ratio of the estimated parameter effect to the estimated parameter standard deviation and a large magnitude *t* with small *p*-value (<0.05) statistically signifies the model coefficient. As it can be seen in the ANOVA result, the linear terms X_T and X_{pH}, all the quadratic terms X^2_T, X^2_{pH}, X^2_D and the interactive term $X_T * X_{pH}$ were the statistically significant model terms. The positive sign of the model coefficients indicates their synergistic effects [41]. Hence, X_T, X_D, $X_T * X_D$, $X_{pH} * X_D$ were seen to influence the adsorption of diclofenac by the nanoadsorbent positively. The per cent contribution (PC) of each model term was computed using the sum of squares (SS) of the corresponding term to that of SS for all the terms:

$$PC = \frac{SS}{\sum SS} \times 100$$

The PC values given in Table 4 show that solution pH contributed the maximum to diclofenac adsorption (47.73%) followed by the quadratic terms, X^2_{pH} (14.99%) and X^2_T (12.71%). Hence, it may be inferred that the application of RSM was most

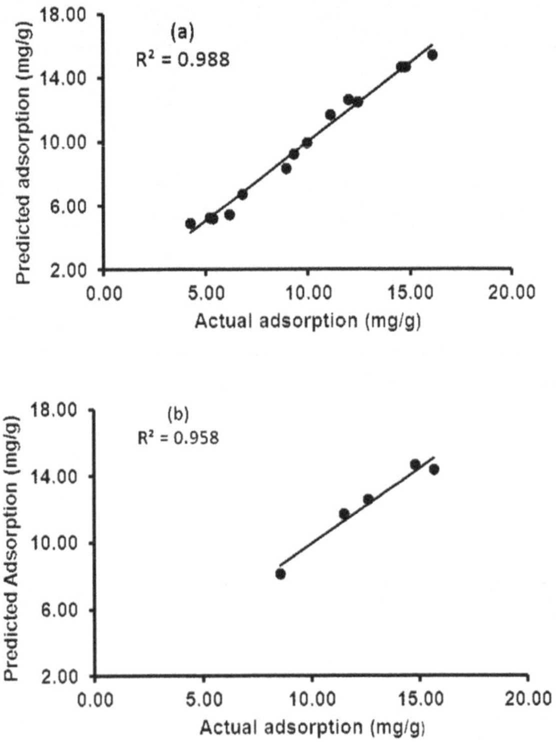

Fig. 7 Plot of the model-predicted and experimental response values from BBD matrix (**a**), validation set (**b**)

suitable to elucidate the quadratic and combined effects of the process parameters in adsorption of diclofenac by the developed magnetic nanoadsorbent.

3.3 RSM Plots for Influence of Process Parameters

The interactive effects of the process parameters have been graphically represented in three-dimensional (3D) response surface plots drawn in the original scale (Fig. 8). A curvature in the RSM plots testifies the significance of all the quadratic model terms during the ANOVA test. The 3D plot in Fig. 8a shows the interactive influence of temperature and pH on diclofenac adsorption by the magnetic nanoadsorbent. The adsorption process was observed to proceed synergistically with an increase in temperature from 20 to 50 °C at the lowest pH (5). At this condition, the maximum adsorption of diclofenac was observed to be 15.89 mg g^{-1} at nanoadsorbent dose of 1 g L^{-1}.

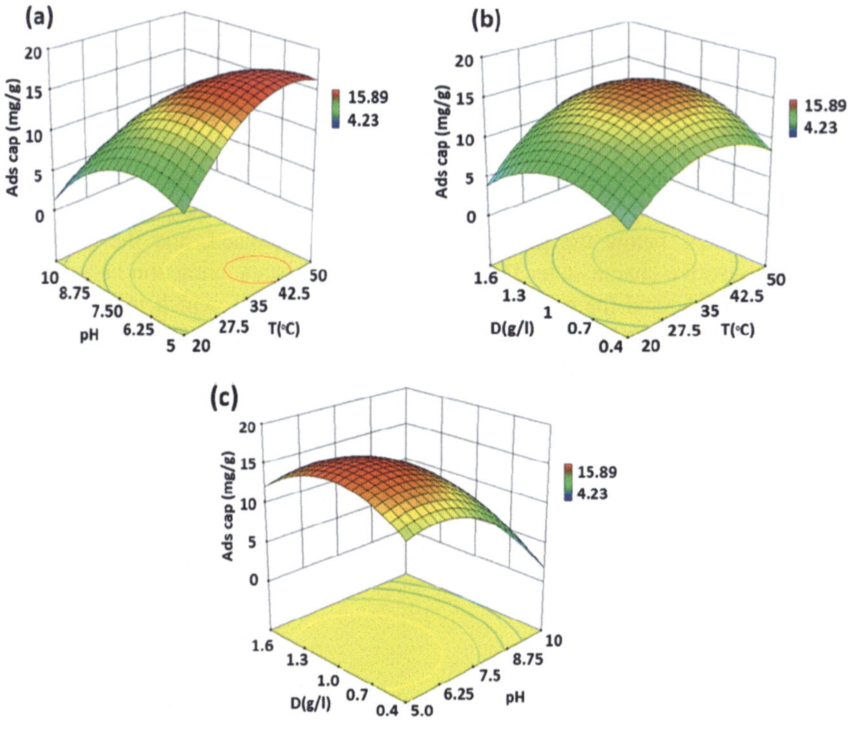

Fig. 8 3D RSM plots showing the interactive effect of process variables: temperature and pH at 1 g L^{-1} MAC dose (**a**), temperature and MAC dose at pH7.5 (**b**), pH and MAC dose at temperature 35 °C (**c**)

Contrary to this observation, a study [18] reported that an increase in temperature caused an increase in diclofenac hydrophilicity which lowered its adsorption affinity towards grape bagasse adsorbent. On the other hand, another study [19] reported an athermic adsorption process where the solvated aqueous diclofenac molecules were not affected by temperature fluctuations for activated carbon sorption system. However, a similar temperature trend in diclofenac uptake was reported [24] where an increase in temperature increased the mobility of diclofenac molecules, enabling them to reach the inner sorption sites of the adsorbent. A similar mechanism may be responsible for the observed endothermic behaviour in diclofenac adsorption by the magnetic nanoadsorbent. The 3D plot in Fig. 8b depicts the trend in diclofenac adsorption influenced by temperature and adsorbent dose at pH 7.5. It can be seen that the diclofenac uptake was again favoured at a higher temperature and a nanoadsorbent dose level of 1 g L^{-1}. The limit in the uptake of diclofenac despite an increase in the nanoadsorbent dose from 1 to 1.6 g L^{-1} could be due to saturation of the sorption sites and fixed availability of diclofenac (60 mg L^{-1}). Moreover, at pH 7.5, the nanoadsorbent develops negative surface charge (pH_{zpc} = 6.8) which does not favour electrostatic interaction with the anionic diclofenac. This may contribute to the limited adsorption despite the increase in the nanoadsorbent dose. The interactive influence of pH and the nanoadsorbent dose is shown in Fig. 8c. It is visible that the sorption process is much favourable at higher nanoadsorbent dose when the solution pH is low 5. Hence, it may be justified that the influence of solution pH was found to be most significant for diclofenac adsorption in the present study. This can be explained by the surface charge states of the adsorbate-adsorbent system as a function of pH [18, 21]. Diclofenac is an acidic drug having pK_a value of 4.08. It exists in the anionic state from the dissociation of its carboxylic group when solution pH exceeds the pK_a value. The pH_{zpc} of the magnetic nanoadsorbent was found to be 6.8. Hence, at pH conditions below the pH_{zpc}, the surface is protonated, and the adsorbent acquires a positive charge state. This induces an electrostatic attraction between the anionic diclofenac and the positively charged nanoadsorbent, which increases their binding affinity. However, at higher pH conditions and particularly when pH exceeds the pH_{zpc}, the nanoadsorbent suffers charge reversal and a drop in adsorption capacity is thus observed. A study conducted on the biosorption of diclofenac onto the biochar was unaffected within the working pH conditions (5–9) [21]. The biosorbent exhibited a high pH_{zpc} of around 9, thereby providing a positive charge to the surface of the biochar in the pH range studied. This favoured electrostatic interactions with the negatively charged diclofenac throughout in addition to the pi-pi dispersive forces between the aromatic ring of the molecule and that of the pi electrons in the sorbent structure.

3.4 *Process Optimization and Model Validation*

Process optimization involves finding the factor settings from the design space, which yields the desired level of response. For optimization of diclofenac adsorption by the magnetic nanoadsorbent, the process parameters as input variables were set to "in range" while the response was set to "maximize" in the input goals. The goals are combined to obtain an overall desirability function, and the optimization algorithm in Design-Expert maximizes this function by starting at random points and proceeding to a maximum via the steepest slope. The optimum process conditions identified by the model were temperature conditions of 42.65 °C, pH of 5.74 and the magnetic nanoadsorbent dose of 1.04 g L^{-1}. Under such optimized conditions, the adsorption capacity of the magnetic nanoadsorbent for 60 mg L^{-1} diclofenac concentration was predicted to be 16.04 mg g^{-1}. This was verified in the laboratory, and corresponding adsorption capacity of 15.31 mg g^{-1} was obtained under experimental conditions.

In order to ensure that the fitted model provides an adequate approximation to the real system, the model was validated by an external dataset. For this, batch experiments comprising five different adsorption assays were performed. The levels of the process variables were chosen at random combinatorial levels within the experimental range and transformed into their respective coded values using Eq. (2). The coded values of the process variables were employed in predicting the response by using the polynomial regression model in Eq. (6). The prediction accuracy was adjudged in terms of R^2, RMSEP and RSEP values determined from the experimental and model-predicted adsorption capacity values. Results showed good agreement between the experimentally obtained and model-predicted datasets (Fig. 7b) with high R^2 and low RMSEP and RSEP values of 0.958, 0.75 and 5.89, respectively. Thus, the selected regression model was suitable for describing the behavioural pattern in the diclofenac-nanoadsorbent sorption system.

3.5 Adsorption Isotherms and Kinetics

The nature of diclofenac adsorption by the magnetic nanoadsorbent was investigated by performing equilibrium studies by the two popular isotherm models, namely the Langmuir and Freundlich isotherms. The Langmuir adsorption isotherm model assumes monolayer adsorption onto the adsorbent surface, having identical and uniform energy sites [52–55]. On the other hand, the Freundlich adsorption isotherm model assumes multilayer adsorption with non-uniform distribution of heat and affinities over heterogeneous surfaces [56–60]. The adsorption isotherm experiments were performed, keeping an initial diclofenac concentration within a range of 20–60 mg L^{-1}. Other process parameters were kept close to the predetermined optimum conditions at a temperature of 40 °C, pH 5 and magnetic nanoadsorbent dose of 1 g L^{-1}. The resultant data were fitted into the two isotherm model equations by linear regression as given in Table 6. The suitability of the models was evaluated based on the regression coefficients (R^2) and average per cent error (APE) as:

Table 6 Isotherm parameters for DIC adsorption by MAC

Isotherm	Equation	Parameters	R^2	APE
Langmuir	$\frac{C_e}{q_e} = \frac{1}{Q_0 b} + \frac{1}{Q_0} C_e$	Q_0 (mg g^{-1}): 26.31 b (L mg^{-1}): 0.04 R_L: 0.33	0.963	2.93
Freundlich	$\log q_e = \log K_F + \frac{1}{n} \log C_e$	K_F (mg g^{-1}): 2.51 n: 1.97	0.949	3.54

$$\mathrm{APE}(\%) = \frac{100}{n}\left[\sum_{i=1}^{n} \frac{y_{i,\mathrm{pred}} - y_{i,\mathrm{exp}}}{y_{i,\mathrm{exp}}}\right] \tag{10}$$

where $y_{i,\mathrm{pred}}$ is the predicted response, $y_{i,\mathrm{exp}}$ is the experimental response and n is the number of data points. The high R^2 (0.967) and low average per cent error (APE = 2.93) observed for the Langmuir isotherm model indicated that diclofenac adsorption by the magnetic nanoadsorbent exhibited monolayer formation and the monolayer adsorption capacity (Q_0) of 26.31 mg g^{-1} was observed. The dimensionless constant separation factor (R_L) obtained as $R_L = 1/(1 + bC_0)$ is an indicator to assess the favorability of adsorption [36, 57, 61, 62]. Since the value of R_L was between 0 and 1, the adsorption of diclofenac by the magnetic nanoadsorbent was a favourable process. The Langmuir and Freundlich adsorption isotherm curves have been given in Fig. 9a, b. The adsorption affinity of other sorbents towards diclofenac reported in the literature has been 7.25 mg g^{-1} in case of biochar from garden waste [21], 100.67 mg g^{-1} in case of biochar obtained from chemically activated moringa seed powder [20] and 653.91 mg g^{-1} by graphene oxide (GO) [63]. In the last study, the high adsorption capacity of GO was observed for a high initial diclofenac concentration of 400 mg L^{-1} and presence of carboxyl, alkoxyl and hydroxyl groups in GO, responsible for its affinity towards the drug molecule. In order to understand the rate of diclofenac uptake by the magnetic nanoadsorbent, kinetic models such as the pseudo-first-order and pseudo-second-order reaction models were applied to the kinetic data. The process conditions for the kinetic experiments were similar to the isotherm study at a fixed DIC concentration of 50 mg L^{-1}. Samples were withdrawn at regular intervals, and the residual DIC concentration was monitored over the intermittent contact period. The parameters of the kinetic models were estimated by fitting the experimental data to respective linearized expressions of pseudo-first-order and pseudo-second-order model as given in Table 7 and curves representing the two kinetic equations have been given in Fig. 9c, d. More excellent agreement given by higher R^2 (0.949) was shown to exist between the experimentally obtained and the pseudo-second-order model derived q_e values. The better fit obtained for the pseudo-second-order kinetic model suggests that the rate-limiting step in the interaction of diclofenac and the magnetic nanoadsorbent was essential of chemical nature. The pseudo-second-order kinetics assumes that interaction between the adsorbate and adsorbent involves

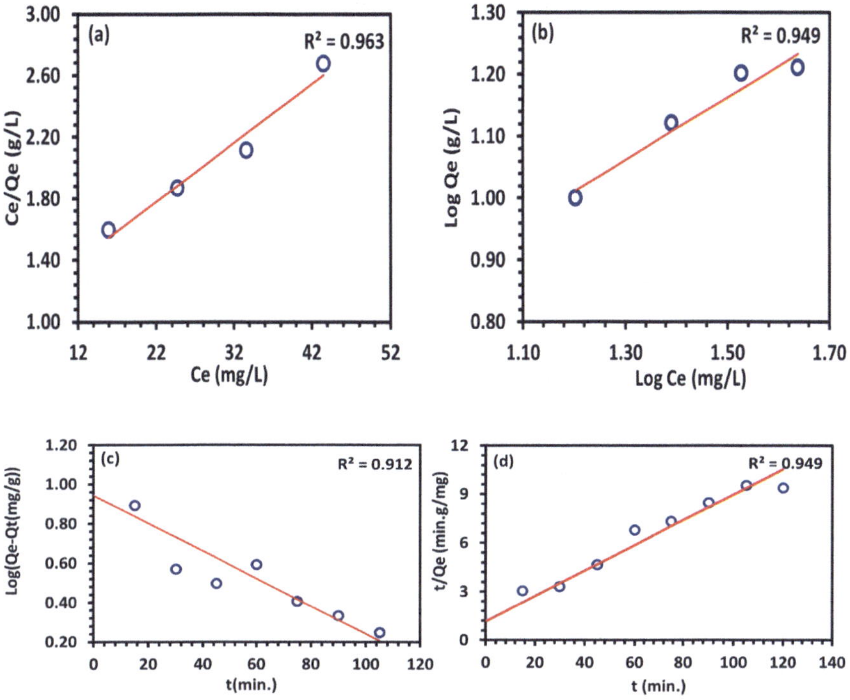

Fig. 9 Adsorption isotherms and adsorption kinetics of DIC onto MAC **a** Langmuir adsorption isotherm, **b** Freundlich adsorption isotherm, **c** pseudo-first-order reaction model, **d** pseudo-second-order reaction model

Table 7 Kinetic parameters for DIC adsorption by MAC

Order of reaction	Equation	Parameters	R^2
		q_{exp} (mg g^{-1}): 12.74	
Pseudo-first-order	$\log(q_e - q_t) = \log_e - \frac{k_1}{2.303}t$	q_e (mg g^{-1}): 9.14 k_1 (min^{-1}): 0.016	0.912
Pseudo-second-order	$\frac{t}{q_t} = \frac{1}{k_2 q_e^2} + \frac{1}{q_e}t$	q_e (mg g^{-1}): 12.82 k_2 (g mg^{-1} min^{-1}): 0.00513 h (mg g^{-1} min^{-1}): 0.8277	0.949

valence forces through sharing or exchange of electrons [36]. The interactive mechanism was investigated through FTIR analysis of the magnetic adsorbent after diclofenac adsorption. Similar sorption mechanism was also seen in recent studies where the adsorption of diclofenac by various adsorbents also followed the Langmuir adsorption isotherm and pseudo-second-order kinetics [21, 63].

3.6 FTIR Analysis of Diclofenac Loaded Magnetic Nanoadsorbent

The FTIR spectrum of the magnetic nanoadsorbent after diclofenac adsorption has been given in Fig. 6c. It can be seen that although the presence of original bands was preserved after adsorption, there was a significant shift in their positions along with a change in intensities of some bands. The broad OH band suffered broadening and was shifted to 3411.62 cm^{-1} from 3418.3 cm^{-1}. This could be possibly due to the displacement of water lattice by the adsorbed diclofenac molecules. A significant shift was observed for the C=C bond from 1634.7 cm^{-1} to higher frequency at 1644.02 cm^{-1} after diclofenac adsorption. The presence of the two chlorine groups in the diclofenac structure induces the electron-withdrawing effect on the adjacent aromatic ring. This could result in $\pi - \pi$ interactions between the electron-deficient aromatic ring in diclofenac and the aromatic rings of the basal graphitic carbon matrix in the magnetic nanoadsorbent. A new and diagnostic peak was formed at 1384.75 cm^{-1} after diclofenac adsorption. The vibration at 1384.75 cm^{-1} is indicative of symmetric stretching from carboxylate group in adsorbed DIC. Hence, it can be inferred that diclofenac was adsorbed onto the magnetic nanoadsorbent mostly by the COO group ligand either by electrostatic interactions or by complexation with Fe since the peak for FeO group at 475.3 cm^{-1} completely disappeared after adsorption by the magnetic nanoadsorbent.

3.7 Desorption Potential

The desorption studies signify the reuse potential of the spent adsorbent. The magnetic nanoadsorbent after adsorption of diclofenac was collected by an external magnet and gently washed with 5 ml Milli-Q water to eliminate the trapped diclofenac molecules loosely impounded in between the particles. The washed adsorbent was contacted with separate solutions of methanol (CH_3OH), Milli-Q water and 0.1 M NaOH. The mixtures were shaken in the water bath shaker at room temperature for 48 h so that maximum desorption took place. The concentration of the desorbed diclofenac in the stripping solvents was analysed by UV–Vis spectrophotometer. The final desorption efficiency (%) of the magnetic nanoadsorbent was calculated using Eq. (11) [34]:

$$\mathrm{DE\%} = \frac{C_0 - C_d}{C_0} \times 100 \tag{11}$$

where C_0 and C_d are the initially adsorbed and residual diclofenac concentration (mg L^{-1}) after desorption, respectively.

Results from desorption experiments indicated that maximum diclofenac desorption took place by 0.1 M NaOH (69.88%) followed by CH_3OH (35.73%) and Milli-Q water (11.4%). A similar trend in desorption was observed in the case of ibuprofen [41]. The highest desorption per cent by NaOH is indicative of the interactive forces being mainly regulated by electrostatic mechanism and hydrophobic interactions between diclofenac and the magnetic nanoadsorbent.

4 Conclusions

PET wastes were successfully converted into activated carbon by CO_2 activation. The developed magnetic nanoadsorbent (carbon/iron composite) showed ferromagnetic behaviour and exhibited an adsorption capacity of 15.89 m g^{-1} (removal percentage of 24.82%) towards diclofenac at 50 mg L^{-1} from water. After process optimization by a second-order response surface model, the adsorption capacity increased to 16.04 mg g^{-1} under process conditions of 42.65 °C temperature, 5.74 pH and magnetic nanoadsorbent dose of 1.04 g L^{-1}. The corresponding adsorption capacity was experimentally determined to be 15.31 mg g^{-1}. The adsorption process was most influenced by solution pH suggesting electrostatic forces being the dominant adsorption mechanism. FTIR analysis of the diclofenac loaded magnetic nanoadsorbent confirmed the adsorption of diclofenac. The magnetic nanoadsorbent showed considerable regeneration potential with good magnetic manoeuvrability and hence can be used as a potential adsorbent in the removal of diclofenac from water.

5 Future Prospectives

In the future, the generation of solid waste is expected to grow in proportion to the increasing population in all the countries across the globe. As waste plastic forms a significant part of the solid waste, recycling technologies are always on the upfront of research. The study herein reveals essential aspects of the potential of waste PET recycling. The waste PET can be suitably modified into a range of highly economical value-added products with alternate applications. For instance, the application of the developed adsorbent can be channelized in the analytical field for the development of inexpensive carbon loaded porous sorbent in extraction techniques. The graphitic nature of the PET activated carbon observed herein indicates its potential for high electrical conductivity, with a good possibility of it to be a substitute to the conventional metal electrodes. Fabrication of activated carbon electrode material from waste PET for detection of emerging contaminants in the environment would be a significant avenue for the application of waste PET activated carbon. The adsorbing potential of the prepared nanomaterial from PET waste can also be tested for sequestration of flue gases, thereby protecting the environment. Hence, the recycled PET nanoadsorbent renders multifaceted propositions for its application.

Acknowledgements The authors would like to express thankfulness to the Director, CSIR-Indian Institute of Toxicology Research, Lucknow (India) for his kind support in this work. The research was a part of Ph.D. work supported by UGC (India)-Senior Research Fellowship.

References

1. Fent K, Weston A, Caminada D (2006) Ecotoxicology of human pharmaceuticals. Aquat Toxicol 76(2):122–159
2. Schwaiger J, Ferling H, Mallow U, Wintermayr H, Negele R (2004) Toxic effects of the non-steroidal anti-inflammatory drug diclofenac: Part I: histopathological alterations and bioaccumulation in rainbow trout. Aquat Toxicol 68(2):141–150
3. Triebskorn R, Casper H, Heyd A, Eikemper R, Köhler H-R, and Schwaiger J (2004) Toxic effects of the non-steroidal anti-inflammatory drug diclofenac: part II. Cytological effects in liver, kidney, gills and intestine of rainbow trout (Oncorhynchus mykiss). Aquatic Toxicol 68 (2):151–166
4. Sousa JCG, Ribeiro AR, Barbosa MO, Pereira MFR, Silva AMT (2018) A review on environmental monitoring of water organic pollutants identified by EU guidelines. J Hazard Mater 344:146–162
5. Zhang J, Chen S, Yang R, Yan Y (2010) Biodiesel production from vegetable oil using heterogenous acid and alkali catalyst. Fuel 89(10):2939–2944
6. Taggart MA, Senacha KR, Green RE, Cuthbert R, Jhala YV, Meharg AA, Mateo R, Pain DJ (2009) Analysis of nine NSAIDs in ungulate tissues available to critically endangered vultures in India. Environ Sci Technol 43(12):4561–4566
7. Krishnaswamy K, Kumar BD (2005) Drug consumption pattern in urban and rural areas of India and their health implications. Citeseer
8. Shanmugam G, Sampath S, Selvaraj KK, Larsson DGJ, Ramaswamy BR (2013) Non-steroidal anti-inflammatory drugs in Indian rivers. Environ Sci Pollut Res 21(2):921–931
9. Singh KP, Rai P, Singh AK, Verma P, Gupta S (2014) Occurrence of pharmaceuticals in urban wastewater of north Indian cities and risk assessment. Environ Monit Assess 186 (10):6663–6682
10. Ziylan A, Ince NH (2011) The occurrence and fate of anti-inflammatory and analgesic pharmaceuticals in sewage and fresh water: treatability by conventional and non-conventional processes. J Hazard Mater 187(1–3):24–36
11. Heberer T (2002) Occurrence, fate, and removal of pharmaceutical residues in the aquatic environment: a review of recent research data. Toxicol Lett 131(1–2):5–17
12. Zhang Y, Geißen S-U, Gal C (2008) Carbamazepine and diclofenac: removal in wastewater treatment plants and occurrence in water bodies. Chemosphere 73(8):1151–1161
13. Ternes TA, Herrmann N, Bonerz M, Knacker T, Siegrist H, Joss A (2004) A rapid method to measure the solid–water distribution coefficient (Kd) for pharmaceuticals and musk fragrances in sewage sludge. Water Res 38(19):4075–4084
14. Mailler R, Gasperi J, Coquet Y, Buleté A, Vulliet E, Deshayes S, Zedek S, Mirande-Bret C, Eudes V, Bressy A, Caupos E, Moilleron R, Chebbo G, Rocher V (2016) Removal of a wide range of emerging pollutants from wastewater treatment plant discharges by micro-grain activated carbon in fluidized bed as tertiary treatment at large pilot scale. Sci Total Environ 542:983–996

15. Habeeb OA, Ramesh K, Ali GAM, Yunus RM (2017) Low-cost and eco-friendly activated carbon from modified palm kernel shell for hydrogen sulfide removal from wastewater: adsorption and kinetic studies. Desalin. Water Treat 84:205–214
16. Habeeb OA, Ramesh K, Ali GAM, Yunus RM, Olalere OA (2017) Kinetic, isotherm and equilibrium study of adsorption capacity of hydrogen sulfide-wastewater system using modified eggshells. IIUM Eng J 18(1):13–25
17. Habeeb OA, Ramesh K, Ali GAM, Yunus RM (2017) Experimental design technique on removal of hydrogen sulfide using CaO-eggshells dispersed onto palm kernel shell activated carbon: experiment, optimization, equilibrium and kinetic studies. J Wuhan Univ Technol-Mater Sci Ed 32(2):305–320
18. Antunes M, Esteves VI, Guégan R, Crespo JS, Fernandes AN, Giovanela M (2012) Removal of diclofenac sodium from aqueous solution by Isabel grape bagasse. Chem Eng J 192:114–121
19. Baccar R, Sarrà M, Bouzid J, Feki M, Blánquez P (2012) Removal of pharmaceutical compounds by activated carbon prepared from agricultural by-product. Chem Eng J 211–212:310–317
20. Bagheri A, Abu-Danso E, Iqbal J, Bhatnagar A (2019) Modified biochar from Moringa seed powder for the removal of diclofenac from aqueous solution. Environ Sci Pollut Res 27 (7):7318–7327
21. Li Y, Taggart MA, McKenzie C, Zhang Z, Lu Y, Pap S, Gibb S (2019) Utilizing low-cost natural waste for the removal of pharmaceuticals from water: Mechanisms, isotherms and kinetics at low concentrations. J Clean Prod 227:88–97
22. Abo El Naga AO, El Saied M, Shaban SA, El Kady FY (2019) Fast removal of diclofenac sodium from aqueous solution using sugar cane bagasse-derived activated carbon. J Mol Liq 285:9–19
23. Sotelo JL, Ovejero G, Rodríguez A, Álvarez S, Galán J, García J (2014) Competitive adsorption studies of caffeine and diclofenac aqueous solutions by activated carbon. Chem Eng J 240:443–453
24. Suriyanon N, Punyapalakul P, Ngamcharussrivichai C (2013) Mechanistic study of diclofenac and carbamazepine adsorption on functionalized silica-based porous materials. Chem Eng J 214:208–218
25. Habeeb OA, Ramesh K, Ali GAM, Yunus RM (2017) Isothermal modelling based experimental study of dissolved hydrogen sulfide adsorption from waste water using eggshell based activated carbon. Malay J Anal Sci 21(2):334–345
26. Habeeb OA, Ramesh K, Ali GAM, Yunus RM, Thanusha TK, Olalere OA (2016) Modeling and optimization for H_2S adsorption from wastewater using coconut shell based activated carbon. Aust J Basic Appl Sci 10(17):136–147
27. Habeeb OA, Ramesh K, Ali GAM, Yunus RM (2017) Optimization of activated carbon synthesis using response surface methodology to enhance H_2S removal from refinery wastewater. J Chem Eng Indus Biotech 1(1):1–17
28. Murugan S, Gu S (2015) Research and development activities in pyrolysis–contributions from Indian scientific community–a review. Renew Sustain Energy Rev 46:282–295
29. de Mello D, Pezzin SH, Amico SC (2009) The effect of post-consumer PET particles on the performance of flexible polyurethane foams. Polym Testing 28(7):702–708
30. Possatto FE, Barletta M, Costa MF, Ivar do Sul JA, and Dantas DV (2011) Plastic debris ingestion by marine catfish: an unexpected fisheries impact. Marine Pollut Bull 62(5):1098–1102
31. Parra JB, Ania CO, Arenillas A, Rubiera F, Pis JJ, Palacios JM (2006) Structural changes in polyethylene terephthalate (PET) waste materials caused by pyrolysis and CO2 activation. Adsorpt Sci Technol 24(5):439–450
32. Rai P, Singh KP (2018) Valorization of Poly (ethylene) terephthalate (PET) wastes into magnetic carbon for adsorption of antibiotic from water: characterization and application. J Environ Manage 207:249–261

33. Patterson AL (1939) The Scherrer formula for X-ray particle size determination. Phys Rev 56 (10):978–982
34. Abdel Ghafar HH, Ali GAM, Fouad OA, Makhlouf SA (2013) Enhancement of adsorption efficiency of methylene blue on Co_3O_4/SiO_2 nanocomposite. Desalin Water Treat 53 (11):2980–2989
35. Franz M, Arafat HA, Pinto NG (2000) Effect of chemical surface heterogeneity on the adsorption mechanism of dissolved aromatics on activated carbon. Carbon 38(13):1807–1819
36. Sadegh H, Ali GAM, Makhlouf ASH, Chong KF, Alharbi NS, Agarwal S, Gupta VK (2018) MWCNTs-Fe_3O_4 nanocomposite for Hg(II) high adsorption efficiency. J Mol Liq 258:345–353
37. Kayan B, Gözmen B (2012) Degradation of Acid Red 274 using H_2O_2 in subcritical water: application of response surface methodology. J Hazard Mater 201–202:100–106
38. Ferreira SLC, Bruns RE, Ferreira HS, Matos GD, David JM, Brandão GC, da Silva EGP, Portugal LA, dos Reis PS, Souza AS, dos Santos WNL (2007) Box-Behnken design: an alternative for the optimization of analytical methods. Anal Chim Acta 597(2):179–186
39. Mourabet M, El Rhilassi A, El Boujaady H, Bennani-Ziatni M, Taitai A (2017) Use of response surface methodology for optimization of fluoride adsorption in an aqueous solution by Brushite. Arab J Chem 10:S3292–S3302
40. Habeeb OA, Kanthasamy R, Ali GAM, Yunus RM (2017) Application of response surface methodology for optimization of palm kernel shell activated carbon preparation factors for removal of H_2S from industrial wastewater. J Teknologi 79(7)
41. Singh KP, Singh AK, Singh UV, Verma P (2011) Optimizing removal of ibuprofen from water by magnetic nanocomposite using box-Behnken design. Environ Sci Pollut Res 19 (3):724–738
42. Mourabet M, El Rhilassi A, El Boujaady H, Bennani-Ziatni M, El Hamri R, Taitai A (2015) Removal of fluoride from aqueous solution by adsorption on hydroxyapatite (HAp) using response surface methodology. J Saudi Chem Soc 19(6):603–615
43. Sing KSW, Everett DH, Haul RAW, Moscou L, Pierotti RA, Rouquerol J, Siemieniewska T (2008) Reporting physisorption data for gas/solid systems. In: Handbook of heterogeneous catalysis, Wiley
44. Lian F, Xing B, Zhu L (2011) Comparative study on composition, structure, and adsorption behavior of activated carbons derived from different synthetic waste polymers. J Colloid Interface Sci 360(2):725–730
45. Bratek W, Świątkowski A, Pakuła M, Biniak S, Bystrzejewski M, Szmigielski R (2013) Characteristics of activated carbon prepared from waste PET by carbon dioxide activation. J Anal Appl Pyrol 100:192–198
46. Asuha S, Zhao S, Wu HY, Song L, and Tegus O (2009) One step synthesis of maghemite nanoparticles by direct thermal decomposition of Fe-urea complex and their properties. Chem Inform 40(24)
47. Cansado IPP, Gonçalves FAMM, Nabais JMV, Ribeiro Carrott MML, Carrott PJM (2009) PEEK: an excellent precursor for activated carbon production for high temperature application. Fuel Process Technol 90(2):232–236
48. Ji Z, Shen X, Song Y, Zhu G (2011) In situ synthesis of graphene/cobalt nanocomposites and their magnetic properties. Mater Sci Eng B 176(9):711–715
49. Da'na E, Sayari A (2011) Optimization of copper removal efficiency by adsorption on amine-modified SBA-15: experimental design methodology. Chem Eng J 167(1):91–98
50. Zhang H, Du Z, Ji Y, Mei M (2013) Simultaneous trace determination of acidic non-steroidal anti-inflammatory drugs in purified water, tap water, juice, soda and energy drink by hollow fiber-based liquid-phase microextraction and ultra-high pressure liquid chromatography coupled to tandem mass spectrometry. Talanta 109:177–184
51. Sahoo C, Gupta AK (2012) Optimization of photocatalytic degradation of methyl blue using silver ion doped titanium dioxide by combination of experimental design and response surface approach. J Hazard Mater 215–216:302–310

52. Langmuir I (1916) The constitution and fundamental properties of solids and liquids. Part I. Solids. J Am Chem Soc 38(11):2221–2295
53. Lee SP, Ali GAM, Algarni H, Chong KF (2019) Flake size-dependent adsorption of graphene oxide aerogel. J Mol Liq 277:175–180
54. Sadegh H, Gama ZA, Nadagoud MN (2017) Adsorption of ammonium ions onto multi-walled carbon nanotubes. Studia Univ Babes-Bolyai Chemia 62(2):233-245
55. Gupta VK, Agarwal S, Sadegh H, Ali GAM, Bharti AK, Hamdy AS (2017) Facile route synthesis of novel graphene oxide-β-cyclodextrin nanocomposite and its application as adsorbent for removal of toxic bisphenol A from the aqueous phase. J Mol Liq 237:466–472
56. Freundlich H (1906) Over the adsorption in solution. J Phys Chem 57(385471):1100–1107
57. Sadegh H, Ali GAM, Nia HJ, Mahmoodi Z (2019) Nanomaterial surface modifications for enhancement of the pollutant adsorption from wastewater, In: Nanotechnology Applications in Environmental Engineering. IGI Global, pp. 28. https://doi.org/10.4018/978-1-5225-5745-6.ch007
58. Seyed Arabi SM, Lalehloo RS, Olyai MRTB, Ali GAM, Sadegh H (2019) Removal of congo red azo dye from aqueous solution by ZnO nanoparticles loaded on multiwall carbon nanotubes. Physica E 106:150–155
59. Sadegh H, Ali GAM, Gupta VK, Makhlouf ASH, Shahryari-ghoshekandi R, Nadagouda MN, Sillanpää M, Megiel E (2017) The role of nanomaterials as effective adsorbents and their applications in wastewater treatment. J Nanostruct Chem 7(1):1–14
60. Sadegh H, Ali GAM, Agarwal S, Gupta VK (2019) Surface Modification of MWCNTs with carboxylic-to-amine and their superb adsorption performance. Int J Environ Res 13(3):523–531
61. Hamidreza S, Gomaa AMA (2019) Potential applications of nanomaterials in wastewater treatment: nanoadsorbents performance. In: Athar H, Sirajuddin A (eds) Advanced treatment techniques for industrial wastewater. IGI Global, Hershey, PA, USA, pp 51–61
62. Maazinejad B, Mohammadnia O, Ali GAM, Makhlouf ASH, Nadagouda MN, Sillanpää M, Asiri AM, Agarwal S, Gupta VK, Sadegh H (2020) Taguchi L9 (34) orthogonal array study based on methylene blue removal by single-walled carbon nanotubes-amine: adsorption optimization using the experimental design method, kinetics, equilibrium and thermodynamics. J Mol Liq 298:112001
63. Hiew BYZ, Lee LY, Lai KC, Gan S, Thangalazhy-Gopakumar S, Pan G-T, Yang TC-K (2019) Adsorptive decontamination of diclofenac by three-dimensional graphene-based adsorbent: Response surface methodology, adsorption equilibrium, kinetic and thermodynamic studies. Environ Res 168:241–253

Waste Plastic-Based Nanomaterials and Their Applications

Kiran Mustafa, Javaria Kanwal, and Sara Musaddiq

Abstract Plastic is one of the most significant hazards to the environment. Plastic is a non-biodegradable material, and several toxic chemicals leach out of it and seep through the soil, water, plants, and animals. This leads to cancer and other serious ailments, aquatic life endangerment, and environmental pollution. The use of waste plastic for the synthesis of different types of nanomaterials can not only save the humans and the environment from the dangers of plastic but also provide beneficial substances for other purposes. Different nanomaterials can be synthesized from the waste plastics, such as polyvinyl chloride plastic is used as the carbon source for the fabrication of MoC_2 nanoparticles. These particles are remarkable electrocatalysts that are involved in the generation of sustainable hydrogen. Similarly, polypropylene plastic waste is used for the synthesis of photoluminescent carbon nanoparticles, which are employed as important bioimaging devices. This chapter will elaborate on the use of different types of polymeric plastic for the synthesis of a variety of nanomaterials. The potential applications of the fabricated materials will also be discussed.

Keyword Plastic · Polystyrene · Nanoparticles · Energy · Pollution control · Waste recycle · Nanomaterials

Abbreviations

AFM	Atomic force microscopy
BET	Brunauer–Emmett–Teller
BPA	Bisphenol A
CNTs	Carbon nanotubes
CVD	Chemical vapor deposition
DDT	Dichlorodiphenyltrichloroethane
DSC	Differential scanning calorimetry
EM	Electromagnetic
FTIR	Fourier transform infrared spectroscopy

K. Mustafa (✉) · J. Kanwal · S. Musaddiq
Department of Chemistry, The Women University Multan, Multan 66000, Pakistan
e-mail: kiranmustafa100@gmail.com

A. S. H. Makhlouf and G. A. M. Ali (eds.), *Waste Recycling Technologies for Nanomaterials Manufacturing*, Topics in Mining, Metallurgy and Materials Engineering, https://doi.org/10.1007/978-3-030-68031-2_27

$g\text{-}C_3N_4$	Graphitic carbon nitride
HfC	Hafnium carbide
Mo_2C	Molybdenum carbide
MRI	Magnetic resonance imaging
MWCNTs	Multi-walled carbon nanotubes
PAHs	Polycyclic aromatic hydrocarbons
PBDEs	Polybrominated diphenyl ethers
PCBs	Polychlorinated biphenyls
PE	Polyethylene
PET	Polyethylene terephthalate
PMMA	Polymethyl methacrylate
PP	Polypropylene
PS	Polystyrene
PTEF	Polytetrafluoroethylene
PU	Polyurethane
PVA	Polyvinyl alcohol
PVC	Polyvinyl chloride
rGO	Reduced graphene oxide
SBR	Styrene-butadiene rubber
SBS	Styrene-butadiene-styrene
SEM	Scanning electron microscopy
SWCNTs	Single-walled carbon nanotubes
TaC	Tantalum carbide
TEM	Transmission electron microscope
TGA	Thermogravimetric analysis
THF	Tetrahydrofuran
XPS	X-ray photoelectron spectroscopy
XRD	X-ray diffraction

1 Introduction

Plastic polymers constitute a significant amount of waste that is produced across the world. 10% of the total household waste consists of plastic, whereas plastic makes up of 60 to 80% of the beaches and oceans waste [1]. The particles of the plastic found in the marine water could release a variety of organic pollutants such as polychlorinated biphenyls (PCBs), polycyclic aromatic hydrocarbons (PAHs), nonylphenol, polybrominated diphenyl ethers (PBDEs), dichlorodiphenyl-trichloroethane (DDT), and bisphenol A (BPA). These chemicals are not only harmful to the aquatic life, but due to the ingestion and biomagnification, they eventually become a serious threat to humans [2–4] A potential route for the conversion of plastic to the microplastics and its absorption by different life forms is depicted in Fig. 1 [5]. Plastic does not undergo degradation and remains a threat to all life forms for several years [6]. Naturally, plastic can be degraded in four different ways, i.e., photodegradation, thermooxidative degradation, hydrolytic degradation, and biodegradation. The process of degradation starts with photodegradation leading to thermooxidative and hydrolytic degradation, eventually completing with biodegradation. The whole process of degradation is prolonged and can take up to 50 or more years for plastic to completely degrade [7, 8].

For several years, scientists have been searching for different ways to minimize the waste plastic polymers. The most commonly used methods are landfills and incineration [2]. Both methods are undesirable in terms of cost, environmental safety, and availability of the space. Only a fraction of the plastic waste is treated by these methods due to economic viability [9].

There is another approach for the reduction of waste plastic from the earth. This approach involves the production of nanomaterials from the waste plastic polymers [10]. Nanomaterials include a variety of substances like nanoparticles, nanotubes, quantum dots, nanosheets, etc. [11–14]. Nanomaterials have multiple applications in industry, food, agriculture, transportation, medicine, housing, communications, etc. [15–19]. All the nanomaterials have attracted a prevalent interest in nearly all

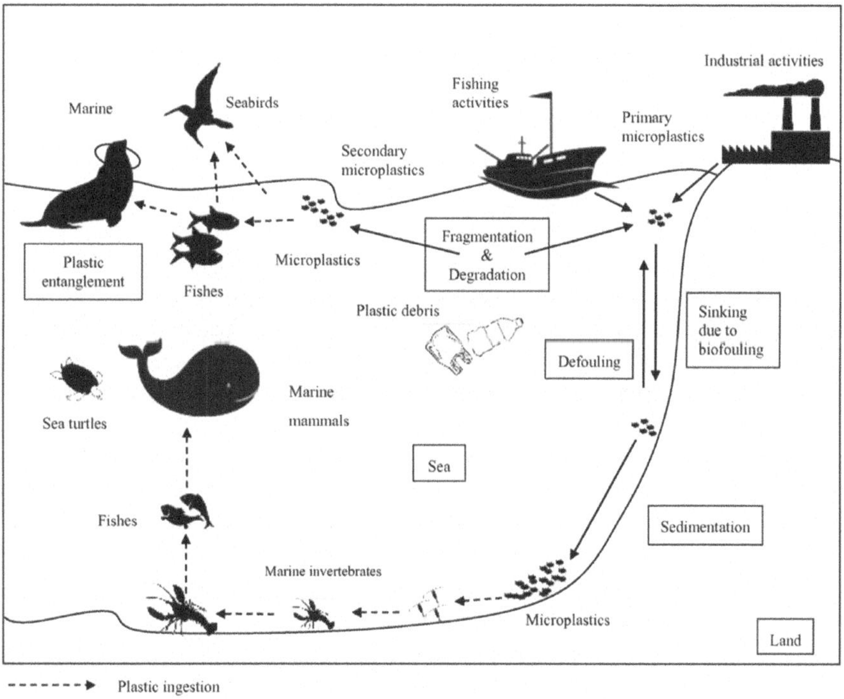

Fig. 1 Potential pathway of plastic debris transportation and its biological interactions. Adapted with permission from Ref. [5] Copyright 2016, Elsevier

domains of science and engineering due to their astonishing physical and chemical properties [20]. Most of the materials and procedures required for the fabrication of the nanomaterials are expensive. The production cost can be reduced by using waste materials for the synthesis of nanomaterials [21–23]. Several waste materials like fruit's peels, spent battery, automotive shredder, crop residue, and building concretes are now used for the synthesis of a variety of nanomaterials [24–28]. Waste plastic can also serve the purpose. It provides important raw material (such as carbonaceous feed) for the synthesis of different nanomaterials. Fabrication of the nanomaterials from waste plastic can serve two important purposes. The plastic pollution can be eliminated, and a useful substance can be obtained from literally the trash. Several researchers have prepared nanoelectrocatalysts by using waste plastic one way or another for the production of sustainable energy [29]. The plastic waste consists of different polymers such as polypropylene (PP), polyethylene (PE), polyvinyl chloride (PVC), polystyrene [30], polyurethane (PU), and polyethylene terephthalate [31, 32]. The present chapter will explain the use of waste plastic polymers for the fabrication of different nanomaterials and the applications of the synthesized nanomaterials because this selective information will be helpful to explore further ventures in the field of recycling of plastic and production of low-cost nanomaterials.

2 Types of Recovered Nanomaterials from Waste Plastic

2.1 Nanoparticles from Waste Plastic and Their Applications

Small scale particles below 100 nm in diameter are referred to as nanoparticles. Based on their characteristics, structural appearance, and shape, they can be categories into various varieties like ceramic nanoparticles, metal nanoparticles, polymeric nanoparticles, and fullerenes. Nanoparticles have peculiar physical and chemical characteristics because of their minimal size and large surface area. Because of absorptivity in the visible region, these nanoparticles show various colors and are considered to be optically active due to their size. The size, structure, and shape of nanoparticles make them very reactive and tough. Nanoparticles are thought to be feasible for widespread applications such as their involvement in catalytic action, energy projects, tomography, therapeutic, and environmental applications [33]. A precise outlook on the widespread applications of nanoparticles is depicted in Fig. 2. Different materials with a variety of origin like biological, organic, inorganic, and the combination of organic and inorganic materials have been used for the synthesis of nanoparticles. Nanoparticles from inorganic materials have been fabricated with variant techniques such as vapor deposition, ball milling, electro-spraying, metallic salts reduction, sol–gel processes, thermos-analysis, and coprecipitation. Organic nanoparticles are usually synthesized by using approaches like dialysis, microemulsion, nanoprecipitation, and via rapid expansion of materials in supercritical solutions. Moreover, biological nanoparticles like viruses, lipoproteins, ferritin, and exosomes are naturally obtained, and the biological nanoparticles can also be prepared from polysaccharides, peptides, and proteins. Similarly, hybrid nanoparticles are fabricated from the organic and inorganic substrates. The coulomb energy, rigidity, superparamagnetism, surface-to-volume ratio, and catalytic action of nanoparticles are usually higher than the respective bulk substances [34]. Tremendous efforts have been made for the fabrication of low-cost, highly effective nanoparticles for various purposes. The use of the plastics for the synthesis of nanoparticles cannot only reduce the cost of processing but will also eradicate the problem of the pollution. Several studies have reported the use of waste plastic for the fabrication of different nanoparticles.

In an investigation, photoluminescent carbon nanoparticles were synthesized from waste plastic bags. The used plastic bags were mainly composed of PP, PVC, or polyvinylidene chloride. These nanoparticles were fabricated in four stages, i.e., thermooxidative degradation, polymerization, carbonization, and passivation. The thermal oxidation of the particles could occur in hydrogen H_2O_2. The oxidized material could accumulate via hydrogen bonding, followed by the subsequent polymerization via aldol condensation or intermolecular dehydration. Intermolecular dehydration led to the formation of hydrophobic carbon cores, and

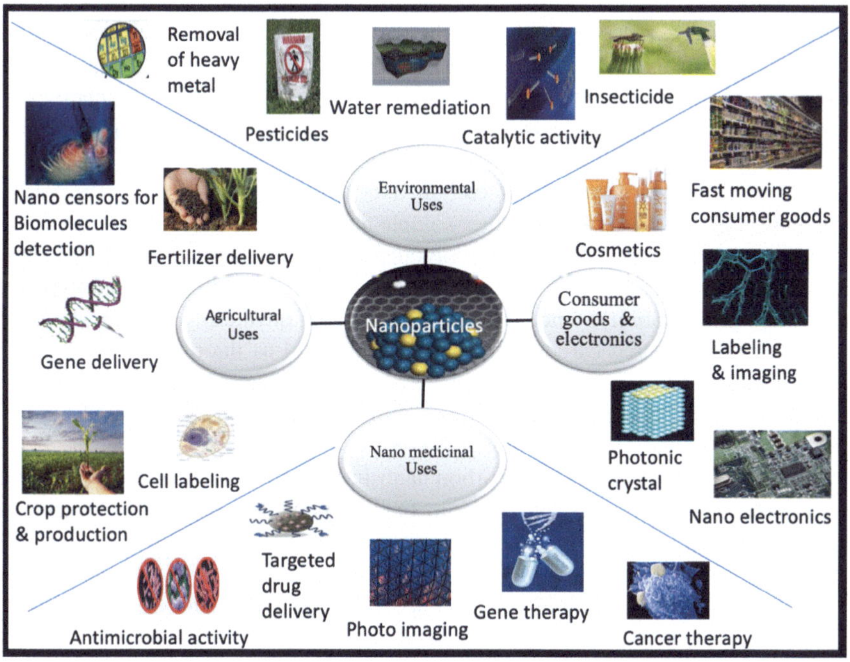

Fig. 2 Applications of nanoparticles

the non-polymerized species in the solution led to the hydrophilic surface passivation, resulting in the formation of oxygen functionalized photoluminescent carbon nanoparticles [35]. The photoluminescent carbon nanoparticles have several applications. They are involved in bioimaging, sensors, drug delivery, optoelectronics, and photocatalysis [36–39].

Expanded PS waste was converted into useful nanoparticles via nanoprecipitation. The method employed for the conversion of PS to nanoparticles was cost-effective and straightforward. In this process, expanded PS was dissolved in THF, which was followed by the addition of water in order to achieve precipitation of PS. Nanoparticles (diameter < 500 nm) thus obtained were later on kept at 50 °C for the evaporation of the excessive solvent. Centrifugation and re-dispersion of the nanoparticles were performed in Milli-Q water in order to achieve the particle separation [40]. PS nanoparticles have several uses, including separation of chemical substances by electrophoresis [41], perturbing of lipid membranes by significantly reducing molecular diffusion in the membranes, hence, serving in the study of cholesterol management, [42] and affecting the coagulation of blood by reducing thrombin generation [43], etc.

Waste PE was employed in an investigation for the fabrication of hafnium carbide (HfC) nanoparticles. The nanoparticles were prepared via the thermal-reduction route in a stainless steel autoclave with waste PE, metallic Li, and hafnium dioxide at 700 °C. XRD analysis of the prepared sample indicated that the nanoparticles of HfC are face-centered cubic. A similar method of preparation was employed for the preparation of HfC with waste PVC and polytetrafluoroethylene (PTFE). The waste plastic was used as the carbonaceous feed in the preparation of the concerned nanoparticles [44]. The HfC has very high melting points and is resistant to corrosion, high phase stability, and low oxygen diffusion coefficient hence is used in aerospace devices [45]. They are also used in spark plasma sintering, which is needed for compacting ceramics [46].

Multifunctionalized fluorescent carbon dots can be prepared from PS waste. Ramanan et al. have synthesized water-soluble nitrogen-doped carbon dots from the waste expanded PS by a single-step solvothermal process. The fabricated carbon dots have shown remarkably small photoluminescence lifetime, low cytotoxicity, promising photostability, and stable luminescence quantum yield in different solutions. The prepared carbon dots were also utilized as fluorescence probes for selective detection of Au^{3+} ions, and 53 n M limit of detection was attained [47]. Carbon dots are frequently used in bioimaging, metallic ion detection (Ag, Hg), optical imaging, and biological works [48–52].

$MgCNi_3$ nanoparticles fixed in carbon nanosheets have been effectively prepared from waste PE in a single-step reaction. For the preparation, these nanoparticles metallic sodium, magnesium chloride, and waste PE were placed in stainless steel autoclave at 800 °C for 10 h. Afterward, the autoclave was cooled to room temperature, and the black product was obtained. TEM and XRD analysis have shown that the $MgCNi_3$ nanoparticles are integrated into the nanosheets, and $MgCNi_3$ has an antiperovskite structure. Magnetization analysis depicts 6.8 K superconducting transition of the synthesized materials [53]. Ternary carbides and nitrides with the antiperovskite structure have several useful properties. They have nearly zero temperature resistivity coefficient, negative thermal expansion, phase separation, and magnetostriction. Owing to these and several other properties, these materials are extensively studied [54, 55].

In a recent study, nanoparticles of molybdenum carbide (Mo_2C) have been prepared with waste PVC. The waste polymer provided the carbonaceous feed as a carbon source, whereas molybdenum sulfide was used for the metallic component. The nanoparticles were prepared at 600 °C in stainless steel autoclave. XRD characterization depicted that the Mo_2C nanoparticles have an orthorhombic crystal structure. SEM and TEM analysis showed the average size of Mo_2C crystalline nanoparticles of 50 nm [56]. Mo_2C are highly active electrocatalysts that are involved in the production of sustainable hydrogen via electrochemical splitting of water. They are also capable of catalyzing hydrodeoxygenation of vanillin and for the production of γ-valerolactone from levulinic acid hydrogenation [57–59].

2.2 *Carbon Nanotubes from Waste Plastic and Their Applications*

Carbon nanotubes are a newly found carbon form consisting of a coiled layer of hexagonally arranged graphite showing as a smooth cylinder. They have a size range between 0.4 and 100 nm while the length is up to 1 mm. Carbon nanotubes are available as single-walled nanotubes (monolayer of the atoms in the above-said arrangement) and as multi-walled nanotubes (multiple layers embedded axially within one another, in the form of matryoshka doll) as shown in Fig. 3. Carbon nanotubes have used a wide range of applications. Particularly, more enormous current bearing potential and mechanical strength of nanotubes as compared to the metals allow them to be used as interconnects in microelectronics. While the considerably high bandwidth of thin single-walled nanotubes proposes their application as small-sized transistor components, besides this, they are considered to be perfectly adapted to low-voltage emitting devices like flat-panel displays because of the small radius curvature at the tips [60].

Carbon nanotubes have enjoyed a prime spot in the different areas of science and engineering since their discovery owing to their remarkable properties such as high tensile strength (100 times more than that of stainless steel), high modulus, low density, high electrical, and thermal conductivities comparable to copper and diamond, respectively [61]. In Fig. 4, the uses of carbon nanotubes in the various field of science are exhibited. Waste plastic of different types has been used for the fabrication of carbon nanotubes. Carbon nanotube has been prepared with vacuum pyrolysis of PP with Fe nanoparticles, where PP, Fe catalyst, and p-xylene were mixed for the formation of the precursor. Precursors were coated with Si wafers, which resulted in the formation of film-shaped species of $\sim$600 nm. Afterward, the prepared specimens were preheated and pyrolyzed in the quartz tube furnace at

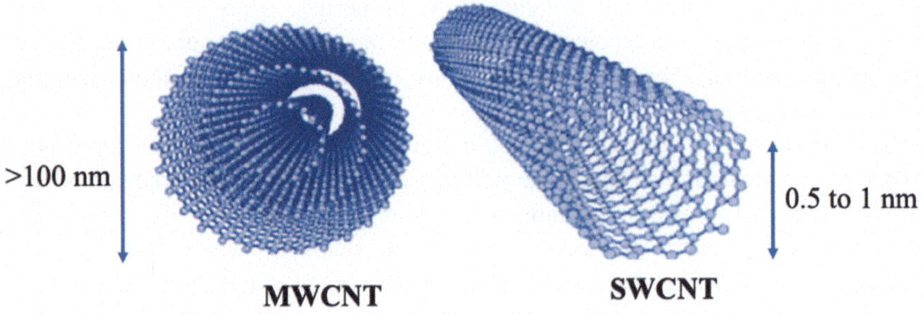

Fig. 3 Graphical comparison of multi-walled carbon nanotubes and single-walled carbon nanotubes. Single-walled carbon nanotubes are monolayered rolled-up graphene, whereas multi-walled carbon nanotubes are several rolled-up graphene layers concentrically

200 °C. The product was collected on formvar covered Cu grids. TEM was used for the characterization of the carbon nanotubes [62].

Thermolytic decomposition of PP and PE produces olefins as the major products, whereas the thermolytic decomposition of PS and PVA yields aromatic hydrocarbons. The chemical reaction between the Fe catalysts and olefins produced from PP and PE is resulted in the formation of long carbon nanotubes. In contrast, the chemical reaction between the Fe catalyst and aromatic hydrocarbons is released

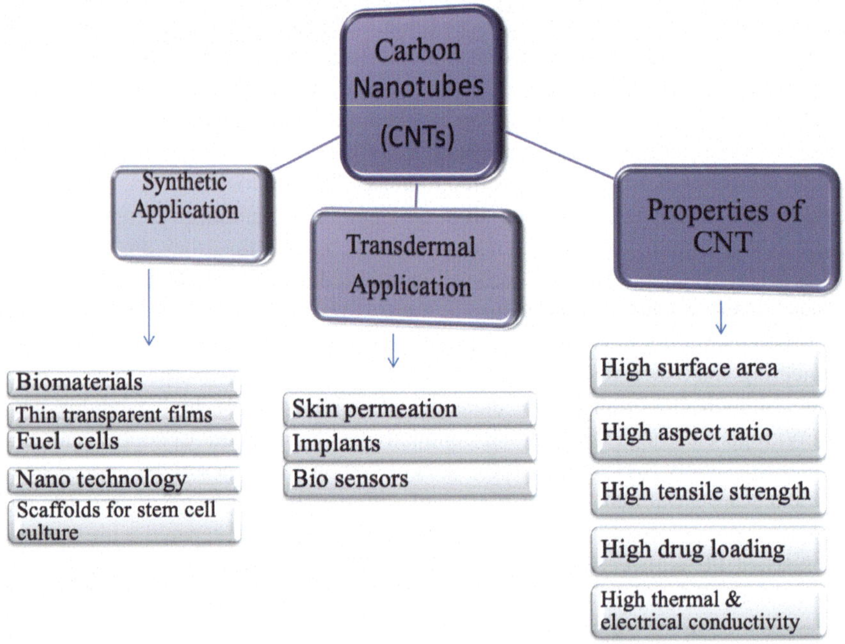

Fig. 4 Some applications and properties of carbon nanotubes (CNT)

from decomposing PS and PVA and resulted in the congealing of the walls of the carbon nanotubes due to the secondary pyrolytic accumulation [62].

Carbon nanotubes could be generated from different waste plastics such as PS, PE, and PP by heating different waste plastics at very high temperatures (700–900 ° C) in an inert nitrogen atmosphere. Afterward, the obtained effluents (both aliphatic and aromatic) were burned with air, but PS residues were not combusted in the air as they have already enriched with oxygen. The burning produced CO and CO_2, which can be termed as the possible carbonaceous sources for the carbon nanotubes. In the end, stainless steel meshes were employed to develop the nanotubes by chemical vapor deposition [63].

Multi-walled carbon nanotubes and sustainable hydrogen were produced from waste (car bumper plastic) as well as virgin PP in a two-step process without the generation of carbon dioxide. In this process, PP was first pyrolyzed in a screw kiln over montmorillonite and zeolites. The effluents thus obtained were decomposed in a moving-bed kiln with Ni catalyst (produced in situ from NiO) for the simultaneous generation of producing multi-walled carbon nanotubes and sustainable hydrogen. An increase in the yield of multi-walled carbon nanotubes was observed with the increase in temperature (from 550 to 750 °C) required for pyrolysis. Also, a significant increase in the yield was observed from the doping of the Ni catalyst with Cu. This revealed that the coupled-catalysts are more efficient for use in the generation of these nanotubes [64].

Bajad et al. reported a process for the synthesis of carbon nanotubes and fuel oil by using waste plastic as a precursor. The nanotubes and the fuel oil were prepared from waste plastic in a locally prepared batch pyrolysis reactor separately. For the preparation of the fuel oil, waste PS granules and mixed waste plastic granules were loaded in the reactor in vacuum at 400 °C, and the effluents were condensed in water condenser at 280 °C. The oil obtained from mixed plastic was greater in quantity as compared to that obtained from waste PS. For the synthesis of nanotubes, the same reactor was employed as an autoclave, and PP bottles were used as precursors. A six trays stainless steel catalyst holder was placed inside the reactor for catalyst holding as well as for the deposition of the carbon nanotubes. The PP and catalyst were placed inside the reactor for 30 min at 800 °C before cooling to the room temperature. Calorific values, FTIR, and density analysis were used to confirm the formation of the fuel oil, whereas the formation of the nanotubes was confirmed with SEM and TEM analysis [65].

Carbon nanotubes have wide-ranging applications, where they are used as a substrate for neuronal growth, and they are involved in the functioning of thin-film transistors. They are conjugated with carbohydrates, proteins, and nucleic acids for various biomedical applications. They are clinically used as biomaterials. They are used in a mixture with metals such as Al for metallurgical purposes and as electrodes in fuel cells [66–71].

2.3 Nanocomposites from Waste Plastic and Their Applications

The nanocomposite is a hard material with multiple phases, where one, two, or three dimensions of one of these phases is below 100 nm [72–78]. They are the substances that integrate nanosized species into some standard material matrix [72–77]. Their structure is a combination of matrix and filler in which nanofillers like fibrils, pieces, and particles encircled and connected as a distinct group in the material matrix. The mixing of the two materials has a drastic enhancement in properties of the substrate materials such as toughness, mechanical strength, thermal, or electrical conductivity. Majority of the nanocomposites with remarkable technical abilities are consisted of two phases and can be categorized into three types, i.e., nano-filamentary composites (consist of specific model nested with nanosized filaments), nanolayered composites (consist of changeable coverings of nanosized dimensions), and nanoparticulate composites (consist of specific form in which nanosized particles are placed) [79, 80]. Figure 5 shows some of the prominent applications of polymeric nanocomposites in the field of medical sciences. Numerous waste plastic polymers have been used for the development of a variety of nanocomposites.

In an environmentally friendly process, sponge-like, Fe/carbon nanotube nanocomposites were prepared by a solvent-free autogenic technique. Waste PP was pyrolyzed at 600 °C in stainless steel autoclave along with equal amounts of ferrocene and NaN_3 for 12 h. Afterward, the mixture was cooled to room temperature, and the product was obtained in the form of black precipitates. The analysis of the nanocomposites by different techniques revealed that both the carbon nanotubes and Fe nanoparticles in the carbon nanotubes were 30 nm in

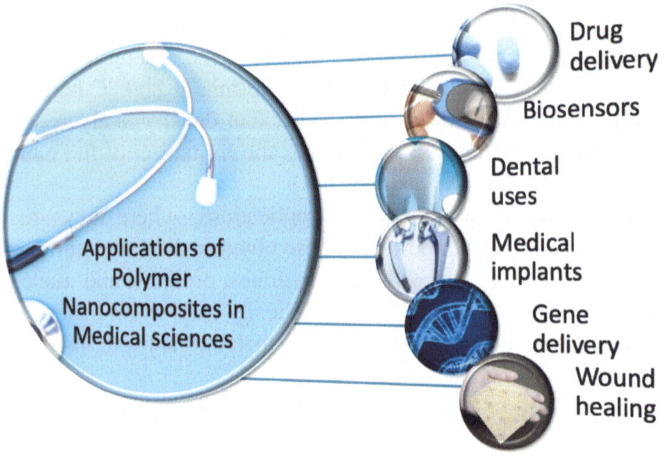

Fig. 5 Applications of polymer nanocomposites in medical sciences

diameter [81]. This type of carbon-encapsulated magnetic nanoparticles has remarkable applications. They are involved in magnetic resonance imaging (MRI), integrated diagnosis and therapeutics, drug delivery, ultra-high-density magnetic recording media, and electromagnetic (EM) wave absorption [82–86].

Waste PVC is also used for the development of useful nanocomposites. One-step conversion of PVC to tantalum carbide/carbon nanocomposite was achieved [87]. Wang et al. used small pieces of waste PVC from the hose and put them in autoclave along with the metallic Li and Ta_2O_5 at 700 °C for 10 h. The raw product thus obtained was washed with hydrochloric acid, distilled water, and absolute alcohol for the removal of residual substances. The product was later vacuum dried and characterized with various techniques including XRD, TEM, thermogravimetric analysis (TGA), and XPS [88]. TaC nanomaterials are considered a remarkable material for cutting tools, rocket nozzles, and other aerospace uses [87].

Fang et al. used waste packaging PE for the synthesis of organic montmorillonite/PE nanocomposites. The nanocomposites were prepared by coextrusion of PE fresh milk packaging bags and nanosized organic montmorillonite. The thermal properties of the waste PE bags and synthesized nanocomposites were also compared. It was found that the melting range of nanocomposites was lowered, whereas the thermal stability was improved. The prepared material was used as the successful asphalt-modifying agent [89]. Asphalt is a common organic binding substance that is used for corrosion resistance, waterproofing, and moisture resistance. Asphalt requires the addition of synthetic polymers such as styrene-butadiene rubber (SBR), styrene–butadiene–styrene (SBS), and PE as the modifiers for proper functioning, and this significantly increases the cost of the material and renders its use [90, 91]. However, the materials mentioned above provide a low-cost solution to the problem, along with the mitigation of undesirable waste.

Nanocomposite photocatalyst consisting of TiO_2 reduced graphene oxide (rGO) and g-C_3N_4 restrained on waste PS were prepared [92]. The nanocomposite was developed in powder as well as in the film form. The facile calcination method was used for the preparation of the nanocomposites. All the raw materials (TiO_2, rGO, g-C_3N_4) other than PS were kept in the furnace at 520 °C for 4 h. The waste PS in the form of local thermocol was dissolved in acetone and dried to result in the transformation of the solid form into small lumps. Afterward, the nanocomposites were immobilized on the PS lumps by the solvent casting method. The mixture of chloroform and ethanol was used as the solvents. The prepared nanocomposites were used in the multiphase airlift reactor for the degradation of remazol turquoise blue (RTB) dye (Cu-phthalocyanine complex). This was done for the evaluation of the photocatalytic behavior of the prepared nanocomposites. 60% and 93% decolorization was achieved with the film and powdered nanocomposites catalysts, respectively, under sunlight irradiation [92]. In the treatment of wastewater, complex dyes are not easily removed. If dyes are left unattended and discharged to the natural water sources, they become a serious threat to aquatic life as well as to humans [93].

Thermoplastic polyurethane (PU) is not only the cause of environmental pollution, but also it is a fire hazard. Ultrathin β-$Co(OH)_2$ nanosheets were fabricated via the surfactant self-assembly process and integrated into thermoplastic PU which reduces the flammability of the material. The integration of ultrathin β-$Co(OH)_2$ nanosheets into the PU matrix efficiently reduced the release of heat and controlled volatiles toxicity [94]. In the same study, the prepared β-$Co(OH)_2$/PU nanocomposites were converted by a green autocatalytic method into high value-added carbon-based materials with 85% yield. This innovative idea helps in the reduction of plastic polymeric waste by its conversion into useful materials via an environmentally friendly process [94].

An investigation has converted waste PET to unsaturated polyester nanocomposites were developed and has also analyzed its technological and economic aspects [95]. In another investigation, PET waste was first depolymerized via diethylene glycol and 1,4-butanediol in the presence of zinc acetate. This resulted in the formation of glycolysis product, which was employed in the synthesis of unsaturated polyester. The unsaturated polyester prepared from the glycolysis products was used for the fabrication of nanocomposites. The unsaturated polyester was melted, and styrene was added to it with vigorous shaking at 60 °C, afterward, different nanofillers such as cloister and modified montmorillonite clay were added to the mixture.

In contrast, hydrogen peroxide and cobalt naphthenate were added as initiators and accelerators, respectively. The obtained nanocomposites were left for curing overnight at room temperature, followed by post-curing for 2 h at 80 °C [96]. The unsaturated polyester resins are the most widely employed thermoset resins because of their proficient physical properties. They are also employed in thermosetting composite polymers as matrix resins [97].

2.4 Graphene-Based Nanomaterials from Waste Plastic and Their Applications

Graphene has appeared as a favorable nanomaterial due to its remarkable characteristics [15, 98–102]. A single two-dimensional sheet of hexagonally arranged carbon is referred to as graphene. In graphene, each carbon is sp^2 hybridized. Of these, the carbon sigma bond that exists in-plane is observed to be the most durable bond, whereas the out of the plane, the π bond, is responsible for delocalized electrons. These delocalized electrons make the graphene active for conductivity and offer weak nuclear forces between graphene and its substrate or among the various layers of graphene. This remarkably unique structure of graphene is responsible for astounding physical properties of the material, making it the epitome of research in the various fields of science and engineering. Among several other properties, graphene also has a promising charge carrying capacity, due to which it acts as light relativist tiny particle or Dirac fermions. Under optimized parameters, they can proceed with excellent dispersion. This property enabled graphene to exhibit a remarkable phenomenon. Graphene is a two-dimensional semiconductor with zero-bandgap. It is also known to have zero band energy and can exhibit shows strong bipolar electric field effect. It also depicts a rare half-integer quantum hall effect for both hole and electron holder when the chemical potential is adjusted via the electric field. Furthermore, graphene exhibits absorptivity of about 2.3% in the visible region and appeared extremely translucent [103]. The heat-conducting power, k, of graphene's monolayer, is about 5000 WmK^{-1} at room temperature. Graphene has excellent mechanical strength. The necessary mechanical qualities of graphene membranes consisting of the independent single sheets were observed by the nano-indentation technique by AFM. Young's modulus and breaking strength of the material are found to be 1.0 T Pa and 42 N m^{-1}, respectively, that make it the strongest materials ever [104]. Graphene has many useful applications; some of them are exhibited in Fig. 6. Several useful graphene-based nanomaterials have been fabricated using waste polymeric plastics.

Graphene foil was synthesized from various plastic waste products, including plastic bottles, i.e., mineral water and cleaning agent bottles, plastic bags, valve bags, protective plastic wrappers, and lunch boxes [105]. These used plastic products are primarily composed of PS, PVC, PP, PET, polymethyl methacrylate (PMMA), and PE. Here, graphene foil has been prepared by solid-state chemical vapor deposition (CVD) technique. In this process, Ni foil was cleaned by using solvents like acetone and alcohol through sonicator and washed with water. Cleaned Ni foil is then subjected to annealing in the presence of Ar/H_2 atmosphere at 1050 °C for 30 min to discharge oxides from the surface of the metal. Afterward, all the chopped plastics were subjected to heat, and the melt was shifted into the

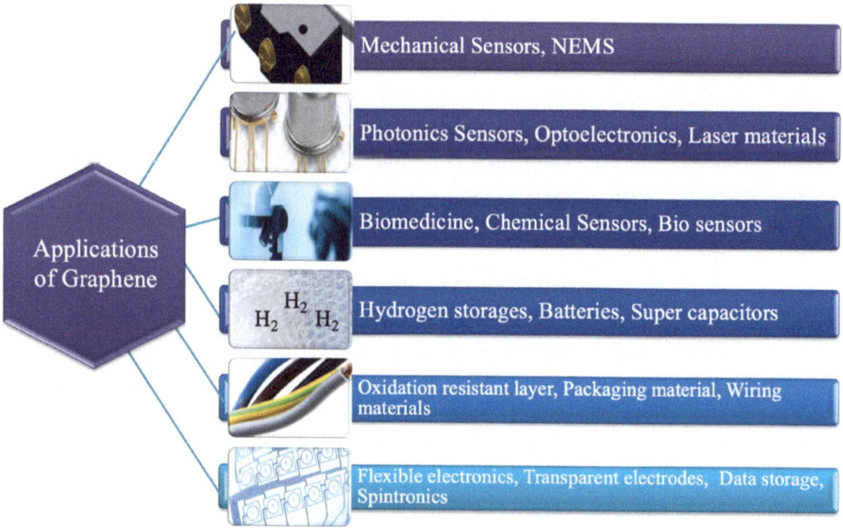

Fig. 6 Applications of graphene in different areas of science and engineering

CVD system. The growth action has occurred at 1050 °C for 120 min. The graphene foil is obtained after the separation of Ni foil through etching in the presence of $FeCl_3/HCl$. Conductivity analysis showed that fabricated graphene foil has a conductance power of 3824 S cm^{-1}. This property enables the material for several practical applications. The prepared foil depicted ultra-low-voltage responsivity and hence was employed in an electro-thermal heater, which heats up to 322.6 °C for the voltage as low as 5 V. Due to the remarkable flexibility of the material, it was also employed as the bendable electrode lithium-ion battery resulting into the high functioning battery [106].

In a recent study, molted salt from PET waste bottles was used for the fabrication of graphene nanostructures. For the preparation of the nanostructures, the plastic was cut into small pieces and was subjected to pyrolysis with sodium chloride into an alumina crucible via an oxidation resistance furnace at 10 °C min^{-1} to 1300 °C, followed by the immediate cooling at room temperature. This led to the formation of solidified salt and the graphene product. The graphene product was separated from the salt by dissolving the mixture in distilled water. The salt gets dissolved in the water, and the graphene formed the suspension, which was separated by vacuum filtration. The formed product was characterized by TEM, XRD, Raman spectroscopy, and differential scanning calorimetry (DSC) [107]. The nanostructured is obtained in the form of crystalline nanosheets of graphite having a thickness of less than 10 nm with a very high electrical conductivity of 1150 S m^{-1}. Graphene-based nanostructure has a wide range of applications because of its high conductivity and surface area. They are used as conductive composites in electronic devices, in energy storage systems, to harvest solar energy, inks conduction, and have other environmental purposes [31, 108–111].

In another study, graphene nanoplatelets were synthesized by using a waste tire as a carbon source. The said nanomaterial was obtained via the chemical treatment of the tires. When the waste tires were subjected to the pyrolysis, carbon black was obtained, which was transformed into graphene. The graphene thus obtained was cheaper in manufacturing than the commercial graphene. TEM and SEM were used for the determination of structure, morphology, and dispersion ability of the prepared nanomaterials. The fabricated nanomaterial was used in the synthesis of polyamide 66-based locomotive pieces. The prepared material was proposed as a novel composite based on thermoplastic for the potential applications in the locomotive industry [112].

In another investigation, PET from the waste bottle was used for the synthesis of graphene via thermal decomposition. The applied method for the conversion of PET to graphene nanoparticles was proven to be facile, reproducible, and cost-effective. In this process, raw PET was torn to pieces and screened to get the appropriate size range between 1 and 3 mm by use of conventional sieve shaker. After that, PET waste was put into a stainless steel autoclave reactor and placed into the furnace at the temperature of about 800 °C for 1 h. The product thus obtained was left to cool for the whole night. Synthesized graphene nanoparticles were analyzed with various characterization techniques, including TEM, SEM, Raman spectroscopy, BET, TGA, and FTIR [113].

Table 1 Nanomaterials from waste plastic and their applications

Waste plastic	Polymer	Nanomaterial	Application	References
Plastic water bottles	PET	Fullerenes	Dermatology and cosmetics	[114, 115]
Plastic foam	PS	Nanofoams	Degradation of dyes	[116]
	PET and PMMA	Nano-SiO_2	Pest control, medicine, concrete mixtures	[117–119]
PVC powder	PVC	Nanoporous carbon	Supercapacitance	[120]
Mixed plastic		Carbon nanosheets	Degradation of dyes	[121]
	PP	Nano-clay/ nanocomposites	Improved mechanical properties of the polymer after recycling	[122]
Woven bags, vessels, foam sheets, bottles, sewage pipes	PP, PE, PET, PVC	Carbon nanosheets	Supercapacitance	[123]
	PTFE	Carbon nanospheres	Energy storage	[124, 125]
Waste plastic	PP	Graphene flakes	Transistors, conductive inks, nanoelectromechanical systems (NEMS), solar cells, smart windows	[126, 127]
Waste PS foam	PS	Superabsorbent nanocomposite		[128]

2.5 Other Nanomaterials from Waste Plastic and Their Applications

Other than the nanomaterials mentioned above, numerous other nanoscale materials are fabricated with a variety of plastics. A brief account is given in Table 1.

3 Conclusion

Plastic waste is one of the biggest modern world problems. Microplastic has adverse effects on both terrestrial and aquatic biomes. In all the ecosystems, the level of microplastic is continuously increasing. The non-biodegradability is allowing the spread of the plastics from municipal sewage to the depths of the oceans. There are several procedures to recycle them, but each has its limitations and disadvantages.

Several attempts have been made to use the waste plastics for the fabrication of a variety of useful nanomaterials. This chapter has elaborated on the synthesis, applications, and characterization approaches of various nanomaterials like nanoparticles, nanotubes, nanocomposites, and graphene-based substances from different waste plastic polymers such as polyvinyl chloride, polyethylene, and polypropylene. The conversion of the waste plastic to the useful nanoparticles is an optimistic approach that cannot only mitigate the pollution but can also provide useful materials at cheaper rates.

4 Future Perspectives

Nanomaterials are still evolving and have deep-rooted applications in modern-day problems. Using the plastic-based nanomaterials to cope with modern problems is a favorable option both in terms of environmental and economic aspects. However, the use of waste plastics has some areas which need the attention of researchers and scientists in future for better productivity and sustainability. For instance, the development of the benchmark process is needed for the conversion of plastics into the nanomaterials as the existing methods are not universal for the wide-ranging polymer of plastic. Different polymers can be converted to the nanomaterials via different methods, which are not feasible on a large scale, as we need separating units as well as the separate treatment plants for variable polymers. A formal protocol of optimized parameters must be developed so that the size and shape of the nanomaterials (particularly carbon nanotubes) can be controlled. Methods for lowering the decomposition temperature of the feed plastic should be determined to cut the energy costs. Several processes of the waste plastic conversion to the nanomaterials involve the generation of undesirable chemicals like dioxins and furans; studies must be made to seek their mitigation or eradication if possible.

References

1. Derraik JGB (2002) The pollution of the marine environment by plastic debris: a review. Mar Pollut Bull 44(9):842–852
2. Hirai H, Takada H, Ogata Y, Yamashita R, Mizukawa K, Saha M, Kwan C, Moore C, Gray H, Laursen D (2011) Organic micropollutants in marine plastics debris from the open ocean and remote and urban beaches. Mar Pollut Bull 62(8):1683–1692
3. Mato Y, Isobe T, Takada H, Kanehiro H, Ohtake C, Kaminuma T (2001) Plastic resin pellets as a transport medium for toxic chemicals in the marine environment. Environ Sci Technol 35(2):318–324
4. Rios LM, Moore C, Jones PR (2007) Persistent organic pollutants carried by synthetic polymers in the ocean environment. Mar Pollut Bull 54(8):1230–1237
5. Li WC, Tse HF, Fok L (2016) Plastic waste in the marine environment: A review of sources, occurrence and effects. Sci Total Environ 566–567:333–349

6. Marqués-Calvo MS, Cerdà-Cuéllar M, Kint DPR, Bou JJ, Muñoz-Guerra S (2006) Enzymatic and microbial biodegradability of poly (ethylene terephthalate) copolymers containing nitrated units. Polym Degrad Stab 91(4):663–671
7. Andrady AL (2011) Microplastics in the marine environment. Mar Pollut Bull 62(8):1596–1605
8. Müller R-J, Kleeberg I, Deckwer W-D (2001) Biodegradation of polyesters containing aromatic constituents. J Biotechnol 86(2):87–95
9. Williams PT (2006) Yield and composition of gases and oils/waxes from the feedstock recycling of waste plastic. Feedstock Recycling and Pyrolysis of Waste Plastics: Converting Waste Plastics into Diesel and Other Fuels:285–313
10. Bazargan A, McKay G (2012) A review – Synthesis of carbon nanotubes from plastic wastes. Chem Eng J 195–196:377–391
11. Ali GAM, Makhlouf SA, Yusoff MM, Chong KF (2015) Structural and electrochemical characteristics of graphene nanosheets as supercapacitor electrodes. Rev Adv Mater Sci 40 (1):35–43
12. Ali GAM, Lih Teo EY, Aboelazm EAA, Sadegh H, Memar AOH, Shahryari-Ghoshekandi R, Chong KF (2017) Capacitive performance of cysteamine functionalized carbon nanotubes. Mater Chem Phys 197:100–104
13. Seyed Arabi SM, Lalehloo RS, Olyai MRTB, Ali GAM, Sadegh H (2019) Removal of congo red azo dye from aqueous solution by ZnO nanoparticles loaded on multiwall carbon nanotubes. Physica E 106:150–155
14. Hegde G, Abdul Manaf SA, Kumar A, Ali GAM, Chong KF, Ngaini Z, Sharma KV (2015) Biowaste sago bark based catalyst free carbon nanospheres: waste to wealth approach. ACS Sustainable Chem Eng 5(9):2247–2253
15. Marina PE, Ali GAM, See LM, Teo EYL, Ng E-P, Chong KF (2019) In situ growth of redox-active iron-centered nanoparticles on graphene sheets for specific capacitance enhancement. Arab J Chem 12(8):3883–3889
16. Ali GAM, Supriya S, Chong KF, Shaaban ER, Algarni H, Maiyalagan T, Hegde G (2019) Superior supercapacitance behavior of oxygen self-doped carbon nanospheres: a conversion of Allium cepa peel to energy storage system. Biomass Conv Bioref https://doi.org/10.1007/s13399-019-00520-3
17. Ali GAM, Divyashree A, Supriya S, Chong KF, Ethiraj AS, Reddy MV, Algarni H, Hegde G (2017) Carbon nanospheres derived from Lablab purpureus for high performance supercapacitor electrodes: a green approach. Dalton Trans 46(40):14034–14044
18. Ghawanmeh AA, Ali GAM, Algarni H, Sarkar SM, Chong KF (2019) Graphene oxide-based hydrogels as a nanocarrier for anticancer drug delivery. Nano Research 12:973–990
19. Abu Bakar NH, Ali GAM, Ismail J, Algarni H, Chong KF (2019) Size-dependent corrosion behavior of graphene oxide coating. Prog Org Coat 134:272–280
20. Fulekar MH, *Nanotechnology: importance and applications*. 2010: IK International Pvt Ltd.
21. Habeeb OA, Ramesh K, Ali GAM, Yunus RM (2017) Low-cost and eco-friendly activated carbon from modified palm kernel shell for hydrogen sulfide removal from wastewater: adsorption and kinetic studies. Desalin Water Treat 84:205–214
22. Ali GAM, Habeeb OA, Algarni H, Chong KF (2018) CaO impregnated highly porous honeycomb activated carbon from agriculture waste: symmetrical supercapacitor study. J Mater Sci 54:683–692
23. Habeeb OA, Ramesh K, Ali GAM, and Yunus RM (2017) Experimental design technique on removal of hydrogen sulfide using CaO-eggshells dispersed onto palm kernel shell activated carbon: Experiment, optimization, equilibrium and kinetic studies. J Wuhan Uni Technol-Mater Sci Ed 32(2):305–320
24. Samaddar P, Ok YS, Kim K-H, Kwon EE, Tsang DCW (2018) Synthesis of nanomaterials from various wastes and their new age applications. Journal of Cleaner Production 197:1190–1209

25. Ali GAM, Yusoff MM, Shaaban ER, Chong KF (2017) High performance MnO_2 nanoflower supercapacitor electrode by electrochemical recycling of spent batteries. Ceram Int 43:8440–8448
26. Aboelazm EAA, Ali GAM, Chong KF (2018) Cobalt oxide supercapacitor electrode recovered from spent lithium-ion battery. Chem Adv Mater 3:67–74
27. Aboelazm EAA, Ali GAM, Algarni H, Yin H, Zhong YL, Chong KF (2018) Magnetic Electrodeposition of the Hierarchical Cobalt Oxide Nanostructure from Spent Lithium-Ion Batteries: Its Application as a Supercapacitor Electrode. J Phys Chem C 122(23):12200–12206
28. Ali GAM, Tan LL, Jose R, Yusoff MM, Chong KF (2014) Electrochemical performance studies of MnO_2 nanoflowers recovered from spent battery. Mater Res Bull 60:5–9
29. Kilic MS, Korkut S, Hazer B, Erhan E (2014) Development and operation of gold and cobalt oxide nanoparticles containing polypropylene based enzymatic fuel cell for renewable fuels. Biosens Bioelectron 61:500–505
30. Evans R, Boyd L, Elam C, Czernik S, French R, Feik C, Philips S, Chaornet E, and Parent Y. *Hydrogen from biomass-catalytic reforming of pyrolysis vapors.*
31. Araby S, Meng Q, Zhang L, Zaman I, Majewski P, Ma J (2015) Elastomeric composites based on carbon nanomaterials. Nanotechnology 26(11):112001
32. Hilado CJ, *Flammability handbook for plastics*. 1998: CRC Press.
33. Khan I, Saeed K, Khan I (2019) Nanoparticles: Properties, applications and toxicities. Arab J Chem 12(7):908–931
34. Moeinzadeh S and Jabbari E, *Nanoparticles and Their Applications*, in *Springer Handbook of Nanotechnology*, B Bhushan, Editor. 2017, Springer Berlin Heidelberg: Berlin, Heidelberg. p. 335–361
35. Hu Y, Yang J, Tian J, Jia L, Yu J-S (2014) Green and size-controllable synthesis of photoluminescent carbon nanoparticles from waste plastic bags. RSC Adv 4(88):47169–47176
36. Cao L, Wang X, Meziani MJ, Lu F, Wang H, Luo PG, Lin Y, Harruff BA, Veca LM, Murray D (2007) Carbon dots for multiphoton bioimaging. J Am Chem Soc 129(37):11318–11319
37. Li H, He X, Kang Z, Huang H, Liu Y, Liu J, Lian S, Tsang CHA, Yang X, Lee ST (2010) Water-soluble fluorescent carbon quantum dots and photocatalyst design. Angew Chem Int Ed 49(26):4430–4434
38. Li Y, Hu Y, Zhao Y, Shi G, Deng L, Hou Y, Qu L (2011) An electrochemical avenue to green-luminescent graphene quantum dots as potential electron-acceptors for photovoltaics. Adv Mater 23(6):776–780
39. Miao P, Han K, Tang Y, Wang B, Lin T, Cheng W (2015) Recent advances in carbon nanodots: synthesis, properties and biomedical applications. Nanoscale 7(5):1586–1595
40. Rajeev A, Erapalapati V, Madhavan N, and Basavaraj MG (2016) Conversion of expanded polystyrene waste to nanoparticles via nanoprecipitation. Journal of Applied Polymer Science 133(4)
41. Na N, Hu Y, Ouyang J, Baeyens WRG, Delanghe JR, De Beer T (2004) Use of polystyrene nanoparticles to enhance enantiomeric separation of propranolol by capillary electrophoresis with Hp-beta-CD as chiral selector. Anal Chim Acta 527(2):139–147
42. Rossi G, Barnoud J, Monticelli L (2014) Polystyrene nanoparticles perturb lipid membranes. the Journal of Physical Chemistry Letters 5(1):241–246
43. Oslakovic C, Cedervall T, Linse S, and Dahlbäck B (2012) Polystyrene nanoparticles affecting blood coagulation. Nanomedicine: Nanotechnology, Biology and Medicine 8 (6):981–986
44. Wang L, Dai W, Zhao D, Zhang F, Zhang K, Cheng Q, Luo Z, He X, Qian Y (2019) A Simple Route for the Direct Conversion of Waste Plastic to Hafnium Carbide Nanoparticles at Low Temperature. Nanoscience and Nanotechnology Letters 11(7):1027–1031

45. Guron MM, Kim MJ, Sneddon LG (2008) A Simple Polymeric Precursor Strategy for the Syntheses of Complex Zirconium and Hafnium-Based Ultra High-Temperature Silicon-Carbide Composite Ceramics. J Am Ceram Soc 91(5):1412–1415
46. Sanders WA and Grisaffe SJ, *The hot-pressing of hafnium carbide (melting point, 7030 F)*. Vol. 303. 1960: National Aeronautics and Space Administration.
47. Ramanan V, Siddaiah B, Raji K, Ramamurthy P (2018) Green Synthesis of Multifunctionalized, Nitrogen-Doped, Highly Fluorescent Carbon Dots from Waste Expanded Polystyrene and Its Application in the Fluorimetric Detection of Au3+ Ions in Aqueous Media. ACS Sustainable Chemistry & Engineering 6(2):1627–1638
48. Dang DK, Sundaram C, Ngo Y-LT, Chung JS, Kim EJ, Hur SH (2018) One pot solid-state synthesis of highly fluorescent N and S co-doped carbon dots and its use as fluorescent probe for Ag+ detection in aqueous solution. Sensors and Actuators B: Chemical 255:3284–3291
49. Wang Y, Kim S-H, Feng L (2015) Highly luminescent N, S-Co-doped carbon dots and their direct use as mercury (II) sensor. Anal Chim Acta 890:134–142
50. Yang C, Thomsen RP, Ogaki R, Kjems J, Teo BM (2015) Ultrastable green fluorescence carbon dots with a high quantum yield for bioimaging and use as theranostic carriers. Journal of Materials Chemistry B 3(22):4577–4584
51. Yang S-T, Cao L, Luo PG, Lu F, Wang X, Wang H, Meziani MJ, Liu Y, Qi G, Sun Y-P (2009) Carbon dots for optical imaging in vivo. J Am Chem Soc 131(32):11308–11309
52. Zheng XT, Ananthanarayanan A, Luo KQ, Chen P (2015) Glowing graphene quantum dots and carbon dots: properties, syntheses, and biological applications. Small 11(14):1620–1636
53. Wang L, Dai W, Yang T, Zhang K, He W, Kong Q, Zhang J, Wang X, Mei T, Hu Y, Shen H, Qian Y (2020) One-step chemical synthesis of $MgCNi_3$ nanoparticles embedded in carbon nanosheets utilizing waste polyethylene as carbon source. Materials Research Express 6(12):126003
54. Yan J, Sun Y, Wu H, Huang Q, Wang C, Shi Z, Deng S, Shi K, Lu H, Chu L (2014) Phase transitions and magnetocaloric effect in $Mn_3Cu_{0.89}N_{0.96}$. Acta Mater 74:58–65
55. Zhang XH, Yuan Q, Han JC, Zhao JG, Jian JK, Zhang ZH, and Song B (2013) Observation of spin-glass behavior in antiperovskite compound Mn3Cu0. 7Ga0. 3N. Applied Physics Letters 103(2):022405
56. Dai W, Lu L, Han Y, Wang L, Wang J, Hu J, Ma C, Zhang K, Mei T (2019) Facile synthesis of Mo2C nanoparticles from waste polyvinyl chloride. ACS Omega 4(3):4896–4900
57. He L, Qin Y, Lou H, Chen P (2015) Highly dispersed molybdenum carbide nanoparticles supported on activated carbon as an efficient catalyst for the hydrodeoxygenation of vanillin. Rsc Advances 5(54):43141–43147
58. Ma R, Zhou Y, Chen Y, Li P, Liu Q, Wang J (2015) Ultrafine molybdenum carbide nanoparticles composited with carbon as a highly active hydrogen-evolution electrocatalyst. Angew Chem Int Ed 54(49):14723–14727
59. Mai EF, Machado MA, Davies TE, Lopez-Sanchez JA, da Silva VT (2014) Molybdenum carbide nanoparticles within carbon nanotubes as superior catalysts for γ-valerolactone production via levulinic acid hydrogenation. Green Chem 16(9):4092–4097
60. Collins PG, Avouris P (2000) Nanotubes for electronics. Sci Am 283(6):62–69
61. Dasgupta K, Joshi JB, Banerjee S (2011) Fluidized bed synthesis of carbon nanotubes–A review. Chem Eng J 171(3):841–869
62. Chung Y-H, Jou S (2005) Carbon nanotubes from catalytic pyrolysis of polypropylene. Mater Chem Phys 92(1):256–259
63. Zhang Q, Huang JQ, Zhao MQ, Qian WZ, Wei F (2011) Carbon nanotube mass production: principles and processes. Chemsuschem 4(7):864–889
64. Liu J, Jiang Z, Yu H, and Tang T. *Production of hydrogen and carbon nanotubes by catalytic pyrolysis of waste polypropylene in a two-step process.*
65. Bajad G, Jain R, Harhare W, Rp V, Bose S (2017) Synthesis of fuel oil and carbon nanotubes in an autoclave using plastic waste as precursor. Mater Manuf Processes 32(5):495–500
66. Bambagioni V, Bianchini C, Marchionni A, Filippi J, Vizza F, Teddy J, Serp P, Zhiani M (2009) Pd and Pt–Ru anode electrocatalysts supported on multi-walled carbon nanotubes and

their use in passive and active direct alcohol fuel cells with an anion-exchange membrane (alcohol= methanol, ethanol, glycerol). J Power Sources 190(2):241–251
67. Kocabas C, Hur SH, Gaur A, Meitl MA, Shim M, Rogers JA (2005) Guided growth of large-scale, horizontally aligned arrays of single-walled carbon nanotubes and their use in thin-film transistors. Small 1(11):1110–1116
68. Liao J, Tan M-J (2011) Mixing of carbon nanotubes (CNTs) and aluminum powder for powder metallurgy use. Powder Technol 208(1):42–48
69. Mattson MP, Haddon RC, Rao AM (2000) Molecular functionalization of carbon nanotubes and use as substrates for neuronal growth. J Mol Neurosci 14(3):175–182
70. Saito N, Haniu H, Usui Y, Aoki K, Hara K, Takanashi S, Shimizu M, Narita N, Okamoto M, Kobayashi S (2014) Safe clinical use of carbon nanotubes as innovative biomaterials. Chem Rev 114(11):6040–6079
71. Yang W, Thordarson P, Gooding JJ, Ringer SP, Braet F (2007) Carbon nanotubes for biological and biomedical applications. Nanotechnology 18(41):412001
72. Ali GAM, Fouad OA, Makhlouf SA, Yusoff MM, Chong KF (2016) Optical and Electrochemical Properties of Co_3O_4/SiO_2 Nanocomposite. Advanced Materials Research 1133:447–451
73. Ali GAM, Yusoff MM, Algarni H, and Chong KF (2018) One-step electrosynthesis of MnO_2/rGO nanocomposite and its enhanced electrochemical performance. Ceramics International
74. Abdel Ghafar HH, Ali GAM, Fouad OA, Makhlouf SA (2015) Enhancement of adsorption efficiency of methylene blue on Co_3O_4/SiO_2 nanocomposite. Desalin Water Treat 53 (11):2980–2989
75. Ali GAM, Fouad OA, Makhlouf SA (2013) Structural, optical and electrical properties of sol-gel prepared mesoporous Co_3O_4/SiO_2 nanocomposites. J Alloy Compd 579:606–611
76. Sadegh H, Ali GAM, Makhlouf ASH, Chong KF, Alharbi NS, Agarwal S, Gupta VK (2018) MWCNTs-Fe_3O_4 nanocomposite for Hg(II) high adsorption efficiency. J Mol Liq 258:345–353
77. Ali GAM, Fouad OA, Makhlouf SA, Yusoff MM, Chong KF (2014) Co_3O_4/SiO_2 nanocomposites for supercapacitor application. J Solid State Electrochem 18(9):2505–2512
78. Fouad OA, Makhlouf SA, Ali GAM, El-Sayed AY (2011) Cobalt/silica nanocomposite via thermal calcination-reduction of gel precursors. Mater Chem Phys 128(1–2):70–76
79. Kumari PVK, Rao YS, Akhila S (2019) Role of nanocomposites in drug delivery. GSC Biological and Pharmaceutical Sciences 8(3):094–103
80. Edited by Pethrick RA, Zaikov JE, and Pielichowski J, *Monomers, oligomers, polymers, composites and nanocomposites research: synthesis, properties and applications*. 2009: Nova.
81. Zhang J, Zhang L, Yang H, Kong Q, Liu Y, Yuan A (2014) Sustainable processing of waste polypropylene to produce high yield valuable Fe/carbon nanotube nanocomposites. CrystEngComm 16(37):8832–8840
82. Choi IA, Li Y, Kim DJ, Pal M, Cho JH, Lee K, Jung MH, Lee C, and Seo WS (2013) Ultra-small, Uniform, and Single bcc-Phased FexCo1-x/Graphitic Shell Nanocrystals for T1 Magnetic Resonance Imaging Contrast Agents. Chemistry–An Asian Journal 8(1):290–295
83. Fan X, Jiao G, Gao L, Jin P, Li X (2013) The preparation and drug delivery of a graphene–carbon nanotube–Fe_3O_4 nanoparticle hybrid. Journal of Materials Chemistry B 1(20):2658–2664
84. Seo WS, Lee JH, Sun X, Suzuki Y, Mann D, Liu Z, Terashima M, Yang PC, McConnell MV, Nishimura DG (2006) FeCo/graphitic-shell nanocrystals as advanced magnetic-resonance-imaging and near-infrared agents. Nat Mater 5(12):971–976
85. Wei X-W, Zhu G-X, Xia C-J, Ye Y (2006) A solution phase fabrication of magnetic nanoparticles encapsulated in carbon. Nanotechnology 17(17):4307
86. Yang Y, Guo Z, Zhang H, Huang D, Gu J, Huang Z, Kang F, Alan Hatton T, Rutledge GC (2013) Electrospun magnetic carbon composite fibers: Synthesis and electromagnetic wave absorption characteristics. J Appl Polym Sci 127(6):4288–4295

87. Wang L, Zhang F, Dai W, Cheng Q, Lu L, Zhang K, Lin M, Shen M, Wang D (2019) One step transformation of waste polyvinyl chloride to tantalum carbide@ carbon nanocomposite at low temperature. J Am Ceram Soc 102(11):6455–6462
88. Alhajri NS, Yoshida H, Anjum DH, Garcia-Esparza AT, Kubota J, Domen K, Takanabe K (2013) Synthesis of tantalum carbide and nitride nanoparticles using a reactive mesoporous template for electrochemical hydrogen evolution. J Mater Chem A 1(40):12606–12616
89. Fang C, Yu R, Li Y, Zhang M, Hu J, Zhang M (2013) Preparation and characterization of an asphalt-modifying agent with waste packaging polyethylene and organic montmorillonite. Polym Testing 32(5):953–960
90. Mouillet V, Farcas F, Besson S (2008) Ageing by UV radiation of an elastomer modified bitumen. Fuel 87(12):2408–2419
91. Yam KL (2010) The Wiley encyclopedia of packaging technology. Wiley
92. Das S, Mahalingam H (2019) Dye degradation studies using immobilized pristine and waste polystyrene-TiO_2/rGO/g-C_3N_4 nanocomposite photocatalytic film in a novel airlift reactor under solar light. J Environ Chem Eng 7(5):103289
93. Schrank SG, dos Santos JNR, Souza DS, Souza EES (2007) Decolourisation effects of Vat Green 01 textile dye and textile wastewater using H_2O_2/UV process. J Photochem Photobiol, a 186(2–3):125–129
94. Zhang J, Kong Q, Yang L, Wang D-Y (2016) Few layered $Co(OH)_2$ ultrathin nanosheet-based polyurethane nanocomposites with reduced fire hazard: from eco-friendly flame retardance to sustainable recycling. Green Chem 18(10):3066–3074
95. Drah A, Rusmirovic J, Milošević M, Kalifa M, Stojiljkovic I, Rančić M, Marinkovic A (2016) Techno-economic analysis of unsaturated polyester production from waste PET. Zastita Materijala 57:605–612
96. Asaad JN, Abd-El-Messieh SL, Ikladious NE (2014) Unsaturated polyester nanocomposites based on poly (ethylene terephthalate) waste using different types of nanofillers. Proc Inst Mech Eng Part N J Nanoeng Nanosyst 228(4):174–183
97. Athawale AA and Pandit JA (2019) Unsaturated polyester resins, blends, interpenetrating polymer networks, composites, and nanocomposites: state of the art and new challenges (Chap. 1). In: Thomas S, Hosur M, Chirayil CJ (eds) Unsaturated polyester resins. Elsevier. pp 1–42
98. CeNeR R, AeK S, KeS S, Govindaraj A (2009) Graphene: the new two-dimensional nanomaterial. Angew Chem Int Ed 48(42):7752–7777
99. Ali GAM, Yusoff MM, Chong KF (2016) Graphene Electrochemical production and its energy storage properties. ARPN J Eng Appl Sci 11(16):9712–9717
100. Gupta VK, Agarwal S, Sadegh H, Ali GAM, Bharti AK, Hamdy AS (2017) Facile route synthesis of novel graphene oxide-β-cyclodextrin nanocomposite and its application as adsorbent for removal of toxic bisphenol a from the aqueous phase. J Mol Liq 237:466–472
101. Ali GAM, Thalji MR, Soh WC, Algarni H, Chong KF (2020) One-step electrochemical synthesis of MoS_2/graphene composite for supercapacitor application. J Solid State Electrochem 24(1):25–34
102. Teo EYL, Ali GAM, Algarni H, Cheewasedtham W, Rujiralai T, Chong KF (2019) One-step production of pyrene-1-boronic acid functionalized graphene for dopamine detection. Mater Chem Phys 231:286–291
103. Wang X, Chen B (2019) Origin of Fresnel problem of two dimensional materials. Sci Rep 9 (1):17825
104. Eigler S (2016) Graphene. An introduction to the fundamentals and industrial applications edited by Madhuri Sharon and Maheshwar Sharon. Angewandte Chemie Int Edn 55 (17):5122–5122
105. Qu L, Cui L, Wang X, Chen N, Ji B (2017) Trash to treasure: converting plastic waste into a useful graphene foil. Nanoscale 9
106. Cui L, Wang X, Chen N, Ji B, Qu L (2017) Trash to treasure: converting plastic waste into a useful graphene foil. Nanoscale 9(26):9089–9094

107. Kamali AR, Yang J, Sun Q (2019) Molten salt conversion of polyethylene terephthalate waste into graphene nanostructures with high surface area and ultra-high electrical conductivity. Appl Surf Sci 476:539–551
108. Hof F, Kampioti K, Huang K, Jaillet C, Derré A, Poulin P, Yusof H, White T, Koziol K, Paukner C (2017) Conductive inks of graphitic nanoparticles from a sustainable carbon feedstock. Carbon 111:142–149
109. Jin S, Jun GH, Jeon S, Hong SH (2016) Design and application of carbon nanomaterials for photoactive and charge transport layers in organic solar cells. Nano Convergence 3(1):8
110. Mauter MS, Elimelech M (2008) Environmental applications of carbon-based nanomaterials. Environ Sci Technol 42(16):5843–5859
111. Ni J, Li Y (2016) Carbon nanomaterials in different dimensions for electrochemical energy storage. Adv Energy Mater 6(17):1600278
112. Okan BS, Menceloğlu Y, Ozunlu BG, Yagci YE (2020) Graphene from waste tire by recycling technique for cost-effective and light-weight automotive plastic part production. AIP Publishing LLC
113. El Essawy NA, Ali SM, Farag HA, Konsowa AH, Elnouby M, Hamad HA (2017) Green synthesis of graphene from recycled PET bottle wastes for use in the adsorption of dyes in aqueous solution. Ecotoxicol Environ Saf 145:57–68
114. Mousavi SZ, Nafisi S, and Maibach HI (2017) Fullerene nanoparticle in dermatological and cosmetic applications. Nanomed Nanotech Biol Med 13(3):1071–1087
115. El Essawy NA, Konsowa AH, Elnouby M, Farag HA (2017) A novel one-step synthesis for carbon-based nanomaterials from polyethylene terephthalate (PET) bottles waste. J Air Waste Manage Assoc 67(3):358–370
116. de Assis GC, Eb S, Leite VD, Rodrigues MO, Galembeck A, Alves MCF, Eastoe J, de Oliveira RJ (2018) Conversion of "Waste Plastic" into photocatalytic nanofoams for environmental remediation. ACS Appl Mater Interfaces 10(9):8077–8085
117. Barik TK, Sahu B, Swain V (2008) Nanosilica-from medicine to pest control. Parasitol Res 103(2):253
118. Quercia G, Brouwers HJH (2010) Application of nano-silica (nS) in concrete mixtures
119. Kerboua N, Cinausero N, Sadoun T, Lopez-Cuesta JM (2010) Effect of organoclay in an immiscible poly (ethylene terephtalate) waste/poly (methyl methacrylate) blend. J Appl Polym Sci 117(1):129–137
120. Cheng LX, Zhang L, Chen XY, Zhang ZJ (2015) Efficient conversion of waste polyvinyl chloride into nanoporous carbon incorporated with MnOx exhibiting superior electrochemical performance for supercapacitor application. Electrochim Acta 176:197–206
121. Gong J, Liu J, Chen X, Jiang Z, Wen X, Mijowska E, Tang T (2015) Converting real-world mixed waste plastics into porous carbon nanosheets with excellent performance in the adsorption of an organic dye from wastewater. J Mater Chem A 3(1):341–351
122. Tri Phuong N, Gilbert V, Chuong B (2008) Preparation of recycled polypropylene/organophilic modified layered silicates nanocomposites part I: the recycling process of polypropylene and the mechanical properties of recycled polypropylene/organoclay nanocomposites. J Reinf Plast Compos 27(18):1983–2000
123. Wen Y, Kierzek K, Chen X, Gong J, Liu J, Niu R, Mijowska E, Tang T (2019) Mass production of hierarchically porous carbon nanosheets by carbonizing "real-world" mixed waste plastics toward excellent-performance supercapacitors. Waste Manage 87:691–700
124. Xu F, Tang Z, Huang S, Chen L, Liang Y, Mai W, Zhong H, Fu R, Wu D (2015) Facile synthesis of ultrahigh-surface-area hollow carbon nanospheres for enhanced adsorption and energy storage. Nature Communications 6(1):1–12
125. Yang X, Li C, Wang W, Yang B, Zhang S, Qian Y (2004) A chemical route from PTFE to amorphous carbon nanospheres in supercritical water. Chem Commun 3:342–343
126. Gong J, Liu J, Wen X, Jiang Z, Chen X, Mijowska E, Tang T (2014) Upcycling Waste Polypropylene into Graphene Flakes on Organically Modified Montmorillonite. Ind Eng Chem Res 53(11):4173–4181

127. Low CTJ, Walsh FC, Chakrabarti MH, Hashim MA, Hussain MA (2013) Electrochemical approaches to the production of graphene flakes and their potential applications. Carbon 54:1–21
128. Liu PS, Li L, Zhou NL, Zhang J, Wei SH, Shen J (2007) Waste polystyrene foam-graft-acrylic acid/montmorillonite superabsorbent nanocomposite. J Appl Polym Sci 104(4):2341–2349

Recycling Nanofibers from Polyethylene Terephthalate Waste Using Electrospinning Technique

Suhad Yasin, Zinab H. Bakr, Gomaa A. M. Ali, and Ibtisam Saeed

Abstract Electrospun nanofibers are a class of nanomaterials appropriate for various applications, such as smart films, filter membranes, catalytic supports, energy generation modules, conversion and storage, photonic and electronic sensors, biomedical scaffolding, and other devices. Electrospinning is a flexible and versatile technique for processing nanofiber materials; it consists of an electrohydrodynamic process in which liquids are electrified to create a beam and then stretched into fibers. The basic set up for electrospinning is relatively simple and, therefore, accessible to almost every laboratory. The main components are a high voltage DC power supply, a syringe device, a spinner, and a conductive collector. Due to the increasing consumption of polyethylene terephthalate-based products and their waste disposal issue, increasing environmental concern has led us to transform this waste into valuable products. This chapter focuses on the research studies of electrospun waste polyethylene terephthalate to produce nanofiber, applied in a different application.

Keywords Nanofibers · Electrospinning · Polyethylene terephthalate · Waste plastic

List of Abbreviations

AMG Aerogel micro granules
AOP Advanced oxidation processes

S. Yasin (✉)
Chemistry Department, College of Science, University of Duhok, Kurdistan Region, Iraq
e-mail: suhad.yasin@uod.ac

Z. H. Bakr
Physics Department, Faculty of Science, Assiut University, Assiut 71516, Egypt

G. A. M. Ali
Chemistry Department, Faculty of Science, Al-Azhar University, Assiut 71524, Egypt
e-mail: gomaasanad@azhar.edu.eg

I. Saeed
College of Science, University of Duhok, Duhok, Iraq

A. S. H. Makhlouf and G. A. M. Ali (eds.), *Waste Recycling Technologies for Nanomaterials Manufacturing*, Topics in Mining, Metallurgy and Materials Engineering, https://doi.org/10.1007/978-3-030-68031-2_28

CD	Cyclodextrin
CNT	Carbon nanotubes
DCM	Dichloromethan
PC	Polycarbonate
PET	Polyethylene terephthalate
PS	Polystyrene
PVP	Polyvinylpyrrolidone
R-PET	Recycled polyethylene terephthalate
SEM	Scanning electron microscopy
TFA	Trifluoroacetic acid

1 Introduction

Nanofibers are a new category of specially designed materials. Materials with sizes from filaments of less than 100–1000 nm are in this classification. Due to their unique design and characteristics, the nanofibers have a high interest in research [1]. In particular, the membrane containing nanofibers provides a high surface area and porosity with extraordinary mechanical strength and sensitivity [2]. Nanomaterials are very attractive in recent decades and have represented most of the research into new materials in the environment, energy, and health [3]. Nanofibers can form particularly porous networks with unique links between their pores. Nanofibers of 5–500 nm in diameter have a surface area of approximately 10,000–1,000,000 $cm^2\ kg^{-1}$, and the high porosity of nanofiber membranes (up to 80%) allows for improved performance in the adsorption of air and water contaminant, as well as a long lifetime [4]. The diameter nanofibers, network composition, and highly specific surfaces make nanofibers necessary for designing the instruments and structures needed, with unique specific features that support our life requirements [5]. The small pores and the high porosity allow nanofibers to filter materials selectively during air and water applications. They can also be used for immobilizing nanoparticles as an ideal anchor fabric for further functioning the composite membrane [6]. It also can be used in battery and power devices as electrode components [7]. Nanofiber membranes are perfect for imitating the original extracellular matrix to regenerate various tissues [8]. In particular, nanofibers' effects in energy conversion (solar cells & fuel cells) and heat and air treatment systems were well understood and promoted [9]. The typical applications of electrospinning polymer nanofiber [10] are shown in Fig. 1.

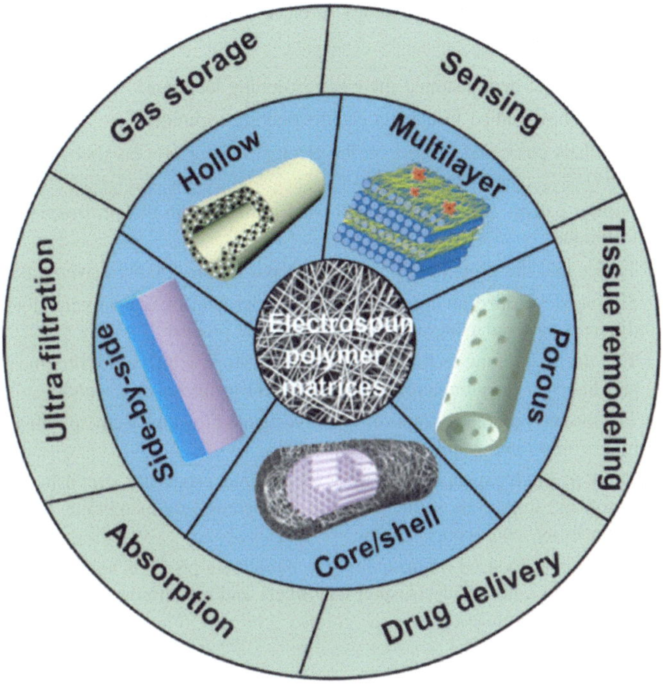

Fig. 1 Various applications of electrospinning polymer nanofiber. Adapted with permission from Ref. [10], Copyright 2012, MDPI

2 Basics of Electrospinning Technique

Because of the simple design and manufacturing of mass production nanofibers, electrospinning is now the most common technique to produce nanofibers. It is nowadays one of the most corporate techniques for manufacturing nanofibers [11, 12]. A variety of manufacturing methods, including drawing, phase separation, template-synthesis, bicomponent extrusion, self-assembly, centrifugal spinning, melt-blown, and electrospinning [13–15]. The electrospinning technique has several advantages over other nanofibers fabrication techniques, as [11, 16, 17] due to its simple instrument, cost-effective comparison to other techniques, continuous process, and fiber diameters of only a few nm to several microns.

The background of the science behind electrospinning stretches back to the earliest times of scientific research. Table 1 shows the milestones of the electrospinning technique [18, 19]. Electrospinning system since the beginning of the 1990s, an essential advancement in the knowledge of this complicated electrospinning system and the control of the creation of nanofibers has been produced [20]. William Gilbert observed a liquid droplet deformation when an electrically loaded piece of amber was brought near the droplet. Electrospinning applies high

Table 1 A critical milestone in the technology of electrospinning [18]

Year	Progress of events
1897	Raleigh first demonstrated the electrospinning technique
1914	Zeleny has studied further on electrospraying technique
1934	Formhals published patents on the electrospinning process (between 1934 and 1944)
1944–2004	Further patents (about 50 patents) were published on the electrospinning technique
1950–1970	Taylor laid the groundwork by his research on electrically driven jets
1966	Simons developed an apparatus for the generation of non-wovens fibers with different patterns using an electrical spinning process
1980	The electrospinning technique in nanotechnology, which used ultrafine nanofibers and fibers in the submicron scale, is subject to intensive research
1981	The molten polymeric solution was investigated for the electrospinning technique by Larrondo and Manley
1990–2000	Many research papers are published in this field, especially after using electrospun nanofibers in various applications, such as composite membrane, tissue engineering, and energy storage
2000–2020	Researchers focused on upscaling the nanofibers products and on extending the application areas of electrospun nanofibers and nanofibrous membranes in all theological applications

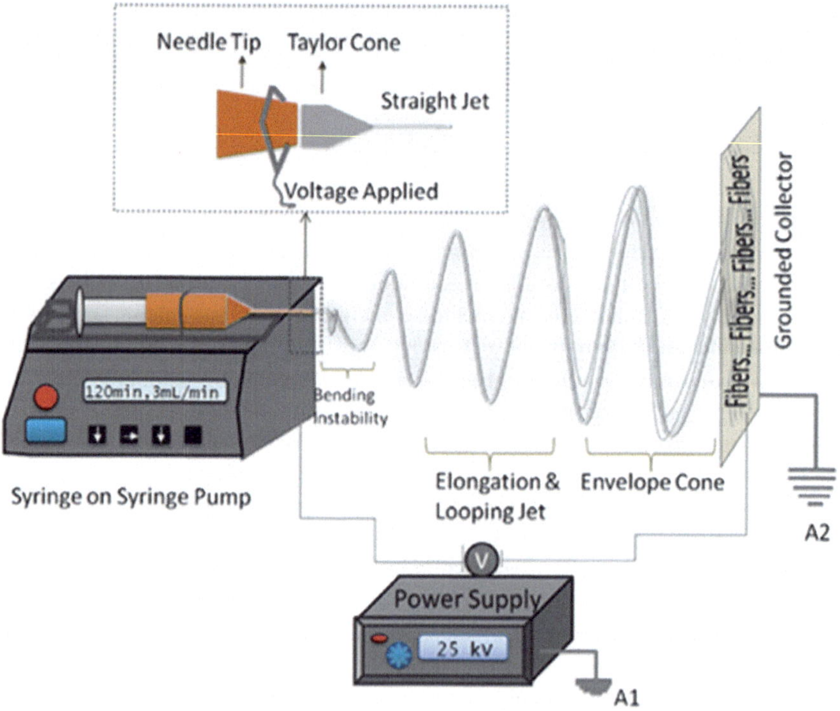

Fig. 2 Electrospinning schematic. Adapted with permission from Ref. [23], Copyright 2012, Elsevier

voltage with the polymer solution or melts on the front end of the nozzle. When the load is concentrated and rejected at the nozzle's end, it drops out of the nozzle into a cone called the Taylor cone [21]. It was the first step in the electrospinning method when the liquid shifts form in the face of an internal electrostatic field.

As soon as the electrostatic strength is big enough to overcome the droplet's surface tension, a finely charged jet ejects from the Taylor cone and performs a sequence of unstable aerial transitions that quickly produce thin fiber after the solvent has been quickly evaporated [22]. The basic design consists of an endless needle with a loaded polymer solution, a spring pump that pushes the solution at a constant speed, a high voltage power supply, and a grounded collector placed at a defined distance from the needle tip. While the solution is pushed off at an individual pace of flow rate, the spinneret produces a droplet of the alternative, as shown in Fig. 2 [23]. Electrospun nanofibers have been prepared for around 100 different polymers [24], of both natural and synthetic origin. Either solvent or melting spinner has been used to prepare these nanofibers. However, this method's knowledge is still very restricted, even with the extensive use of the electrospinning technique [11]. In both theoretical and experimental respects, the parameters that affect the electrospinning process and the resulting nanofibers' morphology are well established [13]. Three parameters affect the electrospinning (1) solution parameters, (2) process parameters, and (3) ambient parameters [11]. The solution parameters include viscosity, conductivity, and surface tension. The process parameters include the applied electric field, the distance between needle and collector, and injecting rate. The ambient parameters include humidity and temperature inside the electrospinning chamber [25, 26].

3 Polyethylene Terephthalate

Polyethylene terephthalate (PET) is a polyester polymer, can be synthesized in two ways. The first procedure is a direct reaction of terephthalic acid with ethylene glycol. This reaction is common esterification of the Fisher type in which an acid interacts with ethanol, following the standard mechanism [27]. The other PET synthesis includes a transesterification reaction in which one ester is transformed into another by reacting dimethyl terephthalate with ethylene glycol [28]. PET seems to be a very popular, low crystallize, high melting point, and fast processing polymer [28, 29]. They are widely utilized in plastic and bottles (food packages and other consumer items, including soft drinks, detergent products, alcoholic beverages, etc.). The PET bottles are light, transparent, chemically resistant, and non-toxic [30]. Due to this functionality, they have a massive presence in the global industries. PET outlets represent 8% by weight and 12% by solid waste worldwide [31]. PET presents no serious threat to the environment but could be considered a pollutant because of its growing plastic waste and its highly resistant biological agents and the environment. PET recycling not only provides PET products but also contributes to a possible partial solution to the production of solid waste [32]. In

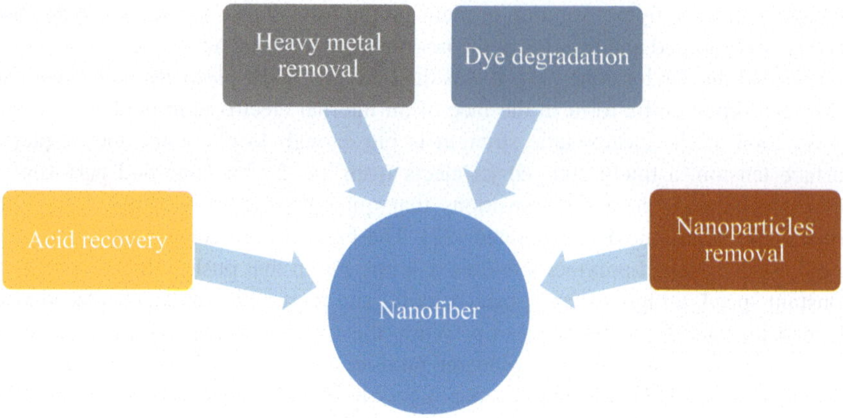

Fig. 3 Applications of nanofibers in wastewater treatment. Adapted with permission from Ref. [46], Copyright 2012, NISCAIR-CSIR

2012, the PET Container Resources Association announced that the US collected 5.6 billion pounds of PET plastic bottles for recycling. The recycling rate was, however, just 30.8%. Consequently, it becomes critical for the environment to find effective ways to recycle PET waste bottles and increase the recycling level [33].

4 Nanofiber Filtration Membranes

Water purification and reuse are significant globally because human life will be at risk without clean water resources [34–37]. Different water purification methods, membrane filtration, are widely applied in recent decades to overcome these challenges [38]. To date, several research and efforts have been conducted to develop new approaches to purify the contaminant and microorganism pollutant water into fresh, safe water [39]. Degradation of harmful organic chemicals and dyes is a critical issue [40–43]. A wide variety of methods, including active carbon adsorption, electrocoagulation, redox treatment, and ultraviolet, are used, but they do not appear to be remarkably cost-effective and efficient. Therefore, there is an immediate demand for new venues for removing pollutants and wastewater treatments [44].

To increase current technological performance and to create a new methodology, nanotechnology has now been actively monitored. Nanomaterial properties for water and wastewater applications cover high surface area for adsorption, high intensity (photo) catalysis, antimicrobial disinfection, and control properties for biofouling and other basic optical and electronic properties for use in advanced water quality control and sensor control processes. Nevertheless, nanotechnology for water purification still has some drawbacks [45]. Electrospun nanofibers and modified electrospun nanofibers use various regeneration pathways, removal, and degradation, as shown in Fig. 3 [46].

5 Applications of Polyethylene Terephthalate Nanofibers

Many studies have focused on using a modification of PET nanofiber in different application fields such as water treatments. The researchers had been used PET as a polymer material in the electrospinning technique. Here is a survey of the studies which started in 2005. Ma et al. [47] studied have been carried out using electrospinning technology to prepare non-woven polyethylene terephthalate nanofiber mats to mimic the native extracellular surface matrix of fibrous proteins towards building a biocompatible surface for the extracellular surface. Veleirinho et al. [48] investigated the electrospun PET nanofibrous matting solutions in trifluoroacetic acid/dichloromethane. The PET and the solvent's concentration have a changeable effect on the morphology and the electrospun nonwovens' mechanical and thermal properties. They found that 10 wt. % of PET giving uniform nanofibers with interconnected pore structure and a diameter of 200–700 nm. Veleirinho and Silva [49] produce good mechanical properties of nanofibrous PET membranes. The use of these membranes in the clarification process for apple juice. An electrospun PET membrane's filtration potential was demonstrated by introducing a new clarifying principle and opening up new venue to the juice processing factories or even to other food industries.

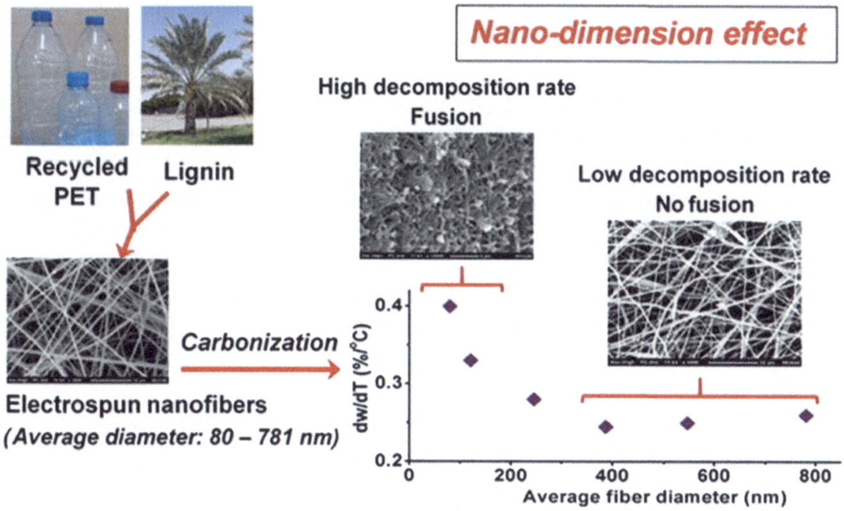

Fig. 4 Carbon nanofibers obtained by combining lignin/recycle PET. Adapted with permission from Ref. [51], Copyright 2019, Elsevier

Mazinani and et al. [50] examined the morphology, structure, and properties of the conductive nanoweb of PET/carbon nanotubes (CNT). Trifluoroacetic acid (TFA)/dichloromethan (DCM) nanocomposite nanofibers have been obtained by electrospinning PET solutions with various CNT concentrations. Smoother nanofibers were seen with less bead structure formation in the morphological test results. Recently, Svinterikos et al. [51] studied the carbonization of electrospun nanofibers consisting of 1/1 mass ratio lignin/recycle PET as shown in Fig. 4. The findings can serve as a road map for the successful development of carbon nanofibers in the sustainability agenda from lignin/recycle PET and other precursor materials.

Zhang et al. [52] studied ZnO precursor/PET hybrid nanofibers fabricated by electrospinning technique. Scanning electron microscopy (SEM) images showed that by increasing the concentration of zinc acetate dihydrate and reduced PET concentration, the obtained nanofibers become smoother and uniformed with lower diameters. Finally, the obtained nanofibers are subject to a mild immersion process to improve the ZnO nanocrystals surface to be a favourable candidate for electrical, photocatalytic, chemical sensors, and solar cell applications. Hadjizadeh et al. [53] presented electrospun PET fibrous with a different diameter from nanometer to micrometers, and with different orientation and mechanical properties by changing the PET concentrations and flow rate. These fibrous could be useful in many applications such as biotechnological and biomedical applications. Kayaci and Uyar [54] investigated the surface modification of electrospun PET nanofibers with cyclodextrin (CD). They find that the nanofibers (PET/CDP) would serve as a perfect candidate for water purification and waste treatment as a filter material for removing volatile organic compounds from air due to its high surface area and the inclusion complex formation of CDP-associated surface.

Santos et al. [55] presented a mat of the ultrathin and nanoscale fiber of lignocellulosic sisal/recycled PET via electrospinning technique one-pot process, which could be an alternative process for those need two-step preparations. Likely, the mechanical properties showed that incorporating sisal with PET improved the moduli value and increased the T_g by up to 20 °C compared to neat PET fibers. Thus, combining lignocellulosic biomass with polymer matrices enhanced and controlled their properties, such as stiffness and hydrophilicity.

Mazrouei-Sebdani et al. [56] studied focused on the possible use of sodium silicate-based aerogels as a filler in PET nanofibers (PNFs) instead of small precursor molecules. Aerogel micro granules (AMG), with successful embedding, have applied to PET's electrospinning solution. The results showed that adding AMG to PET solution yielding, bead aerogel nanofibers with 91.8% porosity, and 815 $m^2\ g^{-1}$ of surface area. Moreover, hydrophobicity, dyeability, and insulation properties have significantly improved.

Nanofibers membranes with a diameter of 100 nm from PET plastic bottles obtained and demonstrated their microfiltration application ability [57]. Latex beads ranging from 30 to 2000 nm in size used to test the filtration efficiency. Gravity filtration helps eliminate more than 99% of the beads with a small size of around 500 nm. Moreover, the mats are functional with quaternary ammonium and biguanide biocides to reduce biofouling, for both gram negatives and gram

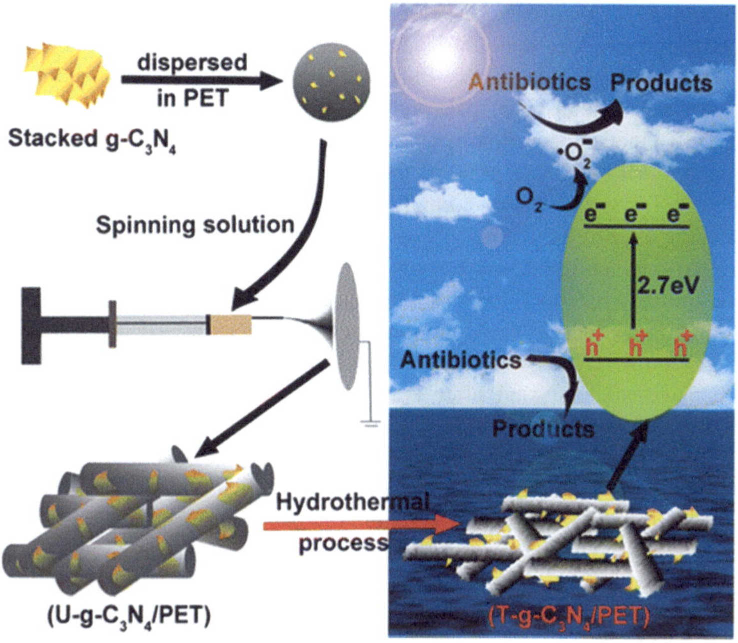

Fig. 5 Illustration of degrading antibiotics using electrospun PAN/GO/Fe_3O_4 under sun illumination. Adapted with permission from Ref. [59], Copyright 2016, American Chemical Society

positives. Thus, recycled PET nanofibers could potentially open the new venue as next-generation filter material.

The ability of PET/PAN/GO/Fe_3O_4 systems to remove two Pb(II) and Cr(VI) ions from both bath and membrane systems was investigated [58]. The PAN/GO/Fe_3O_4 adsorption data for 5% nanofibers from metal ions were best represented in the second and Redlich-Peterson models. Through analysis of the thermodynamic parameters, the endothermic and spontaneous existence of the Pb(II) and Cr(VI) ion adsorption using PAN/GO/Fe_3O_4 (5 wt). Qin et al. [59] showed the simple way to synthesize a recycled photocatalyst with PET was emphasized to disperse carbon graphite nitride (g-C_3N_5) by electrospinning and hydrothermal treatment as shown in Fig. 5. The resulting nanofibers (T-g-C_3N_4/PET) demonstrated strong photocatalytic activity for antibiotic degradation, including sulfaquinoxaline and sulfadiazine, in solar irradiation.

Mahar et al. [60] synthesized for the first time recycled PET nanofibers from waste flasks (R-PET) using the electrospinning technique to check dyeability using two separate disperse dyes, Blue Disperse-56(DB-56) and Disperse Red 167:1(DR 167:1). They suggested that R-PET nanofibers have many advantages, such as superior color rate, high mechanical strength, processing at low temperatures, and short color times compared to other conventional nanofibers. Similarly, Martín et al.

[61] studied PET nanofibers developed by electrospinning and with a simple aminolysis process, i.e. their surface was chemically functional. The resulting product's capacity to adsorb the Pb(II) of synthetic solutions was tested as a function of the contact time, pH, and ion concentration. The adsorbent system provided fast kinetic adsorption, with a total adsorption efficiency of approximately 50 mmol (mmol) Pb(II) per gram of the adsorbent system, reaching 30 min only. Aminated PET nanofibers showed high potential for eliminating heavy metals from contaminated water, eliminating 97% of Pb(II) ions in solution. While Coşkun et al. [62] grafted and modified electrospun PET to produce new fibers with superior properties to remove Cr(VI) ions from the aqueous solution. FT-IR analysis showed the grafting of the acrylonitrile monomer on the PET surface and detected new surface groupings. Moreover, SEM images showed that the PET fiber was more comprehensive and adjusted after grafting with acrylonitrile. luckily, the fiber demonstrates its optimum adsorption capacity in acidic medium (modified fibers). Also, the kinetic results showed that a second-pseudo-kinetic model followed adsorption of Cr(VI) belong to modified fiber. Also, Jin et al. [63] studied wool keratin/PET composite nanofiber membrane preparation for adsorption of Cr(VI). Composite membranes (keratin/PET) maximum adsorption was 75.86 mg g^{-1}. FTIR and XPS analysis showed that disulfide keratin and amino bonds were part of the adsorption process.

Biazar et al. [64] synthesized artificial keratoprostheses from PET treated nanofibers with the CO_2 plasma to modified the surface, then cross-linked with collagen. The development of PET nanofibers with covalent attachment is an effective method to improve keratoprosthesis designs' biocompatibility. Esmaeili et al. [65] investigated the formation of PET, polystyrene (PS), and polycarbonate (PC) nano/microfibers in the corresponding waste plastics. The effect of the needle diameter and the voltage difference used to study the impact of electrospinning's different operating parameters. The data showed that the increase in the voltage applied contributed to the increase in average fibers; however, when preparing the fibers with a voltage difference of 20 kV, the fiber diameters had a smaller distribution. Nevertheless, the decrease in the needle diameter was followed by the creation of PET fibers with lower average fiber diameter. Shahrabi et al. [66] studied PET/polyvinylpyrrolidone (PVP) nanofibers mats prepared by the electrospinning technique. They applied the Taguchi method to determine the optimum values of the parameter. They found that combining the Taguchi method and the linear regression model is an excellent way to predict nanofibers' average diameter. Mohammadkhani et al. [67] showed the use of nanoparticles has always been of great importance on the web, particularly for microwave absorption and photocatalytic operation, for broader and more comfortable use. This work was carried out using a homogeneous sonic RPET solution with NiNPs with a mean diameter of 32.5 nm on average., synthesized via the chemical reduction method, which was used for nickel nano-particulars/recycled PET, web nanofibers (NiNPs/RPET NFs). The use of nanoparticles has always been of great importance on the web, particularly for microwave absorption and photocatalytic operation, for broader and more comfortable use. This work was carried out using a homogeneous sonic RPET solution with NiNPs with a mean diameter of an average of 32.5 nm, synthesized via the chemical reduction method, which was used for nickel nano-particulars/ recycled poly (ethylene terephthalate), web nanofibers (NiNPs/RPET NFs web).

6 Photocatalytic Degradation of Organic Pollutants Using Polyethylene Terephthalate Nanofibers

A photocatalyst is described as an active substance by adsorbing a photon that can accelerate the reaction without being consumed [68]. Typically, these molecules are semiconductors. Thanks to their potential solar energy conversion and environmental purification applications, semiconducting oxide photocatalysts have become highly focused in recent years. Heterogeneous photocatalysis in semiconductors has enormous potential for treating water and air pollutants. This process is known as the Advanced Oxidation Processes (AOP) and is appropriate for oxidation, a wide range of organic compounds. Heterogeneous photocatalysis was developed in the 1970s and was shown to be of interest among AOPs due to its effectiveness in degrading organic recalcitrant. The photoreaction in the presence of a semiconductor catalyst is accelerated. Different semiconductors (TiO_2, ZnO, Fe_2O_3, CdS, ZnS) can be photocatalysts, but TiO_2 has often been observed as an organic pollutant degradation and even complete mineralization [69]. TiO_2, due to its high reactivity, decreased toxicity, chemical stability, and lower costs, TiO_2's photocatalytic and hydrophilic properties make it close to the ideal catalyst [70]. Filtration of the membrane and photodegradation was seen as a positive development to reduce waste and filtration costs in the future. Such new techniques will improve wastewater management at a minimal cost [71].

Abaas et al. [72] studied the impact of various process parameters on the morphology and diameter of the fibers obtained from a solution of PET in trifluoroacetic acid and dichloromethane in a 1:3 ratio. The optimal level factors to obtain small diameter fibers were determined as follows: PET concentration (5%), flow rate (1 mL h^{-1}), applied voltage (15 kV), and distance (15 cm). The aims of the recent study by Yasin et al. [73–75]: (1) reuse PET as a waste material to prepare nanofibers, (2) apply the Tachuqi design for the optimization study of PET nanofibers, (3) composite PET nanofibers mats with CuO nanoparticles, (4) composite PET nanofibers mats with TiO_2 nanoparticles, and (5) applying composite PET nanofibers for photodegradation study of methylene blue in aqueous solution. SEM images of electrospun PET nanofiber morphology are represented in Fig. 6. The different results are shown between runs (designed by the Taguchi method) in surface smoothness, the number of beads, and the uniformity of nanofiber diameter.

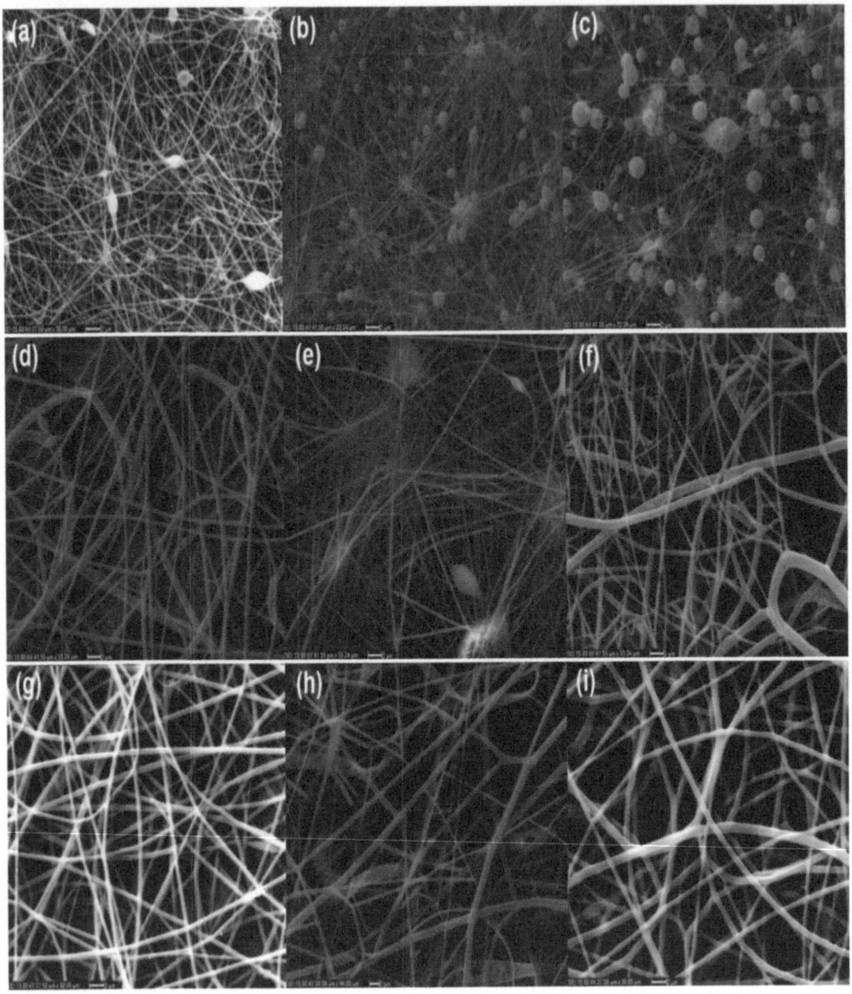

Fig. 6 SEM images of electrospun PET nanofibers. Adapted with permission from Ref. [72], Copyright 2018, IOP Publishing

7 Conclusion

The electrospinning technique is successfully used to produce polyethylene terephthalate nanofibrous mats using different solvents. Polymer concentration, needle tip-to collector distance, feed rate, and applied voltage were studied to investigate the levels of the optimal factors for a thinner fiber diameter during electrospinning. The Taguchi design helps optimize the most important parameters which affect nanofiber results. The concentration of PET solution proved to be the most influential factor in the diameter of PET fibers. The lower concentrations produce the smaller diameters. PET plastic is successful in producing nanofiber with functional characterizations as morphological and thermal behavior. Moreover, modified PET to produce new fibers with superior properties to remove heavy metals such as Cr and Pb from the water could be used as water treatment. Moreover, modified PET nanofibers could be a strong photocatalytic activity for antibiotic degradation to improve the biocompatibility of keratoprosthesis designs and the membrane's filtration photodegradation seen as a positive development to reduce waste and the costs of filtration in the future. Such new techniques will improve wastewater management at a minimal cost.

8 Future Perspectives

Based on the current information and data, it is recommended to use different waste plastic polymers such as polystyrene, polycarbonate, etc. In addition, other types of Design of Experiments could help in comparisons with the Taguchi design. Modification of nanofibers using another nanoparticle could be a promising area of nanocomposite research. Detailed investigations of the filtration system using modified nanofibers, nanofibers membrane to remove heavy metals from wastewater, and photodegradation are needed.

Acknowledgements The authors are grateful for the financial supports by Partnerships for Enhanced Engagement in Research (PEER). It is administered by the U.S. National Academies of Sciences, Engineering, and Medicine (NASEM), PEER/Iraq project/cycle 6.)

References

1. Haider A, Haider S, Kang I-K (2018) A comprehensive review summarizing the effect of electrospinning parameters and potential applications of nanofibers in biomedical and biotechnology. Arab J Chem 11(8):1165–1188
2. Han D, Steckl AJ (2019) Coaxial electrospinning formation of complex polymer fibers and their applications. ChemPlusChem 84(10):1453–1497
3. Huang Y, Song J, Yang C, Long Y, Wu H (2019) Scalable manufacturing and applications of nanofibers. Mater Today 28:98–113
4. Yoon K, Hsiao BS, Chu B (2008) Functional nanofibers for environmental applications. J Mater Chem 18(44):5326
5. Beydoun D, Amal R, Low G, McEvoy S (1999) Role of nanoparticles in photocatalysis. J Nanopart Res 1(4):439–458
6. Tijing LD, Amarjargal A, Jiang Z, Ruelo MTG, Park C-H, Pant HR, Kim D-W, Lee DH, Kim CS (2013) Antibacterial tourmaline nanoparticles/polyurethane hybrid mat decorated with silver nanoparticles prepared by electrospinning and UV photoreduction. Curr Appl Phys 13(1):205–210
7. Cho TH, Sakai T, Tanase S, Kimura K, Kondo Y, Tarao T, Tanaka M (2007) Electrochemical performances of polyacrylonitrile nanofiber-based nonwoven separator for lithium-ion battery. Electrochem Solid-State Lett 10(7):A159

8. Mokhtari F, Salehi M, Zamani F, Hajiani F, Zeighami F, Latifi M (2016) Advances in electrospinning: the production and application of nanofibres and nanofibrous structures. Textile Progress 48(3):119–219
9. Gibson P, Schreuder-Gibson H, Rivin D (2001) Transport properties of porous membranes based on electrospun nanofibers. Colloids Surf A 187–188:469–481
10. Wang C, Wang J, Zeng L, Qiao Z, Liu X, Liu H, Zhang J, Ding J (2019) Fabrication of electrospun polymer nanofibers with diverse morphologies. Molecules 24(5):834
11. Doshi J, Reneker DH (1995) Electrospinning process and applications of electrospun fibers. J Electrostat 35(2–3):151–160
12. Rutledge GC, Fridrikh SV (2007) Formation of fibers by electrospinning. Adv Drug Deliv Rev 59(14):1384–1391
13. Mooney P, Shearer J, Mead J, Barry C, Truong Q, Welsh EA, Pang R, Zukas W, Stenhouse P, Brown P (2018) Bicomponent fiber extraction process for textile applications. J Eng Fibers Fabr 13(1):33–39
14. Krucińska I, Surma B, Chrzanowski M, Skrzetuska E, Puchalski M (2013) Application of melt-blown technology in the manufacturing of a solvent vapor-sensitive, non-woven fabric composed of poly (lactic acid) loaded with multi-walled carbon nanotubes. Text Res J 83 (8):859–870
15. Naeimirad M, Zadhoush A, Kotek R, Esmaeely Neisiany R, Nouri Khorasani S, Ramakrishna S (2018) Recent advances in core/shell bicomponent fibers and nanofibers: a review. J Appl Polym Sci 135(21):46265
16. Almetwally AA, El-Sakhawy M, Elshakankery M, Kasem MH (2017) Technology of nano-fibers: production techniques and properties—critical review. J Textile Assoc 78:5–14
17. Kumbar SG, James R, Nukavarapu SP, Laurencin CT (2008) Electrospun nanofiber scaffolds: engineering soft tissues. Biomed Mater 3(3):034002
18. Ray SS, Chen S-S, Nguyen NC, Nguyen HT (2019) Electrospinning: a versatile fabrication technique for nanofibrous membranes for use in desalination. In: Nanoscale materials in water purification. Elsevier, pp 247–273
19. Tucker N, Stanger JJ, Staiger MP, Razzaq H, Hofman K (2012) The history of the science and technology of electrospinning from 1600 to 1995. J Eng Fibers Fabrics 7(2):63–73
20. Formhals A (1934) Process and apparatus for preparing artificial threads. US Patent: 1975504, in US Patent: 1975504. p. 7
21. Kleivaitė V, Milašius R (2018) Electrospinning—100 years of investigations and still open questions of web structure estimination. Autex Res J 18(4):398–404
22. Ameer PR, Kasoju (2019) Strategies to tune electrospun scaffold porosity for effective cell response in tissue engineering. J Funct Biomater 10(3):30
23. Nataraj SK, Yang KS, Aminabhavi TM (2012) Polyacrylonitrile-based nanofibers—a state-of-the-art review. Prog Polym Sci 37(3):487–513
24. Kulkarni A, Bambole V, Mahanwar P (2010) Electrospinning of polymers, their modeling and applications. Polym-Plast Technol Eng 49(5):427–441
25. Jiang S, Chen Y, Duan G, Mei C, Greiner A, Agarwal S (2018) Electrospun nanofiber reinforced composites: a review. Polym Chem 9(20):2685–2720
26. Li D, Xia Y (2004) Electrospinning of nanofibers: reinventing the wheel? Adv Mater 16 (14):1151–1170
27. Ravindranath K, Mashelkar R (1984) Finishing stages of PET synthesis: a comprehensive model. AIChE J 30(3):415–422
28. Dutt K, Soni RK (2013) A review on synthesis of value added products from polyethylene terephthalate (PET) waste. Polym Sci Ser B 55(7–8):430–452
29. Ravindranath K, Mashelkar R (1986) Polyethylene terephthalate—I. Chemistry, thermodynamics and transport properties. Chem Eng Sci 41(9):2197–2214
30. Strain IN, Wu Q, Pourrahimi AM, Hedenqvist MS, Olsson RT, Andersson RL (2015) Electrospinning of recycled PET to generate tough mesomorphic fibre membranes for smoke filtration. J Mater Chem A 3(4):1632–1640

31. Chinchillas-Chinchillas MJ, Orozco-Carmona VM, Alvarado-Beltrán CG, Almaral-Sánchez JL, Sepulveda-Guzman S, Jasso-Ramos LE, Castro-Beltrán A (2019) Synthesis of recycled poly(ethylene terephthalate)/polyacrylonitrile/styrene composite nanofibers by electrospinning and their mechanical properties evaluation. J Polym Environ 27(3):659–669
32. Choi YJ, Kim SH (2014) Characterization of recycled polyethylene terephthalates and polyethylene terephthalate–nylon6 blend knitted fabrics. Text Res J 85(4):337–345
33. Choi Y-W, Moon D-J, Chung J-S, Cho S-K (2005) Effects of waste PET bottles aggregate on the properties of concrete. Cem Concr Res 35(4):776–781
34. Sadegh H, Ali GAM, Nia HJ, Mahmoodi Z (2019) Nanomaterial surface modifications for enhancement of the pollutant adsorption from wastewater. In Nanotechnology applications in environmental engineering. IGI Global, pp. 143
35. Sadegh H, Ali GAM, Makhlouf ASH, Chong KF, Alharbi NS, Agarwal S, Gupta VK (2018) MWCNTs-Fe_3O_4 nanocomposite for Hg(II) high adsorption efficiency. J Mol Liq 258:345–353
36. Sharifi A, Montazerghaem L, Naeimi A, Abhari AR, Vafaee M, Ali GAM, Sadegh H (2019) Investigation of photocatalytic behavior of modified ZnS:Mn/MWCNTs nanocomposite for organic pollutants effective photodegradation. J Environ Manage 247:624–632
37. Lee SP, Ali GAM, Algarni H, Chong KF (2019) Flake size-dependent adsorption of graphene oxide aerogel. J Mol Liq 277:175–180
38. Strathmann H (2001) Membrane separation processes: current relevance and future opportunities. AIChE J 47(5):1077–1087
39. Bottero J-Y, Rose J, Wiesner MR (2006) Nanotechnologies: tools for sustainability in a new wave of water treatment processes. Integr Environ Assess Manage 2(4):391–395
40. Seyed Arabi SM, Lalehloo RS, Olyai MRTB, Ali GAM, Sadegh H (2019) Removal of congo red azo dye from aqueous solution by ZnO nanoparticles loaded on multiwall carbon nanotubes. Physica E 106:150–155
41. Solehudin M, Sirimahachai U, Ali GAM, Chong KF, Wongnawa S (2020) One-pot synthesis of isotype heterojunction g-C_3N_4-MU photocatalyst for effective tetracycline hydrochloride antibiotic and reactive orange 16 dye removal. Adv Powder Technol 31(5):1891–1902
42. Sadegh H, Ali GAM, Gupta VK, Makhlouf ASH, Shahryari-ghoshekandi R, Nadagouda MN, Sillanpää M, Megiel E (2017) The role of nanomaterials as effective adsorbents and their applications in wastewater treatment. J Nanostruct Chem 7(1):1–14
43. Agarwal S, Sadegh H, Majid M, Makhlouf ASH, Ali GAM, Memar AOH, Shahryari-ghoshekandi R, Tyagi I, Gupta VK (2016) Efficient removal of toxic bromothymol blue and methylene blue from wastewater by polyvinyl alcohol. J Mol Liq 218:191–197
44. K SK, RA, Arumugam P, Berchmans S (2011) Synthesis of gold nanoparticles: an ecofriendly approach using Hansenula anomala. ACS Appl Mater Interf 3(5):1418–1425
45. Gopakumar DA, Pai AR, Pasquini D, Leu S-Y, Abdul Khalil HPS, Thomas S (2019) Nanomaterials-state of art, new challenges, and opportunities. In: Nanoscale Materials in Water Purification. Elsevier. pp 1–24
46. Baskar D, Selvam AK, Ganesan M, and Nallathambi G (2018) Mechanism of nanofibres on removal of water pollutants-a review. Indian J Chem Technol 25(5):451–458
47. Golder A, Hridaya N, Samanta A, Ray S (2005) Electrocoagulation of methylene blue and eosin yellowish using mild steel electrodes. J Hazard Mater 127(1–3):134–140
48. Veleirinho B, Rei MF, Lopes-Da-Silva JA (2008) Solvent and concentration effects on the properties of electrospun poly(ethylene terephthalate) nanofiber mats. J Polym Sci, Part B: Polym Phys 46(5):460–471
49. Veleirinho B, Lopes-da-Silva JA (2009) Application of electrospun poly(ethylene terephthalate) nanofiber mat to apple juice clarification. Process Biochem 44(3):353–356
50. Mazinani S, Ajji A, Dubois C (2010) Fundamental study of crystallization, orientation, and electrical conductivity of electrospun PET/carbon nanotube nanofibers. J Polym Sci, Part B: Polym Phys 48(19):2052–2064

51. Svinterikos E, Zuburtikudis I, Al-Marzouqi M (2019) The nanoscale dimension determines the carbonization outcome of electrospun lignin/recycled-PET fibers. Chem Eng Sci 202:26–35
52. Zhang J, Wen B, Wang F, Ding Y, Zhang S, Yang M (2011) In situ synthesis of ZnO nanocrystal/PET hybrid nanofibers via electrospinning. J Polym Sci Part B Polym Phys 49 (19):1360–1368
53. Hadjizadeh A, Ajji A, Bureau MN (2011) Nano/micro electro-spun polyethylene terephthalate fibrous mat preparation and characterization. J Mech Behav Biomed Mater 4(3):340–351
54. Kayaci F, Uyar T (2014) Electrospun polyester/cyclodextrin nanofibers for entrapment of volatile organic compounds. Polym Eng Sci 54(12):2970–2978
55. Santos RPO, Rodrigues BVM, Ramires EC, Ruvolo-Filho AC, Frollini E (2015) Bio-based materials from the electrospinning of lignocellulosic sisal fibers and recycled PET. Ind Crops Prod 72:69–76
56. Mazrouei-Sebdani Z, Khoddami A, Hadadzadeh H, Zarrebini M (2015) Synthesis and performance evaluation of the aerogel-filled PET nanofiber assemblies prepared by electro-spinning. RSC Adv 5(17):12830–12842
57. Zander N, Gillan M, Sweetser D (2016) Recycled PET nanofibers for water filtration applications. Materials 9(4):247
58. Koushkbaghi S, Jafari P, Rabiei J, Irani M, Aliabadi M (2016) Fabrication of PET/PAN/GO/Fe_3O_4 nanofibrous membrane for the removal of Pb(II) and Cr(VI) ions. Chem Eng J 301:42–50
59. Qin D, Lu W, Wang X, Li N, Chen X, Zhu Z, Chen W (2016) Graphitic carbon nitride from burial to re-emergence on polyethylene terephthalate nanofibers as an easily recycled photocatalyst for degrading antibiotics under solar irradiation. ACS Appl Mater Interf 8 (39):25962–25970
60. Mahar FK, Mehdi M, Qureshi UA, Brohi KM, Zahid B, Ahmed F, Khatri Z (2017) Dyeability of recycled electrospun polyethylene terephthalate (PET) nanofibers: Kinetics and thermodynamic study. J Mol Liq 248:911–919
61. Morillo Martín D, Magdi Ahmed M, Rodríguez M, García M, Faccini M (2017) Aminated polyethylene terephthalate (PET) nanofibers for the selective removal of Pb(II) from polluted water. Materials 10(12):1352
62. Coşkun R, Birgül H, Delibaş A (2017) Synthesis of functionalized PET fibers by grafting and modification and their application for Cr(VI) ion removal. J Polym Res 25(1):29
63. Jin X, Wang H, Jin X, Wang H, Chen L, Wang W, Lin T, Zhu Z (2020) Preparation of keratin/PET nanofiber membrane and its high adsorption performance of Cr(VI). Sci Total Environ 710:135546
64. Biazar E, Ahmadian M, Heidari S, Gazmeh A, Mohammadi S-F, Lashay A, Heidari M, Eslami H, Sahebalzamani M, and Hashemi H (2017) Electro-spun polyethylene terephthalate (PET) mat as a keratoprosthesis skirt and its cellular study. Fibers Polym 18(8):1545–1553
65. Esmaeili E, Deymeh F, Rounaghi SA (2017) Synthesis and characterization of the electrospun fibers prepared from waste polymeric materials. Int J Nano Dimension 8(2):171–181
66. Salehi Shahrabi S, Barzin J, Shokrollahi P (2018) Statistical approach to estimate fiber diameter of PET/PVP blend electrospun using Taguchi method and fitting regression model. Mater Res Express 6(2):025306
67. Mohammadkhani F, Montazer M, Latifi M (2019) Microwave absorption and photocatalytic properties of magnetic nickel nanoparticles/recycled PET nanofibers web. J Text Inst 110 (11):1606–1614
68. Lindner M, Theurich J, Bahnemann DW (1997) Photocatalytic degradation of organic compounds: accelerating the process efficiency. Water Sci Technol 35(4):79–86
69. Gaya UI, Abdullah AH (2008) Heterogeneous photocatalytic degradation of organic contaminants over titanium dioxide: A review of fundamentals, progress and problems. J Photochem Photobiol C 9(1):1–12
70. Faisal M, Abu Tariq M, Muneer M (2007) Photocatalysed degradation of two selected dyes in UV-irradiated aqueous suspensions of titania. Dyes Pigm 72(2):233–239

71. Rajeshwar K, Chenthamarakshan CR, Goeringer S, Djukic M (2001) Titania-based heterogeneous photocatalysis. Materials, mechanistic issues, and implications for environmental remediation. Pure Appl Chem 73(12):1849–1860
72. Abbas JA, Said IA, Mohamed MA, Yasin SA, Ali ZA, Ahmed IH (2018) Electrospinning of polyethylene terephthalate (PET) nanofibers: optimization study using taguchi design of experiment. IOP Conf Ser Mater Sci Eng 454:012130
73. Yasin SA, Abbas JA, Ali MM, Saeed IA, Ahmed IH (2020) Methylene blue photocatalytic degradation by TiO2 nanoparticles supported on PET nanofibres. Materials Today: Proceedings 20:482–487
74. Yasin SA, Abbas JA, Mohamed MA, Saeed IA (2020) Data of characterization of electrospun waste polyethylene terephthalate (PET) nanofibers. Data in Brief 30:105535
75. Yasin SA, Abbas JA, Saeed IA, Ahmed IH (2019) The application of green synthesis of metal oxide nanoparticles embedded in polyethylene terephthalate nanofibers in the study of the photocatalytic degradation of methylene blue. Polym Bull 77(7):3473–3484

Reinforcement of Petroleum Wax By-Product Paraffins as Phase Change Materials for Thermal Energy Storage by Recycled Nanomaterials

Fathi S. Soliman, Heba H. El-Maghrabi, Gomaa A. M. Ali, Mohamed Ayman Kammoun, and Amr A. Nada

Abstract The energy derived from the solar activity is a source that can help solve the high demand that humanity presents in terms of thermal energy, but shows disadvantages due to the changes in prolonged periods and to the variability in very short times. The fundamental thing to take advantage of the higher amount of thermal energy derived from the sun is to count on storage systems that accumulate that energy in the form of latent heat. For this purpose, materials that change from the solid phase to the liquid (Phase Change Materials (PCMs)) are used; this way of storing and reserving energy is beneficial because large amounts of material are available, working isothermally during storage and releasing the energy stored in its solidification process. One of the advantages of latent heat storage is that said energy storage and its consequent delivery are presented in a minimal temperature range, called inter-phase or transition zone. The PCMs have appropriate characteristics for the storage of energy. At present, a diverse range of these materials is known with which it has been experienced, obtaining promising results; The most widely used materials are salts and some organic and inorganic materials; It is important to emphasize that these materials are difficult to regenerate insofar as they are subjected to work cycles, noting the decrease in their storage efficiency and

---

F. S. Soliman (✉) · H. H. El-Maghrabi (✉)
Department of Refining, Egyptian Petroleum Research Institute, Cairo 11727, Egypt
e-mail: fathisamir@gmail.com; fathisamir@epri.sci.eg

H. H. El-Maghrabi
e-mail: hebachem@yahoo.com

G. A. M. Ali
Chemistry Department, Faculty of Science, Al-Azhar University, Assiut 71524, Egypt

M. A. Kammoun
Laboratoire Sciences des Matériaux et Environnement, Faculté des Sciences de Sfax, Université de Sfax, Route. De Soukra Km 4, Sfax, Tunisia

A. A. Nada (✉)
Department of Analysis and Evaluation, Egyptian Petroleum Research Institute, Cairo 11727, Egypt
e-mail: chem_amr@yahoo.com; amr.nada@epri.sci.eg

A. S. H. Makhlouf and G. A. M. Ali (eds.), *Waste Recycling Technologies for Nanomaterials Manufacturing*, Topics in Mining, Metallurgy and Materials Engineering, https://doi.org/10.1007/978-3-030-68031-2_29

consequent dissociation, and on the contrary, paraffins (petroleum by-product) are economical materials with good behavior in storage and with acceptable energy storage ranges.

Keywords Paraffins · Phase change materials · Nanomaterials · Recycling · Petroleum by-products

List of Abbreviations

CNTs	Carbon nanotubes
DSC	Differential scanning calorimeter
NEPCMs	Nanocomposite-enhanced phase-change materials
PCMs	Phase change materials
SEM	Scanning electron microscope
UV	Ultra-violet

1 Introduction

Fossil fuels are the energy source that is most used to boost production processes and transportation. One of the disadvantages of this type of energy is the high degree of generated emissions when burned in the presence of air [1, 2]. This inconvenience has led to the search for energy sources, one of which is solar energy, and this source of energy is abundant. It has no cost, but its variations are very marked, not only during the day, if not throughout the year. These variations depend on climatic conditions typical of the region [3], in addition to the area where the catchment system is circumscribed, as well as the same solar activity (maximum and minimum solar) [4].

Solar energy can be stored thermally sensitive, latent, and thermochemical heat. The most representative systems for sensitive heat storage are water and solid phase matter, such as rocks. The devices for storing and reserving energy in the form of latent heat present highly defined characteristics when they experience a solid phase-to-liquid phase shift. In the thermochemical systems, energy is stored through reversible endothermic reactions [4–6].

The materials that accumulate energy in phase change are characterized by storing a huge amount of heat in the solid–liquid interface and yielding such energy in the solidification processes; They also have excellent efficiencies when they store thermal energy; this is derived from high storage densities, and their isothermal operation during the fusion causes the energy to be stored and released in the solidification phase. The PCMs materials melt and solidify in extensive temperature ranges; these systems have huge latent heat of fusion per unit of volume, large specific heat, the solid and liquid phase are highly conductive, allowing to store and dissociate the energy of the system. The PCMs phase change materials, whose purpose is to store latent heat energy, have probably been used since the Second World War, intended to generate warming of workspaces and homes in cold times. When using energy reserving storage systems in the form of latent heat, two relatively small volumes, their accumulation densities, and their regeneration processes must be taken into account; This does not happen with systems that use sensitive heat [7–9].

2 Classification of the Phase Change Materials

The storage is obtained by maintaining temperatures in specific ranges, and this causes the energy to be absorbed and stored, nowadays, fatty acids, paraffins, salts, and hydrated salts are used as shown in Scheme 1, it is crucial to keep in mind that in the studies made with phase change systems, innumerable substances have been used, however, currently, the most widely used are salts. Throughout development, the technological advances and the diversified use of substances, two groups, the organic components, and the inorganic ones can be identified. Organics do not generate corrosion and are very chemically stable; they are well-matched with many materials and have supreme large heat rates and small vapor pressure. These organic materials are not good thermal conductors; they change their volume and are flammable. The inorganic ones whether salts such as $AlCl_3$, $LiNO_3$, $NaNO_3$, KNO_3, Na_2O_2, KOH, $KClO_4$, LiH, $MgCl_2$, NaCl, Na_2CO_3, KF, LiF, K_2CO_3, NaF, MgF_2 which have melting temperature ranging from 192 to 1271 °C and a heat of fusion ranging from 280 to 936 j g^{-1} [10–13], salts hydrate such as $LiClO_3 . 3H_2O$, $KF . 4H_2O$, $Na_2SO_4 . 10H_2O$, $Na_2P_2O_7 . 10H_2O$, $Na_2HPO_4 . 12H_2O$, which have melting temperature ranging from 8 to 89.3 °C and a heat of fusion ranging from 115.5 to 279.6 j g^{-1} [13–17] over time, have proven to have high latent heat, good thermal conductivity, do not exhibit inflammation, and are very cheap. What puts them at a disadvantage with organic materials, the high degree of corrosion for the metal structures that contain them, break down easily and cool over quickly [18, 19].

The technological choice of the PCMs is carried out in applications by the phase change temperature when several products are in the indicated range, and it is useful to know which is the most convenient and what physical parameters make it possible to have optimal behavior.

3 Properties of Phase Change Materials

A very high melting point characterizes the materials that store energy in phase change, and their operating temperatures generate transitions in their phases, looking for characteristics in these energy storage systems and their consequent release. It has also been shown that systems with organic structure support many fusing and solidification cycles, do not sub-cool [21].

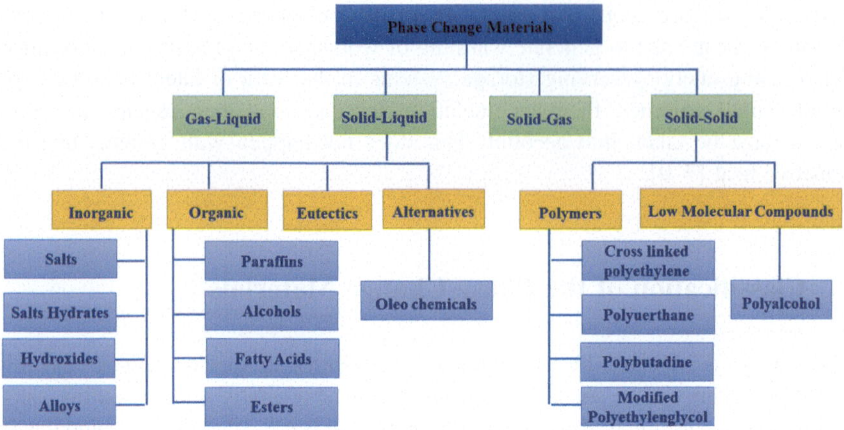

Scheme 1 Classification of the phase change materials. Adapted from Ref. [20], Copyright 2016, Elsevier

The store energy systems based on phase change materials which has high storage density during the phase changing from solid to liquid. In contrast, the stored energy released during the solidification process.

The phase change materials characterized by their temperature nature, high latent heat of fusion, low volumes changes between phases, high specific heat, good thermal conductivities. All of these advantages reinforced the energy transferring to the environmental. Next, a comparison is made between the different means of energy storage. The substances to be used must be chemically stable over time, do not present phase segregation, and be compatible with the security encapsulation. These materials should not be toxic, they should be environmentally friendly in terms of pollution levels, they should not be flammable, and of course, they should also not be at risk of explosion. Finally, they must be abundant and cheap so that they have technical and economic viability [21, 22] (Table 1).

Table 1 Characteristics of phase change materials

Thermal characteristics	Physical characteristics	Chemical characteristics	Economical characteristics	Kinetic characteristics
Accepted phase change temperature, large latent heat of transition, best heat transfer	Accepted phase equilibrium, high density, little change in volume, low vapor pressure	Significant time in chemical stability, well-matched with materials of construction, non-toxic, non-fire hazards	Plentiful, existing, low cost	No supercooling, large nucleation rate

4 Mechanism of the Phase Change Materials

The solid and liquid phases, characterized by cohesive forces derived from the proximity between atoms, and in the solid phase, the molecules vibrate around fixed equilibrium positions. In contrast, in the liquid phase, the molecules must remain between these equilibrium positions. The manifestations of these vibration energies are what we call thermal energy, whose measurement is temperature. Atoms in the liquid phase exhibit higher energy than in the solid phase, so when a solid softens and melts, it is essential to acquire a huge percentage of energy to overcome attraction forces, which keep the solid phase stable. This energy is the latent heat of phase change. It represents a thermal difference or phase enthalpy, liquid–solid, already to solidify. The solid and liquid phases, characterized by cohesive forces derived from the proximity between atoms, and in the solid phase, the molecules vibrate around fixed positions of equilibrium. In contrast, in the liquid phase, the molecules must remain between these equilibrium positions. The manifestations of these vibration energies are what we call thermal energy, whose measurement is temperature, as presented in Fig. 1.

Atoms in the liquid phase exhibit higher energy than in the solid phase, so when a solid softens and melts, it must acquire a large amount of energy to overcome forces of attraction, which keep the solid phase stable. This energy is the latent heat of phase change L. It represents a thermal difference, or what is called phase enthalpy, liquid–solid, already to solidify, it will be necessary to make cadence of all this energy so that the atoms recover its stability of all this energy so that the atoms recover their stability as illustrated in Fig. 2 [24].

5 Paraffins (Petroleum By-Product)

Organic compounds are chemical substances whose structure is based on carbon, hydrogen, and oxygen. In most cases, said compounds contain nitrogen, sulfur, phosphorus, boron, and halogens. Among the organic compounds most commonly

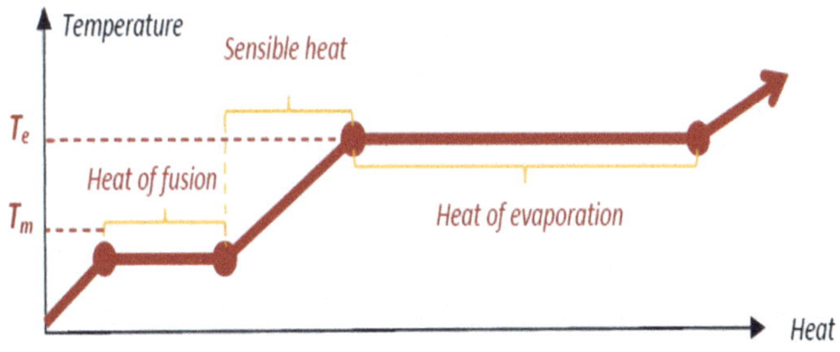

Fig. 1 General heating curve for material with phase changes: temperature as a function of heat added. Tm and Te indicate the melting and evaporation temperatures, respectively. Copied with permission from Ref. [23]

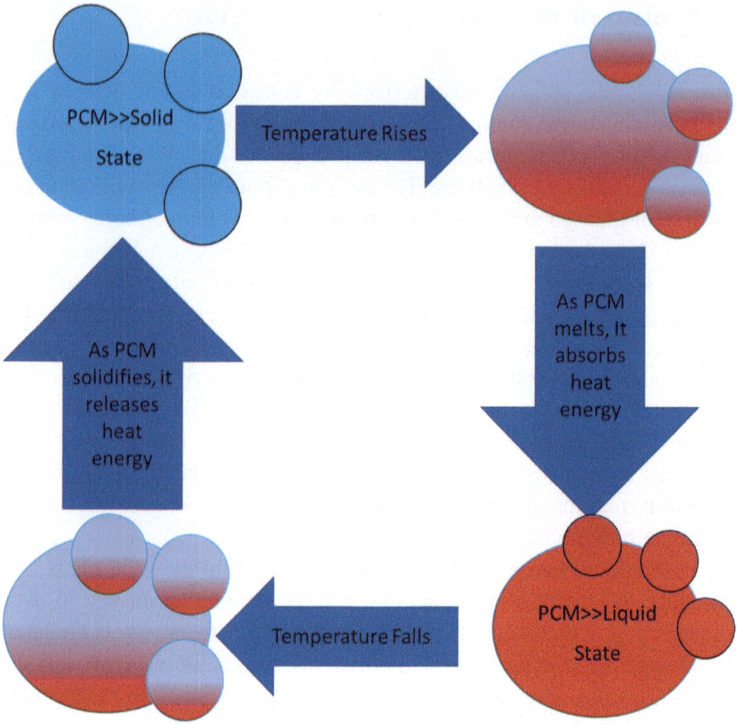

Fig. 2 Atoms recover their stability

used as phase change materials, there are polyols and paraffins. Paraffins are mixed saturated hydrocarbons, which contain alkanes that are obtained during petroleum refining processes; after refining processes are odorless, tasteless, and non-toxic substances; The crystal growth of the paraffins chain frees a high percentage of latent heat; pure paraffins have been presented in Fig. 3 [25, 26].

The increase in the melting point and latent heat of fusion is parallel with the increase in the chain length. Due to its availability in a high-temperature range, paraffins were nominated as the heat of fusion reserving and storage materials. Moreover, according to economic seeing, the only technical grade of paraffins is used as phase change materials in latent heat reserving and storage systems. Paraffins are benign, dependable, expectable, little cost, and do not have a corrosion effect. It is chemically non-reactive and unchanging under 500 °C shows small changes in volume on melting, and has small vapor pressure in the liquid form. Based on these interesting characteristics, systems of paraffins always have a big freeze to melt cycle. The increase in the melting point of alkane is well-matched with the rise in the carbon atoms number depending on the crystal size. There are two types of waxes.

Fig. 3 Pure paraffins material

(i) Macrocrystalline wax, which is generally known as paraffins is a solid and crystalline mixture of hydro-carbons; it is usually obtained in large crystals. It generally consists of normal paraffins ranging from C_{16}–C_{30} and may be higher. Proportions of slightly branched-chain paraffins ranging from C_{18}–C_{36} and naphthenes, especially alkyl-substituted derivatives of cyclopentane and cyclohexane, are also present. The average molecular weight of these paraffins is about 360–420. A paraffins wax melting at 53.5 °C showed a space lattice having C–C bond length of 1.52 Å, a C–C–C bond angle of 110 Å, a C–H bond length of 1.17 Å, and an H–C–H bond angle of 105 Å [25, 27–30].

(ii) Microcrystalline waxes are obtained from the vacuum residue. The source for the production of microcrystalline wax is petrolatum or bright stock [31]. Microcrystalline waxes contain highly branched-chain paraffins in contrast to the macrocrystalline, cyclo-paraffins, and slight percentages of *n*-paraffins and alkylated aromatics. The actual chain length of the *n*-alkanes is approximately C34–C50. Long-chain, branched iso-alkanes predominantly contain chain lengths up to C70 [31]. The branched-chain structures of the composition C_nH_{2n+2} are also found. Isoparaffins of monomethyl alkane and 2-methyl alkanes are being found. As the methyl group's position moves farther from the end of the chain, the corresponding alkane amount becomes smaller. The branched chains in the microcrystalline waxes are presented at random along the carbon chain; meanwhile, in paraffins wax, they are located at the end of the chain. The cyclo-alkanes, however, consist mainly of mono-cyclic systems. Monocyclopentyl, monocyclohexyl, dicyclohexyl paraffins, and polycyclic paraffins are also found. Some microcrystalline waxes are mainly composed of multiple-branched isoparaffins and monocycloparaffins. Moreover, non-hydrogenated microwaxes contain also mainly mono-cyclic and heterocyclic aromatic compounds [31]. Microcrystalline waxes have higher molecular weights (600–800), densities, and refractive indices than paraffins waxes [25–30].

5.1 *Properties of Paraffin Waxes*

5.1.1 Physical Properties

Almost all the physical properties of paraffins are affected by the hydrocarbon chain's diameter, distribution of their components, and degree of branching. Paraffins waxes are composed of 40–90 wt% normal paraffins of about 22–30 carbon atoms and possibly higher, accordingly, differ very little in physical and chemical properties. The remainder is C_{18}–C_{36} iso-alkanes and cycloalkanes. Straight chain alkanes from 20 up to 36 carbon atoms show transition points in the solid phase. Thus, two modifications are known, stable at different temperatures and with different crystal habits. Microcrystalline waxes contain substantial proportions of highly branched or cyclic hydrocarbons ranging from 30 to 75 carbon atoms [26, 30, 32, 33].

Paraffins waxes, relatively simple mixtures, usually have a narrow melting range and are generally lower in melting point than microcrystalline waxes. They usually melt between 46 and 68 °C. The melting point of paraffins waxes increases in parallel with molecular weight. The branching of the carbon chain, at identical molecular weights, results in a decrease in the melting point. Paraffins waxes can be classified according to the melting point to soft (lower m.p.) and hard (higher m.p.) paraffins waxes.

Microcrystalline waxes are more complicated, so it melts over a much wider temperature range. They usually melt between 60 and 93 °C and 38 to 60 °C, respectively. Oil percentage is a fingerprint of the paraffin value. The determination method depends upon the differential solubility of oil and wax in a given solvent [33].

Paraffins wax, microcrystalline wax, and petrolatum have a different degree of affinity for oil content. Paraffins wax has little affinity for oil content. It may be taken as a degree of refinement. Fully refined wax usually has an oil percentage of less than 0.5%. Microcrystalline waxes have a higher tendency to oil than paraffins due to their lesser crystal structure. The oil percentage of microcrystalline wax is 1–4 wt%, depending on wax's grade [26, 29, 30].

5.1.2 Mechanical Properties

The hardness and crystallization behavior of macrocrystalline paraffins waxes are interfered with distinctly by their distribution width, average chain length, and *n*-alkane content [34]. Hardness is the resistance against a body (needle, or cone, or plunger rod) under a defined load; this body is made of a harder material than the substance being tested. To measure the hardness of paraffins waxes, penetration tests are widely accepted. It is a common feature of strength and hardness tests that the test specimens are subjected to short-time stresses. The penetration test is the most widespread technique for determining the hardness and the thermal sensitivity of petroleum waxes. Macrocrystalline waxes change to a greater extent with temperature than that of microcrystalline waxes. An intensification in oil content leads to increased penetration values of both macro-and microcrystalline waxes [26, 34].

5.1.3 Food Grade Properties

Their potential toxicity could be attributed to aromatic residues. The latter is characterized directly by using UV spectra in the spectral zone corresponding to aromatics. Each country has adopted its code governing waxes, which come into contact with food and non-food grade [35].

5.2 *Crystal Structure of Paraffins*

The class of organic crystals represents a broad range of geometries, including needles, plates, cubes, rods, prisms, pentagons, octagons, hexagons, rhomboids, and pyramids. Each of these forms results from crystallization from a solution. The crystals' geometry is determined by the solute/solvent interaction and the physical conditions of the system (e.g., temperature, pressure, and mechanical mixing).

One attractive characteristic of crystals is that they can form various shapes, which are due to the environmental conditions under which they form. They can be large or small, extend long distances or short, be well-defined or diffuse; in short, they can display an impressive array of forms. It is this variety of forms upon which crystal modifiers are intended to take advantage.

All petroleum waxes are crystalline to some degree, and it is possible to classify waxes in terms of the type of crystals formed when the wax crystallizes out of solution [26, 33].

5.2.1 Macrocrystalline Waxes (Paraffins Waxes)

The paraffins crystals appear in three different forms: plates, needles, and mal shapes; the latter are small size, undeveloped crystals, which are often agglomerates. Many researchers have studied the conditions for the formation of these shapes. They have come to the following conclusions:

(1) The three crystal forms of paraffins waxes depend on both the crystallization process's conditions and the chemical composition of the wax.
(2) Plate crystals are obtained from lower boiling points paraffinic distillates, while the needle and mal-shaped crystals are obtained from the higher boiling points ones and vacuum residues.
(3) For a given molecular weight limit, the higher melting point constituents crystallize in plate type in which the crystals are hexagonal plate. The

low-melting ones crystallize in needles while the medium-melting ones crystallize in mal shapes.

(4) Normal paraffins crystallize in plates. Needle crystals contain both aliphatic and cyclic hydrocarbons, while mal-shaped crystals are characterized by their content of branched hydrocarbons.

(5) Low-cooling rates during crystallization will result in large crystals for both plate and needle forms, while the crystal growth formal-shaped crystals are very slight.

(6) The solubility of paraffins in a solvent is inversely proportional to their melting points. In the presence of a solvent, wax mixtures begin to crystallize at relatively low temperatures in plates followed by mal-shaped crystals. However, the constituents crystallizing in needles are more soluble than those crystallizing in plates. Therefore, needles crystals will appear only at lower temperature and higher concentrations

(7) Plate crystals can readily be transformed into needle and mal-shaped crystals. Under appropriate conditions, the needle crystals can be transformed into mal-shaped crystals.

Normal paraffins, C_{17}–C_{34}, may exist in three and possibly four crystal forms. Near the melting point, hexagonal crystals are a stable form. At somewhat lower temperatures, the odd-numbered from C_{19} to C_{29} is orthorhombic, even-numbered ones from C_{18}–C_{26} is triclinic, and those C_{28}–C_{36} is monoclinic [25, 26, 33].

5.2.2 Microcrystalline Waxes

Both *n*-paraffins and isoparaffins crystallize in needle forms; they differ in that the latter does so at all temperatures while higher temperatures are required for the former. The needle form of the isoparaffins differs from that of ceresins or paraffins waxes containing ceresins in that the crystals of the former are large and loose, while those of the latter are extremely small and dense [25, 33].

Microcrystalline waxes may contain substantial percentages up to 30% of paraffins, which crystallize well as high-melting macrocrystalline or paraffins wax when separated. The microcrystalline wax material interferes and imposes its crystallizing habit on the other material. Although the classification of petroleum waxes into macrocrystalline and microcrystalline waxes based on crystal size is valid to a great extent, there is no sharp line separating the two groups. Indeed, a large group of waxes could fall in either class, and these waxes are called intermediate waxes, blended waxes, mal-crystalline waxes, and semi-microcrystalline waxes. However, semi-microcrystalline wax was adopted.

5.3 Manufacture of Paraffins Waxes

The manufacture of paraffins waxes is nearly related to the manufacture of lubricating oils. The raw paraffins distillates and residual oils contain wax, and they are generally solid at ambient temperature. Removal of wax from these fractions is necessary to permit lubricating oil with a good low pour point. Production of paraffins waxes includes the next technological techniques [26]:

(1) Production of slack waxes and petrolatums by de-waxing petroleum products
(2) Refining of the wax products
(3) Deoiling and fractional crystallization
(4) Percolation process
(5) Hydrofinishing process
(6) Acid treatment
(7) Adsorption treatment.

6 Additives to Paraffins

Various ways were applied to improve the thermophysical properties of PCMs, involving inoculation of porous material [36–38], distribution of high thermal conductivity nanoparticles in the phase change material [39], and depositing metal in phase change material [40]. The improving additives with high thermal conductivity, such as metallic nanoparticle [41], carbon material [42], and ceramic particles such as TiO_2, Al_2O_3, CuO, and Fe_2O_3 [43–46] were also used for improvements. Binary metal oxides structures were found to have a high ability to create oxygen vacancies and modify the based material's electronic structure. The creation of an electron- charged space layer and negative oxygen ions on the surface by binary metal oxide improve the thermal conductivity of the based material [47].

6.1 Recycled Nanomaterials

Newly, the managing of waste produced from the industry by benign and economic means is well-thought-out to be one of the key priorities for all the world. Recovering, waste managing, and processing manufacturing waste into novel goods by workable wanted methods are significant problems for scientists and the business sector [48–58]. On a universal scale, almost 30–50 million tons of electronic and electrical wastes are produced annually, with an annual growth percentage of up to 5% [59].

Electronic waste is known to be industrial mines because it is rich in metals. Moreover, electronic waste was a transporter of heavy metals (Cu, Zn, Ni, etc.) and carbon-based wastes. Thus, electronic waste recovery technology is essential to the concept of electronic waste. If recreation technology is peaceful and efficient to the environment, electronic waste is a treasure mine. If recreation technology is primitive, electronic waste can become dangerous for people and the environment [60–62].

Designing and employing end-of-life wastes as a new beginning material for the generation of new high-quality, for instance. Nanoparticles could be a sensible plan to stimulate industrial attention in serious recreation achievements due to potentially acceptable economic returns with state subsidies [63–65]. For example, nanometals like Zn, Mn, Ni, and Li were extracted and purified from all types of batteries (Zn–Mn, Ni-MH, Zn–C batteries, Zn–MnO_2 batteries, Li-ion batteries, and car batteries) by methods such as hydro-metallurgy-precipitation and liquid–liquid extraction, hydrometallurgy, thermal methods-condensation using inert gas and separation by vacuum, sol–gel, leaching, and precipitation [63, 66–80]. Also nanozero-valent Cu, Pb, carbon nanotubes, Zn, Mn, Fe, Ni, Cr, Au, and Ag were extracted and separated from used instruments or engines by different techniques and methods the same as mentioned above [81–90]. Moreover, nanometal oxides such as TiO_2, SiO_2, Fe_3O_4, and mixed metal oxides such as $MnFe_2O_4$ and $CuFe_2O_4$ can be excluded from waste sludge [91–93]. Scheme 2 lists the various spent products that have been treated to date for the recovery of different metals and nanomaterials [64].

6.2 Paraffins Containing Graphene and Carbon Nanotubes

Organic paraffins have drawn important consideration due to their ability to store free thermal energy by large storage cycle through the solid phase and liquid phase within a small temperature range [94, 95]. Besides the high energy storage capacity, considerable thermal and chemical stability, there is no corrosion resistance, and subcooling makes organic paraffins hopeful for reserving latent heat, reservation, and transfer in applications of effectual thermal managing of electrical apparatus, thermal energy reserve, and storage as a heat transference media. The main disadvantage of carbon-based phase change materials is their low thermal conductivity, causing a weak change in temperature and decreased heat transference capability. Customary endeavors to create progressed paraffins center around improving the thermally conductive foams or packed graphite [96–99], or structuring phase change compounds by incorporating the paraffins with thermally conductive fillers of multiple resources and carbon different shapes, and dispersing paraffins into highly thermally conductive porous structures.

The authors studied different forms of graphite on paraffins. Strengthening heat transfer, the results show that different forms of graphite are significantly improved. The phase change heat transference performance of paraffins wax and the strengthening effect of graphite foam are the best. The addition of graphite shows that paraffins wax is fully adsorbed to the cellular micropores of foamed graphite. In

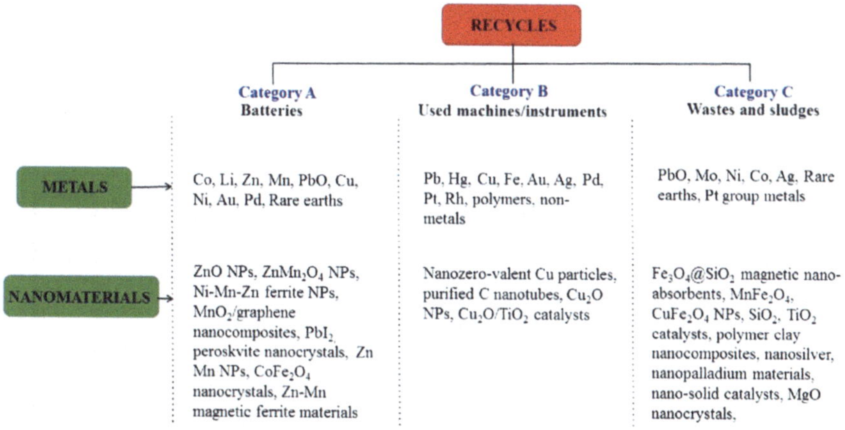

Scheme 2 List the various spent products that have been treated to date for the recovery of different metals and nanomaterials

the middle, the filling of foamed graphite dramatically enhances the phase change material [100]. The phase transition temperature of the composite phase change heat storage material is similar to that of paraffins wax, its phase variable latent heat, and calculated latent heat based on paraffins content in composite materials.

Nevertheless, using solid matrices or combining additives, which have not to phase change, displaces a sufficient percentage of the paraffins capacity and results in the drawback of the heat storage ability. Weakness in heat transfer also occurs through boundaries between nearby nanofills and paraffins matrix. It is still a target in improving refined paraffins to achieve, at the same time, improved thermal transference and energy storage ability [101].

6.3 *Paraffins Containing Nanomaterials*

The most fundamental problem of current PCMs is the weakness in thermal conductivity. To beat this deficiency and gain high thermal characteristics, various techniques have been planned to enhance phase change materials' thermal conductivity, like adding nanoceramics with high thermal conductivity. Teng and Yu reported nanocomposite-enhanced phase change materials (NE PCMs) via the heat conduction method, as demonstrated in Fig. 4 [102]. The thermal conductivity of PCMs can be modified by adding high-conductive compounds, which is a feasible and straightforward method.

Metals or metal oxides such as Al_2O_3 and TiO_2 nanoparticles were usually acclimated to improve heat transference in energy change and managing systems. Some studies have been done to promote latent heat thermal energy storage systems with modified thermal fulfillment. It has been demonstrated that the good thermal

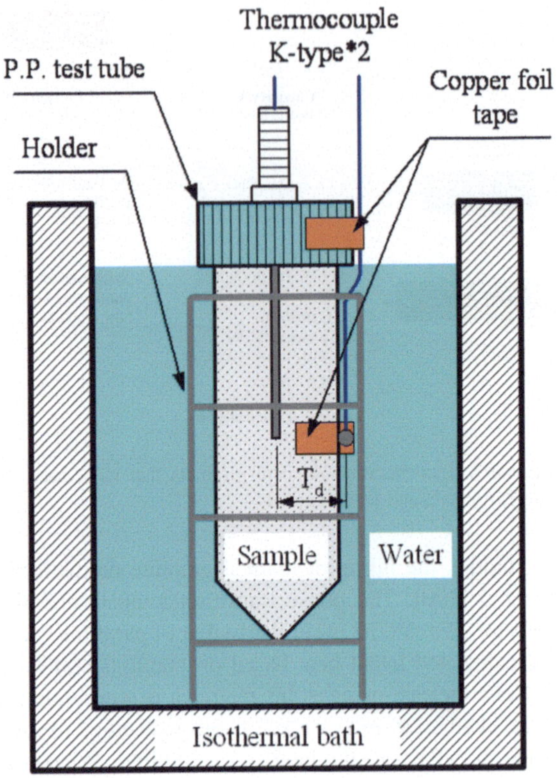

Fig. 4 Test apparatus used for the heat conduction performance of NE PCMs. Reproduced with permission from Ref. [102], Copyright 2012, Springer Nature

conductivity of paraffins latent heat storage can be automatically enlarged using smaller volumetric concentrations of Al_2O_3 nanoparticles [43, 46].

SiO_2 is an inorganic porous compound, which is nonflammable. So, the latent heat of its sturdy shape composite phase change material can be used while the thermal energy loading and disloading cycles. The preparation and characterization of stabilized shape paraffins/SiO_2 composites have been carried out, where paraffins can be applied as the PCM, and SiO_2 is used as the supportive compound [103].

7 Measurement Techniques of Latent Heat of Fusion and Melting Temperature

Analyzing the latent heat of fusion and melting temperature of the phase change materials is done by two techniques: the first one is differential thermal analysis, and the second one is differential scanning calorimeter (DSC), where the target sample and standard compounds are heated from room temperature to the desired temperature at the fixed ramping amount. The difference in temperature between them is proportional to the heat flow difference between the two materials [21, 104].

The most common technique for studying phase change is DSC, where it measures the heat capacities of small samples of materials, whose diameter is commonly 6.3 mm and the length of 6.3 mm. DSC analysis gives the solidification and melting curves and the amount of latent heat stored. The test's principle is to keep a temperature of equilibrium between the tested sample and reference material, usually a certified metal, and to heat or cool at a constant rate. The excess heat absorbed or emitted is then recorded as a function of time and integrated to determine the quantities of total heat exchanged. DSC method is useful for predicting a PCM's performance, mainly when the product is still in development. Once mature, large-scale tests provide a better prediction since it characterizes the material's effective properties, namely MCP and its conditioning [105].

8 Applications of the Phase Change Materials

The storage of energy in the form of latent heat has found several important industrial and domestic applications. Among the most notable are the following: In peak periods of use of electrical energy, since some of the applications of the electrical power in this sense, heating and cooling can be realized by a system of storage of energy with the change of phase. Encapsulation in construction structures; It has become a novel proposal for the conditioning of closed housing spaces since the PCMs directly absorbs the cold or heat without going through an exchange system, as has been commonly analyzed. Currently known: walls impregnated with PCMs; Building blocks for construction that serve as a support matrix for PCMs; The microencapsulation of PCMs in very small structures has been used in solar energy on solar panels and electronic technology [106].

8.1 Textiles

In the mid-1980s, NASA built up the innovation to join phase change materials microcapsules in strands to enhance their exhibition for dealing with the warm obstruction characteristics of a piece of clothing textures, especially space suits. Phase change materials, for example, nonadecane, were exemplified in materials with the expectation of diminishing the effect of outrageous temperature varieties experienced by space travelers during space missions with different levels of temperatures. This was accomplished by drenching empty strands into a phase change material arrangement, for example, a fluid arrangement of hydrated inorganic salt and low atomic weight polyethylene glycol. Despite the fact the temperature-versatile textures presented attractive warm stockpiling and discharge properties, they were questionable and showed poor warm conduct after rehashed warm cycling. Along these lines, different methodologies were grown, for example, smaller scale epitome; to upgrade the exhibition of thermo-controlling fibers, PCM

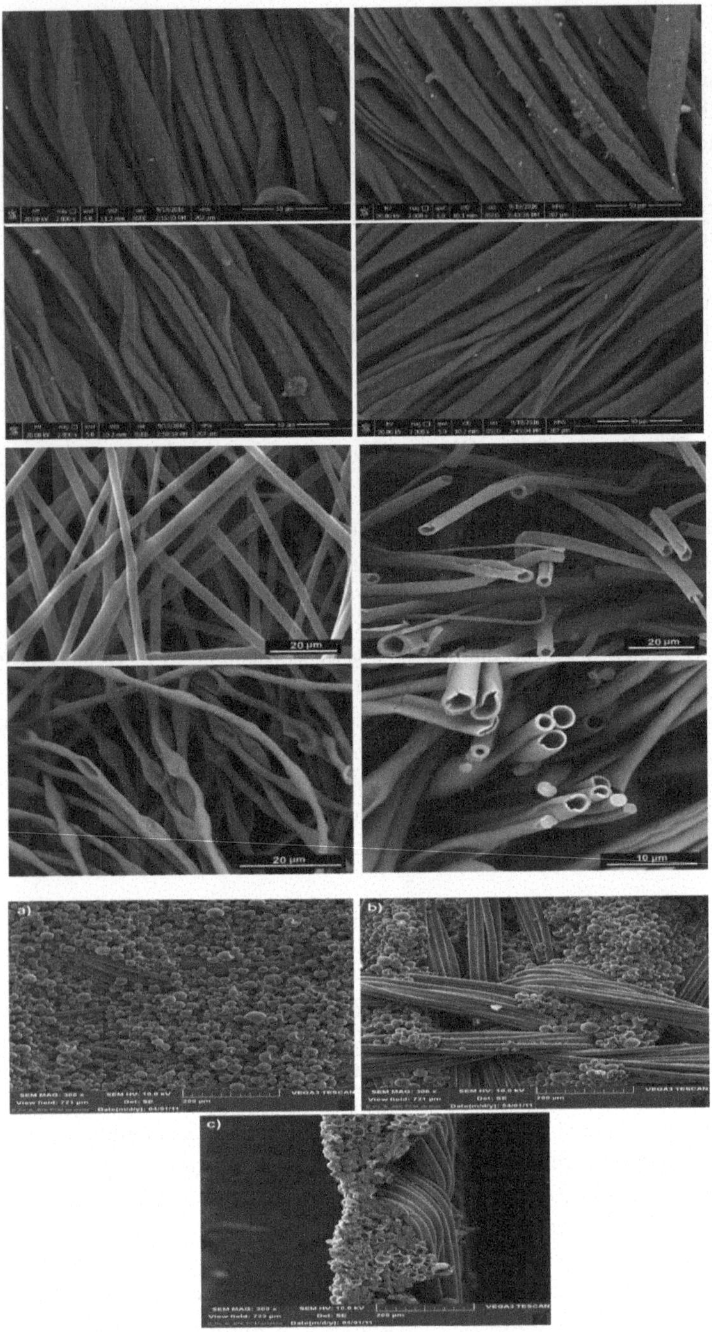

Fig. 5 SEM of improved textile fibers: (**a**) coated side, (**b**) uncoated side, (**c**) cross-section. Reproduced with permission from Refs. [110–112], Copyright 2019, Elsevier

containers are presently utilized monetarily for material fibers and texture coatings. They are used for outside clothes and in blankets, duvets, mattresses, and pillowcases. Moreover, it can be constructed for warmth, textiles containing phase change materials can also stop the overheating, so their effect can act as a thermal regulator (Fig. 5) [4, 107–110].

8.2 Thermal-Chemical Systems

Thermo-compound frameworks are characterized as those administered by dominating qualities of paraffins, for example, non-reactive, non-toxic, beside chemically stable. These characteristics can guard touchy materials against warm stuns just as from organisms for a long time because of cold-stockpiling impacts. Till now, paraffins have been presented as the potential materials dependent on which main thermo-chemical frameworks are expounded in the consequent segments.

8.3 Magnetic Materials and Characterization

Electromagnetic impedances have turned into the worldwide ecological contamination source due to the abundance utilization of electronic gadgets. Accordingly, numerous materials are urged to be manufactured, having a nanoscale thickness and powerful broadband microwave absorption ability to safeguard the electromagnetic waves, as illustrated in Fig. 6 [113]. In any case, when the characterizations identified with electromagnetic conditions of these materials are directed, a few difficulties are experienced during the test planning stage. A composite example is specially arranged in paraffins (nearly 40% by weight), where it can participate as a dielectric host matrix [114, 115].

The fundamental driver of forcing paraffins in such frameworks emerges from its settled and effectively quantifiable permittivity and porousness in the ideal microwave recurrence run, which are exceptionally requested properties to do theoretical demonstrating for magnetized materials. Also, the air hole between the examined test and the holder of the gadget ought to be purposely killed during the portrayal of magnetized properties. What is more, this can be well-guaranteed utilizing paraffins network whose simple embellishment, shape-versatility, and low twisting edge permit to drain airgap uncertainty; therefore, prompting improved trial exactness. Notwithstanding the use of paraffins only to describe the magnetized materials, it is worth referencing that a bi-functional material has been as of late created, containing energy stockpiling just as super-magnetized properties presented in Fig. 7 for SEM and magnetic behavior [116]. Such a material might be better named as an attractive phase change material. It was synthesized through a two-step approach, i.e., Fe_3O_4 was collected onto the *n*-eicosane core, and then, the capsule was formed by SiO_2 (the shell) by sol–gel process [121, 122]. In any case, no huge variety

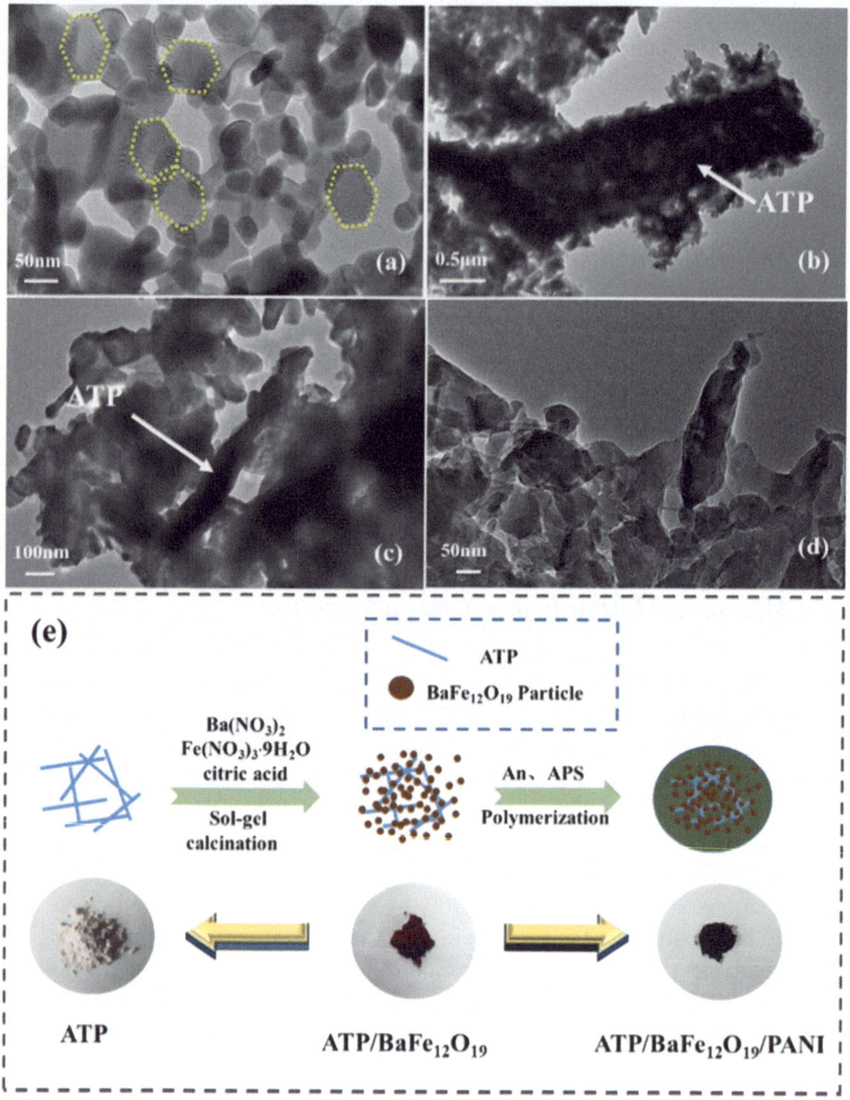

Fig. 6 TEM images of (**a**) $BaFe_{12}O_{19}$, (**b**, **c**) ATP/$BaFe_{12}O_{19}$, and (**d**) ATP/$BaFe_{12}O_{19}$/PANI. (**e**) Schematic illustration of the preparation process of ATP/$BaFe_{12}O_{19}$/PANI. Reproduced with permission from Ref. [113], Copyright 2018, Elsevier

happened in stage change temperature, and the watched attractive hysteresis test made the magnetized phase change material to be exceptionally capable, being referred to with remarkable alluring highlights, for example, low retentivity and low coercivity. Generally speaking, the examinations opens up the fascinating qualities and imaginative uses of PCMs in the space of attraction. From there on,

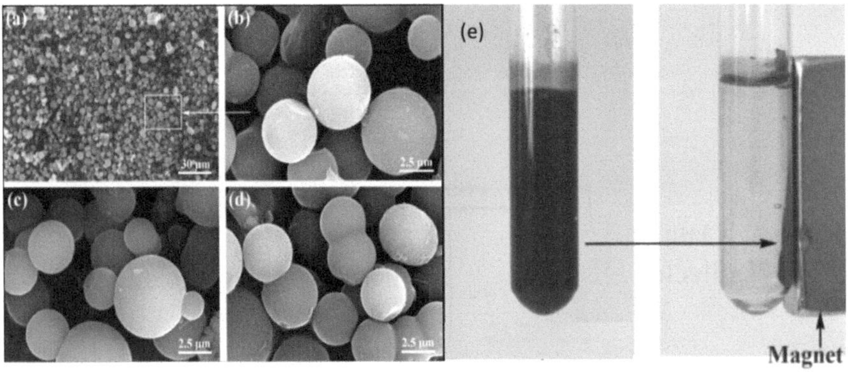

Fig. 7 **(a–d)** SEM micrographs of the magnetic microcapsules synthesized with *n*-eicosane/TEOS with different mass ratios and **(e)** the photographs for the Fe_3O_4 nanoparticles collected by a piece of magnet. Reproduced with permission from Ref. [116], Copyright 2014, Elsevier

investigating the microwave retention properties to diminish the electromagnetic contamination alongside waste warmth stockpiling ought to be the subsequent stage through such magnetized PCMs.

8.4 Thermal Therapy

Thermal medication presented the First World War when French doctor E. B de Sandfort enrolled a patent on inventive consuming treatment. As he alluded his item name, Keritherapy was expected to have paraffins wax and golden resin oil. Afterward, numerous doctors prescribed paraffinic warm treatment as it was beneficial for diminishing joint torments and ailment. Additionally, such wax medicines had demonstrated track records in enhancing the skin [117]. The working instrument behind warm treatment using paraffins wax is not so confounded, i.e., insofar as pieces of the body are submerged in low temperatures liquefied paraffins, or it shall be added straightforwardly onto the pieces of the body, heat move happens from the softened paraffins. In cold areas of the body, accordingly actuating and continuing warm impacts. As it was accepted that warmth medication, as a rule, appease the agonizing body parts, like these warm impacts improved by paraffinic treatment, were medicinally enormous. Also, rhinitis was a typical present situation over the planet, which was perceived by side effects like wheezing and nasal tingling. A stage change-based warm treatment veil has been exhibited to adapt to this sort of unfavorably susceptible issue.

Paraffins wax (softening temperature of 44 °C) was impregnated with non-toxic polypropylene fiber. Along these lines formed into a cuboid block, as demonstrated in Fig. 8, numerical investigation displayed that cuboid square accomplished temperature around 43 °C where it was held for 2250 s in normal air convection, which

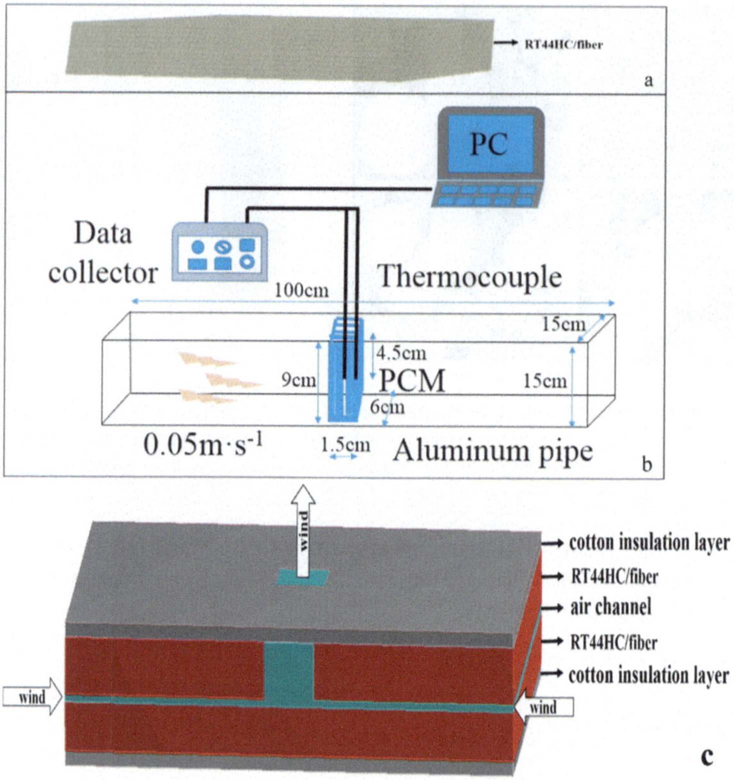

Fig. 8 Schematic diagrams of a RT44HC/fiber composite PCM-based cuboid block (**a**) and set-up for experimentally evaluating the thermal storage performance of the cuboid block (**b**). Channel-type mask based on the RT44HC/fiber composite PCM-based cuboid blocks (**c**). Reproduced with permission from Ref. [118], Copyright 2017, Elsevier

was a reasonable time for warm treatment of nose, particularly experiencing unfavorably susceptible rhinitis. It has also been accounted for that low thermal conductivity of phase change material for warm treatment frameworks is required [118].

8.5 *Biomaterial Storage*

Bio cells much like human cells and tissues should be harmed if the biopsies are taken from the various human body, for example, colon, digestive organs, etc.; in any case, the protection in paraffinic bed mitigates abrupt vulnerabilities. In biobanks, it is a significant typical plan for tissue storage examples in paraffins waxes, where they hold morphology and atomic structures for a long time. What is more, inferable from paraffinic flexibility, those installed tissues may be re-used and

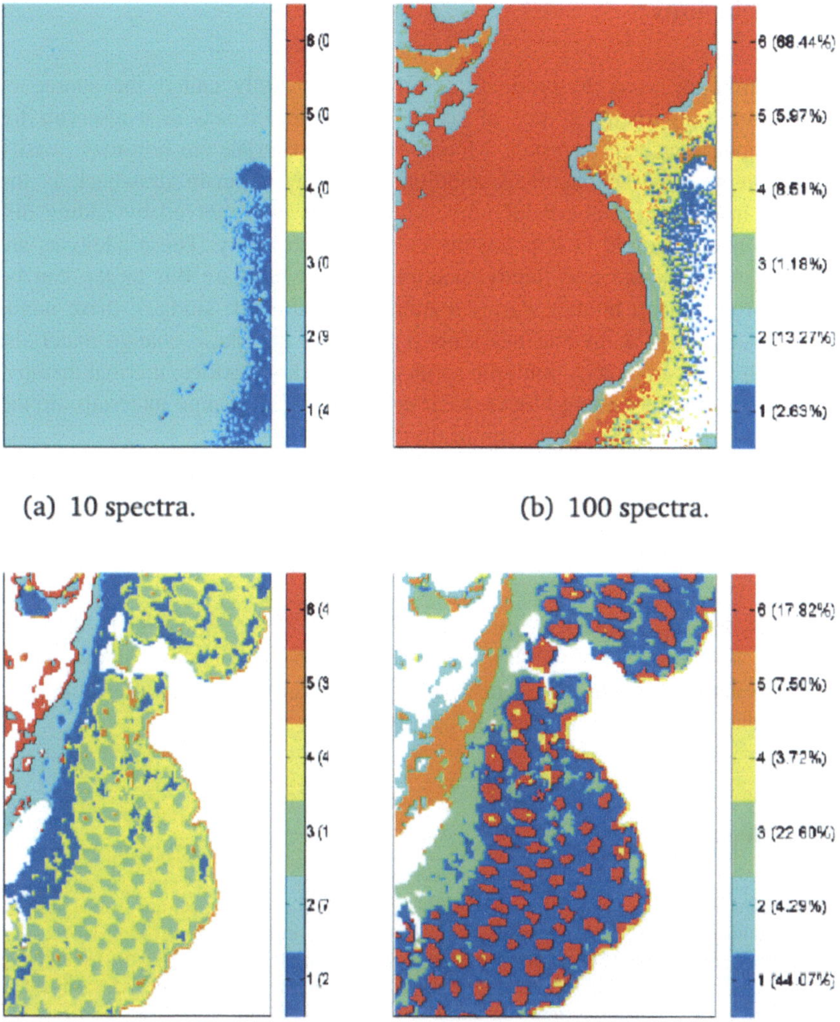

Fig. 9 Influence of the number of spectra of pure paraffin used in the PCA model. Reproduced with permission from Ref. [119], Copyright 2017, Royal Society of Chemistry

reanalyzed the same number of times as required without dealing with paraffins misfortune, as revealed in Fig. 9 [119].

Later the paraffins implanted tissues are de-waxed by multiple methods (chemical, mechanical, and advanced methods preceding) send them for malignant growth determination using infrared spectroscopy. As accomplished within sight of paraffins, the probability of tissue security is credited to the more drawn out substance resistance of paraffins against thermophilic organisms, i.e., microscopic organisms [120].

9 Conclusions

Waste petroleum waxy by-products or what is generally called the source of paraffins are a significant source of latent heat storage. It acts as a reservoir for energy as phase change materials. Moreover, these materials are non-toxic, easily preparation besides the chemical stability. However, its main drawback is the weakness in their thermal conductivity, which needs to be solved by adding different additives required to improve the thermal conductivity. These additives are classified to more than one family, nanomaterials, which are our target, can be extracted from cheap human wastes (batteries, instruments, sludge); these nanomaterials succeeded in freeing heat energy caged in the phase change materials. Phase change materials have multiple applications such as textiles, thermal therapy, biomaterial storage, building blocks, solar cell, etc.; phase change materials are the cheap chance to multiple future applications.

10 Future Perspectives

In the last decade, PCMs have proved their power in the industrial field, while incredible advances in material designing and the physical phenomena' knowledge have been made. In contrast, several challenges remain in terms of alleviation, the reset current, and enhancement the reliability.

References

1. Leonard MD, Michaelides EE, Michaelides DN (2020) Energy storage needs for the substitution of fossil fuel power plants with renewables. Renew Energy 145:951–962
2. Curtin J, McInerney C, Gallachóir BÓ, Hickey C, Deane P, Deeney P (2019) Quantifying stranding risk for fossil fuel assets and implications for renewable energy investment: a review of the literature. Renew Sustain Energy Rev 116:109402
3. Nkwetta DN, Haghighat F (2014) Thermal energy storage with phase change material—a state-of-the art review. Sustain Cities Soc 10:87–100
4. He M, Yang L, Lin W, Chen J, Mao X, Ma Z (2019) Preparation, thermal characterization and examination of phase change materials (PCMs) enhanced by carbon-based nanoparticles for solar thermal energy storage. J Energy Storage 25:100874
5. Prieto C, Cabeza LF (2019) Thermal energy storage (TES) with phase change materials (PCM) in solar power plants (CSP). Concept and plant performance. Appl Energy 254:113646
6. Paksoy H, Sahan N (2012) Thermally enhanced paraffin for solar applications. Energy Proc 30:350–352
7. Lingayat AB, Suple YR (2013) Review on phase change material as thermal energy storage medium: materials, application. Int J Eng Res Appl 3:916–921
8. Cui Y, Xie J, Liu J, Pan S (2015) Review of phase change materials integrated in building walls for energy saving. Proc Eng 121:763–770

9. Mhike W, Focke WW, Mofokeng J, Luyt AS (2012) Thermally conductive phase-change materials for energy storage based on low-density polyethylene, soft Fischer-Tropsch wax and graphite. Thermochim Acta 527:75–82
10. Hasnain S (1998) Review on sustainable thermal energy storage technologies, part I: heat storage materials and techniques. Energy Convers Manage 39(11):1127–1138
11. Heine D (1980) Chemische und physikalische Eigenschaften von Latentwärmespeichermaterialien für Solarkraftwerke
12. Xu B, Li P, Chan C (2015) Application of phase change materials for thermal energy storage in concentrated solar thermal power plants: a review to recent developments. Appl Energy 160:286–307
13. Pielichowska K, Pielichowski K (2014) Phase change materials for thermal energy storage. Prog Mater Sci 65:67–123
14. Khudhair AM, Farid MM (2004) A review on energy conservation in building applications with thermal storage by latent heat using phase change materials. Energy Convers Manage 45(2):263–275
15. Naumann R, Emons H-H (1989) Results of thermal analysis for investigation of salt hydrates as latent heat-storage materials. J Therm Anal 35(3):1009–1031
16. Paris J, Falardeau M, Villeneuve C (1993) Thermal storage by latent heat: a viable option for energy conservation in buildings. Energy Sources 15(1):85–93
17. Nagano K, Mochida T, Takeda S, Domański R, Rebow M (2003) Thermal characteristics of manganese (II) nitrate hexahydrate as a phase change material for cooling systems. Appl Therm Eng 23(2):229–241
18. Kousksou T, Jamil A, El Rhafiki T, Zeraouli Y (2010) Paraffin wax mixtures as phase change materials. Sol Energy Mater Sol Cells 94(12):2158–2165
19. Evers AC, Medina MA, Fang Y (2010) Evaluation of the thermal performance of frame walls enhanced with paraffin and hydrated salt phase change materials using a dynamic wall simulator. Build Environ 45(8):1762–1768
20. Kapsalis V, Karamanis D (2016) Solar thermal energy storage and heat pumps with phase change materials. Appl Therm Eng 99:1212–1224
21. Sharma A, Tyagi VV, Chen C, Buddhi D (2009) Review on thermal energy storage with phase change materials and applications. Renew Sustain Energy Rev 13(2):318–345
22. Kenisarin MM, Kenisarina KM (2012) Form-stable phase change materials for thermal energy storage. Renew Sustain Energy Rev 16(4):1999–2040
23. Sevault A, Kauko H, Bugge M, Banasiak K, Haugen N, Skreiberg Ø (2017) Phase change materials for thermal energy storage in low- and high-temperature applications: a state-of-the-art:1–53
24. Alexiades V, Solomon AD (1993) Mathematical modeling of melting and freezing processes. Hemisphere Publ. Corp., Washington
25. Smith A (1953) The crystal structure of the normal paraffin hydrocarbons. J Chem Phys 21 (12):2229–2231
26. Sequeira A (1994) Lubricant base oil and wax processing. CRC Press
27. Zaky MT, Mohamed NH, Farag AS, Soliman FS (2015) Raising the efficiency of petrolatum deoiling process by using non-polar modifier concentrates separated from paraffin wastes to produce different petroleum products. RSC Adv 5(88):71932–71941
28. Mohamed NH (2012) Competitive study on separation and characterization of microcrystalline waxes using two deoiling techniques. Fuel Process Technol 96:116–122
29. Zaky MT, Mohamed NH, Farag AS (2011) Separation of some paraffin wax grades using solvent extraction technique. Fuel Process Technol 92(10):2024–2029
30. Zaky MT, Mohamed NH (2010) Comparative study on separation and characterization of high melting point macro-and micro-crystalline waxes. J Taiwan Inst Chem Eng 41(3):360–366
31. Meyer G (2009) Thermal properties of micro-crystalline waxes in dependence on the degree of deoiling. SOFW J 135(8):43–50

32. Kuszlik A, Meyer G, Heezen P, Stepanski M (2010) Solvent-free slack wax de-oiling—physical limits. Chem Eng Res Des 88(9):1279–1283
33. Ohlberg SM (1959) The stable crystal structures of pure *n*-paraffins contalmng an even number of carbon atoms in the range C30 to C36. J Phys Chem 63(2):248–250
34. Esquena J, Vilasau J (2013) TF Tadros (ed) Formulation, characterization, and property control of paraffin emulsions. https://doi.org/10.1002/9783527647941
35. Brown W, Marques MR (2013) 14 the United States pharmacopeia/national formulary. In: Generic drug product development: solid oral dosage forms, vol 319
36. Krupa I, Nógellová Z, Špitalský Z, Malíková M, Sobolčiak P, Abdelrazeq HW, Ouederni M, Karkri M, Janigová I, Al-Maadeed MAS (2015) Positive influence of expanded graphite on the physical behavior of phase change materials based on linear low-density polyethylene and paraffin wax. Thermochim Acta 614:218–225
37. Mancin S, Diani A, Doretti L, Hooman K, Rossetto L (2015) Experimental analysis of phase change phenomenon of paraffin waxes embedded in copper foams. Int J Therm Sci 90:79–89
38. Reyes A, Negrete D, Mahn A, Sepúlveda F (2014) Design and evaluation of a heat exchanger that uses paraffin wax and recycled materials as solar energy accumulator. Energy Convers Manage 88:391–398
39. Oya T, Nomura T, Tsubota M, Okinaka N, Akiyama T (2013) Thermal conductivity enhancement of erythritol as PCM by using graphite and nickel particles. Appl Therm Eng 61(2):825–828
40. Yu S, Jeong S-G, Chung O, Kim S (2014) Bio-based PCM/carbon nanomaterials composites with enhanced thermal conductivity. Sol Energy Mater Sol Cells 120:549–554
41. Li M, Wu Z, Kao H, Tan J (2011) Experimental investigation of preparation and thermal performances of paraffin/bentonite composite phase change material. Energy Convers Manage 52(11):3275–3281
42. Xu B, Li Z (2014) Paraffin/diatomite/multi-wall carbon nanotubes composite phase change material tailor-made for thermal energy storage cement-based composites. Energy 72:371–380
43. Wang J, Xie H, Guo Z, Guan L, Li Y (2014) Improved thermal properties of paraffin wax by the addition of TiO_2 nanoparticles. Appl Therm Eng 73(2):1541–1547
44. Jiang X, Luo R, Peng F, Fang Y, Akiyama T, Wang S (2015) Synthesis, characterization and thermal properties of paraffin microcapsules modified with nano-Al_2O_3. Appl Energy 137:731–737
45. Jesumathy S, Udayakumar M, Suresh S (2012) Experimental study of enhanced heat transfer by addition of CuO nanoparticle. Heat Mass Transf 48(6):965–978
46. Mohamed NH, Soliman FS, El Maghraby H, Moustfa Y (2017) Thermal conductivity enhancement of treated petroleum waxes, as phase change material, by α nano alumina: energy storage. Renew Sustain Energy Rev 70:1052–1058
47. Li J, Wang X, Qiao Y, Zhang Y, He Z, Zhang H (2015) High thermal conductivity through interfacial layer optimization in diamond particles dispersed Zr-alloyed Cu matrix composites. Scripta Mater 109:72–75
48. Dhmees AS, Rashad AM, Eliwa AA, Zawrah M (2019) Preparation and characterization of nano SiO_2@ CeO_2 extracted from blast furnace slag and uranium extraction waste for wastewater treatment. Ceram Int 45(6):7309–7317
49. Hegde G, Abdul Manaf SA, Kumar A, Ali GAM, Chong KF, Ngaini Z, Sharma KV (2015) Biowaste sago bark based catalyst free carbon nanospheres: waste to wealth approach. ACS Sustain Chem Eng 5(9):2247–2253
50. Ali GAM, Habeeb OA, Algarni H, Chong KF (2018) CaO impregnated highly porous honeycomb activated carbon from agriculture waste: symmetrical supercapacitor study. J Mater Sci 54:683–692
51. Ali GAM, Divyashree A, Supriya S, Chong KF, Ethiraj AS, Reddy M, Algarni H, Hegde G (2017) Carbon nanospheres derived from Lablab purpureus for high performance supercapacitor electrodes: a green approach. Dalton Trans 46(40):14034–14044

52. Ali GAM, Tan LL, Jose R, Yusoff MM, Chong KF (2014) Electrochemical performance studies of MnO_2 nanoflowers recovered from spent battery. Mater Res Bull 60:5–9
53. Ali GAM, Yusoff MM, Shaaban ER, Chong KF (2017) High performance MnO_2 nanoflower supercapacitor electrode by electrochemical recycling of spent batteries. Ceram Int 43:8440–8448
54. Ali GAM, Abdul Manaf SA, Kumar A, Chong KF, Hegde G (2014) High performance supercapacitor using catalysis free porous carbon nanoparticles. J Phys D-Appl Phys 47 (49):495307–495313
55. Aboelazm EAA, Ali GAM, Algarni H, Yin H, Zhong YL, Chong KF (2018) Magnetic electrodeposition of the hierarchical cobalt oxide nanostructure from spent lithium-ion batteries: its application as a supercapacitor electrode. J Phys Chem C 122(23):12200–12206
56. Ali GAM, Yusoff MM, Algarni H, Chong KF (2018) One-step electrosynthesis of MnO_2/rGO nanocomposite and its enhanced electrochemical performance. Ceram Int 44(7):7799–7807
57. Ali GAM, Supriya S, Chong KF, Shaaban ER, Algarni H, Maiyalagan T, Hegde G (2019) Superior supercapacitance behavior of oxygen self-doped carbon nanospheres: a conversion of allium CEPA peel to energy storage system. Biomass Conv Bioref https://doi.org/10.1007/s13399-019-00520-3
58. Ali GAM, Manaf SAA, Divyashree A, Chong KF, Hegde G (2016) Superior supercapacitive performance in porous nanocarbons. J Energy Chem 25(4):734–739
59. Afroz R, Masud MM, Akhtar R, Duasa JB (2013) Survey and analysis of public knowledge, awareness and willingness to pay in Kuala Lumpur, Malaysia—a case study on household WEEE management. J Clean Product 52:185–193
60. Zeng X, Yang C, Chiang JF, Li J (2017) Innovating e-waste management: from macroscopic to microscopic scales. Sci Total Environ 575:1–5
61. Ruan J, Huang J, Dong L, Huang Z (2017) Environmentally friendly technology of recovering nickel resources and producing nano-Al_2O_3 from waste metal film resistors. ACS Sustain Chem Eng 5(9):8234–8240
62. Ilgin MA, Gupta SM (2010) Environmentally conscious manufacturing and product recovery (ECMPRO): a review of the state of the art. J Environ Manage 91(3):563–591
63. Deep A, Kumar K, Kumar P, Kumar P, Sharma AL, Gupta B, Bharadwaj LM (2011) Recovery of pure ZnO nanoparticles from spent Zn–MnO_2 alkaline batteries. Environ Sci Technol 45(24):10551–10556
64. Dutta T, Kim K-H, Deep A, Szulejko JE, Vellingiri K, Kumar S, Kwon EE, Yun S-T (2018) Recovery of nanomaterials from battery and electronic wastes: a new paradigm of environmental waste management. Renew Sustain Energy Rev 82:3694–3704
65. Li Y, Ye D, Sun Y, Wang Y, Shi B, Liu W, Guo R, Pei H, Zhao H, Zhang J (2020) Recycling materials from degraded lithium-ion batteries for Na-ion storage. Mater Today Energy 15:100368
66. Qu J, Feng Y, Zhang Q, Cong Q, Luo C, Yuan X (2015) A new insight of recycling of spent Zn–Mn alkaline batteries: Synthesis of $Zn_xMn_{1-x}O$ nanoparticles and solar light driven photocatalytic degradation of bisphenol A using them. J Alloy Compd 622:703–707
67. Deep A, Sharma AL, Mohanta GC, Kumar P, Kim K-H (2016) A facile chemical route for recovery of high quality zinc oxide nanoparticles from spent alkaline batteries. Waste Manage 51:190–195
68. Mantuano DP, Dorella G, Elias RCA, Mansur MB (2006) Analysis of a hydrometallurgical route to recover base metals from spent rechargeable batteries by liquid–liquid extraction with Cyanex 272. J Power Sources 159(2):1510–1518
69. Xi G, Li Y, Liu Y (2004) Study on preparation of manganese–zinc ferrites using spent Zn–Mn batteries. Mater Lett 58(7–8):1164–1167
70. Gabal M, Al-Harthy E, Al Angari Y, Salam MA, Asiri A (2016) Synthesis, characterization and magnetic properties of MWCNTs decorated with Zn-substituted $MnFe_2O_4$ nanoparticles using waste batteries extract. J Magn Magn Mater 407:175–181

71. Kim T, Senanayake G, Kang J, Sohn J, Rhee K, Lee S, Shin S (2009) Reductive acid leaching of spent zinc–carbon batteries and oxidative precipitation of Mn–Zn ferrite nanoparticles. Hydrometallurgy 96(1–2):154–158
72. Nan J, Han D, Cui M, Yang M, Pan L (2006) Recycling spent zinc manganese dioxide batteries through synthesizing Zn–Mn ferrite magnetic materials. J Hazard Mater 133(1–3):257–261
73. Duan X, Deng J, Wang X, Guo J, Liu P (2016) Manufacturing conductive polyaniline/graphite nanocomposites with spent battery powder (SBP) for energy storage: a potential approach for sustainable waste management. J Hazard Mater 312:319–328
74. Wen X, Qiao X, Han X, Niu L, Huo L, Bai G (2016) Multifunctional magnetic branched polyethylenimine nanogels with in-situ generated Fe_3O_4 and their applications as dye adsorbent and catalyst support. J Mater Sci 51(6):3170–3181
75. Xiang X, Xia F, Zhan L, Xie B (2015) Preparation of zinc nano structured particles from spent zinc manganese batteries by vacuum separation and inert gas condensation. Sep Purif Technol 142:227–233
76. Xi G-X, Jiao Y-Z, Lu M-X (2008) Preparation of $CoFe_2O_4$ nanocrystal from spent lithium-ion batteries with coprecipitation method. Electron Compon Mater 27(5):19
77. Deng J, Wang X, Duan X, Liu P (2015) Facile preparation of MnO_2/graphene nanocomposites with spent battery powder for electrochemical energy storage. ACS Sustain Chem Eng 3(7):1330–1338
78. Yang L, Xi G, Lou T, Wang X, Wang J, He Y (2016) Preparation and magnetic performance of $Co_{0.8}Fe_{2.2}O_4$ by a sol–gel method using cathode materials of spent Li-ion batteries. Ceram Int 42(1):1897–1902
79. Yao L, Xi Y, Xi G, Feng Y (2016) Synthesis of cobalt ferrite with enhanced magnetostriction properties by the sol−gel−hydrothermal route using spent Li-ion battery. J Alloy Compd 680:73–79
80. Belcher AM, Chen P-Y, Hammond-Cunningham PT, Qi J (2017) Recycling car batteries for perovskite solar cells. Google Patents
81. Singh J, Lee B-K (2016) Recovery of precious metals from low-grade automobile shredder residue: a novel approach for the recovery of nanozero-valent copper particles. Waste Manage 48:353–365
82. Xiu F-R, Zhang F-S (2012) Size-controlled preparation of Cu_2O nanoparticles from waste printed circuit boards by supercritical water combined with electrokinetic process. J Hazard Mater 233:200–206
83. Vermisoglou EC, Giannouri M, Todorova N, Giannakopoulou T, Lekakou C, Trapalis C (2016) Recycling of typical supercapacitor materials. Waste Manage Res 34(4):337–344
84. Zhan L, Xiang X, Xie B, Sun J (2016) A novel method of preparing highly dispersed spherical lead nanoparticles from solders of waste printed circuit boards. Chem Eng J 303:261–267
85. Wu J-Y, Chang F-C (2016) Recovery of nano-Al2O3 from waste aluminum electrolytic solution generated during the manufacturing of capacitors. Desalin Water Treat 57 (60):29479–29487
86. Hu Z, Kurien U, Murwira K, Ghoshdastidar A, Nepotchatykh O, Ariya PA (2016) Development of a green technology for mercury recycling from spent compact fluorescent lamps using iron oxides nanoparticles and electrochemistry. ACS Sustain Chem Eng 4 (4):2150–2157
87. Tunsu C, Petranikova M, Gergorić M, Ekberg C, Retegan T (2015) Reclaiming rare earth elements from end-of-life products: a review of the perspectives for urban mining using hydrometallurgical unit operations. Hydrometallurgy 156:239–258
88. Bandara HD, Field KD, Emmert MH (2016) Rare earth recovery from end-of-life motors employing green chemistry design principles. Green Chem 18(3):753–759
89. Dupont D, Binnemans K (2015) Rare-earth recycling using a functionalized ionic liquid for the selective dissolution and revalorization of Y_2O_3: Eu^{3+} from lamp phosphor waste. Green Chem 17(2):856–868

90. Tan Q, Deng C, Li J (2016) Innovative application of mechanical activation for rare earth elements recovering: process optimization and mechanism exploration. Sci Rep 6:19961
91. Bennett JA, Wilson K, Lee AF (2016) Catalytic applications of waste derived materials. J Mater Chem A 4(10):3617–3637
92. Chen D, Li Q, Shao L, Zhang F, Qian G (2016) Recovery and application of heavy metals from pickling waste liquor (PWL) and electroplating wastewater (EPW) by the combination process of ferrite nanoparticles. Desalin Water Treat 57(60):29264–29273
93. Tang B, Yuan L, Shi T, Yu L, Zhu Y (2009) Preparation of nano-sized magnetic particles from spent pickling liquors by ultrasonic-assisted chemical co-precipitation. J Hazard Mater 163(2–3):1173–1178
94. Nabil M, Khodadadi J (2013) Experimental determination of temperature-dependent thermal conductivity of solid eicosane-based nanostructure-enhanced phase change materials. Int J Heat Mass Transf 67:301–310
95. Khodadadi J, Fan L, Babaei H (2013) Thermal conductivity enhancement of nanostructure-based colloidal suspensions utilized as phase change materials for thermal energy storage: a review. Renew Sustain Energy Rev 24:418–444
96. Chintakrinda K, Weinstein RD, Fleischer AS (2011) A direct comparison of three different material enhancement methods on the transient thermal response of paraffin phase change material exposed to high heat fluxes. Int J Therm Sci 50(9):1639–1647
97. Pincemin S, Py X, Olives R, Christ M, Oettinger O (2008) Elaboration of conductive thermal storage composites made of phase change materials and graphite for solar plant. J Sol Energy Eng 130(1):011005
98. Lafdi K, Mesalhy O, Elgafy A (2008) Merits of employing foam encapsulated phase change materials for pulsed power electronics cooling applications. J Electron Packag 130 (2):021004
99. Sarı A, Karaipekli A (2007) Thermal conductivity and latent heat thermal energy storage characteristics of paraffin/expanded graphite composite as phase change material. Appl Therm Eng 27(8–9):1271–1277
100. Wu J, Feng Y, Liu C, Li H (2018) Heat transfer characteristics of an expanded graphite/ paraffin PCM-heat exchanger used in an instantaneous heat pump water heater. Appl Therm Eng 142:644–655
101. Xin G, Sun H, Scott SM, Yao T, Lu F, Shao D, Hu T, Wang G, Ran G, Lian J (2014) Advanced phase change composite by thermally annealed defect-free graphene for thermal energy storage. ACS Appl Mater Interfaces 6(17):15262–15271
102. Teng T-P, Yu C-C (2012) Characteristics of phase-change materials containing oxide nano-additives for thermal storage. Nanoscale Res Lett 7(1):611
103. Arasu AV, Mujumdar AS (2012) Numerical study on melting of paraffin wax with Al_2O_3 in a square enclosure. Int Commun Heat Mass Transfer 39(1):8–16
104. Buddhi D, Sawhney R, Sehgal P, Bansal N (1987) A simplification of the differential thermal analysis method to determine the latent heat of fusion of phase change materials. J Phys D Appl Phys 20(12):1601
105. Baetens R, Jelle B, Gustavsen A (2010) Phase change materials for building applications: a state-of-the-art review. Energy Build
106. Macía A (2007) Almacenamiento de Energía Solar Térmica Usando Cloruro de Magnesio Hexahidratado. 2007, Tesis de Maestría. Universidad Nacional de Colombia. Medellín
107. Pause B (2010) Phase change materials and their application in coatings and laminates for textiles. In: Smart textile coatings and laminates. Elsevier, pp 236–250
108. Nejman A, Cieślak M (2017) The impact of the heating/cooling rate on the thermoregulating properties of textile materials modified with PCM microcapsules. Appl Therm Eng 127:212–223
109. Bahrar M, Djamai ZI, Mankibi ME, Larbi AS, Salvia M (2018) Numerical and experimental study on the use of microencapsulated phase change materials (PCMs) in textile reinforced concrete panels for energy storage. Sustain Cities Soc 41:455–468

110. Nejman A, Goetzendorf-Grabowska B (2013) Heat balance of textile materials modified with the mixtures of PCM microcapsules. Thermochim Acta 569:144–150
111. Haghighat F, Ravandi SAH, Esfahany MN, Valipouri A, Zarezade Z (2019) Thermal performance of electrospun core-shell phase change fibrous layers at simulated body conditions. Appl Therm Eng 161:113924
112. Hassabo AG, Mohamed AL (2017) Enhancement the thermo-regulating property of cellulosic fabric using encapsulated paraffins in modified pectin. Carbohyd Polym 165:421–428
113. Bai D, Feng H, Chen N, Tan L, Qiu J (2018) Synthesis, characterization and microwave characteristics of ATP/$BaFe_{12}O_{19}$/PANI ternary composites. J Magn Magn Mater 457:75–82
114. Liu P, Yao Z, Zhou J (2016) Fabrication and microwave absorption of reduced graphene oxide/$Ni_{0.4}Zn_{0.4}Co_{0.2}Fe_2O_4$ nanocomposites. Ceram Int 42(7):9241–9249
115. Xu H, Yin X, Li Z, Liu C, Wang Z, Li M, Zhang L, Cheng L (2018) Tunable dielectric properties of mesoporous carbon hollow microspheres via textural properties. Nanotechnology 29(18):184003
116. Jiang F, Wang X, Wu D (2014) Design and synthesis of magnetic microcapsules based on *n*-eicosane core and Fe_3O_4/SiO_2 hybrid shell for dual-functional phase change materials. Appl Energy 134:456–468
117. Vracknos N (2013) Method and apparatus of paraffin treatment of the skin. Google Patents
118. Zhang Q, He Z, Fang X, Zhang X, Zhang Z (2017) Experimental and numerical investigations on a flexible paraffin/fiber composite phase change material for thermal therapy mask. Energy Storage Mater 6:36–45
119. de Lima FA, Gobinet C, Sockalingum G, Garcia SB, Manfait M, Untereiner V, Piot O, Bachmann L (2017) Digital de-waxing on FTIR images. Analyst 142(8):1358–1370
120. Nallala J, Lloyd GR, Stone N (2015) Evaluation of different tissue de-paraffinization procedures for infrared spectral imaging. Analyst 140(7):2369–2375
121. El-Maghrabi HH, Abdelmaged SM, Nada AA, Zahran F, Abd El-Wahab S, Yahea D, Hussein GM, Atrees MS (2017) Magnetic graphene based nanocomposite for uranium scavenging. J Hazardous Mate 322:370–379
122. El-Maghrabi HH, Al-Kahlawy AA, Nada AA, Zakia T (2018) Photocorrosion resistant $Ag_2CO_3@Fe_2O_3/TiO_2$-NT nanocomposite for efficient visible light photocatalytic degradation activities. J Hazardous Mate 360:250–256

Manufacturing of Nanoalumina by Recycling of Aluminium Cans Waste

Aiman Awadh Bin Mokaizh and Jun Haslinda Binti Haji Shariffuddin

Abstract Aluminium drink cans, which are usually disposed of after use, led to sizable land pollution and environmental problems. The main target of this work is to utilize aluminium waste cans to produce nano-alumina that considered a significant oxide that has wide technological-and industrial-applications. The objectives of this study are to synthesize and characterize nano aluminium oxide (alumina) that produced from aluminium waste cans as well as to evaluate the synthesized alumina in the treatment of palm oil mill effluent (POME). In this study, the aluminium waste cans will be transformed into nano and micro-sized alumina by using the sol–gel method. The effect of ageing time and ageing temperature in synthesizing nano Al_2O_3 has been studied. The properties of the synthesized alumina were characterized via FTIR, XRD, SEM, EDX, and BET. The produced alumina was used as an adsorbent in the treatment of POME through adsorption. The experimental outcomes divulged that the powder of the produced alumina at room-temperature has aloft surface areas that being an appropriate property to be applied as an adsorbent in the treatment of POME. AL-6 h-30 °C gave the highest percentage of 29% in COD removal for POME. In short, the utilization of aluminium waste cans to produce nano Al_2O_3 is feasible. This economic and environmental-friendly route turned waste into valuable adsorbent that can be used extensively in the treatment of the POME.

Keywords Aluminium waste cans · Alumina · Sol–gel method; palm oil mill effluent

List of Abbreviations

BET	Brunauer–Emmett–Teller
BJH	Barrett–Joyner–Halenda
CMI	Can manufacturers institute
COD	Chemical oxygen demand

A. A. Bin Mokaizh (✉) · J. H. B. H. Shariffuddin
Faculty of Chemical and Natural Resources Engineering, Universiti Malaysia Pahang, 26300 Gambang, Pahang, Malaysia
e-mail: ay.stmy@gmail.com

A. S. H. Makhlouf and G. A. M. Ali (eds.), *Waste Recycling Technologies for Nanomaterials Manufacturing*, Topics in Mining, Metallurgy and Materials Engineering, https://doi.org/10.1007/978-3-030-68031-2_30

EDX	Energy-dispersive X-ray spectroscopy
FTIR	Fourier transform infrared spectroscopy
OES	Optical emission spectroscopy
POME	Palm oil mill effluent
R&D	Research and development
SEM	Scanning electron microscope
TEM	Transmission electron microscope
TGA	Thermo-gravimetric analysis
XRD	X-ray diffraction

1 Introduction

Recently, Malaysia has been suffering from waste management due-to quick economic growth, people increasing, mainland shortage, needy-infrastructure, and-expertise. Solid waste management is becoming a significant environmental and public health problem in urban areas, as urbanization continues to evolve [1]. Solid waste management is considered a critical issue in Malaysia, which should be taken seriously [1, 2]. With an anticipated populace of more than 31 million in 2016, Malaysia generates more than 25,000 metric tons of domestic waste per day and a total of 7 million tons per year, with an annual increase of 3% [1, 3]. Over the last few decades, intensive Research and Development (R&D) endeavours have been coordinated towards finding cost-effective and practical solutions for waste minimization and utilization [4]. A few of the later patterns in strong squander administration incorporate more current or re-engineered forms coming about in diminishing amounts of emanations, methodologies to maximize squander utilization and improvement of high-value items [4]. One of the most efficient and environmentally ways to handle some categories of hazardous wastes is recycling, which results in decreasing the amount of disposal in landfills [5, 6]. However, according to statistics obtained from the Department of Environment Malaysia, there is only 2% of municipal solid waste that is recycled, 42% of them are incinerated or chemically treated, and the rest is dumped into landfills. Recycling is significant due to several economic and environmental reasons.

There are large amounts of aluminium cans that can be presented in landfills, which are domestic waste, industrial solid waste, and aluminium production waste. Aluminium can be lightweight, strong, easy to package and transport because of using less material and also more valuable than glass or plastic. According to a new report released by the Aluminium Association and Can Manufacturers Institute (ACMI), aluminium cans are the most sustainable beverage package and highest recycling rate that contains 70% recycled content on average more than three times the recycled content of glass or plastic bottles. Although aluminium cans are recyclable, they still carry environmental effects and may pose a health risk to the public [7]. Most drink cans are made of aluminium, while almost all food cans are made of steel because the unique properties of aluminium make it ideal for holding carbonated beverages. Aluminium cans weigh less; however, it can withstand the pressure exerted by the carbon dioxide in the drinks [8]. Aluminium can waste is usually recycled to produce secondary metal, which required 5% of the energy required for primary aluminium production [9]. The production of secondary aluminium cans can be as little as 60 days. Aluminium beverage cans are the most recycled packaging material because of the high value of the scrap and ease of collection. As an addition, aluminium cans retain their properties throughout the process as it is one of the most indestructible materials used two in metal containers. Recycling aluminium can wastes is a common practice since it has a high scrap metal value, can help to reduce production cost and minimize negative environmental effects.

Recycling is considered one of the effective ways of managing solid waste. Recycling provides environmental benefits in terms of energy savings, reduced volumes of waste, and cost-effective. Aluminium refreshment cans, which are by and large disposed of after utilized, have tall esteem of reusing. Recycling of aluminium requires only 5% of the energy for primary aluminium production [9]. Reusing aluminium can squander spare valuable normal assets, vitality, time, and cash [10]. Recycling aluminium waste cans saves up to 95% of the energy needed to produce primary metal and only generates 5% of the greenhouse gas emissions [11]. Recycling of one kilogram of aluminium waste cans saves about 8 kg of bauxite, 4 kg of chemical products, and 14-kWh of electricity [8]. Furthermore, recycling aluminium waste can also help to reduce the volume of waste disposed of in landfills. Recyclable beverage cans are generally composed of aluminium alloy and containing about 1.5 wt% magnesium (Mg) [8]. Table 1 summarizes the composition of components present in the aluminium beverage can.

To address the outbreak of wastes in the environment, the "from waste to wealth" concept had been implemented by transforming recycled aluminium waste cans into aluminium oxide.

Alumina or aluminium oxide is a chemical compound of aluminium and oxygen with the chemical formula Al_2O_3. Alumina is a white solid or nearly colourless crystalline substance and odourless. It serves as the raw material for the broad range of advanced ceramic products and as an active agent in chemical processing. Since the alumina is low cost, has excellent thermal stability, high specific area, surface acidity, and interaction with deposited transitional metals, it is most commonly used

Table 1 Chemical composition of the beverage can (mass %) measured by optical emission spectroscopy (OES) [8]

Component	Composition (%)
Al	97.12
Mn	1.08
Mg	0.863
Fe	0.589
Si	0.18
Others	0.153

as a catalyst, catalytic supports, and adsorbents in the industry [12–14]. Alumina, Al_2O_3, has a molar mass of 101.96 g mol^{-1} and a density of 3.95–4.1 g cm^{-1}. Alumina has high melting and boiling points of 2072 °C and 2977 °C, respectively. Alumina also has a high compression strength, high abrasion resistance, high chemical resistance, high thermal shock resistance, a high degree of refractoriness, and high dielectric strength, which is transparent to microwave radio frequencies and has a low neutron cross section capture [15]. Alumina is used to produce primary aluminium and is extracted from ore bauxite in a refinery process known as Bayer, where the impurities such as iron oxide and silica would be reduced. However, this conventional method of producing alumina is expensive and harmful to the environment because it generates bauxite residue or red mud, which is of insoluble particles. The method of synthesizing alumina from aluminium can waste is a simple, secure, and rapid production process wherein the main returns are energy and time savings, plus it is an environmentally friendly technique. Alumina exists in several different crystalline structure phases due to thermal calcination treatment such as η-, γ-, δ-, θ-, β-, κ-, χ, and α-Alumina [15]. α-Alumina is thermally stable, while the others are in metastable phases. Calcination of aluminium hydroxide ($Al(OH)_3$) at different temperatures results in the variety of crystalline structures of alumina where the increase of temperature leads to transform alumina towards the more stable phase as illustrated in the below process [13] (Fig. 1).

As shown from the above figure, there are several types of alumina; however, the most useful types of alumina are two general classes of alumina, the low surface area, alpha-alumina (α-Al_2O_3), or corundum, and the highly porous alumina, which has widely used called gamma-alumina (γ-Al_2O_3) [13, 15]. Gamma-alumina (γ-Al_2O_3) is highly porous alumina and commonly modified. γ-Alumina has a spinel structure and one of the most important nanomaterials used as an adsorbent or as a support for metal catalysts. Gamma-alumina is one of the greatest extensively utilized in industrial catalysis owing to its relatively enormous surface area, unique surface characteristics, and remarkable structural stability. γ-Alumina is an unstable phase and can be transformed into θ-Alumina at the temperature of 750–1000 °C. γ-Alumina is widely utilized as an adsorbent or a catalyst or catalyst support in many chemical processes such as ammonia synthesis, synthesis gas, and hydrogen production, oils hydrogenation, petroleum refining, and automotive emissions

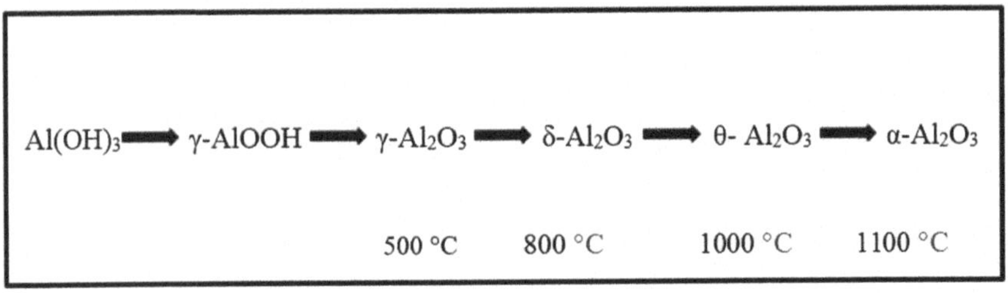

Fig. 1 Alumina phase transformation structures series due to increasing in calcination temperature [13]

control because it has a large specific area, low surface energy, chemically stable, and high mechanical strength [11, 16]. This phase is not used in the high-temperature applications because this phase can transform into the stable phase, which is α-alumina at a higher temperature. Alpha-Alumina (α-Al_2O_3) is the most stable form of alumina. It is used as a structural ceramic due to having good mechanical properties and excellent thermal properties because it is produced during calcination at a high temperature, which is more than 1100 °C [17]. It has a high thermal shock resistance because of the fact of high average grain size for α-Alumina of about 200 μm. The purity of α-Alumina determines the application of the alumina itself. The main application of α-Alumina is to be used in mechanical, electronic, and bio-ceramics applications [18]. In addition, crystalline of α-Al_2O_3 is nanoparticles that have a particle size distribution ranging from 30 to 32 nm, and this nano-alumina structure is extensively used in various fields such as electronics, composites, and bio-ceramics implants applications [19]. Alumina is mainly produced by the Bayer process, a high-energy consumption technology which will lead to high production cost [20]. Therefore, by using aluminium waste cans to produce alumina, it is then a great interest from economic, energetic, and environmental perspectives [21].

The palm oil industry is one of the leading industries in Malaysia, which brings high income to the country. However, the production of crude palm oil results in large amounts of palm oil mill effluent (POME). Discharge of untreated POME will bring a negative impact on the environment. If the untreated palm oil mill effluent is discharged into watercourses, it is sure to cause considerable environmental problems [22]. Thus, this work is pointed to synthesize and describe alumina made from left-over aluminium cans and to assess the synthesized alumina in the treatment of the POME. To achieve the objectives of this study, the identified scopes were to study the effect of reaction time and reaction temperature in producing alumina from aluminium waste cans, physicochemical characterizations of the synthesized alumina, and conducting the treatment of POME using the synthesized alumina that will act as an adsorbent.

Thus, this project can be an efficient solution to the inadequate solid waste management problem in Malaysia, where waste is utilized to produce another useable and beneficial product. From there, we can cut down the amount of solid waste dumped into landfills that will result in a healthier environment and surrounding besides increasing the economic potential. The novelty part of this project is to enhance social, economic, and environmental efficiency. It promotes sustainable development as there is countless pressure from social and ecological policies to cut down the environmental effect of human activities. As an addition, this project will help the government to achieve main policy goals to minimize waste and promote the re-use of materials as well as utilize the produced alumina in the treatment of palm oil mill effluent (POME) as an adsorbent.

2 Experimental Work

2.1 Materials and Methods

Aluminium waste cans were collected around the area of Universiti Malaysia Pahang (UMP). POME samples were obtained from a local palm oil mill based in Lepar, Pahang. The chemicals potassium hydroxide (KOH) and sulphuric acid (H_2SO_4) were from Sigma-Aldrich. Deionized water was obtained from Millipore Ultra Purifier System as a dilution solvent for POME and the preparation of stock solutions.

2.2 Synthesis of Nano γ-Al_2O_3

The strategy of conducting the amalgamation of γ-Alumina was based on the sol–gel approach [23], with a few alterations were done. Aluminium waste cans were severed into little pieces and melted in of 3 M potassium hydroxide (KOH) solution. The blend was sifted, and the pH of the gotten aluminate filtrate solution was adjusted to 9 utilizing 1 M of sulphuric acid (H_2SO_4). The solution was blended until it got to be gel-like. The mixture was aged by varying the ageing time (6 and 24 h) and also at different temperatures (room temperature and 80 °C). The aluminium hydroxide gel was separated by centrifugation and washed several times with deionized water. The obtained gel was placed into the stove at 80 °C for 12 h for the drying preparation. The dry product was calcined in air at 500 °C for 2 h to form porous alumina. The synthesized alumina was denoted as AL-x−y, where x represents reaction time and y reaction temperature. Alumina synthesized at 6 and 24 h ageing time at room temperature (30 °C) was signified as AL-6 h-30 °C and AL-24 h-30 °C, respectively. Alumina synthesized at 6 h reaction time at 80 ° C was denoted as AL-6 h-80 °C.

2.3 Characterization of Nano γ-Al_2O_3

The characterizations of the produced gamma-alumina were explored by utilizing diverse characterization strategies. γ-AL_2O_3 was described via Fourier transform infrared spectroscopy (FTIR), X-ray diffraction (XRD), scanning electron microscope (SEM), energy-dispersive X-ray spectroscopy (EDX) and Brunauer–Emmett–Teller (BET). The functional group of alumina powders was determined by using FTIR. XRD was conducted to confirm the crystalline structure of the synthesized powder. SEM was performed to examine the surface morphology and particle size, while EDX was utilized to decide the natural composition that presents on the catalyst. BET was made to figure out the surface zone, pore volume, and pore estimate of the catalyst.

2.4 Application of Nano γ-Al_2O_3 for POME

Conducting the treatment of POME was based on [13, 24], with some adjustments were done. The type of synthesized aluminas was varied in this experiment based on reaction time and reaction temperature. Before the experiment is conducted, the POME sample collected from the palm oil mill was filtered and diluted with deionized water. For the dilution of the POME sample, the volume ratio of POME was 25% [25]. A concentration of 8 g L^{-1} of alumina was used where 1.6 g of alumina was added into 200 mL of diluted POME sample. Then, the beaker was placed on a magnetic stirrer with a stirrer speed of 300 rpm at room temperature. The samples were stirred continuously for 4 h to generate a homogenous condition. For the first 1 h, 3 mL of samples were withdrawn from the beaker by using a syringe at every 15 min interval. After an hour, the samples were removed at every 30 min interval. The withdrawn sample was transferred into microcentrifuge tubes. Before analysis, the samples were centrifuged at 5000 rpm for 5 min. The concentration of COD of the withdrawn samples was analysed using the COD measurement kit (Hach, DRB200).

3 Results and Discussion

3.1 Characterization of Nano γ-Al_2O_3

3.1.1 FTIR Analysis

Fourier transform infrared (FTIR) examination of the produced alumina was carried out in the wavenumber ranged from 400 to 4000 cm^{-1}. The spectra of AL-6 h-30 °C, AL-24 h-30 °C, AL-6 h-80 °C, and AL-commercial are as indicated in Figs. 2 and 3. The FTIR spectra of AL-6 h-30 °C, AL-24 h-30 °C, AL-6 h-80 °C, and AL-commercial are demonstrated in Figs. 2 and 3. The main peaks at around 485–656 cm^{-1} can be assigned to the Al–O stretching mode of AlO_6. Bands around 715–826 cm^{-1} correspond to Al–O stretching mode in tetrahedron, while bands around 1072 cm^{-1} are related to symmetric bending of Al–O–H. The wide absorbance groups between 900 and 1100 cm^{-1} are relegated to O–H misshapen vibrations [26–28]. From Fig. 2, the groups of AL-6 h-30 °C were around 559 and 1105 cm^{-1}, whereas the groups of AL-24 h-30 °C were about 567 and 1107 cm^{-1}. AL-6 h-80 °C has bands nearly 606 and 1107 cm^{-1}, as illustrated in Fig. 3, while AL-commercial showed the band at 552 cm^{-1}. AL-6 h-30 °C, AL-24 h-30 °C, AL-6 h-80 °C shared similarity with AL-commercial having broadband around 552–567 cm^{-1}, indicating the presence of the Al–O stretching mode of AlO_6. The spectrum of γ-alumina presented a wide pattern with no apparent thin peak extending from 450 to 950 cm^{-1}. The broad extending band shows the presence of amorphous structure or disordered defects. This pattern is typical of a complex

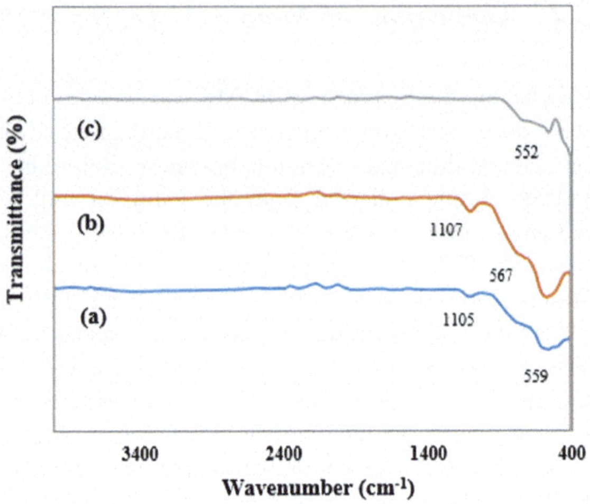

Fig. 2 The cComparison of FTIR spectra highlighting the impact response of ageing time on the produced alumina, **a** AL-6 h-30 °C **b** AL-24 h-30 °C, and **c** AL-commercial

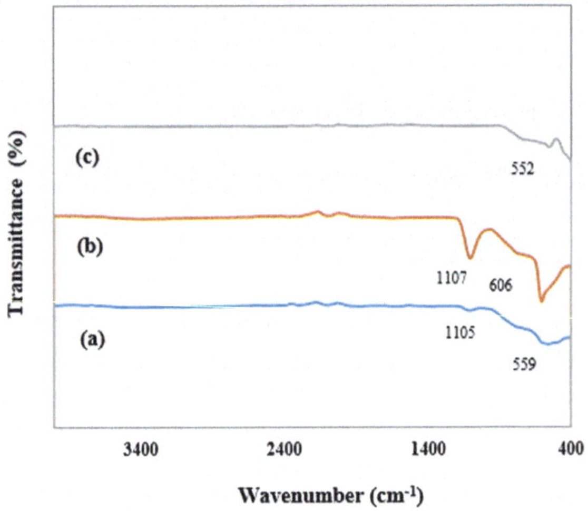

Fig. 3 The comparison of FTIR spectra highlighting the impact response of ageing temperature on the produced alumina, **a** AL-6 h-30 °C **b** AL-6 h-80 °C, and **c** AL-commercial

crystallographic structure, which is the signature of γ-Al_2O_3 [26, 27]. The bands located in 1105 and 1107 cm^{-1} are referred to as the symmetrical deformation vibration of Al–O–H modes [29].

3.1.2 XRD Analysis

XRD analysis was used to analyse the phases present and the crystallinity of the synthesized alumina. Figures 4 and 5 appear the diffraction of the X-ray designs for the synthesized alumina AL-6 h-30 °C, AL-24 h-30 °C, and AL-6 h-80 °C. For

AL-6 h-30 °C, AL-24 h-30 °C, and AL-6 h-80 °C, as shown in Figs. 4 and 5, all the watched crests were distinguished as aluminium oxide or alumina (Al_2O_3) crystalline structure. There were also observed peaks of corundum, which can be defined as a crystalline frame of aluminium oxide, regularly containing follows of iron, titanium, vanadium, and chromium. The XRD designs of synthesized aluminas AL-6 h-30 °C, AL-24 h-30 °C and AL-6 h-80 °C in little point, which is 1–10°, have no XRD crests. This infers that these aluminas have no long-range arranged pore structure. The XRD designs of aluminas within the wide-angle locale proposed the stage of γ-Al_2O_3 at 2θ are 37.5° with wide crests, 46.0°, and 67.0° [30]. Alumina synthesized at 30 °C, as demonstrated in Fig. 4, showed up to be less crystalline in structure with broader crests. The alumina synthesized at 80 °C appeared to be more crystalline in Fig. 5. The crystallite size of the synthesized alumina and commercial alumina, as shown in Table 2, is estimated from X-ray peak broadening using Scherrer formula [31, 32]:

$$D = \frac{k\lambda}{\beta \cos\theta} \tag{1}$$

where D is the average crystallite size, k is the constant with the value of 0.89, λ is the wavelength of the X-ray radiation (0.154 nm), β is the band broadening or full width at half-maximum (FWHM) and θ is the diffraction angle where β and θ are in radians [33]. The size of the obtained γ-Al_2O_3 was found to be ranged from 47 to 55 nm, as listed in Table 2, indicating that the recycled material is in the nanosize range.

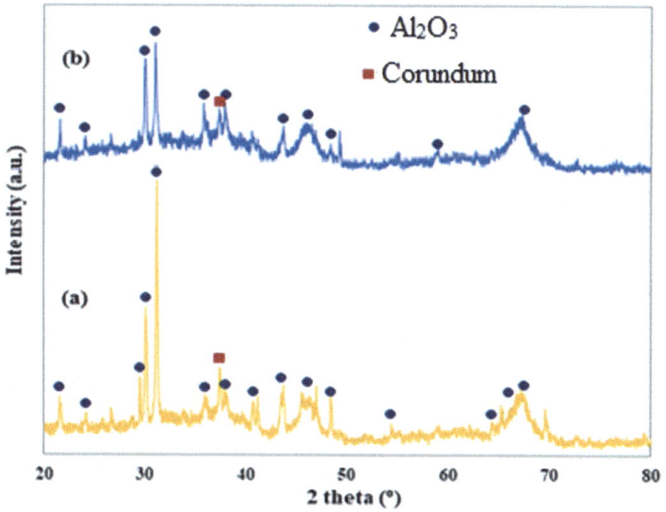

Fig. 4 Comparison of the XRD pattern highlighting the impact response of ageing reaction time on the produced alumina, **a** AL-6 h-30 °C **b** AL-24 h-30 °C

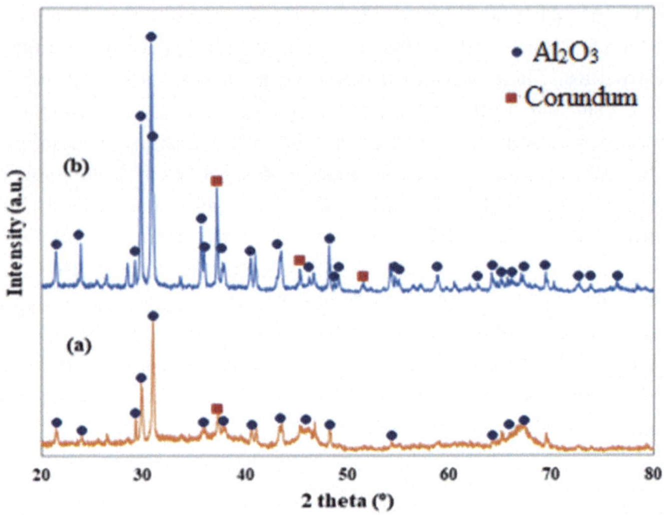

Fig. 5 Comparison of XRD pattern highlighting the impact response of reaction temperature on the produced alumina; **a** AL-6 h-30 °C **b** AL-6 h-80 °C

Table 2 BET analysis of the produced aluminium oxide (alumina)

Sample	Specific surface area (SSA)[a] (m^2 g^{-1})	Pore volume (PV)[b] (cm^3 g^{-1})	Adsorption average pore diameter (4 V/A by BET) (nm)	Crystallite size[c] (nm)
AL-6 h-30 °C	138.5594	0.1285	3.4818	47.12
AL-24 h-30 °C	196.2726	0.1477	3.1412	53.61
AL-6 h-80 °C	89.7032	0.1657	6.5510	59.91
AL-commercial[d]	250.0000	0.8000	9.000	54.89

[a]Use BET method
[b]Calculated by (BJH) method
[c]Use Scherrer formula
[d]Commercial γ-Al_2O_3 [38]

3.1.3 SEM Analysis and EDX

Figure 6 demonstrates the SEM micrographs of AL-6 h-30 °C, AL-6 h-80 °C, AL-24 h-30 °C, and AL-commercial. From Fig. 6a–c, it can be noticed that the solid morphologies for all produced alumina were comparable, with heterogeneous molecule sizes and had a non-spherical shape. They appeared to have hexagonal-like shape particles; these findings were similar to the study by [13, 34]. A few particles with disarranged shapes were got by joining exceptionally little crystallites amid the calcination [8]. Aluminas, AL-6 h-30 °C, AL-6 h-80 °C, AL-24 h-30 °C, showed up to be more disarranged compared to the surface

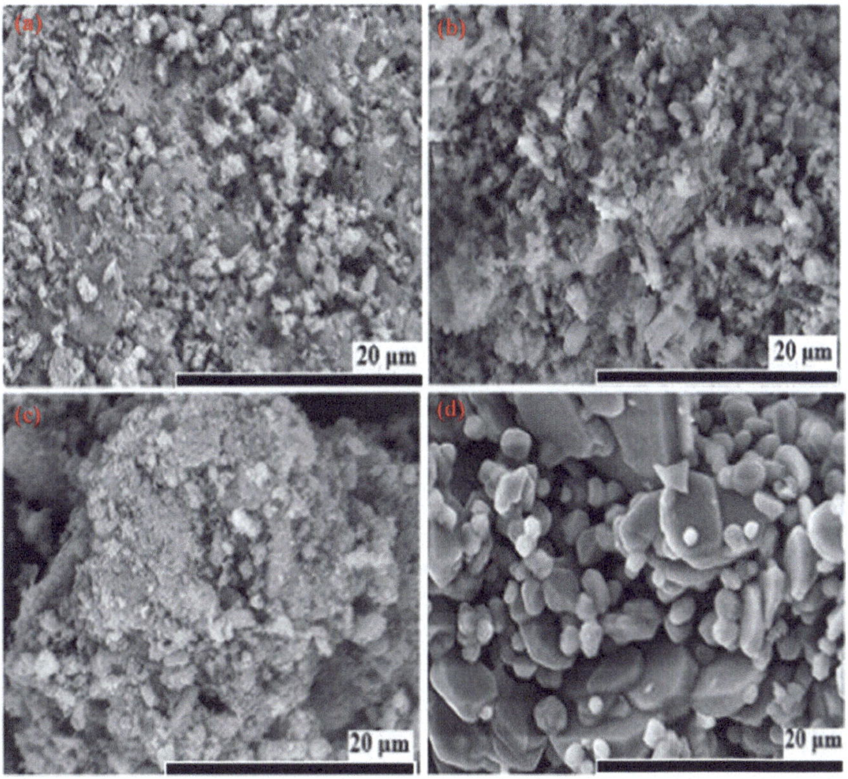

Fig. 6 Comparison of SEM micrographs highlighting the impact response of reaction time and reaction temperature on the produced alumina, **a** AL-6 h-30 °C **b** AL-24 h-30 °C, **c** AL-6 h-80 °C, and **d** AL-commercial (Magnification: 5.0 kx)

morphology of commercial alumina in Fig. 6d. AL-commercial showed a smooth surface and spherical shapes. This may due to different calcination conditions where higher calcination temperature or long calcination time was used in producing the commercial alumina.

EDX analysis in Fig. 7 represents the normal weight and nuclear rate of components that display on the tests of the synthesized aluminas and commercial alumina. The ruddy circle within the figures states to the spot where the EDX test was obtained. AL-commercial is primarily composed of oxygen and aluminium, as displayed in Fig. 7d.

In Fig. 7a–c, the spectrum of the produced alumina AL-6 h-30 °C, AL-6 h-80 °C, and AL-24 h-30 °C was famous that the powders are primarily composed of oxygen, aluminium, sulphur, and potassium. Oxygen has the most noteworthy weight and nuclear rate in all tests. The presence of potassium and sulphuric acid in the synthesized alumina is because of the utilization of potassium hydroxide arrangement in dissolving the aluminium waste cans and the altering the pH,

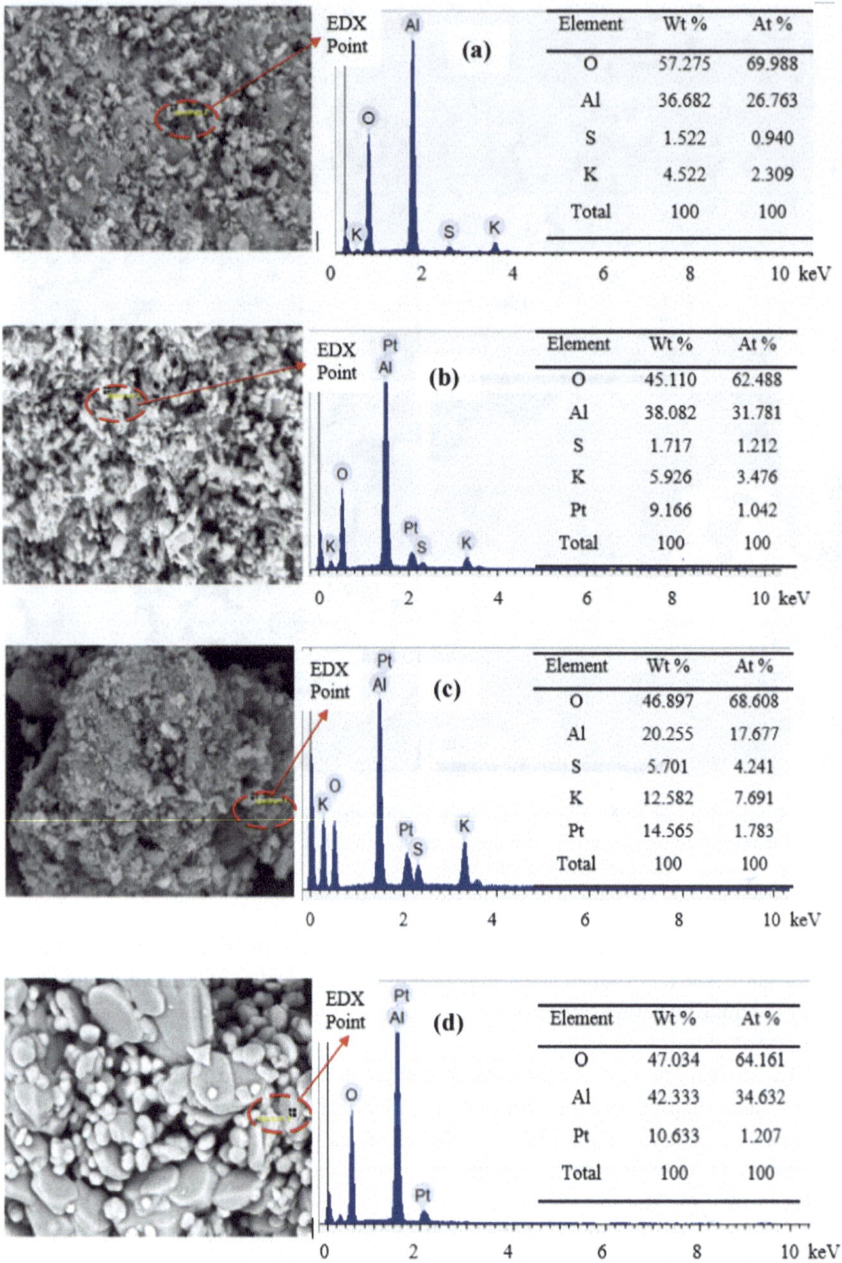

Fig. 7 Comparison of EDX highlighting the impact response of reaction time and reaction temperature on the produced alumina; **a** AL-6 h-30 °C **b** AL-24 h-30 °C, **c** AL-6 h-80 °C, and **d** AL-commercial

respectively. The presence of these after-mentioned chemicals can be reduced by having adequate washing with deionized water throughout forming the aluminium hydroxide gel. Platinum was due to the coating used in the analysis. Regardless of the reaction time and temperature, the synthesized aluminas have a comparative trend within the composition for each component.

3.1.4 BET Analysis

Figure 8 shows the isotherms for the Al_2O_3 tests calcined within the temperature at 500 °C. The N_2 adsorption–desorption isotherms of Al_2O_3 are relegated to sort IV isotherms with the H_2 category of hysteresis agreeing to IUPAC classification, which is characteristic of a mesoporous material. These mesoporous γ-aluminas have moderately limited pore measure dispersions. It was found that expanding response time may broaden the pore estimate conveyance of alumina and leads to the arrangement of bigger pores [35]. In any case, the pore estimate dissemination of γ-aluminas is a contract compared to the commercial alumina. Hence, the main point is to produce the mesoporous gamma-alumina to be used as an adsorbent for palm oil mill effluent (POME).

The specific surface area (SSA), pore volume, and pore diameter of synthesized AL-6 h-30 °C, AL-24 h-30 °C, AL-6 h-80 °C, and AL-commercial are shown in Table 2. The specific surface area (SSA) of AL-6 h-30 °C, AL-24 h-30 °C, and

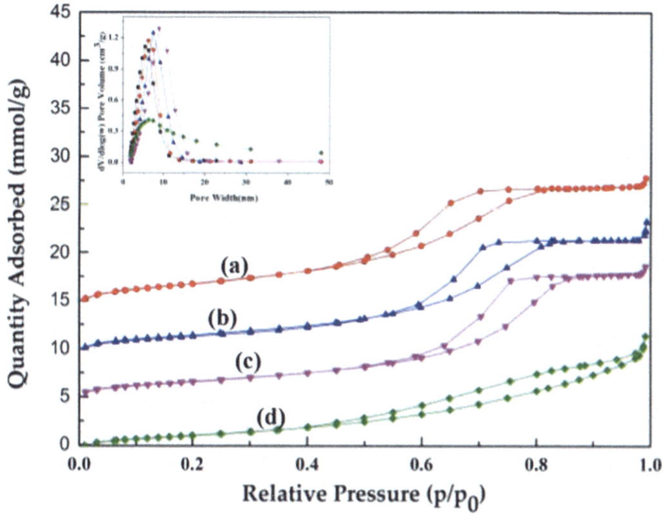

Fig. 8 Comparison of nitrogen adsorption–desorption isotherms curves highlighting the impact response of reaction temperature on the produced alumina; **a** AL-6 h-30 °C **b** AL-24 h-30 °C, **c** AL-6 h-80 °C, and **d** AL-commercial

AL-6 h-80 °C was 138.5594, 196.2726, and 89.7032 $m^2 g^{-1}$, respectively. Alumina, AL-6 h-30 °C and AL-24 h-30 °C, revealed high SSA, whereas their pore volume values were 0.1285 and 0.1477 $cm^3 g^{-1}$, respectively. The pore volume for AL-6 h-80 °C was 0.1657 $cm^3 g^{-1}$. It can be noticed that the values of pore volume are small of the synthesized alumina. The above consequence showed that the sol–gel process was an appropriate process for alumina preparation at calcination temperature 500 °C without the collapse of the mesoporous structure [30].

On the other side, AL-commercial has SSA and PV at 250 $m^2 g^{-1}$ and 0.8 $cm^2 g^{-1}$. The significant conclusion from BET and XRD analysis is that the SSA is dependable on the XRD results; that means, the more crystalline structure, the lower the surface area. Hence, alumina at room temperature for 6 and 24 h has a lesser crystalline structure but higher SSA. However, alumina at 80 °C has the opposite consequence compared to the alumina at room temperature. The surface area of the synthesized alumina at room temperature (30 °C) increased from 138.5594 $m^2 g^{-1}$ to 196.2726 $m^2 g^{-1}$ along with increasing reaction or ageing time (from 6 to 24 h). Ageing/reaction time is considered an essential element to attain higher or lower SSA of the produced alumina [36]. The study extracted by [37] mentioned that the porous materials have a higher SSA around 100 $m^2 g^{-1}$, which could probably be applied as catalyst support. It was observed that the alumina AL-6 h-30 °C and AL-24 h-30 °C had high surface area value (138.5594–196.2726 $m^2 g^{-1}$) and low pore volumes (0.1285–0.1477 $cm^3 g^{-1}$), being suitable to be used as a catalyst support or adsorbent [11, 20]. For the effect of reaction temperature, it is observed that alumina aged at 80 °C (AL-6 h-80 °C) had a larger pore diameter than the alumina aged at room temperature (AL-6 h-30 °C). This trend can be explained by the fact that during ageing at higher temperatures, fewer grains of aluminium hydroxide are formed. This process allowed a more orderly growth of particles and led to the formation of oxides with larger pores and lower surface areas [13].

3.2 *Application of Nano γ-Al_2O_3 as an Adsorbent in the Treatment of POME*

The adsorption performance is measured based on the percentage of removal for chemical oxygen demand (COD) by using different types of synthesized alumina. Figures 9 and 10 show the percentage of COD removal from POME by using 8 g L^{-1} of adsorbents synthesized alumina AL-6 h-30 °C, AL-24 h-30 °C, and AL-6 h-80 °C at room temperature. From Fig. 9a and 10a, it can be seen that the percentage of COD removal for AL-6–30 increased rapidly at the initial stage and begun to slow down after 150 min. It is found that increasing the contact time, the percentage of COD removal increased until adsorption equilibrium is reached. This can be attributed to the availability of sites for the sorbates, and a very high adsorption driving force at the beginning stage resulted in a higher adsorption rate.

After the rapid initial stage, a slower adsorption rate may be explained by the slower diffusion of molecules into the interior pores of the adsorbent, and the molecules subsequently occupy the sites within the adsorbent framework [39]. This observation is similar to the findings by [39, 40]. In Fig. 9b, it is shown that the percentage of COD removal for AL-24 h-30 °C at the initial stage was slow, but there was a sudden rapid increase from 150 to 210 min. This may occur due to the slow absorption of contaminants during the initial stage. However, the percentage of removal started to be constant from 210 min. Based on Fig. 10, both the percentage of COD removal for AL-6 h-30 °C and AL-6 h-80 °C show a similar trend. Therefore, the percentage of COD removal expanded quickly at the starting arrange but moderated within the afterwards organize.

It could be noticed that the highest percentage of COD removal for AL-6 h-30 °C was around 29%, while AL-24 h-30 °C was around 26%. Both AL-6 h-30 °C and AL-24 h-30 °C have high surface areas, and thus, there was no significant difference in the highest percentage of COD removal for both of the alumina. However, the highest percentage of COD removal for AL-6 h-80 °C was only around 7%. The difference in the percentage of COD removal may be due to the difference in the BET surface area. AL-6 h-30 °C has a larger surface area than AL-6 h-80 °C. An increment in the adsorbent surface region and accessibility of more adsorption locales sites result in a higher percentage of COD removal [24].

Figures 11 and 12 indicate the FTIR spectra for the post-reaction powder Spent AL-6 h-30 °C, Spent AL-24 h-30 °C, and AL-6 h-80 °C. Figures 11 and 12 show that Spent AL-6 h-30 °C had a band at 524 cm^{-1}, Spent AL-24 h-30 °C had a band at 533 cm¯1, while Spent AL-6 h-80 °C had a band at 554 cm^{-1}. By comparing to the pre-reaction powder, both of the spent alumina remains the presence of the Al–O stretching mode of AlO_6. This shows that the Spent AL-6 h-30 °C, Spent AL-24 h-30 °C, and AL-6 h-80 °C retain their characteristics as alumina even after the treatment of POME. This also implies that the synthesized alumina remains the same pre- and post-reaction.

Fig. 9 Percentage COD removal highlighting the impact response of the reaction time on the produced alumina; **a** AL-6 h-30 °C and **b** AL-24 h-30 °C

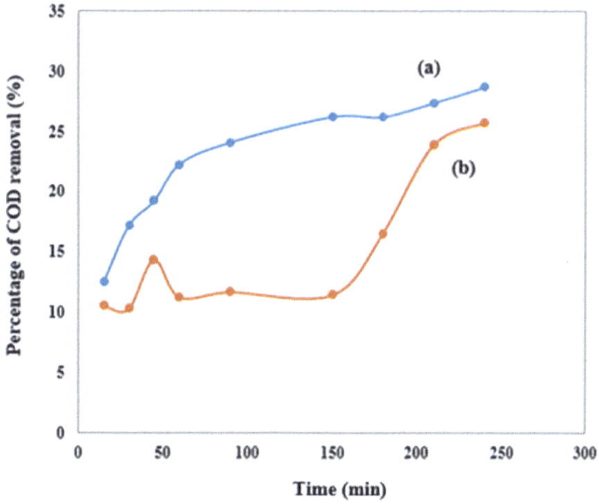

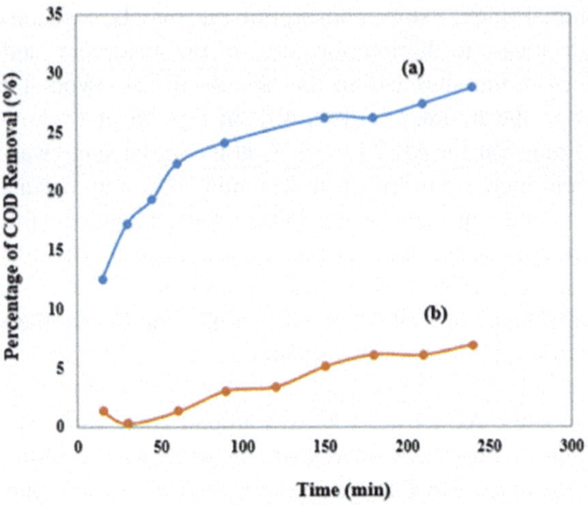

Fig. 10 Percentage COD removal highlighting the impact response of the reaction temperature on the produced alumina; **a** AL-6 h-30 °C and **b** AL-6 h-80 °C

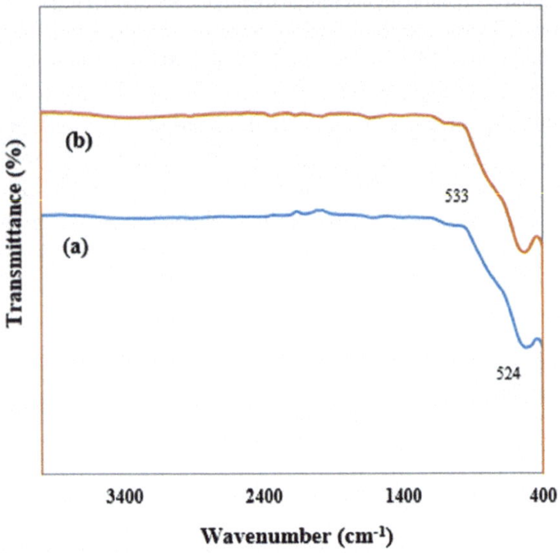

Fig. 11 FTIR spectra comparing the post-reaction alumina for the effect of reaction time; **a** Spent AL-6 h-30 °C and **b** Spent AL-24 h-30 °C

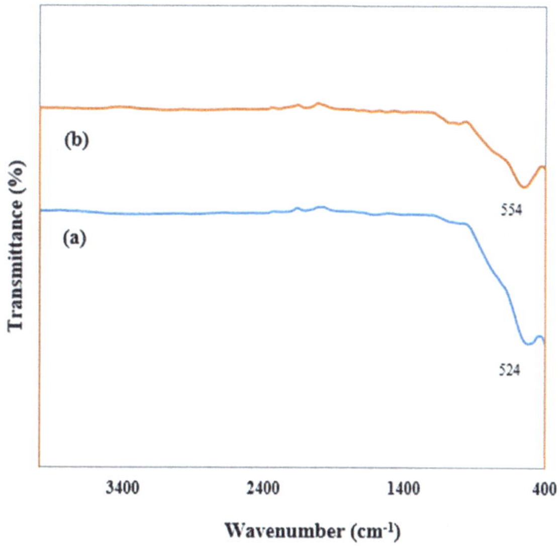

Fig. 12 FTIR spectra comparing the post-reaction alumina for the effect of reaction temperature; **a** Spent AL-6 h-30 °C and **b** Spent AL-6 h-80 °C

4 Conclusions

In short, γ-alumina was properly produced from the aluminium drink cans via sol–gel technique at various ageing times and temperatures. The analysis of FTIR, XRD, SEM, EDX, and BET has been verified the properties of the produced alumina. The synthesized aluminas have high surface areas that are comparable to the commercial alumina. The produced alumina at 30 °C and 24 h had the best state recording the highest surface area at 196.2726 m^2 g^{-1} compared to other produced aluminas to be applied as an adsorbent for POME. The synthesized aluminas were used as an adsorbent in the treatment of POME through adsorption. The percentage of COD removal for POME is influenced by the surface area of alumina. The highest percentage of 29% of COD removal was achieved by using AL-6 30 which has a high BET surface area.

5 Future Prospectives

Based on the results of this research to synthesize γ-alumina from aluminium waste cans and to be applied as an adsorbent in the treatment of POME, a few parameters could be further studied to evaluate the optimal or best conditions in producing high-structured γ-alumina with higher surface areas. Therefore, there are some recommendations for future work as follows: (i) the effect mixing the small pieces of the aluminium waste cans compared to sorting each type of aluminium waste cans on the produced γ-alumina, (ii) the effect of the new technology methods used to synthesize alumina linked to the used method (sol–gel), (iii) the effect of the other chemicals compared to the used one (sulphric acid and potassium hydroxide), (iv) the effect of pH on the synthesis of alumina, (v) the effect of reaction temperatures and ageing times during stirring process to form alumina gel-like, and (vi) the effect of calcination temperatures on the synthesis of alumina as well as the calcination time to produce γ-alumina.

Acknowledgements The authors would like to acknowledge the financial support for this work from the Universiti Malaysia Pahang (UMP) internal grant (RDU180326), Faculty of Chemical and Natural Resources Engineering, UMP.

References

1. Manaf LA, Samah MAA, Zukki NIM (2009) Municipal solid waste management in Malaysia: practices and challenges. Waste Manage 29:2902–2906
2. Aboelazm EAA, Ali GAM, Algarni H, Yin H, Zhong YL, Chong KF (2018) Magnetic electrodeposition of the hierarchical cobalt oxide nanostructure from spent lithium-ion batteries: its application as a supercapacitor electrode. J Phys Chem C 122(23):12200–12206
3. Zainu ZA, Songip AR (2017) Policies, challenges and strategies for municipal waste management in Malaysia. J Sci Technol Innov Policy 3(1):18–22
4. Kumar S, Kumar R, Bandopadhyay A (2006) Innovative methodologies for the utilisation of wastes from metallurgical and allied industries. Resour Conserv Recycl 48:301–314
5. Shinzato MC, Hypolito R (2005) Solid waste from aluminum recycling process: characterization and reuse of its economically valuable constituents. Waste Manag 25(1):37–46
6. Ali GAM, Yusoff MM, Shaaban ER, Chong KF (2017) High performance MnO_2 nanoflower supercapacitor electrode by electrochemical recycling of spent batteries. Ceram Int 43:8440–8448
7. Edwards C (2017) The disadvantages of aluminum can, 18 July 2017. Retrieved from https://www.livestrong.com/article/259503-the-disadvantages-of-aluminumcans/
8. Liu W, Niu T, Yang J, Wang Y, Hu S, Dong Y, Xu H (2011) Preparation of micron-sized alumina powders from aluminium beverage can by means of sol–gel process. Micro Nano Lett 6:852–854
9. Verran G, Kurzawa U (2008) An experimental study of aluminum can recycling using fusion in induction furnace. Resour Conserv Recycl 52:731–736
10. Sheel TK, Poddar P, Murad AW, Neger AT, Chowdhury AS (2016) Preparation of aluminum oxide from industrial waste can available in Bangladesh environment: SEM and EDX analysis. J Adv Chem Eng 06(02)
11. Adans YF, Martins AR, Coelho RE, Das Virgens CF, Ballarini AD, Carvalho LS (2016) A simple way to produce γ-alumina from aluminum cans by precipitation reactions. Mater Res 19:977–982
12. Ahmedzeki NS, Hussein S, Abdulnabi WA (2017) Recycling waste cans to nano gamma alumina effect of the calcination temperature and pH. Int J Curr Eng Technol 7(1):81–88
13. Asencios YJ, Sun-Kou MR (2012) Synthesis of high-surface-area γ-Al_2O_3 from aluminum scrap and its use for the adsorption of metals: Pb (II), Cd (II) and Zn (II). Appl Surf Sci 258:10002–10011
14. Abdelkader A, Osman AI, Halawy SA, Mohamed MA (2018) Preparation and characterization of mesoporous γ-Al_2O_3 recovered from aluminum cans waste and its use in the dehydration of ethanol to dimethyl ether. J Mater Cycles Waste Manage 20(3):1428–1436

15. Matori KA, Wah LC, Hashim M, Ismail I, Zaid MHM (2012) Phase transformations of α-alumina made from waste aluminum via a precipitation technique. Int J Mol Sci 13:16812–16821
16. Aswad MA (2012) Residual stress and fracture in high temperature ceramics. Retrieved from https://www.escholar.manchester.ac.uk/jrul/item/?pid=ukac-manscw:157835
17. Harabi E, Harabi A, Mezahi FZ, Zouai S, Karboua NE, Chehlatt S (2016) Effect of P_2O_5 on mechanical properties of porous natural hydroxyapatite derived from cortical bovine bones sintered at 1,050 °C. Desalin Water Treat 57(12):5297–5302
18. Zhu B, Myat DT, Shin J-W, Na Y-H, Moon I-S, Connor G, Duke M (2015) Application of robust MFI-type zeolite membrane for desalination of saline wastewater. J Membr Sci 475:167–174
19. Hosseini SY, Khosravi-Nikou MR (2016) Synthesis and characterization of nano-sized γ-Al_2O_3 for investigation the effect of temperature on catalytic dehydration of methanol to dimethyl ether. Energy Sources Part A Recov Utiliz Environ Eff 38(7):914–920
20. Adans YF, Martins AR, Coelho RE, Virgens CFD, Ballarini AD, Carvalho LS (2016b) A simple way to produce γ-alumina from aluminum cans by precipitation reactions. Mater Res 19, 977–982
21. Eriksson O, Reich MC, Frostell B, Björklund A, Assefa G, Sundqvist J-O, Granath J, Baky A, Thyselius L (2005) Municipal solid waste management from a systems perspective. J Clean Product 13:241–252
22. Wu TY, Mohammad AW, Jahim JM, Anuar N (2009) A holistic approach to managing palm oil mill effluent (POME): biotechnological advances in the sustainable reuse of POME. Biotechnol Adv 27:40–52
23. Bin Mokaizh AA, Wirman N, Shariffuddin JH (2019) Synthesis of alumina from aluminium can waste to be applied as catalyst support for biodiesel production. IOP Conf Ser Mater Sci Eng 702:012038
24. Mohammed RR, Ketabchi MR, McKay G (2014) Combined magnetic field and adsorption process for treatment of biologically treated palm oil mill effluent (POME). Chem Eng J 243:31–42
25. Ding GT, Yaakob Z, Takriff MS, Salihon J, Rahaman MSA (2016) Biomass production and nutrients removal by a newly-isolated microalgal strain Chlamydomonas sp in palm oil mill effluent (POME). Int J Hydrogen Energy 41:4888–4895
26. Bazyari A, Mortazavi Y, Khodadadi AA, Thompson LT, Tafreshi R, Zaker A, Ajenifujah OT (2016) Effects of alumina phases as nickel supports on deep reactive adsorption of (4, 6-dimethyl) dibenzothiophene: comparison between γ, δ, and θ-alumina. Appl Catal B 180:312–323
27. Djebaili K, Mekhalif Z, Boumaza A, Djelloul A (2015) XPS, FTIR, EDX, and XRD analysis of Al_2O_3 scales grown on PM2000 alloy. J Spectro 2015
28. Jafar Tafreshi M, Masoomi Khanghah Z (2015) Infrared spectroscopy studies on sol–gel prepared alumina powders. Mater Sci 21:28–31
29. Xu N, Liu Z, Bian S, Dong Y, Li W (2016) Template-free synthesis of mesoporous γ-alumina with tunable structural properties. Ceram Int 42:4072–4079
30. Chotisuwan S, Sirirak A, Har-Wae P, Wittayakun J (2012) Mesoporous alumina prepared from waste aluminum cans and used as catalytic support for toluene oxidation. Mater Lett 70:125–127
31. Patterson AL (1939) The Scherrer formula for X-ray particle size determination. Phys Rev 56:978
32. Fouad OA, Makhlouf SA, Ali GAM, El-Sayed AY (2011) Cobalt/silica nanocomposite via thermal calcination-reduction of gel precursors. Mater Chem Phys 128(1–2):70–76
33. Wang X, Shi F, Huang W, Fan C (2012) Synthesis of high quality TiO_2 membranes on alumina supports and their photocatalytic activity. Thin Solid Films 520(7):2488–2492
34. Karim M, Rahman M, Miah M, Ahmad H, Yanagisawa M, Ito M (2011) Synthesis of γ-alumina particles and surface characterization. Open Coll Sci J 4:32–36

35. Ji Y, Zhang F, Yu F, Zhang J, Zhang J (2018) Methyl chloride synthesis over metal chlorides-modified mesoporous alumina catalyst. Catalysts 8(3):99
36. Wu Y, Ma J, Hu F, Li M (2012) Synthesis and characterization of mesoporous alumina via a reverse precipitation method. J Mater Sci Technol 28:572–576
37. Bawa S, Ahmed A, Okonkwo P (2016) The study of thermal effect on the surface properties of gamma-alumina synthesied from kankara kaolin. Niger J Technol 35:66–70
38. Samain L, Jaworski A, Edén M, Ladd DM, Seo D-K, Garcia-Garcia FJ, Häussermann U (2014) Structural analysis of highly porous γ-Al_2O_3. J Solid State Chem 217:1–8
39. Mohammed RR, Chong MF (2014) Treatment and decolorization of biologically treated palm oil mill effluent (POME) using banana peel as novel biosorbent. J Environ Manage 132:237–249
40. Kutty S, Ngatenah S, Johan N, Amat K (2011) Removal of Zn (II), Cu (II), chemical oxygen demand (COD) and colour from anaerobically treated palm oil mill effluent (POME) using microwave incinerated rice husk ash (MIRHA). In: International conference on environmental and industrial innovation IPCBEE, vol 12, pp 90–94

The manufacturer's authorised representative in the EU is Springer Nature Customer Service Centre GmbH, Europaplatz 3, 69115 Heidelberg, Germany. If you have any concerns regarding our products, please contact ProductSafety@springernature.com

Printed and bound by CPI Group (UK) Ltd, Croydon, CR0 4YY
30/09/2026
02225689-0020